Preface

The Fifth International Conference on Computational Science (ICCS 2005) held in Atlanta, Georgia, USA, May 22–25, 2005, continued in the tradition of previous conferences in the series: ICCS 2004 in Krakow, Poland; ICCS 2003 held simultaneously at two locations, in Melbourne, Australia and St. Petersburg, Russia; ICCS 2002 in Amsterdam, The Netherlands; and ICCS 2001 in San Francisco, California, USA.

Computational science is rapidly maturing as a mainstream discipline. It is central to an ever-expanding variety of fields in which computational methods and tools enable new discoveries with greater accuracy and speed. ICCS 2005 was organized as a forum for scientists from the core disciplines of computational science and numerous application areas to discuss and exchange ideas, results, and future directions. ICCS participants included researchers from many application domains, including those interested in advanced computational methods for physics, chemistry, life sciences, engineering, economics and finance, arts and humanities, as well as computer system vendors and software developers. The primary objectives of this conference were to discuss problems and solutions in all areas, to identify new issues, to shape future directions of research, and to help users apply various advanced computational techniques. The event highlighted recent developments in algorithms, computational kernels, next generation computing systems, tools, advanced numerical methods, data-driven systems, and emerging application fields, such as complex systems, finance, bioinformatics, computational aspects of wireless and mobile networks, graphics, and hybrid computation. Keynote lectures were delivered by John Drake – High End Simulation of the Climate and Development of Earth System Models; Marian Bubak – Towards Knowledge – Based Computing: Personal Views on Recent Developments in Computational Science and the CrossGrid Project; Alok Choudhary – Large Scale Scientific Data Management; and David Keyes – Large Scale Scientific Discovery through Advanced Computing.

In addition, four invited presentations were delivered by representatives of industry: David Barkai from Intel Corporation, Mladen Karcic from IBM, Steve Modica from SGI and Dan Fay from Microsoft. Seven tutorials preceded the main technical program of the conference: Tools for Program Analysis in Computational Science by Dieter Kranzlmüller and Andreas Knüpfer; Computer Graphics and Geometric Modeling by Andrés Iglesias; Component Software for High Performance Computing Using the CCA by David Bernholdt; Computational Domains for Explorations in Nanoscience and Technology, by Jun Ni, Deepak Srivastava, Shaoping Xiao and M. Meyyappan; Wireless and Mobile Communications by Tae-Jin Lee and Hyunseung Choo; Biomedical Literature Mining and Its Applications in Bioinformatics by Tony Hu; and Alternative Approaches to

Grids and Metacomputing by Gunther Stuer. We would like to thank all keynote, invited and tutorial speakers for their interesting and inspiring talks.

Aside from the plenary lectures, the conference included 10 parallel oral sessions and 3 poster sessions. Ever since the first meeting in San Francisco, ICCS has attracted an increasing number of researchers involved in the challenging field of computational science. For ICCS 2005, we received 464 contributions for the main track and over 370 contributions for 24 originally-proposed workshops. Of these submissions, 134 were accepted as full papers accompanied by oral presentations, and 89 for posters in the main track, while 241 papers were accepted for presentations at 21 workshops. This selection was possible thanks to the hard work of the 88-member Program Committee and 362 reviewers. The author index contains 1395 names, and over 500 participants from 41 countries and all continents attended the conference. The ICCS 2005 proceedings consists of three volumes. The first volume, LNCS 3514, contains the full papers from the main track of the conference, while volumes 3515 and 3516 contain the papers accepted for the workshops and short papers. The papers cover a wide range of topics in computational science, ranging from numerical methods, algorithms, and computational kernels to programming environments, grids, networking and tools. These contributions, which address foundational and computer science aspects are complemented by papers discussing computational applications in a variety of domains. ICCS continues its tradition of printed proceedings, augmented by CD-ROM versions for the conference participants. We would like to thank Springer for their cooperation and partnership. We hope that the ICCS 2005 proceedings will serve as a major intellectual resource for computational science researchers for many years to come. During the conference the best papers from the main track and workshops as well as the best posters were nominated and commended on the ICCS 2005 Website. A number of papers will also be published in special issues of selected journals.

We owe thanks to all workshop organizers and members of the Program Committee for their diligent work, which led to the very high quality of the event. We would like to express our gratitude to Emory University and Emory College in general, and the Department of Mathematics and Computer Science in particular, for their wholehearted support of ICCS 2005. We are indebted to all the members of the Local Organizing Committee for their enthusiastic work towards the success of ICCS 2005, and to numerous colleagues from various Emory University units for their help in different aspects of organization. We very much appreciate the help of Emory University students during the conference. We owe special thanks to our corporate sponsors: Intel, IBM, Microsoft Research, SGI, and Springer; and to ICIS, Math & Computer Science, Emory College, the Provost's Office, and the Graduate School at Emory University for their generous support. ICCS 2005 was organized by the Distributed Computing Laboratory at the Department of Mathematics and Computer Science at Emory University, with support from the Innovative Computing Laboratory at the University of Tennessee and the Computational Science Section at the University of Amsterdam, in cooperation with the Society for Industrial and Applied Mathematics (SIAM). We invite you to visit the ICCS 2005 Website (http://www.iccs-meeting.org/ICCS2005/) to recount the events leading up to the conference, to view the technical program, and to recall memories of three and a half days of engagement in the interest of fostering and advancing computational science.

June 2005

Vaidy Sunderam
on behalf of
G. Dick van Albada
Jack J. Dongarra
Peter M.A. Sloot

Organization

ICCS 2005 was organized by the Distributed Computing Laboratory, Department of Mathematics and Computer Science, Emory University, Atlanta, GA, USA, in cooperation with Emory College, Emory University (USA), the University of Tennessee (USA), the University of Amsterdam (The Netherlands), and the Society for Industrial and Applied Mathematics (SIAM). The conference took place on the campus of Emory University, in Atlanta, Georgia, USA.

Conference Chairs

Scientific Chair – Vaidy Sunderam (Emory University, USA)
Workshops Chair – Dick van Albada (University of Amsterdam, The Netherlands)
ICCS Series Overall Chair – Peter M.A. Sloot (University of Amsterdam, The Netherlands)
ICCS Series Overall Co-Chair – Jack Dongarra (University of Tennessee, USA)

Local Organizing Committee

Dawid Kurzyniec (Chair)
Piotr Wendykier
Jeri Sandlin
Erin Nagle
Ann Dasher
Sherry Ebrahimi

Sponsoring Institutions

Intel Corporation
IBM Corporation
Microsoft Research
SGI Silicon Graphics Inc.
Emory University, Department of Mathematics and Computer Science
Emory University, Institute for Comparative and International Studies
Emory University, Emory College
Emory University, Office of the Provost
Emory University, Graduate School of Arts and Sciences
Springer

Program Committee

Jemal Abawajy, Deakin University, Australia
David Abramson, Monash University, Australia
Dick van Albada, University of Amsterdam, The Netherlands
Vassil Alexandrov, University of Reading, UK
Srinivas Aluru, Iowa State University, USA
Brian d'Auriol, University of Texas at El Paso, USA
David A. Bader, University of New Mexico, USA
Saeid Belkasim, Georgia State University, USA
Anne Benoit, University of Edinburgh, UK
Michele Benzi, Emory University, USA
Rod Blais, University of Calgary, Canada
Alexander Bogdanov, Institute for High Performance Computing and Information Systems, Russia
Anu Bourgeois, Georgia State University, USA
Jan Broeckhove, University of Antwerp, Belgium
Marian Bubak, Institute of Computer Science and ACC Cyfronet – AGH, Poland
Rajkumar Buyya, University of Melbourne, Australia
Tiziana Calamoneri, University of Rome "La Sapienza", Italy
Serge Chaumette, University of Bordeaux, France
Toni Cortes, Universitat Politecnica de Catalunya, Spain
Yiannis Cotronis, University of Athens, Greece
Jose C. Cunha, New University of Lisbon, Portugal
Pawel Czarnul, Gdansk University of Technology, Poland
Frederic Desprez, INRIA, France
Tom Dhaene, University of Antwerp, Belgium
Hassan Diab, American University of Beirut, Lebanon
Beniamino Di Martino, Second University of Naples, Italy
Jack Dongarra, University of Tennessee, USA
Craig Douglas, University of Kentucky, USA
Edgar Gabriel, University of Stuttgart, Germany
Marina Gavrilova, University of Calgary, Canada
Michael Gerndt, Technical University of Munich, Germany
Yuriy Gorbachev, Institute for High Performance Computing and Information Systems, Russia
Andrzej Goscinski, Deakin University, Australia
Eldad Haber, Emory University, USA
Ladislav Hluchy, Slovak Academy of Science, Slovakia
Alfons Hoekstra, University of Amsterdam, The Netherlands
Yunqing Huang, Xiangtan University, China
Andrés Iglesias, University of Cantabria, Spain
Hai Jin, Huazhong University of Science and Technology, China
Peter Kacsuk, MTA SZTAKI Research Institute, Hungary
Jacek Kitowski, AGH University of Science and Technology, Poland

Lecture Notes in Computer Science 3516

Vaidy S. Sunderam Geert Dick van Albada
Peter M.A. Sloot Jack J. Dongarra (Eds.)

Computational Science – ICCS 2005

5th International Conference
Atlanta, GA, USA, May 22-25, 2005
Proceedings, Part III

Volume Editors

Vaidy S. Sunderam
Emory University
Dept. of Math and Computer Science
400 Dowman Dr, W430, Atlanta, GA 30322, USA
E-mail: vss@mathcs.emory.edu

Geert Dick van Albada
Peter M.A. Sloot
University of Amsterdam
Department of Mathematics and Computer Science
Kruislaan 403, 1098 SJ Amsterdam, The Netherlands
E-mail: {dick,sloot}@science.uva.nl

Jack J. Dongarra
University of Tennessee
Computer Science Department
1122 Volunteer Blvd., Knoxville, TN 37996-3450, USA
E-mail: dongarra@cs.utk.edu

Library of Congress Control Number: 2005925759

CR Subject Classification (1998): D, F, G, H, I, J, C.2-3

ISSN 0302-9743
ISBN-10 3-540-26044-7 Springer Berlin Heidelberg New York
ISBN-13 978-3-540-26044-8 Springer Berlin Heidelberg New York

Springer is a part of Springer Science+Business Media

springeronline.com

Printed in Germany

Typesetting: Camera-ready by author, data conversion by Scientific Publishing Services, Chennai, India
Printed on acid-free paper SPIN: 11428862 06/3142 5 4 3 2 1 0

Krzysztof Zielinski, ICS UST / CYFRONET, Poland
Zahari Zlatev, National Environmental Research Institute, Denmark
Elena Zudilova-Seinstra, University of Amsterdam, The Netherlands

Reviewers

Adrian Kacso
Adrian Sandu
Akshaye Dhawan
Alberto Sanchez-Campos
Alex Tiskin
Alexander Bogdanov
Alexander Zhmakin
Alexandre Dupuis
Alexandre Tiskin
Alexandros Gerbessiotis
Alexey S. Rodionov
Alfons Hoekstra
Alfredo Tirado-Ramos
Ali Haleeb
Alistair Rendell
Ana Ripoll
A. Kalyanaraman
Andre Merzky
Andreas Hoffmann
Andrés Iglesias
Andrew Adamatzky
Andrzej Czygrinow
Andrzej Gościński
Aneta Karaivanova
Anna Morajko
Anne Benoit
Antonio Lagana
Anu G. Bourgeois
Ari Rantanen
Armelle Merlin
Arndt Bode
B. Frankovic
Bahman Javadi
Baowen Xu
Barbara Głut
Bartosz Baliś
Bas van Vlijmen
Bastien Chopard
Behrooz Shirazi
Ben Jackson
Beniamino Di Martino
Benjamin N. Jackson
Benny Cheung
Biju Sayed
Bogdan Lesyng
Bogdan Smolka
Boleslaw Szymanski
Breanndan O'Nuallain
Brian d'Auriol
Brice Goglin
Bruce Boghosian
Casiano Rodrguez León
Charles Shoniregun
Charles Stewart
Chen Lihua
Chris Homescu
Chris R. Kleijn
Christian Glasner
Christian Perez
C. Schaubschlaeger
Christoph Anthes
Clemens Grelck
Colin Enticott
Corrado Zoccolo
Craig C. Douglas
Craig Lee
Cristina Negoita
Dacian Daescu
Daewon W. Byun
Dale Shires
Danica Janglova
Daniel Pressel
Dave Roberts
David Abramson
David A. Bader
David Green
David Lowenthal
David Roberts
Dawid Kurzyniec
Dick van Albada
Diego Javier Mostaccio
Dieter Kranzlmüller
Dirk Deschrijver
Dirk Roekaerts
Domenico Laforenza
Donny Kurniawan
Eddy Caron
Edgar Gabriel
Edith Spiegl
Edward Moreno
Eldad Haber
Elena Zudilova-Seinstra
Elisa Heymann
Emanouil Atanassov
Emilio Luque
Eunjoo Lee
Eunjung Cho
Evarestov
Evghenii Gaburov
Fabrizio Silvestri
Feng Tan
Fethi A. Rabhi
Floros Evangelos
Francesco Moscato
Francis Lau
Francisco J. Rosales
Franck Cappello
Frank Dehne
Frank Dopatka
Frank J. Seinstra
Frantisek Capkovic
Frederic Desprez
Frederic Hancke

Frédéric Gava
Frédéric Loulergue
Frederick T. Sheldon
Gang Kou
Genri Norman
George Athanasopoulos
Greg Watson
Gunther Stuer
Haewon Nam
Hai Jin
Hassan Diab
He Jing
Holger Bischof
Holly Dail
Hongbin Guo
Hongquan Zhu
Hong-Seok Lee
Hui Liu
Hyoung-Key Choi
Hyung-Min Lee
Hyunseung Choo
I.M. Navon
Igor Mokris
Igor Schagaev
Irina Schweigert
Irina Shoshmina
Isabelle Guérin-Lassous
Ivan Dimov
Ivana Budinska
J. Kroc
J.G. Verwer
Jacek Kitowski
Jack Dongarra
Jan Broeckhove
Jan Glasa
Jan Humble
Jean-Luc Falcone
Jean-Yves L'Excellent
Jemal Abawajy
Jens Gustedt
Jens Volkert
Jerzy Waśniewski
Jesus Vigo-Aguiar
Jianping Li
Jing He
Jinling Yang
John Copeland
John Michopoulos
Jonas Latt
Jongpil Jeong
Jose L. Bosque
Jose C. Cunha
Jose Alberto Fernandez
Josep Jorba Esteve
Jun Wu
Jürgen Jähnert
Katarzyna Rycerz
Kawther Rekabi
Ken Nguyen
Ken C.K. Tsang
K.N. Plataniotis
Krzysztof Boryczko
Krzysztof Grzda
Krzysztof Zieliński
Kurt Vanmechelen
Ladislav Hluchy
Laurence T. Yang
Laurent Lefevre
Laurent Philippe
Lean Yu
Leigh Little
Liang Cheng
Lihua Chen
Lijuan Zhu
Luis M. Portela
Luoding Zhu
M. Mat Deris
Maciej Malawski
Magda Sławińska
Marcin Paprzycki
Marcin Radecki
Marcin Smtek
Marco Aldinucci
Marek Gajcki
Maria S. Pérez
Marian Bubak
Marina Gavrilova
Marios Dikaiakos
Martin Polak
Martin Quinson
Massiomo Coppola
Mathilde Romberg
Mathura Gopalan
Matthew Sottile
Matthias Kawski
Matthias Müller
Mauro Iacono
Michał Malafiejski
Michael Gerndt
Michael Mascagni
Michael Navon
Michael Scarpa
Michele Benzi
Mikhail Zatevakhin
Miroslav Dobrucky
Mohammed Yousoof
Moonseong Kim
Moshe Sipper
Nageswara S. V. Rao
Narayana Jayaram
NianYan
Nicola Tonellotto
Nicolas Wicker
Nikolai Simonov
Nisar Hundewale
Osni Marques
Pang Ko
Paul Albuquerque
Paul Evangelista
Paul Gray
Paul Heinzlreiter
Paul Roe
Paula Fritzsche
Paulo Afonso Lopes
Pavel Tvrdik
Paweł Czarnul
Paweł Kaczmarek
Peggy Lindner
Peter Brezany
Peter Hellinckx
Peter Kacsuk
Peter Sloot
Peter H. Welch
Philip Chan
Phillip A. Laplante

Pierre Fraigniaud
Pilar Herrero
Piotr Bala
Piotr Wendykier
Piyush Maheshwari
Porfidio Hernandez
Praveen Madiraju
Putchong Uthayopas
Qiang-Sheng Hua
R. Vollmar
Rafał Wcisło
Rafik Ouared
Rainer Keller
Rajkumar Buyya
Rastislav Lukac
Renata Słota
Rene Kobler
Richard Mason
Richard Ramaroson
Rob H. Bisseling
Robert M. Panoff
Robert Schaefer
Robin Wolff
Rocco Aversa
Rod Blais
Roeland Merks
Roland Wismüller
Rolf Rabenseifner
Rolf Sander
Ron Perrott
Rosemary Renaut
Ryszard Tadeusiewicz
S. Lakshmivarahan
Saeid Belkasim
Salvatore Orlando
Salvatore Venticinque
Sam G. Lambrakos
Samira El Yacoubi
Sang-Hun Cho
Sarah M. Orley
Satoyuki Kawano
Savio Tse
Scott Emrich
Scott Lathrop
Seong-Moo Yoo
Serge Chaumette
Sergei Gorlatch
Seungchan Kim
Shahaan Ayyub
Shanyu Tang
Sibel Adali
Siegfried Benkner
Sridhar Radharkrishnan
Srinivas Aluru
Srinivas Vadrevu
Stefan Marconi
Stefania Bandini
Stefano Marrone
Stephan Olariu
Stephen Gilmore
Steve Chiu
Sudip K. Seal
Sung Y. Shin
Takashi Matsuhisa
Taneli Mielikäinen
Thilo Kielmann
Thomas Ludwig
Thomas Richter
Thomas Worsch
Tianfeng Chai
Timothy Jones
Tiziana Calamoneri
Todor Gurov
Tom Dhaene
Tomasz Gubała
Tomasz Szepieniec
Toni Cortes
Ulrich Brandt-Pollmann
V. Vshivkov
Vaidy Sunderam
Valentina Casola
V. Krzhizhanovskaya
Vassil Alexandrov
Victor Malyshkin
Viet D. Tran
Vladimir K. Popkov
V.V. Shakhov
Włodzimierz Funika
Wai-Kwong Wing
Wei Yin
Wenyuan Liao
Witold Alda
Witold Dzwinel
Wojtek Gościński
Wolfgang E. Nagel
Wouter Hendrickx
Xiaodong Zhang
Yannis Cotronis
Yi Peng
Yong Fang
Yong Shi
Yong Xue
Yumi Choi
Yunqing Huang
Yuriy Gorbachev
Zahari Zlatev
Zaid Zabanoot
Zhenjiang Hu
Zhiming Zhao
Zoltan Juhasz
Zsolt Nemeth

Workshops Organizers

High Performance Computing in Academia: Systems and Applications

Denis Donnelly – Siena College, USA
Ulrich Rüde – Universität Erlangen-Nürnberg

Tools for Program Development and Analysis in Computational Science

Dieter Kranzlmüller – GUP, Joh. Kepler University Linz, Austria
Arndt Bode – Technical University Munich, Germany
Jens Volkert – GUP, Joh. Kepler University Linz, Austria
Roland Wismüller – University of Siegen, Germany

Practical Aspects of High-Level Parallel Programming (PAPP)

Frédéric Loulergue – Université Paris Val de Marne, France

2005 International Workshop on Bioinformatics Research and Applications

Yi Pan – Georgia State University, USA
Alex Zelikovsky – Georgia State University, USA

Computer Graphics and Geometric Modeling, CGGM 2005

Andrés Iglesias – University of Cantabria, Spain

Computer Algebra Systems and Applications, CASA 2005

Andrés Iglesias – University of Cantabria, Spain
Akemi Galvez – University of Cantabria, Spain

Wireless and Mobile Systems

Hyunseung Choo – Sungkyunkwan University, Korea
Eui-Nam Huh Seoul – Womens University, Korea
Hyoung-Kee Choi – Sungkyunkwan University, Korea
Youngsong Mun – Soongsil University, Korea

Intelligent Agents in Computing Systems -The Agent Days 2005 in Atlanta

Krzysztof Cetnarowicz – Academy of Science and Technology AGH, Krakow, Poland
Robert Schaefer – Jagiellonian University, Krakow, Poland

Programming Grids and Metacomputing Systems - PGaMS2005

Maciej Malawski – Institute of Computer Science, Academy of Science and Technology AGH, Krakow, Poland
Gunther Stuer – Universiteit Antwerpen, Belgium

Autonomic Distributed Data and Storage Systems Management – ADSM2005

Jemal H. Abawajy – Deakin University, Australia
M. Mat Deris – College University Tun Hussein Onn, Malaysia

GeoComputation

Yong Xue – London Metropolitan University, UK

Computational Economics and Finance

Yong Shi – University of Nebraska, Omaha, USA
Xiaotie Deng – University of Nebraska, Omaha, USA
Shouyang Wang – University of Nebraska, Omaha, USA

Simulation of Multiphysics Multiscale Systems

Valeria Krzhizhanovskaya – University of Amsterdam, The Netherlands
Bastien Chopard – University of Geneva, Switzerland
Yuriy Gorbachev – Institute for High Performance Computing & Data Bases, Russia

Dynamic Data Driven Application Systems

Frederica Darema – National Science Foundation, USA

2nd International Workshop on Active and Programmable Grids Architectures and Components (APGAC2005)

Alex Galis – University College London, UK

Parallel Monte Carlo Algorithms for Diverse Applications in a Distributed Setting

Vassil Alexandrov – University of Reading, UK
Aneta Karaivanova – Institute for Parallel Processing, Bulgarian Academy of Sciences
Ivan Dimov – Institute for Parallel Processing, Bulgarian Academy of Sciences

Grid Computing Security and Resource Management

Maria Pérez – Universidad Politécnica de Madrid, Spain
Jemal Abawajy – Deakin University, Australia

Modelling of Complex Systems by Cellular Automata

Jiri Kroc – Helsinki School of Economics, Finland
S. El Yacoubi – University of Perpignan, France
M. Sipper – Ben-Gurion University, Israel
R. Vollmar – University of Karlsruhe, Germany

International Workshop on Computational Nanoscience and Technology

Jun Ni – The University of Iowa, USA
Shaoping Xiao – The University of Iowa, USA

New Computational Tools for Advancing Atmospheric and Oceanic Sciences

Adrian Sandu – Virginia Tech, USA

Collaborative and Cooperative Environments

Vassil Alexandrov – University of Reading, UK
Christoph Anthes – GUP, Joh. Kepler University Linz, Austria
David Roberts – University of Salford, UK
Dieter Kranzlmüller – GUP, Joh. Kepler University Linz, Austria
Jens Volkert – GUP, Joh. Kepler University Linz, Austria

Table of Contents – Part III

Workshop on "Simulation of Multiphysics Multiscale Systems"

Workshop on "Grid Computing Security and Resource Management"

2nd International Workshop on Active and Programmable Grids Architectures and Components (APGAC2005)

International Workshop on Computational Nano-Science and Technology

Workshop on "Collaborative and Cooperative Environments"

Workshop on "Autonomic Distributed Data and Storage Systems Management – ADSM2005"

Workshop on "GeoComputation"

Workshop on "Computational Economics and Finance"

Workshop on "Computer Algebra Systems and Applications, CASA 2005"

Workshop on "Intelligent Agents in Computing Systems" – The Agent Days 2005 in Atlanta

Workshop on "Parallel Monte Carlo Algorithms for Diverse Applications in a Distributed Setting"

Poster Session I

Poster Session II

Poster Session III

Table of Contents – Part I

Numerical Methods

Algorithms and Computational Kernels

Nonnumerical Algorithms

Parallel Algorithms

Environments and Libraries

Performance and Scalability

Programming Techniques

Networks and Distributed Algorithms

Parallel and Distributed Computing

Grid Computing

Failure Handling

Optimization

Modeling and Simulation

Image Analysis and Processing

Graphics and Visualization

Computation as a Scientific Paradigm

Hybrid Computational Methods

Complex Systems

Applications

Biomedical Applications

Data Mining and Computation

Education in Computational Science

Emerging Trends

Table of Contents – Part II

Workshop On "High Performance Computing in Academia: Systems and Applications"

Workshop on "Tools for Program Development and Analysis in Computational Science"

Workshop on "Computer Graphics and Geometric Modeling, CGGM 2005"

Workshop on "Modelling of Complex Systems by Cellular Automata"

Workshop on "Wireless and Mobile Systems"

Workshop on "Dynamic Data Driven Application Systems"

Workshop on "Practical Aspects of High-Level Parallel Programming (PAPP)"

Workshop on "New Computational Tools for Advancing Atmospheric and Oceanic Sciences"

2005 International Workshop on Bioinformatics Research and Applications

Workshop on "Programming Grids and Metacomputing Systems – PGaMS2005"

Multiscale Finite Element Modeling of the Coupled Nonlinear Dynamics of Magnetostrictive Composite Thin Film

Debiprosad Roy Mahapatra, Debi Prasad Ghosh, and Gopalakrishnan Srinivasan

ARDB Center for Composite Structures, Department of Aerospace Engineering, Indian Institute of Science, Bangalore 560012, India

Abstract. A multiscale nonlinear finite element model for analysis and design of the deformation gradient and the magnetic field distribution in Terfenol-D/epoxy thin film device under Transverse Magnetic (TM) mode of operation is developed in this work. A phenomenological constitutive model based on the density of domain switching (DDS) of an ellipsoidal inclusion in unit cell of matrix is implemented. A sub-grid scale homogenization technique is employed to upwind the microstructural information. A general procedure to ensure the solution convergence toward an approximate inertial manifold is reported.

1 Introduction

In recent time, multiscale modeling and computation have become powerful tool in solving many problems related to strongly coupled dynamics and multiphysics phenomena in material science and engineering [1, 2]. Such an attempt, with judicious choice of the mathematical models which represent the mutually inseparable physical processes at the molecular scale, the lattice scale, the microstructural scale, and at the continuum scale, can be highly reliable in interpreting the experimental results and also in designing new materials and devices with desired multifunctionality. Many similar problems related to fluidics, transport process, biology and planetary sciences are of immense interest in the research community, which requires the advancement of multiscale computational methodology.

The magnetostrictive materials (a family of rare earth compounds), due to large twinning of their magnetic domains, are ideal candidates for high powered microactuators, tunable microwave devices, shape memory devices and bio-inspired systems. Among the majority of magnetostrictive materials, Tb-Dy-Fe type compounds (industrially known as Terfenol-D) are common and show much reduced macroscopic anisotropy [3]. Promising engineering applications are possible using deposited nanostructured film [4, 5], multi-layers [3] and particulate composite [6]. In magnetostrictive polymer thin film, the magnetic domains are essentially constrained and hence behave differently than their bulk samples. Improving the magnetostrictive performance (e.g. larger magnetostriction, smaller hysteresis, wider range of linearity etc.) by sensitizing the dynamic twinning of

V.S. Sunderam et al. (Eds.): ICCS 2005, LNCS 3516, pp. 1–8, 2005.

these magnetic domains under external field is a complex design task which is far from being straightforward using experiments alone and requires detail physical model and intensive computation. Because of the coupled and nonlinear nature of the dynamics, several computational difficulties exist which require special attention.

Ever since the rationale behind the variationally consistent multiscale finite element method was brought out by Hughes *et al.* [7], a tremendous amount of research has been channelized in that path. Among several approaches to tackle the evolvingly complex multiphysics computational mechanics problems, very few approaches, namely the homogenization method [8, 9], the multi-level hierarchical finite element method [10], the level-set method [11] and few others appear promising. In context of time-dependent problems, the concept of consistent temporal integration [12] is worth noting, which extend itself beyond the classical notion of upwinding. Subsequently, while attempting a strongly nonlinear and coupled problem, the advantage in adopting a low-dimensional manifold based on appropriate error estimate during iteration [2, 13] may be noted. In the present paper we report a new variationally consistent multiscale finite model with the following two features: (1) a sub-grid scale for upwinding the miscrostructural information and (2) an approximate manifold of the original system that retains the desired accuracy in the finite element solution, which is unlike the usual ordering scheme as followed in the conventional asymptotic method of truncation.

Our present problem can be described as the magnetostriction induced transformation process in solid state called magnetic domain switching, which occurs in the microscopic scale. In the macroscopic scale, the dynamics is governed through the coupling between elasticity and electromagnetics. The complexity in mathematical modeling is four-fold: (1) describing an accurate constitutive model using phenomenological framework. Here we use a model based on the density of domain switching (DDS) [14], where the series expansion of the Gibbs free energy functional for cubic non-polar crystalline Laves phase is used along with peak piezomagnetic coefficient on the compressive part of the stress acting along the resultant magnetic field vector (2) including the effect of the volume fraction of the magnetostrictive phase in the dilute composite and microstructural pattern (3) retaining only the essential properties of the small scales and nonlinearity and (4) constructing a variationally consistent finite element model and solving it iteratively.

2 Constitutive Model Based on Density of Domain Switching (DDS)

We consider an unit cell of the two phase composite as shown in Fig. 1, in which the ellipsoidal inclusion (B) in matrix phase (A) is the aggregate of magnetostrictive domains. The effective orientation of the magnetic field in these domains is assumed to be at an angle θ which is also the major axis of the ellipsoid. We

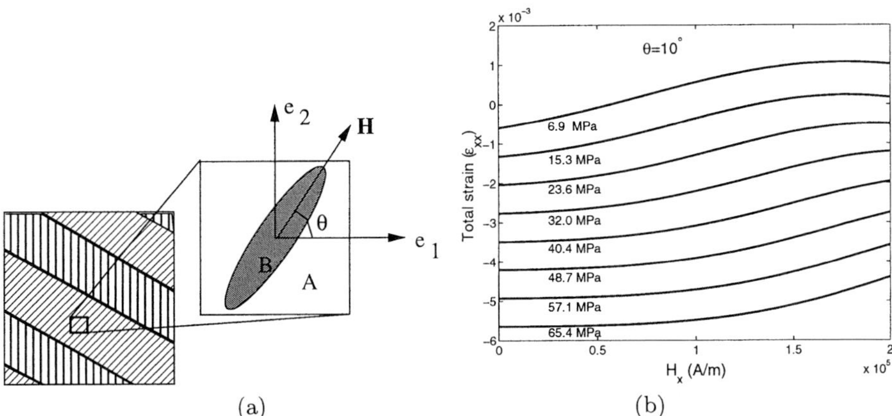

Fig. 1. (a) Unit cell with oriented magnetostrictive phase B in epoxy matrix phase A (b) strain vs. applied magnetic field in the unit cell with $\epsilon = 0.5$, $\theta = 10^o$ under varying stress

consider an effectively one-dimensional constitutive model of A along the major axis of the ellipsoid. Also, this major axis is assumed to be aligned with the resultant magnetic field $\boldsymbol{H}$ under equilibrium. Using the DDS model [14] it is possible to capture the effect of large magnetostriction more effectively than the other models. According to the DDS model, the piezomagnetic coefficient for the phase B can be written as

$$d = \left[\frac{\partial \varepsilon}{\partial H}\right]_\sigma = \tilde{d} e^{-(z-1)^2/A} , \tag{1}$$

where the peak piezomagnetic coefficient $\tilde{d} = \tilde{d}_{cr} + a\Delta\sigma + b(\Delta\sigma)^2$, overstress $\Delta\sigma = \sigma - \sigma_{cr}$, $z = |H|/\tilde{H}$, $\tilde{H} = \tilde{H}_{cr} + \zeta\Delta\sigma$, $A = \sigma_{cr}/\sigma$, σ is the stress, ε is the strain and σ_{cr} is the critical inherent stress of domain switching. a, b, $\tilde{d}_{cr}$, $\tilde{H}_{cr}$ and ζ are material constants obtained by relating the Gibbs free energy and fitting experimental data as discussed in [14]. Let $\epsilon \in (0, 1]$ be the volume fraction of phase B. Applying the mixture rule of rank-one laminate structure under the applied stress components $\{\sigma_{11}, \sigma_{22}, \sigma_{12}\}$ and magnetic field components $\{H_1, H_2\}$ in the unit cell, the constitutive model is obtained, which is expressed in the quasi-linearized form

$$\sigma = \bar{Q}_{AB}\varepsilon - \bar{e}_B H , \quad B = \bar{\mu}_{AB} H + \bar{e}'_B \varepsilon , \tag{2}$$

where ε is the longitudinal strain along the axis of the ellipsoid, B is the magnetic flux density in the film plane, $\bar{Q}_{AB}$ is the effective tangent stiffness, $\bar{e}_B$ and $\bar{e}'_B$ are the tangen coefficients of magnetostriction and $\bar{\mu}_{AB}$ is the effective permeability.

We adopt a quasi-linearization process during time stepping to track the evolution of the morphology in the composite film. Here we first write the coefficients in the constitutive model in Eq. (2) at time $t = t_i$ in terms of the state obtained from the previous time step at $t = t_{i-1} = t_i - \Delta t$ as $\bar{e}^i_B = \bar{e}_B(\epsilon, \sigma^{i-1}, H^{i-1})$,

$\mu^i_{AB} = \mu_{AB}(\epsilon, \sigma^{i-1}, H^{i-1})$. In the subsequent formulation we use the notations: $\boldsymbol{A}.\boldsymbol{B} = A_{ij}B_{jk}$, $\boldsymbol{A} : \boldsymbol{B} = A_{ij}B_{ji}$ as the contraction of tensors over one and two indices, respectively. $|\boldsymbol{A}| := (\boldsymbol{A} : \boldsymbol{A})^{1/2}$. We then rewrite the constitutive model in the coordinate system of the cell (e_1, e_2) as function of the effective angle of orientation $\theta = \theta^i$ of the ellipsoid as

$$\boldsymbol{\sigma} = \boldsymbol{\Gamma}_\varepsilon : \bar{\boldsymbol{Q}}_{AB} : \boldsymbol{\Gamma}_\varepsilon : \boldsymbol{\varepsilon} - \boldsymbol{\Gamma}_\varepsilon : \bar{\boldsymbol{e}}^i_B : \boldsymbol{\Gamma}_H.\boldsymbol{H} =: \bar{\boldsymbol{Q}}^i : \boldsymbol{\varepsilon} - \bar{\boldsymbol{e}}^i.\boldsymbol{H} \;, \tag{3}$$

$$\boldsymbol{B} = \boldsymbol{\Gamma}_H : \bar{\boldsymbol{\mu}}_{AB} : \boldsymbol{\Gamma}_H + \boldsymbol{\Gamma}_H : \bar{\boldsymbol{Q}}_B^{-1} : \boldsymbol{\Gamma}_\varepsilon : \boldsymbol{\sigma} =: \bar{\boldsymbol{e}}^{i^T} : \boldsymbol{\varepsilon} + \bar{\boldsymbol{\mu}}^i.\boldsymbol{H} \;, \tag{4}$$

where $\boldsymbol{\Gamma}_\varepsilon$ is the transformation tensor for strain and stress, $\boldsymbol{\Gamma}_H$ is the transformation tensor for magnetic field vector. Because of the domain switching, the Euler angle θ^i varies over time, which we intend to compute as $\theta^i = \tan^{-1}\left(H_y^{i-1}/H_x^{i-1}\right)$ where $(x, y) \parallel (e_1, e_2)$ and (x, y, z) is the global coordinate system of the composite film.

The electrical source behind the magnetostrictive effect is assumed to be due to the transverse magnetic (TM_z) mode of excitation through array of electrodes parallel to the film plane. We exclude the effect of TE and TEM modes of electromagnetic excitation resulting from any anisotropy in the composite film. Hence $\boldsymbol{H} = \{H_x(x, y, t) \; H_y(x, y, t) \; 0\}^T$ with pointwise prescribed electrical loading $(E_z(z, y, t))$. These should satisfy one of the Maxwell equation for magnetoelectric surface $\nabla \times \boldsymbol{E} = -\dot{\boldsymbol{B}}$, i.e.,

$$\frac{\partial E_z}{\partial x} = -\dot{B}_y \;, \quad \frac{\partial E_z}{\partial y} = \dot{B}_x \;. \tag{5}$$

The deformation can be described by the in-plane displacement and strain

$$\boldsymbol{u} = \{u(x, y, t) \;, v(x, y, t) \;, 0\}^T \;, \quad \boldsymbol{\varepsilon} = \frac{1}{2}(u_{i,j} + u_{j,i}) \;. \tag{6}$$

3 Multiscale Model and Approximate Inertial Manifold

It is important to note that the magnetostriction due to domain switching and the resulting nonlinearity vanishes as $\epsilon \to 0$. In such case, the quasi-linearization of the constitutive law in Eq. (3) results in a linear system whose eigen values are close to that of the slow manifold and a global attractor can be established. Therefore, for $\epsilon \to 0$, it is straightforward to obtain a fairly accurate solution based on center manifold reduction by asymptotic expansion and truncation according to the order of the terms in the PDEs. But in the present case, our objective is to study the film dynamics due to large variation in ϵ spatially or for individual designs with different constituents and manufacturing processes. Therefore, by following the known facts from the reported literature (see [15] and the references therein), we focus our attention on constructing an approximate inertial manifold with global convergence properties in a two-scale finite element framework.

We introduce a slow scale (L_0) and a fast scale (L_1) with the dependent variables defined in them respectively as $(.)_0$ and $(.)_1$, such that

$$\boldsymbol{x}_1 = \boldsymbol{x}/\epsilon\ , \quad t_1 = t/\epsilon\ , \quad \boldsymbol{u} = \boldsymbol{u}_0 + \epsilon \boldsymbol{u}_1\ , \quad \boldsymbol{H} = \boldsymbol{H}_0 + \epsilon \boldsymbol{H}_1 \tag{7}$$

The dynamics of the thin film can now be described using the momentum conservation law

$$\left(\nabla_0 + \frac{1}{\epsilon}\nabla_1\right).\boldsymbol{\sigma} = \bar{\rho}\frac{\partial^2 \boldsymbol{u}_0}{\partial t^2} + \frac{1}{\epsilon}\bar{\rho}\frac{\partial^2 \boldsymbol{u}_1}{\partial t_1^2}\ , \tag{8}$$

and the source free Maxwell's equation

$$\left(\nabla_0 + \frac{1}{\epsilon}\nabla_1\right).\boldsymbol{B} = 0\ . \tag{9}$$

In order to introduce the effect of texture of the composite, we perform homogenization. Simplifying Eq. (8) with the help of Eqs. (3) and (5) and homogenizing over a sub-grid S with nodes $j = 1, \cdots, n$, and a prescribed texture

$$\theta^i(x_s, y_s) = \sum_{j=1}^{n} \psi_{\theta j}(x_s, y_s)\theta^i_j\ , \quad \epsilon = \sum_{j=1}^{n} \psi_{\epsilon j}(x_s, y_s)\epsilon^i_j\ , \tag{10}$$

we get

$$\frac{1}{\Omega_S}\int_{\Omega_S}\left[\bar{\boldsymbol{Q}}^i : \nabla_0.\boldsymbol{\varepsilon} - \nabla_0.\bar{\boldsymbol{e}}^i.\boldsymbol{H} - \bar{\boldsymbol{e}}^i : \nabla_0.\boldsymbol{H} + \frac{1}{\epsilon}\nabla_1.\bar{\boldsymbol{Q}}^i : \boldsymbol{\varepsilon} + \frac{1}{\epsilon}\bar{\boldsymbol{Q}}^i : \nabla_1.\boldsymbol{\varepsilon}\right.$$
$$\left. - \frac{1}{\epsilon}\nabla_1.\bar{\boldsymbol{e}}^i.\boldsymbol{H} - \frac{1}{\epsilon}\bar{\boldsymbol{e}}^i : \nabla_1.\boldsymbol{H}\right] d\Omega_S = \frac{1}{\Omega_S}\int_{\Omega_S}\left[\bar{\rho}\frac{\partial^2 \boldsymbol{u}_0}{\partial t^2} + \frac{1}{\epsilon}\bar{\rho}\frac{\partial^2 \boldsymbol{u}_1}{\partial t_1^2}\right] d\Omega_S \tag{11}$$

Similarly, homogenization of Eq. (9) gives

$$\frac{1}{\Omega_S}\int_{\Omega_S}\left[\nabla_0.\bar{\boldsymbol{e}}^{iT} : \boldsymbol{\varepsilon}_0 + \bar{\boldsymbol{e}}^{iT} : \nabla_0.\boldsymbol{\varepsilon}_0 + \frac{1}{\epsilon}\nabla_1.\bar{\boldsymbol{e}}^{iT} : \boldsymbol{\varepsilon}_0 + \nabla_1.\bar{\boldsymbol{e}}^{iT} : \boldsymbol{\varepsilon}_1 + \frac{1}{\epsilon}\bar{\boldsymbol{e}}^{iT} : \nabla_1.\boldsymbol{\varepsilon}_1\right.$$
$$\left. + \nabla_0.\bar{\boldsymbol{\mu}}^i : \boldsymbol{H}_0 + \bar{\boldsymbol{\mu}}^i : \nabla_0.\boldsymbol{H}_0 + \frac{1}{\epsilon}\nabla_1.\bar{\boldsymbol{\mu}}^i : \boldsymbol{H}_0 + \nabla_1.\bar{\boldsymbol{\mu}}^i : \boldsymbol{H}_1 + \frac{1}{\epsilon}\bar{\boldsymbol{\mu}}^i : \nabla_1.\boldsymbol{H}_1\right] d\Omega_S$$
$$= 0 \tag{12}$$

The property of the texture in Eq. (9) is given in terms of the known distribution function ψ_θ and ψ_ϵ. While evaluating Eqs. (11)-(12) we need projection of the nodal variables on the sub-grid nodes. This is done by a least-square fit over the finite element nodal variables $\boldsymbol{u}^{e(i-1)}$, $\boldsymbol{H}^{e(i-1)}$ at time $t = t_{i-1}$ for the elements overlapping with S and by writing

$$\boldsymbol{u}_k = \sum_{j=1}^{n} \psi_{uj}(x_k, y_k, t_{i-1})\boldsymbol{u}^e_j\ , \quad \boldsymbol{\varepsilon}_k = \sum_{j=1}^{n} \psi_{\varepsilon j}(x_k, y_k)\boldsymbol{\varepsilon}^e_j\ ,$$

$$\boldsymbol{H}_k = \sum_{j=1}^{n} \psi_{Hj}(x_k, y_k, t_{i-1})\boldsymbol{H}^e_j\ , \tag{13}$$

where k is an integration point, superscript e indicates finite element nodal quantities. ψ^e_{uj}, $\psi^e_{\varepsilon j}$ and ψ^e_{Hj} are the functions obtained from the least square fit based on the nodal quantities at previous converged time step $t = t_{i-1}$.

3.1 Finite Element Formulation

So far we have just assumed that the values of the finite element nodal variables $\boldsymbol{q}^e = \{\boldsymbol{u}^{eT}\ \boldsymbol{H}^{eT}\}^T$ from the previous time step $t = t_{i-1}$ are known. In order to solve for the variables at $t = t_i$ the finite element equilibrium is now required. First we rewrite Eqs. (11)-(12) as respectively

$$\mathcal{L}_1(\partial_t, \partial_{t1})\boldsymbol{g}(\boldsymbol{u}_0, \boldsymbol{u}_1) - \mathcal{L}_2(\nabla_0, \nabla_1)\boldsymbol{p}(\boldsymbol{\varepsilon}_0, \boldsymbol{\varepsilon}_1, \boldsymbol{H}_0, \boldsymbol{H}_1) = 0\ , \tag{14}$$

$$\mathcal{L}_3(\nabla_0, \nabla_1)\boldsymbol{g}(\boldsymbol{\varepsilon}_0, \boldsymbol{\varepsilon}_1, \boldsymbol{H}_0, \boldsymbol{H}_1) = 0\ , \tag{15}$$

The final step to obtain our two-scale finite element model is the variational minimization of the weak form, which can be written in the form

$$\delta \int_\Omega \boldsymbol{u}.\left[\mathcal{L}_1\boldsymbol{g} - \mathcal{L}_2\boldsymbol{p}\right] d\Omega + \delta \int_\Omega \boldsymbol{H}.\mathcal{L}_3\boldsymbol{g} d\Omega = 0\ , \tag{16}$$

which leads to the global finite element equilibrium equation

$$\boldsymbol{M}_0\ddot{\boldsymbol{q}}_0^e + \boldsymbol{M}_1\ddot{\boldsymbol{q}}_1^e + \boldsymbol{K}_0\boldsymbol{q}_0^e + \boldsymbol{K}_1\boldsymbol{q}_1^e = \boldsymbol{f}\ . \tag{17}$$

Eq. (17) is further condensed out by eliminating the magnetic field vector $\boldsymbol{H}^e$ by using the pointwise magnetic excitation condition in Eq. (5) rewritten as

$$\left\{ \begin{matrix} H_x \\ H_y \end{matrix} \right\} = (\bar{\boldsymbol{\mu}}^i)^{-1} : \left\{ \begin{matrix} \displaystyle\int_{t_{i-1}}^{t_i} \frac{\partial E_z}{\partial y} dt \\ \displaystyle -\int_{t_{i-1}}^{t_i} \frac{\partial E_z}{\partial x} dt \end{matrix} \right\} - (\bar{\boldsymbol{\mu}}^i)^{-1} : \bar{\boldsymbol{e}}^{iT} : \boldsymbol{\varepsilon}\ . \tag{18}$$

In the present problem when $\epsilon \to 0$, the dynamics is that of semilinear elliptic PDEs in $(\boldsymbol{u}, \boldsymbol{H})$, hence usual bi-linear element with mesh density comparable with the wavenumber of the magnetic excitation is sufficient to produce accurate result. However, such properties no longer hold for $\epsilon \to 1$. Also, $\epsilon\nabla$ is not bounded. A suitable approach in related context of obtaining an approximate manifold (h) is to fix the spatial discretization at a given time step and obtain the approximate solution

$$\boldsymbol{u} = \boldsymbol{u}_0 + \epsilon\boldsymbol{u}_1 = \boldsymbol{u}_0 + \epsilon\boldsymbol{h}(\boldsymbol{u}_0)\ , \tag{19}$$

through iteration. In order to derive the condition to iterate over $\boldsymbol{h}$ while solving Eq. (17), we take help of *apriori* error estimates. Since such an error estimate should ideally be independent of the choice of the scales, i.e. irrespective of the fact that the system is slow (or fast) in that scale [16], we first set $\boldsymbol{u}_0 = \boldsymbol{0}$, $\boldsymbol{H}_0 = \boldsymbol{0}$. Neglecting the inertia terms in the strong form corresponding to Eqs. (11)-(12) and rearranging, we get

$$\nabla_1.\boldsymbol{\varepsilon}_1 = \boldsymbol{R}_1^i : \boldsymbol{\varepsilon} + \boldsymbol{R}_2^i : \boldsymbol{H}_1\ , \quad \nabla_1.\boldsymbol{H}_1 = \boldsymbol{R}_3^i : \boldsymbol{\varepsilon} + \boldsymbol{R}_4^i : \boldsymbol{H}\ , \tag{20}$$

where $\boldsymbol{R}_1^i$, $\boldsymbol{R}_2^i$, $\boldsymbol{R}_3^i$ and $\boldsymbol{R}_4^i$ are functions of $(\epsilon, \theta^i, \boldsymbol{\sigma}^{i-1}, \boldsymbol{H}^{i-1})$ at time $t = t_{i-1}$. We consider a strong estimate

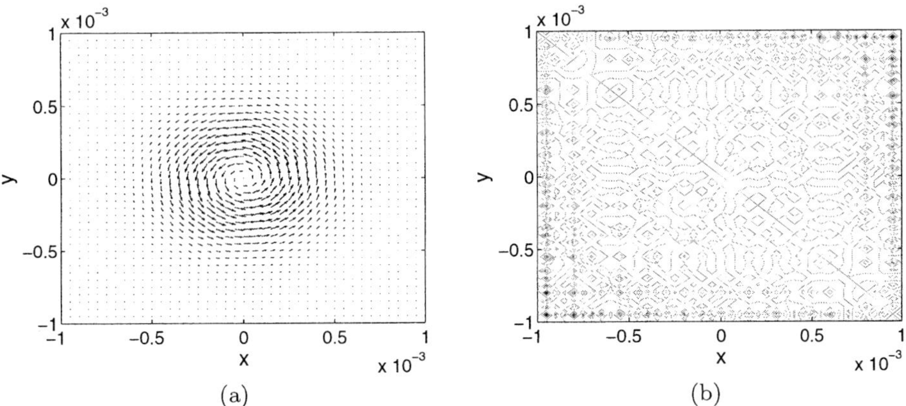

Fig. 2. (a) Orientation of $\boldsymbol{H}$ field vectors and (b) shear strain contour under normally applied short pulse $E_z(t)$ and captured at $t = 50ns$

$$\left(\frac{\partial}{\partial t_1} + \epsilon \nabla_1\right)\left(|\varepsilon_1|^2 + |\boldsymbol{H}_1|^2\right) = 2\varepsilon_1 : \frac{\partial \varepsilon_1}{\partial t_1} + 2\epsilon\varepsilon_1 : \nabla_1.\varepsilon_1$$
$$+2\boldsymbol{H}_1 : \frac{\partial \boldsymbol{H}_1}{\partial t_1} + 2\epsilon \boldsymbol{H}_1 : \nabla_1.\boldsymbol{H}_1 \tag{21}$$

Using Eq. (20) in Eq. (21) and applying the elementary inequality $uv \leq \beta u^2 + v^2/\beta$, finally we get

$$\left(\frac{\partial}{\partial t_1} + \epsilon \nabla_1\right)\left(|\varepsilon_1|^2 + |\boldsymbol{H}_1|^2\right) \leq -2\gamma_1^i|\varepsilon_2|^2 - 2\gamma_2^i|\boldsymbol{H}_1|^2 - 2\gamma_3^i|\dot{\varepsilon}_1|^2 - 2\gamma_4^i|\dot{\boldsymbol{H}}_1|^2 , \tag{22}$$

where $\dot{()} := \partial/\partial t_1$, γ_1^i, γ_2^i, γ_3^i γ_4^i are functions of $(\epsilon, \theta^i, \boldsymbol{\sigma}^{i-1}, \boldsymbol{H}^{i-1})$. Integrating Eq. (22) along the characteristic line $\partial/\partial t_1 + \epsilon\nabla_1 =$ constant, between the interval (t_{i-1}, t_i), and by taking sup on $(x, y) \in S^1$, we have

$$||\varepsilon_1||^2 + ||\boldsymbol{H}_1||^2 \leq e^{-\gamma_1^i \Delta t}||\varepsilon_1^{i-1}||^2 + e^{-\gamma_2^i \Delta t}||\boldsymbol{H}_1^{i-1}||^2$$
$$+e^{-\gamma_3^i \Delta t}||\dot{\varepsilon}_1^{i-1}||^2 + e^{-\gamma_4^i \Delta t}||\dot{\boldsymbol{H}}_1^{i-1}||^2 . \tag{23}$$

We use Eq. (23) and Newmark's finite difference scheme in time to solve Eq. (17) concurrently in $(\boldsymbol{q}_0^e, \boldsymbol{q}_1^e)$. Fig. 2 shows the snap of the evolved field patterns simulated in case of $Tb_{0.27}Dy_{0.73}Fe_{1.95}$-epoxy(30% by vol.) film under nanosecond electrical pulse applied at the center.

4 Concluding Remarks

Based on a phenomenological microscopic model of magnetostrictive polymer composite material, and macroscopic dynamics, a multiscale finite element model with subgrid-scale homogenization is formulated. Error estimates are derived. Numerical simulation of the time-resolved field pattern in Terfenol-D polymer is reported.

References

1. Ghoniem, N.M., Busso, P., Kioussis, N. and Huang, H.: Multiscale modelling of nanomechanics and micromechanics: an overview. Phil. Mag. **83**(31-34) (2003) 3475–3528
2. Melnik, R.V.N. and Roberts, A.H.: Computational models for multi-scale coupled dynamic problems. Future Generation Computer Systems **20** (2004) 453–464
3. Quandt, E., Ludwig, A., Mencik, J. and Nold E.: Giant magnetostrictive TbFe/Fe multilayers. J. Alloys Compounds **258** (1997) 133–137
4. Kumar, D., Narayan, J., Nath, T.K., Sharma, A.K., Kvit, A. and Jin, C.: Tunable magnetic properties of metal ceramic composite thin film. Solid State Communications **119** (2001) 63–66
5. Liu, T., Burger, C. and Chu B.: Nanofabrication in polymer matrices. Prog. Polym. Sci. **2003** 5–26
6. Hommema, J. A.: Magnetomechanical behavior of Terfenol-D particulate composites. MS Thesis, University of Illinois at Urbana-Champaign (1999)
7. Hughes, T.J.R., Feijoo, G.R., Mazzei, L., and Quincy, L.B.: The variational multiscale method – a paradigm for computational mechanics. Computer Methods in Applied Mechanics and Engineering **166** (1998) 3–24
8. Babuska, I.: Homogenization approach in engineering, in: Lions, R., Glowinski (Eds.), Computing Methods in Applied Sciences and Engineering, Lecture Notes in Economics and Mathematical Systems **134** (1976) Springer, Berlin
9. Terada, K. and Kikuchi, N.: A class of general algorithms for multiscale analysis of heterogeneous media. Comput. Methods Appl. Nech. Engrg. **190** (2001) 5427–5464
10. Calgero, C., Laminie, J. and Temam, R.: Dynamical multilevel schemes for the solution of evolution equations by hierarchical finite element discretization. Appl. Numer. Math. **23** (1997) 403–442
11. Chessa, J. and Belytschko, T.: Arbitrary discontinuities in space-time finite elements by level sets and X-FEM. Int. J. Numer. Meth. Engng. **61** (2004) 2595–2614
12. Bottasso, C.L.: Multiscale temporal integration. Comput. Methods Appl. Mech. Engrg. **191** (2002) 2815–2830
13. Margolin, L.G., Titi, E.S. and Wynne, S.: The postprocessing Galerkin and nonlinear galerkin methods - A truncation analysis point of view. SIAM J. Numer. Anal. **41** (2003) 695–714
14. Wan, Y., Fang, D., Hwang, K.-C.: Non-linear constitutive relations for magnetostrictive materials. Int. J. Non-linear Mechanics **38** (2003) 1053–1065
15. Steindl, A. and Troger, H.: Methods for dimension reduction and their application in nonlinear dynamics. Int. J. Solids Struct. **38** (2001) 2131–2147
16. Menon, G. and Haller, G.: Infinite dimensional geometric singular perturbation theory for the Maxwell-Bloch equations. SIAM J. Math. Anal. **33** (2001) 315–346

Large-Scale Fluctuations of Pressure in Fluid Flow Through Porous Medium with Multiscale Log-Stable Permeability

Olga Soboleva

Institute of Computational Mathematics and Mathematical Geophysics, Novosibirsk
630090, pr. Lavrentieva 6, Russia
olga@nmsf.sscc.ru

Abstract. In this paper, we consider subgrid modeling of a filtration flow of a fluid in a nonhomogeneous porous medium. An expression for the effective permeability coefficient for the large-scale component of the flow is derived. The permeability coefficient possesses a log-stable distribution. The obtained formulas are verified by the numerical modeling.

1 Introduction

Field and laboratory observational data imply that such parameters as permeability field and porosity have rather an irregular varying character. In this case small-scale details of permeability and porosity are unknown. The latter are taken into consideration in the statistical models [1],[2], [3],[5], using effective coefficients. In [2], [3], based on the ideas of the renormalized group (RG) by Wilson [4], the subgrid formulas for the effective permeability coefficient were derived, and the diffusion of the interface of liquids in their combined current in a scale-invariant porous medium for log-normal permeability distributions was studied. The Landau-Lifshits formula for the effective permeability within a strict field RG was calculated in Teodorovich [5]. In particular, there are mentioned arguments from monograph [6], according to which the RG methods partially take into account higher orders of the perturbation theory and are to improve the accuracy of the formulas to be obtained. The same arguments are also applicable to the subgrid modeling. If a medium is assumed to satisfy the improved similarity hypothesis by Kolmogorov [2], [7] the RG equations take a very simple form. In this paper, the Wilson RG ideas are used for deriving the subgrid modeling formulas when solving problems of filtration in a multiscale porous medium with a log-stable distribution of permeability. The differential equations for obtaining effective constants were also derived for the media that do not satisfy the improved similarity hypothesis. The formulas obtained are verified by the numerical modeling.

2 Statement of Problem

For small Reynolds' numbers, the filtration velocity v and the pressure are connected by the Darcy law $\mathbf{v} = -\varepsilon(\mathbf{x})\nabla p$, where $\varepsilon(\mathbf{x})$ is a random coordinate

V.S. Sunderam et al. (Eds.): ICCS 2005, LNCS 3516, pp. 9–16, 2005.

function - the permeability coefficient. An incompressible liquid flows through the medium. The incompressibility condition brings about the equation

$$\nabla\left[\varepsilon\left(\mathbf{x}\right)\nabla p\left(\mathbf{x}\right)\right]=0. \tag{1}$$

Let a permeability field be known. This means that at each point $\mathbf{x}$, the permeability is measured by pushing the liquid through a pattern of a very small size l_0 A random function of spatial coordinates $\varepsilon(\mathbf{x})$ is regarded as permeability limit at $l_0 \to 0$, $\varepsilon\left(\mathbf{x}\right)_{l_0} \to \varepsilon\left(\mathbf{x}\right)$. In order to turn to a coarser grid l_1, it is impossible just to smooth the obtained field $\varepsilon(\mathbf{x})_{l_0}$ in the scale $l_1 > l_0$. The field obtained is not a true permeability, which describes filtration in the domain of scales (l_1, L), where L is the largest scale. In order that permeability be defined on a coarser grid, measurements should be taken again by pushing the liquid through larger samples of the size l_1. The necessity of fulfilling this procedure is due to the fact that the permeability fluctuations from the limit (l_0, l_1) have correlations with the pressure fluctuations induced by them. Similar to Kolmogorov [7], we consider the dimensionless field ψ equal to the relation of permeabilities smoothed in two different scales (l, l_1). This approach is described in detail in [2]. Let us denote by $\widetilde{\varepsilon}(\mathbf{x})_l$ the smoothed in scale l the permeability $\varepsilon\left(\mathbf{x}\right)_{l_0}$, then $\psi(\mathbf{x}, l, l_1) = \widetilde{\varepsilon}(\mathbf{x})_{l_1}/\widetilde{\varepsilon}(\mathbf{x})_l$. When $l_1 \to l$, the field $\varphi\left(\mathbf{x}, l\right) = \frac{d\psi(\mathbf{x}, l, l\lambda)}{d\lambda}|_{\lambda=1}$ is obtained that defines all statistical attributes of a porous medium. The relation obtained represents a differential equation, whose solution yields the permeability as function of the field φ

$$\varepsilon\left(\mathbf{x}\right)_{l_0}=\varepsilon_0\exp\left[-\int_{l_0}^{L}\varphi(\mathbf{x},l)\frac{dl}{l}\right]. \tag{2}$$

The permeability is assumed to have inhomogeneities of the scales l from the interval $l_0 < l < L$, and the field φ is isotropic and homogeneous. The fields $\varphi(\mathbf{x}, l)$, $\varphi(\mathbf{y}, l')$, with different scales for any $\mathbf{x}$, $\mathbf{y}$ are statistically independent. This hypothesis is generally assumed to be correct for different models and reflects the fact that the statistical dependence attenuates when fluctuations of scales differ in value [7]. The scale properties of a model are defined from the field $\varphi(\mathbf{x}, l)$. For the scale-invariant models, the condition $\varphi(\mathbf{x}, l) \to \varphi(K\mathbf{x}, Kl)$. should be fulfilled. According to the theorem of sums of independent random fields [8] if the variance $\varphi(\mathbf{x}, l)$ is finite, then for large l. Integral in (2) tends to the field with a normal distribution. However, if the variance is infinite and there exists a non-degenerate, limiting distribution of a sum of random values, then such a distribution will be stable. For simplicity, the field $\varphi(\mathbf{x}, l)$ is assumed to have a stable distribution. With $L/l_0 \gg 1$, it is impossible to calculate the pressure from equation (1) or if it is, then it demands large computer costs. Therefore, we pose the problem to obtain effective coefficients in equations for the large-scale filtration components. For the subgrid modeling we use ideas of the Wilson RG.

3 A Log-Stable Permeability

The growth of irregularity, chaotic state and intermittence in the behavior of physical fields with increasing the scale of measuring made many researchers to reject the log-normal model and consider a general case of log-stable distributions. In [9], distributions of permeability fields and some statistical characteristics were obtained using the experimental borehole data. Also, it was shown that the permeability fields can have log-stable distributions. Stable distributions depend on the four parameters α, β, μ, σ, [10]. The parameter α is within the domain $0 < \alpha \leq 2$, where $\alpha = 2$ corresponds to the Gauss distribution. Statistical moments of order m for $m \geq \alpha$ do not exist except for the case $\alpha = 2$, for which all the statistical moments are determined. Thus, the variance is infinite for $\alpha < 2$, and the mean value - for $\alpha < 1$. For modeling the field $\varphi(\mathbf{x}, l)$, having a stable distribution law, the approach from [11] is used. At the points $(\mathbf{x}_j, l)$, the field φ is defined by the sum of random independent values having stable distributions with the same parameters α, β, $\mu = 0$, $\sigma = 1$ (the form A [10])):

$$\varphi(\mathbf{x}_j, l) = \left(\frac{\Phi_0(l)}{2(\delta\tau ln2)^{\alpha-1}}\right)^{\frac{1}{\alpha}} a^l_{\mathbf{ji}}\zeta^l_{\mathbf{i}} + \varphi_0(l), \tag{3}$$

where $l = 2^\tau$, $\delta\tau$ - is a discretization step, the coefficients $a^l_{\mathbf{ji}}$ have a support of size l^3, depend only on the module of difference of indices (in the sequel the index j can be omitted) $a^l_{\mathbf{ji}} \equiv a^l\left(|\mathbf{i}-\mathbf{j}|\right)$, and for all l the condition $\sum_{k_x}\sum_{k_y}\sum_{k_z}\left(a^l_{k_x k_y k_z}\right)^\alpha = 1$ holds. For $1 \leq \alpha \leq 2$, the thus constructed field φ can be stable, homogeneous and isotropic in spatial variables [11]. If the coefficients $a^l_{\mathbf{ji}}$ satisfy the condition $a^l_{\mathbf{ji}} \equiv a^l\left(\frac{|\mathbf{i}-\mathbf{j}|}{l}\right)$ and the constants $\Phi_0(l)$, $\varphi_0(l)$ are the same for all l, then the field φ will be invariant with respect to the scale transformation. The mean of the field φ exists and is equal to $\varphi_0(l)$. As for the second moments, they are infinite for $\alpha \neq 2$. This complicates carrying out the correlation analysis, applied, for example, in [2] and the approach used in [5]. Nevertheless, for an extreme point $\beta = 1$ the second moments for the permeability field itself exist despite of the absence of variance of the field φ. This case is of interest, because it was experimentally verified [9]. As l_0 is a minimum scale, we can set $\varepsilon(\mathbf{x}) = \varepsilon\left(\mathbf{x}\right)_{l_0}$. Thus, the permeability field $\varepsilon(\mathbf{x})$ within the above-described model has the form:

$$\varepsilon(\mathbf{x}) = \varepsilon_0 \exp\left[-\left(ln2\sum_{\widehat{l_0}}^{\widehat{L}}\varphi(\mathbf{x}, \tau_l)\delta\tau\right)\right], \tag{4}$$

where $L = 2^{\widehat{L}\delta\tau}$, $l_0 = 2^{\widehat{l_0}\delta\tau}$, and the integral in formula (2) is replaced by the sum. For the calculation of moments of first and second orders we use formulas from [11] for $\left\langle e^{-b\zeta}\right\rangle$. For the correlation permeability function we have the estimation:

$$\langle\varepsilon(\mathbf{x})\varepsilon(\mathbf{x}+\mathbf{r})\rangle \simeq C\exp\left[2\delta\tau\ln 2\left(-2^{\alpha-2}\sum_{\widehat{l_r}}^{\widehat{L}}\Phi_0(\widehat{l})\left[\cos\left(\frac{\pi\alpha}{2}\right)\right]^{-1} - \varphi_0(\widehat{l})\right)\right] \tag{5}$$

where $r = 2^{\widehat{l_r \delta\tau}}$. For the self-similar permeability

$$\langle \varepsilon(\mathbf{x})\varepsilon(\mathbf{x}+\mathbf{r})\rangle \simeq C \exp\left[2\left(-2^{\alpha-2}\Phi_0\left[\cos\left(\frac{\pi}{2}\alpha\right)\right]^{-1} - \varphi_0\right)(\ln L - \ln r)\right] \quad (6)$$
$$\simeq C\left(\frac{L}{r}\right)^{-2\left(2^{\alpha-2}\Phi_0\left[\cos\left(\frac{\pi}{2}\alpha\right)\right]^{-1}+\varphi_0\right)}$$

The constant C is not universal, and the exponent for the self-similar permeability in (6) is universal and according to [12] can be measured.

4 A Subgrid Model

Let us divide the permeability function $\varepsilon(\mathbf{x}) = \varepsilon(\mathbf{x})_{l_0}$ into two components with respect to the scale l. The large-scale component $\varepsilon(\mathbf{x}, l)$ is obtained by statistical averaging over all $\varphi(\mathbf{x}, l_1)$ with $l_1 < l$, while the small-scale component is equal to: $\varepsilon'(\mathbf{x}) = \varepsilon(\mathbf{x}) - \varepsilon(\mathbf{x}, l)$:

$$\varepsilon(\mathbf{x}, l) = \varepsilon_0 \exp\left[-\int_l^L \varphi(\mathbf{x}, l_1)\frac{dl_1}{l_1}\right]\left\langle \exp\left[-\int_{l_0}^{l} \varphi(\mathbf{x}, l_1)\frac{dl_1}{l_1}\right]\right\rangle_< \quad (7)$$

$$\varepsilon'(\mathbf{x}) = \varepsilon(\mathbf{x}, l)\left[\frac{\exp\left[-\int_{l_0}^{l}\varphi(\mathbf{x}, l_1)\frac{dl_1}{l_1}\right]}{\left\langle \exp\left[-\int_{l_0}^{l}\varphi(\mathbf{x}, l_1)\frac{dl_1}{l_1}\right]\right\rangle_<} - 1\right], \quad (8)$$

where $<>_<$ means averaging over $\varphi(\mathbf{x}, l_1)$ from the small scale l_1. The large-scale (on-grid) pressure component $p(\mathbf{x}, l)$ is obtained as averaged solution to equation (1), where the large-scale component (1) is fixed, while the small-scale component ε' is random, $p(\mathbf{x}, l) =< p(\mathbf{x}) >_<$. The subgrid component is equal to $p' = p(\mathbf{x}) - p(\mathbf{x}, l)$. Let us substitute the expressions for $p(\mathbf{x})$, $\varepsilon(\mathbf{x})$ in equation (1) with averaging over the component ε':

$$\nabla\left[\varepsilon(\mathbf{x},l)\nabla p(\mathbf{x},l) + < \varepsilon'(\mathbf{x})\nabla p'(\mathbf{x}) >_<\right] = 0. \quad (9)$$

The second term in equation 9) is unknown, but cannot be rejected without preliminary estimation, since the correlation between permeability and the pressure gradient can be essential [1], [5]. The choice of the form of the second term in (9) determines a subgrid model. In order that values of such an expression be estimated, we apply the perturbation theory. In the Wilson RG, the initial value of the scale l is close to that of the least scale l. Subtracting (9) from (1) we obtain the subgrid equation for the pressure $p'(\mathbf{x})$:

$$\nabla\left[\varepsilon(\mathbf{x})\nabla p(\mathbf{x})\right] - \nabla\left[\varepsilon(\mathbf{x},l)\nabla p(\mathbf{x},l) + < \varepsilon'(\mathbf{x})\nabla p'(\mathbf{x}) >_<\right] = 0. \quad (10)$$

Equation (10) is used for finding the pressure $p'(\mathbf{x})$ and cannot be accurately solved. From equation (10), neglecting second order terms of smallness, obtain

$$\Delta p'(\mathbf{x}) = -\frac{1}{\varepsilon(\mathbf{x},l)}\nabla\varepsilon'(\mathbf{x})\nabla p(\mathbf{x},l) \tag{11}$$

According to the Wilson RG, the values $\varepsilon(\mathbf{x},l)$, $p(\mathbf{x},l)$ from the right-hand side equation (11) are considered to be known, their derivatives changing slower than $\varepsilon'(\mathbf{x})$. This corresponds to the idea to obtain a correct on-grid equation in the large-scale limit. Therefore,

$$p'(\mathbf{x}) = \frac{1}{4\pi\varepsilon(\mathbf{x},l)}\int\frac{1}{r}\nabla\varepsilon'(\mathbf{x}')\,d\mathbf{x}'\;\nabla p(\mathbf{x},l), \tag{12}$$

where $r = |\mathbf{x}-\mathbf{x}'|$. we come to the expression for a subgrid term in equation (9) in the large-scale limit:

$$\begin{aligned}\langle\varepsilon'(\mathbf{x})\nabla p'(\mathbf{x})\rangle &\approx \frac{1}{4\pi D\varepsilon(\mathbf{x},l)}\int\Delta\frac{1}{r}\langle\varepsilon'(\mathbf{x})\varepsilon'(\mathbf{x}')\rangle\,d\mathbf{x}'\;\nabla p(\mathbf{x},l)\\ &\approx -\frac{1}{D\varepsilon(\mathbf{x},l)}\langle\varepsilon'(\mathbf{x})\varepsilon'(\mathbf{x})\rangle\nabla p(\mathbf{x},l),\end{aligned}$$

where D is a spatial dimension, here $D = 3$. The model is similar to that mentioned in Landau and Lifshits [13], which is used for effective dielectric permeability of a mixture under simplifying assumptions of small fluctuations of their spatial scale. From (7), (8) and using formulas [11], keeping only first order terms we will have for $\varepsilon'(\mathbf{x})$, obtain

$$\langle\varepsilon'(\mathbf{x})\varepsilon'(\mathbf{x})\rangle \approx \varepsilon^2(\mathbf{x},l)\delta\tau\ln 2\left[\cos\left(\frac{\pi}{2}\alpha\right)\right]^{-1}\left(1-2^{\alpha-1}\right)\Phi_0\left(\tau_{\hat{l}}\right) \tag{13}$$

For the second term in (9) we have

$$\langle\varepsilon'(\mathbf{x})\nabla p'(\mathbf{x})\rangle_< \approx -\frac{\delta\tau\ln 2}{D}\left[\cos\left(\frac{\pi}{2}\alpha\right)\right]^{-1}\left(1-2^{\alpha-1}\right)\Phi_0\left(\tau_{\hat{l}}\right)\varepsilon(\mathbf{x},l)\nabla p(\mathbf{x},l) \tag{14}$$

Substituting (14) in to (9) and keeping only first order terms we find

$$\nabla\left[\left(1-\delta\tau\ln 2\left(\Phi_0\frac{2\left(1-2^{\alpha-1}\right)+D}{2D\cos\left(\frac{\pi}{2}\alpha\right)}+\varphi_0\right)\right)\varepsilon(\mathbf{x},l)\nabla p(\mathbf{x},l)\right] = 0. \tag{15}$$

In the subgrid modeling, the effective permeability coefficient in the scale l must correctly describe the solution to equation (1) within the scales (l, L) and be calculated by a formula of the same form as $\varepsilon(\mathbf{x})_{l_0}$. Thus, the effective permeability is determined from

$$\varepsilon(\mathbf{x}) = \varepsilon_{0l}\exp\left[-\left(ln2\sum_{\hat{l}}^{\hat{L}}\varphi(\mathbf{x},\tau_{\hat{l}})\delta\tau\right)\right]. \tag{16}$$

From (15) for $\delta\tau \to 0$ we obtain that the constant ε_{0l} satisfies the equation

$$\frac{d\ln\varepsilon_{0l}}{d\ln l} = -\Phi_0\left(l\right)\frac{2\left(1-2^{\alpha-1}\right)+D}{2D\cos\left(\frac{\pi}{2}\alpha\right)} - \varphi_0\left(l\right), \quad \varepsilon_{0l}|_{l=L} = \varepsilon_{00} \tag{17}$$

For a self-similar permeability, the solution to equation (17) has the form

$$\varepsilon_{0l} = \varepsilon_{00}\left(\frac{l}{L}\right)^{\Phi_0\frac{2\left(1-2^{\alpha-1}\right)+D}{2D\cos\left(\frac{\pi}{2}\alpha\right)}+\varphi_0}, \tag{18}$$

The constant ε_{00} describes the permeability in the largest possible scale for $l = L$.

5 Numerical Modeling

Equation (1) is numerically solved in the cube with the edge L_0. On the sides of the cube $y = 0$, $y = L_0$ we set the constant pressure $p\left(x,y,z\right)|_{y=0} = p_1$, $p\left(x,y,z\right)|_{y=L_0} = p_2$, $p_1 > p_2$. The pressure on other sides of the cube is set by the linear dependence along y: $p = p_1 + \left(p_2 - p_1\right)/L_0$. . The major filtration flow is directed along the axis y. Here, dimensionless variables are used. The problem is solved in the unit cube, with a unit pressure jump and $\varepsilon_0 = 1$. The permeability field is simulated by (4), $256 \times 256 \times 256$ grid in spatial variables is used, the scale step being $\delta\tau = 1$, $\tau_l = 0,\ldots,-8$.. The coefficients $a^l_{\mathbf{ji}}$ were selected as

$$a^l_{\mathbf{ji}} = \left(\frac{\sqrt{\alpha}}{2^{\tau_l}\sqrt{\pi}}\right)^{3/\alpha} \exp\left(-\frac{\left(\mathbf{x_j}-\mathbf{x_i}\right)^2}{2^{2\tau_l}}\right). \tag{19}$$

The field $\varphi(\mathbf{x},\tau)$ is generated independent for each τ_l. The common exponent in (4) is summed up over statistically independent layers. For an approximate solution, it is possible to use a certain limited number of layers. We have selected the number of layers so that the scale of the largest permeability fluctuations would allow us to approximately change probabilistic mean values averaged over space and the scale of the smallest fluctuations, so that the difference problem will approximate equation (1)sufficiently well, whose grid analog is solved by the iterative method. Independent random values $\zeta^l_{\mathbf{i}}$ in formula (4) were simulated using the algorithm and the program cited [14]. For self-similar media the constants Φ_0, φ_0 can be selected from experimental data for natural porous media. According to [12], the exponent in (6) for some natural media varies within $0.25 - 0.3$.

Fig. 1 shows the lines of the self-similar permeability level in the mid-span section $z = 1/2$ for $\alpha = 2$, which corresponds to the log-normal permeability model and $\alpha = 1,6$, $\beta = 1$, and, respectively, to a stable model. The parameters $\Phi_0 = 0.3$, $\varphi_0 = 0$. A difference between the two models is distinct.

According to the procedure of deriving a subgrid formula, for its verification it is necessary to numerically solve the full problem and to fulfil the probabilistic

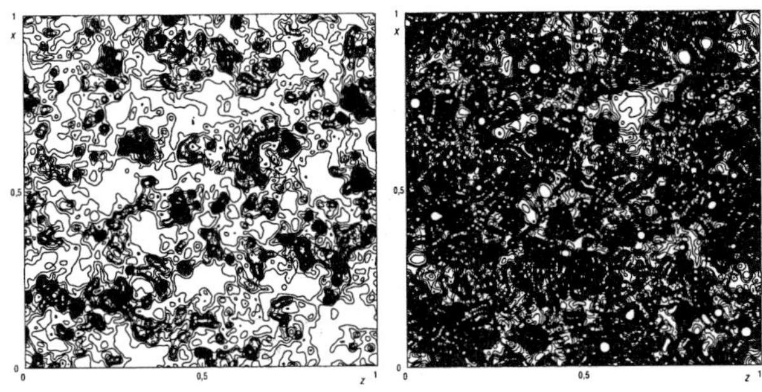

Fig. 1. The lines of levels in the mid-span section $z = 1/2$ of the self-similar permeability for $\alpha = 2$ (on the left), and $\alpha = 1.6$, $\beta = 1$ (on the right)

small-scale fluctuation averaging. As a result we obtain a subgrid term to be compared to a theoretical formula. This paper presents a more efficient version of such a verification that is based on the power dependence of the flow rate on the ratio of a maximum to a minimum scales in the ongrid domain when calculating permeability using (4) provided the contribution of a subgrid domain is not taken into account. The full flow rate of a liquid through a sample should coincide with the true one independent of the scale of the cut off l. The following formula is subject to verification:

$$\left\langle \exp\left[-\left(ln2\sum_{\widehat{l}}^{\widehat{L}}\varphi(\mathbf{x}_j,\tau_l)\delta\tau\right)\right]\nabla p\right\rangle = \frac{p_2-p_1}{y_2-y_1}\left(\frac{l}{L}\right)^{-\chi}, \tag{20}$$

where $\chi = \delta\tau\ln 2\sum_{\widehat{l_1}=\widehat{l}}^{\widehat{L}}\left(\Phi_0\left(\tau_{\widehat{l_1}}\right)\frac{2\left(1-2^{\alpha-1}\right)+D}{2D\cos\left(\frac{\pi}{2}\alpha\right)}+\varphi_0\left(\tau_{\widehat{l_1}}\right)\right)$, the probabilistic mean is replaced for the spatial averaging. This ergodic hypothesis was numerically verified. Then, using the numerical solution to (1), when the fluctuations are $\varepsilon_{-4},...,\varepsilon_{-6}$, we obtain the right-hand side of (20). Fig. 2 shows dependence

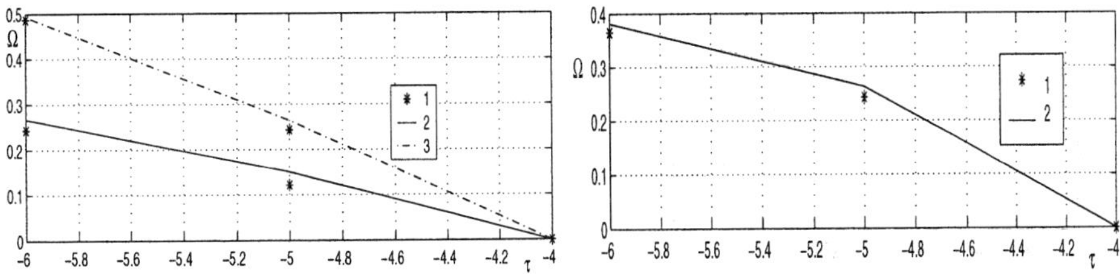

Fig. 2. The dependence of the logarithm of the flow rate Ω on the logarithm of the scale τ for $\alpha = 1,6$, $\beta = 1$, $\varphi_0 = 0$, the numerical simulation is marked stars. On the left - the self-similar model of permeability, 2 - for $\Phi_0 = 0.3$, 3 - for $\Phi_0 = 0.6$. On the right - scale-invariance is absent, $\Phi_0(-4) = 0.3$, $\Phi_0(-5) = 0.6$, $\Phi_0(-6) = 0.3$

of logarithm of the flow rate $\Omega(\tau_{\hat{l}}) = \log_2 \langle \exp\left[-\sum_{\hat{l}}^{\hat{L}} \varphi(\mathbf{x}_j, \tau_{\hat{l}})\delta\tau\right] \frac{\nabla p}{P_2 - P_1} \rangle$ on the logarithm of scale τ for $\alpha = 1,6$, $\beta = 1$, $\varphi_0 = 0$. The results for a self-similar model of permeability are shown on the left Fig. 2. Line 2 corresponds to the value $\Phi_0 = 0.3$, line 3 corresponds to the value $\Phi_0 = 0.6$. The results obtained by numerical simulation are marked stars. On the right side there are shown the results for $\Phi_0(-4) = 0.3$, $\Phi_0(-5) = 0.6$, $\Phi_0(-6) = 0.3$. In this case scale-invariance is absent. The numerical verification with the use of the spatial averaging is in good agreement with theoretical formulas. This paper was partly supported by the Integration Grant SB RAS No 61, and Grant RFBR 04-05-64415.

References

1. Shvidler, M. I.: Statistical hydrodynamics of porous media, Nedra, Moscow (1985) (in Russia)
2. Kuz'min G.A. and Soboleva O.N.: Subgrid modeling of filtration in porous self-similar media. J. Appl. Mech. and Tech. Phys. Vol. 43 (2002) 115–126
3. Kuz'min G.A. and Soboleva O.N.: Displacement of fluid in porous self-similar media. Physical Mesomechanics Vol. 5 (2002) 119–123
4. Wilson K. G. and Kogut J.: The renormalization group and the ϵ-expansion. Physics Reports,12C(2) (1974) 75-199
5. Teodorovich E.V.: The renormalization group method in the problem of effective permeability of stochastically nonhomogeneous porous medium. (in Russian) JETP Vol. 122 (2002) 79–89
6. Bogolubov N. N. and Shirkov D. V.: Introduction of theory of quantum fields, Nauka, Moscow (1976)
7. Kolmogorov A. N.: A refinement of previous hypotheses concerning the local structure of turbulence in a viscous incompressible fluid at high Reynolds number. J. Fluid Mech. Vol. 13 (1962) 82–85
8. Gnedenko B.V. and Kolmogorov A.N.: Limit Distributions for Sums of Independent Random Variables. Addison-Wesley, Cambridge, MA., (1954)
9. Bouffadel M. C., S. Lu et al.: Multifractal scaling of the intrinsic permeability. Physics of the Earth and Planetary Interiors 36(11) (2000) 3211–3222
10. Zolotarev V.M.: One-dimensional Stable Distributions. Amer. Math. Soc., Providence, RI., (1986)
11. Samorodnitsky G. and Taqqu M. S. Stable non-Gaussian random processes. Chapman Hill., New York, London, (1994)
12. Sahimi M.: Flow phenomena in rocks: from continuum models, to fractals,percolation, cellular automata, and simulated annealing. Reviews of Modern Physics 65(4) (1993) 1393–1534
13. Landau L.D. and Lifshitz E.M.: Electrodynamics of Continuous Media. Pergamon Press, Oxford-Elmsford, New York, (1984)
14. Chambers J. M., Mallows C., Stuck B. W. A method for simulating stable random variables. Jornal of the American Statistical Association, 71(354) (1976) 340–344

A Computational Model of Micro-vascular Growth

Dominik Szczerba and Gábor Székely

Computer Vision Lab, ETH, CH-8092 Zürich, Switzerland
{domi, szekely}@vision.ee.ethz.ch
www.vision.ee.ethz.ch

Abstract. In order to supply a growing tissue with oxygen and nutrients and to remove its metabolic wastes, blood vessels penetrating the tissue are formed. Multiple mechanisms are involved in this process ranging over many orders of magnitude: chemical signaling on the bio-molecular level ($10^{-9}m$), genetic program on the protein level ($10^{-7}m$), microscopic mechanical cell interactions ($10^{-5}m$) and external forces and stresses reaching macroscopic scales ($> 10^{-3}m$). Better physiological understanding of this phenomenon could result in many useful medical applications, for example in gene therapy or cancer treatment. We present a simulation framework to study mechanical aspects of the micro-vascular growth using techniques from computational geometry, solid mechanics, computational fluid dynamics and data visualization. Vasculogenesis is modeled as traction driven remodeling of an initially uniform tissue in absence of blood flow. Angiogenesis, the subsequent formation and maturation of blood vessels, is handled as a flow driven remodeling of a porous structure resulting from the preceding stage. The mechanical model of tissue response to the traction forces successfully predicts spontaneous formation of primitive capillary networks. Furthermore, we demonstrate that a shear-stress controlled remodeling of such structures can lead to flow fingering effects observed in real cellular media.

Keywords: angiogenesis, capillary formation, computational model, micro-vascular growth, numerical simulation, shear stress, vascular remodeling, vasculogenesis.

1 Introduction

Experimental studies confirm that physical forces, including tension, stretching or compression, and flow derived quantities like shear stress influence growth and remodeling in living tissues at the cellular level. In particular, it is known that alignment and structural remodeling of endothelial cells can be modulated by mechano-chemical interactions (e.g. [1], [2] and references therein). It is therefore reasonable to assume that physical interactions between blood flow and endothelial cells play an important role in capillary formation during the vascularization of a growing tissue. For example, [3] demonstrated that flow driven remodelling of transcapillary tissue pillars in the absence of cell proliferation is decisively involved in micro-vascular growth in the chick chorio-allantoic membrane. We have recently given a detailed overview of existing mathematical models of vascular growth ([4]). The approaches listed there are generally successful in predicting physiologically realistic systems with relevant biophysical properties,

V.S. Sunderam et al. (Eds.): ICCS 2005, LNCS 3516, pp. 17–24, 2005.

but do not attempt to explain the origin of such structures in a microscopic sense. In addition, they either do not address blood flow effects at all, or only implicitly via Poiseuille's Law. In the same paper we have demonstrated that a simple mechanistic model of an explicit tissue response to blood flow suffices to reproduce bifurcation formation and micro-vessel separation from the underlying capillary bed. Structures resulting from the simulations are comparable to real experimental cases (see e.g. [5]) even though the tissue remodeling procedure used in the model is very primitive. The tissue is treated only in a statistical fashion as a set of independent pixels. This allows a very simple treatment of cellular transport but limits possibilities to model the tissue's elastic properties.

In this study we further investigate mechanical aspects of micro-vessel formation but using more realistic physical modeling of the tissue. In contrary to the previous approach, we now treat the tissue as a deformable object and explicitly address its elastic properties. In addition, we now differentiate between the early capillary plexus formation (vasculogenesis) and its subsequent remodeling (angiogenesis). Vasculogenesis is modeled as traction driven reshaping of an initially uniform tissue represented by a mass-spring mesh. In subsequent steps, as a result of fluid-structure interactions and elastic modeling of the tissue's interior, the emerging clusters elongate, align with the flow and merge with their neighbors, which is a prerequisite for the formation of new generations of blood vessels. In order to provide better solution accuracy at the fluid-structure interfaces, where the shear stress is of primary interest, we now also make use of irregular triangular grids. Such meshes can accurately map complex domains but require more advanced formulation of the flow equations than in case of an isometric Cartesian grid. Moreover, this implies additional difficulties in case of dynamically changing boundaries. Due to permanent modifications of the boundary conditions an efficient re-meshing is necessary for the required adaptation of the domain.

In the following section we will explain how we model the elastic response of the tissue and show that it suffices to explain formation of primitive capillary networks. In the next section the geometrical domain will be defined and the corresponding mesh generation technique will be described. Afterwards we will show how we remodel the system using fluid-structure interactions and demonstrate how this leads to the elongation and alignment of the tissue clusters. Finally, in the concluding section we summarize the approach, indicate its implications and propose future developments.

2 Tissue Modeling

We start by assuming that living cells consist of a soft matrix (cytoplasm) structurally strengthened by stiff elastic fibers (cytoskeleton). We further assume that a growing tissue increases its volume introducing stretching forces in these fibers, which eventually lead to their spontaneous breaking. At least in the first approximation it is natural to model a system of cytoskeleton fibers as a mesh, where uniformly distributed masses are connected by elastic springs. To describe the behavior of such a mass-spring mesh we start with Newton's equation of motion and assume that the external forces are the sum of linear spring deformation forces (the Hook's Law) and non-conservative inter-

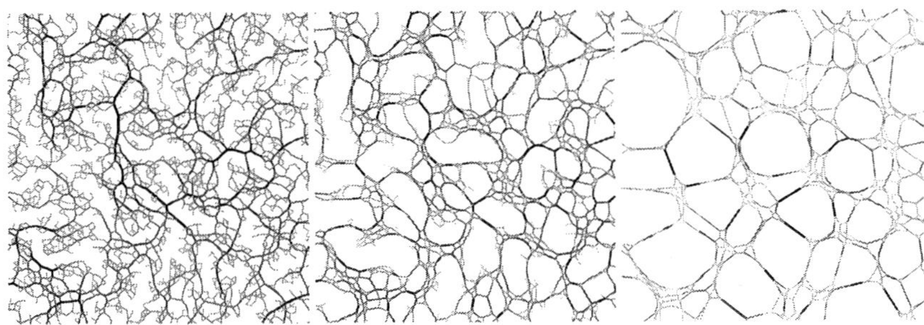

Fig. 1. Computer simulation of a traction-driven remodeling of an initially uniform mesh. The springs are gray- and width- coded according to their stretching forces

nal friction forces due to energy dissipation in cytoskeletal fibers. The resulting motion equation in a matrix form is

$$M\frac{d^2\boldsymbol{r}}{dt^2} + D\frac{d\boldsymbol{r}}{dt} + K\boldsymbol{r} + \boldsymbol{F}^{ext} = 0 , \tag{1}$$

where M is the mass matrix, D damping matrix, K stiffness matrix, $\boldsymbol{r}$ is the vector containing the nodal coordinates and $\boldsymbol{F}^{ext}$ is storing all other external forces (if any). This second order ordinary differential equation is solved numerically using a finite difference scheme. After substituting $d\boldsymbol{r}/dt = \boldsymbol{v}$ and some algebraic rearrangements we obtain the discrete form of the equation, which we use to update the nodal velocities:

$$\boldsymbol{v}^{n+1} = \frac{1}{1 + \frac{D\Delta t}{M}} \left(\boldsymbol{v}^n + \frac{\Delta t}{M}(\boldsymbol{F}^K + \boldsymbol{F}^{ext}) \right) , \tag{2}$$

with $\boldsymbol{v}^t$ being a vector of nodal velocities at iteration t and $\boldsymbol{F}^K$ the spring deformation forces. Now the nodal positions can be obtained using a simple time integration:

$$\boldsymbol{r}^{n+1} = \boldsymbol{r}^n + \boldsymbol{v}^n \Delta t . \tag{3}$$

A detailed description of the mesh generation procedure will be given in the next section. Here we start from a given uniform unstructured triangular grid covering a unit square. As already mentioned before, we assume that a growing tissue increases its volume and thus stretches its structural fibers. If we start to selectively remove springs that are stretched over a given threshold (damage due to exceeding their mechanical resistance), the initially regular mesh remodels into a set of tissue pillars forming a micro-tubular network, exactly as observed in early stages of vasculogenesis. **Fig. 1** shows a few stages of the remodeling process on selected fragments of the whole mesh consisting roughly of 12,000 nodes and 35,000 springs. The springs are width- and gray-coded according to their stretching forces. Our computational model allows to generate results comparable to those presented e.g. in [6] or [7], but as opposed to them, it is not based on a complex mathematical formalism. We do not model cells as scalar fields but base our modeling entirely on physical principles. As we explicitly address the tissue's elastic fibers we do not face numerical difficulties arising from e.g. modeling of surface tension inherent to a density representation. **Fig. 2** shows a comparison of one of our remodeling results with a real experimental case found in [7].

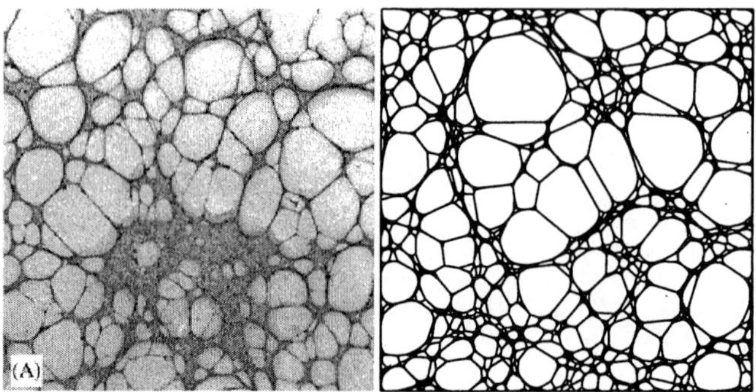

Fig. 2. Left: in vitro angiogenesis experiment from [7] (© Elsevier, reprinted with permission). Right: computer simulation of a traction-driven remodeling of an initially uniform mesh

3 Domain Definition and Mesh Generation

In the previous section we demonstrated how simple mechanistic modeling of cytoskeletal fibers can explain spontaneous formation of micro-channels in an initially homogeneous tissue prior to the actual blood flow. Now we investigate the behavior of the newly created micro-tubules as the blood starts flowing. We note that the resulting network can alternatively be viewed as a set of tissue clusters and the capillaries as narrow passages between them. We now want to approximate the emerging tissue clusters by deformable rings surrounded by an external membrane immersed in the fluid flowing around them. In order to create such rings as well as to generate appropriate meshes necessary to numerically solve the flow equation we need to provide a domain geometry. For that purpose we represent the clusters as polygons, we emit test rays from their centers and run an automatic search for the nearest springs intersecting with them. This information allows us to inspect the topological connections necessary for polygon reconstruction. Unfortunately, the contracting meshes produce many undesired topological artifacts like springs with zero length, degenerated nodal positions or highly non-homogeneous distribution of nodes on the cluster boundary edges. To correctly - and still automatically - reconstruct the polygons these special cases have to be differentiated and handled properly. Additional problems arise from very narrow passages that are often formed between pillars. In an extreme case one tubule is represented by only one spring. This introduces additional difficulties into the mesh generation procedure which we address here either by mesh refinement or by merging the two polygons if refinement leads to excessive mesh sizes.

Once the polygons are available we proceed by generating a mesh. We distribute masses inside the domain boundary and outside the interior polygons. Then we create topological connections between the masses using the Delaunay triangulation. Next, we remove all the triangles that lie outside the domain or inside any polygon. The triangles that are left are broken up into a unique set of springs and each spring is assigned a rest length being a desired mesh edge length at this place. When just one rest length is assigned to all the springs, a uniform (isometric) mesh will form. To achieve refinement near boundaries one can assign different rest lengths for different springs according to e.g. distances to the nearest boundary or local curvature. Once the rest lengths are assigned, the masses are charged electrically and allowed to evolve. Then re-triangulation is performed to correct for any crossing springs, and the same deformation procedure is applied as described in the preceding section. In addition, the masses that come close enough to the boundary segments are snapped to them and not allowed to detach any more. After some iterations this simple method generated triangular meshes of sufficient quality but the time performance turned out to be very low, mostly because of N^2 cost of electromagnetic interactions. The performance was significantly improved when magnetic repulsion between the masses was replaced by a slight increase in the desired spring rest lengths to form an "internal pressure" [8].

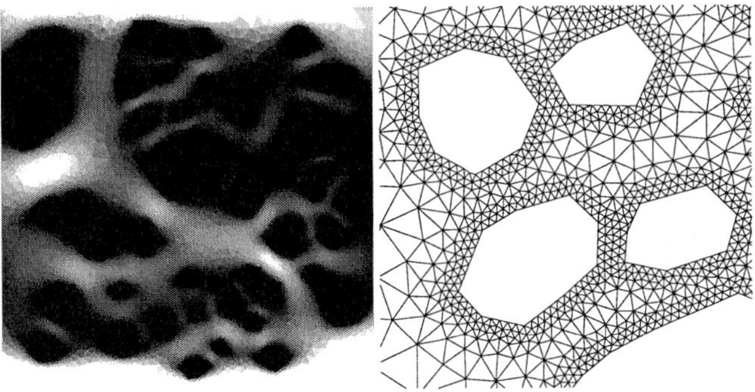

Fig. 3. Left: visualization of velocity in a fragment of a micro-capillary network presented in the previous figures. Right: a fragment of a mesh used to solve the flow equations

4 Flow Equations and Flow-Tissue Interactions

To simulate a structural remodeling of a capillary micro-network the key issue is finding the distribution of forces acting in such a physical domain. To calculate flow conditions we use numerical techniques from computational fluid dynamics. The general form of the Navier-Stokes equation for an incompressible Newtonian fluid is

$$\rho\frac{\partial \boldsymbol{v}}{\partial t} + \rho(\boldsymbol{v}\cdot\nabla)\boldsymbol{v} = \eta\nabla^2\boldsymbol{v} - \nabla p + \boldsymbol{F} \tag{4}$$

$$\nabla\cdot\boldsymbol{v} = 0\,, \tag{5}$$

where ρ is the fluid's density, η the absolute viscosity, p the pressure, $\boldsymbol{v}$ the velocity and $\boldsymbol{F}$ denotes external forces. Eq.4 represents momentum balance and Eq.5 describes fluid continuity. In the case of creeping flow in capillaries the viscous forces are much stronger than the inertial forces and the convective part of the momentum equation does not need to be included. To describe a steady flow in the absence of external forces the time dependent and body force contributions can also be neglected and the Navier-Stokes equations can be rewritten as:

$$\nabla p = \eta\nabla^2\boldsymbol{v} \tag{6}$$

$$\nabla\cdot\boldsymbol{v} = 0\,. \tag{7}$$

The steady-state discretization results in a system of equations with the form:

$$A_P^u u_P + A_i^u u_i = Q_P^u \tag{8}$$

$$A_P^v v_P + A_i^v v_i = Q_P^v \tag{9}$$

$$a_i \boldsymbol{v}_i \cdot \hat{n}_i = 0\ , \tag{10}$$

with A being the domain dependent coefficient matrix, Q the source terms defined by the boundary conditions, a face surface, $\hat{n}$ face normal, u, v velocity components and using the Einstein summation over direct cell neighbors and the compass notation as in [9]. The formulation used above is the same that we already presented for the finite difference discretization [4], but the coefficients are different and have to be derived separately for the finite volume discretization on unstructured triangular meshes. As these equations have to hold true for every volume element in the mesh, we formulate this as a linear problem $Ax = b$, where x is a vector of unknown velocity components and pressure values, A is a domain dependent coefficient matrix and b is defined by the boundary conditions. Unfortunately, Eqs. 8-10 are not suitable for modeling pressure, as pressure is not even present in the continuity equation and is very weakly coupled with the velocity components. Solving such a system of equations in a standard way (i.e. simultaneously) results in oscillatory pressure solutions, therefore this case needs special treatment. To address this problem we adapted the smoothing pressure correction, originally presented for orthogonal grids [10], for the case of unstructured triangular meshes. We continue by solving the partial differential equations on the meshes, generated as described in the previous section (see **Fig. 3**). This results in velocity and pressure values in the center of each volume element. We assume that eventual evolution of the tissue is governed only by shear-induced traction forces at the boundary, which is well justified from experimental observations. We derive now viscous tractions on the boundary walls using the obtained velocity maps and distribute them onto the original mass-spring mesh nodes. Note, that we use two different meshes, a homogeneous one for the mass-spring deformation and an adaptive one for the discretization of the flow equations. This technique is known as a dual mesh approach. Now the original mesh is allowed to deform determining a new geometry and the whole procedure is repeated in a loop. Blood velocity is visualized by reconstructing nodal velocities from the triangle centers (by solving a linear problem) and using OpenGL to plot triangles with linear color interpolation. Unfortunately, simply letting the pillars hang freely, without addressing their internal structure, does not lead to any realistic remodeling. The pillars are only pushed aside all together, somewhat changing shape, but almost entirely preserving distances between them. Even though preferred paths of flow are clearly visible if the velocity is visualized (see gray-coded velocity plot in **Fig. 3**), we found no tendencies for the pillars to elongate and merge to form any kind of corridors or future vessel walls. The alignment of the tissue clusters and their interactions with the neighbors, however, are prerequisites for any vessel formation. Before any large scale simulations can take place, a better model of the dynamic behaviour of single tissue clusters has to be found. We assumed that the mechanical model involved in this case, ignoring the internal structure of the pillars, was not sufficient to correctly reproduce the required local phenomena. We concluded that the pillars were not just holes, but deformable membranes filled with elastic tissue - a structure having certain physical properties.

The springs that we used so far corresponded then to extracellular bindings. Additional springs, much stiffer in order to mimic behavior of solid clusters immersed in a jelly matrix, should now be placed in the pillar interiors, and the whole structure should be separated from the external matrix by a membrane. The structural arrangement of skeletal micro-filaments in cells is not trivial and its proper modeling can be crucial for obtaining correct results. We surround a cell with a membrane that is able to sense shear stress around itself and evaluate its average value. Then we make internal connections between all the membrane's nodes using springs 10 times stiffer than springs in the surrounding matrix. Now, if a part of the membrane senses shear stress higher than average, it will transmit contracting forces to all the internal filaments connected to the activated part. In contrary, if somewhere on the membrane shear stress lower than average is sampled, stretching is induced. Such mechanics successfully predicts elongation of a disk placed in a viscous flow (see **Fig. 4**). The same procedure applied to a bigger mesh is presented in **Fig. 5**. As can be seen from the figures, pillars finally elongated, aligned with the flow and subsequently merged with the neighbors, resulting in a chain which is the basic pre-condition for the formation of a future vessel wall. Based on this more realistic model of single cluster dynamics, further investigations on much bigger domains can be performed. If few pillars demonstrate clear tendencies to align and form chains, chances for micro-tubule formation in large scale simulations (involving e.g. thousands of such clusters) become realistic. This will, however, require computer code parallelization and other super-computing techniques due to excessive demand on computational resources that modeling of such domains would require.

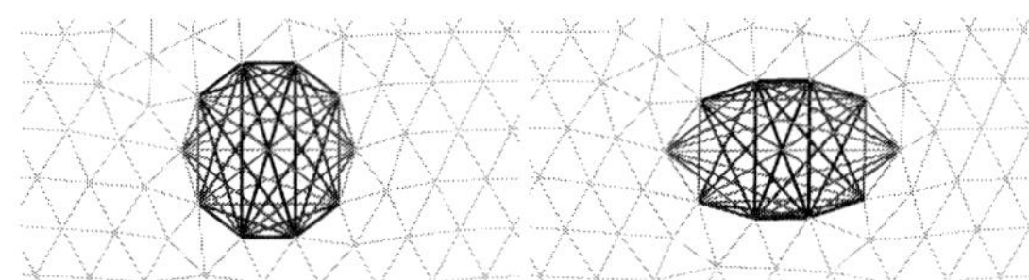

Fig. 4. Deformation of a disk due to its internal mechanics. Flow coming from the left

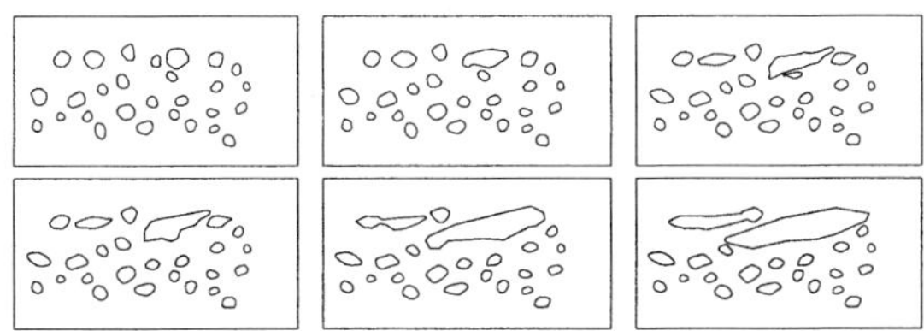

Fig. 5. Elongation of the tissue pillars, alignment with the flow and subsequent merging with neighbors - geometry at different iterations

5 Conclusions

We have presented a multi-physics simulation framework to study mechanical aspects of vascular system formation. We use techniques from solid mechanics to model tissue response, computational geometry to generate meshes needed to solve partial differential equations, computational fluid dynamics to provide flow conditions in the cellular medium and fluid-structure interactions to simulate tissue remodeling. The simulations, visualized using OpenGL, reveal that mechanical traction forces rebuild an initially homogeneous tissue into a micro-tubular network, as observed in early stages of vasculogenesis. By visualizing blood velocity in such networks we find paths of preferred flow, suggesting future formation of some flow corridors or walls. By addressing the internal structure of the emerging cell clusters we demonstrate their alignment in the flow and tendencies to merge with the neighbors to form chains, which is a prerequisite for future micro-tubule formation. Once a realistic model of single cell clusters is available, large scale simulations, using super-computing due to excessive domain sizes, can be performed in future investigations.

Acknowledgments

This work is a part of the Swiss National Center of Competence in Research on Computer Aided and Image Guided Medical Interventions (NCCR Co-Me), supported by the Swiss National Science Foundation.

References

1. Hsiai, T.K., Cho, S.K., Honda, H.M., Hama, S., Navab, M., Demer, L.L., Ho, C.M.: Endothelial cell dynamics under pulsating flows: Significance of high versus low shear stress slew rates. Annals of Biomedical Engineering **30** (2002) 646–656
2. Malek, A.M., Izumo, S.: Mechanism of endothelial cell shape change and cytoskeletal remodeling in response to fluid shear stress. Journal of Cell Science **109** (1996) 713–726
3. Djonov, V., Galli, A., Burri, P.: Intussusceptive arborization contributes to vascular tree formation in the chick chorio-allantoic membrane. Anatomy and Embryology **202** (2000) 347–357
4. Szczerba, D., Székely, G.: Computational model of flow-tissue interactions in intussusceptive angiogenesis. Journal of Theoretical Biology **234** (2005) 87–97
5. Djonov, V., Kurz, H., Burri, P.: Optimality in the developing vascular system: Branching remodeling by means of intussusception as an efficient adaptation mechanism. Developmental Dynamics **224** (2002) 391–402
6. Manoussaki, D.: A mechanochemical model of vasculogenesis and angiogenesis. Mathematical Modelling and Numerical Analysis **37** (2003) 581–599
7. Namy, P., Ohayon, J., Tracqui, P.: Critical conditions for pattern formation and in vitro tubulogenesis driven by cellular traction fields. Journal of Theoretical Biology **227** (2004) 103
8. Persson, P.O., Strang, G.: A simple mesh generator in matlab. SIAM Review, Volume 46 (2), pp. 329-345 (2004)
9. Ferziger, J., Perić, M.: Computational Methods for Fluid Dynamics. Springer-Verlag (2002)
10. Date, A.W.: Complete pressure correction algorithm for solution of incompressible navier-stokes equations on a nonstaggered grid. Numerical Heat Transfer, Part B **29** (1996) 441–458

A Dynamic Model for Phase Transformations in 3D Samples of Shape Memory Alloys

D.R. Mahapatra and R.V.N. Melnik

Mathematical Modelling and Computational Sciences, Wilfrid Laurier University, 75 University Avenue West, Waterloo, Ontario, Canada N2L 3C5

Abstract. Despite recent progress in modelling the shape memory alloy (SMA) behaviour, many difficulties remain due to various limitations of the existing free energy models and strong nonlinearity of these nonlinear materials. Phase kinetics of SMA coupled with thermoelastodynamics is still not fully tractable, and one needs to deal with complicated multiscale character of SMA materials requiring a linkage between their microstructure and macroscopic properties. In this paper we develop a new dynamic model of 3D SMA which employs an improved version of the microscopic Landau theory. Essential properties of the single and multivariant martensitic phase transformations are recovered using consistent steps, which eliminates the problem of non-uniqueness of energy partitioning and relaxes the over-sensitivity of the free energy due to many unknown material constants in previously reported models. We exemplify our results on a model for cubic to tetragonal transformations in a rectangular SMA bar by showing key algorithmic steps which can be extended to more complex cases.

1 Introduction

Modelling of dynamics of phase transformations (PT) in Shape Memory Alloys (SMAs) under the combined effect of stress and temperature is one of the most challenging problems in computational science and engineering. Our better understanding of such dynamics can be achieved with multi-physics multiscale models which assist the researchers in designing new materials and devices by harnessing the shape memory effect. Phenomenological framework based on Landau theory of martensitic transformations (e.g., [1]) has become a convenient choice for basic building blocks in the computational models. However, most of the phenomenological models are applicable at macroscopic and mesoscopic scales as discussed in [2], and the strain field in such model is often beyond the resolution of bain strain. Also it may be noted that a majority of the works (see e.g.[3, 4]) in this direction is based on selectively chosen strain gradient plasticity models that require extensive parameter fitting from experimental data. Also, in such framework, the true nature of nonlinearity and coupling is not fully tractable. For a more detail description of the PT, one requires a microscopic model, such as [5, 6], where the invariance of the free energy with respect to the crystallographic point group symmetry is preserved but the discontinuous

V.S. Sunderam et al. (Eds.): ICCS 2005, LNCS 3516, pp. 25–32, 2005.

nature of the deformation across the individual habit planes is introduced. This provides the most essential information to describe the morphology of the microstructural evolution, whereas the process of reshuffling of the atoms, which actually causes the reconstructive PT, remains beyond such a microstructural scale. The latter process requires the description of chemical interaction energy between atoms [7] and spin stabilized Hamiltonian [8]. Coupling such an atomistic process with the microstructural phenomenology means coupling a molecular dynamic model [9, 10] with a microscopic free energy model [6, 11, 12] through the thermal fluctuation term, strain and the motion of the domain walls.

In the present paper we focus our attention on linking SMA microstructure and macroscopic behaviour by developing the microscopic free energy and the associated coupled field model to simulate the dynamics of 3D macroscopic sample of NiAl undergoing cubic to tetragonal PT. First, we observe that the free energy is highly sensitive to many interdependent constants which are to be obtained experimentally [11]. As a result of this sensitivity and due to a fairly complex polynomial representation of the 3D Landau-Ginzburg free energy function, there may be situations where multidimensional simulations produce unphysical spurious oscillations. In the new model developed in the present paper, the above set of interdependent constants, that need to be determined from experimental data, is reduced to uniquely identifiable set of constants for each transformation surface among the austenitic phase and a particular martensitic variant.

Note that earlier, a new low-dimensional model for the time-dependent dynamics of SMA was derived in [2] and a class of such reduced models was rigorously analyzed in [13]. Computational results presented in [2, 13] confirmed robustness of our modelling approach for a number of practically important cases. The main results of the present paper are pertinent to *the general three-dimensional case.* There are several recent studies where the time-dependent Ginzburg-Landau (TDGL) (Langevin) equation with the correct microscopic description of the Falk-Konopka type free energy function has been analyzed (e.g. [14, 12]). We are not aware of any dynamic model of 3D SMA samples that bases its consideration on the correct Ginzburg-Landau free energy representation in the microscopic scale and the momentum balance and heat conduction in the macroscopic scale. The Ginzburg-Landau free energy function derived in the first part of this paper is incorporated in the coupled dynamic field model. This model is exemplified for cubic to tetragonal PT with single martensitic variant along the axis of a bar with rectangular cross-section under axial stress and heat flux.

2 3D Landau Theory of Martensitic Phase Transformation

It has been demonstrated by Levitas and Preston [6, 12] that the polynomial structures 2-3-4 and 2-4-6 of the Gibbs free energy in order parameter η in Cartesian coordinate can eliminate the problem of unphysical minima and retain all the necessary properties of the Ginzburg-Landau free energy function with respect to point group symmetry of the crystals. Such polynomial structure can be constructed in such a way that the stability of the austenitic phase (A) and martensitic variants (M_j), non-extremal diffusion barrier and nucleation can be described in stress-temperature space. Furthermore, while using such polynomial structure the interfaces (domain walls) $M_j - M_i$ between the martensitic variants (i, j) can be interpreted by using a newly introduced barrierless A nucleation mechanism, i.e. by splitting the original into two simultaneously present interfaces $M_j - A$ and $A - M_i$. In this section a 2-3-4 polynomial structure is constructed by improving upon the model of Levitas and Preston [6, 11].

For analytical clarity we first consider a single variant of martensite and single order parameter $\eta \in [0, 1]$. First we define the Gibbs free energy density in stress-temperature space $(\boldsymbol{\sigma}, \theta)$ as

$$G = -\boldsymbol{\sigma} : \boldsymbol{\lambda} : \boldsymbol{\sigma}/2 - \boldsymbol{\sigma} : \boldsymbol{\varepsilon}_t \varphi(\eta) + f(\theta, \eta) , \tag{1}$$

where $\boldsymbol{\lambda}$ is the constant fourth-rank elastic compliance tensor, $\boldsymbol{\varepsilon}_t$ is the transformation strain tensor at the thermodynamic equilibrium of the martensite (obtained from crystallography), $\varphi(\eta)$ is a monotonic function with $\varphi(0) = 0$ indicating stable A phase and $\varphi(1) = 1$ indicating stable M phase. $f(\theta, \eta)$ is the chemical part of the energy with property: $f(\theta, 1) - f(\theta, 0) = \Delta G^\theta(\theta)$, where ΔG^θ is the difference between the thermal parts of the Gibbs free energy density of the M and A phases, which can be obtained indirectly from experiments [15] The objective now is to obtain the functions φ and f by satisfying their properties mentioned above and the conditions of extremum of the energy for existence of equilibrium of A and M phases: $\partial G/\partial\eta = 0$ at $\eta = 0, 1$.

The new model derived below is based on the assumption that G can be uniquely represented by a polynomial structure in η with the extremum only at $\eta = 0, 1$ and out of these two extremum only one is minimum and the other is maximum for phase transformation (PT) to happen. At equilibrium, we have

$$\frac{\partial G}{\partial \eta} = -\boldsymbol{\sigma} : \boldsymbol{\varepsilon}_t \frac{\partial \varphi(\eta)}{\partial \eta} + \frac{\partial f(\theta, \eta)}{\partial \eta} = 0 , \quad \eta = 0, 1. \tag{2}$$

The total strain tensor ($\boldsymbol{\varepsilon} = -\partial G/\partial \boldsymbol{\sigma}$) is the sum of the elastic strain tensor ($\boldsymbol{\lambda} : \boldsymbol{\sigma}$) and the transformation strain tensor ($\boldsymbol{\varepsilon}_t \varphi(\eta)$). Hence, for reconstructive PT through vanishing misfit strain, the condition

$$\boldsymbol{\sigma} : \boldsymbol{\varepsilon}_t \frac{\partial \varphi(\eta)}{\partial \boldsymbol{\sigma}} - \frac{f(\theta, \eta)}{\partial \boldsymbol{\sigma}} = 0 \quad \forall (\boldsymbol{\sigma}, \eta) \tag{3}$$

must be satisfied. It is observed in the reported results [6] that the transformation barrier is dependent on stress. In the context of interface barrier, Levitas and Preston [12] have treated the associated η to be dependent on $\boldsymbol{\sigma}$. In the present paper, we propose an alternate approach by considering stress-dependent barrier height because the stress $\boldsymbol{\sigma}$ is the only driving factor for PT under isothermal condition. The polynomial structure which satisfies the extremum properties can be expressed as

$$\partial G/\partial \eta = \eta(\eta - 1)(\eta - \eta_b) , \tag{4}$$

so that its roots $\eta = 0, 1$ satisfy Eq. (2) and the root $\eta = \eta_b(\boldsymbol{\sigma}, \theta)$ represents the $A \leftrightarrow M$ PT barrier. Integrating Eq. (4) and imposing the combined properties of $\varphi(\eta)$ and $f(\theta, \eta)$ stated earlier as

$$G(\boldsymbol{\sigma}, \theta, 0) - G(\boldsymbol{\sigma}, \theta, 1) = \boldsymbol{\sigma} : \boldsymbol{\varepsilon}_t - \Delta G^\theta \ , \tag{5}$$

we get that $\eta_b = -6\boldsymbol{\sigma} : \boldsymbol{\varepsilon}_t + 6\Delta G^\theta + 1/2$. Using Eq. (1) in Eq. (4) and by differentiating with respect to $\boldsymbol{\sigma}$, one has

$$-\boldsymbol{\varepsilon}_t \frac{\partial \varphi(\eta)}{\partial \eta} - \boldsymbol{\sigma} : \boldsymbol{\varepsilon}_t \frac{\partial^2 \varphi(\eta)}{\partial \boldsymbol{\sigma} \partial \eta} + \frac{\partial^2 f(\theta, \eta)}{\partial \boldsymbol{\sigma} \partial \eta} = \frac{\partial}{\partial \boldsymbol{\sigma}} \left[\eta_b \eta - (\eta_b + 1)\eta^2 + \eta^3 \right] . \tag{6}$$

The term involving f can be eliminated from Eq. (6) with the help of Eq. (3), and can be expressed as

$$\boldsymbol{\varepsilon}_t \frac{\partial \varphi(\eta)}{\partial \eta} = \eta(\eta - 1) \frac{\partial \eta_b}{\partial \boldsymbol{\sigma}} = \eta(\eta - 1)(-6\boldsymbol{\varepsilon}_t) . \tag{7}$$

Integrating Eq. (7) and following the properties of the transformation strain, i.e. $\varphi(0) = 0$ and $\varphi(1) = 1$, we have $\varphi(\eta) = 3\eta^2 - 2\eta^3 \ , \quad 0 \leq \eta \leq 1$. Substituting this form in Eq. (3) and integrating with respect to η, the chemical part of the free energy density is obtained as

$$f(\theta, \eta) = \boldsymbol{\sigma} : \boldsymbol{\varepsilon}_t (3\eta^2 - 2\eta^3) + \frac{1}{2}\eta_b \eta^2 - \frac{1}{3}(\eta_b + 1)\eta^3 + \frac{1}{4}\eta^4 . \tag{8}$$

For $A \rightarrow M$ PT, the criteria for the loss of stability of A phase is $\partial^2 G / \partial \eta^2 \leq 0$ at $\eta = 0$, which gives the stress driven condition:

$$\boldsymbol{\sigma} : \boldsymbol{\varepsilon}_t \geq \Delta G^\theta + \frac{1}{12} . \tag{9}$$

Similarly, for $M \rightarrow A$ PT, the criteria for the loss of stability is $\partial^2 G / \partial \eta^2 \leq 0$ at $\eta = 1$, which gives the stress driven condition:

$$\boldsymbol{\sigma} : \boldsymbol{\varepsilon}_t \leq \Delta G^\theta - \frac{1}{12} . \tag{10}$$

$M_j \leftrightarrow M_i$ PT or diffused interface can evolve for stresses outside the range obtained by Eqs. (9) and (10). Note that no parameter fitting is required in the present model as opposed to the earlier model [6]. Eqs. (9) and (10) indicate a nonlinear dependence of the transformation surface on the temperature, which can be compared with the experimental data.

2.1 Cubic to Tetragonal Transformation Characteristics

We now consider the cubic to tetragonal PT for single variant martensitic case in absence of the elastic part of the stress. After reducing the stress and strain tensors in 1D, the equilibrium stress-transformation curve is obtained as

$$\eta = \eta_b \Rightarrow \sigma = \frac{1}{\varepsilon_t} \left[\Delta G^\theta + \frac{1 - 2\eta}{12} \right] . \tag{11}$$

Note that the increase in η causes decrease in σ which is consistent. The stress hysteresis (H) is obtained as

$$H = \sigma_{(\eta=0)} - \sigma_{(\eta=1)} = \frac{1}{6\varepsilon_t} , \tag{12}$$

which is independent of the temperature. Eq. (11) also shows nonzero tangent moduli where A and M lose their stability. These observations coincide with the results from the earlier model [6] (Figs.1, 2 and 5).

3 Multivariant Phase Transformation

In order to model realistic situations and macroscopic sample of SMA it is essential to incorporate the effects of (1) martensitic variants (M_k) (2) thermal strain (3) unequal compliances across the interfaces and the resulting inhomogeneity. For cubic to tetragonal transformation there are three variants of martensite according to the point group of crystallographic symmetry. The Gibbs free energy density thus should poses the associated invariance properties. In the mathematical model, this can be cross-checked by interchanging the variant indices (k). In this paper we consider the same order of variation in the compliance tensor and the thermal expansion tensor as in $\varphi(\eta)$ derived in Sec. 2. The Gibbs free energy density for cubic-tetragonal transformation having three variants $k = 1, 2, 3$ is expressed as

$$G = -\boldsymbol{\sigma} : \left[\boldsymbol{\lambda}_0 + \sum_{k=1}^{3} (\boldsymbol{\lambda}_k - \boldsymbol{\lambda}_0)\varphi(\eta_k) \right] : \boldsymbol{\sigma}/2 - \boldsymbol{\sigma} : \sum_{k=1}^{3} \boldsymbol{\varepsilon}_{tk}\varphi(\eta_k)$$

$$-\boldsymbol{\sigma} : \left[\boldsymbol{\varepsilon}_{\theta 0} + \sum_{k=1}^{3} (\boldsymbol{\varepsilon}_{\theta k} - \boldsymbol{\varepsilon}_{\theta 0})\varphi(\eta_k) \right] + \sum_{k=1}^{3} f(\theta, \eta_k) + \sum_{i=1}^{2} \sum_{j=i+1}^{3} F_{ij}(\eta_i, \eta_j) , \tag{13}$$

where $\boldsymbol{\lambda}$ is the second-order forth-rank compliance tensor ($\boldsymbol{\lambda}_0$ is for A phase), $\boldsymbol{\varepsilon}_{\theta 0} = \boldsymbol{\alpha}_0(\theta - \theta_e)$, $\boldsymbol{\varepsilon}_{\theta k} = \boldsymbol{\alpha}_k(\theta - \theta_e)$, $\boldsymbol{\alpha}_0$ and $\boldsymbol{\alpha}_k$ are the thermal expansion tensor of A and M_k. F_{ij} is an interaction potential required to preserve the invariance of G with respect to the point group of symmetry and uniqueness of the multivariant PT at a given material point. The description of PT can now be generalized with three sets of order parameters: $\bar{0} = \{0, \eta_k = 0, 0\}$, $\bar{1} = \{0, \eta_k = 1, 0\}$ and $\bar{\eta}_k = \{0, \eta_k, 0\}$. The extremum property of the free energy density requires

$$\frac{\partial G}{\partial \eta_k} = \eta_k(\eta_k - 1)(\eta_k - \eta_{bk}) = 0 , \quad \eta_k = \bar{0}, \bar{1} , \tag{14}$$

$$\frac{\partial^2 G}{\partial \eta_k^2} \leq 0 , \quad \eta_k = \bar{0} \quad (A \to M_k); \quad \frac{\partial^2 G}{\partial \eta_k^2} \leq 0 , \quad \eta_k = \bar{1} \quad (M_k \to A). \tag{15}$$

The transformation energy associated with $A \leftrightarrow M_k$ is

$$G(\boldsymbol{\sigma}, \theta, \bar{0}) - G(\boldsymbol{\sigma}, \theta, \bar{1}) = \boldsymbol{\sigma} : \boldsymbol{\varepsilon}_{tk} - \Delta G^{\theta} . \tag{16}$$

Combining Eqs. (14) and (16) with similar steps described in Sec. 2, we get

$$\eta_{bk} = -6\boldsymbol{\sigma} : \boldsymbol{\varepsilon}_{tk} + 6\Delta G^\theta + 1/2 \tag{17}$$

Following the steps given in [11], we arrive at the symmetry preserving polynomial structure of the interaction potential

$$F_{ij} = \eta_i\eta_j(1-\eta_i-\eta_j)\left[B\left\{(\eta_i-\eta_j)^2-\eta_i-\eta_j\right\}+D\eta_i\eta_j\right]+\eta_i^2\eta_j^2(\eta_i Z_{ij}+\eta_j Z_{ji}) \tag{18}$$

such that

$$G = \sum_{k=1}^{3}\left[\frac{1}{2}\eta_{bk}\eta_k^2 - \frac{1}{3}(\eta_{bk}+1)\eta_k^3 + \frac{1}{4}\eta_{bk}^4\right] + \sum_{i=1}^{2}\sum_{j=i+1}^{3} F_{ij}(\eta_i,\eta_j) \tag{19}$$

leads to a 2-3-4-5 polynomial structure, where B, D are constants estimated from experiments. The transformation energy associated with $M_i \to M_j$ requires

$$G(\boldsymbol{\sigma},\theta,\bar{\eta}_i) - G(\boldsymbol{\sigma},\theta,\bar{\eta}_j) = \boldsymbol{\sigma} : (\boldsymbol{\varepsilon}_{tj} - \boldsymbol{\varepsilon}_{ti}) \tag{20}$$

which is already satisfied through Eq. (17). The uniqueness of PT at a material point is now imposed through similar steps described in context of Eq. (6).

4 Strongly Coupled Dynamics of the Phase Transformation and Coupled Thermoelasticity in a Rectangular NiAl Bar

The link between microstructure of SMA and its macroscopic behaviour is realized via the coupled field model in which the phase kinetics is governed by the Ginzburg-Landau equation

$$\frac{\eta_k}{\partial t} = -\sum_{p=1}^{3} L_{kp}\left[\frac{\partial G}{\partial \eta_p} - \boldsymbol{\beta}_p : \boldsymbol{\nabla\nabla}\eta_p\right] + \theta\ , \tag{21}$$

where L_{kp} are positive definite kinetic coefficients, $\boldsymbol{\beta}_k$ are positive definite second rank tensor, and the macroscopic energy conservation law is governed by the heat transfer equation

$$\rho\frac{\partial \bar{G}}{\partial t} - \boldsymbol{\sigma} : \boldsymbol{\nabla}\frac{\partial \boldsymbol{u}}{\partial t} + \boldsymbol{\nabla}\cdot\boldsymbol{q} = h_\theta\ , \tag{22}$$

and the momentum balance equation

$$\rho\frac{\partial^2 \boldsymbol{u}}{\partial t^2} = \boldsymbol{\nabla}\cdot\boldsymbol{\sigma} + \boldsymbol{f}\ , \tag{23}$$

where $\bar{G} = G - \theta\partial G/\partial\theta$ is the internal energy, ρ is the mass density, $\boldsymbol{u}$ is the displacement vector, $\boldsymbol{q}$ is the heat flux, h_θ and $\boldsymbol{f}$ are the thermal and mechanical loading, respectively. The displacement is related to the Green strain tensor as $\varepsilon = (1/2)[\boldsymbol{F}^T\boldsymbol{F} - \boldsymbol{I}]$ where $\boldsymbol{F}$ is the deformation gradient. For the purpose of numerical simulation we consider a rectangular NiAl bar with macroscopic domain $([0, L], [y^-, y^+], [z^-, z^+])$ undergoing single variant cubic to tetragonal transformation under uniaxial stress (σ_{11}) due to boundary conditions $f_1(0,t) = (y^+ - y^-)(z^+ - z^-)\sigma_0(t)$, $u_1(0,y,z,t) = 0$, traction-free surfaces, thermal boundary conditions at the ends $\partial\theta(x = 0,t)/\partial x = \bar{\theta}(t)_0$, $\partial\theta(x = L,t)/\partial x = \bar{\theta}(t)_L$, on the surface $(y = y^+, y^-)$, $\partial\theta/\partial y = \bar{\theta}_S$ and on the surface $(z = z^+, z^-)$, $\partial\theta/\partial z = \bar{\theta}_S$. The initial conditions are $\theta(x,0) = \eta_0(x)$, $\theta(x,y,z,0) = \theta_0(x)$, $u(x,y,z,0) = u_0(x)$ and $\partial u(x,y,z,0)/\partial t = 0$. A consistent form of the temperature field is obtained as

$$\theta(x,y,z,t) = \theta_1(x,t) + (y+z)\bar{\theta}_s \; . \tag{24}$$

The longitudinal displacement and strain fields are approximated as

$$u_1 = u(x,t) \; , \quad \varepsilon_{11} = \frac{\partial u}{\partial x} + \frac{1}{2}\left(\frac{\partial u}{\partial x}\right)^2 \; , \quad \varepsilon_{22} = \frac{\partial u_2}{\partial y} \; , \quad \varepsilon_{33} = \frac{\partial u_3}{\partial z} \; . \tag{25}$$

In the present example we assume that the habit planes are normal to e_1, such that $\eta = \eta(x,t)$. For microscopic deformation, the fine graining of strain requires the property of cubic to tetragonal transformation. For this model, we obtain the consistent form of the unknown fields $u_2(x,y,z,t)$ and $u_3(x,y,z,t)$ by imposing the geometric constraints for effectively one-dimensional dynamics, which leads to

$$u_2 = \frac{\mu}{2}y + \frac{\lambda_{21}}{\lambda_{11}}y\varepsilon_{11} - \left(\frac{1}{2} + \frac{\lambda_{21}}{\lambda_{11}}\right)\varepsilon_t\varphi(\eta)y + \alpha\left(1 - \frac{\lambda_{21}}{\lambda_{11}}\right)\int(\theta - \theta_0)dy \; , \tag{26}$$

$$u_3 = \frac{\mu}{2}z + \frac{\lambda_{31}}{\lambda_{31}}y\varepsilon_{11} - \left(\frac{1}{2} + \frac{\lambda_{31}}{\lambda_{11}}\right)\varepsilon_t\varphi(\eta)y + \alpha\left(1 - \frac{\lambda_{31}}{\lambda_{11}}\right)\int(\theta - \theta_0)dz \; , \tag{27}$$

where μ is the prescribed error in the volumetric strain. This quasi-continuum model with discontinuous distribution of $\eta(x,t)$, and continuous fields $u_1(x,t)$, $\theta_1(x,t)$ is then solved by variational minimization of Eqs. (21)-(23).

5 Concluding Remarks

In this paper, a new dynamic multiscale model for simulation of 3D SMA samples has been developed by linking an improved version of the microscopic Landau theory and macroscopic conservation laws. Essential properties of the $A \leftrightarrow M_j$ as well as $M_j \leftrightarrow M_i$ PTs are recovered using consistent steps, which eliminates the problem of non-uniqueness of energy partitioning during PTs and relaxes the over-sensitivity of the free energy due to many unknown material constants in previously reported models. It has been shown how the new 3D model can be reduced to a low dimensional model for simulating the strongly coupled phase kinetics and thermoelasticity in a rectangular bar undergoing cubic to tetragonal phase transformations.

References

1. Falk, F. and Konopka, P.: Three-dimensional Landau theory describing the martensitic phase transformation of shape -memory alloys. J. Phys.: Condens. Matter **2** (1990) 61–77
2. Melnik, R.V.N., Roberts, A.J. and Thomas, K.A.: Computing dynamics of copper-based shape memory alloys via center manifold reduction of 3D models. Computational Materials Science **18** (2000) 255–268
3. Boyd, J.G. and Lagaoudas, D.C.: A thermodynamical constitutive model for for shape memory materials. Part I. The monolithic shape memory alloy. Int. J. Plasticity **12**(9) (1996) 805–842
4. Aurichio, F. and Sacco, E.: A temperature-dependent beam for shape memory alloys: constitutive modelling, finite element implementation and numerical simulations. Comput. Methods Appl. Mech. and Engrg. **174** (1999) 171-190
5. Bhattacharya, K. and Khon, R.V.: Symmetry, texture and the recoverable strain of shape memory polycrystal. Acta Mater. **44**(2) (1996) 529-542
6. Levitas, V.I. and Preston, D.L.: Three-dimensional Landau theory for multivariant stress-induced martensitic phase transformations. I. Austenite $\leftrightarrow$ martensite Physical Rev. B **66** (2002) 134206
7. Chen. L.Q., Wang, Y. and Khachaturyan, A.G.: Kinetics of tweed and twin formation during an ordering transition in a substitutional solid solution. Philos. Mag. Lett. **65**(15) (1992) 15–23
8. Lindgard, P.A. and Mouritsen, O.G.: Theory and model for martensitic transformations. Phy. Rev. Lett. **57**(19) (1986) 2458
9. Clapp, P.C., Besquart, C.S., Shao, Y., Zhao, Y. and Rifkin, J.A.: Transformation toughening explored via molecular dynamics and Monte Carlo simulations. Modelling Simul. Mater. Sci. Eng. **2** (1994) 551–558
10. Rubini, S. and Ballone, P.: Quasiharmonic and molecular-dynamic study of the martensitic transformation in Ni-Al alloys. Phy. Rev. B **48** (1993) 99
11. Levitas, V.I. and Preston, D.L.: Three-dimensional Landau theory for multivariant stress-induced martensitic phase transformations. II. Multivariant phase transformations and stress space analysis. Physical Rev. B **66** (2002) 134207
12. Levitas, V.I. and Preston, D.L.: Three-dimensional Landau theory for multivariant stress-induced martensitic phase transformations. III. Alternative potentials, critical nuclei, kink solutions, and dislocation theory Physical Rev. B **68** (2003) 134201
13. Matus, P., Melnik, R.V.N., Wang, L. and Rybak, I.: Application of fully conservative schemes in nonlinear thermoelasticity: modelling shape memory materials. Mathematics and Computers in Simulation **65** (2004) 489-510
14. Ichitsubo, T., Tanaka, K., Koiwa, M. and Yamazaki, Y.: Kinetics of cubic to tetragonal transformation under external field by the time-dependent Ginzburg-Landau approach. Phy. Rev. B **62** (2000) 5435
15. Fu, S., Huo, Y. and Muller, I.: Thermodynamics of pseudoelasticity - an analytical approach. Acta Mechanica **99** (1993) 1-19

3D Finite Element Modeling of Free-Surface Flows with Efficient $k-\epsilon$ Turbulence Model and Non-hydrostatic Pressure

Célestin Leupi[1] and Mustafa Siddik Altinakar[2]

[1] ISE-STI-LIN, Ecole Polytechnique Fédérale,
Lausanne 1015, Switzerland
Phone: +41.21.693.25.07, Fax + 41.21.693.36.46
celestin.leupi@epfl.ch
[2] NCCHE, The University of Mississipi,
Carrier Hall Room 102 University, MS 38677 USA

Abstract. Validation of 3D finite element model for free-surface flow is conducted using a high quality and high spatial resolution data set. The present research finds its motivation in the increasing need for efficient management of geophysical flows such as estuaries (multiphase fluid flow) or natural rivers with the complicated channel geometry (e.g. strong channel curvature). A numerical solution is based on the unsteady Reynolds-averaged Navier-Stokes equations without conventional assumption of hydrostatic pressure. The model uses implicit fractional step time stepping, with the characteristic method for convections terms . The eddy viscosity is calculated from the efficient $k-\epsilon$ turbulence model. The RANS are solved in the multi-layers system (suitable for the vertical stratified fluid flow) to provide the accurate resolution at the bed and free-surface. The model is applied to the 3D curved open channels flows for which experimental data are available for comparison. Good agreement is found between numerical computations and experiments.

Keywords: Validation; Characteristic method; 3D Curved open channel; secondary currents; Non-hydrostatic pressure.

1 Introduction

Nowadays with the increasing computer power, several 3D computations have been successfully conducted for geophysical flows modeling. Most of these models have used the conventional hydrostatic pressure assumption. However, the natural rivers mostly have complicated geometry with the strong channel curvature. Such flows are of more importance for environmental hydraulic engineering and some related important features such as the secondary flows generated by the channel curvature and the related background turbulence effects, need to be well understood while a 3D description of the velocity field is required. Thus it is useful to resort to a more accurate model in which the hydrostatic assumption

V.S. Sunderam et al. (Eds.): ICCS 2005, LNCS 3516, pp. 33–40, 2005.

is removed. Nevertheless, the importance of non-hydrostatic pressure in computational fluid problems was demonstrated and many researchers have applied 3D non-hydrostatic models to simulate the curved open channel flows. Wu et al. [13] and Olsen [10] used 3D numerical models to study the flow structure and mass transport in curved open channel. Xiabo et al. [2] have simulated the 3D unsteady curved open channel with standard $k - \epsilon$ turbulence model and the non-hydrostatic pressure on the conformal mesh, but conformal mesh could poorly performed for some complicated bathymetry. Lai et al. [4] have used finite volume method on the unstructured grid to simulate 3D flow in meandering channel. For the free surface treatment, most of these 3D models employed the rigid-lid approximation, which have some weaknesses especially in strongly curved open channel flows (see [5]).

Based on the novel approach developed by Leupi et al. [6], the present model adopts the finite element conservative formulation in the multi-layers system for providing an accurate resolution at the bed and the free-surface.

The present work aims at validating the 3D finite element model against well-known non-uniform and unsteady flows in curved open channel flows using a high quality and high spatial resolution data set. The model uses the non-hydrostatic pressure and the state-of-art $k - \epsilon$ turbulence model closure to solve the Reynolds-averaged Navier-Stokes equations (RANS).

In this study, the free-surface movement is controlled through the so-called *integrated continuity* equation. The full 3D governing equations are solved using implicit fractional time marching stepping where final velocity field and pressure term are computed from the hydrodynamic correction. Euler or Runge-Kutta scheme is used to obtain a set of algebraic equations from discretization. An efficient fractional step algorithm from Mohammadi and Pironneau [9] is introduced for the $k - \epsilon$ model. This paper deals with the simulation of the 3D turbulent flow in the open curved channel for which experimental data are available.

2 Governing Equations

Let us consider an incompressible fluid body in a three-dimensional time varying domain $\widehat{\Omega}$ (see also [6]) with Ω the projection of $\widehat{\Omega}$ on the xy horizontal plane. $\widehat{\Omega}$ is bounded by the free-surface Γ_s given by $z = \eta(x, y, t)$, the bottom topography Γ_b given by $z = -\mathfrak{h}(x, y)$, the open boundary denoted by Γ_o. Where $\mathfrak{h}(x, y)$ is the distance between the bottom and the reference plane xy and $\eta(x, y, t)$ the elevation of the free-surface with the respect to the horizontal plane xy. For description of the turbulent motion, the pressure p can be written as the sum of an hydrostatic term p_h and an hydrodynamic correction $p_{nh} = \rho\tilde{p}$,

$$p(\mathbf{x}, t) = p_h + p_{nh} = p_a + g\rho_o(\eta - z) + g\int_z^\eta \Delta\rho \mathrm{d}z + \rho\tilde{p}(\mathbf{x}, t) \tag{1}$$

The 3D non hydrostatic Reynolds Averaged Navier-Stokes (RANS) equations reads

$$\nabla_H \cdot \mathbf{U} + \frac{\partial w}{\partial z} = 0 \tag{2}$$

$$\frac{D\mathbf{U}}{Dt} + g\nabla_H \eta - \nabla_H (\nu_T \nabla_H \mathbf{U}) - \frac{\partial}{\partial z}\left(\nu_T \frac{\partial \mathbf{U}}{\partial z}\right) + g\nabla\left(\int_{\mathfrak{h}}^{\eta} \frac{\Delta\rho}{\rho_o} \mathrm{d}z\right) + \frac{1}{\rho}\nabla\tilde{p} = \mathbf{F}_{\mathrm{xy}} \tag{3}$$

$$\frac{Dw}{Dt} - \nabla_H (\nu_T \nabla_H w) - \frac{\partial}{\partial z}\left(\nu_T \frac{\partial w}{\partial z}\right) + \frac{1}{\rho}\frac{\partial \tilde{p}}{\partial z} = 0 \tag{4}$$

$$\frac{\partial \eta}{\partial t} + \nabla_H \cdot \int_{-\mathfrak{h}}^{\eta} \mathbf{U} \mathrm{d}z = 0 \tag{5}$$

where $\mathbf{U} = (u, v)^T$ is the horizontal velocity vector , $\mathbf{F}_{xy} = (\mathrm{f}v, -\mathrm{f}u)^T$ is vector of body forces with f the Coriolis parameter, g is the gravitational acceleration, ν_T is the eddy viscosity, (see Rodi [12]). $(\nabla\cdot)$ is the $3D$ divergence operator, $\frac{D}{Dt}$ represents the material derivative, and $(\nabla_H\cdot)$ stands as the 2D horizontal divergence operator. $\Delta\rho = \rho - \rho_0$; ρ, ρ_0 are respectively the fluid density and the basic water density.

In eq.(3), the vertical eddy viscosity is defined as

$$\nu_T = \nu + c_\mu \frac{k^2}{\varepsilon} \tag{6}$$

in which ν is the kinematic viscosity.

The $k - \varepsilon$ trubulence equations read ([9]),

$$\frac{Dk}{Dt} - \nabla \cdot \left[c_\mu \frac{k^2}{\varepsilon} \nabla k\right] = c_\mu \frac{k^2}{\varepsilon} G - \varepsilon \tag{7}$$

$$\frac{D\varepsilon}{Dt} - \nabla \cdot \left[c_\varepsilon \frac{k^2}{\varepsilon} \nabla \varepsilon\right] = \frac{c_1}{2} kG - c_2 \frac{\varepsilon^2}{k} \tag{8}$$

The turbulent constants are given: $c_1 = 0.126$, $c_2 = 0.07$, $c_\mu = 0.09$, $c_\varepsilon = 1.92$.

The production is represented by the squared shear frequency, G, (see [7]), such as :

$$G = \frac{1}{2}\left(\|\nabla \mathbf{V}\| + \|\nabla \mathbf{V}\|^T\right)^2 \tag{9}$$

where $\|.\|$ stands as the Euclidian norm, $\mathbf{V} = \mathbf{V}(u, v, w)$ is the 3D velocity vector.

The *depth-integrated continuity* equation eq. (5) allow the model to follow the free-surface position. This equation is obtained by integrating the (local) continuity equation (2) in the z direction using the suitable kinematic free-surface and bottom boundary conditions.

The horizontal velocity is approximated combining the lowest order Raviart-Thomas element ($\mathbb{RT}_0$) in xy plane with the $\mathbb{P}_1$ elements along the vertical direction (see [6], [8], [11]). To discretize the convective term a method based on a Lagrange-Galerkin (or characteristics Galerkin) approach is considered (see [8], [9]) using either Euler scheme or more accurate Runge-Kutta. At each time step it is only required to solve a set of the positive definite symmetric and tridiagonal matrices for the fluxes using the conjugate gradient solver. For the turbulence modeling, the combinaison of the characteristics method with the fractional time stepping algorithm from Mohammadi and Pironneau ([9], [6]) can allow to preserve the positivity of k, ε as well as the stability of the scheme (see [9], [12]). To avoid spurious numerical oscillations, the source term, G, has been discretised explicitly while the sink term has been discretised using the quasi-implicit forms with the consequence that linear terms are linearized (see [9]).

3 Numerical Results

The present model has been applied for simulating a 3D curved open channel flow in Figure 1, for which experimental data are available (see Blanckaert [1]). The

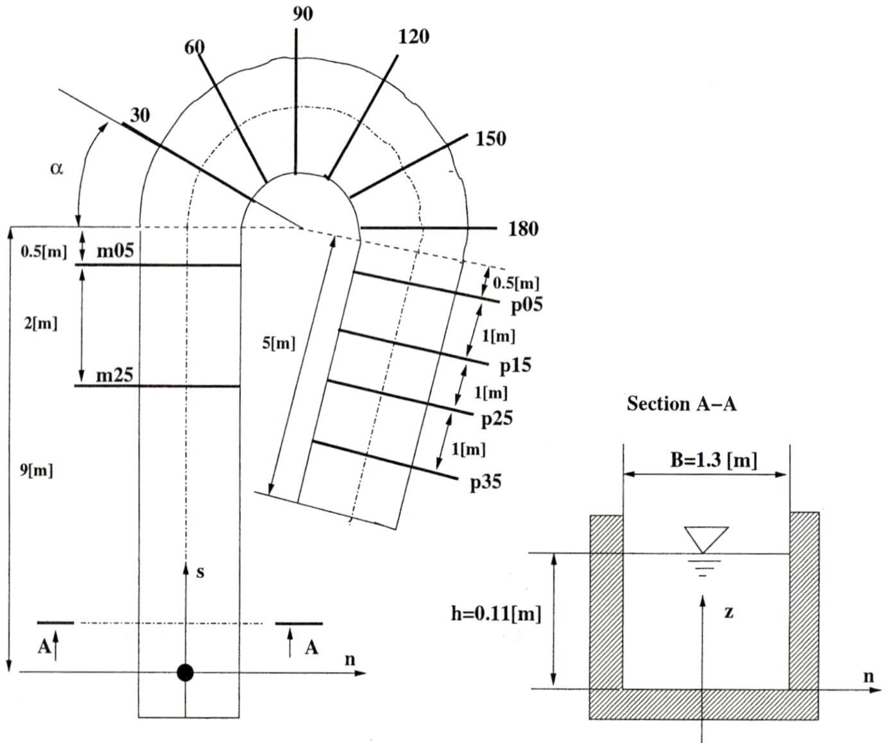

Fig. 1. A 193° Curved open-channel

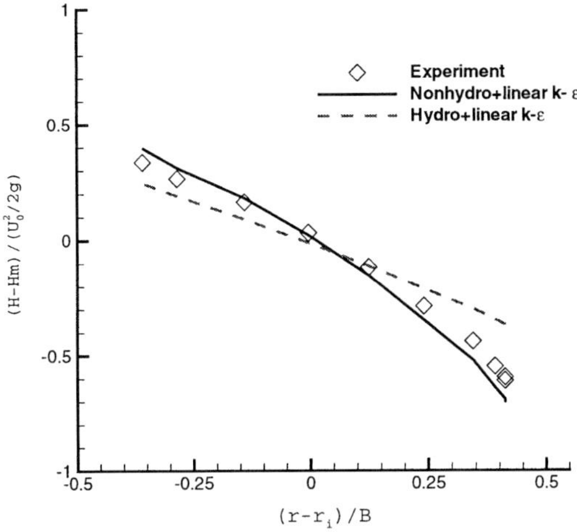

Fig. 2. Experiments versus computed free-surface solutions at section $\alpha = 120^o$ using linear $k - \epsilon$ Turbulence model : Experiment ($\Diamond$) ; Hydrostatic (dashed line) ; Non hydrostatic (solid line)

discharge is set to $Q = 0.089[\mathrm{m}^3/\mathrm{s}]$, the bed slope (i.e. channel bottom slope) is $S_0 = 0.000624$. The rough bed is characterized by an equivalent roughness height, $k_s = 0.0022[\mathrm{m}]$. The flow depth at the outflow (downstream end of the flume) is 0.159[m]. The grid is composed of 50.000 elements and 30.000 nodes. The time step is set to 0.1 [s], and the computation is performed till the flow is well developed at T=1300 [s]. Computations were conducted with the hydrostatic pressure assumption for different cross-sections with the curvature increasing, to determine the conditions where the non-hydrostatic pressure component become significant. Computed solutions predict the gross flow features, whereas the water surface profile is under-estimated at the out bank, and over-estimated at the inner bank.

Figure 2 shows the cross-section at $\alpha = 120^o$, where are found the weaknesses of the hydrostatic pressure solution. Hence agreement with experiments is rather better for non-hydrostatic solutions particularly with the increasing curvature. This suggest that the separation may occur in the vertical flow and the the pressure-driven secondary effects are important. Thus the free surface must be more accurately computed to accounts for its damping effects on the turbulent flow. As observed in figure 3, both hydrostatic and non-hydrostatic pressure solutions show only one secondary flow circulation rotating clockwise from inner bank to outer bank. These predictions do not capture sufficiently the magnitude of the secondary motion. The maximum under prediction in the secondary currents for each vertical examined in this cross-section is ranged between 25 and 95% for the non-hydrostatic solutions and between 30 and 105% for the hydrostatic pressure. In Figure 3 the center of vortex is located at about $z = 0.25[\mathrm{m}]$ for the hydrostatic, about $z = 0.35[\mathrm{m}]$ for the non hydrostatic solutions which is more closed to experiments located at $z = 0.4[\mathrm{m}]$. The predicted secondary currents intensities are weaker than

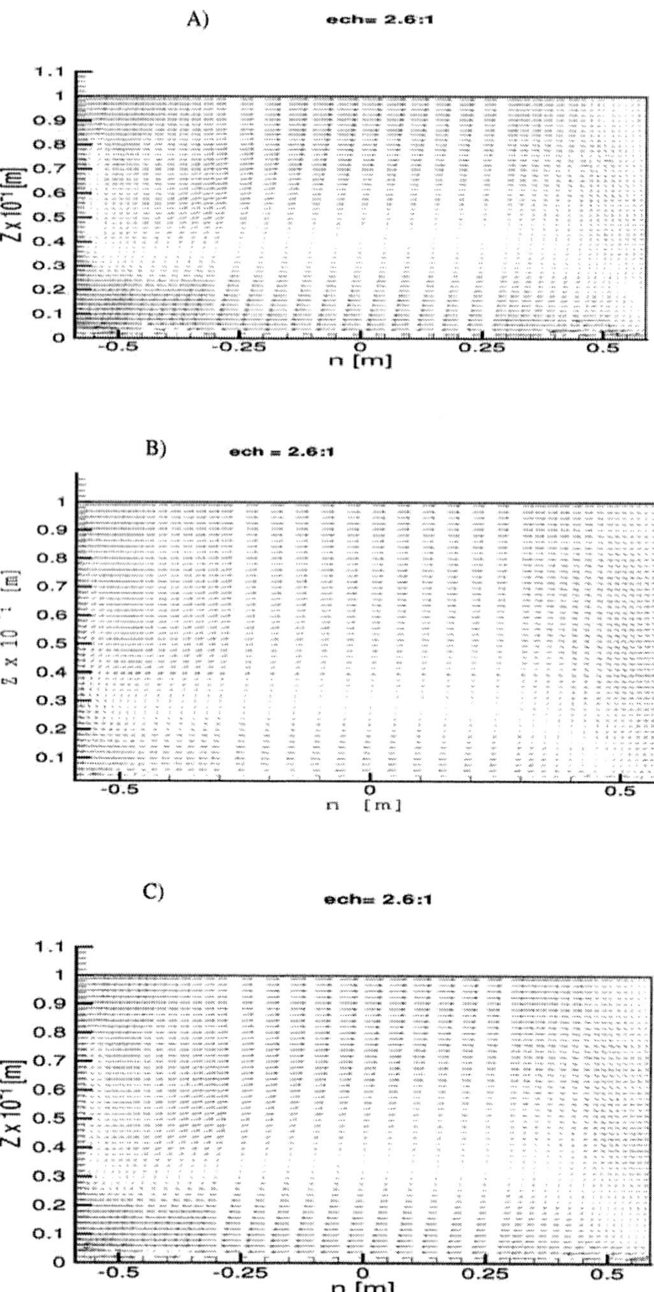

Fig. 3. Experiments versus computed solutions of the of cross-stream velocity vector at section $\alpha = 120^o$: A) Experiment ; B) Hydrostatic ; C) Non hydrostatic

measured, and the transverse velocities are under predicted. The anisotropic stress caused by walls and the junction region is not captured by the model, and the reduced momentum is transfered towards the outer region of the bend as well as the position of the longitudinal velocity maximum. This suggest that the turbulence-driven secondary effects are non-linear. Consequently, as shown by Gatsky et al. [3], the related weaker turbulence-driven secondary motion cannot be reproduced by linear and isotropic eddy-viscosity turbulence models. It should be pointed out that the major flow features such as the presence and rotational sense of the major secondary currents are reproduced and agrees well with experiments. The non-hydrostatic pressure influence is found to be more significant with the increasing curvature region, and although being more expensive, it become useful. This suggest that the pressure-driven secondary currents is relatively important for the accurated description of the velocity field and the use of anisotropic turbulence models is prerequisite to more accurate flow field prediction.

4 Conclusion

Validation of the 3D finite element solver for the RANS equations with the Efficient $k - \epsilon$ turbulence model is conducted successfully using a high quality and high spatial resolution data set. The convection terms have been discretized using the Lagrange-Galerkin approach with advantage that, the CFL restriction is well performed. Moreover, addition of this characteristic method to the conservative form of the PDE and the implicit fractional step time stepping, allow to preserve the mass balance, the positivity of k and ϵ, as well as the stability of the scheme. In computed solutions, the weaker secondary currents were not reproduced, but it shlould be noticed that more refine turbulence modeling can produce improvement for such problem. Computations with and without non-hydrostatic are compared for the same trench to test the validity of the conventional hydrostatic pressure assumption. The model predicts reasonably the complex major features and the 3D flow tests were performed successfully against well-known unsteady non-uniform curved open channel flows.The non-hydrostatic pressure influence is found to be more significant with the increasing curvature region (e.g. cross-section $\alpha = 120^o$). This suggest that the non-hydrostatic pressure may be useful and well suited for complicated geometry flows where its influence is thought to be significant. Further study is needed to improve the general applicability of the model, and the next stage of this work will be focus on the anisotropic turbulence-driven secondary motion.

Acknowledgment

The first author gratefully acknowledged Funding from the Swiss National Science Foundation through grant number 21-65095.01. Michel Deville, Alfio Quarteroni, Edie Miglio and Koen Blanckaert are acknowledged for their fruitfully discussions.

References

1. Koen Blanckaert. *Flow and turbulence in sharp open-channel bends.* PhD thesis, Ecole polytechnique de Lausanne- EPFL, N0. 2545, 2002.
2. Xiabo Chao, Yafei jia, and Sam S. Y. Wang. Three-dimensional simulation of buoyant heat transfer in a curved open channel. In *Proc. Int. Conf. on Advances in hydro-Science and -Engineering, ICHE, NCCHE Mississipi, USA, vol 6*, pages 18–19 & on CDROM, 2004.
3. Thomas B. Gatski, M. Yousuff Hussaini, and John L. Lumley. *Simulation and Modeling of Turbulent Flows.* Oxford University Press, 198 Madison Avenue, New York, New York 10016, 1996.
4. Yong G. Lai, Larry J. Weber, and Virendra C. Patel. Three-dimensional model for hydraulic flow simulation. i: formulation and verification. *ASCE, J. Hydr. Engrg.*, 129(3):196–205, 2003.
5. Michael A. Leschziner and Wolfang Rodi. Calculation of strongly curved open channel flow. *ASCE, J. Hydr. Div.*, 103(10):1297–1314, 1979.
6. Célestin Leupi, Edie Miglio, Mustafa Altinakar, Alfio Quarteroni, and Michel Deville. Quasi-3d finite element shallow-water flow with $k-\varepsilon$ turbulence model. In *Proc. Int. Conf. on Advances in hydro-Science and -Engineering, ICHE, NCCHE Mississipi, USA, vol 6*, pages 400–401 & on CDROM, 2004.
7. Patrick J. Luyten, John Eric Jones, Roger Proctor, Andy Tabor, Paul Tett, and Karen Wild-Allen. Coherens- a coupled hydrodynamical-ecological model for regional and shelf seas. Technical Report MUMM report, Management unit of the Mathematical Models of North Sea, 914pp, COSINUS, 1999.
8. Edie Miglio, Alfio Quarteroni, and Fausto Saleri. Finite element approximation of quasi-3D shallow water equations. *Comput. Methods Appl. Mech. Engrg.*, 174(34):355–369, 1999.
9. Bijan Mohammadi and Olivier Pironneau. *Analysis of k – epsilon Turbulence Model.* Research in Applied Mathematics. John Wiley & Sons, Chichester, 1994.
10. Nils Reidar B. Olsen. Three-dimensional cfd modeling of self-forming meandering channel. *ASCE, J. Hydr. Engrg.*, 129(5):366–372, 2003.
11. Pierre Arnaud Raviart and Jean Marie Thomas. *A mixed finite element method for 2nd order elliptic problems.* Springer-Verlag, Mathematical Aspects of Finite Element Methods, Lecture notes in Methematics.
12. Wolfang Rodi. Turbulence models and their applications in hydraulics. Technical Report 2nd edition, IAHR, Delft, Netherlands, 1984.
13. Weiming Wu, Wolfgang Rodi, and Thomas Wenka. 3D numerical modeling of flow and sediment transport in open channels. *ASCE, J. Hydr. Engrg.*, 126(1):4–15, 2000.

Cluster Computing for Transient Simulations of the Linear Boltzmann Equation on Irregular Three-Dimensional Domains

Matthias K. Gobbert[1], Mark L. Breitenbach[1], and Timothy S. Cale[2]

[1] Department of Mathematics and Statistics,
University of Maryland, Baltimore County,
1000 Hilltop Circle, Baltimore, MD 21250, U.S.A
[2] Focus Center — New York, Rensselaer,
Interconnections for Hyperintegration,
Isermann Department of Chemical and Biological Engineering,
Rensselaer Polytechnic Institute, CII 6015,
110 8th Street, Troy, NY 12180-3590, U.S.A

Abstract. Processes used to manufacture integrated circuits take place at a range of pressures and their models are of interest across a wide range of length scales. We present a kinetic transport and reaction model given by a system of linear Boltzmann equations that is applicable to several important processes that involve contacting in-process wafers with reactive gases. The model is valid for a range of pressures and for length scales from micrometers to centimeters, making it suitable for multiscale models. Since a kinetic model in three dimensions involves discretizations of the three-dimensional position as well as of the three-dimensional velocity space, millions of unknowns result. To efficiently perform transient simulations with many time steps, the size of the problem motivates the use of parallel computing. We present simulation results on an irregular three-dimensional domain that highlights the capabilities of the model and its implementation, as well as parallel performance studies on a distributed-memory cluster show that the computation time scales well with the number of processes.

1 Introduction

Many important manufacturing processes for integrated circuits involve the flow of gaseous reactants at pressures that range from very low to atmospheric. Correspondingly, the mean free path λ (the average distance that a molecule travels before colliding with another molecule) ranges from less than 0.1 micrometers to over 100 micrometers. The typical size of the electronic components (called 'features' during processing) is now below 1 micrometer and the size of the chemical reactor, in which the gas flow takes place, is on the order of decimeters. Thus, models on a range of length scales L^* are of interest, each of which needs to be appropriately selected to be valid on its length scale. These authors have in the

V.S. Sunderam et al. (Eds.): ICCS 2005, LNCS 3516, pp. 41–48, 2005.

past coupled models on several length scales, from feature scale to reactor scale, to form an interactive multiscale reactor simulator [1, 2]. The models used were based on continuum models in all but the feature scale and assumed moderately high pressure to be valid. The current work provides the basis for extending this work to more general pressure regimes. Such multiscale models require well-tested and validated models and numerical methods on every scale of interest. The following results address both the effectiveness of the model and computational efficiency of the numerical method and its parallel implementation.

The appropriate transport model at a given combination of pressure and length scale is determined by the Knudsen number Kn, defined as the ratio of the mean free path and the length scale of interest $\mathrm{Kn} := \lambda/L^*$, which arises as the relevant dimensionless group in kinetic models [3]: (i) For small values $\mathrm{Kn} < 0.01$, continuum models describe the gas flow adequately. (ii) At intermediate values $\mathrm{Kn} \approx 1.0$, kinetic models based on the Boltzmann transport equation capture the influence of both transport of and collisions among the molecules. (iii) For large values $\mathrm{Kn} > 100.0$, kinetic models remain valid with the collision term becoming negligible in the limit as Kn grows.

Our interest includes models on the micro- to meso-scale at a range of pressures, resulting in Knudsen numbers ranging across the wide spectrum from less than $\mathrm{Kn} = 0.1$ to $\mathrm{Kn} \to \infty$, in the regimes (ii) and (iii) above. For flow in a carrier gas, assumed inert, at least an order of magnitude denser than the reactive species, and in spatially uniform steady-state, we have developed a kinetic transport and reaction model (KTRM) [4, 5], given by the system of linear Boltzmann equations for all n_s reactive species in dimensionless form

$$\frac{\partial f^{(i)}}{\partial t} + \mathbf{v} \cdot \nabla_{\mathbf{x}} f^{(i)} = \frac{1}{\mathrm{Kn}} Q_i(f^{(i)}), \quad i = 1, \ldots, n_s, \tag{1}$$

with the linear collision operators

$$Q_i(f^{(i)})(\mathbf{x}, \mathbf{v}, t) = \int_{\mathbb{R}^3} \sigma_i(\mathbf{v}, \mathbf{v}') \left[M_i(\mathbf{v}) f^{(i)}(\mathbf{x}, \mathbf{v}', t) - M_i(\mathbf{v}') f^{(i)}(\mathbf{x}, \mathbf{v}, t) \right] d\mathbf{v}',$$

where $\sigma_i(\mathbf{v}, \mathbf{v}') = \sigma_i(\mathbf{v}', \mathbf{v}) \geq 0$ is a given collision frequency model and $M_i(\mathbf{v})$ denotes the Maxwellian density of species i. The left-hand side of (1) models the advective transport of molecules of species i (local coupling of spatial variations via the spatial derivatives $\nabla_{\mathbf{x}} f^{(i)}$), while the right-hand side models the effect of collisions (global coupling of all velocities in the integral operators Q_i). The unknown functions $f^{(i)}(\mathbf{x}, \mathbf{v}, t)$ in this kinetic model represent the (scaled) probability density that a molecule of species $i = 1, \ldots, n_s$ at position $\mathbf{x} \in \Omega \subset \mathbb{R}^3$ has velocity $\mathbf{v} \in \mathbb{R}^3$ at time t. Its values need to be determined at all points $\mathbf{x}$ in the three-dimensional spatial domain Ω and for all three-dimensional velocity vectors $\mathbf{v}$ at all times $0 < t \leq t_{\mathrm{fin}}$. This high dimensionality of the space of independent variables is responsible for the numerical complexity of kinetic models, as six dimensions need to be discretized, at every time step for transient simulations. Notice that while the equations in (1) appear decoupled, they actually remain coupled through the boundary condition at the wafer surface that models the surface reactions and is of crucial importance for the applications under consideration.

2 The Numerical Method

The numerical method for (1) needs to discretize the spatial domain $\Omega \subset \mathbb{R}^3$ and the (unbounded) velocity space $\mathbb{R}^3$. We start by approximating each $f^{(i)}(\mathbf{x}, \mathbf{v}, t)$ by an expansion $f_K^{(i)}(\mathbf{x}, \mathbf{v}, t) := \sum_{\ell=0}^{K-1} f_\ell^{(i)}(\mathbf{x}, t)\varphi_\ell(\mathbf{v})$. Here, the basis functions $\varphi_\ell(\mathbf{v})$ in velocity space are chosen such that they form an orthogonal set of basis functions in velocity space with respect to an inner product that arises naturally from entropy considerations for the linear Boltzmann equation [6]. Testing (1) successively against $\varphi_k(\mathbf{v})$ with respect to this inner product approximates (1) by a system of linear hyperbolic equations

$$\frac{\partial F^{(i)}}{\partial t} + A^{(1)} \frac{\partial F^{(i)}}{\partial \mathbf{x}_1} + A^{(2)} \frac{\partial F^{(i)}}{\partial \mathbf{x}_2} + A^{(3)} \frac{\partial F^{(i)}}{\partial \mathbf{x}_3} = \frac{1}{\mathrm{Kn}} B^{(i)} F^{(i)}, \quad i = 1, \ldots, n_s, \tag{2}$$

where $F^{(i)}(\mathbf{x}, t) := (f_0^{(i)}(\mathbf{x}, t), ..., f_{K-1}^{(i)}(\mathbf{x}, t))^T$ is the vector of the K coefficient functions in the expansion in velocity space. Here, $A^{(1)}$, $A^{(2)}$, $A^{(3)}$, and $B^{(i)}$ are constant $K \times K$ matrices. Picking specially designed basis functions, the coefficient matrices $A^{(\delta)}$, $\delta = 1, 2, 3$, become diagonal matrices [7, 8]. Note again that the equations for all species remain coupled through the crucial reaction boundary condition at the wafer surface.

The hyperbolic system (2) is now posed in a standard form as a system of partial differential equations on the spatial domain $\Omega \subset \mathbb{R}^3$ and in time t and amenable to solution by various methods. Figure 1 shows two views of a representative domain $\Omega \subset \mathbb{R}^3$; more precisely, the plots show the solid wafer surface consisting of a 0.3 micrometer deep trench, in which is etched another 0.3 micrometer deep via (round hole). The domain Ω for our model is the gaseous region above the solid wafer surface up to the top of the plot box shown at $\mathbf{x}_3 =$ 0.3 micrometers in Figure 1. Since typical domains in our applications such as this one are of irregular shape, we use the discontinuous Galerkin method (DGM) [9], relying on a finite element discretization into tetrahedra of the domain. Explicit time-stepping is used because of its memory efficiency and cheap cost per time step. We accept at this point the large number of time steps that will be required, because we also wish to control the size of the step size to maintain a high level of accuracy.

The degrees of freedom (DOF) of the finite element method are the values of the n_s species' coefficient functions $f_k^{(i)}(\mathbf{x}, t)$ in the Galerkin expansion at K discrete velocities on the 4 vertices of each of the N_e tetrahedra of the three-dimensional mesh. Hence, the complexity of the computational problem is given by $4\, N_e\, K\, n_s$ at every time step. To appreciate the size of the problem, consider that the mesh of the domain in Figure 1 uses $N_e = 7{,}087$ three-dimensional tetrahedral elements; even in the case of a single-species model ($n_s = 1$) and if we use just $K = 4 \times 4 \times 4 = 64$ discrete velocities in three dimensions, as used for the application results in the following section, the total DOF are $N = 1{,}814{,}272$ or nearly 2 million unknowns to be determined at every time step. Then, to reach a (re-dimensionalized) final time of 30.0 nanoseconds, for instance, requires about 60,000 time steps, with time step Δt selected according to a CFL-condition. This size of problem at every time step motivates our interest in parallel computing for this problem. For the parallel computations on a distributed-memory cluster, the spatial domain Ω is partitioned in a pre-processing step, and the disjoint subdomains are distributed to separate parallel processes. The discontinuous Galerkin method for (2) needs the flux through the element faces. At the interface from one subdomain to the next, communications are required among those pairs of parallel processes that share a subdomain boundary. Additionally, a number of global reduce operations are needed to compute inner products, norms, and other diagnostic quantities.

3 Application Results

As an application example, we present a model for chemical vapor deposition. In this process, gaseous chemicals are supplied from the gas-phase interface at $\mathbf{x}_3 = 0.3$ in Figure 1. The gaseous chemicals flow downwards throughout the domain Ω until they reach the solid wafer surface in Figure 1, where some of the molecules react to form a solid deposit. The time scale of our simulations corresponds to forming only a very thin layer; hence the surface is not moved within a simulation. We focus on how the flow behaves when starting from no gas present throughout Ω, modeled by initial condition $f^{(i)} \equiv 0$ at $t = 0$. Here, we use a single-species model with one reactive species ($n_s = 1$). The deposition at the wafer surface can then be modeled using a sticking factor $0 \leq \gamma_0 \leq 1$ that represents the fraction of molecules that are modeled to deposit at ("stick to") the wafer surface. The re-emission into Ω of gaseous molecules from the wafer surface is modeled as re-emission with velocity components in Maxwellian form and proportional to the flux to the surface as well as proportional to $1 - \gamma_0$. The re-emission is scaled to conserve mass in the absence of deposition ($\gamma_0 = 0$). The studies shown use a sticking factor of $\gamma_0 = 0.01$, that is, most molecules re-emit from the surface, which is a realistic condition. The collision operator uses a relaxation time discretization by choosing $\sigma_1(\mathbf{v}, \mathbf{v}') \equiv 1/\tau_1$ with (dimensionless) relaxation time $\tau_1 = 1.0$.

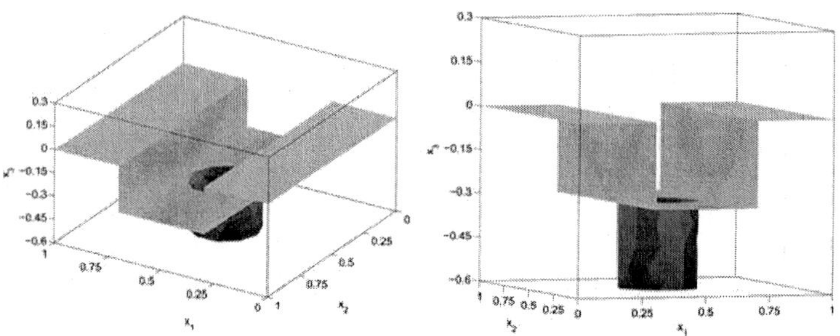

Fig. 1. Two views of the solid wafer surface boundary of the trench/via domain

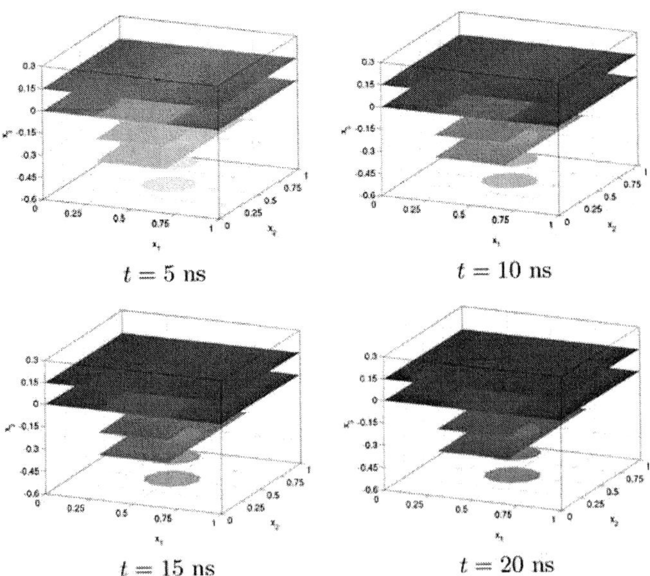

Fig. 2. Slice plots at heights $\mathbf{x}_3 = -0.60, -0.45, -0.30, -0.15, 0.00, 0.15$ at different times t of the dimensionless concentration $c(\mathbf{x}, t)$ for Kn $= 0.1$ throughout domain Ω. Grayscale from light $\Leftrightarrow c = 0$ to dark $\Leftrightarrow c = 1$

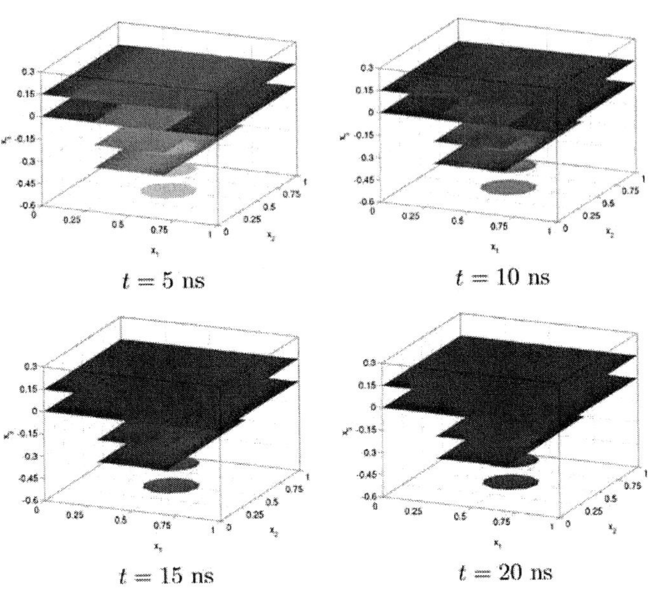

Fig. 3. Slice plots at heights $\mathbf{x}_3 = -0.60, -0.45, -0.30, -0.15, 0.00, 0.15$ at different times t of the dimensionless concentration $c(\mathbf{x}, t)$ for Kn $= 1.0$ throughout domain Ω. Grayscale from light $\Leftrightarrow c = 0$ to dark $\Leftrightarrow c = 1$

Figures 2 and 3 show the results of transient simulations for the values of the Knudsen number Kn = 0.1 and Kn = 1.0, respectively. The quantity plotted for each (re-dimensionalized) time is the (dimensionless) concentration

$$c(\mathbf{x}, t) := \int_{\mathrm{IR}^3} f^{(1)}(\mathbf{x}, \mathbf{v}, t)\, d\mathbf{v}$$

across the domain Ω. The values of the dimensionless concentration $0 \leq c \leq 1$ is represented by the gray-scale color on each of the horizontal slices through Ω at the vertical levels at six values of $\mathbf{x}_3$; the shapes of all slices together indicate the shape of the domain Ω.

At $t = 5$ ns, the top-most slice at $\mathbf{x}_3 = 0.15$ is mostly dark-colored, indicating that a relatively high concentration of molecules have reached this level from the inflow at the top of the domain. The slice at $\mathbf{x}_3 = 0$ shows that the concentration at the flat parts of the wafer surface has reached relatively high values, as well, while the lighter color above the mouth of the trench $(0.3 \leq \mathbf{x}_1 \leq 0.7)$ is explained by the ongoing flow of molecules into the trench. At the slice for $\mathbf{x}_3 = -0.3$, we observe the same phenomenon where the concentration has reached a higher value in the flat areas of the trench bottom as compared to the opening into the via (round hole) below. Finally, not many molecules have reached the via bottom, yet, indicated by the very light color there.

Comparing the plots at $t = 5$ ns in Figures 2 and 3 with each other, the one for the smaller Kn = 0.1 has generally lighter color indicating a slower fill of the feature with gas. The smaller Knudsen number means more collisions among molecules, leading to a less directional flow than for the larger Knudsen number Kn = 1.0. Since the bulk direction of the flow is downward because of the supply at the top with downward velocity, the feature fills faster with molecules in this case.

The following plots in both figures show how the fill of the entire domain with gaseous molecules continues over time. Figure 3 shows that steady-state of complete fill is reached slightly faster for the larger Knudsen number, which is realistic. Consistent results are obtained by considering a wider range of values of Kn than presented here.

The results validate the effectiveness of the model and its numerical method for multiscale simulations for this application.

4 Parallel Performance Results

In this section, parallel performance results are presented for the code running on a 64-processor Beowulf cluster. This system has 32 dual-processor compute

Table 1. Observed wall clock times in minutes up to 64 processes

K	DOF	1	2	4	8	16	32	64
8	226,784	154	93	48	40	21	8	6
64	1,814,272	2,294	1,252	686	323	160	91	50

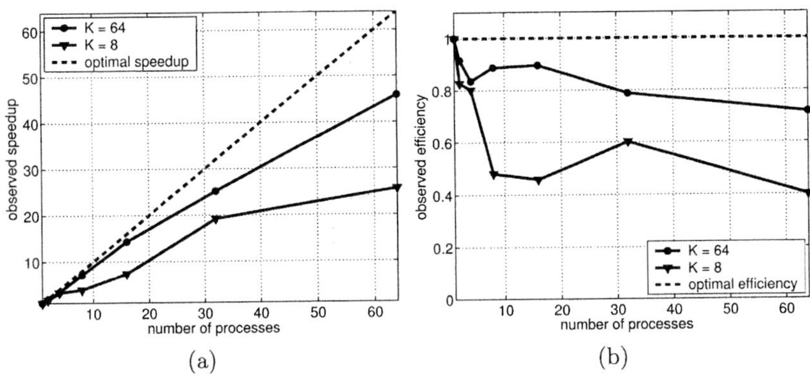

Fig. 4. (a) Observed speedup and (b) observed efficiency up to 64 processes

nodes, each with two Intel Xeon 2.0 GHz (512 kB L2 Cache) chips and 1 GB of memory. The nodes communicate through a high-performance Myrinet interconnect using the Message Passing Interface (MPI).

Table 1 shows observed wall clock times for simulations with Kn = 1.0 up to a final time of 30.0 nanoseconds for the domain with 7,087 elements in Figure 1 for velocity discretizations using $K = 2 \times 2 \times 2 = 8$ and $K = 4 \times 4 \times 4 = 64$ discrete velocities, respectively. The second column in the table lists the number of degrees of freedom (DOF) for reference. The following columns show the observed wall-clock times in minutes for each number of parallel processes 1, 2, 4, …, 64 used; the runs using 1 process use a serial code of the same algorithm. The wall clock times in Table 1 were obtained by computing the difference of the time stamps on the first and last output file written by the code. Thus, these numbers are the most *pessimistic* measure of parallel performance possible, as various extraneous delays like operating system activity, file serving, I/O delay, etc. are included in this measurement in addition to the actual cost of calculations and communications of the numerical algorithm. Clearly, the computation times for three-dimensional models are quite significant, even using the modest resolution of velocity space presented here.

Figure 4 (a) and (b) show the observed speedup and efficiency, respectively, computed from the wall clock times given in Table 1. For $K = 8$, the speedup in Figure 4 (a) appears good up to 4 processes, then deteriorates, but continues to improve again for larger number of processes. This behavior is not unexpected, as 226,784 degrees of freedom are not a particularly large number. The speedup for the more complex case $K = 64$ is clearly significantly better. It is near-optimal up to 16 processes, then drops off slightly. Figure 4 (b) shows the observed efficiency. Both lines reveal a noticeable drop of efficiency from the serial to the 2-process run that was not easily visible in the speedup plot. We explain this drop by overhead of the parallel version of the code as compared to the serial code used as 1-process code. The efficiency shows some additional slight decline up to 4 processes. This poorer efficiency might be due to the particular partitioning of the domain Ω among the processes which is computed independently for each number of processes, thus can result in a particularly inefficient partition in some cases. But it is remarkable that the efficiency does not continue to decrease significantly if more than 4 processes are used. In particular, the more complex case of $K = 64$ maintains its efficiency above 70% nearly all the way to 64 processes. This is a very good result for a distributed-memory cluster and justifies the use of relatively large numbers of parallel processes in more complex cases, e.g., for $K = 8 \times 8 \times 8 = 512$ resulting in 14,514,176 or more than 14.5 million degrees of freedom at every time step.

Acknowledgments

The hardware used in the computational studies was partially supported by the SCREMS grant DMS–0215373 from the U.S. National Science Foundation with additional support from the University of Maryland, Baltimore County. See `www.math.umbc.edu/~gobbert/kali` for more information on the machine and the projects using it. Prof. Gobbert also wishes to thank the Institute for Mathematics and its Applications (IMA) at the University of Minnesota for its hospitality during Fall 2004. The IMA is supported by funds provided by the U.S. National Science Foundation. Prof. Cale acknowledges support from MARCO, DARPA, and NYSTAR through the Interconnect Focus Center. We also wish to thank Max O. Bloomfield for supplying the original mesh of the domain.

References

1. Gobbert, M.K., Merchant, T.P., Borucki, L.J., Cale, T.S.: A multiscale simulator for low pressure chemical vapor deposition. J. Electrochem. Soc. **144** (1997) 3945–3951
2. Merchant, T.P., Gobbert, M.K., Cale, T.S., Borucki, L.J.: Multiple scale integrated modeling of deposition processes. Thin Solid Films **365** (2000) 368–375
3. Kersch, A., Morokoff, W.J.: Transport Simulation in Microelectronics. Volume 3 of Progress in Numerical Simulation for Microelectronics. Birkhäuser Verlag, Basel (1995)
4. Gobbert, M.K., Cale, T.S.: A feature scale transport and reaction model for atomic layer deposition. In Swihart, M.T., Allendorf, M.D., Meyyappan, M., eds.: Fundamental Gas-Phase and Surface Chemistry of Vapor-Phase Deposition II. Volume 2001–13., The Electrochemical Society Proceedings Series (2001) 316–323
5. Gobbert, M.K., Webster, S.G., Cale, T.S.: Transient adsorption and desorption in micrometer scale features. J. Electrochem. Soc. **149** (2002) G461–G473
6. Ringhofer, C., Schmeiser, C., Zwirchmayr, A.: Moment methods for the semiconductor Boltzmann equation on bounded position domains. SIAM J. Numer. Anal. **39** (2001) 1078–1095
7. Webster, S.G.: Stability and convergence of a spectral Galerkin method for the linear Boltzmann equation. Ph.D. thesis, University of Maryland, Baltimore County (2004)
8. Gobbert, M.K., Webster, S.G., Cale, T.S.: A galerkin method for the simulation of the transient 2-D/2-D and 3-D/3-D linear Boltzmann equation. Submitted (2005)
9. Remacle, J.F., Flaherty, J.E., Shephard, M.S.: An adaptive discontinuous Galerkin technique with an orthogonal basis applied to compressible flow problems. SIAM Rev. **45** (2003) 53–72

The Use of Conformal Voxels for Consistent Extractions from Multiple Level-Set Fields

Max O. Bloomfield[1], David F. Richards[2], and Timothy S. Cale[1]

[1] Rensselaer Polytechnic Institute, Dept. of Chemical Engineering, Troy, NY 12180-3590
[2] Synopsys Inc., Mountain View, CA 94043

Abstract. We present and analyze in 2D an algorithm for extracting self-consistent sets of boundary representations of interfaces from level set representations of many phase systems. The conformal voxel algorithm which is presented requires the use of a mesh generator and is found to be robust and effective in producing the extractions in the presence of higher order junctions.

1 Introduction

The level-set method [1] is used in a wide range of simulations in which evolving geometries are important. This method, in which the interface between two phases is represented implicitly as a contour (usually the zero contour) in a scalar field, has many advantages over explicit methods. The most notable advantage is that the geometric and topological complexities associated with updating explicit representations of interfaces are avoided, as level-set evolution is just a matter of solving an equation on the scalar field [1]. If and when an explicit representation is needed in order to perform part of the simulation, the interface can be extracted as a level of the scalar field (usually the zero level). When the geometry contains more than two phases that need to be tracked, such as for thin film grain structures simulations, the method extends quite naturally through the use of multiple scalar fields, identifying each phase with one field [2]. These fields can then be evolved independently using the same level-set equation and be periodically reconciled as shown by Merriman *et al.* [2]. However, the multiple level-set approach suffers from one setback that the standard level-set method does not: Recovering an explicit geometry by extracting the contours is not trivial because each interface is represented multiple times. At points near triple lines and higher order junctions, the explicit representations extracted from each field often disagree. These representations can differ even topologically.

At such points it is difficult or perhaps impossible to construct a consistent boundary mesh without gaps or holes in order to perform a desired simulation on the explicit structure. Consistent meshes are required for an entire class of simulation tasks in which values need to be related across the interfaces, such as mass, momentum, or energy fluxes. This class of simulation includes stress-strain and electromigration calculations on grain structures, which are desireable to link to structure evolution simulations. Bloomfield *et al.* [3] proposed a partial solution for

V.S. Sunderam et al. (Eds.): ICCS 2005, LNCS 3516, pp. 49 – 56, 2005.

this problem which employed rectilinear voxels and robustly arrived at extracted structures that were always consistent. However, these structures (as shown in Figure 1) were problematic in that they had surface triangles with only a small number of normal directions (*i.e.*, the structures were limited to "Manhattan geometries"). In this paper we demonstrate an algorithm that extends the voxel approach and removes the limitation of producing Manhattan geometries. By employing a mesh generator, we produce voxels that conform to the natural interfaces far from higher order junctions, and produce intuitive, topologically consistent, explicit boundary representations at higher-order junctions.

Fig. 1. The voxel extractor reported on by Bloomfield *et al.* [3] solves the consistency problem, but is limited to producing Manhattan geometries

2 Non-consistent Extraction

Although for the purposes of this paper, we represent level-set scalar fields (φ_i) on triangular meshes using linear finite element basis functions, the ideas regarding representation and extraction may be extended to non-simplex and mixed-type meshes, higher-order basis functions, and finite difference grids in both 2 and 3D. We will maintain the convention that $\varphi_i<0$ inside phase i and $\varphi_i>0$ outside phase i and use the signed distance to the interface as the value of φ_i. Note that each phase has a distinct level set function (φ-field). Figure 2 illustrates the problem with representing even simple structures. We begin by initializing three φ-fields to be the signed distances from the explicit starting interfaces (solid lines) and represented the fields on the underlying triangular mesh (dotted lines). Finally, we "recover" the explicit interfaces by extracting line segments (dashed lines) from the φ-fields by interpolating the zeros along the triangles' edges. We see that what would have been a slight rounding of a corner in the standard level-set method [1.], becomes an unphysical situation in the multiple level-set method, leaving a *void* behind in which no single phase is present. It should be noted that the term void as used here does not refer to a physical vacuum, which in this level-set formulation would be treated as a phase and have its own φ-field. Here, void refers to a region of space not associated with any single phase.

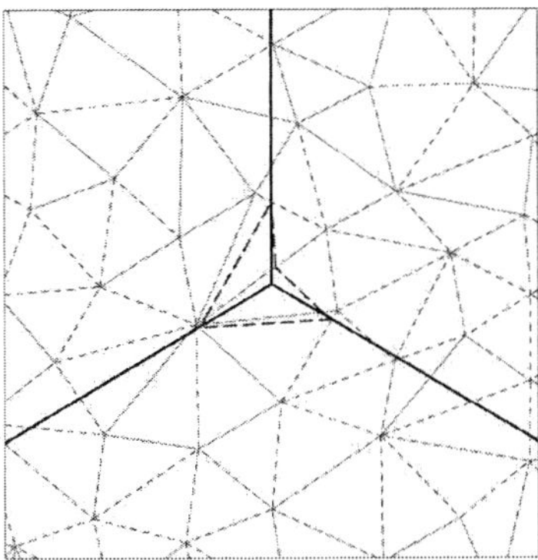

Fig. 2. An explicit set of interfaces (black solid lines) can be represented as contours in scalar fields, themselves represented on a finite element mesh (light dotted lines) or finite difference grid. However when an attempt is made to extract the interfaces, the extracted interfaces (dashed lines) do not come together at triple points and higher order junctions, leaving regions that do not correspond to any physical situation

Each φ-field can be extracted separately, giving smooth boundary representations of each phase in the most of the domain. In these regions, a match for each boundary element can be found among the elements extracted from other fields, allowing the boundary elements to be classified as being a particular type of interface. Constrained meshes may be constructed that include these elements as internal or boundary entities, allowing for further simulations. However, voids of the type shown in Figure 2 are not uncommon, occurring at almost all higher order junctions, and typically are one to two mesh-lengths across. The boundary elements that make up this void do not have matches in the set of elements extracted from other φ-fields; this prevents further execution of a program that requires tracking quantities across interfaces, from one phase to another.

3 Solution Methods

Carefully constructed meshes can potentially capture a single junction, allowing a consistent extraction, but this advantage is lost after the first time step in which the interfaces move. Alternatively, an approach may be taken to "fix" the non-physical situations after they occur, by locating their occurrence, localizing them, and then collapsing vertices and entities in such a way as to achieve a topologically consistent set of boundary representations. This approach, although fraught with special cases and tortuous decision trees, has been successfully implemented for extractions in two-dimensions [4]. We are unaware of a successful attempt to perform the three-dimensional analog. It is worth remarking that in three dimensions, the problem areas are not polygons, but are networks of long, thin "void wires" that meet at higher order junctions. Because of the daunting complexity of the problem in 3D, this approach does not appear promising.

3.1 Voxel Extraction

Voxel extraction has been employed as a partial solution to this dilemma by approaching the problem of extraction by constructing explicit *phases* before attempting to extract the interfaces between them. Bloomfield *et al.* [3] proposed filling the domain with regular rectilinear (*i.e.*, squares or cubes) and using the information embedded in the φ-fields to identify each of there volume elements (voxels) as a part of the explicit representation of the phases. They then looked for neighboring voxels that belonged to different faces and identified the elements between them as boundary elements of both faces. This generated a set of faces in three dimensions and a set of segments in two dimensions that were both consistent and located on or near the intuitively identified interface. Figure 1 shows and example of faces extracted using this voxel method. The results are completely consistent and each boundary element is identified from both sides of the interface.

The clearest drawback of this voxel method is the so-called Manhattan geometry of the results. Because the voxels have a limited set of directions for their faces, the extracted interfaces, which are made up of a subset of these faces, do not reflect the smoothly varying normals associated with most of the systems being represented. Two further improvements can be made to these output. The first approach is to apply a smoothing operator to the results, such as the volume-conserving algorithms reported on by Kuprat *et al.* [5.]. The downsides to this approach are that smoothing operations do not preserve shape global shape, tend to be motivated by aesthetic considerations rather than physical ones, and can be complicated to apply to very unsmooth surfaces. The second approach is to project onto each face a "pseudo-normal" calculated to be the unit direction along the gradient of the appropriate φ-field, similar to the technique used in Phong shading [6]. Such a projection method may be useful for tasks that require normals that more accurately represent the system, such as surface scattering calculations.

Conformal-Voxel Extraction. We demonstrate an extension of the above voxel extraction method that avoids the limitation of producing Manhattan geometries. By extending the concept of a voxelation to include not just tessellations of regular, identical shapes, but *any set of polyhedra that space-fill a domain,* we remove the constraint of having only a small number of distinct normals. With this definition, any geometric mesh that fills a volume (or fills an area in 2D systems) is a voxelation, and we can create a custom voxelation, using a ready-made meshing code and information about the structure in the φ-fields, that will have voxels optimally placed to give extractions that are consistent and are true to the implicit representation away from junctions.

Pulling information from the φ-fields is very important. Without providing guidance to the mesher, the voxel faces are as likely to be transverse to the zero contours in the implicit representation as to conform to them. The simplest way to provide this information is to use the parts of the directly extracted boundaries that do have matches in the other extracted boundaries. In two dimensions this is a set of segments and in three dimensions it is a set of faces, internal to the domain to be meshed. We call these internal elements *model entities*, and provide them to the area or volume mesher as constraints to be included as segments or faces in the resulting voxelation. As is shown below, this will cause the conformal-voxel extracted boundary representation to coincide with the directly extracted version in regions more than a few mesh lengths from triple points and higher order junctions.

The conformal-voxel extraction algorithm can be performed as follows:

1. Begin with discretizations of the n (assuming n phases) φ-fields that implicitly represent the structure to be extracted in a D dimensional domain $\mathbf{\Omega}$.
2. Extract a list $\mathbf{B}_\alpha$ of entities from each field φ_α, $\alpha \in [1,n]$ by interpolating along the level set contour of each.
3. Mark each entity which appears in exactly two of the extracted sets, $\mathbf{B}_\alpha$, as a model entity and add to list $\mathbf{B}'$.
4. Invoke an area (D=2) or volume (D=3) mesher to fill the simulation domain, providing the model entities in $\mathbf{B}'$ as internal constraints. Call this mesh the voxelation, $\mathbf{V}$.
5. Assign each voxel, $V \in \mathbf{V}$, a phase $\alpha(V)$, by an appropriate calculation referencing the fields φ_α.
6. For each voxel V, visit the voxel $U = V.neighbor(k)$ ($k \in [1,D+1]$) on the other side of each segment k (D=2) or face k (D=3). If $\alpha(V) \neq \alpha(U)$, add entity k to the set of conformal voxel extracted boundary entities, $\mathbf{B}''_{\alpha(V)}$.

Each entity in $\mathbf{B}''_\alpha$ should now have a match in some other $\mathbf{B}''_{d \neq \alpha}$ or be on the domain boundary $\partial\mathbf{\Omega}$ and the extraction is complete.

The algorithm is straightforward, with only two tasks left up to the discretion of the implementer. The first is the invocation of the mesher in step 4. Although this point is discussed in detail in section 4 of this work, in this work we use the freeware 2D quality mesh generator `triangle` [7] from the netlib repository [8] and find that no quality constraints on the mesh have to be specified as long as the maximum area of the triangles in the mesh is less than the average area of triangles in the finite element mesh that we use to represent the φ-fields. The second choice that must be made is how the identification of the voxels in done in step 5. Any heuristic that robustly determines in which phase the voxel is located should work; in this work we use the phase associated with the φ-field that has the lowest value at the voxel centroid as the identity of that voxel. This choice is intuitively pleasing, is easy to implement, and empirically works well.

3.2 Examples

In Figure 3, we show the various steps of a conformal voxel extraction of the same φ-fields that produced the void in Figure 2. First, from the directly extracted sets (left), model entities are identified (solid lines). Next, a voxelation (middle) is created using the model entities as internal constraints and each voxel is assigned to a phase (shading). Finally, the conformal voxel extraction (solid blue line, right) is derived by comparing neighboring voxels, and compared to the structure used to initialize the implicit representation (dashed line, right). The resulting extraction is consistent and faithfully captures the normals and positions of the initializing structure.

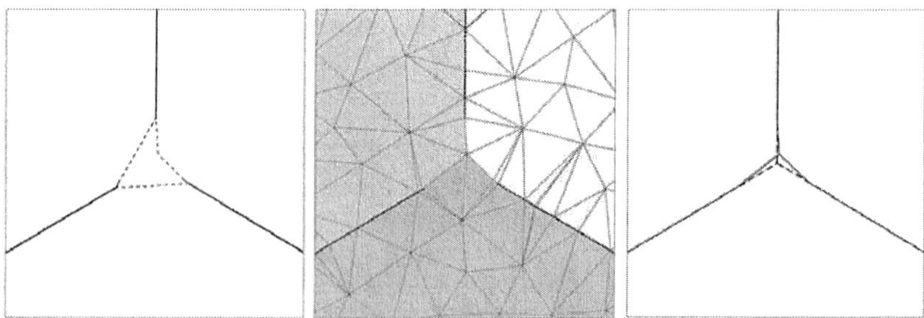

Fig. 3. The conformal voxel extraction procedure applied to the implicit representation shown in Figure 2. First, from the directly extracted sets (left), model entities are identified (solid lines). Next, a voxelization (middle) is created using the model entities as internal constraints and each voxel is assigned to a phase (shading). Finally, the conformal voxel extraction (solid blue line, right) is derived by comparing neighboring voxels, and compared to the structure used to initialize the implicit representation (dashed line, right)

In Figure 4, we show two more examples, the first (left) being a conformal voxel extraction of a higher order junction (a quadruple point). The extraction of the quadruple point indicates one aspect of voxel approaches, conformal or regular, which is that the topology of the extraction can be different that the topology of the structure the implicit representation is meant to represent. Here, the quadruple point present in the initializing structure (solid black lines) is present as a pair of triple points in the extraction. Further discussion of this phenomenon is given in the next section.

The second example in Figure 4 is a conformal voxel extraction (middle) of an 16-phase test structure (right), demonstrating the method's ability to robustly handle

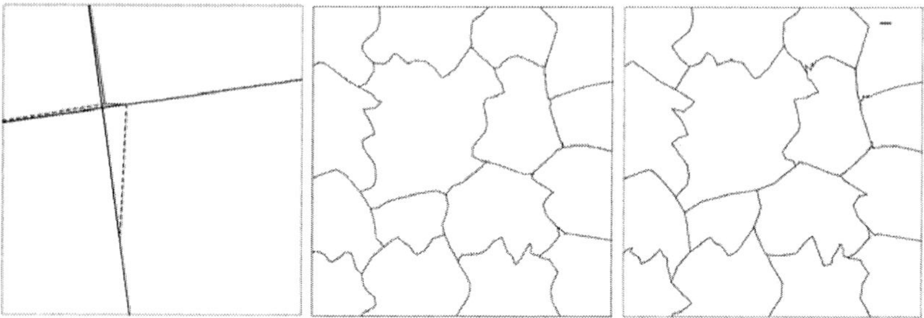

Fig. 4. (left) Drawing showing a conformal voxel extraction (dashed blue line) of a quadruple point (solid black line). Note that the extraction contains a pair of triple points instead of a single quadruple point. (middle) A conformal voxel extraction of (right) a complex 16-phase test structure containing significant fine detail, below the resolution of the underlying finite element mesh used

complex, many-phase structures. In this test structure, there are a variety of small and large geometric features. The extraction faithfully reproduces these geometric features underlying finite element mesh used for the implicit representation. (The input structure is 1x1; the bar in upper right corner of the right figure is 0.025 in length.)

Features smaller than this, such as the oscillations near some of the junctions in the test structure do not appear in the extraction.

4 Discussion and Conclusions

As mentioned in section 3.1, step 4 of the conformal voxel algorithm allows for discretion on the part of the implementer. It is essential that the voxelation include the internal constraints represented by the model entities, or at least alternative constraints based on them. That is, it may be permissible to adjust the coarseness of the model entities using a mesh coarsener to relieve some of the burden on the mesher to be used or to condition the elements in the set of extracted boundary entities. Any changes made to the information represented by the model entities will show up in the extracted boundary entities. However, we have observed in our 2D studies that the mesh quality of the voxelation is not of key importance in obtaining extractions that are true to the input structure. This is true providing that the mesher respects the constraints of the model entities and that the mesh size near the interfaces is less than or equal to the mesh size of the underlying discretizations of the field. In fact, we find that it is the mesh in the underlying finite element representation of the fields that has a determining effect on the amount of detail that is recovered using the conformal voxel extractor. In particular, the locations of triple and higher order junctions produced by the conformal voxel extractor deviate from the location of these junctions in the initializing structure by no more than the characteristic mesh size of the underlying finite element mesh.

This is a powerful result given that there is an implicit upper limit on the amount of detail retained about spatial variations in a field represented on a mesh or a grid with linear interpolants. By making a parallel to Nyquist's sampling theorem [9], we can say that the upper limit on the spatial resolution, *i.e.*, the smallest spatial variation represented is about twice the mesh length. For unstructured grids, this length is someone nebulous, but can be gauged within about a factor of 2, perhaps as a function of position. This agrees with what can be seen in the extraction of the complex test structure show Figure 4, indicating that the conformal voxel method is able to extract a large fraction of the information about the structure implicit in the finite element representation of the φ-fields.

The phenomenon of junction splitting seen in Figure 4 is also a matter of resolution. Although the implicit representation was initialized from an explicit structure with a quadruple point, it cannot be distinguished from the conformal voxel extracted version with two triple points within the resolution of the implicit representation. This situation highlights a characteristic of the comformal voxel method: a particular extraction from a given implicit representation is not unique. By using a voxelation produced by a different mesh algorithm or with different parameters, different extractions can be produced from the same implicit representation. However, in our experience, they will all be the same within the spatial resolution of implicit representation.

The computational complexity of this algorithm is difficult to analyze precisely, because of the use of a separate mesher to produce the voxelation, which will have its own complexity as a function of desired mesh size and number and type of internal constraints. However, if there are N mesh elements in the discretizations of the M input φ-fields, $O(N^{(D-1)/D})$ entities in the union of $\mathbf{B}_\alpha$. Matching up these entities is done pairwise, taking $O(D)$ integer comparisons each, giving an overall complexity of $O(D \cdot N^{2(D-1)/D})$, or $O(N)$ for 2D and $O(N^{4/3})$ for 3D for step 2. The complexity of step 5 is simply $O(N \cdot M)$, leaving the overall complexity for the algorithm system dependent based on the number of phases and the external mesher used.

We have shown several examples of the use of the algorithm in 2D. A demonstration of its use in 3D [10] is beyond the scope of this paper for reasons of brevity. The number of meshers available that can robustly handle the complex internal constraints in an automated fashion is much lower for 3D than for 2D. It should be pointed out that although we use triangular mesh elements for our voxelations, quadrilateral element should work just as well. In 3D, tetrahedral, wedges, and brick elements should be able to be used with equal facility.

This method is highly robust, given a reliable mesher. The only caveat is that if the voxel size is noticeably larger than the resolution of the implicit representation, then the resulting extraction may depend heavily on the voxelation. However, the method will not fail to complete even for very large voxel sizes, happily producing spurious results.

Finally, we note that there is a relationship between the voxelation mesh and the conformal voxel extracted result, namely that the former is a body-fitted mesh for the latter, and potentially can be used as the mesh for further simulation on the extracted system. Should this be desired, more attention should be paid to the quality of the voxelation as it is being produced.

References

1. Osher, S., Sethian, J.A.: Fronts Propogating with Curvature Dependent Speed: Algorithms Based on Hamilton-Jacobi Formulations. J. Comp. Phys. 79 (1988) 12-49.
2. Merriman, B, Bence, J,K, Osher, S.J.: Motion of Multiple Junctions: A Level Set Approach. J. Comp. Phys. 112 (1994) 334-363.
3. Bloomfield, M.O., Richards, D.F., Cale, T.S.: A Computational Framework for Modeling Grain Structure Evolution in Three Dimensions. Phil. Mag. 83(31-34) (2003) 3549-3568.
4. Bloomfield, M.O., Richards, D.F., Cale, T.S.: A Tool for the Representation, Evolution, and Coalescence of 3-Dimensional Grain Structures, presented during the Sixth United States National Congress on Computational Mechanics, August 1-3, 2001, Dearborn, MI.
5. Kuprat, A., Khamayseh, A., George, D., Larkey, L.: Volume Conserving Smoothing for Piecewise Linear Curves, Surfaces, and Triple Lines. J. Comp. Phys. 172 (2001) 99-118.
6. Phong, B.: Illumination for Computer Generated Pictures. Comm. of the ACM, 18(8) (1975) 311-317.
7. A Two-Dimensional Quality Mesh Generator and Delaunay Triangulator: http://www-2.cs.cmu.edu/~quake/triangle.html
8. Netlib Repository at UTK and ORNL: http://www.netlib.org/
9. Papoulis, A.: Signal Processing, McGraw Hill, 1997.
10. Bloomfield, M.O., Richards, D.F., Cale, T.S.: A Conformal Voxel Extraction Method for Extraction of Multiple Level Set Fields in Three Dimensions, *in preparation.*

Nonlinear OIFS for a Hybrid Galerkin Atmospheric Model

Amik St.-Cyr and Stephen J. Thomas

National Center for Atmospheric Research,
1850 Table Mesa Drive, Boulder, 80305 CO, USA
{amik, thomas}@ucar.edu

Abstract. The purpose of this paper is to explore a time-split hybrid Galerkin scheme for the atmospheric shallow water equations. A non-linear variant of operator integration factor splitting is employed as the time-stepping scheme. The hyperbolic system representing slow modes is discretized using the discontinuous Galerkin method. An implicit second order backward differentiation formula is applied to Coriolis and gravity wave terms. The implicit system is then discretised using a continuous Galerkin spectral element method. The advantages of such an approach include improved mass and energy conservation properties. A strong-stability preserving Runge-Kutta scheme is applied for sub-stepping.

1 Introduction

The seminal work of Robert (1981) led to a six-fold increase over the explicit time step for atmospheric general circulation models. To achieve such dramatic gains without recourse to a fully implicit integrator, a semi-Lagrangian treatment of advection was combined with a semi-implicit scheme for the stiff terms responsible for gravity waves. Initially, semi-implicit semi-Lagrangian time-stepping was applied to hyperbolic problems, discretized using low-order finite-differences and finite elements. The traditional semi-Lagrangian algorithm implemented in atmospheric models relies on backward trajectory integration and upstream interpolation, Staniforth and Côté (1991). A potentially lower cost alternative is the operator integrating factor splitting (OIFS) method of Maday et al. (1990) which relies on Eulerian sub-stepping of the advection equation. In contrast with semi-Lagrangian advection, there are no dissipation or dispersion errors associated with upstream interpolation or trajectory integration and the scheme maintains the high-order accuracy of the discrete spatial operators.

A discontinuous Galerkin shallow water model employing a nodal basis and explicit time-stepping is described in Giraldo et al. (2003). Sherwin (2004) demonstrated the advantages of a hybrid Galerkin approach in the context of the incompressible Navier-Stokes equations. Eskilsson and Sherwin (2005) describe a discontinuous Galerkin formulation of the shallow water equations using third-order strong-stability preserving (SSP) Runge-Kutta time-stepping. Here, we

V.S. Sunderam et al. (Eds.): ICCS 2005, LNCS 3516, pp. 57–63, 2005.

investigate a time-split scheme applied to the global shallow water equations in curvilinear coordinates on the cubed-sphere. A second order backward differentiation formula (BDF-2) is combined with SSP-RK sub-stepping of a hyperbolic system. Because the incompressibility constraint has been removed, the fully nonlinear OIFS scheme of St-Cyr and Thomas (2004) is employed. When compared to continuous Galerkin spectral elements, the hybrid scheme results in improved mass and energy conservation properties.

2 Shallow Water Equations

The shallow water equations contain the essential wave propagation mechanisms found in atmospheric general circulation models. These are the fast gravity waves and slow synoptic scale Rossby waves. The latter are important for correctly capturing nonlinear atmospheric dynamics. The governing equations of motion for the inviscid flow of a free surface are

$$\frac{\partial \mathbf{v}}{\partial t} + (f+\zeta)\,\mathbf{k}\times\mathbf{v} + \frac{1}{2}\nabla\,(\mathbf{v}\cdot\mathbf{v}) + \nabla\Phi = 0, \tag{1}$$

$$\frac{\partial \Phi}{\partial t} + (\mathbf{v}\cdot\nabla)\,\Phi + (\Phi_0+\Phi)\;\nabla\cdot\mathbf{v} = 0. \tag{2}$$

h is the height above sea level, $\mathbf{v}$ is the horizontal velocity and $\Phi = gh$ the geopotential height. f is the Coriolis parameter and $\mathbf{k}$ a unit vector in the vertical direction. The geopotential height is decomposed into a perturbation about a constant base state, Φ_0.

To exploit the potential of operator integration factor splitting for systems of time-dependent partial differential equations, St.-Cyr and Thomas (2004) show that a fully nonlinear form of the algorithm is more appropriate. Sub-stepping is applied to

$$\frac{\partial \tilde{\mathbf{v}}}{\partial s} + \tilde{\zeta}\,\mathbf{k}\times\tilde{\mathbf{v}} + \frac{1}{2}\nabla\,(\tilde{\mathbf{v}}\cdot\tilde{\mathbf{v}}) = 0, \tag{3}$$

$$\frac{\partial \tilde{\Phi}}{\partial s} + \nabla\cdot(\tilde{\Phi}\,\tilde{\mathbf{v}}) = 0. \tag{4}$$

with initial conditions $\tilde{\mathbf{v}}(\mathbf{x},t^{n-q}) = \mathbf{v}(\mathbf{x},t^{n-q})$, $\tilde{\Phi}(\mathbf{x},t^{n-q}) = \Phi(\mathbf{x},t^{n-q})$. The integration factor $Q_S^{t^*}(t)$ is applied to the remaining de-coupled system of equations containing the Coriolis and linear gravity wave terms

$$\frac{d}{dt}Q_S^{t^*}(t)\begin{bmatrix}\mathbf{v}\\ \Phi\end{bmatrix} = -Q_S^{t^*}(t)\begin{bmatrix} f\,\mathbf{k}\times\mathbf{v}+\nabla\Phi \\ \Phi_0\,\nabla\cdot\mathbf{v}\end{bmatrix}. \tag{5}$$

An accurate representation of fast-moving gravity waves is not required for large scale atmospheric dynamics and the corresponding terms can be treated implicitly. For a second order BDF-2 scheme, sub-stepping of the rhs terms is not required because $Q_S^{t^n}(t^n) = I$.

The resulting time discretization of (5) is given by

$$\mathbf{v}^n + \frac{2}{3}\Delta t \mathbf{N}\nabla\Phi^n = \frac{4}{3}\mathbf{N}\tilde{\mathbf{v}}^{n-1} - \frac{1}{3}\mathbf{N}\tilde{\mathbf{v}}^{n-2} \tag{6}$$

$$\Phi^n + \frac{2}{3}\Delta t \Phi_0 \nabla\cdot\mathbf{v}^n = \frac{4}{3}\tilde{\Phi}^{n-1} - \frac{1}{3}\tilde{\Phi}^{n-2} \tag{7}$$

where

$$\mathbf{N} = \left(I + \frac{2}{3}\Delta t f \mathbf{M}\right)^{-1}, \quad \mathbf{M} = \begin{bmatrix} 0 & -1 \\ 1 & 0 \end{bmatrix}.$$

The values of the fields $\tilde{\mathbf{v}}$ and $\tilde{\Phi}$ at time levels $n-1$ and $n-2$ are computed by sub-stepping (3) and (4) on the intervals $[t^{n-1}, t^n]$ and $[t^{n-2}, t^n]$. An implicit equation for Φ^n is obtained after space discretization and application of block Gaussian elimination, resulting in a modified Helmholtz problem. The coefficient matrix of this linear system of equations is non-symmetric due to the implicit treatment of the Coriolis terms and is solved using an iterative conjugate-gradient squared (CGS) algorithm with a block-Jacobi preconditioner.

For high-order spectral elements, under-integration using Gaussian quadrature results in discrete operators where the eigenvalues corresponding to high frequency modes are shifted into the right-half plane. A filter is therefore required to stabilize the time-stepping scheme for long time integrations, Fischer and Mullen (2001). Strong-stability preserving (SSP) time integration schemes maintain the non-oscillatory properties of the spatial operator, Gottlieb et al (2001). We next describe the computational domain, high-order Galerkin finite element approximations and SSP-RK sub-stepping.

3 High-Order Galerkin Approximations

The flux form shallow-water equations in curvilinear coordinates are described in Sadourny (1972). Let $\mathbf{a}_1$ and $\mathbf{a}_2$ be the covariant base vectors of the transformation between inscribed cube and spherical surface. The metric tensor is defined as $G_{ij} \equiv \mathbf{a}_i \cdot \mathbf{a}_j$. Covariant and contravariant vectors are related through the metric tensor by $u_i = G_{ij}u^j$, $u^i = G^{ij}u_j$, where $G^{ij} = (G_{ij})^{-1}$ and $G = \det(G_{ij})$. The six local coordinate systems (x^1, x^2) are based on equiangular central projection, $-\pi/4 \le x^1, x^2 \le \pi/4$.

$$G_{ij} = \frac{1}{r^4\cos^2 x^1 \cos^2 x^2} \begin{bmatrix} 1+\tan^2 x^1 & -\tan x^1 \tan x^2 \\ -\tan x^1 \tan x^2 & 1+\tan^2 x^2 \end{bmatrix} \tag{8}$$

where $r = (1 + \tan^2 x^1 + \tan^2 x^2)^{1/2}$ and $\sqrt{G} = 1/r^3 \cos^2 x^1 \cos^2 x^2$.

The time-split hyperbolic system (3) – (4) can be written in as

$$\frac{\partial u_1}{\partial t} + \frac{\partial}{\partial x^1} E = \sqrt{G}\, u^2 (f + \zeta), \tag{9}$$

$$\frac{\partial u_2}{\partial t} + \frac{\partial}{\partial x^2} E = -\sqrt{G}\, u^1 (f + \zeta), \tag{10}$$

$$\frac{\partial}{\partial t}(\sqrt{G}\,\Phi) + \frac{\partial}{\partial x^1}(\sqrt{G}\, u^1 \Phi) + \frac{\partial}{\partial x^2}(\sqrt{G}\, u^2 \Phi) = 0, \tag{11}$$

where

$$E = \frac{1}{2}(u_1 u^1 + u_2 u^2), \quad \zeta = \frac{1}{\sqrt{G}} \left[\frac{\partial u_2}{\partial x^1} - \frac{\partial u_1}{\partial x^2} \right]$$

Consider a scalar component of the hyperbolic system,

$$\frac{\partial u}{\partial t} + \nabla \cdot \mathcal{F} = S.$$

The computational domain Ω is partitioned into elements Ω_k. An approximate solution u_h belongs to the finite dimensional space $\mathcal{V}_h(\Omega)$. Multiplication of by a test function $\varphi_h \in \mathcal{V}_h$ and integration over the element Ω_k results in a weak Galerkin formulation of the problem.

$$\frac{\partial}{\partial t} \int_{\Omega_k} \varphi_h u_h \, d\Omega = \int_{\Omega_k} \varphi_h S \, d\Omega + \int_{\Omega_k} \mathcal{F} \cdot \nabla \varphi_h \, d\Omega - \int_{\partial \Omega_k} \varphi_h \mathcal{F} \cdot \hat{n} \, ds.$$

For a discontinuous Galerkin approximation, a nodal basis for $\mathcal{V}_h$ is employed, consisting of the Legendre cardinal functions. The solutions u_h are expanded in terms of tensor-product basis functions on a Gauss-Lobatto grid.

The flux function $\mathcal{F} \cdot \hat{n}$ is approximated by a Lax-Friedrichs numerical flux

$$\widehat{\mathcal{F}}(u_h^+, u_h^-) = \frac{1}{2} \left[(\mathcal{F}(u_h^+) + \mathcal{F}(u_h^-)) \cdot \hat{n} - \alpha (u_h^+ - u_h^-) \right].$$

Boundary integrals are computed using higher-order Gaussian quadrature. α is the upper bound for the absolute value of eigenvalues of the flux Jacobian $\mathcal{F}'(u)$ in the direction $\hat{n}$. Equations (9) – (11) are written in the semi-discrete form

$$\frac{d\mathbf{u}}{dt} = \mathbf{L}(\mathbf{u}).$$

The second-order three stage SSP RK2-3 integrator of Higueras (2004) is applied to sub-step the above system of ordinary differential equations. This scheme has a CFL number $C = 2$ and efficiency factor $C/3 = 2/3$. The implicit system (6) – (7) is discretized using $\mathbb{P}_N - \mathbb{P}_N$ continuous Galerkin spectral elements.

4 Numerical Experiments

Our numerical experiments are based on the shallow water test suite of Williamson et al (1992). Test case 5 is a zonal flow impinging on an isolated mountain. The center of the mountain is located at $(3\pi/2, \pi/6)$ with height $h_s = 2000\,(1-r/R)$ meters, where $R = \pi/9$ and $r^2 = \min[R^2, (\lambda-3\pi/2)^2+(\theta-\pi/6)^2]$. Initial wind and height fields are

$$
\begin{aligned}
u &= u_0\,(\cos\alpha_0\,\cos\theta + \sin\alpha_0\,\cos\lambda\sin\theta)\\
v &= -u_0\,\sin\alpha_0\sin\lambda\\
g\,h &= g\,h_0 - \frac{u_0}{2}(2\,a\,\Omega + u_0)\,(\sin\theta\,\cos\alpha_0 - \cos\lambda\,\cos\theta\,\sin\alpha_0)^2
\end{aligned}
$$

a is the earth's radius, Ω the rotation rate, $\alpha_0 = 0$, $gh_0 = 5960$ m^2/s^2 and $u_0 = 20$ m/s. The geopotential height field after 15 days of integration is plotted in Fig. 1. The solution is smooth and does not exhibit spurious oscillations.

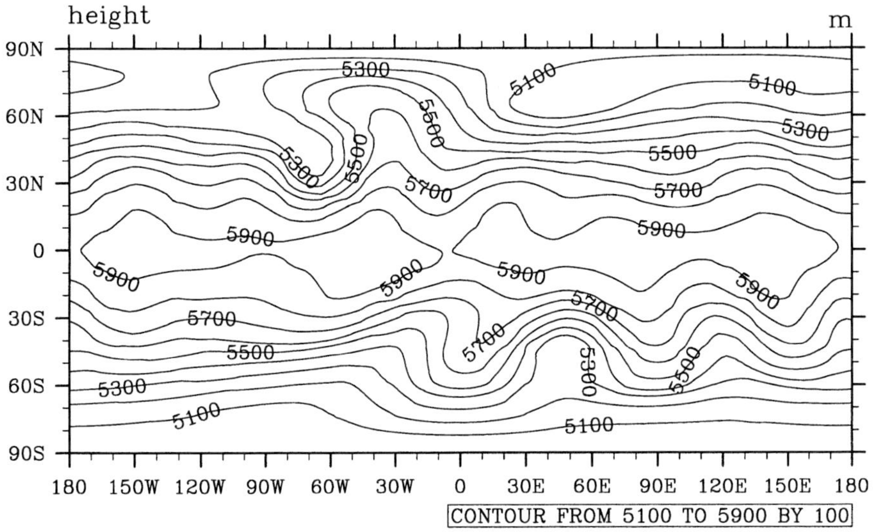

Fig. 1. Shallow water test case 5: Flow impinging on a mountain. Geopotential height field h at fifteen days produced by hybrid scheme. 150 spectral elements, 8×8 Gauss-Legendre Lobatto points per element

The implementation of the BDF-2/RK2-3 scheme is still at an experimental stage. However, it leads to an increase in the model integration rate. Figure 2 is a plot of the ratio of CNLF and BDF-2/RK2-3 execution times for test case 5. The efficiency gain of BDF-2/RK2-3 over the CNLF scheme is close to a factor of two. Thomas and Loft (2002) observed that the CNLF scheme integrated twice as fast as an explicit spectral element shallow water model. Therefore, the total increase in the model integration rate, when compared to the explicit model, is a factor of four. The plateau in the efficiency curve is due to a growing number of solver iterations as the time step size increases.

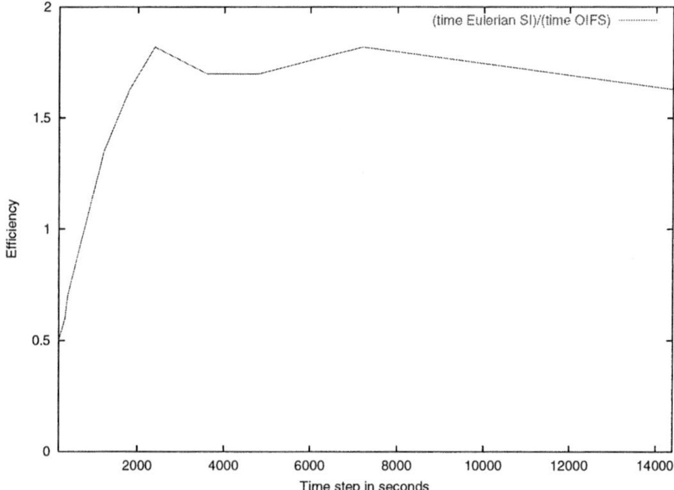

Fig. 2. Shallow water test case 5: Flow impinging on a mountain. Computational efficiency. Ratio of the CNLF and BDF-2/RK2-3 model execution times. $K = 150$ spectral elements, 8×8 Gauss-Legendre points per element

5 Conclusions

The larger time step permitted by the time-split BDF-2/RK2-3 integrator implies that the hybrid Galerkin discretization is well-suited for modeling multiscale atmospheric phenomena. The computation time is currently dominated by an increasing number of solver iterations as the time step size increases. However, a recently developed optimized Schwarz preconditioner maintains a much lower iteration count, St.-Cyr et al (2004). We estimate that this preconditioner could lead to an additional factor of four improvement in the computational efficiency compared to a Crank-Nicholson leap-frog (CNLF) formulation. Although the hybrid scheme is not strictly conservative, the mass loss is significantly less than in a continuous Galerkin spectral element model.

For parallel computation, a clear advantage of the discontinuous Galerkin method is the ability to overlap computation with communication. Specifically, the communication of conserved variables can be overlapped with the computation of the weak divergence and source terms. We are also currently experimenting with a hybrid MPI/OpenMP programming model for SMP cluster computer architectures and plan to report these results in the near future.

References

1. Eskilsson, C. and S. J. Sherwin, 2005: Discontinuous Galerkin spectral/hp element modelling of dispersive shallow water systems. *J. Sci. Comp.*, **22**, 279-298.

2. Fischer, P. F. and J. S. Mullen, 2001: Filter-Based stabilization of spectral element methods. *Comptes Rendus de l'Acadmie des sciences Paris*, t. 332, Série I - Analyse numérique, 265–270.
3. Giraldo, F. X., J. S. Hesthaven, and T. Warburton, 2003: Nodal high-order discontinuous Galerkin methods for spherical shallow water equations. *J. Comput. Phys.*, **181**, 499-525.
4. Gottlieb, S., C. W. Shu, and E. Tadmor, 2001: Strong stability preserving high-order time discretization methods. *SIAM Review*, **43**, 89–112.
5. Higueras, I., 2004: On strong stability preserving time discretization methods. *J. Sci. Comput.*, **21**, 193-223.
6. Maday, Y., A. T. Patera, and E. M. Ronquist, 1990: An operator-integration-factor splitting method for time-dependent problems: Application to incompressible fluid flow. *J. Sci. Comput.*, **5**, 263–292.
7. Robert, A. J., 1981: A stable numerical integration scheme for the primitive meteorological equations. *Atmos.-Ocean*, **19**, 35–46.
8. Sadourny, R., 1972: Conservative finite-difference approximations of the primitive equations on quasi-uniform spherical grids. *Mon. Wea. Rev.*, **100**, 136–144.
9. Sherwin, S. J., 2002: A sub-stepping Navier-Stokes splitting scheme for spectral/hp element discretisations. *Parallel Computational Fluid Dynamics*, North Holland.
10. Staniforth, A., and J. Coté 1991: Semi-Lagrangian integration schemes for atmospheric models – A review. *Mon. Wea. Rev.*, **119**, 2206–223.
11. St-Cyr, A., and S. J. Thomas, 2004: Nonlinear operator integration factor splitting for the shallow water equations. *Appl. Numer. Math.*, to appear.
12. St-Cyr, A., M. Gander and S. J. Thomas, 2004: Optimized RAS preconditioning. Proceedings of the 2004 Copper Mountain Conference on Iterative Methods.
13. Thomas, S. J., and R. D. Loft, 2002: Semi-implicit spectral element atmospheric model. *J. Sci. Comp.*, **17**, 339–350.
14. Williamson, D. L., J. B. Drake, J. J. hack, R. Jakob, P. N. Swarztrauber, 1992: A standard test set for numerical approximations to the shallow water equations in spherical geometry *J. Comp. Phys.*, **102**, 211–224.

Flamelet Analysis of Turbulent Combustion

R.J.M. Bastiaans[1,2], S.M. Martin[1], H. Pitsch[1], J.A. van Oijen[2], and L.P.H. de Goey[2]

[1] Center for Turbulence Research , Stanford University, CA 94305, USA
[2] Eindhoven University of Technology, P.O.Box 513, 5600 MB Eindhoven, The Netherlands
r.j.m.bastiaans@tue.nl

Abstract. Three-dimensional direct numerical simulations are performed of turbulent combustion of initially spherical flame kernels. The chemistry is described by a progress variable which is attached to a flamelet library. The influence of flame stretch and curvature on the local mass burning rate is studied and compared to an analytical model. It is found that there is a good agreement between the simulations and the model. Then approximations to the model are evaluated.

1 Motivation and Objectives

The present research is concerned with the direct numerical simulation (DNS) and analysis of turbulent propagation of premixed flame kernels. The simulations are direct in the sense that the smallest scales of motion are fully resolved, while the chemical kinetics are solved in advance and parameterized in a table by the method of the flamelet generated manifolds (FGM) [8]. The state of the reactions are assumed to be directly linked to a single progress variable. The conservation equation for this progress variable is solved using DNS, with the unclosed terms coming from the table. This allows the use of detailed chemical kinetics without having to solve the individual species conservation equations.

Flame stretch is an important parameter that is recognized to have a determining effect on the burning velocity in premixed flames. In the past this effect has not been taken into account in the flamelet approach for turbulent combustion in a satisfying manner. The laminar burning velocity, which is largely affected by stretch, is an important parameter for modelling turbulent combustion. Flame stretch is also responsible for the creation of flame surface area, affecting the consumption rate as well. In the turbulent case, stretch rates vary significantly in space and time. An expression for the stretch rate is derived directly from its mass-based definition in [4],

$$K = \frac{1}{M}\frac{dM}{dt}, \tag{1}$$

where M is the amount of mass in an arbitrary control volume moving with the flame velocity:

V.S. Sunderam et al. (Eds.): ICCS 2005, LNCS 3516, pp. 64–71, 2005.

$$M = \int_{V(t)} \rho dV. \tag{2}$$

On the basis of this definition, a model for the influence of stretch and curvature on the mass burning rate has been developed. In a numerical study [5], it was shown that this model, with a slight reformulation, shows good agreement with calculations for spherically expanding laminar flames. This formulation, for the ratio of the actual mass burning rate at the inner layer, m_{in}, relative to the unperturbed mass burning rate at the inner layer, m_{in}^0 (for unity Lewis numbers), reads

$$\frac{m_{\text{in}}}{m_{\text{in}}^0} = 1 - \mathcal{K}a_{\text{in}}, \tag{3}$$

with the integral Karlovitz number being a function of flame stretch (1), flame surface area, σ, and a progress variable, $\mathcal{Y}$,

$$\mathcal{K}a_{\text{in}} := \frac{1}{\sigma_{\text{in}} m_{\text{in}}^0} \left(\int_{s_u}^{s_b} \sigma \rho K \mathcal{Y} ds - \int_{s_{\text{in}}}^{s_b} \sigma \rho K ds \right). \tag{4}$$

The integrals have to be taken over paths normal to the flame and s_u, s_b and s_{in} are the positions at the unburned side, the burned side and the inner layer, respectively. The flame surface area, σ, is related to the flame curvature, κ, which is related to the flame normals, n_i on the basis of the progress variable, $\mathcal{Y}$,

$$n_i = -\frac{\partial \mathcal{Y}/\partial x_i}{\sqrt{\partial \mathcal{Y}/\partial x_j \partial \mathcal{Y}/\partial x_j}}, \tag{5}$$

$$\kappa = \frac{\partial n_i}{\partial x_i} = -\frac{1}{\sigma}\frac{\partial \sigma}{\partial s}. \tag{6}$$

In turbulent premixed combustion the total fuel consumption is a result of the combined effect of flame surface increase and local modulation of the mass burning rate. In this study the latter will be investigated on the basis of (3) and possible parameterizations thereof, i.e. models for the Karlovitz integral, (4).

2 Methodology

Freely expanding flames are modelled in a turbulent flow field using DNS. More detailed information about the DNS program can be found in [1]. The fully compressible Navier-Stokes equations are solved supplemented by a conservation equation for the progress variable. For this purpose the mass fraction of carbon dioxide is used, which is monotonically increasing. Unity Lewis numbers are assumed for all species in order to prevent differential diffusion effects from obscuring the direct effects of stretch and curvature on the mass burning rate.

To make the DNS computations affordable, the FGM method of [8] is used to describe the reaction kinetics. FGM can be considered a combination of the flamelet approach and the intrinsic low-dimensional manifold (ILDM) method [7] and is similar to the Flame Prolongation of ILDM (FPI) introduced in [3]. FGM is applied similarly to ILDM. However, the thermo-chemical data-base is not generated by applying the usual steady-state relations, but by solving a set of 1D convection-diffusion-reaction equations describing the internal flamelet structure. The main advantage of FGM is that diffusion processes, which are important near the interface between the preheat zone and the reaction layer, are taken into account. This leads to a very accurate method for (partially) premixed flames that uses fewer controlling variables than ILDM. The manifold used in this paper is based on the GRI3.0 kinetic mechanism with 53 species and 325 reversible reactions [9].

The initial conditions are a laminar spherical flame superimposed on a turbulent field. There is no forcing in the simulation, so the turbulence will decay in time. The chemistry is chosen in relation to the large interest in the power industry in lean premixed combustion engines and there is detailed knowledge of its chemical kinetics. Therefore premixed combustion of a methane/air mixture is used, with an equivalence ratio of $\phi = 0.7$. The evolution of the initial laminar spherical flame kernel to the initial size in the DNS is calculated with detailed chemistry with the CHEM1D code [2].

3 Results

The first simulation, denoted C1, is a lean case with an equivalence ratio of $\phi = 0.7$, domain size of 12 mm, an initial flame kernel radius of approximately 2.9 mm, turbulent fluctuations of $u' = 0.4$ m/s and a turbulence length scale of ℓ_t=1.15 mm. In order to allow for very mild perturbations, initially we study the results at a time equal to 0.026τ, with $\tau = \ell_t/u' = 2.9ms$, taken from the start of the simulation. The time of growth of the laminar flame kernel to the initial DNS size was about 5 ms. The burning velocity of a flat unstretched flame with respect to the unburnt mixture is equal to $s_L^0 = 18.75$ cm/s and the corresponding mass burning rate is $m^0 = 0.213$ kg/m^{2}s. The progress variable is taken to be the carbon dioxide mass fraction, normalized with the maximum adiabatic value. At the left side of figure 1 is a cross section of the field. The contours of the progress variable are deformed only very mildly. It is observed that the scale of the vorticity patches are larger than the integral flame thickness. For this field the mass burning rate is analyzed as a reference case.

Additional analyses are performed in order to assess the basic model (4) under varying physical conditions. The test cases are listed in table 1. In case C2, the effect of grid resolution is investigated. It is assumed that the FGM method is valid in the flamelet regime if the progress variable is approximated with enough accuracy. Since all lengths scales of the gradients of primary

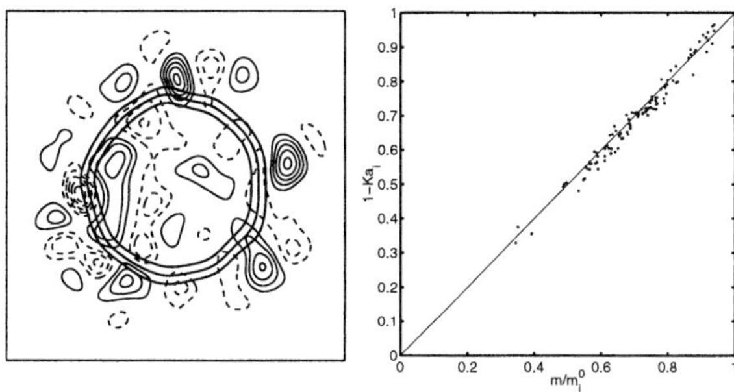

Fig. 1. Case C1, left: Vorticity contours (positive and negative values indicated by solid and dashed lines, respectively) and progress variable (thick lines, values 0.2, 0.5, 0.8), right: Correlation of the actual mass burning rate with the basic model (result of 52000 flamelets found in the domain)

Table 1. Physical properties correspondig to the different simulations

Case	ϕ	u' $[m/s]$	ℓ_t $[mm]$	δ_f $[mm]$	L $[mm]$	r_{ini} $[mm]$	grid	$Re_t = u'\ell_t/s_L^0\delta_f$	tu'/ℓ_t
C1	0.7	0.40	1.15	0.614	12	2.9	254^3	4.0	0.026
C2	0.7	0.40	1.15	0.614	12	2.9	125^3	4.0	0.026
C3	1.0	0.60	0.89	0.475	12	2.9	254^3	4.0	0.026
C4	0.7	0.70	0.77	0.614	20	3.9	254^3	4.7	0.026
C5	0.7	1.31	0.94	0.614	12	2.9	254^3	10.7	0.026
C6	0.7	1.30	0.66	0.614	12	2.9	254^3	7.5	0.026

variables (i.e. the variables that are solved in the present DNS calcualations) are of the same order, this will yield satisfactory solutions. In order to assess the influence of the chemistry a stoichiometric case, C3, is selected, in which the same ratio of the turbulent velocity fluctuations compared to the laminar flame speed, and the turbulent integral length scale compared to the initial flame thickness as used for cases C1 and C2. For the stoichiometric case at unity Lewis numbers the burning velocity is $s_L^0 = 28.17$ cm/s and the corresponding mass burning rate is $m^0 = 0.316$ kg/m^{2}s. An additional case is given by the simulation of an increased initial flame kernel in a larger domain, C4. Here also the effective resolution is decreased. In addition, cases are chosen with increased velocity fluctuations and decreased length scales, cases C5 and C6, respectively.

In the analysis, the stretch rate defined by,

$$\rho K = \frac{\partial}{\partial x_i}(\rho s_L n_i), \tag{7}$$

is evaluated by using the relation for the local burning velocity s_L,

$$s_L = \left(\frac{\partial}{\partial x_i} \left(\frac{\lambda}{Le\bar{c}_p} \frac{\partial \mathcal{Y}}{\partial x_i} \right) + \dot{\rho} \right) \Big/ \left| \frac{\partial \mathcal{Y}}{\partial x_i} \right|, \tag{8}$$

which is a consequence of the combination of the conservation equation for $\mathcal{Y}$ with the kinematic equation for $\mathcal{Y}$. The latter defines the flame speed u_{if} and then the relation for the flame velocity, $u_{if} = u_i + s_L n_i$, can be used to arrive at (8).

Table 2. Differences of the mass burning rate with the basic model

Case	C1	C2	C3	C4	C5	C6
Mean	0.0072	0.0081	0.0075	0.0091	0.0107	0.0094
RMS	0.0215	0.0202	0.0216	0.0236	0.0336	0.0280

Now the actual mass burning rate can be compared to model-values. This is performed by looking for points in the domain that are close to the inner layer and interpolate from there in the direction of positive and negative gradient of the progress variable, with steps of 1/20 times the gridsize. All relevant variables are interpolated over these flamelets and these flamelets are analysed to determine the burning velocity (8) and the model of the mass burning rate given by (4). For the present simulations these analyses lead to lots of starting points (e.g. for case C1: 52000) and thus resulting flamelets. For case C1 the correlation is depicted on the right side of figure 1. This shows that the model is a relatively accurate description of the actual mass burning rate. Deviations of the actual mass burning rate compared to the model (4) are given in table 2 for all six cases. It is seen that the mean error for all cases is about 0.01 or less, with a root mean square value of 0.02 to 0.03 (without normalization). It can be concluded that the model is a good description for all the present cases. Moreover, the grid coarsening shows no real deterioration, indicating that all cases are sufficiently resolved.

Starting from this point approximations to (4) can be considered. First, one can consider the case in which the surface area is taken to be constant, $\sigma = \sigma_{\text{in}}$ as used frequently in the literature,

$$\mathcal{K}a'_{\text{in}} := \frac{1}{m^0_{\text{in}}} \left(\int_{s_u}^{s_b} \rho K \mathcal{Y} ds - \int_{s_{\text{in}}}^{s_b} \rho K ds \right). \tag{9}$$

An improved model can be constructed by assuming that the curvature is not a function of the distance s, but that it remains constant equal to the inner layer value $\kappa = \kappa_{\text{in}}$. By integrating (6) this yields for the surface

$$\sigma = \exp\left(-\kappa_{\text{in}}(s - s_{\text{in}})\right). \tag{10}$$

A third approximation is that the iso-planes of the progress variable are concentric, either cylindrical or spherical yielding

$$\sigma = \left(\frac{\xi/\kappa_{\text{in}} - s}{\xi/\kappa_{\text{in}}} \right)^{\xi}, \tag{11}$$

in which ξ takes the value 2 for spherical curvature and 1 for cylindrical curvature. This has to be limited for distances s beyond the concentric origin, $s > \xi/\kappa_{\text{in}}$, at which $\sigma = 0$.

Table 3. Differences of the mass burning rate determined by the basic model compared to the approximations

Case	C1	C2	C3	C4	C5	C6
$\sigma = \sigma_{\text{in}}$						
Mean	-0.0537	-0.0519	-0.0340	-0.0496	-0.0653	-0.0810
RMS	0.0552	0.0473	0.0373	0.0641	0.0772	0.1004
$\kappa = \kappa_{\text{in}}$						
Mean	0.0062	0.0055	0.0029	0.0026	0.0082	0.0079
RMS	0.0103	0.0085	0.0055	0.0173	0.0186	0.0338
$\xi = 2$						
Mean	-0.0011	-0.0006	-0.0007	-0.0075	-0.0037	-0.0141
RMS	0.0114	0.0101	0.0074	0.0313	0.0224	0.0540
$\xi = 1$						
Mean	-0.0059	-0.0050	-0.0032	-0.0115	-0.0101	-0.0219
RMS	0.0169	0.0142	0.0098	0.0333	0.0281	0.0556

The result of the approximations are given in table 3 for all cases. It is observed that the constant flame surface conjecture gives rise to relatively large error. There is a systematic over-prediction of about 0.05 (without normalization) of the mass burning rate with this model and the fluctuations are of the same order of magnitude. The other approximations give much better results. For the mean differences the spherical approximation, $\xi = 2$, is superior compared to the cylindrical model, $\xi = 1$, and for most cases also compared to the constant curvature model. However, this is not really substantiated when looking at the accompanying fluctuations. For the better resolved cases, C1 and C3, the mean difference is best predicted by the $\xi = 2$ model, but again the accompanying fluctuations are much larger than the model deviation. This suggests that it is not a real improvement. With respect to the fluctuations it seems that constant curvature gives the smallest deviations. Additionally, it can be observed that the constant curvature estimation gives slight under-predictions, whereas the concentric cases give systematic increased values of the mass burning rate. Moreover it can be seen that the stoichiometric case (C3) gives the smallest deviations for any of the present approximations. This indicates that the choice of progress variable for the lean case might not be the best choice.

For closer inspection of all realizations in the field, case C6 is chosen in which the deviations are largest. Correlation plots are shown in figure 2. For this case the basic model does not deviate significantly from the results in figure 1, the only difference being that the range of values is extended more to the origin of the plot. Moreover some features, as indicated above, are clearly reflected like

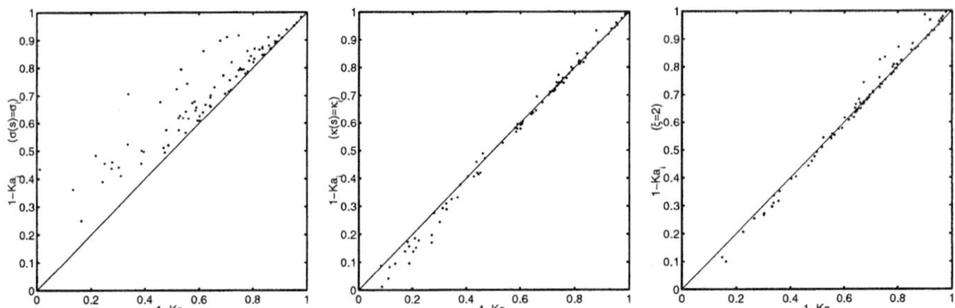

Fig. 2. Case C6, Correlation of the actual mass burning rate with the approximations, left: $\sigma = \sigma_{\mathrm{in}}$, middle: $\kappa = \kappa_{\mathrm{in}}$, right: $\xi = 2$, the case $\xi = 1$ deviates only very little from $\xi = 2$ and the figure is not given (result of 60000 flamelets found in the domain)

the under-prediction of the constant surface case. Furthermore the predictions of the concentric cases are less robust compared to the constant curvature model. The latter however gives deviations at small mass burning rates. This is also observed, to a lesser degree, in the concentric spherical approximation. Near the origin the cylindrical model seems to perform better. This is in agreement with observations in [6], who found that at higher turbulence levels, curvature in premixed turbulent combustion of flame kernels tends to cylindrical modes of deformation of the flame front.

It is obvious that all models do not fit to the true values because no local information on the flame geometry is taken into account in the approximations. If local geometric information is taken into account a much better agreement would be possible and will be a topic of further research. At larger times in the evolution, e.g. case C6, it was found that the basic model (4), gives good correlations (at $t = 0.087\tau$ mean deviation 0.08, rms values of 0.24), see figure 3, whereas all approximations are starting to deteriorate severely. In this case the curvatures have large values, the associated values of radii are within the flame thickness, δ_{f}, as shown in the figure (at the right).

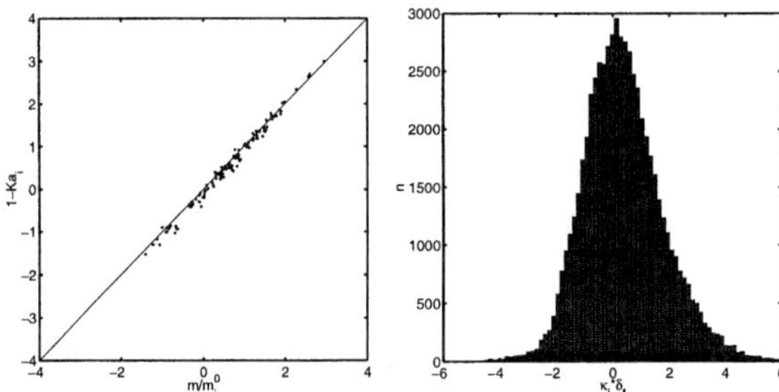

Fig. 3. Results of case C6 at time $t = 0.087\tau$, left: correlation of the actual mass burning rate with the basic model, right: PDF of inner layer curvatures

4 Conclusions

From the previous results it can be concluded that the method of FGM in combination with DNS calculations looks very encouraging. It appears that the FGM is a promising technique to reduce the chemistry and obtain accurate results for the flow, thermodynamics and species. However, apart from a validation in terms of laminar burning velocity, a direct validation is not present for turbulent cases. With respect to this, more validation is needed and the strategy for this will be twofold. By applying a suitable kinetics model with a limited number of species, a DNS can be conducted. This system can be reduced and validated directly against the results of the detailed chemistry calculations. A second method is to increase the dimension of the manifold. It must be studied how many controlling variables are required for a certain accuracy of the predictions. This again can be performed in the framework of the previously mentioned full chemistry DNS.

References

1. BASTIAANS, R. J. M., VAN OIJEN, J. A., MARTIN, S. M., DE GOEY, L. P. H. & PITSCH, H. 2004 DNS of lean premixed turbulent spherical flames with a flamelet generated manifold. In P. Moin, editor, *CTR Annual Research Briefs*, in press.
2. CHEM1D 2002 A one dimensional flame code. Eindhoven University of Technology, http://www.combustion.tue.nl/chem1d.
3. GICQUEL, O., DARABIHA, N. & THEVENIN, D. 2000 Laminar premixed hydrogen/air counter flow flame simulations using flame propagation of ILDM with preferential diffusion. *Proc. Combust. Inst.* **28**, 1901-1908.
4. DE GOEY, L. P. H. & TEN THIJE BOONKKAMP, J. H. M. 1999 A flamelet description of premixed laminar flames and the relation with flame stretch. *Comb. Flame*, **119**, 253-271.
5. GROOT, G. R. A. & DE GOEY, L.P.H. 2002 A computational study of propagating spherical and cylindrical premixed flames. *Proc. Combust. Institute*, **29**, 1445-1451.
6. JENKINS, K.W. & CANT, R.S. 2002 Curvature effects on flame kernels in a turbulent environment. *Proc. Comb. Inst.*, **29**, 2023-2029.
7. MAAS, U. & POPE, S. B. 1992 Simplifying Chemical Kinetics: Intrinsic Low-Dimensional Manifolds in Composition Space. *Combust. Flame* **88**, 239-264.
8. VAN OIJEN, J.A. 2002 Flamelet-Generated Manifolds: Development and Application to Premixed Laminar Flames. Ph.D. thesis, Eindhoven University of Technology, The Netherlands.
9. SMITH, G. P., GOLDEN, D. M., FRENKLACH, M., MORIARTY, N. W., EITENEER, B., GOLDENBERG, M., BOWMAN, C.T., HANSON, R. K., SONG, S., GARDINER JR., W. C., LISSIANSKI, V.V. & Z. QIN, Z. 1999 http://www.me.berkeley.edu/gri_mech/

Entropic Lattice Boltzmann Method on Non-uniform Grids

C. Shyam Sunder and V. Babu

Department of Mechanical Engineering,
Indian Institute of Technology,
Madras, India 600 036.
vbabu@iitm.ac.in

Abstract. The entropic lattice Boltzmann method (ELBM) has recently been shown to be capable of simulating incompressible flows with good accuracy and enhanced stability. However, the method requires that uniform grids be used, which makes it prohibitively expensive for simulating large Reynolds number flows. In this paper, an interpolation scheme is proposed to extend the applicability of this method to arbitrary non-uniform meshes, while retaining the salient features of ELBM such as stability and computational efficiency. The new scheme is used to simulate flow in a lid driven cavity as well as flow past a square cylinder and is shown to largely reduce the grid requirements. The results of the simulation agree very well with other numerical experimental data available in the literature

Keywords: Entropic Lattice Boltzmann ISLB.

1 Introduction

In the last decade the Lattice Boltzmann method (LBM) has attracted a lot of attention and is fast emerging as an alternative to finite volume and finite element techniques. Lattice Boltzmann methods have been successful in simulating many simple and complex hydrodynamics including turbulent flows[1]. The main attractions of LBM are ease of implementation, high computational efficiency and easy parallelizability. There are many variations of LBM in use, among which are finite volume Lattice Boltzmann method, multiple relaxation Lattice Boltzmann methods, interpolation supplemented Lattice Boltzmann method, and entropic Lattice Boltzmann method.

The Lattice Boltzmann Equation (LBE) initially originated as a floating point recast of the evolution equation of lattice-gas cellular automaton dynamics. The simplest form of LBE, namely, the Lattice Bhatnagar Gross Krook (LBGK) form, can be written as follows.

$$f_i(x + c_i\delta_i, t + \delta_t) - f_i(x,t) = -\frac{1}{\tau}\{f_i(x,t) - f_i^{eq}(x,t)\} \tag{1}$$

V.S. Sunderam et al. (Eds.): ICCS 2005, LNCS 3516, pp. 72–79, 2005.

where

$$c_i = \begin{cases} 0, & i = 0, \\ c\{cos((i-1)\pi/2), sin((i-1)\pi/2)\}, & i = 1,2,3,4 \\ \sqrt{2}c\{cos[(i-5)\pi/2+\pi/4], sin((i-5)\pi/2+\pi/4)\}, & i = 5,6,7,8 \end{cases} \quad (2)$$

where $c = \partial x/\partial t$ is the lattice speed, $f_i(x,t)$ represents the probability for a particle moving in the direction c_i to reside at the location x, at time t and f_i^{eq} is the equilibrium distribution corresponding to the particular velocity and density. The right hand side of the equation represents the single relaxation time model collision process with τ representing the dimensionless relaxation time. The local hydrodynamic quantities are given by

$$\rho = \sum f_i, \quad \rho u = \sum f_i c_i \quad (3)$$

One of the shortcomings of the LBM is numerical instability. The reason for the instability is that no bounds are imposed on the values f_i^{eq} and f_i during the collision process making it possible for f_i to take negative values depriving it of any physical sense [2]. One of the ways to ensure the positivity of f_i is to define the corresponding equilibrium value as a minimum of a convex function, known as the H function, under the constraint of the local conservation laws [2].

Apart from the stability issue, another well known problem associated with the discrete velocity models is non-adherence to the equation of the state [3],[4]. In these models, the local equilibrium entropy does not obey the thermodynamic definition of the temperature being a function of entropy and energy [4]. These issues were addressed in the formulation of the ELBM [4],[5],[6],[7]. This is discussed in the next section.

As described by Eq. 1, particles at a lattice site undergo collision followed by advection. The left hand side of Eq. 1 can be split into two parts viz. calculation of $f_i(x, t+\delta_t)$ (updating post collision values) and calculation of $f_i(x+c_i\delta_t, t+\delta_t)$ (advection in the direction c_i). This advection however can be done only to the neighboring lattice sites at a distance of $c_i\delta_t$ which constrains the lattice used for the simulation to be a uniform square mesh. This issue was addressed by He et al [8] for a 9 velocity LBE model, wherein the extension of LBE for nonuniform meshes was outlined. The objective of the present work is to extend the ELBM method to non-uniform meshes also using the methodology outline by He et al [8].

2 Entropic Lattice Boltzmann Method

The construction of the entropic Lattice Boltzmann method (ELBM) is based on a good choice of the discrete velocities, the H function and an expression for the equilibrium values[9]. The discrete form of the H function is derived from the continuous Boltzmann H function given by $\int F \ln F dc$ where $F(\boldsymbol{x}, c)$ is the single particle distribution function, $\boldsymbol{x}$ is the position vector and c is the continuous velocity. For 2D athermal cases, the discrete form the H function can be written as

$$H_{(W_i,f_i)} = \sum_{i=0}^{8} f_i \ln\left(\frac{f_i}{W_i}\right) \tag{4}$$

where f_i represents discrete velocities and W_i the weights associated with each direction. The weights in one dimension are $\{1/6, 2/3, 1/6\}$ for the directions 'Right', 'Zero' and 'Left' respectively and the weights for higher dimensions can be constructed by multiplying the weights associated with each component direction[10]. The equilibrium value of the discrete velocity is the minimizer of the corresponding H function under the constraints of local conservation laws given by Eq. 3. The explicit solution for the f_i^{eq} in D dimensions is

$$f_i^{eq} = \rho W_i \prod_{j=1}^{D} (2u_j - \left(\sqrt{1+3u_j^2}\right)) \left(\frac{2u_j + \sqrt{1+3u_j^2}}{1-u_j}\right)^{c_{ij}/c_s} \tag{5}$$

where j is the index for spatial directions, c_s is the velocity of sound and the exponent c_{ij}/c_s can only take the values +1, 0, -1. H can be guaranteed to remain non-increasing everywhere by a two step process explained in Fig. 1. The population is changed first by keeping the H constant. In the second step dissipation is introduced to decrease H. It can be guessed that in the steady state i.e. when $f_i^{eq} = f_i$, value of H remains constant. The BGK form of the collision in ELBM is

$$f_i(x + c_i\delta_t, t + \delta_t) - f_i(x,t) = \alpha\beta\left[f_i(x,t) - f_i^{eq}(x,t)\right] \tag{6}$$

where $\beta = 6/(1+6\nu)$ is the relaxation frequency with ν being the kinematic viscosity. The parameter α can be calculated by solving the following nonlinear equation

$$H(f) = H(f + \alpha\Delta) \tag{7}$$

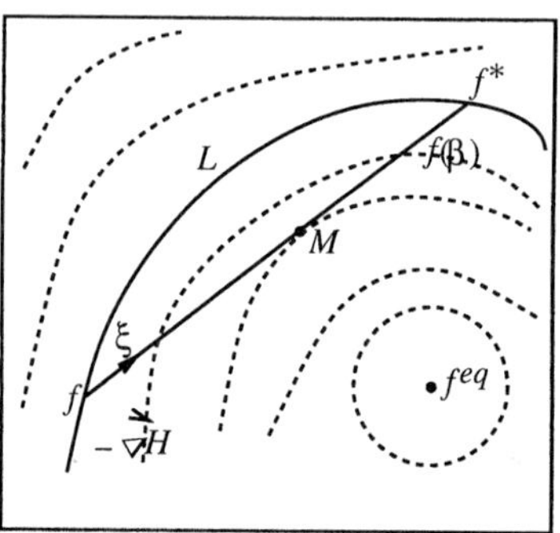

Fig. 1. Graphical representation of the collision process. The curves shown represent lines of constant entropy and ξ represents the collision integral. The first step of the collision process is the calculation of the point f^*, by solving equation Eq. 7. Then $f(\beta)$ is determined using 'over-relaxation' based on the value of β[9]. In case of BGK collision, the point M coincides with f^{eq}

where Δ represents the bare departure from the local equilibrium i.e. $\Delta = f^{eq} - f$. The value of α is usually close to 2 (and equal to 2 when $f = f^{eq}$). The procedure for explicit calculation of α was described by Ansumali and Karlin [10]. By using different value for α at each lattice site, the relaxation time is adjusted locally to ensure compliance with the H theorem. This guarantees the positivity of the distribution function, which gives the scheme good non-linear stability.

3 Entropic Lattice Boltzmann Method on Non-uniform Grids

The ELBM as outlined above is restricted to uniform square meshes. With increasing Reynolds number, this results in a tremendous increase in the total number of points required and hence an increase in computation time. Moreover, ELBM is computationally more expensive than LBM because of multiple calculations of $'\ln'$ function at every lattice point. The present work aims at extending the applicability of ELBM to arbitrary nonuniform meshes which can help reduce the total number of grid points required for a simulation.

The distribution functions (f) in LB models are continuous functions in both space and time, although they are used to represent population of particles locally. This gives us scope for interpolation schemes to be used in the interior of the computational domain and extrapolation schemes at the boundary.

In short, the implementation of ELBM on nonuniform grids can be explained in the following steps:

1. Start with an initially guessed flow field.
2. For each lattice site, calculate f^{eq} according to Eq. 5 and hence find Δ $(= f^{eq} - f)$.
3. Calculate α by solving the nonlinear equation 7.
4. Perform collision according to equation 6.
5. Advect the post collision values to appropriate neighboring nodes (imaginary square lattice on which the LB automaton resides).
6. Calculate the values on the 'actual' lattice sites (mesh used in the simulation) by means of quadratic interpolation.
7. Apply the necessary boundary conditions.
8. Goto step 2.

It should be noted that step 5 can be combined with step 6, eliminating the need to store and maintain the values of f_i on the nodes of the imaginary lattice. Computationally ELBM can be divided into three kernels viz. collision (steps 2, 3, 4), advection and interpolation (steps 5 and 6), boundary conditions (step 7). In general, collision accounts for about 90% of the total computational load, and advection and interpolation about 8%.

In the case of ELBM on square grids i.e. without interpolation being done on the grid, the time spent in collision accounts for almost 98% of the computational load. This shows that the overhead associated with the introduction of the interpolation routine is not very high. It will be shown in the coming sections that the savings incurred in terms of grids is enormous and more than compensates for the extra calculation being done in interpolation.

4 Enhancement in Reynolds Number

Another advantage of using nonuniform or rectangular meshes is the possibility of increasing the Reynolds number of the simulation for a given number of grid points and without any loss of stability. The Reynolds number of the simulation, Re is equal to UL_{char}/ν where U is the reference velocity, ν is kinematic viscosity in lattice units and L_{char} is the characteristic length for the flow in grid units and is equal to $N_{char}\ \delta^{mesh}/\delta^{automaton}$. Here δ^{mesh} is the spacing of the computational grid and $\delta^{automaton}$ is the spacing of the grid on which the automaton resides. N_{char} is the number of grid points per characteristic length measured on the latter grid. Now for square uniform meshes, $\delta^{automaton} = \delta^{mesh}$. However for rectangular and nonuniform meshes, δ^{mesh} can be chosen to be greater than $\delta^{automaton}$ resulting in increase in the Reynolds number of the simulation by a factor of $\delta^{mesh}/\delta^{automaton}$. The use of nonuniform and rectangular grids does not affect the stability of the system (or the LB automaton) as the $\delta^{automaton}$ is not being altered. Hence for a given number of grid points, the Reynolds number is enhanced. Moreover, by choosing different levels of grid spacing it is possible to greatly reduce the number grid points required for the simulation.

5 Results and Discussion

In this section, results from numerical simulations using ELBM on non-uniform grids are presented. Both steady (lid driven cavity flow) and unsteady flows (flow past a square) are simulated to demonstrate the suitability of the methodology.

5.1 Lid Driven Cavity Flow

In this classical test problem, fluid is contained within three stationary walls while the fourth (top) wall moves with a certain velocity. This results in the formation of a large primary vortex near the center of the cavity and smaller vortices (two or three, depending on the Reynolds numbers) near the corners. A non-uniform mesh with about 60-100 grid points along each side of the cavity has been used for the simulations. Diffusive boundary condition [11] has been used for all the four walls. In the diffusive boundary condition, the corners can be associated with either of the walls or the normal can be made to be inclined at a 45° angle. In the current implementation, the corners are treated as points belonging to the side walls (left or right). The other choice for the association of corner points is also possible and would not affect the flow by a great extent as three of the four diagonal speeds (f) do not propagate into the domain. The results of the simulation compare well with those available in the literature as shown in Table 1.

Table 1. Comparison of results for the lid driven cavity problem

Re	Reference	Primary vortex (Ψ_{max}, X, Y)	Lower left vortex $(\Psi_{max} \times 10^{-5}, X, Y)$	Lower right vortex $(\Psi_{max} \times 10^{-4}, X, Y)$
400	[13]	(0.1121, 0.5608, 0.607)	(1.30, 0.0549, 0.0510)	(6.19, 0.8902, 0.1255)
	[14]	(0.1130, 0.5571, 0.607)	(1.42, 0.0508, 0.0469)	(6.42, 0.8906, 0.1250)
	current	(0.1120, 0.548, 0.596)	(1.25, 0.04, 0.036)	(5.92, 0.88, 0.112)
1000	[13]	(0.1178, 0.5333, 0.564)	(22.2, 0.0902, 0.0784)	(16.9, 0.8667, 0.1137)
	[14]	(0.1179, 0.5313, 0.562)	(23.1, 0.0859, 0.0781)	(17.5, 0.8594, 0.1094)
	current	(0.1175, 0.536, 0.564)	(18.2, 0.052, 0.040)	(17.1, 0.852, 0.108)
5000	[13]	(0.1214, 0.5176, 0.537)	(135, 0.0784, 0.1373)	(30.3, 0.8075, 0.0745)
	[15]	(0.1190, 0.5117, 0.535)	(136, 0.0703, 0.1367)	(30.8, 0.8086, 0.0742)
	current	(0.095, 0.508, 0.528)	(132, 0.066, 0.128)	(30.2, 0.8025, 0.068)

5.2 Flow Past a Square

Here, flow past a square immersed in a stream of fluid is considered. As shown in the figure, a square of side L_{char} is placed symmetrically at a location $10L_{char}$ down stream of the inlet AB. The size of the computational domain is $45L_{char}$ in the direction of the flow and $25L_{char}$ perpendicular to the flow (resulting in a blockage ratio of 4%). The side CD represents the outlet boundary, and

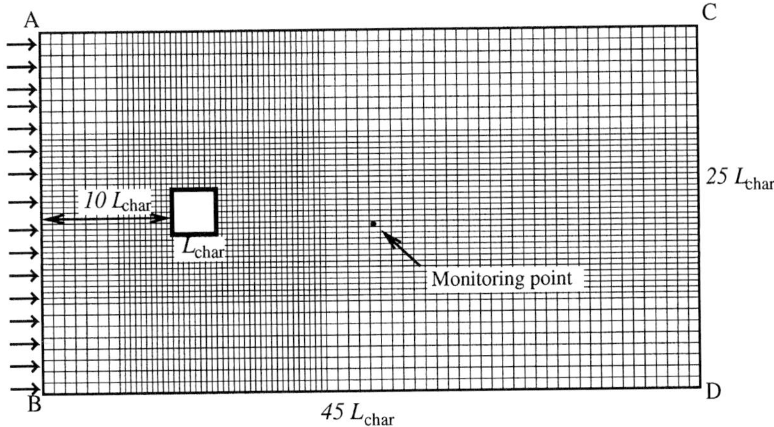

Fig. 2. Flow past a Square

the sides BC and DA are free stream boundaries. This flow becomes unsteady for Reynolds numbers above 40 or so. If the dominant frequency of the vortex shedding is denoted by n_f then the Strouhal number of the flow is $n_f L_{char}/U_\infty$, where U_∞ is the free stream velocity. In the past, several techniques such as the Finite Difference, Finite Volume and Lattice Boltzmann methods, have successfully predicted the Strouhal number - Reynolds number relation at low and moderate Reynolds numbers [17],[18]. At high Reynolds numbers, conventional techniques face difficulties such as enormous grid requirements, lack of stability and lack of a good exit boundary condition [1]. The difficulty with the exit boundary condition can be handled by increasing the domain size in the direction of flow, albeit, at a tremendous increase in the computational cost. In contrast, the non-linear stability of the entropic Lattice Boltzmann method together with lesser grid requirements gives an added advantage over the normal LB methods. Thus, the usage of non-uniform grids with ELBM makes it possible to simulate the flow at high Reynolds numbers and with a reduced computational cost. Table 2 shows the variation of Strouhal number with Reynolds number as predicted by experiments and numerical calculations. It is evident that the present calculations are able to predict the Strouhal number very well and the agreement is better than that of the results available in the literature so far. The grid sizes used in the present case are much smaller than the grid requirements of normal LBM. At high Reynolds numbers, the N_{char} used in the current case ranges from 30-50, whereas the N_{char} required for normal LBM would be around 128-256 [19] resulting in more than an order of magnitude savings in the grid requirement (Table 2).

Table 2. Comparison of Strouhal number and grid sizes for different Reynolds numbers

Re	Strouhal Number				Grid Size	
	Expt[16]	Current	FV[17]	FD[18]	Current	LBM [19]
250	0.140-0.145	0.136	0.17	0.165	60000	112500
400	0.127-0.136	0.132	0.145	0.161	"	"
500	0.126-0.133	0.132	0.174	-	240000	288000
800	0.121-0.126	0.129	-	-	"	"
1000	0.120-0.122	0.130	-	0.142	$4.5 - 5.4 \times 10^5$	$18 - 73 \times 10^6$
2000	0.120-0.129	0.131	-	0.142	"	"
3000	0.123-0.130	0.133	-	0.145	"	"

6 Conclusion

An interpolation scheme to enable the use of non-uniform meshes is proposed for the entropic lattice Boltzmann method. The method has been successfully used for simulating the steady flow in a lid driven cavity and the unsteady flow over a rectangular cylinder for Reynolds number up to 3000. For the former problem, the location as well as the value of the streamfunction at the center of the vortices is predicted very well by the present calculations, when compared with data available in the literature. For the latter problem, the unsteady vortex shedding frequency predicted by the method compares well with the experimental values reported by Okajima [16]. The proposed interpolation scheme preserves all the salient features of ELBM, like stability, and locality of the collision operator.

Moreover, the interpolation itself is a very small addition to the overall computational cost. Considerable savings in grid sizes are shown to be possible by comparing with earlier implementations of LBM. This interpolation scheme is quite general and can be easily implemented on different types of grids.

References

1. S. Succi, The Lattice Boltzmannn Equation for Fluid Dynamics and Beyond, Oxford university press, Oxford, 2001.
2. S. Succi, I. V. Karlin, H. Chen, Colloquium: Role of the H theorem in lattice Boltzmann hydrodynamic simulations, Rev. Modern Physics 74 (2002).
3. M. Ernst, Discrete Models of Fluid Dynamics, edited by Alves A. S. (World Scientific, Singapore) 1991.
4. S. Ansumali, I. V. Karlin, H. C. Ottinger, Minimal entropic kinetic models for hydrodynamics, Europhys. Lett. 63 (6) (2003) 798-804.
5. B. M. Boghosian, P. J. Love, P. V. Coveney, Iliya V. Karlin, Sauro Succi and Jeffrey Yepez, Galilean-invariant lattice-Boltzmann models with H theorem, Phys. Rev. E 68, (2003) 025103.
6. B. M. Boghosian, J. Yepez, P. V. Coveney, A. Wagner, Entropic lattice Boltzmann method, Royal Society of London Proceedings Series A, vol. 457, Issue 2007, p.717
7. H. Chen, C. Teixeira, H-theorem and origins of instability in thermal lattice Boltzmann models, Computer Phy. Comm., vol. 129, Issue 1-3, pp.21-31
8. X. He, L. S. Luo, M. Dembo, Some progress in Lattice Boltzmann method :Part 1. Nonuniform Mesh Grids, J. Comp. Phys. 129 (1996) 357-363.
9. I. V. Karlin, A. Ferrante, H. C. Ottinger, Perfect entropy functions of the Lattice Boltzmann method, Europhys. Lett. 47 (2) (1999) 182-188.
10. S. Ansumali, I V. Karlin, Single relaxation time model for entropic lattice Boltzmann methods, Phys. Rev. E 65 (2002) 056312.
11. S. Ansumali, I. V. Karlin, Kinetic boundary conditions in the lattice Boltzmann method, Phys. Rev E 66 (2002) 026311.
12. X. He, L. S. Luo, M. Dembo, Some Progress in Lattice Boltzmann method: Enhancement of Reynolds number in simulation, Physica A 239 (1997) 276-285.
13. S. Hou, Q. Zou, S. Chen, G.D. Doolen, A.C. Cogley, Simulation of Cavity Flow by the Lattice Boltzmann Method, J. Comp. Phys. 118 (1995) 329.
14. R. Schreiber, H.B. Keller, Driven cavity flows by efficient numerical techniques, J. Comp. Phy 49 (1983) 310.
15. U. Ghia, K.N. Ghia, C.Y. Shin, High-Re solutions for incompressible flow using the N.S. equations and a multigrid method, J. Comp. Phys. 48 (1982) 387.
16. A. Okajima, Strouhal numbers of retangular cylinders, J. Fluid Mech. 123 (1982) 379-398.
17. A. Sohankar, C. Norberg, L. Davidson, Simulation of three-dimensional flow around a square cylinder at moderate Reynolds numbers, Phys. Fluids 11 (1999) 288.
18. R. W. Davis, E. F. Moore, Vortex shedding process behind two dimensional buff bodies, J. Fluid Mech. 116 (1982) 475.
19. G. Baskar, V. Babu, Simulation of the Flow Around Rectangular Cylinders Using the ISLB Method, AIAA-2004-2651 (2004).

A Data-Driven Multi-field Analysis of Nanocomposites for Hydrogen Storage

John Michopoulos, Nick Tran, and Sam Lambrakos

Materials Science and Component Technology Directorate,
U.S. Naval Research Laboratory,
Washington DC, 20375,U.S.A
{john.michopoulos, nick.tran, sam.lambrakos}@nrl.navy.mil

Abstract. This paper focuses on computational parameter identification associated with heat and mass diffusion macro-behavioral models of hydrogen storage systems from a continuum multiphysics perspective. A single wall nanotube (SWNT) based composite pellet is considered as our representative finite continuum system. The corresponding partial differential equations (PDEs) governing the spatio-temporal distribution of temperature and hydrogen concentration are formulated. Analytical solutions of the system of coupled PDEs are constructed and utilized in the context of inverse analysis. The corresponding non-linear optimization problem is formulated in order to determine the unknown parameters of the model, based on an objective function and constraints consistent with experimentally acquired data along with the physical and utilization requirements of the problem. Behavioral simulation results are presented in an effort to demonstrate the applicability of the methodology. Finally, we indicate potential extensions of this methodology to multi-scale and manufacturing process optimization.

1 Introduction

The activities described herein are a part of a larger effort associated with the development of a data-driven environment for multiphysics applications (DDEMA) [1, 2, 3, 4]. The analytical methodology for approaching system identification problems is based on establishing a corresponding inverse problem that can be solved by means of global optimization as shown earlier [5, 6, 7] for various problems ranging from material nonlinear constitutive response to welding characterizations.

The recent utilization of SWNTs as an essential component for nanocomposites for hydrogen storage has lead to the direct need for modeling the behavioral characteristics of such systems during both the hydriding and dehydriding stages. The desire for manufacturing highly efficient, inexpensive and long lasting hydrogen storage systems underscores our motivation for modeling such system as a special case of our general effort.

In order to achieve these goals, we proceeded with the modeling of a nanocomposite pellet's hydriding and dehydriding response coupled with heat conduction

V.S. Sunderam et al. (Eds.): ICCS 2005, LNCS 3516, pp. 80–87, 2005.

and chemical reactivity under the framework of an inverse problem setting applied to preliminary experiments that generated the data to drive our modeling. Continuum multi-field modeling efforts have been based on continuum thermodynamics and conservation theorems [8, 9, 10]. Here we present the results of our multi-field modeling effort [5] for the problem at hand along with a description of the associated solutions.

2 Behavioral Modeling of Hydrogen Storage Pellet

Consider a medium with isotropic and homogeneous aggregate properties at the macro-length scale. Furthermore, consider that this medium is exposed to temperature and multi-species concentration boundary conditions. As heat diffusion proceeds in the medium so does multi-species mass diffusion. The species may or may not be chemically reactive with the host medium or with each other. A medium under such a multi-field excitation includes as a special case the SWNT-enhanced materials considered for hydrogen storage. The general procedure for deriving the continuum multiphysics model for this system is analogous to the one followed for hygrothermoelastic composites elsewhere [10]. The resulting general system of PDEs [11] describes all open continuum systems under the influence of temperature, and multi-species diffusion with chemical reactions among the species. However, for the case of the hydrogen storage continua of the type used here we do not have cross-species reactivity since hydrogen is the only species involved. The diffused hydrogen only reacts with the matrix and the embedded SWNTs of the composite. This process can be abstracted as an absorption/desorption diffusion. Therefore, as a first approximation we will assume a single component system (i.e. we will consider the concentration of hydrogen C as our mass concentration variable) with no chemical reaction involved and T as our temperature field state variable. In this case the following pair of coupled PDEs is valid:

$$\frac{\partial C}{\partial t} = D_m \nabla^2 C + \lambda D_h \nabla^2 T, \tag{1a}$$

$$\frac{\partial T}{\partial t} = \nu D_m \nabla^2 C + D_h \nabla^2 T. \tag{1b}$$

Here, D_m and D_h are the mass and heat diffusivities, respectively, and λ, ν are both coupling coefficients. This heat-mass diffusion system is completed by the generalized boundary conditions (of the third kind),

$$T(\mathbf{x}, t) + \frac{k_h(\mathbf{x})}{h_T} \frac{\partial T(\mathbf{x}, t)}{\partial \mathbf{n}} = \phi_T(\mathbf{x}, t), \mathbf{x} \in S \tag{2a}$$

$$C(\mathbf{x}, t) + \frac{k_m(\mathbf{x})}{h_C} \frac{\partial C(\mathbf{x}, t)}{\partial \mathbf{n}} = \phi_C(\mathbf{x}, t), \mathbf{x} \in S \tag{2b}$$

where S represents the boundary, $\mathbf{x}$ and $\mathbf{n}$ the position and outward normal vectors on the boundary, respectively, k_h, k_m are the heat and mass conductivities, respectively, h_T, h_C are the heat and mass transfer coefficients respectively, and finally $\phi_T(\mathbf{x},t), \phi_C(\mathbf{x},t)$ are the prescribed distributions of temperature and mass concentration on the boundary, respectively. Boundary conditions of the first and second kinds can be produced by appropriate omission of the second terms (first kind) and the first terms (second kind) of these equations. The cylindrical symmetry of the SWNT composite pellet suggests transforming the governing equations from the cartesian coordinate system to a cylindrical frame of reference ($\{x,y,z,t\} \to \{r,\vartheta,z,t\}$). In addition, the axisymmetric character of the applied boundary conditions further simplify the system of Eqs. (1) by coordinate transformation of the nabla operators. It has been shown elsewhere [10] that it is possible to uncouple the system of Eqs. (1) by using the solutions of the uncoupled PDEs if λ, ν are constants via the method of normal coordinates. One form of the uncoupled solution is given by

$$\begin{aligned} T = T_0 + (T_f - T_0)F_1(r/r_o, z/l_o, D_h t/r_o^2, D_h t/l_o^2, u_d) \\ +\nu u_d(C_f - C_0)F_2(r/r_o, z/l_o, D_m t/r_o^2, D_m t/l_o^2, 1/u_d) \end{aligned} \tag{3a}$$

$$\begin{aligned} C = C_0 + (C_f - C_0)F_1(r/r_o, z/l_o, D_m t/r_o^2, D_m t/l_o^2, 1/u_d) \\ +\lambda(T_f - T_0)F_2(r/r_o, z/l_o, D_h t/r_o^2, D_h t/l_o^2, u_d) \end{aligned} \tag{3b}$$

where T_f, C_f are the final values of the temperature and hydrogen concentrations, respectively, that are applied on the boundary and reached by the continuum at equilibrium conditions and where,

$$F_i(r,z,\tau_r,\tau_z,u) = \Psi_i^c(r,\tau_r,u)\Psi_i^s(z,\tau_z,u). \tag{4}$$

The solutions F_i for the finite cylinder geometry have been constructed here as the product of the two one dimensional solutions corresponding to an infinite cylinder of radius r_o and an infinite slab of thickness $2l_o$ with appropriate scaling of time via usage of τ_r and τ_z, respectively. The functions Ψ_i^j with $j = c, s$ and $i = 1, 2$ are defined by

$$\Psi_1^j(x_j,\tau_j,u) = [1 - H_1^j(u)]\Psi^j(x_j, D_2(u)\tau_j) + H_1^j(u)\Psi^j(x_j, D_1(u)\tau_j) \tag{5a}$$

$$\Psi_2^j(x_j,\tau_j,u) = H_2^j(u)[\Psi^j(x_j, D_2(u)\tau_j) - \Psi^j(x_j, D_1(u)\tau_j)], \tag{5b}$$

where we have used the contractions,

$$H_1(u_d) = \frac{1}{2}[1 - (1 - u_d)H_2(u_d) \tag{6a}$$

$$H_2(u_d) = \frac{D_1 D_2}{(D_2 - D_1)u_d}, \tag{6b}$$

where $u_d = D_m/D_h$ and

$$D_1 = 2u_d/[1 + u_d + \sqrt{(1-u_d)^2 - 4u_d(1-\lambda\nu)}] \tag{7a}$$

$$D_2 = 2u_d/[1 + u_d - \sqrt{(1-u_d)^2 - 4u_d(1-\lambda\nu)}]. \tag{7b}$$

Functions Ψ^j are the solutions of the normalized one-dimensional problems for the cylinder $m = 0$ and the slab $m = 1/2$ and are defined by

$$\Psi^j(x,\tau) = 1 - 2\sum_{i=1}^{\infty} \frac{1}{\mu_i}[1 + \frac{2m}{Bi} + (\frac{\mu_i}{Bi})^2]^{-1} x^m \frac{J_{-m}(\mu_i x)}{J_{1-m}(\mu_i)} e^{-\mu_i^2 \tau}, \quad (8)$$

in terms of the Bessel functions J_i. The coefficients μ_i are the roots of the transcendental equation

$$Bi J_{-m}(\mu) - \mu J_{1-m} = 0. \quad (9)$$

3 Inverse Problem Setup

The set of diffusivities (D_h, D_m), the coupling constants (λ, ν) and the four instances of the Biot number Bi ($(Bi)_T^c = h_T r_o/k_h, (Bi)_C^c = h_C r_o/k_m, (Bi)_T^s = h_T l_o/k_h, (Bi)_C^s = h_C l_o/k_m$) determine the behavioral model completely when the geometry and initial and boundary conditions are known. Determination of these constants from experimental data constitutes the inverse problem formulation.

3.1 Experimental Setup and Associated Data (NT)

The pellet samples for the hydridingdehydriding study were made by uniaxially cold pressing the Mgx wt.% Mm powder. The details of manufacturing and processing the pellets are given elsewhere [12, 13]. Figure (1) shows the pellet and the experimental setup for hydriding and dehydriding the pellet. Hydriding and dehydriding of Mgx wt.% Mm pellets were carried out inside a dual reactor made of seamless inconel tubes. Two thermocouples were attached to the outside of each tube and were calibrated to reflect the temperature of the reaction zone. All data from the hydriding and dehydriding process were digitally captured by

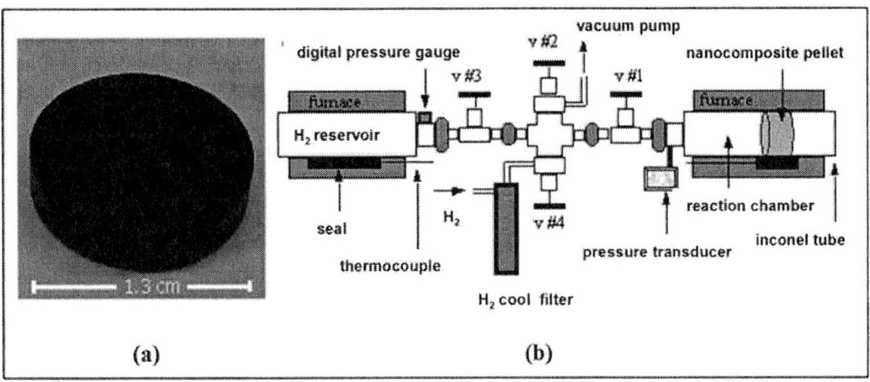

Fig. 1. Hydrorgen storage pellet (a) and experimental setup schematic for hydriding and dehydriding of thc pellet (b)

a data acquisition board, which was controlled by our custom code [13]. The hydrogen uptake or release of the Mgx wt.% Mm pellets was determined by measuring the pressure changes in the reaction chamber before and after each hydriding or dehydriding treatment.

3.2 Nonlinear Optimization

The general outline of using design optimization for parameter identification is a well established discipline. It usually involves the usage of an optimization module that utilizes behavioral data from the actual physical system (collected experimentally) as well as simulated data from a potential model. The design variables to be identified are the constants of the model. Ultimately, the simulated behavior has to reproduce the experimentally observed behavior and therefore an objective function must be minimized. For an objective function $f = f(\mathbf{X})$ with $\mathbf{X}$ being the resultant vector defined by the design variable component vectors such as $x_1\mathbf{i}_1, \ldots, x_n\mathbf{i}_n \in \mathbf{X}^n$, where $\mathbf{X}^n$ is the vector space spanned by the basis vectors $\mathbf{i}_1, \ldots, \mathbf{i}_n$. For this general case the optimization problem can be expressed by

$$min[f(x_1, \ldots, x_n)]_{x_i \in \Re^n} \tag{10a}$$

$$c_j^{eq}(x_1, \ldots, x_n) = 0, c_i^{ineq}(x_1, \ldots, x_n) \geq 0, j = 1, \ldots, p \tag{10b}$$

Relations (10b) express the equality and inequality constraints.

In our analysis the behavior of the system is described by the time evolutions of the total temperature and hydrogen absorption. The stimulation of the system is expressed by the multi-field boundary conditions that define the temperature and hydrogen mass concentration on the boundary. We define $T_i^{sim}(t_i), C_i^{sim}(t_i)$ and $T_i^{exp}(t_i), C_i^{exp}(t_i)$ to be the simulated and experimental values, respectively, of the temperature and hydrogen concentration at a discrete evaluation point $i = 1. \ldots, p$ where p is the total number of evaluations. We can now form the objective function as the sum of the squares of their differences in the least square sense, as follows:

$$\begin{aligned} f(x_1, \ldots, x_n) = & \sum_{i=1}^{p} [T_i^{exp}(t_i) - T_i^{sim}(x_1, \ldots, x_n, t_i)]^2 + \\ & + \sum_{i=1}^{p} [C_i^{exp}(t_i) - C_i^{sim}(x_1, \ldots, x_n, t_i)]^2 \end{aligned} \tag{11}$$

We set the simulated values to be the average values of the solutions of the PDEs along the edge of the pellet for the temperature and over the entire volume for the hydrogen concentration according to

$$T_i^{sim}(x_1, \ldots, x_n, t_i) = \int_{-l_0}^{l_0} T(r_0, z, t_i; x_1, \ldots, x_8) dz \tag{12a}$$

$$C_i^{sim}(x_1, \ldots, x_n, t_i) = 2\pi \int_0^{r_0} \int_{-l_0}^{l_0} C(r, z, t_i; x_1, \ldots, x_8) dr dz. \tag{12b}$$

The integrands in these relations are the ones defined in Eqs. (5) with the special one-to-one substitution $\{x_1, \ldots, x_8\} \rightarrow \{D_h, D_m, \lambda, \nu, (Bi)^c_T, (Bi)^s_T, (Bi)^c_C, (Bi)^s_C\}$.

4 Preliminary Results

The implementation of the optimization procedure was formulated in Mathematica [14] with various global nonlinear optimization algorithms encoded in the package Global Optimization "GO-5.0" [15]. The function "GlobalSearch" that yielded the quickest results utilizes a generalized hill climbing technique that is based on Newton's method but uses a generalized gradient rather than a derivative, and allows for analytic linear or nonlinear constraints. Multiple starts are used to test for the existence of multiple solutions. The multiple starts are generated randomly from the region defined by the range of parameter values input by the user. Feasible starting regions are not needed, but it is assumed that objective function values in this region are Real.

The initial and boundary conditions for the computation of the solutions of the PDEs governing the system as required by the simulated values within the objective function, were chosen to correspond to the experimental procedure. Specifically they were $T_0 = 27^oC, T_f = 200^oC, C_0 = 0$ and $C_f = 1.1$.

Dehydriding data were also considered along with the hydriding data but have not effected the identified parameters.

Figure (2) shows the hydriding time evolution observed experimentally along with the one computed after parameter identification took place. The identified model fits the hydriding experimental data very well and similar performance is observed (but not included here) for the dehydriding data. Therefore, the model may be adopted as a "calibrated" model for simulation of the behavior of similar systems in many other situations and applications, where the shape of specimen varies but the nanocomposite material remains the same.

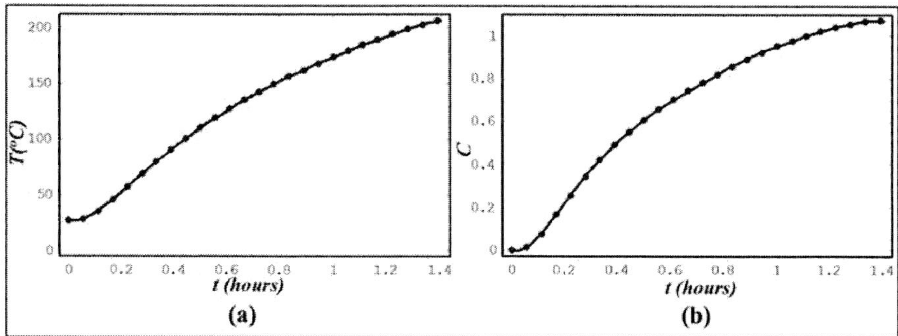

Fig. 2. Experimental (dots) and simulated (continuous line) after the parameter identification has been conducted, for the time evolution of the temperature (a) and Hydrogen concentration (b) time evolutions during the hydriding process

5 Potential Extensions

Careful consideration of the identified parameters implies two distinct correlations. The subset $\{D_h, D_m, \lambda, \nu\}$ is related to the characteristic behavior of the nanocomposite system and the macroscopic length scale, while the subset $\{(Bi)_T^c, (Bi)_T^s, (Bi)_C^c, (Bi)_C^s\}$ is related to the characteristic behavior of the reactor-nanocomposite system owing to the fact that Biot numbers are functions of the heat and mass transfer coefficients that in turn depend on the fluid properties of the gas (hydrogen) according to the Dittus-Boelter equation. These are controllable by the reactor/furnace setup. This suggests that the above referenced optimization procedure is capable of identifying all parameters of interest to the technology practitioner. Therefore, a natural extension of this technique could include any parameters that participate in a given utility (efficiency of performance), aging, or economic model of the system. This extension is well posed as long as the parameters of these models are related to any of the behavioral parameters of the system and an appropriate objective function can be constructed.

In addition to optimizing the usability aspects of such a system one can also optimize the condition of manufacturing related to desirable performance (from efficiency, economic, aging, etc perspectives) by establishing the process control models associated with the manufacturing of such systems, with the lower length scale behavioral models that may be derived from *ab initio* modeling methodologies among many others.

6 Conclusions

We have utilized general coupled mass and heat diffusion, to construct a PDE model to capture the hydrogen storage behavior of a nanocomposite pellet system. We have formulated the inverse problem approach for utilizing data to create a procedure for the data-driven identification of the free parameters participating in the behavior model. We implemented the computational equivalent of the procedure and performed the identification based on experimental data collected for the hydriding cycle of the nanocomposite pellet system. Comparison of the identified system to the experimental data, demonstrated how well the identification has performed. Finally, two specific motivating extensions were discussed to define the context of our future intensions.

Acknowledgements

The authors acknowledge the support by the National Science Foundation under grant ITR-0205663. In addition, the authors would like to extend their special thanks to Dr. F. Darema for her constant availability and encouragement throughout this effort.

References

1. Michopoulos, J., Tsompanopoulou, P., Houstis, E., Rice, J., Farhat, C., Lesoinne, M., Lechenault, F., DDEMA: A Data Driven Environment for Multiphysics Applications,in: Proceedings of International Conference of Computational Science - ICCS'03, Sloot, P.M.A., et al. (Eds.), LNCS 2660, Part IV, Springer-Verlag, Haidelberg, (2003) 309-318.
2. Michopoulos, J., Tsompanopoulou, P., Houstis, E., Rice, J., Farhat, C., Lesoinne, M., Lechenault, F., Design Architecture of a Data Driven Environment for Multiphysics Applications, in: Proceedings of DETC'03, Paper No DETC2003/CIE-48268, (2003).
3. Michopoulos, J., Tsompanopoulou, P., Houstis, E., Farhat, C., Lesoinne, M., Rice, J., Joshi, A., On a Data Driven Environment for Multiphysics Applications, Future Generation Computer Systems, in-print (2005).
4. Michopoulos,J., Tsompanopoulou, P., Houstis, E., Joshi A., Agent-based Simulation of Data-Driven Fire Propagation Dynamics, in: Proceedings of International Conference of Computational Science - ICCS'04, Bubak, M., et al. (Eds.), LNCS 3038, Part III, Springer-Verlag, Haidelberg, (2004) 732-739.
5. Michopoulos, J., Computational Modeling of Multi-Field Ionic Continuum Systems, in: Proceedings of International Conference of Computational Science - ICCS'04, Bubak, M., et al. (Eds., LNCS 3039, Part IV, Springer-Verlag, Haidelberg, (2004) 621-628.
6. Michopoulos, J., Automated Characterization of Material Constitutive Response, in: Proceedings of Sixth World Congress of Computational Mechanics (WCCM-VI 2004), Tsinghua University Press & Springer-Verlag, Haidelberg, (2004) 486-491.
7. Lambrakos, S.G., Milewski, J.O., Analysis of Processes Involving Heat Deposition Using Constrained Optimization, Sci. and Tech. of Weldong and Joining, 7(3), (2002), 137.
8. Truesdell, C., Toupin, R., 1960, "The Classical Field Theories", in Handbuch der Physik (Herausgegeben con S. Flugge) **III/1**, Springer-Verlag, Berlin.
9. Green, A. E., Naghdi, P. M., 1995, "A unified procedure for construction of theories of deformable media. I. Classical continuum physics", Proc. Roy. Soc. London Ser. **A 448** (1934), pp. 335-356.
10. Sih, G.C., Michopoulos, J.G., Chou S.C., "Hygrothermoelasticity", Martinus Nijhoff Publishers (now Kluwer Academic), Dordrecht, (1986).
11. Michopoulos, J.,Lambrakos, S.G., Tran, N. E., Multi-Field Characterization of Single Wall Nano-Tube Composites for Hydrogen Storage, ASME-CIE conference, Long Beach, CA, September 24-28, 2005, To appear, (2005).
12. Tran, N. E. , Imam, M. A., Feng, C. R., Evaluation of Hydrogen Storage Characteristics of Magnesium-misch Metal Alloys. J Alloys Comp ,359, (2003), 225-229.
13. Tran, N. E., Lambrakos, S.G., Purification and Defects Elimination of Single-Walled Carbon Nanotubes by Thermal Reduction Technique. To appear.
14. Wolfram, S., The Mathematica Book,Wolfram Media, (2004).
15. Loehle Enterprises, Global Optimization 5.0 for Mathematica, Loehle Enterprises, (2004).

Plug and Play Approach to Validation of Particle-Based Algorithms

Giovanni Lapenta and Stefano Markidis

Los Alamos National Laboratory, Los Alamos, NM 87545, USA
{lapenta, markidis}@lanl.gov

Abstract. We present a new approach for code validation. The approach is based on using particle-based algorithms to simulate different levels of physical complexity. We consider here heat and mass transfer in a multicomponent plasma at the kinetic and fluid level. By representing both levels using particle methods we can design a component based software package, Parsek, to conduct validation using a plug and play approach. With the plug and play paradigm, different components representing different physical descriptions but all based on a common particle algorithm can be interchanged without altering the overall software architecture and the overall algorithm. The advantage of the plug and play approach is that validation can be conducted for each component and switching between physical descriptions requires the validation of just the affected components, not entire codes.

1 Introduction

There exist Particle-based algorithms for every level of description, from the quantum ab-initio molecular dynamics (MD), to classic MD, to kinetic particle in cell (PIC) to fluid particle-based algorithms (e.g. fluid PIC, SPH, vortex methods). While other methods compete with particle-based methods at the macroscopic fluid level, particle-based methods are nearly universally used at the kinetic and MD level.

The availability of a common class of algorithms at all levels of description provides a unique possibility inaccessible to other classes of algorithms. All particle-based algorithms share a common software infrastructure made of particles and interactions (either pair-interactions or interactions mediated by a field discretized on a grid). Using a particle-based approach at all levels it is possible to design a single software package based on a component architecture. All the levels of description are formulated using the same classes of software components (e.g. particle movers or components for particle-particle interactions). Within a single software package it is possible to go across all levels of description simply activating a component in place of another, for example a kinetic particle mover (based on Newton's equations for the particle position and velocity) can be replaced by a fluid particle mover (where the motion is computed according to the local flow speed rather than the particle velocity).

V.S. Sunderam et al. (Eds.): ICCS 2005, LNCS 3516, pp. 88–95, 2005.

We term the approach *plug and play validation* because it allows to validate coarser levels of description with finer levels of description by plugging in components and testing them in use.

In the present work, we follow the standard nomenclature [2] of terming the process of *verification* as the task of making sure that the equations of the model are solved correctly and the process of *validation* as the task that the model equations represent with fidelity the actual physics of the system under investigation. Our focus here is on validation alone.

The approach proposed here brings validation to a new level because it not only validates the prediction of coarse levels by testing their results. It also validates coarse levels by directly testing the assumptions behind them (e.g. equations of state or closure relations) validating their assumptions using the information from finer levels of description.

While our focus is on validation, we remark that the approach proposed has also the advantage of simplifying verification because all levels are present within a common software infrastructure and verification can be done component by component. The alternative of using totally different codes and algorithms for the different levels clearly requires a much more extensive effort of verification.

We present a specific example. We consider the heat conduction in a system of multiple ionized gases. The fluids are not in local thermodynamic equilibrium and each has its own temperature and velocity. The fluid closures for this case are not consolidated. We test a model we recently developed based on the 13 moment approach:

$$\mathbf{q}_s = \sum_t \lambda_{st} \nabla T_t \tag{1}$$

where $\mathbf{q}_s$ is the heat flux in species s due to temperature gradients in all other species T_t. The coefficients λ_{st} can be computed from the kinetic model [1]. We will simulate the system at the kinetic and at the fluid level using a single pilot code based on particle methods both for the kinetic and the fluid model. The plug and play approach is used in the design of the pilot code and in the execution of the validation protocol. The process of validation proposed here would determine the validity of the fluid model used (eq. 1) and of the formulation of the transport coefficients λ_{st}.

2 The Physical Challenge

Plasma physics provides an ideal test-bed for developing and testing multiscale, multiphysics algorithms. Plasma physics intrinsically includes two level of multiplicity. First, the plasma species have intrinsically different scales. The electrons are lighter (in a typical hydrogen plasma, the ion mass is approximately 1836 times larger than the electron mass). Second, two types of interactions occur in a plasma: short range and long range. While in a typical gas the interactions are only short range (collisions), in plasmas the presence of charge in the components introduces also a long range force due to collective electromagnetic interactions.

Plasma physics encompasses the whole range of physical scales and models known to mankind. Even not including the topic of quark-gluon plasmas, proper plasma physics ranges from quantum many-body systems to cosmological general relativity models on the structure of the universe. We propose that the approach described here can indeed in principle cover the whole spectrum of physics at these two ends and everything in between. Of course realistically, we need to limit the scope here. And we decide to limit the scope to just two levels of description: the kinetic level and the fluid level.

At the kinetic level, a N-body system is described through the evolution of the distribution function in phase space. The classical formalism for this level of description is the Boltzmann equation.

At the fluid level, a N-body system is described through the evolution of the moments of the distribution function, such as density, velocity and temperature. A great variety of fluid models can be derived from the kinetic level taking a different number of moments of the distribution.

A well known feature of fluid models and their derivation from kinetic models is the need for closure relations. The moment formulation does not form a closed system of equations per se. The equation for the evolution of a moment of any order generally involves higher order moments. Relationships expressing higher order moments in terms of lower order moments are called *closure relations* and are needed to turn the fluid model into a well posed closed problem.

Examples of closure relations are the equations of state: $p = p(n,T)$ and relationships expressing the heat flux or the stress tensor in terms of lower order moments. A classic example is the Fourier law: $q = -k\nabla T$, generalized in our case to eq. (1).

In the present work we consider an initially uniform system with density $n_s = 1$ for all species s, at rest $\mathbf{v}_s = 0$. But we assume that the system is not in thermodynamic equilibrium and each species is initially set up with a uniform temperature gradient. The electrons represent a neutralizing background.

A complete description of the physics and of the mathematical models used can be found in Ref. [1]. The kinetic model can be formulated with the Boltzmann equation:

$$\frac{\partial f_s}{\partial t} + \mathbf{v} \cdot \frac{\partial f_s}{\partial \mathbf{x}} + \frac{q_s}{m_s}\mathbf{E} \cdot \frac{\partial f_s}{\partial \mathbf{v}} = St(f_s) \tag{2}$$

for each species s, including electrons and ions. Charge imbalance could be created in the transient evolution, but we assume that magnetic fields are neglected. This latter assumption is not realistic in most laboratory experiments but it is acceptable within the scope of the present work.

The fluid model is based on a continuity, momentum and energy equation for each species. We use the standard textbook formulation of multiple fluid models [3]. However, we investigate the issue of how many moments are required to represent the physics correctly and what closure relations can be used. We base our fluid analysis on our recent work on this topic [1] and on the classic 13-moment approach [3].

The model poses a formidable challenge, a challenge that we try to address with particle-based algorithms.

3 Particle-Based Method Meets the Challenge

We consider three classes of particle-based methods. We categorize them according to the level of physical description.

First, we consider molecular dynamics (MD), MonteCarlo (MC), or generally particle-particle (PP) methods, typically based on a direct mathematical representation of the physical system. In this class of methods, the particles represent a physical entity: a particle, a molecule or a cluster of atoms and molecules (such as a nanoparticle or a subunit of it). The interactions among the computational particles mimic as best we know the actual physical interaction among the real physical objects.

Second, we consider kinetic particle-mesh (PM) methods, often referred to as particle in cell (PIC) methods. This class differs from the PP approach in two ways: 1) While the PP approach uses the direct simulation of N-bodies through their fundamental laws of motion, the PM approach studies the N-body problem at the kinetic level, as described by the Boltzmann transport theory of classical or relativistic N-body systems. In PM, the computational particle represents an element of phase space. 2) While the PP approach handles only particles and their pair interactions, the PM approach introduces average fields and for their discretization introduces a computational mesh.

Third, we consider fluid particle-based methods (fluid PIC, SPH, vortex methods, ...). While the data structure and the software engineering issues typical of fluid PIC are similar to kinetic PIC, the mathematical and physical bases for it are entirely different. In fluid PIC, the computational particle represents an element of the fluid (or solid) in the coordinate space (rather than of the phase space) and its significance is similar to that of a fluid element in a common Lagrangian scheme. Indeed, the fluid PIC method can be viewed as a peculiar implementation of the arbitrary Lagrangian Eulerian (ALE) approach, widely used in industrial engineering applications.

In our study PP approaches are used to handle short range interactions at the kinetic level, PM methods are used to treat long range interactions at the kinetic level. The combined used of PP and PM methods is often referred to as PPPM or P^3M. Fluid PIC is used to handle the fluid description.

4 Plug and Play Approach to Validation

We can identify two primary advantages of particle-based methods in addressing the issue of validation.

First, there is a direct relationship of particle-based algorithms with the physical system under investigation. Computational particles can be directly linked with physical observables (particle themselves in PP, elements of the phase space in PM and parcels of fluid in fluid PIC). Furthermore, the interactions among the particles (either pair interactions or field-mediated interactions) are direct mathematical constructs of actual measurable interactions in the real world.

Second, particle-based algorithms use the same data structure and the same software architecture to treat all different levels of description, from the fundamental quantum level of ab-initio quantum MD, to classical PP methods, to kinetic PIC, to fluid PIC. All levels of description of N-body systems are covered by particle-based algorithms. No other class of numerical algorithms spans the same range. Based on the two properties above, we can propose that particle-based algorithms provide a unique prototyping environment for Validation & Verification (V&V).

We propose a line of investigation closed on the particle-based algorithms themselves and investigates for a specific system the limits of validity, the fidelity and the predictive potency of different levels of description. Having, as particle-based algorithm do have, the most fundamental level of description (kinetic for weakly coupled systems and molecular dynamics for strongly coupled systems), one has the exact correct theoretical answer based on our most complete and modern understanding of the physics world. By reducing the complexity and stepping down the ladder of further approximations and simplifications, we can address the question of "what is the simplest model to capture a given process".

Within the particle-based approaches, this can be accomplished without changing the overall software architecture, with as simple a switch as changing for example the equations of motion from relativistic to classic, or the field equations from fully electromagnetic to electrostatic. We propose here a plug and play component based software architecture where the properties of the particles and of their interactions through pair forces or through fields can be exchanged by replacing a software component with a compatible one based on a different level of sophistication.

No doubt, this has a strong impact on the cost of the simulations or on the time scales and length scales involved. But if a system is chosen where the most fundamental approach is feasible, stepping down the ladder of approximate models will be progressively easier and more feasible. We note that the plug and play approach just outlined is not feasible with other non particle-based approaches. For example Eulerian methods would not have an equivalent component to be replaced with at the kinetic level (at least a mature one) and even less so at the molecular dynamics level. In that case, the plug and play approach would not be feasible and different codes would need to be used instead making the V&V much more difficult and time consuming, as all parts of the code would need V&V, not just the new component plugged in. Different software architectures would be involved, adding a level of uncertainty and clouding the issue. Particle-based methods allow the plug and play approach primarily because they work as the physics system work allowing a direct comparison with them.

4.1 Specific Example

Let us consider an example: the formulation of heat conduction in hydrodynamics simulations. We propose to start at the fundamental kinetic level with a PIC simulation (assuming a weakly coupled system, but the same approach would function by starting from a PP approach for high energy density systems) where a first principle approach provides the exact heat transfer by taking the appropriate moments of the particle distribution function.

The heat conduction is defined as the third order moment of the particle distribution function:

$$q_{hs} = \frac{m_s}{2} \sum_{k=1}^{3} \int c_{sh} c_{sk} c_{sk} f_s d\mathbf{v}_s \tag{3}$$

where $\mathbf{c}_s$ is the random velocity (difference between actual velocity and average fluid velocity) for species s, h and k label the coordinates.

From a kinetic description, the heat flux can be computed directly using eq. (3).

In the fluid model, heat flux is computed using the 13-moment fluid closure and is defined as in eq. (1) where the transport conefficients λ_{st} are provided by theoretical estimates [1].

In our plug and play validation approach, we replace the components for kinetic description with components for fluid descriptions based on different level of sophistication and we test not only whether the predictions are still valid, but also whether the assumptions behind each reduced model is correct. Moreover, this can be done one component at a time, focusing on the effect of only replacing one component within a common code without worrying at the same time with having replaced a whole code with another. The approach proposed is a striking alternative to the classic approach of just validating closure relations by testing their predictions when applied in existing fluid codes. Here besides doing that, we also test directly the assumptions behind the closures, thanks to the availability of the more fundamental kinetic (or MD) level. This provides information on what is not valid in certain closures allowing us to improve them.

5 Implementation in Parsek

To implement the plan discussed above and demonstrate the feasibility of the plug and play approach for the validation of particle-based algorithms, we have developed a component based package called Parsek.

Parsek uses components for each of the four classes of components: particles, fields, interactions between particle and interactions between fields.

Each component is implemented in an object oriented language. We considered here Java and C++. Two approaches have been tested for object orientation: coarse grained or fine grained [4].

In the fine grained approach each object corresponds to the smallest physical entity: particle or cell. In the coarse grained approach each object represents a population of the smallest physical entities: particle species or grid. The fine grained approach is composed of small objects and uses arrays of objects (or pointers to them) to describe the whole population of them (species for the particles and grid for the cells). The coarse grained approach, instead, uses arrays of native double precision numbers to represent the population and the whole population is the object. Figure 1-a illustrates the difference between these two approaches. The great advantage of the latter approach is the avoidance of arrays of objects (or pointers to them), resulting in a more efficient use of the cache on processors such as the Pentium family.

Figure 1-b proves indeed that a significant advantage can be gained by using a coarse grained approach. A similar study has been conducted in a previous study using similar Java implementations [4] and is reported in Fig. 1-b. For our pilot code Parsek, we settled on the coarse grained approach.

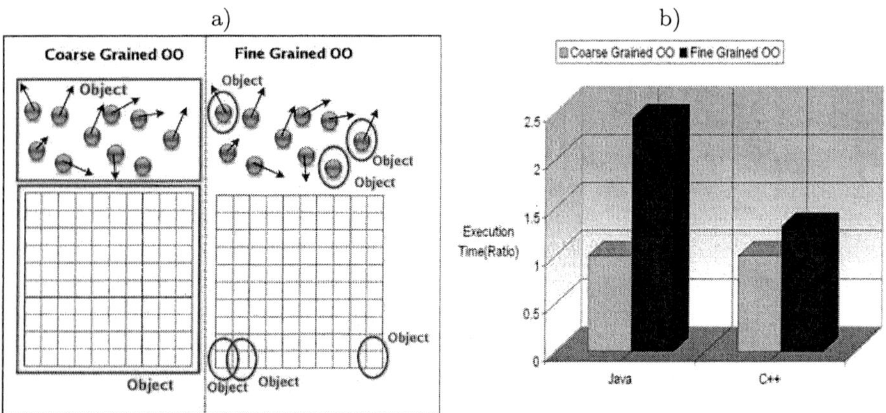

Fig. 1. Coarse grained and a fine grained object oriented versions (a). Comparison of fine and coarse object orientation performance in Java and C++ (b)

The computational challenge posed by the task under investigation (and particularly by the kinetic level) requires the use of a fully parallelized version of the algorithms. We adopted a 3D domain decomposition where portions of the systems including the cells and the particles in them are apportioned to a processor. The requirement to keep the particles and the cells belonging to the same physical domain in the same processor is due to the need to interpolate the information between particles and cells to compute the interactions. A consequence of this choice is the need to dynamically allocate the particles to the processors as the particles cross domain boundaries. In the examples considered here the system is 1D and is obtained as a limit of the 3D case by taking only once cell in the y and z direction and applying periodic boundary conditions.

We have considered two parallelization environments: message passing and multithreading. Message passing is a widely used programming paradigm and it is based on the use of libraries such as MPI. We have developed a version of our software package, Parsek, in C++ using the MPI library to handle the communication. Figure 2 shows the scaling obtained in a typical problem by the C++/MPI version of Parsek. The test were conducted on a distributed memory cluster of INTEL Xeon processors.

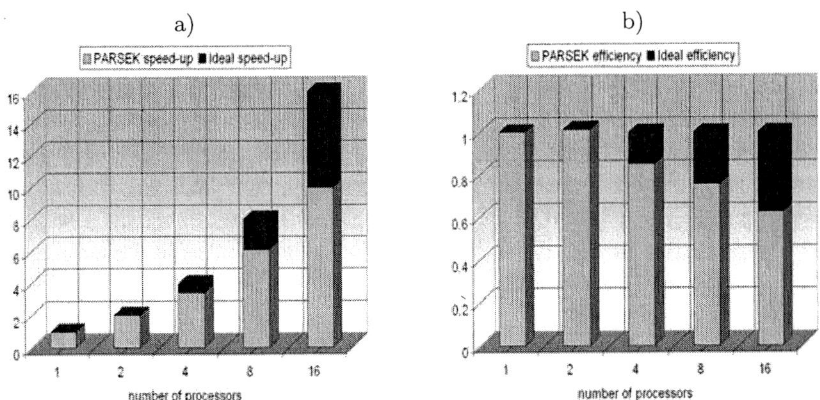

Fig. 2. Parallel speed-up (a) and parallel efficiency (b) of a C++ version with parallelization based on the MPI library. Study conducted on a distributed memory cluster

As an alternative, we considered also the multithreading approach typical of grid-based languages such as Java. The tests (not reported here) were conducted on a shared memory four processor SUN workstation and the observed scaling were very efficient. However, when the same approach is attempted on a cluster machine, native Java multithreading cannot be used and tools such as Java Party need to be used [5].

References

1. L. Abrardi, G. Lapenta, C. Chang, *Kinetic theory of a gas mixture far from thermodynamic equilibrium: derivation of the transport coefficients*, LANL Report, LA-UR-03-8334.
2. S. Schleisinger, R.E. Crosbie, R.E. Gagne, G.S. Innis, C.S. Lalwani, J. Loch, R.J. Sylvester, R.D. Wright, N. Kheir, D. Bartos, *Terminology for Model Credibility*, **Simulation**, 32:103, 1997.
3. J.M. Burgers, *Flow equations for composite gases*, Academic Press, New York, 1969.
4. S. Markidis, G. Lapenta, W.B. VanderHeyden, Z. Budimlić, *Implementation and Performance of a Particle In Cell Code Written in Java*, **Concurrency and Computation: Practice and Experience**, to appear.
5. N. T. Padial-Collins, W. B. VanderHeyden, D. Z. Zhang, NM E. D. Dendy, D. Livescu ,*Parallel operation of cartablanca on shared and distributed memory computers*, **Concurrency and Computation: Practice and Experience**, 16(1):61-77, 2004.

Multiscale Angiogenesis Modeling

Shuyu Sun[1], Mary F. Wheeler[1],
Mandri Obeyesekere[2], and Charles Patrick Jr[2]

[1] The Institute for Computational Engineering and Sciences (ICES), The University of Texas at Austin, Austin, Texas 78712, USA
[2] The University of Texas (UT) M.D. Anderson Cancer Center, and The UT Center for Biomedical Engineering, Houston, Texas 77030, USA

Abstract. We propose a deterministic two-scale tissue-cellular approach for modeling growth factor-induced angiogenesis. The bioreaction and diffusion of capillary growth factors (CGF) are modeled at a tissue scale, whereas capillary extension, branching and anastomosis are modeled at a cellular scale. The capillary indicator function is used to bridge these two scales. To solve the equation system numerically, we construct a two-grid algorithm that involves applying a mixed finite element method to approximate concentrations of CGF on a coarse mesh and a point-to-point tracking method to simulate sprout branching and anastomosis on a fine grid. An analysis of the algorithm establishes optimal error bounds for each of the processes – CGF reaction-diffusion, capillary extension, sprout branching and anastomosis – and overall error bounds for their coupled nonlinear interactions.

1 Introduction

Angiogenesis is the outgrowth of new vessels from pre-existing vasculature, and it plays an important role in numerous clinical indications, including wound healing, tissue regeneration and cancer. A deep understanding of angiogenesis is critical for reparative strategies since the capillary network dictates tissue survival, hemodynamics, and mass transport. The angiogenic system is strongly nonlinear, possessing multiple, integrated modulators and feedback loops. This complexity limits the *in vitro* and *in vivo* experiments that may be designed and the amount of non-confounding information that can be gleaned. Consequently, computational models simulating the intercellular growth patterns of capillaries within a tissue are essential to understanding and analyzing these phenomena. However, most angiogenesis modeling approaches in the literature have been restricted to a single scale (e.g., see [1, 2, 3]), even though, in fact, the genetic, biochemical, cellular, biophysical and physiological processes in angiogenesis are intimately and tightly coupled across spatial and temporal dimensions.

2 Multiscale Angiogenesis Modeling Equations

In this section, we develop a deterministic two-scale tissue-cellular angiogenesis model. Let Ω be a bounded domain in $\mathbb{R}^d$ ($d = 1$, 2 or 3) and T the final

V.S. Sunderam et al. (Eds.): ICCS 2005, LNCS 3516, pp. 96–103, 2005.

simulation time. The capillary network is represented by an indicator binary function $n = n(\mathbf{x}, t)$. Denote by c_i, $i = 1, 2, \cdots, N_{CGF}$, the concentration of a species of capillary growth factors (CGF), where N_{CGF} is the number of CGF components.

2.1 A Tissue Scale Model for CGF Behaviors

We assume that CGF component j is released in the extracellular matrix at a rate α_j. The diffusivity of CGF is denoted by $\mathbf{D}_j$, a general second order tensor, and the diffusive flux of CGF is $\mathbf{q}_j = -\mathbf{D}_j \nabla c_j$. The consumption (binding) of CGF by endothelial cells occurs only in the place where $n = 1$, and its rate is assumed to be proportional to the CGF concentration. Thus the consumption rate is $\lambda_j n c_j$, where λ_j is the consumption parameter of CGF j. We model the natural decay of CGF by a rate $\lambda_j^* c_j$. The mass balance of CGF yields the following equation: $\partial c_j / \partial t = \nabla \cdot (\mathbf{D}_j \nabla c_j) + \alpha_j (1 - n) - \lambda_j n c_j - \lambda_j^* c_j$.

2.2 A Cellular Scale Model for Capillary Dynamics

We model sprout extension by tracking the trajectory of individual capillary tips. We denote by $\mathbf{p}_i(t) \in \mathbb{R}^d$ the position of capillary tip i at time t. Under certain biological conditions, cells behind the sprout tips undergo mitosis, and sprout extension subsequently occurs. The movement of an individual sprout tip during proliferation depends on the direction and speed of the sprout extension: $d\mathbf{p}_i(t)/dt = \sum_j k_{p,j}(c_j)\mathbf{u}_{0,j}(\mathbf{q}_j)$, where $k_{p,j}$, a function of CGF concentration c_j, represents the cell proliferation rate and $\mathbf{u}_{0,j}$, a normalized vector specifying the capillary outgrowth direction, is a function of the corresponding diffusive flux. We consider general functions $k_{p,j}(\cdot)$ and $\mathbf{u}_{0,j}(\cdot)$ in this paper.

We denote by S the set of all active capillary tips. The behaviors of capillaries are described by the movement of the tips, which includes sprout extension and the changes of the tip set by branching and anastomosis. The tip set S remains unchanged during capillary extension, but it is modified at branching or anastomosis because these events change the number of elements in S.

In capillary branching, we terminate the parent capillary tip label and start two new labels for the two resultant daughter tips. We denote the branching trigger function by $f_{BT}(\tau, c_1, c_2, \cdots, c_{N_{CGF}})$ and assume that the sprout branches as soon as $f_{BT}(\tau, c_1, c_2, \cdots, c_{N_{CGF}}) \geq 0$, where τ is the age of the capillary tip and $c_i = c_i(\mathbf{p}, t)$ is the CGF concentration at the location occupied by the capillary tip. For example, $f_{BT}(\tau, c_1, c_2, \cdots, c_{N_{CGF}}) = \tau - \tau_a$ specifies uniform sprout branching, where every sprout performs branching after maturing for the length of time τ_a. Mathematically, sprout branching increases the number of elements in the tip set S and is denoted by $S(t^+) = B\left(S(t^-)\right)$.

Anastomosis, the fusion of capillary sprouts, is assumed to occur when a sprout tip meets another sprout tip physically (tip-to-tip anastomosis) or a sprout tip joins another sprout physically (tip-to-sprout anastomosis). After a tip-to-sprout anastomosis, the tip cell forms a part of the loop and no longer undergoes sprout extension, i.e. the tip no longer exists. We distinguish two types of tip-to-tip anastomosis: in a "head-on-head" anastomosis, both tips become inactive, whereas in a "shoulder-on-shoulder" anastomosis, only one of the two tips becomes inactive. Mathematically, the anastomosis mechanism decreases the number of elements in the tip set S and is written as $S(t^+) = A(S(t^-))$.

2.3 Bridging Cellular and Tissue Scales

The CGF concentration profile strongly influences sprout extension, branching and anastomosis, all of which control the development of a capillary network. The capillary network, in turn, affects the bioreaction and diffusion of CGF. The capillary indicator function n is determined by the history of sprout tip positions. We define the set $N_C(t)$ occupied by the capillary network at time t as $N_C(t) = \bigcup_i \bigcup_{\tau \le t} \mathbf{B}_{r_{EC}}(\mathbf{p}_i(\tau))$, where $\mathbf{B}_{r_{EC}}(\mathbf{x}) = \{\hat{\mathbf{x}} : |\hat{\mathbf{x}} - \mathbf{x}| \le r_{EC}\}$. The radius of an endothelial cell r_{EC} is assumed to be constant. The capillary indicator function may be written as $n = \chi_{N_C}$, where χ_E is the standard set characteristic function, i.e. $\chi_E(\mathbf{x}) = 1$ for $\mathbf{x} \in E$ and $\chi_E(\mathbf{x}) = 0$ otherwise.

2.4 A Modified Model Based on Cell Level Averaging

With initial and boundary conditions, the previous equations in this section represent a mathematical system for two-scale modeling of angiogenesis. In this paper, we analyze a modified system, which is based on cell level averaging:

$$\frac{\partial c}{\partial t} = -\nabla \cdot (\mathbf{q}) + \alpha(1-n) - \lambda n c - \lambda^* c, \tag{1}$$

$$\mathbf{q} = -\mathbf{D}\nabla c, \tag{2}$$

$$\frac{d\mathbf{p}_i}{dt} = \mathcal{M}_S\left(k_p(c)\mathbf{u}_0(\mathbf{q})\right), \qquad \forall i \in S, \tag{3}$$

$$n = \mathcal{M}_S\left(\chi_{N_C}\left(\{\mathbf{p}_i : i \in S\}\right)\right), \tag{4}$$

$$S(t^+) = A\left(B_{\mathcal{M}}\left(S(t^-)\right)\right). \tag{5}$$

We only consider a single CGF species here for simplicity of presentation, though the analysis of multiple CGF species is a straightforward extension. The averaging operator (or the mollifier) $\mathcal{M}_S$ is defined for $f \in L^1(0,T;L^1(\Omega))$ by $\mathcal{M}_S(f)(\mathbf{x},t) = \int_{\mathbf{B}_{r_{EC}}(\mathbf{x})\cap\Omega} f(\hat{\mathbf{x}},t)d\hat{\mathbf{x}}/\mathrm{meas}(\mathbf{B}_{r_{EC}}(\mathbf{x})\cap\Omega)$, where $\mathrm{meas}(\cdot)$ denotes the Lebesgue measure. The stabilized branching operator $B_{\mathcal{M}}$ is formed from the original branching operator B by replacing c by $\mathcal{M}_S c$ in $f_{BT}(\tau, c)$. We note that $\mathcal{M}_S$ may be viewed as a modification operator to reflect averaged information collected by a tip cell.

We consider the boundary condition $\mathbf{q} = 0$ on $\partial\Omega$ and the initial condition $c = c_0$ at $t = 0$ for the CGF concentration. We impose the following initial conditions for capillary tips: $S = S_0$ and $\mathbf{p}_i = \mathbf{p}_{i,0}$ at $t = 0$. We note that, by using algebraic equations, anastomosis and branching are described as instantaneous events, whereas sprout extension is modeled as a continuous-in-time process using ordinary differential equations (ODEs). Since the number of elements in the set S changes with time, the number of unknowns in the system varies with time.

3 A Two-Grid Algorithm

A mixed finite element (MFE) method is employed to approximate the CGF diffusion-reaction equation on a concentration grid. We trace the trajectory of each capillary tip using a standard ODE solver. A point-to-point tracking method is proposed to simulate sprout branching and anastomosis, where algebraic conditions are applied for branching, and geometric conditions are checked for anastomosis. While the concentration grid is a mesh at the tissue scale, the capillary network forms a secondary grid at the cellular scale. Locally conservative L^2 projections are used for data transfer between the two grids when needed.

Let $(\cdot,\cdot)_D$ denote the $L^2(D)$ inner product over a domain $D \subset \mathbb{R}^d$ for scalar functions or the $\left(L^2(D)\right)^d$ inner product for vector functions, and, when $D = \Omega$, we drop the subscript. Let $\|\cdot\|_{L^p(D)}$, $1 \le p \le \infty$, be the $L^p(D)$ norm for a scalar function or the $\left(L^p(D)\right)^d$ norm for a vector function. Similarly, let $\|\cdot\|_{H^s(D)}$ be the standard $H^s(D)$ norm or the $\left(H^s(D)\right)^d$ norm. Throughout this paper, we denote by C a generic positive constant that is independent of h and by ϵ a fixed positive constant that may be chosen to be arbitrarily small. We define the following standard spaces: (1) $W = L^2(\Omega)$; (2) $\mathbf{V} = H(\text{div};\Omega) = \left\{\mathbf{v} \in (L^2(\Omega))^d : \nabla\cdot\mathbf{v} \in L^2(\Omega)\right\}$; (3) $\mathbf{V}^0 = \{\mathbf{v} \in H(\text{div};\Omega) : \mathbf{v}\cdot\nu = 0 \text{ on } \partial\Omega\}$, where ν denotes the outward unit normal vector on $\partial\Omega$. The weak formulation of the CGF diffusion-reaction equation is to find $c \in W$ and $\mathbf{q} \in \mathbf{V}^0$ such that $c(x,0) = c_0(x)$ and $\forall w \in W$, $\forall \mathbf{v} \in \mathbf{V}^0$, $\forall t \in (0,T]$:

$$\left(\frac{\partial c}{\partial t}, w\right) = (-\nabla\cdot\mathbf{q}, w) + (\alpha(1-n) - (\lambda n + \lambda^*)c, w)\,, \tag{6}$$

$$\left(\mathbf{D}^{-1}\mathbf{q}, \mathbf{v}\right) = (c, \nabla\cdot\mathbf{v})\,. \tag{7}$$

We let $\mathcal{E}_h = \{E_i\}$ denote a partition of $\overline{\Omega}$ into elements E_i (for example, triangles or parallelograms if $d = 2$) whose diameters are less than or equal to h. Let $(W_h, \mathbf{V}_h) \subset (W, \mathbf{V})$ be a mixed finite element space of order r that possesses an associated projection operator $\Pi_h : \mathbf{V} \to \mathbf{V}_h$ satisfying: (1) $\nabla\cdot\mathbf{V}_h = W_h$; (2) $(\nabla\cdot\Pi_h\mathbf{q}, w) = (\nabla\cdot\mathbf{q}, w)$, $\forall \mathbf{q} \in \mathbf{V}$, $\forall w \in W_h$; (3) $\|\Pi_h\mathbf{q} - \mathbf{q}\|_{L^2(\Omega)} \le C\|\mathbf{q}\|_{H^s(\Omega)} h^{\min(r+1,s)}$; (4) $\|\mathcal{P}_h c - c\|_{L^2(\Omega)} \le C\|c\|_{H^s(\Omega)} h^{\min(r+1,s)}$, where $\mathcal{P}_h$ is the L^2 projection from W onto W_h: $(\mathcal{P}_h c, w) = (c, w)$, $\forall c \in W$, $\forall w \in W_h$. Obviously, we have $\nabla\cdot\Pi_h\mathbf{q} = \mathcal{P}_h\nabla\cdot\mathbf{q}$, $\forall \mathbf{q} \in \mathbf{V}$. The continuous-in-time mixed finite element method for approximating the CGF diffusion-reaction equation is to find $c_h(t) \in W_h$ and $\mathbf{q}_h(t) \in \mathbf{V}_h^0 = \mathbf{V}^0 \cap \mathbf{V}_h$ such that $\forall w \in W_h$, $\forall \mathbf{v} \in \mathbf{V}_h^0$, $\forall t \in (0,T]$:

$$\left(\frac{\partial c_h}{\partial t}, w\right) = (-\nabla\cdot\mathbf{q}_h, w) + (\alpha(1-n_h) - (\lambda n_h + \lambda^*)c_h, w)\,, \tag{8}$$

$$\left(\mathbf{D}^{-1}\mathbf{q}_h, \mathbf{v}\right) = (c_h, \nabla\cdot\mathbf{v})\,, \tag{9}$$

$$(c_h(\cdot,0), w) = (c_0(\cdot), w)\,. \tag{10}$$

We denote by c_h, $\mathbf{q}_h$, $\mathbf{p}_h$ and n_h the finite element solutions for the CGF concentration, the CGF diffusive flux, the capillary tip positions and the capillary indicator function respectively. We first analyze the error of the CGF concentration assuming that the error of the capillary indicator function is given:

Theorem 1. (CGF bioreaction-diffusion) *We assume that* $c \in L^2(0,T;H^s(\Omega))$, $\partial c/\partial t \in L^2(0,T;H^s(\Omega))$ *and* $\mathbf{q} \in (H^s(\Omega))^d$. *We further assume that the diffusivity tensor* $\mathbf{D}$ *is uniformly symmetric positive definite and bounded from above, that* c *is essentially bounded, that parameters* λ *and* λ^* *are nonnegative, and that parameters* α *and* λ *are bounded. Then, for any given* $\epsilon > 0$, *there exists a constant* C *independent of the mesh size* h *such that*

$$\begin{aligned}&\|c_h - c\|_{L^\infty(0,T;L^2(\Omega))} + \|\mathbf{q}_h - \mathbf{q}\|_{L^2(0,T;L^2(\Omega))} \\ &\leq C\left(\|c\|_{L^2(0,T;H^s(\Omega))} + \left\|\frac{\partial c}{\partial t}\right\|_{L^2(0,T;H^s(\Omega))} + \|\mathbf{q}\|_{L^2(0,T;H^s(\Omega))}\right) h^{\min(r+1,s)} \\ &+\epsilon \|n_h - n\|_{L^2(0,T;L^2(\Omega))}.\end{aligned} \tag{11}$$

Proof. We let $c_I = \mathcal{P}_h c$ and $\mathbf{q}_I = \Pi_h \mathbf{q}$, and define the finite element error $E_c = c_h - c$, the projection error $E_c^I = c_I - c$, and the auxiliary error $E_c^A = c_h - c_I$. Similarly, $E_\mathbf{q} = \mathbf{q}_h - \mathbf{q}$, $E_\mathbf{q}^I = \mathbf{q}_I - \mathbf{q}$, $E_\mathbf{q}^A = \mathbf{q}_h - \mathbf{q}_I$. We also define $E_n = n_h - n$.

Subtracting (6) and (7) from (8) and (9) respectively, splitting E_c and $E_\mathbf{q}$ according to $E_c = E_c^I + E_c^A$ and $E_\mathbf{q} = E_\mathbf{q}^I + E_\mathbf{q}^A$, and choosing $w = E_c^A$ and $\mathbf{v} = E_\mathbf{q}^A$, we observe

$$\begin{aligned}\left(\frac{\partial E_c^A}{\partial t}, E_c^A\right) &= (-\nabla \cdot E_\mathbf{q}, E_c^A) - ((\lambda c + \alpha)E_n, E_c^A) \\ &\quad - ((\lambda n_h + \lambda^*)E_c, E_c^A) - \left(\frac{\partial E_c^I}{\partial t}, E_c^A\right), \quad t \in (0,T],\end{aligned} \tag{12}$$

$$(\mathbf{D}^{-1}E_\mathbf{q}^A, E_\mathbf{q}^A) = (E_c, \nabla \cdot E_\mathbf{q}^A) - (\mathbf{D}^{-1}E_\mathbf{q}^I, E_\mathbf{q}^A), \qquad t \in (0,T]. \tag{13}$$

Recalling the orthogonality of projections $\mathcal{P}_h$ and Π_h, we add the two error equations (12) and (13) to obtain, for any $t \in (0,T]$,

$$\begin{aligned}&\left(\frac{\partial E_c^A}{\partial t}, E_c^A\right) + (\mathbf{D}^{-1}E_\mathbf{q}^A, E_\mathbf{q}^A) + ((\lambda n_h + \lambda^*)E_c^A, E_c^A) \\ &= -((\lambda c + \alpha)E_n, E_c^A) - ((\lambda n_h + \lambda^*)E_c^I, E_c^A) - \left(\frac{\partial E_c^I}{\partial t}, E_c^A\right) - (\mathbf{D}^{-1}E_\mathbf{q}^I, E_\mathbf{q}^A).\end{aligned} \tag{14}$$

We note that, as a binary function, n_h must be non-negative and bounded. We bound the right hand side of (14) using Cauchy-Schwarz inequality, the boundedness of λ, c, α and n_h, and the orthogonality, and then we rewrite the left hand side of (14) in forms of the L^2 norm to conclude:

$$\begin{aligned}&\frac{1}{2}\frac{d}{dt}\|E_c^A\|_{L^2(\Omega)}^2 + \left\|\mathbf{D}^{-\frac{1}{2}}E_\mathbf{q}^A\right\|_{L^2(\Omega)}^2 + \left\|(\lambda n_h + \lambda^*)^{\frac{1}{2}}E_c^A\right\|_{L^2(\Omega)}^2 \\ &\leq \epsilon\|E_n\|_{L^2(\Omega)}^2 + C\|E_c^A\|_{L^2(\Omega)}^2 + \epsilon\left\|\mathbf{D}^{-\frac{1}{2}}E_\mathbf{q}^A\right\|_{L^2(\Omega)}^2 \\ &+C\left(\|c\|_{H^s(\Omega)}^2 + \|\mathbf{q}\|_{H^s(\Omega)}^2\right) h^{2\min(r+1,s)}.\end{aligned} \tag{15}$$

Noting that the finite element solution at $t = 0$ is an L^2 projection, i.e. $c_h(0) = \mathcal{P}_h c_0$, we have $E_c^A(0) = 0$. We first integrate (15) with respect to the time t, and apply Gronwall's inequality. Then recalling the triangle inequality and the approximation results of projection operators Π_h and $\mathcal{P}_h$, we establish the theorem. □

We now estimate the error on capillary networks. Denote by $J = (t_0, t_F]$ the capillary tip life time period, where the final time of the capillary tip t_F is the time when the tip becomes inactive due to either branching or anastomosis, or is the final simulation time T, whichever is shorter. Due to page limitation, we list Lemmas 1–4 below without detailed proofs. Applications of the triangle inequality and Gronwall's inequality give the error bound on capillary extension (Lemma 1). Exploration of the sprout branching stability condition leads to its error estimate (Lemma 2). Geometric analysis results in the error bound on anastomosis (Lemma 3). A straightforward manipulation of the set of the capillary network establishes an L^∞ upper bound on the error of the capillary indicator function (Lemma 4). Based on the four lemmas, the approximations of overall capillary behaviors are concluded in Theorem 2. The error estimate for the coupled system is finally obtained in Theorem 3 as a consequence of individual error bounds on the CGF concentration and capillary behaviors.

Lemma 1. **(Capillary extension)** *We assume that c is bounded, that $k_p(c)$ is a Lipschitz continuous function of c, and that $\mathbf{u}_0(\mathbf{q})$ is a Lipschitz continuous function of $\mathbf{q}$. Then, there exists a constant C independent of the mesh size h such that $\|\mathbf{p}_h - \mathbf{p}\|_{L^\infty(J)} \le C\,|\mathbf{p}_{0,h} - \mathbf{p}_0| + C\,|t_{0,h} - t_0| + C\,\|\mathbf{q}_h - \mathbf{q}\|_{L^1(0,T;L^1(\Omega))} + C\,\|c_h - c\|_{L^1(0,T;L^1(\Omega))}$, where $\mathbf{p}$ is the position of an individual capillary tip, $\mathbf{p}_0$ and t_0 are the initial position and initial time when the tip is formed, and $\mathbf{p}_h$, $\mathbf{p}_{0,h}$ and $t_{0,h}$ are their approximate solutions respectively.*

Lemma 2. **(Sprout branching)** *We assume that the branching trigger function $f_{BT}(\tau, c)$ is a uniformly Lipschitz function of c with a Lipschitz constant C_L, and there is a positive constant such that $\partial f_{BT}/\partial\tau \ge C_0 > 0$. We further assume that $\partial c/\partial t \in L^\infty(0,T;L^1(\Omega))$, $\nabla c \in L^\infty(0,T;L^1(\Omega))$ and $d\mathbf{p}/dt \in L^\infty(J)$. If C_L/C_0 is sufficiently small, there exists a constant C independent of the mesh size h such that $|t_{B,h} - t_B| + |\mathbf{p}_{B,h} - \mathbf{p}_B| \le C\,|t_{0,h} - t_0| + C\,\|\mathbf{p}_h - \mathbf{p}\|_{L^\infty(J)} + C\,\|c_h - c\|_{L^\infty(0,T;L^1(\Omega))}$, where t_B and t_0 are the sprout branching time and the tip birth time of the capillary under consideration, and $t_{B,h}$ and $t_{0,h}$ are their approximate solutions respectively.*

Lemma 3. **(Anastomosis)** *We assume that the intersecting angle θ_A of the two parent capillaries i and j at the location of anastomosis satisfies $C_0 \le |\sin\theta_A| \le 1 - C_0$, where $0 < C_0 < 1/2$ is a fixed small constant. We further assume that there exists a constant C_1 such that $|d\mathbf{p}/dt| \ge C_1 > 0$ in the neighborhoods of the anastomosis location. In addition, we assume that $d\mathbf{p}_i/dt$ and $d\mathbf{p}_j/dt$ are bounded. Then there exists a constant C independent of the mesh size h such that $|t_{A,h} - t_A| + |\mathbf{p}_{A,h} - \mathbf{p}_A| \le C\,\|\mathbf{p}_{i,h} - \mathbf{p}_i\|_{L^\infty(J_i)} + C\,\|\mathbf{p}_{j,h} - \mathbf{p}_j\|_{L^\infty(J_j)}$, where t_A is the time of anastomosis, and $\mathbf{p}_A$ the location of anastomosis.*

Lemma 4. (Capillary indicator function) *We assume that the exact total capillary length is bounded. Then there exists a constant C independent of the mesh size h such that $\|n_h - n\|_{L^\infty(0,T;L^\infty(\Omega))} \leq C \sum_i \|\mathbf{p}_{i,h} - \mathbf{p}_i\|_{L^\infty(J_i)}$.*

Theorem 2. (Capillary behavior) *Let the assumptions in Lemmas 1–4 hold. In addition, we assume that the number of capillaries from the exact solution is bounded. Then there exists a constant C independent of the mesh size h such that*

$$\|n_h - n\|_{L^\infty(0,T;L^\infty(\Omega))} \leq C \left(\|\mathbf{q}_h - \mathbf{q}\|_{L^1(0,T;L^1(\Omega))} + \|c_h - c\|_{L^\infty(0,T;L^1(\Omega))} \right). \tag{16}$$

Proof. Combining Lemmas 1, 2 and 3, using the assumption on the boundedness of the capillary number, and noting that a $L^1(0,T;L^1(\Omega))$ norm of a function is no greater than its $L^\infty(0,T;L^1(\Omega))$ norm, we obtain $\sum_i \|\mathbf{p}_{i,h} - \mathbf{p}_i\|_{L^\infty(J_i)} \leq C \|\mathbf{q}_h - \mathbf{q}\|_{L^1(0,T;L^1(\Omega))} + C \|c_h - c\|_{L^\infty(0,T;L^1(\Omega))}$. The theorem follows from this inequality and Lemma 4. □

Theorem 3. (Final result on overall error bound) *Let the assumptions in Theorems 1 and 2 hold. Then there exists a constant C independent of the mesh size h such that*

$$\|c_h - c\|_{L^\infty(0,T;L^2(\Omega))} + \|\mathbf{q}_h - \mathbf{q}\|_{L^2(0,T;L^2(\Omega))} + \|n_h - n\|_{L^\infty(0,T;L^\infty(\Omega))} \leq C h^{\min(r+1,s)}. \tag{17}$$

Proof. Recalling that the constant ϵ in (11) may be chosen to be arbitrarily small, we let $\epsilon = 1/(2C)$, where C is the constant in (16). Observe that the $L^2(0,T;L^2(\Omega))$, $L^1(0,T;L^1(\Omega))$, and $L^\infty(0,T;L^1(\Omega))$ norms of a function are less than or equal to its $L^\infty(0,T;L^\infty(\Omega))$, $L^2(0,T;L^2(\Omega))$ and $L^\infty(0,T;L^2(\Omega))$ norms, respectively. The overall error bound follows from Theorems 1 and 2. □

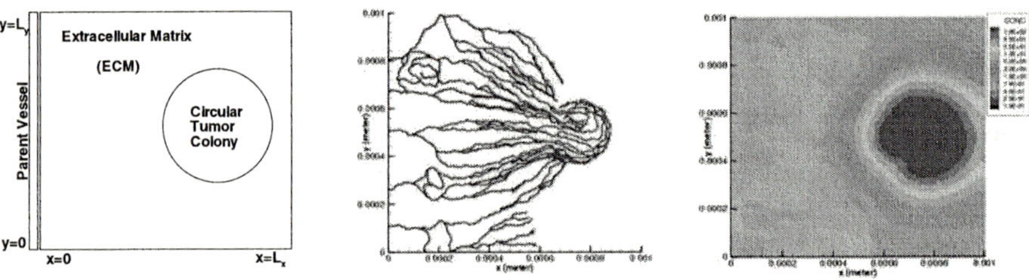

Fig. 1. An angiogenesis simulation example. Left: domain geometry; Middle: capillary network at 28 days; Right: CGF concentration map ($\times 10^{-16}$mol/m^2) at 28 days

We now present a numerical example simulating tumor-induced angiogenesis (illustrated in Fig. 1). All modeling parameters are taken from [4] except that the tumor colony is represented by a large CGF release rate: $\alpha = 6.42 \times 10^{-18}\text{mol/m}^2\text{s}$ in the tumor area and $\alpha = 6.42 \times 10^{-21}\text{mol/m}^2\text{s}$ in the remaining extracellular matrix (ECM). The CGF is released from both the ECM and the tumor cells, but substantially more so from the latter. Clearly, endothelial cells at the capillary sprout tips are migrating from the parent vessel to the right, toward the tumor colony, as induced by the CGF concentration gradient. The simulation produces capillary networks with realistic structures and morphologies. Readers are referred to [4, 5, 6] for more simulation results.

4 Conclusions

We have presented a deterministic two-scale model to simulate angiogenesis, where CGF bioreaction and diffusion at a tissue scale are coupled with capillary extension, branching and anastomosis at a cellular scale. To solve the coupled angiogenesis system, we have proposed a two-grid numerical algorithm based on MFE on a coarse mesh and a point-to-point tracking method on a fine grid. Optimal order error estimates have been derived for the proposed scheme. The model generates an overall dendritic structure of the capillary network morphologically similar to those observed *in vivo*, and captures significant vascular patterning, such as vascular loops and backward growth. Clearly, experimental work is needed to enhance and verify the model.

Acknowledgment

Supported in part by a grant from the University of Texas Center for Biomedical Engineering.

References

1. A. R. A. Anderson and M. A. J. Chaplain. Continuous and discrete mathematical models of tumor-induced angiogenesis. *Bull. Math. Biol.*, 60:857–900, 1998.
2. M. A. J. Chaplain. The mathematical modelling of tumour angiogenesis and invasion. *Acta Biotheor.*, 43:387–402, 1995.
3. H. A. Levine, S. Pamuk, B. D. Sleeman, and M. Nilsen-Hamilton. Mathematical modeling of capillary formation and development in tumor angiogenesis: Penetration into stroma. *Bull. Math. Biol.*, 63:801–863, 2001.
4. S. Sun, M. F. Wheeler, M. Obeyesekere, and C. W. Patrick Jr. A deterministic model of growth factor-induced angiogenesis. *Bull. Math. Biol.*, 67(2):313–337, 2005.
5. S. Sun, M. F. Wheeler, M. Obeyesekere, and C. W. Patrick Jr. Deterministic simulation of growth factor-induced angiogenesis. In *Proceedings of American Institute of Chemical Engineers 2004 Annual Meeting*, Austin, Texas, November 7-12, 2004.
6. S. Sun, M. F. Wheeler, M. Obeyesekere, and C. W. Patrick Jr. Nonlinear behavior of capillary formation in a deterministic angiogenesis model. In *Proceedings of the Fourth World Congress of Nonlinear Analysts*, Orlando, Florida, June 30 - July 7, 2004.

The Simulation of a PEMFC with an Interdigitated Flow Field Design

S.M. Guo

Dept. Mechanical Engineering,
Louisiana State University,
Baton Rouge, LA 70803, USA
+1 225 578 7619
sguo2@lsu.edu

Abstract. This paper presents the simulations for a two-dimensional PEMFC with an interdigitated flow channel design using FEMLAB. The multi-species flow of O_2, H_2, H_2O and inert N_2 is examined over the entire fuel cell working range. The transportations of these gases in the porous anode/cathode are predicted using the Maxwell-Stefan transport equations and the Fick's law; the flow field is predicted using the Darcy's law; and the electrical field is simulated using a conductive media model. The standard current-voltage performance curve, the species concentration of O_2, H_2, H_2O and inert N_2, mass fluxes, electrical current and potential distributions have been obtained.

1 Introduction

Fuel cells are electrochemical devices, which utilize fuel and oxidant to produce electricity and heat [1]. Because the fuel is converted chemically to electricity, fuel cells may operate at higher efficiencies than conventional internal combustion engines. Fuel cells are classified by the electrolytes. The common electrolyte of a Proton Exchange Membrane Fuel Cell (PEMFC) is a thin layer of proton permeable polymer membrane. In a PEMFC system, the coupled physical and electrochemical processes take place. Transport resistances of the gaseous species in the feeding channels and the porous electrodes lead to the so-called concentration over-potential; the transportation of H^+ in the electrolyte layer forms a large portion of ohmic loss; the activation energy barriers for the electrochemical reactions are related to the charge-transfer processes at the electrode-electrolyte interfaces. The theoretical potential of a hydrogen/oxygen cell, operating under standard conditions of 1 bar and 25°C, is about 1.2 volt. However, due to losses, the voltage of a single cell is much less than its theoretical value.

The PEMFC flow-field, including the feeding channels and the porous electrodes, has a significant influence to the cell performance. The performance of a fuel cell is often described by its current-voltage (*I-E*) relationship. Fuel, such as hydrogen, and small amount of water vapor are supplied to the PEMFC anode side. Due to the electrochemical reactions, steam forms at the cathode. For a typical planar PEMFC design, feeding channels are provided by the bipolar plates with a typical serpentine

V.S. Sunderam et al. (Eds.): ICCS 2005, LNCS 3516, pp. 104 – 111, 2005.

or parallel channel structure. Due to the long channels, serpentine flow-fields have large pressure losses between the inlet and the outlet. Although straight parallel design exhibits lower pressure differences, inhomogeneous reactant gas distribution can easily occur. These flow channels must distribute fuel and air over the reaction sites and remove the products. Recently, a number of novel flow channels have been proposed for Proton Exchange Membrane Fuel Cells, in order to address the PEMFC mass transport and water management problems. The most promising design is the interdigitated flow field design [2,3,4]. Figure 1 shows the schematic drawing of this design. The flow channels are dead-ended, forcing gas to flow through the porous diffusion layer. Comparing to the flow in a conventional straight parallel flow field design, the use of the interdigitated fuel/air distributors imposes a pressure gradient between the inlet and the outlet channels, forcing the convective flow of the fuel/oxidant through the porous electrodes. Thus the interdigitated design in effect converts the transport of gases to/from the Triple Phase Boundary (TPB) active sites, along the interface between the electrodes and electrolyte, from a diffusion dominated mechanism to a forced convection dominated mechanism. The convective flow through the porous electrode reduces the gas diffusion distance to and from the reaction sites. By having fuel/oxidant flow over the shoulders of the gas distributor, the electrode active area over the shoulder is used more effectively. The shear force of the gas stream helps removing the water condensate, which is entrapped in the electrode layer, thereby reducing the flooding problem. This design has been proven to be very effective by some experimental studies. Wang et al. [3] presented an experimental study of PEM fuel cells with interdigitated flow fields under different operation parameters. Nguyen [4] presented a comparative experimental study of fuel cells with interdigitated flow fields and parallel straight channel flow fields. They reported that the interdigitated flow design could extend the PEMFC operable regime to higher current densities and consequently, a 50-100% increase in the fuel-cell performance could be obtained as a result of the use of interdigitated fuel/air distributors. To study a fuel cell performance mathematically, especially for a design with thick porous electrodes and a strong convective flow, a proper mass transport model must be applied. At high current density, the losses are dominated by the limitation of transport the fuel/oxidant to the reaction sites, the so-called concentration or mass transport over-potential. Mass transport in the porous electrodes depends on the structure of the porous electrodes, such as the porosity, tortuosity and mean pore size. Washak et al. [5] and Suwanwarangku et al. [6] conducted comparative studies using the Fick's law and the Dusty Gas Model for a Solid Oxide Fuel Cell. They found that the current density, the reactant concentration and the pore size were the three key parameters for choosing a proper porous media simulation model. The Dusty Gas Model works better for the H_2–H_2O and CO–CO_2 systems, especially under high operating current densities, low reactant concentrations and small pore sizes.

To perform a parametrical optimization for the flow field design, multiphysics based numerical simulation offers many advantages comparing to the experimental approach.

The aim of this paper is to study the effect of using different mass transport models on the fuel cell performance simulation. Two models are examined in this paper. The first one is the Fick's model, and the second model applies the Stefan-Maxwell diffusion equations to the mixture. The performance of a PEMFC with an interdigitated flow channel design was simulated using FEMLAB. The simulation includes the multi species transportation in the porous electrodes and the coupled electrical current and potential distributions.

2 The Simulation Model

FEMLAB is a commercial Partial Differential Equation (PDE) solver, which can solve coupled multi-physical problems. Partial differential equations are the governing equations for most physical phenomena and provide the foundation for modeling a wide range of scientific and engineering problems. There are three ways of describing PDEs in FEMLAB, coefficient form, general form and the weak form. The coefficient form is suitable for linear or nearly linear models and the rest two are suitable for nonlinear models. FEMLAB runs finite element analysis to solve the PDEs, together with adaptive meshing and error controls.

The simulation domain used in this paper is a 2-dimensional cross section of an interdigitated PEMFC flow field, shown in Figure 1. The oxidation of hydrogen and the reduction of oxygen take place at the anode and cathode side reaction boundaries respectively. Electrons are transported to an outer circuit at the anode and received at the cathode because only proton ions can pass through the electrolyte membrane.

$H_2 \rightarrow 2H^+ + 2e^-$ at anode side reaction boundary

$O_2 + 4H^+ + 4e^- \rightarrow 2H_2O$ at the cathode side reaction boundary

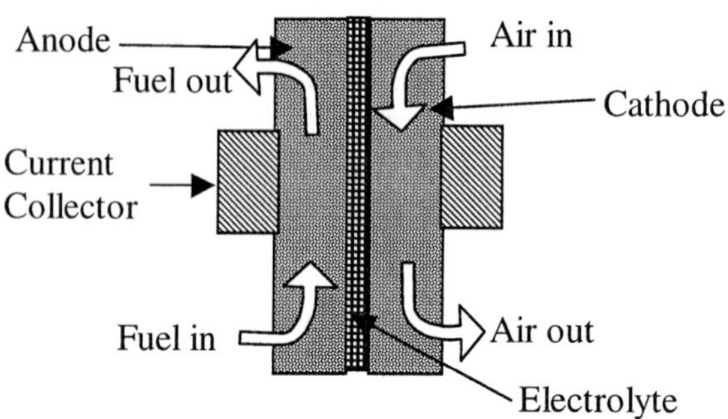

Fig. 1. Interdigitated Flow Field Design

The thickness of the anode and cathode is set to be 0.25 mm and the height is set to be 2 mm; the electrolyte layer has a thickness of 0.1 mm. Along the height direction, current collectors of 1 mm in length are in contact with the outer surface of the anode and cathode in a symmetrical manner, leaving the inlet and outlet ports a dimension of 0.5 mm. Hydrogen and an inert gas mixture is fed into the anode side inlet port while oxygen, inert nitrogen and water are fed into the cathode side. The gas mixtures are treated as incompressible due to the low flow velocities. In the simulation, the hydrogen is oxidized along the electrode/electrolyte interface, which has a zero thickness. The electrical potential of the PEMFC is specified as an input parameter over the entire working range.

A set of PEMFC governing equations is specified in FEMLAB. Equation 1 is the steady state continuity equation.

$$\nabla \bullet (CV) = 0 \tag{1}$$

Where C is the total gas mixture molar concentration; V is the velocity; CV is the total molar flux. For continues gas phase flow in the porous electrodes, Darcy's law, Equation 2, is used.

$$V = -\frac{k_p}{\mu} \nabla P \tag{2}$$

Where, k_p is the permeability of the medium, μ is the dynamic viscosity of the gas mixture, ∇P gives the pressure gradient.

For the multi species mass transfer in the electrodes, both Fick's law and the Maxwell-Stefan diffusion and convection mass transfer models were tested. Using the Fick's law, the diffusion flux in the porous electrodes is calculated using equation 3.

$$N^d = -D^e \nabla C \tag{3}$$

D^e is the effective diffusivity of the gas, C is the concentration. By assuming equal counter-current molar fluxes, according to Chan et al. [7], the composition independent D^e can be found in Eq. 4

$$D_1^e = \left(\frac{1}{D_{12}^e} + \frac{1}{D_{1k}^e}\right)^{-1} \tag{4}$$

D_{12}^e is the effective binary bulk diffusion coefficient. D_{1k}^e is the effective Knudsen diffusion coefficient, which depends on temperature and structure of the porous material. For a convective flow in a porous medium, combining the diffusive and convective fluxes, the flux equation can be written as equation 5.

$$N = -\left(\frac{1}{D_{12}^e} + \frac{1}{D_{1k}^e}\right)^{-1} \nabla C - \frac{X_1}{RT}\frac{k_p P}{\mu} \nabla P \tag{5}$$

For the simulation using the Maxwell-Stefan diffusion and convection model. The Maxwell-Stefan multi-component diffusion is given in equation 6.

$$\frac{\partial}{\partial t}\rho\omega_i + \nabla \bullet \left[-\rho\omega_i \sum_{j=1}^{N} D_{ij} \left\{ \begin{array}{l} \frac{M}{M_j}\left(\nabla\omega_j + \omega_j \frac{\nabla M}{M}\right) \\ +\left(x_j - \omega_j\right)\frac{\nabla P}{p} \end{array} \right\} + \omega_i \rho u + D_i^T \frac{\nabla T}{T} \right] = R_i \tag{6}$$

where D_{ij} is the diffusion coefficient (m^2/s), P the pressure (Pa), T is the temperature (K), u the velocity vector (m/s), x and ω are mole and mass fractions. The density, ρ (kg/m^3), is calculated based on the mole fractions and mole masses of gas species.

At inlet, pressure and the feeding gas mass fractions are specified. At the outlets, a convective flux boundary condition is applied. The local current density is a function of the local species concentration, physical structure of the electrodes and the specified cell output electrical potential. For steady state calculations, due to the conservation of current, the anode side current density is the same as the current density at the cathode. At the anode and cathode reaction boundary, the species mass transfer is related to the local current density according to: $-n \bullet n_{H_2} = -\frac{i_a}{2F}$ at anode reaction boundary and $-n \bullet n_{O_2} = \frac{i_C}{4F}$ at the cathode reaction boundary.

The potential difference between the cathode and anode current collectors represents the cell voltage. In the simulation, the potential at the anode current collector was arbitrarily chosen to be zero, while the cell voltage at the cathode current collector is set as a fixed boundary condition. The potential distributions in the anode, cathode and the electrolyte are modeled as conductive media using equation7.

$$\nabla \bullet (-k \nabla E) = 0 \tag{7}$$

where k is the effective conductivity (S/m) and E is the potentials in the calculation domain. The rest of the boundaries were set as the electrical insulators or as a symmetrical boundary. Normal to the reaction boundary, $n \bullet (-k \nabla E)$ gives the current density.

3 Results and Discussions

The *I-E* performance curve of a PEMFC has been obtained using two different mass transfer models, Fick's model and the Maxwell-Stefan model. Figure 2 shows a typical fuel and oxidant distribution inside a PEMFC with an interdigitated flow field design (see Fig.1 for feeding directions). A 60% H_2 is fed to the anode side (left half) and a 21% O_2 is fed to the cathode (right half) inlet port.

As expected, the Fick's model results show discrepancies to the Maxwell-Stefan model results in the high current density region, see Fig. 3. Start at about 3000A/m^2, Fick's law started to over-predict the current density at fixed cell output electrical potentials. At the high current end, the difference between these two models is about 15%. This is mainly caused by the over prediction of the oxygen concentration at the cathode side reaction boundary. The oxygen concentration at the inlet is 21%. Due to the electrical chemical reactions along the cathode/electrolyte boundary, the oxygen concentration at the reaction sites is far less than the inlet value, see Fig. 2. Figure 4 shows the predicted oxygen concentration average along the cathode reaction boundary under different current densities using those two models. The difference between the two models is obvious at the high current region, where the Fick's model predicts a higher O_2 concentration. Figure 5 shows the production of water due to the electrochemical reactions. The curve shows a strong linear relationship between the water formation and the current density.

The definition of anode side H_2 concentration is based on the hydrogen partial pressure. Because of the electrochemical reactions, hydrogen depletes along the flow direction. The mass transport process affects the local concentration of fuel. Hydrogen concentration reduces almost linearly with the increase of current density. At about 6500 A/m^2, the right end of the curve, the hydrogen mass transfer in the porous anode reaches its limiting current density.

The commercial FEMLAB software has many built-in multi physics models. However, the advantage of a general software package, which claims to solve "many" physical problems, is also likely to be its weakness. The build-in models are generally

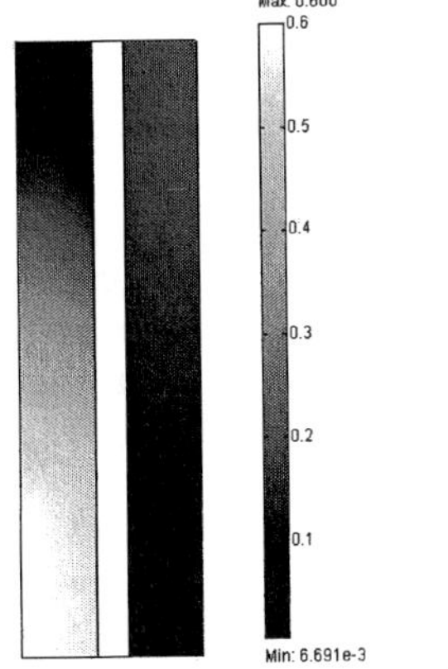

Fig. 2. A typical H_2 (left) and O_2 (right) concentration in the PEMFC

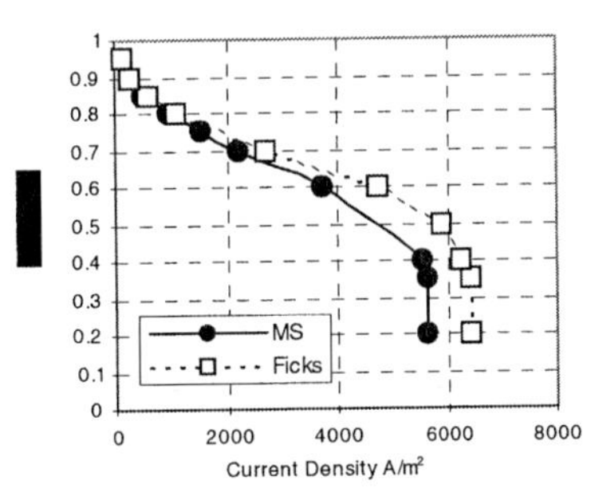

Fig. 3. PEMFC *I-E* curves predicted using the Fick's model and the Maxwell-Stefan model

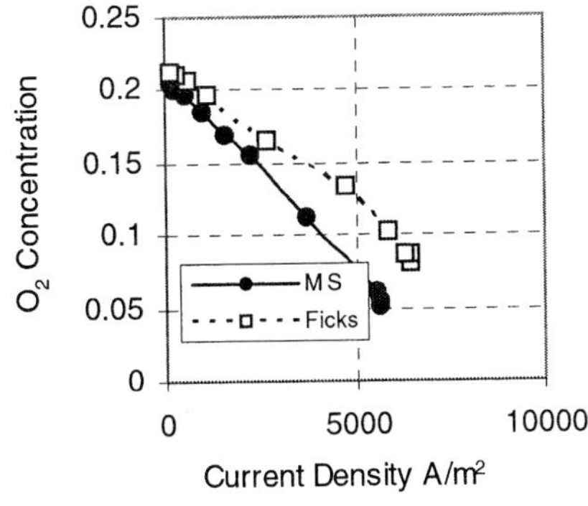

Fig. 4. PEMFC cathode reaction boundary O_2 concentration predicted using the Fick's model and the Maxwell-Stefan model

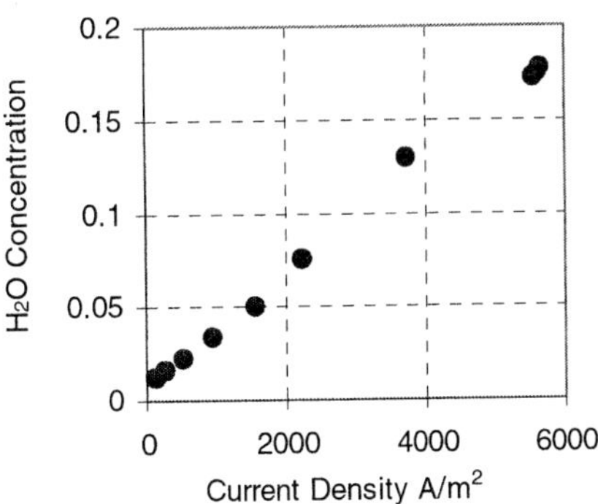

Fig. 5. PEMFC cathode reaction boundary H_2O concentration as a function of cell current density

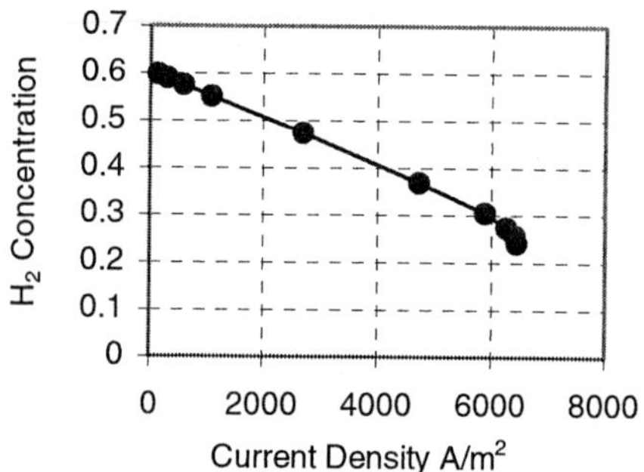

Fig. 6. PEMFC anode reaction boundary H_2 concentration

to be the basic physical and mathematical models for a particular physical problem. For example, because only Darcy's law was used in the simulation to predict the fluid flow in the porous electrodes, the accuracy near the reaction boundaries is not expected to match the accuracy of a well-tuned CFD case, which uses Navier-Stokes equations with proper turbulent models. To solve a real multi-physical problem, a balanced approach must be carefully taken. The number of physical phenomena, to be modelled in software, must be considered in conjunction with the expected overall accuracy and the available computing power.

4 Conclusions

In the past two decades, due to the energy shortage and the environmental concerns, great efforts have been put into fuel cell research. To improve the fuel cell performance, multiphysics analysis could be employed. Using proper fuel cell models, fuel cell simulations could provide detailed understanding and solutions on how to reduce the losses and how to improve the operating efficiency. Comparing with an experimental based approach, computational simulations are low cost and could quickly provide the physical understandings to a particular fuel cell design. The PEMFC flow-field, including feeding channels and porous electrodes, has significant influence to the cell performance. The local concentration of fuel and oxidant is affected by the mass transport processes, which include the mixing of multiple gases in the porous anode/cathode, the reactions of fuel and oxygen and the transportation of fuel and electrochemical products near the reaction sites. This paper presents the numerical simulations of gas transport in the porous electrodes of a PEMFC with an interdigitated flow channel design using FEMLAB. The simulation is a coupled fluid dynamics and electrical potential field problem with multi species mass transfer and chemical reactions. This complicated PEMFC problem has been solved using FEMLAB over the entire working range. Two porous media mass transport models, the Fick's Model and the Maxwell-Stefan model were used in the simulations for multi-species of O_2, H_2, H_2O and inert N_2. The standard current-voltage performance curve and the detailed species concentration, mass fluxes, electrical current and potential distributions have been obtained. The simple but less accurate Fick's model is validated against the Maxwell-Stefan model. Fick's model was found to over-predict the oxygen concentration along the cathode side reaction boundary and thus over-predict the cell performance in the high current density region.

References

1. Larminie, J., Dicks, A., Fuel cell systems explained, Wiley, ISBN: 047084857x, (2003)
2. Jung, S. Y., Nguyen, T.V., Multicomponent Transport in Porous Electrodes of Proton Exchange Membrane Fuel Cells Using the Interdigitated Gas Distributors, Journal of The Electrochemical Society, v 146, n1, 1999, p 38-45
3. Wang, L., Liu, H., Performance studies of PEM fuel cells with interdigitated flow fields, Journal of Power Sources, 134, p185-196, (2004)
4. Nguyen, T.V., Gas distributor design for proton-exchange-membrane fuel cells, J. Electrochem. Soc. 143, n 5, p L103-L105, (1996)
5. Washak, H., Guo, S.M., Turan, A., 2005, Gas Transport in Porous Electrodes of Solid Oxide Fuel Cells, the Ninth International Symposium on Solid Oxide Fuel Cells (SOFC IX), The 207th Meeting of The Electrochemical Society, Québec City Convention Centre, May 15-20, 2005
6. Suwanwarangkul, R., Croiset, E., Fowler, M.W., Douglas, P.L., Entchev, E., Douglas, M.A., Performance comparison of Fick's, dusty-gas and Stefan–Maxwell models to predict the concentration overpotential of a SOFC anode, Journal of Power Sources **122** 9-18, (2003)
7. Chan, S.H., Khor, K.A., Xia, Z.T., A complete polarization model of a solid oxide fuel cell and its sensitivity to the change of cell component thickness, Journal of Power Sources, **93**, 130-140, (2001)

Multiscale Modelling of Bubbly Systems Using Wavelet-Based Mesh Adaptation

Tom Liu and Phil Schwarz

CSIRO Minerals, Clayton, VIC 3168, Australia
{Tom.Liu, Phil.Schwarz}@CSIRO.AU

Abstract. Since typical industrial-scale reactors may contain many millions of bubbles, the extension of direct free-surface modelling techniques to resolve every bubble in the vessel would require far more computational power than will be available. A more immediate solution is to couple macro-scale reactor models to micro-scale models of individual bubbles and collections of a small number of bubbles. In this paper, a micro-scale modelling technique was presented and tested on the situation of a single rising bubble. The micro-scale model was based on the Volume-of-Fluid (VOF) technique combined with a dynamic mesh adaptation based on wavelet analysis to ensure a sufficient resolution at the gas-liquid interfaces. The method was based on a multi-block parallel scheme with mesh adaptivity facilitated by wavelet analysis embedded into a commercial CFD package CFX. Examples of the performance of the scheme for a bubble rising problem are given.

1 Introduction

Bubbly systems are used widely in the minerals and process industries to increase agitation, supply reactants (e.g. oxygen), or to effect particle separations (as in froth flotation). Effective two-fluid (or phase-averaged) modelling techniques have been developed to simulate the multi-phase fluid flow, and heat and mass transfer in such reactors, and are now widely used to assist vessel design [1,2]. These methods do not seek to resolve each bubble, but are based volume-averaged equations in which the physics of bubble-fluid interactions is modelled though constitutive terms analogous to the Reynolds stresses that appear in the RANS equations for a single phase turbulent flow. This necessarily reduces the accuracy and predictive capability of such models, and increases their dependence on empirical data and fitting parameters, particularly in cases where bubble-bubble interactions such as coalescence are important. However, in practice, it is difficult to develop accurate constitutive relations for complex systems valid over wide operating ranges using physical experiments.

A computationally efficient solution to improving the macro-scale multi-fluid models is to couple them to micro-scale models of individual bubbles and collections of a small number of bubbles. The strategy is to determine improved closure relationships for the multi-fluid model through the analysis of the micro-scale models. Where phenomena occur at widely differing scales, the numerical analysis may be employed to analyse some particular processes at fine scales, for instance, the breakup of a single bubble,

V.S. Sunderam et al. (Eds.): ICCS 2005, LNCS 3516, pp. 112–119, 2005.

coalescence of two or more bubbles, or bubble-particle interactions. The information obtained at fine scales can be applied in macroscale CFD or other models.

This approach has advantages over the experimental determination of constitutive relationships because physical information on the processes, e.g. shear rate dependency of the viscosity or stress field, can be obtained locally and analysed in detail. The numerical simulations used in this way may be called numerical experiments [3], though they should be viewed as the complementary to physical experiments, rather than a substitute.

The fine scale simulations differ from traditional macroscale CFD simulations used in reactor design in that more attention is paid to local information and a very fine mesh must be used to resolve full detail of interactions between the flow and the phase boundaries. If fixed meshes are employed in a traditional CFD simulation for solving practical engineering problems, the full representations of fine scale phenomena are difficult or even impossible for the foreseeable future because of the overwhelming computational costs [4]. Therefore an alternative approach that is more efficient must be developed.

Micro-scale modellings (or numerical experiments) for multiphase problems generally involve moving interfaces between immiscible fluids. The accurate simulation of fluid flows with sharp fronts presents a problem with considerable difficulties [5]. Many types of interface tracking methods have been developed, but they generally fall into two categories. In the first type, a deformable finite volume or finite element mesh is used such that a particular mesh boundary moves so that it always coincides with the interface [6]. The other strategy is to keep the mesh fixed and use a separate procedure to describe the position of the interface. These methods are reviewed in [7]. The interface can be represented on a fixed grid in a variety of ways, either explicitly or implicitly. The Volume of Fluid method (VOF) is one of the most popular implicit interface tracking schemes [5]. Physical problems which are considered in practice require three-dimensional calculations with surface tension, non-catastrophic breakage and reconnection of the interface. The VOF technique naturally allows for the latter, and has been modified by various workers to include surface tension.

Commercial software packages such as CFX and FLUENT are usually employed to model practical multiphase flows due to the complexities of the phenomena and the geometries. Although mesh adaptation has been provided by the commercial CFD packages, they generally do not allow for dynamic mesh adaptation [8]. If a mesh sufficiently fine to resolve detail near the interface is applied on the whole geometry, the computational requirements can be huge. Such a fine mesh is not necessary on most of the geometry, but because of the movement of the interface, the localized mesh refinement requires a dynamic mesh adaptation.

The largest difficulty with adaptive methods is to determine the mobility of grids. Adaptive wavelet methods have been developed to solve Navier-Stokes equation at high Reynolds numbers [9]. Wavelets have the ability to accurately and efficiently represent strongly inhomogeneous piecewise continuous functions [10]. Using such techniques, the numerical resolution can be naturally adapted to intermittent the structures of flows at fine scales with significant decrease of computational efforts and memory requirements. However, efforts have been mainly made around using wavelets as an orthogonal and complete basis, spanning a space in which to seek approximate solutions satisfying the equation in a Galerkin or collocation sense [4, 9]. To apply such methods however would require the development of a new program for each particular flow-related problem, whereas general-purpose commercial CFD packages are preferred.

Hestheven and Jameson [4] developed a different approach to utilise the unique properties of wavelets, and this approach can be applied in a grid-based method utilized by CFD packages. In this method, wavelets were employed to detect the existence of high frequency information and supplied spatial locations of strongly inhomogeneous regions. Very fine grids were used only in these regions. In this method, wavelets were used for grid generation and order selection only, whilst the scheme for solving the partial differential equation was based on conventional finite difference/element schemes, albeit defined on variable grids. The method provides a possibility to embed the wavelet-based grid generation into current commercial CFD packages to reduce computational costs by using adaptive meshes.

In this paper, a multi-block (domain) parallel scheme with the adaptivity facilitated by wavelet analysis was proposed and embedded into commercial CFD package CFX to track moving free surfaces efficiently.

2 Wavelet Analysis and Wavelet-Based Grid Adaptation

Wavelet analysis is an emerging field of applied mathematics that provides tools and algorithms suited to the type of problems encountered in multiphase process simulations. It allows one to represent a function in terms of a set of base functions, called wavelets.

Wavelet transform involves representing general functions in terms of simple, fixed building blocks at different scales and positions. These building blocks, which are actually a family of wavelets, are generated from a single fixed function called the "mother wavelet" by translation and dilation (scaling) operations. In contrast to the traditional trigonometric basis functions which have an infinite support, wavelets have a compact support. Therefore wavelets are able to approximate a function without cancellation. In the basic work of Daubechies [10], a family of compactly supported orthonormal wavelets is constructed. Each wavelet number is governed by a set of L (an even integer) coefficients $\{p_k : k = 0, 1, \cdots, L-1\}$ through the two-scale relation:

$$\phi(x) = \sum_{k=0}^{L-1} p_k \phi(2x - k) \tag{1}$$

and the equation:

$$\psi(x) = \sum_{k=2-L}^{1} (-1)^k p_{1-k} \phi(2x - k) \tag{2}$$

where functions $\phi(x)$ and $\psi(x)$ are called scaling function and wavelet, respectively. The fundamental support of the scaling function $\phi(x)$ is in the interval $[0, L-1]$ while that of the corresponding wavelet $\psi(x)$ is in the interval $[1 - L/2, L/2]$. The coefficients p_k appearing in the two-scale relation (1) are called wavelet filter coefficients. All wavelet properties are specified through these coefficients. For more details see Daubechies' paper [10].

Interpolating functions are generated by the autocorrelation of the usual compactly supported Daubechies scaling functions [11]. Such an autocorrelation function $\theta(\cdot)$ verifies trivially the equality $\theta(n)=\delta_{0n}$, and generates a multi-resolution analysis. The approximate solution of the problem $u_j(\cdot)$ defined on interval [0, 1] is written in terms of its values in the dyadic points:

$$u_j(x)=\sum_n u_j(2^{-j}n)\theta(2^j x-n) \tag{3}$$

and such a function is exact at the dyadic points.

Consider the function:

$$\theta(x)=\int_{-\infty}^{+\infty}\phi(y)\phi(y-x)dy \tag{4}$$

Denote V_j the linear span of the set $\{\theta(2^j x-k), k\in Z\}$. It can be proven that V_j forms a multi-resolution analysis where $\theta(\cdot)$ plays the role of a scaling function (nonorthonormal). In this paper, a modified interpolating function was constructed in order to achieve an interpolating operator on the interval with the same accuracy as the counterpart on the line. Such functions were introduced by Bertoluzza and Nald [11].

$$\widetilde{\theta}_{j0}=\sum_{k=-L+1}^{0}\theta_{jk},\quad \widetilde{\theta}_{j,2^j}=\sum_{k=2^j}^{2^j+L+1}\theta_{jk}$$

Consider $I_j u$ as the form:

$$\mathrm{I}_j u=u(0)\widetilde{\theta}_{j0}+\sum_{k=1}^{2^j-1}u(x_k)\theta_{jk}+u(1)\widetilde{\theta}_{j2^j} \tag{5}$$

The sparse point representation (SPR) and the grid generation technique based on Holmstorm's work [12] were employed for grid adaptation. The SPR was based on interpolating wavelet transform (IWT) on dyadic grids. A feature of the basis is the one-to-one correspondence between point values and wavelet coefficients. The interpolating subdivision scheme recursively generated the function values on a fine grid from the given values on a coarse grid. At each level, for odd-numbered grid points, the differences between the known function values and the function values predicted by the interpolation from the coarser grid were calculated. These differences were termed as wavelet coefficients d_k^j, which gave the information about the irregularity of the function: $d_k^j=u(x_{2k+1}^{j+1})-I^j u(x_{2k+1}^{j+1})$.

3 Problem Definition and Method Formulation

Wavelet-based adaptive methods have been successfully applied to solve many problems [13, 9]. The grids were adapted based on the change of the solutions. Fine meshes were assigned only on the regions where the solutions changed sharply. However, for current CFD commercial packages, e.g. CFX, there is a restriction to the dynamic mesh adaptation: the topology of the mesh must remain fixed. Therefore, the advantages of wavelet-based grid adaptation cannot be fully utilised at this stage. A modified wavelet-based mesh adaptation methodology was proposed in this section.

Wavelet-based grid generation supposes a calculation which begins with evenly spaced samples of a function. It has usually been applied to simple geometries and structured grids [11]. To combine wavelet analysis with CFX commercial CFD package, a multi-block (domain) formulation was proposed. A moving cubic subdomain or block with fine mesh was designed to track the free surface and combined the geometric flexibility and computational efficiency of a multi-block (domain) scheme with the wavelet-based mesh adaptivity. The moving block was like a "microscope" - it was used to track local information required. Coarse mesh was employed outside the moving block (subdomain). Therefore the scheme was computationally efficient without loss of accuracy.

To clearly describe the scheme, a simple case, a single rising bubble, was considered. Two algorithms were proposed as follows: wavelet-based adaptive structured grid using junction box routines; wavelet-based adaptive unstructured grid using CFX expression language and junction box routines

3.1 Problem Definition

A numerical experiment was designed by Krishna and Baten [14] to simulate the single bubble rising process using CFX 4. A 2D rectangular column (25mm × 90mm) involving 144000 grid cells was used. It took about two weeks using six R8000 processors for a simulation run.

In this paper, the same configuration was modelled using CFX 5.7. The only difference was that a wavelet-based mesh adaptation algorithm was embedded into the CFX package so that a moving fine-mesh block was used to track the bubbles. This technique used only 10% of the number of grid cells that would be required for a uniform mesh to obtain similar accuracy.

The simulation was carried out in a rectangular column using 2D Cartesian coordinate grid. No-slip wall boundary condition was imposed, and the column was modelled as an open system. The VOF model was employed to describe transient motion of the gas and liquid phases using the Navier-Stokes equations. Surface tension was included. For the convective terms, high resolution differencing was used. First order backward Euler differencing was used for time integration. The time step used in the simulation was 0.00003s or smaller.

3.2 Wavelet-Based Adaptive Structured Grid Using Junction Box Routines

CFX-MESHBUILD was used to generate the initial structured mesh consisting of five blocks. To utilise wavelet analysis, the proposed moving fine-mesh block must be either a cube or a rectangle. The initial size and position of the block depends on those of the bubble at initial time, see Table 1. Therefore the proposed mesh adaptation method can be applied in a complex geometry. The mesh distribution of block B1 was chosen as uniform grid $(2^7+1) \times (2^6+1)$ to apply wavelet analysis. Because the bubble extended in x-direction significantly during the process, higher resolution was used in this direction. The mesh distributions of other blocks can be different. For simplicity, the same distributions were given for the shared edges and uniform grids were chosen, see Figure 1 and Table 2.

Table 1. Initial bubble/block size and position

Initial Bubble (Diameter) / Block	Initial Bubble / Block Positions (Centre)
4mm / (6mm×6mm)	(12mm, 4mm) / (12mm, 4mm)
8mm / (10mm×10mm)	(12.mm, 6mm) / (12mm, 6mm)

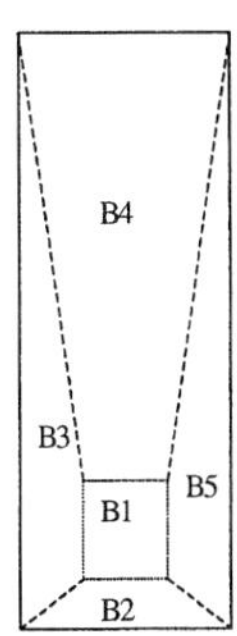

Fig. 1. Geometry and blocks of the bubble rising column

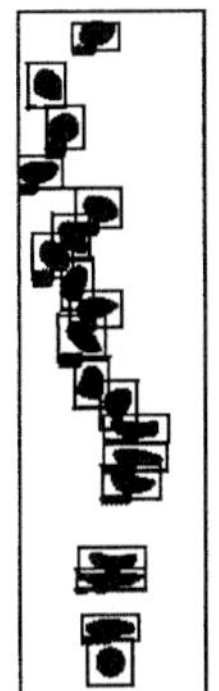

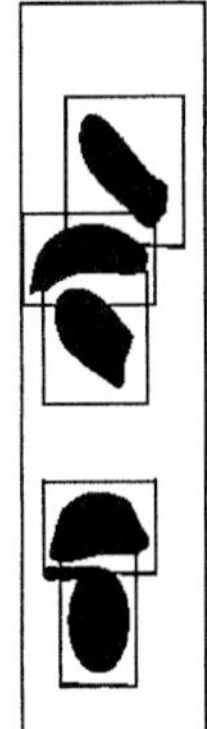

Fig. 2. Snapshots of typical rising trajectorie of bubbles of 4 and 8 mm diameter and B1

Table 2. Distributions of blocks

	Distribution of B1	Distributions of B2 and B4	Distributions of B3 and B5	Ratio of adaptive mesh to full mesh
4mm bubble	$(2^7+1)\times(2^6+1)$	$2^7\times 2^5$	$2^6\times 2^5$	0.08
8mm bubble	$(2^7+1)\times(2^6+1)$	$2^7\times 2^5$	$2^6\times 2^5$	0.08

Using CFX 5.7, the volume fractions on the vertices of each cell can be obtained. Due to the limitation of CFX solver, the topology of the mesh must remain fixed. Points cannot be added to or removed from the grid. Only the position of free interface was detected using IWT and wavelet coefficients. At each mesh adaptation time step, the minimum distances between the interface and the boundaries of block B1 was calculated to decide if the position and size of B1 should be changed and a new grid was generated. The algorithm is as follows:

Step 1. Calculate the wavelet coefficients of volume fraction of gas in x-direction from the top and bottom rows of B1 to determine the nearest position between the interface and the top or bottom boundaries of B1. The IWT and wavelet coefficients are calculated from level 6 to level 7 on each row. If most of the wavelet coefficients on one row are larger than the threshold ε, e.g. 80% of the total number on one row, this means the interface is on or near the position.

Step 2. Calculate the wavelet coefficients of volume fraction of gas in y-direction from the left and right columns of B1 to determine the nearest position between the interface and the left or right boundaries of B1 using the same method in Step 1.

Step3. If all the distances between the interface and the boundaries of B1 are larger than given minimum distance d_p, the mesh remains fixed; Otherwise, move the boundaries of B1 to the positions until the distances between the interface and the boundaries are not less than d_p.

Step 4. Set limits to the position of block B1 to avoid mesh folding. There should be a given minimum distance between B1 and the boundaries of the column.

Step 5. Generate a new uniform mesh based on the new position of B1 for each block using the same distribution as that of the initial mesh and pass the new grid to CFX.

The block B1 was moved based on the wavelet analysis of the volume fraction function of gas. Therefore, the bubble was always in B1. There is a correlation between the frequency of the grid adaptation and the tolerance distance between the interfaces and the boundaries of B1 when the adaptation is performed. If the grid adaptation is not done often enough, a larger tolerance distance must be given so that the front does not move out of B1 between the adaptations. Conversely, more frequent adaptations demand a small distance.

It is natural to define the parallel computation partitions based on five blocks. In this work, three partitions were defined, partition 1 including B1, partition 2 including B2 and B3, partition 3 including B3 and B4. Partition 1 was calculated on master machine. The whole grid was determined based on partition1 and passed to slave machines.

In practice, unstructured grids are often used in complex geometries. The proposed method can also be applied to unstructured grids using CFX expression language and junction box routines together. The method has been used for the same simulation as described above.

4 Simulation Results

The simulations were run on a Linux system. Snapshots of typical bubble trajectories are shown in Fig 2. They were similar to the results of Krishna and Baten [13]. Figure 2 clearly showed how the fine mesh moves with the bubble. It confirmed the ability to use the wavelet analysis within a multi-block framework to achieve considerable savings in computing time without loss of accuracy. The number of grid cells was reduced by 90%. The CPU time of the adaptive mesh method was only 1/8 that of the method using the fully fine mesh.

5 Conclusions

A wavelet-based adaptive mesh scheme was embedded into a commercial CFX package to track moving free surfaces. It significantly saves computational cost and memory requirements. The scheme was demonstrated in a single bubble rising simulation. A moving "microscope" was designed to reveal details of local information with a realistic computing time. The scheme provides the possibility to design some numerical experiments to analyse processes at a fine scale and utilise the local information obtained in macroscale models. Further work to apply the method in such a scheme is underway.

References

1. Lane, G.L, Schwarz, M.P., Evans, G.M.: Predicting gas-liquid flow in a mechanically stirred tank. Applied Mathematical Modelling 26 (2002) 223-235
2. Schwarz, M.P.: Simulation of gas injection into liquid melts. Applied Mathematical Modelling 20 (1995) 41-51
3. Ohta, M., Iwasaki, E., Obata, E., Yoshida, Y.: A numerical study of the motion of a spherical drop rising in shear-thinning fluid systems. Journal of Non-Newtonian Fluid Mechanics 116 (2003) 95-111
4. Hesthaven, J.S., Jameson, L.M.: A wavelet optimized adaptive multi-domain method. ICASE Report No.97-52, NASA Langley Research Centre, Hampton, 1997
5. Unverdi, S.O., Tryggvason, G.: A front-tracking method for viscous, incompressible multi-fluid follows. Journal of Computational Physics 100 (1992) 125-137
6. Drew, D.A.: Mathematical modelling of two-phase flow. Annual Review of Fluid Mechanics 15 (1983) 261-271
7. Sethian, J.A.: Level Set Methods. Cambridge University Press, Cambridge, UK 1996.
8. CFX User Guide. Release 5.7, 2004, ANSYS Canada Ltd, Waterloo, Canada
9. Vasilyev, O., Paolucci, S.: A fast adaptive wavelet collocation algorithm for multidimensional PDEs. Journal of Computational Physics 138 (1997) 16-56
10. Daubechies, I.: Orthonormal bases of compactly supported wavelets. Communication on Pure and Applied Mathematics 41 (1988) 909-996
11. Bertoluzza, S., Naldi, G.: A wavelet collocation method for the numerical solution of partial differential equations. Applied and Computational Harmonic Analysis 3 (1996) 1-9
12. Holmstrom, M.: Solving hyperbolic PDEs using interpolating wavelets. SIAM Journal on Scientific Computing 21 (1999) 405-420
13. Santos, P.C., Cruz, P., Magalhaes, F.D., Mendes, A.: 2-D wavelet-based adaptive-grid method for the resolution of PDEs. AIChE Journal 49 (2003) 706-717
14. Krishna, R., Baten, J.M.: Rise characteristics of gas bubble in a 2D rectangular column: VOF simulations vs experiments. Int. Commn. Heat Mass Transfer 26 (1999) 965-974

Computational Study on the Effect of Turbulence Intensity and Pulse Frequency in Soot Concentration in an Acetylene Diffusion Flame

Fernando Lopez-Parra and Ali Turan

School of Mechanical, Aerospace and Civil Engineering,
The University of Manchester, Po Box 88,
M60 1QD, Manchester, UK
F.Lopez-Parra@postgrad.umist.ac.uk

Abstract. A computational investigation of the effect of turbulence structure in the formation and depletion of soot in non-premixed acetylene turbulent diffusion flames is presented. Two separate modelling approaches are investigated: 1. Realizable k-ε turbulence model combined with non-adiabatic strained laminar flamelets to solve the reaction mechanisms accounting for the effect of non-equilibrium species and, 2. Standard k-ε turbulence model and Eddy-Break-Up –EBU-with volumetric global reaction and Eddy-Dissipation model for chemistry. In both cases the results obtained show that increments in the input Reynolds number yield lower concentrations of soot. It was also found that low frequency sinusoidal pulse in the fuel inlet velocity can contribute to further reduce the soot concentration in the flame. The soot and nuclei source codes were solved as post-processed scalars and considered to be "passive" species.

1 Introduction

The main objective of the present work is the creation of a mathematical subroutine that would reproduce a soot model based on the eddy dissipation concept of Magnussen[1-5] and that can be implemented into a commercial solver. The performance of the sub-routine is tested with a 2-dimensional axi-symmetric diffusion acetylene flame. Over the last few decades, there has been a growing interest in the modeling of particulates from combustion systems. Although the mechanisms of soot formation are not completely understood, the behavior of this carcinogenic[6,7] specie has been computationally reproduced in the past with an acceptable degree of accuracy[1-5,8-10]. There are two differentiated stages in which soot forms[11]: the inception of the particles for which soot will ultimately form *-nucleation-* and the subsequent growth of these particles. The growth of the carbonaceous particles takes place in two different stages, referred to as the agglomeration and the surface growth. Soot originates from the C_2 radicals that form as a consequence of the break down *–pyrolysis-* of the fuel molecule. In turbulent chemical reactions, there is a strong influence of the flow parameters in the performance of the combustion. As a result, an inhomogeneous structure of the appearance of reacting species will develop. In these situations, the molecular mixing of the fuel and oxidant, which is highly intermittent, takes inside the fine

V.S. Sunderam et al. (Eds.): ICCS 2005, LNCS 3516, pp. 120 – 128, 2005.

structures, which occupy a fraction of the total volume of the domain. These are believed to be three-dimensional vortex tubes of very small dimensions in one or two directions, but not in the other one. These are vortex are of the same characteristic dimensions as the Kolmogorov structures[13]. These regions for the last link in the turbulence energy transfer cascade, and in modeling environments, it is assumed that fuel and oxidant are perfectly mixed. In general, high turbulent flows of high Reynolds number would present a spectrum of eddies of different sizes. These eddies transfer mechanical energy to their immediate neighbors and the interaction between the larger and the smaller eddies represents the main source of production of turbulence kinetic energy. On the other hand, the dissipation of kinetic energy into heat, due to the work done by molecular forces on the eddies, takes place in the smallest eddies.

The mass fraction occupied by the fine structures is given by

$$\gamma^* = 9.7 \cdot \left(\frac{\nu \cdot \varepsilon}{k^2} \right)^{0.75} \tag{1}$$

where k is the turbulent kinetic energy –TKE-, where ν is the kinematic viscosity and ε is the turbulence dissipation rate -TDR .

The mass transfer per unit mass and unit time between the fine structures and the surrounding fluid is expressed by

$$\dot{m} = 23.6 \left(\frac{\nu \cdot \varepsilon}{k^2} \right)^{0.25} \frac{\varepsilon}{k} \tag{2}$$

Due to the reaction taking place within the fine structures, these will have a higher temperature with respect to the local mean temperature. The increment of temperature, ΔT, is computed as

$$\Delta T = \frac{\Delta H_R \cdot c_{\min}}{\rho \cdot c_p} \tag{3}$$

where ΔH_R is the heat of reaction of the fuel, ρ is the density, c_p is the local specific heat capacity of the mixture and c_{min} is the minimum of c_{fu} and c_{o2}/r_{fu}, where c_{fu} and c_{o2} are the gravimetric concentrations of fuel and oxygen respectively and r_{fu} is the gravimetric oxygen requirement to burn 1kg of fuel.

In the Magnussen model, the fine structures and the surrounding fluid are assumed to be in local equilibrium and the concentration of species in the fine structures and the surrounding fluid are related to the local mean concentrations by

$$\frac{c_i}{\rho} = \frac{c_i^*}{\rho^*} \gamma^* \chi + \frac{c_i^o}{\rho^o} \left(1 - \gamma^* \chi\right) \tag{4}$$

where χ is a local variable that accounts for the fraction of fine structures that are actually heated enough to react and c is the concentration of species.

After defining all the preliminary and more relevant variables in the model, the mean nuclei and soot formation rates are expressed as follows

$$R_{n,f} = \left(n_o^* \frac{\gamma^* \chi}{\rho^*} + n_o^o \frac{1-\gamma^* \chi}{\rho^o} + g_o n^* \frac{\gamma^* \chi}{\rho^*} \left(N^o - N^* \right) \right) \rho + (f-g) \cdot Y_{nuc} \cdot \rho - g_0 \rho N^o Y_{nuc} \tag{5}$$

$$R_{s,f} = m_p \left(a \cdot Y_{nuc} \cdot \rho + b N^* \frac{\gamma^* \chi}{\rho^*} \rho \left(n^o - n^* \right) \right) - \rho b n^o Y_{soot} \tag{6}$$

where Y_{nuc} and Y_{soot} are the mass fractions of nuclei and soot, N is the soot particle concentration and (f-g) is the nuclei branching-termination coefficient and a, b and g_o are model constants.

The mean rates of nuclei and soot oxidation are comparable to the rate of fuel combustion.

$$R_{pollut,c} = \dot{m} \cdot \chi \cdot c_{\min} \frac{Y_{pollut}}{Y_{fu}} \tag{7}$$

The net rates of nuclei and soot formation were computed by subtracting the rate of combustion from the rate of formation. The resulting equation was used as the source term for soot and nuclei in the user-defined function written for FLUENT. One of the major challenges presented in was the appropriate selection of a linearization term. Source terms in FLUENT are written as a two-part variable, with well differentiated implicit and explicit sides, e.g. $S_\phi = A + B\phi$, where ϕ would be the dependent variable and A and $B\phi$ would be the explicit and implicit parts respectively. FLUENT automatically determines whether the given expression for B enhances the stability and if this is not the case, the source term is handled explicitly. In current work, algebraic manipulation of the expressions given above yielded the separate explicit and implicit terms required for the simulation.

This soot model also relies in the correct specification of the model constants. Whereas some of these are rather universal, the soot particle diameter, d_p, and the pre-exponential constant for nucleation, a_o need to be adjusted for different flow conditions. In the present work, the two following expressions have been used, combined with the experimental measurements of Magnussen[5] to determine these parameters for each case:

$$\dot{m} \cdot a_o = \text{constant} \tag{8}$$

$$d_p = \frac{D \cdot c_{fu,o}^{1.6}}{U_o^{0.6}} \left(\frac{\rho_{fl}}{\rho_o} \right) \frac{1}{\nu_{fl}^{0.2}} \,.\, \text{constant} \tag{9}$$

where sub-index *o* and *fl* refer to inlet and flame conditions respectively.

The constants employed in the turbulence model are given in Table I, and the turbulence viscosity was user-defined as

$$\mu_t = \frac{\rho \cdot C_D \cdot k^2}{\varepsilon} \tag{10}$$

Table 1. Turbulence model constants

C_D	C_1	C_2	σ_h	σ_k	σ_ε	σ_{fu}	σ_f	σ_s	σ_{nuc}
0.09	1.44	1.79	0.7	1.0	1.3	0.7	0.7	0.7	0.7

The basic transport equations solved in these simulations can be written in a general for as follows

$$\frac{\partial}{\partial x}\left(\bar{\rho}\bar{U}\right)+\frac{1}{y}\frac{\partial}{\partial y}\left(y\bar{\rho}\bar{V}\right)=0\text{, for mass} \tag{11}$$

and for momentum and scalar quantities:

$$\bar{\rho}\bar{U}\frac{\partial \phi}{\partial x}+\bar{\rho}\bar{V}\frac{\partial \phi}{\partial y}=\frac{1}{y}\frac{\partial}{\partial y}\left(y\frac{\mu_t}{\sigma_\phi}\frac{\partial \phi}{\partial y}\right)+R_\phi \tag{12}$$

All transport equations were solved using a second order upwind discretization scheme, except the turbulence dissipation rate. In the case of soot and nuclei, QUICK discretization scheme was employed. The convergence criteria were 1E-3 for most parameters, apart from the energy and soot, for which it was set to 1E-6 and 1E-5 respectively. In reacting flows, the fact that the flow has reached convergence does not necessarily imply that an equilibrium state has been reached. Thus, the concentrations of some product species were monitored at points near the exit of the combustor.

2 Procedure

There are three wee differentiated areas in the present work: the implementation of the Magnussen model using both flamelet and EBU chemistry models, the analysis of the effect of Reynolds number in the final soot concentration and the effect of sinusoidal fuel pulses in the in also the soot concentration.

Nowadays there is a varied choice for turbulence and combustion models, and it cannot be said that one outperforms the others in all flow conditions. It is normally good practice to decide the models to employ by solving a flame or combustion environment that closely resembles the one subject to investigation. In this case, a turbulent piloted diffusion flame[13-15] –referred to as flame D- was reproduced in order to evaluate the performance of the turbulence model and the combustion model. In order to improve the predictions of the spreading of the diffusion jet, realizable k-ε was chosen to model the momentum transport equations. Based on the good agreement of the reproduced flame D in FLUENT with the results obtained with conditional moment closure[13,14] and the experimental data[15], this methodology was assumed to be suited for the simulation of 2-dimensional axi-symmetric diffusion flames. Despite this work being based on the findings of Magnussen, the boundary conditions were modified due to a lack of boundary condition information such as the shape of the issuing nozzle or the TKE and TDR –which can be critical for combustion simulation-, the grid employed, or any discussion as to how the flame can be stable at the given fuel inlet velocity. Therefore, the fuel inlet velocities were reduced with respect to those employed in references[1-5]. In the present study, jet Reynolds numbers of 3300, 16500, 20000 and 25000 – U_{jet}=10, 50, 60 and 75m/s respectively- were tested using the flamelet model and Reynolds numbers of 8300, 11500 and 16500 for when the EBU was employed. A sinusoidal fuel inlet delivery was also analysed, with frequencies 50, 100 and 200Hz and amplitude of 25m/s. The time steps were adjusted in each case to ensure 20 divisions in each complete cycle and 2nd Order Implicit was employed for the unsteady discretization.

3 Results

The first part of the study was the simulation of flame D using the flamelet model in order to see how it compares to the CHEMKIN mechanisms used in the literature. Figure 1 depicts the values of mixture fraction at the x/d = 15 station of the piloted flame D. This result is in close agreement with the experimental data[15] and predictions made using detailed chemical mechanisms[13,14] Likewise, the temperature estimation and the main species mass fractions –figure 3- are within reasonable agreement with experiments. The axial velocity was also compared at the different stations, again showing good agreement with the sources of data. As a result of this study, it was concluded that the flamelet model and the selected realizable k-ε turbulence model were suited to the simulation of turbulent diffusion flames, and therefore were also used to simulate the acetylene flame. The flame D simulation was also employed as an initial step into the grid-independence study, based on both temperatures and species concentrations mainly, and is later applied to the acetylene investigation.

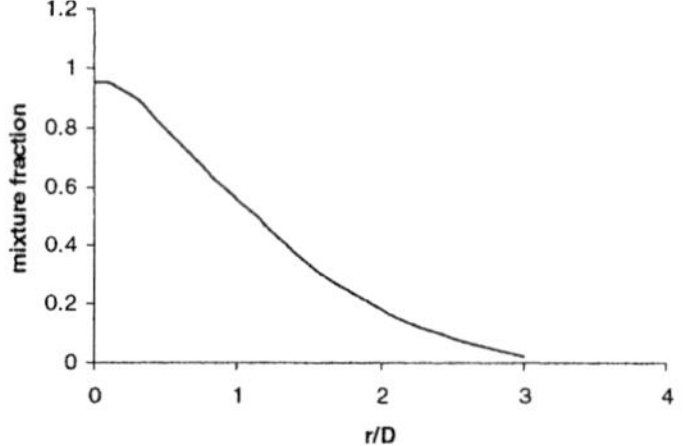

Fig. 1. Flame D, mixture fraction

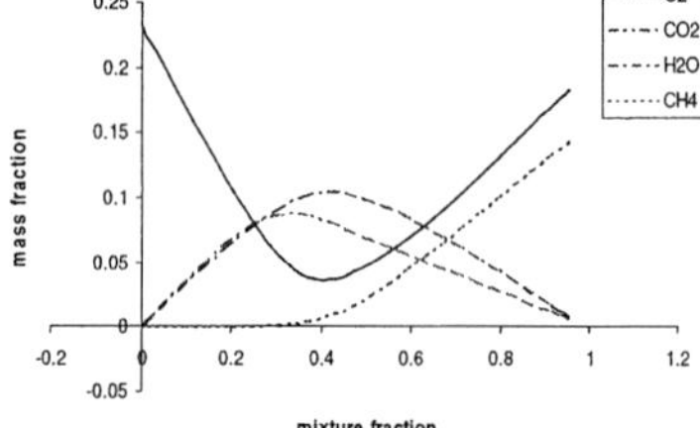

Fig. 2. Flame D, species mass fraction

In the case of the acetylene combustion, the geometry differed slightly from what it can be inferred from the literature. The major differences are that in the present case a converging nozzle is not employed and the issuing jet is not placed in the centre of a circular plate, and instead is placed in an unconfined environment. Figure 3 depicts the turbulence dissipation rate near the jet exit, where a large difference in the dissipation values is observed. These variations of turbulence dissipation will have an effect on the soot and nuclei formation rates, but this effect is more clearly seen when combined with the turbulence kinetic energy in order to monitor the turbulence time scale, as depicted in figure 4. In this figure a large reduction in the time scale is observed between the low Re and the other higher Re conditions. In can also be appreciated that the decrease of the eddy lifetime is also non-linear, since the decrease between

Re=16500 and Re=20000 is larger than that between Re=20000 and Re=25000. These results imply that the residence time allowed for soot to form and growth in the higher Reynolds number conditions is reduced compared to those lower Reynolds. The effect of the turbulence intensity on the formation and combustion of soot is depicted in figure 5. It can be seen that for lower Re, the peak soot concentration is closer to the nozzle that for the higher Re. However, one other feature of this graph is the fact that the Reynolds number does not have a strong effect on the location of the peak soot concentration for higher Re. On the other hand, it does have a strong impact on the maximum soot mass fraction, which reduces as Re increases. Consequently, it can be said that the input boundary conditions do have a significant influence on the soot concentration in turbulent flames. However, it can also be seen that there is a deviation of the peak of the soot concentration towards the exit of the combustor. These differences in are thought to be due to an inappropriate linearization of the scalar transport equations and are currently under study.

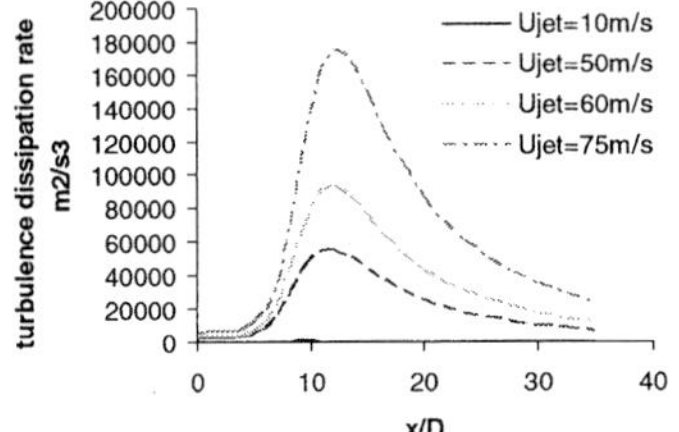

Fig. 3. Acetylene, turbulence dissipation rate

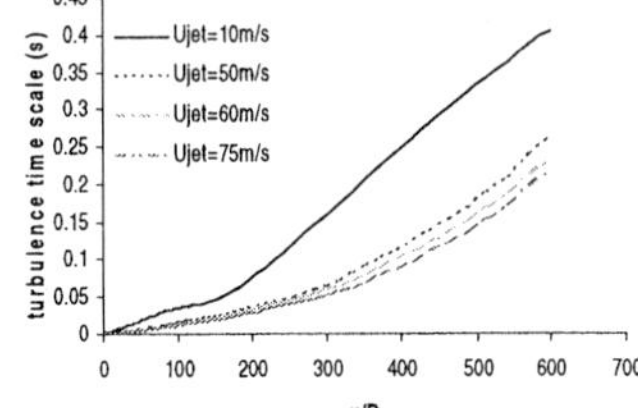

Fig. 4. Acetylene, turbulence time scale

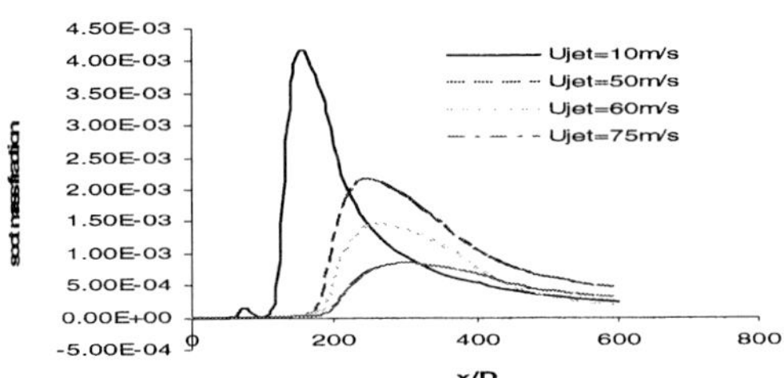

Fig. 5. Acetylene, soot mass fraction at different Re

A similar effect of the Reynold number on the final concentration of soot was observed when the EBU was employed. Comparing the results directly with those obtained with the flamelet model, the EBU presents the peak soot concentrations in the areas of the flame where it could be expected, unlike the flamelet model. This is probably due to the incompatibility of the flamelet model, which assumes that fuel and oxidizer cannot co-exist, unless the strain rate extinguishes the flame, and Magnussen soot model assumes that the fine structures and the surrounding flow are in equilibrium, which seems to be contradictory with the flamelet. Figure 6 depicts the soot concentration along the symmetry axis for the three fuel inlet velocities studied.

This trend also seen along the radial lines, like the example take from a vertical station 0.3m downstream the jet exit.

These results indicate that increased levels of turbulence can, indeed, reduce the soot emissions from a combustion reaction. One other factor that can enhance the depletion of soot particles is the entrainment of oxygen into the core of the diffusion flame, i.e. the fuel rich area. This could be achieved by pulsing the fuel, thus lengthening and twisting the shear layer through which air and fuel ultimately mix. Figures 8, 9 and 10 depict a comparison between the root-mean-square –rms- of the soot concentrations for the three frequencies investigated and the steady state concentration along the symmetry axis and two vertical stations located 0.2 and 0.3m downstream the nozzle exit. The pulsed-fuel cases were run for a total flow time of about 2.7 seconds to allow the pulse to stabilize. Thus we assumed that the quantities seen in one pulse would be repeated in the next one.

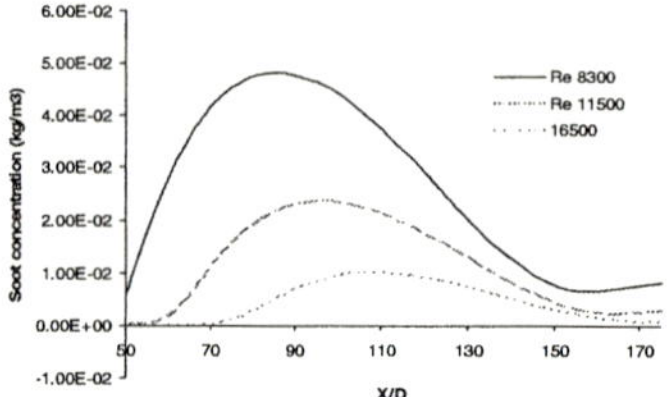

Fig. 6. Axial Soot concentration

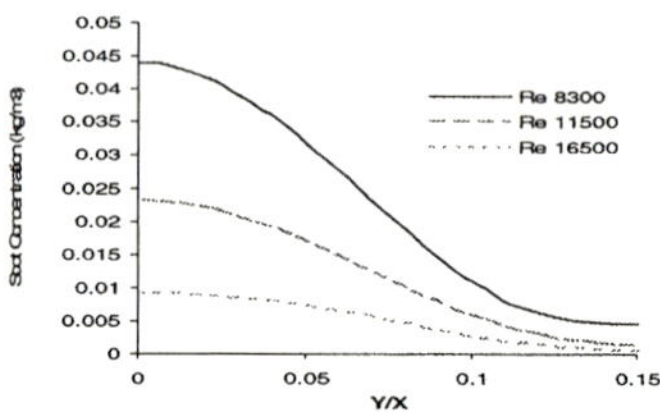

Fig. 7. Radial soot concentration at X=0.3m

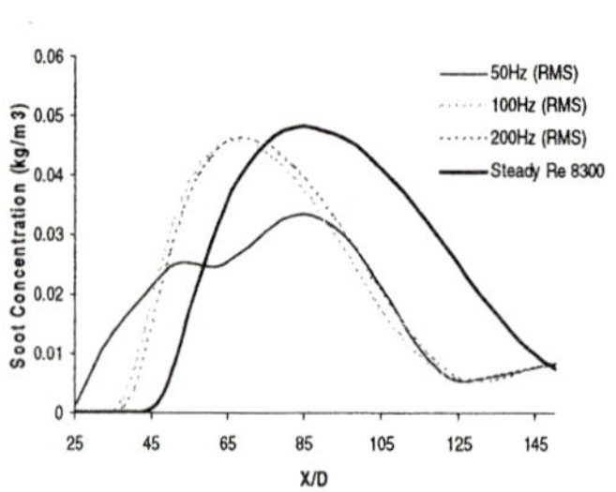

Fig. 8. Soot rms along axis

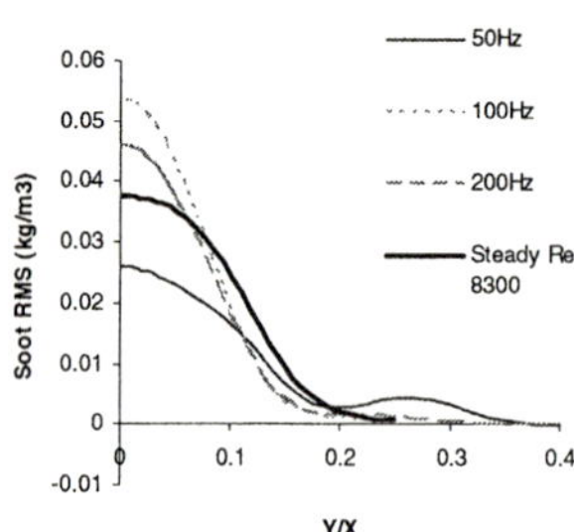

Fig. 9. Soot rms at X=0.2m

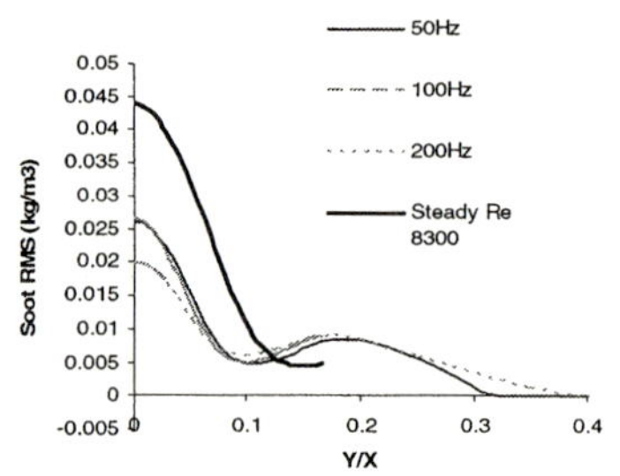

Fig. 10. Soot rms at X=0.3m

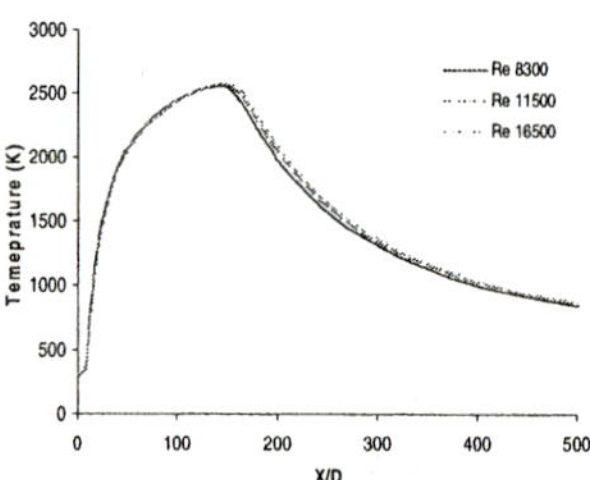

Fig. 11. Axial temperature, steady state

Figure 11 depicts the temperature predicted during the steady state simulations at different Reynolds numbers. Because the temperature is very similar in each case, the reduction in soot concentration must come from the difference in the turbulent kinetic energy and dissipation, as seen in the flamelet case. It is not surprising to find a higher concentration of soot with the pulsed fuel respect to the steady state in Figure 9 because, as can be appreciated in Figure 10, the peak of soot concentration is moves towards the jet exit when the fuel is pulsed. Thus, a more realistic reading of the effect of pulsing in the soot concentration can be made from Figure 10.

4 Conclusions

A soot model with turbulence interaction has been satisfactorily implemented onto a commercial code. At present work is being done to treat soot as"active" specie.

The flamelet model has proved to produce very good thermodynamic solutions, but seems to be incompatible with the soot model.

Both flamelet and EBU simulations have shown a decreasing concentration of soot with increasing flow Reynolds number, whilst temperature distribution has remained fairly unchanged.

Pulsing fuel has shown a decrease in the final concentration of soot based on stabilized pulsed flames and compared to a steady velocity equal to the amplitude of the pulse. Should it be compared to the rms of the inlet velocity, these results would be even larger.

References

1. Magnussen, B.F. *On the structure of turbulence and generalized eddy dissipation concept-for chemical reaction in turbulent flow.* 19th AIAA Science Meeting, 1981.
2. Magnussen, B.F. *Modelling of NOx and soot formation by the eddy dissipation concept.* International Flame Foundation First Topic Oriented Technical Meeting, 1989.
3. Magnussen, B.F. *15th Symposium (Int) on Combustion.* The Combustion Institute, Pittsburgh, 1974.
4. Magnussen, B.F., Hjertager, B.H. *16th Symposium (Int) on Combustion.* The Combustion Institute, Pittsburgh, 1976.
5. Magnussen, B.F., Hjertager, B.H, Olsen, J.G, Bhaduri, D. *17th Symposium (Int) on Combustion.* The Combustion Institute, Pittsburgh, 1979.
6. Thilly, W.G. Soot Combustion Systems and its Toxic Properties, pp1-12, Plenum Press, New York, 1983.
7. Boyland, E. Soot Combustion Systems and its Toxic Properties, pp13-24, Plenum Press, New York, 1983.
8. Tesner, P.A., Snegiriova, T.D., Knorre, V.G. *Combustion and Flame,* 17, pp253-260, 1971.
9. Tesner, P.A., Tsygankova, E.I., Guilazetdinov, L.P., Zuyev, V.P., Loshakova, G.V. *Combustion and Flame,* 17, pp 279-285, 1971.
10. Srivatsa, S.K. *NASA-Lewis Research Center, NAS3-22542,* NASA CR-167930, Garrett 21-4309, 1982.

11. Prado, G., Lahaye, J., Haynes, B.S. Soot Combustion Systems and its Toxic Properties, pp145-162, Plenum Press, New York, 1983.
12. Kolmogorov, A.N., *Journal of Fluid Mechanics,* 13, 82, 1962.
13. Fairweather, M., Woolley, R.M., *Combustion and Flame,* 138, pp3-19, 2004.
14. Fairweather, M., Woolley, R.M. *Combustion and Flame,* 133, pp393, 2003.
15. Information available at http://www.ca.sandia.gov/TNF

Application Benefits of Advanced Equation-Based Multiphysics Modeling

Lars Langemyr and Nils Malm

COMSOL AB, Tegnérgatan 23, SE-111 40 Stockholm, Sweden
Contact author: Nils Malm, telephone +46 8 412 95 291
nils@comsol.se

Abstract. In just the past few years, the field of mathematical modeling with equation-based tools has seen a number of advances and innovations that allow software developers to create programs that are at once far more powerful yet easier to use. To illustrate the benefits that users gain from the latest features and capabilities, this paper examines a problem where software that supports extended multiphysics calculates the applied voltage needed to produce a predefined thermally induced bending of a microscale device. The package used in this problem is FEMLAB, an equation-based environment for solving a large class of systems of coupled partial differential equations.

The software relies on a Galerkin discretization of a weak formulation of the PDEs using generic shape functions. Arbitrary couplings between equations are possible, and a correct discrete residual and Jacobian can be obtained for both linear and nonlinear problems. Therefore, the software can handle tightly coupled systems using an efficient all-at-once Newton solver or an implicit time-stepping method.

Linear systems are described by giving coefficient values for a generic PDE form. More general systems are specified as conservation laws where the user defines an arbitrary flux vector and source for each equation. In addition, the user has direct access to the underlying weak form, which adds additional flexibility to the system. For the less mathematically oriented user, predefined application templates set up the equations based on physical properties. The resulting equations can be viewed, modified, and arbitrarily coupled to other physics.

Direct access to the weak form allows fully integrated multidimensional modeling. Separate equations can be modeled on the boundary of a domain, or contributions can be added to the main equations. In particular, the weak form provides for the straightforward implementation of non-standard boundary conditions and those conditions connecting different types of physics, as is the case when doing, for example, fluid-structure interaction.

In traditional multiphysics problems, the interaction between fields is local in space. In addition to such local couplings, FEMLAB supports the use of non-local variables in any equation. We refer to that capability as *extended multiphysics*, and it can be used to model control systems, couple multidimensional models, couple discrete and semi-analytical models, solve inverse and optimization problems, and more. Non-local coupling variables can be defined through arbitrary coordinate mappings, as integrals over selected domains, or as integrals along a given direction. FEMLAB internally knows the sensitivity of the coupling variables with respect to the degrees of freedom, and therefore it can accurately compute the full Jacobian for extended multiphysics model.

V.S. Sunderam et al. (Eds.): ICCS 2005, LNCS 3516, p. 129, 2005.

Large Eddy Simulation of Spanwise Rotating Turbulent Channel and Duct Flows by a Finite Volume Code at Low Reynolds Numbers

Kursad Melih Guleren* and Ali Turan

University of Manchester, School of Mechanical, Aerospace and Civil Engineering, M60 1QD Manchester, UK
M.Guleren@postgrad.manchester.ac.uk
A.Turan@manchester.ac.uk

Abstract. The objective of this study is to show the highly complex features of rotational turbulent flow using a widely known finite volume code. The flow subjected to an orthogonal rotation is investigated both qualitatively and quantitatively in a three-dimensional channel and a duct using FLUENT. The predictions of rotational flow calculations, presented for low Reynolds numbers, both in channel and duct are in good agreement with the DNS predictions. It is of interest to present the capability of the code for capturing the multi-physics of internal flow phenomena and to discuss the Coriolis effects for two rotational rates. The results show that FLUENT is able to predict accurately first and second order turbulent statistics and it also captures the proper secondary flow physics which occur due to rotation and the geometry itself. These results are very encouraging for the simulation of the flow in a centrifugal compressor, which is the main goal of the authors in the long term.

1 Introduction

It is well known that investigation of the turbulent fluid motion is a challenging research area; neither an analytical solution exists nor it can be exactly defined mathematically. Its complexity is generally explained with its unsteadiness, three-dimensionality, dissipative and diffusive features. In addition, it contains a broad spectrum, which is formed by various size of eddies. For example, scales of these eddies can be of the order of the size of the flow geometry and of the size of Kolmogorov scale, which is known as the smallest scale. Even without rotation, turbulent is certainly a multiscale process. Combining the effects of rotation with turbulence makes the flow physics more interesting, however more complex, and difficult to analyze either experimentally or numerically. It was confirmed by previous studies that rotation changes not only the mean flow but also the turbulence field itself. Although, there exist a wide range of studies in literature as to how and why the multi-physics of these flows are affected depending on the Reynolds and rotation numbers,

* Permanent address: Cumhuriyet University, Dept. of Mech. Eng., 58140, Sivas, Turkey.

V.S. Sunderam et al. (Eds.): ICCS 2005, LNCS 3516, pp. 130 – 133, 2005.

criteria still remains to be formulated clearly for practical industrial flow applications including the centrifugal compressor.

The present study analyzes a turbulent rotating channel flow at a low Reynolds number of Re=2800 for two rotation numbers of Ro=0.1 and Ro=0.5 ($Re=U_bh/\nu$, $Ro=2\Omega h/U_b$, where U_b is the bulk velocity, h is the half width of the channel, ν is the kinematic viscosity and Ω is the rotation rate of the flow). In addition, the turbulent duct flow is investigated at a low Reynolds number of Re=4410 for two rotation numbers of Ro=0.013 and Ro=0.053 ($Re=U_bD/\nu$, $Ro=2\Omega D/U_b$, where D stands for the hydraulic diameter of the duct).

2 The Model

For the channel flow, dimensions of the geometry are L_x=6.4h, L_y=2h, L_z=3.2h. For the duct flow, spanwise and radial lengths are set to be equal to $L_y=L_z=D$ while the streamwise length is taken as L_x=6.28D. The flow is assumed to be fully developed, isothermal, incompressible and rotating at a fixed positive angular velocity parallel to the spanwise direction, $\mathbf{\Omega}=(0,0,\Omega)$.

The numerical calculations were performed using the development version of the general-purpose code FLUENT V6.2 [1] using the Dynamic Smagorinsly-Lilly Model [2],[3]. The code is based on a finite-volume method with an unstructured grid algorithm. The LES incorporates 2nd order central differencing for the diffusive and convective terms for the channel flow calculations and 3rd order MUSCL for duct flow calculations. A fully second-order implicit scheme is applied for temporal discretization while the PISO algorithm and PRESTO! scheme are employed for the velocity-pressure coupling and pressure interpolation, respectively.

The computational domain is formed by $66 \times 66 \times 66$ and $66 \times 60 \times 60$ cells (in the x,y and z-directions) for channel flow and duct flow, respectively. The computational grid is equally spaced along the homogenous directions (x and z-directions for channel and x-direction for the duct) and stretched non-uniformly between the solid walls [from y=0 (bottom wall) to y=2h (top wall) in the channel, from y=0 (bottom wall) to y=D (top wall) and from z=0 (lateral wall) to z=D (lateral wall) in the duct]. Non-slip boundary conditions and periodic boundary conditions were applied for the walls and homogenous directions, respectively. Constant mass flow rate was assumed in the flow directions rather than constant pressure drop.

3 Results and Discussion

Fig. 1. shows the distribution of mean and turbulent intensities for low and high rotational rates. For both cases, excellent agreement for the mean velocity was found with the DNS data [4] except for a slight increase near the pressure side (y=0) at the low rotation rate. For this case, radial and spanwise turbulent intensities are remarkably under-predicted near the pressure side, but they gradually approach the DNS data through center of the channel. Spanwise intensity, which is mainly responsible for turbulent kinetic energy, is under- and over-predicted near the pressure and suctions sides, respectively. For the high rotational case, similar trends are obtained however the discrepancies and the range that is affected are smaller in this case.

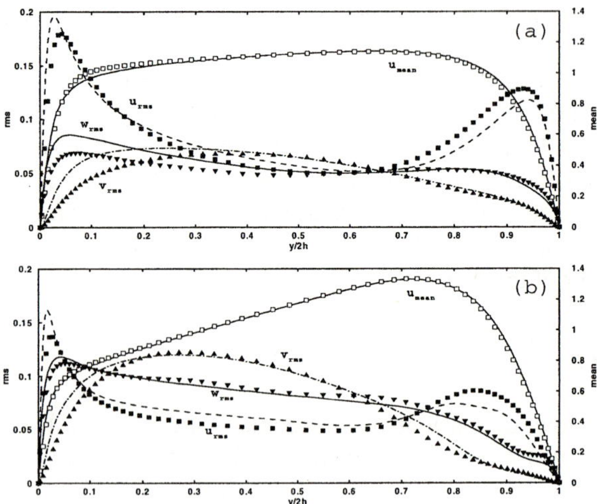

Fig. 1. Mean velocity and turbulent intensity profiles for the rotating channel flow at Ro=0.1 (a) and at Ro=0.5 (b). Present LES results, shown with symbols, are compared with DNS results of Kristoffersen and Andersson [2], shown with lines. Values are normalized by bulk velocities

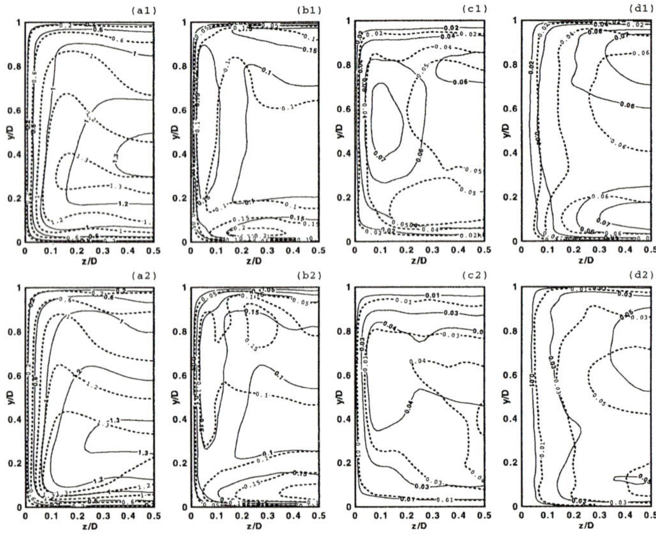

Fig. 2. Mean velocity (a1,a2) and turbulent contours (b1,b2 for streamwise, c1,c2 for normal-wise, d1,d2 for spanwise directions) for rotating duct flow at Ro=0.013 shown with lines and at Ro=0.053 shown width dashes lines. While the top figures represent the DNS results of Gavralakis[16], the bottom figures represent the present LES results. Due to symmetry, half of the duct is shown

Spatial distribution of the mean velocity and turbulent intensities for half of the square duct is also shown for two rotational cases in Fig. 2. The peak region of the predicted mean velocity (a2) has a tendency to shift towards the bottom corner; otherwise, the LES predictions are in good agreement with the DNS data [5] for both rotational cases. Streamwise turbulent intensity (b2) seems to be consistent with the DNS, but the remaining intensities (c2,d2) are remarkably under-predicted. However, the under-prediction for Ro=0.053 is less than that for Ro=0.013. Notwithstanding these discrepancies, our results are similar to those of Palleres and Davidson [6]. Considering the increase in rotational number, mean velocity and turbulent intensity decrease near the suction side and increases towards the pressure and lateral side of the duct. The normal and spanwise turbulent intensities are observed to be reduced near the suction side and become enhanced near the pressure side.

4 Conclusion

Spanwise rotating channel and duct flow were investigated at low Reynolds numbers using LES. Although there are some discrepancies at low rotational numbers, the results are generally in good agreement with DNS predictions. These discrepancies are thought to be caused primarily by the SGS model incorporated in FLUENT. Additionally, the highest accuracy for the numerical schemes in the code is third order: it is well known that such attributes might provide another source for discrepancies.

Concerning future studies, plans are already underway to test different SGS models, including dynamic kinetic energy SGS model [7] and the wall-adapting local eddy-viscosity (WALE) model [8], in order to understand the performances of these SGS models for rotating channel and duct flow problems.

References

1. FLUENT 6.1 USER GUIDE Fluent Inc. Lebanon, USA (2001)
2. Germano, M., Piomelli, U., Moin, P. & Cabot, W. H. A dynamic subgrid-scale eddy viscosity. Phys. Fluids 7 (1991) 1760
3. Lilly, D. K. A proposed modification of the Germano subgrid-scale closure method. Phys. Fluids 4 (1992) 633
4. Kristoffersen, R. & Andersson, H. I. Direct simulation of low-Reynolds-number turbulent flow in a rotating channel. J. Fluid Mech. 256 (1993) 163
5. Gavralakis, S. Direct numerical simulation (DNS) of the Rotating Square Duct flow at a low turbulent Reynolds number. http://lin.epfl.ch/index2.php/link/staff/id/43 (results not published yet)
6. Palleres, J. & Davidson, L. Large-eddy simulations of turbulent flow in a rotating square duct. Phys. Fluids 12 (2000) 2878
7. Kim, W. and Menon, S. Application of the localized dynamic subgrid-scale model to turbulent wall-bounded flows. J.Fluid Mech. 35th Aerospace Sciences Meeting & Exhibit, Reno, NV, (1997) AIAA Paper 97-0210
8. Nicoud, F. & Ducros, F. Subgrid-scale stress modeling based on the square of velocity gradient tensor. Flow, Turb. Comb. 62 (1999) 183

Modelling Dynamics of Genetic Networks as a Multiscale Process

Xilin Wei, Roderick V.N. Melnik, and Gabriel Moreno-Hagelsieb

Mathematical Modelling and Computational Sciences,
Wilfrid Laurier University,
75 University Ave W, Waterloo, Ontario, N2L 3C5, Canada

Abstract. A key phenomenon in the dynamics of genetic networks is the cell cycle. In the study of this phenomenon, an important task is to understand how many processes, acting on different temporal and spatial scales, interact in the cell.

In this paper we deal with the problem of modelling cell cycles. We start our analysis from the Novak-Tyson model and apply this deterministic model to simulate relative protein concentrations in several different living systems, including Schixosaccharomyces pombe to validate the results. Then we generalize the model to account for the nonlinear dynamics of a cell division cycle, and in particular for special events of cell cycles. We discuss the obtained results and their implications on designing engineered regulatory genetic networks and new biological technologies.

1 Introduction

Cells process information in complex ways. During the cell cycle, an eukaryotic cell duplicates all of its components and separates them into two daughter cells. This process is composed of four phases: G1 phase in which size of the cell is increased by producing RVA and synthesizing protein, S phase in which DNA are replicated, G2 phase in which the cell continues to produce new proteins and grows in size, and M (mitosis) phase in which DNA are separated and cell division takes place [1], [3]. From the outset, we are in a situation where we have to deal with different biological events with different spatial and temporal scales.

The problem of modelling dynamics of genetic networks, including those for cell cycles, has been actively addressed in the past decades [2]. New improved models have been recently developed with increasing capability to predict competitively experimental results [3]. The Novak-Tyson model for a cell cycle in [3] contains over 40 parameters that are of the same units but vary from less than 10^{-2} to 35. A stochastic generalization of that model was presented in [4].

In the present work, we start our analysis from the Novak-Tyson model and apply this deterministic model to simulate relative protein concentrations in several different living systems. Then, we generalize the model to account for the nonlinear dynamics of a cell division cycle, and in particular for special

V.S. Sunderam et al. (Eds.): ICCS 2005, LNCS 3516, pp. 134–138, 2005.

events of cell cycles. We show that the effects of such fluctuations may have important implications on designing engineered regulatory genetic networks due to the sensitivity of the model to parametrization processes.

2 Mathematical Models of Cell Cycles

Based on the original Novak-Tyson model, in this section, we develop a new model that accounts for fluctuations of concentrations in response to the multi-scale character of cellular activities.

2.1 The Novak-Tyson Model

With $x_1(t) = Cdc13_T(t), x_2(t) = preMPF(t), x_3(t) = Ste9(t), x_4(t) = Slp1_T(t)$, $x_5(t) = Slp1(t)$, $x_6(t) = IEP(t)$, $x_7(t) = Rum1_T(t)$, $x_8(t) = SK(t)$ and $MPF(t)$ denoting the relative concentrations of the corresponding proteins, and $x_9(t) = M(t)$ the mass of the cell in the cell cycle, the equations and parameters in the Novak-Tyson model are given in Table 1 where the time t for variables x_i, $i = 1, 2, \ldots, 9$; MPF, TF, $Trimer$ and Σ is dropped.

Table 1. The Novak-Tyson Model. All constants have units $\min^{-1}$, except the J's and K_{diss} which are dimensionless

Equation	No.	Parameters
$\frac{d}{dt}x_1 = k_1 x_9 - (k_2' + k_2'' x_3 + k_2''' x_5)x_1$	(1)	$k_1 = k_2' = 0.03$, $k_2'' = 1.0$, $k_2''' = 0.1$;
$\frac{d}{dt}x_2 = k_{wee}(x_1 - x_2) - k_{25}x_2$		$k_3' = 1.0$, $k_3'' = 10.0$, $J_3 = 0.01$,
$-(k_2' + k_2'' x_3 + k_2''' x_5)x_2$	(2)	$k_4' = 2.0$, $k_4 = 35.0$, $J_4 = 0.01$;
$\frac{d}{dt}x_3 = (k_3' + k_3'' x_5)\frac{1-x_3}{J_3+1-x_3}$		$k_5' = 0.005$, $k_5'' = 0.3$, $J_5 = 0.3$,
$-(k_4' x_8 + k_4 MPF)\frac{x_3}{J_4+x_3}$	(3)	$k_6 = 0.1$, $k_7 = 1.0$, $k_8 = 0.25$,
$\frac{d}{dt}x_4 = k_5' + k_5'' \frac{MPF^4}{J_5^4+MPF^4} - k_6 x_4$	(4)	$J_7 = J_8 = 0.001$; $k_9 = 0.1$, $k_{10} = 0.04$,
$\frac{d}{dt}x_5 = k_7 x_6 \frac{x_4-x_5}{J_7+x_4+x_5} - k_8 \frac{x_5}{J_8+x_5} - k_6 x_5$	(5)	$J_9 = J_{10} = 0.01$; $k_{11} = 0.1$, $k_{12} = 0.01$,
$\frac{d}{dt}x_6 = k_9 MPF \frac{1-x_6}{J_9+1-x_6} - k_{10}\frac{x_6}{J_{10}+x_6}$	(6)	$k_{12}' = 1$, $k_{12}'' = 3$, $K_{diss} = 0.001$;
$\frac{d}{dt}x_7 = k_{11} - (k_{12} + k_{12}' x_8 + k_{12}'' MPF)x_7$	(7)	$k_{13} = k_{14} = 0.1$; $k_{15} = 1.5$, $k_{16}' = 1$,
$\frac{d}{dt}x_8 = k_{13}TF - k_{14}x_8$	(8)	$k_{16}'' = 2$, $J_{15} = J_{16} = 0.01$;
$\frac{d}{dt}x_9 = \mu x_9$	(9)	$V_{awee} = 0.25$, $V_{iwee} = 1$,
$Trimer = \frac{2x_1x_7}{\Sigma+\sqrt{\Sigma^2-4x_1x_7}}$	(10)	$J_{awee} = J_{iwee} = 0.01$;
$MPF = \frac{(x_1-x_2)(x_1-Trimer)}{x_1}$	(11)	$V_{a25} = 1$, $V_{i25} = 0.25$,
$TF = GK(k_{15}x_9, k_{16}' + k_{16}''MPF, J_{15}, J_{16})$	(12)	$J_{a25} = J_{i25} = 0.01$; $k_{wee}' = 0.15$,
$k_{wee} = k_{wee}' + (k_{wee}'' - k_{wee}')GK(V_{awee}, V_{iwee}MPF, J_{awee}, J_{iwee})$	(13)	
$k_{25} = k_{25}' + (k_{25}'' - k_{25}')GK(V_{a25}MPF, V_{i25}, J_{a25}, J_{i25})$	(14)	
where $\Sigma = x_1 + x_7 + K_{diss}$ and		$k_{wee}'' = 1.3$, $k_{25}' = 0.05$, $k_{25}'' = 5$;
$GK(a,b,c,d) = \frac{2ad}{b-a+bc+ad+\sqrt{(b-a+bc+ad)^2-4ad(b-a)}}$.		$\mu = 0.005$.

2.2 The Generalized Model with Fluctuations

Since a cell cycle involves nonlinear changes of the protein concentrations related to multiple spatial and temporal scales, the regulation of cellular activities contains a degree of uncertainty [3], [4]. Specifically, at the G1 phase, $Ste9$ and $Rum1$ are activated while $Slp1$ and $Cdc13_T$ are reducing rapidly. From the results of the deterministic model and experimental observations, the magnitudes of $Ste9$, $Cdc13_T$ and $Slp1$ are large enough to introduce fluctuations and the fluctuations of their derivatives are expected. SK is also active at the latter part of the G1 phase. During the S phase which is shorter than G1 and G2 phases but much longer than M phase, the magnitudes of $Cdc13_T$ and $preMPF$ are large enough to generate fluctuations of their changing rates. During the G2 phase, the magnitudes of $Cdc13_T$ and $preMPF$ continue to increase. In the M phase, the magnitudes of $Cdc13_T$, $preMPF$ and $slp1$ changes rapidly and are large enough to introduce fluctuations. IEP is also active in the M phase.

If the magnitude of the relative concentration of a protein $x_i(t)$ is beyond certain value (we use 0.3 for such a value in this paper), we need to modify the right hand sides (RHSs) of equations (1)–(9). Based on the experimental results (see Fig. 1) and taking into account that the period of the cell cycle is about $T = 138.63$ minutes [3], we suggest to multiply the RHSs of equations (1)–(9) by the functions $f_1(t)$, $f_2(t)$, ..., $f_9(t)$ respectively, where

$$f_j(t) = \begin{cases} 1+r, & kT \le t \le kT + \alpha_j \text{ or } kT + \beta_j \le t \le (k+1)T; \\ 1.0, & \text{otherwise}, \end{cases}, \quad j = 1,\ 5; \tag{15}$$

$$f_\ell(t) = \begin{cases} 1+r, & kT + \gamma_\ell \le t \le kT + \lambda_\ell; \\ 1.0, & \text{otherwise}, \end{cases}, \quad \ell = 2,\ 3,\ 6,\ 8; \tag{16}$$

$$f_4(t) = f_7(t) = 1.0; \; f_9(t) = 1 + r, \tag{17}$$

k is a nonnegative integer, r is a control parameter that provides us with the amplitude of fluctuations, $\alpha_1 = 3$, $\beta_1 = 20$, $\alpha_5 = 15$, $\beta_5 = T - 5$, $\gamma_2 = 10$, $\lambda_2 = T$, $\gamma_3 = 0$, $\lambda_3 = 20$, $\gamma_6 = T - 10$, $\lambda_6 = T$, $\gamma_8 = 10$ and $\lambda_8 = 20$. Note that the choice of $f_i(t)$ for $i = 1, \ldots, 9$ is not unique, but the above choice for $r = 0$ is consistent with experimentally confirmed results of [3].

3 Computational Experiments

Both models, described in the previous section, have been implemented in MATLAB. We applied stiff solvers to deal efficiently with numerical difficulties caused by variability of model parameters. The initial conditions in all experiments are $x(0) = (0.45, 0, 1.0, 0, 2.1, 0, 0.05, 0, 1.0)$. In our first set of experiments, we use the deterministic Novak-Tyson model. The results with this model are presented in Fig. 1. Here and in all figures that follow we present two cycles. We observe that the relative concentrations of proteins are qualitatively the same as those obtained in [3], given differences of initial conditions. Replacing k''_{wee} (parameters k''_{wee} and k''_{25} are responsible for rate of tyr-phosphorylation and dephosphorylation) by 0.3 in the above model as suggested in [3], we get a model for the cell cycle of $Wee1^-$ mutants. The results obtained in this case are presented in Fig. 2. We can see that the relative concentrations of $Cdc13_T$, MPF and $preMPF$

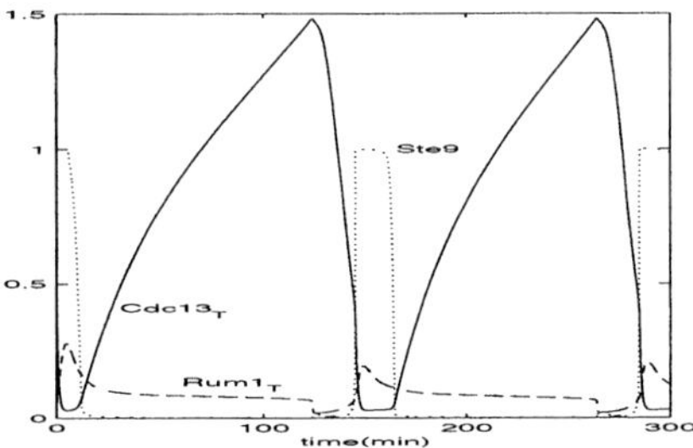

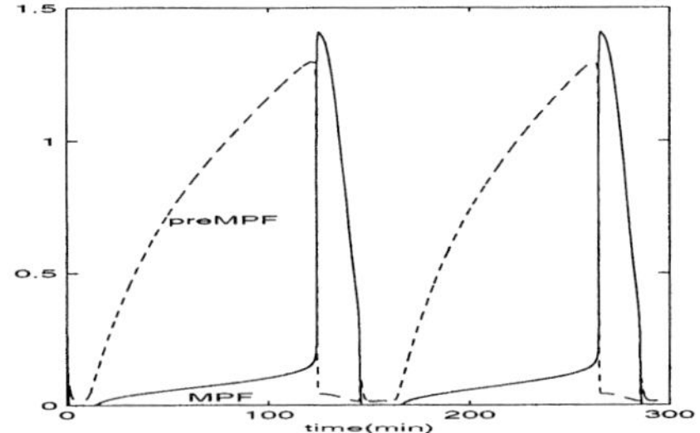

Fig. 1. Numerical simulation of the model in Section 2.1

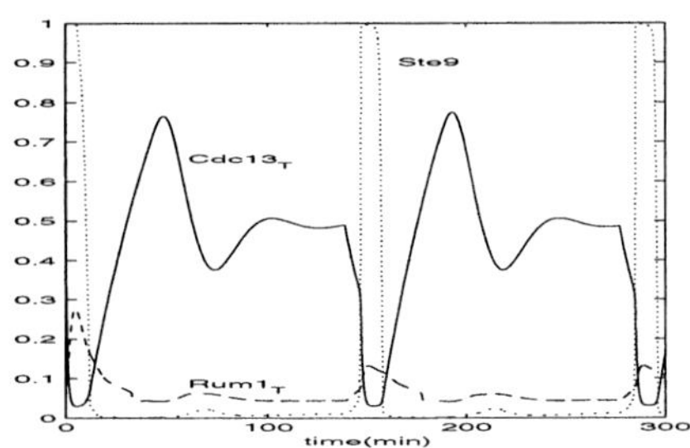

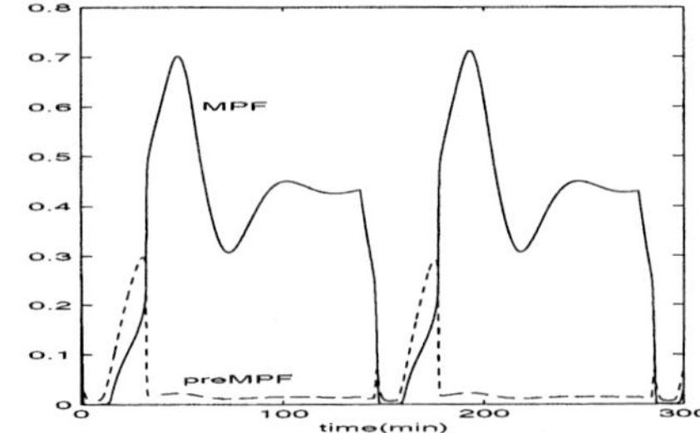

Fig. 2. Numerical simulation of the model with $k''_{wee} = 0.3$

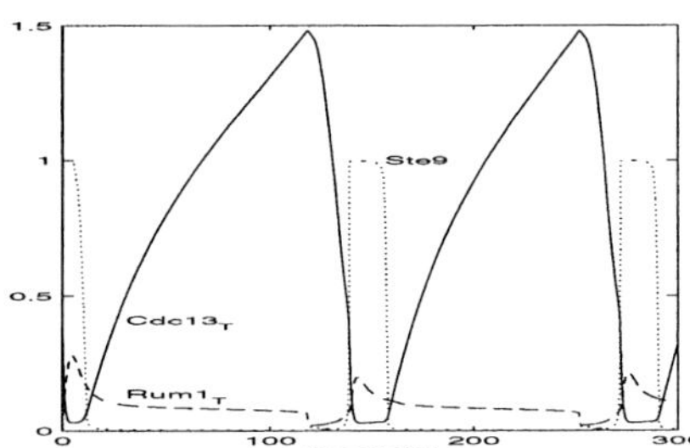

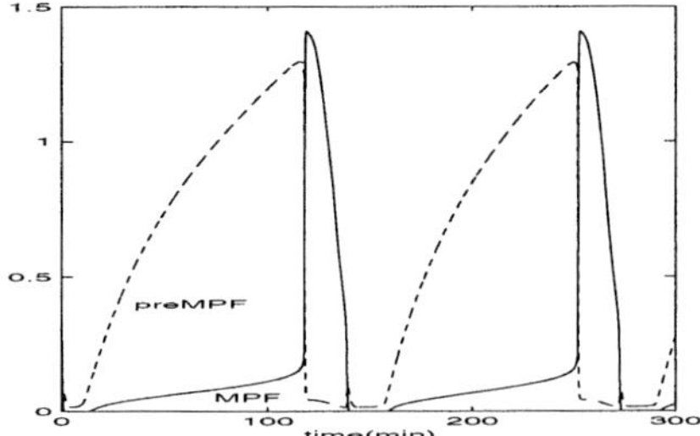

Fig. 3. Numerical simulation of the generalized model with $r = 0.05$

in Fig. 2 are quite different from those in Fig. 1. We have also analyzed the situations with $k''_{wee} = 0.3$ and $k''_{25} = 0.02$, as well as with $k''_{25} = 0.02$, keeping k''_{wee} the same as in our first model. In both cases, noticeable changes in relative MPF were observed.

In our second set of experiments, we use the generalized model given in Section 2.2. Setting sequentially $r = 0.001$, $r = 0.005$, $r = 0.01$ and $r = 0.05$ in (15)–(17), we obtained cell cycles with reduced cycle times. The results for two cycles for $r = 0.05$ are shown in Fig. 3. They demonstrate that it is possible to regulate the cell cycle by adjusting the perturbation control parameter r.

4 Conclusions

In this paper, we proposed a new model of cell cycle processes. The model takes into account special events during the cell cycle. The developed methodology can also be used to guide investigations on multiscale phenomena in designing engineered regulatory genetic networks and new biological technologies.

References

1. Chen, L., Wang R., Kobayashi, T. J. and Aihara K.: Dynamics of Gene Regulatory Networks with Cell Division Cycle, Phys. Rev. E, **70** (2004), 011909.
2. Jong, H.D.: Modeling and Simulation of Genetic Regulatory Systems: A Literature Review, J. of Computational Biology, **9(1)** (2002), 67–103.
3. Novak, B., Pataki, Z., Ciliberto, A. and Tyson, J. J.: Mathematical Model of the Cell Division Cycle of Fission Yeast, CHAOS, **11(1)** (2001), 277-286.
4. Steuer, R.: Effects of Stochasticity in Models of the Cell Cycle: from Quantized Cycle Times to Noise-induced Oscillations, J. of Theoretical Biology, **228** (2004), 293–301.

Mathematical Model of Environmental Pollution by Motorcar in an Urban Area

Valeriy Perminov

Belovo Branch of Kemerovo State University,
652600 Belovo, Kemerovo region, Russia
pva@belovo.kemsu.ru

Abstract. In the present paper it is developed mathematical model for description of heat and mass transfer processes and predicting velocity, temperature and pollution concentrations near roadway. To describe convective transfer controlled by the wind and gravity, we use Reynolds equations for turbulent flow. The boundary value problem was solved numerically. A discrete analog for equations was obtained by means of the control volume method. Methods of predicting concentrations of automobile exhaust gases near roadways are needed for the planning and design of roads and nearby structures.

1 Introduction

Mathematical model for description of heat and mass transfer processes and predicting velocity, temperature and pollution concentrations near roadway is constructed as a result of an analysis of known experimental data and using concept and methods from reactive media mechanics [1] and existing environmental pollution models [2,3]. It is considered that 1) the flow has a developed turbulent nature, molecular transfer being neglected, 2) gaseous phase density doesn't depend on the pressure because of the low velocities of the flow in comparison with the velocity of the sound, 3) the traffic is uniformly distributed over all lanes, 4) two dimensional model used to predict the concentrations along a line normal to highway. The forest in forest belt represents a non-deformable porous-dispersed medium [4].

2 Problem Formulation

Let the coordinate reference point x_1, $x_2 = 0$ be situated at the center of the road surface source at the height of the roughness level, axis Ox_2 directed upward, axis Ox_1 directed parallel to the ground's surface to the right in the direction (Fig. 1).

The problem formulated above is reduced to a solution of the Reynolds and transport equations for turbulent flow:

$$\frac{\partial \rho}{\partial t} + \frac{\partial}{\partial x_j}(\rho v_j) = 0, \; j = 1,2, \; i = 1,2; \tag{1}$$

V.S. Sunderam et al. (Eds.): ICCS 2005, LNCS 3516, pp. 139 – 142, 2005.

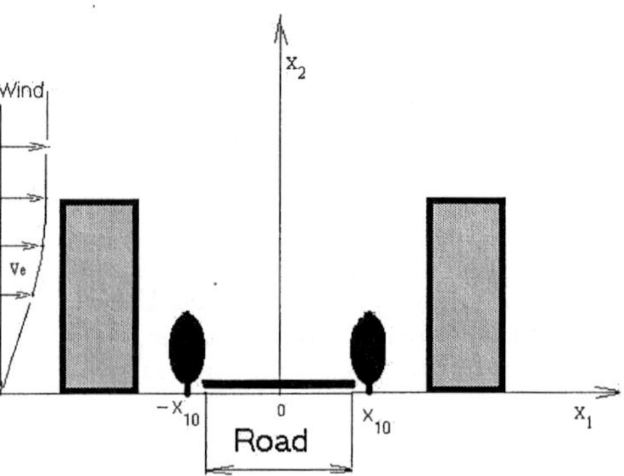

Fig. 1. Street canyon

$$\rho \frac{dv_i}{dt} = -\frac{\partial P}{\partial x_i} + \frac{\partial}{\partial x_j}(-\rho \overline{v_i' v_j'}) - \rho s c_d v_i \,|\vec{v}\,| - \rho g_i ; \tag{2}$$

$$\rho c_p \frac{dT}{dt} = \frac{\partial}{\partial x_j}(-\rho c_p \overline{v_j' T'}) - v_2 (\rho_e g + c_p \rho \frac{dT_e}{dx_2}); \tag{3}$$

$$\rho \frac{dc_\alpha}{dt} = \frac{\partial}{\partial x_j}(-\rho \overline{v_j' c_\alpha'}), \; \alpha = 1,4; \tag{4}$$

$$\sum_{\alpha=1}^{5} c_\alpha = 1, \; P_e = \rho R T \sum_{\alpha=1}^{5} \frac{c_\alpha}{M_\alpha}, \; \frac{\partial T_e}{\partial x_2} = \gamma_e , \vec{v} = (v_1, v_2), \vec{g} = (0, g). \tag{5}$$

The system of equations (1)–(5) must be solved taking into account the following initial and boundary conditions:

$$t = 0: v_1 = 0, v_2 = 0, T = T_e, c_\alpha = c_{\alpha e}, T_s = T_e; \tag{6}$$

$$x_1 = -x_{1e} : v_1 = V_e(x_2), v_2 = 0, T = T_e, c_\alpha = c_{\alpha e}; \tag{7}$$

$$x_1 = x_{1e} : \frac{\partial v_1}{\partial x_1} = 0, \frac{\partial v_2}{\partial x_1} = 0, \frac{\partial c_\alpha}{\partial x_1} = 0, \frac{\partial T}{\partial x_1} = 0; \tag{8}$$

$$\begin{gathered} x_2 = 0: v_1 = 0, \; v_2 = V_{20}, T = T_0, c_\alpha = c_{\alpha 0}, |x_1| \le x_{10}, \\ v_1 = 0, \; v_2 = 0, T = T_e, c_\alpha = c_{\alpha e}, |x_1| > x_{10}; \end{gathered} \tag{9}$$

$$x_2 = x_{2e} : \frac{\partial v_1}{\partial x_2} = 0, \; \frac{\partial v_2}{\partial x_2} = 0, \; \frac{\partial c_\alpha}{\partial x_2} = 0, \; \frac{\partial T}{\partial x_2} = 0. \tag{10}$$

Here and above $\frac{d}{dt}$ is the symbol of the total (substantial) derivative; t is time; x_i, v_i, (i = 1, 2) are the Cartesian coordinates and the velocity components; ρ, T - density and temperature of air, P - pressure; c_P – constant pressure specific heat of the gas, c_α- mass concentrations (α=1 – CO, 2 –CH_x, 3 – NO_x, 4 – soot, 5 – inert components of air); V_e – wind speed at the height *1,5 m*, M_α - molecular mass of α -components of the gas phase; V_{20} - velocity of automobile exhaust gases, γ_e – gradient of stratification of temperature, c_d is an empirical coefficient of the resistance of the vegetation, s is the specific surface of the forest elements; index e corresponds to the unperturbed parameters of the medium. The components of the tensor of turbulent stresses and the turbulent fluxes of heat and mass are written in terms of the gradients of the average flow [4]. The road is modeled as a plane source of burning motorway fuels with known temperature and concentrations of automobile exhaust gases.

3 Numerical Simulation and Results

The boundary value problem (1) – (10) was solved numerically. A discrete analog for equations was obtained by means of the control volume method using the SIMPLE algorithm [5]. The accuracy of the program was checked by the method of inserted analytical solutions. Analytical expressions for the unknown functions were substituted in (1)–(10) and the closure of the equations were calculated. Next, the values of the functions were inferred with an accuracy of not less than 1%. Our practical situation requires calculation domain with obstacles in the middle of the flow region. It is convenient to employ a Cartesian coordinates but to modify the difference equations using method of fiction domain [5]. In the present calculations the results are obtained by using data: T=*300K*, V_e =*2 m/sec*, the width of highway is 12 m, the number of cars passing per unit time (hour) is 3000, percentage structure of traffic (cars - 56%, lorries and buses – 19%, automobiles with diesel engine – 25%). The distribution of

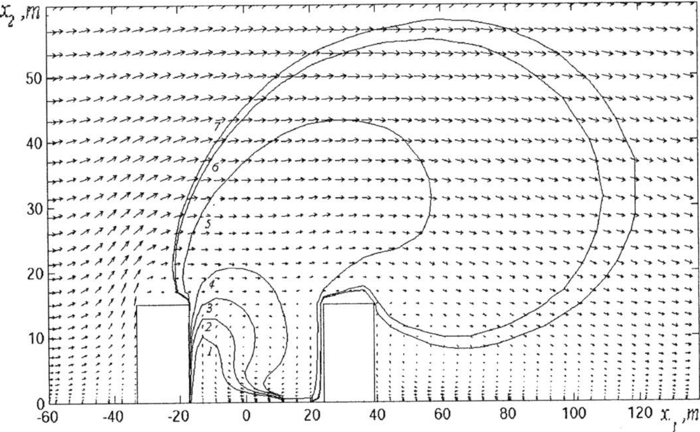

Fig. 2. Distributions of the vector velocity field and the concentration of carbon monoxide; 1 – 1 – 5,0, 2 – 2,0, 3 – 1,0, 4 – 0,5, 5 – 0,1, 6 – 0,05, 7 – 0.04 mg/m^3

temperature of air, velocity, concentrations of components were obtained at different instants of time. The fields of mass concentrations of automobile emissions *CO* and vectorial fields of velocity (Fig. 2) were presented at the moment $t=6\ min$ when the steady situation is realized. The wind field under the street canyon interacts with the gas-jet obstacle that forms from the surface source of heated air masses of automobile emissions. Recirculating flow forms beyond the zone of heat and mass release, i.e. street canyon. We can note that the distribution of velocity and concentration are deformed in the domain by the action of wind, which interacts with the buildings. Similarly, the others fields of component concentrations of pollutants are deformed. It allows investigating dynamics of environmental pollution under influence of various conditions: meteorology conditions and parameters of traffic flow (the number of cars of different types, traffic density and etc.). By increasing the wind velocity, the results show that the concentration of *CO* is drop around the street canyon more intensive.

4 Conclusions

The obtained results are agreed with the laws of physics and experimental data obtained near highway Moscow-Petersburg (Russia). Mathematical model and the results of the calculation give an opportunity to evaluate critical levels of environmental pollution and the damage from motorcar.

References

1. Sedov, L.G.: Mechanics of Continuous Medium. Science. Moscow (1976) (in Russian)
2. Maddukuri, C.S.: A Numerical Model of Diffusion of Carbon Monoxide Near Highways //Journal of Air Pollution Control Association, Vol. 32, 8 (1982) 834-836
3. Perminov, V.A.: Mathematical Modeling of Environmental Pollution by the Action of Motor Transport // Advances in Scientific Computing and Application, Science Press, Being/New York, (2004) 341-346
4. Grishin, A.M.: Mathematical Modeling Forest Fire and New Methods Fighting Them, F.Albini (ed.), Publishing House of Tomsk University, Tomsk (Russia), (1997)
5. Patankar, S.: Numerical Heat Transfer and Fluid Flow, Hemisphere Publ. Co., New York (1980)

The Monte Carlo and Molecular Dynamics Simulation of Gas-Surface Interaction

Sergey Borisov, Oleg Sazhin, and Olesya Gerasimova

Department of Physics, Ural State University, 620083, Ekaterinburg, Russia
sergei.borisov@usu.ru

Abstract. A testing procedure and a program product for modeling gas-surface scattering process have been developed. Using the developed product the numerical simulation of the thermal transpiration phenomenon at free molecular conditions of the gas flow in channels with the use of different scattering kernels has been carried out. The surface structure influence on energy and momentum exchange in a gas-surface system has been studied by the use of Molecular Dynamics method.

1 Thermal Transpiration Phenomenon Study by Monte Carlo Method Realized for Different Gas-Surface Scattering Kernels

The most well-known diffuse-specular scheme of the boundary conditions to the heat and mass transfer equations of rarefied gas dynamics developed by Maxwell is successfully used for the majority of practical calculations (see, for instance, [1]). But some experimental data and corresponding theoretical calculations based on this scheme come into conflict with each other. As an example, one of the results of such calculations affirms that the thermal transpiration phenomenon (or "thermo molecular pressure difference effect" as it appears in scientific papers) does not depend on the kind of the gas and the surface state [2]. Such result contradicts the rather reliable experiments, for instance [3]. Apparently, the diffuse-specular scheme is not suitable for the correct description of the gas-surface scattering process at the non-isothermal rarefied gas flow, the striking example of which is the thermal transpiration phenomenon.

The use of the diffuse-specular scheme does not provide the dependence of scattering process on the gas molecule state that leads to contradiction between the theory and the experiment especially for non-isothermal gas flow. To eliminate such contradiction the boundary conditions that include certain data about the state of gas molecules interacting with the surface must be applied.

Nowadays, besides the diffuse-specular scheme other boundary conditions based on scattering kernels developed by Epstein [4] and Cercignani-Lampis [5,6] are widely recognized. The mathematical forms of these kernels contain certain expressions where the velocity of a gas molecules incident on the surface and the surface temperature are included. The Cercignani-Lampis and the Epstein scattering kernels are based on a certain physical ground and they satisfy all the requirements established for a scattering kernel [7].

V.S. Sunderam et al. (Eds.): ICCS 2005, LNCS 3516, pp. 143 – 146, 2005.

To test the correctness of modeling the gas-surface scattering the program product that provides simulating the behavior of non-interacting molecules in finite space while changing the shape of the limited surface as well as the modeling method of the scattering process, the surface temperature distribution and the initial gas state has been developed. The free molecular version of the Monte Carlo direct simulation method [8] has been realized. The efficiency of the program product has been demonstrated on the example of reaching the equilibrium state of the gas in the bulbs of various forms at the temperature perturbation of the surface. On the base of this product the results that do not contradict the principal postulates of the gas kinetic theory have been achieved. This fact has initiated our interest to apply the developed approach for studying the thermal transpiration phenomenon in rarefied gas at non-isothermal conditions that meets the problem in description while using the scheme of boundary conditions based on the Maxwell scattering kernel.

To understand the problem let us consider the free molecular stationary gas flow in cylindrical channel connecting two bulbs where the gas is in the equilibrium with the "hot" and the "cold" bulb at the temperature T_h and T_c accordingly. The main equation for the gas pressure reached in each bulb is

$$\frac{P_h}{P_c} = \left(\frac{T_h}{T_c}\right)^{\gamma}, \tag{1}$$

where P_h – the gas pressure in the hot bulb, P_c – the gas pressure in the cold bulb, γ – a so called thermal transpiration coefficient.

The γ value is observed close to ½ in all simulation procedures that use the Maxwell kernel with any kernel parameter. The simulation of the thermal transpiration phenomenon with the use of Cercignani-Lampis and Epstein kernels demonstrates significant dependence of the thermal transpiration coefficient γ on the kernel parameters.

It has been shown that with the use of both the Cercignani-Lampis and the Epstein kernels the thermal transpiration effect depends on the channel's length/radius ratio, the surface temperature distribution along the channel and does not depend on the bulbs' temperature ratio. The stationary gas temperature distribution inside the channel depends on the channel's length/radius ratio and practically does not depend on the kernel parameters. The stationary gas concentration distribution depends both on the channel's length/radius ratio and the kernel parameters.

The comparison with the most reliable experiments shows that the simulation based on the use of the Cercignani-Lampis scattering kernels provides satisfactory description of the gas-surface scattering at non isothermal rarefied gas flow conditions at all. Due to strong dependence of the thermal transpiration coefficient on kernel parameters one can expect similar result while using the Epstein kernel.

2 Molecular Dynamics Simulation of Energy and Momentum Transfer in a Gas/Solids System

A great number of structural models describing current gas dynamics experiments and forecasting and momentum exchange in a "gas – rough surface" system have been developed. Every model corresponds to definite material, grain orientation and surface structural phase. An attempt to build an adequate model of surface structure and to describe the real experiment for rarefied gas flow in a rectangular channel with the rough walls has been realized with the use of Monte Carlo Test Particle Simulation Method [9]. Other approach for statistical modeling the roughness proposed in [10] is based on the assumption that the separate elements of the surface microstructure are the cones of the same height and top angle. These approaches for simulation of the surface structure as similar ones are not adequate completely to the real situation because of their "artificial" character based on "imagination" but not on the topography of the real surface.

The methods of scanning probe microscopy, in particular, atomic force microscopy that are developed intensively last years give on opportunity to get an information on specific features of the surface structure and to develop boundary conditions adequate to the real situation. The attempt to simulate the surface structure with the use of such approach has been realized recently [11].

In this study we investigate the topography of platinum plate used in gas dynamics experiments to estimate the roughness of the real surface. The surface structure has been studied with the use of AFM Explorer in a contact regime of scanning. Using the obtained data the main parameters characterizing surface microstructure have been determined.

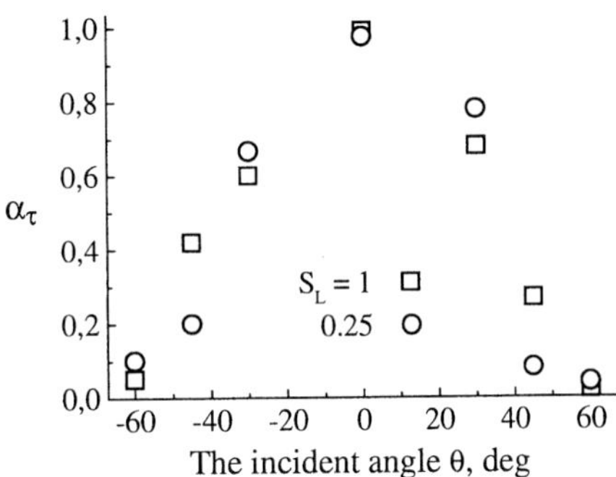

Fig. 1. The tangential momentum accommodation coefficient α_τ for xenon/platinum system

To simulate the gas-solids interaction the classical molecular dynamics method is applied. The Knudsen's accommodation coefficients for tangential and normal momentum, as well as for kinetic energy have been calculated taking into account the gas nature and the surface structure. The results of the tangential momentum accommodation calculation for xenon/platinum system as a function of an incident angle θ for two values of the velocity ratio S_L are presented in figure 1. S_L is introduced as a ratio of the surface movement velocity to the most probable gas molecule velocity.

3 Conclusions

The results of numerical simulation of the thermal transpiration phenomenon at free molecular conditions of the gas flow in channels with the use of the Maxwell, the Cercignani-Lampis and the Epstein scattering kernels are presented. The principal outcome of the study is the statement that in contrast to the Maxwell scheme of boundary conditions the use of the Cercignani-Lampis and the Epstein kernels permits to describe more correctly the non-isothermal internal rarefied gas flow. The obtained results show that there are no principle problems for gas-surface interaction description using numerical simulation procedures, in particular DMCS and molecular dynamics method. Some technical problems could be met under way of the AFM data use in simulations and finding acceptable form of interaction potentials as well as their parameters.

Acknowledgments

The research described in this publication was made possible by Awards No: 03-53-5117 of INTAS and No: REC-005 (EK-005-X1), Y2-P-05-15 of U.S. Civilian Research & Development Foundation for the Independent States of the Former Soviet Union (CRDF).

References

1. Siewert, C. E.: Poiseuille and thermal-creep flow in cylindrical tube. J. Comp. Physics 160 (2000) 470-480.
2. Sharipov F., Seleznev V.: Data on Internal Rarefied Gas Flow. J. Phys. Chem. 27 (3) (1998) 657-706.
3. Edmonds T., Hobson J.P.: A study of thermal transpiration using ultra high vacuum techniques. J. Vac. Sci. Technol. 2 (1965) 182-197.
4. Epstein M.: A model of the wall boundary condition in kinetic theory. J. AIAA 5(10) (1967) 1797-1800.
5. Cercignani C., Lampis M.: Kinetic model for gas-surface interaction. Transp. J. Theory and Stat. Phys. 1 (1971) 101-114.
6. Lord R.G.: Some further extensions of Cercignani-Lampis gas-surface interaction model. J. Phys. Fluids 7 (1995) 1159-1161.
7. Cercignani C.: The Boltzmann Equation and its Application. Springer, New York (1988).
8. Bird G.A.: Molecular Gas Dynamics and Direct Simulation of Gas Flows. Oxford University Press, Oxford (1996).
9. Sazhin O.V., Kulev A.N., Borisov S.F.: The role of the surface structure in formation of an ultra rarefied gas flow in a channel. J. ThermoPhysics & Aeromechanics, 8(3) (2001) 391-399.
10. Sawada T., Horie B.Y., Sugiyama W. J. Vacuum, 47(6-8) (1996) 795-797.
11. Gerasimova O.E., Borisov S.F., Boragno C., Valbusa U.: Modeling of the surface structure in gas dynamic problems with the use of the data of atomic force microscopy. J. Eng. Phys. & Thermophys., 76(2) (2003) 413-416.

GIVS: Integrity Validation for Grid Security

Giuliano Casale and Stefano Zanero*

Dipartimento di Elettronica e Informazione,
Politecnico di Milano - via Ponzio 34/5 - 20133 Milano Italy
{casale, zanero}@elet.polimi.it

Abstract. In this paper we address the problem of granting the correctness of Grid computations. We introduce a Grid Integrity Validation Scheme (GIVS) that may reveal the presence of malicious hosts by statistical sampling of computations results. Performance overheads of GIVS strategies are evaluated using statistical models and simulation.

1 Introduction

Computational grids are touted as the next paradigm of computation [1]. Since they provide access to the resources needed for computational intensive applications, a Grid enviroment usually spans a heterogeneous set of machines.

When a Grid extends beyond the systems of a single administrative authority it suddenly becomes a network of potentially untrusted nodes, on which a remote user submits a potentially harmful piece of code together with sensitive data, as already happens in generic peer-to-peer systems [2]. The ensuing problems of security and privacy have not been fully explored in literature yet.

We identify at least four different problems that arise when extending Grid computing beyond a trusted network:

1. Defending each participant from the effects of potentially malicious code executed on the Grid [3, 4].
2. Avoiding unfair parties which tap computational resources without sharing their own [5, 6].
3. The privacy of data submitted for computation should be adequately protected, depending on their sensitivity.
4. The effect of the presence of one or more peers that are interested in making the overall computation fail.

Our paper will deal in particular with the last problem. We propose a general scheme, named Grid Integrity Validation Scheme (GIVS), based on problem replication and submission of test problems to the Grid hosts.We evaluate the effectiveness and the overhead of the proposed solution both analitically and through simulation.

* This work has been partially supported by the Italian FIRB-Perf project.

V.S. Sunderam et al. (Eds.): ICCS 2005, LNCS 3516, pp. 147–154, 2005.

The paper is organized as follows: in Section 2 we analyze the problem and we introduce our integrity scheme. Section 3 describes a statistical model for evaluating the security level and performance trade-offs of alternative validation scheme within GIVS. Finally, in Section 4 we draw conclusions.

2 The Grid Integrity Validation Scheme

The problem of detecting when remote untrusted machine is running a particular piece of mobile code is difficult to solve in the general case. Checksums and digital signatures can be used to verify the identity of a piece of code we have received, but cannot be used to prove to a remote party that we are actually executing it. Cryptographic protocols would require that the mobile code is endowed with a secret that cannot be accessed by someone controlling the host on which it is running, a constraint that looks impossible to achieve.

A possible solution would be to design the code in such a way that tampering with results in a non-detectable way would require a much longer time than calculating the real solution. Nevertheless, this would require a correct estimate of the real computation times, and thus would implicitly rely on the fact that the malicious user will correctly declare the computational power he has available.

Unless the particular problem we are dealing with has some sort of checksum property, which allows to check the correctness of the results without having to rerun the program, the only feasible approach is to check the correctness of the results by sampling. Specifically, we propose to use some *test problems* to check the correctness of the results. The idea resembles a blind signature protocol [7], in which the submitter prepares N similar documents, the signer opens $(N-1)$ documents checking they are correct, and then blindly signs the last document, with a probability $p = 1/N$ to be cheated. In our case, documents are replaced by computational tasks. We also require to perform our controls by harnessing the computational power of the Grid as much as possible. This is the key idea of our Scheme for Integrity Validation of Grid computation, or GIVS.

In order to generate the test problems required for the integrity validation, we propose a *bootstrap* phase. During the bootstrap, one or more trusted machines compute correct solutions to a subset of test problems, chosen among the available ones according to some criteria (e.g. difficulty, sensitivity to errors, ...). These problems are then used during normal Grid activity to perform security controls, since when a test problem is scheduled to a user, the computated solution is compared to the correct solution to detect whether the user is malicious. Hence, users trying to corrupt the solution of a test problem would be easily identified.

While test problems seem a simple solution to the Grid integrity validation problem, some drawbacks can be identified. First, test problems require a bootstrap phase that slows down the computation startup and waste part of resources. Moreover, increasing the number of test problems could turn into a continuous performance overhead that may become unacceptable.

An alternative solution, consists in replicating a set of problems of unknown solution throughout the Grid, that we call *replicated problems* or just *replicas.* The integrity validation is performed in this case by comparing the different results provided by the untrusted hosts to the same replicated problem. If a conflict is detected, a trusted machine is asked to compute the correct result, so that the malicious hosts can be identified. Compared to the test problems approach, we are now accepting a trade-off between the performance overheads imposed by the integrity validation and the degree to which we can rely on our own security controls.

Notice that if a subset of M machines are cooperating to cheat, if they receive any test problem o replica in common, they can detect it and cheat. Moreover, if any of the results previously computed by any of the M machines is given to another machine as a test problem, it can be detected. The risk connected to this situation are evaluated in the next sections.

3 Performance-Security Trade-Offs

3.1 Notation and Basic Definitions

Let us begin to model a Grid composed of H hosts, with $M \leq H$ malicious nodes that are cooperatively trying to disrupt the computation. In normal conditions we expect $M << H$.

Let T be a period of activity of the Grid after the bootstrap phase. We discretize T in the K intervals $T_1, \ldots, T_K$, thus $\sum T_k = T$. In each period T_k, every host h receives a collection $\mathbf{P_h^k} = (p_1^k, \ldots, p_P^k)$ of P_h problems[1]. We assume the solution of each problem p_i takes the same computational effort on all machines of the Grid. We can classify the problems $p_m \in \mathbf{P_h}$ in several classes: we denote test problems with q_m, replicas with r_m, and generic problems, which don't have any role in security controls, are denoted with g_m. Thus, for instance, the set $\mathbf{P_h} = (c_1, \ldots, c_{C_h}, r_1, \ldots, r_{R_h}, g_1, \ldots, g_{G_h})$ is composed of C_h test problems, R_h replicas and G_h generic problems with $P_h = C_h + R_h + G_h$.

In the rest of this section we explore the two different strategies for GIVS described in Section 2: in the first one we use only test problems, while in the second all test problems are replaced with replicas. Our aim is to study, for each strategy, the performance overhead OH introduced by the security controls, and the probability p_{CF} of accepting a corrupted final value as a correct result.

3.2 First Strategy: Test Problems

We now consider that each host receives a constant number of problems P_h, among which G_h are generic problems, while $C_h = P_h - G_h$ are test problems. Then, no problem replicas are present in the sets $\mathbf{P_h}$ ($R_h = 0$). We denote with

[1] For clarity, we will omit in the rest of the paper the indices k when we do not need to refer to different time slices.

$\mathbf{q_h}$ the subset of test problems of $\mathbf{P_h}$ and with C_h the cardinality of $\mathbf{q_h}$. All test problems $q_m \in \mathbf{q_h}$ are distinct and extracted randomly from a set of C elements.

Under these assumptions, the performance overhead OH of this scheme is given by $OH = H \cdot C_h$ since the solution of the test problems does not give any useful result for the applications running on the Grid.

We observe that the probability that two machines share the same control set is negligible. In fact, given that the probability of receiving a particular control set is the inverse of the number of possible distinct sets $\mathbf{q_h}$, we have

$$p(\mathbf{q_{h_k}}) = \binom{C}{C_h}^{-1} = \frac{C_h!\,(C - C_h)!}{C_h!}.$$

Now, observing that two random extractions $\mathbf{q_1}$ and $\mathbf{q_2}$ of the control sets are independent, the probability of two hosts h_1 and h_2 of receiving the same control set $\mathbf{q_{h_k}}$ is

$$p(\mathbf{q_{h_1}} \equiv \mathbf{q_{h_2}}) = p(\mathbf{q_{h_k}})^2 = \left(\frac{C_h!\,(C - C_h)!}{C!}\right)^2.$$

In non-trivial cases this probability is extremely low. As an example, with $C = 5$ and $C_h = 3$, it narrows down to a modest 0.25%. As C grows, this probability tends quickly to zero.

We now consider the probability $p(q_m \in \mathbf{q_h})$ that a particular problem q_m belongs to a control set $\mathbf{q_h}$. This probability is modeled using an hypergeometric distribution[2], since it reduces to the probability of extracting a winning ball from a set of C, in C_h extractions without replacement. Then we have

$$p(q_m \in \mathbf{q_h}) = hypergeom(C, 1, C_h, 1) = \frac{(C-1)!}{C!}\frac{C_h!(C - C_h)!}{(C_h - 1)!(C - C_h)!} = \frac{C_h}{C}$$

It is interesting to note that such probability is the same as if we would have dropped the no-replacement condition in the extraction, and so it suggest that, from a statistical point of view, the condition of uniqueness of the test problems in the same $\mathbf{q_h}$ can be dropped.

We now turn our attention to p_{CF}. In order to quantify such probability, we first need to estimate the mean number $\bar{s}$ of test problems that each user shares with the other malicious hosts. We can try to model such quantity as the probability $P_M(k)$ that each host of a group of M has exactly k overlaps with the others. For the case $M = 2$, host i has probability $P_2(k) = hypergeom(C, C_h, C_h, k)$, since the winning problems are exactly the C_h that belong to the control set $\mathbf{q_j}$ of host j. Unfortunately, if we try to extend this idea to the case with three or more hosts, we find that we cannot simply add together the pairwise probabilities, because we must account for the overlap between all three sets. This extension is quite difficult to derive analitically with a closed form, and due to its combinatorial nature, it raises some concerns about the computational complexity of

[2] We denote with $hypergeom(n + m,\ n,\ N,\ i)$ the probability of having i successful selections from a hypergeometric distribution with n winning balls out of $n + m$ and after N extractions.

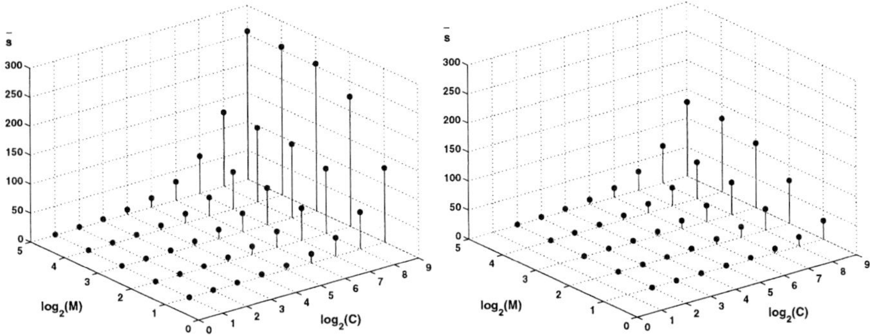

Fig. 1. Expected number of overlappings for different ratios of C_h versus C. In the left figure, $C_h/C = 1/2$. On the right, $C_h/C = 1/4$

a possible iterative description. Moreover, what we actually need is the mean number of overlaps $\bar{s}$, rather than the complete probability distribution, and so a simulation can provide accurate estimates of $\bar{s}$ at a reasonable computational cost and with limited modelling efforts.

In Figure 1 are shown the estimated $\bar{s}$, as a function of the number of malicious hosts M and of the total number of test problems C for the cases $C_h/C = 1/2$ and $C_h/C = 1/4$. All the experiments have been conducted using an Independent Replications Estimate of the mean $\bar{s}$ values with 50 replication of a total of over 1500 different samples. Confidence intervals have been computed for a 95% confidence level; however, since such intervals have resulted to be extremely tight to the estimated $\bar{s}$, we have omitted them from all figures.

Assuming now to know the value of $\bar{s}$, we can model the success probability p_{CF} of a malicious user using an hypergeometric distribution. We consider a malicious user h who gives b wrong answers to the problems of $\mathbf{P_h}$ and answers correctly to the $\bar{s}$ test problems that he has in common with other malicious users. In this case its probability of success p_{CF} is the probability of never giving a wrong answer a test problem, i.e.

$$p_{CF} \cong hypergeom(P_h - \bar{s}_+, C_h - \bar{s}_+, b, 0) = \frac{(P_h - C_h)!(P_h - \bar{s}_+ - b)!}{(P - C_h - b)!(P_h - \bar{s}_+)!}$$

where $\bar{s}_+ = ceil(\bar{s})$ is used instead of $\bar{s}$ since the formulas require integer values. Please note that if $\bar{s}$ is integer the previous formula holds with the equality. The previous formula is defined only for $b < C_h - \bar{s}_+$, otherwise we have trivially that the user always corrupts a test problems and so $p_{CF} = 0$.

The qualitative behavior of p_{CF} is shown in Figure 2, where the probability is plotted against the ratio of undetected test problems $(C_h - \bar{s}_+)$ to the number of problems P_h. The figure is plotted with $\bar{s} = 0$, since this term simply shifts the origin of the x-axis of a corresponding quantity (e.g. with $\bar{s} = 10$ p_{CF} is 1 for $x = C_h/P_h = 10$). Several observations can be drawn from the shape of the p_{CF}. First, as intuitive, if a host tries to cheat on a single problem, its probability of success decreases linearly with the number of test problems. This has a very important consequence on every integrity validation schema: if the correctness of the global computation requires all subtasks to be computed correctly, a severe performance overhead is required to grant to grant the correctness of the results. Then, according to the model, a Grid provider would require a strong additional cost to grant the security of such applications. However, in many cases of interest a malicious user would need to affect a large number of tasks in to affect the global solution. For example, affecting a multiobjective optimization of a set of functions over a numerical domain would probably require the simultaneous corruption of the results on many different subdomains for each of the objective functions. In the case of multiple corruptions ($b >> 1$), the probability p_{CF} drops quickly as the number of test problems grows. This suggests, if possible, to split the functional domain in the highest possible number of regions, although a trade-off with the involved communication overhead should be also taken into account [8]. However, even in this case, a large waste of computational resources is required to grant low values of p_{CF}. Therefore, using replicas could be a suitable way to handle these problems.

3.3 Second Strategy: Validation Using Replicated Problems

We are now going to extend the model developed for the first strategy to the case where replicated problems replace test problems. Problem sets have now the general form $\mathbf{P_h} = (r_1, \ldots, r_{R_h}, g_1, \ldots, g_{G_h}) = \mathbf{r_h} \cup \mathbf{g_h}$, and the replicas $r_m \in \mathbf{r_h}$ are drawn from a set of U problems of unknown solution, in such a way that ρ replicas of each problem are scheduled on the Grid. As explained before, the main advantage here is that, as a side effect of the security controls, the solution of U problems is computed, and so less computational resources are wasted for security purposes. In fact, the minimum performance overhead when ρ replicas of each of the U problems are sent on the Grid is $OH^- = (\rho - 1)U$ that accounts for the fact that replicas also yield the solution of the unknown problems. However, we must handle the case where m problems have been recognized as potentially corrupted, and so the performance overhead becomes $OH = (\rho - 1)U + 2m$ since we lack m solutions and these must be recomputed by trusted machines. However, in normal conditions we expect $OH \cong OH^-$.

As observed for the first strategy, finding an analytical expression for the average number of overlappings $\bar{s}$ is quite difficult. Furthermore, the model of the second strategy should include additional constraints and problems that are difficult to be handled by means of a statistical model. In fact, if we want to ensure that exactly ρ replicas of each of the U problems are computed on the H hosts, during the assignment of the replicas we must satisfy the constraint $C_h H = \rho U$.

Moreover, we observe that the constraint excludes the possibility of using a completely random assignment of the replicas: let us consider, for example, Table 1, where is shown a possible random assignment of $U = 10$ problems with a replication factor $\rho = 2$ (i.e. $\rho U = 20$). In this case we still need to assign two replicated problems to host h_5, but only the replicas of problems 2 and 9 are

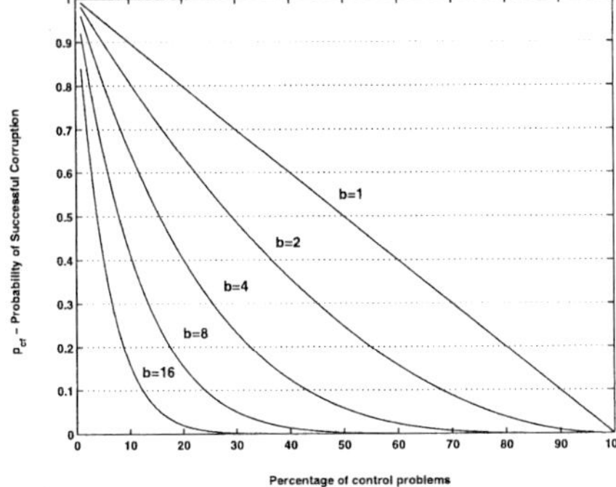

Fig. 2. Probability p_{CF} of accepting a corrupted result as function of C_h/P_h (here is $\bar{s}_+ = 0$) and of the number of simultaneous wrong answers b

Host	Replicated problems	
h_1	1 5 7	10
h_2	3 7 8	10
h_3	3 4 6	8
h_4	1 4 5	6
h_5	2 9 ?	?
Unassigned replicated problems		
Avail	2 9	

Table 1. An example of bad random assignment: any further assignment would alert h_5 that he is subject to a security control

still unassigned. But this assignment is unacceptable, because it would break the condition that a host must receive distinct test problems. In this case, a random strategy would lock searching for an impossible assignment of the replicas.

A simple assignment scheme that avoids locking is the following: we assign randomly, until possible, the replicas. When a lock condition is detected, we change some of the previous assignment in order to solve correctly the assignment. For instance, in Table 1 we could easily avoid locking assigning replicas 2 and 9 to h_4 and moving 5 and 6 to h_5. It is clear that such heuristics is quite complex to be modeled analitically, while using simulation we can easily estimate the involved overlappings $\bar{s}$. In Figure 3 we show in logarithmic scale some numerical results from a simulation with $H = 48$ and $\rho = 2$: we can identify an exponential growth of the overlapping as the number of malicious users M and the unknown problems U grow. This turns to be a linear growth in linear scale, and so we conjecture an approximation $\bar{s} \cong f(M, U)$, where $f(\cdot, \cdot)$ is a linear function.

Given that $\bar{s}$ are known, the analysis of the probability p_{CF} for the second strategy results in the same formula derived for the first strategy, where the quantities C_h and C are replaced respectively by R_h and U. We also observe that, in most cases, the average overlappings $\bar{s}$ for the second strategy are generally lower that those of the first one, but this is affected by the presence of $H - M$ non-malicious hosts. The simulation of such users is a requirement for modeling the second strategy, since otherwise we would have trivially that $\bar{s} = R_h$. Such hosts, instead, are not considered in the simulations of the first strategy.

It is also interesting to note that U is in general bigger than C even with the optimal replication factor $\rho = 2$. This observation suggests that choosing between the two strategies should be done after a critical comparison of both the performance overheads OH and the p_{CF} for the two presented schemes, according to the specific characteristics of the Grid and its workload.

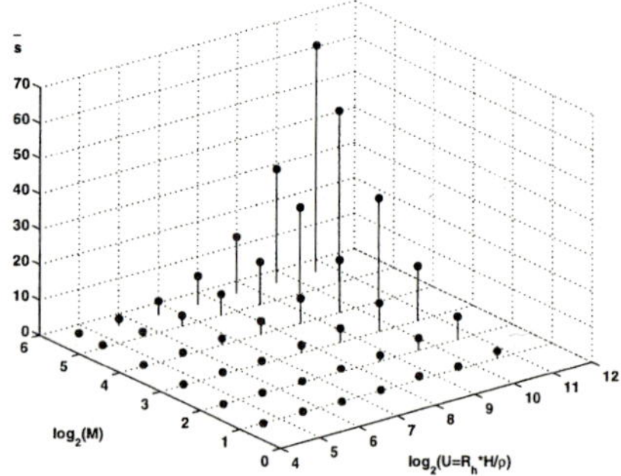

Fig. 3. Expected number of overlaps for the second strategy, with $H = 48$ and $\rho = 2$

4 Conclusions and Future Works

In this paper we have introduced GIVS, a Grid Integrity Validation Scheme, based on the idea of using test problems in order to detect the presence of malicious users. We have defined two different strategies for generating and distributing these problems, studying performance overhead and success probabilities of each using statistical models and simulation. We have shown that the cost for ensuring integrity is far from negligible. Thus, future extensions of our work include optimization techniques such as the introduction of trust and reputation mechanisms.

References

1. Foster, I., Kesselman, C.: The grid: blueprint for a new computing infrastructure. Morgan Kaufmann Publishers Inc. (1999)
2. Oram, A., ed.: Peer-to-Peer: Harnessing the Power of Disruptive Technologies. O'Reilly & Associates, Inc. (2001)
3. Wahbe, R., Lucco, S., Anderson, T.E., Graham, S.L.: Efficient software-based fault isolation. In: Proc. of the 14th ACM Symp. on O.S. princ., ACM Press (1993) 203–216
4. Necula, G.C.: Proof-carrying code. In: Proc. of the 24th ACM SIGPLAN-SIGACT Symp. on Principles of programming languages, ACM Press (1997) 106–119
5. Damiani, E., di Vimercati, D.C., Paraboschi, S., Samarati, P., Violante, F.: A reputation-based approach for choosing reliable resources in peer-to-peer networks. In: Proc. of the 9th ACM CCS, ACM Press (2002) 207–216
6. Aberer, K., Despotovic, Z.: Managing trust in a peer-2-peer information system. In: Proc. of the 10th Int'l Conf. on Inf. and Knowledge Manag., ACM Press (2001) 310–317
7. Chaum, D.: Blind signatures for untraceable payments. In: Advances in Cryptology - Crypto '82, Springer-Verlag (1983) 199–203
8. Muttoni, L., Casale, G., Granata, F., Zanero, S.: Optimal number of nodes for computations in a grid environment. In: 12th EuroMicro Conf. PDP04. (2004)

On the Impact of Reservations from the Grid on Planning-Based Resource Management

Felix Heine[1], Matthias Hovestadt[1], Odej Kao[1], and Achim Streit[2]

[1] Paderborn Center for Parallel Computing (PC^2),
Paderborn University, 33102 Paderborn, Germany
{fh, maho, okao}@upb.de
[2] Zentralinstitut fuer Angewandte Mathematik (ZAM),
Forschungszentrum Juelich (FZJ), 52425 Juelich, Germany
a.streit@fz-juelich.de

Abstract. Advance Reservations are an important concept to support QoS and Workflow Scheduling in Grid environments. However, the impact of reservations from the Grid on the performance of local schedulers is not yet known. Using discrete event simulations we evaluate the impact of reservations on planning-based resource management of standard batch jobs. Our simulations are based on a real trace from the parallel workload archive. By introducing a new option for scheduling reservations in planning-based resource management, less reservation requests are rejected. Our results are important for increasing the acceptability of the Grid technology. We show, that a limited number of additional resource reservations from the Grid have only a limited impact on the performance of the traditionally submitted batch jobs.

1 Introduction

Currently, a gap [2] exists between the demands of Grid middleware and the capabilities of the underlying resource management systems (RMS). While Grid middleware systems provide a certain amount of support for Quality of Service (QoS), underlying RMSs offer only limited functionality in this aspect. Service Level Agreements (SLA) [7] are powerful instruments for assuring QoS, but simply adding SLA negotiation mechanisms to an existing RMS is not sufficient. A demand for advanced abilities in runtime management and control exists. A basic mechanism for accomplishing a guaranteed resource usage are *advance reservations.* They denote the reservation of a fixed amount of resources for a given time span. In the context of Grid Computing, reservations are indispensable to realize the simultaneous availability of distributed resources and to orchestrate Grid workflows. In queuing-based systems reservations are complex to realize, as scheduling focuses only on the present resource usage. In contrast, planning-based resource management systems [3] do resource planning for the present and future. Their design implicitly supports advance reservations.

In this paper we evaluate the impact of reservations on the schedule quality of planning-based resource management systems. This impact is important to

V.S. Sunderam et al. (Eds.): ICCS 2005, LNCS 3516, pp. 155–162, 2005.

evaluate, as reservations affect the scheduling of the traditionally existing workload of batch jobs on supercomputers. Reservations occupy resources for a fixed time span, which gives the scheduler less opportunities to schedule batch jobs. With only some reservations in the system, this impact can be invisible, as the scheduler has still some free slots in the schedule. However, with an increased usage of Grid technology more reservations will be submitted to the underlying resource management systems. At a certain point, the impact will become visible and traditional batch jobs have to wait longer for execution.

We evaluate this impact by using discrete event simulations and generating a fixed amount of batch jobs from a workload trace. We increase the amount of submitted reservation requests and measure the average slowdown of batch jobs. Additionally, we measure the rejection rate of the submitted reservation requests and the utilization of the system.

The intuitive approach for scheduling reservations is to reject them, if the requested resources are already scheduled to batch jobs. We introduce a new approach for scheduling reservations, where such interfering batch jobs are scheduled to a later time and the reservation can be accepted.

The remainder of this paper is structured as follows: we first give an overview on related work, following an overview on the characteristics of planning-based scheduling. In section 4 we present the results of our simulations, thus measuring the impact of advance reservations. A brief conclusion closes this paper.

2 Related Work

Many publications deal with the implementation and applications of advance reservations. However, not much publications have been written about the impact of reservations on resource management.

In 2000, Smith, Foster and Taylor published their results on scheduling with advance reservations [8]. At this, they proposed and evaluated several algorithms, regarding utilization, mean wait time and the mean offset from requested reservation time, which is the time between the reservation request and the receipt of the reservation. All measurements were generated with an Argonne National Laboratory (ANL) workload trace. A basic assumption in their simulations is, that running jobs that have been reserved are not terminated to start other jobs. It is also assumed that reserved jobs will always start on time. These assumptions imply the necessity of run time estimates for each job to assure that resources are free on time. In fact, extending this approach to all types of jobs would lead to planning-based scheduling.

For systems which do not support resumable jobs (i. e. intermediate results of a job are not saved, so that a job can be restarted after termination), measurements show an superproportional increase of waiting time by increasing the percentage of reservations. Likewise, the mean offset from requested reservation time is increasing superproportional. By decreasing the percentage of queued jobs that can not be delayed by a reservation, waiting time as well as offset decrease. For resumable jobs, which can be terminated and restarted at a later time without losing results, the jobs to terminate can be determined using an equation regarding number of nodes and used computing time. For this type of jobs, further decrease of waiting time and offset can be measured.

In [1] a queuing scheduler is enhanced to support advanced reservations. The authors state that users with lower priorities may gain start time advantages by using advanced reservations. To prevent this, the authors applied a shortest notice time for every reservation. It is defined using a predictive waiting time as if the reservation would have been submitted as a queued job. The prediction is based on historical data. A performance evaluation shows that advanced reservations prolong the queue waiting time, although no start time advantages are taken. Hence, longer shortest notice times have to be applied to prevent this.

Similar as before, this work reveals the restrictions of queuing-based scheduling. As only the present resource usage is scheduled, decisions on future reservations can not be based on future load information, instead predictions have to be done. With applying notice times to reservations, users can not specify arbitrary desired start times for their reservations. This is possible in planning-based scheduling presented in this paper.

3 Scheduling Reservations in Planning-Based Systems

If we look in depth at the way current resource management systems work, we find two different approaches: *queuing-based* and *planning-based* systems. In [3] we give a detailed overview on the two types of scheduling approaches. Examples for planning-based systems are the Maui scheduler [4] or CCS [5]. Choosing one of these approaches has a huge impact on the provided features and in particular on the possibility to support advance reservations.

Due to their design queuing-based systems do not support reservations by nature. Hence, work-arounds have to be used, e. g. the insertion of dummy-jobs in high-priority queues, or suspend and resume of low-priority jobs, but they all have drawbacks. Either, the number of reservations is limited, or in fact a second, planning-based scheduler is implemented on top of the queuing-based resource management system. In planning-based systems, on the other hand, reservations are implicitly available in a trivial and easy way.

We call standard batch jobs *variable jobs* as they can move on the time axis until they are started. The scheduler places such jobs at the earliest possible free slot in the schedule. If prior jobs finish earlier than expected, or get aborted, the scheduler might re-schedule the job to an earlier time. Depending on the scheduling policy (e. g. SJF, LJF), jobs may also be moved to a later point in time. *Reservations* have a user specified start time and the scheduler can either accept or reject the reservation for the given start time. Reservations limit the scheduler in placing variable jobs, as resources are occupied for a fixed time.

In case the scheduler cannot place a reservation due to a conflict with a not yet running variable job, we have two options. In the first one, which we call reject, a new reservation is only accepted, if the resources requested for the given time frame are available. The reservation is rejected, if the resources are occupied by already running jobs, by previously accepted reservations, or by planned variable jobs.

With the new variable jobs move option the scheduler additionally accepts those reservations, for which the requested resources are only occupied by variable jobs which are planned, but not yet running. The scheduler plans the overlapping variable jobs at a later time, so that the reservations can be accepted. Still, a reservation is rejected, which collides with already running jobs or other reservations.

Note, starvation of variable jobs is possible with the variable jobs move option, in case newly submitted reservations constantly delay variable jobs. Appropriate mechanisms have to be implemented in this case to prevent starvation.

4 Evaluation

It is common practise to use discrete event simulations for the evaluation of job-scheduling strategies. For this purpose we implemented the reject and variable jobs move options in our MuPSiE system (Multi Purpose Scheduling Simulation Environment).

For evaluating the simulated schedules we use different metrics. We are focusing on the slowdown metric s_i (= response time divided by run time) and weight it by the job area a_i (= run time · requested resources), resulting in the $SLDwA$ metric defined as $SLDwA = \sum a_i \cdot s_i / \sum a_i$.

For measuring reservations other metrics have to be used than for variable jobs, as reservations do not have a waiting time which depends on the scheduler's performance and workload. As already mentioned in Section 3 reservations are either started at their requested start time or they are rejected. Hence, the total number of rejected reservations and respectively the rejection rate in percent of all submitted reservations is measured in our evaluations.

Finally, the overall utilization of the machine is used to measure the efficiency of the scheduler to place all simulated requests, variable jobs and accepted reservations. This information about utilization is of particular interest for system owners.

4.1 Workload

An evaluation of job scheduling policies requires to have job input. In this work a variable job (Section 3) is defined by the submission time, the number of requested resources, and the estimated run time, which is mandatory for planning-based resource management system. The use of a discrete event simulation environment for the evaluation requires additional data about the actual run time of a job. A reservation is defined by the same four parameters like a variable job, but additionally a requested start time is specified.

The variable jobs in this work are derived from a trace of the Parallel Workload Archive [6]. We are using the CTC trace from the Cornell Theory Center. It was taken from a 512-node IBM SP-2 machine during July 1996 to May 1997. It contains 79,279 valid jobs and only 430 nodes of the 512 are available for batch processing.

Unfortunately, no traces are available which contain the appropriate data about reservations required for our simulations as described above. Hence, reservations have to be generated synthetically. In our work this is a two step process and we use the variable jobs from the CTC trace as a basis.

In a first step variable jobs are randomly chosen to get their properties (submission time, number of requested resources, estimated and actual run time). These properties are the basis for newly generated reservations. An input parameter defines, in percent of all variable jobs, how many reservations should be generated in total. If set to 100%, every variable job is used to copy its properties and generate a new reservation, hence the amount of submitted variable jobs and reservation is 79,279 each. With reservations added to the set of variable jobs, the overall workload submitted to the scheduler is increased.

The most important property of a reservation, its requested start time, is generated in the second step. At first a start delay is computed as follows: $start\ delay = rand[0,1] \cdot estimated\ run\ time \cdot start\ factor$. Assume that the start factor is initially set to 1, then the start delay is a random number between 0 and the estimated run time. This random start delay is added to the original submission time of the variable job and the resulting value is the requested start time for the reservation. The input parameter *start factor* is used to modify the start delay, so that the requested start time of reservations is potentially further (start factor > 1.0) or sooner (start factor < 1.0) in the future. With these two steps all necessary properties of a new reservation are copied and generated.

4.2 Results

Figure 1 and Figure 2 show the results for the two different scheduling options, three different start factors, and the three used performance metrics. Note that we repeated each simulation ten times and took the average result to exclude singular effects from generating random jobs.

With the variable jobs move option the scheduler moves interfering variable jobs to a future start time. This implies, that less reservations are rejected,

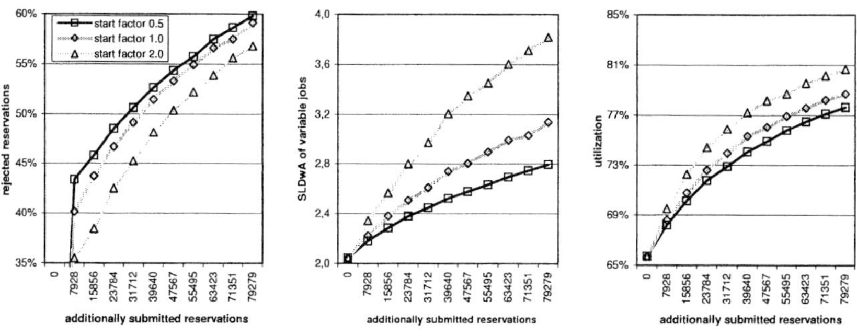

Fig. 1. Results with start factor 0.5, 1.0, and 2.0 for the reject option

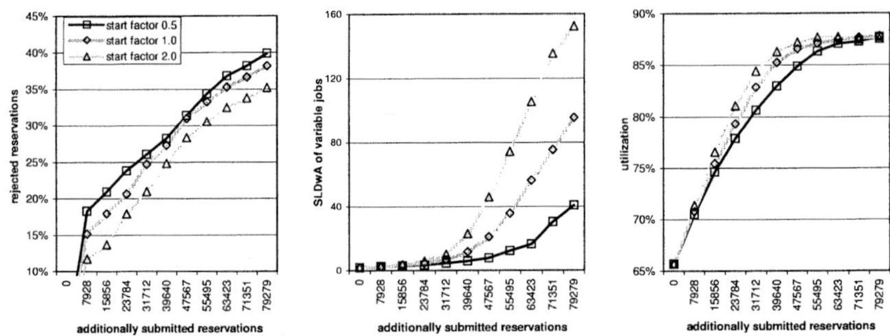

Fig. 2. Results with start factor 0.5, 1.0, and 2.0 for the variable jobs move option

but variable jobs have to wait longer for a start, i. e. their slowdown (SLDwA) increases. Still, if resources are occupied by previously accepted reservations, or by already started reservations, or variable jobs, new reservation requests are rejected with the variable jobs move option.

Focusing on the percentage of rejected reservation and the SLDwA metrics for the two scheduling options in Figures 1 and 2 shows this behavior. The more reservation requests are submitted, the more reservations the scheduler has to reject, as these additional reservations increase the workload and potentially more resources are occupied by other reservations or running jobs. This is the case for both scheduling options. Independent of how many reservations are submitted, the scheduler has to reject more reservations with the reject option, than with the variable jobs move option. For different amounts of submitted reservations the difference between both options is at least 20 percentage-points. With only some reservations submitted, this constant difference means, that more than twice as much reservations are rejected with the reject option. Note, with no reservations submitted at all, obviously no reservations are rejected (or accepted) by both options.

Taking a look at the SLDwA values for all scheduled variable jobs confirms the above made assumption. With the reject option, the scheduler does not replace variable jobs for new reservation requests. They are started at their originally assigned start time and they are not further delayed. However, made reservations may occupy resources for variable jobs submitted in the future, so that their (and thereby the overall) slowdown increases slightly. The curve for variable jobs move clearly shows, that moving variable jobs further in the future and thereby accepting more reservations, significantly increases the slowdown of variable jobs. The results become worse, the more reservations are submitted.

With additionally submitted reservations the workload increases and thereby the utilization, too. However, the utilization can not increase unbounded, which can be seen for the variable jobs move option in particular. Not more than approx. 88% is possible. The planned schedule is so compact, that additional reservations would only increase the slowdown (SLDwA) of variable jobs. Comparing the two scheduling options shows, that with the variable jobs move option the utilization is higher than with the reject option. This is based on the fact, that more reservations can be accepted (cf. the percentages of rejected reservations) and that the scheduler has more freedom in planning variable jobs by placing them further in the future.

The start factor is used to modify the start delay, so that the requested start time of reservations is potentially further (start factor > 1.0) or sooner (start factor < 1.0). Increasing the start factor from to 0.5 to 1.0 and 2.0 means that the time difference between the submission and average requested start time of reservation requests gets larger. Combined with the random computation of the requested start times, they are equally distributed over a larger time span. This reduces the limitations for the scheduler to accept reservations and to (re)place variable jobs. Thus the scheduler can accept more reservations. This can be seen in the diagrams for both scheduling options. With a start factor of 2.0 less reservations are rejected, independent of the total amount of submitted reservations.

With a similar argumentation one could expect that the SLDwA values of variable jobs should decrease with increasing start factors: requested start times are further in the future and variable jobs might be placed at an earlier time in the schedule. However, the diagrams show different results. Increasing the start factor leads to increased SLDwA values for both scheduling options. On a first look the made assumption is correct and variable jobs are potentially placed earlier in the schedule. However, the reservations are still in the schedule, which means that the scheduler is still limited in placing variable jobs. As previously mentioned more reservations are accepted with increased start factors and this additionally limits the scheduler in placing variable jobs. In the end, the scheduler is more limited in placing variable jobs with an increased start factor, so that the SLDwA of variable jobs increases for both options.

As less reservations are rejected with increased start factors, the workload increases and the generated utilization is higher. This can be seen for both scheduling options, but observing the utilization diagram for the variable jobs move option in detail shows, that at the beginning the difference becomes larger and at the end almost the same utilization is achieved. With the variable jobs move option the system runs in a saturated state with many reservations submitted. Accepting more reservations as a result of increased start factors does only mean that the SLDwA values increase and variable jobs have to wait longer. Hence, the differences in the percentage of rejected reservations are smaller for the variable jobs move option than for the reject option. The saturated state is not reached with the reject option, so that a higher utilization is generated by the scheduler with different amounts of reservations submitted and the achieved utilization values are smaller.

5 Conclusion

At the beginning of this paper we revealed the demand for QoS and guaranteed resource usage support in modern resource management systems. To this end, planning systems are well suited for this endeavor due to their design. Planning the present and future resource usage by assigning proposed start times to all waiting jobs makes reservations trivial to schedule. Although reservations are well-known, their impact on planning-based scheduling was not yet evaluated.

We introduced two scheduling options, called reject and variable jobs move, and evaluated the impact of different numbers of reservations on the performance of these options. As no real workload trace contains reservation requests, they were added synthetically. Based on the properties of randomly chosen jobs from the trace, we generated several job sets with different amounts of added reservations. Furthermore, we generated different start delays. The performance was measured by: the percentage of rejected reservations, the average slowdown of variable jobs weighted by their area (SLDwA), and the utilization of the system.

Our results show, that both scheduling options have their benefits. With reject the SLDwA of variable jobs does not increase much due to the additionally accepted reservations, but many reservations are rejected by the scheduler. With the variable jobs move option more reservations are accepted by the scheduler, as variable jobs are moved to a future start time. Hence, the result is an increased SLDwA of variable jobs. Generating reservations, which are further in the future, means that the scheduler can accept more reservations, so that the utilization increases. As more reservations are accepted the scheduler is more limited in placing variable jobs, hence their SLDwA increases.

In contrast to a queuing system, scheduling reservations in a planning system does not mean to terminate running jobs to guarantee the requested start time of the reservation. This is a direct consequence of planning the present and future resource usage not only of reservations, but also of all waiting variable jobs.

References

1. J. Cao and F. Zimmermann. Queue scheduling and advance reservations with cosy. In *Proc. of 18th Intl. Parallel and Distributed Processing Symposium*, 2004.
2. G. Fox and D. Walker. e-Science Gap Analysis. Technical report, Indiana University, USA, 2003.
3. M. Hovestadt, O. Kao, A. Keller, and A. Streit. Scheduling in HPC Resource Management Systems: Queuing vs. Planning. In *Proc. of the 9th Workshop on Job Scheduling Strategies for Parallel Processing*, 2003.
4. D. Jackson, Q. Snell, and M. Clement. Core Algorithms of the Maui Scheduler. In *Proc. of 7th Workshop on Job Scheduling Strategies for Parallel Processing*, 2001.
5. A. Keller and A. Reinefeld. Anatomy of a Resource Management System for HPC Clusters. In *Annual Review of Scalable Computing, Singapore University Press*, 2001.
6. Parallel Workloads Archive. `http://www.cs.huji.ac.il/labs/parallel/workload`.
7. A. Sahai, S. Graupner, V. Machiraju, and A. v. Moorsel. Specifying and Monitoring Guarantees in Commercial Grids trough SLA. Technical Report HPL-2002-324, Internet Systems and Storage Laboratory, HP Laboratories Palo Alto, 2002.
8. W. Smith, I. Foster, and V. Taylor. Scheduling with Advanced Reservations. In *Proc. of the 14th Intl. Conference on Parallel and Distributed Processing Symposium*, 2000.

Genius: Peer-to-Peer Location-Aware Gossip Using Network Coordinates[1]

Ning Ning, Dongsheng Wang, Yongquan Ma, Jinfeng Hu, Jing Sun, Chongnan Gao, and Weiming Zheng

Department of Computer Science and Tecohnology, Tsinghua University, Beijing, P.R.C
{nn02, myq02, hujinfeng00, gcn03}@mails.tsinghua.edu.cn
{wds, zwm-dcs}@tsinghua.edu.cn

Abstract. The gossip mechanism could support reliable and scalable communication in large-scale settings. In large-scale peer-to-peer environment, however, each node could only have partial knowledge of the group membership. More seriously, because the node has no global knowledge about the underlying topology, gossip mechanism incurs much unnecessary network overhead on the Internet. In this paper, we present Genius, a novel peer-to-peer location-aware gossip. Unlike many previous location-aware techniques which utilize BGP or other router-level topology information, Genius uses the network coordinates map produced by Vivaldi as the underlying topology information. By utilizing the information, Genius could execute near-preferential gossip, that is, the node should be told the gossip message by nodes as close as possible, through which much unnecessary 'long-range' communication cost could be reduced. Further, the node direction information inherited in the coordinate space is exploited. We present preliminary experimental results which prove the feasibility of our scheme.

1 Introduction

Peer-to-Peer applications have been popularized in the Internet in recent years. Building reliable, scalable, robust and efficient group communication mechanism on top of peer-to peer-overlay is an important research topic. Such mechanism in peer-to-peer environment much meet following three requirements: the first one is scalability; the second one is reliability, robustness and decentralized operation; the third one is efficiency. The gossip communication mechanism pioneered by [2] emerged as a good candidate which has the potential to satisfy the requirements mentioned above.

The work in this paper deals with the unstructured system which is developed in the previous Scamp[4] protocol. However, the Scamp protocol does not take the underlying topology into account, that is, it is not location-aware. This causes much unneces-

[1] This work is supported by National Natural Science Foundation of China (60273006).

V.S. Sunderam et al. (Eds.): ICCS 2005, LNCS 3516, pp. 163 – 170, 2005.

sary network overhead on internet. The waste of network resources is needless and should be reduced.

Many previous works [6][7][8] address the location-aware issue from various aspects in different settings. Due to shortcomings of these schemes explained at length in Section 4, we propose Genius, a novel peer-to-peer location-aware scheme. In Genius, a network coordinates map which indicates the coordinate of each node is produced by Vivaldi[1] when nodes join the system gradually. Our scheme obtains topology information from this map without the help of network-layer information provided by domain administrators.

In Genius, individual node gains the rather global view of the whole system by exchanging its local information with other nodes in the system. After gathering node information over the whole system, the individual node analyzes them and adjusts its local view so that each individual node has a balanced view of the whole system. Based on these adjusted location-aware local views, Genius could execute location-aware gossip which enables more nodes to be told the gossip message by more nearby nodes, and accordingly reduces the network overhead.

The primary contributions of this paper are the following:

1. It presents Genius, a peer-to-peer location- aware gossip using network coordinates without the help of network-layer information.
2. It proposes an approach to cluster nodes in the coordinate space in a decentralized manner.
3. Genius is the first system to exploit the node direction information in the coordinates map to our knowledge.

The rest of the paper is organized as follows. Section 2 describes the Genius design. Section 3 presents results. Section 4 discusses related work. Finally, Section 5 summarizes our conclusion and future work.

2 Genius Design

There are several steps for node in Genius to obtain location-aware local view.

2.1 Node Arrival and the Network Coordinates Map Construction

In Genius, new node joins the system according to the process specified in the subscription section of Scamp[4]. After collecting latency information from nodes in its *InView*[4], the new node computes good coordinates for itself using Vivaldi[1] which does not depend on the selection of landmarks[3]. When the coordinates converge, one global network coordinates map is formed and Genius enters into nearby nodes exchange stage.

According to [1], coordinates drawn from a suitable model can accurately predict latency between hosts and inter-host RTT is dominated by geographic distance. Thus, the node coordinates reflect its geographic location and it is reasonable to infer the geographic relation of nodes based on the coordinates map.

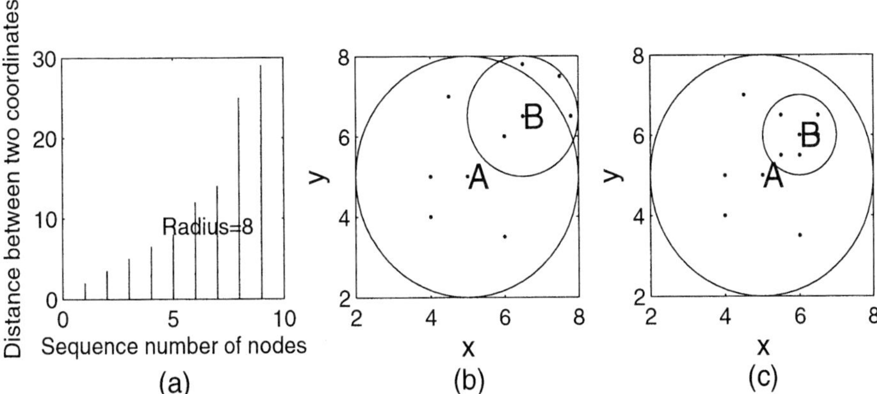

Fig. 1. Illustration of nearby nodes exchange and Clique formation. (a) is the selection of the Clique radius.(b) is the increasing of the Clique radius, and (c) is the decreasing of the Clique radius

2.2 Nearby Nodes Exchange Process and Faraway Nodes Exchange Process

The nearby nodes exchange action is to exchange topology information among the close-by nodes called "Clique" which is formed in the exchanging process simultaneously. The process is actually to cluster nodes in the coordinate space in a decentralized manner. Every node in the system executes the process. The process is described as follows.

Firstly, each node orders the distances between the coordinates of itself and the nodes in its *PartialView* (i.e. the latency between them) in the order of increasing distance. Find the largest difference between the adjacent distances from the smallest one to largest one, which is larger than the preceding difference and the succeeding difference. Define the smaller one of the two distances between which is the largest difference as radius r. This is illustrated by Figure 1(a). The largest difference in Figure 1(a) is the one between the distance of node #5 and that of node #6 which is 4(i.e.12-8) and larger than 1.5(i.e. 8-6.5) and 2(i.e. 14-12). The nodes in its *PartialView* whose distance is smaller than r are considered to belong to the same Clique as it does and r is the radius of the Clique.

According to the Clique radius, there are two sets of nodes in the *PartialView*: nodes belonging to its Clique are called 'Clique contact' and all the nodes except Clique nodes are called 'long-range' contact. The intuition behind the selection of Clique radius r is that the nodes in the same Clique should have relative similar distance from the source node and the nodes not belonging to the same Clique should have relative larger distance.

For acquiring more information about the Clique and preventing the Clique radius being excessively inaccurate just depending on its own *PartialView*, the node needs exchange its information about the Clique, that is Clique membership and Clique radius, with other nodes in its Clique. Thus, it will ask other nodes in its Clique about

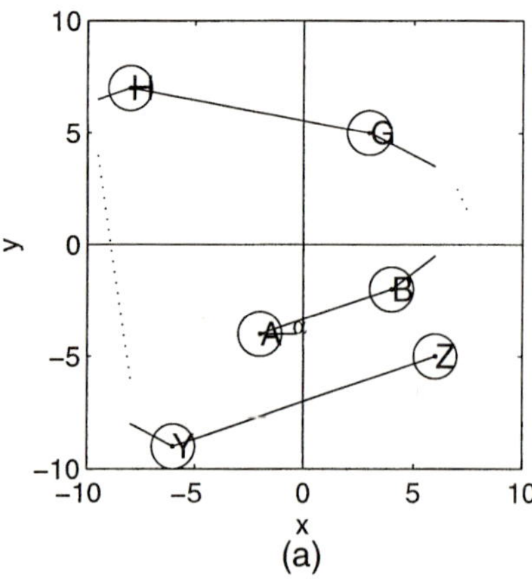

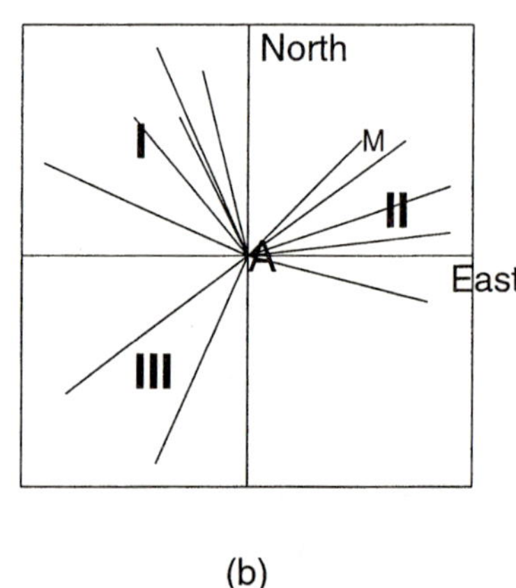

Fig. 2. Illustration of faraway nodes exchange and the 'long–range'contact set adjustment. In (a), node A traverses the whole system from node B to node Z. The circle centered at each node represents its Clique.'long–range' contacts of node A in different directions are shown in (b). They form areas of different angulardensity, such as area I,II,III

which nodes are in their Clique and their Clique radius. For the exchanging process progressing in the right direction and converging to the right state reflecting the real topology, we should force the node exchanging information with nodes whose information is more accurate than that of itself. We define the Clique radius of the node as its credit. Smaller the radius, more credit should be given to the node. The node only trusts nodes whose radius is smaller than its own and exchanges information with them.

Exchanging action is described as follows. As illustrated in Figure 1(b), node B is in the Clique of node A. The radius of Clique of node A is 3 and that of Clique B is 2. So, node A trusts node B and adds the nodes in the Clique of node B into its own 'Clique node' set of its *PartialView*. But the size of original *PartialView* is maintained for future use. This intuition behind the augment of membership is the transitivity property, that is, node A and node B belong to a same Clique, and that, nodes in the Clique of node B belong to the same Clique as node B do, so, nodes in the Clique of node B should belong to the Clique of node A. After adding more nodes into its own 'Clique node' set, node A starts modifying its Clique radius. Then, node A uses similar method described above to determine the new radius based on the result of the augment of Clique membership. This example illustrates the increasing of Clique radius. The example of deceasing of excessively large Clique radius of node A is shown in Figure 1(c). Through repeatedly exchanging information with trusted nodes in its own Clique, the node modifies its Clique radius in every step. The process converges until the variance of radius per step is smaller than one small percent of the radius, say 1%, in several successive steps.

After the node has exchanged information among nodes in its Clique, it needs execute the faraway nodes exchange process which is used to exchange information about 'long-range' contacts to gain a rather complete view of the whole system. We propose a method exploiting the node direction information in the coordinates map.

We observe that nodes with the same latency from the node, but in different direction mean differently to the node. Thus, nodes should not only be classified in the measure of latency and node direction is also an important parameter to be considered. For the node to be able to obtain a summary view of the whole system, our method works as follows. In Figure 2(a), node A wants to exchange information. Firstly, node A selects the longest 'long-range' contact node B which is of longest distance from node A in its 'long-range' contact set. The vector AB makes an angle of α with axis x. It then visits node B, gets one copy of 'long-range' contacts of node B. Next, node B finds the set of its 'long-range' contacts which satisfy that vector BC makes an angle of $\alpha+\beta$ with axis x (suppose node C is a random element of the set), then selects a node C randomly from the set as the next stop of the traverse. Similarly, node A visits node C, and gets one copy of its 'long-range' contacts. After that, node C selects a node D randomly from its 'long-range' contacts satisfying that vector CD makes an angle of $\alpha+2\beta$ with axis x as the next stop. Analogically, node E, F, and G are selected as the next stops. The process terminates when the last vector which is formed by the last two stops, such as YZ, makes an angle between $\alpha-\beta$ and $\alpha+\beta$ with axis x. This termination condition implies that node A traverses a circle on the whole system. When no node in the 'long-range' contacts satisfies the above requirement, the node most approximately meeting the requirement is selected.

The selection of β is determined as follows. Preliminary experimental results from Section 3 show that traversing about 35 nodes are enough to obtain a rather complete view of the system. So, it is suitable to let β be 360/35, approximately 10 degree. Making β positive means counterclockwise traverse and making β negative means clockwise traverse.

When node A traverses the system, it aggregates the copies of 'long-range' contacts of visited nodes in the network by processing over the data when it flows through the nodes, discarding irrelevant data and combining relevant data when possible. The goal of our aggregation of 'long-range' contacts is to identify a set of contacts in which only one contact belongs to every distinct Clique, that is, this set contains just one representative contact for each Clique.

The aggregation at each traversed node is described as follows. When node A visits node B, if the distance between node A and the node in the 'long-range' contact set of node B (for instance, node s) is smaller than the Clique radius of node s, node s is considered to belong to the same Clique of the 'long-range' contact of node A. When the traverse is complete, node A obtains the summary knowledge about other Cliques in the system. Every node in the system traverses the system and gets its own 'long-range' contact set. Although working in this way seems too costly, this kind of traverse happens only once when the system initializes.

2.3 The *PartialView* Adjustment and Maintenance in Dynamic Environment

After the detailed knowledge about its own Clique and a summary knowledge about other Cliques in the system are obtained by each node, the *PartialView* of each node should be adjusted so as to the size of the *PartialView* is $O(log(N))$, where N is the size of the system.

Firstly, we determine the size of the result *PartialView* of node A as the mean size of all original *PartialView* size of visited nodes in the previous traverse. This size is

O(*log*(N)). Secondly, we compute the proportion between the size of the Clique of node A and the size of the 'long-range' contact set. Then, we could determine the size of the result Clique contact set and that of the result 'long-range' contact set.

The next job is to select determined size of nodes from the two former sets. Basically, the selection strategy lies on the most concern of the system designer. There is an obvious tradeoff between network overhead (i.e. message redundancy) and reliability in gossip mechanism. For the Clique contacts, if the designer most concerns about reducing network overhead, those nodes which are nearest should be kept in the result Clique contact set. On the contrary, if the designer most concerns about gossip reliability, the result Clique contacts should be selected randomly from the former Clique contact set of node A. For the 'long-range' contacts (see Figure 2(b)), suppose a 'long-range' node M, node A computes the angle formed by vector AM and axis x. Node A computes the angle for each node in the 'long-range' contact set. If reducing network overhead is the most concern, contacts are selected to be kept according to the density of distribution of angles of contacts. If the most concern is gossip reliability, contacts in sparse area, such as area III, encompassing few nodes should be given priority to be selected. Given the location-aware *PartialView*, more nodes could be told the gossip message by closer neighbors. Thus, location-aware gossip is achieved.

Nodes leaves the system according to the process specified in the unsubscription section of Scamp[4].Because Vivaldi[1] naturally adapts to network changes, the node coordinates remain accurate.

3 Results

Preliminary experiments mostly concentrate on the faraway nodes exchange process. We use a data set which involves 1740 DNS servers mentioned in [1] as the input to

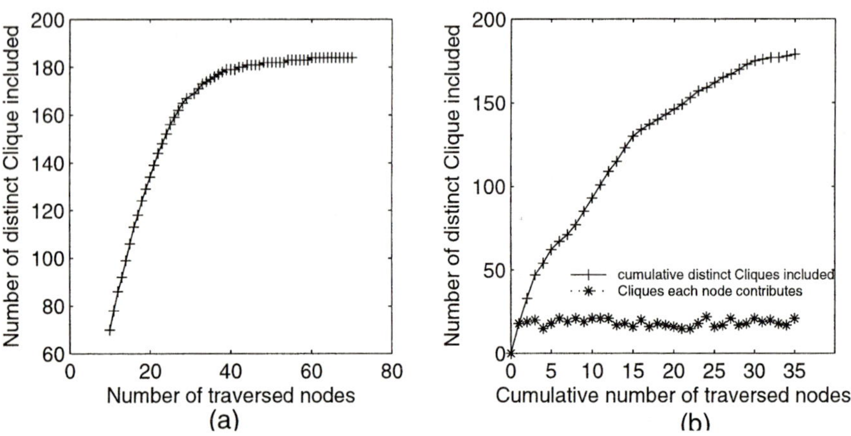

Fig. 3. Preliminary experimental results focusing on faraway nodes exchange process. (a) is the number of distinct Cliques included as a function of number of traversed nodes. (b) is the number of distinct Cliques included as a function of cumulative number of traversed nodes in one experiment in which 35 nodes are traversed

Vivaldi. The node coordinates output by Vivaldi is taken as the input to our simulator. Our simulator implements the logic in the paper.

Figure 3(a) shows that when more than 35 nodes are traversed, the number of distinct Cliques collected increases very slowly during the faraway nodes exchange process. The experiment suggests suitable number of nodes required to be traversed for obtaining a rather complete view of the system is 35.

Figure 3(b) shows that in one experiment in which 35 nodes are traversed, how the number of distinct Cliques included increases along with the progression of the traverse.

4 Related Work

Many location-aware techniques are proposed to achieve location-aware operations in different environment. LTM[6] is a measurement-based technique designed to do location-aware query in unstructured peer to peer systems. It builds a more efficient overlay by cutting low productive connections which is detected by TTL2 detector and choosing physically closer nodes as logical neighbors. Routing underlay[8] provides underlying network topology for overlay services by aggregating BGP routing information from several nearby routers. However, this approach is somewhat too ideal because not all the administrators of domain are willing to provide a feed from their BGP router to the underlay. In contrast, our network coordinates-based approach will not encounter this problem since coordinates are computed by participants cooperatively with no relation to BGP router.

As to the clustering of nodes in coordinate space, Sylvia Ratnasamy *et. al.*[9] proposed a binning scheme. A node measures its round-trip-time to each of d landmarks and orders the landmarks in order of increasing RTT. Thus, every node has an associated ordering of landmarks. The nodes having the same ordering of landmarks belong to the same bin. But, unfortunately, the approach corresponds to halve the right angle in the d-D Euclidean space, thus it does not reflect the node partition based on distance. Some research works also deal with the location-aware gossip. Localiser [7] chooses an energy function over all the nodes and edges in the system incorporating its locality-related objectives and then the function is minimized by simulated annealing through decentralized operation. However, Localiser depends on the choosing of the parameter sensitive energy function too heavily and it is possible for Localiser to be trapped at a local minimum.

5 Conclusions and Future Work

In this paper, we present Genius, a peer to peer location-aware gossip using network coordinates. In contrast to other approaches, it does not rely on any kind of network-layer information which is probably not available. It also does not depend on certain energy function very much. In the future, more experiments about Clique formation and gossip overhead, reliability should be conducted.

References

[1] Frank Dabek, Russ Cox, Frans Kaashoek, Robert Morris Vivaldi: A Decentralized Network Coordinate System In Proceedings of ACM SIGCOMM'04, Portland, Oregon, Aug, 2004

[2] A. Demers, D. Greene, C. Hauser, W. Irish, J. Larson, S. Shenker, H. Stuygis, D. Swinehart, and D. Terry. Epidemic algorithms for replicated database maintenance. In Proceedings of 7th ACM Symp. on Operating Systems Principles, 1987.

[3] T. S. Eugene Ng and Hui Zhang Predicting Internet Network Distance with Coordinates-Based Approaches In Proceedings of IEEE INFOCOM'02 New York, June 2002.

[4] A. Ganesh, A.-M. Kermarrec, and L. Massouli´e. Peer to-peer membership management for gossip-based protocols. In IEEE Transactions on Computers, 52(2), February 2003.

[5] A.-M.Kermarrec, L.Massouli'e, and A.J. Ganesh Probabilistic reliable dissemination in large-scale systems. IEEE Transactions on Parallel and Distributed Systems, 14(3), March 2003

[6] Yunhao Liu, Xiaomei Liu, Li Xiao, Lionel Ni, Xiaodong Zhang Location-aware Topology Matching in P2P Systems. In Proceedings of IEEE INFOCOM'04 Hong Kong, March, 2004

[7] Laurent Massouli´, Anne-Marie Kermarrec, Ayalvadi J. Ganesh Network Awareness and Failure Resilience in Self-Organising Overlay Networks In IEEE Symposium on Reliable and Distributed Systems, Florence, 2003

[8] Akihiro Nakao, Larry Peterson and Andy Bavier A Routing Underlay for Overlay Networks In Proceedings of ACM SIGCOMM'03 Karlsruhe, Germany, August, 2003

[9] Sylvia Ratnasamy, Mark Handley, Richard Karp, Scott Shenker Topologically-Aware Overlay Construction and Server Selection. In Proceedings of IEEE INFOCOM'02 New York, June 2002.

DCP-Grid, a Framework for Conversational Distributed Transactions on Grid Environments

Manuel Salvadores[1], Pilar Herrero[2], María S. Pérez[2], and Víctor Robles[2]

[1] IMCS, Imbert Management Consulting Solutions,
C/ Fray Juan Gil 7, 28002 Madrid , Spain
[2] Facultad de Informática – Universidad Politécnica de Madrid,
Campus de Montegancedo S/N,
28.660 Boadilla del Monte, Madrid, Spain
mso@imcs.es
{pherrero, mperez, vrobles}@fi.upm.es

Abstract. This paper presents a Framework for Distribute Transaction processing over Grid Environment, called DCP-Grid. DCP-Grid complements *Web Services* with some OGSI functionalities to implement the *Two Phase Commit* (2-PC) protocol to manage two types of Distribute Transactions, Concurrent and Conversational transactions, properly in this kind of environment. Although DCP-Grid is still under development at the Universidad Politécnica de Madrid, in this paper, we present the design and the general characteristics associated to the implementation of our proposed Framework.

1 Introduction

The introduction of *Services Oriented Architectures* (SOA) [1] [2], in the last few years, has increased the use of new distributed technologies based on *Web Services* (WS) [3]. In fact, e-science and e-business processes have adopted this technology to improve the integration of some applications. The coordination of this type of processes, based on WS, needs the transactional capability to ensure the consistency of those data that are being handled by this kind of applications.

A transaction could be defined as the sequence of actions to be executed in an atomic way. This means that all the actions should finish - correctly or incorrectly- at the same time as if they were an unique action.

The four key properties associated to the transactions processing are known as the ACID properties - *Atomicity, Consistency, Isolation, y Durability*[4]. The aim of our proposal is to build a Framework, based on grid technologies, to coordinate distributed transactions that are handling operations deployed as Web Services.

The Grid Technology, which was born at the beginning of the 90's , is based on providing an infrastructure to share and coordinate the resources through the dynamic organizations which are virtually distributed[5] [6].

In order to make possible the development of DCP-Grid, we will take into account the *Grid Web Services* (GWS) characteristics. The GWS, defined in the *Open Grid Service Infrastructure* (OGSI) [7], could be considered as an extension of the WS. The GWS introduce some improvement on WS, which are necessary to the

V.S. Sunderam et al. (Eds.): ICCS 2005, LNCS 3516, pp. 171 – 178, 2005.

construction of standard, heterogeneous and open Grid Systems. The OGSI characteristics on which DCP-Grid has being designed and built are: Statefull and potentially transient services; Service Data; Notifications; portType extension; Grid Service Handle (GSH) and Grid Service Reference (GSR).

OGSI is just an specification, not a software platform, and therefore, we need a middleware platform, supporting this specification, in order to deploy the DCP-Grid Framework. From all the possible platforms to be used, we have decided to use the Globus Toolkit [8] platform for this Project because in the most extended nowadays. More specifically, we have being working with GT3 (version 3.2) for DCP-Grid due to its stability.

In this paper we will start describing the state of the art in the dealing area as well as their contributions to the DCP-Grid design, subsequently we will move to the architectural design of our proposal and we will give some details related to the framework implementation to finish with some conclusions, ongoing directions and future work.

2 Related Work

The standard distributed transactional processing model more extended is the X/Open [14] model, which defines three rolls (Resource Manager RM, Transaction Processing Manager TPM and Application Program AP) [9] [14].

Based on the Web Service technology two specifications to standardize the handling of transactions through open environments have arisen. These specifications are WS-Coordination [10] and WS-Transaction [13], developed by IBM, Bea and Microsoft. In them, the way to group multiple Web Services as a transaction is exposed, but the form of coordination of the transactions is not specified. On the other hand, the Business Transaction Protocol specification (BTP) [13], proposed by OASIS, defines a transactional coordination based on workflows. This specification is complex to handle and integrate [12]. Based on GT3 [8] we try to construct a simple proposal for the implementation of a transactional manager adopting the X/Open model. In the proposed design, the analogies are visible.

2.1 Two Phase Commit Protocol

The *Two phase commit* (2-PC) protocol is an ACID compliant protocol to manage DTs. How 2-PC works is easy to explain. A Transaction with a 2-PC protocol does not commit all the actions if not all of them are ready to be committed. This process works in two phases, as its names indicates, first phase called *Voting Phase* and second phase called *Commit Phase*.

During de *Voting Phase,* each and every action notifies to the system their intentions to commit theirs operation. This phase terminate when all the actions are ready to be committed. Then starts the *Commit Phase*, during this phase the system notifies to each and every action to be finished, the conclusion of the operation takes place when the commit message arrives from the system to each and every action.

These two components work in two different phases, during the *Voting Phase* when an application requests to the DTC to commit a Transaction, the DTC sends PREPARE_TO_COMMIT to all the RM that have uncommitted actions of the Transaction, then the DTC waits for a period to receive all the RMs responses.

There are two possible responses READY_TO_COMMIT or UNABLE_TO_COMMIT. If all the RMs responses to the DTC are READY_TO_COMMIT message then the DTC sends a COMMIT message to all RMs, but if any of the resource managers sends a UNABLE_TO_COMMIT or not response in a limit time then the DTC sends a ROLLBACK message to all the RM. In Figure 1 we can appreciate the successfully scenario of a commit transaction over 2-PC protocol.

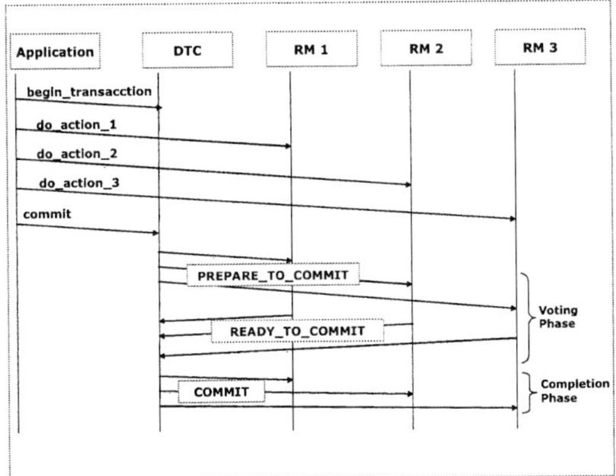

Fig. 1. 2-PC messages intereaction

Distributed Transactions and 2-PC protocol will be the pillars of our Framework because DCP-Grid will provide 2-PC protocol to support transaction processing.

2.2 Classification of Distributed Transactions

We assume two categories *Concurrent Distributed Transactions* and *Conversational Distributed Transactions*:

- *Concurrent Distributed Transactions*: are the transactions formed by actions that have not dependencies between. In this case, the different actions can be sending by the application layer in a concurrent way improving the service time processing.
- *Conversational Distributed Transactions*: are the transactions composed by dependent actions. An example of this kind of transaction could be the following. This scenario refers for any transaction in which action N depends on, at least, one or more actions previously executed, being N-i the maximum number of actions to be included in this dependence, and i a natural number representing the position.

3 Scenarios

As we mentioned in previous section of this paper, there is two different kinds of categories regarding to each proposed scenario.

3.1 Scenario 1: Concurrent Distributed Transactions

In this scenario we will concentrate on the logical operation of the Framework regarding transactions composed by independent actions.

Let's imagine a scenario composed by two actions A and B, and a client application which wants execute these actions in a Transactional way. In this case it would necessary define a transaction like Tx {idTx, coordinator, A, B} where idTx is transaction id and A, B are the actions that compose the transaction. The element coordinator references to the DTC process that coordinates the phases of the transaction.

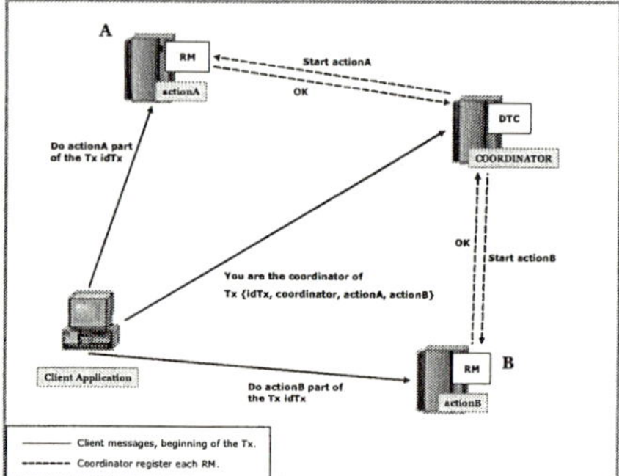

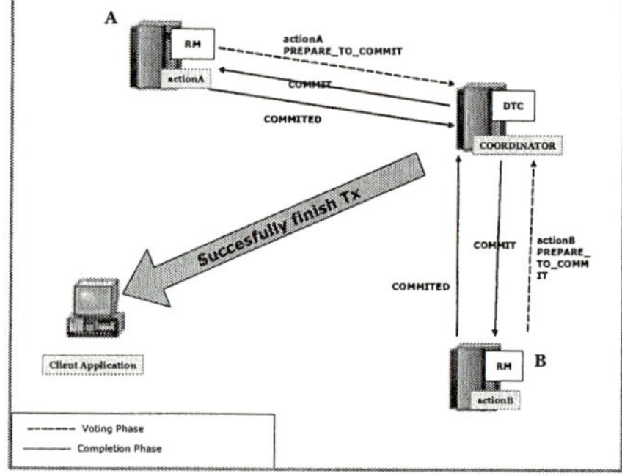

Fig. 2. Scenario for Concurrent Distributed Transactions

Figure 2 shows how the transaction starts, first the client applications sends to each RM the respective action and the idTx associated with the Tx. At same time, the client, sends to the coordinator all the information of the Tx, with this information the coordinator register in the DTC all the actions of the Tx and sends a message to each RM for start their associated actions.

Once each and every action is ready to be committed, it will send a message to the DTC notifying the current state (PREPARE_TO_COMMIT), as it were mentioned previously this is the *Voting Phase*. After all RMs have sent their status message the DTC will decide about finish the transaction sending a COMMIT message to each RM. The stage meanwhile the DTC is sending the COMMIT message is the *Commit Phase* of the 2-PC protocol.

The case of ROLLBACK in this scenario will given by the overcoming of period (TIMEOUT) or if any RM sends a fail message (UNABLE_TO_COMMIT) to the DTC, in this situation a ROLLBACK message will be send to each RM.

It is easy to see a small modification over the previous explication of 2-PC, in this scenario due to the concurrency, are the RM's who sends the message PREPARE_TO_COMMIT not the DTC.

3.2 Scenario 2: Conversational Distributed Transactions

This scenario describes a DT composed by the same two previous actions A and B which can't working currently, due to this, these actions are sent sequentially because of its dependencies, the client recovers the partial results and it invokes other actions with these results. Due to this, is the client who decides when the transaction must finalize, and when the 2-PC protocol must begin their commit phases (*Voting* and *Completion*).

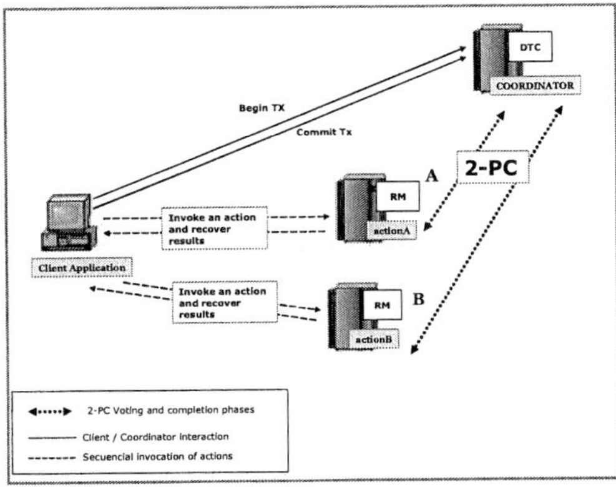

Fig. 3. Scenario for Conversational Distributed Transactions

The *Figure 3* shows the process in this scenario. First the client application invoke the coordinator to start the transaction Tx, after, call sequentially the necessary actions and to finish the transaction invoke the coordinator with a COMMIT message. At this moment, the DTC manages the 2-PC protocol to cross the *Voting* and *Completion* phases like in scenario 1. In case of failure of some action, the DTC will send a ROLLBACK message to the rest of them.

In addition, if some RM does not respond in a time limit then a ROLLBACK message will be sent by the DTC to each RM. By this way, we avoided blockades of long duration. The problem of blockades and concurrency access will be explained in more detail in section 4.1.

4 Our Approach to Grid Environments

Taking advantage of the OGSI characteristics, we propose a DCP-Grid to be introduced in a Grid Environment. As our first approach at the *Universidad Politécnica de Madrid* (UPM), we have decided to introduce a new interface which we have called *ITransactionSupport.* This new interface provides to Grid Services with operations that provides "rollback" and "commit" functionality. Every Grid Service which wants take part of a transactional execution will extend this interface, this solution could be achived thanks to the *PortType Extension*, provided by OGSI.

On the other hand, we have defined the element *Distributed Transaction Coordinator (DTC)*, key concept for DCP-Grid. Our solution proposes to the DTC like a *Grid Service*, we assigned to this service the name of *TXCoordinationService*, with this design we took advantage of all features of *Grid Service* to develop the DTC.

In order to separate the interface of the logical behavior our proposal introduce a new component, the engine that manages all the process around the commit protocol. This component is the *TXManagementEngine.* We can look inside the coordinator in the next figure:

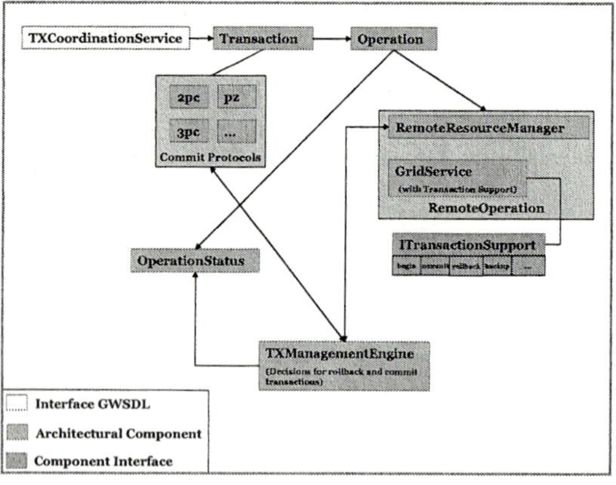

Fig. 4. DTC building blocks Architecture

The TXManegementEngine makes its decisions in function from the information updated by the *TXCoordinationService*, for example, when the TXCoordinationService receives a message PREPARE_TO_COMMIT it update the correspondent *OperationStatus.* When the *TXManagementEngine* detects that all *OperationStatus* are in PREPARE_TO_COMMIT then sends a COMMIT message to all remote resource managers.

Each remote RM can commit their operations because it implements *ITransacctionSupport.* In case of some message received by the *TXCoordinationStatus* will be a UNABLE_TO_COMMIT message then the *TXManegementEngine* will make a rollback sending the corresponding message to each remote RM.

The RM is another *Grid Service* deployed in each system that contains transactional *Grid Services* due to this the TXManagementEngine contained in the DTC can invoke the *TXResourceManagerService* for the commit or rollback interface operations and a situation of blockade will occur.

4.1 Solutions to the Situations of Blockade

A key point to be taken into account is the resolution of the interblockades; this factor could be critical in some scenarios as the following: a client application invokes a service and this execution is part of a transaction. At a certain moment, this service not finished yet because of another action associated to the same transaction is not in the commit phase. At this same moment, another client application sends the execution of a transaction in which there is the same action blocked before. If this action is not released, another execution will not be able to enter and a situation of blockade will occur.

In order to solve this situation our proposal establishes a time limits of delay to process commit once the service has been processed, this timeout will be a parameter to be established by each and every service. In future versions of DCP-Grid, we will tackle the problem deeply.

4.2 Solutions to Concurrent Access Situations

During the investigation of DCP-Grid a problem with the concurrent access appear. What happens if an RM has deployed a service that allows parallel access to different client execution, if two clients invoke the service simultaneously when the respective DTCs want to process COMMIT or ROLLBACK? How the RM knows which is the message (COMMIT / ROLLBACK) associated to each execution?

To solve this situation our proposal generates a unique ID that will be propagated in corresponding messages to the respective DTCs and RMs. With this ID the two components (DTCs and RMs) will be able to associate the message received with the corresponding operation. This session ID identifies uniquely each transaction. Another problem appear with this solution How can generate unique IDs across different RMs in separated system?

To solve this new problem DCP-Grid will negotiate the session ID between the RMs. The internal operation is as it is described to continuation, the DTC associated to the transaction generate the transaction ID based on the system address, after the DTC sends the start message to each RM. If some RM detects that it has an active id with same value, then this RM will send a CHANGE_ID message to the DTC, the DTC will generate a new ID only for this RM. Internally the DTC will store this information for future messages.

5 Conclusions and Future Work

In this paper we have presented our approach to implement an architecture supporting transactional Grid WS execution. This approach is based on some of the main properties of OGSI specification [7]. As ongoing work, currently we are developing a similar framework for concurrent distributed transactions on grid environments which will be presented in the Workshop on KDMG'05. So many future research lines has been opened for DCP-Grid but maybe the most interesting would be the building of an environment to support transactions on distributed and heterogeneous databases based on the concepts and ideas that we have presented in this paper.

References

[1] Douglas K. Barry, Web Services and Service-Oriented Architecture: The Savvy Manager's Guide, Morgan Kaufmann Publishers 2003

[2] Mark Endrei, Jenny Ang, Ali Arsanjani, Sook Chua, Philippe Comte, Pal Krogdahl, Min Luo, Tony Newling, "Patterns: Service Oriented Architecture", IBM RedBook SG24-6303-00

[3] "Web Services Main Page at W3C", http://www.w3.org/2002/ws/, Worl Wide Web Consortium, (consultado en dic/2004)

[4] Berntein, New Comer "Principles of Transaction Processing", Editorial: Kaufman, 1997

[5] I. Foster, C. Kesselman. The Physiology of the Grid: An Open Grid Services Arquitecture for Distributed System Integration . 2002. http://www.globus.org/research/papers/ogsa.pdf

[6] Miguel L. Bote-Lorenzo, Yannis A. Dimitriadis, Eduardo Gómez-Sánchez "Grid Characteristics and Uses: A Grid Definition", LNCS 2970, 291-298

[7] S. Tuecke, K. Czajkowski, I. Foster. Grid Service Specification. Technical Report. Jun 2003. www-unix.globus.org/toolkit/draft-ggf-ogsi-gridservice-33_2003-06-27.pdf

[8] "Globus Toolkit Project", The Globus Alliance, http://www.globus.org (consulted 2004/12)

[9] I. C. Jeong, Y. C. Lew. DCE "Distributed Computing Environment" based DTP "Distributed Transaction Processing" Information Networking (ICOIN-12) Jan. 1998

[10] F. Cabrera et al., "Web Services Coordination (WS-Coordination)" Aug. 2002, ww.ibm.com/developerworks/library/ws-coor/

[11] F. Cabrera et al., "Web Services Transaction (WS-Transaction)" Aug. 2002, www.ibm.com/developerworks/library/ws-transpec/.

[12] Feilong Tang, Minglu Li, Jian Cao, Qianni Deng, Coordination Business Transaction for Grid Service. LNCS3032 pag. 108-114 (Related Work Section)

[13] OASIS BTP Committee Specification 1.0, 3 June 2002, Business Transaction Protocol, http://www.choreology.com/downloads/2002-06-03.BTP.Committee.spec.1.0.pdf

[14] X/Open Specification, 1988, 1989, February 1992, Commands and Utilities, Issue 3 (ISBN: 1-872630-36-7, C211); this specification was formerly X/Open Portability Guide, Volume 1, January 1989 XSI Commands and Utilities(ISBN: 0-13-685835-X, XO/XPG/89/002).

Dynamic and Fine-Grained Authentication and Authorization Architecture for Grid Computing

Hyunjoon Jung, Hyuck Han, Hyungsoo Jung, and Heon Y. Yeom

School of Computer Science and Engineering, Seoul National University,
Seoul, 151-744, South Korea
{hjjung, hhyuck, jhs, yeom}@dcslab.snu.ac.kr

Abstract. The Globus Toolkit makes it very easy and comfortable for grid users to develop and deploy grid service. As for the security mechanism, however, only static authentication and coarse-grained authorization mechanism is provided in current Globus Toolkit. In this paper we address the limitations of current security mechanism in the Globus Toolkit and propose a new architecture which provides fine-grained and flexible security mechanism. To implement this without modifying existing components, we make use of the Aspect-Oriented Programming technique.

1 Introduction

With the advent of grid computing, Globus Toolkit[8] have been playing an important role for making grid systems. Thanks to the support of various components in the Globus Toolkit, many grid developers have been able to develop their own grid systems very conveniently. When the grid service administrator - in this paper, we would like to divide grid users into grid service administrator and general grid users for explaining more precisely - make their own grid systems, the grid computing environment usually consists of many service nodes which are managed correlatively. Namely, many nodes can be separated logically into specific domains which can be under the same policy, so called virtual organization [11]. The grid service administrator should manage various kinds of services according to the policy of each domain.

Although it is possible in current Globus Toolkit, there are some limitations. If a new user wants to use some grid service, the grid service administrator must change the policy of the service which might be currently running. However, to change the security policy of a specific grid service, it is required to stop the grid services related to the altered security policy. After the policy change, the grid services are redeployed according to the new security policy.

Besides, current security scheme of Globus Toolkit supports only service level authorization. In other words, method level authorization is not supported. If the grid service administrator wants to configure different authorization level for each method, he should make extra grid service under different security policy. Consequently, current security configuration scheme in Globus Toolkit is static and coarse-grained, which results in extra jobs for grid service administrator.

V.S. Sunderam et al. (Eds.): ICCS 2005, LNCS 3516, pp. 179–186, 2005.

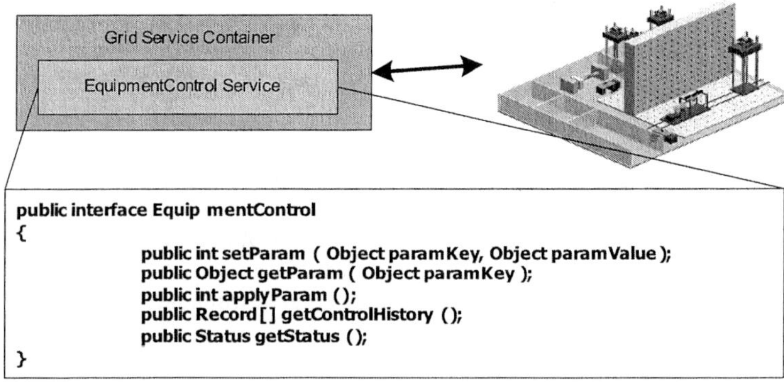

Fig. 1. Motivation Example

To mitigate these limitations, we present a dynamic authentication and authorization scheme. Our scheme also allows fine-grained method level security scheme.In designing and implementating this scheme, we have used *Aspect-Oriented Programming* [9, 10, 14] so that existing Globus Toolkit is left unmodified.

The rest of this paper is organized as follows. We discuss the motivation of this project in detail in section 2 and introduce our design goals and strategies in section 3. And then we propose new architecture in section 4. In section 5, we explain how to implement our scheme with *Aspect-Oriented Programming.* The section 6 addresses issues of grid security related to our work and future works. In the end, we conclude our paper with summary and contribution.

2 Motivation

In this section, we describe the limitations of security scheme in current Globus Toolkit. And then our focuses of this paper will be introduced.

In grid computing, there are many users who try to access some services. And there are also many kinds of grid services being executed by user's request. To manage users and services securely, grid computing have adopted security policies for each user and service. In practice, Globus Toolkit ensures the security of grid service and user with exploiting grid security infrastructure (GSI) [12].

With the advent of web-based architecture, grid service also have evolved into conforming architecture[5]. However, current Globus Toolkit supports only static management of security policy in web-based architecture. In other words, the security policy of specific service would be set before deploying. Besides, while that service is running, security policy cannot be reconfigured and applied instantly to that service. Owing to this kinds of limitations, grid service administrator should first stop the service, reconfigure specific security policy and then redeploy that service. Since this reconfiguration and redeplyment should be done for each small change, the burden on the part of the administrator could be huge.

In addition to the static management of security policy, security level in authorization can be also considered as a limitation. When the grid service administrator deploys some services with current Globus Toolkit, the level of authorization is the service instance. Figure 1 describes the motivating example. When *EquipmentControl* service is deployed, grid administrators configure security policies for each user level. If *EquipmentControl* grid service would be subject to the policy of table1, it is impossible to provide service to each user's permission using existing Globus Toolkit.

Table 1. The list of user's permission of each service method

user group	defined security policy
experiment, operator	permit all methods
other reseacher	permit getXXX() methods
guest	permit getControlHistory() method

In other words, current Globus Toolkit cannot support method-level authorization scheme. Since only service-level authorization scheme is provided, the grid service administrator ought to make each service separately according to each user's request and authorization policy. Consequently, it leads to the inefficient utilization of resources.

3 Design Goals and Strategy

As stated in the previous section, our goal in this paper is to ameliorate the current Globus Toolkit as follows:

- To provide dynamic authentication and authorization
- To support fine-grained authorization
- To implement this scheme as an extra module without modifying the Globus Toolkit.

To provide dynamic authentication and authorization means that grid service administrators would be able to reconfigure the security policy of running services without suspending them. Supporting fine-grained authorization presents enhancements of security granularity and removes the redundancy of grid services. For instance, if some grid administrator would like to change security policy of some grid services, they would be troubled like the example of the section2. It is an inefficient utilization of resources due to multiple security policies. Therefore, method level authorization with dynamicity would be a better alternative to solve this kind of problem. In order to support these two security schemes in the Globus Toolkit, it would be easy to implement if we modify security part in Globus Toolkit.

In that case, however, Globus Toolkit should be recompiled and then it causes undesirable extra needs. For these challenges of implementation, we adopt gracious programming skill that is *Aspect-Oriented Programming*. It ensures that any existing program does not need to be modified. By exploiting this way, we would present more compact and modularized architecture. Aspect-oriented programming would be explained down to the minutest details in the implementation part.

4 Proposed Architecture

To achieve our goals, we propose the architecture composed of three components in accordance with each role as shown in figure2.

4.1 FSecurity Manager

Originally, security mechanism of Globus Toolkit is mainly divided into authentication part which verifies someone and authorization part which checks the right of a user in grid service. And the policy of the authentication and authorization can be set only ahead of the deployment of grid service. Hence, to achieve the dynamicity of security scheme, *FSecurity Manager* intercepts the request of authentication and authorization in current Globus Toolkit when the request is invoked. And then, that request is processed in *FSecurity Manager* with Security Configuration Data.

For instance, we assume that the grid administrator expects 10 users would use his grid service A before deployment. But, after some time, 5 new users would want to use the grid service A. In that case, the grid administrator must stop the grid service A to modify new security policy for the 5 new users. And after applying new security policy, grid service will be executed again correctly. However, this kind of process would be changed with *FSecurity Manager* as follows. Above all, the most troubled part which do not check the security policy at running time is changed with intercepting request of *FSecurity Manager*.

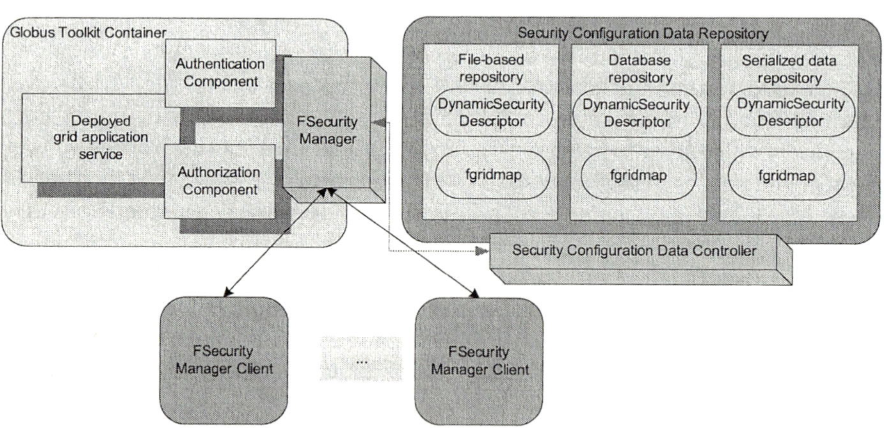

Fig. 2. Proposed architecture

And then *FSecurity Manager* reads the security policy at that time, applied that scheme to authentication and authorization instantly. Therefore, *FSecurity Manager* supports the reconfigurable security processing.

Moreover, *FSecurity Manager* provides a fine-grained authorization processing with adding the element which describes the relation between user and grid service method. In current Globus Toolkit, before the service deployment, grid service administrator used to configure the gridmap file which describes simply those who can use specific grid service. However, that mechanism is changed into supporting the dynamic management and the fine-grained description with *FSecurity Manager*. In the former case, grid service administrator can reconfigure gridmap file even though service is executing. There is no need to stop the service and redeploy it later. And as for the latter, the description which some user have rights to use some grid service method is added into new type gridmap file, which is called fgridmap.

4.2 Security Configuration Data Controller

Security Configuration Data Controller is an extra module in order to manage security policy information. Security policy information can be stored as a XML file, serialized file or database records. To support this kind of function, this controller is responsible for connecting, storing, and loading specific security policy information with a data repository.In real grid computing, grid administrator can install this part in a machine which Globus Toolkit is executed or another machine which may be dedicated to data repository.

4.3 FSecurity Manager Client

This component supports that grid service administrator controls and manages the security policy of grid service from a remote or local node. Client program can be installed in a ordinary computing node or mobile node such as PDA, mobile phone. Also the grid service administrator can manage a *FSecurity Manager* via web interface.

5 Implementation

As we mentioned above, our implementation without the change of Globus Tookits is done with a refined programming skill which is Aspect-Oriented Programming. In this part, we would describe the reason why we have adopt Aspect-Oriented Programming and how it was applied. And then the operation flow of the proposed architecture would be explained with a structural perspective.

5.1 Aspect-Oriented Programming and Our Architecture

Aspect-Oriented Programming (AOP) complements *Object-Oriented programming* by allowing the developer to dynamically modify the static OO model to create a system that can grow to meet new requirements. The reason why we use *Aspect-Oriented Programming* technology is to separate the core part of program from the extra part of program because there is some kinds of different properties between the former and the latter. For instance, many of us have developed simple web applications that use servlets as the entry point, where a servlet accepts the values of a HTML form, binds them to an object, passes them into the application to be processed, and then returns a response to the user.The core part of the servlet may be very simple, with only the minimum amount of code required to fulfill the use-case being modeled. The code, however, often inflates to three to four times its original size by the time extra requirements such as exception handling, security, and logging have been implemented.

Therefore, AOP allows us to dynamically modify our static model to include the code required to satisfy the extra requirements without having to modify the original static model - in fact, we don't even need to have the original code. Better still, we can often keep this additional code in a single location rather than having to scatter it across the existing model, as we would have to if we were using OO on its own.

The Globus Toolkit 3 is also implemented with Object-Oriented programming. Therefore, the supplement of Globus Toolkit should conform to the Object Oriented model. However, Globus Toolkit is so complicated and complex. And the security part of Globus Toolkit are involved with almost area. On the ground of that, this case would be modeled as the cross-cutting concerns which are named in aspect-oriented programming.

Whenever grid users try to access some grid services, security component must be activated and then perform the process of authentication and authorization. Hence, the invocation point of security component in Globus Toolkit should be defined as point-cut in aspect-oriented programming which is the term given to the point of execution in the application at which cross-cutting concern needs to be applied. And then to perform some predefined action, code will be implemented as an advice which is the additional code that someone wants to apply to the existing model. As a result, aspect-oriented programming is exploited to instantiate proposed architecture using point-cut and advice which is named in that field. Besides, it is possible to implement the proposed architecture without any modification of Globus Toolkit thanks to this programming technique.

5.2 Operational Flow

In this section, we describe the implementation details of the proposed architecture. The implementation was done using AspectJ[13].

Let us first look at the left side of figure3. It shows how the authentication is processed in our architecture. Originally, initSecurityDescriptor and getSecurityDescriptor are performed in current Globus Toolkit. However, there's only one time check which means getSecurityDescriptor does not check SecurityDescriptor at running time. Hence, we defined two point-cuts at the invocation time of initSecurityDescriptor and getSecutityDescriptor. In this way, any invocation of SecurityDescriptor can be substituted DynamicSecurityConfig which always

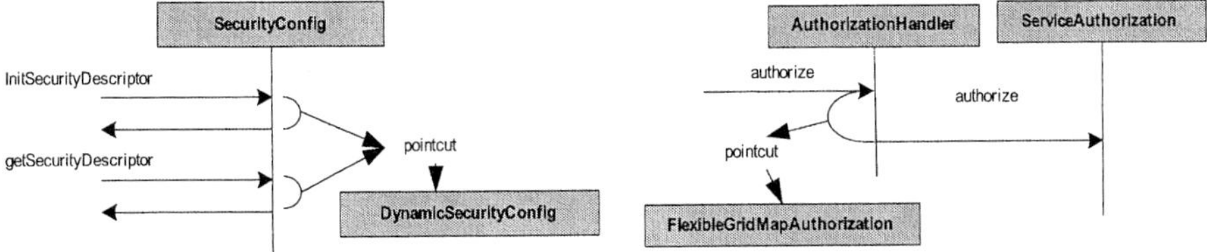

Fig. 3. The operational flow of authentication part(left) and authorization part(right)

checks the SecurityDescriptor and applies up-to-the-minute information to the specific grid service at running time.

In authorization, FlexibleGridMap Authorization (fgridmap authorization) functions as the core. Whenever AuthorizationHandler is invocated, the pointcut leads the execution flow to fgridmap authorization. And then operation would be continued with the authorization which conform the defined policy between the user and grid service method. The style of fgridmap is illustrated in the right side of figure3. This authorization part can be performed using the latest authorization information for the specific grid service no matter when the update has been made. As was done in authentication part, the processing of authorization is performed at running time with the help of aspect-oriented programming.

To explain the whole flow of operation with authentication and authorization, let us assume that there are grid user 1, grid user 2, and grid user 3. Also, there is a grid service S1 which can provide methodA, methodB, and methodC. If user 1 who is set to use methodA and methodB, the request for methodC by user 1 shuld be denied by the authorization part. Even if new grid user 4 who is not set on DynamicSecurityDescriptor at deployment time, grid user 4 may be added according to the decision of grid service administrator. And then the request by grid user 4 in compliance with the DynamicSecurityDescriptor and the FlexibleGridmap may be accepted or denied.

To sum up, with the proposed architecture, grid service administrator can configure the authorization scheme up to method level. Moreover, grid service administrator can reconfigure the authentication and authorization policy at service running time. It enhances the limitation of Globus Toolkit by using aspect-oriented programming.

6 Conclusion

In this paper, we present a dynamic and fine-grained security architecture for grid service administrators who use Globus Toolkit. With this component, the extra needs of many grid administrators who frequently have to reconfigure security policies and redeploy grid services can be alleviated very simply.

To make current Globus Toolkit more efficient, we have studied the limitation of security part of current Globus Toolkit and then proposed some refinements of this research. The proposed architecture has been implemented using Aspect-

Oriented Programming technique without modyfing Globus Toolkit. Therefore, in real grid computing, grid service administrator can use this component very conveniently with existing Globus Toolkit.

References

1. Von Welchr, Ian Foster, Carl Kesselman, Olle Mulmo, Laura Pearlman, Steven Tuecke, Jarek Gawor, Sam Meder and Frank Siebenlist "X.509 Proxy Certificates for Dynamic Delegation", In Annual PKI R&D workshop, April,2004
2. Von Welch, Frank Siebenlis, Ian Foster, John Bresnaban, Karl Czajkowski, Jarek Gawor, Carl Kesselman, Sam Meder, Laura Pearlman and Steven Tuecke , "Security for Grid Services", In IEEE Symposium on High Performance and Distributed Computing ,June,2003.
3. Ian Foster, Carl Kesselman, Cene Tsudik and Steven Tuecke "A Security Architecture for Computational Grids", In 5th ACM Conference on Computer and Communication Security, 1998.
4. "Security in a Web Services World: A Proposed Architecture and Roadmap", A Joint White Paper from IBM Corporation and Microsoft Corporation, April 2002.
5. Tuecke S. and Czajkowski K. and Foster I. and Frey J and Graham S. and Kesselman C. and Maguire T. and Sandholm T. and Snelling D. and Vanderbilt P "Open Grid Services Infrastructure(OGSI) Version 1.0", Global Grid Forum, June 2003.
6. Nataraj Nagaratnam, Philippe Janson, John Dayka, Anthony Nadalin, Frank Siebenlist, Von Welch, Ian Foster and Steve Tuecke "The Security Architecture for Open Grid Services", July 2002.
7. Foster, I., Kesselman, C., Nick, J. and Tuecke, S. "The Physiology of the Grid: An Open Grid Services Architecture for Distributed Systems Integration " , Globus Project, 2002. http://www.globus.org/research/papers/ogsa.pdf.
8. Jarek Gawor, Sam Meder, Frank Siebenlist and Von Welch, "GT3 Grid Security Infrastructure Overview", In ,Globus Project 2003.
9. the AspectJ Team, "The AspectJTM Programming Guide", Copyright (c) 1998-2001 Xerox Corporation, 2002-2003 Palo Alto Research Center, Incorporated. All rights reserved.
10. Markus Voelter, "Aspectj-Oriented Programming in Java" in the January 2000 issue of the Java Report
11. Foster, I. and Kesselman, C. and Tuecke, S., "The anatomy of the grid : Enabling scalable virtual organizations", Intl. J. Supercomputer Applications, 2001
12. I. Foster, C. Kesselman, G. Tsudik, S. Tuecke., "Security Architecture for Computational Grids", 5th ACM Conference on Computer and Communications Security Conference, pp. 83-92, 1998.
13. Palo Alto Research Center, "The AspectJ(TM) Programming Guide", http://eclipse.org/aspectj/
14. Tzilla Elrad ,Robert E. Filman and Atef Bader "Aspect-oriented programming: Introduction", In Communications of the ACM , 2001

GridSec: Trusted Grid Computing with Security Binding and Self-defense Against Network Worms and DDoS Attacks*

Kai Hwang, Yu-Kwong Kwok, Shanshan Song, Min Cai Yu Chen, Ying Chen, Runfang Zhou, and Xiaosong Lou

Internet and Grid Computing Laboratory, University of Southern California, 3740 McClintock Ave., EEB 212, Los Angeles, CA 90089-2562, USA
{kaihwang, yukwong, shanshas, mincai, rzhou, cheny, chen2, xlou}@usc.edu
http://GridSec.usc.edu

Abstract. The USC GridSec project develops distributed security infrastructure and self-defense capabilities to secure wide-area networked resource sites participating in a Grid application. We report new developments in trust modeling, security-binding methodology, and defense architecture against intrusions, worms, and flooding attacks. We propose a novel architectural design of Grid security infrastructure, security binding for enhanced Grid efficiency, distributed collaborative IDS and alert correlation, DHT-based overlay networks for worm containment, and pushback of DDoS attacks. Specifically, we present a new pushback scheme for tracking attack-transit routers and for cutting malicious flows carrying DDoS attacks. We discuss challenging research issues to achieve secure Grid computing effectively in an open Internet environment.

1 Introduction

Over the last few years, a new breed of network worms like the *CodeRed, Nimda, SQL Slammer*, and *love-bug* have launched widespread attacks on the Whitehouse, CNN, Hotmail, Yahoo, Amazon, and eBay, etc. These incidents created worm epidemic [8] by which many Internet routers and user machines were pulled down in a short time period. These attacks had caused billions of dollars loss in business, government, and services. Open resource sites in information or computational Grids could well be the next wave of targets. Now more than ever, we need to provide a secure Grid computing environment over the omni-present Internet [6].

---

* The paper was presented in the *International Workshop on Grid Computing Security and Resource Management* (GSRM'05) in conjunction with the *International Conference on Computational Science* (ICCS 2005), Emory University, Atlanta, May 22-25, 2005. The research reported here was fully supported by an NSF ITR Grant 0325409. Corresponding author: Kai Hwang, USC Internet and Grid Computing Lab, EEB 212, Los Angeles, CA 90089. E-mail: kaihwang@usc.edu, Tel.: (213) 740-4470. Y.-K. Kwok participated in this project when he was a visiting associate professor at USC on sabbatical leave from HKU.

V.S. Sunderam et al. (Eds.): ICCS 2005, LNCS 3516, pp. 187 – 195, 2005.

Network-centric computing systems manifest as Grids, Intranets, clusters, P2P systems, etc. Malicious intrusions to these systems may destroy valuable hosts, network, and storage resources. Network anomalies may appear in many Internet connections for *telnet, http, ftp, smtp, Email,* and *authentication* services. These anomalies cause even more damages. Internet anomalies found in routers, gateways, and distributed hosts may hinder the acceptance of Grids, clusters, and public-resource networks [10]. Our work is meant to remove this barrier from Grid insecurity. This article reports our latest research findings in advancing security binding and building self-defense systems tailored for protecting Grid resource sites.

- Architectural design of the Grid security infrastructure in Section 2
- Security binding for trusted resource allocation in Grid job scheduling [12] in Section 3.
- The CAIDS distributed IDS and alert correlation system in Section 4
- The salient features of a DHT (*distributed hash table*) overlay [1, 13] for supporting distributed worm containment [1, 8] in Section 5
- A real-time pushback scheme to combat DDoS (*Distributed Denial of Service*) attacks [2, 3, 9] in Section 6.

2 GridSec Security Infrastructure Architecture

Our GridSec security architecture is designed to be a wide-area defense system that enables high degree of trust [7] among the Grid sites in collaborative computing over the Internet. As illustrated in Fig. 1, GridSec adopts DHT-based overlay architecture

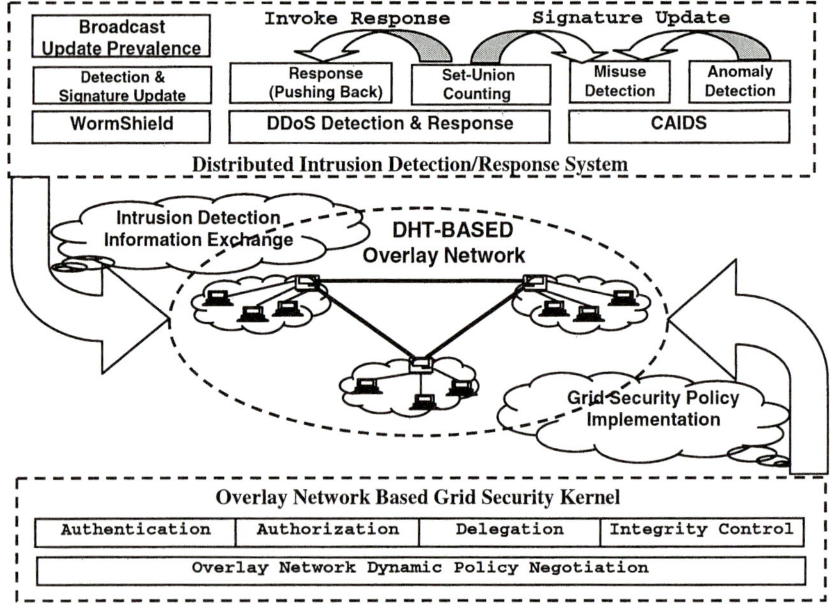

Fig. 1. GridSec infrastructure for building self-defense capabilities to protect Grid sites

as its backbone. As a virtual communication structure lay logically on top of physical networks, our overlay network maintains a robust virtual inter-networking topology. Through this topology, trusted direct application level functionalities facilitates inter-site policy negotiation and management functions such as authentication, authorization, delegation, policy exchange, malicious node control, job scheduling, resource discovery and management, etc.

The GridSec system functions as a *cooperative anomaly and intrusion detection system* (CAIDS) [6]. Intrusion information is efficiently exchanged by the overlay topology with confidentiality and integrity. Each local IDS is autonomous, and new algorithms can be added easily due to the high scalability of the overlay. Each node may work as agent for others and varies security models/policies can be implemented. As shown in Fig. 1, currently available functional blocks include the WormShield [1], CAIDS [6] and DDoS pushback scheme [2]. We are currently integrating our newly developed worm and flooding defense algorithms into the GridSec *NetShield* system .

3 Security-Binding for Trusted Resource Allocation

The *reputation* of each site is an aggregation of four major attributes: *prior job execution success rate, cumulative site utilization, job turnaround time,* and *job slowdown ratio.* These are behavioral attributes accumulated from historical performance of a site [12]. The defense capability of a resource site is attributed to *intrusion detection, firewall, anti-virus/worm,* and *attack response capabilities.* Both site reputation and defense capability jointly determine the *trust index* (TI) of a resource site. In [12], we have suggested a novel fuzzy-logic approach to generating the local trust index from the above-mentioned attributes.

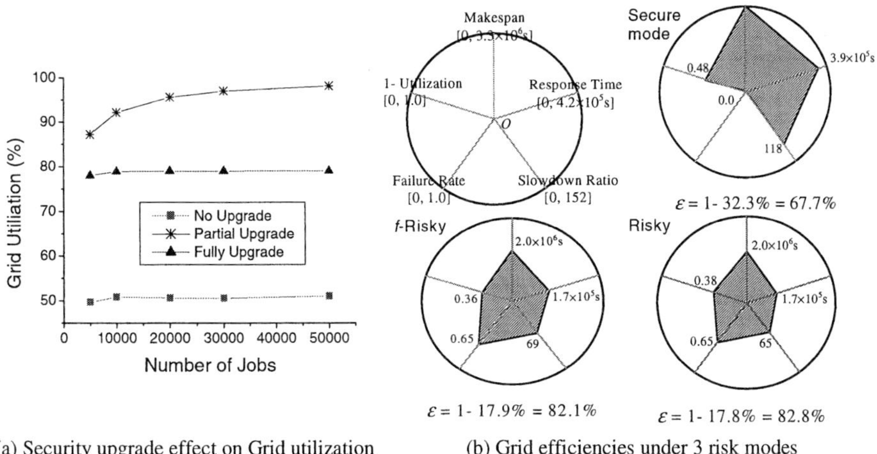

(a) Security upgrade effect on Grid utilization (b) Grid efficiencies under 3 risk modes

Fig. 2. Effects on Grid utilization and Grid efficiency by security enhancement

On the other hand, user jobs provide their *security demand* (*SD*) from resource site. A *trusted resource allocation* (TRA) scheme must satisfy a *security-assurance condition*: $TI \geq SD$ during mapping jobs to resource sites. We evaluate the effectiveness of a TRA scheme by considering five performance metrics: *makespan*, *job failure rate*, *site utilization*, *response time*, and *slowdown ratio*, etc. We report in Fig. 2(a) the effects of trust integration towards total Grid utilization. *No upgrade* policy corresponds to resource allocation without trust integration. While *full upgrade* and *partial upgrade* entail a full scale and a resource-constrained trust integration, respectively. Higher Grid utilization is observed after the integration process.

This security-binding scheme is effective in mapping large-scale workloads in NAS and PSA benchmark experiments [12]. New performance metrics are developed to assess the effects of trust integration and secure allocation of trusted resources to enormous Grid jobs. Our secure binding scheme scales well with both job number and Grid size. Trusted job outsourcing makes it possible to use open Grid resources with confidence and calculated risks. We consider three risk conditions in remote job executions, namely, *conservative* mode, *f-risky* and *risky* mode representing various levels of risk the jobs may experience.

The cumulative Grid performance of these three modes is shown in Fig.2(b) by three 5-D Kiviat diagrams under 3 risk conditions. The five dimensions correspond to five performance metrics. The smaller is the shaded polygon at the center of the Kiviat diagram, the better is the *Grid efficiency,* defined by $\varepsilon = (1 - A_{shaded} / A_{circle})$. This implies that more efficient Grid has *shorter makespan* and *response time* and *lower slowdown, failure rate,* and *under-utilization rate* (1- *utilization rate*). Our NAS simulation results shows that it is more resilient for the global job scheduler to tolerate job delays introduced by calculated risky conditions, instead of resorting to job preemption, replication, or unrealistic risk-free demand.

4 Distributed Intrusion Detection/Alert Correlation

The CAIDS we built [6] can be deployed at various Grid sites to form a *distributed IDS* (DIDS) supported by alert correlation sensors. These sensors are scattered around the computing Grid. They generate a large amount of low-level alerts. These alerts are transmitted to the alert correlation modules to generate high-level intrusion reports,

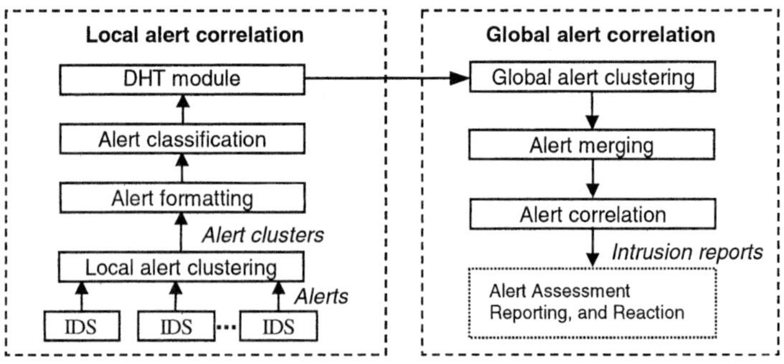

Fig. 3. Alert operations performed in local Grid sites and correlated globally

which can provide a broader detection coverage and lower false alarm rate than the localized alerts generated by single IDS. Figure 3 shows the alert operations performed by various functional modules locally and globally [4].

Similar to a major earthquake, one large attack has a series of after attacks. The global alert correlation is to detect the relationship among the attacks. We need a high-level view of attacks. The system detects the intention and behavior of attackers. An early detection report can be generated to minimize the damages. We have tested the CAIDS system at USC with an Internet trace of 23.35 millions of traffic packets, intermixed with 200 attacks from the Lincoln Lab IDS dataset.

In Fig.4, we plot the ROC curves corresponding to 4 attack classes. The detection rate grows quickly to its peak value within a small increase of false alarm rate. To achieve a total detection rate above 75% of DoS attacks, we have to tolerate 5% or more false alarms. The R2L (*root-to-local*) attacks have the second best performance. The port-scanning Probe attacks perform about the same as R2L attacks. The U2R (*user-to-root*) attacks have the lowest detection rate of 25% at 10 % false alarms, due to the stealthy nature of those attacks. When the false alarm rate exceeds 5%, all attacks reaches their saturated performance.

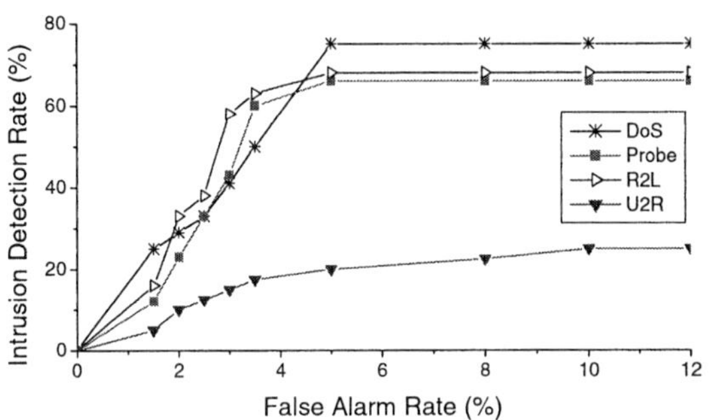

Fig. 4. Intrusion detection rate versus false alarm rate in using the CAIDS (Cooperative Anomaly and Intrusion Detection System) developed at USC [6]

5 DHT-Based Overlay for Worm Containment

We build a scalable DHT overlay to cover a large number of autonomous domains in edge networks. Our *WormShield* system [1] consists of a set of geographically distributed monitors located in multiple administrative domains (Fig.5). They are self-organize into a structured P2P overlay ring network based on the Chord algorithm [13]. Each monitor is deployed on the DMZ (*Demilitarized Zone*) of the edge network and analyzes all packets passing through it.

In *WormShield,* each monitor *i* remembers the set of source addresses *S(i,j)* and the set of destination addresses *D(i,j)* for each substring *j*. When the global prevalence of substring *j* is greater than the prevalence threshold T_p, each monitor will send their locally maintained source and destination addresses to the root monitor root *j*. The

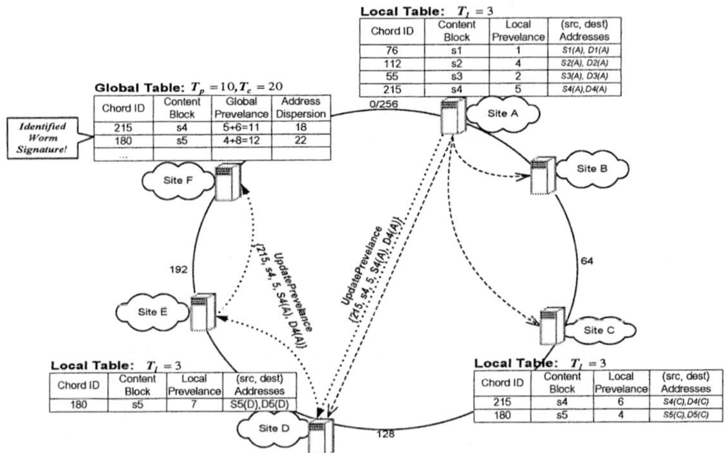

Fig. 5. An example of WormShield system with six sites for worm control

root monitor then compute the global address dispersion for the substring *j*. If *C(j)* is greater then an address dispersion threshold T_c, the substring *j* will be identified as a potential worm signature [11]. The root monitor will construct a multicast tree on the overlay network and disseminate the signature to all monitors participated.

For each monitor *i*, we use Rabin footprint algorithm to compute the substrings for each packet payload. Then it computes the local prevalence *L(i,j)* for each substring *j*. After a predefined interval *t* or *L(i,j)* is greater than a local prevalence threshold T_l, monitor *i* will update the global prevalence *P(j)* for substring *j* that tracks all prevalence seen in the network with *WormShield* monitors deployed. A selected monitor is assigned to maintain the global prevalence for a substring *j* using consistent hashing as in Chord [13].

6 Tracking and Pushback DDoS Attacks

We tackle two issues towards effective DDoS defense: (1) accurately identifying the ingress routers (i.e., the edge routers of the domain to be protected) that unknowingly participate in the forwarding of malicious DDoS attack flows; and (2) identifying the malicious flows and incisively cutting such flows at these *Attack-Transit Routers* (ATRs) [2].

Real-Time Traffic Matrix Tracking: We propose a low-complexity traffic monitoring technique that is based on measuring both the *packet-level* and *flow-level* traffic matrices among routers in real-time. Our proposed technique is based on accumulating very lightweight statistics for packets or flows at each router within the domain. When huge volumes of packets or flows arrive at a particular last-hop router as depicted in Fig.6, this victim router identifies the ATRs with very high accuracy using the lightweight statistics exchanged among the routers. It only requires *O(log log N)* storage capacity for *N* packets or flows on each router [5, 9], compared with the *O(N)* complexity in using a Bloom filter [2].

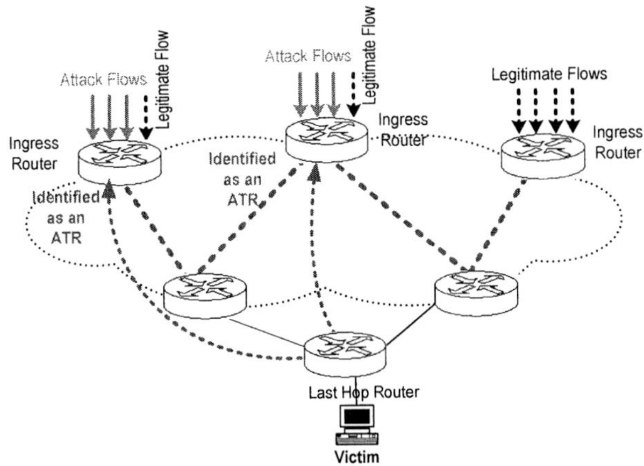

Fig. 6. The pushback scheme for identifying the attack-transit routers and blocking the malicious flows with spoofed source IP addresses

Packet- and Flow-Level Counting: Define S_i as the set of packets that enter the domain from an ingress router r_i, and D_j as the set of packets that leave the domain from an egress router r_j. To compute packet-level traffic matrix, we use the first 28-byte invariant bytes of a packet (20-byte IP header with 4 bytes masked out plus the first 12 bytes of payload). This will result in a very small collision rate. We compute the packet-level traffic matrix $A = \{a_{ij}\}$, where $a_{ij} = |S_i \cap D_j| = |S_i| + |D_j| - |S_i \cup D_j|$ [9]. Here, we can easily compute $|S_i|$ and $|D_j|$ at each router.

For the term $|S_i \cup D_j|$, we use two probabilistic counting techniques, namely the *stochastic averaging algorithm* and *distributed max-merge algorithm* [5]. which require only O (*log log N*) storage space for N packets in the set. For flow-level traffic matrix, we use the 5-tuple {*source IP, source port, destination IP, destination port, protocol*} as the identifier for each packet. The flow-level traffic matrix $B = \{b_{ij}\}$, where $b_{ij} = |S_i^F \cap D_j^F|$ is computed in a similar fashion. The counting complexity is $O(N)$, where N is the number of packets in a set.

Even with the ATR identification issue efficiently solved by our novel traffic tracking technique, the second issue is also a daunting challenge because IP addresses are commonly spoofed, that making correct identification of malicious flows very difficult. We propose a new MAFIC algorithm to support the adaptive packet dropping policy at the identified ATRs [3]. Through probing, MAFIC would drop malicious attack packets with very high accuracy while minimizes the loss on legitimate traffic flows [3].

Our NS-2 simulation indicates that the traffic tracking and flow cutting by dropping attacking packets are up to 90% accurate, as revealed in Fig.7(a). The scheme reduces the loss of legitimate flows to less than 5%. Figure 7(b) shows the false positive rates are quite robust and scalable under increasing domain sizes. The dup-ACK based probing is quite accurate in identifying the attack flows as the identification errors are mostly below 1%. The false alarm rates are less than 0.05% for a TCP flow rates from 35% to 95% of the total traffic volume.

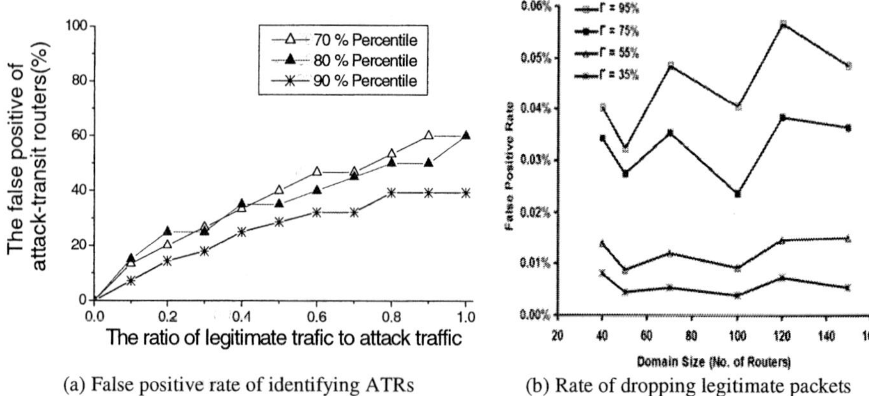

(a) False positive rate of identifying ATRs (b) Rate of dropping legitimate packets

Fig. 7. The false positive rate of identifying ATRs and of dropping legitimate packets at the identified ATRs with different percentages (70%, 80%, and 90%) of traffic flows actually passing through the identified ATRs (*attack-transit routers*)

7 Conclusions

The NSF/ITR-supported GridSec project at its second year has made encouraging progress in trust management, security-driven job scheduling, trusted resource allocation, distributed IDS, collaborative alert correlation, worm containment, and distributed DDoS pushback. We offer a scalable security overlay architecture, experimental validation of distributed IDS design, and new schemes to capture network worms and pushback DDoS attacks. The GridSec system offers early warning of Internet worm spreading and launching effective pushback operations to protect Grid resources.

In the research front, we suggest several meaningful challenges for further work. The major threats come from software vulnerability and naïve users. Today's Windows, Unix and Linux variants are by no means immune from worm attacks, let alone free from DDoS flood attacks. Outbreaks must be dealt with immune response swiftly. The major research challenge lies still in the containment area. In particular, we need automated signature generation and fast suppression of malicious flows.

Internet outbreak detection and monitory are other big challenges. The reaction time, containment strategies, deployment scenarios are all yet to be worked out. We have identified the requirements of *robustness, resilience, cooperativeness, responsiveness, efficiency*, and *scalability.* The DHT-base security overlays offer a viable approach towards a fast cybersecurity solution. Of course, further advances in operating-system security, active networks, and trust management are also important.

References

[1] M. Cai, K. Hwang, Y.-K. Kwok, Y. Chen, and S. Song," Fast Conatinment of Internet Worms for Epidemic Defense using Distributed-Hashing Overlays", *IEEE Security and Privacy,* submitted July 2004 and revised March 6, 2005, to appear Nov/Dec. 2005.

[2] M. Cai, Y. K.-Kwok and K. Hwang, "Inferring Network Anomalies from Mices: A Low-Complexity Traffic Monitoring Approach", in preparation for submission to *ACM SIGCOMM Workshop on Mining Network Data*, 2005
[3] Y. Chen, Y.-K. Kwok, and K. Hwang, "MAFIC: Adaptive Packet Dropping for Cutting Malicious Flows to Pushback DDoS Attacks," *Proc. Int'l Workshop on Security in Distributed Systems* (SDCS-2005), in conjunction with ICDCS 2005, Columbus, Ohio, USA, June 2005.
[4] F. Cuppens and A. Miege, "Alert Correlation in a Cooperative Intrusion Detection Framework," *IEEE Symposium on Security and Privacy*, 2002, pp.187-200.
[5] M. Durand and P. Flajolet, "LogLog Counting of Large Cardinalities," *Proc. European Symp. on Algorithms*, 2003.
[6] K. Hwang, Y. Chen, and H. Liu, "Protecting Network-Centric Computing System from Intrusive and Anomalous Attacks," *Proc. IEEE Workshop on Security in Systems and Networks* (SSN'05), in conjunction with *IPDPS 2005*, April 8, 2005.
[7] S. Kamvar, M. Schlosser, and H. Garcia-Molina, "The EigenTrust Algorithm for Reputation Management in P2P Networks," *Proc. of WWW,* 2003.
[8] H. A. Kim and B. Karp, "Autograph: Toward Automated Distributed Worm Signature Detection," *Proc. USENIX Security Symposium*, 2004.
[9] M. Kodialam, T. V. Lakshman, and W. C. Lau, "High-speed Traffic Measurement and Analysis Methodologies and Protocols," Bell Labs Technical Memo, Aug. 2004.
[10] N. Nagaratnam, P. Janson, J. Dayka, A. Nadalin, F. Siebenlist, V. Welch, S. Tuecke, and I. Foster, "Security Architecture for Open Grid Services," http://www.ggf.org/ogsa-sec-wg
[11] S. Singh, C. Estan, G. Varghese and S. Savage, "Automated Worm Fingerprinting," *Proc. of the USENIX Symp.on Operating System Design and Implementation*, S.F., Dec. 2004.
[12] S. Song, K. Hwang, and Y.-K. Kwok, "Security Binding for Trusted Job Outsourcing in Open Computational Grids," *IEEE Trans. Parallel and Dist. Systems*, revised Dec. 2004.
[13] I. Stoica, R. Morris, D. Karger, M. F. Kaashoek, H. Balakrishnan, "Chord: A P2P Lookup Protocol for Internet Applications," *Proc. ACM SIGCOMM*, 2001.

Design and Implementation of DAG-Based Co-scheduling of RPC in the Grid

JiHyun Choi[1], DongWoo Lee[2], R.S. Ramakrishna[3], Michael Thomas[4], and Harvey Newman[5]

[1,2,3]Department of Information and Communication, Gwangju Institute of Science and Technology, Republic of Korea
{jhchoi80, leepro, rsr}@gist.ac.kr
[1,4,5] California Institute of Technology, Pasadena, CA91125, USA
{jchoi, thomas, newman}@hep.caltech.edu

Abstract. Effective scheduling in the Grid consisting of heterogeneous and distributed resources is imperative in order to counter unacceptably large overheads of the Grid. We proposed the grid middleware (pyBubble) supporting the DAG based co-scheduling for improving the performance of the RPC mechanism. DAG based co-scheduling reduces redundant transmission of input and output data from execution of the sequence of related client requests, thereby decongesting the network. We demonstrate the efficiency of DAG based co-scheduled RPC in experiments compared with the overhead of the traditional RPC mechanism.

1 Introduction

The Grid is the Internet-connected computing and data management infrastructure. Computing and data resources are geographically dispersed in different administrative domains with different policies for security and resource usage. The computing resources are highly heterogeneous, ranging from single PCs and workstations, cluster of workstations, to large supercomputers[5]. With the technology of the Grid we can construct large-scale and scientific applications over these distributed and heterogeneous resources. There are many critical issues that need to be efficiently resolved to support the ever-increasing number of applications that can benefit from the Grid Computing infrastructure.

GridRPC[8] is a programming model based on client-server remote procedure call(RPC), with features added to allow easy programming and maintenance of code for scientific applications on the Grid. Application programmers write parallelized client programs using simple and intuitive GridRPC APIs that hide most of the complexities involving Grid programming. As a result, programmers lacking experience in parallel programming, let alone the Grid, can still construct Grid Applications effortlessly[3]. Most applications in Grid Computing generally have large input data sets and intricate data dependency. Moreover, data transfer in large distributed systems can add an unacceptable amount of overhead. The

V.S. Sunderam et al. (Eds.): ICCS 2005, LNCS 3516, pp. 196–204, 2005.

goal of this research is to devise simple and effective strategies for dealing with these issues.

2 Related Works

Grid programming involves the Interface and the Run-time system. WS-Resource Framework defines Web service convention to enable the discovery of, introspection on, and interaction with stateful resources in standard and interoperability ways. Many enhanced interfaces for grid computing such as Grid MPI and Grid RPC have been developed. With regard to Message Passing Interface(MPI), several systems, notably, MPICH-G2, MPI_Connect[1], PACX-MPI, MagPIe[2] and Stampi[3] have been studied for connecting different MPI implementations. Conventional parallel programming is implemented by these MPI-like systems in tightly coupled high performance parallel computing networks or network of workstations. MPI needs middleware to interact with the global grid.

The other programming paradigm of importance is RPC (Remote Procedure Call)[2]. It is used for calling remote functions through a simple programming interface. To a user, RPC presents a transparent interface to a remote function. RPC is the most promising candidate for Grid programming interface due to its simplicity and user friendliness. With this interface, the grid middleware can invoke a remote grid resource via RPC calls. Ninf-g[9], NetSolve-G(Grid Solve)[4], OmniRPC[7] and so forth provide Grid RPC interface to their systems.

pyBubble[6] provides DAG(Directed Acyclic Graph)-based co-scheduling as a mechanism to minimize repeated interactions among resources. Unlike Ninf and NetSolve, our system can store the applications task dependencies in DAGs. These DAGs allow pyBubble to schedule communications with a functionality-based scheduling algorithm. Our DAG-based Grid Runtime with RPC programming interface exhibits substantial performance improvements in terms of the execution time of related RPC requests. This efficiency increases in larger distributed systems that suffer from large data transfer overheads.

3 Motivation

In the traditional RPC paradigm, individual RPCs are processed independently. The actual scheduling of the remote invocation is unilaterally determined by the remote service receiving the RPC request. Clients, however, may have to meet scheduling constraints. If a remote call entails submitting a batch job or a sequence of related tasks, the client may at least want to know what the queue

[1] MPI_Connect, http://icl.cs.utk.edu/projects/mpi-connect

[2] MagPIe, http://www.cs.vu.nl/albatros

[3] Stampi, http://ssp.koma.jaeri.go.jp/en/index.html

[4] GridSolve, http://www.nsf-middleware.org/documentation/NMI-R4/0/gridsolve

length is, or have some notion of the expected time of completion. Clients may also need to co-schedule multiple RPCs if the input and output parameters of these multiple RPCs are interrelated. Co-scheduling will help avoid the transmission of redundant data, resulting in an overall shortened response time and reduced network congestion. The total processing time can also be shortened by executing modules concurrently whenever possible. However, GridRPC systems do not support any mechanism to co-schedule GridRPC that targets heterogeneous and loosely-coupled systems over wide-area networks. This work is an attempt to fill this gap.

Figure1[1] illustrates two kinds of data flow involving multiple RPCs. The client invokes the services of three servers for processing a job consisting of three tasks. There is ample opportunity to reduce redundant network traffic between clients and servers when we execute this series of related RPCs. In the left diagram, servers always return the result to the client after the execution of their tasks. But, in the co-scheduled system, servers don't need to return the intermediate results to the client, but instead send the intermediate results to other servers directly as their inputs. In the right diagram, server3 executes the final task, and sends only the final result back to the client.

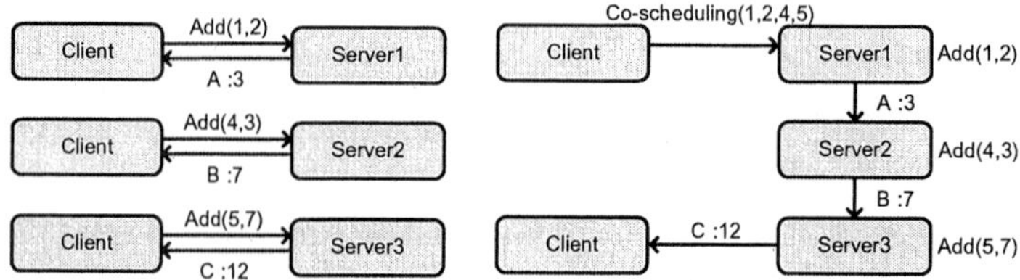

Fig. 1. Data flow compared with co-scheduling

Grid Computing usually involves the processing of very large amounts of data using distributed resources over wide area. When we take the overhead of the data transfer time into consideration, decongestion of network traffic by co-scheduling in the Grid can substantially contribute toward reducing the overall response time of multiple RPCs.

4 Framework: The pyBubble

Our system, pyBubble, is a web service-based Grid middleware for parallel and distributed computation. This system intends to be a GridRPC system that uses XML-based RPC for the interactions between the client application and remote computing resources. pyBubble is written in the Python programming language to support portability across multiple platforms. pyBubble uses SOAP as the transport encoding and supports DAG based co-scheduling and a restart protocol for improving the dependability of the system.

4.1 SOAP-based Implementation

pyBubble[6] uses SOAP for performing remote procedure calls. SOAP provides an envelope that encapsulates XML data for transfer through the Web infrastructure with a convention for Remote Procedure Call (RPCs) and a serialization mechanism based on XML Schema data type. We note that other RPC mechanisms for Grids are possible, including XML-RPC[5] which also uses XML over HTTP. While XML provides tremendous flexibility, it currently has poor support for binary data due to a significant encoding cost[4]. Therefore, we compress the xml documents before they are transferred in order to reduce the overhead caused by the substantial encoding overhead of XML.

4.2 pyBubble Architecture and General Scenario

Figure 2 shows each of these components of pyBubble and illustrates the relationship and data flow between the components. pyBubble consists of the client, the resource broker, and resource servers. We can get metadata of available servers by intergrating monitoring services, but we can use specific host configuration information collected from servers in this work.

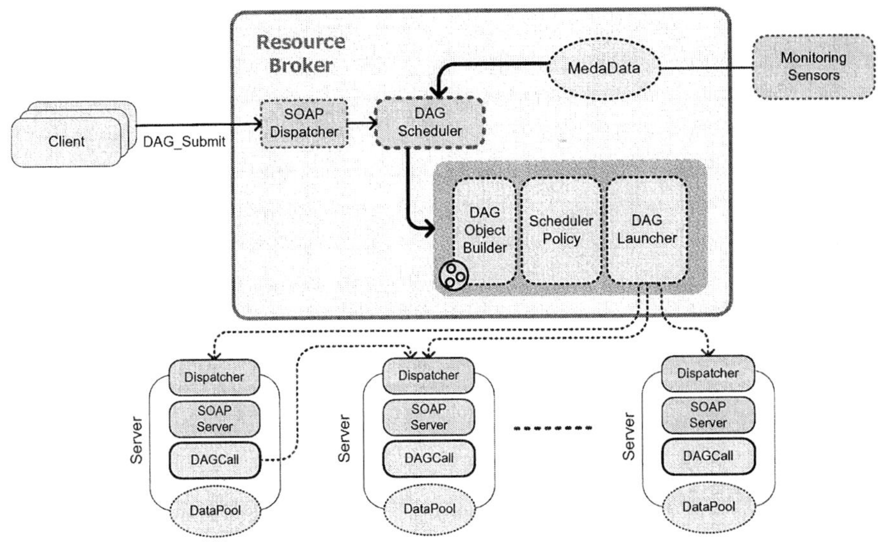

Fig. 2. The architecture of pyBubble consisting of client, broker and servers

We can assume that the user has several tasks targeted to distributed servers. First, the user sends the tasks to the pyBubble broker with the programming interface for DAG-based co-scheduling which constructs the DAG from client's tasks, subject to constraints. The client specifies the input data, the result labels and the function names with the programming interface. The tasks to be sent to

[5] XML-RPC, http://www.xml-rpc.com

the broker should be a sequence of interrelated requests. The broker ranalyzes the relationships of input and output data within tasks, and then checks the precedence or dependency between tasks. It then constructs the DAG. The DAG has the information about intra-task relationships.

Tasks are assigned to the right server based on the scheduling algorithm because the broker has the information about available servers. To execute the tasks, first, the broker submits root tasks in the DAG. The root tasks are assigned to multiple servers to be executed in parallel. They call the respective child tasks in sequence according to the DAG. They also send the result, the scheduling information for the child task, and the DAG to the child tasks. After all the tasks are completed, the server executing the final task sends its result to the broker.

5 DAG-Based Co-scheduling of RPC

5.1 The DAG Model and Application Encoding Using DAG

The DAG is a generic model of a parallel program consisting of a set of interdependent processes (nodes). A node in the DAG represents a task. The graph also has directed edges representing a partial order among the tasks. The partial order creates a precedence-constrained directed acyclic graph[20]. For scheduling purposes, tasks are described by a DAG. A DAG incorporates all the precedence relationships among tasks and information about the assigned task in order to make scheduling decisions. After scheduling, the DAG includes information about which task is assigned to which server.

5.2 DAG-Based Co-scheduling Algorithm with Function Availability

After constructing the DAG, the broker schedules the DAG tasks based on function availability since the broker can get information of configuration files specified on each server and knows which resource can execute which function. Figure 3 shows the pseudocode of DAG-based co-scheduling by functional availability. The function definition procedure assists in finding the available resources offering the requested function and this returns the candidate set of the resources. The variable `DAG` is the collection of the requests described in DAG, and `Rmax` is the number of resources. `ResourcePool` includes the information about available resource to which the broker assigns some task. Each task in the DAG finds the cadidate resources set offering the function called by the method `FindResHavingFunc`. In the candidate resource set, one resource is selected for assignment to the task. After finding the root tasks, they are executed with the `Execute` method.

```
Scheduling Definition:
DAG: collection of the request described in DAG
Rmax: the number of resources
ResourcePool: {R1,R2,   Ri, Rmax}
Ri: { F1, F2,  , Fn}

Begin
    for each T in DAG
    Do
        CandidateResourceSet :=
FindResHavingFunc(T,ResourcePool)
        AssignedRes := RandomSelect(CandidateResourceSet)
        MappingTaskResource(AssignedRes, T)
    Done
    RootTask := FindRootTask(Dag)
    Execute(RootTask)  // Root Tasks execute in parallel
    End

Function Definition:
FindResHavingFunc(Task, ResourcePool)
    CandidateSet = {}
    For eash R in ResourcePool
    Do
        IF Task.Func R.func then
             CandidateSet := R
    Done
    Return CandidateSet
```

Fig. 3. Pseudocode of scheduling by functional availability

6 Experiment Results

The experiments compare co-scheduled RPC with conventional RPC. The performance criteria are data size, number of processors, and the CPU power. The efficiency of DAG-based co-scheduling can vary with these factors.

6.1 Application: Image Processing

Image Processing is appropriate for studying the efficiency of co-scheduling. In this experiment, images can be very large, on the order of several gigabytes in size. The execution of a series of transformations - in a specific order - on an image is common in image processing. This experiment shows the improved efficiency of DAG-based co-scheduling. The results of experiments depend of the combination of the function set and the DAG construction.

Comparison with increasing image size. Figure 4 shows that the execution time of conventional RPC increases sharply when the data size increases due to network traffic overhead. But the increase in the execution time of co-scheduled RPC is not as drastic. Parallel processing of concurrent tasks and the resulting reduction in redundant network traffic contribute to this improved performance.

Comparison on the basics of CPU Performance. This experiment uses CPU power as the performance criterion: the high performance group and the low

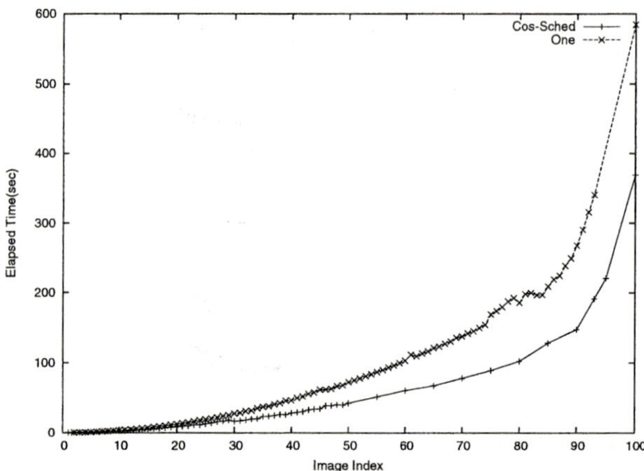

Fig. 4. Execution time of Co-scheduled RPC and Conventional RPC

performance group. Figure 5 compares co-scheduled RPC with conventional RPC in both the server groups. the performance of the low performance group is not improved significantly with co-scheduled RPC because that has large overhead for co-scheduling tasks and is not less afftected by reducing network overhead.

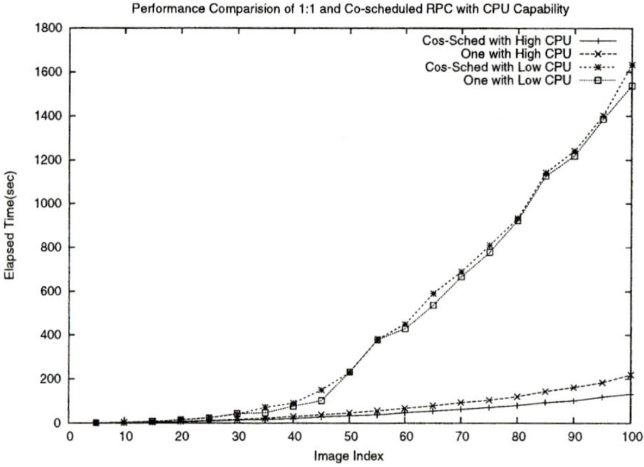

Fig. 5. Comparison of conventional and co-scheduled RPC in low performance CPU group and high performance RPC group

Comparison with the different number of processors. In figure 6, the single processor records the worst performance as expected. Two and three processors achieve the best performance. The DAG has three root tasks and two child tasks and hence two or three are just the right numbers. Two processors return a performance below that of three processors. This is understandable in light of the fact that there are three (concurrent) root tasks in the DAG. Four and five processors also exhibit good performance until the number of images reaches 80. Thereafter, the performance degrades due to heavy network traffic induced by large sized images.

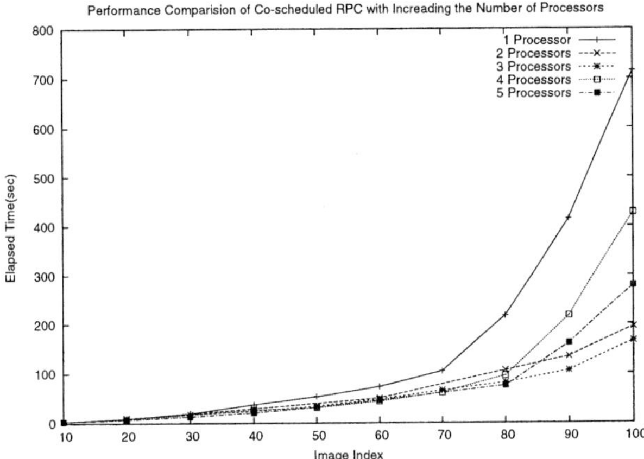

Fig. 6. Performance comparison of co-scheduled RPC with increasing of processor

7 Conclusion

In this paper we have proposed a DAG based co-scheduling technique as a tool for efficient RPC programming in the Grid. Co-scheduling intends to avoid redundant transmission of inputs and outputs in order to reduce network traffic congestion. The system supports a portable programming interface for DAG based co-scheduling as a user facility. DAG-based applications are scheduled using functionality- and input output data location-based co-scheduling algorithm. Image processing applications were used for test purposes and have proved that DAG based co-scheduling exhibits considerable performance improvements over conventional RPC.

References

1. Dorian C. Arnold, Dieter Bachmann, and Jack Dongarra. Request sequencing: Optimizing communication for the grid. In *Euro-Par*, pages 1213–1222, 2000.
2. Gregory L. Field Lola Gunter Thomas Klejna Shankar Lakshman Alexia Prendergast Mark C. Reynolds David Gunter, Steven Burnett and Marcia E. Roland. *Client/Server Programming With Rpc and Dce.* Que, 1995.
3. GlobalGridForum. http://www.ggf.org.
4. M. Govindaraju, A. Slomenski, V. Choppella, R. Bramley, and D. Gannon. Requirements for and evaluation of rmi protocols for on the performance of remote method invocation for scientific computing. In *Proc. of the IEEE/ACM International Conference on Supercomputing (SC 2000)*, November 2000.
5. Ian Foster, Carl Kesselman, editor. *The GRID2: Blueprint for New Computing InfrastructureMeyers.* Morgan Kaufmann, 2003.
6. DongWoo Lee and JiHyun Choi. http://pybubble.sourceforge.net, 2004.
7. Mitsuhisa Sato, Taisuke Boku, and Daisuke Takahashi. Omnirpc: a grid rpc ystem for parallel programming in cluster and grid environment. In *CCGRID*, pages 206–, 2003.

8. Keith Seymour, Hidemoto Nakada, Satoshi Matsuoka, Jack Dongarra, Craig A. Lee, and Henri Casanova. Overview of gridrpc: A remote procedure call api for grid computing. In *GRID*, pages 274–278, 2002.
9. Y. Tanaka, Hidemoto Nakada, Satoshi Sekiguchi, Toyotaro Suzumura, and Satoshi Matsuoka. Ninf-g: A reference implementation of rpc-based programming middleware for grid computing. *J. Grid Comput.*, 1(1):41–51, 2003.

Performance Analysis of Interconnection Networks for Multi-cluster Systems

Bahman Javadi[1], J.H. Abawajy[2], and Mohammad K. Akbari[1]

[1] Department of Computer Eng. and Information Technology,
Amirkabir University of Technology, Hafez Ave.,Tehran, Iran
{javadi, akbari}@ce.aut.ac.ir
[2] School of Information Technology, Deakin University,
Geelong, VIC 3217, Australia
jemal@deakin.edu.au

Abstract. With the current popularity of cluster computing systems, it is increasingly important to understand the capabilities and potential performance of various interconnection networks. In this paper, we propose an analytical model for studying the capabilities and potential performance of interconnection networks for multi-cluster systems. The model takes into account stochastic quantities as well as network heterogeneity in bandwidth and latency in each cluster. Also, blocking and non-blocking network architecture model is proposed and are used in performance analysis of the system. The model is validated by constructing a set of simulators to simulate different types of clusters, and by comparing the modeled results with the simulated ones.

1 Introduction

Advances in computational and communication technologies has made it economically feasible to conglomerate multiple clusters of heterogeneous networked resources leading to the development of large-scale distributed systems known as multi-cluster systems. Performance analysis and evaluation of multi-cluster systems in general and interconnection networks in particular is needed for understanding system behavior and the analysis of innovative proposals. However, performance analysis in such systems has proven to be a challenging task that requires the innovative performance analysis tools and methods to keep up with the rapid evolution and ever increasing complexity of such systems.

This paper addresses the network interconnects performance analysis problem for multi-cluster computing systems. The motivation for considering this problem is that multi-cluster systems are gaining more importance in practice and a wide variety of parallel applications are being hosted on such systems as well [1]. Also, many recent cluster builders are concerned with two primary factors: cost and performance [17].

While cost is easily determined and compared, performance is more difficult to assess particularly for users who may be new to cluster computing. Moreover, with the current popularity of cluster computing, it is increasingly important to understand the capabilities and potential performance of various network interconnects for cluster

V.S. Sunderam et al. (Eds.): ICCS 2005, LNCS 3516, pp. 205–212, 2005.

computing systems [18]. In addition, performance analysis in such systems has proven to be a challenging task that requires the innovative performance analysis tools and methods to keep up with the rapid evolution and ever increasing complexity of such systems [16].

In this paper, we present a new methodology that is based on Jackson network technique to analytically evaluate the performance of network interconnects for multi-cluster systems. The model takes into account stochastic quantities as well as network heterogeneity in bandwidth and latency in each cluster. Bandwidth is the amount of data that can be transmitted over the interconnect hardware in a fixed period of time, while latency is the time to prepare and transmit data from a source node to a destination node. Also, blocking and non-blocking network architecture model is proposed and are used in performance analysis of the system. The message latency is used as the primary performance metric. The model is validated by constructing a set of simulators to simulate different types of clusters, and by comparing the modeled results with the simulated ones.

The rest of the paper is organized as follows. In Section 2, related work is discussed. In Section 3, we describe the proposed analytical model. We present the model validation experiments, in Section 4. Finally, Section 5 summarizes our findings and concludes the paper.

2 Related Work

Generally, multi-cluster systems can be classified into *Super-Cluster* and *Cluster-of-Cluster*. A good example of Super-Cluster systems is DAS-2 [5], which is characterized by large number of homogenous processors and heterogeneity in communication networks. In contrast, Cluster-of-Clusters are constructed by interconnecting multiple single cluster systems thus heterogeneity may be observed in communication networks as well as processors. The LLNL multi-cluster system which is built in by interconnecting of four single clusters, MCR, ALC, Thunder, and PVC [6] is an example of cluster-of-cluster system. In this paper, we will focus our discursion on the Super-Cluster system with homogenous processors and heterogeneous communication networks.

Currently, there are three possible ways to address this problem – simulation, prediction and analytical modeling. The limitations of simulation-based solutions are that it is highly time-consuming and expensive. Similarly, techniques based on predictions from measurements on existing clusters would be impractical. An alternative to simulation and prediction approaches is an analytical model, which is the focus of this paper. An accurate analytical model can provide quick performance estimates and will be a valuable design tool. However, there is very little research addressing analytical model for interconnects in multi-cluster systems. The few results that exist are based on homogenous cluster systems and the evaluations are confined to a single cluster [2, 3, 4, 19]. With all probability, multiple cluster systems would be configured from heterogeneous components, rendering exiting optimization solutions unusable in heterogeneous multi-cluster environment. In contrast, our work focuses on heterogeneous multi-cluster computing systems. To this end, we present a generic model to analytically evaluate the performance of multi-cluster systems. We believe that our work is the first to deal with heterogeneous multi-cluster environments.

3 Proposed Analytical Model

The architectural model of the system assumed in this paper similar to [12]. It is made up of C clusters, each cluster i is composed of N_i processors of type T_i, i=1,...,C. Also, each cluster has two communication networks, an Intra-Communication Network ($ICN1_i$), which is used for the purpose of message passing between processors, and an intEr-Communication Network ($ECN1_i$), which is used to transmit messages between clusters, management and also for the expansion of system. Note that, ECN can be accessed directly by the processors of a cluster without going through the ICN.

The proposed model is based on the following assumptions that are widely used in similar study [3, 4, 12, 14]:

1. Each processor generates packets independently which follows a Poisson process with a mean rate of λ and inter-arrival times are exponentially distributed.
2. The arrival process at a given communication network is approximated by an independent Poisson process. This approximation has often been invoked to determine the arrival process in store-and-forward networks [13]. In this paper we apply the store-and-forward network, e.g., Ethernet-based networks. Therefore, the rate of process arrival at a communication network can be calculated using Jackson's queuing networks formula [7].
3. Each processor granted the network as a packet transmission.
4. The destination of each request would be any node in the system with uniform distribution.
5. The processors which are source of request must be waiting until they get service and they cannot generate any other request in wait state.
6. The number of processors in all clusters are equal ($N_1=N_2=\ldots=N_C=N_0$) with homogenous type of ($T_1=T_2=\ldots T_C=T_0$).
7. Message length is fixed and equal to M bytes.

A packet is never lost in the network. Also, the terms "request" and "packet" are used interchangeably throughout this paper.

3.1 Queuing Network Model

Based on the characteristics of the system model, each communication network can be considered as service center. The queuing network model of system is shown in Fig. 1, where the path of a packet through various queuing centers is illustrated. As is shown in the model, the processor requests will be directed to service center ICN1 and ECN1 by probability *1-P* and *P*, respectively. According to assumption 1, the request rate of a processor is λ, so the input rate of ICN1 and ECN1 which feed from that processor will be $\lambda(1-P)$ and λP, respectively. The additional inputs at these service centers, γ_{I1} and γ_{E1}, are due to the requests generated by other processors of the same cluster. The output of ICN1 is feedback to the same processor, and also ε_{I1} represents the response to other processors in the same cluster.

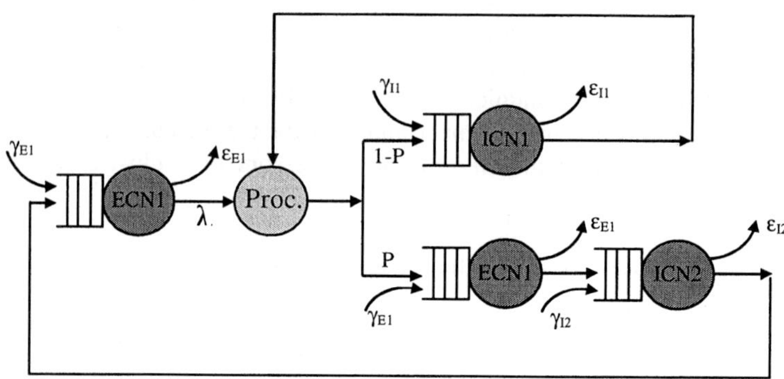

Fig. 1. Queuing Network Model of a Super-Cluster System

The external request (out of cluster) of a cluster goes through the ECN1 with probability *P* and then ICN2. In the return path, it again accesses the ECN1 to get back to the node, which initiated the request. As mentioned before, ε_{E1} and ε_{I2} are responses to the other requests except the one under consideration. So, the total requests of the processors received by service centers in the first stage can be calculated as follows:

$$\lambda_{I1} = (1-P)\lambda + \gamma_{I1} = (1-P)\lambda + (N_0-1)(1-P)\lambda \\ = N_0\,(1-P)\lambda \tag{1}$$

$$\lambda_{E1(1)} = P\lambda + \gamma_{E1} = P\lambda + (N_0-1)P\lambda \\ = N_0\,P\lambda \tag{2}$$

where $\lambda_{E1(1)}$ is the input rate of ECN1, the one which is feed by the processor.

In the second stage, the input request rate of ICN2 in forward path and ECN1in feedback path can be computed by following equations:

$$\lambda_{I2} = \lambda_{E1(1)} + \gamma_{I2} = N_0\,P\lambda + (C-1)\lambda_{E1(1)} \\ = N_0\,P\lambda + (C-1)\,N_0\,P\lambda = C\,N_0\,P\lambda \tag{3}$$

$$\lambda_{E1(2)} = \lambda_{I2}/C = N_0\,P\lambda \tag{4}$$

where $\lambda_{E1(2)}$ is the input rate of ECN1 from feedback path. According to equations (2) and (4), the input rate of ECN1 is:

$$\lambda_{E1} = \lambda_{E1(1)} + \lambda_{E1(2)} = 2N_0\,P\lambda \tag{5}$$

The average number of waiting processors in each service center can be computed through queue length of each center. So, the average of total waiting processors in the system will be:

$$L = C\,(2.L_{E1} + L_{I1}) + L_{I2} \tag{6}$$

which L is denoting the queue length of each service center. As mentioned in the assumption 5, the waiting processors would not be able to generate new requests, so the effective request rate of the processor would be less than λ. Applying the method described in [8] to find the effective request rate of a processor, it is directly dependent to the ratio of number of active processors to total number of processors. Therefore, L and λ are computed iteratively based on following equation, until no considerable change is observed between two consecutive steps:

$$\lambda_{eff} = \frac{N-L}{N} \times \lambda \tag{7}$$

As it can be seen in the previous equations, the probability P has been used as the probability of outgoing request within a cluster. According to assumption 4, this parameter is computed base on structure of system [20] by the following equation:

$$P = \frac{(C-1) \times N_0}{(C \times N_0) - 1} \tag{8}$$

In this paper, message latency is selected as a primary performance metric. However, most of the other performance metrics for the queuing network model of a multi-cluster system are related to the message latency with simple equations [12]. To model the mean message latency, we consider effective parameters as follows. In such systems, the mean network latency, that is the time to cross the network, is the most important part of the message latency. Other parameters such as protocol latency can be negligible.

Since the system under study is symmetric, averaging the network latencies seen by message generated by only one node for all other nodes gives the mean message latency in the network. Let S be the source node and D denotes a destination node such that $D \in A\text{-}\{S\}$ where A is the set of all nodes in the network. The network latency, T_C, seen by the message crossing from node S to node D consist of two parts: one is the delay due to the physical message transmission time, T_W, and the other is due to the blocking time in the network, T_B. Therefore, T_C can be written as:

$$T_C = T_W + T_B \tag{9}$$

These parameters are strongly depended on the characteristics of the communication network which is used in the system. Of this, we take into account two different networks in our model as following.

3.2 Blocking and Non-blocking Network Model

For non-blocking architecture, we use the *Multi-Stage Fat-Tree* topology, which is used in some cluster systems such as Thunder [9]. For modeling blocking interconnect architecture, a *Linear Array* of switches is used. Due to space limitation, we have not included the details of blocking and non-blocking analysis here. Interested reader can refer to [20].

4 Performance Evaluation

In order to validate the technique and justify the approximations, the model was simulated using the OMNeT++ [15]. Requests are generated randomly by each processor with an exponential distribution of inter-arrival time with a mean of $1/\lambda$ where is λ is fixed to 0.25 msg/sec in all experiments. The destination node is determined by using a uniform random number generator. Each packet is time-stamped after its generation. The request completion time is checked in to compute the message latency in a "*sink*" module. For each simulation experiment, statistics were gathered for a total number of 10,000 messages. In our study, we used two well-known network technologies, Gigabit Ethernet (GE) and Fast Ethernet (FE), which are widely used in cluster systems. We also used the same value for the latency and bandwidth of each network as reported by [10]. Two different communication network scenarios for network heterogeneity were simulated. However, due to space limitations, a subset of the results is presented here.

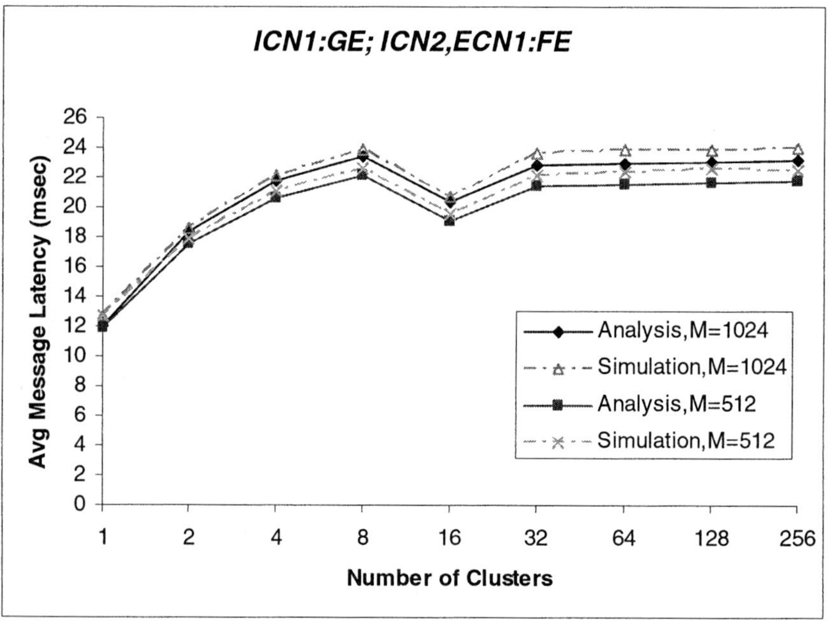

Fig. 2. Average Message Latency vs. Number of Clusters for Non-blocking Networks

Fig. 2 shows the average message latency in a multi-cluster system with $N=256$ nodes and non-blocking communication network against those provided by the simulator for the message size of 1024 and 512 bytes. The horizontal axis in the figures represents the number of clusters in the system. To have a better performance analysis of the system, we used the blocking communication networks, with the same parameters. As we expected, the average message latency in this system is much larger than in the system with non-blocking networks. These results reveal that average message latency in blocking network with uniform traffic pattern is 1.4 to 3.1 times larger than non-blocking network.

5 Conclusion and Future Directions

A performance model is an essential tool for behavior prediction of a system. It is used to analyze intricate details of the system and various design optimization issues. One such model based on queuing networks is presented in this study to predict the message latency of multi-cluster systems. Two different networks, blocking and non-blocking, were used in our modeling of the system. The analysis captures the effect of communication network architecture on the system performance. The model is validated by constructing a set of simulators to simulate different types of clusters, and by comparing the modeled results with the simulated ones. The future works focus on improving the analytical model to tack into account more effective parameters, modeling of communication networks with technology heterogeneity and propose a similar model to another class of multi-cluster systems, Cluster-of-Clusters.

References

1. J. H. Abawajy and S. P. Dandamudi, "Parallel Job Scheduling on Multi-Cluster Computing Systems," In P*roceedings of the IEEE international Conference on Cluster Computing (CLUSTER'03)*, Dec. 1-4, 2003, Hong Kong.
2. X. Du, X. Zhang, Z. Zhu, "Memory Hierarchy Consideration for Cost-Effective Cluster Computing," *IEEE Transaction on Computers*, Vol. 49, No.5, pp. 915-933, Sept. 2000.
3. B. Javadi, S. Khorsandi, and M. K. Akbari, "Study of Cluster-based Parallel Systems using Analytical Modeling and Simulation" *International Conference on Computer Science and its Applications (ICCSA 2004)*, May 2004, Perugia, Italy.
4. B. Javadi, S. Khorsandi, and M. K. Akbari, "Queuing Network Modeling of a Cluster-based Parallel Systems", *7th International Conference on High Performance Computing and Grids (HPC ASIA 2004)*, July 2004, Tokyo, Japan.
5. The DAS-2 Supercomputer. http://www.cs.vu.nl/das2
6. B. Boas, "Storage on the Lunatic Fringe", Lawrence Livermore National Laboratory, Panel at SC2003, Nov. 2003, Arizona, USA.
7. D. Bertsekas, R. Gallager., *Data Networks*, Prentice Hall Publishers, New Jersey, 1992.
8. H. S. Shahhoseini, M. Naderi, "Design Trade off on Shared Memory Clustered Massively Parallel Processing Systems", *The 10th International Conference on Computing and Information (ICCI '2000)*, Nov. 2000, Kuwait.
9. "Thunder Statement of Work", University of California, Lawrence Livermore National Laboratory, Sept. 2003.
10. M. Lobosco, and L. de Amorim, "Performance Evaluation of Fast Ethernet, Giganet and Myrinet on a Cluster", *Lecture Notes in Computer Science*, volume 2329, pp. 296-305, 2002.

11. J. H. Abawajy, "Taxonomy of Job Scheduling Approaches in Cluster Computing Systems," Technical Report, Deakin University, 2004.
12. B. Javadi, M. K. Akbari, J.H. Abawajy, "Performance Analysis of Multi-Cluster Systems Using Analytical Modeling", *International Conference on Modeling, Simulation and Applied Optimization*, Sharjah, United Arab Emirates, Feb. 2005.
13. L. Kleinrock, Queuing System: Computer Applications, Part 2, John Wily Publisher, New York, 1975.
14. H. Sarbazi-Azad, A. Khonsari, M. Ould-Khaoua, "Analysis of k-ary n-cubes with Dimension-order Routing", *Journal of Future Generation Computer Systems*, pp. 493-502, 2003.
15. Nicky van Foreest. Simulation Queuing Networks with OMNet++, in Tutorial of OMNnet++ Simulator, Department of Telecommunications, Budapest University of Technology and Economics, Apr. 2002.
16. J. H. Abawajy, "Dynamic Parallel Job Scheduling in Multi-cluster Computing Systems," *4th International Conference on Computational Science*, Kraków, Poland, pp. 27-34, 2004.
17. Chee Shin Yeo, Rajkumar Buyya, Hossein Pourreza, Rasit Eskicioglu, Peter Graham, Frank Sommers, "Cluster Computing: High-Performance, High-Availability, and High-Throughput Processing on a Network of Computers", Handbook of Innovative Computing, Albert Zomaya (editor), Springer Verlag, 2005.
18. H. Chen, P. Wyckoff, and K. Moor, "Cost/Performance Evaluation of Gigabit Ethernet and Myrinet as Cluster Interconnects," *Proc. 2000 Conference on Network and Application Performance (OPNETWORK 2000)*, Washington, USA, Aug. 2000.
19. J. Hsieh, T. Leng, V. Mashayekhi, and R. Rooholamini, "Architectural and Performance Evaluation of GigaNet and Myrinet Interconnects on Clusters of Small-Scale SMP Servers," *Proc. 2000 ACM/IEEE conference on Supercomputing (SC2000)*, Dallas, USA, Nov. 2000.
20. Bahman Javadi, J. H. Abawajy, Mohammad K. Akbari, "Performance Analysis of Interconnection Networks for Multi-Cluster Systems," Technical paper, School of Information Technology, Deakin University, Geelong, VIC 3217, Australia, 2005.

Autonomic Job Scheduling Policy for Grid Computing

J.H. Abawajy

School of Information technology,
Deakin University, Geelong, VIC, 3217, Australia

Abstract. Autonomic middleware services will play an important role in the management of resources and distributed workloads in emerging distributed computing environments. In this paper, we address the problem of autonomic grid resource scheduling and propose a scheduling infrastructure that is capable of self-management in the face of dynamic behavior inherent to this kind of systems.

1 Introduction

Grid computing is a network of geographically distributed heterogeneous and dynamic resources spanning multiple administrative domains [5]. Grids can potentially furnish large computational and storage resources to solve large-scale problems. However, the need to integrate many independent and heterogeneous subsystems into a well-organized virtual distributed systems introduces new levels of complexity making the underlying Grid environment inherently large, complex, heterogeneous and dynamic. Also it makes the systems highly susceptible to a variety of failures. Some of these failures include node failure, interconnection network failure, scheduling middleware failure, and application failure. Due to the inherent complexity and vulnerabilities, achieving large-scale distributed computing in a seamless manner on Grid computing systems introduces not only the problem of efficient utilization and satisfactory response time but also the problem of fault-tolerance [1].

One way to address these problems is to make Grid middleware technologies to embrace the concept of self-configuring systems that are able to act autonomously and adapt to changes in application or user needs. There is a synergy towards designing and building computing systems that are capable of running themselves, adjusting to varying circumstances, and preparing their resources to handle most efficiently the workloads we put upon them [7]. This new frontier of designing and building next generation computing systems has become known as autonomic computing [7]. The overall goal of autonomic computing is to make systems anticipate needs and allow users to concentrate on what they want to accomplish. The question is then how to design and develop autonomic grid resource management infrastructure that is capable of self-management (i.e., self-control, self-healing, self-configuration, self-optimization, and self-protection) in the face of dynamic behavior inherent to this kind of systems?

V.S. Sunderam et al. (Eds.): ICCS 2005, LNCS 3516, pp. 213–220, 2005.

Within this broad problem area, we address the problem of autonmous Grid resource scheduling in the presence of a variety of failures. By autonomous Grid resource scheduling we mean a schediling infrastructure that is capable of acting autonomously and adapt to changes in application or resource failures. We propose an autonomic scheduling infrastructure that is capable of proactively detect and rectify potential faults as applications are executing.

The rest of the paper is organized as follows. In Section 2, related work is presented. This section also establishes the fact that, to a large extent, the problem considered in this paper has not been fully addressed in the literature. Section 3 presents the proposed autonomic scheduling policy. Finally, Section 4 presents the conclusion and future directions.

2 Related Work

The human body is self-healing. For example, broken bones mend, and cuts heal. The concept of developing the next generation of computing systems should be driven by the conceptual similarity between biological systems and digital computing systems [6]. Hence, the objectives of autonomic computing is to build computing systems and services that are capable of managing themselves; that can anticipate their workloads and adapt their resources to optimize their performance. [6][7][10].

Enhancing the core services of a Grid middleware technologies with autonomic capabilities so that the functions are self-managing is an important research area. There has been some work towards autonomic grid computing [8][11][6] [9]. Liu, and Parashar [11] present an environment that supports the development of self-managed autonomic components, dynamic and opportunistic composition of these components using high-level policies to realize autonomic applications, and provides runtime services for policy definition, deployment and execution. In [6], an autonomic architecture to achieve automated control and management of networked applications and their infrastructure based on XML format specification is presented.

Our main focus in this paper is on autonomic grid resource scheduling middleware. Existing approaches statically allocate or release resources to Grid applications. Moreover, although fault-tolerance is one of the desirable properties of any grid scheduling algorithm, unfortunately it has not been factored into the design of most existing scheduling strategies for Grid computing systems. Research coverage of fault tolerant scheduling is limited as the primary goal for nearly all scheduling algorithms developed so far has been high performance by exploiting as much parallelism as possible. Achieving integrated scheduling and fault-tolerance goal is a difficult proposition as the job scheduling and fault-tolerance are difficult problems to solve in their own right.

Clearly, there is a need for a fundamental change in how scheduling middleware for the next generation Grid computing are developed and managed. We believe that the ability to self-manage while effectively exploiting the variably sized pools of resources in an scalable and transparent manner must be an integral part of Grid computing scheduling middleware. We are not aware of any work that is currently concentrate on autonomic Grid resource management in general and scheduling middleware in particular. To this end, we propose a scheduling infrastructure that is capable of dynamically scheduling resources while at the same time capable of self-heal to various types of failures.

3 Autonomic Scheduling Middleware Infrastructure

In this section, we discuss the proposed autonomic scheduling policy based on a hierarchical approach shown in Figure 1. Note that hierarchical scheduling policies have been used in various platforms such as cluster computing [2] and grid computing [4].

3.1 System Architecture

The core system architecture is designed around **L**-levels of virtual hierarchy, which we refer to as a *virtual organization tree*, as shown in Figure 1. At the top of the virtual organization tree, there is a *system scheduler* while at the leaf level there is a *local scheduler (LS)* for each node. In between the *system scheduler* and the *local schedulers*, there exists a hierarchy of *middle schedulers(CS)*.

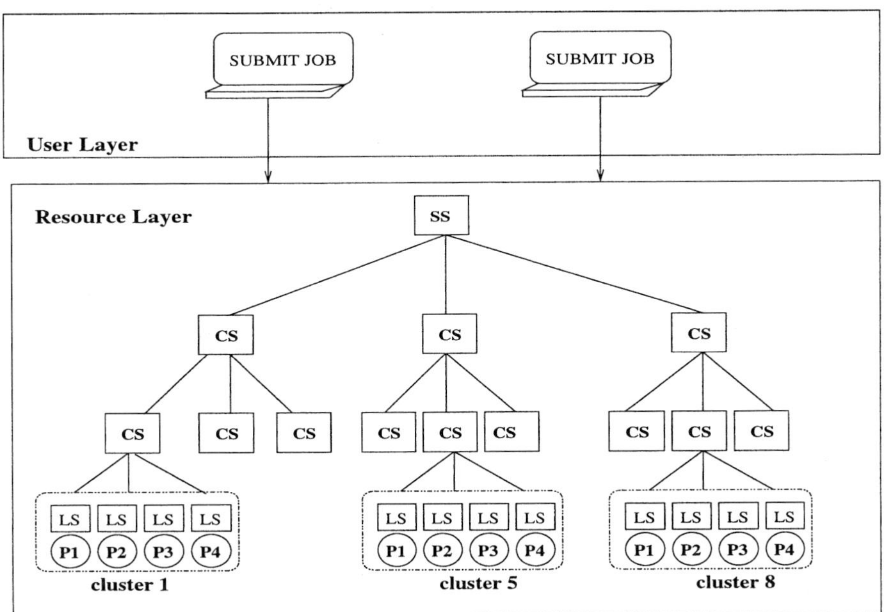

Fig. 1. An example of virtual organization tree (SS: System scheduler; CS: middle scheduler; LS: Local scheduler; and P_i: node i)

We refer to all processors reachable from a given node in the virtual organization tree as its *partition-reach.* We associate a parameter called *base load level* with each node in the virtual organization tree. For non-leaf nodes, the base load level is set to zero. For all the leaf-level nodes, the base load level is the same as the multiprogramming level (MPL) of the node. The MPL parameter of a node controls the maximum number of tasks that can concurrently execute at any given time [2].

3.2 Autonomic Scheduling Middleware

Within the hierarchical structure shown in Figure 1, an autonomic Grid resource scheduler can be viewed as a collection of autonomic schedulers that can manage their internal behaviors and their relationships and interactions with other schedulers and the system. The *Autonomic Scheduling (ASP)* policy automatically replicats jobs and tasks over several grids and processors, keep track of the number of replicas, instantiate them on-demand and delete the replicas when the primary copies of the jobs and tasks successfully complete execution. In this paper, we assume that every middle scheduler in the system is reachable from any other middle scheduler unless there is a failure in the network or the node housing the cluster scheduler. A scheme to deal with node, scheduler and link failures is discussed in [3] [4]. Also, without lose of generality, we assume that all incoming jobs are submitted to the system scheduler where they are placed in the *job wait queue* until a placement decision is made. Figure 2 shows the pseudo-code of the ASP policy. As shown in Figure 2, the ASP policy has three main components namely; *Self-scheduling*, *Job and Task Scheduling* and *Fault Management* components. These components are described in detail in the following subsection.

3.3 Self-scheduling

ASP is demand-driven where nodes in the system look for work when their load is below a given threshold. Specifically, whenever the *current load level* of a non-root node in the cluster tree falls below its *base load level*, the node sends a *Request for Computation (RFC)* message asking for $\mathbf{T_{req}}$ units of computation to its parent. After sending RFC message to its parent, the node updates its *base load level* to ensure that it can have only one outstanding RFC at any given time. When a parent with no unscheduled work receives a RFC from a child, if there is a pending RFC at the parent, the new RFC is backlogged and processed when work becomes available. Otherwise, the RFC recursively ascends the cluster tree until the RFC reaches either the *system scheduler* or a node that has unassigned jobs. In the later case, a set of jobs/tasks are transferred down the hierarchy along the path the RFC has traveled. This amount is determined dynamically during parent and child negotiations and the number of unscheduled jobs as described in the following section.

SHS algorithm

1. Self-Scheduling
   (a) IF Current load level falls below base load level

$$T_{req} = \text{base load level} - \text{current load}. \tag{1}$$

   (b) Update base load level
   ENDIF
2. Job/Task Assignment
   (a) Determine ideal number of jobs/tasks that can be scheduled

$$T_{target} = \lceil T_r \times \text{number of tasks queued} \rceil \tag{2}$$

   where T_r is the *transfer factor* and is computed as follows:

$$T_r = \frac{\text{partition-reach of the child node}}{\text{partition-reach of the parent node}} \tag{3}$$

   (b) Adjust the number of jobs that will actually be transferred down one level to the child as follows:

$$T_{target} = \begin{cases} min(T_{req}, \text{number of tasks queued}) & \text{if} T_{req} > T_{target} \\ min(T_{req}, \text{child partition-reach}) & \text{Otherwise.} \end{cases} \tag{4}$$

   where T_{req} is the is the number of computations requested by the child scheduler.
   (c) Finally, if the target size determined in Equation 2 is smaller than the T_{req} from the child, then the parent will send the lesser of T_{req} and the *number of tasks queued*. Otherwise, the set of tasks sent to the child is the lesser of T_{target} and *child partition-reach*.
   (d) Send a replica of the job to each site and update job table and backup scheduler.
3. Monitor the job execution
   FOR replica-interval DO
   (a) Prompt all remote SMs for job status
   (b) Determine the number of healthy replicas (i.e., h)
   (c) IF any replica is done THEN
      i. Tell all remote SMs to terminate their replica
      ii. Update job table
   (d) ELSEIF $(h > n)$ THEN
      i. Select last replica from set to terminate
      ii. Update job table
   (e) ELSE
      i. Pick next site from set to terminate
      ii. Inform the remote SM to execute the job
      iii. Update job table
   ENDIF
   ENDFOR

Fig. 2. Self-healing scheduling algorithm

3.4 Job and Task Placement

First, we determine an ideal number of jobs/tasks that can possible be sent to a child scheduler as shown in Equation 2. Once the target number, T_{target}, of jobs that can possibly be transferred to a child is determined, the algorithm then considers the size of the RFC from the child as a hint to adjust the number of jobs that will actually be transferred down one level to the child. Finally, if the target size is smaller than the T_{req} from the child, then the parent will send the lesser of T_{req} and the *number of tasks queued.* Otherwise, the set of tasks sent to the child is the lesser of T_{target} and *child partition-reach.* Note that this is a dynamic load distribution algorithm that changes the size of the batch at run time, allocating large size to larger clusters while smaller clusters are allocated small size.

With respect to which jobs are dispatched, ASP favors jobs that have their replicas within the partition reach of the requesting schedulers. If there are no such jobs, then jobs belonging to the left sibling of the requesting node is searched. If this fails the jobs of the right sibling of the requesting node are selected. This process continues until the exact number of jobs to be sent to the requesting node is reached. The motivation for this job selection scheme is that we minimize replica management overheads (e.g., the replica instantiation latency) in case the original job fails. We also reduce the job transfer latency as we have to only send control messages to the child scheduler if the replica is already located there. Finally, it reduces the time that a child scheduler waits for the jobs to arrive, which increases system utilization. After dispatching the jobs to a child, the parent informs the backup scheduler about the assignment and then updates the application status table (AST) to reflect the new assignment.

3.5 Failure Management

A fail-over strategy is used when a link or a node failure is detected. Link failure is addressed by rerouting traffic from the failed part of the network to another portion of the network. For non-root nodes, the child scheduler is informed to communicate through its closest sibling of the parent. If the failed node is the root, then we choose the closest functional and reachable sibling. When the failed link is repaired, traffic is rerouted over the primary route. Similarly, when a node is repaired, if the node used to be the primary scheduler, then the node that replaced it is told to send to the recovered node and all former children are also notified to update their parent id. The node then rejoins the system and provides services immediately after recovery. If it was a backup node, recovery is not necessary. A detailed discussion of the fail-over strategy is given in [3][4].

Job Replication. The replica creation and placement is to ensures that a job and its constituent task are stored in a number of locations in the cluster tree. The policy maintains some state information for failure and recovery detections in Application Status Table (AST). Jobs are replicted over clusters while tasks are replicated over processors. Specifically, when a job with fault-tolerance requirement arrives into the system, SHS undertakes the following steps: (1) create a replica of the job; (2) keep the replica and send the original job to a child that is alive and reachable; and (3) update the application status table (AST) to reflect where the job replicas are located. This process recursively follows down the cluster tree until we reach the lowest level cluster scheduler (LCS) at which point the replica placement process terminates.

Replica Monitoring. The SHS monitors applications at job-level (between non-leaf nodes and their parents) and at task-level (between leaf nodes and their parents). A monitoring message exchanged between a parent and a leaf-level node is called a *report* while that between non-leaf nodes is called a *summary.* A report message contains status information of a particular task running on a particular node and sent every *REPORT-INTERVAL* time units. In contrast, the *summary* message contains a collection of many reports and sent every *SUMMARY-INTERVAL* time periods such that *REPORT-INTERVAL* < *SUMMARY-INTERVAL.*

When a processor completes execution of a task, the report message contains a *FINISH* message. In this case, the receiving scheduler deletes the corresponding replica and informs the backup scheduler to do the same. When the last replica of a given job is deleted, the job is declared as successfully completed. In this case, the cluster scheduler immediately sends a summary message that contains the *COMPLETED* message to the parent scheduler, which deletes the copy of the job and forward the same message to its parent. This process continues recursively until all replicas of the job are deleted.

Failure Detection and Recovery. After each assignment, the children periodically inform their parents the health of the computations as discussed above. If the parent does not receive any such message from a particular child in a given amount of time, then the parent suspects that the child has failed. In this case, it notes this fact in the AST and sends a request for report message to the child. If a reply from the child has not been received within a specific time frame, the child is declared dead. When a failure is detected, a recovery procedure is initiated to handle the failure. The recovery mechanism restarts a replica of the failed primary task as soon as possible.

4 Conclusion and Future Directions

In this paper, we presented a scalable framework that loosely couples the dynamic job scheduling approach with the hybrid (i.e., passive and active replications) approach to schedule jobs efficiently while at the same time providing fault-tolerance. The main advantage of the proposed approach is that fail-soft behaviour (i.e., graceful degradation) is achieved in a user-transparent manner. Furthermore, being a dynamic algorithm estimations of execution or communication times are not required. An important characteristic of our algorithm is that it makes use of some local knowledge like faulty/intact or busy/idle states of nodes and about the execution location of jobs. Another important characteristic of the proposed approach is that they are applicable for a wide variety of target machines including Grid computing.

We are currently implementing and studying the performance of the proposed policy. In the proposed self-healing distributed framework, the latency of detecting the errors might be affected by message traffic in the communication network. To address this problem, we intend to develop an on-line mechanism to dynamically measure the round-trip time of the underlying network and calculate the error latency accordingly.

Acknowledgement. I appreciate the help of Maliha Omar without whose help this paper would not have been realized.

References

1. J. H. Abawajy. Fault-tolerant scheduling policy for grid computing systems. In *Proceedings of IEEE 18th International Parallel & Distributed Processing Symposium (IPDPS04)*, pages 238–146, 2004.
2. Jemal H. Abawajy and Sivarama P. Dandamudi. Parallel job scheduling on multi-cluster computing systems. In *Proceedings of IEEE International Conference on Cluster Computing (CLUSTER'03)*, pages 11–21, 2003.
3. Jemal H. Abawajy and Sivarama P. Dandamudi. A reconfigurable multi-layered grid scheduling infrastructure. In Hamid R. Arabnia and Youngsong Mun, editors, *Proceedings of the International Conference on Parallel and Distributed Processing Techniques and Applications, PDPTA '03, June 23 - 26, 2003, Las Vegas, Nevada, USA, Volume 1*, pages 138–144. CSREA Press, 2003.
4. Jemal H. Abawajy and Sivarama P. Dandamudi. Fault-tolerant grid resource management infrastructure. *Journal of Neural, Parallel and Scientific Computations*, 12:208–220, 2004.
5. Ian T. Foster, Carl Kesselman, and Steven Tuecke. The anatomy of the grid - enabling scalable virtual organizations. *CoRR*, cs.AR/0103025, 2001.
6. Salim Hariri, Lizhi Xue, Huoping Chen, Ming Zhang, Sathija Pavuluri, and Soujanya Rao. Autonomia: An autonomic computing environment. In *IEEE International Performance Computing and Communications Conference*, 2003.
7. Paul Horn. Autonomic computing. *The Economist print edition*, September 19, 2002.
8. Gang HUANG, Tiancheng LIU, Hong MEI, Zizhan ZHENG, Zhao LIU, and Gang FAN. Towards autonomic computing middleware via reflection. In *COMPSAC*, 2004.
9. Gail Kaiser, Phil Gross, Gaurav Kc, Janak Parekh, and Giuseppe Valetto. An approach to autonomizing legacy system. In *IBM Almaden Institute Symposium*, 2002.
10. Jeffrey O. Kephart and David M.Chess. The vision of autonomic computing. *IEEE Computer*, pages 41–50, 2003.
11. H. Liu and M. Parashar. A component based programming framework for autonomic applications. In *In Proceedings of the International Conference on Autonomic Computing*, 2004.

A New Trust Framework for Resource-Sharing in the Grid Environment

Hualiang Hu, Deren Chen, and Changqin Huang

College of Computer Science, Zhejiang University, Hangzhou, 310027, P.R. China
huhualiang@163.net

Abstract. The open and anonymous of grid make the task of controlling access to sharing information more difficult, which cannot be addressed by traditional access control methods. In this paper, we identify access control requirements in such environments and propose a trust based access control framework for grid resource sharing. The framework is an integrated solution involving aspects of trust and recommendation models, based the discretionary access control (DAC), and are applied to grid resource-sharing systems. In this paper, we integrate technology of web services into idea of trust for describing resources.

1 Introduction

Grid applications are distinguished form traditional client-server applications by their simultaneous use of large numbers of resources, dynamic resource requirements, use of resources from multiple administrative domains, complex communication structures, and stringent performance requirements, among others [1].

Although grid security infrastructure (GSI) has been widely adopted as the core component of grid applications, GSI, which provides a basic secure and reliable grid-computing environment, is still at its early stage of development. Since GSI is built upon PKI, risks factors due to the use of PKI have to be considered carefully such as compromising of private keys or theft of certificates for the following reasons:

1. Parallel computations that acquire multiple computational resources introduce the need to establish security relationships not simple between a client and s server, but among potentially hundreds of processes that collectively span many administrative domains.
2. The inter-domain security solutions used for grids must be able to interoperate with, rather than replace, the diverse intra-domain access control technologies inevitably encountered in individual domains.
3. In such a distributed system, a huge set of entities cannot be known in advance.
4. Authentication alone is sometimes not enough to make one confident about allowing a requested access or action rather, a kind of trust is also along with authentication.
5. In order to increase the scalability of a distributed system, it should be possible to delegate the authority to issue access certificates.

V.S. Sunderam et al. (Eds.): ICCS 2005, LNCS 3516, pp. 221 – 228, 2005.

6. In the traditional security systems, an access control mechanism such as Access Control List (ACL) is not expressible and extensible. Whenever new or diverse conditions and restrictions arise, the application is required to change or rebuild [1, 2]. At present, trust for grid is solely built on authentication of identity certificates. As authentication is not insufficient for establishing strong security, it is critical that a proper trust evaluation model for grid is needed. In this paper, we present an access control framework for grid resource-sharing systems, which provides grid users better access control services whilst preserving the decentralized structure of the grid environment. The framework extends a traditional access control model to meet the requirements of grid resource- sharing. The paper is organized as follow. Section 2 identifies the requirements of an access control model for grid environment. Section 3 discusses the characteristics of grid environment. Section 4 explains our access control framework in detail, including the overall architecture, authentication process, scoring scheme. Section 5 gives our concluding remarks.

2 Access Control Requirements

We have identified main requirements that an access control model for grid resource-sharing system should support:

1. Single sign-on: A user should be able to authenticate once.
2. Protection of credentials: User's credentials must be protected.
3. Interoperability with local security solutions: while our security solutions may provide inter-domain access mechanisms, access to local resources will typically be determined by a local security policy that is enforced by local security mechanisms. It is impractical to modify every local resource to accommodate inter-domain. Access; instead, one or more entities in a domain must act as agents of remote clients/users for local resources.
4. Resources may require different authentication and authorization mechanisms and policies, which we will have limited ability to change.
5. Limits are placed on the overall amount of resource consumed by particular groups or users.
6. Decentralized control: The centralized access control authority does not exist in a grid environment. We must take this decentralization of the access control model for grid environment must into account.
7. Node classification: one attribute of grid environment is uncertainty. Their interacting partners are mostly unknown, unlike most other systems, where the users are known. Previously unknown users, who request access to the system, may contact entities. Hence, the framework proposed must provide a mechanism for entities to classify users and assign each user to access rights accordingly.
8. Encourage sharing resources: Sharing is another grid's property. The framework proposed does not only protect the resources, but also provides technologies to grid for resources sharing. This means giving entities the ability to control access to their resources, at the same time, entities must be confident that participation in the system will give them better chance to access to the resources they want.

3 Grid Resource Sharing Systems

Grid resource sharing allows any two nodes in the system to directly access resources from each other systems. To achieve this flexible direct access, grid environment has to support three kinds of operations:

1. Publish: The resources should firstly be described and published in the grid. The provider must pass the identity verification of the resources.
2. Find: There are two ways, which the resources requester finds information. One of them is browse pattern, in the case the result is not unique; the other is drill down pattern, in the case, the result is unique.
3. Bind: By analyzing the binding information gathering from the Services, including the access path of resources, the invoking parameters of resources, the return value, the transmission protocols and the requirements of security; the resource Requester deploys its own system and then invokes the remote resources provided by the resource provider.

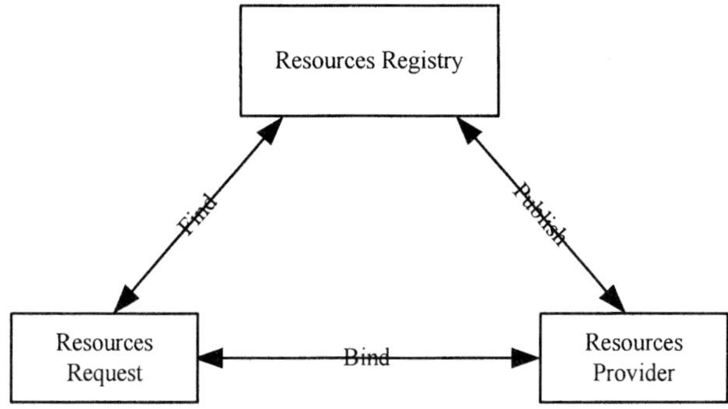

Fig. 1. Relationship of three roles [3]

After three steps, resource is ready. During three steps, there exists following technologies:

1. UDDI (Universal Discovery, Description, Integration). The UDDI specification was first defined by IBM, Microsoft and Ariba in July 2000. It was developed from the DISCO (Discovery of web services) of Microsoft and ADS (Advertisement and Discovery of services) of IBM.UDDI is the registry specification of web services. It defines a way to publish and discover information about web services so that the users who need the service can find and use it conveniently. This specification consists of a core information model provides four aspects of basic information about web services: By means of UDDI, components of web services can complete "register once, access anywhere".
2. WSDL (web service description language). WSDL provides an abstract language for defining the published operations of a service with their respective parameters and data types. It also addresses the definition of the location and binding details of

the service and describes the network into the set of communication end-point, which expresses what a component of web services can do k where it is and how service requester invokes it. Otherwise, WSDL also provides the standard format of describing request for a service requester.

4 Access Control Framework

We propose an access control framework based on the discretionary access control (DAC) model [4]. The basic behind discretionary access control is that the owner of an object should be trusted to manage its security. More specifically, owners are granted full access rights to objects under their control, and are allowed to decide whether access rights to their objects should be passed to other subjects or groups of subjects at their own discretion. In the DAC, a discretional owner of resources has the right for the control of access discretion. Due to anonymousness for grid, we cannot pre-assign access rights to users.

4.1 Terms and Definitions

Following definitions can be found in [5]

Definition 1 (Entities): The entities are all the components involved with the operation of a grid computing system, which can be related to each other via certain trust relationships. These include the user, host (resource provider) node and trusted third parties such as the resource management system (RMS) or a CA.

Definition 2 (Direct Trust): Direct Trust is the belief that one entity holds in another entity in its relevant capacity with reference to a given trust class.

Definition 3 (Indirect Trust): A host node often encounters a client node that it has never met. Therefore, the host has to estimate the client's trustworthiness using recommendations, which a client submits other nodes'.

Definition 4 (Recommended Trust): Recommended Trust expresses the belief in the capacity of an entity to decide whether another entity is reliable in the given trust class and in its honesty when recommending third entities.

Definition 5 (Hard Trust): Hard Trust is trust is the trust derived from cryptographic based mechanisms. This can be treated as a meta-level attribute.

Definition 6 (Soft Trust): Soft Trust is the trust relationship derives from non-cryptographic based mechanisms which employ methods like recommendation protocol, observations, interactions or combination of them.

In this paper, we only discuss Soft Trust, which can address the uncertainty for grid.

4.2 Overall Architecture

Our framework contains client (user) node, service provider node and trust third parties such as the resource management system (RMS) or a CA. Shared resources in framework are rated depending on their size and content; each resource being assigned two thresholds which capture two access aspects. Only if the request end's two access values both equal to and greater than the corresponding thresholds of the resource. It can access resources that it wants. The request end is responsible to collect recommendations that contain the information needed to evaluate its access values for a particular host. After each transaction, direct trust and direct contribution of both the client node and host nodes are updated accordingly to the satisfaction level of the transaction, which then affect the future evaluation of the access values between these two nodes.

4.3 Authentication

An issue in determining trust is how the user authenticated to the current session. Hence, the authentication, in our framework, must be mutual. This means both a client node and a host node need to be authenticated with each other. The authentication process is initialized by the client node that wishes to make contact.

A user may have an X.509 certificate that he uses from certain machines, a password that he may use from other sites, and may not even want to type in a password from a very un-trusted site. The user may also not yet have some of the authentication methods available [7].

The host node' local database tracks client nodes' records, for authentication purposes and access control purposes such as trust and contribution calculation. In summary, the authentication information that is saved per session consists of a 128-bit GUID (global unique identifier [8]) number and a pair of public/private keys. The node gives out the GUID and the public key as its identity and uses the private key for authentication, the name (nickname) of the user, the authentication method (certificate, password, null), and fields that other users might have added after the user authenticated into the session.

4.4 Evaluating System

A host node classifies its client nodes based on their records in grid environment. After authentication process with the host, a client node is required to send its rating certificates to the host to calculate the client's relative access values. The host node perceives the client's trustworthiness and contribution level, based on the client's relative access values (trust value and contribution value). The trust value is to ensure the node is trusted to interact with. The contribution value is to promote contribution for the host in grid environment.

In our framework, the trust value is key factors for making access control decision can address uncertainty factor in grid. By the trust value, a host node can determine whether it should trust that client to allow it to access to the local resources.

4.5 Rating Certificate Management

A rating certificate contains the direct trust value and the direct contribution score of the recommending node on the recommended node and contains the new trust value and the updated contribution score. After transaction, system automatically updates trust value and contribution value.

4.6 Interaction Procedure

We introduce definition for describing entities' interaction procedure [6]. The interaction procedure is shown as figure 2.

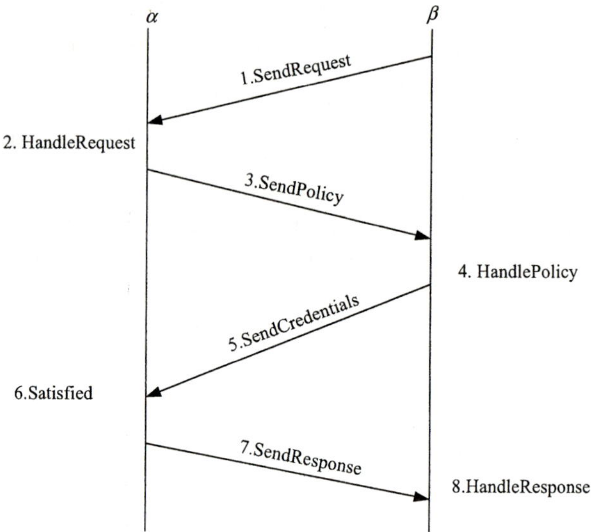

Fig. 2. Interact Procedure

Entities include α and β.
Definition 1 α.Trust(β): α trustsβ;
Definition 2 α.MakeCommentβ): αcomments on β;
Definition 3 α.SendRequest(β): α sends an request to β;
Definition 4 α.HandleRequest(β.Request): αhandles with request which β sends to α;
Definition 5 α.SendPolicy(β): αsends policy toβ;
Definition 6 α.SendPolicy(β. Policy): αhandles with β'policy;
Definition 7 α.SendCredentials(β): αsends Credentials to β;
Definition 8 α.Satistied(β.Credentials): αhandles withβ's message for credentials, to judge whether ß satisfies à'trust-policy;
Definition 9 α.SendResponse(β): αsends the result to β;
Definition 10 α.HandleResponse(β.Response): αhandles with the result for β.Response;

5 Relational Works

In this section, we examine several papers that examine issues that are peripherally related. A model for supposed in [9]. This trust-based model allows entities to tune their understanding of another entity's recommendations. A model for supporting behavior trust based on experience and reputation is proposed in [10]. The Community Authorization Service (CAS) in [11]. A design and implementation of a secure service discovery service (SDS) is in [12]. The SDS can be used by service providers as well as clients. Formalizations of trust in computing systems were done by Marsh [13]. He attempted to integrate the various facets of trust from the disciplines of economics, psychology, philosophy and sociology.

6 Concluding Remarks

We propose a novel trust framework based on the DAC model for resources-sharing in grid environment.

Our proposed trust model is optimistic in that the majority of nodes start with access value, while a small group of nodes have overdue access value. The concept of trust value can address uncertainty problem. Hence, host nodes can assigns appropriate access privileges to each visitor accordingly.

The trust framework we proposed can also provide trust guarantees to each node. We provided rating certificate that stores and updates the trust value and the contribution value. Involvement values, which are associated with each community within which a node is a member, play an important role in determining accurate trust value of nodes.

Finally we explained the interact procedure between entities.

Our future challenge ahead is to refinement the proposed framework based access scheme in grid environment and how to compute trust value efficiently over the grid.

References

1. I. Foster and C. Kesselman. A Security Architecture for Computational Grids. In Proceeding of the 5th ACM Conference on Computer and Communication Security, November 2-5, 1998.
2. H. Lei, G. C. Shoja. Dynamic Distributed Trust Model to Control Access to Resources over the Internet, Communications, Computers and Signal Processing. August 28-30, 2003.
3. Dr. Liang-Jie (L.J) Zhang. On Demand Business Collaboration With Services Computing.
4. J. McLean. The Specification and Modeling of Computer Security. IEEE Computer, Jan 1990.
5. L. Ching, V. Varadharajan and W. Yan. Enhancing Grid Security With Trust Management. In proceedings of the 2004 (IEEE) international Conference on Services Computing (SCC 04).
6. N.YuPeng and C. YuanDa. Automation Trust Establishment in Open Network Environment. Computer Engineering 2004; 30(16);124-125.
7. D. Agarwal, M. Thompson, M. Perry and M. lorch, A New Security Model for Collaborative Environments.http://dsd.lbl.gov/Collaboratories/Publications/WACE-IncTrust-final-2003.pdf.
8. D. Box, Essential COM, Addison Wesley, 1998.
9. Karl Aberer & Zoran Despotovic, Managing Trust in a Node-2-Node Information System, ACM CIKM, Nov 2001.

10. A. Abdul-Rahman and S. Hail. Supporting Trust in Virtual Communities. Hawaii Int'I Conference on System Sciences, 2000.
11. L. Pearlman et al. A Community Authorization Service for Group Collaboration. IEEE Workshop on Policies for Distributed Systems and Networks, 2002.
12. S. E. Czerwinski, B. Y. Zhao et al. An Architecture for a Secure Service Discovery Service. 5th Annual Int'I Conference on Mobile Computing and Networks(Mobicom'99).
13. S. Marsh. Formalizing Trust as a Computational Concept. Ph.D. Thesis, University of Stirling, 1994.

An Intrusion-Resilient Authorization and Authentication Framework for Grid Computing Infrastructure

Yuanbo Guo[1,2], Jianfeng Ma[2], and Yadi Wang[1]

[1] School of Electronic Technology, Information Engineering, University, Zhengzhou, Henan 450004, China
yuanbo_g@hotmail.com

[2] The Ministry of Education Key Laboratory of Computer Networks and Information Security, Xidian University, Xi'an, 710071, China

Abstract. A correctly and continually working authorization and authentication service is essential for the grid computing system, so it is very necessary to maintain efficient this service with high availability and integrity in the face of a variety of attacks. An intrusion-resilient framework of authorization and authentication service for grid computing system is presented in this paper. This service is able to provide fault tolerance and security even in the presence of a fraction of corrupted authorization and authentication servers, avoiding any single point of failure. We use a cryptographic (*f*, *n*) secret sharing scheme to distribute parts of the clients' proxy certificates and use a secure multi-party computation scheme to perform the signatures such that the proxy certificate can be issued in a distributed fashion without reassembly when a legal client registrant at the Globus host. By using Non-Malleable Proof, the "man-in-the-middle attack" can be prevented; by distributing the secret data across several authorization and authentication servers, the compromise of a few servers will not compromise the availability of data. And, under the assumption of a Diffie-Hellman decisional problem, a passive adversary gets zero knowledge about the system's private key *X*, and so cannot to issue the certification for any client, neither to impersonate a legal authorization and authentication server.

1 Introduction

At the base of any grid environment, there must be mechanisms to provide security, including authentication, authorization, data encryption, and so on. The Grid Security Infrastructure (GSI) component of the Globus Toolkit provides robust security mechanisms. Especially, it provides a single sign-on mechanism, so that once a client is authenticated, the certificate authority in the GSI will generate a proxy certificate and send it to the client, then the client can use the proxy certificate to perform actions within the grid computing system. But this mechanism has its own drawback. For example, since all clients' certificates are maintained by the certificate authority, it requires the certificate authority to be trusted and so introduces dependence upon the certificate authority and thus creates a single point of failure, and, if successfully compromised, would immediately lead to the exposure of all clients' confidentiality of certificates, interruption communication, or other forms of denial of service.

V.S. Sunderam et al. (Eds.): ICCS 2005, LNCS 3516, pp. 229 – 236, 2005.

To address this issue, in this paper we present an intrusion-resilient authorization and authentication framework for grid computing infrastructure, that removes this single point of failure by distributing the authorization and authentication service among n authorization and authentication servers (which will be denoted as *AASs* briefly in the remaining of the paper). The framework uses a verifiable secret sharing scheme to distribute shares of the system's secret to *AASs* and use a secure multi-party computation scheme to perform the signatures such that the proxy certificate can be signed in a distributed fashion without reassembly. When a legal client want to sign-on the grid computing system, he/she sends a sign-on request to the Globus host firstly, the Globus host will forward this request to a his-own-choosing subset of the n *AASs*, and the contacted *AASs* answer with some information enabling the client to compute his own proxy certificate and verify its correctness.

Our framework is intended to tolerate the compromise of up to f *AASs*, where $3f+1 \leq n$. Compromised *AASs* are assumed to be under the full control of an attacker and to have Byzantine behavior[1], but it is also assumed that the attacker cannot break the cryptographic algorithms used. Under these assumptions, our service ensures the confidentiality, integrity and availability of authorization and authentication service for grid computing infrastructure. Also, our system is designed to be easy to embed into existing grid computing environment. For example, the *AASs* are built with commodity PCs, an administrator can just swap a failed host out with another.

2 System Architecture

Let *Client* = $\{C_1, \ldots, C_m\}$ be a set of m clients and let $AAS = \{AAS_1, \ldots, AAS_n\}$ be a set of n *AAS*s(authorization and authentication servers for grid computing infrastructure). All of them share a global clock (i.e., the system is synchronous). Each client connects himself or herself with $2f + 1$ *AAS*s by point-to-point encrypted and authenticated channels, where f *is* the threshold of secret sharing scheme we used. The member in the *AAS*s can be good (i.e., it honestly executes the protocol) or bad (i.e., it is controlled by an adversary and can deviate from the protocol in arbitrary ways), but a majority of good *AAS*s are always present across the system.

The service works as follows. A **dealer**, which is needed only in the system initialization phase, distributes system's private key X among all n *AAS*s by using Feldman's $(f+1, n)$ VSS scheme [3] straight forward. So along with the secret shares, he also generates a verification share to go along with each key share, and broadcast the verification shares to all clients.

Each secret share is then distributed secretly to a different *AAS*, and all corresponding verification shares are broadcast to all of the clients. No *AAS* knows the whole X, but only knows his own assigned share.

The architecture of our intrusion-resilient authorization and authentication service can be illustrated as Figure 1.

---

[1] A compromised host deviates arbitrarily from its specified protocols. It is also called arbitrary behavior or malicious behavior.

Each *AAS* maintains a configuration file which specifies what types of resource are allowed to be access by which client. When a client wants to apply for some services of or execute a task on the grid computing system, he/she registers himself/herself at the Globus hosts firstly, with the submission of his/her password or some others type of credential and the requested services or tasks. Then, Globus host will performs the initial validation; if success, it sends an authentication and authorization request to the *AAS*s on behalf of the client. Based on the identification of the client, each *AAS* consults its configuration files to determine what resources in the grid environment are authorized to the client, and then uses its share of the system's private key to compute and send a share of proxy certificate to the client, which we called sub-certificate in the remaining paper. It also uses its verification share to compute a zero-knowledge proof which will prove the correctness of its sub-certificate.

These sub-certificate and zero-knowledge proofs are then collected by the Globus host. Then, the host chooses f of them. After verifying the correctness of all of the proofs of these sub-certificates, it can combine them to form the actual proxy certificate and send it to the client. Afterwards, the client can request the needed resources to the grid computing system by using this proxy certificate.

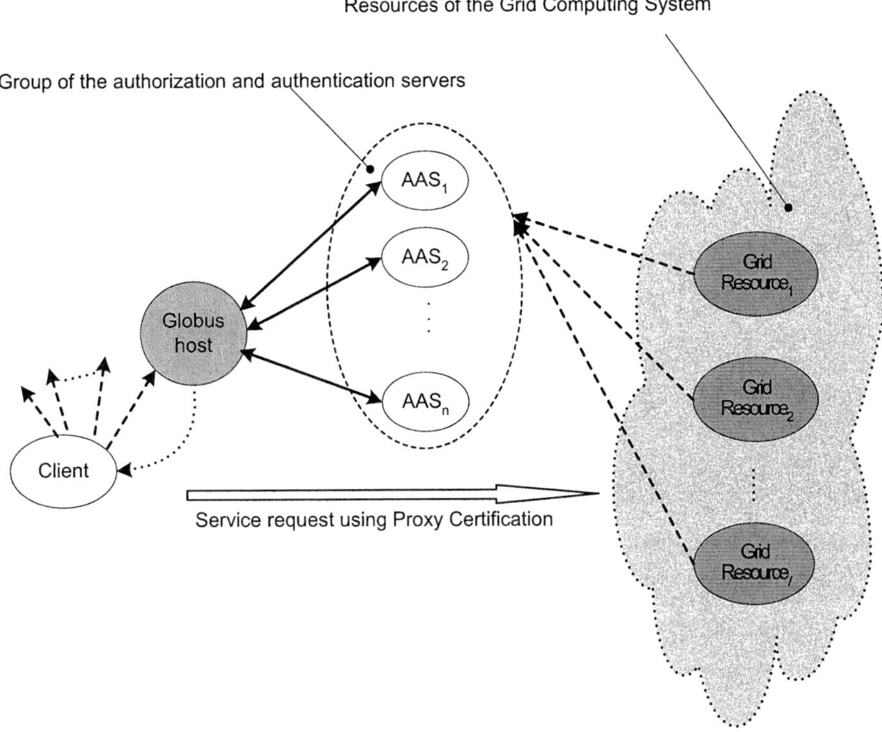

Fig. 1. Architecture of intrusion-resilient authorization and authentication service for grid computing environment

3 The Details of the Framework

Denote the system's private key as X_{AAS} and the corresponding public key Y_{AAS},. Denote a client C's X_C and Y_C. Our scheme starts with a large prime $p = 2q+1$, where q is also a large prime, and a generator g for the subgroup $G_q= \{i^2 \mid i \in Z_p^*\}$ of quadratic residues of Z_p^*. A collision-free one-way hash function $H(x)$ is also needs to be made public. Then, our framework can be divided into 3 phase:

Initialization phase: The dealer distribute the system's private key $X_{AAS} \in Z_q^*$ among all AAS by using Feldman's $(f+1, n)$ VSS scheme [3] straight forward.

1) Select randomly f elements $a_1, \ldots, a_f$ of Z_q^*. These coefficients define a polynomial of degree t in $Z_q^*[X]$:

$$F(X)= X_{AAS} + a_1X + a_2X^2 + \ldots + a_f X^f$$

2) Compute $X_1, \ldots, X_n$ of Z_q and $g_1, \ldots, g_n$ of Z_p as follows:

$$X_i = F(i) \bmod q \text{ and } g_i = g^{X_i} \bmod p$$

The shares $X_1, \ldots, X_n$ must then be distributed secretly to the n authorization and authentication servers $AAS_1, \ldots, AAS_n$, respectively. The generator g and the elements $g_1, \ldots, g_n$ are made public. They must be known by all clients and all AASs. Each AAS_i can verifies the share X_i he or she received by using g and g_i.

As in Shamir's secret sharing scheme[4], any subset of f+1 values among $X_1, \ldots, X_n$ allows one to reconstruct F by interpolation, and then to compute the value $X_{AAS} = F(0)$. For example, given $X_{i_1}, \ldots, X_{i_{f+1}}$, one has $X_{AAS} = \sum_{s=1}^{f+1} b_{i_s} X_{i_s} \bmod q$, where

$$b_{i_s} = \prod_{j=1, j\neq s}^{f+1} \frac{i_j}{i_s - i_j} \bmod q.$$

Authorization Phase: The grid computing system authorizes AASs to authenticate the clients and issue the corresponding proxy certificates. Each grid resources GR_i, where i = 1, 2, …, l, is associated with an access control table CL_i, which is used to specifying what types of access to this resources of which client are allowed. After authorization, all GR_i's access control tables are maintained by each AAS.

Registration and Proxy Certificate Generation Phase: When a client C, whose ID is ID_C, wants to apply for some services or execute some tasks, he/she registers at the Globus host firstly. The client should submit to Globus host his/her password or other credential and the requested services or tasks. On receiving these information, the Globus host will then performs the initial validation of the client; if successes, it sends an authentication and authorization request to a robust subset of AASs (in our case, the servers of size at least $2f$+1, where f is the threshold of the secure secret sharing scheme we used) on behalf of the client. Each AAS does the following:

1) Compute sub-signature of client C's private key $SC_i = X_C^{X_i} \; mod\, p$;
2) Randomly select $r \in Z_q^*$ and compute

$$u = g^r \; mod\, p \text{ and } v = X_C{}^r \; mod\, p$$

3) Compute $s_i = H(g, g_i, ID_C, SC_i, u, v)$ and $t_i = r + s_i \cdot X_i \; mod\, q$.
4) Based on the identification of the client, each AAS consults its configuration files to determine what resources in the grid environment are authorized to the client, and then generate a sub-certificate (SC_i, s_i, t_i, C) for client C (denoted as $Cert_C^i$), and then send it to the Globus host.

In step 3) and 4), both s_i and t_i are used here to prove the validity of SC_i without revealing X_i. This uses a technique of non-malleable proof proposed by [5].

After receiving $Cert_C^i$ send by S_i, the Globus host does following:

1) Compute $u' = g^{t_i} / g_i^{S_i} \bmod p$ and $v' = X_C{}^{t_i} / SC_i^{S_i} \bmod p$
2) compute $s_i' = H(g, g_i, ID_C, SC_i, u', v')$

If $s_i' = s_i$, then the client accept the proof and get the sub-certificate $Cert_C^i$; else, Globus host can raise an alarm to the administrator of system.

After receiving and checking f+1 sub-certificates $Cert_C^i, \ldots, Cert_C^{i+f+1}$ from different f+1 $AASs$, the Globus host compute $SC = \prod_{s=1}^{f+1} SC_{i_s}{}^{b_{i_s}} \; mod\, p$, where b_{i_s} is obtained from $S_{i_1}, \ldots, S_{i_{f+1}}$ and $b_{i_s} = \prod_{j=1, j \neq s}^{f+1} \frac{i_j}{i_s - i_j} \bmod q$. By this interpolation method, Globus host can obtain $SC = X_C^X \; mod\, p$.

Then, a proxy certificate containing the signature SC for client C can be generated and send to client C by Globus host.

Afterwards, the proxy certificate can be used by the client to access the grid resources he/she needs.

4 Security Analysis

4.1 Correctness and Robustness

By correctness and robustness, we mean that all legal clients can get a correct proxy certificate if there are no more than f dishonest $AASs$.

Theorem 1: If there are no more than f dishonest $AASs$, then all legal clients can get a correct proxy certificate.

Proof: In our case, we only need to show that Globus host will compute a correct $SC = X_C^X$ for legal client C.

From the details of the framework we know that, before Globus host compute and recover SC, it has received and verified $f+1$ sub-certificates $Cert_C^i, \ldots, Cert_C^{i+f+1}$

from different f+1 *AASs*. Then by using interpolation method, Globus host can compute $SC = \prod_{s=1}^{f+1} SC_{i_s}^{\,b_{i_s}} = \prod_{s=1}^{f+1} (X_C^{X_i})^{b_{i_s}} = (X_C^{X_i})^{\sum_{s=1}^{f+1} b_{i_s} X_{i_s}} = X_C^X$.

4.2 Security Against Impersonators Attack

Impersonations means that server AAS_j impersonate server AAS_i to send message (SC_i, s_i, t_i, C) to the Globus host.

Theorem 2: In the random oracle model, a dishonest *AAS* cannot forge, with a non negligible probability, a valid proof for a valid sub-certificate.

Proof: From the above scheme we know that SC_i, s_i, t_i satisfy the following equations:

$$g^{t_i} = g^{r+s_i \cdot X_i} = g^r \cdot (g^{X_i})^{s_i} = u \cdot g_i^{s_i} \tag{4.1}$$

$$X_C^{t_i} = X_C^{r+s_i \cdot X_i} = X_C^r \cdot (X_C^{X_i})^{s_i} = v \cdot SC_i^{s_i} \tag{4.2}$$

Since g^{t_i} and $g_i^{s_i}$ are both in G, by the closure of G under multiplication we know that Equation (4.1) yields $u \in G$. This means that it exists λ such that $u = g^{\lambda}$. So we have $g^{t_i} = g^{\lambda} \cdot g_i^{s_i} = g^{\lambda} \cdot g^{X_i s_i}$, which implies: $t_i = \lambda + s_i \cdot X_i$. Now Equation (4.2) becomes:

$$X_C^{t_i} = vSC_i^{s_i} \Leftrightarrow X_C^{\lambda + S_i \cdot X_i} = vSC_i^{s_i} \Leftrightarrow X_C^{\lambda} v^{-1} = (X_C^{-X_i} \cdot SC_i)^{s_i}$$

This yields two possible cases:

(1) $SC_i = X_C^{X_i}$. In this case, the sub-certificate is correct. $v = X_C^{\lambda}$ and for all $s_i \in Z_q$, the verifier equations trivially hold.

(2) $SC_i \neq X_C^{X_i}$. In this case, we must have

$$s_i = \log_{X_C^{-X_i} \cdot SC_i} X_C^{\lambda} v^{-1} \tag{4.3}$$

From equation (4.3), one can see that, if SC_i is not a valid sub-certificate, once the triplet (SC_i, u, v) is chosen, then there exists a unique $s_i \in Z_q$ that satisfies the verifier equations. In the random oracle model, the hash function H is assumed to be perfectly random. Therefore, the probability that H (g, g_i, ID_C, SC_i, u, v) equals s_i is $\frac{1}{q}$, once (SC_i, u, v) fixed.

On the other hand, if the attacker perform an adaptively chosen message attack by querying an oracle N times, the probability for the attacker to find a triplet (SC_i, u, v), such that H (g, g_i, ID_C, SC_i, u, v) is $P_{success} = 1-(1-\frac{1}{q})^N \approx \frac{N}{q}$, for large q and N. Now if k is the number of bits in the binary representation of q, then $P_{success} \le \frac{N}{2^k}$. Since a computationally bounded server can only try a polynomial number of triplets, then when k is large, the probability of success, which is $P_{success} \le \frac{N}{2^k} << 1$, is negligible.

4.3 Security Against Secret Compromise

The most serious attack is for an attacker to compromise f *AASs* and reconstruct the system's private key X, and then to issue the certification for any client, whether legal or illegal. To combat this, f should be chosen to be sufficiently large to make compromising f machines exceedingly difficult.

Arbitrarily setting f to be very large, however, has reliability concerns. In the extreme case, if $f = n$, then the system has n points of failure—where a proxy certificate cannot be issued. f must be chosen so that $n - f - 1$ is sufficiently large to handle node failures.

Secondly, each machine will, ideally, be running a different operating system with different versions of software to make compromising a machine more difficult.

4.4 Security Against Denial of Service

Since f machines are necessary to successfully compute a correct proxy certificate for the legal client and there are n machines in total, if $n - f - 1$ of them crashes or fails, the key will not be able to be computed. To prevent attackers from compromising or flooding $n - f - 1$ machines, n and f will need to be appropriately chosen to make this probabilistically low. In the case where machines crash because they are faulty, we can bound the probability of failure.

4.5 Failure Handling

Consider that each *AAS* fails—crashes or does not respond—with probability r. We assume r is independent. The probability that a key cannot be computed is:

$$p_f = r^f$$

That is, p_f is the probability that the system is unable to perform useful work; a state of failure.

Solving for r, we find: $$r = p_f^{1/f}$$

Obviously, any system will need to set n and f such to maximize r.

5 Discussions and Conclusion

A combination of fairly standard secure multi-party computation algorithm, secret sharing scheme, zero knowledge proof and traditional authorization protocols was used to develop an intrusion-resilient framework of authorization and authentication for grid computing environment, which can provide correct services even in the presence of faulty insiders of *AASs*. Its security requirements can be shown to theoretically hold if no more than f *AAS*s are compromised, the attacker does not break the cryptographic algorithms, and the network assumptions are satisfied. But it is obviously that the claim of no more than f *AAS*s are compromised relies on the assumption that it is substantially harder for an attacker to penetrate several *AAS*s than a single one. In fact, with enough time and resources, a determined attacker can compromise a sufficient number of *AAS*s to break the system. To justify this assumption we rely on two complementary approaches: diversity and proactive updates. With diversity, an attacker has to devise different penetration mechanisms for each server, so that it is impossible for him/her to work to kill other *AAS*s at once when he/she kills one. Diversity can be achieved by running a different version of our software on each of the *AAS*s, running on different hardware and operating systems, and putting the different *AAS*s under the responsibility of different administrator.

Proactive security is the ability to gracefully recover from compromises in such a way that any past compromises don't assist a current attacker. Proactive update implies that once every time period (say once an hour) the *AAS*s compute a new sharing of the same system's privacy key. Consequently, not only does the attacker have to penetrate multiple *AAS*s, he/she has to do so within a single time period. Proactive update is normally implemented by re-sharing the private key in such a way that the new shares have no relationship to the old shares, and therefore if an attacker compromises some *AASs*, these *AASs* are recovered, and then he/she compromises more *AASs*, the information he gains isn't enough to compute the key.

However, due to the organization's budget and security constraints, the parameters of the system's diversity and proactive should be set accordingly.

Finally, it should be noted that, even with above measures, resilience is not absolute in the real world. Our scheme can also be vulnerable to network-based denial-of-service attack based on flooding, though this attack is not specific to our system, as current TCP/IP protocols make it difficult to defend against such attacks in any system. Intrusion resilient can be effective only if combined with traditional measures to avoid and remove vulnerabilities in system components, and with intrusion detection mechanisms to help respond to attacks before it is too late.

References

1. Hung-Yu Chien, New spproach to authorization and authentication in distributed environments, Communications of the CCISA, vol.9 no.3, pp.63-69, Jun. 2003
2. Mambo, M., Usuda, K., and Okamoto, E., Proxy signatures for delegating signing operation, Proc. 3rd ACM conf. on Computer and Communications Security, 1996.
3. P. Feldman, A Practical Scheme for Non-interactive Verifiable Secret Sharing, Proc. 28th IEEE Annual Symposium on Foundations of Computer Science, pp. 427-437, Oct. 1987
4. A. Shamir. How to Share a Secret. Communications of the ACM, vol. 22, no. 11, pp. 612–613, Nov. 1979.
5. Chaum D. and Pedersen. T. Wallet Databases with Observers. In Crypto'92, number 740 in Lecture Notes in Computer Science, Santa Barbara, CA, Springer-Verlag, 1992, pp. 89–105
6. Yair Frankel, Peter Gemmell, Philip D. MacKenzie, and Moti Yung. Optimal resilience proactive publickey cryptosystems. In IEEE Symposium on Foundations of Computer Science, pp 384–393, 1997.
7. Yair Frankel, Peter Gemmell, Phillip D MacKenzie, and Moti Yung. Proactive RSA. In Advanced In Cryptology – Crypto 97, volume 1294 of Lecture Notes in Computer Science, pp 440–454 Springer-Verlag, 1997.
8. Globus OGSA Home Page. http://www.globus.org/ogsa/
9. Foster I, Kesselman C. The Grid: Blueprint for a new computing infrastructure. San Fransisco: Morgan Kaufmann Publisher, 1999.

An Active Platform as Middleware for Services and Communities Discovery

Sylvain Martin* and Guy Leduc

Research Unit in Networking, Université de Liège,
Institut Montefiore B28, 4000 Liège 1, Belgium
{martin,leduc}@run.montefiore.ulg.ac.be
http://www.run.montefiore.ulg.ac.be/

Abstract. In an increasing number of cases, network hosts need to locate a machine based on its *role* in a service or community rather than based on a well-known address. We propose and evaluate WASP, a lightweight active platform where ephemeral state left in the network can help locate service providers such as request dispatchers or computation aggregators. In an active grid architecture, WASP can also help locate participants, build and manage overlays.

1 Introduction

In peer-to-peer systems in general and grid computing as a specific case, we are facing the problem of communities of machines that need to cooperate together without having any *a priori* knowledge of their respective existence. Existing schemes often rely on the existence of some centralizing machine that swaps peers addresses. With the increase of service popularity and the decrease of average connection time, keeping these so-called *pong servers* scalable is a real challenge.

In this paper, we study a solution based on active networks that will help network operators and peer-to-peer application designers to introduce aggregators and caches of *pong servers* transparently in the network. Our active platform also allows discovery of more generic services and could be used to locate computation dispatchers and aggregators as well.

1.1 Why Yet Another Active Platform?

Despite an important amount of active network platforms has been proposed and studied, none has really successfully reached the real-world deployment level. In response to this reality, WASP (*Weightless Active packets for ephemeral State Processing*) is made lightweight enough and resource-friendly so that its presence in the network does not become a nuisance for network managers. WASP is not built from scratch: it is an effort to merge the advantage of two recently proposed platforms ESP (*Ephemeral State Processing*) and SNAP (*Safe and Nimble Active Packets*) [5]) that also focus on

* Sylvain Martin is a Research Fellow of the Belgian National Fund for Scientific Research (FNRS).

V.S. Sunderam et al. (Eds.): ICCS 2005, LNCS 3516, pp. 237–245, 2005.

safe, flexible and efficient – in a word, *practical* (cf Moore et *al.* [2]) – active networks. Section 2 shows how both computation time, storage and network bandwidth are kept under low and predictable limits for WASP packets.

Unlike fully-featured active platforms, WASP alone is not able to offer complex services such as flow transcoding or realtime auction though it can help *advertise* and *locate* such services, helping machines that offer such services (hereafter called *service providers*) and machines that use such services (hereafter called *end-systems*) to build and manage adequate *tunnels* and *overlays* in an automated way, thus providing an effective support for active grid architectures. We will show in Sect. 3 how WASP can be used to build such solutions.

The design of our proposed platform is detailed in Sect. 4. In order to evaluate its performance, we translated ESP instructions into WASP code and compared execution timings. The results of this evaluation are presented in Sect. 5

1.2 What Slows Down Active Packets

The speed at which our active platform will be able to process active packets will define the network locations where it can actually be deployed. Our goal is to offer a solution that can be running even on a border router in a transit domain, taking advantage of *network processor* technologies to achieve high throughput. A simple *active packet* crossing an active node will incur different types of time-consuming operations:

Classification. Filters should be applied so that the active packet is recognized as such. Depending on the architecture, this filtering may be based on a dedicated *transport protocol type*, the presence of an *IP header option*, etc.

Delivery. After classification, we still have to make the packet available to the component that will process it. In the worst case, the whole packet should be copied into user-space before applicative decoding could be applied. Most performance-targetted platforms ([3, 4]) thus decode and process packets at kernel level.

Decoding. Some platforms like ESP [4] and SNAP [5] are able to process the operations for the packet as soon as it is delivered, but in most of the other frameworks, a data decoding phase is needed to deserialize objects, strings, lists, etc.

Processing. By interpretation or execution of compiled code from a cache. We chose interpretation since it better fits inherent characteristics of network processors [9].

2 Resource Aware System

2.1 Execution Time

Even if untrusted code cannot perform harmful operations thanks to sandboxing, it is impractical to detect malicious code that will run silly loop, consuming available CPU time on routers without performing any 'useful' job.

Like in SNAP[5], WASP avoids this by prohibiting *backward jumps* and providing only instructions with predictable execution time. Together with in-place processing of packets at kernel level, this allows our lightweight control tasks to be performed at the lowest cost for the router.

2.2 Memory and Storage

Most active protocols will need information to be stored temporarily on intermediate nodes, so that it can be later retrieved by other active packets. It is important for network availability and performance that this local storage remains easy to manage and can automatically discard information that is no longer pertinent.

ANTS [1] and many other platforms use *soft-state* based memory management to release memory that has not been used by packets for a given amount of time. Unfortunately, soft-state based managers make it hard for the access control to define if there will be sufficient memory to accept the flow.

Alternatively, ESP [4] proposes the *ephemeral state* approach, where data are kept for a constant period, *regardless of how frequent the data is referenced during that period.* If, in addition, all the data slots in the store have the same size, collecting free-for-reuse slots becomes simple enough to execute without disturbing packet forwarding tasks on the router, and flow access checking simply requires that the router checks how many different slots are used by the flow. We will call *tags* these fixed-size data that the node associates with a key for a fixed amount of time, like in the ESP terminology.

2.3 Network Bandwidth

Taking care of local resources is required to achieve platform *safety* but not sufficient. If no restriction is enforced, an ill-intentioned active packet could easily create clones of itself will all the allowed execution times, and clones of the clones at the next router so that a single emitted packet will overload the destination (in addition of network links close to that destination). In the case of the *WASP* platform, packets do not have the ability to create child packets unless they are targetted at a multicast address on a multicast router. All it can do is block the packet or send it back to its source.

3 Service Discovery with WASP

Literature has presented a number of active network-based solutions to various problems such as real-time auction, hierarchical web caches, video stream filters, reliable multicast and many more. All these applications can be viewed as applying a custom *service* (filtering, merging, splitting of packets) at some strategic points in the network. It is somehow expected that, sooner or later, network operators will integrate such services in their network as their presence could help reduce the required bandwidth. The behaviour of an active service can thus be seen as follows:

1. identify anchor points in the network (*i.e.* machines able to host the service),
2. detect at which anchor point(s) service deployment is strategically most useful,
3. route packets requiring the service towards deployed *service provider(s)*[1],
4. apply merge/split/filter service on packets received by the service provider(s).

In a grid computing environment, those services could for instance consist of collecting available computations sites, routing computations requests towards the closest

[1] *i.e.* "nodes offering the packet filter/split/merge service", not Internet Service Providers.

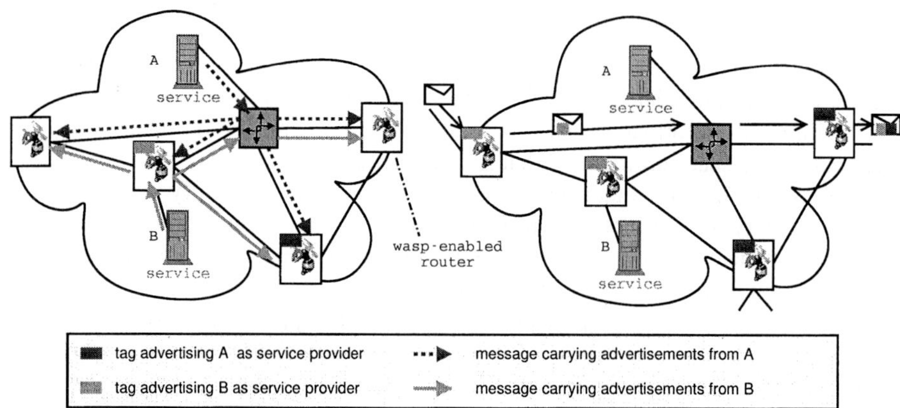

Fig. 1. Advertising (left-side) and looking for (right-side) service

(or less loaded) point of presence or even perform distribution/aggregation of computation requests hierarchically.

3.1 The Managed Dynamic Overlay Alternative

Most existing solutions for such problems, like OPUS [7] or the X-Bone [8], set up an infrastructure that dynamically creates *overlay networks* interconnecting end-systems and intermediate service providers with *tunnels* in order to obtain the desired logical topology. Unfortunately, none of these works have suggested a truly scalable and completely decentralised method for identifying available anchor points in a very large-scale network. Moreover, maintaining the infrastructure, monitoring the available resources and expressing applications' needs in a generic fashion remains a resource-intensive activity even when hierarchically divided such as in OPUS.

In the following, we will show how, with sensibly less support from all sides, WASP manages to offer enough information to *end-systems* and *service providers* so that they can take strategic decisions themselves.

3.2 Basis of Service Discovery with WASP

The idea behind the WASP platform is to provide a lightweight environment that can be used for locating the most interesting service provider(s) independently of what the service will actually do. Once service-providing node(s) have been located, the end-system can adjust application behaviour so that the applicative flow go through the discovered provider(s).

When a new applicative flow is initiated, small active packets are used to probe the network on the route to be taken. Each time such probe crosses a WASP node, it will lookup the node store to see if it can find *advertisements* of the expected service, consisting of the provider address and cost for reaching that provider from the local node. Depending on the application needs, probes will perform some pre-filtering of the collected advertisements or simply store them all. Based on this collected knowledge of the network, the end-system can evaluate the different options and enforce the one that will result in the best utility.

The same kind of active packets can also be used by the service providers to install advertisements in routers of the local domain. Again, the programmability of the advertisement packets allows expression of simple policies such as only keeping the advertisement(s) of the (k) nearest service provider(s) in a router, and report other service providers to the advertiser.

Figure 1 illustrates that two-phase process: servers A and B first flood the domain with WASP packets advertising their presence, avoiding to re-install a tag in a router that already contains a better tag (e.g. advertising a closer or less loaded service provider). A source S can then use another WASP packet to record those tags as a list of provider P and branchpoint X information: $(P_{addr}, X_{addr}, cost(S, X), cost(X, P))$. Note that by simply changing the program in service advertisement and lookup packets, we are able to select the service provider that is closest to the source or to the destination, or to keep only the provider that will lead to smallest path deviation in each domain and leave final selection to the end systems.

3.3 Flooding Locally

In order to advertise the service, providers have to locate WASP routers in the local domain and send them WASP packets that will install advertisement tags. We benefit here from the fact that WASP processing is *optional* so no overlay of WASP-enabled routers need to be pre-established. In our previous work [6], we show how knowing the routing table of the local domain suffices to discover all the active routers of that domain.

A particularity of *ephemeral* storage is that the advertising tag will be deleted after a fixed period τ, regardless of any refresh we could try to perform. Therefore, there may be a small delay between the moment where a WASP router decides to remove an advertisement tag and the moment where an advertisement refresh comes. Even if the server manages to learn precisely the tag's lifetime τ it cannot completely avoid the risk that client packets may not see any advertisement. If this risk cannot be afforded, it is still possible for a service to use two separate keys k_1 and k_2 that will be refreshed with a period $\tau + \epsilon$ but such as advertisements of k_1 and k_2 are separated by a delay of e.g. $\tau/2$. A client that doesn't find the "primary tag" (referenced by k_1) can then check the "backup tag" (referenced by k_2) to see whether the service is really missing.

4 The WASP Platform

As we previously said, WASP is derived from ESP router[4]. An ESP node consists of *Ephemeral State Stores* (ESS) containing *tags* that packets access based on 64 bit *keys*. Each packet requests the execution of one of the pre-defined operations on certain tags. Despite operations can be modified a bit (e.g. changing threshold values, selecting operators, etc), they remain thighly bound to multicast-related applications. It is also a bit disappointing to see that as soon as one wishes to implement more complex feature such as *reliable* flows merging, other very-specific operations need to be added.

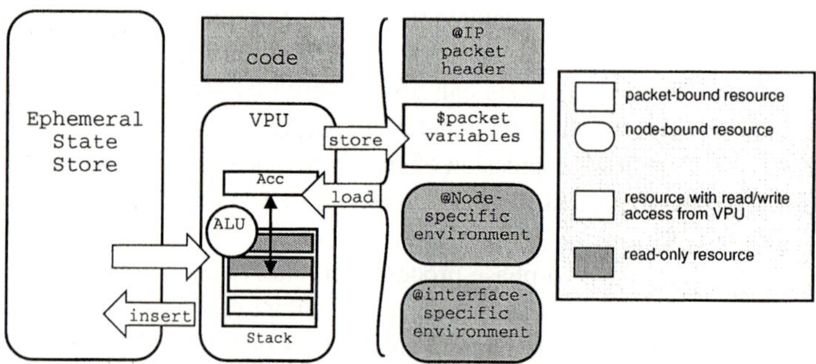

Fig. 2. WASP Execution Environment

The WASP platform thus keeps the overall design of ESP but replaces pre-defined operations by a *virtual processor* interpreting a bytecode language inspirated by SNAP [5]. Another couple of extensions have been brought to ESP in order to allow more efficient solutions to services and communities discovery and flow management tasks in general, like the "return" behaviour and protected tags, explained later.

4.1 WASP Packets

WASP uses the *active packets* paradigm: each packet contains its own code and the data on which it can operate. WASP code and data can be stored in the payload of an IP packet or it can be piggybacked on another packet as an IP header extension. The packet's code consists of up to 256 *microbytes* for a WASP *Virtual Processing Unit* that will eventually lead to a packet control instruction telling whether the packet should continue towards its destination, return to its source or be dropped. The data part of the WASP packet is available as a 128 byte of RAM to the VPU during packet interpretation. Other parts of the packet (code, IP header, payload) are not alterable by the WASP code and only the IP header is readable. Figure 2 shows how node and packet storage areas are viewed by WASP code.

4.2 The WASP Node

A WASP node has several *Ephemeral State Stores* that associate 64 bit keys with small, fixed-size data into *tags*. Each ESS is bound to a *Virtual Processing Unit* that processes the WASP packets on a given *location* like e.g. "incoming on eth0". The VPU state is reset everytime a new packet is processed, which means that all the communications and exchanges between packets will occur *in the ESS* associated with the VPU.

Each VPU on the node exports a few information for WASP packets, like its IP address, netmask, the local node time, etc. Outgoing VPU will also export interface-related information like queue length and number of packets sent. These environment variables appear as a bank of read-only memory for the WASP VPU and allow the WASP programmer to design various monitoring or self-adapting services.

Table 1. Relative timings for ESP operations processing in CPU cycles, sorted by ESS accesses

operation	ESS	WASP, nocache	WASP, cached	WASP, MAPping	native ESP
forward	0	129	123	-	-
compare	2	549	543	-	316
count	2	721	592	586	349
collect	4	1245	958	842	633
rchild	6	2058	1845	1509	775
rcollect	8	2980	2394	2020	1091

4.3 Super Packets and Protected Tags

In existing applications using ESP router, there's no need for access control to tags. It is simply assumed that the each source picks up a random 64 bit word and uses it as a key. Chances that two sources randomly pick the same (publicly available) tag and send packets over routes that cross the same router (otherwise no collision occurs) are virtually nul. When using WASP to advertise services, however, participants are required to use a *well-known* tag value that both service advertiser and service user will put in WASP packets. Unfortunately, we cannot safely use public well-known tags as an attacker could hijack the traffic of a given operator to its own network by advertising his machine as a proxy on the operator's network.

To solve such problems, WASP introduces *protected tags* that can only be modified by *super packets*. The node tells whether a tag is protected or not by checking its key against a specific pattern, and will allow writes to such tags only to packets that are marked 'trusted' in their WASP header. All a network manager will have to do in this case is (1) filter out WASP super packets at ingress nodes from the outside and (2) use super packets to advertise services within his own network.

5 Performance Evaluation

These tests are based on the linux module version of ESP software, running on a 1GHz Pentium 3 machine with default compilation options. Timings were measured using the internal *time stamp counter* of the processor, averaging on 1000 tests to avoid any unwanted side effects of caches, etc. The virtual node state is maintained such that the same (longest) code sequence is evaluated at each iteration. For each of the five ESP instructions on that distribution (*compare*, *count*, *collect*, *rchild* and *rcollect*), we wrote an equivalent *WASP* packet. Note that in order to achieve good performance, it is usually required to tune packet code so that the data organization better suits the instruction flow, as illustrated by Table 5.

One key feature for fast interpretation of WASP code will be how good the interpreter is at avoiding repetitive access to the ESS, and one way to achieve this is by *caching* intermediate results. Tests carried were in favour of very small caches (1 entry) since more complex policies tend to eat all the cache benefit in their initialization.

Alternatively, we can offer larger values for each key in the ESS. Instead of having a single 64-bit word, we now allow a whole memory bank of 32 bytes which can be *mapped* in the VPU's memory. In complex operations like *rcollect/rchild*, the state we process is no longer atomic, but instead consists of tuples. While ESP then requires one key per field (and thus one ESS lookup at least per field), WASP allows a few fields to be grouped together as long as they fit one 32 byte bank.

Our tests with the Pentium-based implementation shows an improvement of 18% (two variables per bank) to 35% (four variables per bank) in the processing time as soon as several variables need to be updated, and we expect improvement to be even more important on saturated ESS storage (e.g. when collisions occur inside of the ESS hash table).

6 Conclusion and Future Work

We have presented a lightweight active platform that combines advantages of ESP's per-node storage and SNAP's safe and efficient language. Despite its use of a bytecode interpreter instead of native code, our work still shows execution performance of only 150% to 200% of corresponding native code and is much more generic than the existing ESP framework.

The proposed platform elegantly solves the problem of locating available third-party service providers. We also expect that it could also help peers of a community to find each other, even without the help of a "pong server" and we are investigating the possibility of having *private* tags that would be restricted to packets from the same 'protocol'.

Acknowledgment

We would like to address special thanks to Jiangbo Li from Kenneth L. Calvert's team for having so kindly replied to all our questions related to ESP.

This work has been partially supported by the Belgian Science Policy in the framework of the IAP program (Motion PS/11 project) and by the E-Next European Network of Excellence.

References

1. D. Wetherall, A. Whitaker: *ANTS - an Active Node Transfer System. version 2.0* http://www.cs.washington.edu/research/networking/ants/
2. J. Moore and S. Nettles: *Towards Practical Programmable Packets*, In Proc. of the 20th IEEE INFOCOM. Anchorage, Alaska, April 2001.
3. E. Nygren, S. Garland, and M. Kaashoek: *PAN: A High-Performance Active Network Node Supporting Multiple Mobile Code Systems*, In Proc. of IEEE OPENARCH, pp. 78-89, New York, March 1999.
4. K. Calvert, J. Griffioen and S. Wen: *Lightweight Network Support for Scalable End-to-End Services*, in Proc. of ACM SIGCOMM, pp. 265-278 Pittsburg, PA. August 2002.
5. Jonathan T. Moore: *Safe and Efficient Active Packets*, Technical Report MS-CIS-99-24, University of Pennsylvania, October 1999.
6. S. Martin and G. Leduc: *A Dynamic Neighbourhood Discovery Protocol for Active Overlay Networks*, in Proc. of IWAN, pp. 151-162, Kyoto, Japan, December 2003.

7. R. Braynard, D. Kostić et al. *Opus: an Overlay Peer Utility Service*, in Proc. of the 5th IEEE OPENARCH, pp 168-178, New York, June 2002.
8. J. Touch and S. Hotz, *Dynamic Internet Overlay Deployment and Management Using the X-Bone* in Proc. of ICNP 2000, Osaka Japan, pp. 59-68.
9. Intel Corporation *The IXP1200 Hardware Reference Manual*, August 2001.

p2pCM: A Structured Peer-to-Peer Grid Component Model

Carles Pairot[1], Pedro García[1], Rubén Mondéjar, and Antonio F. Gómez Skarmeta[2]

[1] Department of Computer Science and Mathematics, Universitat Rovira i Virgili
Avinguda dels Països Catalans 26, 43007 Tarragona, Spain
{carles.pairot, pedro.garcia}@urv.net
[2] Department of Computer Engineering, Universidad de Murcia
Apartado 4021, 30001 Murcia, Spain
skarmeta@fcu.um.es

Abstract. In this paper we present p2pCM, a new distributed component-oriented model aimed to structured peer-to-peer grid environments. Our model offers innovative contributions like a lightweight distributed container model, an adaptive component activation mechanism, which takes into account network proximity, and a decentralized component location and deployment service. We believe that all of the features our component-oriented model provides can be very promising for the development of future wide-area distributed applications.

1 Introduction

Developers face an important problem when willing to build wide-area component-based distributed applications. Practically there are no frameworks oriented to such target market. In this paper we present **p2pCM**: a peer-to-peer component model for building distributed component-based applications on top of a **structured peer-to-peer grid**. Our model offers traditional and novel component services by means of an underlying object middleware (Dermi [3]). To the best of our knowledge, this is the first component model built on top of a structured P2P grid infrastructure.

The main contributions of this paper are the design of a new wide-area component model that runs on top of a structured P2P grid infrastructure; the utilization of a decentralized component location and deployment facility; and the mplementation of a resilient and autonomous lightweight container model, which provides component's life cycle services, and many others. Component instantiation is of special interest because it takes into account network locality properties.

The rest of the paper is structured as follows, in Section 2 we give an overview of p2pCM's architecture and services, in Section 3 we analyze possible usage scenarios for our component model, and in Section 4 we explore related work. We conclude in Section 5 by providing an outline of future work.

2 p2pCM Architecture and Services

The decentralized component model we have designed (p2pCM) runs on top of a structured peer-to-peer grid. We have implemented the majority of traditional compo-

V.S. Sunderam et al. (Eds.): ICCS 2005, LNCS 3516, pp. 246–249, 2005.

nent models services, and adapted them to the underlying topology. These include a **decentralized component location and deployment facility**, and a **decentralized lightweight container model**, which provides a component persistence and life cycle service, as well as an adaptive activation policy.

2.1 Decentralized Component Location and Deployment

All components must be previously registered into the system so as they can be used by any client. This *deployment* phase is found in all traditional component models, as well as in ours. In our model, we use a decentralized naming service which stores all component's metadata (read in *deployment* time), as well as the component's class files. This is done to allow dynamic component class loading in clients that do not have the necessary component classes. Our naming service benefits from the efficient overlay network routing properties thus hashing the component's identifier, and storing the values into the node whose key is closer to this hash. Fault tolerance is provided by he underlying Dermi [3] middleware in a transparent way to the developer.

2.2 Decentralized Container Model

In p2pCM we have tried to avoid heavy and monolithic containers used by traditional component models (e.g. J2EE or CORBA), and opted for designing a **decentralized lightweight container model**. In our case, all of the nodes that belong to the network are containers, and as such, they can house components. The idea is that any component can be run in any node (unless restricted by security constraints), because each node runs a lightweight container. Our containers are fault resilient and autonomous: components are replicated all along the network using the underlying Dermi [3] middleware. If a container fails, surely other containers housing those components will exist in the network. Now we are going to briefly describe the different services our decentralized container model offers to components.

2.2.1 Component Life Cycle Service

The container is responsible for activation and passivation of components. When creating a component's instance, the following may happen:

- *No other component instance is already active in the network.* In this case, the instance will be activated on our local container, and a local reference is returned so as we can interact with it.
- *Other component instances are already active in the network.* In such case, we will get a reference to the **closest instance** (basically, a Dermi *anycall* is done). However, if this closest instance informs us that it cannot accept more requests (it may be overwhelmed), a local copy is activated.

Component instances are passivated when they are not called during a certain amount of time in order to save resources on the node.

2.2.2 Component Persistence

Stateful components need to be ensured state persistence. p2pCM allows this by providing a persistence mechanism in order to take care of total passivation of component instances. If all component instances are passivated, their shared state is lost, and this situation is not desired. The persistence strategy is chosen by the component itself by means of the *persistence* metadata tag in the deployment phase. Persistence itself is managed by the decentralized object replication mechanisms of Dermi.

2.2.3 Adaptive Component Activation

In wide-area environments, some component instances could become temporarily overwhelmed with requests of nearby clients. To avoid such problem, we use an **adaptive component activation** mechanism.

The idea is to know which immediate node is delivering most messages to the overwhelmed node (via the *getPreviousHopHandle()* method), and start an algorithm which keeps activating new component instances all along the overwhelming request path, until stress is relieved.

We are currently working in this adaptive activation scheme in order to fine tune several parameters, and as future work we will perform extensive simulations to validate our approach.

3 p2pCM Usage Scenario

One scenario where our model could be successfully applied would be in an application similar to a SETI@Home or United Devices Cancer Research Project model. Such applications normally require a central server which distributes computing units to home computers for analyzing. Our component model could be used to build an application which would efficiently be fault tolerant and resilient to high request peaks.

The idea would be that interested nodes would activate a *Processing* component on their local containers. This component could thus be receptive to a *DataFeed* component's calls. The *Processing* components would be aligned into a multicast group, so as the data feeder requested data unit analysis by *anycalling* or *multicalling* [3] to the *Processing* component group.

Once each *Processing* component finished its data unit analysis, instead of returning data back to the central *DataFeed* component (which would possibly create an important bottleneck there), it would pass data to another *Results* component by means of an event. This component would manage the results persistence by replicating itself and storing state information in the P2P grid infrastructure. Whenever the *DataFeed* component wished to obtain the results (cyclically every x hours), it would *anycall* to its closest *Result* component instance, which would provide it with the latest gathered results.

4 Related Work

In the Grid world there exist solutions like Fractal/ProActive [1], which is a hierarchical and dynamic component model. Nevertheless, its approach is different form p2pCM, in the sense that virtual node and virtual machine mapping is performed on the component's deployment descriptor, thus not allowing self adaptation to node failures.

P2PComp [2] is built on top of an unstructured P2P network, and it is mainly aimed to address the development needs for mobile P2P applications. It features a lightweight container model as well, and provides many services, including synchronous / asynchronous remote invocations, hot swapping, service fetching and ranking. All these services are basically oriented for highly mobile and dynamic applications.

5 Conclusions and Future Work

In this paper we have presented p2pCM, a decentralized component model built on top of a structured P2P Grid. Our approach manages to provide many interesting services to component developers. Our main contributions include a **decentralized component location and deployment facility**, and the adoption of a **decentralized lightweight and fault resilient container model**. We have shown as well, possible usage scenarios of our system. However, we still do not provide other services found in commercial component-oriented models, like security or transactions, which need to be further investigated.

A first prototype of p2pCM is available through http://ants.etse.urv.es/p2pcm. We believe that the features p2pCM offers can be very promising for the development of future wide-area distributed applications.

This work has been partially funded by the Spanish Ministry of Science and Technology through project TIC-2003-09288-C02-00.

References

1. Baude, F., Caromel, D., and Morel, M., "From Distributed Objects to Hierarchical Grid Components", *Proc. DOA 2003,* LNCS, pp. 1226-1242, Nov. 2003.
2. Ferscha, A., Hechinger, M., et al, "A Light-Weight Component Model for Peer-to-Peer Applications", *Proc. of ICDCS 2004*, pp. 520-527.
3. Pairot, C., García, P., and Gómez Skarmeta, A. F., "Dermi: A New Distributed Hash Table-based Middleware Framework". *IEEE Internet Computing*. Vol 8, No. 3, 2004.

Resource Partitioning Algorithms in a Programmable Service Grid Architecture

Pieter Thysebaert, Bruno Volckaert, Marc De Leenheer, Filip De Turck, Bart Dhoedt, and Piet Demeester

Department of Information Technology, Ghent University - IMEC, Sint-Pietersnieuwstraat 41, B-9000 Gent, Belgium
{pieter.thysebaert, bruno.volckaert}@intec.ugent.be

Abstract. We propose the use of programmable Grid resource partitioning heuristics in the context of a distributed service Grid management architecture. The architecture is capable of performing automated and exclusive resource-to-service assignations based on Grid resource status/properties and monitored service demand. We present two distinct approaches for the partitioning problem, the first based on Divisible Load Theory and the second built on Genetic Algorithms. Advantages and drawbacks of each approach are discussed and their performance is evaluated using NSGrid. Results show that automated resource-to-service partitioning simplifies scheduling decisions, improves service QoS support and allows efficient computational/network resource usage.

1 Introduction

As more and more application-types are being ported to Grid environments, an evolution from pure computational and/or data Grids to full-scale service Grids [1] is taking place. A "service Grid" denotes a Grid infrastructure capable of supporting a multitude of *application types* with varying QoS levels. With widespread Grid adoption also comes the need for automated distributed management of Grids, as the number of resources offered on these Grids rises dramatically. Automated self-configuration/optimization of Grid resource usage can greatly reduce management complexity, and at the same time achieve better resource utilization [2]. In this paper, the focus is on the automated deployment of resource partitioning algorithms, which intelligently (i.e. based on current service needs and Grid status) assign Grid resources (network, computing and data/storage resources) to a particular service class for exclusive use during a specified time frame. In doing so, we wish to improve service class priority support and Grid resource utilisation while at the same time simplifying scheduling decisions. Well-known service-driven Grid scheduling frameworks such as AppLeS [4] and GrADS [5] differ from our approach in that we use a Service Management Architecture which operates independent of the scheduling system and actively monitors application behaviour at runtime.

In order to compare the performance of a service managed Grid versus a non-service managed Grid we use NSGrid (reported upon in [3]), an ns-2 based Grid

V.S. Sunderam et al. (Eds.): ICCS 2005, LNCS 3516, pp. 250–258, 2005.

simulator capable of accurately modeling different Grid resources, management components and network interconnections. More specifically, we evaluated both GA-based and DLT-based resource partitioning strategies, both when network aware and when network unaware scheduling algorithms are used.

The remainder of this paper is structured as follows: section 2 gives an overview of the service management architecture. Section 3 elaborates on the different resource partitioning strategies, while their evaluation in a typical Grid topology is discussed in section 4. Concluding remarks are presented in section 5.

2 Service Management Architecture

We regard a Grid as a collection of *Grid Sites* interconnected by WAN links. Each Grid Site has its own resources (computational, storage and data resources abbreviated as CR, SR and DR respectively) and a set of management components, all of which are interconnected by means of LAN links. Management components include a *Connection Manager* (capable of bandwidth reservation support, and responsible for monitoring available link bandwidth), an *Information Service* (IS) (storing the properties and status of the registered resources) and a *Scheduler*. Every resource in our model is given an associated service class property (stored in the Information Services). The basic unit of work in our model is a *job*, which can roughly be characterised by its length (execute time on a reference processor), required input data, the amount of output data and the service class to which it belongs. Each Grid Site has one or more *Grid Portals* through which users can submit their jobs. Once submitted, a job gets queued at the *Scheduler*, which in turn queries both local and foreign ISs for resources adhering to the job's requirements. Once the results of those queries are returned, the Scheduler applies one of its scheduling algorithms and (if possible) selects one or more DRs (for input data), together with one or more SRs (for storing output data) and a CR (for processing). If the scheduling algorithm is network aware, the Connection Manager is queried for information about available bandwidth on paths between resources and, once a scheduling decision is made, attempts to make connection reservations between the selected resources.

A distributed service management architecture was implemented in NSGrid in order to evaluate the performance of different resource partitioning strategies. Each Grid Site has a local *Service Manager* interacting with the local IS, Connection Manager and *Service Monitor*. The Service Monitor component monitors local characteristics of each service class; it stores inter-arrival times, I/O data requirements and processing length of jobs. At specified intervals, the Service Monitor sends the collected information to all known foreign Service Monitors, so they can keep their 'foreign service characteristics' information up-to-date.

The Service Manager queries the local Service Monitor for information regarding the different services. When the monitored service characteristics do not differ (with regard to a certain threshold) from the ones used to partition the Grid resources in a previous run, no repartitioning will occur. If this is not the case, or if no partitioning has been done yet, the Service Manager will query the

ISs for Grid resource properties/status. Once the answer to these queries has been received, one of the resource partitioning algorithms (detailed in section 3) is applied to the resource set, and the resulting solution is sent back to the ISs, which in turn change the service-exclusive attribute of their stored resources.

3 Partitioning Strategies

Recall that we are trying to partition resources into service class resource pools. A solution to this problem is a mapping from resource to a particular service type, and this for all resources returned from the Service Manager - IS queries. A resource can also be assigned service type '0', meaning it can be used by any service type. Exhaustively searching for a cost function optimum quickly becomes infeasible, as the number of solutions that needs to be evaluated is $(\#servicetypes+1)^{\#resources}$. To find a suitable solution in reasonable time, we used two distinct approaches: one uses Divisible Load Theory while the other uses a Genetic Algorithm to obtain a resource-to-service mapping.

3.1 DLT-Based Partitioning

Whenever a Grid reaches a steady state (e.g. a Grid processing a periodic load), stochastic parameters regarding the distributions of job IAT, duration and I/O-needs can be derived for each Service Type by the Service Monitoring Architecture. These parameters can then be used to fuel an ILP designed to

1. Assign an exclusive Service Type to each Computational Resource.
2. Determine the optimal *schedule* of the periodic workload over the Grid's resources, taking into account the Service Type assignation.

An approximation used to limit the number of integer variables in this problem is to treat the aggregate workload as arbitrarily divisible (hence the name "Divisible Load Theory") [6]. In this context, values of interest are $arrivals_s^n$ - the load per time unit arriving at site s and belonging to service type n, $Sets_n$ and $Size_n$ - the datasets available to service type n jobs and their respective sizes. Main decision variables in the problem are $x_{c,n}$ (binary, assigning resource type n to CR c) and $\alpha_{i,n}^c$ (real-valued, amount of service type n load per time unit processed at CR c which arrived at site i). Auxiliary variables needed to fulfill routing constraints on the input datasets and generated output data have been dubbed $in_{n,j}^l$ (bandwidth needed on link l for transport of dataset j of service type n) and out_s^l (bandwidth needed on link l for transport of output to SR s).

Using the Divisible Load approach, the resource-to-service assignation can now be modeled as a cost minimization problem with several classes of constraints[1]. The capacity constraints to be observed are

$$\forall c \in CR. \sum_{i \in Sites} \sum_{n \in ST} \alpha_{i,n}^c \leq Cap_c \tag{1}$$

[1] Abbreviations used:GW = Gateways, L^+ = outgoing links, L^- = incoming links.

$$\forall l \in L. \sum_{n \in ST} \sum_{j \in Sets_n} in^l_{n,j} + \sum_{s \in SR} out^l_s \leq Cap_l \tag{2}$$

Network traffic is routed according to following constraints:

$$\forall n \in ST, j \in Sets_n. \sum_{d \in DR: j \in Sets_d} \sum_{l \in L^+_d} in^l_{n,j} = \frac{\sum_{s \in Sites} arrivals^n_s \times Size_n}{|Sets_n|} \tag{3}$$

$$\forall c \in CR, n \in ST, j \in Sets_n. \sum_{l \in L^-_c} in^l_{n,j} = \frac{\sum_{i \in Sites} \alpha^c_{i,n} \times Size_n}{|Sets_n|} \tag{4}$$

$$\forall c \in CR, s \in SR. \sum_{l \in L^+_c} out^l_s = \sum_{n \in ST} \alpha^c_{Sites_s,n} \times Size_n \tag{5}$$

$$\forall s \in SR. \sum_{l \in L^-_s} out^l_s = \sum_{n \in ST} arrivals^n_{Sites_s} \times Size_n \tag{6}$$

$$\forall g \in GW, n \in ST, j \in Sets_n. \sum_{l \in L^-_g} in^l_{n,j} = \sum_{l \in L^+_g} in^l_{n,j} \tag{7}$$

$$\forall g \in GW, s \in SR. \sum_{l \in L^-_g} out^l_s = \sum_{l \in L^+_g} out^l_s \tag{8}$$

A feasible schedule is obtained by

$$\forall i \in sites, n \in ST. \sum_{c \in CR} \alpha^c_{i,n} = arrivals^n_i \tag{9}$$

Constraints concerning the exclusive reservation of each CR:

$$\forall c \in CR. \sum_{n \in ST} x_{c,n} = 1 \tag{10}$$

$$\forall c \in CR, n \in ST. \sum_{i \in Sites} \alpha^c_{i,n} \leq x_{c,n} \times Cap_c \tag{11}$$

The "cost" to be minimized can take on several forms; for instance, the total amount of data traveling over network links per unit of time (in the steady-state Grid) can be described in terms of problem variables as

$$\sum_{l \in L} \left(\sum_{n \in ST, j \in Sets_n} in^l_{n,j} + \sum_{s \in SR} out^l_s \right) \tag{12}$$

Using this cost function in the ILP results in a workload schedule and Service Type assignation yielding minimal aggregate network load for a given arrival process. Alternatively, one can choose to minimize the maximal unused CR fraction, which results in an "even" workload distribution across all CRs according to their respective capacities. This can be modeled by adding the constraints

$$\forall c \in CR, n \in ST. cost \geq \frac{\left(x_{c,n} \times Cap_c - \sum_{i \in Sites} \alpha^c_{i,n} \right)}{Cap_c} \tag{13}$$

3.2 GA-Based Partitioning

The resource type assignment can easily be encoded into an n-tuple of service type IDs, where n equals the number of resources. These *chromosomes* can then be fed to a Genetic Algorithm which evaluates the fitness of each chromosome (i.e. possible service type assignment) w.r.t. a cost function (see algorithm 3.1). Unlike an ILP, this cost function need not be "linear" in the decision variables, giving this approach more expressive power than the DLT-based partitioning.

Algorithm 3.1: GENETIC ALGORITHM(*resources*)

```
population_initial ← (b(1,0),...,b(m,0)), t←0
while stopcondition false
  do {
    for i←1 to m   comment: proportional selection
      do {
        x←rand[0,1], k←1
        while k<m and x < Σ_{j=1}^{k} f(b_{j,t}) / Σ_{j=1}^{m} f(b_{j,t})
          do k←k+1
        b_{i,t+1}←b_{k,t}
      }
    for i←1 to m−1 step i+2   comment: two-point crossover
      do {
        if rand[0,1] ≤ ρ_C
          then {
            pos1←rand[1,m], pos2←rand[1,m]
            if pos1>pos2
              then switch(pos1,pos2)
            for k←pos1 to pos2
              do switch(b_{i,t+1}[k], b_{i+1,t+1}[k])
          }
      }
    for i←1 to m   comment: mutation
      do {
        for k←1 to m
          do {
            if rand[0,1] < ρ_M
              then b_{i,t+1}[k]←rand[0,#ST]
          }
      }
    t←t+1
  }
```

Global Service CR Partitioning. This cost function takes into account the computational processing needs and priority of the different service types (ST). The Service Manager queries the ISs for all local CRs and calculates average service processing needs $\forall ST \cdot ppower_{req_{ST}} = sites_{ST} \times \frac{ptime_{ref_{ST}}}{IAT_{ST}}$. Average processing time of a job from service class ST on a reference processor is denoted by $ptime_{ref_{ST}}$, while $sites_{ST}$ denotes the amount of Grid portals launching jobs from this service class. The relative processing power assigned to a service type is then given by: $\forall ST \cdot ppower_{asg_{ST}} = \sum_{\forall CR \in ST} \frac{speed_{CR}}{speed_{CR_{ref}}} \times ptime_{ref_{ST}}$. The importance of assigning resources to foreign service types can be adjusted by the local Service Manager by tweaking the foreign service policy $\rho_{ST_{foreign}}$. Once CR query answers have been received, GA 3.1 will be started with cost function 3.2 (GA equivalent of equation 13), with an objective to assign each service type a same amount of processing power *relative* to their requested processing power.

Network Partitioning. Since the Service Monitor keeps track of I/O data characteristics of each service, data intensiveness relative to the other services can be calculated. This in turn can be used to perform per-service network bandwidth reservations. We have implemented a proof-of-concept network partitioning strategy, in which the Service Manager calculates average data requirement percentages for each service class $bw_{req_i} = \frac{\frac{bw_{input_i} + bw_{output_i}}{IAT_i}}{\sum_{\forall j \in ST} \frac{bw_{input_j} + bw_{output_j}}{IAT_j}}$ and passes this information to the Connection Manager, which will make service type bandwidth reservations on all network links for which it is responsible.

Algorithm 3.2: $f_{CRpart_{global}}(x)$

$result \leftarrow \frac{ppower_{asg_0}}{2}$, $maxAlloc_{over} \leftarrow 0$, $maxAlloc_{under} \leftarrow 0$

for $i \in ST_{local} \cup ST_{foreign}$

$$\textbf{do} \begin{cases} aux \leftarrow ppower_{req_i} - ppower_{asg_i} \\ \textbf{if } aux < 0 \\ \quad \textbf{then} \begin{cases} \textbf{if } -aux > maxAlloc_{over} \\ \quad \textbf{then } maxAlloc_{over} \leftarrow -aux \\ aux \leftarrow ppower_{asg_i} \end{cases} \\ \quad \textbf{else} \begin{cases} \textbf{if } \frac{aux}{ppower_{req_i}} > maxAlloc_{under} \\ \quad \textbf{then } maxAlloc_{under} \leftarrow \frac{aux}{ppower_{req_i}} \\ aux \leftarrow ppower_{asg_i} - aux \end{cases} \\ \textbf{if } i \in ST_{foreign} \\ \quad \textbf{then } aux \leftarrow aux \times \rho_{ST_{foreign}} \\ result \overset{+}{\leftarrow} \frac{priority_i}{(\sum_{j \in ST} priority_j)} \times aux \end{cases}$$

$result \overset{-}{\leftarrow} maxAlloc_{over} + maxAlloc_{under}$

4 Performance Evaluation

4.1 Simulation Setup

A fixed Grid topology was used for all simulations. First, a WAN topology (containing 8 core routers with an average out-degree of 3) was instantiated. Among the edge LANs, we chose 12 of them to represent a Grid site. Each site's resources and management components are connected through 1Gbps LAN links, while Grid site interconnections consist of dedicated 10Mbps WAN links. We have assigned 3 CRs to each Grid Site. To reflect the use of different tiers in existing operational Grids, not all CRs are equivalent: the least powerful CR has two processors (operating at reference speed). A second class of CRs has four processors, and each processor operates at twice the reference speed. The third CR type contains 6 processors, each operating at 3 times the reference speed. Conversely, the least powerful CR is 3 times as common as the most powerful CR, and twice as common as the middle one. We have assumed that SRs offer "unlimited" disk space. Each site has at its disposal exactly one such SR. Each site's DR contains 6 out of 12 possible data sets. These data sets are distributed in such a way that 50% of the jobs can have local access to their needed data.

We used 2 equal-priority service classes (each accounting for half of the total job load); one is more data-intensive, while the other is more CPU-intensive (see table 4.2). Jobs were scheduled using one of two scheduling algorithms; the first algorithm only uses available CR capacities to make decisions, while the second also takes into account network link loads [7]. Once sufficient statistical data about the job parameters had been gathered by the Service Monitor, the Service Manager was instructed to apply a partitioning algorithm to the Grid's resources. We measured average job response time (JRT) and network usage.

4.2 Comparison of DLT- and GA-Based Partitioning

In general, our GA-based partitioning strategy provides more functionality, as it is able to support different priority schemes, shared resource and local vs. foreign service differentiation. Its main drawback is the time needed to complete a GA run (with reasonable results); on our sample scenario, this takes about $800s$, while the DLT-based approach needs only $10s$. For the GA approach, we used Grefenstette's settings, with an initial population of 30, $\rho_C = 0.9$ and $\rho_M = 0.01$, and a stop condition of 100 runs.

	CPU-Job	Data-Job
Input(GB)	0.01-0.02	1-2
Output(GB)	0.01-0.02	1-2
IAT(s)	30-40	30-40
Ref. run time(s)	100-200	40-60

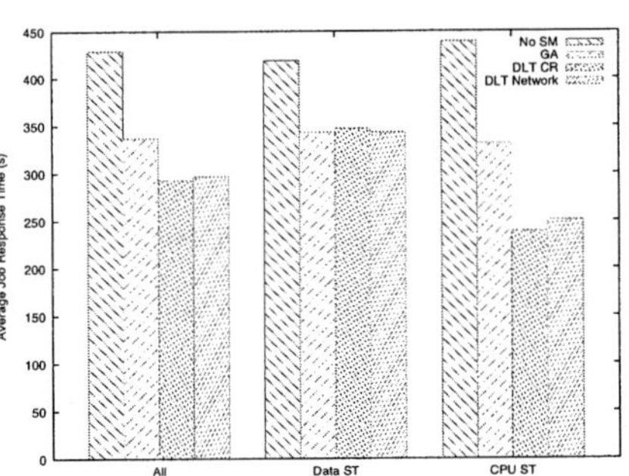

Fig. 1. Job Class Properties & Non-Network Aware Scheduling: Metrics

4.3 Non-network Aware Scheduling

Using a non-network aware scheduling algorithm over all Grid partition strategies results in average JRTs as shown in figure 4.2. Clearly, the use of Grid partitioning based on accurate job characteristic predictions has a positive influence. This behavior is due to resources being reserved for exclusive use by a service class, forcing the scheduler not to assign jobs to less-optimal resources (e.g. non-local access to needed input data), but to keep the job in the scheduling queue until a service-assigned resource becomes available. Optimizing the Grid partitions for minimal network usage does not yield a significant improvement, as the scheduling algorithm does not take into account network loads and diverges from the workload distribution as proposed by the DLT ILP.

4.4 Network Aware Scheduling

Values for average JRT and network usage observed when jobs are scheduled using a network aware algorithm are shown in figure 2. Again, when partitioning for optimal CPU usage, average JRTs are improved when compared to JRTs obtained when no partitioning strategy is used[2]. In analogy, after partitioning for minimal network utilization, network resources are less loaded.

[2] GA stands for GA-based computational partitioning, while GA-CONN denotes GA-based computational partitioning + network partitioning.

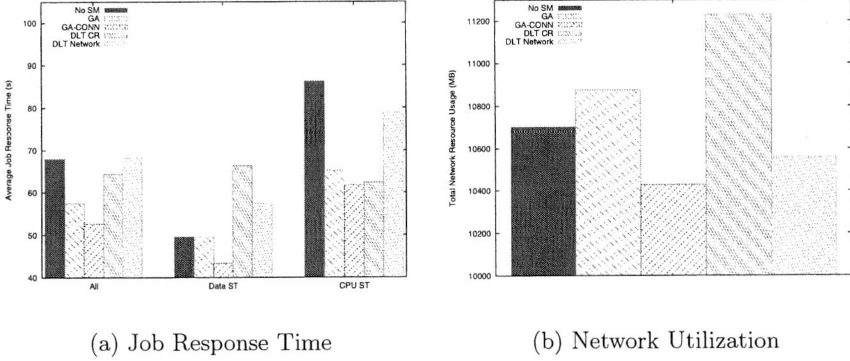

(a) Job Response Time

(b) Network Utilization

Fig. 2. Network Aware Scheduling: metrics

We measured the time it takes to calculate a scheduling decision and noticed a decrease in scheduling time of 28% when comparing the service managed Grid to the non-service managed Grid (i.e. from an average 12.71s in the non service managed case to 9.13s in the service managed Grid). This can be explained by the fact that when resources are partitioned among services, less resource query results will be returned to the scheduler, allowing easier scheduling decisions.

5 Conclusions

In this paper we proposed a distributed service management architecture, capable of monitoring service characteristics at run-time and partitioning Grid resources among different priority service classes. Two specific partitioning algorithms were discussed, and we indicated how our architecture dynamically invokes each algorithm with suitable parameters. We evaluated these algorithms using NSGrid: besides easing the process of schedule making decisions, Service Partitioning does not lead to a deterioration of Grid performance, both when job response times or network resource utilization is measured. A possible preference of one algorithm over the other depends on the trade-off between the size of the needed feature set and the algorithm's computational complexity.

References

1. I. Foster, C. Kesselman, J.M. Nick, S. Tuecke, "*Grid services for distributed system integration*", IEEE Computer, Vol. 35-6, pp. 37-46, 2002
2. J.O. Kephart, D.M. Chess, "*The vision of autonomic computing*", IEEE Computer, Vol. 36-1, pp. 41-50, 2003
3. B. Volckaert, P. Thysebaert, F. De Turck, P. Demeester, B. Dhoedt, "*Evaluation of Grid Scheduling Strategies through a Network-aware Grid Simulator*", Proc. of PDPTA 2003, Vol. 1, pp. 31-35, 2003

4. F. Berman et al., "*Adaptive Computing on the Grid Using AppLeS*", IEEE Transactions on Parallel and Distributed Systems, Vol. 14-4, pp. 369-382, 2003
5. H. Dail, F. Berman, H. Casanova, "*A Decoupled Scheduling Aproach for Grid Application Development Environments*", Journal of Parallel and Distributed Computing, Vol. 63-5, pp. 505-524, 2003
6. P. Thysebaert, F. De Turck, B. Dhoedt, P. Demeester, "*Using Divisible Load Theory to Dimension Optical Transport Networks for Computational Grids*", in Proc. of OFC/NFOEC 2005
7. P. Thysebaert, B. Volckaert, F. De Turck, B. Dhoedt, P. Demeester, "*Network Aspects of Grid Scheduling Algorithms*", Proc. of PDCS 11, pp. , 2004

Triggering Network Services Through Context-Tagged Flows

Roel Ocampo[1,2], Alex Galis[2], and Chris Todd[2]

[1] Department of Electrical and Electronics Engineering, University of the Philippines, Diliman, Quezon City, 1101 Philippines

[2] Department of Electronic and Electrical Engineering, University College London, Torrington Place, London WC1E 7JE

{r.ocampo, a.galis, c.todd}@ee.ucl.ac.uk

Abstract. Next-generation Grids will require flexible and adaptive network infrastructures that would be able to provide the requisite quality of service for computational flows. We discuss a mechanism where network flows are tagged with network context information, triggering dynamic network services, including QoS and adaptation services. An incremental deployment strategy and some initial results are presented.

1 Introduction

Due to the nature of the computational tasks that grids handle, the underlying resources they use are required to deliver nontrivial qualities of service [1]. In networked environments, where resources such as bandwidth are often shared, there may be situations where computational flows would have to compete for the use of resources along with flows from less critical applications. To deal with this, a pragmatic approach might be to differentiate the various network flows and set an adaptation policy P on a particular group of flows based on the class of users U generating the flow, for a particular type of usage or activity A, under a certain situation or set of network conditions C. In other words, an adaptation P is triggered by the general parameters (U, A, C).

Currently the mechanisms that exist to differentiate network flows provide very limited information about the flow, and are usually non-extensible. In this work, we propose the use of rich and extensible information in the form of network context as a means of differentiating flows and triggering network services, particularly adaptation services. In Sect. 2 we argue that end-hosts are rich sources of information needed in triggering adaptation services, and that a mechanism for sharing such information with the rest of the network is needed. In Sect. 3 we present the concept and design for such a mechanism, called context tagging, and Sect. 4 briefly discusses an incremental deployment strategy. Section 5 discusses some initial results. Section 6 briefly discusses efforts that are related to our current work, and finally Sect. 7 concludes and describes the work ahead.

V.S. Sunderam et al. (Eds.): ICCS 2005, LNCS 3516, pp. 259–266, 2005.

2 End-Hosts and Flow Context

As a consequence of the end-to-end principle [2], nodes such as routers often have a very limited view of the state of the network. While they often have a global view of the network's topological state as a result of routing exchanges, routers may have a limited, per-node awareness of other parameters such as traffic levels, congestion, or bit error rates. Other information that may characterize the state of the network, such as path loss, delay and jitter are usually sensed and processed at end-hosts.

Aside from end-to-end traffic-related parameters, end-hosts are rich sources of other types of information such as the characteristics and capabilities of end-devices, the applications generating the traffic, and the identities and activities of users such as their movement and location. All of these pieces of information are relevant in (a) determining whether a certain condition C exists within the network or user environment, (b) in identifying whether the traffic belongs to the class of users U, and (c) if the traffic is the result of usage activity A. These parameters also help in the design of appropriate adaptation policies applicable to the user group under these circumstances, or within that particular *context.*

2.1 The Context of a Flow

Context is defined in the New Oxford Dictionary of English as "the circumstances that form the setting for an event ... and in terms of which it can be fully understood and assessed." Dey, Salber and Abowd further define context as "any information that can be used to characterize the situation of entities ... that are considered relevant to the interaction between a user and an application, including the user and the applications themselves" [3]. Although the latter definition is tailored to interactions between users and applications, it can still be used as a template to describe the notion of context relevant to the interaction between a user application and the network, or between a user and the network.

One entity we consider relevant to this interaction is the *flow*, defined as distinguishable streams of related datagrams, typically resulting from the activity of a single entity [4]. The fact that flows are attributable to specific user groups and their activities makes them ideal sources of adaptation triggers U and A. We define the context C_f of a network flow as any information that can be used to characterize its situation, including information pertaining to other entities and circumstances that give rise to or accompany its generation at the source, affect its transmission through the network, and influence its use at its destination. This collectively includes not only the intrinsic, "low-level" characteristics of a flow, such as its traffic profile, but also the nature of the applications, devices, and the activities and identities of the users that produce or consume the flow.

We envision flow context to be used within the network in the following ways:

- to trigger adaptation directly on the flow itself. For example, the content of a flow may be compressed, transcoded, encrypted, classified, assigned a particular QoS treatment, marked, or rerouted in response to link constraints, network state, or application requirements [5].

- to trigger immediate or future network-wide management actions that are not necessarily targeted to exclusively affect the context-carrying flow. For example, a new flow may trigger the activation of a reserve link in order to guarantee sufficient bandwidth for future flows.
- to trigger network services that are specific or personalized for a user or group of users. A context tag may carry information about the user generating the flow, such as her profile or preferences. The network's response may be to offer these specific services to the user, or to automatically provide them in a transparent way. These new services may or may not directly affect the current flow.
- as a means to collect information for long-term purposes, without triggering any immediate adaptation or service, such as for traffic engineering or network optimization, or for consumption by cognitive or knowledge-based entities within the network [6].

Our approach involves tagging network flows with context obtained from sensors located mainly at end-hosts, by injecting packets that carry the flow's context within the flow itself. These context tags are preferably located at the start or within the first few packets of a flow, although there may be times where they may be reinserted anytime during the flow lifetime, especially to signal changes in the flow's context.

3 Context-Tagging Architecture

A diagram illustrating the main functional components in our scheme is show in Figure 1. Some of the components are based on a cognitive framework previously described in [7].

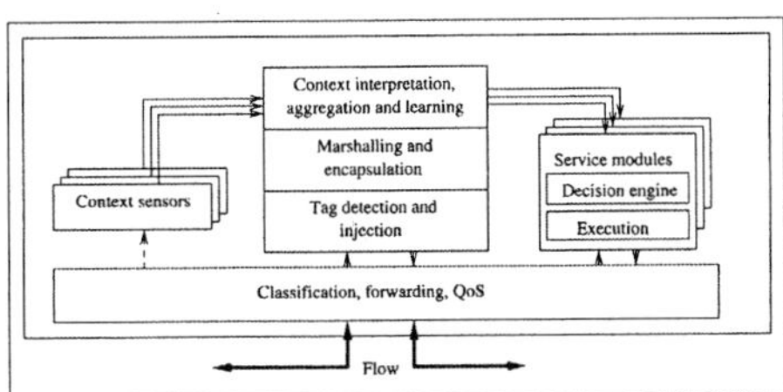

Fig. 1. Context tagging framework

3.1 Tag Creation and Injection

Sensors collect low-level context information, usually measurable data or low-level events. These are transformed into useful form through the process of interpretation, which may involve the application of an algorithm, comparison with a model, or the application of reasoning or other cognitive processes. Aggregation involves the examination of separate pieces of context information to generate new contextual knowledge.

The context information is then passed down to the marshalling and encapsulation stage, where it is transformed into an Extensible Markup Language (XML) document [8]. The use of XML provides a rich and extensible means of representing context information, and allows the formal specification of languages governing the interpretation of such context. It is further encapsulated in a User Datagram Protocol (UDP) transport-layer datagram [9]. The use of UDP allows other context-"interested" hosts, including the end-destination of the flow, to detect or demultiplex out the context tag packet from the rest of the flow.

Further down, the tag injection and detection stage encapsulates the UDP datagram in an IP packet whose header contains the IP Router Alert Option as described in RFC 2113 for IPv4 [10], and RFC 2711 for IPv6 [11]. This option has the semantic "routers should examine this packet more closely."

Routers within the network may also inject context tags to describe information aggregated from multiple tags within a single flow, or to describe flow aggregates called *macroflows*. In some cases, such as within the Internet core, routers may be more interested in context descriptions of large macroflows rather than the individual constituent flows, for reasons of scalability. It may be also necessary for routers to inject context tags to describe any adaptation they may have performed on a flow or macroflows.

3.2 Service Triggering

Routers within the network detect the context tag by virtue of the Router Alert option in the IP header. Routers that either do not support the option, or do not recognize the context payload simply forward the packet to the next hop. At end-hosts, the context tag may be demultiplexed out of the flow by virtue of the UDP port number. If no equivalent context-processing process exists at the destination host the tag is either silently dropped, or an error message may be returned.

Once received, the context payload is extracted and sent up the stack, to the context interpretation and aggregation module. The context may either be used for long-term information gathering or learning, or may result in a specific adaptation being triggered. Service (including adaptation) modules subscribe to certain context values or events within the context interpretation module. When these events occur or the values are presented to the service module, a decision engine determines if the service or adaptation is executed. In the case of routers, the adaptation may be applied directly to the flow carrying the tag, or to a group of flows, or a larger-scale adaptation or management function may be invoked. Service modules on different nodes may communicate with each other and may operate in a coordinated fashion; however, the details of this operation is beyond the scope of this paper.

The architecture also allows end-hosts to receive and process context tags, and to contain adaptation components. In this case, the adaptation could be on the incoming or outgoing network flow, or on a user application, or influence some operation of the receiving device.

4 Deployment Strategy

Our context-tagging scheme can be incrementally deployed on networks by progressively adding components that either support flow context sensing and tag generation, or provide context-triggered services and adaptation, or both. Context sensing functionality may be added to end-hosts or incrementally on network nodes such as routers, or dedicated boxes may be inserted within the network in order to inspect flows and inject context tags. For nodes that will provide context-triggered services, the service modules and the core router functionalities (classification, forwarding, QoS) do not necessarily have to be closely coupled; the context-related and adaptation functions could reside on a separate device "bolted" onto a conventional router, and SNMP [12] may be used to effect service execution. While there are obvious advantages in terms of performance and a wider range of functionalities are possible in the closely-coupled case, the loosely-coupled example is given here to illustrate that the scheme may be incrementally deployed on the existing Internet infrastructure.

5 Initial Results

A simple flow-generating application using *libnet* [13], a toolkit that allows low-level packet construction, handling and injection was designed and used to simulate the operation of our system and perform initial validation on some of its components. Our application generated a flow containing exactly one context tag packet carrying the Router Alert Option in the IP header and a rudimentary XML-formatted context payload encapsulated in UDP. The rest of the flow consisted of UDP datagrams carrying dummy payloads. The context tag packet was positioned well within the flow, rather than among the first few packets, so that the adaptation could be well-observed. The flow was sent through a Linux-based router that in turn ran a process that detected the context-tagged packet.

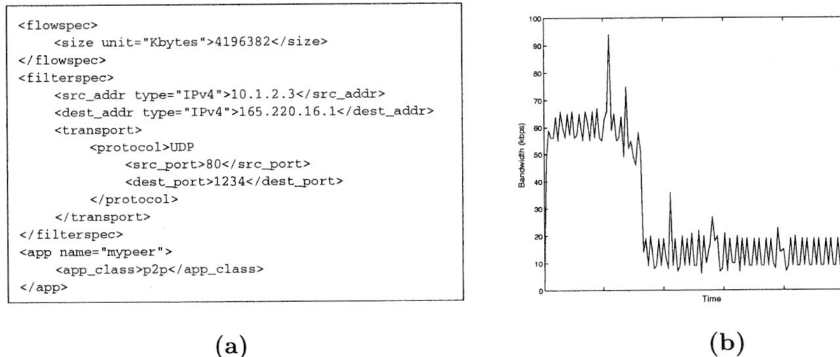

```
<flowspec>
    <size unit="Kbytes">4196382</size>
</flowspec>
<filterspec>
    <src_addr type="IPv4">10.1.2.3</src_addr>
    <dest_addr type="IPv4">165.220.16.1</dest_addr>
    <transport>
        <protocol>UDP
            <src_port>80</src_port>
            <dest_port>1234</dest_port>
        </protocol>
    </transport>
</filterspec>
<app name="mypeer">
    <app_class>p2p</app_class>
</app>
```


Fig. 2. (a) Context tag fragment. (b) Adaptation response on router

On the Linux router, the tag was processed, triggering a simple adaptation. In this example, the specific adaptation was to map the flow into an appropriate traffic class. Within the context tag a component called the *filterspec* provided a function similar to a similarly-named component in the Resource Reservation Protocol (RSVP): a means to define the set of packets that constitute the flow described by the tag [14]. In this experiment our filterspec used a Linux `u32` filter [15] descriptor that was directly parsed by the router and passed on to the adaptation module. In future implementations we expect to use a wider variety of mechanisms to specify filterspecs, including high-level descriptions that could be mapped by the context interpreter and service modules to specific classification mechanisms within the router. The adaptation module used the Linux `iptables` and `tc` mechanisms [15] to map the flow to a traffic class.

A fragment of the context tag is shown in Fig. 2a. The router's response to the context tag is shown in Fig. 2b. Prior to receiving the tag, the network accommodated the flow's full offered load (60 kbps average). Upon receiving the tag, the router effectively limited the bandwidth allocated to the flow to an average of 14 kbps.

6 Related Work

The network edge is a rich source of context information. End-hosts are in the best position to provide information on the applications that generate or consume traffic, the users whose activities drive these applications, and the computational and networking capabilities of the devices on which these applications run. Currently no widely deployed and sufficiently flexible mechanism exists for the edge to share context information within the network. Many context-aware systems employ publish-subscribe mechanisms that allow clients to obtain context information from sources [3, 16]. Our approach however contemplates a more loosely-coupled, connectionless communications model between end-hosts and nodes.

While network- and transport-layer protocol headers contain rudimentary tags (fields) that provide some hints on the type of applications that generate the flow, or to some extent, the nature and characteristics of the flow's content, these tags are not expressive enough to encapsulate the rich context information available at the edge. Schemes that provide flow descriptions often focus exclusively on their QoS characteristics, rather than a more general description of flow context. Resource reservation protocols such as RSVP [14] and the experimental ST-II [17] use a *flowspec* to describe the characteristics of a flow, usually in terms of bandwidth, delay, and other QoS-related characteristics [18]. Other schemes such as the Session Description Protocol (SDP) [19] and Multipurpose Internet Mail Extensions (MIME) [20] deliver flow or session context to end-hosts rather than network nodes, and are limited to very specific application domains.

7 Conclusion and Future Work

A key component needed to enable the deployment of context-aware networks is a method for sharing network context information. We have discussed a simple yet extensible and semantically-rich mechanism that tags flows through the network with context, delivering context directly to nodes along the unicast path or multicast tree. This allows network services, particularly adaptation services, to be triggered based on flow context. Our on-going work is focused on further validating our architecture, evaluating its performance, and exploring other application scenarios.

We are likewise currently exploring a parallel implementation of our components on an active networking [21] platform. We believe that such an approach will allow us to achieve a greater amount of flexibility through an on-demand deployment of context sensing, tagging, interpretation and adaptation components within the network.

Routers and end-hosts would be able to efficiently share context only if they subscribe to a common information model. An important component of our ongoing work is the development of an ontology and appropriate context-sharing languages.

Finally, as with any system that senses and distributes context information, there are concerns on security and privacy that we also hope to address in the near future.

Acknowledgment. This paper describes work partially undertaken in the context of the E-NEXT - Information Society Technologies (IST) FP6-506869 project. The IST program is partially funded by the Commission of the European Union. The views and conclusions contained herein are those of the authors and should not be interpreted as necessarily representing the E-NEXT project. Roel Ocampo acknowledges support by the Doctoral Studies Fellowship program of the University of the Philippines.

References

1. I. Foster. What is the Grid? A Three Point Checklist. *Grid Today*, July 2002.
2. J. Saltzer, D. Reed, and D. Clark. End-to-End Arguments in System Design. *ACM Transactions on Computer Systems*, November 1984.
3. A. K. Dey, D. Salber, and G. D. Abowd. A Conceptual Framework and a Toolkit for Supporting the Rapid Prototyping of Context-Aware Applications. *Human-Computer Interaction (HCI) Journal*, 16 (2-4), 2001
4. R. Braden, D. Clark, and S. Shenker. Integrated Services Architecture in the Internet: an Overview. *Request for Comments 1633*, June 1944.
5. M. Yarvis, P. Reiher and G. Popek. Conductor: A Framework for Distributed Adaptation. *Proc. 7th Workshop on Hot Topics in Operating Systems*, March 1999.
6. D. Clark, C. Partridge, J. Ramming, and J. Wroclawski. A Knowledge Plane for the Internet. *Proc. ACM SIGCOMM 2003 Conference on Applications, Technologies, Architectures and Protocols for Computer Communications*, August 2003.

7. R. Ocampo and H. De Meer. Smart Wireless Access Points for Pervasive Computing. *Proc. First Working Conference on Wireless On Demand Network Systems (WONS '04)*, Lecture Notes in Computer Science LNCS 2928, January 2004.
8. T. Bray, J. Paoli, C.M. Sperberg-McQueen, E. Maler, and F. Yergeau (editors). Extensible Markup Language 1.0 (Third Edition). *W3C Recommendation 04 February 2004*, http://www.w3.org/TR/REC-xml
9. J. Postel. User Datagram Protocol. *Request for Comments 768*, August 1980.
10. D. Katz. IP Router Alert Option. *Request for Comments 2113*, February 1997.
11. C. Partridge and A. Jackson. IPv6 Router Alert Option. *Request for Comments 2711*, October 1999.
12. J. Case, M. Fedor, M. Schoffstall, J. Davin. A Simple Network Management Protocol (SNMP). *Request for Comments 1157*, May 1990.
13. M. Schiffman. Libnet. http://www.packetfactory.net/ Projects/Libnet/
14. L. Zhang, S. Deering, D. Estrin, S. Shenker, and D. Zappala. RSVP: A New Resource ReSerVation Protocol. *IEEE Network*, September 1993.
15. B. Hubert, T. Graf, G. Maxwell, R. van Mook, M. van Oosterhout, P. Schroeder, J. Spaans, P. Larroy (editors). Linux Advanced Routing and Traffic Control. http://www.lartc.org.
16. T. Kanter. A Service Architecture, Test Bed and Application for Extensible and Adaptive Mobile Communication. *Proc. Personal Computing and Communication Workshop 2001 (PCC'2001)*, April 2001.
17. C. Topolcic. Internet Stream Protocol Version 2 (ST-II). *Request for Comments 1190*, October 1990.
18. C. Partridge. A Proposed Flow Specification. *Request for Comments 1363*, September 1992.
19. M. Handley, V. Jacobson. SDP: Session Description Protocol. *Request for Comments 2327*, April 1998.
20. N. Freed and N. Borenstein. Multipurpose Internet Mail Extensions Part Two: Media Types. *Request for Comments 2046*, November 1996.
21. D. L. Tennenhouse and D. J. Wetherall. Towards an Active Network Architecture. *Computer Communication Review*, 26(2), April 1996.

Dependable Execution of Workflow Activities on a Virtual Private Grid Middleware

A. Machì, F. Collura, and S. Lombardo

ICAR/CNR Department of Palermo
{machi, s.lombardo, f.collura}@pa.icar.cnr.it

Abstract. In this paper we relate QoS management to Workflow management and discuss dependable execution on a middleware layer, named Virtual Private Grid (VPG), of sets of processes performing a workflow activity. We propose two patterns of interaction between a Workflow Management System and the distributed workflow management support. The patterns monitor resource availability & connectivity, and, in case of fault of any resource, ensure job completion by re-mapping a process graph and restarting it. We also describe current implementation of the patterns and of the run-time support.

1 Introduction

Quality of Service (QoS) can be defined as "the set of those quantitative and qualitative characteristics which are necessary in order to achieve the required functionality of an application" [1]. We can suppose that QoS could be specified as a set of *non-functional* attributes some of which are not expressible numerically.

One category of QoS qualitative characteristics is *dependability* that characterizes the degree of certainty that an activity is performed [2].

Dependability is of main concern in grid environments where applications are ran by coordinating processes mapped on a set of computational resources co-allocated but just virtually co-reserved. In fact, according to the Virtual Organization paradigm, control of grid shareable resources is maintained by Local Organizations and no insurance is given to clients on persistence of their availability.

Even if a number of tools and high level collective services have been developed in the grid community for supporting co-reservation [3], up to now other aspects of grid-awareness as completion insurance, fault-tolerance and resources management are still in charge of user control code. The most popular grid technology for e-science, namely the Globus Toolkit 2 (GT2), and most projects based on it [4][5], offer adequate APIs for *staging* of processes and data, and for process *enactment*, but offer very limited support for process *monitoring* and just a kill mechanism to support workflow *enforcement*. Process monitoring and control are in charge of user who interleaves management code to business code to implement grid-awareness or self-adaptiveness.

The Web Services Resource Framework, emerging from the OGSA (GT3) experience and based on Web Services technology, defines a few models and patterns specifying non-functional requirements of Web Services life cycle and composition [6]. Namely, WS-ResourceLifetime [7] defines mechanisms for managing WS-Resources,

V.S. Sunderam et al. (Eds.): ICCS 2005, LNCS 3516, pp. 267 – 274, 2005.

WS-ServiceGroups [8] describes how collections of services can be represented and managed. WSBaseFaults [9] defines a standard format for reporting exceptions.

But WSRF specifics are limited and focused to the implementation of the Service Oriented Architecture (SOA) paradigm [10] and are scarcely useful to implement QoS control policies and patterns in e-science grid-enabled applications still using legacy code structured as a set of cooperative processes. In this paper we describe a workflow management support service and propose two patterns for supporting dependability of execution.

In section 2 we state equivalence between execution of a process graph and execution of a workflow activity. In section 3 we describe two patterns for management of dependable execution on the grid of sets of concurrent processes described by direct graphs.

Finally, in section 4, we describe a component implementation over GT2 of a grid middleware layer supporting non-functional aspects of workflow management through services for process life cycle monitoring and an asynchronous event bus service.

2 Process Graphs and Workflow Activities

The Workflow Management Coalition organization (WFMC) [11] defines a "*business process*" as "a set of one or more linked procedures or activities which collectively realise a business objective or policy goal" [12]. It defines an *activity* as "a description of a piece of work that forms one logical step within a process. An activity is typically the smallest unit of work that is scheduled by a workflow engine during process enactment. An activity typically generates one or more work items. A work item is a representation of the work to be processed (by a workflow participant) in the context of an activity within a process instance."

We recognize in these definitions a basic model useful for partitioning legacy applications whose business logic can be expressed as a workflow in a set of activities some of which can be separately enacted and managed on the grid. An activity may be carried out by a single workflow participant or by a set of participants organized in a workflow sub-graph. It can be hosted in a SOA service and managed honoring a Service Level Agreement (SLA) [13].

In the following we restrict our considerations to activities that may be described by a direct graph of virtual processes where one virtual node operates as an activity front-end. It coordinates the routing of work-items to other virtual processes participants to activity expressed by workflow sub-graph.

Controlled execution of an activity consists then in an atomic execution on the grid of an instance of a process graph mapped onto grid resources.

Adhering to the terminology commonly used by the WFMC community we can recognize two main control activities:

- *Routing*: the direction of work-items processing flow which is implicitly defined by the process graph and expresses part of the business logic;
- *Management*, which consists in *deployment* of virtual processes onto grid resources, staging of application data, *enactment* (initialization) of activity procedures, *monitoring* of QoS respect, *enforcement* or rerouting of processing flow in case of fault or not compliance with expected QoS.

In the WFMC scenario, management activities are assigned to a WorkFlow Management (WFM) engine holding mechanisms able to enforce workflow rerouting. Present grid toolboxes offer tools for deployment [14], or enactment and monitoring of resource status [15] but do not support collective activity monitoring and re-routing. This is mainly due to their relying on a synchronous communication paradigm (rsh-like) used both for functional and non-functional control.

The Grid Application Development System (GrADS) helps building natively self-consistent applications in which configurable objects, named COPs (Configurable Object Programs), embody at compile time intimate knowledge of application while objects drivers (application manager and executor) embody grid-awareness [16]. GrADS relies on a proprietary run-time system to support QoS monitoring and reconfiguration of COPs through application specific sensors and actuators.

In the framework of the Grid.it project, a new grid-aware programming paradigm has been proposed [17] and the ASSIST environment is being implemented for assisting native development of components configurable in respect to performance attributes. The key point for enabling flexible control is wrapping executive code in components including drivers of standard routing patterns (parmods) and a local proactive executor serving external enforcement through a non-functional interface.

Neither grid-aware programming paradigm defines patterns for integrating control of external activities in theirs application *Managers*.

We propose a grid management support engine that allows application or workflow *Managers* to extend their control on activities on the grid.

In the following we describe a grid middleware offering services for management of the graph of processes on the grid and two patterns implementing dependable execution of workflow actions.

3 Patterns for Dependable Execution of Process Graphs

Fig.1 shows a workflow scenario where two steps of a pipe (activities) execute on resources obtained on-demand from the grid (grid-enabled). Activities are performed by graphs of cooperating processes.

Dependable activity execution is required to avoid overrun of data-item flow-control. If any resource faults, activities need to be remapped and restarted.

Monitoring of stable execution of process activities is committed to a grid middleware. A *grid-front-end* process coordinates a network of sensor/actuator *demons* installed on grid nodes. It offers to the (activity or workflow) *Manager* a high level synchronous RPC functional interface for mapping process graphs, starting them, polling status of their processes, aborting them.

Two interaction patterns allow demons to monitor the regular execution of any process activated on grid resources and *Manager* to restart the graph in case of fault of a grid node or connection. While executing these patterns the *grid-front-end* process and the *demons* interact synchronously via RPC calls and asynchronously via signals, the *Manager* and the grid front-end just via RPC calls.

The ***resource availability & connectivity check*** pattern verifies, at regular time interval, activity of graph processes and functionality of significant connections among them. Exchanging of *ping messages* between demons at each connection end tests connectivity. A *demon* notifies to the *grid-front-end* a *missing-connection signal* when unable to reach its peer at end of any significant connection.

Furthermore, each demon emits regularly a *heartbeat* signal to notify the *grid-front-end* that node it monitors is operative. In case of extended absence of the *heartbeat* signal, the *grid-front-end* infers a fault condition, due to the death of the demon, fault of related connection or host shutdown. Successful result of connection test leads to infer a demon fault and register a *process fault* status, unsuccessful result to register a *node fault* status.

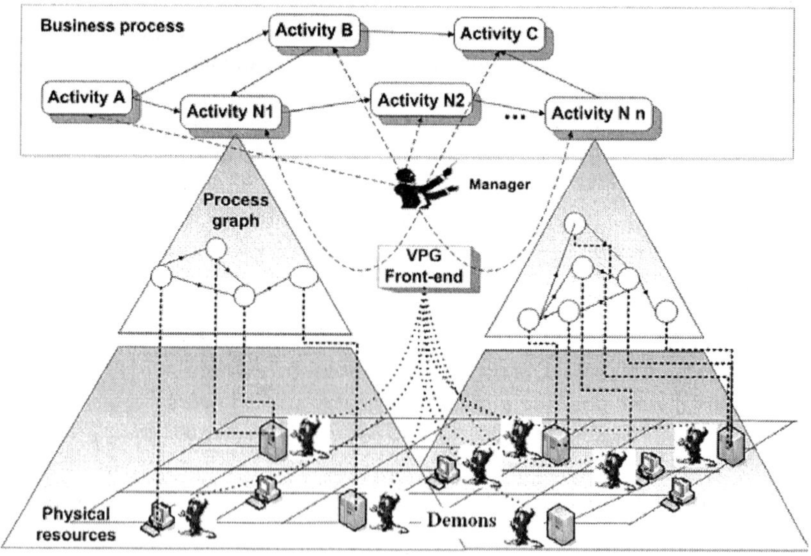

Fig. 1. Actors playing management of workflow activities on the grid

The ***insured completion*** pattern guarantees that a process graph comes to a known end (completed or terminated). *Manager* enacts graph through *grid-front-end* commands and polls the collective status of its process. In case of any node failure it restarts graph execution. Each *demon* monitors processes on its own node. It catches signals emitted by the local operating system and notifies them asynchronously to the *grid-front-end,* that updates graph status on behalf of *Manager.*

The following mechanisms ensure valid knowledge of process vitality and ensured completion:

- *Local operating system signals* inform the *demon* about normal process termination or abort. These signals are notified to the *grid-front-end.*
- *Keep-alive signal*: it is notified regularly from *grid-front-end* to each slave *demon* to maintain its service alive.
- *Automatic shutdown*: a *demon* not receiving for a while the *keep-alive signal* infers a fault of connection with *grid-front-end* node or failure of front-end itself. It kills

any process activated on the node, performs a garbage collection procedure that cleans-up any software element deployed and terminates.

Fig. 2 shows an UML sequence diagram that details an occurrence of the ***insured completion*** pattern after a process fault. Before graph restart, processes on live nodes are explicitly killed. Release of resources allocated on the fault node by the automatic shutdown procedure is omitted.

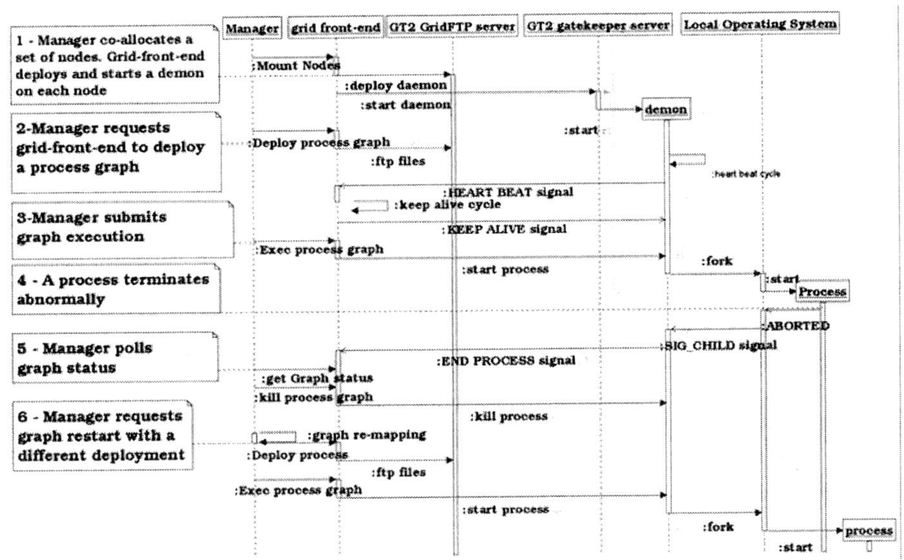

Fig. 2. UML sequence diagram illustrating events and actions in a sample occurrence of the insured completion pattern

4 The Virtual Private Grid Middleware

In the framework of the Grid.it Project we have developed a grid support engine, named Virtual Private Grid (VPG), implementing the roles of *grid-front-end* and *demon* organized as a middleware layer over GT2. VPG manages and monitors the life cycle of process graphs over a pool of grid resources. It offers to a client *Manager* a distributed operating system façade, making grid resources to appear virtually as belonging to a heterogeneous private cluster of processors.

VPG manages multiple process graphs described by lists of processes and arcs connecting them. Attributes of a process are the name of the node on which it is mapped and the list of the software elements required for its execution (executables, libraries and data files). Attributes of the link are the identifiers of processes connected.

Current middleware implementation (VPG 1.2) [18] provides the following functionalities:

- Grid nodes management:
  - ***Mount/Unmount*** of a grid node;
  - ***Retrieve status:*** return activity status of a grid node (disconnected, connected, activated, active).
- Process graph lifecycle primitives:
  - ***Staging*** of a set of files (executable, libraries, etc) on grid nodes. For each file the (redundant) deployment node list is specified. An identifier is assigned to the file set.
  - ***Garbage collection***: removal of any files belonging to a file set from grid nodes where they have been staged.
  - ***Enactment*** of the process graph. An identifier is assigned to each process according to its rank in the process list.
  - ***Kill*** of graph processes included in a list of identifiers.
  - ***Restart*** of graph processes included in a list of identifiers.
  - ***Retrieve status***: return activity status of processes belonging to a process graph (inited, aborted, running, done). Activity progress status is unsupported.
- Event services: an asynchronous event bus is provided for exchanging signals among VPG components according to the event subscribe/notify pattern.

Fig 3 shows, through an UML deployment diagram, VPG components and actors interacting with them. The VPG Master acts as a grid front-end for the *Manager* while slave demons, called Remote Engine run on each node of the grid.

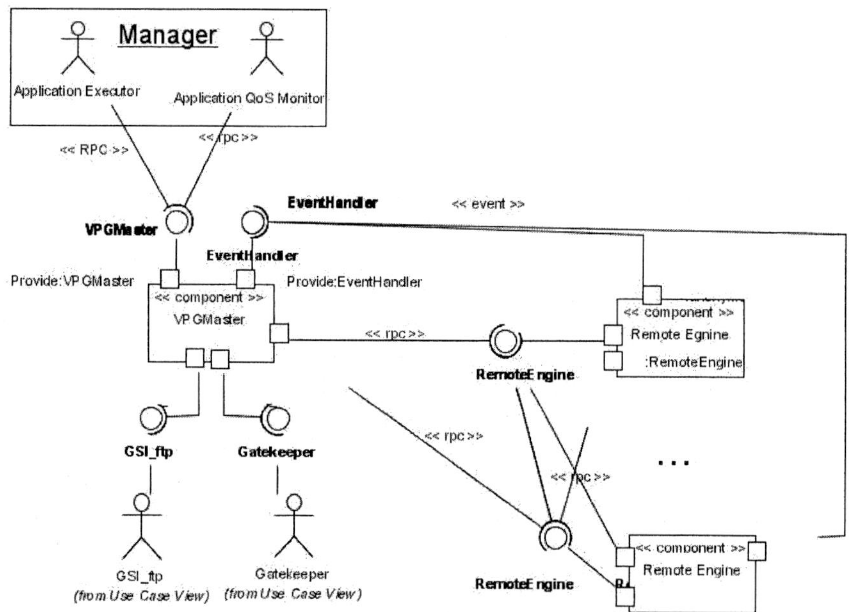

Fig. 3. Interaction among VPG components and actors shown through an UML deployment diagram

VPG Master: provides a functional interface that exposes every of the over mentioned functionalities. It maintains and updates in a register the status of each process controlled. It provides the asynchronous event bus and registers itself for any events related to remote process status changes. The Remote Engine provides local processes lifecycle primitives (run, kill, status, clean), and sends to master events about processes status changing or fault of connections among meaningful neighbor nodes. This entity is automatically deployed during node mount and it is removed during unmount.

Present VPG implementation is based on the usage of several design patterns: acceptor-connector, reactor, proxy, wrapper and adapter provided by the open-source object-oriented framework ACE [19].

Grid nodes management and file transfer are implemented over Globus Toolkit2 tools and services: GRAM for start-up of the Remote Engine, GridFTP for file staging.

APIs for calling VPG Master through synchronous Remote Procedure Calls (RPC) formatted according to the XML-RPC specifications or for direct integration in the *Manager* code, are provided in the VPG 1.2 SDK.

5 Conclusions and Future Work

In the paper we have introduced a methodology for considering QoS management of grid activities in the context of workflow management in SOA architectures. In particular we have discussed the simple case of dependable execution of graphs of processes performing a self-consistent activity.

We have shown that, in this case, two control patterns can be implemented without intervention over the application code, by exploiting an external run-time support offering collective control services to the WFM system.

Control rights over graph processes are obtained by spawning them through remote engines. Restart of the entire process graph has been indicated as a minimal solution for insured execution in case of fault of any grid resources.

A more interesting pattern for insured execution should support assisted recover from a node fault, or at least from a slave node fault.

Design of a pattern for joining or disjoining slaves from a master-slave graph routed by a parallel skeleton is a topic of current research.

Acknowledgements

Work supported by Italian Ministry of Scientific Research: , Project FIRB ***Grid.it.***

References

[1] Vogel A., Kerhervé B., Von Bochman G. Ad Gecsei J. (1195). Distributed Multimedia and QoS: A Survey, IEEE Multimedia Vol. 2, No. 2, p10-19

[2] ITU-T Rec. I.350: General aspects of Network Performance and Quality of Service in Digital Networks, including ISDN.

[3] I. I. Foster, M. Fidler, A. Roy, V, Sander, L. Winkler. End-to-End Quality of Service for High-end Applications. Computer Communications, 27(14):1375-1388, 2004.
[4] The Condor® Project , http://www.cs.wisc.edu/condor/
[5] The DataGrid Project, http://eu-datagrid.web.cern.ch/eu-datagrid/
[6] From Open Grid Services Infrastructure to WS-Resource Framework: Refactoring & Evolution. K. Czajkowski, D. Ferguson, I. Foster, J. Frey, S. Graham, T. Maguire, D. Snelling, S. Tuecke, March 5, 2004.
[7] Frey, J., Graham, S., Czajkowski, C., Ferguson, D., Foster, I., Leymann, F., Maguire, T., Nagaratnam, N., Nally, M., Storey, T., Sedukhin, I., Snelling, D., Tuecke, S., Vambenepe, W., and Weerawarana, S. 2004. WS-ResourceLifetime. http://www-106.ibm.com/developerworks/library/ws-resource/wsresourcelifetime.pdf.
[8] Graham, S., Maguire, T., Frey, J., Nagaratnam, N., Sedukhin, I., Snelling, D., Czajkowski, K., Tuecke, S., and Vambenepe, W. 2004. WS-ServiceGroups. http://www-ibm.com/developerworks/library/ws-resource/ws-servicegroup.pdf.
[9] Teucke, S., Czajkowski, K., Frey, J., Foster, I., Graham, S., Maguire, T., Sedukhin, I., Snelling, D., Vambenepe, W. 2004. WS-BaseFaults. http://www-106.ibm.com/developerworks/library/wsresource/ws-basefaults.pdf.
[10] K.Channabasavaiah, K.Holley, E.M.Tuggle, "Migrating to a service-oriented architecture", http://www-06.ibm.com/developerworks/webservices/library/ws-migratesoa/
[11] The Workflow Management Coalition, www.wfmc.org
[12] Reference Model - The Workflow Reference Model (WFMC-TC-1003, 19-Jan-95, 1.1)
[13] Web Service Level Agreements (WSLA) Project - SLA Compliance Monitoring for e-Business on demand http://www.research.ibm.com/wsla/
[14] R. Baraglia, M. Danelutto, D. Laforenza, S. Orlando, P. Palmerini, P. Pesciullesi, R. Perego, M. Vanneschi "AssistConf: a Grid configuration tool for the ASSIST parallel programming environment", Eleventh Euromicro Conference on Parallel, Distributed and Network-Based Processing February 05 - 07, 2003 Genova, Italy
[15] WS GRAM Docs, http://www-unix.globus.org/toolkit/docs/3.2/gram/ws/ index.html
[16] The Grid Application Development Software Project (GrADS), http://hipersoft.cs.rice.edu/grads/index.htm
[17] M. Aldinucci, S. Campa, M. Coppola, M. Danelutto, D. Laforenza, D. Puppin, L. Scarponi, M. Vanneschi, C. Zoccolo "Components for high performance Grid programming in the Grid.it Project". In Proc. Of Intl. Workshop on Component Models and Systems for Grid Applications.
[18] S. Lombardo, A. Machì. "Virtual Private Grid (VPG 1.2) Un middleware a supporto del ciclo di vita e del monitoraggio dell'esecuzione grafi di processi su griglia computazionale", Technical report RT-ICAR-PA-11 -2004, October 2004
[19] The ADAPTIVE Communication Environment (ACE) http://www.cs.wustl.edu/ ~schmidt/ACE.ht

Cost Model and Adaptive Scheme for Publish/Subscribe Systems on Mobile Grid Environments

Sangyoon Oh[1,2], Sangmi Lee Pallickara[2], Sunghoon Ko[1], Jai-Hoon Kim[1,3], and Geoffrey Fox[1,2]

[1] Community Grids Computing Laboratory, Indiana University, Bloomington, IN. USA
{ohsangy, leesangm, suko, jaikim, gcf}@indiana.edu
[2] Department of Computer Science, Indiana University, Bloomington, IN. USA
[3] Graduate School of Information and Communications, Ajou University, Suwon, S. Korea
jaikim@ajou.ac.kr

Abstract. Publish/subscribe model is appropriate in many push based data dissemination applications. This paper presents cost model for publish/subscribe systems, analyze its performance, and compare to other interaction-based models such as the client-server model and the polling model. Based on the cost analysis, we have proposed an adaptive model which can dynamically select an appropriate model for each client independently.

1 Introduction

Publish/subscribe system [1] have been widely used in many applications [2], [3]. Publish/subscribe system consists of publishers (ES: Event Source), servers (EBS: Event Brokering System), and subscribers (ED: Event Displayer). After a publisher publishes data (events) asynchronously to a server, the server disseminates the data (events) to subscribers which registered their interest on the server. Thus publish/subscribe model is appropriate in many applications such as data dissemination services, information sharing, service discovery, etc. Fig. 1 depicts system configurations.

In this paper, we present cost model for publish/subscribe systems, analyze its performance, and compare to other interaction based models such as a client-server model and a polling models. We can estimate performance and adopt publish/subscribe systems effectively by using our proposed cost model and analysis of publish/subscribe systems. Based on the cost analysis, we propose adaptive model which can dynamically select an appropriate model for each client independently. We believe the adaptive scheme we introduce here is very useful for the mobile and ubiquitous services where characteristics of device and networks are diverse and dynamically changing. We also experimentally measured and compared performance of publish/subscribe model to client/server model

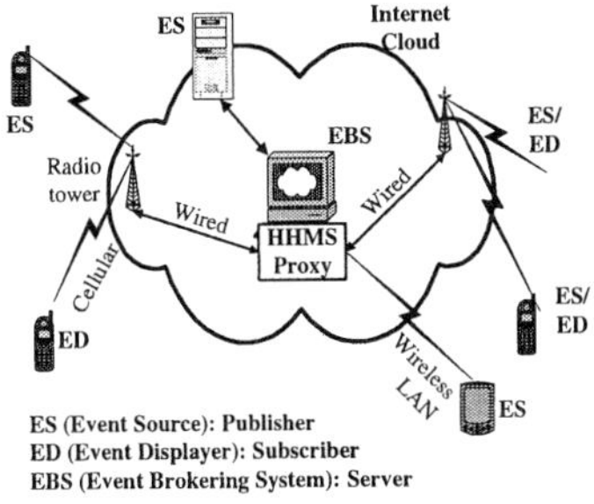

Fig. 1. Pub/Sub System Configurations

V.S. Sunderam et al. (Eds.): ICCS 2005, LNCS 3516, pp. 275 – 278, 2005.

on our test bed including mobile device and NaradaBrokering [4] (our publish/subscribe based message brokering system) to verify correctness of our model on the real systems. Our cost analysis model is simple but accordant with experimental results.

2 Cost Model

System Models

To evaluate the cost model for different systems, we assume following system parameters to analyze cost: α, publish rate of event; β, subscriber's access rate of published events or request rate of client in the client/server models; $c_{ps}(\alpha)$, publish/subscribe cost per event, $c_{pub} + c_{sub}$; $c_{rr}(\beta)$, cost per request and reply; $c_{poll}(\alpha,T)$, cost of periodic publish or polling; $c_{delay}(\alpha,T)$, cost of delaying publish; $s(n)$, effect of sharing among n subscribers; t_{ps}, time delay for publish/subscribe, $t_{pub} + t_{sub}$; t_{rr}, time delay for request and reply; $t_{poll}(\alpha, T)$, time delay for periodic publish.

Cost Analysis

In this analysis, we analyze cost of three different models without any failure of communication link or node. We consider (1) conceptual total cost (e.g., the number of message, amount of message, or time delay) per unit time for each model, (2) cost for each access by client (or subscriber), (3) time delay for access after subscriber's (or client's) intention, and (4) time delay between event occurrence and notification to subscriber (or recognition by client). Cost can be the number of message, amount of message, or time delay. Table 2 shows the summary of the cost for each model analyzed in this paper. Please refer to [5] for detailed analysis.

Table 1. The cost of the selected model

Model	Publish/Subscribe	Request/Reply	Polling
conceptual total cost per time unit	$\alpha\,(c_{pub} + n\,s(n)c_{sub})$	$\beta\,n\,c_{rr}$	$(c_{poll}(\alpha,T) + c_{delay}(\alpha,T))\,/\,T$
cost for each access	$\frac{\alpha}{\beta}\,(\frac{c_{pub}}{n} + c_{sub})$	c_{rr}	$c_{poll}(\alpha,T) + c_{delay}(\alpha,T)$
time delay between intention and access	0	t_{rr}	T/2
time delay between event occurrence and notification/recognition (access)	$t_{ps} = t_{pub} + t_{sub}$ $(t_{ps} = t_{pub} + t_{sub} + \frac{1}{\beta})$	$\frac{1}{2\beta}$	T/2

Adaptive Scheme

Adaptive scheme can choose an appropriate model among publish/subscribe and request/reply models. Each client node can select its own model independently (hybrid model) and change its model during its service (dynamic model).

In this paper, we consider cost per each client's access as a cost metric. During a period of time, the average number of events occurred per client's access is measured for each client. At the end of the period, the average cost for each client's access is computed using the analysis in section 2, which is $\frac{\alpha}{\beta}(\frac{c_{pub}}{n}+c_{sub})$, where $\frac{\alpha}{\beta}$ is average number of event occurred per client's access and n is the number of subscriber. In our adaptive scheme, average number of event and the number of subscriber are obtained experimentally during the execution of application. At the end of the period, the model that is expected to require less cost than the other model during the following period is selected independently for each client. Fig.2 shows that publish/subscribe model is appropriate when the number of client is large and/or the number of event per client's access is small.

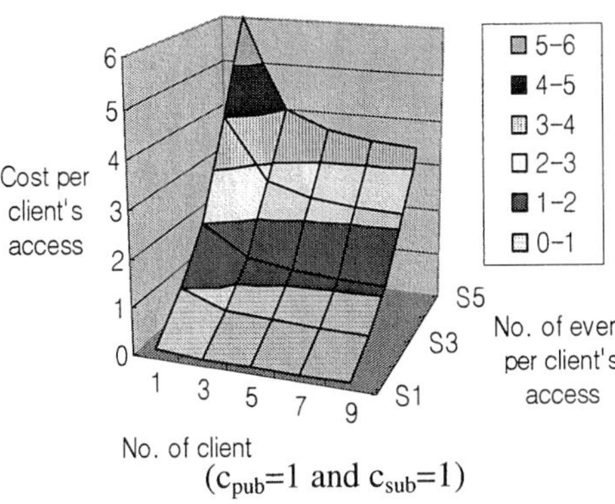

Fig. 2. Cost per client's access of publish / subscribe model

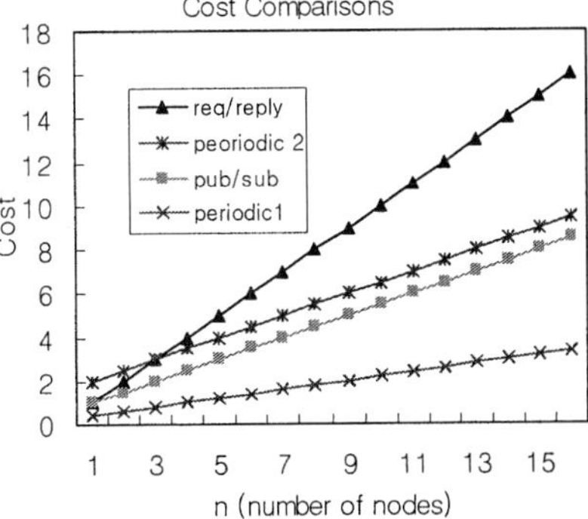

(α = 0.5, s(n)=1, c_{ps} = 2, and c_{rr} = 2; $c_{pub}(\alpha,T)$= c_{pub}, $c_{sub}(\alpha,T)$= c_{sub}, and $c_{delay}(\alpha,T)$= 0 for periodic1; $c_{pub}(\alpha,T)$= $\alpha T c_{pub}$, $c_{sub}(\alpha,T)$= $\alpha T c_{sub}$, $c_{delay}(\alpha,T)$= $2\alpha T c_{dealy}$ for periodic2)

Fig. 3. Communication cost by varying number of clients

3 Performance Comparisons

Parametric Analysis

In this section, we describe performance comparisons by parametric analysis. We set system parameters as shown in Table 2. Fig.3 shows performance comparisons between publish/subscribe, request/reply, and polling systems. Since publish/subscriber system disseminates data via server instead of individually for each client, it requires less cost than request/reply system. As the number of client node increases, the cost gap between two systems increases. Periodic polling system saves cost by transferring data once per period when delay cost is negligible. However, cost increases as delay cost increases. Polling system is viable approach when data delay is allowed and cost is negligible.

Table 2. Parameters

Param.	Values
α, β	0.5
c_{ps}, c_{rr}	2
c_{pub}, c_{sub}	1
$c_{poll}(\alpha, T)$	1 or αT
$c_{delay}(\alpha, T)$	0, T, or αT
$s(n)$	$1/n$ - 1
t_{ps}, t_{rr}	1
t_{proc}	1 or 5
$t_{poll}(\alpha, T)$	1, T, or αT

Experimental Results

The performance framework consists of NaradaBrokering System and Handheld Message Service (HHMS) [6]. The framework executes test applications – written in J2ME MIPD 2.0 and J2SE. We wrote two applications for two different communication paradigms – pub/sub and Remote Procedure Call (RPC). A client of the first application (ED) echoes message back to the sender (ES). With them, we experimented to perform the test for a cost of message. By varying a size of message, we measured RTT of different size messages. The result is shown in Fig.4. A client of the second application replies ACK back to the sender to measure RTT of different number of mobile clients. The second test was executed with mobile client emulators that come with Sun Microsystems' J2ME WirelessToolkit. This is a limited configuration, but it is still enough to exemplify the analysis we've made in Section 2. We define the data transition time of publish/subscribe and RPC as RTT/2 and RTT respectively from the semantics of each messaging scheme. Experiments are performed on mobile devices – a Samsung SPH-I300 phone and a Treo 600 with Sprint PCS Vision service and Linux machine.

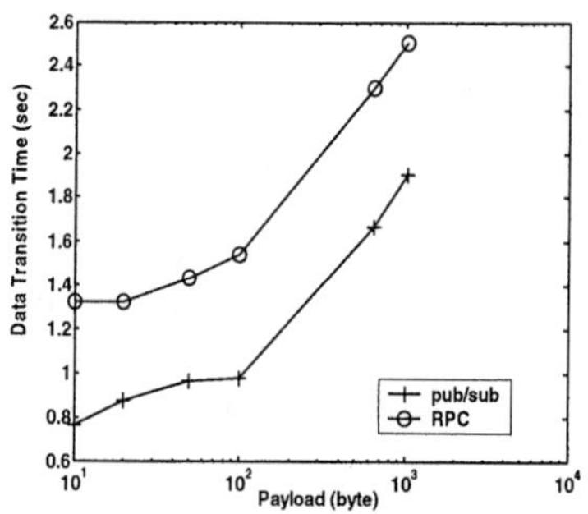

Fig. 4. Latency by Payload

4 Conclusions

We presented cost analysis model for publish/subscribe systems. Based on the cost analysis, we proposed an adaptive scheme which can dynamically select an appropriate model for each client independently. Experimental results (delay time) from our test bed were quite similar to our cost analysis models, which verifies that our cost model is useful to select proper model and to design adaptive schemes.

References

1. P. Eugster, P. Felber, R. Guerraoui, and A. Kermarrec, "The Many Faces of Publish/Subscribe", ACM Computing Surveys, vol. 35, no. 2, pp. 114-131, Jun. 2003.
2. A. Rowstron, A. Kermarrec, M. Castro, and P. Druschel, "SCRIBE: The design of a large-scale event notification infrastructure", Networked Group Communication, 2001.
3. A. Uyar, S. Pallickara and G. Fox, "Audio Video Conferencing in Distributed Brokering Systems", Proc. of the International Conf. on Communications in Computing, June 2003.
4. S. Pallickara and G. C. Fox, "NaradaBrokering: A Middleware Framework and Architecture for Enabling Durable Peer-to-Peer Grids", Proc. of ACM/IFIP/USENIX International Middleware Conference Middleware, pp 41-61, 2003.
5. S Oh, S. Lee Pallickara, S. Ko, J. Kim, G. Fox, "Cost Model and Adaptive Scheme for Publish/ Subscribe Systems on Mobile Environments", Community Grids Lab Technical Report, Dec. 2004. (http://grids.ucs.indiana.edu/ptliupages/publications/)
6. S. Oh, G. C. Fox, S. Ko, "GMSME: An Architecture for Heterogeneous Collaboration with Mobile Devices", Proc. of the Fifth IEEE/IFIP MWCN 2003, Singapore, Oct. 2003.

Near-Optimal Algorithm for Self-configuration of Ad-hoc Wireless Networks

Sung-Eok Jeon and Chuanyi Ji

Georgia Tech, Atlanta GA 30332, USA

Abstract. To quantify the goodness of a configuration, we develop a probabilistic model of network configuration. A probabilistic graph then represents the statistical dependence in network configuration, and shows that self-configuration can be optimized if the graph has nested local dependence.

1 Introduction

Self-configuration is for the network to achieve a desired network configuration by end-users in a distributed fashion, where each node adjusts its local configuration by following a local algorithm based on information from neighbors. A key requirement is that the local algorithm is optimal so that the desired global configuration can be achieved. The desired "globally optimal configuration" is considered to the one that makes trade-off between spatial channel-reuse maximization and constraint on both topology and reconfiguration cost. We consider topology formation and re-covery from failures. As many algorithms and protocols have provided promising results for self-configuration, it is not clear when self-configuration is implementable in a fully distributed fashion. This work intends to address the following questions: (a) How to quantify the goodness (i.e., optimality) of a network configuration? (b) When and how can local adaptation of nodes result in a globally optimal configuration?

2 Problem Formulation

Consider an ad-hoc wireless network with N nodes, with positions, $\underline{X} = \{X_1, \cdots, X_N\}$. Let σ_{ij} be random activity of link (i, j), referred to as a "communication dipole," $\sigma_{ij} = 1$ if node i is transmitting to node j; and $\sigma_{ij} = -1$, otherwise. A network configuration is a combination of both the topology and link activities, G=$(\underline{\sigma}, \underline{X})$; and an optimal configuration includes an optimal physical topology $(\underline{X})$ with maximal channel reuse $(\underline{\sigma})$ under management constraints. The management constraints considered in this work are for (a) link quality (e.g. on signal to interference + noise (SINR)), (b) maximized geographical coverage of nodes while maintaining 1-connectivity, and (c) minimized cost in re-adjusting positions for failure recovery. The optimal configuration $(\underline{\sigma}^*, \underline{X}^*)$ is defined as

V.S. Sunderam et al. (Eds.): ICCS 2005, LNCS 3516, pp. 279–283, 2005.

the most likely configuration that maximizes the likelihood function, $\arg\max_{(\underline{\sigma},\underline{X})} P(\underline{\sigma},\underline{X}|\underline{\sigma_0},\underline{X_0})$, where $(\underline{\sigma_0},\underline{X_0})$ is the initial configuration.

Self-organization is to obtain optimal distributed algorithms characterized by local rules, $g_i(\ )$, for $1 \le i \le N$, so that node positions can be adapted by end-users,

$$(\hat{X_i}(t+1), \hat{\sigma_{ij}}(t+1)) = \underset{(X_i(t+1),\sigma_{ij}(t+1))}{\arg\max} \quad g_i(X_i(t+1), \sigma_{ij}(t+1)|X_{N_i}(t), \sigma_{N_i}(t)), \quad (1)$$

where N_i and σ_{N_i} are the neighbors of node i and dipole σ_{ij}.

3 Cross-Layer Model of Network Configuration

To quantify the goodness of a network configuration, we obtain the likelihood $P(\underline{\sigma},\underline{X}|\ \underline{\sigma_0},\underline{X_0})$ which is from $P(\underline{\sigma}|\underline{X},\underline{\sigma_0},\underline{X_0})$ and $P(\underline{X}|\underline{\sigma_0},\underline{X_0})$.

3.1 Link Activities

Assuming the traffic demand is all-to-all for simplicity, the management constraint for $\underline{\sigma}$ is then to maximize the spatial channel reuse. Thus, $P(\underline{\sigma}|\underline{X},\underline{\sigma_0},\underline{X_0}) = P(\underline{\sigma}|\underline{X})$. Each feasible communication configuration of a network can be represented with the total energy of the network, which is referred to as "configuration Hamiltonian."

• Configuration Hamiltonian : The configuration Hamiltonian of a network is then characterized by the total negative power $-\sum_{ij} P_j$ and denoted as $U'(\underline{\sigma}|\underline{X})$, which is $-\sum_{ij} P_i l_{ij}^{-4}\eta_{ij} + \sum_{ij}\sum_{mn\in N_{ij}^I} (2\sqrt{P_i P_m} l_{ij}^{-2} l_{mj}^{-2} - P_m l_{mj}^{-4})\eta_{ij}\eta_{mn} - \sum_{ij}\sum_{mn\in N_{ij}^I}\sum_{uv\in\{N_{ij}^I, N_{mn}^I\}} 2\sqrt{P_m P_u} l_{mj}^{-2} l_{uj}^{-2}\ \eta_{ij}\eta_{mn}\eta_{uv} + R_I(\underline{\sigma},\underline{X}) + R_3(\underline{\sigma},\underline{X}) + \beta\cdot\sum_{ij} \|\text{SINR}_{ij} - \text{SINR}_{th}\|$, where $\eta_{ij} = (\sigma_{ij}+1)/2$, N_{ij}^M is the neighboring dipoles within MAC range of an active dipole σ_{ij}; and the dipoles outside the MAC range, denoted as N_{ij}^I, are allowed to be active concurrently, resulting in interference. Relevant interference neighbors are those whose power exceeds a threshold, i.e., $P_{th} \le P_m \cdot l_{mj}^{-4}$. The minimum region that covers all relevant interference neighbors is referred to as the "interference range."

The interference outside the inference range is denoted with a remainder $R_{I_{ij}}(\underline{\sigma},\underline{X})$. With $R_I(\underline{\sigma},\underline{X}) = \sum_{ij} R_{I_{ij}}(\underline{\sigma},\underline{X})$, being the total contribution due to interference outside the interference range.

• Boltzmann Distribution : The total energy $U'(\underline{\sigma}|\underline{X})$ can now be related to probability $P(\underline{\sigma}|\underline{X})$ using an analogy between communication activities in ad-hoc wireless networks and particles in statistical physics [2]. The probability distribution of particle systems obeys the Maxwell-Boltzmann distribution [2]. As a result, for a configuration Hamiltonian, the corresponding Boltzmann distribution is $P(\underline{\sigma}|\underline{X}) = \exp\frac{-U'(\underline{\sigma}|\underline{X})}{T} / \sum_{\underline{\sigma}} \exp\frac{-U'(\underline{\sigma}|\underline{X})}{T}$.

3.2 Random Position of Nodes

We now obtain the probability distribution of node positions $P(\underline{X}|\underline{\sigma_0}, \underline{X_0})$, assuming that current node positions are conditionally independent of initial conditions, i.e., $P(\underline{X}|\underline{\sigma_0}, \underline{X_0}) = P(\underline{X}|\underline{X_0})$.

With no management purposes, nodes' movement can be characterized by a two-dimensional random-walk around fixed positions, where $P(\underline{X}|\underline{X_0})$ is a multivariate Gaussian distribution with an exponent $U(\underline{X}|\underline{X_0}) = (\underline{X} - \underline{X_0})^T \cdot (\underline{X} - \underline{X_0})/2\sigma^2$, and variance of node movement σ^2. The Hamiltonian $U(\underline{X}|\underline{X_0})$ shows that aimless motions are penalized.

Management constraints make nodes move cooperatively to achieve a pre-defined constraint, e.g., 1-connected topology. The 1-connectivity can be achieved by a Yao-like graph [3], which can be implemented with

$$C(X_i, X_j) = \begin{cases} 0 & , \frac{|l_{ij} - l_{th}|}{l_{th}} < \epsilon_0 \text{ or } j \notin N_i \\ |l_{ij} - l_{th}| & , \text{otherwise} \end{cases} \tag{2}$$

where ϵ_0 is a small constant, l_{th} is a threshold, l_{ij} is the distance between nodes i and j.

The extended Hamiltonian for the topology is $U'(\underline{X}|\underline{X_0}) = U(\underline{X}|\underline{X_0}) + \sum_i \sum_{j \in N_i} \zeta \cdot C(X_i, X_j)$, where $U(\underline{X}|\underline{X_0})$ is due to free movements, ζ is a weighting constant, and N_i is the set of the nearest neighboring nodes of node i for every angle θ (θ=90° in this work).

For faults recovery, the propagation of configuration changes may need to be minimized across the entire network, and thus an additional penalty function can be introduced as the cost of reconfiguration, e.g., $f_{P_1}(\underline{\sigma}, \underline{X}) = \|(\underline{\sigma}, \underline{X}) - (\underline{\sigma_0}, \underline{X_0})\|$, where $\| \ \|$ denotes a vector norm. The resulting topology constraint is then $C'(X_i, X_j) = C(X_i, X_j) + C_1 \cdot f_{P_1}(\underline{\sigma}, \underline{X})$, where C_1 is constant, weighting the trade-off between global optimization and local recovery.

3.3 Graphical Representation of Network Configuration

To determine the optimality of distributed self-configuration, it suffices to examine with graphical models [1] whether the obtained likelihood function is factorizable. A set of random variables, e.g., $\underline{X}$, is called a Gibbs Random Field (GRF) if it obeys Gibbs distribution, $Z_0^{-1} \exp\frac{-U(\underline{X})}{T}$ [1]. Hammersley-Clifford Theorem [1] shows that a GRF is is equivalent with an MRF, which shows the spatial Markov dependence among nodes.

Since the Boltzmann distribution of a network configuration obeys a Gibbs form, a network configuration can be modeled with an MRF. First, consider the random link activities given node positions. Since link activities for dipoles are all dependent, the graph is fully connected. However, $R_I(\underline{\sigma}, \underline{X})$ and $R_3(\underline{\sigma}, \underline{X})$ are negligible in Hamiltonian. The remaining two terms define an approximation $P^l(\underline{\sigma}|\underline{X})$, to the original Boltzmann distribution, where $P^l(\underline{\sigma}|\underline{X}) = \frac{1}{Z_0} \cdot \exp(-\sum_{ij} \alpha_{ij}(\underline{X}) \cdot \frac{\sigma_{ij}+1}{2} - \sum_{mn \in N_{ij}^I} \alpha_{ij,mn}(\underline{X}) \cdot \frac{\sigma_{mn}+1}{2} \frac{\sigma_{ij}+1}{2})/T$. The approximated MRF, $(\underline{\sigma}|\underline{X})$, is now the well-known second-order Ising model [1].

Next, for the 1-connectivity topology, $U^{'}(\underline{X}|\underline{X_0})$ shows that the random field, $(\underline{X}|\underline{X_0})$, is a Potts model with random bonds [1] with the first-order neighbors. As a result, the coupled MRF $(\underline{\sigma}, \underline{X}|\ \underline{\sigma_0}, \underline{X_0})$ can be fully described by an Ising model and a Potts model together, representing the nested local dependence with random bonds. Especially, for the special case of free node-movement, the coupled random field becomes a Random Bond Ising Model (RBIM) [1].

4 Self-configuration: Distributed Algorithm and Failure-Recovery

The nearly optimal configuration is the one that maximizes the approximated likelihood function. Maximizing the global likelihood function reduces to maximizing local likelihood at cliques, i.e., for $1 \leq i, j \leq N$, $(\hat{X_i}(t+1), \hat{\sigma_{ij}}(t+1)) = \arg\max_{(X_i(t+1), \sigma_{ij}(t+1))} P^l(X_i(t+1), \sigma_{ij}(t+1)\ |X_{N_i}(t), \sigma_{N_i}(t))$. Therefore, the nearly optimal local rule $g_i() = P^l()$. Stochastic relaxation is an optimal distributed algorithm through Gibbs sampling [1]. Now self configuration is viewed as local optimizations of Hamiltonian on cliques.

An important application of self-configuration is adaptive recovery from failures. Consider node failures from an optimal configuration. There is a tradeoff between global optimization of network configuration and failure localization. Figure 1 (a) shows the self-configured topology, where the failed nodes are

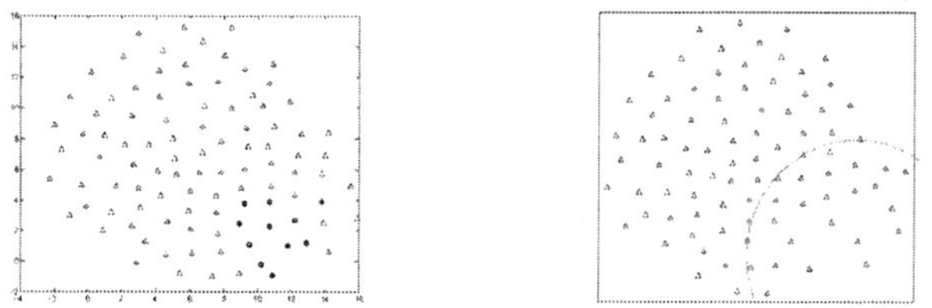

(a) Self-configuration and Failure Event (b) After Self-Recovery of $\underline{X}$

Fig. 1. Self-Recovery with Local Optimal Algorithm

marked by stars. Due to failure localization, Figure 1 (b) shows that only nodes within the arc are involved in the recovery.

5 Conclusion

We find that a random-bond model is the probability distribution of the network configuration where link activities are coupled by random coefficients due to node positions. Probability graphs show the conditions on when distributed self-configuration is optimal: (a) interference from far away nodes is negligible, and (b) management constraints are local on both connectivity and reconfiguration cost.

Acknowledgement. This work is supported by NSF ECS-0300605 (ECS-9908578). The authors would like to thank Mo Li, John Barry and Guanglei Liu for many useful discussions.

References

1. S. Geman, and D. Geman, Stochastic Relaxation, Gibbs Distributions, and the Bayesian Restoration of Images. IEEE Trans. PAMI vol. 6, 1984.
2. K. Huang, Statistical Mechanics. John Wiley & Sons.
3. R. Wattenhofer, L. Li, P. Bahl, and Y. Wang, Distributed topology control for power efficient operation in multihop wireless networks. In Proc. IEEE Infocom, 2001

The Applications of Meshfree Particle Methods at the Nanoscale

Weixuan Yang and Shaoping Xiao

Department of Mechanical and Industrial Engineering and Center for Computer-Aided Design, The University of Iowa, 3131 Seamans Center, Iowa City, IA, 52241-1527, USA
{weixuan-yang, shaoping-xiao}@uiowa.edu

Abstract. Since meshfree particle methods are beneficial in simulating the problems involving extremely large deformations, fractures, etc., these methods become attractive options in multiscale modeling, especially when approaching a large number of atoms. In this paper, we propose preliminary research on applying meshfree particle methods to solve nanoscale problems. A quasicontinuum technique, i.e. the Cauchy-Born rule, is implemented into the meshfree particle methods so continuum approaches for large deformation problems or fracture problems at the nanoscale can be performed. Furthermore, the meshfree particle methods can be coupled with molecular dynamics via the bridging domain coupling technique. The examples show that the meshfree particle methods can benefit either hierarchical or concurrent multiscale modeling at the nanoscale.

1 Introduction

With the development of nanotechnology, numerical simulation plays an important role in nanoscale material and device design. To develop a potential numerical method, which can efficiently model micro/nano systems, has been one of the forefront research topics of computational nanotechnology.

Among a variety of numerical simulation techniques, molecular dynamics (MD) has become a powerful tool to elucidate complex physical phenomena [1-2]. Up to billions of atoms can be simulated by MD when studying the crack propagation [2] at the atomistic level with parallel computing techniques. However, most MD simulations are still restricted on both small length and short time scales. Therefore, multiscale methods have been of more and more interest to simulate large nanoscale systems. The recently developed multiscale methods can be divided into two classes: hierarchical multiscale methods [3] and concurrent multiscale methods [4-7]. In hierarchical multiscale modeling, the continuum approximation is based on the properties of a subscale model, such as a MD model. The intrinsic properties of materials are sought at the atomic level and embedded in the continuum model according to the quasicontinuum technique, which is also called the Cauchy-Born rule [8-9]. The Cauchy-Born rule states that the deformation is locally homogeneous.

Concurrent multiscale methods use an appropriate model to solve each length scale simultaneously. Recently, some concurrent multiscale techniques [4-6], particularly coupling methods between the continuum model and the molecular model, have been

V.S. Sunderam et al. (Eds.): ICCS 2005, LNCS 3516, pp. 284 – 291, 2005.

developed. One of the key issues the concurrent multiscale methods must overcome is the occurrence of spurious numerical phenomena, such as non-physical reflections on the interfaces between the molecular and continuum models. Most researchers use the Langevin equation [6] or other filtering processes to eliminate spurious reflections. Xiao and Belytschko developed a bridging domain coupling method [7], which can eliminate the spurious wave reflection automatically.

Mostly, finite element methods are used in the hierarchical or concurrent multiscale methods with the implementation of the quasicontinuum technique. It is known that the meshfree particle methods [10] are more attractive for a variety of problems with moving boundaries, discontinuities, and extremely large deformations. Therefore, the incorporation of the meshfree particle methods and the quasicontinuum technique will have much potential to solve the above problems at the nanoscale. Belytschko and Xiao [11] found that the meshfree particle methods with Lagrangian kernels are more stable than those with Eulerian kernels. In this paper, only the meshfree particle methods with Lagrangian kernels are considered. With the implementation of the quasicontinuum method, the meshfree particle methods can be used to simulate large nano systems. Furthermore, based on the idea of the bridging domain coupling method [7], the meshfree particle methods can be coupled with molecular dynamics to accomplish a multiscale modeling for large nano systems.

The outline of this paper is as follows: We will introduce the meshfree particle methods; The Cauchy-Born rule will then be implemented into the meshfree particle methods, which can also be coupled with molecular dynamics; Several examples are studied in the following section and the last section presents the conclusions.

2 Meshfree Particle Methods at the Nanoscale

2.1 Discrete Equations

The physical princ iples governing the continuum are the conservation of mass, momentum and energy. A so-called total Lagrangian description is employed (see Belytschko, Liu and Moran [12]); therefore, the linear momentum equations are

$$\frac{\partial P_{ji}}{\partial X_j} + \rho_0 b_i = \rho_0 \ddot{u}_i \ . \tag{1}$$

where ρ_0 is the initial density, $\mathbf{P}$ is the first Piola-Kirchhoff stress tensor, $\mathbf{X}$ is the reference coordinates, $\mathbf{b}$ is the body force per unit mass, $\mathbf{u}$ is the displacement and the superposed dots denote material time derivatives. The weak form of the momentum conservation equation is

$$\int_{\Omega_0} \delta u_i \rho_0 \ddot{u}_i d\Omega_0 = \int_{\Omega_0} \delta u_i \rho_0 b_i d\Omega_0 - \int_{\Omega_0} \delta F_{ij} P_{ji} d\Omega_0 + \int_{\Gamma_0} \delta u_i \bar{t}_i d\Gamma_0 \tag{2}$$

where Ω_0 is the reference configuration, δu_i is the test function, F_{ij} is the gradient of deformation and $\bar{t}_i$ is the prescribed boundary traction. The particle approximation is

$$u_i(\mathbf{X},t) = \sum_I w_I(\mathbf{X}) u_{iI}(t) \ . \tag{3}$$

where $w_I(\mathbf{X})$ is a Lagrangian kernel function, which is the function of reference coordinates. With a similar expansion for $\delta\mathbf{u}(\mathbf{X})$, the following discrete equations can be obtained:

$$M_I \ddot{u}_{iI} = F_{iI}^{ext} - F_{iI}^{int} \ , \quad M_I = \rho_0 V_I^o \ . \tag{4}$$

where V_I^0 is the volume associated with particle I in the reference configuration. F_{iI}^{ext} and F_{iI}^{int} are the external and internal nodal forces, respectively, given by

$$F_{iI}^{ext} = \int_{\Omega_0} \rho_0 w_I b_i d\Omega_0 + \int_{\Gamma_0^t} N_I \bar{t}_i d\Gamma_0 \ , \quad F_{iI}^{int} = \int_{\Omega_0} \frac{\partial w_I(\mathbf{X})}{\partial X_j} P_{ji} d\Omega_0 \tag{5}$$

If the nodal integration scheme [11] is used in the meshfree particle methods, the internal nodal forces in (5) can be calculated by

$$F_{iI}^{int} = \sum_J V_J^0 \frac{\partial w_I(\mathbf{X_J})}{\partial X_j} P_{ji}(\mathbf{X_J}) \ . \tag{6}$$

The Nodal integration scheme may result in one of instabilities due to the rank deficiency. A stress point integration scheme [11] can be used to stabilize it.

2.2 Implementation of the Quasicontinuum Technique

In a continuum model, the potential energy depends on the elongations and angle changes of the bonds at the atomistic level. The total potential of the continuum model can be written as

$$W^C = \int_{\Omega_0} w_C d\Omega \tag{7}$$

where w_C is the potential energy per unit volume. Then, the first Piola-Kirchhoff stress can be obtained from the potential of the continuum by

$$\mathbf{P} = \frac{\partial w_C(\mathbf{F})}{\partial \mathbf{F}} \ . \tag{8}$$

where $\mathbf{F}$ is the deformation gradient. In this paper, it is assumed the molecular structure in the volume associated with each particle is under a constant deformation gradient. Therefore, the first Piola-Kirchhoff stress at each particle can be evaluated through (8). In other words, (8) serves as the constitutive equation for meshfree particle methods at the nanoscale. For curved monolayer crystalline membranes such as nanotubes, an extension of the Cauchy-Born rule, called the exponential Cauchy-Born rule, can be used (see Arroyo and Belytschko [9]).

2.3 Coupling with Molecular Dynamics

Belytschko and Xiao [7] proposed a multiscale method called the bridging domain coupling method, in which, the molecular model and the continuum model overlap at their junctions in a bridging domain.

In this paper, molecular dynamics and the meshfree particle method are coupled via the bridging domain coupling technique. The complete domain in the initial configuration is denoted by Ω_0. The domain is subdivided into the subdomain treated by continuum mechanics, Ω_0^C, and the one treated by molecular dynamics, Ω_0^M. The intersection of these two subdomains is called the bridging domain denoted by Ω_0^{int} in the initial configuration. The bridging domain multiscale modeling of a molecular chain is shown in Figure 1.

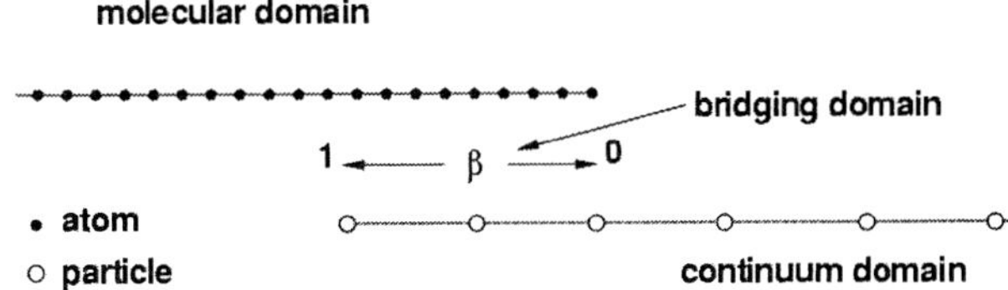

Fig. 1. A Bridging domain coupling model for a molecular chain

In expressing the total Hamiltonian of the system we employ a scaling parameter β in the bridging domain as shown in Figure 1. The scaling parameter β vanishes at one end of the bridging domain and is unity at another end. Therefore, the Hamiltonian for the complete domain is taken to be a linear combination of the molecular and continuum Hamiltonians

$$\begin{aligned} H = \beta H^M + (1-\beta)H^C = \sum_I \beta(\mathbf{X}_I)\frac{\mathbf{p}_I^M \cdot \mathbf{p}_I^M}{2m_I} + \beta W^M \\ + \sum_I (1-\beta(\mathbf{X}_I))\frac{\mathbf{p}_I^C \cdot \mathbf{p}_I^C}{2M_I} + (1-\beta)W^C \end{aligned} . \quad (9)$$

where W^M is the potential in the molecular model, and W^C is the strain energy in the continuum model. The discrete equations can be obtained via the classical Hamiltonian mechanics. The details can be found in [7].

3 Examples

3.1 Bending of a Nano Beam

The bending of a nano cantilever beam is considered in this example. The nano beam contains 5,140 atoms and the dimensions are: length $L = 270nm$ and height $H = 15.6nm$.

A pair potential function is used to approximate the interaction between nearest atoms,

$$U(l) = 0.5k(l - l_0)^2 \tag{10}$$

where $k = 10000N/m$ and $l_0 = 1nm$.

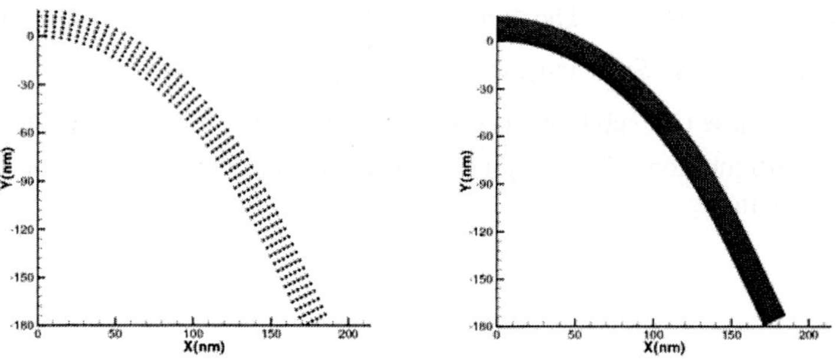

(a) The meshfree particle method (b) The molecular mechanics calculation

Fig. 2. Deformed configurations of the nanobeam

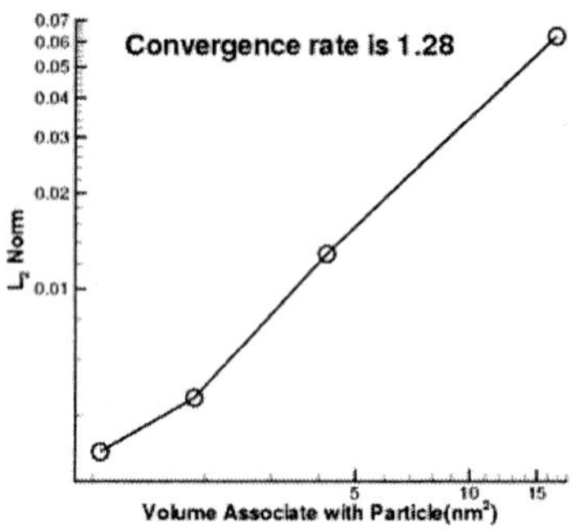

Fig. 3. Convergence of the nanoscale meshfree particle method

We use the meshfree particle method with 250 particles to simulate the bending of this nano beam. The prescribed displacement is applied on the right upper corner of the beam. The final configuration of the nano beam is shown in Figure 2(a). Figure 2(b) shows the deformed beam when performing the molecular mechanics calculation, and it supports the meshfree particle method result. As shown in figure 3, the convergence is also studied by using the l_2 error in displacement for the meshfree particle method.

3.2 A Nano Plate with a Central Crack

Meshfree particle methods are advantageous to simulate fracture problems. In this example, the meshfree particle method is used to study the stress concentration of a

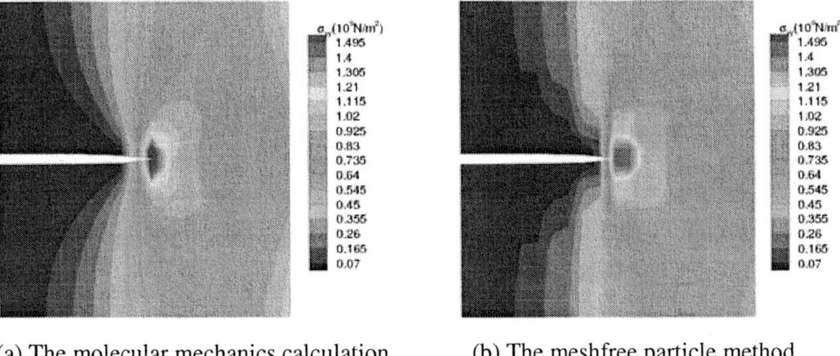

(a) The molecular mechanics calculation (b) The meshfree particle method

Fig. 4. Comparison of stress concentration at the crack tip

nano plate containing an central crack. This nano plate contains 86,915 atoms with the triangular molecular structure. The dimensions are: $L = 270nm$ and $M = 280nm$, and the crack length is $135nm$. The crack is initialized by taking a number of bonds out. The meshfree particle model has 400 particles, and a visibility criterion is used to construct the kernel functions for the particles near the crack or around the crack tip. Figure 4 shows the comparison of the stress (σ_{yy}) contour obtained from the molecular mechanics calculation with the one from the meshfree particle method. It can be seen that they are in accord.

3.3 Wave Propagation in a Molecular Chain

In this example, the wave propagation in a molecular chain, which contains 2001 atoms, is simulated. The LJ 6-12 potential function is used as the interatomic potential function between the nearest atoms, and it is

$$w_M = 4\varepsilon\left[(\sigma / r)^{12} - (\sigma / r)^6\right]. \tag{11}$$

where the constants are chosen as: $\sigma = 3.4e^{-10}m$ and $\varepsilon = 1.65e^{-21}J$. The mass of each atom is set to be $3.8\times 10^{-10}\,kg$.

In the bridging domain coupling modeling of this molecule chain, there are 1001 atoms in the molecular domain and 200 particles in the continuum domain. The initial wave is the combination of high frequency and low frequency waves and starts to propagate from the molecular domain. A non-physical phenomenon, shown in Figure 5(a), can be observed if using a handshake coupling technique [4] without the application of the artificial viscosity. It is possible to see that the high frequency wave is reflected while the low frequency wave passes the continuum domain. Such a phenomenon is also called the spurious wave reflection. However, with the bridging domain coupling technique, the spurious wave reflection can be eliminated as shown in Figure 5(b).

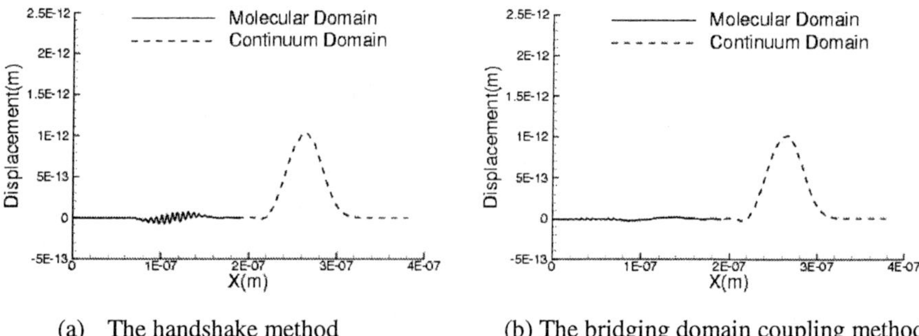

(a) The handshake method (b) The bridging domain coupling method

Fig. 5. Multiscale simulations on the wave propagation in a molecular chain

4 Conclusions

In this paper, the quasicontinuum technique (the Cauchy-Born rule) was implemented into the meshfree particle methods. Therefore, numerical simulations in nanotechnology can be valuable in regards to the meshfree particle methods. This progress makes it possible to treat extremely large deformation problems and the problems involving discontinuities, such as fractures, at the nanoscale. The examples showed that the nanoscale meshfree particle methods can give accurate results when compared with the molecular mechanics calculation outcomes. In addition, the meshfree particle methods can be coupled with molecular dynamics via the bridging domain coupling technique. The spurious wave reflection can be eliminated without any additional filtering processes.

References

1. Rountree, C. L., Kalia, R. K., Lidorikis, E., Nakano, A., Van, B. L., Vashishta, P.: Atomistic aspects of crack propagation in brittle materials: Multimillion atom molecular dynamics simulation. Annu. Rev. Mater. Res. 32(2002) 377-400
2. Abraham, F. F., Gao, H.: How fast can crack move. Phys. Rev. Lett. 84(2000) 3113-3116
3. Tadmor, E. B., Phillips, R., Ortiz, M.: Hierarchical modeling in the mechanics of materials. Int. J. Solids Struct. 37 (2000) 379-389
4. Abraham, F., Broughton, J., Bernstein, N., Kaxiras, E.: Spanning the continuum to quantum length scales in a dynamic simulation of brittle fracture. Europhys. Lett. 44(1998), 783-787
5. Rudd, R. E., Broughton, J. Q.: Coarse-grained molecular dynamics and the atomic limit of finite elements. Phys. Rev. B. 58(1998) R5893-R5896
6. Wagner, G. J., Liu, W. K.: Coupling of atomic and continuum simulations using a bridging scale decomposition. J. Comp. Phys. 190(2003) 249-274
7. Xiao, S., Belytschko, T.: A bridging domain method for coupling continua with molecular dynamics. Comput. Method Appl. M. (2004) in press
8. Tadmor, E. B., Ortiz, M., Phillips, R.: Quasicontinuum analysis of defects in solids. Philos. Mag. A 73(1996) 1529-1563

9. Arroyo, M., Belytschko, T.: A finite deformation membrane based on inter-atomic potentials for the transverse mechanics of nanotubes. Mech. Mater. 35(2003) 193-215
10. Belytschko, T., Krongauz, K., Organ, D., Fleming, M., Krysl, P.: Meshless methods: An overview and recent developments. Comput. Method Appl. M. 139(1996) 3-47
11. Belytschko, T., Xiao, S.: Stability analysis of particle methods with corrected derivatives. Comput. Math. Appl. 43(2002) 329-350
12. Belytschko, T., Liu, W. K., Moran, B.: Nonlinear Finite Elements for Continua and Structures. Wiley, New York (2000)

Numerical Simulation of Self-heating InGaP/GaAs Heterojunction Bipolar Transistors

Yiming Li[1,2] and Kuen-Yu Huang[3]

[1] Department of Computational Nanoelectronics, National Nano Device Laboratories, Hsinchu 300, Taiwan
[2] Microelectronics and Information Systems Research Center, National Chiao Tung University, Hsinchu 300, Taiwan
ymli@faculty.nctu.edu.tw
[3] Institute of Electronics, National Chiao Tung University, Hsinchu 300, Taiwan

Abstract. We numerically simulate effects of the self-heating on the current-voltage characteristics of InGaP/GaAs heterojunction bipolar transistors (HBTs). A set of coupled nonlinear ordinary differential equations (ODEs) of the equivalent circuit of HBT is formed and solved numerically in the large-signal time domain. We decouple the corresponding ODEs using the waveform relaxation method and solve them with the monotone iterative method. The temperature-dependent energy band gap, the current gain, the saturation current, and the thermal conductivity are considered in the model formulation. The power-added efficiency and the 1-dB compression point of a three-finger HBT are calculated. This approach successfully explores the self-heating and the thermal coupling phenomena of the three-finger transistors under high power and high frequency conditions. The numerical algorithm reported here can be incorporated into electronic computer-aided design software to simulate ultra-large scale integrated and radio frequency circuits.

1 Introduction

High power heterojunction bipolar transistors (HBTs) operated at micrometer- and millimeter-wave frequencies have been of great interest; in particular, for the advanced wireless and fiber communication [1], [2], [3], [4], [5]. These transistors, fabricated for high power applications, usually have a structure of multiple fingers to spread the current and the dissipated heat. Therefore, effects of the self-heating and the thermal coupling among fingers is one of the important issues for design and fabrication of advanced radio frequency (RF) circuits. Electrical characteristics depend on the surrounding temperature, so the thermal effect significantly influences the linearity of operating transistors [6]. Therefore, a self-heating model should be considered when exploring power dissipations of InGaP/GaAs devices and related RF circuits.

In this paper we examine effects of the self-heating on a three-finger InGaP/GaAs HBT operated at high frequency. Using the waveform relaxation

V.S. Sunderam et al. (Eds.): ICCS 2005, LNCS 3516, pp. 292–299, 2005.

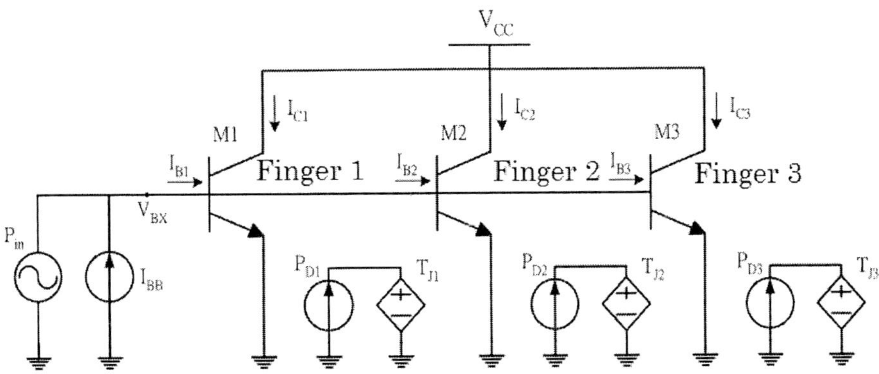

Fig. 1. An equivalent circuit of the simulated three-finger HBT under constant current and high frequency excitations

(WR), the monotone iterative (MI), and Runge-Kutta methods [7], [8], [9], [10], a set of nonlinear ordinary differential equations (ODEs) with thermal models is solved numerically in the time domain. The governing ODEs of the equivalent circuit of HBTs are formulated by the Kirchhoff's current law [11]. This solution technique has recently been developed by us for circuit simulations [7], [8], [9], [10]. The temperature-dependent physical quantities are modeled and several important engineering factors, such as the power-added efficiency and the 1-dB compression point of the simulated three-finger HBT are calculated. Our modeling and simulation successfully explores the self-heating and the thermal coupling phenomena of the studied three-finger transistors circuit operated under high power and high frequency conditions.

This paper is organized as follows. In Sec. 2, we state the physical model and numerical method. In Sec. 3, we present the simulation results. Finally, we draw conclusions.

2 Self-heating Modeling and Numerical Method

A multifinger HBT is formed by several sub-HBTs with their own collector and emitter, where their base is connected together. As shown in Fig. 1, an equivalent circuit with a thermal network of the three-finger HBT is studied in this work, where each finger is theoretically assumed to be identical. A thermal model that describes the relation between the power dissipation and the junction temperature is adopted. The electrical model of HBT considered in our simulation is based on the Gummel-Poon (GP) large signal model [7], [8], [9], [10]. For the thermal-electrical feedback mechanism, the temperature-dependant equations are introduced to the GP model

$$E_g(T_J) = E_g(T_A) + \frac{E_a \cdot T_A^2}{T_A + E_b} + \frac{E_a \cdot T_J^2}{T_J + E_b} , \tag{1}$$

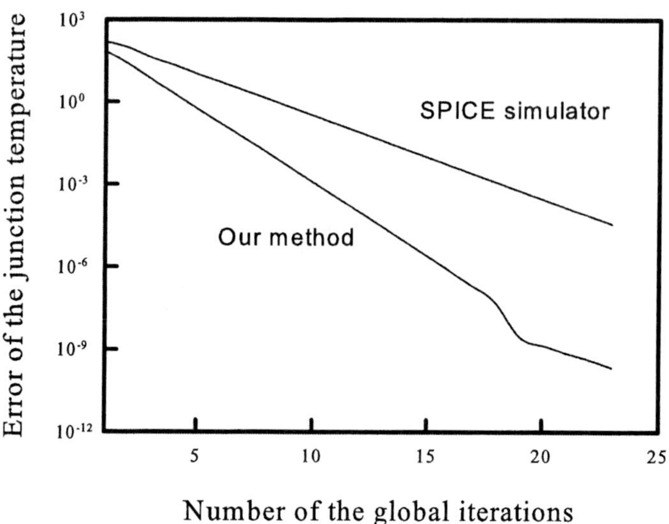

Fig. 2. A plot of the maximum norm error of the junction temperature versus the number of iterations of the WR loop

$$IS(T_J) = IS \cdot (\frac{T_J}{T_A})^{XTI} \cdot exp[(\frac{E_g(T_A)}{k \cdot T_A}) - (\frac{E_g(T_J)}{k \cdot T_J})], \tag{2}$$

$$ISE(T_J) = ISE \cdot (\frac{T_J}{T_A})^{\frac{XTI}{NE} - XTB} \cdot exp[(\frac{E_g(T_A)}{NE \cdot k \cdot T_A}) - (\frac{E_g(T_J)}{NE \cdot k \cdot T_J})], \tag{3}$$

$$ISC(T_J) = ISC \cdot (\frac{T_J}{T_A})^{\frac{XTI}{NC} - XTB} \cdot exp[(\frac{E_g(T_A)}{NC \cdot k \cdot T_A}) - (\frac{E_g(T_J)}{NC \cdot k \cdot T_J})], \tag{4}$$

$$BF(T_J) = BF \cdot (\frac{T_J}{T_A})^{XTB}, \tag{5}$$

$$BR(T_J) = BR \cdot (\frac{T_J}{T_A})^{XTB}, \tag{6}$$

where T_J and T_A are the junction and the ambient temperature, respectively. We note that for high power devices, T_A is the temperature on the back of the substrate. Above equations include the temperature dependance of energy band gap (E_g), the saturation current (IS), the collector and emitter leakage current (ISC and ISE), and the current gain (BF and BR). The thermal model expresses the relation between the power dissipation and the junction temperature. The junction temperature with considering the temperature-dependent thermal conductivity for the three-finger HBT is given by

$$\mathbf{T}_J = T_A\{1 - \frac{(BB-1)}{T_A}[\mathbf{R}_{TH} \cdot \mathbf{P}_D]\}^{\frac{-1}{BB-1}} = \begin{bmatrix} T_{J1} \\ T_{J2} \\ T_{J3} \end{bmatrix}, \tag{7}$$

where T_{Jn} is the junction temperature of the nth finger, $n = 1, 2, 3$. $\mathbf{R}_{TH} \cdot \mathbf{P}_D$ is given by

$$\mathbf{R}_{TH} \cdot \mathbf{P}_D = \begin{bmatrix} R_{T11} & R_{T12} & R_{T13} \\ R_{T21} & R_{T22} & R_{T23} \\ R_{T31} & R_{T32} & R_{T33} \end{bmatrix} \cdot \begin{bmatrix} P_{D1} \\ P_{D2} \\ P_{D3} \end{bmatrix}. \tag{8}$$

R_{Tnn} and R_{Tnm} denote the self-heating thermal resistance of the nth finger and the coupling thermal resistance which counts the coupled heat from the mth finger to the nth finger, respectively. Furthermore, the power dissipation of the nth finger is denoted by P_{Dn}. Equivalent circuit of the three-finger HBT is shown in Fig. 1. The finger 1 of this HBT is represented by $M1$, and $M2$ and $M3$ are the fingers 2 and 3, respectively. I_{BB} denotes the constant bias current at the base and P_{in} is the power of RF input signal. The behavior of Fingers 1 and 3 is the same for the identical fingers assumption. By considering the models above, the electrical and thermal feedback equations for the power HBTs are achieved and solved self-consistently in the time domain, where the temperature-dependent thermal conductivity is also included in our numerical solution.

The corresponding coupled nonlinear ODEs of the equivalent circuit of HBTs above is first decoupled using the WR method and then solved with the MI and Runge-Kutta methods. Compared with the conventional method, the Netwon's iteration method, used in the well-known SPICE circuit simulator [8], [9] our approach in solving the self-heating model is robust and cost-effective for the large-scale time domain circuit simulation of HBTs.

3 Results and Discussion

First of all, we verify the convergence of the solution method. The testing case is the device with the collector voltage $V_{CC} = 5V$, the input current bias $I_{BB} = 0.5mA$, the frequency is centralized at $1.8GHz$, and the convergence criterion is that the maximum norm error of the output voltage is less than $10^{-11}V$. Figure 2 shows a plot of the maximum norm error of the junction temperature versus the number of iterations. Simulation with our method demonstrates better convergence property than the result of the SPICE circuit simulator [9]. However, the simulation with the SPICE circuit simulator takes more than 100 iterations to meet the specified stopping criterion. For each time step, the CPU time of our method is of the order of 10^2 sec. running on a HP workstation XW 8000. The convergence of SPICE circuit simulator depends upon the selection of initial guess and the time steps which complicates the solution procedure for practical engineering application. Our solution algorithm converges monotonically and the DC solutions, in general, are used as the starting solution.

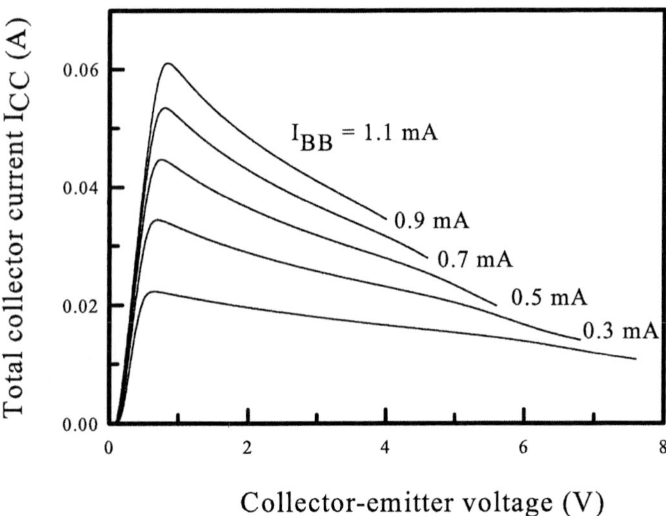

Fig. 3. The curves of I_{CC}-V_{CE} with respect to different I_{BB} for the simulated three-finger InGaP/GaAs HBT circuit. It is significantly different from the result without considering the self-heating [7], [8], [9]

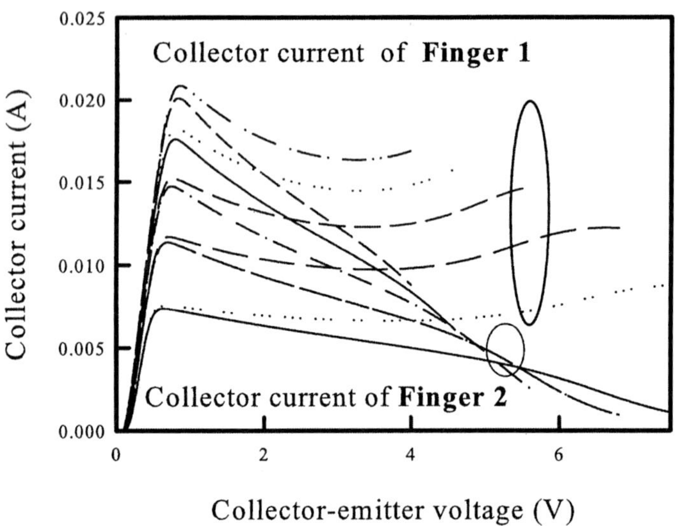

Fig. 4. The simulated I_{CC}-V_{CE} curves with respect to different I_{BB} for the fingers 1 and 2 of the HBT. The result of Finger 3 is omitted according to the property of symmetry

The simulated I-V characteristics of the three-finger InGaP/GaAs HBT circuit are shown in Figs. 3 and 4. Each curve in these figures represents the col-

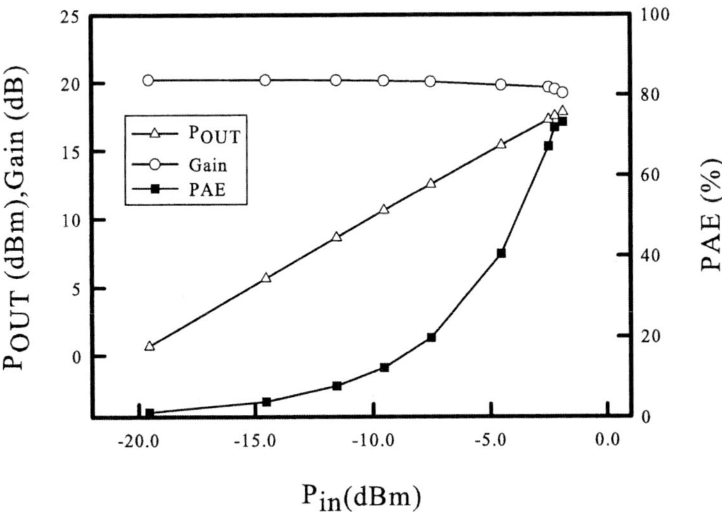

Fig. 5. The computed P_{OUT}, PAE, and Gain versus P_{in}

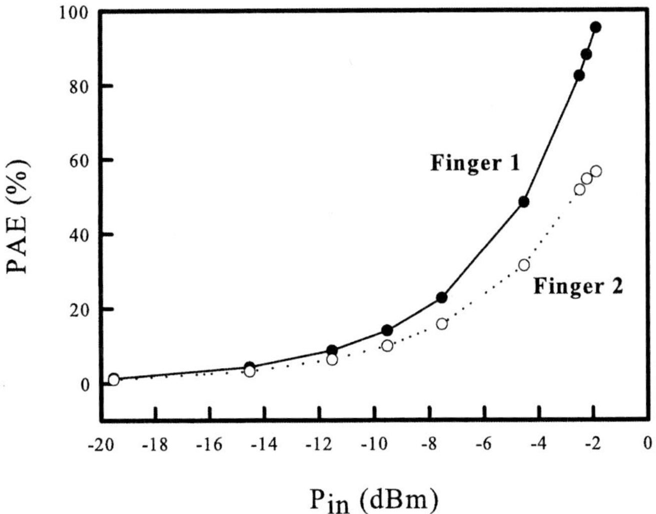

Fig. 6. Plots of the PAE of Fingers 1 and 2 versus the input power

lector current under constant I_{BB}. Due to the effect of self-heating, the total collector current decreases when the collector-emitter voltage increases, shown in Fig. 3. It results in a negative differential resistance region in the I-V characteristics and a collapse of current gain in the three-finger HBT circuit. As shown in Fig. 4, the collector current of the central finger (Finger 2) decreases rapidly, compared with the results of neighbor figures, such as Finger 1. It is due to the strongly coupled heat from its neighbor two fingers, Finger 1 and Finger 3. As the collector-emitter voltage increases even more ($V_{CC} > 4V$), an abrupt reduction of the collector current of Finger 2 occurs. With the electrical and thermal interaction, our modeling and simulation can explore the phenomenon of collapse for the multifinger HBTs under high voltage and high current bias.

As shown in Fig. 5, it is the calculated output power (P_{OUT}), the power-added efficiency (PAE), and the power gain (Gain) versus the different values of the input power (P_{in}). The input excitation is a single tone signal at 1.8 GHz. The bias condition of this single tone simulation is with $V_{CC} = 3.6$ V and $I_{BB} =$ 0.6 mA. In this simulation, we have taken the effect of heating of the input high frequency signal into consideration. It is found that, shown in Fig. 5, the Gain and PAE degrade as P_{in} increases, and the $1 - dB$ compression point (P_{1-dB}) is -2.45 dBm. The effect of thermal coupling among fingers also influences the performance of the three-finger device structure. As shown in Fig. 6, the PAE of the central finger (Finger 2) is reduced and degraded when $P_{in} > -3dBm$. In the meanwhile, the PAE of the neighbor finger (Finger 1) still increases as P_{in} increases. This phenomenon illustrates that the performance degradation of the whole transistor is mainly dominated by the hotter central finger.

4 Conclusions

In this paper, we have numerically solved a set of nonlinear and self-heating ODEs for exploring the electrical characteristics of InGaP/GaAs device. The solution approach is mainly based on the WR and the MI methods. Compared with the well-known SPICE circuit simulator, our approach has successfully shown its robustness. Different electrical characteristics have been calculated to examine effects of self-heating on the simulated three-finger HBT circuit. We believe that the solution method will benefit the community of electronic computer-aided design; in particular, for modern RF circuit simulation. To perform optimal designs for specified circuits, we are currently implementing intelligent algorithms with the developed method.

Acknowledgments

This work is supported in part by the National Science Council (NSC) of TAIWAN under contracts NSC-93-2215-E-429-008 and NSC 93-2752-E-009-002-PAE, and the grant of the Ministry of Economic Affairs, TAIWAN under contracts No. 92-EC-17-A-07-S1-0011 and No. 93-EC-17-A-07-S1-0011.

References

1. Yanagihara, M., Sakai, H., Ota, Y., Tamura, A: High fmax AlGaAs/GaAs HBT with L-shaped base electrode and its application to 50 GHz amplifier. Solid-State Electron. 41 (1997) 1615-1620

2. Oka, T., Hirata, K., Suzuki, H., Ouchi, K., Uchiyama, H., Taniguchi, T., Mochizuki, K., Nakamura, T: High-speed small-scale InGaP/GaAs HBT technology and its application to integrated circuits. IEEE Trans. Electron Devices. 48 (2001) 2625-2630
3. Troyanovsky, B., Yu, Z., Dutton, R. W.: Physics-based simulation of nonlinear distortion in semiconductor devices using the harmonic balance method. Comput. Methods Appl. Mech. Engrg. 181 (2000) 467-482
4. Zhu, Y., Twynam, J. K., Yagura, M., Hasegawa, M., Hasegawa, T., Eguchi, Y., Amano, Y., Suematsu, E., Sakuno, K., Matsumoto, N., Sato, H., Hashizume, N.: Self-heating effect compensation in HBTs and its analysis and simulation. IEEE Trans. Electron Devices. 48 (2001) 2640-2646
5. Heckmann, S., Sommet, R., Nebus, J.-M., Jacquet, J.-C., Floriot, D., Auxemery, P., Quere, R.: Characterization and modeling of bias dependent breakdown and self-heating in GaInP/GaAs power HBT to improve high power amplifier design. IEEE Trans. Microwave Theory and Techniques 50 (2002) 2811-2819
6. Park, H.-M., Hong, S.: A novel temperature-dependent large-signal model of heterojunction bipolar transistor with a unified approach for self-heating and ambient temperature effects. IEEE Trans. Electron Devices. 49 (2002) 2099-2106
7. Li, Y., Cho, Y.-Y., Wang, C.-S., Hung, K.-Y.: A Genetic Algorithm Approach to InGaP/GaAs HBT Parameters Extraction and RF Characterization. Jpn. J. Appl. Phys. 42 (2003) 2371-2374
8. Huang, K.-Y., Li, Y., Lee, C.-P.: A Time Domain Approach to Simulation and Characterization of RF HBT Two-Tone Intermodulation Distortion. IEEE Trans. Microwave Theory and Techniques. 51 (2003) 2055-2062
9. Li, Y., Kuang, K.-Y.: A Novel Numerical Approach to Heterojunction Bipolar Transistor Circuit Simulation. Comput. Phys. Commun. 152 (2003) 307-316
10. Li, Y.: A Monotone Iterative Method for Bipolar Junction Transistor Circuit Simulation. WSEAS Trans. Mathematics. 1 (2002) 159-164
11. Liu, W.: Handbook of III-V Heterojunction Bipolar Transistor. John Wiley & Sons (1998)

Adaptive Finite Volume Simulation of Electrical Characteristics of Organic Light Emitting Diodes

Yiming Li[1,2] and Pu Chen[2]

[1] Department of Computational Nanoelectronics, National Nano Device Laboratories, Hsinchu 300, Taiwan
[2] Microelectronics and Information Systems Research Center, National Chiao Tung University, Hsinchu 300, Taiwan
ymli@faculty.nctu.edu.tw

Abstract. In this paper a two-dimensional simulation of organic light emitting devices (OLEDs) using an adaptive computing technique is presented. A set of drift-diffusion equations including models of interface traps is solved numerically to explore the transport property of OLED structures. The adaptive simulation technique is mainly based on the Gummel's decoupling algorithm, a finite volume approximation, a monotone iterative method, a posteriori error estimation, and an unstructured meshing scheme. With this computational approach, we investigate the intrinsic and terminal voltage-current characteristics of OLEDs with respect to different material parameters, thickness of materials, and length of structure.

1 Introduction

Organic electroluminescence has been of great interest in various display applications. Organic light emitting diode (OLED) displays are lightweight, durable, power efficient and ideal for portable applications [1]. They have lower material costs and fewer processing steps than their liquid crystal display (LCD) counterparts. As such, the OLED display appears to be a strong candidate as a replacement technology in a variety of mobile application areas. OLEDs with different thin-film structures, consisting of emitter and carrier transport layers, have recently been reported [2], [3], [4], [5], [6]. According to the simple device geometry in OLEDs, one-dimensional (1D) transport model, the drift-diffusion model, has generally been solved along the transport direction for studying the electrical properties of OLEDs [4], [5], [6]. However, a multidimensional modeling and simulation plays a crucial role for exploring the effect of device structure and material on the electrical characteristics of OLEDs.

In this paper a set of drift-diffusion (DD) equations is solved with an adaptive computing technique [7], [8], [9], [10] for a two-dimensional (2D) simulation of OLEDs. For the simulation of OLEDs, the DD equations consist of the Poisson equation, the current continuity equation of electron, the current continuity equation of hole, and models of interface traps. First of all we decouple the three partial differential equations (PDEs) according to the Gummel's procedure. Based on adaptive unstructured mesh and finite volume (FV) approximation, each decoupled PDE is discretized and then solved by means of the monotone iterative (MI) method instead of

V.S. Sunderam et al. (Eds.): ICCS 2005, LNCS 3516, pp. 300 – 308, 2005.

Newton's iteration (NI) method. The method of monotone iteration is a constructive alternative for numerical solutions of PDEs. It has been reported that, compared with the NI method, the major features of the MI method are (1) it converges globally with any arbitrary initial guesses; (2) its implementation is much easier than NI method; and (3) it is inherently ready for parallelization [8]. Furthermore, due to the efficient posteriori error estimation, the variation of physical quantities, such as the gradients of potential and current density, can be automatically tracked. Therefore, the terminal characteristics are accurately calculated. The proposed adaptive computing technique shows the simulation accuracy and numerical robustness for the simulation of 2D OLEDs. Effects of geometry, the trap density and the Schottky barrier height [11] on the current-voltage (I-V) curves of the simulated 2D OLED are examined using the developed 2D simulation program.

This paper is organized as follows. In the section 2, we state the transport model and the adaptive computing technique for the 2D simulation of OLED. In the section 3, the results of numerical simulation are discussed. Finally we draw the conclusions.

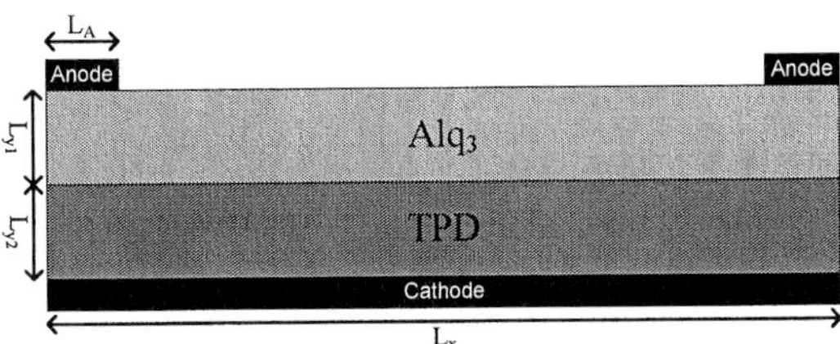

Fig. 1. A cross-sectional view of the studied OLED structure, where L_A is the length of contacts of anode and L_x is the length of contact of cathode. The L_{y1} is the width of material Alq3 which is the layer of electron transport and the L_{y2} is the width of material TPD which is the layer of hole transport

2 Mathematical Model and Computational Methodology

Based on the well-known inorganic charge transport continuum model [9], the drift-diffusion model, electron and hole transport in the OLED is described using the current continuity equations coupled to the Poisson equation [11]. Along with the appropriate boundary conditions, which for OLEDs require appropriate formalisms for current injection at each of the contacts, these equations are solved to obtain solutions for the electrostatic potential, electric field, carrier densities, and current densities for electron and hole, respectively. The investigated 2D structure of OLED, shown in Fig. 1, is a TPD/Alq_3 two-layer device. Modeling of traps in numerical simulation of OLEDs is widely debated [3]. The electron-hole mobility taking the field-dependent Poole-Frenkel form in the current continuity equations of electron and hole is for shallow trapping of carriers. For deep traps, an extra recombination term and inclusion of their charge in the Poisson equation are considered. Deep traps can be described by a discrete and exponential distribution. The simulated OLED is based on the 2D structure of the tris-(8-hydoroxyquinoline)-aluminum (Alq_3) for the layer of electron transport and the triphenyl-diamine (TPD) for the layer of hole transport. As shown in Fig. 1, we assume the length of anode (L_A) is equal to 20 nm, the length of cathode (L_X) is equal to 400 nm, and the thicknesses of Alq_3 layer (L_{y1}) and TPD layer (L_{y2}) are equal to 40 nm, respectively. We solve the steady-state DD model [11-14], which consists of

$$\Delta\psi = -\frac{q}{\varepsilon}(p - n - N_A + N_D - n_t + p_t), \tag{1}$$

$$\nabla J_n = -q(G - R) \text{, and} \tag{2}$$

$$\nabla J_p = -q(G - R). \tag{3}$$

The current equations, shown in Eqs. (2) and (3), for electron and hole are given by

$$J_n = -q\mu_n(nE - \frac{K_B T}{q}\nabla n) \text{ and} \tag{4}$$

$$J_p = -q\mu_p(pE + \frac{K_B T}{q}\nabla p). \tag{5}$$

In Eq. (1), ψ is the electrostatic potential, ε is the dielectric constant, N_A and N_D are the densities of acceptor and donor, n_t and p_t are the densities of trapped electrons and holes, respectively. Maxwell-Boltzmann statistics is adopted for the electron and hole densities. In Eqs. (2) and (3) G is the electron and hole generation rate and the carrier generation by thermal excitation across the gap is assumed. Two carrier recombination rates, the optical recombination rate R_{opt} and the Shockley-Read-Hall recombination rate R_{srh} are assumed in the simulation [11]. We consider here the densities of trapped electrons and holes for the jth trap level

$$n_t = \frac{N_{tj}}{1 + \frac{1}{g} e^{\frac{E_{tj} - E_{fn}}{K_B T}}} \text{ and } p_t = \frac{P_{tj}}{1 + \frac{1}{g} e^{\frac{E_{tj} - E_{fp}}{K_B T}}},$$

where N_{tj} (P_{tj}) is the electron (hole) trap density, E_{tj} is the trap energy relative to the conduction band edge, g is the trap degeneracy, and E_{fn} (E_{fp}) is the electron (hole) quasi-Fermi level. Boundary conditions are assumed for the DD model above [11-14].

In the solution procedure, the adaptive computing is mainly based on the Gummel's decoupling algorithm, the FV approximation, the MI method, a posteriori error estimation, and an unstructured meshing technique. This simulation methodology has been developed in our recent work for semiconductor device simulation [7], [8], [9], [10]. Each Gummel decoupled PDE is approximated with the FV method over unstructured meshes. The corresponding system of nonlinear algebraic equations of the FV approximated PDE is solved with the MI method and the posteriori error estimation scheme is applied to assess the quality of computed solutions. It has been shown that the method of MI converges monotonically [8]. The adaptive mechanism is based on an estimation of the gradient of computed solutions, such as the electrostatic potential, the carrier density, and the current density. A posteriori error estimation is applied to provide local error indicators for incorporation into the mesh refinement strategy. The local error indicators guide the adaptive refinement process.

3 Results and Discussion

In this section we first present the computational efficiency of the method by solving the OLED under a given biasing condition. The applied anode voltage and cathode voltage on the OLED are 10 V and 0 V, respectively, and the barrier height on the contact is assumed to be 0.3 eV. For this testing case, the obtained initial and final refined meshes are shown in Fig. 2. The stopping criteria for the MI and Gummel's loops are 10^{-6} and 10^{-3}, respectively. The initial mesh has 153 nodes and the final one consists of 1681 nodes. The adaptive computing process includes 5 refinement levels, which shows the computational efficiency. Different simulation cases are further performed to explore the intrinsic and terminal electrical characteristics of the OLEDs.

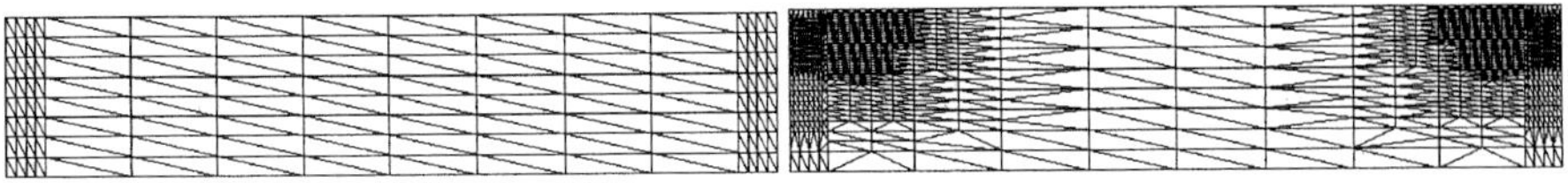

Fig. 2. The left figure is the initial mesh and the right one the 5th refined final mesh

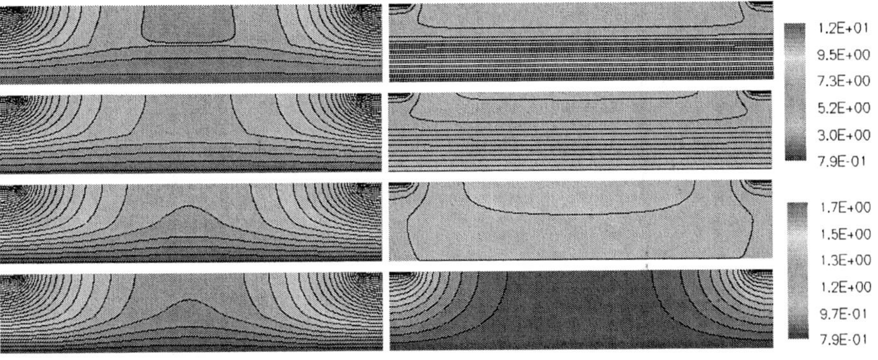

Fig. 3. The simulated potential of the OLED under the 10 V (left column) and the zero bias (right one). The barrier height is 0.3 eV (top figure), 0.6 eV, 0.9 eV, and 1.2 eV (bottom one)

We note that the simulation of OLEDs require the input of accurate material parameters, such as density of state, barrier height, and carrier mobility. However, unlike the case for inorganic semiconductor devices, these material parameters are often poorly characterized and are strongly dependent upon the fabricated samples. Our selection of the material parameters provides only the starting point for the electrical simulation of OLEDs. Good agreement between simulation and measurement should be subject to further calibration. Contour plots of the electrostatic potential and the electron density are shown in Figs. 3 and 4, respectively, where the OLED under two different biasing conditions the anode voltage is equal to 10.0V and 0 V, are simulated. The cathode voltage is fixed at 0.0V and the barrier height of contact varies from 0.3 eV to 1.2 eV with step 0.3 eV. The unit of the color bars, shown in Fig. 3, is in Volt. The computed electrostatic potential significantly shows the importance of selection of barrier height. Different barrier heights imply different current injection. The unit of the color bars of electron density is with per meter cubic. Figures 3 and 4 demonstrate the necessary for the advanced design of OLED structures by using a 2D simulation.

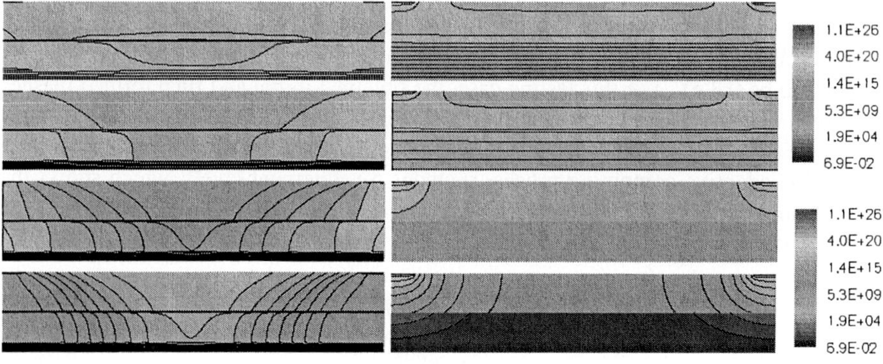

Fig. 4. The contour plots of the simulated electron density. The setting is the same with Fig. 3

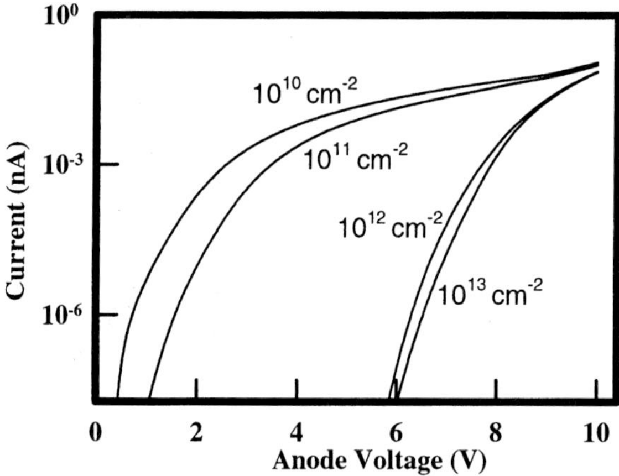

Fig. 5. The I-V curves of the simulated OLED with different trap densities of electron

To explore the effect of trap density on the transport characteristics, we simulate the current-voltage (I-V) curves with respect to different electron trap densities, where the hole trap density is neglected. As shown in Fig. 5, the I-V curves per unit area are calculated with different densities of traps which range from 10^{10} cm^{-2} to 10^{13} cm^{-2}. The anode voltage is equal to 10 V and the barrier height is fixed at 0.3 eV. It is found that the higher electron trap densities get no significant benefit to improve the electrical performance of the OLED. The choice of OLED's contacts is crucial, and it affects the minority and majority currents, the recombination rates, and the efficiency. At the interface of metal and semiconductor of OLED, the variation of barrier heights is wide. The operation of OLED depends upon the asymmetry of the barrier heights at the two contacts, shown in Fig. 1; ITO is one of the preferred anode materials due to the transparency and relatively high work function. Metal, such as Al, Ca, or Mg, with low work functions is selected as cathode material. To explore the effect of the Schottky barrier height on the I-V relationships [11], the Schottky barrier height on the contacts is simply varied from 0.0 to 1.2 eV, shown in Fig. 6.

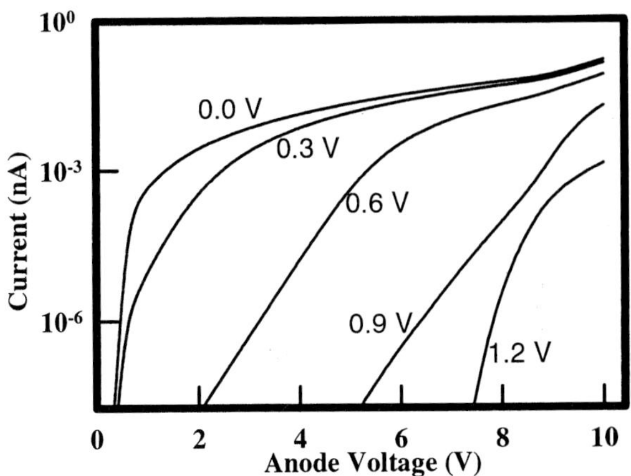

Fig. 6. The I-V curves of the simulated OLED with different Schottky barrier heights

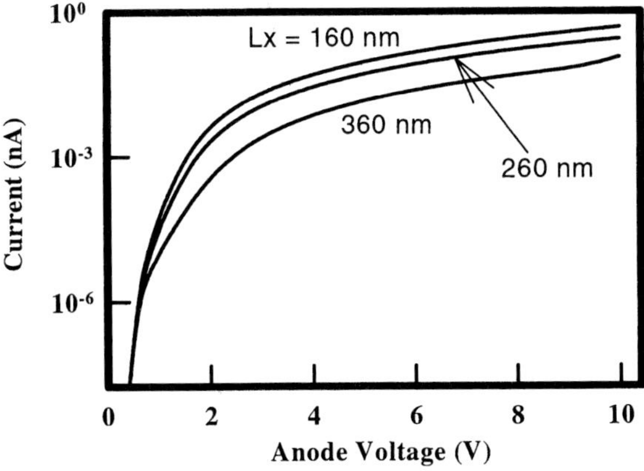

Fig. 7. The I-V curves of the simulated OLED with different lateral length L_x

With the 2D simulation, the lateral length and thickness of transport layer are changed for exploring the I-V curves. It is found that small L_x produces high current level when the L_{y1} and L_{y2} are fixed, shown in Fig. 7. Due to uniform distribution of electric fields, the small variation on L_{y1} and L_{y2} does not significantly alter the level of current.

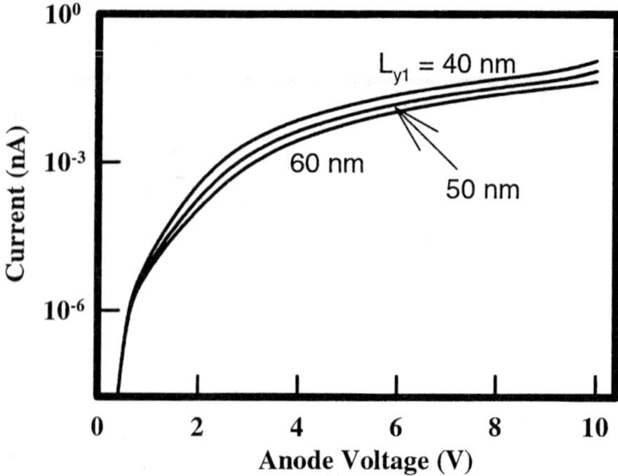

Fig. 8. The I-V curves with respect to different thicknesses of electron transport layer L_{y1}

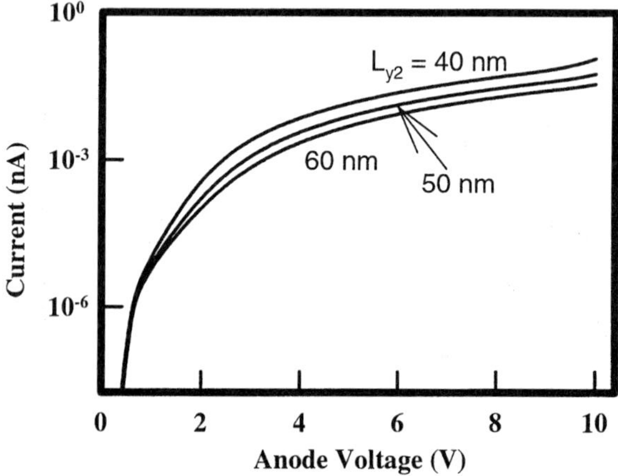

Fig. 9. The I-V curves with respect to different thicknesses of hole transport layer L_{y2}

4 Conclusions

In this paper we have successfully applied adaptive computing technique to solve the 2D DD model for the simulation of electrical characteristics of OLEDs. Effects of the carrier traps and device geometry on the transport properties have been studied. The 2D macroscopic modeling and simulation is a starting point for theoretical investigation of electrical characteristics of OLEDs. It benefits the design of structure and optimization of characteristics. Application of technology computer-aided design software to modern display industry requires accurate modeling and calibration of material parameters. In contrast to the macroscopic simulation, we believe that microscopic approaches will physically provide accurate calculation.

Acknowledgments

This work is supported in part by the National Science Council (NSC) of TAIWAN under contracts NSC-93-2215-E-429-008 and NSC 93-2752-E-009-002-PAE, and the grant of the Ministry of Economic Affairs, Taiwan under contracts 92-EC-17-A-07-S1-0011 and 93-EC-17-A-07-S1-0011.

References

1. Goh, J.-C., Chung, H.-J., Jang, J., Han, C.-H.: A New Pixel Circuit for Active Matrix Organic Light Emitting Diodes. IEEE Elec. Dev. Lett. 23 (2002) 544-546
2. Ruhstaller, B. Carter, S.A., Barth, S., Riel, H., Riess, W., Scott J. C.: Transient and Steady-State Behavior of Space Charges in Multilayer Organic Light-Emitting Diodes. J. Appl. Phys. 89 (2001) 4575-4586
3. Waler, A. B., Kambili A., Martin S. J.: Electrical Transport Modelling in Organic Electroluminescent Devices. J. Phys.: Condens. Matter 14 (2002) 9825-9876
4. Ruhstaller, B., Beierlein, T., Riel, H., Karg, S., Scott, J. C., Riess W.: Simulating Electronic and Optical Processes in Multilayer Organic Light-Emitting Devices. IEEE J. Sel. Topics Quantum Elec. 9 (2003) 723-731
5. Blades, C. D. J., Walker, A. B.: Simulation of Organic Light Emitting Diodes. Synth. Met. 111-112 (2000) 335-340
6. Barth, S., Müller, P., Riel, H., Seidler, P. F., Riess, W., Vestweber, H., Bässler, H.: Electron Mobility in Alq Thin Films Determined via Transient Electroluminescence from Single- and Multilayer Organic Light-Emitting Diodes. J. Appl. Phys. 89 (2001) 3711
7. Li, Y., Yu, S.-M.: A Two-Dimensional Quantum Transport Simulation of Nanoscale Double-Gate MOSFETs using Parallel Adaptive Technique. IEICE Trans. Info. & Sys. E87-D (2004) 1751-1758
8. Li, Y.: A Parallel Monotone Iterative Method for the Numerical Solution of Multidimensional Semiconductor Poisson Equation. Comput. Phys. Commun. 153 (2003) 359-372
9. Li, Y., Sze, S, M., Chao, T.-S.: A Practical Implementation of Parallel Dynamic Load Balancing for Adaptive Computing in VLSI Device Simulation. Eng. Comput. 18 (2002) 124-137

10. Li, Y., Liu, J.-L., Chao, T.-S., Sze, S. M.: A New Parallel Adaptive Finite Volume Method for the Numerical Simulation of Semiconductor Devices. Comput. Phys. Commun. 142 (2001) 285-289
11. Sze, S. M.: Physics of semiconductor devices. Wiley-Interscience, New York (1981)
12. Crone, B. K., Davids. P. S., Campbell. I. H., Smith, D. L.: Device Model Investigation of Bilayer Organic Light Emitting Diodes. J. Appl. Phys. 87 (2000) 1974-1982
13. Lupton, J. M., Samuel, I. D. W.: Temperature-Dependent Single Carrier Device Model for Polymeric Light Emitting Diodes. J. Phys. D: Appl. Phys. 32 (1999) 2973-2984
14. Kawabe, Y., Jabbour, G. E., Shaheen, S. E., Kippelen, B., Peyghambarian, N.: A Model for the Current–Voltage Characteristics and the Quantum Efficiency of Single-Layer Organic Light Emitting Diodes. Appl. Phys. Lett. 71 (1997) 1290-1292

Characterization of a Solid State DNA Nanopore Sequencer Using Multi-scale (Nano-to-Device) Modeling

Jerry Jenkins, Debasis Sengupta, and Shankar Sundaram

CFD Research Corporation 215 Wynn Drive,
Huntsville, AL 35805
{jwj, dxs, sxs}@cfdrc.com

Abstract. Nanobiotechnology is a rapidly advancing frontier of science with great potential for beneficial impact on society. Unfortunately, design of integrated nano-bio systems is a complex, laborious task with large failure rates. Current models describing molecular level behavior are expensive, while device design codes lack the necessary nanophysics. The objective of this work is to demonstrate multiscale, multiphysics modeling of an integrated nanobio device, where nanoscale effects are efficiently integrated with a continuum model. A three-level modeling paradigm was developed for this purpose. The feasibility of this approach is demonstrated by characterizing a nanopore-based DNA sequencing device. In the demonstration calculations, the dependence of the device performance on the nucleotide sequence, pore diameter, and applied voltage was determined. Extension of the approach for describing biomolecular processes in other commercial nanobiosystems is discussed. The main conclusions of the device level simulations are presented along with an overview of future work.

1 Introduction

Research efforts directed at the problem of integrating nanotechnology and biology to form integrated nano-bio systems are becoming a priority. Integrated nano-bio systems have emerged as strong candidates for single molecule detection and genomic sequencing. Understanding the relationship between molecular behavior and system/device response, requires the development of modeling and simulation tools, which can simulate phenomena spanning a wide spectrum of length and time scales (Figure 1(a)). We have developed a hierarchical theoretical and computational framework (Figure 1(b)) that addresses the problem of coupling molecular level information into a continuum level device simulation. Each modeling and simulation level in the hierarchy is chosen with careful attention to a cost-benefit analysis.

The first level in the hierarchy is an atomistically detailed model of the nanoscale phenomena of interest. Molecular simulations enable a physical understanding of the fundamental mechanisms that underlie nanoscale phenomena. Results from the molecular level simulations are used to parameterize the second level stochastic

V.S. Sunderam et al. (Eds.): ICCS 2005, LNCS 3516, pp. 309 – 316, 2005.

model. Stochastic models (Master Equation or Fokker-Planck Equation) are utilized to bridge molecular and continuum levels because they are able to correctly predict the average properties of a system as well as the fluctuations inherent in the nanoscale. The final level is a continuum model of the system of interest, which couples information from the preceding two levels. Stochastic models can be tightly integrated into the continuum model for high-fidelity design calculations. Insights obtained using this modeling hierarchy can guide the choice of candidate design strategies for specific integrated nanobiosystems.

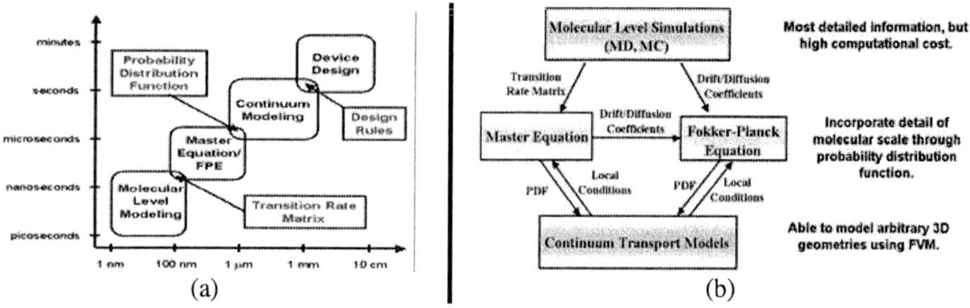

Fig. 1. (a) Applicable range of scales for each modeling level. (b) Illustration of the different levels of modeling. The text next to the arrows is the information transferred at each level

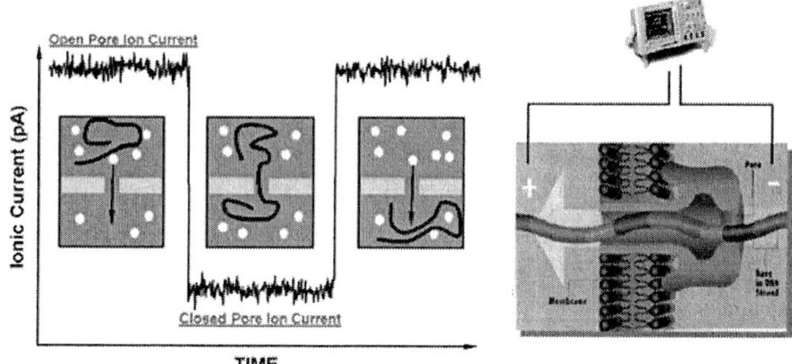

Fig. 2. As negatively charged nucleic acids pass through the nanopore in single file, they block the flow of current in a manner characteristic of the polymer's length and sequence [2]

The demonstration case we have chosen (Figure 2) is the problem of DNA translocation through nanpores. This area has come under a great deal of investigation recently due to its promise for inexpensive ultra high-throughput sequencing of DNA. A voltage bias applied to a nanopore has been shown to induce charged single stranded DNA and RNA molecules to translocate through the pore. Each translocating molecule blocks the open pore ionic current, providing an electrical signal that depends on several characteristics of the molecule [1]. The natural system has limitations due to the fixed size of the pore, stability, and noise characteristics. These difficulties can be overcome using a solid state nanopore [2].

The paper begins with a description of the molecular modeling in §2, followed by the development of the stochastic model in §3. This information is then coupled to the device level calculations to evaluate a proposed ultrahigh throughput DNA sequencing device (§4). The main conclusions of the device level simulations are presented along with an overview of future work.

2 Molecular Level Modeling

Translocation times for DNA to cross a nanopore are strongly dependent on the pore diameter [1]. Nanopores with diameters of 1.5nm and 2.0nm, were constructed based on experiments presented in the literature [3]. The pores were created using an in-house software that constructs a cube of silicone dioxide using the method of Brodka [4]. The cube is truncated to form a solid block of with dimensions of 5nm x 5nm x 5nm [5]. An overhead view of the 2nm nanopore is shown in Figure 3(a).

Experimental work has also revealed that homopolymers of adenine and cytosine have significant differences in their translocation times [2]. For this reason, poly$(dC)_4$ and poly$(dA)_4$ were chosen for this computation with poly$(dC)_4$ shown in Figure 3(b). Once generated, the nucleotides were placed in the nanopore and aligned with the axis of the pore with the 5' end pointing in the positive z-direction. Water and KCl were added to the system using tcl scripting language built into the VMD molecular graphics software [6]. The final system consists of a 6.1nm layer of water on top of the 5nm crystal structure and a 2nm layer of water on the bottom of the crystal for a total height of 13.1nm (Figure 3(c)). The total number of atoms for each system is approximately 29,000.

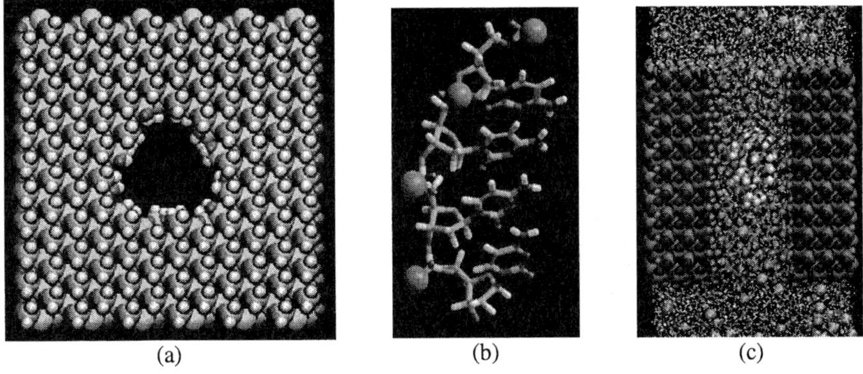

Fig. 3. Generation of (a) the nanopore in a Block of SiO_2 with dimensions of 5x5x5 nm. (b) 4-mer of Cytosine with potassium counter ions. (c) Final system (cutaway view) with nucleotide in the nanopore including counter-ions and solvent water

Molecular Dynamics (MD) simulations were performed in three phases following the work of Auffinger [7]. The phases consist of: (a) the initial heating of the water and ions, (b) equilibration of the entire system, and (c) production. The NAMD software was used for all molecular simulations. (NAMD was developed by the Theoretical and Computational Biophysics Group in the Beckman Institute for Advanced Science and Technology at the University of Illinois at Urbana-Champaign) [8]. The Verlet-I or r-RESPA Multiple Time Stepping (MTS) algorithm was used in order to increase the time step to 2fs [9]. During the production simulations a Langevin based temperature controller was used to maintain the temperature [10], but was not coupled to hydrogen atoms. The Particle Mesh Ewald (PME) algorithm was used to model the periodic boundary conditions, with grid spacing of 64, 64, and 128 in the x, y, and z directions, respectively. All forcefield parameter information was taken from the CHARMM27 all hydrogen nucleic acid forcefield, derived from the work of [11] and [12]. Parameters for the potassium and chlorine ions were taken from [13]. The charge and Lennard-Jones parameters for the SiO_2 were taken from [14].

Entropic affects can contribute significantly to the free energy, especially if a molecule is moving from solution into a confined space. The method of Schlitter [15] was used to compute the state entropy from the production phase of the molecular dynamics trajectories. The method is uses an approximate Gaussian form to approximate the canonical configurational distribution function. This method has been widely used to determine the entropy change for protein stability [16], and the stability of double stranded DNA [17]. A striking feature (Figure 4(a)) is that the 1.5nm pores have larger entropy changes relative to the surroundings when compared to the 2.0 nm pores. The largest entropic penalty was felt by poly(Cytosine)$_4$ in the 1.5nm pore. This contribution to the free energy is approximately 25 kcal/mol, while the 2nm pore case for cytosine only has a 11 kcal/mol entropic barrier. The adenine shows a similar pattern of higher entropy outside of the pore and higher entropy change.

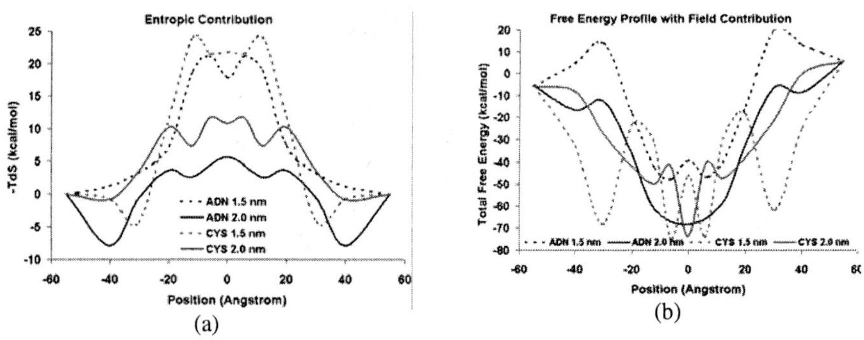

Fig. 4. (a) Prediction of the entropic component of the overall free energy of the system. (b) Free Energy Profiles (ΔU-$T\Delta S$) of translocation for poly(dC)$_4$ and poly(dA)$_4$ translocating through the 1.5 and 2.0nm pores

Figure 4(b) illustrates that the free energy profiles are similar for all four cases. Focusing in on the portion of the graph that contains the pore (–25 to 25 on the x-axis), we can draw some conclusions pertaining to the contribution to translocation times from the free energy. The relative free energy profiles show that Cytosine has a lower free energy relative to Adenine within the pore, for both pore diameters, having the lowest free energy in the 1.5nm pore. Adenine shows a lower free energy within the 2.0nm pore relative to the 1.5nm pore, while Cytosine shows a similar free energy for both pores. The barriers extracted from the nanopores are very similar to the shape of the energy profiles extracted from ion channels [18]. The contribution to the electric field tilts the free energy profiles11.5 kcal/mol assuming a 120 mV bias across the 5nm thickness of the nanopore.

3 Stochastic Model of DNA Translocation

The stochastic model forms the second modeling scale (complementary to the molecular modeling described above). The inputs to this level are the translocation free energy profiles and the dependence of the profile on applied voltage. The goal is to compute the distribution of translocation times and the mean translocation time. The distribution of translocation times was computed using the method of Lubensky [19], and are shown in Figure 5(a). The mean translocation times were then computed from the distribution and checked against the method of Schulten [20].

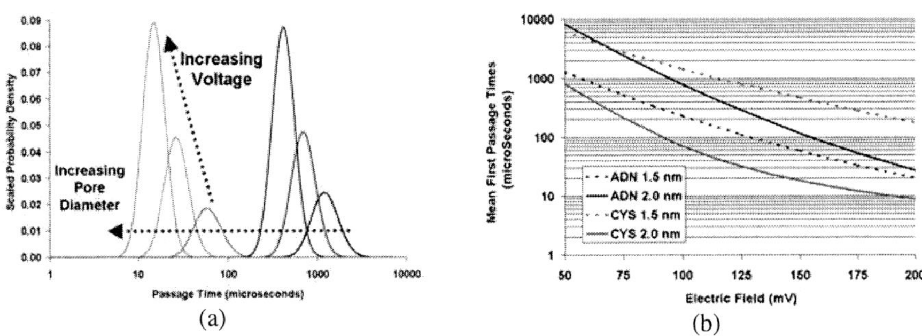

Fig. 5. (a) Variation in the distribution of passage times with applied field and pore diameter for cytosine. (b) Mean passage time dependence on applied voltage for all four cases

For all four cases, Figure 5(b) shows an exponential drop in the mean passage time with applied voltage, in agreement with several other researchers [21, 22]. Looking at the individual nucleotides, the cytosine shows a longer passage time with the 1.5 nm pore as opposed to the 2.0 nm pore. However, the Adenine chain shows a longer passage time with the 2.0 nm pore as opposed to the 1.5 nm pore. Also to note is the fact that there is a much larger difference in passage time comparing pore diameters with cytosine than with adenine. The reason for the differences is not immediately clear until we look at the free energy profiles (Figure 4(b)). The Adenine profiles show a large difference in the depth of the minimum on the order of 20% of the total depth. While the cytosine nucleotides do not show the same dependence. The information computed in this section will be coupled in directly with the continuum level model presented in the next section.

4 Continuum Level Model

This forms the final element in the three-tiered paradigm A pore model is constructed that takes into account the arrival rate of DNA in the proper conformation, and the translocation of DNA across the pore. A boundary condition was implemented in the multiphysics modeling software CFD-ACE+ (marketed by ESI Group), which accounts for the rate of capture as well as the rate of translocation (Figure 2) The translocation model contains two steps, with the first step being the rate of capture of a DNA (**R**) chain into the nanopore. This rate was not computed, but was taken from the literature [2]. The second step is the translocation of the captured DNA across the pore to free up the pore. This is assumed to occur at a rate of $1/\tau$ where, τ is the mean translocation time computed using the free energy profiles (Figure 5(b)). Both τ and **R** are dependent on the applied voltage.

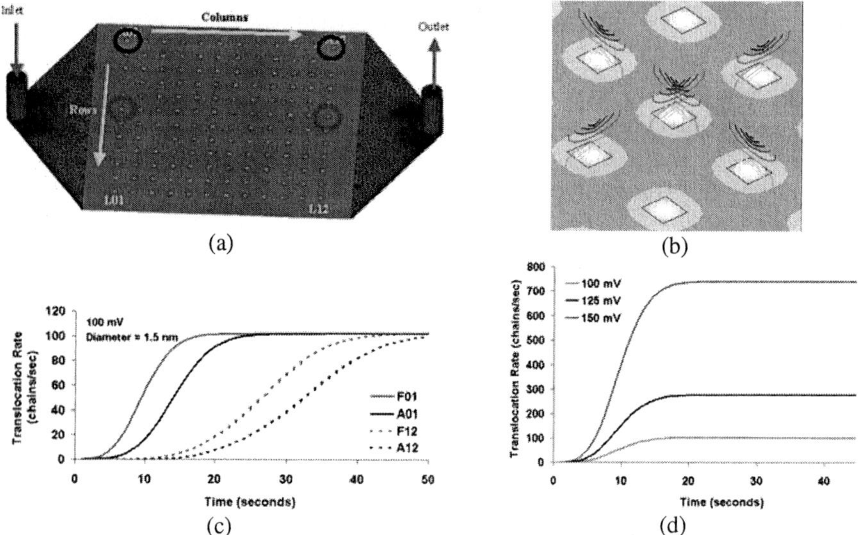

Fig. 6. (a) Three dimensional geometry of the notional device showing the labeling of the nanopore containing patches. The patches are shown in yellow. (b) Zoom in view of the electric field on each patch optimized to minimize cross talk. (c) Dependence on the Translocation Rates on the Placement. For the case of CYS 1.5 nm at 100, 125, and 150 mV. (d) Dependence on the Translocation Rates on the Voltage for the Case of F01 Spot

The device chosen for demonstration (Figure 6(a)) is a 21 mm by 7mm by 0.25 mm thick microfluidic chip containing 144 nanopore bearing patches. Each patch is uniquely identified (similar to a standard spreadsheet) with a letter and a number. Within the computational domain of the chip, the Navier-Stokes equation for an incompressible fluid is solved along with the equation of mass transport for DNA. A sink term is added at each patch based on the rate of translocation. Initially there is assumed to be no DNA in the chip. At time zero, DNA begins to flow into the chip at a concentration of 10^{-6} Molar being carried by the fluid that is being pumped at a rate of 100 microL/min. The total simulated time is 50 seconds, long enough for the pore translocation to reach a steady state. Also, a constant voltage was held across each patch, and the solution for the electric field was computed during the simulation. This calculation allowed for the determination of the optimal spacing of the pore containing patches to minimize electronic cross talk (Figure 6(b)).

To understand the dependence of the output signals upon pore placement, four pores were monitored. A01, A12, F01, and F12 (see Figure 6(a) for locations). These represent extremes in terms of placement within the flow field and placement along the line of flow. Figure 6(c) shows the evolution of the translocation rate for three voltages and the four candidate spots. All three voltages show a similar trend of F01 and A01 being the first to reach steady state, while F12 and A12 show a nearly 20 second lag. This lag time will be strongly dependent upon the inlet flowrate of the device, with an increasing flowrate decreasing the lag time. An increase in the applied voltage increases the point at which the patch translocation rate reaches a steady state. Figure 6(d) shows the dependence of the DNA translocation rates on the applied pore voltage. The translocation rates are shown to increase exponentially in agreement with [21].

5 Summary and Conclusions

The design of devices that incorporate both nanoscale and continuum scale physics is challenging due to the nature of simulation techniques for each scale. This paper presents a general three-tiered modeling paradigm for use in simulating devices containing a nanoscale component. The models presented will be incorporated into a continuum code and ultimately used to simulate integrated nanobio devices. Future objectives for this research include the development of a generic framework for simulation of systems of Master Equations and Fokker-Planck Equations. The ME and FPE solver will be designed as a separate module to be integrated with a continuum code for simulations of integrated nano-bio systems. Future work will focus on the demonstration of the feasibility and efficiency of the proposed approach on alternative test systems.

The example system is one of the simulation of a nanopore-based DNA sequencing device. The results from the device level calculation show that the overall translocation rate is exponentially dependent on the applied voltage. The placement of the patches of nanopores affects the rate at which they reach a steady translocation rate. The final result is that the nanopore diameter had only a marginal affect on the overall translocation rates. The main conclusion from the device level modeling is that in order to improve the performance of this type of a device it is desirable to increase the rate at which DNA is captured. This can be accomplished via the addition of a surface treatment around the pore via functionalization.

Acknowledgement

The authors gratefully acknowledge NASA-AMES (NNA04CB26C, Technical Monitor: Dr. Harry Partridge) for funding, and thank Dr. Harry Partridge and Dr. Jun Li for their many helpful discussions.

References

1. Stolc, V., Cozmuta I., Brock, M., O'Keeffe, J.: Presentation given at the DARPA Joint BIOFLIPS/SIMBIOSYS Principal Investigators Meeting, Monterey, CA (2003).
2. (a)Meller, A., Branton D.: Electrophoresis, 23 (2002) 2583-2591 and (b) Meller, A., Nivon L., Branton D.: Physical Review Letters, 86 (2001) 3435-38.
3. Li, J., Stein, D., McMullan, C., Branton, D., Aziz, M., and Golovchenko, J.: Nature 412 (2001) 166-8.
4. Brodka, A. and Zerda, T.: Journal of Chemical Physics 104 (1996) 6319-6326.
5. Li, J., Gershow, M., Stein, D., Brandin, E., Golovchenko, J.: Nature Materials, 2 (2003) 611-15.
6. Humphrey, W., Dalke, A. and Schulten, K.: "VMD - Visual Molecular Dynamics", J. Molec. Graphics, 14 (1996) 33-38.
7. Auffinger, P. and Westhof, E.: Journal of Molecular Biology, 269 (1997) 326-341.
8. Kalé, L., Skeel, R., Bhandarkar, M., Brunner, R., Gursoy, A., Krawetz, N., Phillips, J., Shinozaki, A., Varadarajan, K., and Schulten, K.: NAMD2: Greater scalability for parallel molecular dynamics. Journal of Computational Physics, 151 (1999) 283-312.
9. Izaguirre, J., Reich, S. Skeel, R.: Journal of Chemical Physics, 110 (1999) 9853-64.
10. Skeel, R., Izaguirre, J.: Molecular Physics 100 (2002) 3885-91.
11. Foloppe, N. and MacKerell, Jr., A.D.: Journal of Computational Chemistry, 21 (2000) 86-104.
12. MacKerell, Jr., A.D. and Banavali, N.: Journal of Computational Chemistry, 21 (2000) 105-120.
13. Beglov, D. and Roux, B.: Journal of Chemical Physics, 100 (1994) 9050-9063.
14. Rovere, M., Ricci, M., Vellati, D., and Bruni, F.: Journal of Chemical Physics, 108 (1998) 9859-9867.
15. Schlitter, J.: Chemical Physics Letters, 215 (1993) 617-21.
16. Malliavin, T., Jocelyn, G. Snoussi, K. and Leroy, J.: Biophysical Journal, 84 (2003) 1-10.
17. Rueda, M., Kalko, S., Luque, J., and Orozco, M.: JACS, 125 (2003) 8007-14.
18. Edwards, S., Corry, B., Kuyucak, S., Chung, S.: Biophysical Journal, 83 (2002) 1348-60.
19. Lubensky, D. and Nelson, D.: "Driven Polymer Translocation Through a Narrow Pore" Biophys. J. 77 (1999) 1824-38.
20. Park, S., Sener, M., Lu, D., Schulten, K.: J. Chem. Phys. 119 (2003) 1313:19.
21. Ambjornsson, T. Apell, S. Konkoli, Z. Marzio, E. Kasianowica, J.: Journal of Chemical Physics. 117 (2002) 4063-73.
22. Chuang, J., Kantor, Y., Kardar, M.: Phys. Rev. E. E65 (2002) 11802(1-8).

Comparison of Nonlinear Conjugate-Gradient Methods for Computing the Electronic Properties of Nanostructure Architectures*

Stanimire Tomov[1], Julien Langou[1], Andrew Canning[2], Lin-Wang Wang[2], and Jack Dongarra[1]

[1] Innovative Computing Laboratory, The University of Tennessee, Knoxville, TN 37996-3450

[2] Lawrence Berkeley National Laboratory, Computational Research Division, Berkeley, CA 94720

Abstract. In this article we report on our efforts to test and expand the current state-of-the-art in eigenvalue solvers applied to the field of nanotechnology. We singled out the nonlinear conjugate gradients (CG) methods as the backbone of our efforts for their previous success in predicting the electronic properties of large nanostructures and made a library of three different solvers (two recent and one new) that we integrated into the parallel PESCAN (Parallel Energy SCAN) code [3] to perform a comparison.

1 Introduction

First-principles electronic structure calculations are typically carried out by minimizing the quantum-mechanical total energy with respect to its electronic and atomic degrees of freedom. Subject to various assumptions and simplifications [5], the electronic part of this minimization problem is equivalent to solving the single particle Schrödinger-type equations (called Kohn-Sham equations)

$$\hat{H}\psi_i(r) = \epsilon_i \psi_i(r), \tag{1}$$

$$\hat{H} = -\frac{1}{2}\nabla^2 + V$$

where $\psi_i(r)$ are the single particle wave functions (of electronic state i) that minimize the total energy, and V is the total potential of the system. The wave functions are most commonly expanded in plane-waves (Fourier components) up to some cut-off energy which discretizes equation (1). In this approach the

* This work was supported by the US Department of Energy, Office of Science, Office of Advanced Scientific Computing (MICS) and Basic Energy Science under LAB03-17 initiative, contract Nos. DE-FG02-03ER25584 and DE-AC03-76SF00098.

V.S. Sunderam et al. (Eds.): ICCS 2005, LNCS 3516, pp. 317–325, 2005.

lowest eigen-pairs are calculated for $\hat{H}$ and the Kohn-Sham equations are solved self-consistently. For a review of this approach see reference [5] and the references therein. The computational cost of this approach scales as the cube of the number of atoms and the maximum system size that can be studied is of the order of hundreds of atoms. In the approach used in PESCAN developed by L-W. Wang and A. Zunger [9] a semi-empirical potential or a charge patching method [7] is used to construct V and only the eigenstates of interest around a given energy are calculated, allowing the study of large nanosystems (up to a million atoms). The problem then becomes: find ψ and E close to a given E_{ref} such that

$$H\psi = E\psi, \tag{2}$$

where H represents the Hamiltonian matrix, which is Hermitian with dimension equal to the number of Fourier components used to expand ψ. The dimension of H may be of the order of a million for large nanosystems. The eigenvalues E (energy of state ψ) are real, and the eigenvectors ψ are orthonormal.

In many cases, like semiconductor quantum dots, the spectrum of H has energy gaps and of particular interest to physicists is to find a few, approximately 4 to 10, of the interior eigenvalues on either side of the gap which determines many of the electronic properties of the system. Due to its large size H is never explicitly computed. We calculate the kinetic energy part in Fourier space, where it is diagonal, and the potential energy part in real space so that the number of calculations used to construct the matrix-vector product scales as $n \log n$ rather than n^2 where n is the dimension of H. Three dimensional FFTs are used to move between Fourier and real space. H is therefore available as a procedure for computing Hx for a given vector x. Thus one more requirement is that the solver is matrix free. Finally, repeated eigenvalues (degeneracy) of approximately 3 maximum are possible for the problems discussed and we need to be able to resolve such cases to fully understand the electronic properties of our systems.

Currently, equation (2) is solved by a CG method as coded in the PESCAN package [9]. While this program works well for 1000 atom systems with a sizable band gap (e.g., 1 eV), it becomes increasingly difficult to solve for systems with (1) large number of atoms (e.g, more than 1 million); (2) small band gap, and where (3) many eigenstates need to be computed (e.g, more than 100), or to solve eigenstates when there is no band gap (e.g, for Auger or transport calculations). Thus, new algorithm to solve this problem is greatly needed.

The focus of this paper is on nonlinear CG methods with folded spectrum. The goal is to solve the interior eigenstates. Alternative for the folded spectrum transformation are shift-and-invert or fixed-polynomial [6]. Our choice of method is based on the highly successful current scheme [3] which has been proven to be efficient and practical for the physical problems we are solving. It will be the subject of other studies to further investigate the applicability of other alternatives like the Lanczos and Jacobi-Davidson method.

In Section 2 we describe the three eigensolvers investigated in the paper. We give our numerical results in Section 3, and finally, in Section 4, we give some concluding remarks.

2 Nonlinear CG Method for Eigenvalue Problems

The conventional approach for problems of very large matrix size is to use iterative projection methods where at every step one extracts eigenvalue approximations from a given subspace S of small dimension (see e.g. [2]). Nonlinear CG methods belong to this class of methods. Let us assume for now that we are looking for the smallest eigenvalue of the Hermitian operator A.

This eigenvalue problem can be expressed in terms of function minimization as: find the variational minimum of $F(x) = \ < x, Ax >$, under the constraint of $x^T x = I$, on which a nonlinear CG method is performed. The orthonormal constraint $x^T x = I$ makes the problem nonlinear.

In this section, we first give a description of the algorithms that we have implemented in our library, namely: the preconditioned conjugate gradient method (PCG), the PCG with $S = \mathtt{span}\{X, R\}$ method (PCG-XR), and the locally optimal PCG method (LOPBCG). Finally, we describe the spectral transformation that we use to get the interior eigen-values of interest.

2.1 PCG Method

In Table 1, we give a pseudo-code of the PCG algorithm for eigen-problems. This is the algorithm originally implemented in the PESCAN code (see also [5, 8]).

Here `P` is a preconditioner for the operator A, X is the block of `blockSize` column eigenvectors sought, and λ is the corresponding block of eigenvalues. In

Table 1. PCG algorithm

```
do i = 1, niter
    do m = 1, blockSize
        orthonormalize X(m) to X(1 : m − 1)
        ax = A X(m)
        do j = 1, nline
            λ(m) = X(m) · ax
            if (||ax − λ(m) X(m)||_2 < tol .or. j == nline) exit
            r_{j+1} = (I − X X^T) ax
            β = ((r_{j+1} − r_j)·P r_{j+1}) / (r_j·P r_j)
            d_{j+1} = −P r_{j+1} + β d_j
            d_{j+1} = (I − XX^T) d_{j+1}
            γ = ||d_{j+1}||_2^{−1}
            θ = 0.5 |atan (2 γ d_{j+1}·ax) / (λ(m) − γ^2 d_{j+1}·A d_{j+1})|
            X(m) = cos(θ) X(m) + sin(θ) γ d_{j+1}
            ax = cos(θ) ax + sin(θ) γ A d_{j+1}
        enddo
    enddo
    [X, λ] = Rayleigh − Ritz on span{X}
enddo
```

Table 2. LOBPCG algorithm

```
do i = 1,niter
   R = P (A X_i − λ X_i)
   check convergence criteria
   [X_i, λ] = Rayleigh − Ritz on span{X_i, X_{i−1}, R}
enddo
```

the above procedure, $X^T X = I$ is satisfied throughout the process. $(I - XX^T)$ is a projection operator, which when applied to y deflates $\mathtt{span}\{X\}$ from y, thus making the resulting vector orthogonal to $\mathtt{span}\{X\}$. The matrix-vector multiplication happens at line 15. Thus there is one matrix-vector multiplication in each j iteration. The above procedure converges each eigen-vector separately in a sequential way. It is also called state-by-state (or band-by-band in the physics community) method, in contrast to the Block method to be introduced next.

2.2 LOBPCG Method

Briefly, the LOBPCG method can be described with the pseudo-code in Table 2. Note that the difference with the PCG is that the `m` and `j` loops are replaced with just the blocked computation of the preconditioned residual, and the Rayleigh-Ritz on $\mathtt{span}\{\mathtt{X_i}\}$ with Rayleigh-Ritz on $\mathtt{span}\{\mathtt{X_{i-1}}, \mathtt{X_i}, \mathtt{R}\}$ (in the physics community Rayleigh-Ritz is known as the process of diagonalizing A within the spanned subspace, and taking the "blocksize" lowest eigen vectors). The direct implementation of this algorithm becomes unstable as X_{i-1} and X_i become closer and closer, and therefore special care and modifications have to be taken (see [4]).

2.3 PCG-XR Method

PCG-XR is a new algorithm that we derived from the PCG algorithm by replacing line 18 in Table 1 with

```
18      [X, λ] = Rayleigh − Ritz on span{X,R}
```

The idea, as in the LOBPCG, is to use the vectors R to perform a more efficient Rayleigh-Ritz step.

2.4 Folded Spectrum

Projection methods are good at finding well separated (non-clustered) extremal eigenvalues. In our case, we are seeking for interior eigenvalues and thus we have to use a spectral transformation, the goal being to map the sought eigenvalue of our operator to extremal eigenvalues of another one.

To do so we use the folded spectrum method. The interior eigenvalue problem $Hx = \lambda x$ is transformed to find the smallest eigenvalues of $(H - E_{ref} I)^2 x = \mu x$. The eigenvalues of the original problem are given back by $\mu = (\lambda - s)^2$.

The PCG algorithm in its folded form (FS-PCG) is described in [8]. To adapt the folded spectrum to LOBPCG (FS-LOBPCG), we have added three more block vectors that store the matrix-vector products of the blocks X, R, and P with the matrix H. This enables us to control the magnitude of the residuals for an overhead of a few more axpy operations (otherwise we just have access to the magnitude of the residuals of the squared operator). Also the deflation strategy of LOBPCG is adapted in FS-LOBPCG, as the vectors are deflated when the residual relative to H has converged (not H^2).

3 Numerical Results

3.1 Software Package

We implemented the LOBPCG [1], PCG, and PCG-XR methods in a software library. Currently, it has single/double precision for real/complex arithmetic and both parallel (MPI) and sequential versions. The library is written in Fortran 90. The folded spectrum spectral transformation is optional. The implementation is stand alone and meant to be easily integrated in various physics codes.

A test case is provided with the software. It represents a 5-point operator where the coefficients a (diagonal) and b (for the connections with the 4 closest neighbors on a regular 2D mesh) can be changed. In Table 1 the output of the test is presented. It is performed on a Linux Intel Pentium IV with Intel Fortran compiler and parameters ($a = 8, b = -1 - i$). We are looking for the 10 smallest eigenstates, the matrix size is $20,000$, and the iterations are stopped when all the eigencouples (x, λ) satisfy $\|Hx - x\lambda\| \leq \mathtt{tol}\|x\|$, with $\mathtt{tol} = 10^{-8}$.

Table 3. Comparison of the PCG, PCG-XR and LOBPCG methods in finding 10 eigenstates on a problem of size $20,000 \times 20,000$

	PCG	LOBPCG	PCG-XR
time (s)	37.1	61.7	20.2
matvecs	3,555	1,679	1,760
dotprds	68,245	137,400	37,248
axpys	66,340	158,261	36,608
copys	6,190	9,976	3,560

In general LOBPCG always performs less iterations (i.e. less matrix-vector products) than PCG. This advantage comes to the cost of more vector operations (axpys and dot products) and more memory requirements. In this case, LOBPCG performs approximately 2 times more dot products for 2 times less matrix vector

[1] http://www-math.cudenver.edu/ aknyazev/software/CG/latest/lobpcg.m (revision 4.10 written in Matlab, with some slight modifications).

Table 4. Comparison of FS-PCG, FS-PCG-XR and FS-LOBPCG methods in finding 10 eigenstates around the gap of quantum dots of increasing size

(20Cd, 19Se) n = 11,331	# matvec	outer it	time
FS-PCG(50)	4898	(8)	50.4s
FS-PCG-XR(50)	4740	(6)	49.1s
FS-LOBPCG	4576		52.0s

(83Cd, 81Se) n = 34,143	# matvec	outer it	time
FS-PCG(200)	15096	(11)	264 s
FS-PCG-XR(200)	12174	(5)	209 s
FS-LOBPCG	10688		210 s

(232Cd, 235Se) n = 75,645	# matvec	outer it	time
FS-PCG(200)	15754	(8)	513 s
FS-PCG-XR(200)	15716	(6)	508 s
FS-LOBPCG	11864		458 s

(534Cd, 527Se) n = 141,625	# matvec	outer it	time
FS-PCG(500)	22400	(6)	1406 s
FS-PCG-RX(500)	21928	(4)	1374 s
FS-LOBPCG	17554		1399 s

products than the PCG method, since the 5-point matrix-vector product takes approximately the time of 7 dot products, PCG gives a better timing.

The CG-XR method represents for this test case an interesting alternative for those two methods: it inherits the low number of matrix vector products from the LOBPCG and the low number of dot products from the PCG method.

3.2 Numerical Results on Some Quantum Dots

In this section we present numerical results on quantum dots up to thousand of atoms. The experiments are performed on the IBM-SP `seaborg` at NERSC.

For all the experiments we are looking for `mx` = 10 interior eigenvalues around $E_{ref} = -4.8$eV, where the band gap is about 1.5 to 3 eV. We have calculated 4 typical quantum dots: (20Cd,19Se), (83Cd,81Se), (232Cd,235Se), (534Cd,527Se). These are real physical systems which can be experimentally synthesized and have been studied previously using the PCG method [10]. Nonlocal pseudopotential is used for the potential term in equation (1), and spin-orbit interaction is also included. The cutoff energy for the plane-wave basis set is 6.8 Ryd. The stopping criterion for the eigenvector is $\|Hx - x\lambda\| \leq \texttt{tol}\|x\|$ where $\texttt{tol} = 10^{-6}$. All the runs are performed on one node with 16 processors except for the smallest case (20Cd,19Se) which is run on 8 processors. All the solvers are started with the same initial guess.

The preconditioner is the one given in [8]: diagonal with diagonal elements

$$p_i = \frac{E_k^2}{(\frac{1}{2}q_i^2 + V_0 - E_{ref})^2 + E_k^2)},$$

where q_i is the diagonal term of the Laplacian, V_0 is the average potential and E_k is the average kinetic energy of the wave function ψ. It is meant to be an approximation of the inverse of $(H - E_{ref})^2$.

A notable fact is that all the solvers find the same 10 eigenvalues with the correct accuracy for all the runs. Therefore they are all robust.

The timing results are given in Table 4. For each test case the number of atoms of the quantum dot and the order n of the corresponding matrix is given. The parameter for the number of iterations in the inner loop ($nline$) for FS-PCG and FS-PCG-XR is chosen to be the optimal one among the values 20, 50, 100, 200, and 500 and is given in brackets after the solver.

From Table 4, we observe that the three methods behave almost the same. The best method (in term of time) being either FS-PCG-XR or FS-LOBPCG.

FS-LOBPCG should also benefit in speed over FS-PCG and FS-PCG-XR from the fact that the matrix-vector products are performed by block. This is not the case in the version of the code used for this paper where the experiments are performed on a single node. The blocked implementation of FS-LOBPCG in PESCAN should run faster and also scale to larger processor counts as latency is less of an issue in the communications part of the code.

Another feature of FS-LOBPCG that is not stressed in Table 4 is its overwhelming superiority over FS-PCG when no preconditioner is available. In Table 5 Left, we illustrate this later feature. For the quantum dot (83Cd,81Se), FS-LOBPCG runs 4 times faster than FS-PCG without preconditioner whereas it runs only 1.4 times faster with. For the four experiments presented in Table 4, the number of inner iteration that gives the minimum total time is always attained for a small number of outer iteration, this is illustrated in Table 5 Right for (232Cd, 235Se) where the minimum time is obtained for 6 outer iterations. Another and more practical way of stopping the inner iteration is in fixing the requested tolerance reached at the end of the inner loop. We call FS-PCG(k) FS-PCG where the inner loop is stopped when the accuracy is less than $k^{n_{outer}}$, where n_{outer} is number of the corresponding outer iteration. In Table 5 Right, we give the results for FS-PCG(10^{-1}) and (223Cd,235Se). It comes without a surprise that this solver converge in 6 outer steps. This scheme looks promising. It also allows a synchronized convergence of the block vectors.

Table 5. Left: Comparison of FS-PCG and FS-LOBPCG with and without preconditioner to find mx = 10 eigenvalues of the quantum dots (83Cd,81Se); **Right:** The problem of finding the best inner length for FS-PCG can be avoided by fixing a tolerance as stopping criterion in the inner loop

(83Cd, 81Se) n = 34,143		# matvec	time
FS-PCG(200)	precond	15096	264 s
FS-LOBPCG	precond	10688	210 s
FS-PCG(200)	no precond	71768	1274 s
FS-LOBPCG	no precond	17810	341 s

(232Cd, 235Se) n = 75,645	# matvec	outer it	time
FS-PCG(100)	17062	(15)	577 s
FS-PCG(200)	15716	(6)	508 s
FS-PCG(300)	15990	(4)	517 s
FS-PCG(10^{-1})	15076	(6)	497 s

4 Conclusions

In this paper, we described and compared 3 nonlinear CG methods with folded spectrum to find a small amount of interior eigenvalues around a given point. The application is to make a computational prediction of the electronic properties of quantum nanostructures. The methods were specifically selected and tuned for computing the electronic properties of large nanostructures. There is need for such methods in the community and the success of doing large numerical simulations in the field depend on them. All three methods are similar and thus often the results are close; a general ranking being: FS-LOBPCG is the fastest, next FS-PCG-XR and finally FS-PCG. In terms of memory requirement the three methods are ranked in the same way: FS-LOBPCG/FS-PCG-XR requires four/two times as much memory as FS-PCG. As our problem scales up the memory has not shown up as a bottleneck yet, i.e. using FS-LOBPCG is affordable.

The main drawback of FS-PCG and FS-PCG-XR is their sensitivity to the parameter `nline` (the number of iterations in the inner loop). In order to get rid of this parameter one can instead have a fixed residual reduction to be achieved on each step of the outer loop.

On other applications, the performance of FS-LOBPCG would be still better than FS-PCG if a fast block matrix-vector product and an accommodating preconditioner are available.

Finally, based on our results, if memory is not a problem and block version of the matrix-vector multiplication can be efficiently implemented, the FS-LOBPCG will be the method of choice for the type of problems discussed.

Acknowledgements

This research used resources of the National Energy Research Scientific Computing Center, which is supported by the Office of Science of the U.S. DOE.

References

1. Arbenz, P., Hetmaniuk, U.L., Lehoucq, R.B., Tuminaro, R.S.: A comparison of eigensolvers for large-scale 3D modal analysis using AMG-preconditioned iterative methods. Int. J. Numer. Meth. Engng (to appear)
2. Bai, Z., Demmel, J., Dongarra, J., Ruhe, A., van der Vorst, H. (Editors): *Templates for the solution of Algebraic Eigenvalue Problems: A Practical Guide*, SIAM, Philadelphia (2000)
3. Canning, A., Wang, L.W., Williamson, A., Zunger, A.: Parallel empirical pseudopotential electronic structure calculations for million atom systems. J. Comp. Phys. **160** (2000) 29–41
4. Knyazev, A.: Toward the optimal preconditioned eigensolver: locally optimal block preconditioned conjugate gradient method. SIAM J. on Scientific Computing **23(2)** (2001) 517–541
5. Payne, M.C., Teter, M.P., Allan, D.C., Arias, T.A., Joannopoulos, J.D.: Iterative minimization techniques for ab initio total-energy calculations: molecular dynamics and conjugate gradients. Rev. Mod. Phys. **64** (1992) 1045–1097

6. Thornquist, H.: Fixed-Polynomial Approximate Spectral Transformations for Preconditioning the Eigenvalue Problem. Masters thesis, Rice University, Department of Computational and Applied Mathematics (2003)
7. Wang, L.W., Li, J.: First principle thousand atom quantum dot calculations. Phys. Rev. B **69** (2004) 153302
8. Wang, L.W., Zunger, A.: *Pseudopotential Theory of Nanometer Silicon Quantum Dots application to silicon quantum dots.* In Kamat, P.V., Meisel, D.(Editors): *Semiconductor Nanoclusters* (1996) 161–207
9. Wang, L.W., Zunger, A.: Solving Schrodinger's equation around a desired energy: application to silicon quantum dots. J. Chem. Phys. **100(3)** (1994) 2394–2397
10. Wang, L.W., Zunger, A.: Pseudopotential calculations of nanoscale CdSe quantum dots. Phys. Rev. B **53** (1996) 9579

A Grid-Based Bridging Domain Multiple-Scale Method for Computational Nanotechnology

Shaowen Wang[1], Shaoping Xiao[2,3], and Jun Ni[1,2,4]

[1]Academic Technology -- Research Service, Information Technology Services
[2] Department of Mechanical and Industrial Engineering
[3] Center for Computer-Aided Design
[4] Department of Computer Science,
The University of Iowa, Iowa city, IA, 52242, USA
{Shaowen-wang, shaoping-xiao, jun-ni}@uiowa.edu

Abstract. This paper presents an application-level Grid middleware framework to support a bridging domain multi-scale method fornumerical modeling and simulation in nanotechnology. The framework considers a multiple-length-scale model by designing specific domain-decomposition and communication algorithms based on Grid computing. The framework is designed to enable researchers to conductlarge-scale computing in computational nanotechnology through the use of Grid resources for exploring microscopic physical properties of materials without losing details of nanoscale physical phenomena.

1 Introduction

In the study of nano-materials or nano-devices, a molecular dynamics (MD) model [1] plays an important role. However, many existing MD algorithms have limitations either on length or on time scales due to the lack of enough compute power. Even the most powerful high performance computing (HPC) system is still not powerful enough to perform a complete MD simulation [2]. Therefore, an innovative computational methodology for MD simulations is urgently needed.

Recently-developed concurrent multiscale methods [3] take into consideration of multiple length scales in predicting macroscopic properties. For example, a bridging domain coupling method [4] performs a linkage between the continuum and molecular models. However, if the ratio of mesh size in the continuum domain to the lattice space in the molecular domain is too large, there is a difficult to eliminate a spurious phenomenon in which a nonphysical wave reflection occurs at the interface of different length scales. Therefore, concurrent multiscale methods also require a tremendous computing power. In addition, a small time step (the order of femtosecond) must be selected to meet numerical stability of MD stimulation at the molecular model. Such requirement causes to waste compute time during simulations. A feasible and efficient method is needed to conduct a complete modeling and simulation of large nano systems within long time scales. This requirement stimulates the authors to develop a Grid-based Bridging Domain Multiscale Method (GBDMM).

V.S. Sunderam et al. (Eds.): ICCS 2005, LNCS 3516, pp. 326–333, 2005.

Grid computing [5-7] enables users to assemble large-scale, geographically-distributed, heterogeneous computational resources together for intensive MD computations. This assemblage is dynamically orchestrated using a set of protocols as well as specialized software referred to as Grid middleware. This coordinated sharing of resources takes place within formal or informal consortia of individuals and/or institutions that are often called Virtual Organizations (VO) [8]. In a similar way, Grid application-specific middleware must be developed for MD related multi-scale methods. We intend to develop a micro/macro coupling method and Grid application middleware for multiscale computation in nano-science and technology.

Section 2 reviews the bridging domain coupling method and describes its extensions in the GBDMM. Section 3 addresses domain-decomposition and communication algorithms for the GBDMM. Section 4 presents the framework of Grid nano-middleware for GBDMM's implementation.

2 Grid-Enhanced Bridging Domain Multiscale Method (GBDMM)

2.1 Bridging Domain Coupling Method

The bridging domain coupling method contains a continuum domain, Ω^C, which is overlapped with a molecular domain, Ω^M, through a bridging domain, Ω^{int} (Fig. 1). The scaling of the molecular and continuum Hamiltonians is performed in the bridging domain. The Lagrange multiplier method is used to conjunct the molecular domain with the continuum domain via constraint condition(s), $\mathbf{u}^C(\mathbf{X}_I,t)-\mathbf{u}_I^M(t)=0$, where $\mathbf{u}$ is the displacement. The total Hamiltonian of the system can be written as

$$\begin{aligned} \mathrm{H} = & \int_{\Omega^C} \beta(\mathbf{X})\left(\mathrm{K}^C+\mathrm{U}^C\right)d\Omega^C + \sum_{I}\left(1-\beta(\mathbf{X}_I)\right)\left(\mathrm{K}_I^M+\mathrm{U}_I^M\right) + \\ & \sum_{I\in\Omega^{\text{int}}} \lambda(\mathbf{X}_I)\left(\mathbf{u}^C(\mathbf{X}_I,t)-\mathbf{u}_I^M(t)\right) \end{aligned} \tag{1}$$

where $\beta(\mathbf{X})$ equals zero in Ω^M and unity in Ω^C except the bridging domain. Within the bridging domain, Ω^{int}, it smoothly varies from zero to one. If an irregular bridging domain is considered, a nonlinear function $\beta(\mathbf{X})$ can be introduced. The function can be derived from a signed distance function, or defined by radial basis functions from a set of points. K and U are the kinetic and potential energies. $\lambda(\mathbf{X})$ is a Lagrange multiplier field that can be obtained from the finite element approximation as shown in Fig. 1. Based on the classical Hamiltonian mechanics, the discrete equations of Equation (1) can be expressed as

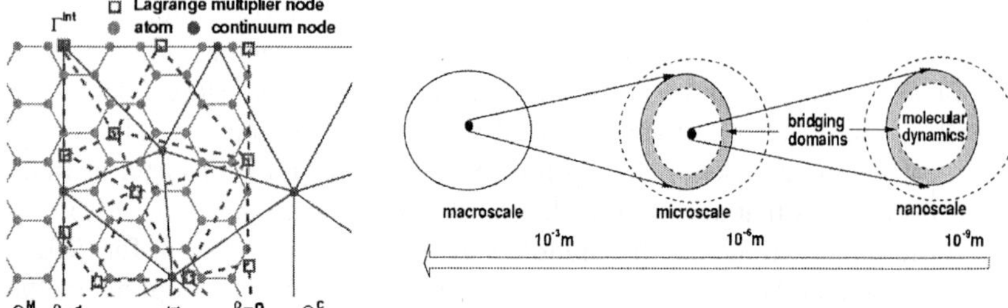

Fig. 1. The bridging domain coupling model for a graphite sheet

Fig. 2. Length scales covered in the BDMM

$$
\begin{aligned}
\overline{\mathrm{M}}_{\mathrm{I}}\ddot{\mathbf{u}}_{\mathrm{I}}^{\mathrm{C}} &= \mathbf{f}_{\mathrm{I}}^{\mathrm{extC}} - \mathbf{f}_{\mathrm{I}}^{\mathrm{intC}} - \mathbf{f}_{\mathrm{I}}^{\mathrm{LC}} \quad \text{in} \quad \Omega^{\mathrm{C}}, \\
\overline{\mathrm{m}}_{\mathrm{I}}\ddot{\mathrm{u}}_{\mathrm{I}}^{\mathrm{M}} &= \mathbf{f}_{\mathrm{I}}^{\mathrm{extM}} - \mathbf{f}_{\mathrm{I}}^{\mathrm{intM}} - \mathbf{f}_{\mathrm{I}}^{\mathrm{LM}} \quad \text{in} \quad \Omega^{\mathrm{M}}
\end{aligned}
\tag{2}
$$

where $\overline{M}_I$ and $\overline{m}_I$ are the modified nodal and atom masses; $\mathbf{f}_I^{ext}$ and $\mathbf{f}_I^{\mathrm{int}}$ are the external and internal forces, independently; $\mathbf{f}_I^{L}$ is the force due to constraints in the bridging domain. During simulation, velocities are obtained independently in the continuum and molecular domains from Equation (2) without the consideration of constraint forces. Constraint conditions are then used to calculate the Lagrange multipliers. Finally, the constraint forces in Equation (2) are used to correct the nodal/atom velocities in the bridging domain, thus eliminating the wave reflection automatically.

2.2 Multiscale Considerations in GBDMM

Based on the bridging domain coupling method, we proposed a Bridging Domain Multiscale Method (BDMM), which bridges from nanoscale to microscale to macroscale (Fig. 2). A bridging domain multiscale model contains a macro-scale domain (~10^{-3} m) as a linear elasticity domain in which linear finite element methods (FEM) can be used. The material properties can be obtained from the Representative Volume Element (RVE) technique with molecular dynamics simulation. The BDMM embeds a micro-scale domain (~10^{-6} m) in the macro-scale domain. It models physical behaviors of materials using either nonlinear FEM [9] or extended FEM [10] (if crack propagation is considered at the microscale). Furthermore, a quasi-continuum technique [11] is implemented with the FE methods at the microscale, so that the constitutive equations can be constructed based on the atomic level energy. A sub-domain in a microscale domain can be treated as a nanoscale (molecular) domain (~10^{-9} m) in which molecular dynamics is employed. The BDMM couples various scales without grading down mesh sizes. Such coupling leads to a straightforward implementation of different time steps via a multiple-time-step algorithm.

3 Domain Decomposition and Communication Algorithms

To perform simulations using GBDMM, an efficient algorithm of domain decomposition must be developed. An entire physical domain with multiple length scales specified can be hierarchically divided into multiple sub-domains. The first generation (FG) sub-domains, such as Ω_{01}^{M} (see Fig3), are generated based on variation in length scales. These sub-domains are allocated on heterogeneous Grid computing resources. Such first-generational domain decomposition is designed to avoid high network latency, and hence produce a loosely-networked computation on Grids.

Each FG sub-domain can further be divided into a number of second generation (SG) sub-domains being parallel-processed on different computational nodes within a Grid resource. A bridging sub-domain is only shared between two SG sub-domains, which belong to two different length scales, respectively. Although one SG sub-domain can be overlapped with two or more SG sub-domains, the bridging sub-domains are not overlapped with each other.

There are two types of communications in the GBDMM (see Fig. 4). The inter-domain communication among the SG sub-domains takes place within an interconnection within a Grid resource, performing a traditional parallel computation. This computation guarantees the consistency and integrity of SG sub-domains and boundary information for solving the governing equations of atom motions. The procedure for solving equations of atom motion on each Grid cluster is independent. After solving the equations at each time step, communication takes place between two clusters. The bridging coupling techniques are then applied to correct the trial velocities of nodes or atoms in each bridging sub-domain independently.

For example, a simple GBDMM have two sub-domains Ω_A and Ω_B in the molecular domain, allocated to a single Grid resource (often a computing cluster), and Ω_C and Ω_D in the continuum domain, allocated to another Grid resource. The communication between the group of Ω_A and Ω_B, and the group of Ω_C and Ω_D takes place on networks in a Grid environment. Ω_A and Ω_B share an atom E. Bridging domains Ω_A^{int} and Ω_B^{int} perform energy and force calculations. The results from Ω_B^{int} are used to calculate the body potential functions and inter-atomic forces in Ω_A, since Ω_A and Ω_B^{int} share with same atoms. The size of bridging domains depends on the selected potential functions, especially the cutoff distance for Van der Waals potential functions.

The continuum sub-domains, Ω_C and Ω_D share a boundary node F as shown in Fig. 4. F^C and F^D represent the same node F in different sub-domains. Unlike inter-domain communications (which do not require bridging domains), the communication between continuum sub-domains, Ω_C and Ω_D exchange internal forces of boundary nodes through the networks among Grid resources. For instance, the internal force of node F^C, calculated in the sub-domain Ω_C is fetched to the sub-domain Ω_D to form an external force on the node of F^D, but in an opposite direction. A similar procedure is conducted to pass the internal force of node F^D as the negatively external force of node F^C. Therefore, the motions of node F, updated from Ω_C and Ω_D are consistent.

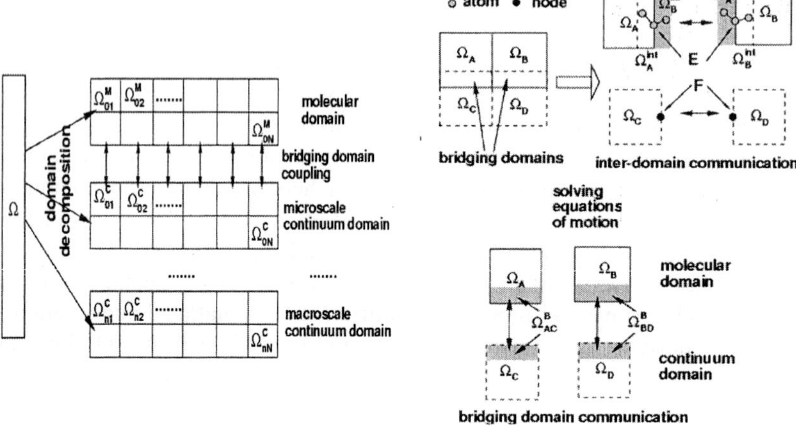

Fig. 3. Domain decomposition for the GBDMM method

Fig. 4. A demonstration for domain communications

After equations of motion are solved independently on each cluster, bridging-domain communication takes place independently in each bridging sub-domain, such as Ω^B_{AC} and Ω^B_{BD}. For instance, the trial velocities of atoms in Ω^B_{BD} of Ω_B are transferred to Ω_D, while the trial velocities of nodes in Ω^B_{BD} of Ω_D are transferred to Ω_B. The bridging domain coupling technique is then applied to correct the trial velocities of atoms in Ω^B_{BD} of Ω_B as well as in Ω^B_{BD} of Ω_D, respectively on each mater nodes.

4 Framework of Grid Middleware for GBDMM Implementation

An application-level Grid middleware framework is designed (in Fig. 5) to enable a GBDMM implementation. In this framework, middleware manages available computing resources, schedules decomposed domains to appropriate Grid resources (i.e. clusters). It contains (1) a task scheduling advisor (TSA) that takes the result of domain decomposition as input to produce scheduling plans and achieve high-performance simulation through load-balancing; and (2) an information broker (IB) that leverages Grid information services to provide the task scheduling advisor with a resource discovery query functions [12].

The framework is centered on the TSA that is used to schedule sub-domains to an appropriate set of Grid resources discovered to achieve optimal performance: tasks are allocated to balance computations across the available set of resources. The sub-domains are converted to the tasks that are placed in Grid-resource queues. The TSA is designed to achieve high levels of performance by balancing tasks across available resources. It determines the correspondence between tasks and the available Grid resources. Within the TSA, several static scheduling strategies [13] such as Min-min and Max-min have been implemented. The fine granularity of the domain decomposition in GBDMM is designed to achieve high level parallelism. Static scheduling strategies are developed to assign tasks based on computational intensity information for each sub-domain, as well as the variability in the computing capacity of each Grid resource. The workflow of the nano-middleware enhanced GBDMM with Grid computing is shown as Fig. 6.

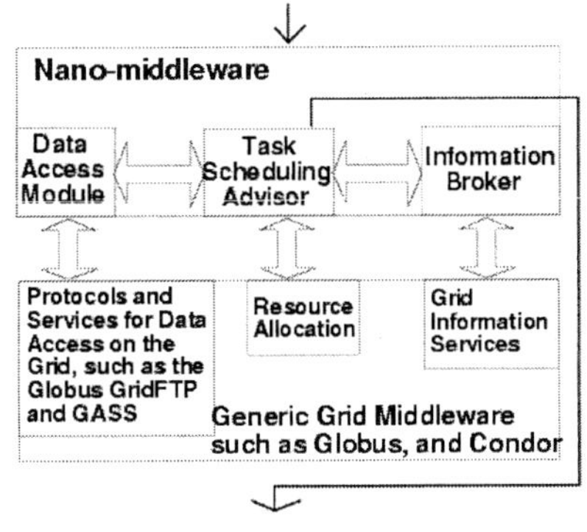

Fig. 5. The Grid nano-middleware architecture. Note: Globus is a software project, the purpose of which is to develop protocols and services for computational grids. Condor is a high throughput computing system

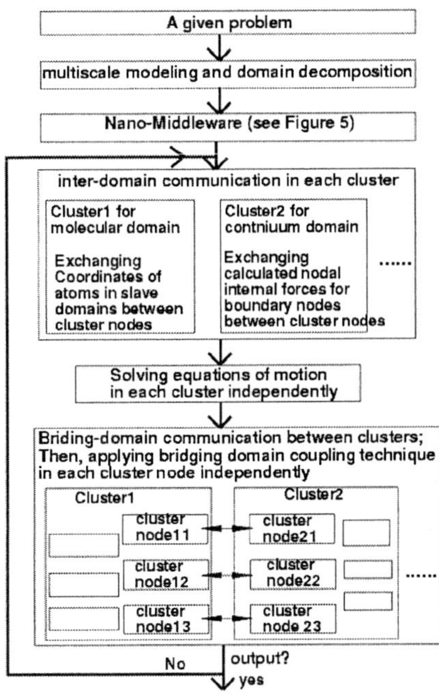

Fig. 6. The work flow of GBDMM

TSA is used to schedule sub-domains to an appropriate set of Grid resources discovered to achieve optimal performance: tasks are allocated in a way that balances computations across the available selection of resources. The sub-domains information is converted to the tasks that are placed in Grid-resource queues. The TSA is designed to achieve high levels of performance by balancing tasks across available resources. It determines the correspondence between tasks and the available Grid resources. The TSA deploys both static and dynamic scheduling, planning to operate static and dynamic scheduling on tasks upon the computing recourse available. The TSA dynamically facilitate the swap of tasks between computing resources according to a dynamic performance evaluation. The fine granularity of the domain decomposition in GBDMM is carefully design to achieve high level parallelism. Static scheduling strategies are developed to assign tasks based on computational intensity information for each sub-domain, as well as the variability in the computing capacity of each Grid resource.

Since Grid resources is very a complex set (resources and network complexities), an IB is needed to perform a self-organized-resource discovery. The IB for the nano-middleware is designed based on a self-organized grouping (SOG) method that manages the Grid complexity. Therefore, the IB contains three strategies: 1) to control resource heterogeneity during a certain period of time; 2) to capture resource dynamics through a publishing mechanism and aggregating mechanism that handles resource dynamically; (3) to enable the assemblage of large number of resources for applications across VO boundaries.

The developed information broker is based on a self-organized resource discovery method. It is well known that currently, the Grid represents an extremely complex distributed computing environment for developing applications because: 1) Grid resources (e.g., CPU, network, storage, and special-purpose instrument) are heterogeneous; 2) Grid resources are dynamic, and they tend to have faults that may not be predictable; 3) Grids are often distributed across security domains with large number of resources involved. Due to this complexity, it has been a great challenge for developing efficient methods for Grid resource discovery that refers to the process of locating satisfactory resources based on user requests. Grid resource discovery must handle the search for desirable subsets of large number of accessible resources, the status and availability of which dynamically change.

The workflow of the nano-middleware enhanced GBDMM with Grid computing is shown as Fig. 6. The workflow demonstrates the integration of computational nanotechnology and computer science in this multi-disciplinary project.

5 Conclusion and Future Work

The GBDMM is an extension of the bridging domain coupling method that allows multiple length/time scales to be treated. The GBDMM's domain decomposition and communication are designed to be applicable to distributed Grid resources. The Grid nano-application-middleware includes a task scheduling advisor and an information broker. This middleware enables the GBDM to schedule resources, distribute tasks, and achieve load balancing. Our ongoing work is to implement and apply GBDMM to study mechanical properties of nanostructured materials.

References

1. Rountree C. L., Kalia R. K., Lidorikis E., Nakano A., Van B. L., Vashishta P: Atomistic aspects of crack propagation in brittle materials: Multimillion atom molecular dynamics simulation. Annual Review of Materials Research 32 (2002) 377-400
2. Abraham F. F., Walkup R., Gao H., Duchaineau M., Rubia T. D., Seager M.: Simulating materials failure by using up to one billion atoms and the world's fastest computer: brittle fracture, Proceedings of the National Academy of Science of the United States of America. 99(9) (2002) 5777-5782
3. Muralidharan K., Deymier P. A., Simmons J. H.: A concurrent multiscale finite difference time domain/molecular dynamics method for bridging an elastic continuum to an atomic system, Modelling and Simulation in Materials Science and Engineering, 11(4) (2003) 487-501

4. Xiao S. P., Belytschko T.: A bridging domain method for coupling continua with molecular dynamics, Computer Methods in Applied Mechanics and Engineering, 193 (2004) 1645-1669
5. Foster I., Kesselman C.: The Grid: Blue Print for a New Computing Infrastructure. Morgan Kaufmann Publishers, Inc. San Francisco, CA (1999)
6. Foster I.: The Grid: A new infrastructure for 21st century science, Physics Today (2003) February, 42-47
7. Berman F., Fox G., Hey T.: The Grid, past, present, future, Grid Computing, Making the Global Infrastructure a Reality. edited by R. Berman, G. Fox, and T. Hey John Wiley & Sons, West Sussex, England (2003)
8. Foster I., Kesselman C., Tuecke S., The anatomy of the grid: enabling scalable virtual organizations, International Journal Supercomputer Applications. 15(3) (2002)
9. Belytschko T., Liu W. K., Moran B.: Nonlinear Finite Elements for Continua and Structures, Wiley, New York (2000)
10. Belytschko T., Parimi C., Moës N., Sukumar N., Usui S.: Structured extended finite element methods for solids defined by implicit surfaces, International Journal for Numerical Methods in Engineering. 56 (2003) 609-635
11. Tadmor E. B., Ortiz M., Phillips R.: Quasicontinuum analysis of defects in solids, Philosophy Magazine A. 73 (1996) 1529-1563
12. Wang, S., Padmanabhan, A., Liu, Y., Briggs, R., Ni, J., He, T., Knosp, B. M., Onel Y.: A multi-agent system framework for end-user level Grid monitoring using geographical information systems (MAGGIS): architecture and implementation. In: Proceedings of Lecture Notes in Computer Science. 3032 (2004) 536-543
13. Braun, T. D., Siegel, H. J., Beck, N., Boloni, L. L., Maheswaran, M., Reuther, A. I., Robertson, J. P., Theys, M. D., Yao, B., Hensgen, D., and Freund, R. F.: A comparison of eleven static heuristics for mapping a class of independent tasks onto heterogeneous distributed computing systems. Journal of Parallel and Distributed Computing. 61(2001) 810-837

Signal Cascades Analysis in Nanoprocesses with Distributed Database System

Dariusz Mrozek, Bożena Małysiak, Jacek Frączek, and Paweł Kasprowski

Silesian University of Technology, Department of Computer Science,
ul. Akademicka 16, 44-100 Gliwice, Poland
{Mrozek, malysiak, kasprowski}@polsl.pl,
jacekf@polsl.gliwice.pl

Abstract. The signal cascades are a number of successive biochemical reactions, occurring in the cells. In these reactions take part many proteins (often enzymes) and the entire process may be compared to the dominoes effect. The common term used for define a varied biochemical mechanisms regulating processes in the nanonetworks is signal transduction executed in the signal cascades. These processes can be realized in a closed area of space which contains proper quantities of substrates and a set of control molecules working in a predefined manner which was determined by their chemical construction, including changes of chemical activity reached usually by the conformational changes. Information about the signal cascades that happen in the various type of cells for given processes can be retrieved from the biochemical research accessible in the biochemical databases. In this work, the simulation process of the signal transduction in bio-nanoprocesses using the distributed database environment is presented and illustrated by the biological example.

1 Introduction

In outward things development of nanotechnology is in initial stage. Nowadays, manufacturing methods are very simple and basic at the molecular level. Manufactured products are made from cells. The properties of those products depend on how those cells are arranged. The first approach to implement the nanoprocesses would be coping biocompatible standards in design and implementation phase of such processes. The signal cascades are a number of successive biochemical reactions, occurring in the cells. In these reactions take part many proteins (often enzymes) and the entire process may be compared to the dominoes effect – when the first protein of the cascade is modified, such modified protein has an effect on the next protein modifying it in some direction. Thanks to the signal cascades diverse organisms can live and develop. Furthermore, if some protein take part in signal cascade incorrectly (as a result of mutation in gene coding this protein), effect of this single change can result in the tumor development.

The common term used for define a varied biochemical mechanisms regulating processes in the nanonetworks is signal transduction which is executed in the signal cascades. These processes (coping biocompatible solutions) can be realized in a closed area of space which contains appropriate quantities of substrates and a set of con-

V.S. Sunderam et al. (Eds.): ICCS 2005, LNCS 3516, pp. 334–341, 2005.

trol molecules working in a predefined manner which was determined by their chemical construction, including changes of chemical activity reached usually by the conformational changes. These molecules have a decisive role of a control system. The system is stimulated by external signal molecules coming from outside of the closed area of space to its surface, meeting the function of nanoprocess border. The signal cascades are the fixed cycles of transformations in which the molecular control system performs its control tasks. Actually, this kind of system performing nanoprocess ought to be called as a nanonetworks. The selected substrates (inorganic ions and most of metabolites e.g. sugars, amino acids, and nucleotides) and final products can penetrate through the border of the process in determined conditions. The external stimulating signal molecules only activate the receptors placed in the border of a nanonetwork, excluding selected ions fulfilling the control functions in the signal cascades. The signal transduction [4, 21] is the process of internal transformations and conversions of control molecules in signal cascades expressed as a control of nanoprocesses in a nanonetwork. Generally, one can talk about wide area and local nanonetworks. In living organisms wide area nanonetworks [13, 19] is represented by an electro-chemical net (e.g. nervous system) and extracellular communication net (e.g. immune and hormone systems). The local area nano-network is represented by a cell. Exploitation of a living cell is one of the most natural (but not basic) approach to build the local nanonetwork performing a given production process, which can be modified through the external control. An assurance of survival requirements is the basic condition, which have to be satisfied for the cell. Stimulation of inputs of the signal cascades in a nanoprocess cell from its environment is necessary for the control. Information about the signal cascades that happen in the various type of cells for given processes can be retrieved from the biochemical research available in the biochemical databases. Finally, a sequence of stimuli signals for given process can be determined.

2 Architecture of the Simulation System Implementation

The mirror-based architecture presented in Fig. 1 defines the foundation of the simulation system used in the signal cascade analysis. The approach arose as the consequence of the distribution of domain-specific information. Nowadays, the growing number of central repositories for collecting biochemical and biomedical data is observed and many organizations, research institutes and university laboratories around the world lead their projects in order to understand the mysterious nature of living cells. This cause the situation the particular information is distributed in many databases managed by different institutions connected to specific domain, e.g. the huge amount of biomolecular structural data is stored in public databases of the PDB (Protein Data Bank [1]) managed by the RCSB[1] or NCBI's[2] MMDB [20], amino acid sequence information stored in the UniProt[3] [3, 11] and NCBI's databases, human gene data stored in the H-Invitational DB provided by JBIRC[4], protein interaction data

[1] Research Collaboratory for Structural Bioinformatics (http://www.rcsb.org).
[2] National Center for Biotechnology Information (http://www.ncbi.nlm.nih.gov).
[3] UniProt Consortium (http:// www.uniprot.org).
[4] Japan Biological Information Research Center (http://www.jbirc.aist.go.jp).

managed by BIND[5] [12], etc. Fortunately for the community, all the data is available from these institution's FTP sites in the form of text files (e.g. PDB [17], mmCIF [5] formats), XML-structured files or using dedicated application programming interfaces (OMG's Macromolecular Structure Specification and Biomolecular Sequence Analysis Specification [15]) and loaders (OpenMMS [2, 18] toolkit). Searching information across distributed heterogeneous database systems that store all the enormous volumes of data has become a very important aspect of the scientific research and may be a great challenge.

In the mirror-based approach in Fig. 1, during the query process, data does not come directly from the distributed data sets. The complete sets or subsets of databases are first mirrored to the local server(s) and then queries are submitted to local replicas. In this architecture the source databases are called the primary databases and the local servers are called the secondary databases.

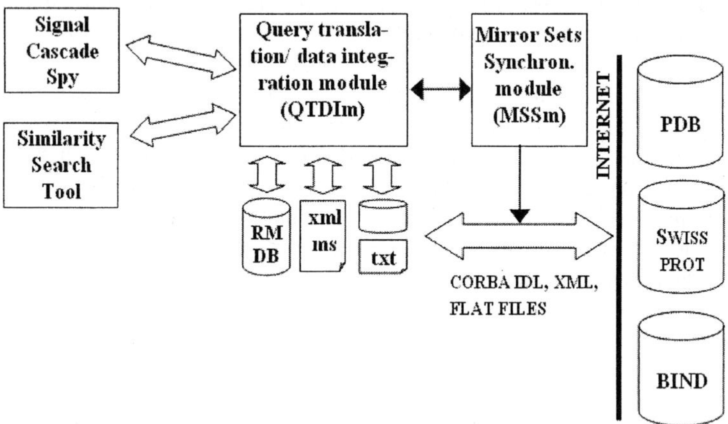

Fig. 1. The architecture with mirrored data sets synchronized with the primary databases

The main components of the architecture are: mirrored datasets in the form of relational databases (RMDB icon – Relational Mirrored Database), XML- and TXT- format mirror sets (XML-MS, TXT), the Query Translation/Data Integration module (QTDIm), the Mirror Sets Synchronization module (MSSm) and End-User tools. User queries, similarity searching and advanced computing are triggered from End-User Tools through the QTDIm which is a kind of controller in the process of information extraction. The synchronization process is coordinated by the MSSm, which can be forced to make replicas by the QTDIm, if needed.

3 Approximate Methods of Similarity Searching

At present, the afford of implementing the structure similarity search has been made with approximate methods. The main assumption of approximate retrieval methods is

[5] Biomolecular Interaction Network Database of Blueprint Initiative (http://www.bind.ca).

that for the answer on the given query we can obtain a set of the objects from database which are consistent with criteria defined in the query with given degree. Queries of this type require defining characteristic function, which determines in what degree the searched object is consistent with criteria defined in the query and threshold value, which allows to qualify the objects that should occur in the answer. Existing retrieval algorithms for the biological databases are based on principles valid in the objects approximate retrieval methods. In the biological databases two trends are separated:

- similarity searching by a protein sequences alignment,
- similarity searching by alignment of a three-dimensional protein structures.

During the research (example introduced in section 5) the PDB [1], BIND [12] and BioCarta [9] datasets were exploited but there is a possibility to access other mirrored databases like UniProt/Swissprot [11] and GenBank [8]. Some features of developed simulation software are presented in Fig. 2.

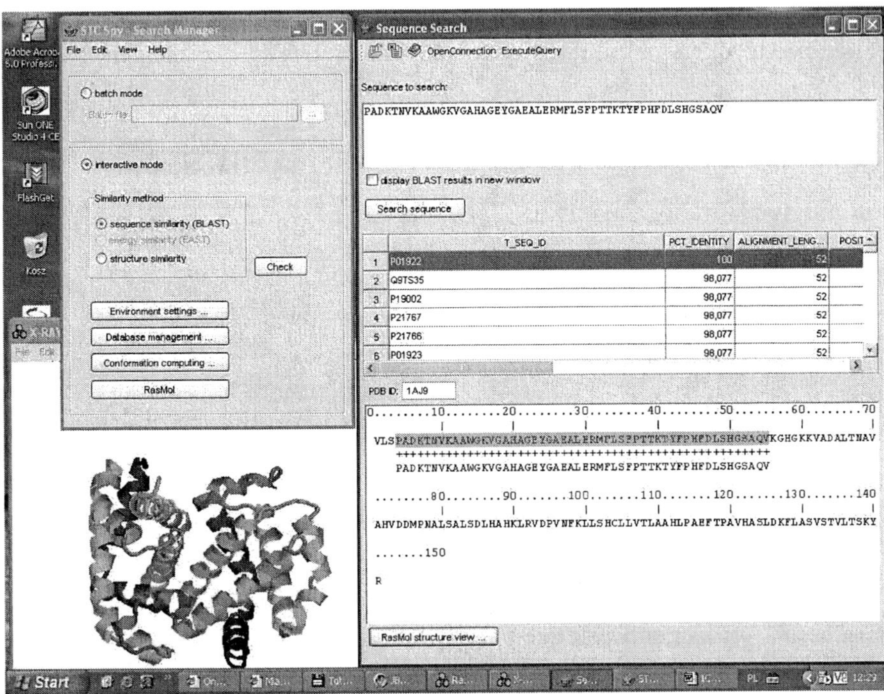

Fig. 2. Signal Transduction Spy – the Search Manager

Two modes of preparing experiment remain feasible: batch mode and interactive step-by-step mode (left window). Processing data in the batch mode seems to be better because it is going without any control and doesn't affect the user attention – the experiment may take a long time especially during all the processes that use protein conformation computation module to investigate changes in protein 3D structure as a result of environmental parameters changes (e.g. temperature, pH). However, the batch process is supposed to be described in the special purpose process-description files containing all steps of the computation. The mode is in the development phase.

The interactive mode provides users to see what changed during each step of the experiment and to make corrections or to change the way the process goes (e.g. see each step of the signal transduction and the substrates of the cascade, see the network of interactions, explore of protein conformation, change the similarity search method, etc.). In the mode users make use of RasMol [10] to display semi-result structures of their activities in graphical form.

4 Processing of a Single Step of the Signal Cascade

Processing of the signal cascade may be divided on processing respective steps of the cascade, represented by reactions on each level. Fig. 3a presents a sample network of the reactions between nanostructures (e.g. proteins, polypeptides). Each reaction results with a creation of a final product. The product may be a protein in the appropriate state and conformation, which can be the activator or inhibitor of the other reactions which take place in the living cell or nanoprocess. Fig. 3b shows the algorithm of tracking the signal cascade step – e.g. 1.7.2.24 reaction [24] marked in the Fig. 3a.

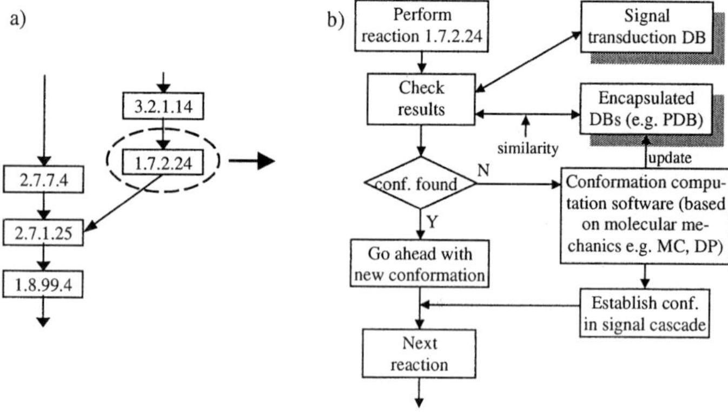

Fig. 3. a) Sample network of reactions, b) Processing of a single step of the cascade

During the process of tracking cascade's steps it is possible to check the conformational results (of nanostructure like polypeptide) with the databases consisting of the structural information (e.g PDB [1]). These databases work as an integral parts of the entire system. The retrieval of similar polypeptides is based on the approximate methods (section 3). If the structure or similar structure (with similarity determined) is found in the database, the system retrieves appropriate information and tracks to the next step of the cascade with the retrieved polypeptide. The information about cascade steps comes from other databases (e.g. BioCarta [9], BIND [12], aMAZE [24]). If the resultant structure is not found in the database, it must be computed with the use of molecular mechanics methods and software, with the assumption of getting minimal potential energy. We are strongly convinced that minimizing of the energy expression describing a conformational energy of structure [7, 22, 23] leads to the problem related with multiple local minima or saddle points and so simple algorithms of optimization e.g. gradient-based, Newton's or Fletcher-Powell have to be modified to find the global minimum. The implementation of algorithms of dynamic programming (DP) [25] or Monte Carlo (MC) methods [16, 26] is better solution. Moreover, these methods are time- and computer's memory consuming. Once the new conformation is established, it is possible to look for the interactions with other proteins and reactions the new protein can participate and meanwhile the database is being updated.

5 Example of Selected Process in the Signal Cascade Analysis

The dataflow in the signal transduction simulation system using the distributed databases environment [14] is presented in Fig. 4. In the system the three databases work as encapsulated parts of the computing process visible through the interfaces A and B.

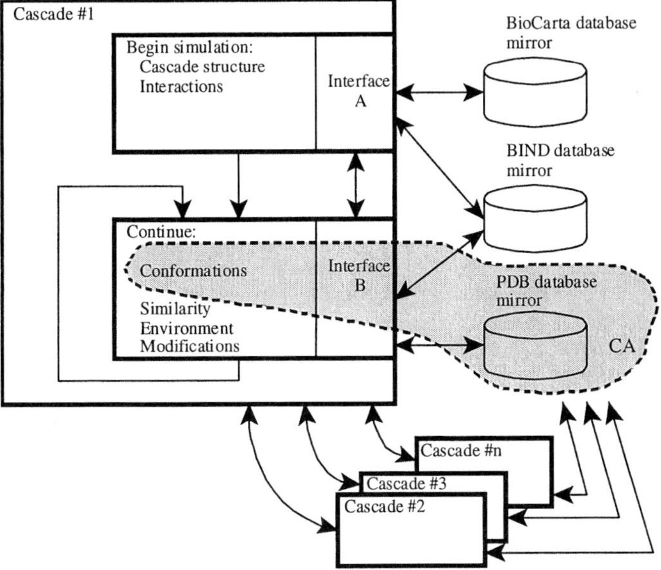

Fig. 4. Dataflow in simulation process

The BioCarta database mirror allows the cascade structures data [9], the BIND database mirror allows nanostructures interactions data [12], and the PDB database mirror allows the molecule structures data [1].

The example presents the approach of conformation determination in signal cascade including the human CDK2 kinase during the phosphorylation process [6] using proper database. In inactive form the CDK2 kinase can be retrieved from the PDB database as molecule signed by 1B38 [1]. During the phosphorylation process the threonine 160 in main chain of CDK2 is modified. To find this conformation change is very difficult by computation. The better solution is to search the proper, existing information from database. In Fig. 4 the shadowed area CA presents this simulation/database operation. To make more visible the critical place responsible for the conformation of kinase CDK2 (in the ATP complex) described by the 1B38 file was processed to extract the threonine 160 neighborhood [6]. Retrieved from 1B38 file the middle part of the main chain of amino acids of the CDK2 is presented in Fig. 5a. The same part of the main chain of CDK2 after the phosphorylation (extracted from 1B39 file from PDB database) is presented in Fig. 5b. The changes in the conformation can be found basing on molecular mechanics approach, however, the results obtained from operation CA using selected database seem to be more practical.

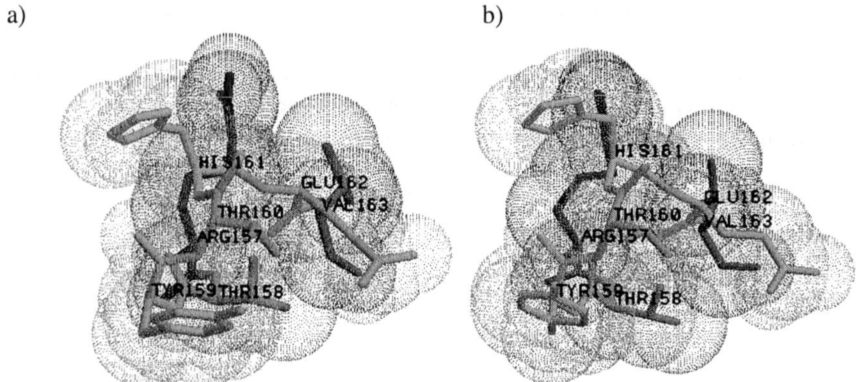

Fig. 5. a) Sticks representation with the van der Waal's dot surfaces of the conformation in a middle part (residues 157-163) of the non-phosphorylated CDK2 (the threonine 160 in the center of view), b) the same view of CDK2 in a middle part of the amino acids chain (residues 157-163) presenting the results of phosphorylation of threonine 160 on the conformation switching of the whole CDK2 structure

6 Evolution of the System and Future Research

Searching information in the distributed bioinformatics databases takes integral part of the signal cascade analysis process. The paper presents the simulation system under development in the Institute of Informatics, Silesian University of Technology, Gliwice, Poland intended to signal transduction simulation using the data from BIND, BioCarta, PDB (Rutgers) databases, and others. The considered system includes the algorithms of fused signal cascades analysis, the nanostructures conformation determination, the sequence (amino acids chain) and shape (conformation) similarity searching and the solution of the encapsulation of databases (DB mirrors, interfaces and control) enables to use the proper database in the on-line computing.

References

1. Berman, H.M., Westbrook, J., Feng, Z., Gilliland, G., Bhat, T.N., Weissig, H., Shindyalov, I.N., Bourne, P.E.: The Protein Data Bank. Nucleic Acids Res. 28 (2000), 235–242
2. Greer, D.S., Westbrook, J.D., Bourne, P.E.: An ontology driven architecture for derived representations of macromolecular structure. Bioinformatics. 18 (2002) 1280-1281
3. Apweiler, R., Bairoch, A., Wu, C.H., et al.: UniProt: the Universal Protein knowledgebase. Nucleic Acids Research. 32 (2004) D115-D119

4. Berridge, M.J.: The Molecular Basis of Communication within the Cell. Scientific American. 253 (4) (1985) 142-152
5. Bourne, P.E., Berman, H.M., Watenpaugh, K., Westbrook, J.D., Fitzgerald, P.M.D.: The macromolecular Crystallographic Information File (mmCIF). Methods Enzymol., 277 (1997) 571–590
6. Brown, N.R. et al.: Effects of Phosphorylation of Threonine 160 on Cyclin-dependent Kinase 2 Structure and Activity. J. of Biol. Chem. 274(13) (1999) 8746-8756
7. Znamirowski, L., Zukowska, E.D.: Simulation of Post-translational Conformations in the Ribosomal Polypeptide Synthesis. Proc. of the IASTED Intern. Conf. Modeling and Simulation, Marina del Rey, California, ACTA Press, Anaheim-Calgary-Zurich (2002) 97-102
8. Benson, D.A., Karsch-Mizrachi, I., Lipman, D.J., Ostell, J., Wheeler, D.L.: GenBank: update. Nucleic Acids Res. 32(Database issue) (2004) D23-6
9. BioCarta: Charting Pathways of Life. http://www.biocarta.com/genes/
10. Sayle R. RasMol: Molecular Graphics Visualization Tool. Biomolecular Structures Group, Glaxo Wellcome Research & Development, Stevenage, Hartfordshire 1998, H. J. Bernstein. v.2.7.1.1, rasmol@bernstein-plus-sons.com
11. Boeckmann, B., Bairoch, A., Apweiler, R., et al.: The SWISS-PROT protein knowledgebase and its supplement TrEMBL in 2003. Nucleic Acids Res. 31 (2003) 365-370
12. Bader, G.D., Betel, D., Hogue, C.W.V.: BIND: the Biomolecular Interaction Network Database. Nucleic Acids Research. Vol. 31(1) (2003) 248-250
13. Snyder S.H.: The Molecular Basis of Communication between Cells. Scientific American, 253 (4) (1985) 132-141, (1985)
14. Signal Transduction Simulation System Using the Distributed Databases Environment. Research Project BW/2004/05, Institute of Informatics, Silesian University of Technology, Gliwice (2005)
15. http://www.omg.org
16. Warecki, S., Znamirowski, L.: Random Simulation of the Nanostructures Conformations, Intern. Conference on Computing, Communication and Control Technology, Proceedings Volume I, The Intern. Institute of Informatics and Systemics, Austin, Texas, August 14-17, p. 388-393, (2004)
17. Callaway, J., Cummings, M., et al.: Protein Data Bank Contents: Atomic Coordinate Entry Format Description, Federal Govern. Agency. (1996) http://www.rcsb.org/pdb/docs/format/
18. http://openmms.sdsc.edu
19. Tonegawa, S.: The Molecules of the Immune System, Scientific American, 253 (4) (1985) 122-131
20. Wang, Y., Addess, K.J., Geer, L., Madej, T., Marchler-Bauer, A., Zimmerman, D., Bryant S.H.: MMDB: 3D structure data in Entrez. Nucleic Acids Res. 28 (2000) 243-245
21. Ray, L.B.,: The Science of Signal Transduction. Science. 284 (1999) 755-756
22. Znamirowski, L.: Switching VLSI Structures. Reprogrammable FPAA Structures. Nanostructures. Studia Informatica. Vol. 25 (4A) (60) (2004) 1-236
23. Ponder, J.: Tinker – Software Tools for Molecular Design. Dept. of Biochemistry & Molecular Biophysics, Washington University, School of Medicine, St. Louis (2001)
24. van Helden, J., Naim, A., Mancuso, R., et al.: Representing and analysing molecular and cellular function using the computer. Biol Chem. 381(9-10) (2000) 921-35
25. Bellman, R.: Dynamic Programming. Princeton University Press, Princeton, N. J. (1957)
26. Metropolis, N., Ulam, S.: The Monte Carlo Method, Journal of the American Stat. Assoc., 44 (247) (1949) 335-341, (1949)

Virtual States and Transitions, Virtual Sessions and Collaboration

Dimitri Bourilkov

University of Florida, Gainesville, FL 32611, USA
bourilkov@phys.ufl.edu

Abstract. A key feature of collaboration is having a *log* of what and how is being done - for private use/reuse and for sharing selected parts with collaborators in today's complex, large scale scientific/software environments. Even better if this log is *automatic*, created on the fly while a scientist or software developer is working in a habitual way, without the need for extra efforts. The `CAVES` (Collaborative Analysis Versioning Environment System) and `CODESH` (COllaborative DEvelopment SHell) projects address this problem in a novel way, building on the concepts of *virtual state* and *virtual transition* to provide an automatic persistent logbook for sessions of data analysis or software development in a collaborating group. Repositories of sessions can be configured dynamically to record and make available in a controlled way the knowledge accumulated in the course of a scientific or software endeavor.

1 Introduction

Have you sifted through paper folders or directories on your computer, trying to find out how you produced a result a couple of months (or years) ago? Or having to answer a question while traveling, with your folders safely stored in your office? Or a desperate collaborator trying to reach you about a project detail while you are hiking in the mountains? It happened many times to me, and there must be a better way.

2 Automatic Logbooks

We explore the possibility to create an automated system for recording and making available to collaborators and peers the knowledge accumulated in the course of a project. The work in a project is atomized as *sessions*. A session is defined as a transition T from an initial $|I>$ to a final $|F>$ state: $|F>= T|I>$. The work done during a session is represented by the transition operator T. The length of a session is limited by the decision of the user to record a chunk of activity. A state can be recorded with different level of detail determined by the users: $|State>=|Logbookpart, Environment>$. The system records automatically, in a persistent way, the logbook part for future use. The environment is considered as available or provided by the collaborating group. The splitting between

V.S. Sunderam et al. (Eds.): ICCS 2005, LNCS 3516, pp. 342–345, 2005.

logbook and environment parts is to some extent arbitrary. We will call it a *collaborating contract*, in the sense that it defines the responsibilities of the parties involved in a project. A good splitting is characterized by the transition operator T acting in a meaningful way on the logged part without affecting substantially the environment.

At sufficient level of detail, each final state $|F>$ can be reproduced at will from the log of the initial state $|I>$ and the transition T. The ability to reproduce states bring us to the notion of *virtual state* and *virtual transition*, in the sense that we can record states that existed in the past and can recreate them on demand in the future. Virtual states serve as *virtual checkpoints*, or delimiters for *virtual sessions*. In a traditional programming language, the user typically codes the transition T, i.e. provides in advance the source of the program for producing a given final state, following the language syntax. The key idea in our approach is that the user works in a habitual way and the log for the session is created *automatically, on the fly,* while the session is progressing. There is no need per se to provide any code in advance, but the user can execute preexisting programs if desired. When a piece of work is worth recording, the user logs it in the persistent session repository with a unique identifier [1].

Let us illustrate this general framework with an example. We want to log the activities of users doing any kind of work on the command line e.g. in a `UNIX` shell. At the start of a session the system records details about the initial state. Then the user gives commands to the shell (e.g. ls, cd, cp, find etc). During the session users could run some existing user code, possibly providing input/output parameters when invoking it. The system collects and stores all commands and the source code of all executed scripts as used during the session, automatically producing a log for the transition T. Later users may change any of the used scripts, delete or accidentally lose some of them, forget which parameters were used or why. When the same or a new user wants to reproduce a session, a new sandbox (clean slate) is created and the log for a given unique identifier is downloaded from the repository. In this way both the commands and the scripts, as well as the input/output parameters, "frozen" as they were at the time of the session, are available, and the results can be reproduced.

3 The CODESH/CAVES Projects - Working Implementations of Automatic Logbooks

The `CODESH` [2] and `CAVES` [3] projects take a pragmatic approach in assessing the needs of a community of scientists or software developers by building series of working prototypes with increasing sophistication. By extending with automatic logbook capabilities the functionality of a typical `UNIX` shell (like tcsh or bash) - the `CODESH` project, or a popular analysis package as `ROOT` [4] - the `CAVES` project,

---

[1] The virtual data idea explores similar concepts, putting more emphasis on the transition - abstract transformation and concrete derivations, coded by the users in a virtual data language [1].

these prototypes provide an easy and habitual entry point for researchers to explore new concepts in real life applications and to give valuable feedback for refining the system design.

Both projects use a three-tier architecture, with the users sitting at the top tier and running what looks very much like a normal shell or identical to a ROOT session, and having extended capabilities, which are provided by the middle layer. This layer is coded in Python for CODESH or in C++ for CAVES, by inheriting from the class which handles the user input on the command line. The implemented capabilities extend the example in Sect. 2. The command set is extensible, and users can easily write additional commands. The lower tier provides the persistent back-end. The first implementations use a well established source code management system - the Concurrent Versions System CVS, ideal for rapid developments by teams of any size. The CVS tags assume the role of unique identifiers for virtual sessions, making it possible to extract session information identical to the one when the session log was first produced.

The CAVES/CODESH systems are used as building blocks for collaborative analysis/development environments, providing "virtual logbook" capabilities and the ability to explore the metadata associated with different sessions. A key aspect of the project is the distributed nature of the input data, the analysis/development process and the user base. This is addressed from the earliest stages. Our systems are fully functional both in local and remote modes, provided that the necessary repositories are operational and the datasets available. This allows the users to work on their laptops (maybe handhelds tomorrow) even without a network connection, or just to store intermediate steps in the course of an active analysis/development session for their private consumption, only publishing a sufficiently polished result. This design has the additional benefit of utilizing efficiently the local CPU and storage resources of the users, reducing the load on the distributed services (e.g. Grid) system. The users have the ability to replicate, move, archive and delete session logs. Gaining experience in running the system will help to strike the right balance between local and remote usage.

A possible scenario is that a user chooses during a session to browse through the tags of other users to see what work was done, and selects the log/session of interest by extracting the peers log. Here two modes of operation are possible: the user may want to reproduce a result by extracting and executing the commands and programs associated with a selected tag, or just extract the history of a given session in order to inspect it, possibly modify the code or the inputs and produce new results.

Our first releases are based on a quite complete set of commands providing interesting functionality. Screenshots from a real session are shown in Fig. 1.

In this paper we outline the main ideas driving the CODESH/CAVES projects. The history mechanism or the script utility of UNIX shells can just log the commands and standard output of a working session in a private file, leaving all the rest to the users. Our automatic logbook/virtual session approach does much more by managing *private and shared repositories of complete session logs*, including the commands *and* the programs, and the ability to reproduce the re-

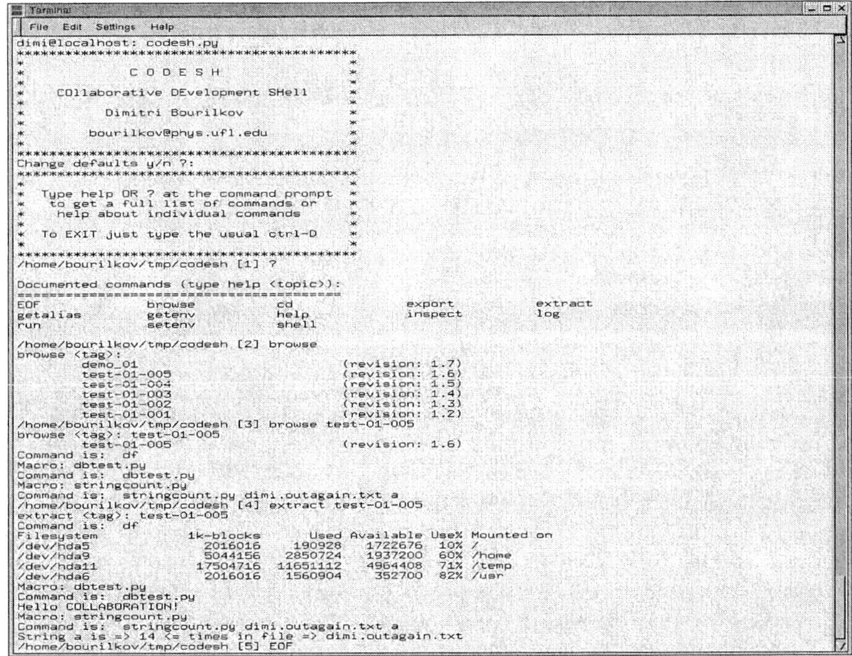

Fig. 1. Example of a `CODESH` session

sults or reuse them for future developments. Public releases of the first functional systems for automatic logging in a typical working session are available for interested users. Possible directions for future work include adding different back-ends - SQL, web or grid service oriented [5], using GSI security, automatically converting session logs to workflows, the ability to develop locally and seamlessly schedule more CPU/data intensive tasks on grid infrastructure.

The study is supported in part by the United States National Science Foundation under grants NSF ITR-0086044 (GriPhyN) and NSF 0427110 (UltraLight).

References

1. Foster, I. *et al.*: Chimera: A Virtual Data System for Representing, Querying, and Automating Data Derivation. In: 14th International Conference on Scientific and Statistical Database Management (SSDBM 2002), Edinburgh, 2002
2. Bourilkov, D.: THE CAVES Project - Collaborative Analysis Versioning Environment System; THE CODESH Project - Collaborative Development Shell. arXiv:physics/0410226 ; http://xxx.lanl.gov/abs/physics/0410226
3. Bourilkov, D.: The CAVES Project: Exploring Virtual Data Concepts for Data Analysis. arXiv:physics/0401007, and references therein ; http://xxx.lanl.gov/abs/physics/0401007
4. Brun, R. and Rademakers, F.: ROOT - An Object Oriented Data Analysis Framework. Nucl. Inst. & Meth. in Phys. Res. **A 389** (1997) 81–86
5. Grid-enabled Analysis Environment project: http://ultralight.caltech.edu/gaeweb/

A Secure Peer-to-Peer Group Collaboration Scheme for Healthcare System*

Byong-In Lim, Kee-Hyun Choi, and Dong-Ryeol Shin

School of Information and Communication Engineering,
Sungkyunkwan University,
440-746, Suwon, Korea, +82-31-290-7125
{lbi77, gyunee, drshin}@ece.skku.ac.kr

Abstract. P2P (Peer-to-Peer) applications allow flexible organization, distribution of role to participating peers and the ability to share information with the other peers for group collaborations. As a result, P2P systems are not only gaining importance, but also becoming ubiquitous media for information exchange. JXTA is a P2P application development infrastructure that enables developers to easily create service-oriented software. This paper presents a low-cost, patient-friendly JXTA-based healthcare system, which is comprised of medical sensor modules in conjunction with wireless communication technology. Among the most important activities associated with the healthcare system are the sharing of medical information and collaborative medical work. The JXTA grouping service facilitates to share data in a secure manner under P2P environments. Through JXTA grouping service, we implemented a prototyped system, which improves the ability of the healthcare workers to cope with dynamic situation, which in turn makes it possible to offer more efficient medical services.

1 Introduction

Advancements in sensor, information and communication technology can play an important role in achieving cost reduction and efficiency improvement in healthcare delivery systems, to the extent that this offers high-quality medical service anytime and anywhere. However, conventional healthcare systems, including the Mobihealth [1] project, utilize a central server for the look up of information. Furthermore, the possibility of offering seamless service is limited by the network connectivity of wireless devices or resource-constraints. In addition, traditional systems only focus on organic communication and service utilization between patients and hospital, whereas they ignore the systematic communication and sharing of information between healthcare workers in the hospital. For these reasons, these systems cannot cope with acute situations dynamically and promptly. Also, privacy and security are potential problems. Patient's data should be available irrespective of their location, but only to authorized healthcare workers.

---

* This research was partially supported by a grant from the CUCN, Korea and Korea Science & Engineering Foundation (R01-2004-000-10755-0)

V.S. Sunderam et al. (Eds.): ICCS 2005, LNCS 3516, pp. 346 – 349, 2005.

To resolve these problems, we propose a JXTA-based healthcare system, which operates in a peer-to-peer (P2P) environment so as to offer seamless service by distributing the healthcare services among the healthcare workers. This sharing of information about the medical treatment that patients receive between the healthcare workers in the JXTA environment improves their ability to cope with dynamic situations, which in turn makes it possible to offer more efficient medical services. Furthermore, the JXTA grouping service makes it possible to share data in a secure manner under the P2P environment. So, our system protects patient's data. In this paper, we focus on the system architecture design in hospital region which is based on JXTA grouping mechanisms.

This paper is structured as follows: Section 2 discusses system architecture for healthcare services and the system implementation is presented in section 3. Finally, this paper is concluded in section 4.

2 System Architecture for Healthcare Services

2.1 Architecture Overview

Our system consists of two regions, viz. the Healthcare region and the Hospital region, as shown in Figure 1.

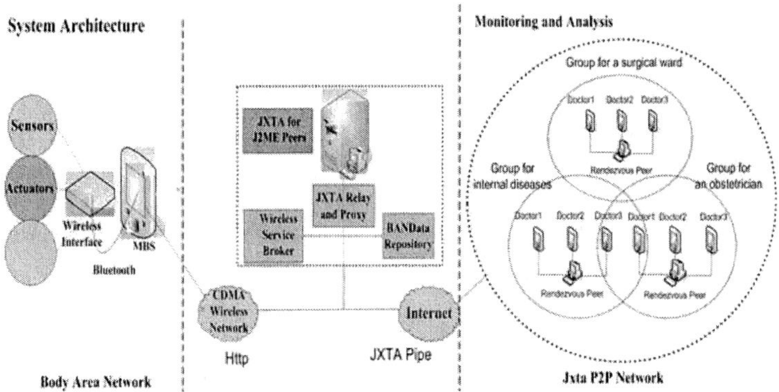

Fig. 1. System Architecture

The healthcare BAN consists of sensors, a Wireless Interface, PDA communication, and various facilities. The Hospital region is composed of the JXTA P2P network that supports the doctor's mobility and dynamic service management modules. A P2P network distributes the information among the member nodes, instead of concentrating it at a single server. A JXTA relay peer is defined among the two regions, in order to provide a systematic computing environment between the Healthcare region and the Hospital region. JXTA relays act as proxies for the individual patients' PDAs, as well as taking care of all the heavier tasks on their behalf. For more details, refer to [2]. In next section, we describe how healthcare services with collaboration of JXTA secure group environment are formed and managed in a hospital region.

2.2 Secure Group Collaboration Scheme in Hospital

Among the most time consuming activities associated with the public health sector are documentation and data exchange. In a hospital, the healthcare workers form a peer group within the institution. Together they can create a secure peer group with various access rights, so as to allow the sharing of the patient's data (e.g. his or her patient UUID, patient sensor data, etc) and other kind of information (e.g. a first-aid, drugs to prescribe, etc) under the secure environment. Figure 2 shows a sequence diagram of group collaboration scheme in hospital domains.

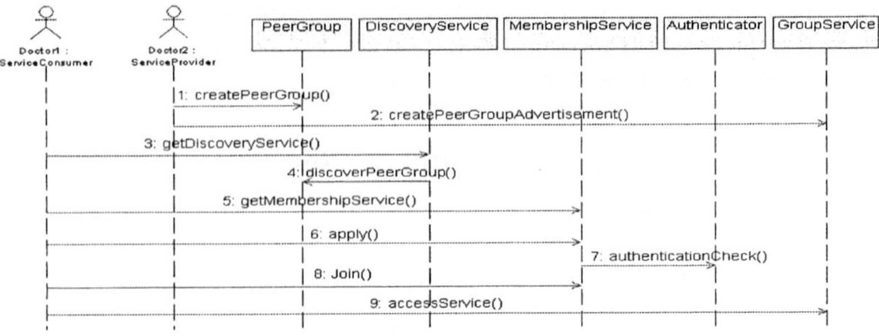

Fig. 2. Sequence Diagram of Group Collaboration Scheme

JXTA combines the overlay network with a namespace abstraction referred to as a peer group [3]. Any peer may create a new peer group at any time for any reason to communicate with any set of peers. The only requirement is that all peers have agreed upon a common set of services and protocols for implementing secure group collaboration (authentication, authorization, message propagation, discovery service, etc). Consequently, JXTA grouping service is an effective means to implement P2P collaboration system under secure environment. This is the main reason for which the healthcare system is based on the JXTA platform. In our system, the healthcare region consists of several groups (e.g. groups corresponding to specific surgical wards, dental treatment, etc). The collaboration scheme of secure peer group is as follows. Before healthcare workers can interact with the group, they need to create a peer group. When creating the peer group, the healthcare workers involved initialize the local secure environment, and then publish a peer group advertisement, which includes the peer group ID and peer group medical service. After the secure group is created, they can perform collaborative service and functionality of the peer. Searching and sharing are done on the peer group level, i.e., shared contents are only available to the peer group. In order for service consumers to utilize medical service, they must firstly achieve their guaranteed passwords as well as the information of secure peer groups which provide medical service in the JXTA network. In JXTA, the membership service is used to apply for peer group membership, join a peer group, and exit from a peer group. The membership service allows a peer to establish an identity within a peer group. Once an identity has been established, a credential is available, which allows the peer to prove that they have obtained the identities rightfully. In our system, the security is ensured by using JXTA grouping mechanism to implement group level security policies for membership and service access. Also, the JXTA grouping mechanism supports the sharing of medical information and collaborative medical work, which in turn makes it possible to offer more efficient medical services.

3 Implementation

In this study, we utilize the JXTA Platform to develop a medical-service management tool as well as secure group collaborations in a hospital domain. We also implemented a software for monitoring the patient's status, in the BAN as well as in

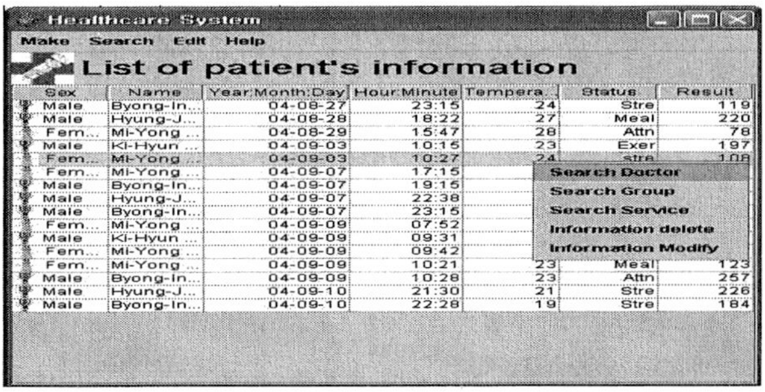

Fig. 3. Data Management JXTA Application

the JXTA application used for managing the patient's information; medical status and personal information, as shown in Figure 3. The use of the JXTA Platform with its discovery and grouping mechanisms enables us to offer efficient and adaptive medical services.

4 Conclusion

In this paper we have presented the design of a JXTA-based healthcare system which enables healthcare workers to share data in a secure manner under the P2P environment. This paradigm offers significant advantages in healthcare workers to cope with dynamic situations, thus improving the quality of the medical service that they can provide. Future work includes the detailed simulations and performance evaluations for the developed systems.

References

1. Nikolay Dokovsky, Aart van Halteren, Ing Widya, "BANip: enabling remote healthcare monitoring with Body Area Networks", International Workshop on scientific engineering of Distributed Java applications, November 27-28, 2003.
2. Byongin Lim, et al, "A JXTA-based Architecture for Efficient and Adaptive Healthcare Services", ICOIN 2005, LNCS 3391, pp. 776-785, January 31- February 2, 2005.
3. PKI Security for JXTA Overlay Network, Jeff Altman, February 2003.

Tools for Collaborative VR Application Development

Adrian Haffegee[1], Ronan Jamieson[1], Christoph Anthes[2], and Vassil Alexandrov[1]

[1] Centre for Advanced Computing and Emerging Technologies, The University of Reading, Reading, RG6 6AY, United Kingdom
sir04amh@reading.ac.uk
[2] GUP Linz, Johannes Kepler University Linz, Altenbergerstraße 69, A-4040 Linz, Austria

Abstract. This paper introduces a tool set consisting of open source libraries that are being developed to facilitate the quick and easy implementation of collaborative VR applications. It describes functionality that can be used for generating and displaying a Virtual Environment (VE) on varied VR platforms. This is enhanced to provide collaboration support through additional modules such as networking. Two existing VR applications which make use of these tools are described. Both were developed effortlessly over a short period of time, and demonstrate the power of these tools for implementing a diverse range of applications.

1 Introduction

With the growing range and power of VR systems, increasing numbers of users are gaining access to VEs. These are virtual spaces created within the system, that a user can enter and then interact with. Linking these using various technologies expand the virtual domain, and bestow the users with quasi person to person interaction.

A potentially limiting factor in the development of collaborative VR applications is the time, effort and expense of their implementation. A new tool set consisting of the CAVE Scene Manager (CSM), networking utilities and additional component libraries have been designed and implemented. These aid developers in creating their applications by providing the key functionality of generating, displaying and enhancing a VE.

Work relating to this development will be discussed in Section 2. This will be followed by Sections 3 and 4 respectively; first with a description of the CSM, and then the additional tools. Two sample applications currently using this tool set are then mentioned in Section 5, before the paper concludes in Section 6.

2 Related Work

Designing tools for collaborative VR application development requires the consideration of several key areas, most of which have existing implementations.

V.S. Sunderam et al. (Eds.): ICCS 2005, LNCS 3516, pp. 350–358, 2005.

Networked VR viewers, such as Coanim which comes with the CAVERNsoft toolkit [1], allow the display of geometries for different remote users in a basic Networked Virtual Environment (NVE). In this type of application users can modify attributes in the shared environment (e.g. transformations for geometries) or trigger animation events. Further descriptions of this particular type of VE can be found in [2, 3].

Scenegraph libraries (OpenGL Performer [4], OpenSG [5]), provide a structured database of entities and their relationships in the scene. Typically these entities are rendered graphically enabling representation of potentially complex environments.

Navigation and interaction within the VE is required to allow users to interface with the environment. A good overview of different navigation methods can be found in [6].

Avatars, which are used to graphically represent remote users, both in their position and actions, are a highly important feature for collaborative environments. A description of how these can be implemented is found in [7].

Many different types of VR hardware such as CAVEs [8] or head mounted displays can be used for the display of VEs. A generic application should allow operation with many of these differing systems, providing hardware independence. Offerings providing similar functionality include those from CAVELib [9], and VRJuggler [10].

3 The CAVE Scene Manager (CSM)

The CSM handles the VE's scenegraph representation (described below), as well as any display's video rendering. This module is built on top of a scenegraph library, forming an integral part of the tool set.

Although scenegraph independence would ideally be desired for the tool set, OpenSG [5] was chosen for the initial design and implementation due to its cluster support, structure and for being open source.

3.1 Configuration

Despite its name, the CSM can be used for displaying on a wide range of VR systems and not just CAVEs. Many parameters of these systems, for example sizes, graphics configuration, or tracking dimensions, change with each installation. To allow for these differences the CSM uses a configuration file which describes the particular setup. With this approach, by just referencing a different configuration file the same application can be run on any VR platform.

To allow simple integration with existing VR installations, the configuration file format was made compatible with that used for CAVELib configuration. While the CSM does not utilize all the CAVELib parameters, it uses those sufficient for generating multi-wall stereo images, ignoring those it does not require.

In addition to the standard CAVELib configuration options, additional optional parameters have been included for advanced CSM configuration.

3.2 Internal Scenegraph Representation

The internal structure of the CSM contains a node tree for the positioning of the user within the VE. Figure 1 shows the OpenSG representation of these nodes within the CSM. The circles represent the nodes which are used to describe the graph hierarchy, while the rectangles are node cores determining the type of node functionality.

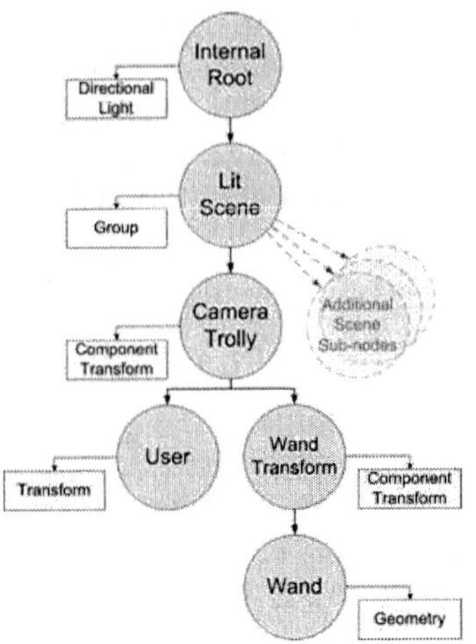

Fig. 1. Node tree for the CAVE Scene Manager

The *internal root* node has a directional light source core for a headlight, with its beacon being the *lit scene*. This ensures every node below the internal root will be lit if the headlight is on.

An OpenSG component transform is used to position a *camera trolley* node. This represents the navigated position in the scene of a trolley upon which the cameras are mounted. Two child nodes from their camera trolley parent are used to show the positions of the two tracking devices; the user position from head tracking, and also the wand position as generated by the CAVE. Finally a basic geometry node is added to display the default wand within the environment. The user, wand and camera trolley can be moved independently, and the CSM has functions for updating these positions as required. A common technique in the CAVE is to move the camera trolley using the joystick on the controller, with user and wand positions coming from the tracking devices. For desktop operation all these movements would come from a keyboard/mouse combination.

A VE application passes the root node of the its scene to the CSM, where it gets added as a child node of the internal root. This is useful for setting a scene that has a fixed location and does not move, such as a street or a room. However, often additional entities are needed to represent objects or remote users within the scene. Functionality has been provided to add and remove the node tree representations of these, and also to change their base position and orientation as required. These functions reference the entity through its root node without the CSM needing any knowledge of the entities' composition. This allows complex animated geometries to be easily incorporated into the scene.

In a similar way the default wand geometry can be overridden to display any node tree while still tracking any movement of the wand. This could be used to model a particular type of pointer, a sophisticated moving hand, or even a hand using or manipulating some additional object.

3.3 Display Walls

The CSM's main task is to display images, optionally stereo, on one or more screens (monitors or projection walls). Virtual cameras are placed in the scene, and are used to determine the images being rendered.

The CSM configuration determines how many cameras exist in the scene. It has fields for display wall configuration which describe the position of the viewing device relative to the user's position. Defaults have been provided for the regular CAVE walls which form a cube around the user. When these are not sufficient, such as for a curved screen, the system is configured with the corner coordinates of the screen or projection plane. One camera is used for each wall being rendered, facing in the appropriate direction for that wall.

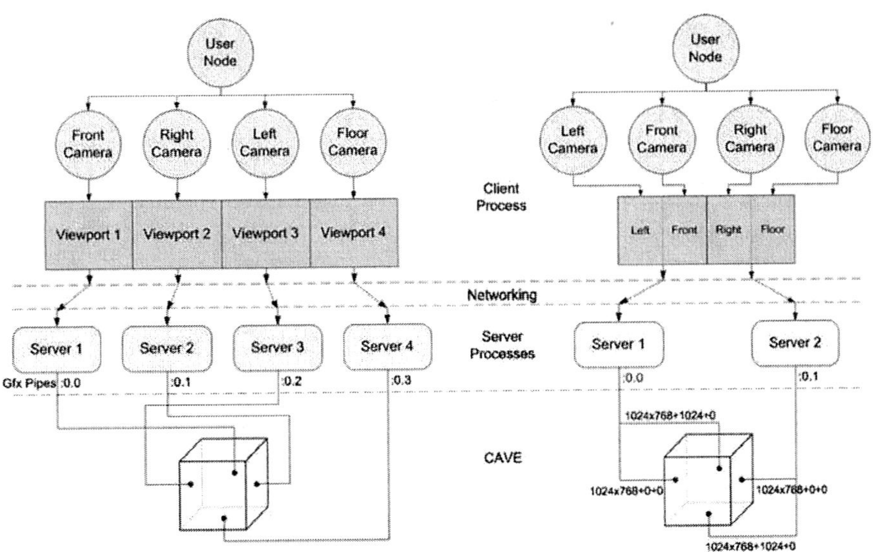

Fig. 2. Client/Server Cluster Window for a 4 pipe (left) and 2 pipe (right) CAVE

3.4 Client/Server Clustering Windows

The OpenGL Utility Toolkit (GLUT), which is used for displaying the graphics windows can only output to one graphics pipe per process. This is an issue when multiple displays are required. An alternative to a multiple process scheme uses clusters, with the full environment being constructed through a network of individual computers, each responsible for its own portion.

Both of these approaches are supported in the CSM through OpenSG's client/server clustering. A client process handles the scenegraph manipulations, and one or more server processes render the graphics. These processes can exist on the same system, operate remotely, or a combination of the two. This is achieved through CSM configuration.

When clustering, the views generated by the client are split over a grid with as many cells as there are servers. Each server only requires the data for its cell, which it renders for its connected display.

Figure 2 shows two possible approaches for clustering for a four wall CAVE. In the architecture on the left each wall is driven through its own pipe. For this the client window is split into 4 viewports, with each camera placing its output into one of these. Each server receives a quarter of the client window, therefore receiving the data from the camera for the wall it represents.

The architecture on the right is slightly different in that the system only has two pipes, with each of these driving two walls. In this case the client window is split in half, with each half being sent to a different server. Two viewports are combined onto each of these halves to display the correct image split over two walls. In this example the split is vertically down the middle of each graphic pipe.

4 Additional Functionality

In addition to the scene and display functionality provided by the CSM, several other useful tools are also included. These include a command parser for accepting user input, which would then be used to activate certain commands or functionality. Individual application configuration files can also be utilized, and work along the same lines as for CSM configuration.

Three of the more important tools are avatars, networking and navigation/interaction, which we describe below.

Avatars. Two types of avatar currently exist. Simple avatars have been included with limited realism, however they do make use of range based level of detail, allowing a potentially large number of them to be simultaneously active with little processing overhead. Alternatively, highly graphical avatars with realistic motion can also be used. These increase observer immersion as they move and perform pre-set actions in highly realistic manner.

Networking. A networking infrastructure has been developed providing low latency, and reliability. This has been integrated into a topology based on the descriptions from [11], making it possible for remote clients to reliably communicate. Internally, the networking tools control default network handling, including the necessary messaging for remote avatar representation. To this, functionality has been provided to allow sharing of user defined messages. As will be seen in the sample applications these messages can be used for advanced collaboration.

Network simulators and logging/debug routines have also been developed for assistance in application development.

Navigation and Interaction. Each time the displayed frame is updated the application polls the tracking information from shared memory. While the head and wand tracking data are passed to the CSM, the controller information is used for navigation and interaction. Current navigation makes use of the joystick and wand orientation to move within the scene. It is anticipated that this could be further extended through the integration of a navigation framework such as that described in [12].

Basic collision detection between the user and the scene has also been provided for when one of the wands buttons had been activated. Either as a result of this, or directly from the interaction with the wand it is possible to transmit events to any remote users, thereby forming the basis of user interaction and collaboration beyond that achieved from a simple visual response.

5 Applications

5.1 Flooding Visualisation

This application demonstrates the benefits gained through visualization in an immersive environment. As part of the EU CROSSGRID project, it uses the GRID for landscape flooding simulation [13]. The CSM provides an integral part of this application, allowing easy viewing on the various types of VR hardware.

Fig. 3. Interaction in the CAVE, Curved Screen and on a desktop system

The data to be simulated is calculated with Globus middleware on the GRID, and then transferred to the display clients through GRIDFTP. The simulation makes use of two types of VRML 2.0 models. The first makes up the landscape and consists of a height map geometry of the terrain, upon which a photographic image has been superimposed. This provides a realistic environment with good spatial orientation.

The second group of models, are used for representing the flood water. A separate set of triangle meshes are provided for each time stamp during the flooding process, and these are applied in sequence to the terrain to simulate the flood.

For efficiency in maintaining a realistic frame rate (>30fps), all geometry data is preloaded into the scene on start-up. The scene root has two children; one for the terrain, and the second a switch node which parents each individual flooding time slice and controls which image is being rendered. The switch node steps through these slices in order, providing smooth animation. On a desktop system the simulation is controlled via a keyboard, whereas fully immersive systems will also use their wand. A flying navigation mode is used, allowing the scene to be viewed from any angle. The keyboard or wand is also used for animating forward or backwards at a pre-defined speed, or for pausing the simulation. Figure 3 demonstrates the simulation being used on three different VR platforms.

5.2 Virtual House Application

The second application has been developed as an easily extendible VE that allows remote users to interact collaboratively with objects within a virtual house environment. It is built on top of the VR tool set which manage display and navigation, both in the CAVE and on the desktop. It uses the networking functionality to distribute each user's motion and event data, and the avatars to display these locally.

Current functionality shares tasks associated with opening and closing a door in the virtual house. Demonstration of this can be seen in Figure 4, which show a CAVE user about to manipulate a moving door that has been opened by a desktop user. In each case the remote user position is depicted by an avatar. Virtual objects that can be interacted with in this environment are defined as

Fig. 4. Collaborative door opening in the Virtual House application

Dynamic Interaction Objects (DIO). They contain attributes and functionality relative to how they can perform. For instance, the door DIO has attributes declaring its axis of rotation and the degree of rotation allowed. For ease of management and flexibility these attributes are contained in a text file which is read during execution.

On initialization the scenegraph is traversed, and DIOs are created for those defined from the text file. Simple collision detection is used to determine if a user is interacting with one of these DIOs, and if so the relevant attributes are modified. Additionally, an Object Manager component continuously services the

DIOs, allowing them to modify the VE as required. Events, such as the door starting to open, are sent over the network for remote notification.

Future improvements will increase functionality of the DIO, allowing it to respond to multiple input events, thereby developing realistic multi-user concurrent collaboration. Further study is also required of potential latency issues, both in terms of avatar movement and event passing.

6 Conclusion

This work describes a useful set of tools that have successfully been used to implement different types of collaborative VR environments. However it should be noted that these form the basis of a potentially larger and more advanced offering. Additional work could see advanced networking, the integration of a navigation library, improved avatars and further collaboration support, all which would allow equally simple, but more advanced application development.

References

1. Park, K., Cho, Y., Krishnaprasad, N., Scharver, C., Lewis, M., Leigh, J., Johnson, A.: CAVERNsoft G2: A toolkit for high performance tele-immersive collaboration. In: VRST, Seoul, Korea, ACM Press (2000) 8–15
2. Macedonia, M.R., Zyda, M.J.: A taxonomy for networked virtual environments. IEEE MultiMedia **4** (1997) 48–56
3. Matijasevic, M.: A review of networked multi-user virtual environments. Technical report tr97-8-1, Center for Advanced Computer Studies, Virtual Reality and Multimedia Laboratory, University of Southwestern Lousiana, USA (1997)
4. Rohlf, J., Helman, J.: IRIS Performer: A high performance multiprocessing toolkit for real-time 3D graphics. In: SIGGRAPH, ACM Press (1994) 381–394
5. Reiners, D.: OpenSG: A scene graph system for flexible and efficient realtime rendering for virtual and augmented reality applications. PhD thesis, Technische Universität Darmstadt (2002)
6. Bowman, D.A., Koller, D., Hodges, L.F.: Travel in immersive virtual environments: An evaluation of viewpoint motion control techniques. In: VRAIS (1997) 45–52
7. Badler, N.I., Phillips, C.B., Webber, B.L.: Simulating Humans: Computer Graphics Animation and Control. Oxford University Press, New York, NY, USA (1992)
8. Cruz-Neira, C., Sandin, D.J., Defanti, T.A., Kenyon, R.V., Hart, J.C.: The CAVE: Audio Visual Experience automatic virtual environment. Communications of the ACM **35** (1992) 64–72
9. VRCO website. http://www.vrco.com/ (2004)
10. Bierbaum, A.D.: VRJuggler: A virtual platform for virtual reality application development. Master's thesis, Iowa State University, Ames, Iowa (2000)
11. Anthes, C., Heinzlreiter, P., Volkert, J.: An adaptive network architecture for close-coupled collaboration in distributed virtual environments. In: VRCAI, Singapore (2004) 382–385

12. Anthes, C., Heinzlreiter, P., Kurka, G., Volkert, J.: Navigation models for a flexible, multi-mode VR navigation framework. In: VRCAI, Singapore (2004) 476–479
13. Hluchy, L., Habala, O., Tran, V.D., Simo, B., Astalos, J., Dobrucky, M.: Infrastructure for grid-based virtual organizations. In: Proceedings of the International Conference on Computational Science, Part III, LNCS 3038, Springer Verlag (2004) 124–131

Multicast Application Sharing Tool – Facilitating the eMinerals Virtual Organisation

Gareth J. Lewis[1], S. Mehmood Hasan[1], Vassil N. Alexandrov[1], Martin T. Dove[2], and Mark Calleja[2]

[1] Advanced Computing and Emerging Technologies Centre, School of Systems Engineering, University of Reading, Whiteknights, P.O. Box 225, Reading, RG6 6AY, United Kingdom
{g.j.lewis, s.m.hasan, v.n.alexandrov}@rdg.ac.uk
http://acet.rdg.ac.uk

[2] Department of Earth Sciences, University of Cambridge, Downing Street, Cambridge, CB2 3EQ, United Kingdom
{martin, mcal00}@esc.cam.ac.uk

Abstract. The eMinerals Virtual Organisation consists of a consortium of individuals affiliated to geographically distributed academic institutions. Collaborative tools are essential in order to facilitate cooperative work within this Virtual Organisation. The Access Grid Toolkit has been widely adopted for this purpose, delivering high quality group-to-group video and audio conferencing. We briefly mention this technology and describe the development of a Multicast Application Sharing Tool designed specifically for this environment.

1 Introduction

The eMinerals project [1] is a NERC funded test-bed project, whose primary aim is to use grid computing methods to facilitate simulations of environmental processes at the molecular level with increased levels of realism. The project involves partners from seven academic institutions throughout the United Kingdom. In this setting, there is a clear need for inter-institute collaboration to collectively achieve the aims and objectives of the project.

Collaborative Computing systems aim to complement human face-to-face communication by providing various tools which enhance users' experience. The emergence of multicast triggered a rush in the development of group communication software. Collaborative tools have also become popular with the widespread availability of broadband. The Access Grid [2] has become well known for high quality group-to-group collaboration across the Internet. It has been widely adopted by the academic community and uses multicast for streaming audio and video. Unfortunately, multicast is still not available to many institutions

V.S. Sunderam et al. (Eds.): ICCS 2005, LNCS 3516, pp. 359–366, 2005.

and home users (due to the reluctance of the Internet Service Providers (ISPs) in adopting this technology).

Person-to-person communication is enriched by an ability to share, modify, or collaboratively create data and information. Our aim at The University of Reading is to provide an Application Sharing tool which allows effortless sharing of legacy applications. The Application Sharing tool is developed by the authors, specifically to be used in a group communication environment. This tool will be integrated with Access Grid 2.x to provide enhanced functionality. We are also interested in developing an inclusive collaborative system, allowing unicast participants to interact with multicast groups in a dynamically changing environment.

2 The eMinerals Virtual Organisation

The eMinerals consortium consists of collaborators from seven different academic institutions UK-wide. The project has three components: the science driving the project, the development of simulation codes, and setting up a grid infrastructure for this work. Each of these components are dealt with by the scientists, application developers and the grid experts respectively.

The eMinerals Virtual Organisation consists of individuals affiliated to different institutions. These individual bring with them their own ideas, expertise (in their respective fields), and resources which are bound by their local administrative rules. The eMinerals mini-grid provides the grid infrastructure necessary to manage these resources in order to support the Virtual Organisation (VO).

Collaborative tools are essential in facilitating cooperative work within a Virtual Organisation. These tools provide synchronous and asynchronous communication between geographically distributed participants. The Access Grid currently allows synchronous collaboration using video, audio and other services. Multicast Application Sharing Tool (MAST) complements the available Access Grid services, allowing sharing of arbitrary legacy applications.

3 The Access Grid

The Access Grid is an advanced collaborative environment, which is used for group-to-group collaboration. To achieve high quality collaboration with the Access Grid, a specialised suite can be constructed. A typical Access Grid suite consists of a machine for the audio, video and for the display. The video streams are displayed through several high quality projectors and specialist sound equipment to enhance the sound quality and to reduce the echo. Multicast is used for the transport of video and audio data to multiple hosts. Access Grid uses the concept of Virtual Venues (VV) to allow groups with similar interests to interact, an example of this is The University of Reading VV. Each of the VV's has associated multicast addresses and ports over which streams the video and the audio. The VV uses different ports and sometimes different multicast addresses to distinguish between the audio and video streams. This provides flexibility for the user to decide whether to receive audio, video or both. Unicasting and broadcasting provide extremes in addressing - unicast uses a single IP address and port as an endpoint, and broadcasting propagates to all IP addresses on a subnet. Multicast provides an intermediate solution, allowing a set of IP addresses to be identified and ensures that datagrams are only received by interested participants.

There are two major versions of the Access Grid Toolkit. The initial release AG 1.x used the Robust Audio Tool (RAT) [4] for audio and the Video Conferencing tool (VIC) [3] for video. Recently, the 2.x version of the Access Grid Toolkit was released. The new version still uses the stand-alone tools VIC and RAT for the audio and video conferencing, but also includes enhanced features, such as, a new Venue Client and Certificate Management functionality.

4 Application Sharing

Video and audio are essential for an effective collaborative experience, allowing participants to mimic natural human interaction. Another major component involves the sharing of material between participants. This includes activities, such as collaborative viewing of visual data, document editing by various colleagues and collaborative code development between geographically distributed peers. Application sharing is the ability to share and manipulate desktop applications between multiple participants, thus facilitating the aforementioned activities.

There are two approaches to implementing Application Sharing. The first, namely colloaboration-unaware, involves sharing of applications in a transparent manner to the application developers. This method does not require application developers to adapt their applications to be shared by distributed participants. The method generally works by sharing a visual representation of the legacy application, without sharing application specific data structures. The second approach, namely collaboration-aware, relies on the applications having an awareness of the collaborative functionality.

The decision upon which of the methods to use depends on the specific scenario. The collaboration-unaware model allows any legacy application to be shared without modification of the legacy applications source code. The collaboration-aware model requires the legacy application to be adapted to include the functionality necessary for collaboration. An advantage of the collaboration-aware method is that the amount of data that has to be transferred between the participants is significantly lower and so the response time is comparatively high.

In a situation where a single application was to be shared between many participants, (and the functionality of the application was known and could be modified) the collaboration-aware model would probably be preferred. The functionality is known and so the code could be modified to include the necessary collaborative features. This model has been used in another project involving the authors, namely The Collaborative P-GRADE Portal [6], where a workflow editor is shared between several participants. By providing collaborative functionality and ensuring synchronisation between participants, one can allow distributed participants to work collaboratively on a work-flow.

When designing a tool to share many different, legacy applications across different platforms, the collaboration-unaware model is usually preferred. It is not possible to adapt every legacy application to be used collaboratively, and so in this situation the Application Sharing tools tend to transfer the visual data between participants. The advantage of this is that any application could theoretically be shared, and the application only needs to be running on one of the participants machines. The main disadvantage, as mentioned earlier, is the relatively slow response times due to the amount of visual data being transferred.

4.1 Related Work

There are several applications sharing tools available, which are currently used within the Access Grid. The main examples are; IGPix [7], Distributed PowerPoint (DPPT) [8], and Virtual Network Computing (VNC) [5].

DPPT is one of the most widely-used applications for sharing slide presentations across the Access Grid. It provides a mechanism by which a presenter can control a PowerPoint slide show on multiple sites from a single machine. DPPT is specifically designed for slide shows and users are unable to share any other application. Each participant must also obtain a copy of the slides prior to a session. IGPix overcomes some of the problems associated with DPPT, specifically that, slides do not have to be circulated amongst the group members. IGPix is being used extensively with AG Toolkit 2.x for sharing PowerPoint slides. However, IGPix does not have a scalable model since it uses a client/server topology, with multiple participants connecting to a central point. Another problem is that only one node is able to present via IGPix at any given time.

Virtual Network Computing (VNC) was the most popular tool to be used in Access Grid 1.x in order to share applications during meeting. VNC shares the entire desktop with the participants in the group. However, it can be easily modified to share a specific application.

VNC has become the leading solution for desktop sharing. It is designed for point to point communication, consisting of a server (Xvnc) and a light-weight viewer (vncviewer). A participant wishing to share a particular application runs the application and allows the rest of the group to make individual TCP connections to his/her machine.

The main disadvantage associated with the aforementioned application sharing solutions (which makes them unsuitable for use within the scalable Access Grid) is their use of Unicast protocols for the transfer of the visual data. Sending multiple packets of the same data over the network is not scalable and can lead to performance degradation.

4.2 Multicast Application Sharing Software (MAST)

Application Sharing tools fall into one of the two categories, those that share collaboration-aware applications and those that share collaboration-unaware applications. As previously discussed, there are advantages and disadvantages associated with both approaches, which makes them best suited for different scenarios. In the case of MAST, the main objective is to be able to share a range of applications between groups of distributed participants. To achieve this main objective, MAST is designed to share legacy applications without any modification to the shared applications source code.

There are two versions of MAST, one for Microsoft Windows and the other for Linux, which can be used in conjunction with the Access Grid Toolkit to enhance the group-to-group collaborative experience. There are several factors that had to be considered when designing MAST:

- Allows Multicasting and Unicasting: To to be in-line with VIC and RAT, MAST allows the data to sent to the participants using multicast or unicast. If multicast is enabled on the network, then the application can send multicast packets. However if the multicast is not enabled, the application can be sent unicast to a multicast bridge.
- Simple Configuration: MAST has a settings menu which allows a participant to select whether the data is sent multicast or unicast. The user can also add details about themselves to be seen by other participants.
- Using a Single Multicast group: Each Virtual Venue within the Access Grid has a unique multicast address and port which is used for video (VIC) and audio (RAT) streams. A similar idea is used for MAST so that participants in a particular VV can share applications on the multicast address associated with the VV using a unique port. Since the multicast address and port combination is well-known within the VV, participants are able to collaborate with ease. The same multicast address and port combination must be used for all the shared applications within a VV. To enable this, MAST has to be able to deal with multiple streams of graphical data being transferred in a single multicast group. MAST uniquely identifies each of the shared application streams and lists these applications with the participants name.
- Reducing Screen wastage: Due to the lack of screen space when using the desktop Access Grid tools, such as PIG [9]/PIGLET [10], we felt that it was important to reduce the screen area required by MAST. The GUI of MAST resembles that of many Instant Messengers, with a simple list of participants, that can be expanded to show all the applications currently being shared by a participant. The user can enlarge a particular application to its normal size within the sharing window. There is only one sharing window, as it was felt that having multiple sharing windows would cause wastage of valuable screen space.

Image Capture. The main goal of MAST is to get the visual data associated with a single application and send this data to other participants within multicast group. An important issue involves obtaining the graphical data. The visual data for the whole application could be acquired and sent to other participants each time an event occurs or after a set interval. Sending visual data associated with the whole application would be inefficient. If only a small proportion of the screen is changing, it would make more sense to send only the changes. MAST achieves this by splitting the visual representation of the legacy application into sections. When an event occurs or the timer expires, each section is checked to see if there is a change. If the section has changed, it is sent to the other participants. The receiving participants identify the section that has changed and update their local view of the application.

Changes to the visual representation of an application can occur due to hardware or software events. If the user clicks a button on the application, the view of the application will change in response to the event. Similarly if an application initialises an event, such as, an animation, or a progress bar, the visual representation of the application will change. There are two possible methods for reacting to these events - the first is to check the visual representation after a default interval. This method works well for software events, by updating the screen to show changes not induced by external events, such as mouse or keyboard events. The interval between checking for updates is extremely important. The default interval value must attempt to provide good responsiveness, whilst ensuring relatively low overhead. To reduce wastage of processor time, the interval must be relatively high. This makes using the interval method unacceptable for changes due to external events. In a standard operating system users would expect the visual response to an event within several hundred milliseconds. If the interval is set to one second, then the shared application will appear "sluggish" to the user. The alternative to setting the interval is to check sections after an external event. It is sensible to assume that the visual representation of the shared application will change once it receives an external event. Using this method ensures that the shared application will remain responsive.

MAST combines the two methods, the interval can be set relatively high to avoid undue wastage of processor time, whilst still capturing changes that are not due to external events. Checking the segments after each external event (associated with the shared applications window) means that visual changes due to external events are processed quickly, improving the interactive performance.

Transport Design. The visual data must be transferred to each of the participants within a multicast group. Ideally, the data would be sent to each of the participants using multicast. As with the two main AG tools, VIC and RAT, the data can be sent via multicast, if it is enabled on the local network, or by unicast to a bridge. The transport system has been designed to allow the transport of any type of data and this has been achieved by creating template transport classes. The advantage of this is that other applications can easily be developed to use the same transport system.

To be used successfully within a group-to-group collaboration, MAST uses multicast to send the visual data. IP Multicast is an unreliable protocol meaning that packets of data could be lost or received out of order. If some packets are lost then the remote participants view of the shared application could be incomplete. MAST does not have any knowledge of packet loss and therefore it will assume that the section is updated and will not resend the section. To overcome this problem, and the problem of latecomers, MAST must resend sections periodically even if they appear to be unchanged since the previous send. Re-sending sections could be done all at once after a set number of intervals, at that moment it could take a relatively long time to obtain visual data for the entire application not to mention the increased load on the network. During this high overhead period, external events from the user could take longer to process and so the responsiveness of the shared application would be effected. MAST attempts to balance this load by re-sending a few sections after each interval, this reduces the overhead associated with refreshing the shared application, and maintains the responsiveness to the user.

Session Management. An important aspect of the transport system is the identification of the separate participants and their shared application streams. MAST does not use a central server, which means that the participant list cannot be managed centrally. Each participant must manage their own participant and application lists - this is achieved by uniquely identifying each participant and application stream. When a new application stream is received, MAST checks if the owner of the shared application is present in its own participant list. If the participant is present, then this application is added to the list. If the participant is not present, then the participant and the application name is added to the list. Each instance of MAST must be responsible for detecting participants that leave the Virtual Venue or applications that are no longer being shared. Detecting leaving participants is relatively simple - while a stream is being received a flag is set to indicate that a participants application is active. After a set interval, the flags are cleared - if any streams have a cleared flag at the next interval the application name or the participant will be removed from the list.

5 Conclusion

In this paper, we introduced the eMinerals Virtual Organisation and described the factors that motivated us to develop the Multicast Application Sharing Tool. We described the collaborative model used along with its advantages and disadvantages. The paper gives a detailed description of some of the design and implementation issues associated with the development of MAST.

Our work in the immediate future will be to complete integration of the two versions of MAST to provide a single tool that can be used between the two platforms, and the integration with the current AG Toolkit.

References

1. Dove, M.T., Calleja, M., Wakelin, J., Trachenko, K., Ferlat, G., Murray-Rust, P., H De Leeuw, N., Du, Z., Price, G.D., Wilson, P.B., Brodholt, J.P., Alfredsson, M., Marmier, A., Ptyer, R., Blanshard, L.J., Allan, R.J., Van Dam, K.K., Todorov, I.T., Smith, W., Alexandrov, V.N., Lewis, G.J., Thandavan, A., Hasan, S.M.: Environment from the molecular level: an escience testbed project. AHM 2003 (Nottingham 2-4/9/2003)

2. The Access Grid Project website. Available on: http://www.accessgrid.org.
3. Videoconferencing Tool (VIC) website. Available on: http://www-mice.cs.ucl.ac.uk/multimedia/software/vic/
4. Robust Audio Tool (RAT) website. Available on: http://www-mice.cs.ucl.ac.uk/multimedia/software/rat/
5. Richardson, T., Stafford-Fraser, Q., Kenneth, Wood, R., Hopper, A.: Virtual Network Computing. IEEE Internet Computing, Volume 2, Number 1 January/February 1998
6. Nemeth, C., Dozsa, G., Lovas, R., Kascuk, P.:The P-GRADE Grid Portal. Computational Science and Its Applications - ICCSA 2004 International Conference Assisi, Italy, LNCS 3044, pp. 10-19
7. The IGPix website. Available on: http://www.insors.com
8. The DPPT website. Available on: http://accessgrid.org/agdp/guide/dppt.html
9. Personal Interface to Access Grid (PIG) website. Available on: http://www.cascv.brown.edu/pig.html
10. Personal Interface To Access Grid, Linux Exclusively Thanks (PIGLET) website. Available on: http://www.ap-accessgrid.org/linux/piglet.html

The Collaborative P-GRADE Grid Portal

Gareth J. Lewis[1], Gergely Sipos[2], Florian Urmetzer[1], Vassil N. Alexandrov[1], and Peter Kacsuk[2]

[1] Advanced Computing and Emerging Technologies Centre, School of Systems Engineering, University of Reading, WhiteKnights, P.O. Box 225, Reading, RG6 6AY, United Kingdom
{g.j.lewis, f.urmetzer, v.n.alexandrov}@rdg.ac.uk
http://www.acet.rdg.ac.uk

[2] MTA SZTAKI, Laboratory of Parallel and Distributed Systems, H-1518 Budapest, Hungary
{sipos, kacsuk}@sztaki.hu
http://www.lpds.sztaki.hu

Abstract. Grid portals are increasingly used to provide uniform access to the grid infrastructure. This paper describes how the P-GRADE Grid Portal could be used in a collaborative manner to facilitate group work and support the notion of Virtual Organisations. We describe the development issues involved in the construction of a collaborative portal, including ensuring a consistent view between participants of a collaborative workflow and management of proxy credentials to allow separate nodes of the workflow to be submitted to different grids.

1 Introduction

The Grid Infrastructure is essential in supporting the development of Virtual Organisations. The Grid enables the sharing of resources dynamically and in a secure manner. Grid Portals are increasingly used as a convenient interface to the Grid by providing uniform access to grid resources.

P-GRADE [1] is a graphical programming environment, used in the development of parallel applications. P-GRADE incorporates GRAPNEL (GRAphical Process Net Language), which is a graphical parallel programming language and GRED (Graphical Editor), which can be used to write parallel applications. Along with several other components, P-GRADE provides an abstraction from the low level details associated with the message passing. P-Grade has several layers which are used in the development and execution of parallel programs. The layer of particular interest, for the purposes of this paper, is the Workflow layer. A Workflow is a set of consecutive and parallel jobs, which are cooperating in the execution of a parallel program [2]. Different nodes of a certain workflow can be executed on different resources within a grid. The Workflow layer allows users to design and execute workflows, specifying properties of individual nodes, and specifying whether the output data of one node is used as the input of another node in the workflow.

V.S. Sunderam et al. (Eds.): ICCS 2005, LNCS 3516, pp. 367–374, 2005.

The P-GRADE Grid Portal [2] provides uniform access to underlying grid resources. It has two main components, the Portal Server, and the Workflow Editor (A Java Webstart application). The Portal Server was developed using Gridsphere, a grid portal development framework. It is responsible for managing several aspects, such as, security, monitoring and execution visualisation. The Workflow editor allows a user to design a workflow and provides graphical feedback on its progress during the execution.

The objective of the project outlined in this paper, was the development of a collaborative P-GRADE Portal, which could be used by several participants in "real-time". The motivation for this collaborative ability was two-fold; firstly, it would allow different users to collaboratively construct workflows and secondly, participants could use their personal certificates to submit nodes of the workflow to different grids.

2 The P-GRADE Portal

As previously mentioned, the P-GRADE Portal consists of two major components, the Portal Server and the Workflow editor. Together these components facilitate the development of workflows and their execution within a grid. The Workflow Editor allows the end-user to construct workflow graphs whilst, the Portal Server is responsible for managing the security as well as the visualisation and monitoring of the execution. The Workflow Editor is separate from the Portal and runs on the end users local machine. It communicates with the Portal Server to provide information about the current workflow. The user of the workflow editor can create a workflow graph consisting multiple jobs, which are either sequential, PVM or MPI programs.

The Grid Security Infrastructure (GSI) [4] (based on X.509 certificates) is used to provide secure authentication and access to the grid resources. Portals are becoming popular for providing a convenient interface to these computational grids. The grid portals rely on the MyProxy [5] repository to enable the integration of the portal with the GSI. The MyProxy server allows the grid portal to use the GSI to interact with the grid resources. The P-GRADE portal is responsible for managing the downloading of proxy credentials from the MyProxy Server, and allowing users to view the lifetime of the proxies. Several different proxies can be downloaded but only a single proxy can be active, as each workflow utilises the resources from a single grid.

The Portal must be able to monitor the execution of the workflow and provide visual feedback to the end-user. The P-GRADE portal allows the user to monitor the workflow in real-time, using the Mercury monitoring tool [2]. It represents the progress of the execution, both in the workflow editor and online within the portal. The user can monitor the entire workflow and the individual jobs. The portal shows the communication between the workflow jobs and the processes within each job.

3 Collaborative Version of the P-GRADE Portal

A driving force in the development of the Grid Infrastructure is the collaboration of distributed partners (forming Virtual Organisations) interacting over the grid. The collaborative P-GRADE portal supports the idea of distributed groups working toward a common goal using the underlying grid infrastructure. The main motivations for the development of a collaborative P-Grade Portal, were to allow workflows to be constructed collaboratively, and facilitate submission of different parts of the workflow to several grids. There are two approaches to implementing a collaborative application such as the workflow editor. The first involves sharing of the application in a transparent manner, and does not require it to be adapted to be aware of the collaboration. Collaborative-unaware sharing would be achieved by sharing a visual representation of the same workflow editor, without sharing any application specific data structures. The second approach depends upon the application being aware of its collaborative functionality. In the case of the P-GRADE Portal it is sensible to adapt the workflow editor and the portal server to work in a collaborative manner. The portal server provides a central point to which each workflow editor can connect and be synchronised. The server also provides the solution to transferring data between different grids (as described in the following sections).

The collaborative version of the workflow editor must allow workflows to be constructed in a collaborative manner. The end-users must be able to add and remove nodes of the workflow with these changes being visible to each participant. For successful collaboration, between the users cooperating in the construction of a collaborative workflow, the portal must ensure that each participant has the same view. Simultaneous editing of the workflow could lead to inconsistencies, where participants have a different view of the same data. The client/server network topology assists the implementation of floor control. A single copy of the workflow data can be contained at the portal server and its consistency assured by guaranteeing that access to the central workflow data is mutually exclusive. The Workflow editor must also provide a visual representation of the workflows execution - currently this functionality is provided by representing different states of execution in different colours. The execution functionality of the Workflow editor must evolve to work within a collaborative environment.

The Portal Server must also be adapted to include the necessary collaborative functionality. An important issue in the development of the collaborative P-GRADE portal, is management of the proxy credentials, which are downloaded from the MyProxy server. The current P-GRADE Portal allows different proxy credentials to be downloaded, but as a workflow can only be submitted to a single grid, only one of the proxies is active. A major aim of the shared P-GRADE portal was to allow separate parts of collaborative workflows to be submitted to different grids. To facilitate this objective The Credential Manager [2], must be adapted to be able to use multiple proxy credentials concurrently. The visualisation and monitoring of an executing workflow is an important consideration within the collaborative setting. The portal server must ensure that all participants of a collaborative workflow receive the same visual information, and that the central workflow object is not modified by any workflow editor during the execution period.

3.1 Collaborative P-GRADE Portal Design

This section expands upon the issues described above and gives details on specific design decisions. There are three main stages in the development and execution of parallel applications using the P-GRADE portal. The initial step is the construction of the Workflow graph, this involves constructing how the consecutive and parallel jobs cooperate, and where output from one job can be used as input to another. Before the workflow can be submitted to the grid the portal server must ensure that the necessary proxy credentials are available. This step is managed by the Credential Manager, which downloads the proxy credentials from the MyProxy server. The final stage is the execution of the workflow including its monitoring and visualisation. The current P-GRADE portal will have to be adapted at each of these stages to be utilised successfully within a collaborative setting.

3.2 Collaborative Workflow Construction

The construction of the workflow graph is achieved in the workflow editor which, runs on the users local machine. Much of the development associated with the construction of the workflow graph will be within the workflow editor. However, the portal server is important in this step as it contains the stored workflows and will be responsible for ensuring a consistent view of the workflow between the distributed participants. In the current version of the P-GRADE portal, the workflow editor runs on the users local machine and connects to the central P-GRADE portal server to exchange information related to the workflow graph. This graph is then used by the server to submit the workflow to the grid infrastructure. The workflow editors have no reason to send information between themselves as a workflow is "owned" by a single user. In the collaborative version, the editors working on the same workflow will have to communicate with each other to ensure the workflows consistency. Having the workflow editors connect directly to each other would not be sensible and would lead to problems seen in many peer-to-peer applications. The central portal server allows the editors to indirectly communicate with each other. By containing a central workflow object at the server, and ensuring mutual exclusive access to all or part of this object, we can achieve a consistent view between the distributed editors.

The collaborative P-GRADE portal will be implemented with several workflow objects. The workflow editor will contain workflow objects called Local Workflows Objects (LWO), these will contain the data present within the stored workflows and additional information concerning the locking and un-saved changes. The portal server will have two types of workflow objects, the first will be known as the Dynamic Global Workflow Object (DGWO) and will contain the same information as local workflow objects. The DGWO objects will be updated by each of the workflow editors when they perform a local-to-global update. The LWO for each of the workflow editors will be updated by a global-to-local update. The second workflow object at the portal server will be the Static Global Workflow Object (SGWO). This object will contain the data stored in the saved collaborative workflow and the locking information, the SGWO will have no data associated with components that are not currently saved (Fig. 1).

The major interfaces for the project lay between these different workflow objects. The first interface is between the SGWO and the DGWO - this interface represents the point between the collaborative workflow object, where the changes to local workflows are stored, and the workflow that mirrors the data contained on disk. There are several functions that must be present in this interface to enable interactions between these two objects. The SGWO contains the workflow data from disk and the locking information for the entire collaborative workflow. The interface must allow the stored collaborative workflows to be retrieved from the portal server, the locking information to be synchronised between the two global workflow objects and the unsaved changes from the DGWO to be incorporated into the SGWO during the saving process. The SGWO can be used specifically by the portal server for submitting the workflow to the grid resources. Another advantage of the SGWO is that updates to the object are more efficient than saving changes to disk when saving or when updating locking information.

The second major interface lays between the portal server and the workflow editors. The workflow editors must be able to communicate with the portal server to maintain the consistency of the collaborative workflow between the participants. There is some additional functionality that must be available to ensure consistency, including: local-to-global updates, global-to-local updates, locking nodes and unlocking nodes. The workflow editor communicates with the portal server via HTTP, and so cannot receive messages asynchronously from the server. The workflow editor must be responsible for polling the server to receive the latest global view. When the workflow editor polls the server it will be involved in the local-to-global synchronisation. The server can respond to this request by issuing a response in the form of a global-to-local update. Obviously the polling interval is crucial, the interval should be small to ensure a more consistent workflow between the participants, however, if the interval is too small, then it could lead to an undue load on the network.

The locking of the workflow is also an important consideration at this interface. The DGWO contains data that is shared between several different editors concurrently. As with all shared data, there is a distinct possibility of race conditions due to concurrent access to the data. To overcome this problem the portal server must ensure that access to the DGWO is mutually exclusive. The portal must provide a locking mechanism by which participants are able to lock a part of the workflow to be modified. In the interface there must be two functions available to the workflow editor. The first is the lock function, which will allow participants to lock a node and associated nodes. If participants had to acquire a lock for the entire workflow, it would place a large constraint on effective collaboration. The alternative is that the participants are able to lock

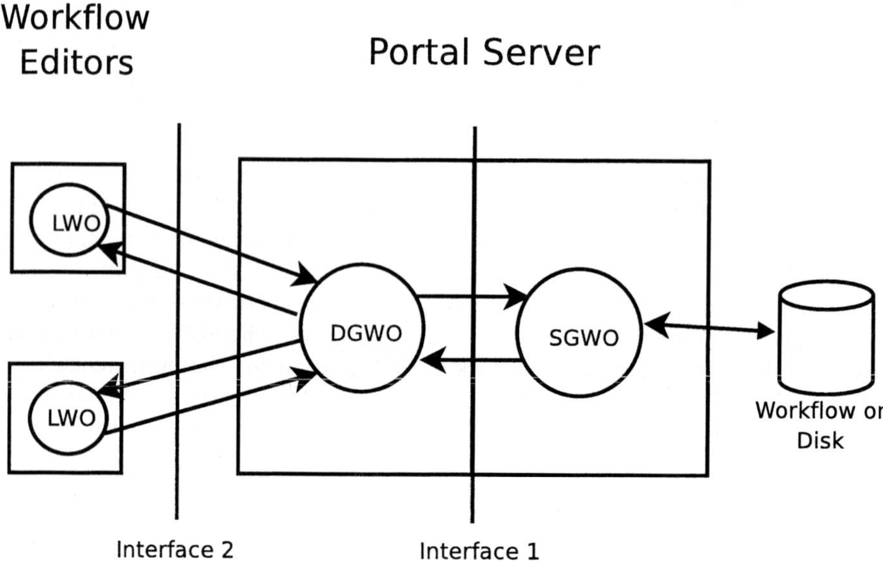

Fig. 1. Overview of the Collaborative P-GRADE Portal showing workflow objects

single jobs and associated jobs of a workflow. The portal server must manage the locking mechanism to make sure that multiple editors cannot lock the same nodes. The management of the locking will take place at the SGWO level. The unlock function must also be available to the workflow editors. Once a participant has completed the modification to part of the workflow, these nodes should be released. There must also be a set time interval for how long a participant can hold a lock before it is forcibly removed by the system. A single participant must not be able to monopolise a particular part of the workflow and any failing workflow editors should not be able to hold locks indefinitely.

3.3 Execution and Management of a Collaborative Workflow

Once the collaborative workflow has been constructed it is ready to be executed on a grid or over several grids. There are several issues that have to be considered when submitting the collaborative workflow. The aim is to have parts of the collaborative workflow submitted to different grids. To achieve this, the Portal must be responsible for effectively managing the proxy credentials, ensuring locking of the workflow during the execution period and facilitating the transfer of intermediate data between the different grids.

Credential Management. In the current version of the P-GRADE portal, the Credential Manager is responsible for managing the proxy credentials. This involves allowing users to download an appropriate proxy certificate from the MyProxy server, and providing information such as its lifetime. The current version allows multiple proxies to be downloaded, but allows only one of these to be actively used for submission of the workflow. A non-collaborative workflow is submitted to a single grid and so it is sensible to be able to use only one proxy. An aim of the collaborative portal is to allow parts of a single workflow to be submitted on different grids using different participants proxy credentials. To achieve this aim, the Credential Manager must be adapted to be capable of managing several active proxies for a single workflow.

To manage several active proxies, the portal server must be capable of associating active proxies with different nodes of the workflow graph. The user must be able to specify within the editor which proxy will be used for which node. The current version of the portal, is not capable of automatically downloading proxies - this would be necessary in the collaborative version. The portal must keep track of the relation between the users, the MyProxy server and the available grid resources.

Workflow Locking. Once a collaborative workflow has been submitted by one of the collaborative users, the portal must ensure that all editors, are locked and set to an execution state. This feature is currently available within the portal, but consideration must be given to ensuring that all workflow editors are within this locked state before submission to the grid can proceed.

Transfer of Intermediate Data. Different nodes of the workflow could be submitted to different grids within the proposed collaborative P-GRADE portal. The major issue with this design is that there is generally no facility available for the data to be transfered between the different grids. Thus in the workflow, outputs from some jobs will be needed as inputs to other jobs on different grids. The central portal server can facilitate the indirect transfer of the output data (from one node) for use as input data to another node (submitted on a different grid). The portal server must be able to handle the transfer of this data identifying from which data the node has been received and to which node the data must be delivered. The management of the Proxy Credentials is essential in facilitating the indirect transfer of the intermediate data. Once execution completes, the output data from the collaborative workflow must accessible to each participant of the collaborative session.

4 Conclusion

In this paper, we have discussed the issues and motivations involved in the development of a collaborative P-GRADE grid portal. The goal of this collaborative portal was to allow distributed participants to work on "collaborative workflows", run workflows, and monitor the progress of the nodes' execution across different grids. The paper describes how the workflow editor and portal server must be adapted to be used in a collaborative manner, ensuring a consistent view of the workflow by providing synchronisation and locking mechanisms. Management of the proxy credentials by the portal server has been discussed to allow the automatic download of proxies by monitoring the relationship between users, and available grid resources.

References

1. Peter Kacsuk: Visual Parallel Programming on SGI Machines. The SGI Users' Conference, Krakow, Poland, pp. 37-56, 2000
2. Csaba Nemeth, Gabor Dozsa, Robert Lovas and Peter Kascuk: The P-GRADE Grid Portal. Computational Science and Its Applications - ICCSA 2004: International Conference Assisi, Italy, LNCS 3044, pp. 10-19
3. Joshy Joseph, and Craig Fellenstein: Grid Computing. On-Demand Series, Prentice Hall, 2004
4. R. Butler, D. Engert, I. Foster, C. Kesselman, S.Tuecke, J.Volmer and V. Welch: Design and Deployment of a National-Scale Authentication Infrastructure. IEEE Computer, 33(12):60-66, 2000
5. J. Novotny, S. Tuecke, and V. Welch: An Online Credential Repository for the Grid: MyProxy. In Symposium on High Performance Distributed Computing, San Francisco, Aug. 2001.
6. J. Chris Lauwers and Keith A. Lantz: Collaboration awareness in support of collaboration transparency: Requirements for the next generation of shared window systems. In Proceedings of ACM CHI90 Conference on Human Factors in Computing Systems, pages 303-311, 1990.
7. G J. Lewis, S. Mehmood Hasan, Vassil N. Alexandrov: Building Collaborative Environments for Advanced Computing. In the proceedings of the 17th International Conference on Parallel and Distributed Systems (ISCA), pp. 497-502, San Francisco, 2004.

An Approach for Collaboration and Annotation in Video Post-production

Karsten Morisse[1] and Thomas Sempf[2]

[1] University of Applied Sciences Osnabrück, D-49076 Osnabrück, Germany
kamo@fhos.de
[2] Nero AG, D-76307 Karlsbad, Germany
thomas.sempf@gmx.net

Abstract. In the different phases of a movie production, there is a need for intensive discussion about the produced sequences. Especially if the discussion members are located at different places they will have problems to precisely describe their critics for the video production. This paper describes an approach and a tool for video collaboration to simplify the video production process for video editing service providers and their customers.

1 Introduction

The process of making a movie, be it just a commercial or a full-time film, is a long and difficult task with many different stages, persons and companies involved. During each step in this process, several different persons will collaborate and information, media assets, ideas, and management decisions for the upcoming phases of the production process will be made and have to be stored for discussion either in electronic form, like emails or word-processing documents, or in paper form, like printed documents or pictures. These created documents will be used in subsequent phases of the production process and will be duplicated and delivered to other companies and persons in charge of the tasks to be done later on.

It is usual that even short productions, like a 60 second commercial, are international productions with people involved from locations scattered all over the world. This might cause problems in the production process.

The organization of this article is as follows. In section 2 we describe common problems in the video post-production process. Functional requirements for a collaboration process in video production are characterized in section 3. Related work is considered in section 4 and, finally in section 5 we present a collaboration tool for some of the mentioned tasks.

2 Problems in Video Production Process

After finalization, but also during a video production, there is always a demand for discussion about the processed scenes. However, the necessary information

V.S. Sunderam et al. (Eds.): ICCS 2005, LNCS 3516, pp. 375–382, 2005.

exchange between the persons involved is difficult if the service provider (the sfx expert) and the customer (director or producer) are at different locations.

In general, the verbal discussion supported by sketches about scene cut, picture elements, movements or texturing is often difficult, because verbal description can be inaccurate or misleading. For example, the description of a position of an object or an actor is difficult. Verbal constructs like "right of the chair", "he moves fast through the scene" are ambiguous. What exactly means "right of the chair"? How far to the right has it to be positioned? If the persons involved have a common monitor or screen at which they are looking, the verbal description can be supported by hand or by a pointer. But how to do this without having a common screen to look at?

Problems in coloring and color design are special for the area of animation and comic productions. If the blue color of an object does not satisfy the demands, what is the right "blue" and how "blue" shall it be? Particularly colors are hard to describe verbally. How to do that if there is no common view on a screen?

A discussion using collaboration media like telephone, audio or video conference is not the solution, because even with these media there is no common view on a single screen and thus, verbal descriptions will be ambiguous. The discussion about a single scene requires each participant to know the exact position of that scene (e.g. time code) and to navigate to that position synchronously.

Particularly the production of computer generated scenes is often an iterative process. After the screening of a single scene it would be helpful to continue the scene in a sense of digital story-board development. For distributed participants this could be done verbally and each has to sketch the ideas individually. Of course the same ambiguities as described above can be foreseen. In general, the necessary duplication and delivering of these documents to all involved persons is a troublesome task. Even in times of electronic groupware and document management systems this can take a lot of time and is always a reason for delay and misunderstanding because papers are missing and most important, the given information and instructions are inaccurate.

One solution to overcome all the mentioned problems is to travel and come together to have the necessary common look on the screen. However, to simplify the video production process it would be helpful to have an online collaboration tool which supports at least the remote discussion about video content with a common look on a video screen for all the distributed participants.

3 Requirements for Collaboration in Video Post-production

What are the requirements for a tool supporting the video production process?

It would be helpful for spatially distributed participants to have a common look on the screen with the possibility of a synchronized navigation through the video sequence. This common screen must be the base for the collaboration process and a tool supporting collaboration for video processing should provide all or at least some of the following features:

Synchronized Navigation and a Common Video Screen: For efficient collaboration on video processing a common screen with the possibility to navigate in a synchronized way through a scene is a must. It has to be possible for each participant to navigate to a specific time code without the compulsion for the others to navigate manually to the same scene. There must be one navigation process for all participants.

Graphic and Text Annotation: It must be possible to annotate single frames and complete sequences. An annotation should be defined as a text or a graphic object directly on the video layer, e.g. a sketch like a circle to emphasize a particular effect in the picture. Moreover, it should be possible to give separate comments to a single annotation as a text or an audio sequence.

Multi User Approach: It is desirable that more than two participants can join a collaboration session in a way, that each of them can follow the discussion process actively and passively.

Archiving of Annotations: Given annotations must be storable, i.e. it must be possible to recall an review them later on during the production process.

Categorized Annotation: It would be helpful if annotations could be grouped in categories, e.g. annotations for texturing or annotations for actor movements. This is necessary for a structured discussion process.

High-Quality Video Material: The common video screen is the base for a collaboration tool. It would not be sufficient, if the provided video quality is the typical thumb nail quality known from video streaming in the internet. To support discussions about production details high quality video material is absolute necessary.

Story Board Development: It would be very useful and convenient if a system could assist the document development process, e.g. treatment or story board, by collecting the information, making it available to all involved persons and companies, and connecting all persons, information, and media assets online for collaboration. Such a distributed video production system can offer many new and helpful possibilities for streamlining and simplifying discussions and organizations tasks in the video production process.

Video Content Processing: Support of typical video processing tasks would also be helpful too, e.g. automatic scene detection and/or connection to NLE systems with import facility of edit decision lists (EDL). Another interesting feature would be the synchronized simulation of special video effects or video filters.

4 Related Work

A search for systems with the mentioned abilities revealed that there are some systems that enable the user to do some of the mentioned tasks, but there is no system that supports the case of distributed users. All the systems are for single user usage and not intended for collaboration with several persons involved. Systems for single user experience in the storyboard design are for example Storyboard Artist [7], FrameForge 3D Studio [4], Storyboard Lite [8].

Video annotation systems are not known to assist the video production process. However, there are some systems for video description. These systems supports the task to describe the actual plot of a video. Most of these systems keep ready a set of tools for navigating through the video sequence and adding metadata to it. This metadata is then stored on a server for later retrieval. The evolving standard for this metadata description is MPEG-7 [6], which is mainly used to search in video databases for specific topics. An example for such a software is the IBM MPEG-7 Annotation Tool [5]. It enables the user to annotate MPEG-1 and MPEG-2 based video sequences. The user is supported with an automatic shot detection, which detects the different scenes in the video file. Based on that information the software tool creates a navigation structure, so the user can easily navigate through the different scenes. To define an annotation the user can choose from a set of keywords, which can be extended, to describe the scene. Again no distributed collaboration with several users is possible, thus the participants have to meet personally to discus the video in front of a video screen.

Another interesting project is the filmEd [1] research project of the University of Queensland, Australia. The project was started to create tools for annotating, discussing, and resuming videos and to develop multimedia rich learning environments. They also analyzed and discovered the need for software tools to discuss and annotate video sequences in real-time over the Internet in a distributed manner. They developed a prototype called Vannotea, which enables the collaborative indexing, annotation and discussion of audiovisual content over high bandwidth networks. It enables geographically distributed groups connected across broadband networks to perform real time collaborative discussion and annotation of high quality digital film/video and images. Based on the GrangeNet, which connects four Australian cities with a 10 Gigabit line, it enables the exchange of high quality video in MPEG-2 format between the participants of a meeting. A streaming server supports the participants with the needed video data and a special annotation server stores and hosts all annotation information. To reduce the latency during a session, more then one streaming server is used and the video is transmitted as a multicast stream. The annotation information is stored in XML format, which is based on a combination of elements of the Dublin Core [3] and the MPEG-7 standard. At the point of the writing only text information can be stored in the system which can be bound to rectangular areas in the video. The navigation system is master/client based, that means that only one person can navigate through the video sequence and all clients will be automatically synchronized.

In [9] a hint could be found to Quicktime Synchro, a tool from Hybride which was used during production of the movie Spy Kids II and which might fulfill some of the requirements given in section 3. But even on request no further information was provided about it.

5 DiVA – An Approach for Video Collaboration and Annotation in Video Post-production

DiVA[1] [2], developed at the University of Applied Sciences Osnabrück, is a collaboration tool which supports some of the requirements mentioned in section 3. Annotations can be given as a text or a graphic object directly on the video layer. Based on a plugin architectural approach it could be extended to satisfy different demands. Figure 1 shows an annotated video sequence with textual and graphical annotation divided in several annotation categories.

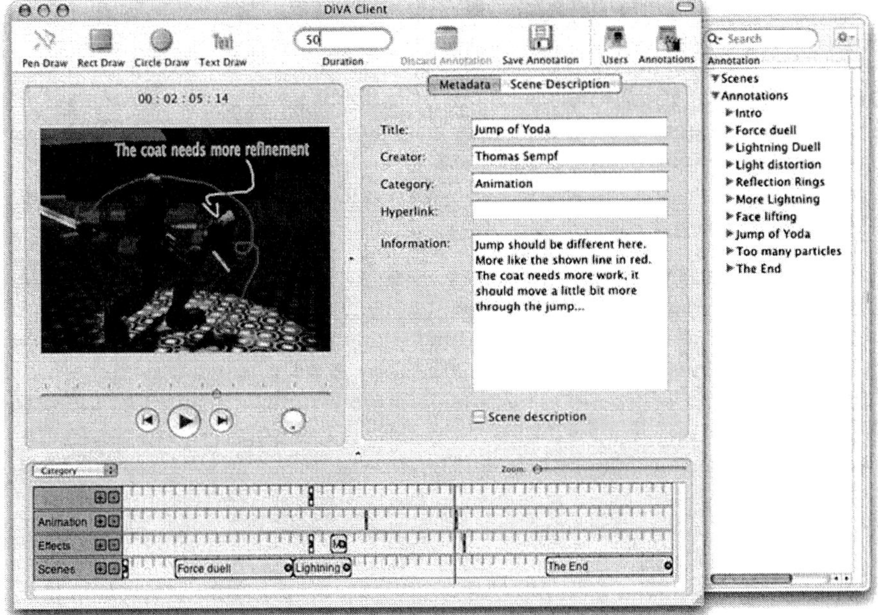

Fig. 1. DiVA - Scene annotation with graphic and text objects, synchronized navigation and annotation archiving

5.1 Multi User Approach

DiVA is a client-server based system approach for the collaboration process in video production. Annotations are defined on the base of sessions. Sessions run under the control of a server, which allows several users to join a session concurrently. The video content of a collaborative session is defined with a URL and can be loaded automatically from the clients. If necessary, the video content can be loaded from the local file system of the clients[2]. Moreover, streaming

[1] DiVA - **Di**stributed **V**ideo **A**nnotation.

[2] This might be an option for long video sequences with an huge amount of data where the video content is distributed in advance via DVD.

content is also possible but for practical reasons this will not be used due to the minor video quality. This could change in the near future with the emerging of high efficiency video codecs like MPEG-4 part 10/H.264 or VC-1, which enables the streaming of video content, even in high definition quality, over a small bandwidth.

5.2 Navigation and Common Video Screen

Video sequences can be navigated in several ways. Every position can be accessed directly by a timeline. Forward and backward movement from frame to frame is possible. For fast navigation a jog shuttle is available. Each of the navigation approaches is done concurrently for all participants of the collaboration session. I.e. if one user navigates his DiVA-client to a particular position, each of the other clients of that session follows the navigation commands. A blocking mechanism to prevent from navigation deadlocks is provided. Only one participant can navigate actively, the others are following this navigation in a passive mode and are blocked for active navigation.

5.3 Annotations

DiVA provides a set of possibilities to annotate a video sequence and single video frames. Different tools like circles, rectangles and a drawing pencil can be used to annotate the video content. Annotations from different users are drawn in different colors and they can be defined for exactly one single frame or an arbitrary sequence of frames. Therefore it is possible to define annotations for object movements over a set of video frames. For detailed annotations within that sequence, further annotations can be defined.

If one user has started an annotation, the navigation functionality is blocked for the other users. Annotations are directly drawn on the video layer and are immediately visible for all others users of that session. In combination with a discussion via telephone or AV-conferencing this is an effective visual support of the discussion process about the video content. If the user has finished his annotation, the other users can annotate the same or other scenes of the video sequence in the same way. Thus, all users have equal rights, there is no privileged user. Therefore the discussion process is supported in a visual way, combining the video content with the given annotations to highlight errors, misunderstandings and optimization possibilities of the content.

For a comfortable access there are different views for the annotations. A list view lists all annotations in a alphabetic order. Annotations can be reviewed or deleted. It is also possible to search annotations by keywords. A timeline based view gives an chronological view of the annotations sorted by categories. In the timeline, the duration of annotations can be changed or they can be moved.

Graphic and Text Annotation. Annotations can be defined as elementary graphic objects like circles, rectangles and lines or as text objects directly in the video layer. Different users are annotating in different colors, thus comments can be separated easily. The technical realization of the annotation tools is done via a plugin architecture so that the system is open for further tools.

Categorized Annotation. DiVA supports different categories of annotations. Examples are general scene descriptions, color and shading errors or wrong object positions. Each category will be displayed in an individual timeline, thus there is an easy way to navigate from one annotation to others of the same category. This is important in reviewing recorded annotation sessions.

Archiving of Annotations. After the discussion about a single scene the annotation can be stored on a server. A set of metadata can be defined for each annotation. This could be additional remarks or even a URL as a hyperlink. The stored annotations are bundled in a session archive on the server so that they can be accessed later on during the production process.

The categorization of annotations with a corresponding keyword based search mechanism allows a direct access to relevant annotations. This might be helpful in the following production steps or even in other video productions.

5.4 Network Requirements and Video Quality

If the video content is available at client side, the network requirements are rather low since there is only a flow of control messages over the network between clients and server.

DiVA is developed under Apple OS X and fully supports the Quicktime platform so that each Quicktime codec can be used for the video layer. Thus low-bandwidth video codecs, like Sorensen or MPEG-4 are supported as well as HD production codecs, like DVCPro-HD.

6 Summary and Further Work

DiVA, a collaboration platform for video production, supports different steps for a distributed video production process. Particularly it provides a common view on a video screen for spatially distributed users with the possibility for each participant to navigate and annotate the video content and to archive the given annotations on a server. Right now, there is no support in DiVA for story-board development or advanced video content processing. Also missing is a direct integration of video or audio conferencing facilities. This and an automatic scene detection mechanism is under implementation or will be implemented in the future . The distributed approach for navigation and annotation of video content makes DiVA unique.

A desired feature would be an object-oriented approach for video based annotations, i.e. coloring or texturing an object in the video layer with a single mouse click. However, that requires an integration of the collaboration platform in specialized video compositing systems or, an object-oriented approach for video compression[3].

Of course, DiVA will not supersede screening during the production process with all relevant members of the production team at a single location. However, it might be an useful collaboration tool for the intermediate discussion process.

Another application area for video annotation systems could be the usage as a distributed video teaching system. Here the teacher could easily visualize the activities in a scene, mark important characteristics, and explain everything at the same time to the participants of the session. This would improve the understanding of the scene and the plot dramatically, additionally all explanations could be archived, so that a later review would be possible for the students. Sequentially a whole archive of video scenes could be created. To make this archive more useful, those annotations could be extended with recorded audio and video comments and explanations. As mentioned before, students could use this archive for studying but also for meeting again to discuss some scenes, which they did not understand or which they want to explain to other students. These possibilities would create a new learning environment and could be especially helpful for universities teaching video production.

References

1. R. Schroeter, J. Hunter, D. Kosovic: FilmEd - Collaborative Video Indexing, Annotation and Discussion Tools Over Broadband Networks. International Conference on Multi-Media Modeling, Brisbane, Australia, January 2004.
2. Sempf, T., Morisse, K.: Video Annotation in der Postproduktion. Digital Production. **1** (2005) 103-105.
3. Dublin Core Metadata Initiative: http://www.dublincore.org/ (Feb 19, 2005)
4. FrameForge 3D Studio: http://www.frameforge3d.com/ (Feb 19, 2005)
5. IBM MPEG-7 Annotation Tool: http://www.alphaworks.ibm.com/tech/videoannex (Feb 19, 2005)
6. MPEG: http://www.chiariglione.org/mpeg/ (Feb 19, 2005)
7. Storyboard Artist: http://www.storyboardartist.com/artist.html (Feb 19, 2005)
8. Storyboard Lite: http://www.zebradevelopment.com/ (Feb 19, 2005)
9. Tetiker, T.: Spy Kids 2 Digital Production. **2** (2003) 26-34.

[3] MPEG-4 provides this kind of object-based video compression by using the Binary Format for Scene Description (BIFS). However, several video object planes foreseen in the MPEG-4 standard are usually not used today. In the future an integration of MPEG-4 object based coding could be integrate, so that not only whole frames in a scene can be annotated, but rather the different video sprites in the movie. This would open the possibility to exchange video objects in the scene to demonstrate how a plot could look like if for example a video sprite or the background is exchanged. All those changes could be stored with a reference to the used sprites and later again reconstructed.

A Toolbox Supporting Collaboration in Networked Virtual Environments

Christoph Anthes and Jens Volkert

GUP, Institute of Graphics and Parallel Programming,
Johannes Kepler University, Altenbergerstrasse 69, A-4040 Linz, Austria
canthes@gup.uni-linz.ac.at

Abstract. A growing interest in Collaborative Virtual Environments (CVEs) can be observed over the last few years. Geographically dislocated users share a common virtual space as if they were at the same physical location. Although Virtual Reality (VR) is heading more and more in the direction of creating lifelike environments and stimulating all of the users senses the technology does not yet allow communication and interaction as it is in the real world. A more abstract representation is sufficient in most CVEs. This paper provides an overview on tools which can be used to enhance communication and interaction in CVEs by visualising behaviour. Not only is a set of tools presented and classified, an implementation approach on how to use these tools in a structured way in form of a framework is also given.

1 Introduction

CVEs have emerged in various forms in the recent years. Interaction and communication in these environments are realised in many different ways. The users can send text messages, use audio and video communication; they can change attributes of the simulation, can share data and might even be able to collaboratively manipulate this data. Our research concentrates on environments using desktop VR or immersive VR systems such as Head Mounted Displays (HMDs) or CAVEs [4]. In these environments the users can manipulate objects with natural interaction techniques and experience the feeling of co-presence.

Performing collaborative tasks in Networked Virtual Environments (NVEs) is more difficult than performing them in a real world environment through the limitations of the interfaces. Different factors such as latencies, abstract graphical representation, limited field of view (FOV) and precision of input devices make natural interaction a challenging task. VR technology on the other hand provides many possibilities to enhance or even to completely change the users perception of the Virtual Environment (VE). Collaboration to the level that two users can manipulate the same object simultaneously [3] combined with natural interaction offers new challenges in the area of computer supported cooperative work (CSCW).

Although some systems support this collaboration on a technical level, i.e. they provide networking, allow for concurrent object manipulation or provide

V.S. Sunderam et al. (Eds.): ICCS 2005, LNCS 3516, pp. 383–390, 2005.

audio communication. However no tools exist so far which allow a rich enhancement of the VE to support collaboration on a more abstract level by visualising additional information about the users and the interaction. Worlds can be shared and manipulated, but an abstract toolset visualising collaboration is not available yet. This paper presents a set of tools which can be used to support collaborative work in immersive NVEs. The tools are described in their functionality as well as their benefit for collaboration. All of these tools augment the environment in a way which is only available through VR technology. Communication and interaction could be improved through the use of the these tools. This approach tries not to overcome technical issues such as lag or limited FOV, it rather enhances the VE with additional information which is not available in a real world environment. The toolbox is integrated into a CVE framework, which provides support for collaboration on a technical level.

This section has given an introduction on the topic of NVEs and collaboration. The following section shows an insight into the area of related work. Section three presents different tools which have been identified as supportive for collaborative tasks. The fourth section will give an overview on the architecture of the collaborative toolbox. The last section concludes the paper and shows some future work in the area.

2 Related Work

Numerous research has been done in the area of NVEs and CVEs. Good overviews on these topics are given in [8], [9] and [11] on a technical level. They describe how VEs have to be designed to be scalable and interactive. Psychological aspects of collaborative tasks in NVEs have been examined [16]. Users perform placing tasks to build up cube, while their behaviour is analysed. Aspects of collaboration have been researched in a medical visualisation in [6]. Audio and video communication are used in this scenario with several different projection VR systems. Advantages and disadvantages of different VE systems and different communication methods are analysed.

Cooperative object manipulation has been researched by many authors. Good examples are given in [14] where different attributes of single objects can be changed simultaneously. In other scenarios such as the piano carriers [15] the users are able to manipulate the same object simultaneously in a way that they manipulate the same attribute concurrently. In these examples the position of the manipulated object results from a transformation based on both user inputs. This is concurrent object manipulation in a three dimensional environment is described in [10] as the highest level of collaboration. All these scenarios incorporate complex communication and interaction, but none of them provides abstract tools in form visualisation of collaborative activities.

Many NVEs exit but none exploits the support of collaboration the level desired. CAVERNSoft G2 [5] provides a collaborative geometry viewer. The users have the possibility to share a scene in a VE. While discussing the scene they are able to manipulate it sequentially or they can indicate areas of the shared world using a virtual pointing device.

CVEs usually have avatar representations which represent an embodiment of the user. One example approach using avatars is given in [13]. There is full anatomically based body articulation possible which allows the use of postures and gestures in a CVE. Avatars allow dislocated users in a VE the feeling of co-presence. Good overviews on the design of avatars are given in [1].

3 Tools

Different categories or classes of tools can be identified for the support of collaboration in immersive VEs. They range from visual to audio support. The field of haptics becomes more and more important in the area of collaboration but is not researched in this project. The tools presented are categorized into 5 different areas. Gaze visualisation enhances the local users information about the remote users perception of the world. Object higlighting allows users to clearly show interest on certain objects in the VE to remote users. An overview of the whole VE is shown to the users by the map tool and complex movements can be stressed by the use of the movement visualisation tool. Sonfication is another aspect which is important especially when feedback is needed and textual messages would decrease the feeling of immersion.

3.1 Gaze Visualisation

Gaze visualisation helps the collaborating partners to see of what the dislocated users are possibly aware of. This tool is especially useful, when the FOV of the users is limited, which is the case when HMDs or desktop VR systems are used.

Since most VR systems do not analyse the actual gaze of the user a rough approximation is used. It is based on the gaze vector of the avatar. On desktop systems the centre of the screen is assumed as the centre of interest of the user. In immersive environments, where head tracking is available the gaze vector is based on orientation and position of the head sensor. On a desktop system the centre of the screen is used for approximation of the users gaze. Locally the gaze centre can be displayed by a crosshair.

Frustum Visualisation. This type of gaze visualisation enables other participants in the environment to see what parts of the scene the remote user could see. The frustum is displayed with a very high transparency value not to irritate the observing users or to hide objects from them. In the case of desktop VR or HMDs one frustum is displayed. If several camera orientations are used for the different display walls as in the case of CAVEs or multiple screen installations the setup data of the display walls is transferred to the remote users and several frustums, one for each used screen, are displayed. Using stereo display the frustums of the left and the right eye are merged together.

Gaze Vector Visualisation. The gaze vector indicates not only the direction the user is looking; it indicates as well the distance of the far clipping plane from the viewing frustum. A line is displayed from the avatars gaze centre in the direction of the gaze. The line stops at the far clipping plane.

Parameters of the gaze visualisation are color, opacity and color range which will allow different frustum or vector visualisations for different remote users.

3.2 Object Highlighting

The possibility to highlight objects in a NVE allows users more easily to show to each other about which object they are discussing or with what they intend to interact. In [12] several interaction techniques were suggested for object selection. These interaction techniques could easily be used for displaying an interest in certain objects. CVEs could profit from the use of two particular ways for highlighting of objects.

At-a-Distance by Pointing. The user points at a certain object with his input device. The position and orientation of the wand are measured, a ray is cast in the scene and the first object of which the bounding volume is hit is highlighted. On a desktop VR environment either a picking algorithm is applied to check where the user points with the mouse or the avatars hand movement is implemented by emulating a tracked users hand.

Gaze Directed. The user looks at a certain object which will be highlighted. For this the gaze vector calculated by the mechanisms described in the previous section are be used. The first bounding volume which is hit by the gaze vector is highlighted.

These selection techniques can be used continuously during the whole simulation or they can be toggled by input signals from a wand, a mouse or a keyboard. Several possibilities arise as well in the way how the highlighting is done. The whole object or it's bounding volume can be highlighted by a colour change or a change of the objects opacity. Instead of color change flashing of the color can be applied to get more attention. If flashing of an object is enabled, the color change is activated after a the object is selected for certain amount of time.

Parameters of the object highlighting are similar to the ones given in gaze visualisation. Color, opacity and color range highlight the objects geometry or the objects bounding volume. The time to activate flashing after selection has to be set as well.

3.3 Movement Visualisation

Complex movements can be visualised in many ways. In the area of cartoons which represent their own reality in a more abstract way complex motion is visualised by the use of motion lines. The support of this framework is given by motion lines and shadow objects. Considering two dimensional CSCW applications the use of motion lines has been suggested by [7].

Two different types of motion visualisation are given by the framework. Lines at object vertices can be drawn as well as shadow objects can be displayed.

Motion Lines. Motion Lines are drawn between two positions in the time line. A point taken into account for representation of a motion line is a vertex from a moved object. Using the motion lines it is not only possible to display the area where the object was moved. By changing color, opacity or length of the motion line it is possible to display the speed as well.

Shadow Objects. A shadow object is a copy of the geometry of the actual object with a higher transparency value. Options in this tool allow the amount of shadow objects displayed in a set timeframe with constant transparency or increasing transparency over the time. Shadow objects can be connected with motion lines from one time frame to another.

Since the use of shadow objects and motion lines can produce a high amount of geometry data displayed it is possible to display bounding volumes or draw lines between the different bounding volumes instead of using the whole geometries. Less computational power is required in that case especially if the original object consists of a high amount of polygons.

Movement visualisation is not only important for moved objects; it can be a powerful tool to display avatar movement as well. The movement of the whole avatar as a single object can be displayed, in the case of immersive VEs as the movement of its hand can be displayed as well.

The parameters for the movement visualisation describe how long a line or a shadow object shall be displayed. A fading factor specifies the opacity of the fading object.

3.4 Maps

Maps in NVEs can support spatial orientation and allow users to easily navigate to their desired location in the world [2]. Using landmarks in larger environments or floor plans in smaller environments allows users to discuss where they meet in the VE.

The map has different levels of detail depending on the current zoom factor. The smaller the displayed area is the more detail is given. An abstract representation of large geometries is given from a top perspective. This is realised by projecting the bounding volumes of large objects like buildings and landmarks on the map. The next level of detail shows the geometries of the large objects. Additionally it is possible to enhance the map with textual information. If the VE simulates a city, street names and names of buildings can be given.

Supporting collaboration an avatar radar is implemented which displays all surrounding avatars. If one avatar speaks to the client via voice connection the avatar can be highlighted on the map for a set period of time. The names of the avatars are displayed on the map as well. If a users avatar is interacting with objects in the scene a zoom on the map should automatically indicate an interaction. The focus on the map displays the interacting avatar form a top view with the object he is interacting with. Different states of the user in the environments can be visualised by different colours or textual output. Avatars can be highlighted in a different way when they are working with objects, when they are talking, when they are moving or when they are idle.

Parameters of the map include initial scale, level of detail, avatar visualisation and additional textual information.

3.5 Audio

Sonification of additional information can be useful in NVEs. If avatars are entering or leaving the environment the users should be notified. A voice notification that a user has entered the environment is given. In order to not to reduce the immersion, status messages can be given via audio rather than being displayed in a status window. If a collision with an object has happened or a user has picked up or selected an object audio feedback provides a useful addition to the visual cue. The volume of the audio can be set as a parameter. Different aspects of the sonification can be turned on and off, some of these values are user entry notification, interaction messages, system messages and collision messages.

4 Toolbox Architecture

The core of the toolbox is formed by the toolbox object which contains a list of available tools. Depending on the tool changes are performed once, like gaze visualisation where the frustum or the gaze vector are added to the scene graph, or several times like object highlighting, where it has to be constantly checked whether a new object has been selected or not. Each tool uses an update method which will be called once or every frame. This is defined by an update flag which can be set. Desiging your own tools becomes a very easy task, by deriving from existing tool classes and overriding the update method. All tools in the framework can be activated or deactivated individually. The configuration file

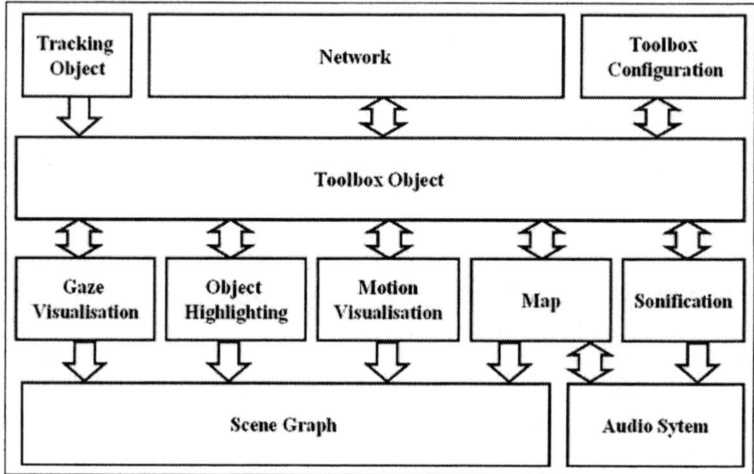

Fig. 1. Toolbox Architecture

is stored locally in an XML format providing the intial setup. The different attributes of the tools can be changed dynamically during runtime. Additional tools can be activated or already active tools can be turned off. The users have to be able to choose which tools suit them best.

The tool setup can affect the VE in three different ways. The tools can be activated for the avatars representing the remote users locally, which allows enhancements of the perception of what the remote users are doing. Examples were discussed in section three. The second possibility allows the users to enable some of their tools locally, an example would be in the area of sonification if a user is notified if he picks up or uses an object in the VE. The last possibility allows the user to share his configuration file with other users. They could get information which tools the user has activated. Figure 1 gives a rough overview of the communication between the different components of the toolbox and other modules in the CVE.

5 Conclusions and Future Work

The paper has described a selection of tools which can be used to improve collaborative activities in a NVE by augmenting the VE. The users working with these tool are able to perceive the environment with additional information such as the gazing direction, viewing area, movement visualisation of the dislocated user. The tools represented were explained in functionality as well as in their meaning for collaboration. Modular design allows users to select the type of additional information they want to have displayed depending on their needs. The framework including this toolbox can be used as a testbed for the analysis of human behaviour in CVEs. Additional logging mechanisms which record tracking data and other user input can be used to replay the whole simulation after the collaborative task is completed. For analysis it can be heplful to enable the frustum visualisation afterwards for example.

References

1. Norman I. Badler, Cary B. Phillips, and Bonnie L. Webber. *Simulating Humans: Computer Graphics Animation and Control.* Oxford University Press, New York, NY, USA, 1992.
2. Douglas A. Bowman, Elizabeth T. Davis, Larry F. Hodges, and Albert N. Badre. Maintaining spatial orientation during travel in an immersive virtual environment. *Presence: Teleoperators and Virtual Environments*, 8(6):618–631, 1999.
3. Wolfgang Broll. Interacting in distributed collaborative virtual environments. In *Virtual Reality Annual International Symposium (VRAIS)*, pages 148–155, Los Alamitos, March 1995. IEEE.
4. Carolina Cruz-Neira, Daniel J. Sandin, Thomas A. Defanti, Robert V. Kenyon, and John C. Hart. The cave: Audio visual experience automatic virtual environment. *Communications of the ACM*, 35(6):64–72, June 1992.

5. CAVERNsoft G2: A Toolkit for High Performance Tele-Immersive Collaboration. Kyoung park and yong cho and naveen krishnaprasad and chris scharver and michael lewis and jason leigh and andrew johnson. In *Virtual Reality Software and Technology (VRST)*, pages 8–15, Seoul, Korea, 2000. ACM Press.
6. Gernot Goebbels and Vali Lalioti. Co-presence and co-working in distributed collaborative virtual environments. In *Proceedings of the 1st international conference on Computer graphics, virtual reality and visualisation*, pages 109–114, Camps Bay, Cape Town, South Africa, 2001. ACM Press.
7. Carl Gutwin. Traces: Visualizing the immediate past to support group interaction. In *Conference on Human-Computer Interaction and Computer Graphics*, pages 43–50, May 2002.
8. Chris Joslin, Igor S. Pandzic, and Nadia Magnenat Thalmann. Trends in networked collaborative virtual environments. *Computer Communications*, 26(5):430–437, March 2003.
9. Michael R. Macedonia and Michael J. Zyda. A taxonomy for networked virtual environments. *IEEE MultiMedia*, 4(1):48–56, Jan-Mar 1997.
10. David Margery, Bruno Arnaldi, and Noel Plouzeau. A general framework for cooperative manipulation in virtual environments. In M. Gervautz, A. Hildebrand, and D. Schmalstieg, editors, *Virtual Environments*, pages 169–178. Eurographics, Springer, June 1999.
11. Maja Matijasevic. A review of networked multi-user virtual environments. Technical report tr97-8-1, Center for Advanced Computer Studies, Virtual Reality and Multimedia Laboratory, University of Southwestern Lousiana, USA, 1997.
12. Mark R. Mine. Virtual environment interaction techniques. Tr95-018, University of North Carolina, Chapel Hill, NC 27599-3175, 1995.
13. Igor Pandzic, Tolga Capin, Elwin Lee, Nadia Magnenat Thalmann, and Daniel Thalmann. A flexible architecture for virtual humans in networked collaborative virtual environments. *Computer Graphics Forum*, 16(3), September 1997.
14. Marcio S. Pinho, Doug A. Bowman, and Carla M.D.S. Freitas. Cooperative object manipulation in immersive virtual environments: Framework and techniques. In *Virtual Reality Software and Technology (VRST)*, pages 171–178, Hong Kong, November 2002.
15. Roy A. Ruddle, Justin C. D. Savage, and Dylan M Jones. Symmetric and asymmetric action integration during cooperative object manipulation in virtual environments. *ACM Transactions on Computer-Human Interaction*, 9:285–308, 2002.
16. Josef Wideström, Ann-Sofie Axelsson, Ralph Schroeder, Alexander Nilsson, and Åsa Abelin. The collaborative cube puzzle: A comparison of virtual and real environments. In *Collaborative Virtual Environments (CVE)*, pages 165–171, New York, NY, USA, 2000. ACM Press.

A Peer-to-Peer Approach to Content Dissemination and Search in Collaborative Networks

Ismail Bhana and David Johnson

Advanced Computing and Emerging Technologies Centre,
School of Systems Engineering,
The University of Reading,
Reading, RG6 6AY, United Kingdom
{i.m.bhana, d.johnson}@reading.ac.uk

Abstract. There are three key driving forces behind the development of Internet Content Management Systems (CMS) – a desire to manage the explosion of content, a desire to provide structure and meaning to content in order to make it accessible, and a desire to work collaboratively to manipulate content in some meaningful way. Yet the traditional CMS has been unable to meet the latter of these requirements, often failing to provide sufficient tools for collaboration in a distributed context. Peer-to-Peer (P2P) systems are networks in which every node is an equal participant (whether transmitting data, exchanging content, or invoking services) and there is an absence of any centralised administrative or coordinating authorities. P2P systems are inherently more scalable than equivalent client-server implementations as they tend to use resources at the edge of the network much more effectively. This paper details the rationale and design of a P2P middleware for collaborative content management.

1 Introduction

There are three key driving forces behind the development of Internet Content Management Systems (CMS) – a desire to manage the explosion of information (or content), a desire to provide structure and meaning to content in order to make it accessible, and a desire to work collaboratively to manipulate content in some meaningful way. Yet the traditional CMS has been unable to meet the latter of these requirements, often failing to provide sufficient tools for collaboration in a distributed context. The distributed CMS addresses the need to delegate control of resources and serves as a more natural paradigm for the collaboration in the CMS. However, with the burgeoning mobile market and an increasing need to support a range of end-user devices for content authoring, sharing, and manipulation has lead to a new requirement for meaningful collaborative tools that are able to deal with the complexity and heterogeneity in the network.

Most of current popular open source and commercial CMS implementations (e.g. Zope [1], Cocoon [2], and Magnolia [3]) are based on the client-server model. The client-server model has many obvious advantages in terms of familiarity (amongst developers, administrators and users), ease of deployment and administration, simplified version control and archiving, manageability in access control, security and data

V.S. Sunderam et al. (Eds.): ICCS 2005, LNCS 3516, pp. 391 – 398, 2005.

consistency. However, the relative lack of complexity in these systems results in a number of limitations in scalability and reliability, particularly where there is a rapidly fluctuating user base or changing network, as is common in mobile networks. The client-server model is essentially static and does not scale well as the number of clients increases, both because of limitations on the server and limitations in bandwidth around a heavily loaded server (the congestion zone). Server clusters, load balancing, and edge caches (as used in Zope) lessen the problem in some circumstances but are a costly solution and cannot overcome the problem entirely.

In contrast, the P2P approach restores an element of balance to the network. Firstly, whilst servers are still a central element of the network there is no steadfast reliance on a particular set of central provides. P2P systems are thus much more scalable. P2P systems are also in many circumstances much more fault-tolerant (i.e. resources and services have high availability) due to the potential for replication redundancy (for resources that are replicated amongst peers). Moreover, P2P systems can be more efficient in bandwidth utilisation because they tend to spread the load of network traffic more evenly over the network.

These properties are highly significant in relation to the design of a collaborative CMS, particularly in a heterogeneous context (i.e. spanning operating system, network, and mobile boundaries). However, due to increased complexity the P2P approach also presents us with a number of challenges – particularly in ensuring consistency, security, access control and accountability. The JXTA CMS [4] the Edutella project [5], and the Hausheer and Stiller approach [6] are attempts to tackle the content problem from a P2P perspective.

Building on the traditional strengths of the CMS, the possible additional elements of fault tolerance, availability, flexibility and a sufficient set of collaborative tools are critical in ensuring the future success of the CMS. The following sections of this paper give details of the rationale and design of a P2P middleware for mobile and adhoc collaborative computing (known as Coco) that includes services to support collaborative content management.

2 Our Approach

Our goal, as described in [7], is to develop a framework that supports collaboration in a way that enables users to self-organise and communicate, share tasks, workloads, and content, and interact across multiple different computing platforms. The rationale for designing a collaborative content system on P2P networks is based on the desire to achieve scalability, enabling a collaborative system to scale with a dynamically changing user base, and resilience. Our goal is also to support self-organisation and dynamic behaviour by developing systems and services that support the organisation of individuals into groups with shared interests and allowing the formation of dynamic collaborations. As a starting point, our model builds on the general CMS lifecycle depicted in figure 1. This model places collaboration at the heart of content management. The figure illustrates that content management is a continual process of creation, collaboration, and dissemination.

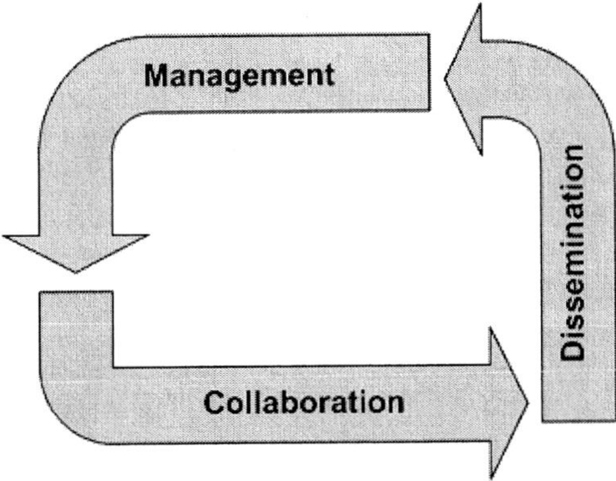

Fig. 1. The CMS lifecycle: a process of content creation, collaboration, and dissemination

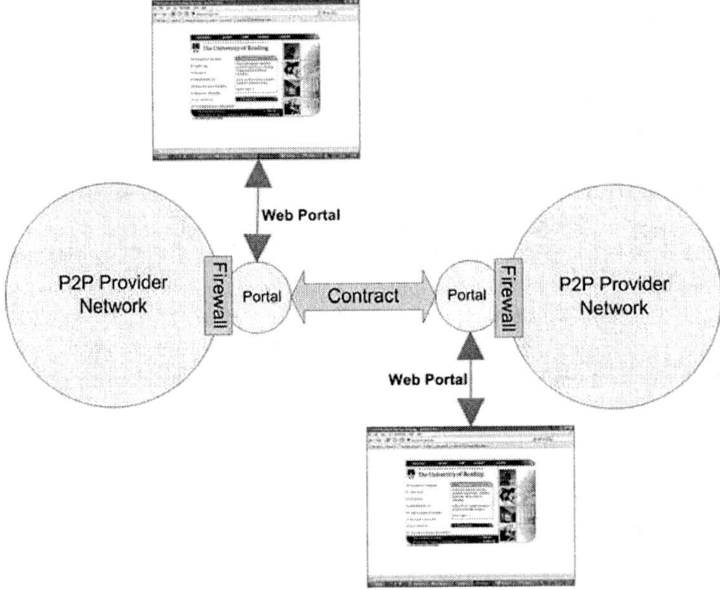

Fig. 2. Accessing Live P2P Content; Provider Networks that will typically represent some sort of real world enterprise, such as a university, local education authority, company, or organisation, but may be any arbitrary collection of peers with a shared interest

The Coco content network can be viewed as a hybrid P2P system built above a pure P2P JXTA network. The network consists of self-regulating regions called *provider networks* that will typically represent some sort of real world enterprise, such as a university, local education authority, company, or organisation, but may be any arbitrary collection of peers with a shared interest. The provider networks act as a trusted region (or a secure domain of trust). Peers are not limited to the same network, they may be geographically dispersed, or behind firewalls or routers, or may be mobile devices, as illustrated by figure 2.

Whilst peers within provider networks are able to interact freely in a pure P2P manner, each provider network consists of at least one peer (the portal peer) that acts as a gateway and (Web) portal into the provider network for external peers. It is this peer that enables internal peers to interact with external peers and assists in the process of authentication and access control. The portal peer is also able to act as a Web proxy to the P2P network residing within the institution, enabling users to search and retrieve content over the Web (using the company or university website, for instance) without requiring them to install the relevant P2P software. This content is live, meaning that the state of the network is continually changing as peers appear and disappear at will. The system also enables agreements to be formed between provider networks supporting (in future) logging and reporting. For networks to communicate in such a way it is important to define a set of interoperability standards – standard ways of representing resources as well as invoking operations on remote peers in the network.

2.1 Content as a Resource

Content, in this context, is defined as the set of resources available for consumption within the network. This definition ranges from the obvious, such as files and documents, to the less intuitive, such as services and computing resources, to things that do not generally have an opaque representation, such people (represented, for example, using VCards [8]). This formulation has much in common with the ethos of the Resource Description Format (RDF) [9] and it is, in fact, RDF that is used in our implementation as the language (rather than framework, as RDF is essentially Web-based) of resource description.

In order for non-Web resources to be described in a P2P context they are represented using a unique Universal Resource Name (URN). The content system uses an URN notation to form a unique *content identifier* for each unit of content generated using a cryptographic digest. Whilst it is normal to uniquely reference an RDF resource using a URL, there may be many situations in a replicated P2P network in which a given resource is duplicated across many network nodes or devices and hence a location-dependent representation is inadequate. The given representation allows a resource to be referenced without an *a priori* knowledge of its location within the network.

Metadata describing content is cached by peers in the network to ensure high availability and performance of search queries. Each peer is responsible for managing its cache and stale resources may be purged at regular intervals.

2.2 Service Invocation

The CMS is deployed as a P2P Web service using open protocols such as WSDL [10] and SOAP [11]. Search queries are submitted using Web service invocations (although an API is required to deal with P2P interactions, including dynamically discovering peers and peer services). Using open standards such as SOAP provides us with enormous flexibility as it abstracts away the service interfaces from the underlying transport or middleware. Content can therefore be searched and retrieved over the

JXTA network as well as over the Web as a standard Web service – where the individual peer acts as a web server and is able to tunnel requests through the P2P network (this is essentially what the portal peer does). As figure 3 illustrates the invocation process consists of three steps:

- Service Advertisement & Discovery – the service descriptor (WSDL) for the peer hosting an instance of the content service is propagated to peers using the JXTA discovery mechanism. This enables peers to dynamically discover new services as they appear on the network.
- Authentication & Authorisation – the next step (if applicable) is for the consumer peer to authenticate with the relevant authority that will grant access to the peer. This process is optional and allows a peer to delegate the authorisation process as might be desirable in an enterprise or educational context. Fail-over mechanisms will be in place in future if the portal peer is temporarily unavailable. This step may also be used to add additional support for logging, versioning, or charging for the service provided by a particular peer within a provider network.
- Invocation – once the peer has the relevant service description and authorisation it is able to query the service-hosting peer directly. Search queries are normally answered with RDF containing information about resources available on the peer, as well as cached information about resources available on other peers (if specified in the request).

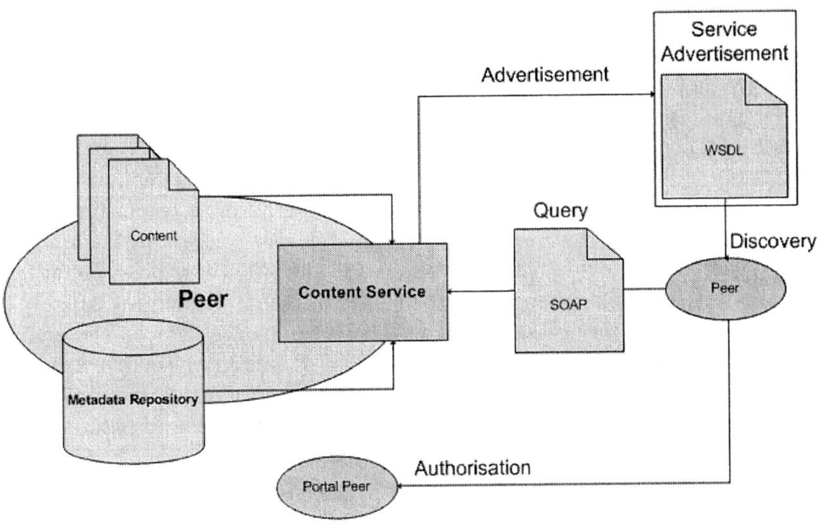

Fig. 3. Invocation of a P2P Web Service: each peer acts as a web server and is able to propagate Web service invocations through the P2P network

An XML metadata repository (using Xindice [12]) is used to store and retrieve resource entries. The advantage of using an open source XML database is that we don't need to worry about storage and retrieval issues and developing database optimisations. RDF resources are simply added to database and can retrieved later using XPath [13]. Metadata is normally stored using some form of formal schema, such as Dublin Core [14].

2.3 Enabling Mobile Devices

In the mobile arena, we are building on the Java 2 Platform Micro Edition (J2ME) [15] and JXTA for J2ME (JXME) [16] to allow mobile devices, such as phones and PDAs, to participate in the content transactions. Mobile devices have significant hardware constraints compared to desktop and enterprise machines. These limitations include:

- Significantly less processing power
- Limited runtime memory
- Little or no persistent memory
- Very small screens with limited modes of interaction
- Lower network bandwidth and higher network latencies

By basing software for mobile devices on the J2ME platform, the range of device capabilities in the mobile computing market is accounted for through the use of different J2ME device configurations and profiles. A configuration defines the features of a Java Virtual Machine (JVM) that a device can use, and a profile is the definition of the set of Java API's that are available to the developer for that class of device. JXME has been implemented for the Connected, Limited Device Configuration [17] and the Mobile Information Device Profile [18] (CLDC/MIDP) that is also the most widely used configuration and profile combination.

Coco for mobile devices (MicroCoco) is being built on J2ME and JXME. MicroCoco includes services to consume the collaborative content management services provided by Coco. Services that require searching textual data are ideally suited to J2ME based applications, because the restrictive modes of interaction imposed by mobile computing devices are tailored to text input and J2ME accounts for this limitation in its standard interface components. The mobile device peers will not function as content providers, but only consumers because even though it is possible for devices such as PDAs and mobile phones to host and share content, it is highly unlikely that a user will have a mobile device as their primary computing device. The amount of persistent memory is limited in comparison to that of a desktop machine and we have assumed that users will not want to keep a great number of documents on the mobile device. Many mobile devices also do not have the appropriate third party software to view documents (such as Adobe PDF or Microsoft Word files).

However, a user may wish to search for and record their search results whilst on the move. By having a mobile application that can search the content network for resources, users are given the facility to participate in research whilst on the move. Search results can be stored locally on the mobile device. To facilitate sending search results to a desktop peer that a user may also be running, the mobile peer can be linked with a desktop peer in a similar manner in which Bluetooth devices are paired. The user can then send the search results from the mobile device to a desktop machine where the user can then download the documents, all of which occurs in a purely P2P manner.

3 Conclusions

Our experience indicates that the decentralised (P2P) approach works very well for content distribution. Our rationale for designing a collaborative content system on P2P networks was out of a desire to achieve scalability, as well as to enable a diverse range of devices to participate in collaborative processes. We wanted to provide a framework that supports the interaction of groups or individuals with shared interests.

The difficulty in taking the P2P approach is that there is an inevitable increase in the design complexity of a CMS and it makes it difficult to achieve many of the things that traditional CMSs do well. For instance, for version control and archiving, strong consistency is required to ensure that all elements of a version history are always accessible. Workflow is another area that can be complicated with a decentralised model – it requires flexible organisational models that can be easily customised, which in turn rely on security and access control mechanisms. Logging and reporting is another key area where flexible mechanisms must be in place to facilitate accountability.

Our intention in the near future is to take the development of the Coco content service forward through a series of alpha and beta releases. In future, we intend to make developments in the areas of pricing/charging, replication and versioning, logging, authentication and access control, privacy and accountability, and security.

References

1. Zope: Open Source Application Server for Content Management Systems; Version 2.7.4, http://www.zope.org/Products/, (2005)
2. Cocoon: XML publishing framework; Version 2.0.4, Apache Software Foundation, http://xml.apache.org/cocoon/, (2005)
3. Magnolia: Open Source Content Management; www.oscom.org/matrix/magnolia.html, 2005.
4. Project JXTA, CMS; http://cms.jxta.org/servlets/ProjectHome, (2005)
5. EDUTELLA: A P2P Networking Infrastructure Based on RDF; http://edutella.jxta.org/, (2005)
6. Hausheer, D., Stiller, B., Design of a Distributed P2P-based Content Management Middleware; In Proceedings 29th Euromicro Conference, IEEE Computer Society Press, Antalya, Turkey, September 1-6, (2003)
7. Bhana, I., Johnson, D., Alexandrov, V.N., Supporting Ad Hoc Collaborations in Peer-to-Peer Networks; PCDS04, San Francisco, (2004)
8. VCard Overview; http://www.imc.org/pdi/vcardoverview.html, (2005)
9. Resource Description Framework (RDF), W3C Semantic Web; http://www.w3.org/RDF/, (2005)
10. Web Services Description Language (WSDL) 1.1, http://www.w3.org/TR/wsdl, (2005)
11. W3C SOAP Specification; http://www.w3.org/TR/soap/, (2005)
12. Apache Xindice; Apache Software Foundation, http://xml.apache.org/xindice/, (2005)
13. XML Path Language (XPath), W3C XML Path Language (XPath); Version 1.0, http://www.w3.org/TR/xpath, (2005)
14. Dublin Core Metadata Initiative (DCMI), Interoperable online metadata standards; http://dublincore.org/, (2005)

15. Java 2 Platform Micro Edition (J2ME), http://java.sun.com/j2me/, (2005)
16. JXME: JXTA Platform Project, http://jxme.jxta.org/proxied.html, (2005)
17. Connected Limited Device Configuration (CLDC);JSR 30,JSR139, http://java.sun.com/products/cldc/, (2005)
18. Mobile Information Device Profile (MIDP); JSR 37, JSR 118, http://java.sun.com/products/midp/, (2005)

TH-VSS: An Asymmetric Storage Virtualization System for the SAN Environment

Da Xiao, Jiwu Shu, Wei Xue, and Weimin Zheng

Department of Computer Science and Technology, Tsinghua University,
100084 Beijing, China
xiaoda99@mails.tsinghua.edu.cn

Abstract. Storage virtualization is a key technique to exploit the potential of SANs. This paper describes the design and implementation of a storage virtualization system for the SAN environment. This system has asymmetric architecture, and virtualization operations are done by a metadata server allowing management to be accomplished at a single point. The system has better scalability compared to symmetric systems and can support heterogeneous platforms of hosts. Agent software was implemented in the volume management layer of hosts so that any standard HBA card can be used. The metadata server manages storage resources automatically and configuration of storage pools can be changed dynamically. The metadata server is also used for system monitoring, which is the basis of dynamic storage resource management and automatic failure recovery. Test results showed that the overhead introduced by our virtualization layer is negligible, and the performance of storage system was enhanced effectively by the striping strategy in organizing storage pools. The number of seeks per second to a logical volume allocated from a pool of four disks was increased by 55.2% compared to a plain FC disk.

1 Introduction

The introduction of storage area networks (SANs) can significantly improve the reliability, availability, and performance of storage systems. However, SANs must be managed effectively so that their potential can be fully exploited. Storage virtualization is often seen as the key technology in SAN management. Storage virtualization separates physical storage from the server's operating system and provides storage users with unified storage pools and logical volumes. It can provide users with much larger storage space than a single physical disk and a much better utilization of disk capacity. Furthermore, virtualization offers a new level of flexibility. Storage systems can be added to or removed from storage pools without downtime, thus enabling fast adaptation to new requirements.

The first approach to storage virtualization in a cluster environment is host-based. Two typical examples of this approach are CLVM[3] and the EVMS cluster[5]. CLVM is an extension of Linux LVM in a single system environment to add cluster support. The EVMS cluster is a framework proposed by IBM which

V.S. Sunderam et al. (Eds.): ICCS 2005, LNCS 3516, pp. 399–406, 2005.

also supports virtualization in cluster environments. Both of these approaches adopt a host-based symmetric architecture. Each node in the cluster has the right to perform virtualization management tasks. The consistency of metadata is maintained by communications between nodes. This puts a heavy burden on hosts and reduces the scalability of the system. Besides, heterogeneous platforms of hosts are not supported by systems with this symmetric architecture.

Another approach is network-based in-band virtualization. This approach calls for an appliance to be installed between the hosts and the storage devices, which redirects all I/Os between hosts and storage devices. Representatives of this approach are DataCore Software's SANsymphony[6] and HP's StorageApps[4]. The advantage of this approach is that agent software need not be installed on hosts, so transparency to hosts is achieved. Its main drawback is that the appliance may become a SAN bottleneck, thus limiting the SAN performance and scalability and significantly complicating the design of large-scale highly available configurations.

A third approach is network-based out-of-band virtualization. In this approach, the handling of metadata is separated from the data path and is done by a dedicated appliance. It enables direct data transfer between hosts and storage subsystems. StoreAge's SVM[7] adopts such an architecture.

Moreover, the main consideration in all the approaches mentioned above is storage capacity when grouping physical disks into storage pools. These approaches ignore other properties of virtual storage such as bandwidth, latency, and reliability.

This paper describes the design and implementation of a storage virtualization system for SAN environments called the Tsinghua virtualization storage system (TH-VSS). In this system, an asymmetric architecture is adopted. Metadata management tasks are done by a metadata server so that single-point management is achieved. The system has better scalability compared to symmetric virtualization architecture and can support heterogeneous platforms in hosts. Agent software is implemented in the volume management layer of hosts, which results in improved flexibility. Any standard HBA card can be used. Storage resources are managed automatically by the metadata server, and the configuration of storage pools can be changed dynamically. The metadata server is also used for system monitoring, which is the basis of dynamic storage resource management and automatic failure recovery.

2 TH-VSS Design

2.1 Architecture of TH-VSS

TH-VSS consists of three parts: the virtualization agents on the hosts, a meta server, and storage devices. The hosts and meta server are attached to the SAN via a FC HBA on the motherboard. Storage devices are also attached to the SAN via their fibre channel interfaces. The meta server and agent are connected to each other through Ethernet. The TH-VSS was designed for a heterogeneous

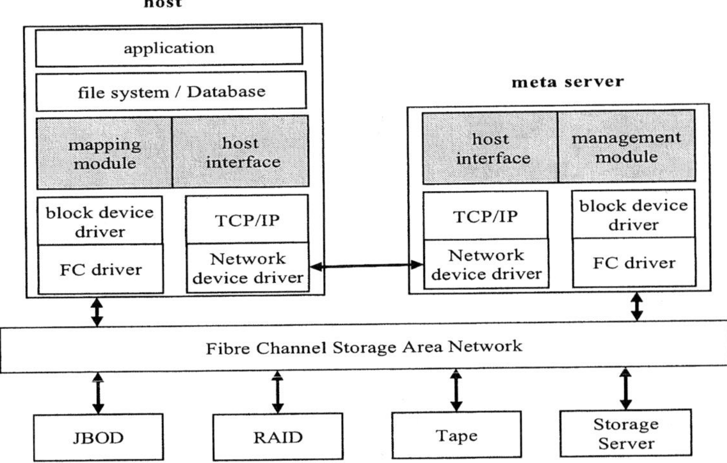

Fig. 1. Overall architecture of the TH-VSS

SAN environment. Hosts may run different OSs, and storage devices include JBOD, RAID subsystems, tape drives, and storage servers, as well as others. The metadata management module on the meta server performs the metadata management tasks, and logical to physical address mapping of I/O requests are handled by the mapping layer of the agent on the host. The software architecture of the system is shown in Fig. 1 (gray boxes are software modules implemented in the TH-VSS).

2.2 Virtualization Process

The system works as shown in Fig. 2. On startup, the system monitor on the meta server scans all physical storage devices attached to the SAN. For each device, the system monitor writes a label and a UUID on the device to create a physical volume on it. Then the system monitor groups the storage devices into different storage pools according their detected properties. Next, the administrator issues a command to create a logical volume through the administration interface. The administration interface tells the metadata manager to allocate space for the logical volume from the storage pool and create it. Then the administrator issues another command to assign the volume to the Solaris server. On receiving such a command, the meta server sends the UUIDs of the physical volumes in the storage pool and the mapping table of the volume to the agent on the Solaris server through host interfaces. The agent locates the correct disks according to the UUIDs and creates the logical volume in a kernel for further use. Then it sends a response to the meta server to tell it that the volume has successfully been created. Finally, applications on the Solaris server can access the logical volume directly without consulting the meta server.

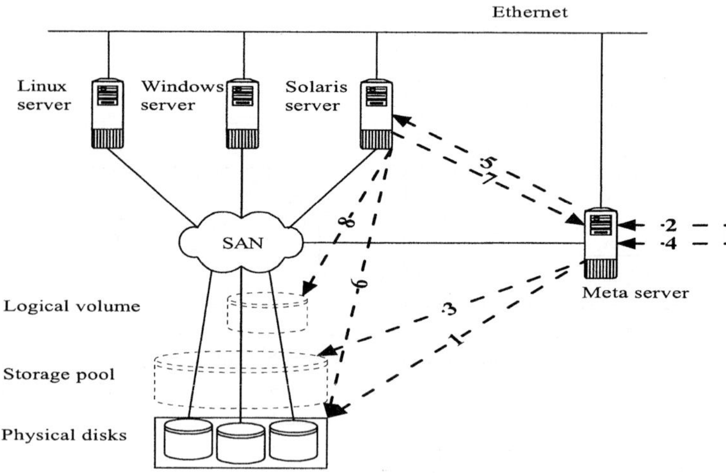

Fig. 2. Virtualization process

3 Key Techniques in TH-VSS Implementation

3.1 Storage Resource Management

The lowest level in the TH-VSS storage hierarchy is the physical volume. Multiple physical volumes are merged into a storage pool, from which logical volumes can be allocated. The TH-VSS supports concatenation (linear), stripe (RAID-0) and mirror (RAID-1) schemes.

In commercial data storage environments, various kinds of storage systems can be attached to the SAN, including the JBOD system, the RAID subsystem, tape drives and storage servers. Different kinds of storage systems have different properties in bandwidth, latency, and reliability. When the meta server starts up, it scans all the storage devices attached to the SAN and writes a label and a UUID at the head of each detected device to mark it as a physical volume. Then it puts the physical volume into the appropriate storage pool according to its property. The composition of a storage pool can be changed dynamically. When a new storage device is added to the SAN, the system monitor on the meta server will detect the addition of the new device, query its type, create a physical volume on it and put it into the appropriate storage pool.

When a physical volume is added to a striped storage pool, the data of the logical volumes must be re-striped among physical volumes to improve performance. A mirroring mechanism is used to ensure that the data of the logical volumes is not corrupted during the re-striping process. Fig. 3 shows the process of re-striping. Data movement can be interrupted at any time due to errors or other reasons. LV1 can continue to function normally because it contains the newest copy of data.

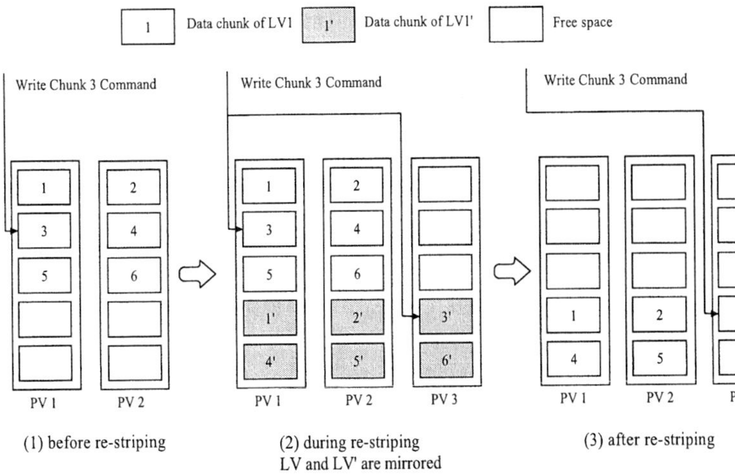

Fig. 3. Process of re-striping

3.2 Online Change of Mapping Table

In order to meet the requirements for uninterrupted service, some management tasks need to change mapping tables of logical volumes online. In SAN environments, where storage is shared among multiple hosts, the access of hosts to logical volumes must be controlled so that the data of logical volumes is consistent. A locking mechanism is used to ensure data consistency when the mapping table of logical volume is changed. A meta server may send lock and unlock requests of a particular logical volume to the agent. Upon the agent's receiving a lock request, any I/O request that has already been mapped by the mapping table of a logical volume but has not yet completed will be flushed. Any subsequent I/O request to that logical volume will be postponed for as long as the volume is locked. On receiving an unlock request, any postponed I/O request will be mapped by the new mapping table and gets re-queued for processing.

The process by which mapping tables are changed online is as follows. First, the meta server identifies the hosts to which this logical volume has been assigned, and sends a LOCK_LV request to agents on these hosts. On receiving the request, the agent performs the locking operation and then sends a LOCK_LV response to the meta server. After receiving the responses from the agents to the LOCK_LV request, the metadata manager generates a new mapping table. Then it sends a RELOAD_TABLE request to the agents. The agents replace their old mapping tables of the logical volume with the new ones and send a response to the meta server. After all the old mapping tables are replaced with the new ones, the meta server sends an UNLOCK_LV request to the agents. The agents map the postponed I/O requests with the new mapping table for processing and send an UNLOCK_LV response. Finally, the metadata manager of the meta server writes the updated metadata back to the heads of the physical volumes.

In this process, data inconsistency due to hosts' access to a logical volume with different mapping tables is avoided by having the meta server send an UNLOCK_LV request after all the old mapping tables have been replaced with new ones successfully.

4 Experimental Results

In this section we will present the experimental results of our virtualization system. The test system consisted of two Linux servers, two Windows servers, a meta server and an FC disk array, all of which were connected to a 2 Gigabit FC switch. Their configurations are shown in Table 1.

Table 1. Test configuration

Machine	Linux server	Windows server	Meta server
CPU	Intel Xeon 2.4GHz x 2	Intel Itanium2 1GHz x 2	Intel Xeon 2.4GHz x 2
Memory	1GB	1GB	1GB
OS	Linux(kernel: 2.4.26)	Windows Server 2003	Linux(kernel: 2.4.26)
FC HBA	Emulex LP982(2Gb/s)	Emulex LP982(2Gb/s)	Emulex LP9802
FC Disk	Seagate ST3146807FC x 5		

4.1 Overhead of Virtualization Layer

We used the Intel Company's IOMeter test program to evaluate the overhead introduced by the virtualization layer. First we tested the performance of a plain FC disk. Then we created a linear storage group on the disk and allocated a logical volume from it. To derive the overhead, we compared the performance of the VSS volume to the performance of the physical disk. The access pattern was sequential reading in 4-KB blocks. The comparison of average response times for different block sizes is shown in Fig. 4.

We can see from the results that the impact of the virtualization layer on the I/O performance was negligible in respect to bandwidth and average response time.

4.2 Impact of Stripe Mapping Strategy

In order to evaluate the performance improvement when using striped storage pools, first we created a striped pool with one disk and allocated a logical volume LV1 from it. We tested the performance of LV1 using Bonnie Benchmark. We created a file system on LV1 and allowed random access to fixed size files. We measured the number of random seeks per second on the LV1. Then we added another disk to the pool and repeated the test. The process was repeated until

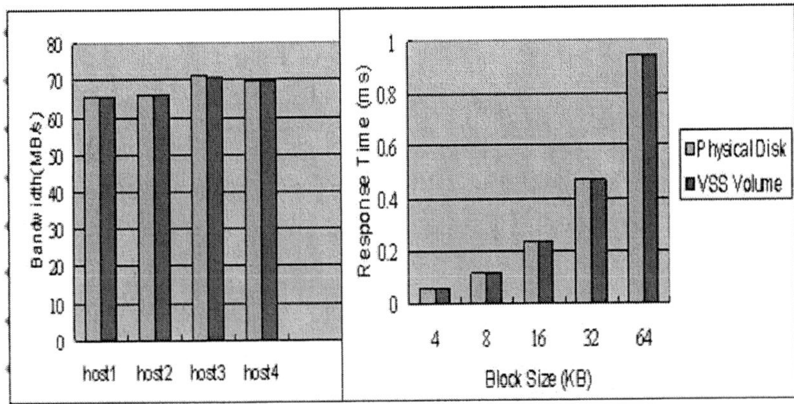

Fig. 4. Overhead of TH-VSS. First figure shows the comparison of bandwidth. Second figure shows the comparison of average response time

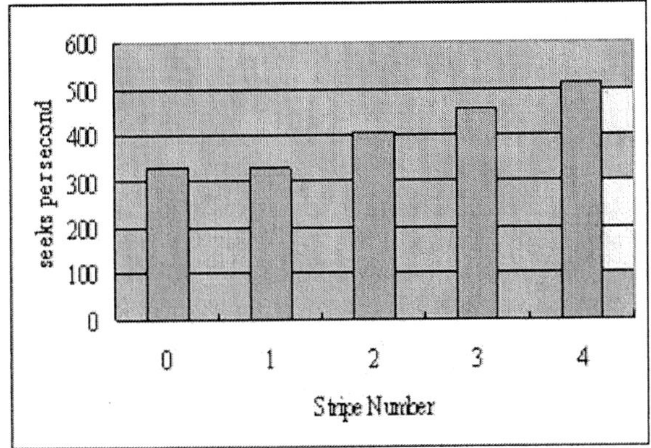

Fig. 5. Impact of stripe mapping strategy on performance

four disks had been added to the pool. The results are shown in Fig. 5, including a comparison of the results with the case of a plain disk labeled with 0.

The figure shows that the striped logical volume can provide better I/O capacity than that of a single disk, and the more disks in the striped storage pool, the better the performance of the LV1. When the LV1 is striped over 4 disks, the number of seeks per second is increased by 55.2%.

5 Conclusion

In this study, we designed and implemented a storage virtualization system for the SAN environment. An asymmetric architecture was adopted to achieve a single point of management, and metadata consistency was ensured in the SAN environment where storage is shared by multiple hosts. The new system's meta server implements the function of dynamic management of storage resources and system monitoring. Storage pools with different properties can be provided. The test results showed that the addition of our virtualization layer introduces little overhead to the system, and the striped storage pool can improve I/O performance effectively.

Acknowledgements

The work described in this paper was supported by the National Natural Science Foundation of China (No. 60473101) and the National High-Tech Research and Development Plan of China (No. 2004AA111120).

References

1. David, T., Heinz, M.: Volume Managers in Linux. In Proceedings of the FREENIX Track:2001 USENIX Annual Technical Conference, Boston, Massachusetts, USA (2001)
2. Sistina Software, Inc.: Logical Volume Manager. http://www.sistina.com
3. Heinz, M.: Linux Cluster Logical Volume Manager. In Proceedings of the 11th Linux Kongress , Erlangen, Germany (2004)
4. Hewlett-Packard Company: HP StorageApps sv3000 White Paper. (2002)
5. Ram, P.: EVMS Cluster Design Document. http://evms.sourceforge.net/clustering/
6. SAN Symphony version 5 datasheet. http://www.datacore.com (2002)
7. StoreAge White Paper. High-Performance Storage Virtualization Architecture. http://www.storeage.com
8. André, B., Michael, H.: V:Drive - Costs and Benefits of an Out-of-Band Storage Virtualization System. In Proceedings of the 12th NASA Goddard, 21st IEEE Conference on Mass Storage Systems and Technologies, College Park, Maryland, USA (2004)
9. Chang Soo Kim, Gyoung Bae Kim, Bum Joo Shin: Volume Management in SAN Environment. In Proceedings of 2001 International Conference on Parallel And Distributed Systems, KyongJu City, Korea (2001)
10. Common Information Model (CIM) Specification, v2.2. http://www.dmtf.org/standards/cim
11. Friedhelm, S.: The SCSI Bus & D E Interface. Addison-Wesley, second edition (1998)

Design and Implementation of the Home-Based Cooperative Cache for PVFS

In-Chul Hwang, Hanjo Jung, Seung-Ryoul Maeng, and Jung-Wan Cho

Division of Computer Science, Dept. of Electrical Engineering & Computer Science, KAIST, 373-1 Kusung-dong Yusong-gu, Taejon, 305-701, Republic of Korea
{ichwang, hanjo, maeng, jwcho}@calab.kaist.ac.kr

Abstract. Recently, there has been much research about cluster computing to get high performance using low-cost PCs connected with high-speed interconnection networks. In many research areas, many distributed file systems have been developed. In many distributed file systems, PVFS (Parallel Virtual File System) provides users with high bandwidth by stripping data over I/O servers. In PVFS, there is no file system cache. For a new file system cache for PVFS, we designed and implemented cooperative cache for PVFS (Coopc-PVFS). Because the previous developed Coopc-PVFS is a hint-based cooperative cache, a cache manager reads/writes files using approximately correct information so that it has a little read/write overhead. And previous studies about cooperative cache are only focused on shared read data and don't care about write performance. In this paper, we describe the design and implementation of the home-based Coopc-PVFS to improve read/write performance. Also, we evaluate and analysis the performance of the home-based Coopc-PVFS in comparison to PVFS and to the hint-based Coopc-PVFS.

1 Introduction

Recently, there has been much research about cluster computing to get high performance using low-cost PCs connected with high-speed inter-connection networks. For an efficient usage of cluster computing, many efficient components of cluster is necessary - efficient interconnection networks, Operating System supports and etc. Among many research areas, many distributed file systems which access disks slower than any other component in cluster computing have been developed.

Among research areas of distributed file system, cooperative cache [3, 4, 5] was proposed to reduce servers' load and to get high performance. Because the access time of other clients' memories is faster than that of servers' disks, to get a block from other clients' memories is faster than to get the block from servers' disk. In cooperative cache, a client finds a block first from its own file system cache, and then other clients' file system caches before getting the block from servers' disks.

In many distributed file systems, PVFS (Parallel Virtual File System) [1] provides users with high bandwidth by stripping data over I/O servers. There is no file system cache in PVFS - PVFS just supports data transfers from/to I/O servers. For a new file system cache for PVFS, we designed and implemented cooperative cache for PVFS (Coopc-PVFS) [9].

V.S. Sunderam et al. (Eds.): ICCS 2005, LNCS 3516, pp. 407 – 414, 2005.

Because the previous developed Coopc-PVFS is a hint-based cooperative cache, a cache manager reads/writes files using approximately correct information. If a hint is incorrect, read/write overhead can be large. And previous studies about cooperative cache are only focused on shared read data and don't care about write performance. To solve this problem, we design and implement the home-based Coopc-PVFS. In the home-based Coopc-PVFS, every cache manager manages exact information with little management overhead and written data can be buffered in home nodes. We also evaluate and analysis the performance of the home-based Coopc-PVFS in comparison to PVFS and to the hint-based Coopc-PVFS.

This paper is organized as follows. In the next section, we present about PVFS and cooperative cache. In section 3, we describe about Coopc-PVFS and the previous developed hint-based Coopc-PVFS. In section 4, we present the design and implementation of the home-based Coopc-PVFS. In section 5, we evaluate and analysis the performance of the home-based Coopc-PVFS in comparison to PVFS and to the hint-based Coopc-PVFS. Finally, we summarize major contributions of this work and discuss future work in section 6.

2 Related Work

2.1 PVFS (Parallel Virtual File System)

PVFS consists of compute nodes, a metadata manager and I/O servers. The compute nodes are clients that use PVFS services. A metadata manager manages metadata of PVFS files. The I/O servers store actual data of PVFS files. In PVFS, a file data is stripped over I/O servers.

There was a study for the file system caching effect of PVFS. Vilayannur et al. [2] designed and implemented a file system cache of a client. They showed that a file system cache in a client is efficient if many applications in the client share files among them. But their research was limited to a file system cache in a single node.

2.2 Cooperative Cache

Cooperative cache [3, 4, 5] was proposed to reduce servers' load and to get high performance. In cooperative cache, if a file system cache in a client doesn't handle a request to a file, the client sends the request to the other client's cache that caches the file rather than to the server because the access time of another client's memory is faster than that of the server's disk. Servers' load can be reduced in cooperative cache so that it is scalable as the number of clients increases. Because there is much more memory in the cooperative cache than in a single file system cache, the cooperative cache can handle more requests and improve overall system performance.

There have been many studies about cooperative caching. Dahlin et al. [3] suggested the efficient cache management scheme called N-chance algorithm, Feeley et al. [4] suggested another efficient cache management scheme called modified N-chance algorithm in GMS (Global Memory Service). Sarkar et al. [5] suggested the hint-based cooperative caching to reduce the management overhead using hint. Thus, the hint-based cooperative cache is scalable and can be adopted in the large-scale system such as cluster computer.

3 Cooperative Cache for PVFS

3.1 Overview of Cooperative Cache for PVFS

In the PVFS, an application reads/writes a file through I/O servers without a file system caching facility. We made Coopc-PVFS as we added a cache manager to a PVFS client.

In figure 1, we present the workflow of Coopc-PVFS added to a PVFS client.

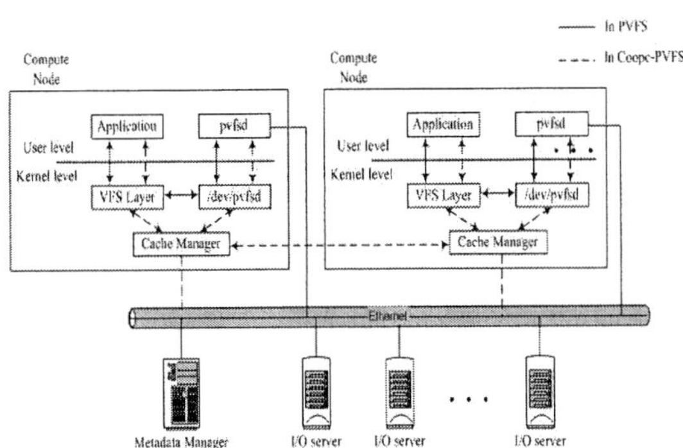

Fig. 1. Workflow of Coopc-PVFS

After we added a cache manager to a PVFS client, a cache manager can cache data and does cooperation with other clients' cache managers.

3.2 Hint-Based Cooperative Cache for PVFS

Because of large overhead to maintain accurate information about cached blocks, we designed Coopc-PVFS as a hint-based cooperative cache [9]. To maintain the hint – opened clients list, we added new function to the metadata manager to keep the clients list that contains information of clients that opened the file before. Whenever a client opens a file, the client gets both the metadata and the opened clients list of the file from the metadata manager. To accurately look up where is a block, a client must manage information about cached blocks in its own cache and other clients. To maintain this information, each cache manager manages its own bitmap and exchanges its own information with each other when it gets the block from another client.

Unlike previous hint-based cooperative cache research [5], we managed information and cached blocks per block, not per file. Because many clients share large files among them in a parallel file system, it is more adaptable to manage information and to cache per block than per file in Coopc-PVFS.

In PVFS, all the accesses to files go through the I/O servers. To conserve the consistency the same as in PVFS, the cache manager must invalidate blocks cached in other clients before writing the block to the I/O server in Coopc-PVFS. To do so, whenever an application writes a block, the cache manager sends cache invalidation messages to the others through the metadata manager before sending the written block to the I/O server.

4 Design and Implementation of the Home-Based Cooperative Cache for PVFS

In the home-based Coopc-PVFS, every cached block has its own 'home node'. The home node of a block is statically determined and is not changed forever. When a client reads/writes a file, a cache manager finds a block first in its own cache and sends block request to the home node if a block is not in its own cache. Every read/write request to a block missed in every client' file system cache goes to the home node so that there is cache locality in the home-based Coopc-PVFS. Cache locality and little information management overhead in the home-based Coopc-PVFS can improve more read performance than the hint-based Coopc-PVFS. Also, a home node can buffer written data from clients and all write operations to a block can be gathered and buffered in a home node. Therefore, the home-based cooperative cache can improve more write performance than the hint-based Coopc-PVFS and other previous cooperative caches - only focused on shared read data - because of write buffering and gathering. In the hint-based Coopc-PVFS, there is much penalty of false information and information management overhead so that read/write performance could be worse than the home-based Coopc-PVFS.

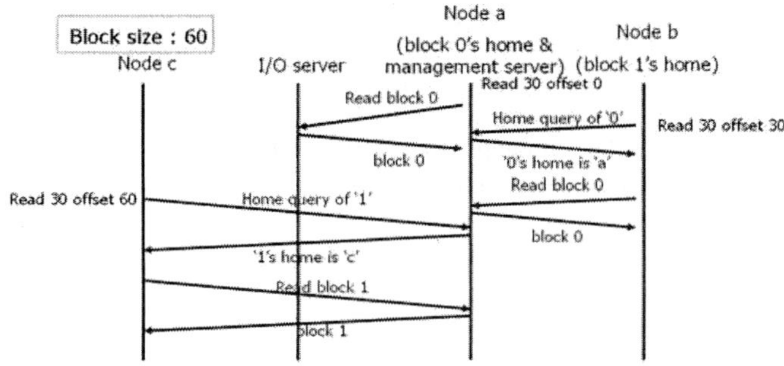

Fig. 2. Message transfers when read operations occur in Home-based Coopc-PVFS

For scalability of home-based Coopc-PVFS, we manage servers per file to reduce home node management overhead: every file has its management server. The management server allocates and manages home nodes of a file. Therefore, we can distribute home node management overhead over nodes. The management server of a file is determined at first file open time in the system and this management server information is saved in the metadata. Therefore, every client can know the management server of a file.

Figure 2 shows message transfers when read operations occur.

When a client reads a block, a cache manager in the client finds a block from its own cache first and read operation is over if there is a block in its own cache. If there is not in its own cache, a cache manager finds a home node of the block. If the home node is not allocated, a cache manager gets the home node from a management server of the file. After knowing the home node, a cache manager gets the block from the home node and caches it so that read operation is over. When a home node gets block requests from other nodes, it transfers the block if there is the block in its own cache or it transfers the block to other clients' cache managers after it gets the block from I/O servers if there is not. Also, the home node remembers nodes that read the block so that the home node can invalidate all cached blocks when there are write operations to the block.

Figure 3 shows message transfers when write operations occur.

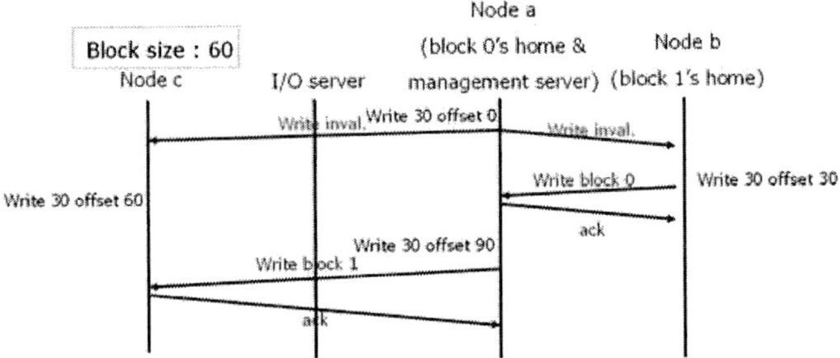

Fig. 3. Message transfers when write operations occur in Home-based Coopc-PVFS

When a client writes a block, a cache manager looks up which node is the home node. If there is no home node, a cache manager gets the home node from a manager node.

If the home node is itself, a cache manager invalidates all cached blocks in others' cache before writing a block and buffers written data to its own cached block. If the home node is another node, a cache manager sends written data to the home node and the home node cache manager invalidates all cached blocks in others' caches before buffering written data to its own cache. Therefore, all written data are buffering in cooperative cache and write performance can be improved in home-based cooperative cache. Also, all written data for a block can be buffering in only one node-home node and all write operations to a block are gathered in the home node so that we can reduce I/O servers' overhead.

The home-based Coopc-PVFS is implemented like the previous hint-based cooperative cache. A cache manager is not using Linux page cache system so that a replacement manager in a cache manager manages cached blocks according to the amount of free memory in the system using LRU list. For managing home nodes, a management server per file manages home lists. To manage management servers of files in a metadata manager, we added management servers' lists to a metadata manager. After a client opens a file, a cache manager gets home nodes of a file from a management server. Using this information, a cache manager can read/write blocks to/from home nodes of blocks.

5 Performance Evaluation

We used CAN cluster [6] in KAIST to evaluate the performance of Coopc-PVFS. The system configuration of CAN cluster is presented in table 1.

Table 1. System configuration

CPU	Pentium IV 1.8GHz
Memory	512MByte 266MHz DDR
Disk	IBM 60G 7200rpm
Network	3c996B-T(Gigabit Ethernet) 3c17701-ME(24port Gigabit Ethernet Switch)
OS , PVFS	Linux(Kernel version 2.4.18) , 1.6.0

The metadata manager was allocated in one node and the I/O server was allocated in another node. And four other clients were used to execute the test applications –a simple matrix multiplication program and BTIO benchmark programs. Each program operates like below:

◦ Matrix multiplication program: Each process in four nodes reads each part of two input files of 1024*1024 matrix and does matrix multiplication, and then writes results to a output file. Using the MPI library [8], applications are executed in four nodes.

◦ BTIO benchmark programs: BTIO is a parallel file system benchmark. It is one of NAS Parallel Benchmark [7]. BTIO contains four programs. In table 2, we present each program. We can evaluate the performance using four nodes with smallest size -class s- in BTIO.

Table 2. BTIO benchmark programs

Full (*mpi_io_full*)	MPI I/O with collective buffering
Simple (*mpi_io_simple*)	MPI I/O without collective buffering
Fortran(*fortran_io*)	Fortran 77 file operations used
Epi (*ep_io*)	Each process writes the data belonging to its part of the domain to a separate file

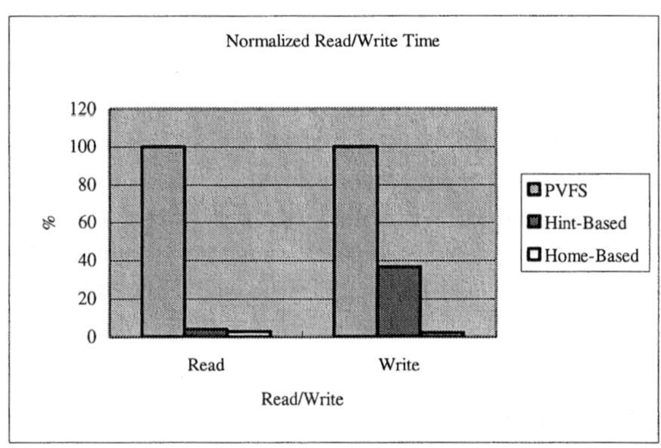

Fig. 4. Normalized read/write time in matrix multiplication program

5.1 Execution Time of Matrix Multiplication Program

In figure4, we present normalized total read/write time based on total read/ write time in PVFS when we execute matrix multiplication program.

Matrix multiplication program is a read-dominant program and total read time in PVFS is about 12 seconds. In the hint-based Coopc-PVFS and in the home-based Coopc-PVFS, total read time is about 0.5 seconds and 0.3 seconds. In this program, read files (input files) are always read and written file (output file) is always written. So, total read time in the home-based Coopc-PVFS is similar to in the hint-based Coopc-PVFS. Total write time in home-based Coopc-PVFS decreases about 94% of total write time in hint-based Coopc-PVFS because there is no write overhead and home nodes do write-buffering in home-based Coopc-PVFS.

5.2 Execution Times of BTIO Benchmark Programs

In figure 5, we present normalized total read/write time based on total read/write time in PVFS when we execute BTIO benchmark programs.

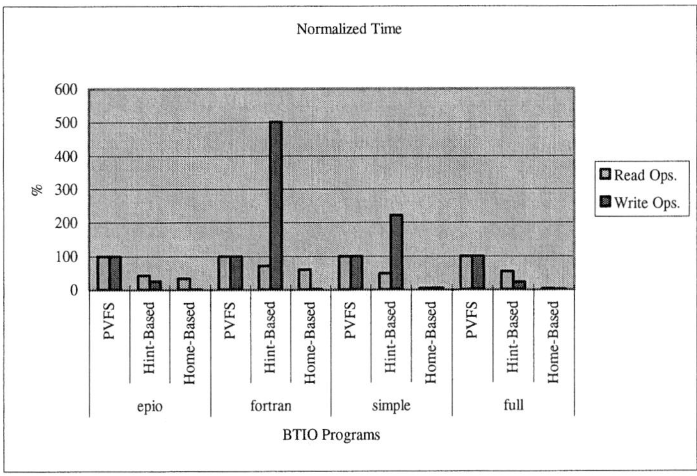

Fig. 5. Normalized read/write time in BTIO benchmark programs

Total read time in the home-based Coopc-PVFS is shorter than in the hint-based Coopc-PVFS and in PVFS because of caching effects and small management overhead. When read/write are all mixed up (fortran and simple cases), total write time in the hint-based Coopc-PVFS is much worst than in PVFS. In all programs, total write time in the home-based Coopc-PVFS is much shorter than in the hint-based Coopc-PVFS and in PVFS because all write operation is buffered in home nodes.

6 Conclusion and Future Work

In this paper, we present the home-based Coopc-PVFS. Also, we evaluate and analysis the performance of the home-based Coopc-PVFS in comparison to PVFS and to the hint-based Coopc-PVFS. The home-based Coopc-PVFS improve read/write performance more than the hint-based Coopc-PVFS. When read operation occurs, a client gets a block just from the home node in the home-based Coopc-PVFS with less overhead than in the hint-based Coopc-PVFS so that there is cache locality in the home-based Coopc-PVFS. When a client write a block, the home-based Coopc-PVFS does write buffering in the home node and gathers all write operations to a block in the home node so that write performance in the home-based Coopc-PVFS is much better than in the Coopc-PVFS and in PVFS.

In the future, we will evaluate the performance of Coopc-PVFS with real parallel programs and various benchmarks. We will do more research about cache management schemes and home node management schemes in the home-based cooperative cache.

References

[1] P. H. Carns, W. B. Ligon III, R. B. Ross, and R. Thakur, "PVFS: A Parallel File System For Linux Clusters", Proceedings of the 4th Annual Linux Showcase and Conference, Atlanta, GA, October 2000, pp. 317-327

[2] M.Vilayannur,M.Kandemir, A.Sivasubramaniam, "Kernel-Level Caching for Optimizing I/O by Exploiting Inter-Application Data Sharing", IEEE International Conference on Cluster Computing (CLUSTER'02),September 2002

[3] Dahlin, M., Wang, R., Anderson, T., and Patterson, D. 1994. "Cooperative Caching: Using remote client memory to improve file system performance", In Proceedings of the First USENIX Symposium on Operating Systems Design and Implemntation. USENIX Assoc., Berkeley, CA, 267-280

[4] Feeley, M. J., Morgan, W. E., Pighin, F. H., Karlin, A. R., and Levy, H. M. 1995. "Implementing global memory management in a workstation cluster", In Proceedings of the 15th symposium on Operating System Principles (SOSP). ACM Press, New york, NY, 201-212

[5] Prasenjit Sarkar , John Hartman, "Efficient cooperative caching using hints", Proceedings of the second USENIX symposium on Operating systems design and implementation, p.35-46, October 29-November 01, 1996, Seattle, Washington, United States

[6] Can cluster, http://camars.kaist.ac.kr/~nrl

[7] Parkson Wong, Rob F. Van der Wijngaart, NAS Parallel Benchmark I/O Version 2.4, NAS Technical Report NAS-03-002, NASA Ames Research Center, Moffett Field, CA 94035-1000

[8] W. Gropp, S. Huss-Lenderman, A. Lumsdaine, E. Lusk, W. Nitzberg, W. Saphir, M. Snir. MPI: The Complete Reference (vold.2). MIT Press, 1998

[9] In-Chul Hwang, Hojoong Kim, Hanjo Jung, Dong-Hwan Kim, Hojin Ghim, Seung-Ryoul Maeng and Jung-Wan Cho, Design and Implementation of the Coopeartive Cache for PVFS, International Conference on Computational Science 2004, Krakow, Poland, June 2004.

Improving the Data Placement Algorithm of Randomization in SAN

Nianmin Yao[1], Jiwu Shu[2], and Weimin Zheng[3]

[1] Department of computer science and technology, tsinghua university
lucos@126.com
[2] shujw@tsinghua.edu.cn
[3] zwm-dcs@tsinghua.edu.cn

Abstract. Using the randomization as the data placement algorithm has many advantages such as simple computation, long term load balancing, and little costs. Especially, some latest works have improved it to make it scale well while adding or deleting disks in large storage systems such as SAN (Storage Area Network). But it still has a shortcoming that it can not ensure load balancing in the short term when there are some very hot data blocks accessed frequently. This situation can often be met in Web environments. To solve the problem, based on the algorithm of randomization, an algorithm to select the hot-spot data blocks and a data placement scheme based on the algorithm are presented in this paper. The difference is that it redistributes a few very hot data blocks to make load balanced in any short time. Using this method, we only need to maintain a few blocks status information about their access frequency and more than that it is easy to implement and costs little. A simulation model is implemented to test the data placement methods of our new one and the one just using randomization. The real Web log is used to simulate the load and the results show that the new distributing method can make disks' load more balanced and get a performance increased by at most 100 percent. The new data placement algorithm will be more efficient in the storage system of a busy Web server.

1 Introduction

With the developing of the Internet, a dramatic growth of enterprise data storage capacity can be observed in the last couple of years. Many things including a lot of enterprise data coming onto the internet; data warehouse, e-business and especially the Web contents contribute the growth of storage. So the performance and capacity of the storage are becoming the bottleneck of IT's progress. Now, the SAN (Storage Area Network) is a popular technology to solve the problem. It can ensure the reliability, serviceability, scalability and availability of the storage. Though the SAN technology has so many virtues, there are also a few hard technical problems which need to be solved. One of these problems is how to place the data among the disks, e.g. the data placement algorithm.

V.S. Sunderam et al. (Eds.): ICCS 2005, LNCS 3516, pp. 415–422, 2005.

The good data placement algorithm can greatly improve the performance of the storage. More than that, it can also decrease the costs of reorganizing the data among the disks in SAN while adding or deleting a disk which is a very common operation in managing the storage system.

2 Related Work

An algorithm commonly being used to place data blocks is RAID[3]. It has been an industrial standard being applied in most of commercial products. This method has many virtues such as fault tolerance, high performance and so on. But it has a shortcoming that all the data must be reconstructed while adding or deleting disks. Additionally, the RAID 5 can not be used in large-scale storage system such as SAN because with the number of disks in the storage system increasing, its writing performance is becoming lower. So RAID is often used in one disk array and it should not be a good data placement algorithm in the whole SAN.

Another simple and effective method to distribute the data blocks among the SAN is using the randomization which is often used now. In this algorithm, every data block is put into a randomly selected physical disk block. But in fact, the physical disk block is not completely randomly selected. A pseudo-random function is often used because the right place for a data block can be easily found through a simple computation and need not to maintain a very large map list describing the relationship between the data blocks and the physical disk blocks. Randomly selecting the location of the data block is good at keeping the load balancing among all the disks in SAN in a long run. But as the RAID technology, while adding or deleting the disks, the whole data must be reorganized. And more than that, it can't ensure that load is balanced in any short term, for example, when there are hot data accessed frequently during a short time.

To decrease the costs of reorganizing the data when adding or deleting disks, an improved randomization algorithm was presented in [2]. The main work of [2] focused on the hash function being used in the randomization algorithm. Once the number of disks is changed, it only needs to change the hash function and move a little data in order to reconstruct the data. Similar works have been done in [1, 9].

Being improved, the algorithm of randomization now has so many advantages such as simple computation, long term load balancing, little costs and high scalability. But all the methods described above have the same shortcoming that they all can not ensure load is balanced when there is hot data during some short time. Because if there are some data blocks which are accessed very frequently, then the load is more often focused on a few disks where these hot data blocks are saved. In this situation which will often be met in Web environment, the load is not balanced and the storage resources can not be fully utilized while using the algorithms discussed above to distribute the data blocks. Intuitively, if the information of the access frequency of the data blocks is applied into the distributing scheme, we can make disks loads more balanced. But all the work above did not use this kind of information, because it will cost a lot of resources if the information of all the data blocks access frequencies are to be maintained. Some data placement algorithms which take the access frequency of blocks into account have been given, for example [5] , but they can only be used in a small storage system such as a disk array, because they totally ignore the high costs of maintaining all the information of access frequency of blocks. The data placement algorithm must be simple and effective enough to be used. We will give a new method based on the randomization in this paper to distribute the data blocks which can keep short term load balancing and costs little.

3 The Algorithm to Select Hot Data Blocks

The network traffic follows the self-similarity model, e.g. it can be bursty on many or all time scales[6]. So, there will often be a few hot spot data blocks which burden most loads in the servers. As discussed, the above distributing methods obviously can not deal with this situation well and make full use of the storage resources. Because not until the information of data blocks access frequency is concerned in distributing data blocks, we can not really keep the load balancing. But how to decrease the costs of maintaining the information of all the data blocks access frequency? In fact, for some servers on the Internet, we only need to maintain the access frequency of a small part of all the data blocks, because there are only a small part of files of the servers on the Internet which can be accessed frequently during a short time. In the study of distributed Web servers, many works make use of this characteristic of servers [2, 3]. For example, study of the world cup 98s Web log indicates that only 3MB space covers 99% requests in some short time. So the critical point to solve the problem is how to effectively select these hot data blocks.

In this section, we will give an algorithm to select the hot data blocks. Tts data structures are two lists which have fixed length. One called "HL" which will hold the hot data blocks and the other called "CL" which contains candidate hot data blocks. Their length is decided by the storage space and it can be 1% of the whole space in blocks or lower. HL's elements are ordered by blocks' access frequency and CL's elements are ordered approximately by their insert sequence. The elements' data structure contained in the list has information about address or access frequency such as frequency and hot_levle. When we say $D1 > D2$, it means that $D1.hot\_levle > D2.hot\_level$ or else $D1.hot\_levle = D2.hot\_level$ and $D1.frequency > D2.frequency$.

The algorithm is described below. In the algorithm, "HOT_LEVEL", "UPGRADE_LEVEL" and "CYCLE" are the predefined parameters. CYCLE is the interval of the number of time units to check and organize the two lists. "Time" always denotes the current time. The text behind "//" is comment.

```
(Step 1) Old_time=time;
         make all data blocks' frequency in HL and CL be 0;
(Step 2) A request of data block D comes;
```

```
(Step 3) if(D in HL)
         //D' in HL and D'.logical_address==D.logical_address
         then D'.frequency++;
         elseif(D in CL)
           then {D'.frequency++; move D' to CL's tail;}
         else {
           if (CL is full)  then remove the head of CL;
           insert D into CL's tail;
         }
(Step 4) if((time - old_time)<CYCLE) then goto (Step 2);
(Step 5) if((time - old_time)>(1+1/3)*CYCLE)  // *1
         then goto (Step 1);
(Step 6) for(all d in HL){  //deal with all the elements in HL
           if(d.frequency > HOT_LEVEL) then d.hot_level++;
           else{
             d.hot_level= d.hot_level/2;  //*2
             if(d.hot_level < 1) then remove d from HL;
           }
         }
         reorder elements in HL;
(Step 7) for(all d in CL){//deal with all the elements in CL
           if(d.frequency > HOT_LEVEL) then {
             d.hot_level++
             if (d.hot_level>UPGRADE_LEVEL and ((HL is not full)
                   or (HL is full and d>HL's tail) )then {
               if (HL is full) then remove HL's tail element;
               insert d into HL;
             }
           }
         }
(Step 8) goto (step 1);
```

We can see that when the algorithm is functioning, the elements in HL are the hottest data blocks whose access count is more than HOT_LEVEL in about a time unit for several continuous CYCLEs. The sentence marked by "*1" ensure that the computing error of frequency is not too big and when this sentence is true, we can believe that the system is much idle because there is no request in 1/3(can be changed according to the status of load) time unit. The sentence marked by "*2" makes the data block's hot_level decrease more quickly than it increases. Each time when an element in CL is accessed, it will be moved to the tail of CL so that hot data blocks are not easy to be squeezed out of the CL because we always squeeze the head element out of the CL. So when running for a while, elements in the CL must be relatively hot and can be candidates to be selected into the HL.

Obviously, this algorithm can make sure that all the hottest data blocks will be put in the HL and its costs are very low because there are only a little simple computations in it and it needs not to access the disks and maintain all the blocks access frequencies.

4 A New Data Placement Scheme

Now, we can get the information about the hottest data blocks by the algorithm given in section 3, then the next question is how to use it. We will present a new data placement scheme in this section based on the algorithm of randomization and the algorithm to select the hottest data blocks we have given.

SAN systems usually provide storage virtualization functions. Its architecture is described in fig. 1. The meta data server maintains all the information about the data distribution. Before a server reads a data block, it must first get the real location of the data from the meta data server. So we can implement the function of distributing data blocks in the meta server. By the way, some meta data can be cached in the server and the actual data accesses need not pass through the meta data server, so the meta data server usually can not become the bottleneck of the system.

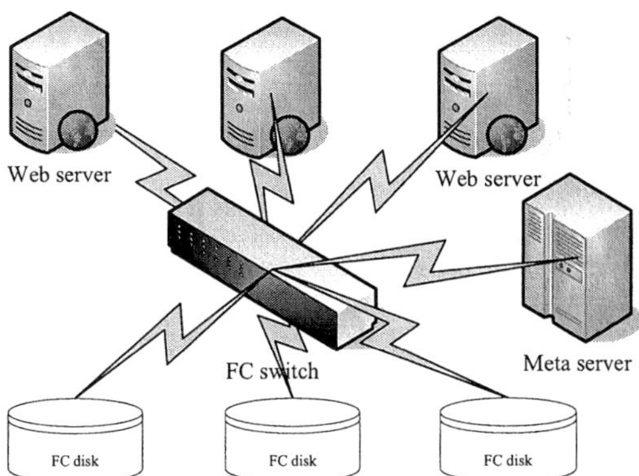

Fig. 1. Virtual Storage Architecture in SAN

In the meta server, we can implement the data placement function like this:

We always assign MAXQL as the command queue length of the busiest disk, MINQL as the command queue length of the idlest disk. In the following description, MAX_QUEUE_LENGTH and DIFF_QUEUE_LENGTH are pre-defined parameters. Initially, the data placement follows the randomization algorithm, for example, using some pseudo-random hash function. When the system is running, how our data placement scheme is functioning can be described below. These steps can be inserted after the algorithm's (Step 7) and before its (Step 8) described in section 3.

```
(Step 1) Get the correct value of MAXQL and MINQL.
(Step 2) if not (MAXQL> MAX_QUEUE_LENGTH) then goto (Step 9);
(Step 3) if not (MAXQL-MINQL)>DIFF_QUEUE_LEN then goto (Step 9);
(Step 4) if (There aren't any hot spot data blocks in the
            most busy disk)
         then goto (Step 9);
(Step 5) Copy the most hot block to the most idle disk;
(Step 6) Change the physical address of the moved data block
         in meta server;
(Step 7) if (There aren't any data blocks whose physical
             addresses were changed and now are not hot)
         then goto (Step 9);
(Step 8) Delete the redundant copies of the ever hot data blocks
         and change their physical addresses to their original
         addresses in meta server;
(Step 9) end;
```

In the scheme, we do not delete the original moved data block, and then the hot data block has two redundant copies, so we can also schedule a few requests to the hot data blocks for better balancing the load. In the scheme, some disks' space is spent for the redundant copy of hot data blocks. But it is so small that it can be ignored. For example, if the scheme is used in the Web server of World Cup 98, the wasted disk space is no more than 3M bytes.

In this distributing scheme, most of the data blocks' placement follows the randomization method. It only needs to process a little hot spot data blocks which have much effect on the storage performance during a short time. So, its costs are low enough to be applied in the practical storage system.

5 Experiments

To prove our design's efficiency, we implemented a simulation model depicted in fig. 1 using the CSIM18[10] which is a famous tool to simulate the discrete events. The model we are using is similar to the one described in [2] . In the model, one "ROUND" is defined as the time needed by a disk to process a data block request. The servers continuously send requests to the disks and record their response time. The difference between our model and the one in [2] is that in our model the disks' requests queue is long enough to hold all the requests, e.g. the requests will not be discarded. Especially, we use the real Web access log to drive the model, so it can truly reflect the SAN environment. The 1000,000 requests we used to drive the model are extracted from the world cup 98's web log at 1998.5.1. For simplification, every data block request represents a file request in the log.

In the first experiment, we set the number of disks as 5 and change the arriving rate of the requests. Its test results are drawn in fig. 2. In fig. 2, the X axis represents the interval of incoming requests in percentile of ROUND. So with the number decreasing, the load is heavier. We can see that in most of the situations, our distributing method is much better than the one using randomization except that when system is too idle or too busy. We deem that in the two extreme situations, the load balancing function of the new distributing method doesn't work well. The reason of the phenomena can be gotton in our data placement method such as (Step 5) in section 3 and (Step 2), (Step 3) in section 4. Because when the storage system are too idle or busy, it will not need to balance the load what is just our data placement method does.

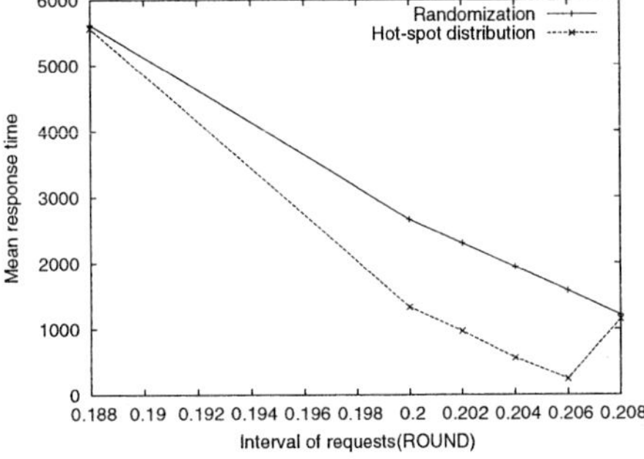

Fig. 2. Tests results of two data placement methods

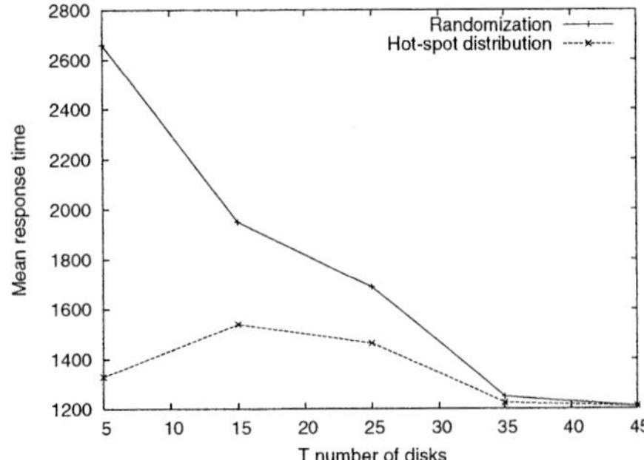

Fig. 3. Test results while changing the number of disks

In the testing, we found that the parameters of the model especially the HOT_LEVEL have very big effect on the performance. For example, when we set the HOT_LEVEL as 8, then when the interval of requests is 20% of a ROUND, the mean response time can reach at 880 ROUNDs which is much better than before. So if the parameters of the model can be adjusted dynamically according to the load of the system, we can get more performance enhancements. This will be our future work.

In our second experiment, we change the number of disks in the model and get the results described in fig. 3. In order to let the system be not too idle, we set the interval of requests as 1/n ROUND, in which n denotes the number of disks. Obviously, we can see from fig. 3 that our data placement scheme is much better than the other. It can be noticed that when the number of disks is 45, the two methods get the same results. We believe that it is because that the system is too idle in this configuration. The followed test proved our conjecure that when we increase the load, we get much better results than the randomization method. For simplification, we would not show this results here.

6 Conclusions

In this paper, some data placement algorithms are analyzed and they all can not keep disks' load balanced when there are hot data accessed frequently which have much effect on the performance of the system. The reason is that they do not take the access frequency of disks into consideration because they may think it will cost too much. So we present a new data placement method based on the randomization algorithm which can evenly distributing the hottest data blocks among disks and is also very easy to implement and costs little.

We implement a simulation model to test the new data placement algorithm and it proves that it can better balance the load of disks and get much higher performance. We will implement it in a true SAN system with the virtual storage function in the future. We can say that it will be very promising while applying the data placement algorithm into the storage system of a busy Web server.

Acknowledgements. This research is supported by China Postdoctor foundation(No.023240008), the National Natural Science Foundation of China (No: 60473101) and the National High-Tech Research and Develop Program of China (No: 2004AA111120).

References

1. A. Brickmann, K. Salzwedel, and C. Scheideler: Compact, adaptive placement schemes for non-uniform capacities. In Proceedings of the 14th ACM Symposium on Parallel Algorithms and Architectures (SPAA), Winnipeg, Manitoba, Canada (Aug, 2002) 53–62
2. Cherkasova L, Karlsson M: Scalable web server cluster design with workload-aware request distribution strategy with WARD. In: Proceedings of the 3rd International Workshop on Advanced Issues of E-Commerce and Web-Based Information Systems. Los Alamitos: IEEE Computer Society Press (2001) 212–221
3. D. A. Patterson, G. Gibson and R. H. Katz: A case for Redundant Arrays of Inexpensive Disks(RAID). In Proceedings of the 1988 ACM Conference on Management of Data(SIGMOD) (June 1988) 109–116
4. Du ZK, Zheng MY, Ju JB: A distributed algorithm for content-aware Web server clusters. Journal of Software, 2003,14(12) 2068–2073
5. G. Weikum, P. Zabback, and P. Scheuermann: Dynamic File Allocation in Disk Arrays. In Proc. of ACM SIGMOD, (May 1991) 406–415
6. J. C. Mogul: Network behavior of a busy web server and its clients, Technical Report WRL 95/5, DEC Western Research Laboratory, Palo Alto, CA (1995)
7. M. E. Crovella and A. Bestavros: Self-similarity in world wide web traffic: Evidence and possible causes, In Proceedings of the 1996 ACM SIGMETRICS International Conference on Measurement and Modeling of Computer Systems (May 1996)
8. P. Berenbrink, A. Brinkmann, and C. Scheideler: Design of the PRESTO multimedia storage network. In International Workshop on Communication and Data Management in Large Networks, Paderborn, Germany (October 5 1999)
9. R. J. Honicky, E. L. Miller: A fast algorithm for online placement and reorganization of replicated data. 17th International Parallel and Distributed Processing Symposium (IPDPS) (2003)
10. Schwetman, H.: Object-oriented simulation modeling with C++/CSIM17. In Proceedings of the 1995 Winter Simulation Conference. ed. C. Alexopoulos, K. Kang, W. Lilegdon, D. Goldsman. Washington, D.C. (1995) 529–533

Safety of a Server-Based Version Vector Protocol Implementing Session Guarantees*

Jerzy Brzeziński, Cezary Sobaniec, and Dariusz Wawrzyniak

Institute of Computing Science,
Poznań University of Technology, Poland
{Jerzy.Brzezinski, Cezary.Sobaniec, Dariusz.Wawrzyniak}@cs.put.poznan.pl

Abstract. Session guarantees are used to manage replica consistency of a distributed system from the client perspective. This paper defines formally the guarantees, presents and proves safety of a protocol implementing session guarantees using server-based version vectors.

1 Introduction

Replication is a key concept in providing high performance and availability of data and services in a distributed system. However, replication introduces the problem of data consistency that arises when replicas are modified. Required properties of distributed system with respect to consistency depend in general on application and are formally specified by *consistency models*. There are numerous consistency models developed for *Distributed Shared Memory* systems. These models, called *data-centric* consistency models [1], assume that servers replicating data are also accessing the data for processing purposes. In a mobile environment, however, clients accessing the data are not bound to particular servers, they can switch from one server to another. This switching adds a new dimension of complexity to the problem of consistency. *Session guarantees* [2], called also *client-centric* consistency models [1], have been proposed to define required properties of the system regarding consistency from the client's point of view. Intuitively: the client wants to continue processing after a switch to another server so that new operations will remain consistent with previously issued operations within a *session*. The relationships between data-centric and client-centric consistency models have been analyzed in [3, 4]. Protocols implementing session guarantees must efficiently represent sets of operations performed in the system. Version vectors based on vector clocks [5] may be used for this purpose. This paper presents a protocol, called VsSG, implementing session guarantees, that uses server-based version vectors. The protocol is based on a concept presented in [2]. The main contribution of this paper is a formal proof of safety of the VsSG protocol. The formalism introduced in this paper includes also formal definitions of session guarantees.

* This work was supported in part by the State Committee for Scientific Research (KBN), Poland, under grant KBN 3 T11C 073 28.

V.S. Sunderam et al. (Eds.): ICCS 2005, LNCS 3516, pp. 423–430, 2005.

2 Session Guarantees

In this paper we consider a weakly consistent replicated storage system. The system consists of a number of *servers* holding a full copy of a set of *shared objects*, and *clients* running applications that access the objects. Clients are separated from servers, i.e. a client application may run on a separate computer than the server. A client may access a shared object after selecting a single server and sending a direct request to the server. Clients are mobile, i.e. they can switch from one server to another during application execution. Session guarantees are expected to take care of data consistency observed by a migrating client. The set of shared objects replicated by the servers does not imply any particular data model or organization. Operations performed on shared objects are divided into *reads* and *writes*. A read does not change states of the shared objects, while a write does. A write may cause an update of an object, it may create a new object, or delete an existing one. A write may also atomically update states of several objects.

Operations on shared objects issued by a client C_i are ordered by a relation $\overset{C_i}{\rightarrow}$ called *client issue order*. A server S_j performs operations in an order represented by a relation $\overset{S_j}{\rightarrowtail}$. Writes and reads on objects will be denoted by w and r, respectively. An operation performed by a server S_j will be denoted by $w|_{S_j}$ or $r|_{S_j}$.

Definition 1. *Relevant writes $RW(r)$ of a read operation r is a set of writes that has influenced the current state of objects observed by the read r.*

The exact meaning of *relevant writes* will strongly depend on the characteristics of a given system or application. For example, in case of simple isolated objects (i.e. objects with methods that access only their internal fields), relevant writes of a read on object x may be represented by all previous writes on object x.

Session guarantees have been defined in [2]. The following more formal definitions are based on those concepts. The definitions assume that operations are unique, i.e. they are labeled by some internal unique identifiers.

Definition 2. *Read Your Writes (RYW) session guarantee is defined as follows:*

$$\forall C_i \forall S_j \left[w \overset{C_i}{\rightarrow} r|_{S_j} \Rightarrow w \overset{S_j}{\rightarrowtail} r \right]$$

Definition 3. *Monotonic Writes (MW) session guarantee is defined as follows:*

$$\forall C_i \forall S_j \left[w_1 \overset{C_i}{\rightarrow} w_2|_{S_j} \Rightarrow w_1 \overset{S_j}{\rightarrowtail} w_2 \right]$$

Definition 4. *Monotonic Reads (MR) session guarantee is defined as follows:*

$$\forall C_i \forall S_j \left[r_1 \overset{C_i}{\rightarrow} r_2|_{S_j} \Rightarrow \forall w_k \in RW(r_1) : w_k \overset{S_j}{\rightarrowtail} r_2 \right]$$

Definition 5. *Writes Follow Reads (WFR) session guarantee is defined as follows:*

$$\forall C_i \forall S_j \left[r \overset{C_i}{\rightarrow} w|_{S_j} \Rightarrow \forall w_k \in RW(r) : w_k \overset{S_j}{\rightarrowtail} w \right]$$

3 The VsSG Protocol Implementing Session Guarantees

The proposed VsSG protocol implementing session guarantees intercepts communication between clients and servers; at the client side before sending a request, at the server side after receiving the request and before sending a reply, and at the client side after receiving the reply. These interceptions are used to exchange and maintain additional data structures necessary to preserve appropriate session guarantees. After receipt of a new request a server checks whether its state is sufficiently up to date to satisfy client's requirements. If the server's state is outdated then the request is postponed and will be resumed after updating the server.

Servers occasionally exchange information about writes performed in the past in order to synchronize the states of replicas. This synchronization procedure eventually causes total propagation of all writes directly submitted by clients. It does not influence safety of the VsSG protocol but rather its liveness, therefore it will not be discussed in this paper (example procedure is presented in [6]). In contrast with [2] we do not assume total ordering of non-commutative writes which is treated by us as an orthogonal problem.

Every server S_j records all writes performed locally in a history. The writes result from direct client requests, or are incorporated from other servers during synchronization procedure. The writes are performed sequentially, therefore the history is totally ordered. Formally, histories are defined as follows:

Definition 6. *A history H_{S_j} is a linearly ordered set $\left(\mathcal{O}_{S_j}, \overset{S_j}{\rightarrowtail} \right)$ where $\mathcal{O}_{S_j}$ is a set of writes performed by a server S_j, and relation $\overset{S_j}{\rightarrowtail}$ represents an execution order of the writes.*

During synchronization of servers the histories are *concatenated.* Intuitively, a concatenation of two histories is constructed as a sum of the first history, and new writes found in the second history. The orderings of respective histories are preserved, and new writes are added at the end of the first history.

Server-based version vectors. Version vectors are used for efficient representation of sets of writes required by clients and necessary to check at the server side. Version vectors used in this paper have the following form: $\left[v_1 \; v_2 \; \ldots \; v_{N_S} \right]$, where N_S is the total number of servers in the system. A single position v_j denotes the number of writes performed by a server S_j, and changes whenever a new write request is performed by the server. Because every server increments the version vector for every write, and the changes are done at different positions, the values of version vectors at servers during execution of writes are unique. A write in VsSG protocol is labeled with a vector timestamp set to the current value of the version vector V_{S_j} of the server S_j performing the write for the first time. In the presentation of the VsSG protocol the vector timestamp of a write w is returned by a function $T : \mathcal{O} \mapsto V$. A single i-th position of the version vector timestamp associated with a write will be denoted by $T(w)[i]$.

The VsSG protocol (presented in Alg. 1) interacts with requests sent from clients to servers and with replies sent from servers to clients. A request is a couple $\langle op, SG \rangle$, where op is an operation to be performed, and SG is a set of session guarantees required for this operation. Before sending to the server, the request is supplemented with a vector W representing the client's requirements. A reply is a triple $\langle op, res, W \rangle$ where op is the operation just performed, res represents the results of the operation (delivered to the application), and W is a vector representing the state of the server just after performing the operation.

Before sending a request by a client C_i, a vector W representing its requirements is calculated based on the type of operation, and the set SG of session guarantees required for the operation. The vector W is set to either $\mathbf{0}$, or W_{C_i} — a vector representing writes issued by the client C_i, or R_{C_i} — a vector representing writes relevant to reads issued by the client, or to a maximum of these two vector (lines 1, 3 and 6). The maximum of two vectors V_1 and V_2 is a vector $V = \max(V_1, V_2)$, such that $V[i] = \max(V_1[i], V_2[i])$.

On receipt of a new request a server S_j checks whether its local version vector V_{S_j} *dominates* the vector W sent by the client (line 9), which is expected to be sufficient for providing appropriate session guarantees. A version vector V_1 dominates a version vector V_2, which is denoted by $V_1 \geq V_2$, when $\forall i : V_1[i] \geq V_2[i]$. If the state of the server is not sufficiently up to date, the request is postponed (line 10), and will be resumed after synchronization with another server (line 32). As a result of writes performed by a server S_j, its version vector V_{S_j} is incremented at position j (line 14), and a timestamped operation is recorded in history H_{S_j} (lines 15 and 16). The current value of the server version vector V_{S_j} is returned to the client (line 18) and updates the client's vector W_{C_i} in case of writes (line 20), or R_{C_i} in case of reads (line 22).

4 Safety of the VsSG Protocol

Definition 7. *A supremum of a set of writes O_{S_j}, denoted by $\overline{V}(O_{S_j})$, is a vector that is set to $\mathbf{0}$ for an empty set, and for nonempty sets its i-th position is defined as $\overline{V}(O_{S_j})[i] = \max_{w \in O_{S_j}} T(w)[i]$.*

Lemma 1. *For every server S_j running VsSG protocol at every moment $\overline{V}(O_{S_j}) = V_{S_j}$.*

Proof. By induction. 1) Basis. At the very beginning $V_{S_j} = \mathbf{0}$, and the set of writes $O_{S_j} = \emptyset$, therefore $\overline{V}(O_{S_j}) = \mathbf{0}$, hence $\overline{V}(O_{S_j}) = V_{S_j}$. 2) Induction step.

On send of request message $\langle op, SG\rangle$ from C_i to S_j

$W \leftarrow$ 0
if (iswrite(op) **and** MW $\in SG$) **or** (**not** iswrite(op) **and** RYW $\in SG$) **then**
    $W \leftarrow \max(W, W_{C_i})$
end if
if (iswrite(op) **and** WFR $\in SG$) **or** (**not** iswrite(op) **and** MR $\in SG$) **then**
    $W \leftarrow \max(W, R_{C_i})$
end if
send $\langle op, W\rangle$ to S_j

On receipt of request message $\langle op, W\rangle$ from client C_i at server S_j

while $\left(V_{S_j} \not\geq W\right)$ **do**
    wait
end while
perform op and store results in res
if iswrite(op) **then**
    $V_{S_j}[j] \leftarrow V_{S_j}[j] + 1$
    timestamp op with V_{S_j}
    $H_{S_j} \leftarrow H_{S_j} \cup \{op\}$
end if
send $\left\langle op, res, V_{S_j}\right\rangle$ to C_i

On receipt of reply message $\langle op, res, W\rangle$ from server S_j at client C_i

if iswrite(op) **then**
    $W_{C_i} \leftarrow \max(W_{C_i}, W)$
else
    $R_{C_i} \leftarrow \max(R_{C_i}, W)$
end if
deliver $\langle res\rangle$

On receipt of update message $\langle S_k, H\rangle$ at server S_j

foreach $w_i \in H$ **do**
    if $V_{S_j} \not\geq T(w_i)$ **then**
        perform w_i
        $V_{S_j} \leftarrow \max\left(V_{S_j}, T(w_i)\right)$
        $H_{S_j} \leftarrow H_{S_j} \cup \{w_i\}$
    end if
end for
signal

Every Δt at server S_j

foreach $S_k \neq S_j$ **do**
    send $\left\langle S_j, H_{S_j}\right\rangle$ to S_k
end for

Algorithm 1. VsSG protocol implementing session guarantees

Let us assume a state where condition $\overline{V}(O_{S_j}) = V_{S_j}$ holds. The set O_{S_j} and the version vector V_{S_j} can change only in the following two situations:

- The server S_j accepts a new write requested by a client. This causes the value of $V_{S_j}[j]$ to be incremented by 1, next the write is timestamped with the current value of vector V_{S_j}, and the write is added to O_{S_j}. This causes $\overline{V}(O_{S_j})$ to be also incremented at position j by 1 (lines 14 and 16 of Alg. 1). As a result the condition $\overline{V}(O_{S_j}) = V_{S_j}$ still holds.
- The server S_j incorporates a write w received from another server. This causes the current value of V_{S_j} to be maximized with the vector $T(w)$ of the write being added (line 28). The new write is then added to O_{S_j} (line 29). As a result values of V_{S_j} and $\overline{V}(O_{S_j})$ will be incremented at the same positions by the same values, therefore the condition $\overline{V}(O_{S_j}) = V_{S_j}$ still holds. □

Definition 8. *A write-set $WS(V)$ of a given version vector V is defined as $WS(V) = \bigcup_{j=1}^{N_S} \{w \in O_{S_j} : T(w) \leq V\}$.*

Lemma 2. *For any two vectors V_1 and V_2 used by servers and clients of the VsSG protocol $V_1 \geq V_2 \Leftrightarrow WS(V_1) \supseteq WS(V_2)$.*

Proof. 1) Sufficient condition. By contradiction, let us assume that $V_1 \geq V_2 \wedge WS(V_1) \not\supseteq WS(V_2)$, which means that $\exists w\, [w \notin WS(V_1) \wedge w \in WS(V_2)]$ and, according to Definition 8: $\exists j\, (T(w)[j] > V_1[j] \wedge T(w)[j] \leq V_2[j]) \Rightarrow V_1[j] < V_2[j] \Rightarrow V_1 \ngeq V_2$. 2) Necessary condition. By contradiction, let us assume that $WS(V_1) \supseteq WS(V_2) \wedge V_1 \ngeq V_2$, which means that $\exists j : V_1[j] < V_2[j]$. Version vectors at position j are only incremented when a new write is performed by a server S_j (line 14). It means that $\exists w \in O_{S_j}\, [w \in WS(V_2) \wedge w \notin WS(V_1)]$ and hence $WS(V_1) \not\supseteq WS(V_2)$. □

Lemma 3. *At any time during execution of VsSG protocol $O_{S_j} = WS(V_{S_j})$.*

Proof. By contradiction: 1) Let us assume that $\exists w \in O_{S_j} : w \notin WS(V_{S_j})$. According to Definition 8, a write w does not belong to $WS(V_{S_j})$ when $T(w) \nleq V_{S_j}$. This implies that $\exists k : T(w)[k] > V_{S_j}[k]$, and, according to Lemma 1, $T(w)[k] > \overline{V}(O_{S_j})[k]$, which implies $\overline{V}(O_{S_j}) \ngeq T(w)$. Based on Definition 7, $w \notin O_{S_j}$ — a contradiction. 2) Let us assume that $\exists w \in WS(V_{S_j}) : w \notin O_{S_j}$. According to Definition 7, a write w does not belong to O_{S_j} when $\overline{V}(O_{S_j}) \ngeq T(w)$. This implies that $\exists k : T(w)[k] > \overline{V}(O_{S_j})$, and, according to Lemma 1, $T(w)[k] > V_{S_j}[k]$, which implies $T(w) \nleq V_{S_j}$. Based on Definition 8, $w \notin WS(V_{S_j})$ — a contradiction. □

Lemma 4. *At any time during execution of VsSG protocol $WS(W_{C_i})$ contains all writes issued by a client C_i.*

Proof. A write issued by a client C_i and performed by a server S_j updates the client's vector W_{C_i} by calculating a maximum of its current value and value of the server version vector V_{S_j} (lines 18 and 20). Hence, after performing the write $W_{C_i} \geq V_{S_j}$, and (according to Lemma 2) $WS(W_{C_i}) \supseteq WS(V_{S_j})$, and (according to Lemma 3) $WS(W_{C_i}) \supseteq O_{S_j}$. It means that the write-set $WS(W_{C_i})$ contains all writes requested directly at server S_j, including also writes requested by the client C_i at server S_j. The vector W_{C_i} monotonically increases, therefore no past write is lost in case of a migration to another server. □

Lemma 5. *At any time during execution of VsSG protocol $WS(R_{C_i})$ contains all writes relevant to reads issued by a client C_i.*

Proof. A read issued by a client C_i and performed by a server S_j updates the client's vector R_{C_i} by calculating a maximum of its current value and value of the server version vector V_{S_j} (lines 18 and 22). Hence (according to Lemmata 2 and 3 $R_{C_i} \geq V_{S_j} \Rightarrow WS(R_{C_i}) \supseteq WS(V_{S_j}) = O_{S_j}$. It means that the write-set $WS(R_{C_i})$ contains all writes performed at server S_j, therefore also writes relevant to reads requested by the client C_i at server S_j. The vector R_{C_i} monotonically increases, therefore no past write is lost in case of a migration to another server. □

Theorem 1. *RYW session guarantee is preserved by VsSG protocol for clients requesting it.*

Proof. Let us consider two operations w and r, issued by a client C_i requiring RYW session guarantee. Let the read follow the write in the client's issue order, and let the read be performed by a server S_j, i.e. $w \overset{C_i}{\rightarrow} r|_{S_j}$. After performing w we have (according to Lemma 4) $w \in WS(W_{C_i})$. Because $V_{S_j} \geq W_{C_i}$ is fulfilled before performing r (lines 3 and 9), we get (according to Lemma 2) $WS(V_{S_j}) \supseteq WS(W_{C_i}) \Rightarrow w \in WS(V_{S_j})$. Because local operations at servers are totally ordered, we get $w \overset{S_j}{\rightarrowtail} r$. This will happen for any client C_i requiring RYW and any server S_j, so $\forall C_i \forall S_j \left[w \overset{C_i}{\rightarrow} r|_{S_j} \Rightarrow w \overset{S_j}{\rightarrowtail} r \right]$, which means that RYW session guarantee is preserved. □

Theorem 2. *MR session guarantee is preserved by VsSG protocol for clients requesting it.*

Proof. Let us consider two reads r_1 and r_2, issued by a client C_i requiring MR session guarantee. Let the second read follow the first read in the client's issue order, and let the second read be performed by a server S_j, i.e. $r_1 \overset{C_i}{\rightarrow} r_2|_{S_j}$. After performing r_1 we have (according to Lemma 5) $\forall w_k \in RW(r_1) : w_k \in WS(R_{C_i})$. Because $V_{S_j} \geq R_{C_i}$ is fulfilled before performing r_2 (lines 6 and 9), we get (according to Lemma 2) $WS(V_{S_j}) \supseteq WS(R_{C_i}) \Rightarrow \forall w_k \in RW(r_1) : w_k \in WS(V_{S_j})$. Because local operations at servers are totally ordered, we get $\forall w_k \in RW(r_1) : w_k \overset{S_j}{\rightarrowtail} r_2$. This will happen for any client C_i and any server S_j, so $\forall C_i \forall S_j \left[r_1 \overset{C_i}{\rightarrow} r_2|_{S_j} \Rightarrow \forall w_k \in RW(r_1) : w_k \overset{S_j}{\rightarrowtail} r_2 \right]$, which means that MR session guarantee is preserved. □

A theorem and a proof for MW are analogous to RYW. A theorem and a proof for WFR are analogous to MR. Full versions of the theorems and proofs can be found in [7].

5 Conclusions

This paper has presented formal definitions of session guarantees, a VsSG protocol implementing session guarantees, and finally a correctness proof showing that the protocol is safe, i.e. appropriate guarantees are provided. It is worth mentioning, however, that though the server-based version vectors used in the VsSG protocol are sufficient for fulfilling session guarantees, they are not necessary. Thus, other approaches are also possible, and they have been discussed in [8]. The sets of writes represented by version vectors are supersets of the exact sets resulting from appropriate definitions. The accuracy of the write-set representation is therefore an important factor of a protocol implementing session guarantees influencing its performance. This problem is currently considered, and appropriate simulation experiments are being prepared.

References

1. Tanenbaum, A.S., van Steen, M.: Distributed Systems — Principles and Paradigms. Prentice Hall, New Jersey (2002)
2. Terry, D.B., Demers, A.J., Petersen, K., Spreitzer, M., Theimer, M., Welch, B.W.: Session guarantees for weakly consistent replicated data. In: Proceedings of the Third International Conference on Parallel and Distributed Information Systems (PDIS 94), Austin, Texas, September 28-30, 1994, IEEE Computer Society (1994) 140–149
3. Brzeziński, J., Sobaniec, C., Wawrzyniak, D.: Session guarantees to achieve PRAM consistency of replicated shared objects. In: Proc. of Fifth Int. Conference on Parallel Processing and Applied Mathematics (PPAM'2003), LNCS 3019, CzOstochowa, Poland (2003) 1–8
4. Brzeziński, J., Sobaniec, C., Wawrzyniak, D.: From session causality to causal consistency. In: Proc. of 12th Euromicro Conference on Parallel, Distributed and Network-Based Processing (PDP2004), A Coruña, Spain (2004) 152–158
5. Mattern, F.: Virtual time and global states of distributed systems. In Cosnard, Quinton, Raynal, Robert, eds.: Proc. of the Int'l. Conf. on Parallel and Distributed Algorithms, Elsevier Science Publishers B. V. (1988) 215–226
6. Petersen, K., Spreitzer, M.J., Terry, D.B., Theimer, M.M., Demers, A.J.: Flexible update propagation for weakly consistent replication. In: Proc. of the 16th ACM Symposium on Operating Systems Principles (SOSP-16), Saint Malo, France (1997) 288–301
7. Brzeziński, J., Sobaniec, C., Wawrzyniak, D.: Safety of VsSG protocol implementing session guarantees. Technical Report RA-003/05, Institute of Computing Science, Poznań University of Technology (2005)
8. Kobusińska, A., Libuda, M., Sobaniec, C., Wawrzyniak, D.: Version vector protocols implementing session guarantees. In: Proc. of Int'l Symposium on Cluster Computing and the Grid (CCGrid 2005), Cardiff, UK (2005)

Scalable Hybrid Search on Distributed Databases

Jungkee Kim[1,2] and Geoffrey Fox[2]

[1] Department of Computer Science, Florida State University, Tallahassee FL 32306, U.S.A
jungkkim@cs.fsu.edu,
[2] Community Grids Laboratory, Indiana University, Bloomington IN 47404, U.S.A
gcf@indiana.edu

Abstract. We have previously described a hybrid keyword search that combines metadata search with a traditional keyword search over unstructured context data. This hybrid search paradigm provides the inquirer additional options to narrow the search with some semantic aspect from the XML metadata query. But in earlier work, we experienced the scalability limitations of a single-machine implementation. In this paper, we describe a scalable hybrid search on distributed databases. This scalable hybrid search provides a total query result from the collection of individual inquiries against independent data fragments distributed in a computer cluster. We demonstrate our architecture extends the scalability of a native XML query limited in a single machine and improves the performance for some queries.

1 Introduction

With the popularity of computer communication networks, there have been many efforts to share and exchange information between resources on the Net. There are two main approaches to search on networks—searching over structured data or searching over unstructured data. An archetypal example representing structured data is a relational database, while information retrieval represents search over the unstructured data. Extending these approaches to the Internet environment uncovers new research areas. The theories of SQL queries over structured data give way to XML queries—searching over semistructured data—while newly developed Web search engines are based on information retrieval technologies.

Our hybrid keyword search paradigm lies between those two approaches. The hybrid search is fundamentally based on keyword search for the unstructured data, and adds supplemental search on metadata attached to each unstructured document. We adopt XML—the de facto standard format for information exchange between machines—as a metalanguage for the metadata. We demonstrated the practicality of the hybrid keyword search in [8, 9], but we also experienced the scalability limitations of a single-machine implementation. Particularly, the native XML database we used had very limited scalability. In this

V.S. Sunderam et al. (Eds.): ICCS 2005, LNCS 3516, pp. 431–438, 2005.

paper, we will adopt a distributed strategy to improve the scalability of hybrid keyword search. We will experimentally illustrate performance improvement and larger scalability of the new architecture.

The rest of this paper is organized as follows. In the next section we describe our architecture and compare to a grid system. Section 3 describes our search architecture on distributed databases. We illustrate a query processing architecture on a distributed database in Section 4. In Section 5, we evaluate our hybrid search over a distributed database. We summarize and conclude in Section 6.

2 Hybrid Keyword Search on Distributed Databases

A distributed database is a collection of several or many databases stored in different computers and *logically interrelated* through a network. A distributed database management system (DDBMS) can be defined as a software application that manages a distributed database system so that to users it seems like a single machine—it provides transparency. Let us describe our distributed architecture according to the transparency criteria of [11].

- Data Independence: the hybrid searches on each machine are logically and physically independent each other. The data of each database have the same schema and the schema do not change. Each database has its own management system and only returns query results against the user inquiry. So our architecture has physical data independence due to data encapsulation in each machine.
- Network Transparency: the network connection of our system depends on message-oriented middleware or peer-to-peer middleware, and the middleware administrator or agent is only concerned with the connection. The end-user does not perceive the detailed network operation of the distributed hybrid search inquiry.
- Replication Transparency: we assume our distributed architecture is restricted to a computer cluster, and no replication exists in the cluster. Simply: our system is replication transparent. Data replication is usually necessary to increase the locality of data reference in a distributed environment. Actually, the locality is guaranteed in a clustering architecture.
- Fragmentation Transparency: our experiments with hybrid search on a distributed database will show that the data can be partitioned into each database within the limitation of the local database. Full partition is the easiest case for the distributed database management system. The data are partitioned into a chunk of XML instances with their associated unstructured data, and this type of fragmentation is *horizontal.*

We can summarize that in our architecture each database is totally independent. The query result for the distributed databases is the collection of query results from individual database queries.

Data independence and horizontal fragmentation features make our architecture different from other general architectures for federated database systems surveyed in [13]. If data are not independent and user inquiries require joining query data from different machines with various schemas, we can consider a distributed querying framework based on recently emerging Grid infrastructure [5, 4]. OGSA-DQP (Open Grid Services Architecture; Distributed Query Processing) [1] is an example of such framework.

3 Distributed Database Architecture

The architecture of the hybrid search service on each local machine depends on the local service provider who joins the distributed database, and we demonstrated such architectures in [8, 9]. One is utilizing an XML-enabled relational DBMS with nested subqueries to implement the combination of query results against unstructured documents and semistructured metadata. The other is based on a native XML database and a text search library. To associate metadata with unstructured documents, we assign the file name for the document as the key of the metadata. A hash table is used for a temporary storage of metadata query results and the keyword search maps to the table subsequently for joining.

The remaining issue for organizing the distributed database is the computer network technology that connects the databases, and provides the network transparency to the user. Tanenbaum and Steen [14] suggested that message-oriented middleware is one of best application tools for integrating a collection of databases into a multidatabase system. We utilize a message-oriented middleware implementation—NaradaBrokering [12, 6]. It includes JMS compliant topic-based communication, which meets the minimum requirements for the network transparency in a distributed database. NaradaBrokering also provides a cooperating broker network for increased network scalability.

Figure 1 summarizes our general architecture. The search client is a publisher for a query topic, and a group of search services subscribe on the same topic. When the client publishes the query, the query message is broadcast to the search service subscribers. The query results from the search services are returned back to the client by publishing the message on a dynamic virtual channel—a *temporary topic* whose session object was attached to the query message originally delivered to the search service.

The architecture can be grown to a cooperating broker network. One or more heterogeneous search services could be attached to each broker and each broker can relay messages to other brokers. The cooperative network usually follows a hierarchical structure, for better performance.

4 Query Processing

The query processing of each database (fragment) on our distributed architecture depends on the local database or the search library. Due to the full partition in our database distribution, the query processing in the DDBMS is simple, and query propagation and result collections are the only interesting processes.

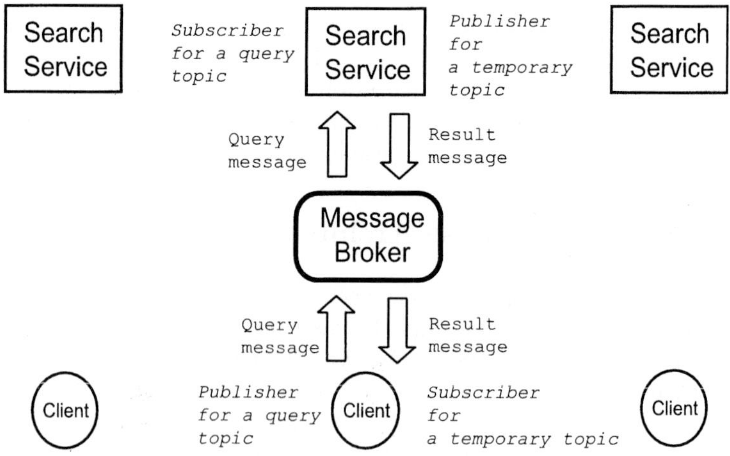

Fig. 1. Scalable Hybrid Search Architecture on Distributed Databases

A query processing architecture on a distributed database is shown in figure 2. The search client is a query publisher and the search service is a query subscriber. The user types query parameters through a user interface, and they are amalgamated into a message along with a job property and a temporary topic. The job property in this case is an integer property assigned to the "QUERY" final variable. The JMS `MapMessage` message type is used for the query message in our Java program.

The content of this message is a set of *name-value* pairs. The name is always a string and the value can have one of several allowed types. We used a string and an integer type. Those values could be empty for the string, and zero for the integer, if there is no user input. The `JMSReplyTo` property is a message header that contains a temporary topic.

The message broker delivers query messages to search services that have already subscribed to the same topic. The listener in the search service captures the query message, which includes a property header filled with a temporary topic. The extracted query parameters are passed to the local query processing service, which produces query results.

Those query results are returned back only to the client that published the query message. This works because the temporary topic is unique to this client. The inquirer client listens for the returned messages, and displays query results to the user.

5 Experimental Performance

In this section, we evaluate our hybrid search over a distributed database. A cluster of 8 computers forms a distributed environment, and all computers are located within a local network whose network bandwidth is very high.

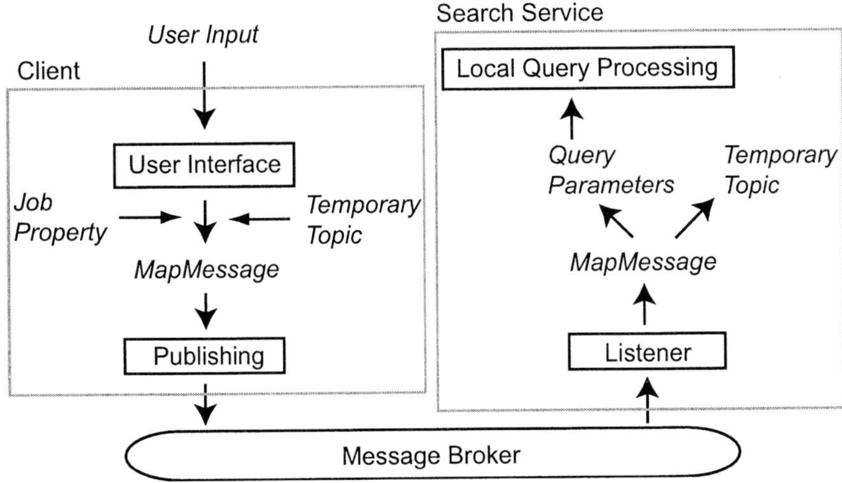

Fig. 2. A Query Processing Architecture on a Distributed Database

In these experiments we use 100,000 XML instances extracted from the *DataBase systems and Logic Programming* (DBLP) XML records [10] and 100,000 abstract text files from the TREC collection, OHSUMED [7]. We select article XML instances only, and non-ASCII characters are converted to ASCII characters. We cut off embedded metadata in the collection and extract the abstract part only. The data set is horizontally partitioned into 8 fragments, and each fragment on each machine has 12,500 XML instances and 12,500 text files. All eight computers have the same specification—2.4 GHz Intel Xeon CPU with 2 GB of memory, running a Linux 2.4 kernel and Java Hotspot VM 1.4.2. Those machines are connected with each other by 1 Gbps links, but the switch for the outside connections has 100 Mbps bandwidth. Apache Xindice 1.1b4 [3] is used as a native XML database, and Jakarta Lucene 1.3 [2] is utilized for text management.

We use two different communication middlewares—JXTA version 2.3 and NaradaBrokering version 0.96rc2—for the performance comparison, and three experimental architectures for 8 nodes as follows:

- JXTA: 1 node acts a rendezvous node and all other 7 nodes are connected to the rendezvous node. All of the nodes are located in the same subnet and all the query propagations are broadcast only within the subnet. This is because the current JXTA implementation does not meet the specification for selective query propagation from the rendezvous peer.
- Single NaradaBrokering: 1 node has a server and a search service, and the other nodes only have search services. All the search services are clients for the communication.
- Multiple NaradaBrokering Cluster: 1 node has a root server and the other nodes have second-level servers. Each node has a NaradaBrokering server

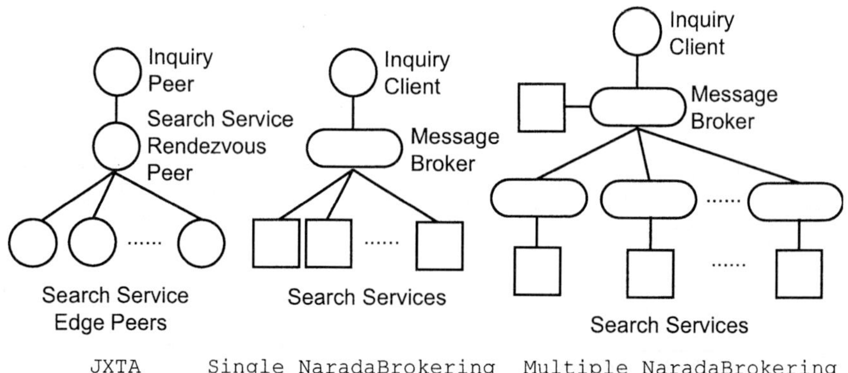

Fig. 3. Examples of communication middleware architectures

and a search service. This architecture gives us an idea of the performance difference between a cooperative network and a single message broker.

Examples of these architectures are shown in figure 3.

Figure 4(a) shows the average response time for an author exact match query over 8 search services using the three approaches. We choose 8 queries that combine an author and a keyword, and the databases are indexed against the author element in the XML instances and keywords for the unstructured data. Each query matches to only one of eight search services. The number of matches is between 1 and 3. We present the graphs separately for the average response time of no match and match result cases. We can interpret the difference between matched and non-matched query time to mean that the additional overhead for processing matched results is more than half of the total query time. The local processing time for non-matched query is very short as a result of the indexing against the author name. But the matched query takes long for joining the query results, and it is larger than the communication time. The query time of NaradaBrokering connections is shorter than that of JXTA connections. The time for eight connections of NaradaBrokering is a little shorter than 1 broker connection. But considering standard deviations—more than 70 ms—the two NaradaBrokering cases are statistically indistinguishable. NaradaBrokering uses Java NIO—the new I/O for better performance provided in JDK 1.4. JXTA version 2.3 does not use the Java NIO features.

We evaluate further hybrid search queries for the year match. In these queries, there are two different keyword selections. One has a few keyword matches—only 4 documents in 100,000 unstructured data documents, and the other has many keyword matches—41,889 documents out of 100,000. The year in the query is "2002" and it is hit in 7,752 out of 100,000 XML instances. We only show single NaradaBrokering case because multiple NaradaBrokering cluster had little difference. Figure 4(b) shows the average response time for the year match hybrid query over 8 search services in the cluster. Compared with the per-

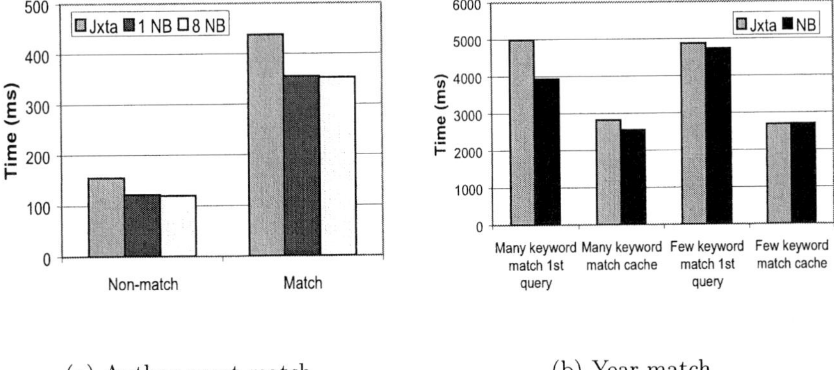

(a) Author exact match

(b) Year match

Fig. 4. Average response time for queries over 8 search services

formance experiment results on a single machine [1] running an XML-enabled commercial database, the year match queries on the distributed database show dramatic improvement in performance. The query time is improved from 82.9 seconds to less than 0.3 second for a few keyword matches, and from half an hour to less than 0.3 second for many keyword matches. This big performance improvement derives from *horizontal partitioning*, which reduces data retrieving time from disk with the relation partitioning. We divide those data into each machine by continuous ranges, and it is called *range partitioning*. The second and later repeats of the same inquiry take a shorter time than the first attempt. This is because the internal cache on the databases improves the query response time for recently answered inquiries. The query time in NaradaBrokering middleware is faster than that in JXTA, similar to the author matches.

Through experimental performance tests, we established that a distributed database can overcome the limit in target size of XML query when Xindice is run on a single machine. There was no problem up to 100,000 target XML instances in the query over the distributed database. The local native XML database could not produce accurate query results when the number of stored XML instances was over 16,000. Another contribution from the distributed query processing is the performance improvement for some queries over large target results. Our experiment was focused on the exact match, because Xindice does not provide context-based indexing on XML elements and attributes—there was no performance improvement from XML element indexing for an approximate match search.

[1] This machine was used for performance tests in [9]. We should add that since this paper was originally submitted we evaluated newer version of the commercial database in one of the cluster computers with large resource allocations. With this modified configuration, the query time was less than 1 second for the many keyword matches.

6 Conclusion

In this paper we described a scalable hybrid search on distributed databases. Those distributed architectures are mainly based on several or many computer network connections utilizing a message-oriented middleware or a peer-to-peer network framework. The hybrid search provides a total query result generated from a union of queries against data fragments in a computer cluster. The aspect of horizontal partitioning for our architecture contributed a performance improvement for some queries comparing to those on a single machine. Furthermore our new architecture extended the scalability of Xindice XML query, limited to a small size on a single machine.

References

1. N. Alpdemir, A. Mukherjee, N. Paton, and P. Watson. Service-Based Distributed Querying on the Grid. In *Proceedings of International Conference on Service Oriented Computing (ICSOC)*, December 2003.
2. Apache Software Foundation. Jakarta Lucene. World Wide Web. http://jakarta.apache.org/lucene/.
3. Apache Software Foundation. Xindice. World Wide Web. http://xml.apache.org/xindice/.
4. F. Berman, G. Fox, and A. Hey, editors. *Grid Computing: Making The Global Infrastructure a Reality.* John Wiley & Sons, 2003.
5. I. Foster and C. Kesselman, editors. *The Grid 2: Blueprint for a New Computing Infrastructure.* Morgan Kaufmann, 2003.
6. G. Fox, S. Pallickara, and S. Parastatidis. Towards Flexible Messaging for SOAP Based Services. In *Proceedings of International Conference for High Performance Computing and Communications(SC)*, November 2004.
7. W. Hersh, C. Buckley, T. Leone, and D. Hickam. OHSUMED: An interactive retrieval evaluation and new large test collection for research. In *Proceedings of the 17th Annual ACM SIGIR Conference*, 1994.
8. J. Kim, O. Balsoy, M. Pierce, and G. Fox. Design of a Hybrid Search in the Online Knowledge Center. In *Proceedings of the IASTED International Conference on Information and Knowledge Sharing*, November 2002.
9. J. Kim and G. Fox. A Hybrid Keyword Search across Peer-to-Peer Federated Databases. In *Proceedings of East-European Conference on Advances in Databases and Information Systems (ADBIS)*, September 2004.
10. M. Ley. Computer Science Bibliography. World Wide Web. http://www.informatik.uni-trier.de/~ley/db/.
11. T. Ozsu and P. Valduriez. *Principles of Distributed Database Systems.* Prentice Hall, 1999.
12. S. Pallickara and G. C. Fox. NaradaBrokering: A Distributed Middleware Framework and Architecture for Enabling Durable Peer-to-Peer Grids. In *Proceedings of International Middleware Conference*, June 2003.
13. A. Sheth and J. Larson. Federated Database Systems for Managing Distributed, Heterogeneous, and Autonomous Databases. *ACM Computing Surveys*, 22(3):183—236, September 1990.
14. A. Tanenbaum and M. Steen. *Distributed Systems: Principles and Paradigms.* Prentice Hall, 2002.

Storage QoS Control with Adaptive I/O Deadline Assignment and Slack-Stealing EDF

Young Jin Nam and Chanik Park[†]

School of Computer and Information Technology,
Daegu University,
Kyungbuk, Republic of Korea
yjnam@daegu.ac.kr
[†]Department of Computer Science and Engineering/PIRL,
Pohang University of Science and Technology,
Kyungbuk, Republic of Korea
cipark@postech.ac.kr

Abstract. Storage QoS control enforces a given storage QoS requirement for each I/O request from different storage clients that share an underlying storage system. This paper proposes an efficient storage QoS control scheme that features adaptive I/O deadline assignment and slack-stealing EDF scheduling. Simulation results with various I/O workloads show that the proposed scheme outperforms previous approaches in terms of response time variation, average response times, and miss ratio of the target response time.

1 Introduction

Embedding QoS feature into a storage system needs to define storage QoS specifications, map the storage QoS specifications (requirements) onto the underlying storage resources, and enforce the storage QoS requirements for each I/O request from different virtual disks (storage clients). This paper mainly emphasizes the storage QoS enforcement, also called the real-time QoS control (briefly QoS control). It is generally accepted that a storage system is characterized by its I/O performance; that is, the IOPS and RT relationship that depicts the variation of an average response time as a function of I/O requests per second (briefly IOPS). Thus, our QoS specification should capture this basic feature of a storage system as the first step. While the storage QoS specification in a broad sense may encompass other features of a storage system, such as data reliability, system costs, our QoS specification focuses mainly on the aspect of storage I/O performance that includes an average request size, a target response time, and a target IOPS. We define a QoS requirement of the virtual disk as a storage service required from a virtual disk in terms of QoS specification. The QoS requirement from a virtual disk i (VD_i) is represented as ($SZ_i, IOPS_i^{targ}, RT_i^{targ}$), where SZ_i represents an average I/O request size(KB), $IOPS_i^{targ}$ represents a target IOPS, and RT_i^{targ} represents a target response time(msec) [1]. The QoS requirement can be easily expanded to support a storage cluster environment and a more detailed specification having multiple pairs of a target IOPS and a target response time [1].

A few QoS control schemes for storage resources have been introduced [1, 2, 3, 5]. We can categorize the characteristics of the previous schemes into three classes. Class 1

V.S. Sunderam et al. (Eds.): ICCS 2005, LNCS 3516, pp. 439–446, 2005.

includes the derivatives of packet-based fair queuing schemes for network resources [3]. It proportionates the entire storage bandwidth according to a given set of resource weights allotted to each virtual disk that shares the same storage system. In addition, it attempts to reorder a given I/O sequence in order to reduce overhead caused by disk head movements. Note that they do not directly take control of the demanded response time; instead, they control only the storage bandwidth. Class 2 operates mainly on a rate-based QoS control using a leaky bucket [5]. It attempts to guarantee a given QoS requirement simply by throttling the IOPS of the incoming I/O requests from each virtual disk. Class 2 is expected to have the same drawbacks as Class 1. Class 3 guarantees the target response time by assigning a deadline time to each incoming I/O request only if the current IOPS is not greater than its target IOPS and then scheduling the pending I/O requests according to the EDF (Earliest Deadline First) scheme [2]. If the current IOPS is greater than its target IOPS, I/O requests have no deadline. Otherwise, the deadline is set by adding the target response time to the current time. Let us call this type of I/O deadline assignment target-IOPS-based I/O assignment. In contrast to Class 1 and 2, this approach directly controls the target response time for a given target IOPS. It can also support a QoS requirement with multiple pairs of a target IOPS and a target response time.

2 The Proposed Scheme

The key features of the proposed QoS control scheme include the adaptive I/O deadline assignment based on the current IOPS and the current queue depth of a virtual disk, and slack-stealing EDF scheduling that exploits any available slack between the current time and the earliest deadline time to minimize the underlying storage overhead. Note that the proposed scheme falls into Class 3 that takes control of both the target IOPS and the target RT for each I/O request from different virtual disks.

Adaptive I/O Deadline Assignment. The key idea is to adaptively determine a deadline time of each I/O request according to the current IOPS and the current queue depth for each virtual disk. First, the proposed assignment scheme obtains an actual target response time denoted by $act\_RT_i^{targ}$ of VD_i according to the current IOPS condition with respect to its target IOPS, as given in Algorithm 1. If the current IOPS is higher than its target IOPS, an I/O request is served as if it were a best-effort I/O request having no deadline. If the current IOPS is equal to the target IOPS, the actual target RT is the same as the original target RT. If the current IOPS is lower than its target IOPS, its actual target RT decreases in proportion to the ratio of the current IOPS to the target IOPS. Second, the proposed assignment scheme empirically measures the current average queue depth denoted by $qdepth_i^{targ}$ for each VD_i. It computes a unit target response time denoted by $unit\_rt_i^{targ}$ for $act\_RT_i^{targ}$, as given in Algorithm 1. Note that the $unit\_rt_i^{targ}$ is meaningful only if the current IOPS is not greater than the target IOPS. Finally, the proposed assignment scheme assigns the deadline of r_i^k as a function of the current position in the queue denoted by $qpos_i^{cur}(r_i^k)$; that is, the deadline of an I/O request increases from the queue head in a piece-wise linear manner up to its target RT by $unit\_rt_i^{targ}$. In consequence, the proposed assignment scheme can avoid delays with the processing of I/O requests from a virtual disk having a larger target RT until all the I/O requests from virtual disks having a smaller target RT are processed.

Algorithm 1: Adaptive I/O deadline assignment

input : r_i^k, $IOPS_i^{cur}$, T_{cur}
output : $deadline(r_i^k)$
begin
  if $(IOPS_i^{cur} > IOPS_i^{targ})$ **then**
    $deadline(r_i^k) \leftarrow T_{be}$;
  else

$$act\_RT_i^{targ} \leftarrow \begin{cases} T_{be} & \text{if } IOPS_i^{cur} > IOPS_i^{targ} \\ RT_i^{targ}\left(\frac{IOPS_i^{cur}}{IOPS_i^{targ}}\right) & \text{if } IOPS_i^{cur} \leq IOPS_i^{targ} \end{cases};$$

$$unit\_rt_i^{targ} \leftarrow \frac{act\_RT_i^{targ}}{qdepth_i^{targ}};$$

$$deadline(r_i^k) \leftarrow T_{cur} + min\{RT_i^{targ}, unit\_rt_i^{targ} \times qpos_i(r_i^k)\};$$

  end
end

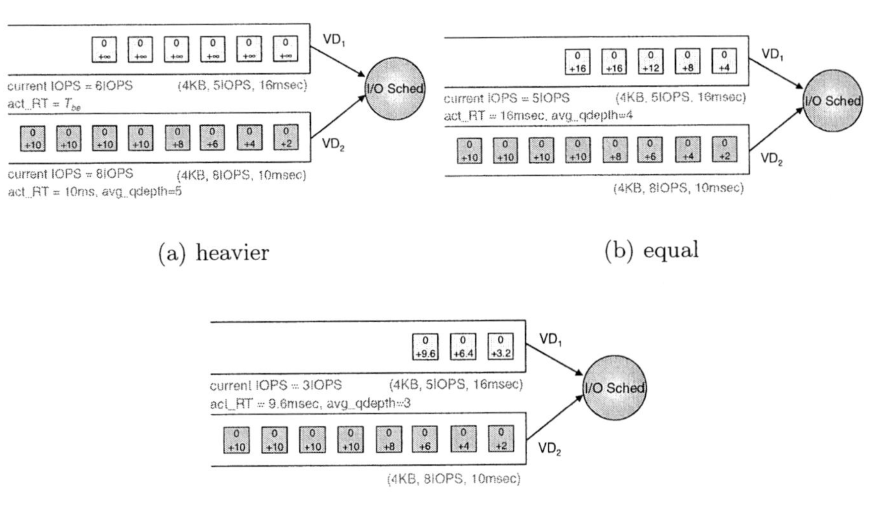

(a) heavier

(b) equal

(c) lighter

Fig. 1. Examples of the adaptive I/O deadline assignment with $IOPS_2^{cur} = IOPS_2^{targ}$: (a) heavier ($IOPS_1^{cur} > IOPS_1^{targ}$), (b) equal ($IOPS_1^{cur} = IOPS_1^{targ}$), (c) lighter ($IOPS_1^{cur} = IOPS_1^{targ}$)

Figure 1 show three examples of the adaptive I/O deadline assignment scheme. Each example assumes that Q_1 and Q_2 for VD_1 and VD_2 are (4KB, 5IOPS, 16msec) and (4KB, 8IOPS, 10msec), respectively. First, Figure 1 presents an example for the condition that $IOPS_1^{cur} > IOPS_1^{targ}$ and $IOPS_2^{cur} = IOPS_2^{targ}$. Since the current IOPS of VD_1 is higher than its target IOPS, the I/O deadline of each I/O request for VD_1 is set to T_{be}, implying no deadline. By contrast, since the current IOPS of VD_2 is equal to its target IOPS, its actual target RT of $act\_RT_2^{targ}$ is 10msec, as with its original target RT. Assuming that the observed average queue depth is 5, we obtain that $unit\_rt_2^{targ} = 2$ msec. Finally, the deadline of each I/O request from the queue head increases from 2msec up to 10msec by 2msec.

Second, Figure 1(b) shows an example of the proposed deadline assignment when the current IOPS of both storage clients are the same as their target IOPS; that is, $IOPS_1^{cur} = IOPS_1^{targ}$ and $IOPS_2^{cur} = IOPS_2^{targ}$. The I/O deadline assignment of VD_1 is performed as with VD_2 in the previous example. Since the current IOPS of VD_1 is equal to its target IOPS, its actual target RT of $act\_RT_1^{targ}$ is 20msec. Assuming that the observed average queue depth is 4, we obtain that $unit\_rt_1^{targ} = 5$ msec. Finally, the deadline of each I/O request from the queue head starts from 5msec and increases up to 20msec by 5msec. The I/O deadline assignments by the proposed scheme for VD_1 and VD_2 in Figure 1(b) reveals that the I/O scheduler can choose the I/O requests from each queue in a fair manner. Third, Figure 1(c) presents an illustrative example for the condition that the current IOPS of VD_1 is smaller than its target IOPS. To begin, the actual target RT of $act\_RT_1^{targ}$ decreases to 9.6msec in proportion to the ratio of the current IOPS to its target IOPS. Assuming that the observed average queue depth is 3, we obtain that $unit\_rt_1^{targ} = 3.2$ msec. Finally, the deadline of each I/O request from the queue head starts from 3.2msec and increases up to 9.6msec by 3.2msec. Finally, observe that the proposed assignment scheme determines a deadline time of each I/O request adaptively to the changing IOPS by adjusting its target response time and by increasing the I/O deadline by its unit target response time from the queue head. Consequently, we expect that the adaptive I/O deadline assigner will be able to give better fairness in servicing I/O requests with low RT variations in response times and provide better performance isolation to different virtual disks that share the same storage system.

Slack-Stealing EDF Scheduling: Another feature of the proposed QoS control scheme is to exploit the available slack between the current time and the earliest deadline time statistically in order to minimize storage overhead. The proposed scheduling algorithm selects an I/O request that not only minimizes the underlying storage overhead when scheduled, but also causes no deadline miss for the I/O request having the earliest deadline time. The proposed scheduling algorithm operates in two steps. First, it determines a set of I/O requests that entail no deadline miss for the I/O request with the earliest deadline, denoted by $R_{eligible}$. Next, it selects an I/O request that is likely to minimize the underlying storage overhead caused mainly by mechanical disk head movements. Then, the design of the proposed scheduling algorithm raises the following two issues: how to predict the service time of an I/O request and how to estimate storage overhead caused by scheduling the I/O request.

The proposed scheduling algorithm needs to compute the service time of a given I/O request in order to exploit or steal any existing slack time between the current time and the time when the service of the I/O request having the earliest deadline should be started to meet its deadline. Unfortunately, it is not possible to precisely predict the I/O service time under either a disk or a storage system. Previous research to estimate the I/O service time exists based on a theoretical model of disks or storage systems [6]. However, this approach has the following drawbacks. First, it cannot capture the feature of a time-varying service time according to changes in I/O workload patterns. Second, modeling a disk or a storage system requires understanding the detailed architectures for the disk or the storage system that are generally unavailable. Thus, the proposed scheduling algorithm assumes that the I/O service time is time-variant, and it measures the time-varying I/O service time by monitoring the service time of each I/O request as used in [4]. That is, it collects the service time of each I/O request during a given monitoring interval and then averages out the service times during the interval. We denote with $serv\_time(t)$ the current I/O service time at time t. Note that a single I/O service time is used for all virtual disks that share the underlying storage, because the service time does not include the queuing delay in the pending queues. Storage overhead of a single disk is equal to the overhead time to move from the current head position to the position to serve the I/O request. An exact computation of overhead time demands to estimate a seek time and a rotational delay between two I/O requests. However, in large-scale storage systems typically equipped with a large non-volatile cache memory, the exact estimation of storage overhead becomes almost impossible. Thus, the proposed scheduling algorithm simply estimates the overhead time between two I/O requests as the absolute distance between the start block address of the given I/O request and that of its previous I/O request.

Algorithm 2 gives the description on the proposed scheduling algorithm. We assume that N virtual disks share the same storage system. Recall that PQ_i represents the I/O pending queue dedicated to VD_i. The notation of r_i^h represents the I/O request at the head of PQ_i. Denote with $addr(r_i^k)$ the start block address of r_i^k. We define a set of eligible I/O requests $R_{eligible}$ that resides at the head of the I/O pending queue and entails no deadline miss for the I/O request that has the earliest deadline time.

Algorithm 2: Slack-stealing EDF scheduling

$T_{cur} \leftarrow$ current time;
$serv\_time(t) \leftarrow$ current I/O service time;
assume that $deadline_{earliest} = deadline(r_e^h)$;
$addr_{last} \leftarrow$ the start block address scheduled at last;

// determine $R_{eligible}$ set
if $T_{cur} + serv\_time(t) \leq deadline_{earliest} - serv\_time(t)$ **then**
  $R_{eligible} \leftarrow \{r_i^h | r_i^h \in PQ_i \text{ and } PQ_i \neq \emptyset\}$;
  find r_s^h, $abs(addr(r_s^h), addr_{last}) = min_{r_i^h \in R_{eligible}}\{abs(addr(r_i^h), addr_{last})\}$;
else
  $r_s^h \leftarrow r_e^h$;
endif
remove r_s^h from PQ_s and schedule it to the underlying storage;

In order to determine $R_{eligible}$, the proposed scheduling algorithm examines only the first I/O request at each queue in order to minimize the I/O scheduling overhead. When the proposed scheduling algorithm inspects all the pending I/O requests within all the queues, its time complexity becomes $O(N + M)$, where N is the number of the virtual disks that share the storage system and M is the maximum number of I/O requests within the pending queues. Considering $M >> N$, the approach that checks all pending I/O requests is expected to cause a considerable amount of overhead for I/O scheduling with the increase of the number of the pending I/O requests and the number of virtual disks that share the storage system. By contrast, the time complexity of the proposed scheduling algorithm is only $O(N)$.

3 Performance Evaluations

We evaluate the performance of the proposed QoS control scheme on our storage simulator that consists of an I/O workload generator, a set of virtual disks (storage clients), an underlying storage system. (See [1] for more details.) We also assume the following simulation environments. Two virtual disks named VD_1 and VD_2 issue only read I/O requests. The QoS requirement of each virtual disk is defined as $(SZ_1^{targ}, IOPS_1^{targ}, RT_1^{targ})$ = (4KB, 45IOPS, 70msec) and $(SZ_2^{targ}, IOPS_2^{targ}, RT_2^{targ})$ = (4KB, 45IOPS, 100msec). Note that the underlying storage system can satisfy the given QoS requirements properly. We determined the QoS requirements empirically by measuring the I/O performance of the underlying storage system for the given I/O workloads.

Adaptive I/O Deadline Assignment: To begin, we focus on evaluating the effectiveness of the proposed assignment scheme by disabling the slack-stealing EDF scheme; instead, the EDF scheduler is employed as an I/O scheduler. We use four workload sets: WS_1^d, WS_2^d, WS_3^d, and WS_4^d. Each workload set is characterized by different I/O traffic intensity from VD_2; that is, the current IOPS of VD_2 becomes higher than its target IOPS in WS_1^d $(IOPS_2^{cur} > IOPS_2^{targ})$, the current IOPS of VD_2 is equal to its target IOPS in WS_2^d $(IOPS_2^{cur} = IOPS_2^{targ})$, the current IOPS of VD_2 becomes lower than its target IOPS in WS_3^d $(IOPS_2^{cur} < IOPS_2^{targ})$, and the current IOPS of VD_2 becomes much lower than its target IOPS in WS_4^d $(IOPS_2^{cur} << IOPS_2^{targ})$. We can expect that the proposed assignment scheme will outperform the previous target-IOPS based I/O assignment under the workload sets of WS_3^d and WS_4^d, where the current IOPS of VD_2 becomes smaller than its target IOPS. Table 1 compares the performance results of the target-IOPS-based I/O deadline assignment and the proposed assignment scheme. As expected, the proposed assignment scheme reduced the variation of response times of VD_2 in WS_3^d and WS_4^d On the contrary, it increased the variation of response times of VD_1, whereas the target RT miss ratio still remains zero. In summary, the simulations results verified that the proposed assignment scheme could overcome the drawbacks of the target IOPS-based I/O deadline assignment; that is, *a high RT variation of response times due to unfairness in the processing of I/O requests from virtual disks having a larger target response time, and poor performance isolation by assigning its deadline based on its original target RT regardless of the current IOPS.*

Slack-Stealing EDF Scheduling: Recall that the proposed scheduling algorithm selects an I/O request that not only minimizes the underlying storage overhead when scheduled, but also causes no deadline miss for the I/O request having the earliest deadline time. Note that the scheduling algorithms under test will employ the adaptive I/O deadline assigner. Three workload sets, WS_1^s, WS_2^s, and WS_3^s, are used for this performance evaluation. The I/O traffic intensity for each virtual disk in WS_1^s is equal to the given QoS requirements; that is, $IOPS_1^{targ}$ and $IOPS_2^{targ}$. However, the I/O traffic intensity for each virtual disk in WS_2^s and WS_3^s is increased by 10% and 20% respectively, compared with its target IOPS. Table 2 compares the performance results of the EDF scheduling and the proposed scheduling scheme. Observe that the proposed scheduling scheme reduced the target response time(RT) miss ratio of VD_1 and VD_2 in WS_2^s and WS_3^s by improving the response times of each virtual disk. To summarize, the simulation results verified that *the slack-stealing EDF scheduling algorithm could reduce storage overhead by reordering the I/O requests as long as the deadline times of I/O requests are not missed, resulting in better target RT miss ratios and average response times, compared with the EDF scheduling algorithm.*

Table 1. Summary of the performance results of the adaptive I/O deadline assignment

		Avg. IOPS (IOPS)		Avg. RT (msec)		Var. RT		Targ. RT miss ratio	
		targ-IOPS	prop	targ-IOPS	prop	targ-IOPS	prop	targ-IOPS	prop
WS_1^d	VD_1	43.5	43.4	41.4	40.8	–	–	0.21	0.19
	VD_2	67.7	68.0	122.1	118.8	–	–	0.58	0.58
WS_2^d	VD_1	44.8	44.0	22.8	22.1	105.0	73.6	0	0
	VD_2	44.0	44.7	26.9	27.5	159.5	161.4	0	0
WS_3^d	VD_1	44.6	44.2	16.4	17.0	**14.6**	**12.7**	0	0
	VD_2	24.7	25.0	16.8	15.9	**20.8**	**17.3**	0	0
WS_4^d	VD_1	44.5	44.5	12.4	12.6	**2.2**	**2.6**	0	0
	VD_2	5.0	5.1	14.0	12.5	**17.9**	**6.5**	0	0

Table 2. Summary of the performance results of the slack-stealing I/O scheduling

		Avg. IOPS (IOPS)		Avg. RT (msec)		Targ. RT miss ratio	
		EDF	prop	EDF	prop	EDF	prop
WS_1^s	VD_1	44.0	44.5	22.1	22.4	0.00	0.00
	VD_2	44.7	44.4	27.5	23.6	0.00	0.00
WS_2^s	VD_1	48.7	49.2	28.2	28.5	**0.03**	**0.02**
	VD_2	49.1	49.1	40.8	29.8	**0.02**	**0.00**
WS_3^s	VD_1	53.4	53.8	45.1	39.3	**0.16**	**0.07**
	VD_2	52.2	53.1	58.8	41.0	**0.10**	**0.01**

4 Conclusion and Future Work

This paper proposed an efficient QoS control scheme that enforces the QoS requirements of multiple virtual disks (or storage clients) that share the same storage system. The proposed QoS control scheme consists of two key components: the adaptive I/O deadline assignment and the slack-stealing EDF scheduling for storage systems. The key of the adaptive I/O deadline assignment is to adaptively determine the deadline time of each I/O request according to the current IOPS and the current queue depth. Thus, it could overcome the drawbacks of the target IOPS-based deadline assignment: a high RT variation of response times due to unfairness in the processing of I/O requests from virtual disks having a larger target response time, and poor performance isolation by assigning its deadline based on its original target RT regardless of the current IOPS. The key of the slack-stealing EDF scheduling is to steal any available slack between the current time and the earliest deadline time in order to minimize the underlying storage overhead. The proposed scheduling algorithm selects an I/O request that minimizes the underlying storage overhead when scheduled, while causing no deadline miss for the I/O request having the earliest deadline time. We raised two design issues concerning how to predict the I/O service time of an I/O request and how to estimate storage overhead (disk head movement) caused by scheduling an I/O request, and provided reasonable solutions. We implemented the proposed QoS control scheme on our storage simulator. The simulation results for the adaptive I/O deadline assignment under various competing I/O workload sets showed that the proposed assignment scheme not only provides better fairness with lower RT variation, but also assures a better performance isolation for each virtual disk. Performance evaluations for the slack-stealing EDF scheduling revealed that the proposed scheduling scheme could provide better target RT miss ratios and response times by reducing storage overhead under various I/O workload sets.

In future work, we plan to implement the proposed QoS control scheme on top of an actual storage system and evaluate its performance with actual I/O traffic. In addition, we need to evaluate different techniques to predict an I/O service time and storage overhead for the slack-stealing EDF scheduling.

Acknowledgments

This research was supported by the Daegu University Research Grant, No 20040825. This research has been also supported in part by the Ministry of Education of Korea for its support toward the Electrical and Computer Engineering Division at POSTECH through its BK21 program, in part by HY-SDR IT Research Center, and in part by grant No. R01-2003-000-10739-0 from the Basic Research Program of the Korea Science and Engineering Foundation.

References

1. Y. J. Nam, *Dynamic Storage QoS Control for Storage Cluster and RAID Performance Enhancement Techniques.* Ph.D Dissertation, POSTECH, February 2004.
2. C. Lumb, A. Merchant, and G. Alvarez, "Facade: Virtual storage devices with performance guarantees," in *Proceedings of Conference on File and Storage Technologies*, March 2003.
3. Y. Nam and C. Park, "A new proportional-share disk scheduling algorithm: Trading-off I/O throughput and qos guarantees," *Lecture Notes in Computer Science*, vol. 1067, pp. 257–266, June 2003.
4. A. Chandra, W. Gong, and P. Shenoy, "Dynamic resource allocation for shared data centers using online measurements," in *Proceedings of the 11th International Workshop on Quality of Service*, June 2003.
5. H. Lee, Y. Nam, and C. Park, "Regulating I/O performance of shared storage with a control theoretical approach," in *Proceedings of the 21st IEEE Mass Storage Systems Symposium/12th NASA Goddard Conference on Mass Storage Systems and Technologies (MSST2004)*, April 2004.
6. M. Uysal, G. Alvarez, and A. Merchant, "A modular, analytical throughput model for modern disk arrays," in *Proceedings of the Ninth International Symposium on Modeling, Analysis and Simulation of Computer and Telecommunications Systems*, pp. 183–192, August 2001.

High Reliability Replication Technique for Web-Server Cluster Systems

M. Mat Deris[1], J.H. Abawajy[2], M. Zarina[3], and R. Mamat[4]

[1] Faculty of Information Technology and Multimedia,
College University Tun Hussein Onn
Batu Pahat, Johor, Malaysia
mustafa@kustem.edu.my
[2] Deakin University, School of Information Technology,
Geelong, VIC, Australia
jemal@deakin.edu.au
[3] KUSZA College,Kuala Terengganu, Malaysia
zarina@kusza.edu.my
[4] Department of Computer Science,
Faculty of Science and Technology, KUSTEM, K.Terengganu, Malaysia
rab@kustem.edu.my

Abstract. Providing reliable and efficient services are primary goals in designing a web server system. Data replication can be used to improve the reliability of the system. However, mapping mechanism is one of the primary concerns to data replication. In this paper, we propose a mapping mechanism model called enhanced domain name server (E-DNS) that dispatches the user requests through the URL-name to IP-address under Neighbor Replica Distribution Technique (NRDT) to improve the reliability of the system.

1 Introduction

With the ever increasing applications in the world wide web (WWW) such as distance learning education and e-commerce, the need for a reliable web server is likely to increase [6]. Thus, providing reliable and efficient services are primary goals in designing a web server cluster (WSC) system. This is due to the constraints of the eventual failure of hardware or/and software components. In order to provide reliable services, a WSC needs to maintain the availability of some data replicas while preserving one-copy consistency among all replicas [4]. Therefore, data replication plays an important role in a WSC as a highly reliable system.

The most common approaches in the replication techniques are the synchronous and asynchronous replications. The former provides a 'tight consistency' between data stores. This means that the latency between the data consistency achieved will be zero. Data in all nodes are always the same, no matter from which replica the updated originated. However, synchronous replication has drawbacks in practice [5]. The major argument is that, the response time to execute an operation is high because the time taken for all nodes that have the same copy to 'agree' to execute an operation is high. Whilst the asynchronous replication provides a 'loose consistency' between data

V.S. Sunderam et al. (Eds.): ICCS 2005, LNCS 3516, pp. 447 – 456, 2005.

stores. The replication process occurs asynchronous to originating transaction. In other words, there is always some degree of lag between the time the originating transaction is committed and the effects of the transaction are available at any replica(s) [1]. Nevertheless, the response time is lower than that of the synchronous technique.

Reliability refers to the probability that the system under consideration does not experience any failure in a given time interval. Thus, a reliable WSC is one that can continue to process user's requests even when the underlying system is unreliable [3]. When components fail, it should still be able to continue executing the requests without violating the database consistency.

In this paper we propose the enhanced domain name server (E-DNS) algorithm as a mapping mechanism under the Neighbor Replica Distribution Technique (NRDT) to improve the reliability of the system. In case of server failure, the server will be switched to neighboring IP address.

The paper is organized as follows. Section 2 reviews NRDT model. Section 3 describes typical DNS. Section 4 describes the concept of E-DNS. Section 5 concludes the proposed work.

2 Neighbor-Replica Distribution Technique (NRDT)

2.1 The Model

In a replicated database, copies of a data object may be stored at several sites in the network. Multiple copies of a data object must appear as a single logical data object to the transactions. This is termed as one-copy equivalence and is enforced by the replica control technique. The correctness criteria for replicated database is one-copy serializability [11], which ensures both one-copy equivalence and the serializable execution of transactions. In order to ensure the one-copy serializability, a replicated data object may be read by reading a quorum of copies, and it may be written by writing a quorum of copies. The selection of a quorum is restricted by the quorum intersection property to ensure one-copy equivalence: For any two operations o[x] and o'[x] on a data object x, where at least one of them is a write, the quorum must have a non-empty intersection. The quorum for an operation is defined as a set of copies whose number is sufficient to execute that operation.

Briefly, a site *i* initiates a NRDT transaction to update its data object. For all accessible data objects, a NRG transaction attempts to access a NRDT quorum. If a NRDT transaction gets a write quorum with non-empty intersection, it is accepted for execution and completion, otherwise it is rejected. We assume for the read quorum, if two transactions attempt to read a common data object, read operations do not change the values of the data object. Since read and write quorums must intersect and any two NRDT quorums must also intersect, then all transaction executions are one-copy serializable.

In the design of the WSC, a client on the Internet will notice only one IP address coming from the cluster, not those individual servers in the cluster. The cluster (with only one IP address visible to the public) is composed of a node called Request Distributor Agent (RDA) and a group of servers. The servers are logically connected to each other in the form of a grid structure, each of which is connected to RDA. Figure 1 shows architecture of the cluster-server system with nine servers. The RDA will forward legitimate Internet requests to the appropriate servers in the cluster. It returns any replies from the servers back to the clients.

Each node has the premier data (eg. *file a* will be located to server A, *file b* will be located to server B, and so on). The RDA will forward any request to the appropriate server with the premier data file. This is done by maintaining a server-service table that contains all the services provided by the cluster together with the corresponding addresses and their neighbors.

One advantage of a server cluster over a single server is its high security. If a single server is used, it is reachable from the Internet and therefore vulnerable for vicious [3]. Only the RDA has the IP address visible from the Internet, and all other stations of the cluster bear only private IP address. Therefore, all cluster-server stations are not reachable directly from the outside. A firewall system may be installed on the RDA to protect the whole cluster. To attack one of the cluster server stations, one has to first land on RDA and launch an attack from there. Network Address Translation is used on RDA to translate the destination IP address of incoming packets to an internal IP address, and that of the outgoing packets to the IP address on Internet where the requests.

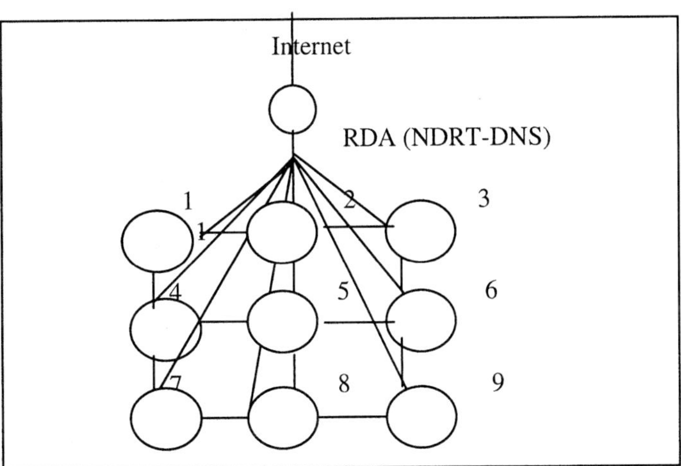

Fig. 1. A Cluster with 9 servers

2.2 The NRDT Technique

In NRDT, all sites are logically organized in the form of a two-dimensional grid structure. For example, if a NRDT consists of nine nodes, it will be logically organized in the form of 3 x 3 grid as shown in Figure. 1. Each node has a *master* data file. In the remainder of this paper, we assume that replica copies are data files. A node is either operational or failed and the state (operational or failed) of each node is statistically independent to the others. When a node is operational, the copy at the node is available; otherwise it is unavailable.

***Definition 2.2.1*:** A node X is a neighbor to node Y, if X is logically-located adjacent to Y.

A data will replicate to the neighboring nodes from its primary node. The number of data replication, *d*, can be calculated using Property 2.2, as described below.

***Property 2.2*:** The number of data replication from each node, $d \leq 5$.

Proof: Let n be a set of all nodes that are logically organized in a two-dimensional grid structure form. Then n nodes are labeled m(i,j), $1 \leq i < \sqrt{n}$, $1 \leq j < \sqrt{n}$. Two way links will connect nodes n(i,j) with its four neighbors, nodes m(i±1,j) and m(i,j±1), as long as there are nodes in the grid. Note that, four nodes on the corners of the grid, have only two adjacent nodes, and other nodes on the boundaries have only three neighbors. Thus the number of neighbors of each node is less than or equal to 4. Since the data will be replicated to neighbors, then the number of data replication from each node, *d*, is:

$$d \leq \text{the number of neighbors} + \text{a data from node itself} = 4+1 = 5.$$

For example, from Figure. 1, data from node 1 will replicate to node 2 and node 4 which are its neighbors. Node 5 has four neighbors, which are nodes 2, 4, 6, and 8. As such, node 5 has five replicas.

For simplicity, the primary node of any data file and its neighbors are assigned with vote one and vote zero otherwise. This vote assignment is called neighbor binary vote assignment on grid. A neighbor binary vote assignment on grid, B, is a function such that

$$B(i) \in \{0,1\},\ 1 \leq i \leq n$$

where B(*i*) is the vote assigned to node *i*. This assignment is treated as an allocation of replicated copies and a vote assigned to the node results in a copy allocated at the neighbor. That is,

$$1 \text{ vote} \equiv 1 \text{ copy}.$$

Let
$$L_B = \sum_{i=1}^{n} B(i)$$

where, L_B is the total number of votes assigned to the primary node and its neighbors and it also equals to the number of copies of a file allocated in the system. Thus, $L_B = d$.

Let r and w denote the read quorum and write quorum, respectively. To ensure that the read operation always gets up-to-date values, r + w must be greater than the total number of copies (votes) assigned to all nodes. The following conditions are used to ensure consistency:

1) $1 \leq r \leq L_B$, $1 \leq w \leq L_B$,
2) $r + w = L_B + 1$.

Conditions (1) and (2) ensure that there is a nonempty intersection of copies between every pair of read and write operations. Thus, the conditions ensure that a read operation can access the most recently updated copy of the replicated data. Timestamps can be used to determine which copies are most recently updated.

Let S(B) be the set of nodes at which replicated copies are stored corresponding to the assignment B. Then

$$S(B) = \{i| B(i) = 1, 1\leq i\leq n\}.$$

Definition 2.2.2: For a quorum q, a *quorum group* is any subset of S(B) whose size is greater than or equal to q. The collection of quorum group is defined as the *quorum set*.

Let Q(B,q) be the quorum set with respect to the assignment B and quorum q, then

$$Q(B,q) = \{ G| G\subseteq S(B) \text{ and } |G| \geq q\}$$

For example, from Figure. 1, let node 5 be the primary node of the master data file x. Its neighbors are nodes 2, 4, 6 and 8. Consider an assignment B for the data file x, such that

$$B_x(5)=B_x(2)=B_x(4)=B_x(6)=B_x(8) = 1$$

and $L_{B_x} = B_x(5)+ B_x(2)+ B_x(4)+ B_x(6)+ B_x(8) = 5$.

Therefore, $S(B_x) = \{5,2,4,6,8\}$.

If a read quorum for data file x, r =2 and a write quorum $w = L_{Bx}-r+1 = 4$, then the quorum sets for read and write operations are $Q(B_x,2)$ and $Q(B_x,4)$, respectively, where

$$Q(B_x,2)=\{\{5,2\},\{5,4\},\{5,6\},\{5,8\},\{5,2,4\},\{5,2,6\},\{5,2,8\},\{5,4,6\},\{5,4,8\},\{5,6,8\},\{2,4,6\},\{2,4,8\},\{2,6,8\},\{4,6,8\},\{5,2,4,6\},\{5,2,4,8\},\{5,2,6,8\},\{5,4,6,8\},\{2,4,6,8\},\{5,2,4,6,8\}\}$$

and

$$Q(B_x,4) = \{\{5,2,4,6\},\{5,2,4,8\},\{5,2,6,8\},\{5,4,6,8\},\{2,4,6,8\},\{5,2,4,6,8\}\}$$

2.3 The Correctness of NRDT

In this section, we will show that the NRDT technique is one-copy serializable. We start by defining sets of groups called *coterie* [12] and to avoid confusion we refer the sets of copies as *groups*. Thus, sets of *groups* are sets of sets of copies.

***Definition* 2.3.1.** *Coterie*. Let **U** be a set of *groups* that compose the system. A set of *groups* **T** is a coterie under **U** iff

i) $G \in \mathbf{T}$ implies that $G \neq \emptyset$ and $G \subseteq \mathbf{U}$.
ii) If $G, H \in \mathbf{T}$ then $G \cap H \neq \emptyset$ (intersection property)
iii) There are no $G, H \in \mathbf{T}$ such that $G \subset H$ (minimality).

By definition of coterie and from Definition 2.2.2, then Q(B,w) is a coterie, because it satisfies all coterie's properties.

Since read operations do not change the value of the accessed data object, a read quorum does not need to satisfy the intersection property. To ensure that a read operation can access the most recently updated copy of the replicated data, two conditions in sub-section 2.2 must be conformed. While a write quorum needs to satisfy read-write and write-write intersection properties. The correct criterion for replicated database is one-copy serializable. The next theorem provides us with a mechanism to check whether NRDT is correct.

Theorem 2.3.1. The NRDT technique is one-copy serializable.

Proof: The theorem holds on condition that the NRDT technique satisfies the quorum intersection properties, i.e., write-write and read-write intersections. Since Q(B,w) is a coterie then it satisfies the write-write intersection. However, for the case of a read-write intersection, it can be easily shown that $\forall G \in Q(B,r)$ and $\forall H \in Q(B,w)$, then $G \cap H \neq \varnothing$.

3 Overview of DNS

At its most basic level, the DNS provides distributed database of name-to-address mappings spread across a hierarchy of nameservers. The namespace is partitioned into a hierarchy of domains and subdomains with each domain administered independently by an authoritative nameserver. Nameservers store the mapping of names to address in resource records, each having an associated time to live (TTL) field that determines how long the entry can be cached by other nameserver in the system. A large TTL value reduces the load on the nameservers but limits the frequency of update propagation through the system. The different types of resource records and additional details about the DNS are described in [9]. The most widely used nameserver implementation in the DNS is the Berkeley Internet Name Domain (BIND) [2].

Nameserver can be implemented in the form of iterative or recursive queries. In an iterative query, the nameserver returns either an answer to the query from its local database, or a referral to another nameserver that may be able to answer the query. In handling a recursive query, the nameserver return a final answer, querying any other nameserver necessary to resolve the name. Most nameserver within the hierarchy are configured to send and accept only iterative queries. Local nameserver, however, typically accept recursive queries from clients.

4 E-DNS

E-DNS is proposed to enhance DNS according to the NRDT requirements. The function is similar to RDA in NRDT model. In case of server(s) fails, user cannot access data due to no other server take over the service. All clients or user requests from Internet are cross through to DNS to decide the right server to replay the request. When a server fails, the client who maps the name to IP address will find out that the server is down. The problem still unsolved although the client may press 'reload' or 'refresh' button in their browsers. This problem is unacceptable and unreliable. E-DNS is a viable alternative or solution to this problem because it is developed to take over the service in case of server(s) failure. To do this we used the complete DNS BIND software and only nameserver (it is the text file) in DNS will be updated according to match NRDT requirement. NRDT program will check the server fail/up every *t* second (time can be configured). E-DNS centralizes dispatching host name into single IP address and update it according to NRDT pattern. In case of server failures, the server will be switched to its neighboring IP address. One of the main attractions of this approach is its ease of deployment.

4.1 E-DNS Algorithm

We design an algorithm for E-DNS. There are three important modules for E-DNS algorithms: communication module, recovering module and neighbor module. The communication module will check the server either up or down/fail, while the neighbor module will seek the appropriate neighbors when the primary server fails and the recovering module will update the failed of nameserver in DNS.

```
Main
 Read time.conf  #user define the time
 Do while time true
  Call communication-module
  If died
   Call neighbor-module until true
  If true
   Call recovering-module
  End do
End Main

Procedure communication-module
 Create sockets
 Assign port to sockets
 Get service
 Get IP
 Ping IP
End procedure

Procedure Neighbor-Module
 Read from NRDT able (have logic structure by metric)
 Insert into array (#row,#cow)
 For i = 1 to 2
  For j = 1 to 2
   If i = 1
    If j = 1
     If row !=1
     New-metric = cow – 1,row
   If j = 2
    New-metric = cow + 1,row
  End if #i = 1

  If i =2
   If j = 1
     New-metric = cow, row – 1
   If j = 2
```

```
        New-metric = cow, row +1
    End if # i= 2
  End for #i
End for #j
End procedure
```

```
Procedure recovering-module
  Read DNS text fail
  Seek IP dead (#IP take from communication module)
  Replace new IP (#IP take from neighbor module)
  Re-run DNS software
End Procedure
```

4.2 Simulation of E-DNS

Assume that in a web sever cluster, there are six servers with the IP-address shown below:

1) p1.project.com is primary mail server with IP 999.999.9.111
2) p2.project.com is primary ftp server with IP 999.999.9.222
3) p3.project.com is primary web server with IP 999.999.9.333
4) p4.project.com is primary distance learning server with IP 999.999.9.444
5) p5.project.com is primary telnet server with IP 999.999.9.555
6) p6.project.com is primary other server with IP 999.999.9.666

1. Under normal *conFiguration* on BIND nameserver file, mapping address is simply done through the name to the IP address of single server as show below;

p1.project.com	IN	A	999.999.9.111
p2.project.com	IN	A	999.999.9.222
p3.project.com	IN	A	999.999.9.333
p4.project.com	IN	A	999.999.9.444
p5.project.com	IN	A	999.999.9.555
p6.project.com	IN	A	999.999.9.666

2. *ConFiguration* of the NRDT table consists of port, server name, IP and logical structure for every node (metric). The logical table *conFiguration* based NRDT is shown as.

80	p1.project.com	999.999.9.111	11
21	p2.project.com	999.999.9.222	12
23	p3.project.com	999.999.9.333	13
80	p4.project.com	999.999.9.444	21
80	p5.project.com	999.999.9.444	23
80	p6.project.com	999.999.9.444	24

3. The NRDT program will check the server either it is up or down/fail. For example, if the server with 999.999.9.222 (metric 12) fails, then its neighbors (could be with the

metrics 11,13,22) will be replaced (for example with IP address 999.999.9.333). The nameserver update becomes;

p1.project.com	IN	A	999.999.9.111
p2.project.com	IN	A	999.999.9.333
p3.project.com	IN	A	999.999.9.333
p4.project.com	IN	A	999.999.9.444
p5.project.com	IN	A	999.999.9.555
p6.project.com	IN	A	999.999.9.666

When requests come from Internet to browser ftp server, with the primary IP address 999.999.9.222, it will automatically mapped to the IP address 999.999.333. Thus, services on Internet are accessible at any point of time. Consequently, the reliability of the server is increased.

5 Conclusion

Web server cluster is a popular architecture used to improve the reliability and availability of the systems. However, with the current DNS algorithm, it is unreliable due to the fact that in a case of server(s) fails, users cannot access the data since no other server take over the service. In this paper, an enhanced domain name server (E-DNS) algorithm on NRDT technique, has been proposed to improve the reliability of the WSC. The algorithm is used as a mapping mechanism to dispatch the user requests through the URL-name to IP-address. It centralizes dispatching host name into single IP address and update it according to NRDT pattern. In case of server failures, the server will be switched to its neighboring IP address. The simulation showed that the proposed algorithm works well and provides a convenient approach to increase the reliability of WSC.

References

[1] M. Buretta, "*Data replication: Tools and Techniques for Managing Distributed Information*", John Wiley, New York, (1997).

[2] P. Albitz, and C.Liu, "DNS and BIND." *O'Reilly and Associates*, inc.,(2001).

[3] J. Liu, L Xu, B. Gu, J. Zhang, "Scalable, High Performance Internet Cluster Server", *IEEE 4th Int'l Conf.* HPC-ASIA, Beijing, pp. 941-944, (2000)

[4] M. Mat Deris, A. Mamat, P. C. Seng, H. Ibrahim, "Three Dimensional Grid Structure for Efficient Access of Replication Data", *Int'l Journal of interconnection Network, World Scientific*, Vol. 2, No. 3, pp 317-329, (2001).

[5] A. Moissis, "*Sybase Replication Server: A practical Architecture for Distributing and Sharinf Information*", Technical Document, Sybase Inc, (1996).

[6] E. Pacitti and E. Simon, " Update Propagation Strategies to Improve Fresh LazyMaster Replicated Databases", *Journal* VLDB, Vol. 8, no. 4, (2000).

[7] H. H. Shen, S. M. Chen, and W. M. Zheng,"Reserch on Data Replication Distribution Technique for Web Server Cluster", *IEEE Proc. 4th Int'l. Conference on Performance Computing,* Beijing, pp. 966-968, (2000).

[8] W. Zhou and A. Goscinski, "Managing Replicated Remote Procedure Call Transactions", *The Computer Journal*, Vol. 42, no. 7, pp592-608, (1999).
[9] P. Mockapetris , " Domain names-implementation and specication ," Internet Request for Commnets (RFC 1035), (November 1987).
[10] Internet Software Consortium, " Berkeley Internet domain (BIND)",http://www.isc.org /product /BIND, (June 2000).
[11] P.A. Bernstein and N.Goodman, " An Algorithm for Concurrency Control and Recovery in Replicated Distributed Databases,"*ACM Trans. Database Systems*,vol 9,no. 4(1994),pp.596-615.
[12] M. Maekawa," A $\sqrt{n}$ Algorithm for Mutual Exclusion in Decentralized Systems,"*ACM Trans. Computer Systems*,vol. 3,no. 2(1992), pp. 145-159.

An Efficient Replicated Data Management Approach for Peer-to-Peer Systems

J.H. Abawajy

Deakin University,
School of Information technology,
Geelong, Victoria, 3217 Australia

Abstract. The availability of critical services and their data can be significantly increased by replicating them on multiple systems connected with each other, even in the face of system and network failures. In some platforms such as peer-to-peer (P2P) systems, their inherent characteristic mandates the employment of some form of replication to provide acceptable service to their users. However, the problem of how best to replicate data to build highly available peer-to-peer systems is still an open problem. In this paper, we propose an approach to address the data replication problem on P2P systems. The proposed scheme is compared with other techniques and is shown to require less communication cost for an operation as well as provide higher degree of data availability.

1 Introduction

Peer-to-peer (P2P) network systems are one of the important and rapidly growing distributed system paradigms in which participants (the peers) rely on one another for service, rather than solely relying on dedicated and often centralized servers. The relationships among the nodes in the network are equal, nodes may join and leave the network in an ad-hoc manner and communication and exchange of information is performed directly between the participating peers. There is a growing research and industrial interest on peer-to-peer (P2P) systems. The success of P2P systems is due to many potential benefits such as fault-tolerance through massive replication; scale-up to very large numbers of peers, dynamic self-organization, load balancing, and parallel processing. Examples of P2P systems include Napster [3], Gnutella [4], and KaZaA[5] and Freenet [6].

Recently, peer-to-peer systems have become popular mechanism for large-scale content sharing. It is well known that techniques to increase the resilience and availability of stored data are fundamental to building dependable distributed systems. However, unlike traditional client-server applications that centralize the management of data in a few highly reliable servers, peer-to-peer systems distribute the burden of data storage, computation, communications and administration among thousands of individual nodes. Data management in this context offers new research opportunities since traditional distributed database techniques need to scale up while supporting high data autonomy, heterogeneity, and dynamicity.

V.S. Sunderam et al. (Eds.): ICCS 2005, LNCS 3516, pp. 457 – 463, 2005.

Generally P2P systems consider the data they offer to be very static or even read-only [9]. However, advanced peer-to-peer applications are likely to need more general replication capabilities. For example, a patient record may be replicated at several medical doctors and updated by any of them during a visit of the patient, e.g. to reflect the patient's new weight [8]. Other typical applications where new data items are added, deleted, or updated frequently by multiple users are bulletin-board systems, shared calendars or address books, e-commerce catalogues, and project management information [9]. Also, the inherent characteristics of peer-to-peer systems, require them to employ some form of replication to provide acceptable service to their users. For example, the erratic behaviour of online availability and the complete lack of global knowledge coupled with the absence of any centralisation makes P2P environments unreliable [9]. Ulike traditional distributed systems, the individual components of a peer-to-peer system experience an order of magnitude worse availability. This is because peer-to-peer systems are characterized by susceptibility to failure (e.g., node may be switched off), join and leave the system, have intermittent connectivity, and are constructed from low-cost low-reliability components. The study of a popular peer-to-peer file sharing system found that the majority of peers had availability rates of under 20% [1].

While much of the attention in the peer-to-peer systems research has been focused on the issues of providing scalability, free-rider problem or routing mechanisms within P2P networks, the resilience and availability of the data has so far seldom been mentioned. Therefore, how best to replicate data to build highly available peer-to-peer systems is still an open problem. In most peer-to-peer (P2P) systems data is assumed to be rather static and updates occur very infrequently. For application domains beyond mere file sharing such assumptions do not hold and updates in fact may occur frequently. Therefore, data replication in the presence of updates and transactions remains an open issue as well.

In this paper, we discuss an extension of our previous work on replica placement and management [2] to handle a system that support data replication in a transactional framework for weakly connected environments such as P2P systems. The proposed scheme uses quorum-based protocol for maintaining replicated data and shown to provide both high data availability and low response time. The proposed approach imposes a logical three dimensional grid structure on data objects based on a box shape organization and uses a sense-of-direction approach (SODA) for both read and write operations. We show that the proposed approach presents better average quorum size, high data availability, low bandwidth consumption, increased fault-tolerance and improved scalability of the overall system as compared to standard replica control protocols.

The rest of the paper is organized as follows. Section 2 presents related work. Section 3 presents our approach. Section 3 presents the proposed replica management protocol. In order to show the merits of the proposed approach, we present comparative analysis of the proposed approach against an existing approach in Section 4. Concluding remarks and future directions is reported in Section 5.

2 Related Work

The rapid popularization of Internet-based P2P applications such as Napster [3], Gnutella [4], and KaZaA[5] has inspired the research and development of technologies for P2P services and systems. An efficient data replication management (DRM) technique is one of the important P2P technologies. Through an efficient DRM, the availability of P2P services and their data can be significantly increased by replicating them on multiple systems connected with each other, even in the face of system and network failures [2]. From the viewpoint of data management these systems should address two critical areas [9]:

1. Efficient, scalable data access which is provided more or less by all approaches, and
2. Updates to the data stored, especially with respect to replication and low online probabilities.

However, the data sharing P2P systems like Gnutella and Kaaza deal with static, read-only files (e.g. music files) for which update is not an issue. Also, in systems such as Napster and Gnutella, replication occurs implicitly as each file downloaded by a user is implicitly replicated at the user's workstation. However, since these systems do not explicitly manage replication or mask failures, the availability of an object is fundamentally linked to its popularity and users have to repeatedly obtain the data. Also, if an update of a data item occurs this means that the peer that holds the item changes it. Subsequent requests would get the new version. However, updates are not propagated to other peers which replicate the item. As a result multiple versions under the same identifier (filename) may co-exist and it depends on the peer that a user contacts whether the latest version is accessed. The same holds true for most decentralised systems such as Gnutella [4].

ActiveXML [11] is a declarative framework that harnesses web services for data integration, and is put to work in a peer-to-peer architecture. It supports the definition of replicated XML fragments as Web service calls but does not address update propagation. Update is addressed in P-Grid [1], a structured network that supports self-organization. The update algorithm uses rumour spreading to scale and provides probabilistic guarantees for replica consistency. However, it only considers updates at the file level in a mono-master mode, i.e. only one (master) peer can update a file and changes are propagated to other (read-only) replicas.

Freenet [6] partially addresses updates which are propagated from the updating peer downward to close peers that are connected. Freenet uses a heuristic strategy to route updates to replicas which is uncertain to guarantee eventual consistency. Searches replicate data along query paths ("upstream"). In the case of an update (which can only be done by the data's owner) the update is routed "downstream" based on a key-closeness relation. Since the routing is heuristic, the network may change, and no precautions are taken to notify peers that come online after an update has occurred, consistency guarantees are limited. Also, peers that are disconnected do not get updated.

In OceanStore [10] every update creates a new version of the data object (versioning). Consistency is achieved by a two-tiered architecture: A client sends an update to the object's "inner ring" (some replicas who are the primary storage of the object and perform a Byzantine agreement protocol to achieve fault-tolerance and consistency) and some secondary replicas that are mere data caches in parallel. The inner ring commits the update and in parallel an epidemic algorithm distributes the tentative update among the secondary replicas. Once the update is committed, the inner ring multicasts the result of the update down the dissemination tree. To our knowledge analysis of the latency and consistency guarantees for this update scheme has not been published yet.

3 Replica Management Approach

Regardless of the underlying system topology, P2P systems need some form of replication to achieve good query latencies, load balance, and reliability. We now briefly describe the architecture of the proposed data replication system. Our system assumes an infrastructure-less peer-to-peer system, i.e., all peers are equal and no specialised infrastructure, e.g., hierarchy, exists. No peer has a global view of the system but base their behaviour on local knowledge, i.e., its routing tables, replica list, etc. The peers can go offline at any time according to a random process that models the behaviour when peers are online.

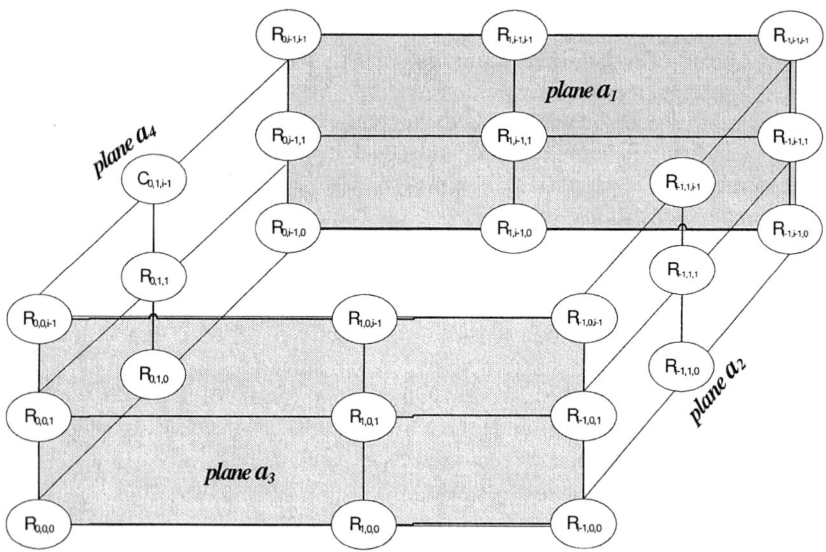

Fig. 1. The organization of replicas with four planes (i.e., α_i) and the circles in the grid represent the sites

3.1 System Architecture

Given N copies of a data object, we logically organize the N copies into a box-shape structure with four planes (i.e., α_1, α_2, α_3, and α_4) as shown in Fig. 1. Each copy of the object (circles in Fig. 1) is located at x, y, z coordinate ($C_{x,y,z}$) in a given plane (e.g., $C_{0,0,0}$, $C_{0,0,1}$, ..., $C_{l-1,l-1,l-1}$). We define a pair of copies that can be constructed from a hypotenuse edge in a box-shape structure is called hypotenuse copies.

3.2 Operations

The basic architecture of Fig. 1 supports operations for creating objects, creating and deleting object replicas, and performing reads and writes on the shared objects in a transactional framework. Read operations on an object are executed by acquiring a read quorum that consists of any hypotenuse copies. In Fig. 1, copies {$R_{0,0,0}$, $R_{l-1,l-1,l-1}$}, { $R_{0,0,l-1}$, $R_{l-1,l-1,0}$}, { $R_{0,l-1,l-1}$, $R_{l-1,0,0}$}, or { $R_{l-1,0,l-1}$, $R_{0,l-1,0}$} are hypotenuse copies and any one pair of which is sufficient to execute a read operation. Since each pair of them is hypotenuse copies, it is clear that, read operation can be executed if one of them is accessible, thus increasing the fault-tolerance of this protocol.

In contrast, write operations are executed by acquiring a write quorum from any plane that consists of: (1) hypotenuse copies; and (2) all vertices copies. For example, if the hypotenuse copies, say { $R_{0,0,0}$, $R_{l-1,l-1,l-1}$} are required to execute a read operation, then copies {$R_{0,0,0}$, $R_{l-1,l-1,l-1}$, $R_{l-1,l-1,0}$, $R_{0,l-1,l-1}$, $R_{0,l-1,0}$} are sufficient to execute a write operation, since one possible set of copies of vertices that correspond to { $R_{0,0,0}$, $R_{l-1,l-1,l-1}$} is {$R_{l-1,l-1,l-1}$, $R_{l-1,l-1,0}$, $R_{0,l-1,l-1}$, $R_{0,l-1,0}$}. Other possible write quorums are {$R_{0,0,0}$, $R_{l-1,l-1,l-1}$, $R_{l-1,l-1,0}$, $R_{l-1,0,l-1}$, $R_{l-1,0,0}$}, {$R_{l-1,l-1,l-1}$, $R_{0,0,0}$, $R_{0,0,l-1}$, $R_{l-1,0,l-1}$, $R_{l-1,0,0}$}, {$R_{l-1,l-1,l-1}$, $R_{0,0,0}$, $R_{0,0,l-1}$, $R_{0,l-1,l-1}$, $R_{0,l-1,0}$}, etc. It can be easily shown that a write quorum intersects with both read and write quorums in this protocol.

3.3 Advantages

The advantage of the proposed approach is that it tolerates the failure of more than three quarter of the copies. This is because the proposed protocol allows us to construct a write quorum even if three out of four planes are unavailable as long as the hypotenuse copies are accessible. To show this, consider the case when only one plane which consists of four copies of vertices and hypotenuse copies are available, e.g., the set {$R_{l-1,l-1,l-1}$, $R_{0,0,0}$, $R_{0,0,l-1}$, $R_{l-1,0,l-1}$, $R_{l-1,0,0}$} is available as shown in Fig. 1. A transaction in the proposed can be executed successfully by accessing those copies in a quorum. Hence the write quorum in the proposed protocol is formed by accessing those available copies. Read operations, on the other hand, need to access the available hypotenuse copies.

Thus the proposed protocol enhances the fault-tolerance in write operations compared to the grid configuration protocol. Moreover, proposed protocol ensures that read operations have a significantly lower cost, i.e., two copies, and have a high degree of availability, since they are not vulnerable to the failure of more than three quarter of the copies. Write operations, on the other hand, are more available than the grid-based configuration protocol since only five copies are needed to execute write operations.

4 Performance Analysis

In the analysis of the update algorithm we focus on the amount of communication required to achieve consistency and provide probabilistic guarantees for successful and appropriate results for queries. To this end, we compared the proposed scheme with ROWA when the number of copies is set to 16 (i.e., N= 16). We also assumed that all copies have the same availability.

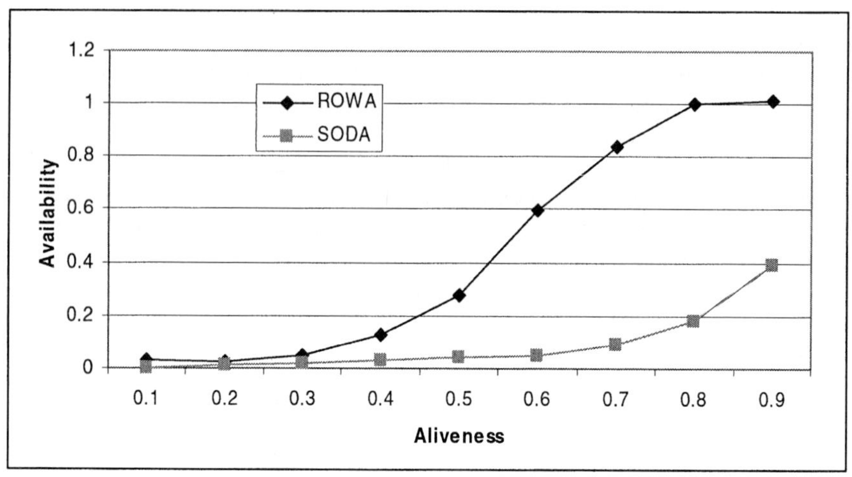

Fig. 2. Comparison of the write availability between SODA and ROWA

The result of the experiment is shown in Fig. 2. As shown, the proposed scheme has the lowest cost for write operation compared to ROWA protocol. This is because of the fact that in the ROWA protocol, an update operation needs to access all the replicas of the file in the system. Thus, the communication cost of an update operation in ROWA protocol is directly proportional to the number of replicas. In contrast, the proposed protocol needs only 5 copies at most, which results in significantly lower communication costs for comparable data availability.

5 Conclusions and Future Directions

The problem of protocol for maintaining replicated data has been widely studied in the distributed database systems. Recently, the need to support replication over wide-area networks and to use the Internet infrastructure as the basis for building a perpetual data store has spawned new research directions. Until recently, the challenges in providing high availability to P2P systems is poorly understood and only now being studied. Existing protocols are designed primarily to achieve high availability by updating a large fraction of the copies which provides some (although not significant) load sharing. We presented a new quorum-based protocol for maintaining replicated data across distributed P2P systems. The proposed approach is constructed on the organization of data in a box shape. We presented an analysis of the overhead and availability of the new protocol and showed that it performs better than the ROWA protocols. We are planning to implement the proposed protocol on various architectures including Data Grid, Peer-to-Peer, and Mobile systems.

Acknowledgement. The help of Maliha Omar is greatly appreciated. Without her kind support, this paper would not have been completed. The financial support of Deakin University is also gratefully acknowledged.

References

1. S. Saroiu, P. K. Gummadi, and S. D. Gribble. A measurement study of peer-to-peer file sharing systems. In *MMCN*, 2002.
2. M. Mat Deris, J. Abawajy and H.M. Suzuri, "*An Efficient Replicated Data Access Approach for Large-Scale Distributed Systems*", In Proceedings of IEEE International Conference on Cluster and Grid Computing (CCGRID 2004).
3. Napster. http://www.napster.com/.
4. Gnutella. http://www.gnutella.com/.
5. KaZaA[KaZaA. http://www.kazaa.com/.
6. I. Clarke et al. Protecting Free Expression Online with Freenet. *IEEE Internet Computing, 6(1),* 2002.
7. Q. Lv, P. Cao, E. Cohen, K. Li, and S. Shenker. "Search and replication in unstructured peer-to-peer networks." In *Proc. of the 16th annual ACM International Conf. on Supercomputing (ICS'02),* New York, USA, June 2002.
8. R. Akbarinia, V. Martins, E. Pacitti, P. Valduriez.*Replication and Query Processing in the APPA Data Management System*, in: "Int. Workshop on Distributed Data and Structures (WDAS'2004), Lausanne", 2004.
9. Anwitaman Datta, Manfred Hauswirth, Karl Aberer,Updates in Highly Unreliable, Replicated Peer-to-Peer Systems,
10. S. Rhea, C. Wells, P. Eaton, D. Geels, B. Zhao, H. Weatherspoon, and J. Kubiatowicz. Maintenance-free global data storage. *IEEE Internet Computing*, 5(5), 2001.
11. Serge Abiteboul, Angela Bonifati, Gregory Cobena, Ioana Manolescu, Tova Milo: Dynamic XML Documents with Distribution and Replication, SIGMOD 2003

Explore Disease Mapping of Hepatitis B Using Geostatistical Analysis Techniques

Shaobo Zhong[1], Yong Xue[1,2,*], Chunxiang Cao[1], Wuchun Cao[3], Xiaowen Li[1], Jianping Guo[1], and Liqun Fang[3]

[1] State Key Laboratory of Remote Sensing Science, Jointly Sponsored by the Institute of Remote Sensing Applications of Chinese Academy of Sciences and Beijing Normal University, Institute of Remote Sensing Applications, Chinese Academy of Sciences, P.O. Box 9718, Beijing 100101, China

[2] Department of Computing, London Metropolitan University, 166-220 Holloway Road, London N7 8DB, UK

[3] Institute of Microbiological Epidemiology, Academy of Military Medical Sciences, Beijing, China

zhongshaobo163@163.com, y.xue@londonmet.ac.uk

Abstract. This paper presents the application of Exploratory Spatial Data Analysis (ESDA) and Kriging from GIS (ArcGIS8.3) in disease mapping through the analysis of hepatitis B in China. The research shows that geostatistical analysis techniques such as Kriging and ESDA have a good effect in disease mapping. Kriging methods can express properly the spatial correlation. Furthermore, unlike model-based methods, which largely depend on assumption for disease data, the Kriging method is more robust for the data. So it can be used more widely and is more operational. What's more, the Kriging method may be adapted to interpolate nonstationary spatial structure. This can expand its application more largely. At last, the Kriging method can estimate the uncertainty of prediction while many deterministic methods cannot do so. In conclusion, it is an effective operational procedure to gain a deep insight into the disease data through ESDA before mapping disease using the Kriging method.

1 Introduction

When we practise analysis and surveillance of diseases, the use made of maps of disease incidence are many and various. Disease maps can be used to assess the need for geographical variation in health resource allocation, or could be useful in research studies of the relation of incidence to exploratory variables (Lawson *et al.* 1999). Disease mapping usually chooses certain spatial interpolation method(s), and then creates a continuous surface of disease distribution according to geographically distributed sampling data of disease. There are all kinds of spatial interpolation methods, which include Inverse Distance Weighted (IDW), global polynomial, local polynomial, and Kriging etc. IDW, global polynomial and local polynomial belong to

[*] Corresponding author.

V.S. Sunderam et al. (Eds.): ICCS 2005, LNCS 3516, pp. 464–471, 2005.

deterministic spatial interpolation methods. This is, random effect is not taken into consideration for these methods. However, the Kriging method differs from these methods. It addresses that spatial data not only exist trend but also spatial variation. Differing from the deterministic interpolation methods, it introduces random factors and correlation factors, which are reasonable for many infectious diseases, into its activities (Carrat *et al.* 1992, Torok *et al.* 1997, Kleinschmidt *et al.* 2000, Croner *et al.* 2001). Thus, the result of disease mapping is more correct.

Exploratory Data Analysis (EDA) is an approach/philosophy for data analysis that employs a variety of techniques (mostly graphical) to insight into data itself such as uncover underlying structure, extract important variables, detect outliers and anomalies, test underlying assumptions, develop parsimonious models and determine optimal factor settings (http://www.itl.nist.gov/div898/handbook/eda/). Exploratory Spatial Data Analysis (ESDA) is EDA related to spatial data. Before Kriging the disease data, we often want to learn about the data first (e.g. whether the data exist spatial trend, which is important for selecting a proper Kriging) and we can preprocess data for meet certain needs (Johnston *et al.* 2001). ESDA enables us to gain a deeper understanding of disease data so that we can take correct action and make better decisions on issues relating to our data.

This paper first introduces EDA and Kriging. Then it presents an application of Kriging in disease mapping according to a case study of Hepatitis B in China. We conclude that geostatistical analysis techniques such as Kriging and ESDA have a good prospect in disease mapping. It is an effective operational procedure to gain a deep insight into the disease data through ESDA before mapping disease using Kriging methods.

2 Materials and Methods

2.1 Exploratory Spatial Data Analysis Techniques

EDA is an approach/philosophy for data analysis that employs a variety of techniques (mostly graphical) to gain deep insight into the data (http://www.itl.nist.gov/div898/handbook/eda/). It is different from classical analysis or Bayesian analysis in the procedure.

- For classical analysis, the sequence is:
  Problem, Data, Model, Analysis, Conclusions
- For EDA, the sequence is:
  Problem, Data, Analysis, Model, Conclusions
- For Bayesian, the sequence is:

  Problem, Data, Model, Prior Distribution, Analysis, Conclusions.

For EDA, clearly seen from the above, the data collection is not followed by a model imposition, rather it is followed immediately by analysis with a goal of inferring what model would be appropriate. EDA tries to uncover and understand the data from the data itself. So it has unique advantages over others.

ESDA is a special EDA, which aims at the spatial data. With ESDA, we can:

- Examining the distribution of spatial data,
- Looking for global and local outlier,
- Looking for global trend,
- Examining semivariogram/covariance and spatial structure and directional variation of the data.

In epidemiological study, all kinds of disease data hold their own characteristics (e.g. spatial structure and distribution etc.). In order to ascertain or verify these characteristics, ESDA provides us with some good approaches. Now, almost all the mainstream GIS software packages are equipped with geostatistical analysis tools. For instance, ArcGIS developed by ESRI provides Geostatistical Analyst as an extension. It includes these ESDA tools: Histogram, QQPlot, Voronoi map, Trend Analysis and Semivariogram/Covariance cloud etc (Johnston *et al.* 2001).

2.2 Kriging Method

The kriging method was put forward by a South African mining engineer D.G. Krige in 1951 and was developed by a famous French geographer G. Matheron (Matheron 1963). The method absorbs the concept of geostatistics, which addresses that any continuous properties in geo-space are quite irregular and cannot be modeled using simple smoothing mathematical function, but can be depicted properly with random surface. The property changing continuously along with the geo-space is called as "regionalized variable". It can be used in depicting continuous index variables such as air pressure, elevation and so on. Though the disease rates, unlike air pressure or elevation, holds obvious continuously distributed characteristics, health data are better examined by methods that assume that disease rates are spatially continuous (Rushton 1998). In practice, this assumption is reasonable for most widely distributed diseases.

The interpolation model of Kriging can be expressed as follows:

$$Z(s) = m(s) + \varepsilon_1(s) + \varepsilon_2 \tag{1}$$

where $Z(s)$ denotes the interpolation value, which is composed of three parts: $m(s)$ reflects the spatial trend of the data, $\varepsilon_1(s)$ is the variation relevant to the change of spatial location s. And ε_2 is the residual (Gaussian noise), which has mean 0 and covariance σ^2 in space and has nothing to do with the change of the spatial location. s, which is looked on as the position expressed with x(longitude), y(latitude) ordinates, denotes the location of sampling points.

In generally, $\varepsilon_1(s)$ is presented with semiviaogram $\gamma(h)$ in Kriging Model. Semiviogram, only relevant to the distance h of sample point pairs (this characteristic is called as stationary), reflects the effect on which the change of spatial location have and be expressed as the function of h. Usually, semivariogram is monofonic increasing function of h and reflects the change of spatial correlation along with h, i.e. the closer the distance is, the stronger the spatial correlation is. Versus, the weaker the correlation is.

All kinds of Kriging models, including Simple kriging, Ordinary kriging, Universal kriging, Indicator Kriging, Probability Kriging etc., are based on the above formula (1). Each Kriging model is suited for specific situation and should be chosen seriously on purpose of best prediction.

2.3 Description of the Data and Processing

China is a high prevalence area of Hepatitis B. Several nation-wide censuses of hepatitis B show the proportion of the HBsAg carriers is over 10% on average and the prevalence of HBV over 60%. Furthermore, reported acute hepatitis cases are 2,700,000 every year according to the statistics of CDC in China. Hepatitis B has been one of the most serious problems of public health in China.

The data for this research were obtained from the statistics of Hepatitis B between 1994 ~ 1998 when the nation-wide screening was carried out thoroughly. Thus, there is a great deal of high quality and detailed sample data. Nevertheless, west and north of China have less sample data than other places such as middle, south and east (Fig. 1) due to different administrative region size and imbalance of socioeconomic development.

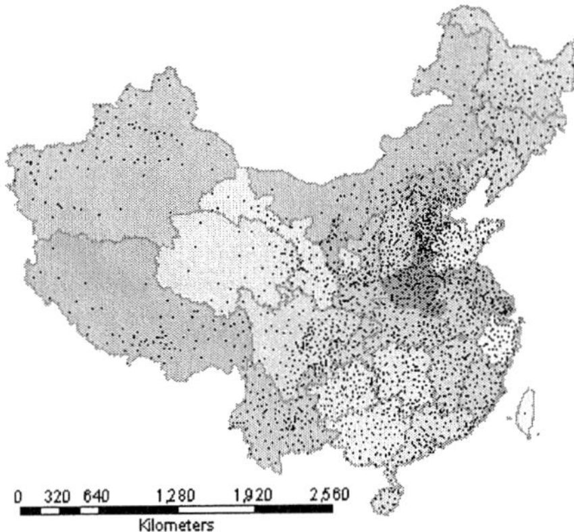

Fig. 1. The distribution of sampling points all over the nation without some islands in Southern Sea

In order to carry out the preprocessing and mapping, we have mainly collected the following data: 1) 1994-1998 county level new cases data of hepatitis B, 2) 1994-1998 county level age-grouped census and 3) 1994 screening data of hepatitis B all over the nation. We first calculate Standardized Mortality Ratio (SMR) of Hepatitis B from the above data (http://www.paho.org/English/SHA/be_v23n3- standardization.htm), then mapping SMR of Hepatitis B in China using Geostatistical techniques including ESDA, Kriging etc.

3 Result and Discussion

3.1 Trend in the Data

Trend analysis is very important for the subsequent choice of Kriging model. In ArcGIS, Trend analysis tool provides a three-dimensional perspective of the data. The locations of the sampling points are plotted on the x, y plane. Above each sampling point, the value is given by the height of a stick in the z dimension. Especially noteworthy, the values of z are projected onto the x, z plane and the y, z plane, which makes users more easily find out the trend of the data.

Fig. 2, plotted from the disease data of 1994, shows the space distribution of the sampling points of the disease through the Trend Analysis tool. The population data are derived from the 1994~1998 censuses. The disease data were provided by Institute of Microbiological Epidemiology, Academy of Military Medical Sciences, China. From the graph, we can find out clearly that there is nonlinear trend (e.g. quadratic) from the data set. Thus, we may choose Ordinary Kriging or Universal Kriging to interpolate the data because they both can detrend the data.

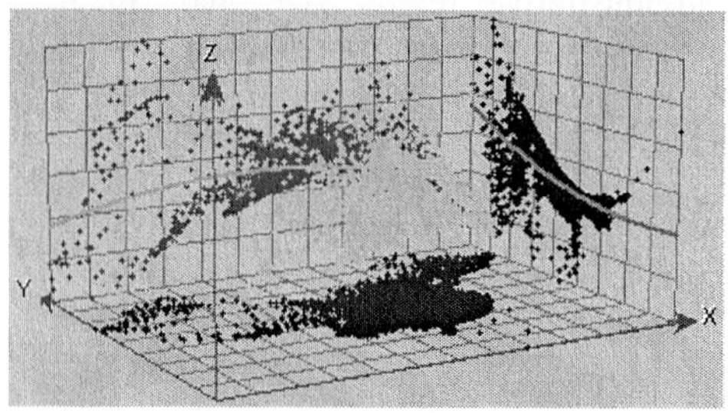

Fig. 2. Disease data of 1994 visualization in 3D space, x, y represent the locations of the disease sampling points and z is the SMR of the disease

3.2 Distribution of the Data Set

Certain Kriging methods work best if the data is approximately normally distributed (a bell-shaped curve). Furthermore, Kriging also relies on the assumption of stationary. In many cases, the distribution of the original data maybe not meet the need of the Kriging model and it is necessary to transform the data (e.g. Logarithmic transformation, Box-Cox transformation and Arcsine transformation). Histogram tool of geostatistical analyst extension in ArcGIS can explore the distribution of the data and make transformation, if necessary, for the data.

Before transformation, quite obviously, the distribution has a positive skewness from the histogram of the SMR data of 1994. By performing the Box-Cox transformation and adjusting the parameter, we produce a fairly good-shaped distribution.

3.3 Empirical Semivariogram

Semivariogram is a key component in a Kriging model. Exploring of the semivariogram mainly is to fit its model. There are many commonly used models for semivariogram (Edward *et al.* 1990). Figure 3 is the empirical semivariogram of the SMR of 1994, which is plotted using ArcGIS. For a large number of data, empirical semivariogram is an effective approach to fit the semivariogram. From the graph (a) and (b), we infer that the semivariogram of the data exists obvious anisotropy. The anisotropy is presented in (a) (spherical models with different parameters) and (b) (the ellipse). Here, we chose the spherical model to fit the scatter points. For more information about empirical semivariogram, and fitting models, please refer to ArcGIS.

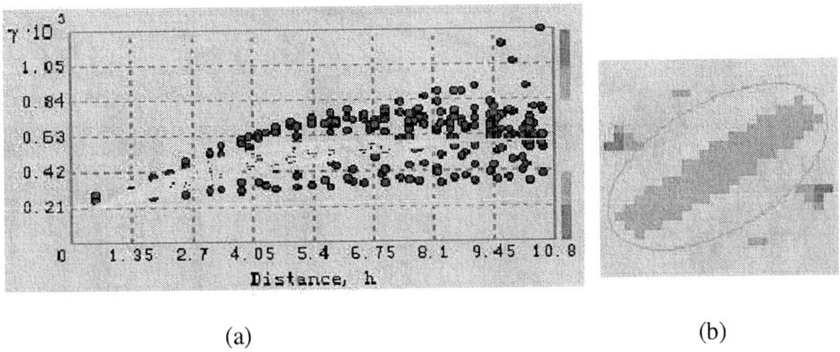

(a) (b)

Fig. 3. Empirical semivariogram of the SMR of 1994. (a) presents different fit models changing with direction, and (b) presents the semivariogram surface

3.4 Mapping Disease and Results Validation

From sections 3.1, 3.2 and 3.3, finally, we can choose a proper Kriging model to interpolate the data, here Universal Kriging, semivariogram model is spherical, exists anisotropy and the data are transformed by Box-Cox with parameter 0.65 at first. With the Kriging tool from ArcGIS, we get the interpolation map. Figure 4 is produced from the data of 1994.

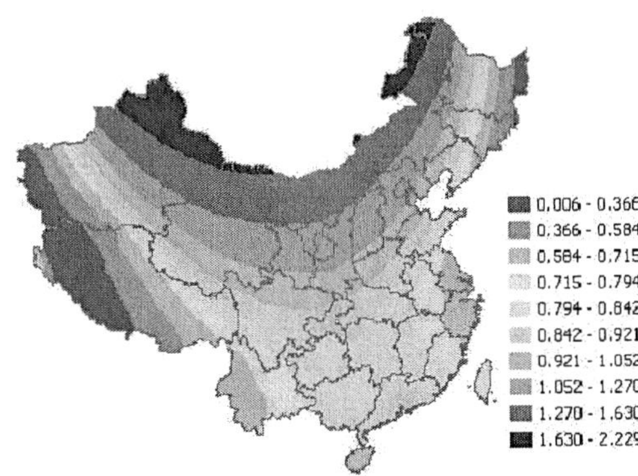

Fig. 4. SMR mapping result of hepatitis B of 1994 using Universal Kriging

When producing the result map using ArcGIS, simultaneously we get the validation result (Figure 5). The Kriging method allows us to validate the predicted values. There are two main validation methods: Cross-validation and Validation. Figure 5 is the validating result of predicted SMR of 1994 using Cross-validation. From this figure, we can see the precision of mapping is quite high. This implies that Kriging methods are suitable for the disease mapping.

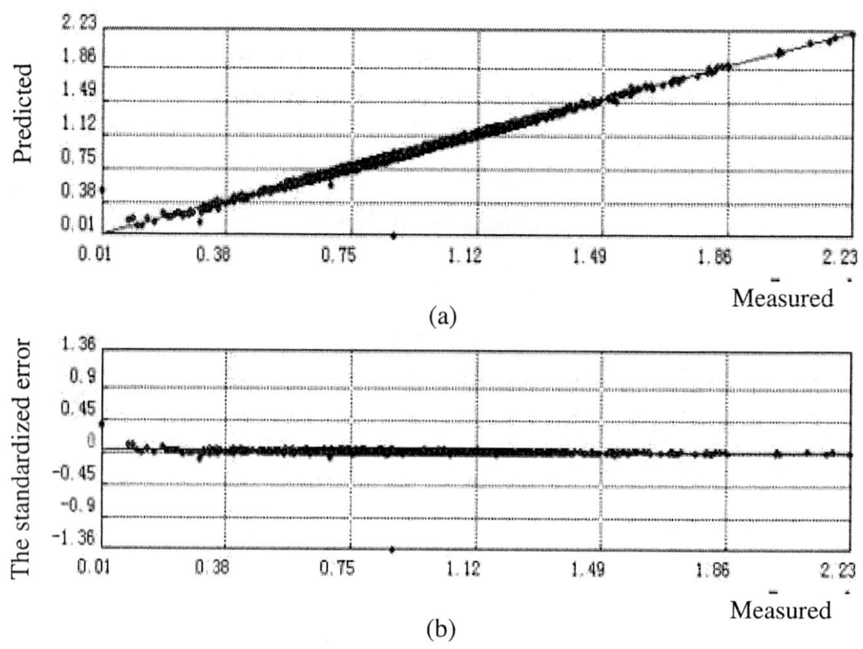

Fig. 5. Cross validation result. (a) is the comparison of predicted values and observed values and (b) is the standardized error, which is the square root of what is called Kriging variance and is a statistical measure of uncertainty for the prediction

4 Conclusions

Research on disease mapping covered a long history. All kinds of methods from simple ones such as point map, choropleth map and so on to complex ones like model-based methods, Kriging etc. provide analysis and surveillance of disease with strong measures. In general, complex model-based methods are more effective than the simple ones and produce more accurate results. This paper presents the application of ESDA and Kriging in disease mapping through the analysis of hepatitis B in China. The research shows that geostatistical techniques such as Kriging and ESDA have a good effect in disease mapping. Kriging methods can expresses properly the spatial correlation. Furthermore, unlike model-based methods, which largely depend on assumption for disease data (Bailey 2001), the Kriging method is more robust for the data. So it can be used more widely and is more operational. What's more, the Kriging method may be adapted to interpolate nonstationary spatial structure (Brenning *et al.* 2001). This can expand its application more largely. At last, the Kriging method can estimate the uncertainty of prediction while many deterministic methods cannot do so. In conclusion, it is an effective operational procedure to gain a deep insight into the disease data through ESDA before mapping disease using the Kriging method.

Acknowledgement

This publication is an output from the research projects "CAS Hundred Talents Program" and "Innovation Project, Institute of Remote Sensing Applications" funded by Chinese Academy of Sciences. The authors wish to acknowledge related departments and individuals.

References

1. Bailey, T.C., 2001, Spatial statistical methods in health. Cad Saúde Públ, 17: 1083-1098.
2. Brenning, A., and Boogaart, K.G.v.d., 2001, Geostatistics without stationary assumptions within GIS. Proceedings of 2001 Annual Conference of the International Association for Mathematical Geology, Cancun, Mexico, September 6-12, 2001.
3. Carrat, F., and Valleron, A.J., 1992, Epidemiologic mapping using the "kriging" method: application to an influenza-like illness epidemic in France. American Journal of Epidemiology, 135(11):1293-1300.
4. Croner, C.M., and Cola, L.D., 2001, Visualization of Disease Surveillance Data with Geostatistics. Presented at UNECE(United Nations Economic Commission for Europe) work session on methodological issues involving integration of statistics and geography, Tallinn, September 2001, available at http://www.unece.org/stats/documents/2001/09/gis/25.e.pdf.
5. Edward, H., Isaaks, R., and Mohan, S., 1990, Applied Geostatistics, Oxford University Press.
6. Johnston K., Ver Hoef, J.M., Krivoruchko, K., and Lucas N., 2001, Exploratory Spatial Data Analysis. In Use ArcGIS Geostatistical Analyst (digital book), edited by ESRI (San Diego: ESRI), pp. 81-112.
7. Kleinschmidt, I., Bagayoko, M., Clarke, G.P.Y., Craig, M., and Le Sueur, D., 2000, A spatial statistical approach to malaria mapping. International Journal of Epidemiology, 29(2):355-361.
8. Lawson, A.B., Böhning, D., Biggeri, A., Lesaffre, E., and Viel, J.F., 1999, Disease Mapping and Its Uses. In Disease Mapping and Risk Assessment for Public Health, edited by A.B. Lawson (New York: John Wiley & Sons Ltd.), pp. 3-13
9. Matheron, G.., 1963, Principles of geostatistics. Economic Geology, 58: 1246-66.
10. NIST/SEMATECH, e-Handbook of Statistical Methods, 2004, http://www.itl.nist.gov/div898/handbook/eda/section1/eda11.htm, accessed on November 5, 2004.
11. Rushton, G., 1998, Improving the geographic basis of health surveillance using GIS. In GIS and Health, edited by A. Gatrell and M. Loytonen(Philadelphia: Taylor and Francis, Inc.), pp. 63-80.
12. Standardization: A Classic Epidemiological Method for the Comparison of Rates, http://www.paho.org/English/SHA/be_v23n3-standardization.htm, accessed on November 5, 2004.
13. Torok, T.J., Kilgore, P.E., Clarke, M.J., Holman, R.C., Bresee, J.S., and Glass, 1997, Visualizing geographic and temporal trends in rotavirus activity in the United States, 1991 to 1996. Pediatric Infectious Disease Journal, 16(10):941-46.

eMicrob: A Grid-Based Spatial Epidemiology Application

Jianping Guo[1], Yong Xue[1,2,*], Chunxiang Cao[1], Wuchun Cao[3], Xiaowen Li[1], Jianqin Wang[1], and Liqun Fang[3]

[1] State Key Laboratory of Remote Sensing Science, Jointly Sponsored by the Institute of Remote Sensing Applications of Chinese Academy of Sciences, and Beijing Normal University, Institute of Remote Sensing Applications, Chinese Academy of Sciences, P.O. Box 9718, Beijing 100101, China

[2] Department of Computing, London Metropolitan University, 166-220 Holloway Road, London N7 8DB, UK

[3] Institute of Microbiology and Epidemiology, the Academy of Military Medical Sciences, the Chinese PLA, Beijing 100071, PR China

gjpgis@163.com, y.xue@londonmet.ac.uk

Abstract. The use of Grid technologies allows us to make progress in the prediction accuracy of epidemiological patterns, epidemiological modeling, risk predictions of infectious diseases etc by combining the geo-information and molecular simulation analysis methods. In this paper, we mainly design the eMicrob, in particular, build up the e-Microbe miniGrid deployed in IRSA, CAS and IME, the Chinese PLA. The architecture is as follows: Firstly we review related grid applications that are motivating widespread interest in Grid concepts within the scientific and engineering communities. Secondly we talk about the key methodologies and strategies involved in the construction of eMicrob. In the third section, the system design of the eMicrob, in particular about the architecture of the eMicrob miniGrid is discussed. Finally, we draw some conclusion in the process of the building of eMicrob and make some discussion about the challenges. It has been proven that the methods based on the Grid technologies are revolutionary and high efficient through the experience of the establishment and deployment of the e-Microbe miniGrid.

1 Introduction

Epidemiologists study a diverse range of health conditions as well as the impact that various exposures have on the manifestation of disease. However, we have done some new researches in the epidemiology (science of the public health, http://www.acepideiology2.org/aboutepid/whatis.asp) based on the new emerging technology — Grid computing, integrated with geographical information system (GIS) and remote sensing (RS), which will establish a virtual organization comprised of scientists and engineers from many different fields and discipline. Grid computing, as a new technology, is mainly concerned with "coordinated resource sharing and

* Corresponding author.

V.S. Sunderam et al. (Eds.): ICCS 2005, LNCS 3516, pp. 472–475, 2005.

problem solving in dynamic, multi-institutional virtual organizations." (Foster *et al.* 2001). A principal objective in the development of Grid computing is to make such resources available to users, regardless of their geographical location or institutional affiliation. While this goal is some way from being achieved, we have developed eMicrob miniGrid in the use of Grid tools, which allows us to make progress in the prediction accuracy of epidemiological patterns, epidemiological modeling, risk predictions of infectious diseases etc by combining the geo-information and molecular simulation.

Now through the eMicrob project we know that patterns and risk predictions can be made through GIS, by which we can study the dynamics of the epidemiological diseases based on above-mentioned environment factors achieved from remote sensing based on the grid platform. The undergoing eMicrob project benefits from grid and spatial technologies in various ways.

In this paper, we firstly introduce the related Grid projects, as good paradigms, which will help us to design the eMicrob system more precisely. The design of the eMicrob miniGrid is outlined and the architecture of the eMicrob miniGrid is discussed in details. At the end of this paper, we draw some conclusions and make discussion about the challenges eMicrob is encountering.

2 Methodology and Strategy

The eMicrob aims at building up a Grid platform (eMicrob miniGrid) in the long run, which will provide secure access to heterogeneous data and expensive resources in different locations, facilitate the study of transmission pattern, spatial distribution patterns, epidemiological trends, even advance the risk prediction precision about the infectious diseases i.e. Severe Acute Respiratory Syndrome (SARS), schistosomiasis (Savioli *et al.* 1997), avian influenza, malaria and cholera, etc.

We take advance of the following methodologies and strategies to design and construct the eMicrob. First, we build up a science team comprising of 3 kinds of technician, i.e. developer of code, epidemiologist and specialist from spatial domains. Second, we build our different database based on the SRB (http://www.npaci.edu/DICE/SRB/) technology. Moreover, we deploy the miniGrid on the platform – HIT-SIP developed by us .now we will expound on it. HIT-SIP (Wang *et. al.*, 2004) Grid platform in Institute of Remote Sensing Applications, Chinese Academy of Science is an advanced High-Throughput Computing Grid testbed using Condor developed by Department of Computer Sciences, University of Wisconsin-Madison. Heterogeneous computing nodes include two sets of Linux computers and WIN 2000 professional computers and one set of WIN XP computer provide stable computing power. The Grid pool uses java universally to screen heterogeneous characters. In general, users can use the heterogeneous Grid and share its strong computing power to process remote sensing images with middlewares as if in one supercomputer. Across the HIT-SIP platform, we can process and analyze the remote sensed images, for example, supervised classification and unsupervised classification and retrieval of various environment factors, which will be used in such discipline as biology, public health, and other spatially related disciplines.

3 Architecture of the eMicrob MiniGrid

To build the e-Microbe miniGrid, we deployed the miniGrid between IME Grid center (8 nodes) and IRSA Grid center (10 nodes). Both centers are required to use such universal grid platform as Globus, Condor or Condor-G as a starting point in order to develop the web data portals and computing portals, moreover, the epidemiologist and RS and GIS scientists can use them easily.

The framework consists of the following components:

- eMicrob User Interface and Web Portal Service, including data portal
- All kinds of databases and Model Pools
- Data Cache
- Application Modules

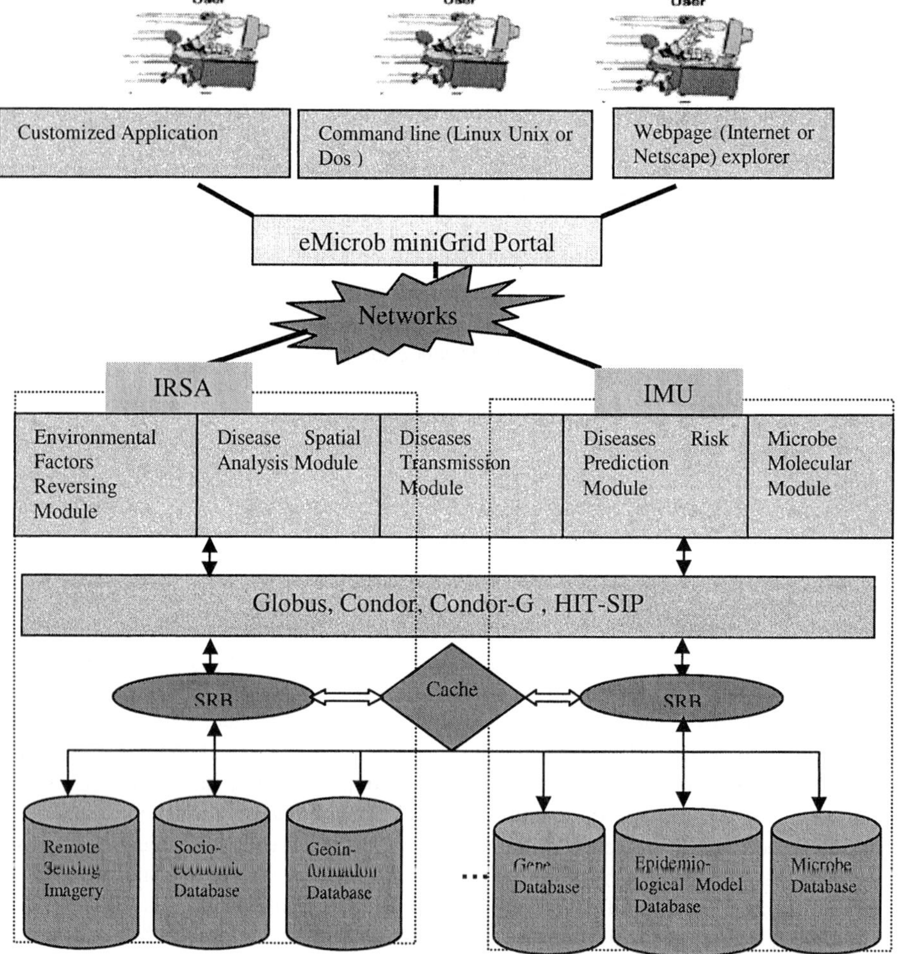

Fig. 1. The architecture o the eMicrob miniGrid

Based on the above-mentioned database components and computer tools, the architecture of the eMicrob miniGrid is given in Figure 1.

4 Conclusions and Results

The eMicrob miniGrid has provided a new platform of epidemiology investigation on which we collect remotely sensed data and other related data. Then integrate them into epidemiology applications researches, based on which analyses of microbe and/or virus combined with the real-time experimental data, are made and visualization are resulted in.

The science outcome based on the eMicrob miniGrid not only include the outcomes of spatial domain, such as we can disclose potential links between remotely sensed factors and specialized diseases (Beck *et al.* 2000), which has been proven prevalent in the research of epidemiological diseases, but also include ones of epidemiological disease such as disease controlling and prediction, and rules of the microorganism variances, the character of hepatitis B virus or the coronavirus causing SARS, taking it for example, can be analyzed on such eMicrob miniGrid platform.

We think that the scientist, both from spatial information and epidemiological fields, will greatly promote their researches in the epidemiological disease and in the end will provide more scientific information to the government officials to do correct decision-making. Meanwhile, through the experience of the design of the eMicrob and the implementation of the miniGrid, we recognize that the method based on the eMicrob miniGrid has made the study of the epidemiological diseases change deeply and will have a great potential impact for it can make the epidemiology and microbe discipline to make great stride in the near future.

Acknowledgements

This publication is an output from the research projects "CAS Hundred Talents Program" and "Innovation Program" funded by IRSA, CAS. The authors wish to acknowledge related departments and individuals.

References

Foster I., Kesselman C., Tuecke S., 2001, "The Anatomy of the Grid: Enabling Scalable Virtual Organizations". *International Journal of Super-computer Applica*tions, 15(3),: 200-222.

Beck L. R., Bradley M, Lobitz Byron L. Wood, 2000, Remote Sensing and Human Health: New Sensors and New Opportunities. *Emerging Infectious Diseases*, Vol. 6, No. 3, May–June 2000:217-226.

Savioli, L., Renganathan, E., Montresor, A., Davis, A., Behbehani, K., 1997, Control of schistosomiasis – a global picture. *Parasitology Today*, 11, 444-448.

Jianqin Wang, Xiaosong Sun, Yong Xue, Yanguang Wang, Ying Luo, Guoyin Cai, Shaobo Zhong, and Jiakui Tang, 2004, Preliminary Study on Unsupervised Classification of Remotely Sensed Images on the Grid. *Lecture Notes in Computer Science,* Vol. 3039, pp.995-1002.

Self-organizing Maps as Substitutes for K-Means Clustering

Fernando Bação[1], Victor Lobo[1,2], and Marco Painho[1]

[1] ISEGI/UNL, Campus de Campolide, 1070-312 LISBOA, Portugal
bacao@isegi.unl.pt
[2] Portuguese Naval Academy, Alfeite, 2810-001 ALMADA, Portugal
vlobo@isegi.unl.pt

Abstract. One of the most widely used clustering techniques used in GISc problems is the k-means algorithm. One of the most important issues in the correct use of k-means is the initialization procedure that ultimately determines which part of the solution space will be searched. In this paper we briefly review different initialization procedures, and propose Kohonen's Self-Organizing Maps as the most convenient method, given the proper training parameters. Furthermore, we show that in the final stages of its training procedure the Self-Organizing Map algorithms is rigorously the same as the k-means algorithm. Thus we propose the use of Self-Organizing Maps as possible substitutes for the more classical k-means clustering algorithms.

1 Introduction

The widespread use of computers and Geographical Information Systems (GIS) made available a huge volume of digital geo-referenced data (Batty and Longley 1996). This growth in the amount of data made multivariate data analysis techniques a central problem in Geographical Information Science (GISc). Amongst these techniques, cluster analysis (Jain, Murty et al. 1999) is one of the most used, and it is usually done using the popular k-means algorithm. Research on geodemographics (Openshaw, Blake et al. 1995; Birkin and Clarke 1998; Feng and Flowerdew 1998; Openshaw and Wymer 1994), urban research (Plane and Rogerson 1994; Han, Kamber et al. 2001), identification of deprived areas (Fahmy, Gordon et al. 2002), and social services provision (Birkin, Clarke et al. 1999) are examples of the relevance that clustering algorithms have within today's GISc research.

There have been a number of tests comparing SOM's with k-means (Balakrishnan, Cooper et al.1994; Openshaw and Openshaw 1997; Waller, Kaiser et al. 1998). Conclusions seem to be ambivalent as different authors point to different conclusions, and no definitive results have emerged. Some authors (Flexer 1999; Balakrishnan, Cooper et al. 1994; Waller, Kaiser et al. 1998) suggest that SOM performs equal or worst than statistical approaches, while other authors conclude the opposite (Openshaw and Openshaw 1997; Openshaw, Blake et al. 1995).

The main objective of this paper is to analyze the performance of the SOM and k-means in clustering problems, and evaluate them under specific conditions. We

V.S. Sunderam et al. (Eds.): ICCS 2005, LNCS 3516, pp. 476–483, 2005.

review both algorithms and then compare their performance on specific problems, using two synthetic datasets, and three real-world datasets.

2 K-Means Algorithm and Its Initialization

The k-means algorithm is widely known and used so only a brief outline is presented (for a thorough review see (Kaufman and Rousseeuw 1990; Fukunaga 1990; Duda, Hart et al. 2001)). K-means is an iterative procedure, to place cluster centers, which quickly converges to a local minimum of its objective function (Bradley and Fayyad 1998; Kanungo, Mount et al. 2002). This objective function is the sum of the squared Euclidean distance (L2) between each data point and its nearest cluster center (Selim and Ismail 1984; Bradley and Fayyad 1998). This is also known as "square-error distortion" (Jain and Dubes 1988). It has been shown that k-means is basically a gradient algorithm (Selim and Ismail 1984; Bottou and Bengio 1995) which justifies the convergence properties of the algorithm. he original online algorithm (MacQueen 1967) is as follows:

```
Let       k  be the predefined number of centroids
          n  be the number of training patterns
          X  be the set of training patterns x_1, x_2,..x_n
          P  be the set of k initial centroids μ_1, μ_2,... μ_k taken from X
          η  be the learning rate, initialized to a value in ]0,1[
  Repeat
          For i=1 to n
             Find centroid μ_j∈P that is closer to x_i
             Update μ_j by adding to it Δμ_j = η(x_i - μ_j)
          Decrease η
          Until η reaches 0
```

There are a large number of variants of the k-means algorithm. In this study we use the generalized Loyd's algorithm (Duda, Hart et al. 2001), which yields the same results as the algorithm above (Bottou and Bengio 1995). The popularity of this variant in statistical analysis is due to its simplicity and flexibility. It does not, however, specify how the initial centroids should be selected. Due to the gradient nature of the algorithms, these initial centroids have a decisive effect on which areas of the solution space can be searched. In all but the simplest cases the solution space contains many local optima to which the k-means algorithm may converge. To guarantee that a good solution will be found, multiple initializations of the algorithms are usually tested, and only the best final solution is kept.

By far the most common initialization, called "Forgy Approach" (Peña, Lozano et al. 1999), consists on randomly selecting k of available data patterns as centroids. This method has as main advantage it's simplicity: the selection requires no prior knowledge or computational effort, and multiple initializations will usually cover rather well the solution space. This is the initialization procedure used by default by software packages such as SAS Enterprise Miner, Matlab, and Clementine.

Instead of choosing k samples, we may divide the dataset into k subsets, and then use the centroids of these sets as seeds. We wil call this the "random selection method" (Peña, Lozano et al. 1999). It is similar to calculating the centroid of the whole dataset and then obtaining k perturbations of this point (Thiesson, Meek et al. 1999). More random selection based algorithms have been proposed (Kaufman and Rousseeuw 1990; Cano, Cordón et al. 2002; MacQueen 1967) each with specific strengths and weaknesses. The order by which the initial seeds are presented may influence the final outcome, so re-ordering techniques have been used in (Fisher, Xu et al. 1992) and (Roure and Talavera 1998). Sensitivity to outliers is another important problem that can be minimized by repeating random selection. Various methods can be used to repeat selection and clustering on smaller datasets, such as proposed by by (Bradley and Fayyad 1998).

Genetic algorithms are a well established technique to "guide randomness", and thus can be used to generate successive random selections. This approach is followed by (Peña, Lozano et al. 1999). Several attempts have been made to avoid randomness in the selection of seeds by using deterministic density estimation methods, and selecting the points of higher density as seeds. Such is the approach followed in (Bradley and Fayyad 1998), (Fukunaga 1990). Another family of initialization methods comes from heuristics that use the distance between candidate seeds as a guide for their selection (Katsavounidis, Jay Kuo et al. 1994;Al-Daoud and Roberts 1994;Tou and González 1974).

Hierarchical clustering algorithms are widely used and can produce meaningful clusters of data, but they usually do not minimize the objective function of the k-means algorithm. They may however be used to obtain a good approximation that can be used as seed for it's initialization. This approach was proposed in (Fisher 1987), and is used under different forms by (Higgs, Bemis et al. 1997; Snarey, Terrett et al. 1997;Meila and Heckerman 2001). The major drawback of these types of initializations is that they require a lot of computational effort.

Comparing results with all these initialization techniques is a Herculean task, and since simple random selection (or "Forgy Approach") is the most common and simple, we will use it in the comparisons with SOM.

3 Self-organizing Maps and Their Use in Obtaining K-Clusters

Although the term "Self-Organizing Map" could be applied to a number of different approaches, we shall use it as a synonym of Kohonen's Self Organizing Map (Kohonen 1982;Kohonen 2001), or SOM for short, also known as Kohonen Neural Networks.

The basic idea of a SOM is to map the data patterns onto a n-dimensional grid of neurons or units. That grid forms what is known as the output space, as opposed to the input space where the data patterns are. This mapping tries to preserve topological relations, i.e., patterns that are close in the input space will be mapped to units that are close in the output space, and vice-versa. So as to allow an easy visualization, the output space is usually 1 or 2 dimensional. The basic SOM training algorithm can be described as follows:

```
Let X  be the set of n training patterns x1, x2,..xn
    W  be a pxq grid of units wij where i and j are their
       coordinates on that grid
    α  be the learning rate, assuming values in ]0,1[, initialized
       to a given initial learning rate
    r  be the radius of the neighborhood function h(wij,wmn,r),
       initialized to a given initial radius
1  Repeat
2    For k=1 to n
3      For all wij∈W, calculate dij = || xk - wij ||
4      Select the unit that minimizes dij as the winner wwinner
5      Upadate each unit wij∈W: wij = wij + α h(wwinner,wij,r) || xk - wij ||
6    Decrease the value of α and r
7  Until α reaches 0
```

The neighborhood function h is usually a function that decreases with the distance (in the output space) to the winning unit, and is responsible for the interactions between different units. During training, the radius of this function will usually decrease, so that each unit will become more isolated from the effects of its neighbors. It is important to note that many implementations of SOM decrease this radius to 1, meaning that even in the final stages of training each unit will have an effect on its nearest neighbors, while other implementations allow this parameter to decrease to zero.

SOMs can be used in many different ways, even within clustering tasks (Bação, Lobo et al. 2005). In this paper we will assume that each SOM unit is a cluster center, and thus a k-unit SOM will perform a task similar to k-means. It must be noted that SOM and k-means algorithms are rigorously identical when the radius of the neighborhood function in the SOM equals zero (Bodt, Verleysen et al. 1997). In this case the update only occurs in the winning unit just as happens in k-means (step 4).

4 Experimental Setting

4.1 Datasets Used

The data used in the tests is composed of 4 basic datasets, two synthetic and two real-world. The real-world datasets used are the well known iris dataset (Fisher 1936) and sonar dataset (Sejnowski and Gorman 1988). The iris dataset has 150 observations with 4 attributes and 3 classes, while the sonar dataset has 208 observations with 60 attributes and 2 classes. Two synthetic datasets were created. The first dataset, DS1, comprises 400 observations in two-dimensions with 4 clusters. Each of these clusters has 100 observations with a Gaussian distribution around a fixed center. The variance of these Gaussians was gradually increased during our experiments. The second data set, DS2, consists of 750 observations with 5 clusters with Gaussian distributions defined in a 16 dimensional space.

4.2 Robustness Assessment Measures

In order to access the performance of the two methods a set of three measurements was used. The first one is the quadratic error i.e., the sum of the squared distances of each point to the centroid of its cluster. This error is divided by the total dispersion of each cluster so as to obtain a relative measure. This measure is particularly relevant as it is the objective function of the k-means algorithm. Additionally, the standard deviation of the mean quantization error is calculated in order to evaluate the stability of the results found in the different trials. The second measure used to evaluate the clustering is the mean classification error. This measure is only valid in the case of classification problems and is the number of observations attributed to a cluster where they do not belong. Finally, a structural measurement is used in order to understand if the structural coherence of the groups is preserved by the clustering method. This measure is obtained by attributing to each cluster center a label based on the labels of the observations which belong to its Voronoi polygon. If more than one centroid receives a given label (and thus at least one of the labels is not attributed) then the partition is considered to be structurally damaged.

5 Results

Each dataset was processed 100 times by each algorithm, and the results presented in table 1 constitute counts or means. Table 1 presents a summary of the most relevant results. A general analysis of table 1 shows a tendency for SOM to outperform k-means. The mean quadratic error over all the datasets used is always smaller in the case of the SOM, although in some cases the difference is not sufficiently large to allow conclusions. The standard deviation of the quadratic error is quite enlightening showing smaller variations in the performance of the SOM algorithms. The class error indicator reveals a behavior similar to the mean quadratic error. Finally, the structural error is quite explicit making the case that SOM robustness is superior to k-means.

Looking closer at the results in different datasets, there is only one data set in which k-means is not affected by structural errors. The reason for this is related with the configuration of the solution space. In the sonar dataset the starting positions of the k-means algorithm are less relevant than in the other 3 datasets.

Table 1. Comparison of SOM and k-means on different datasets, using the average quadratic error. its standard deviation, average classification error, and average structural error, over 100 independent initializations

Dataset	Method	Quadratic error	Std(Qerr)	ClassErr	Struct Err
IRIS	SOM	86.67	0.33	9.22	0
	k-means	91.35	25.76	15.23	18
SONAR	SOM	280.80	0.10	45.12	0
	k-means	280.98	3.18	45.34	0
DS1	SOM	9651.46	470.36	1.01	0
	k-means	11341.49	2320.27	12.77	58
DS2	SOM	27116.40	21.60	7.40	0
	k-means	27807.97	763.22	15.51	49

The real-world dataset refers to enumeration districts (ED) of the Lisbon Metropolitan Area and includes 3968 ED's which are characterized based on 65 variables, from the Portuguese census of 2001. Exploratory analysis of this dataset using large size SOMs and U-Matrices suggests that we should consider 6 clusters within this dataset. To find the exact locations and members of these 6 clusters we applied a batch k-means algorithm to this data, and compared the results with those obtained with a 6x1 SOM. In both cases we repeated the experiment 100 times with random initializations. The quadratic error obtained with k-means was 3543 ± 23 with a minimum of 3528, whereas with SOM we obtained 3533 ± 6 with a minimum of 3529. These results show that the best clustering obtained with each method is practically the same, but on average SOM outperforms k-means and has far less variation in it's results.

6 Conclusions

The first and most important conclusion that can be drawn from this study is that SOM is less prone to local optima than k-means. During our tests it is quite evident that the search space is better explored by SOM. This is due to the effect of the neighborhood parameter which forces units to move according to each other in the early stages of the process. This characteristic can be seen as an "annealing schedule" which provides an early exploration of the search space (Bodt, Cottrell et al. 1999). On the other hand, k-means gradient orientation forces a premature convergence which, depending on the initialization, may frequently yield local optimum solutions.

It is important to note that there are certain conditions that must be observed in order to render robust performances from SOM. First it is important to start the process using a high learning rate and neighborhood radius, and progressively reduce both parameters to zero. SOM's dimensionality is also an issue, as our tests indicate that 1-dimensional SOM will outperform 2-dimensional matrices. This can be explained by the fact that the "tension" exerted in each unit by the neighboring units is much higher in the case of the matrix configuration. This tension limits the plasticity of the SOM to adapt to the particular distribution of the dataset. Clearly, when using a small number of units it is easier to adapt a line than a matrix.

These results support Openshaw's claim which points to the superiority of SOM when dealing with problems having multiple optima. Basically, SOM offers the opportunity for an early exploration of the search space, and as the process continues it gradually narrows the search. By the end of the search process (providing the neighborhood radius decreases to zero) the SOM is exactly the same as k-means, which allows for a minimization of the distances between the observations and the cluster centers.

References

Al-Daoud, M. and S. Roberts (1994). New Methods for the Initialisation of Clusters. Leeds, University of Leeds: 14.

Bação, F., V. Lobo, M. Painho (2005). "The Self-Organizing Map, Geo-SOM, and relevant variants for GeoSciences." Computers & Geosciences, Vol. 31, Elsevier, pp. 155-163.

Balakrishnan, P. V., M. C. Cooper, V.S. Jacob, P.A. Lewis (1994). "A study of the classification capabilities of neural networks using unsupervised learning: a comparison with k-means clustering." Psychometrika 59(4): 509-525.

Batty, M. and P. Longley (1996). Analytical GIS: The Future. Spatial Analysis: Modelling in a GIS Environment. P. Longley and M. Batty. Cambridge, Geoinformation International: 345-352.

Birkin, M. and G. Clarke (1998). "GIS, geodemographics and spatial modeling in the UK financial service industry." Journal of Housing Research 9: 87-111.

Birkin, M., G. Clarke, M. Clarke (1999). GIS for Business and Service Planning. Geographical Information Systems. M. Goodchild, P. Longley, D. Maguire and D. Rhind. Cambridge, Geoinformation.

Bishop, C. M. (1995). Neural Networks for Pattern Recognition, Oxford University Press.

Bodt, E. d., M. Cottrell, M. Verleysen (1999). Using the Kohonen Algorithm for Quick Initialization of Simple Competitive Learning Algorithms. ESANN'1999, Bruges.

Bodt, E. d., M. Verleysen, M. Cottrell (1997). Kohonen Maps versus Vector Quantization for Data Analysis. ESANN'1997, Bruges.

Bottou, L. and Y. Bengio (1995). Convergence Properties of the K-Means Algorithms. Advances in Neural Information Processing System. Cambridge, MA, MIT Press. 7 G: 585-592.

Bradley, P. and U. Fayyad (1998). Refining initial points for K-means clustering. International Conference on Machine Learning (ICML-98).

Cano, J. R., O. Cordón, F. Herrera, L. Sánchez (2002). "A Greedy Randomized Adaptive Search Procedure Applied to the Clustering Problem as an Initialization Process Using K-Means as a Local Search Procedure." International Journal of Intelligent and Fuzzy Systems 12: 235-242.

Duda, R. O., P. E. Hart, D. Stork (2001). Pattern Classification, Wiley-Interscience.

Fahmy, E., D. Gordon, S. Cemlyn (2002). Poverty and Neighbourhood Renewal in West Cornwall. Social Policy Association Annual Conference, Nottingham, UK.

Feng, Z. and R. Flowerdew (1998). Fuzzy geodemographics: a contribution from fuzzy clustering methods. Innovations in GIS 5. S. Carver. London, Taylor & Francis: 119-127.

Fisher, D. H. (1987). "Knowledge Acquisition Via Incremental Conceptual Clustering." Machine Learning 2: 139--172.

Fisher, D. H., L. Xu, N. Zard (1992). Ordering effects in clustering. Ninth International Conference on Machine Learning, San Mateo, CA.

Fisher, R. A. (1936). "The use of Multiple Measurements in Taxonomic Problems." Annals of Eugenics VII(II): 179-188.

Flexer, A. (1999). On the use of self-organizing maps for clustering and visualization. Principles of Data Mining and Knowledge Discovery. Z. J.M. and R. J., Springer. 1704: 80-88.

Fukunaga, K. (1990). Introduction to statistical patterns recognition, Academic Press Inc.

Han, J., M. Kamber, A. Tung (2001). Spatial clustering methods in data mining. Geographic Data Mining and Knowledge Discovery. H. Miller and J. Han. Londor, Taylor & Fancis: 188-217.

Higgs, R. E., K. G. Bemis, I. Watson, J. Wikel (1997). "Experimental Designs for Selecting Molecules from Large Chemical Databases." Journal of Chemical Information and Computer Sciences 37(5): 861-870.

Jain, A. K. and R. C. Dubes (1988). Algorithms for clustering data, Prentice Hall.

Jain, A. K., M. N. Murty, P. Flynn (1999). "Data Clustering: A review." ACM Computing Surveys 31(3): 264-323.

Kanungo, T., D. M. Mount, N. Netanyahu, C. Piatko, R. Silverman, A. Wu (2002). "An efficient k-means clustering algorithm: analysis and implementation." IEEE Transactions on Pattern Analysis and Machine Intelligence 24(7): 881-892.

Katsavounidis, I., C.-C. Jay Kuo, Z. Zhang (1994). "A new initialization technique for generalized Lloyd iteration." IEEE Signal Processing Letters 1(10): 144 - 146.

Kaufman, L. and P. J. Rousseeuw (1990). Finding groups in data : an introduction to cluster analysis. New York, John Wiley & Sons.

Kohonen, T. (1982). Clustering, Taxonomy, and Topological Maps of Patterns. Proceedings of the 6th International Conference on Pattern Recognition.

Kohonen, T. (2001). Self-Organizing Maps. Berlin-Heidelberg, Springer.

MacQueen, J. (1967). Some methods for classification and analysis of multivariate observation. 5th Berkeley Symposium on Mathematical Statistics and Probability, University of California Press.

Meila, M. and D. Heckerman (2001). "An Experimental Comparison of Several Clustering and Initialization Methods." Machine Learning 42: 9-29.

Openshaw, S., M. Blake, C. Wymer (1995). Using neurocomputing methods to classify Britain's residential areas. Innovations in GIS. P. Fisher, Taylor and Francis. 2: 97-111.

Openshaw, S. and C. Openshaw (1997). Artificial intelligence in geography. Chichester, John Wiley & Sons.

Openshaw, S. and C. Wymer (1994). Classifying and regionalizing census data. Census Users Handbook. S. Openshaw. Cambridge, UK, Geo Information International: 239-270.

Peña, J. M., J. A. Lozano, P. Larrañaga (1999). "An empirical comparison of four initialization methods for the k-means algorithm." Pattern recognition letters 20: 1027-1040.

Plane, D. A. and P. A. Rogerson (1994). The Geographical Analysis of Population: With Applications to Planning and Business. New York, John Wiley & Sons.

Roure, J. and L. Talavera (1998). Robust incremental clustering with bad instance orderings: a new strategy. IBERAMIA 98 - Sixth Iberoamerican Conference on Artificial Intelligence, Lisbon, Springer Verlag.

Sejnowski, T. J. and P. Gorman (1988). "Learned Classification of Sonar Targets Using a Massively Parallel Network." IEEE Transactions on Acoustics, Speech, and Signal Processing 36(7): 1135 -1140.

Selim, S. Z. and M. A. Ismail (1984). "k-means type algorithms: a generalized convergence theorem and characterization of local optimality." IEEE Trans. Pattern Analysis and Machine Intelligence 6: 81-87.

Snarey, M., N. K. Terrett, P. Willett, D. Wilton (1997). "Comparison of algorithms for dissimilarity-based compound selection." Journal of Molecular Graphics and Modelling 15(6): 372-385.

Thiesson, B., C. Meek, D. Chickering, D. Heckerman (1999). Computationally Efficient Methods for Selecting Among Mixtures of Graphical Models. Bayesian Statistics. J. M. Bernardo, J. O. Berger, A. P. Dawid and A. F. M. Smith. Oxford, UK, Oxford University Press. 6.

Tou, J. and R. González (1974). Pattern Recognition Principals. Reading, MA, Addison Wesley Publishing Company.

Waller, N. G., H. A. Kaiser, J. Illian, M. Manry (1998). "A comparison of the classification capabilities of the 1-dimensional Kohonen neural network with two partitioning and three hierarchical cluster analysis algorithms." Psychometrika 63(1): 5-22.

Key Technologies Research on Building a Cluster-Based Parallel Computing System for Remote Sensing

Guoqing Li and Dingsheng Liu

45 BeiSanHuanXi Road, P.O. Box 2434, Beijing, 100086,China
Open laboratory, Remote Sensing Satellite Ground Station, China Academic of Sciences
{gqli,dsliu}@ne.rsgs.ac.cn

Abstract. Remote sensing image processing needs high performance computing to answer the fast growing data and requirement. Cluster-based parallel remote sensing image processing shows an effective way to overcome it. With an example of PIPS, paper gives basic theory of it, such as system structure, parallel model, and data distribution strategy and software integration and so on. Many experiments have proved that such technology can afford a receivable parallel efficiency with low cost hardware equipment. Moreover, it is friendly for experts who know remote sensing applications well and parallel computing less in developing their own parallel application implementations.

1 Introduction

In our information era, earth observation information has been the chief resource of information we have to handle. Remote sensing data is the mean part of EO information which comes from plane and satellite platforms. Remote sensing technology is fast developing on the filed of spatial resolution, spectrum resolution, time resolution, with which the data scale is expanding rapidly. A single scene of TM image with 7 bands can reach 280M disk storage. A global and big area geo-information always reaches to TB scale with in a single project. This information gives us more detail knowledge about our living environment.

However, the limit of processing speed and processing scale has been the bottleneck for remote sensing development and application. Most desktop workstations do not have sufficient computing power to perform such image processing. Generally large super-computers such as MPP are the main frameworks for such processing [1], while new researches on cluster system have shown the potential power and good performance to overcome such problem. Cluster system is based on current normal software and hardware desktops, such as linux and ANSC C/C++ . It gives a low cost, high operation possibility and expandability [2]-[3]. In some degree, it can be a perfect alternate plan of expansive Super Computers.

In this paper, we describe the design and implementation of Cluster-based Parallel Remote Sensing Processing System (PIPS). It is not only a general image processing parallel system, with the optimization of large image, but a special remote sensing image processing system. PIPS uses a message passing model based on MPICH standard which is designed to be the parallel desktop of PIPS. At the same time, PIPS is a

V.S. Sunderam et al. (Eds.): ICCS 2005, LNCS 3516, pp. 484–491, 2005.

developing desktop for some experts who know remote sensing applications well and know parallel technology less. With the develop model, they can design their own parallel applications by easy methods within very short time, and finally they can reassemble a special parallel remote sensing image processing desktop.

2 System Structure

Mainly, the processing of remote sensing image is data parallel model. Large data quantity can cover up the profile of arithmetic parallel. In cluster, we want to take computation extension profile from the multi CPU and multi I/O capability. Data is not processing locally, which is send to distributed computer nodes and processed there. Finally, the results data will be send back and new image be created. This course is a representative data parallel structure. The most time cost is happened in the network transmission for large EO data.

2.1 Basic Structure

2.1.1 Structure

In general, a whole running cluster processing system on remote sensing includes three chief modules, Initialization, Arithmetic Implementation and Finalization, shown in Fig. 1 as PIPS system. Firstly, Initialization module PIPSInit contains the following tasks: setting system parameters (run-mode XT/VT), starting related processes (MPI and process control), initializing temporary files, parameter files and PIP image files. Second, PIPSRun, the Arithmetic Implementation module, is the main part of a PIPS routine. The difference of different algorithms can bring different routines. However, it always contains three steps, image sending (distribution), processing and collection. Lastly, Finalization is running by PIPSEnd module. Corresponding to PIPSInit, it closes files, finishes processes and deletes rubbish.

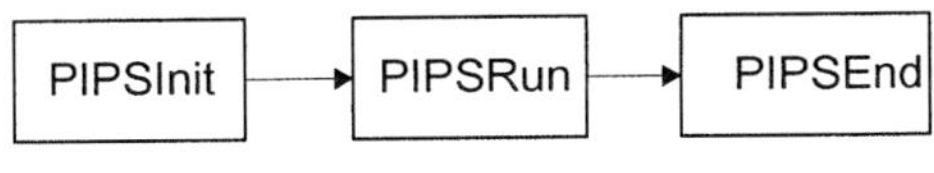

Fig. 1. structure

From such structure we can find that most initialization and finalization works is similar and which can be build into common modules. The difference of every processing function is realized in the part of PIPSRun, especially in some short subroutines.

2.1.2 System Frame

Application Routines use library function to finish certain processing function and function library contains basic image processing routines. Generally, they have been designed to be parallel routines, and some of them are full processing units. Some routines, however, are much basically and generally, such as I/O operation, having been assembled to be a low level library Kernel Function Library. Some advance function such as load balance, dynamic process management etc. MPI is the lowest support of PIPS system, which is invisible for application users and developer.

2.1.3 Multi-layer Parallel Model

There are three main parallel modes in remote sensing field, message-passing mode [4], share memory mode [5] and parallel file system mode [6]. Most parallel arithmetic in recent works all can be found form one of three modes. Each parallel mode above can be used to build a whole parallel processing system. Coming from different special research environment and research organization, such mode all have their superiority and inferiority. As many experimentation shown, in remote sensing processing, individual mode always cannot bring a perfect result mostly.

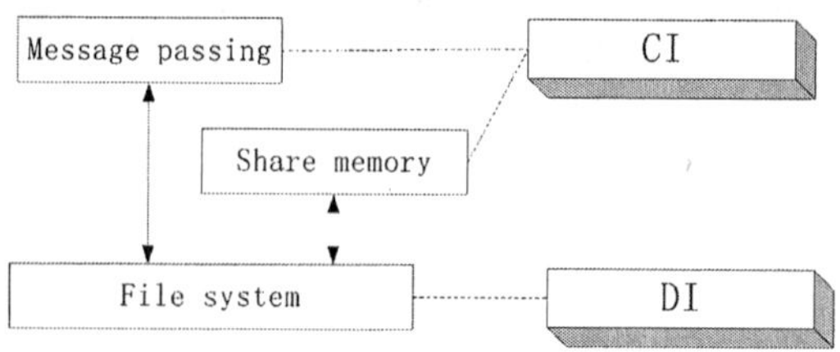

Fig. 2. Multi-layer parallel structure

Message-passing is most fast and steadily parallel mode, which has perfect standards system (PVM and MPI) and has been supported wildly by most parallel computer producers. Program based on Message-passing mode is clearly and readable, is also easily to build special parallel programming tools with their debugging and watching environment. However, such mode also brings the block of buffer and net transfer, especially for fine granularity parallel strategy. Share memory mode and Parallel file system mode can make the programming simply, reduce the changing of serial program. However, the parallel task control is too difficult to manage. Such modes is still in the research stage without stable and authoritative publish version. Uniform technical standard have not been confirmed.

However, such parallel modes can work together, and they do not exclude each other. There are two types of data transfer in the view of parallel architecture, CI (control information) and DI (data information), shown in Fig.2. Message-passing mode is fit for the CI, whereas parallel file system is fit for DI. Combining such three parallel modes, multi-layer parallel mode [9] can give better performance. There are two combinations: message-passing with parallel file system and share memory with parallel file system. In our next version of PIPS, we improve the parallel architecture by combining message-passing with parallel file system.

2.2 Data Distribution

As a basic model of message passing structure, large EO data have to be handled with data distribution operations, which can move target file data from one node to many nodes, and then collect them back as a whole. Data distribution is finalized by parallel data I/O library. This library is the base of module programming of data parallel, which can simplify different, reduce time and improve effective. According to given image data, they can create distribute strategy with interior dispatching engine. The strategy can reach load balance, reduce communication and speed transfer.

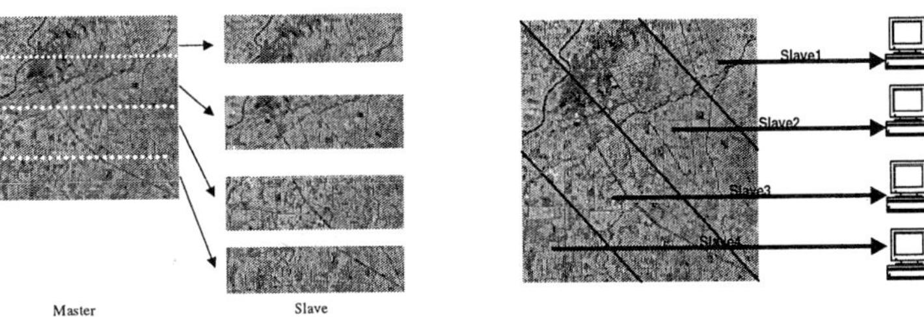

Fig. 3. whole image to line sequence image **Fig. 4.** whole image to specially type

With parallel file system, such parallel I/O has been building in the parallel read and writes operation as a global file object, which can reduce 30% transmission time.

The general rule of data distribution is reducing communication and offering complete data for certain application function. It is not only a course of space distribution but a re-assembling course of data structure.

In remote sensing, distribution modes is following: whole image to whole image, whole image to line sequence image, whole image to column sequence image, whole image to cross sequence image, whole image to matrix and whole image to specially type. (As Fig. 3 and Fig. 4)

2.3 Implementation

2.3.1 Data Parallel

In large remote sensing image processing, the representative parallel mode is data parallel. The FOR sentence and DO sentence in sequence programs always means a lot of calculation and can cause parallel possibility, which means FOR and DO sentences can be done parallel by many computer nodes. In PIPS, a whole processing can be divided into three steps of image sending, image processing and image collection. Modularization is a good developing method for users and developers, because they need not consider the details of distribution and collection while keeping good parallel performance.

2.3.2 Algorithmic Parallel

Not all operations in remote sensing reach good parallel performance only with data parallel, especially the operations reacted to GIS. On the other hand, algorithmic parallel sometime can get better efficiency however. For instance, DEM establishment and wavelet transfer are typical algorithmic parallel implementations. PIPS provides **complex mode** to fit these conditions. Developers have to write PIPSRun part as a whole by themselves, which contributes to the customized function library.

3 Developing Toolkit

PIPS has powerful developing ability. Many remote sensing experts are very familiar with the theories and methods of remote sensing technology but they have less knowledge on parallel technology. The developing toolkit of PIPS can help them transfer their sequence program expediently.

3.1 Function Library

In PIPS the functions are independently and the process loading is the main way of software integration. As shown in Fig. 5, Function library includes full program structure, the main part and other routines. What users should do is only little part of program routines, which have relationship with algorithm he want to carry out. The other part of routines, including complex control and interface function, should not be considered again. So the users can build up their whole parallel applications in a short time.

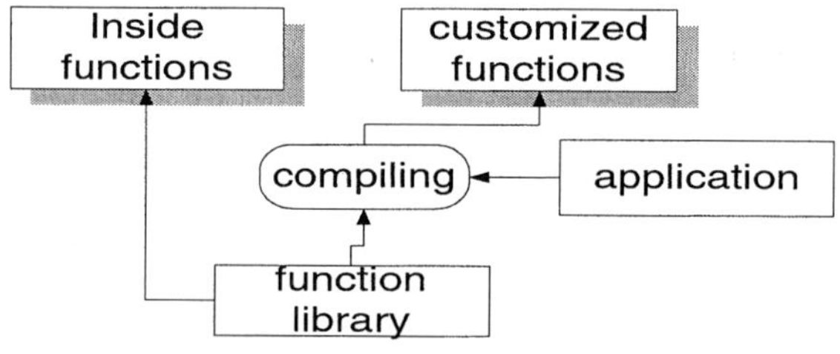

Fig. 5. function library

3.2 Develop Mode

Different algorithms can be realized with different parallel mode, the simple mode and the complex mode, seen in Fig.. Most of applications, which can be simplified to be three steps, distribution, processing and collection, belongs to simple mode.

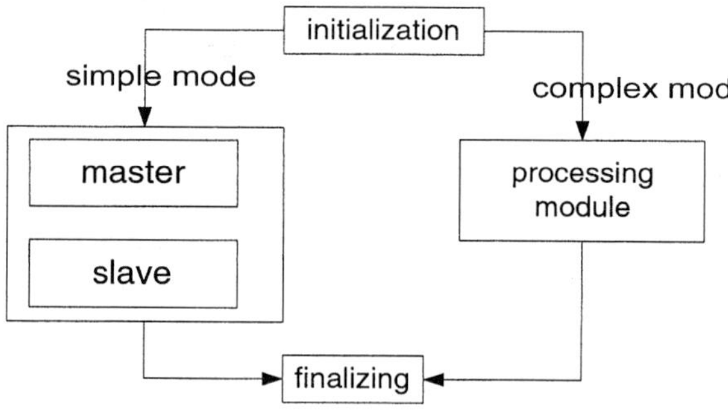

Fig. 6. developing mode

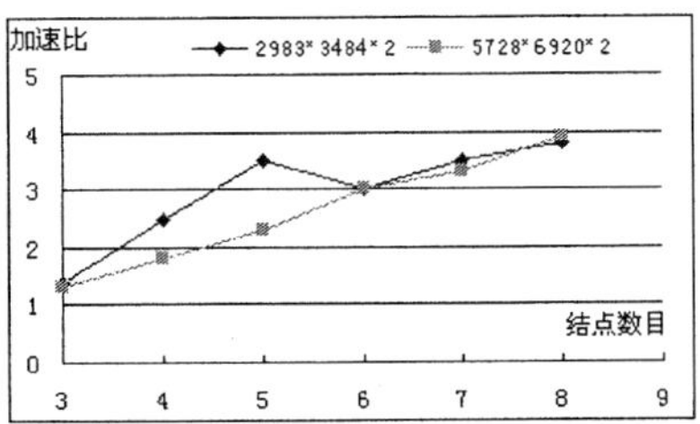

Fig. 7. parallel efficiency of Rotate

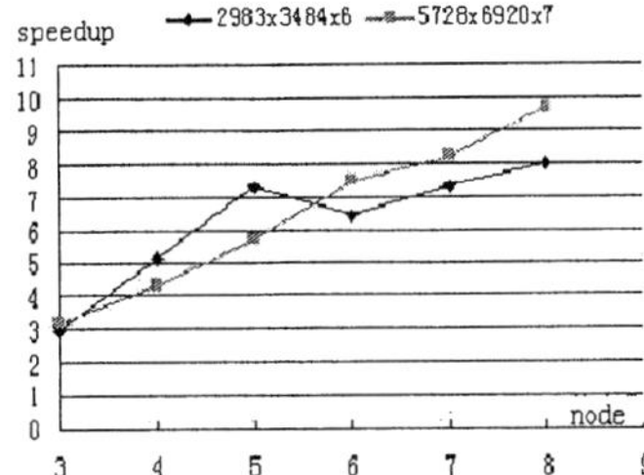

Fig. 8. Walsh transfer parallel efficiency

In this mode, what users should modify is only routines **pipsmaster** in master node and **pipsslave** in slave node. The other modules have been designed fitting this change automatically. For example, the K-L transfer [7], FFT transfer and classification operations are typical application of simple mode. The programming of this mode is clear, simple and easy to compare and debug.

However, there are still some applications of remote sensing cannot be formed with simple mode. To these applications, experts have to use special optimized algorithm. The complex mode of PIPS let it also can be involved in the whole structure of PIPS. Some applications include wavelet, DEM, SAR pre-processing [8], rotate, mosaic are typical complex mode.

3.3 Integrated Running Environment

PIPS is running on LINUX operation system with MPI parallel tools on cluster hardware. For an integration running system, many factors should be taken into account, auto detecting the running mode (VT/XT), static load balance, compute node selection, controlling the disk space, version conflicting and error control etc. Moreover, the running mode of MPI is very complex also. PIPS has build an integrated running environment, which can do above things at same time and automatically.

4 Experimental Results

In this section we present experiment results using PIPS system to perform a variety of image processing applications tasks. These experiments were done using a cluster of 9 PCs connected with 100M Ethernet. Experimental images were offered by RSGS and had three types. 7 band small TM images (2983 x 3484 x 7 band) , 7 band whole TM images (5728 x 6920 x 7 band) and 7 band mosaic TM images (22000 x 27000 x 7 band).

Parallel efficiency in speedup index has three conditions also. To most remote sensing applications, standard PIPS parallel mode (simple mode) can do well enough, such as WALSH transfer.

Complex mode can bring enough good result for some special tasks, sometime they can reach linear speedup. For example, rotate operation with slant strip algorithm can do 20000 x 20000 or more large image with parallel efficiency great than 50% (parallel efficiency means here speedup/nodes).

When we use parallel file system replaces MPI parallel IO routines, the performance reaches a new standard. As an example, above rotate operation-computing speed reaches 1.70 times faster than MPI mode.

In the theories of classical parallel computing, the speedup is always less than the parallel nodes number. However, in the field of data complexity applications, speedup index sometime can large than parallel nodes number. The parallel cluster is not only increasing the CPU number but increasing the available hard disk space and the total memory size. Some problems are too large to processing without using swap memory space in single node, but they are still less than the real total memory of cluster. We have taken a mosaic example for AnHui Prov. China, which include 9 scenes TM images with bands of 5/4/3 together and PIPS costs less than 7 minutes, which cannot be successfully running on the normal PC platform.

5 Conclusions

In general, the problems of remote sensing image processing are data parallelism problems. Course-grain parallelism is the main method for them. The overall system is based on such parallelism. We have been doing some researches on fine-course type, for example, the operator of mosaic and wavelet.

As a typical cluster-based remote sensing data processing system, PIPS is successful in transferring the classical sequence programming to parallel programming. PIPS have been developed form 1996, which has three main versions, PIPS, PIPS2000, and PIPS2005. New version PIPS is adopted with multi parallel theory and reaches a high performance. It shows rather better result not only in realizing algorithms, parallel efficiency, and integration condition but also in developing desktop. An experiment system on whole satellite ground processing station is being built on the base of PIPS, which is the first taste with cluster parallel computation environment.

Acknowledgement

This paper is fund by China National Key Research Project (863) "research on fast processing technology for large remote sensing information" and finished at open laboratory, RSGS, CAS.

References

1. K.A.Hawick, H.A.James, Distributed High-Performance Computation for Remote Sensing, Proc. Of Supercomputing '97,San Jose, Nov. 1997, Technical report.
2. Chao-Tung Yang, Chi-Chu Hung, Parallel Computing in Remote Sensing Data Processing.
3. Chao-Tung Yang,Chih-Li Chang, Using a Beowulf Cluster For a Remote Sensing Application, 22nd Asian Conference on Remote Sensing, Nov, 2001
4. Marc Snir, MPI: The Complete Reference MIT Press 1996
5. http://www.ict.ac.cn/chpc/dsm/dist.html

6. Zhu Yaofei, Li Guoqing, The research and experimentation of parallel file system in remote sensing image parallel processing system, Master's thesis, China remote sensing satellite ground station.
7. Li Guoqing ,Li Xia, Liu Dingsheng parallel processing on K-L transfer(in Chinese) '98 academic conference of China remote sensing at Dalian
8. Chen Lin, Liu Dingsheng parallel processing research on SAR imaging(in Chinese) '98 academic conference of China remote sensing at Dalian
9. Li Guoqing,, Liu Dingsheng, Research work on parallel mode of geo-image processing, Journal of Image and Graphics, Vol. 8. Spec 2003

Grid Research on Desktop Type Software for Spatial Information Processing

Guoqing Li, Dingsheng Liu, and Yi Sun

45 BeiSanHuanXi Road, P.O. Box 2434, Beijing, 100086, China
Open laboratory, Remote Sensing Satellite Ground Station, China Academic of Sciences
{gqli,dsliu,ysun}@ne.rsgs.ac.cn

Abstract. In the research works of spatial information grid, the grided modification of desktop style processing software is very important. Based on the analysis of such software structure, paper gives its technical strategy. Function Grided modifying is the acceptable method instead of whole package grided. Within the research, PCI and PIPS are brought as two kinds examples. Some key technology is followed being discussed and an experiment is given to show an interested result.

1 Desktop Processing Resources and PCI Software

In spatial information processing field, workstations are the main computing platform. With them, desktop style processing is using widely. Desktop software can work in single powerful workstation with support of other peripheral equipments, such as plotter, printer and tape reader etc. Users can do their simple processing works in such machine. With the development of user's requirement, computer has to handle large sale of data, high complex arithmetic within a limited time. In addition, local data and software and computing capability are found not sufficient for use, network cooperation and resource sharing is need.

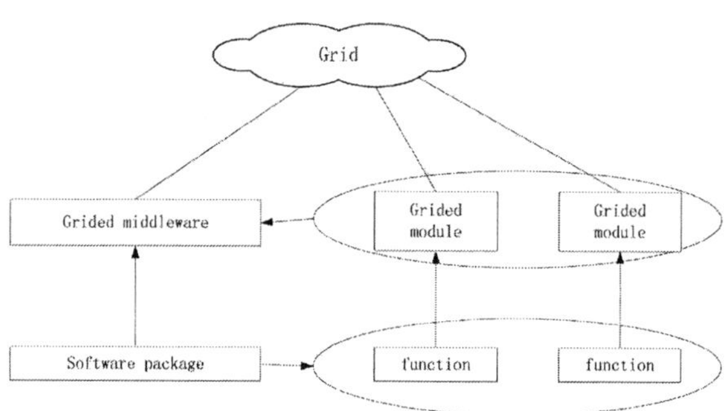

Fig. 1. How to grid a desktop software package

Grid is a method to realize such resource cooperation. With grid, users hope they can use a processing function only with their requirement, not to care of where the

V.S. Sunderam et al. (Eds.): ICCS 2005, LNCS 3516, pp. 492–495, 2005.

function is, which platform the function hoisted, how to learn the interface parameters and how to code a program to quote such function. Such grid resource is simple and standard [1]. However, each desktop software is very different with others.

Desktop processing software package is formed by many functions, which have been integrated and can be use by integration interface or can be use directly with command line mode. The grid transfer of desktop software package is a course of reconstruct, shown in Fig. 1. We think it is no sense to say software package grid modify. Within the grid, a whole application software package cannot be a useful and searchable grid resource to users, because any gird resource should be registered with their resource description according to formal attributions. A large software package has to be separated into function pieces, which can be recognized with their resource registration.

Granularity is real a problem we have to discuss. However, there is less research works has been taken on it. Generally, there are three-type function in software package, processing type, engine type and interface type. Processing type function can realize a certain operation according to arithmetic coding, which always can be run independent. Engine type builds the framework of package and supports common basic function, such as runtime environment, file system and display, while engine type function cannot be use independent. Interface type function contributes to the data format transfer and develops API, some of which can be use as a single function. Within these three types, only processing type and part interface type functions are valuable to grided, which can be mapped to resource tree and managed by grid resource register services.

However, such natural granularity separation is not certain to fit the need of grid service; some special research should be taken for it. But clearly, the fine granularity services can combine into a coarse size granularity service. When considering granularity question, we should pay more attention to the separating difficulty and network performance [2].

The following work is taken with legal PCI develop tools without accessing PCI source code. Such develop tools is also been offered with the purchase of whole package.

2 Mainframe

As above said, what can be managed in gird is only the fine-granularity services which separated from PCI package. Such fine-granularity services combine to be a whole middleware layer of PCI, which can be named the gird version of PCI. So, the research work on PCI package grid research can bring a normal method for other desktop remote sensing and GIS processing software package.

In general, PCI hosted workstation can use Windows or UNIX platform. Being a grid node, the server should install basic grid protocols and support services, which bring this node to be a grid node, such as a Globus GT3 environment. But if a PCI hosted workstation is configured as a standard grid node, this grid support services will heavily increase the load, which will affect the normal processing performance of PCI function. So, an independence agent node is needed to be a bridge of grid world and PCI node. Agent node locates in the same LAN with PCI node, they communicate with a couple of socket program. Agent can be a less capability computer, which installed as a grid node in the same time. As shown in Fig. 2, the resource registered in grid is such agent node instead of the real PCI node. If a grid user wants to use a PCI services, what he quotes is the agent shell services, which will transfer real operation command to the PCI background.

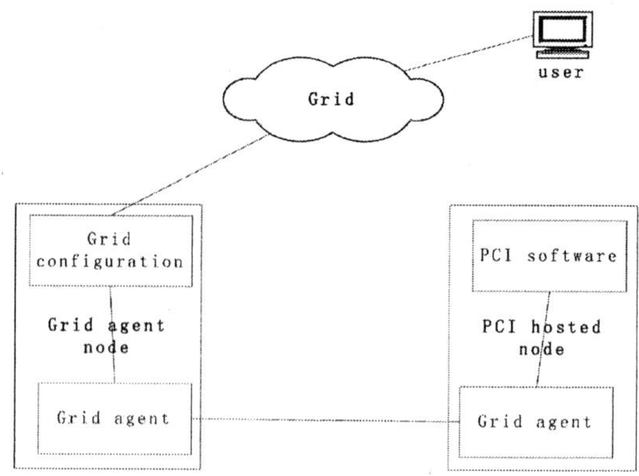

Fig. 2. PCI function grided frame structure

In this structure, PCI node is not changed for the temporary using of grid, such resource can be protecting as independence whole. In fact, it is the real value of grid. Out-grid-sharing will not bring security problem; the upgrade of PCI version also can be hidden to gird user if only keeps the PCI services interface not changed.

The realization of such desktop software function, all can be separated into the following five steps: Enter-parameters formatting (EPF), Enter-image formatting (EIF), quoting PCI function (QF), Outer-image formatting (OIF) and Out-parameters formatting (OPF), shown in Fig. 3. There are two type of information exchanged within grid and desktop software function, operation parameters and image data files' URL. Such information is designed in format of XML. EPF transfers the parameters in XML stream into format of PCI accepted, while EIF will access the source data files (or sub files or database) and replica them to local node via agent node, then EIF also need to change such files into the file format which PCI can handle, such pix file format. OPF and OIF give a reversed operation, which transfer result parameters and files into grid accessible format and return to grid runtime environment. Middle part is QF, which is a standard command line quoting of PCI functions.

3 Implementation and Experiment

We designed some experiments about PCI functions in Globus/OGSA environment with application server as JDK. PCI software hosted in a PC workstation configures with Pentium4 2.0GHz and 256MB memory. The agent node is a normal PC computer, which connects PCI node with 100M LAN. Fuzzy classification has been taken as an example with a whole scene TM data, about 40MB, transferred from another grid node. Experiment shows that the cost in EPF, EIF, OPF and OIF is 45sec., and QF costs 220min.

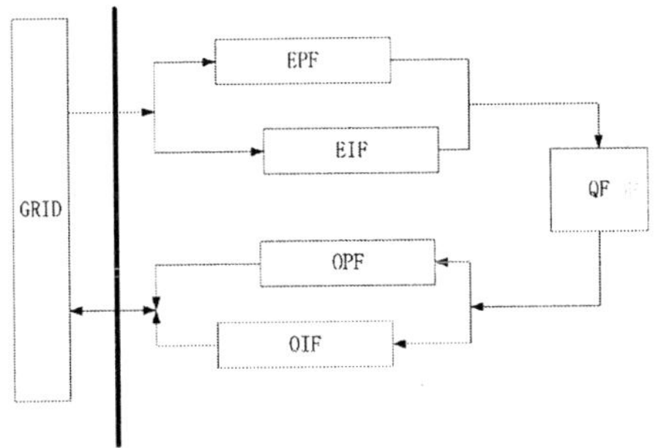

Fig. 3. Five steps of function grided realization

4 Conclusion

Above all, desktop spatial software package can be grided modification successfully. Such modification of software package is realized in form of function grided, instead of whole package grided. Grided software can reach resource sharing and operation control. The additional cost is acceptable and processing efficiency is similar between classical method and grid method.

Acknowledgement

This paper is fund by China National Key Research Project (863) "spatial information grid key technology research and application" and finished at the open laboratory, RSGS, CAS.

References

1. Bart Jacob Taking advantage of Grid computing for application enablement, IBM technical report. http://www-900.ibm.com/developerWorks/cn/grid/gr-overview/index_eng.shtml
2. Li Guoqing, Research work on spatial information application grid technology, doctoral dissertation, RSGS, CAS, 2004

Java-Based Grid Service Spread and Implementation in Remote Sensing Applications

Yanguang Wang[1], Yong Xue[1,2,*], Jianqin Wang[1], Chaolin Wu[1], Yincui Hu[1], Ying Luo[1], Shaobo Zhong[1], Jiakui Tang[1], and Guoyin Cai[1]

[1] State Key Laboratory of Remote Sensing Science, Jointly Sponsored by the Institute of Remote Sensing Applications of Chinese Academy of Sciences and Beijing Normal University, Institute of Remote Sensing Applications, Chinese Academy of Sciences, P.O. Box 9718, Beijing 100101, China

[2] Department of Computing, London Metropolitan University, 166-220 Holloway Road, London N7 8DB, UK

{wyg_nju@hotmail.com, y.xue@londonmet.ac.uk}

Abstract. Remote sensing applications often concern very large volumes of spatio-temporal data, the emerging Grid computing technologies bring an effective solution to this problem. The Open Grid Services Architecture (OGSA) treats Grid as the aggregate of Grid service, which is extension of Web Service. It defines standard mechanisms for creating, naming, and discovering transient Grid service instances; provides location transparency and multiple protocol bindings for service instances; and supports integration with underlying native platform facilities. It is not effective used in data-intensive computing such as remote sensing applications because its foundation, Web Service, is not efficient in scientific computing. How to increase the efficiency of the grid services for a scientific computing? This paper proposes a mechanism Grid service spread (GSS), which dynamically replant a Grid service from a Grid node to the others. We have more computers to provide the same function, so less time can be spent completing a problem than original Grid system. This paper also provides the solution how to adept the service duplicate for the destination node's Grid environment; how each service duplicate communicates with each other; how to manage the lifecycle of services spread etc. The efficiency of this solution through a remote sensing application of NDVI computing is demonstrated. It shows that this method is more efficient for processing huge amount of remotely sensed data.

1 Introduction

With the development of modern space remote sensing technology, the sensors have got a great increment in spatio-resolution and spectrum-resolution, and have made huge volumes of data for our remote sensing applications. While today's PC is faster than the Cray supercomputer of 10 years ago, it is still often inadequate to provide a

* Corresponding author.

V.S. Sunderam et al. (Eds.): ICCS 2005, LNCS 3516, pp. 496 – 503, 2005.

satisfying computing power, and at the same time the resources of supercomputers are very precious and expensive. Grid computing can give a good solution (Foster and Kesselman 1999). Grid computing technology is being developed to solve two kinds of problems. First, there is much resource wasting in the Internet. Such resources include processing cycles, disk space, data and network. Second, the integration of different systems deployed in a large company tends to be difficult. We need standard technology and platform to support such integration.

To solve this problem, Grid computing considers all the available resource in the network as a "super computer". User can transparently use and manage all these resources. Grid computing also provides a series of standard to integrate heterogeneous systems.

Since 2002, Open Grid Service Architecture (OGSA) which integrates the Globus has stood with the Web Service standard, and is going to be the unified standard for the Grid computing. The basic concept of OGSA is essentially a Web service with improved functionalities and behaviors. Web service is selected because it is a more suitable candidate for Internet scale application compared to other distributed computing technologies, such as CORBA, RMI and EJB (Ceram 2002). First, web service is based on a collection of open standards, such as XML, SOAP, WSDL and UDDI. It is platform independent and programming language independent because it uses standard XML language. Second, web service uses HTTP as the communication protocol. That is a big advantage because most of the Internet's proxies and firewalls will not mess with HTTP traffic. However, Web service is not powerful to build complex applications. It lacks some functionality, such as lifecycle management, notification and persistency. And web service is stateless which means it can't remember what has been done from one invocation to another.

Grid service provides more versatile functionality than web service. We will cover these functionalities in Section 2. Although Grid services get many advantages from Web Services, at the same time they also inherit the low efficiency in scientific computing of Web Service. This paper will propose a mechanism – Grid Services Spread (GSS). It allows the Grid services on a Grid node to extend dynamically to the others in order to increase the number of the nodes serving for the same task. We will focus on this mechanism in details in Section 3. At present, the implementation is based on Java. In Section 4, we will apply this mechanism for a remote sensing application to demonstrate how it works and what it can give us.

2 Background About Grid Service and Globus Toolkit 3

A computational Grid is a hardware and software infrastructure that provides dependable, consistent, pervasive, and inexpensive access to high-end computational capabilities. And the latest development of Grid technology gives some progress, that Grid computing is concerned with coordinated resource sharing and problem solving in dynamic, multi-institutional virtual organizations, to the definition.

Comparing Web Services, Grid services has been created with the advatanges including transient service which make the life of the serive can be managed by users and service data which indicates the characters of the service, lifecycly management and notification. These adtantages are the tools which we use for building the Grid services spread.

The Globus Toolkit (http://www.globus.org) developed by Argon National Laboratory, University of Chicago and University of Southern California, has become the industrial standard Grid middleware. The famous projects, which used and are using Globus Toolkit, include SF-Express, NASA OVERFLOW-D2, X-ray CMT, Cactus, MM5, National Technology Grid, The European DataGrid, NASA Information Power Grid, ASCI Distributed Resource Management (DRM) Testbed, etc. Globus Toolkit 3 (GT3) is the integrate of the original Globus Toolkit and Web Services, and it implements The Open Grid Services Infrastructure (OGSI) specification, which is the technical specification of Grid services, as an extension to Apache Axis (an implementation of SOAP).

GT3 core implements OGSI, it is a very important part in GT3. GT3 Base Services layer maily include Globus Resource Allocating Mangagement (GRAM in short), Index Service and Reliable File Transfer Service (RFT) which are very important to Globus Toolkit and GSS. GT3 Data Service layer includes Replica Management, which is very useful in applications that have to deal with very big sets of data. When working with large amount of data, we're usually not interested in downloading the whole thing, we just want to work with a small part of all that data. Replica Management keeps track of those subsets of data we will be working with. GRAM handles job submission and management. Index Services are used in discovering services like UDDI in Web Services. RFT allows us to perform large file transfer between the client and the Grid Service. In GT3 core, user does not subscribe to the service instance, but to each service data element. This fine-grained notification may reduce network traffic and improve the system performance.

Grid Service has plenty of contents, but the purpose of this paper is not describing Grid Service in details. More knowleges about Grid Service can be obtained in the The Globus Toolkit 3 Programmer's Tutorial (Sotomayor 2004).

Other Grid Service layer is maintained for the services made by user, for example remote sensing application. These services are built on the Core, Base Service and the Data Service in general. Security is an important factor in Grid-based applications,but this paper doesn't concern the security subject at this stage of study.

3 Problems and Solution

The biggest problem of Grid service solving data-intensive computing is that the overhead of remote procedure call is very high. The main reason is that Web Services are based on those standard interoperation technologies,such as XML, SOAP, UDDI which use XML text datastreams.They need plenty of serialization and unserialization. Web Services are necessary to Grid service for the privilege of platform independent, programming language independent and going through Internet proxies and firewalls.This paper will propose a method — Grid Services Spread, which allow a Grid service extend to the other Grid nodes so that more PCs provide their computing resource to the same client request.The request naturally will be completed in shorter time,especially for the application using vast volumes of data.

3.1 Design

The main purpose of GSS is that let more Grid nodes dynamically join to serving for one application. The GSS consists of, in fact, two Grid services source GSS (SGSS) and destination GSS (DGSS). SGSS is in charge of packing all binary code of the service spread, and then sending it to DGSS on the other Grid nodes. DGSS unpacks the bundle from the SGSS and deploys all binary codes to correct paths as the information provided by SGSS. RS Grid Service (Remote Sensing Grid Service) on Node A is called Source Grid Service in GSS, and RS Grid Service Duplicate is called Service Duplicate. Service Duplicate is, in fact, a grid service which have the same functionalities with its Source Grid Service and more characters, such as transience that its lifecycle depends on requests from clients.

More computing resource can been used dynamically with GSS. In Figure 1, Node A only use NDVI services on Node B without GSS, while with GSS Node A can ask SGSS on Node B to copy NDVI Grid service to the other nodes, such as Node C, Node D, and then the NDVI Grid Service Client on Node A can send NDVI Service request to the NDVI Service duplicates on Node C and D by the spread information from SGSS on Node B. Three services will be working for the client at a same time, at last the client will integrate the result from each service.

3.2 Challenges and Solutions

A Grid service need user to write the following three parts, Service interface which are GWSDL file, Service implementation which are based on Java in this paper, and the Deployment Description (WSDD). With help of them and ANT, which is a java build tool, WSDD and the final java classes are encapsulated to a GAR file (Grid Archive). ANT deploys this GAR file to the Grid services container. So, how to realize GSS? We meet the following challenges.

Which files should be packed by SGSS?
Before copying a Grid service to the other node, the necessary files about this service should be collected. Which files should be selected? This problem involves two conditions. One is the original GAR file for the service exists. In this case, it is only required to send this GAR file to DGSS as the pack. This situation is very realistic and important. The other is the GAR file does not exist or we do not know where the GAR file is. We find there are some information about the service in *undeploy* directory. For a simple application, it is possible to make out GAR file because we can find where the class files and the WSDL files are, but for a complex one, it may not be so. The best method is maintaining GAR file in a specific position and using a file to keep the mapping between the GSH of a Grid service and where its GAR file is.

What kind are those services duplicate on the other nodes? Persistent or Transient?
A Grid service duplicate should be destroyed when the application it serves for is finished, or there are more and more Grid service duplicates running on the node and holding plenty of resources. Therefore those service duplicates should be transient. Who decide the lifecycles of the service duplicates? The service duplicates can be destroyed only when the service client receives satisfactory result.

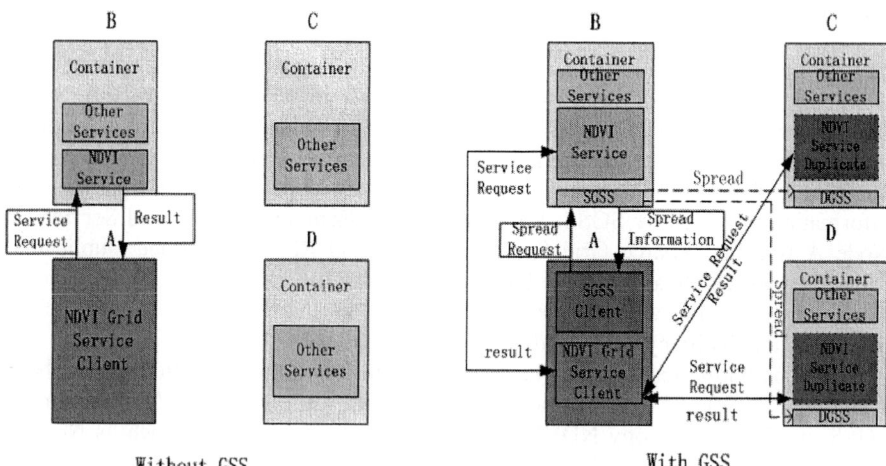

Fig. 1. The contrast of NDVI service without and with GSS

How to deploy and undeploy the service duplicates dynamically without restarting the container?

We find that GT3 does not support dynamic service deployment at present through our experiments. A new Grid service deployment is always available only by restarting the container. Our experiments indicate that it's mainly an Axis (a SOAP engine) problem. We can realize dynamic Grid service deployment in GT3 if Axis notices the new services and makes them available to the Grid. Each WAR (Web Archive) contains a fresh copy of the axis/ogsa web-app and the new services are made available when this web-app starts up. We can stop and start this web-app to make our services reinitialize and re-read their configuration files.

4 Using GSS in an Application of Remote Sensing – NDVI

Our goal is to test the efficiency of using GSS for Earth surface geophysical parameters determination form remote sensing data. Vegetation indices (VIs) are spectral transformations of two or more bands designed to enhance the contribution of vegetation properties and allow reliable spatial and temporal inter-comparisons of terrestrial photosynthetic activity and canopy structural variations. As a simple transformation of spectral bands, they are computed directly without any bias or assumptions regarding plant physiognomy, land cover class, soil type, or climatic conditions. They allow us to monitor seasonal, inter-annual, and long term variations of vegetation structural, phenological, and biophysical parameters and are important parameter to various kinds of local, regional, and global scale models, including general circulation and biogeochemical models. Numerous studies have shown the utility of the Vegetation Indices (VIs) from climate studies to famine early warning detection, epidemiology and renewable natural resources management. The NDVI has also been shown to be highly correlated with net primary production (NPP).

The Normalized Difference Vegetation Index (NDVI) (Tucker 1979, Jackson *et al.* 1983), which is related to the proportion of photo synthetically absorbed radiation, is calculated from atmospherically corrected reflectance from the visible and near infrared remote sensing sensor channels as:

$$(CH2 - CH1) / (CH2 + CH1)$$

Where the reflectance values are the surface bidirectional reflectance factors for MODIS bands 1 (620 - 670 nm) (CH1) and 2 (841 - 876 nm) (CH2). The NDVI is successful as a vegetation measure in that it is sufficiently stable to permit meaningful comparisons of seasonal and inter-annual changes in vegetation growth and activity.

4.1 Experiment

In this experiment, we used 1000 MODIS images (1000x1200 pixels), and each image is about 1.1M. We calculate the NDVI as batches using these MODIS images through different approaches: single PC without NDVI grid service, NDVI grid service on only one node and NDVI grid service spreading to 5 PCs with GSS.

Our test bed consists of 1 PC running Linux9.0 and 5 PCs running Windows 2000 operating system. It makes a good advantage of the individual PCs, especially in a work group, to constitute a Grid system based on GT3 core and GT3 Gars, which include RFT, MMJFS, GRAM and MDS etc. Table 1 gives the configuration information of our test bed on GT 3.

NDVI Service Client program on Tgp2.tgp requests the SGSS Service on wang.tgp to spread the NDVI Service. Then the SGSS returns GSHs of the NDVI Service duplicates that has been spread to Cai.tgp, Ly.tgp, Zhong.tgp and Hu.tgp. When getting the information, Client program divides remotely sensed data (MODIS data in the experiment) in geometric parallelism and send request to all the NDVI Services including original one and duplicates.

4.2 Result and Analysis

This experiment involved three cases:

1) We calculate NDVI with single PC without Globus Toolkit, using varying amount of data from 11 MB to 1100MB (form 10 images to 1000 images).
2) We do that with Grid Service, but without Grid Service Spread mechanism. NDVI Grid Service only runs on one PC. The amount of data still is varied from 11MB to 1100MB.
3) We use Grid Service Spread. The situation of utilizing data still is so.

This experiment was done for 10 times successively and we give the average value in Table 2, which is the statistics of the results of the three cases. We define the efficiency of GSS is the ratio of these two execution times:

$$\text{Efficiency} = \text{Execution of case 1} / \text{Execution of case 3}$$

Table 1. The configurations of GT 3 test bed

Host name	IP address	CPU	OS	Globus Toolkit	Description
Tgp2.tgp	192.168.0.200	1.7GHz	Win 2000	GT 3 core and Gars	Service Client
Wang.tgp	192.168.0.1 159.226.117.121	2.0GHz	Linux9.0	GT 3 core and Gars	NDVI Service SGSS
Hu.tgp	192.168.0.119	2.6GHz	Win2000	GT3 core and Gars	Service Duplicate DGSS
Ly.tgp	192.168.0.120	2.0GHz	Win2000	GT3 core and Gars	Service Duplicate DGSS
Cai.tgp	192.168.0.110	2.6GHz	Win 2000	GT 3 core and Gars	Service Duplicate DGSS
Zhong.tgp	192.168.0.19	2.6GHz	Win 2000	GT 3 core and Gars	Service Duplicate DGSS

Table 2. The results of the three cases

Cases \ Amount of Data	10 Images (11MB)	100 Images (110MB)	500 Images (550MB)	1000 Images (1100MB)
Without Grid Service (1 PC)	3.5 s	42.9 s	321.8 s	1093.8 s
1 PC providing Grid Service without GSS	8.4 s	49.2 s	342.4 s	1128.4 s
NDVI Grid Service spreading to 5 PCs with GSS	15.6 s	24.8 s	84.3 s	253.7 s

Case 2 is little slower than case 1 because of the serialization and unserialization of Grid service, but the difference varies little with the increase of the amount of data. In case 3, GSS even use more time to calculate NDVI when the amount is 11M for it is also a Grid service in fact. And the advantage of GSS is not huge when 100 images are used, but it is very surprising effective when it comes to 550 M (efficiency = 3.7) and 1100 M (efficiency = 4.3). We are convinced that we will have a better efficiency if the service can be spread to a bigger range. For a specific range of GSS, the efficiency will be better if the amount of data processed is larger.

5 Conclusion

Grid Service, the combination of Grid Technologies and Web Service, is a trend because it can really change the whole Internet into a huge real Grid system. But its performance is not satisfying in scientific computing, especially in data-intensive. Grid Service Spread mechanism can help us use more computing resource to service for our applications dynamically. With the help of GSS, we will have more powerful computing abilities and Throughput. It shows that this method is more efficient for processing huge amount of remotely sensed data.

Acknowledgements

This publication is an output from the research projects "CAS Hundred Talents Program" and "Monitoring of Beijing Olympic Environment" (2002BA904B07-2) and "Remote Sensing Information Processing and Service Node" funded by the MOST, China and "Aerosol fast monitoring modeling using MODIS data and middlewares development" (40471091) funded by NSFC, China.

References

Ceram, E., 2002, *Web Services Essentials,* (Sebastopol: O'Reilly & Associates, Inc), pp, 7-24.

Foster, I., Kesselman, C. 1999, *The Grid: Blueprint for a New Computing Infrastructure* (San Francisco: Morgan Kaufmann), pp, 16-50

Foster, I., Kesselman, C. and Tuecke, S., 2001, The Anatomy of the Grid: Enable Scalable Virtual Organizations. Available online at http://www.globus.org/reserch/papers.html (accessed 10 Month 2004)

Foster, I., Kesselman, C., Nick, J., Tuecke, S., 2002, The Physiology of the Grid. Available online at http://www.Gridforum.org/ogsi-wg/drafts/ogsa_draft2.9_2002-06-22.pdf (accessed 10 Month 2004)

Jackson, R.D., P.N. Slater, and P.J. Pinter, 1983, Discrimination of growth and water stress in wheat by various vegetation indices through clear and turbid atmospheres. *Remote Sensing of the Environment*, **15**, 187-208.

Sotomayor, B., 2004, The Globus Toolkit 3 Programmer's Tutorial, Available online at http://www.casa-sot\ omayor.net/gt3-tutorial (accessed 10Month 2004)

Tucker, C. J., 1979, Red and photographic infrared linear combinations for monitoring vegetation. *Remote Sensing of the Environment*, **8**, pp, 127-150.

Tuecke, S., Czajkowski, K., Foster, I., Frey, J., Graham, S., and C. Kesselman, C., 2002, Grid Service Specification. Available online at http://www.Gridforum.org/ogsi-wg/drafts/GS_Spec_draft03_2002\-07-17.pdf (accessed 10 Month 2004)

Modern Computational Techniques for Environmental Data; Application to the Global Ozone Layer

Costas Varotsos*

University of Athens, Department of Applied Physics,
Bldg. Phys. 5, Panepistimiopolis, GR-157 84 Athens, GR
covar@phys.uoa.gr
* Currently with the University of Maryland, Department of Meteorology,
3417 Computer and Space Science Bldg., College Park, MD 20742
covar@atmos.umd.edu

Abstract. The physics laws, which govern the atmospheric phenomena, are mostly non-linear and therefore the application of the conventional Fourier spectral analysis on the time series of the atmospheric quantities reveals that these are usually non-stationary. Quite often these non-stationarities conceals the existing correlations and therefore new analytical techniques capable to eliminate non-stationarities in the data should be employed. The most recent analytical methods used along these lines are the wavelet techniques and the detrended fluctuation analysis. Much attention has been paid recently to the latter technique, which has already proved its usefulness in a large variety of complex systems. As a paradigm, the detrended fluctuation analysis is applied to the column ozone data. Specifically the zonally and globally averaged column ozone observations conducted by ground-based (1964-2004) and satellite-borne (1979-2003) instrumentation are employed to detect long-range correlations in column ozone time series. The results show that column ozone fluctuations exhibit persistent long-range power-law correlations for all time lags between 4 months - 11 years.

1 Introduction

Trends in total ozone content (TOZ) are caused by external effects and they are usually supposed to have a smooth and monotonous or slowly oscillating behavior. Therefore, for the reliable detection of long-range correlations, it is essential to distinguish trends from the long-range fluctuations intrinsic in the data. Usually, the short-range correlations are described by the autocorrelation function, which declines exponentially with a certain decay time. For the long-range correlations, however, the autocorrelation function declines as a power-law. However, the direct calculation of the autocorrelation function is usually not appropriate due to noise superimposed on the collected data and due to underlying trends of unknown origin. The detrended fluctuation analysis (DFA), which will be discussed later, is a well established method for determining the scaling behavior of noisy data in the presence of trends without knowing their origin and shape.

V.S. Sunderam et al. (Eds.): ICCS 2005, LNCS 3516, pp. 504–510, 2005.

Very recently, *Varotsos* [2005] showed that the amplitudes of large TOZ fluctuations (in seasons of enhanced ozone depletion) obey a power-law scaling. This means that correlations between these points decrease according to a power law and are therefore scale-invariant. He also suggested that the Arctic and Antarctic TOZ fluctuations exhibit persistent long-range correlations for all time lags between 4 days - 2.5 years. This means that TOZ fluctuations at different times are correlated and the corresponding correlation function decays much slower than the exponential decay, i.e. a power-law decay. In other words persistence refers to the "memory" or internal correlation within the TOZ time series. For example, there is a tendency an increase in TOZ to be followed by another increase in TOZ at a different time. Furthermore he demonstrated the crucial role of the planetary waves to the scaling dynamics of TOZ over the high latitudes in both hemispheres, since the elimination of the TOZ long-term trend leads to persistent (antipersistent) long-range power-law correlations for time lags shorter (longer) than 10 days. It is worth noting here that a series is persistent if adjacent values are positively correlated, whereas a series is antipersistent if adjacent values are inversely correlated.

To reach to the aforementioned conclusions *Varotsos* [2005] employed the DFA method, which as mentioned above, allows the detection of long-range power-law correlations in a time-series with noise that often can mask true correlations. In this respect, due to the fact that the physics laws, which govern the atmospheric phenomena, are mostly non-linear, the application of the conventional Fourier spectral analysis on the time series of the atmospheric quantities reveals that these are usually non-stationary (the correlation functions are not invariant under time translation) (Chen et al. 2002, and references therein). Quite often these non-stationarities (e.g. trends and cycles) conceals the existing correlations (or other intrinsic properties) and therefore new analytical techniques capable to eliminate non-stationarities in the data should be employed (Hu et al. 2001).

The most recent methods used along these lines are the wavelet techniques (e.g. Koscielny-Bunde et al. 1998) and the DFA that introduced by Peng et al. (1994). Much attention has been paid recently to the latter technique, which has already proved its usefulness in a large variety of complex systems, for example, in southern oscillation index, in turbulence, in biology, in financial analysis and in other self-organizing critical systems (e.g. Ausloos and Ivanova, 2001; Weber, and Talkner, 2001; Chen et al. 2002; Varotsos et al. 2003ab; Collette and Ausloos 2004).

The present paper examines the time scaling of the TOZ fluctuations over the tropical and mid-latitudinal zones of both hemispheres and globally, thus contributing to the attempt for a selected choice between the proposed climate models for a projection of the TOZ levels in the future, taking into account the feedback between climate change and the ozone layer.

2 The Time Scaling and Correlations of the TOZ Fluctuations

The zonally and globally averaged TOZ observations performed by ground-based (1964-2004) and satellite-borne (1979-2003) instrumentation is used in order to efficiently search for time scaling, by adopting, however, a data analysis technique, which is not debatable due to the non-stationarity of the data. Thus, to study the temporal correlations of TOZ fluctuations the method of DFA with acceptable error bars is herewith used. This method stems from random walk theory, and permits the detection of intrinsic self-similarity in non-stationary time series (Talkner and Weber 2000). Therefore, this method has the advantage of avoiding seasonal-like trends and non-stationarity effects. According to DFA method, the time series is first integrated and then it is divided into boxes of equal length, Δt. In each box, a least squares line (or polynomial curve of order l, DFA-l) is then fitted, in order to detrend the integrated time series by subtracting the locally fitted trend in each box. The root-mean-square (rms) fluctuations $F_d(\Delta t)$ of this integrated and detrended time series is calculated over all time scales (box sizes). More specifically, the detrended fluctuation function $F(\tau)$ is calculated as follows (Kantelhardt et al. 2002):

$$F^2(\tau) = \frac{1}{\tau} \sum_{t=k\tau+1}^{(k+1)\tau} [y(t) - z(t)]^2 \,, \quad k=0, 1, 2, \ldots, \left(\frac{N}{\tau} - 1\right) \tag{1}$$

where $z(t) = at+b$ is the linear least-square fit to the τ data points contained into a class.

For scaling dynamics, the averaged $F^2(\tau)$ over the N/τ intervals with length τ is expected to obey a power-law, notably:

$$\langle F^2(\tau) \rangle \sim \tau^{2\alpha} \tag{2}$$

and the power spectrum function scales with $1/f^{\beta}$, with $\beta=2\alpha-1$.

An exponent $\alpha \neq 1/2$ in a certain range of τ values implies the existence of long-range correlations in that time interval as, for example, in fractional Brownian motion, while $\alpha = 1/2$ corresponds to the classical random walk (white noise). If $0 < \alpha < 0.5$, power-law anticorrelations are present (antipersistence). If $0.5 < \alpha \leq 1.0$, then $0 < \beta \leq 1$, and persistent long-range power-law correlations prevail; the case $\beta=1$ ($\alpha=1$) corresponds to the so-called $1/f$ noise. In addition, when $1< \alpha <1.5$, then long-range correlations are again present (but are stronger than in the previous case); the value $\alpha=1.5$ corresponds to the Brownian noise (e.g., *Talkner and Weber*, 2000).

It is worth noting that since the time series is first summed the noise level due to imperfect observations is reduced.

It has recently been recognized (Hu et al, 2001) that the existence of long-term trends in a time series may influence the results of the correlation analysis. Therefore, the effects of TOZ trends have to be distinguished from TOZ intrinsic fluctuations. To this end, before applying scaling analysis to the TOZ time-series, all TOZ data were deseasonalized and detrended.

In Figure 1, a log-log plot of the function $F_d=\sqrt{\langle F^2(\tau) \rangle}$ is shown, by employing the DFA-1 to the deaseasonalised and detrended monthly mean values of the ground-based TOZ values, during 1964-2003, over the belt 90°S-90°N. Since $\alpha = 1.1$ ($\pm$ 0.13) for the interval time ranging from about 4 months to 11 years long-range correlations are present. This suggests persistent long-range power-law correlations in global TOZ fluctuations. This persistence suggests that an anomaly in global TOZ in one time frame continues into the next.

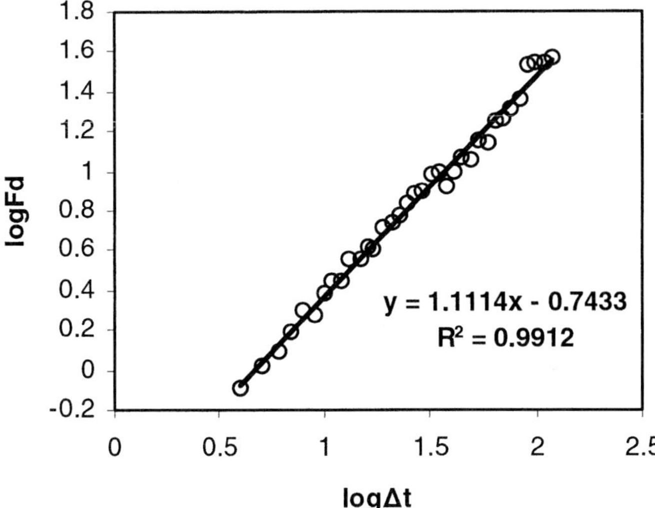

Fig. 1. DFA-function in log-log plot for the deseasonalised and detrended TOZ, during 1964-2003, over the belt 90°S-90°N

In the following, the temporal correlations of the deaseasonalised and detrended monthly mean values of TOZ are also examined for the belt 25°S-25°N during 1964-2003. The result obtained from the application of the DFA-1 to the aforementioned TOZ values over tropics is presented at Figure 2. The finding drawn from this figure (α = 1.1 ± 0.11) is that TOZ over tropics exhibits persistent long-range power-law correlations for the interval time ranging from about 4 months to 11 years.

It is worth noting that similar to the above-discussed results are also found by applying the DFA-1 method to the monthly zonal mean V8 TOZ values over tropics and globally.

We now turn to the extra-tropics and specifically to the ground-based TOZ data for the latitude belt 25°N-60°N during 1964-2003. The application of the DFA-1 method to these deseasonalised and detrended monthly mean values of TOZ reveals that long-range power-law correlations exist for all time scales (fig.3a). Since α_1 = 1.2 ± 0.13 (for time scales shorter than 2 years) the long-range correlations in TOZ exhibit "stronger memory" (the process forgets more slowly its past behavior) compared to that of α_2 = 0.6 ± 0.09 (for time scales longer than 2 years). Higher persistence implies a stronger correlation between successive data points. It is also worth noting that "memory" or correlations exist at all time scales over which the power law is valid.

Finally, the DFA-1 method is also applied on the deseasonalised and detrended monthly mean values of TOZ in the latitude belt 25°S-60°S during 1964-2003. The results obtained are depicted in Fig.3b, where persistency of TOZ fluctuations is observed. In particular, for time scales shorter (longer) than 2 years $\alpha_1 = 1.1 \pm 0.11$ ($\alpha_2 = 0.6 \pm 0.07$).

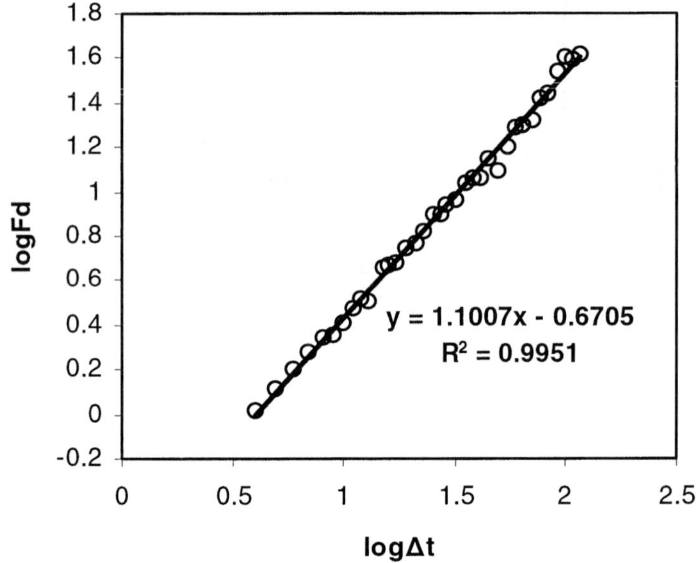

Fig. 2. DFA-function in log-log plot for the deseasonalised and detrended TOZ during 1964-2003, over the belt 25°S-25°N

It is worth mentioning that similar to the above-discussed results for the TOZ variability over extra-tropics are also found by applying the DFA-1 method to the monthly zonal mean V8 TOZ values.

The results obtained clearly show that the monthly mean column ozone fluctuations over tropics, extra-tropics and globally exhibit persistent long-range correlations for all time lags between 4 months - 11 years, which correspond to the 1/f noise. Over extra-tropics, this persistency becomes weaker for time lags between 2-11 years. It is well known that the 1/f noise is one of the most common features in nature. Superposition of effects giving rise to signals with scale invariant distributions of correlation times, the so-called scale similarity, could be on the basis of the observed behavior. However, a proper explanation for such a behavior is still lacking and for this reason the physical origin of the 1/f noise is pretty much an open question (Fanchiotti et al, 2004).

The aforementioned findings seem to favor specific models for the description of the ozone depletion and may lead to better predictability on the global TOZ evolution along the lines of nonlinear dynamics.

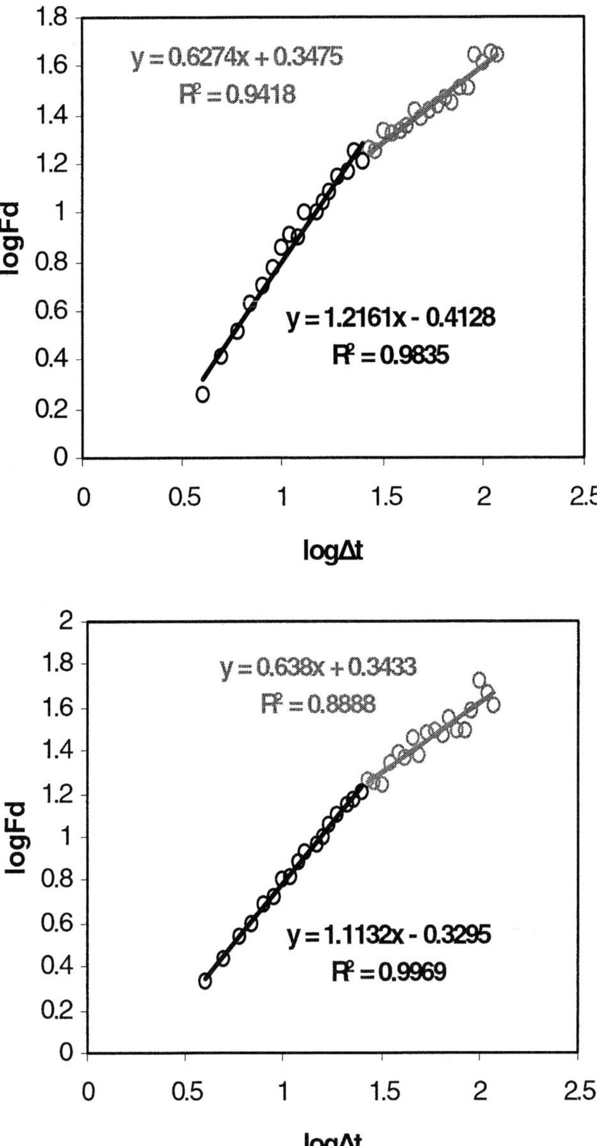

Fig. 3. DFA-function in log-log plot for the deseasonalised and detrended TOZ during 1964-2003, over the belt : (a) 25°N-60°N, and (b) 25°S-60°S

3 Conclusions

The investigation of the existence of the long-range correlations to the zonally and globally averaged column ozone data derived from observations performed by ground-based (1964-2004) and satellite-borne (1979-2003) instrumentation shows the following:

The monthly mean column ozone fluctuations over tropics, extra-tropics and globally exhibit persistent long-range correlations for all time lags between 4 months - 11 years, which correspond to the 1/f noise. Over extra-tropics, this persistency becomes weaker for time lags between 2-11 years.

The above-mentioned findings demonstrate the nature of specific atmospheric mechanisms that operate and affect the ozone layer in a power law fashion. These could also be a good test of atmospheric chemistry-transport models. Apart from reproducing instantaneous absolute values, model results should also demonstrate the scaling behavior.

Acknowledgments

TOMS data were produced by the Ozone Processing Team at NASA's Goddard Space Flight Center. The ground-based data are credited to Vitaly Fioletov, Experimental Studies Division, Air Quality Research Meteorological Service of Canada.

References

Ausloos, M., and K. Ivanova, 2001: Power-law correlations in the southern-oscillation-index fluctuations characterizing El Nino. *Phys. Rev. E,* **63** (4): art. no. 047201.

Chen, Z., P. C. Ivanov, K. Hu, and H. E. Stanley, 2002: Effect of nonstationarities on detrended fluctuation analysis. *Phys. Rev. E,* **65** (4), art. no. 041107.

Collette, C., and M. Ausloos, 2004: Scaling analysis and evolution equation of the north atlantic oscillation index fluctuations. ArXiv:nlin.CD/0406068 vl 29 June.

Fanchiotti, H., S.J. Sciutto, C.A. Garcia, and C. Hojuat, 2004: Analysis of sunspot number fluctuations. ArXiv:nlin.AO/0403032 vl 16 March.

Hu, K., P. C.Ivanov, Z. Chen, P. Carpena, and H. E. Stanley, 2001: Effect of trends on detrended fluctuation analysis. *Phys. Rev. E* **64** (1): art. no. 011114.

Kantelhardt, J.W., S. A. Zschiegner, E. Koscielny-Bunde, S. Havlin, A. Bunde, and H. E. Stanley, 2002: Multifractal detrended fluctuation analysis of nonstationary time series. *Physica A* **316** *(*1-4): 87-114.

Koscielny-Bunde, E., A. Bunde, S. Havlin, H. E. Roman, Y. Goldreich, H. J. Schellnhuber, 1998: Indication of a universal persistence law governing atmospheric variability. *Phys. Rev. Lett.* **81** (3), 729-732.

Peng, C. K., S. V. Buldyrev, S. Havlin, M. Simons, H. E. Stanley, and A. L. Goldberger, 1994: Mosaic organization of DNA nucleotides. *Phys. Rev. E* **49** (2), 1685-1689.

Talkner, P., and R. O. Weber: 2000: Power spectrum and detrended fluctuation analysis: Application to daily temperatures. *Phys. Rev. E* **62** (1): 150-160 Part A.

Varotsos, C., 2005: Power-law correlations in column ozone over Antarctica. *Int. J. Rem. Sensing* (in press).

Varotsos, P. A., N. V. Sarlis, and E. S. Skordas, 2003a: Long-range correlations in the electric signals that precede rupture: Further investigations. *Phys. Rev. E,* **67**, 21109-21121.

------------, ------------, and -------------, 2003b: Attempt to distinguish electric signals of a dichotomous nature. *Phys. Rev. E* **68** (3): art. no. 031106.

Weber, R. O., and P. Talkner, 2001: Spectra and correlations of climate data from days to decades. *J. Geophys. Res.,* **106**, 20131-20144.

PK+ Tree: An Improved Spatial Index Structure of PK Tree

Xiaolin Wang, Yingwei Luo*, Lishan Yu, and Zhuoqun Xu

Dept. of Computer Science and Technology, Peking University, Beijing, P.R.China, 100871
lyw@pku.edu.cn

Abstract. Spatial index is very important in GIS. PK tree and Hilbert R tree are two well-known spatial index structures. Comparison operations are very little in PK tree, while disk I/O operations are quite little in Hilbert R tree. PK+ Tree is an improved spatial index structure from PK tree. In PK+ tree, Comparison operations are less than in Hilbert R tree, while disk I/O operations are almost the same as in Hilbert R tree.

1 Introduction

With the development of information technology about data mining and multimedia, massive data processing gets more and more important in spatial processing. Spatial indexing is a key issue for massive spatial data processing.

Many kinds of spatial index structures exist nowadays. They can be divided into two main categories: partition tree and R tree. These two categories of spatial index structure have their own weakness and strongpoint. PK tree and Hilbert R tree are two efficient spatial index structures of each category.

In this paper, a new spatial index structure, PK+ tree, which are from PK tree, is introduced in order to improve storage usage, so that it may be more efficient for high dimensional data. In the next section, the problems of PK tree are analyzed. In the third section, PK+ tree is discussed in details. In the fourth section, the efficiency of storage and query of PK+ tree are compared with of PK tree and Hilbert R tree.

2 PK Tree and Its Weakness

PK tree is an efficient spatial index structure of partition tree category. In PK tree, an initial space cell C_0 are required to enclosing all the data objects to be indexed. C_0 can be split into $R=(r_x, r_y)$ sub cells C_1, and so on, so that the width and height of each C_i is $1/r_x$ and $1/r_y$ of C_{i-1}. Only those K-instantialable cells have their corresponding nodes in PK tree.

A key issue of PK tree is that it's poor efficiency of storage. The average node usage (ANU) is the percentage of valid entries to the capacity of entries in the tree. For data points in two dimensional space, each node has the capacity to hold $R(K\text{-}1)$ en-

---

* Corresponding author: LUO Yingwei, lyw@pku.edu.cn.

V.S. Sunderam et al. (Eds.): ICCS 2005, LNCS 3516, pp. 511 – 514, 2005.

tries, but the ANU is only 40%. For the same data set, the ANU of a Hilbert R tree may reach to 84% or 87%. When processing data object other than point, for example curves or surfaces, the ANU of a PK tree can reach only 12%, while Hilbert R tree may also reach around 85%. The poor efficiency of storage leads to more disk I/O operations while access the index tree. For data point, the ANU of PK tree is half of the ANU of Hilbert R tree, but PK tree yells more efficient query performance than Hilbert R tree. But for curves or surfaces, the query performance of PK tree is much poorer than Hilbert R tree. The main reason is that the ANU of PK tree is far more lees than that of Hilbert R tree. With such poor ANU, PK tree are limited to be used for high dimensional data object. To solve this problem, PK+ tree is introduced in this paper aiming at improving the ANU.

3 PK+ Tree

The ANU of PK tree drops with large $R=(r_x, r_y)$. If R can be limited to the smallest value, then the ANU might be greatly increased. In PK+ tree, R is restricted to be no more than 3. To partition a given space cell C in two-dimensional space, firstly C is split into three cells at x dimension, then at y dimension, and then at x dimension, and so on. The three cells are not symmetrical. Two cells of them are the two half of C, while the third cell is C itself! But the third cell is not the same as C, since is hold all data objects sits in C but not in the other two half cells. The third cell is called middle cell, and the two half-cells are called left cell and right cell (or bottom cell and top cell) correspondingly, as shown in figure 1.

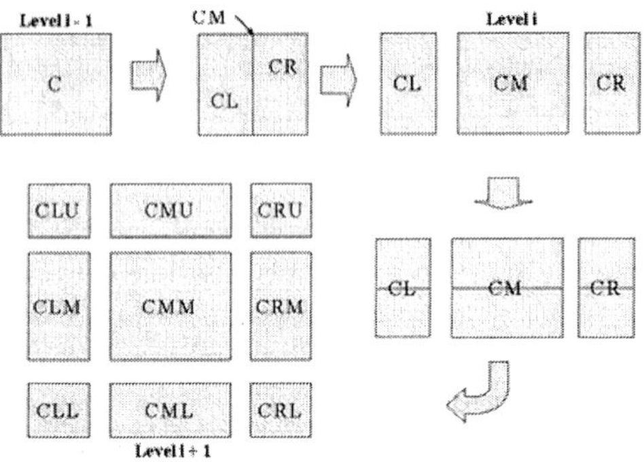

Fig. 1. PK+ Tree Cell Splitting

PK+ tree also requires that any node in the tree is K-intantialable. Insertion or deletion is implemented with two steps. First, search from the root to find the proper node to be modified, and modify it; then, trace back to the root and checking each node on the path if it's K-intantialable, if not, some promotion, division or deletion operations will be done to keep all nodes K-intantialable. The process of K-intantialable checking in PK+ tree is the same as in PK tree, except that they apply different kinds of splitting rules.

PK+ tree split the space cell at each dimension in turn, which makes a lower degree of regularity of partition than PK tree, but the ANU is greatly increased. In PK+ tree, the extent of a middle node overlaps with its sibling left (right) node. The overlapping of nodes might reduce some query's performance.

4 Performance Comparison

In the section, a series of experiments are done to examine the query performance of PK+ tree. Two dimensional data objects in two-dimensional space are being used in our experiment. A cluster data set from *SEQUOIA 2000 Benchmark* and a random data set are used. In these experiments, we compare PK+ tree with corresponding PK tree and 3-2 Hilbert R tree.

We count disk I/O operations and compare operations instead of measure the time cost in queries as the measurement of performance, thus we avoid the effects of implementation issues to our experiments.

Since in PK tree and PK+ tree, theoretically, the ANU will never be large than 50%. To make the storage efficiently, two kinds of methods might be applied. One method is to store each node in one or more disk pages according to the amount of valid entries in each node (SNMP, Single-node-multi-pages). The other method is to store several nodes in the same disk page if the total valid entries can be hold in one disk page (MNSP, Multi-nodes-single-page).

4.1 Storage Usage

For a given data set, the storage usage is the total valid entries vs. the total entries capacity of the index structure. ANU is used to describe the storage usage.

Table 1 shows the comparison of storage usage of PK tree, PK+ tree and Hilbert R tree. SNMP is compared with MNSP to show the effect of MNSP. Since Hilbert R tree is storage optimized, no more optimized method is considered in the comparison.

Table 1. ANU of PK+ tree, PK tree and Hilbert R tree

	SNMP		MNSP	
	Cluster data	Random data	Cluster data	Random data
PK+ tree	46.37%	50.83%	67.62%	66.78%
PK tree	12.95%	12.69%	25.49%	38.15%
Hilbert R tree	88.17%	85.52%	□	□

Table 1 shows that, MNSP outperforms SNMP, but does not make the ANU of PK+ tree reach the ANU of Hilbert R tree.

4.2 Query Performance

Our experiments show that, for both cluster data set and random data set, PK+ tree greatly reduces disk I/O operations per query than PK tree, but has almost the same disk I/O operations per query as Hilbert R tree. Since the ANU of PK+ tree is approximately 75% of Hilbert R tree, it is obvious that PK+ tree requires much less compare operations per query than that of Hilbert R tree. In more details, for small query regions, PK+ tree yells less disk I/O operations than Hilbert R tree; for query of inside, PK+ tree yells much less disk I/O operations than Hilbert R tree.

5 Conclusion

In this paper, an improved spatial index structure based on PK tree - PK+ tree is introduced. PK+ tree outperforms PK tree when used for high dimensional object indexing. PK+ tree has a higher ANU and hence requires less disk I/O operations in queries. Though disk I/O operations in queries of PK+ tree is almost the same as that of Hilbert R tree, but Hilbert R tree yells a higher ANU, compare operations in queries is more than that of PK+ tree. PK+ tree is perfect for application that cares both disk I/O performance and CPU performance.

Acknowledgement

This work is supported by the National Research Foundation for the Doctoral Program of Higher Education of China under Grant No. 20020001015; the National Grand Fundamental Research 973 Program of China under Grant No.2002CB312000; the National Science Foundation of China under Grant No.60203002; the National High Technology Development 863 Program under Grant No. 2002AA135330 and No. 2002AA134030; the Beijing Science Foundation under Grant No.4012007.

References

1. Wei Wang, Jiong Yang and Richard Muntz: PK-tree: A Dynamic Spatial Index Structure for Large Data Sets, Kluwer Academic Publishers (1997)
2. Wei Wang, Jiong Yang and Richard Muntz: Pk-Tree: A Spatial Index Structure For High Dimensional Point Data, Kluwer Academic Publishers (2000).
3. Ibrahim Kamel and Christos Faloutsos: Hilbert R-tree: An improved R-tree Using Fractals, Morgan Kaufmann Publishers Inc (1994).
4. Bernhard Seeger and Hans-Peter Kriegel: Techniques for Design and Implementation of Efficient Spatial Access Methods, Morgan Kaufmann Publishers Inc (1988).
5. Andreas Henrich, Hanas-Werner Six and Peter Widmayer: The LSD Tree: Spatial Access to Multidimensional Point and Non-point Objects, Morgan Kaufmann Publishers Inc (1989).

Design Hierarchical Component-Based WebGIS

Yingwei Luo, Xiaolin Wang, Guomin Xiong, and Zhuoqun Xu

Dept. of Computer Science and Technology, Peking University, Beijing, P.R.China, 100871
lyw@pku.edu.cn

Abstract. A practical component-based WebGIS named as Geo-Union is presented. Geo-Union consists of four layers: storage layer, service layer, component layer and application layer. Service layer can be partitioned into another two layers: client service layer and server service layer. The architectures and object constitutions of each layer in Geo-Union are discussed in details. The Web application model of Geo-Union is also presented. At last, some future works in WebGIS, such as interoperability, security, distributed computing and intelligent computing, are indicated and simply explored.

1 Introduction

Geographical Information System, GIS, is an effective tool that digitally reflects the geometry spatial situation on which human society lives and the various transitional spatial data. GIS describes the attributes of these spatial data and simulates the action of geospatial objects in a model way. Under the supports of software and hardware, GIS uses the given formats to support input/output, memory and display. It also provides the service of inquiring geometry spatial information, doing compositive analyses and making assistant decision. After several years' development, GIS is being wildly used in every aspect and plays an important role.

WebGIS is the Internet GIS that has Browse/Server architecture. Recently, WebGIS application has become more and more popular in many GIS user communities [1] because of the fascinating development of computer networks as well as the more and more popular use of the Internet. The key steps to promote WebGIS to a more practical situation are rational adjusting computation functions and enhancing performance. In this paper, a practical multi-layer component-based WebGIS model Geo-Union and its Web application model are discussed.

2 Multi-layer Component-Based WebGIS Model Geo-Union

Component modeling is one of the main approaches to enhance functions of WebGIS. Geo-Union system is based on component technique and Client/Server architecture. The system includes four layers: storage layer, service layer, component layer and application layer, with client and server existing in service layer [2-4], shown in Figure 1. The multi-layer component-based model enables GIS functions to be distributed in network effectively and brings high reusability of the system. Furthermore, it provides effective functions for further development and integration with other systems.

V.S. Sunderam et al. (Eds.): ICCS 2005, LNCS 3516, pp. 515 – 522, 2005.

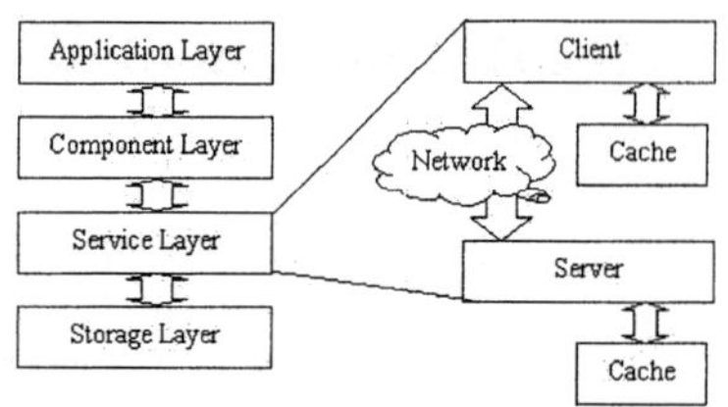

Fig. 1. Architecture of Geo-Union

2.1 Geo-Union Storage Layer

Storage Layer is the fundamental part in Geo-Union architecture, responsible for storage management of GIS data. With the help of ORDB, the layer stores and manages geography spatial and non-spatial data. Main objective of the layer includes how to present and store GIS data and how to maintain relations of these data. Below is type of GIS data: (1) Layer: Collection of spatial entities with the same type. (2) Entity: Spatial object composed of geometry and attribute data. Geometry data represents geometrical location of spatial object while attribute data describes society data. (3) Legend: A method to visualize Spatial Entity. (4) Legend library: Composed of 0~n Legends. (5) Reference system: Reference frame and Attitude frame of the Layer. (6) Display Setting Item: Mapping relation between Spatial Entity and Legend. (7) Display Setting: Composed of Display Setting Items.

Figure 2 shows the relations among GIS data. Encapsulation and management of GIS data exist in all the layers: server, client and component layer, and the relation in each layer are the same.

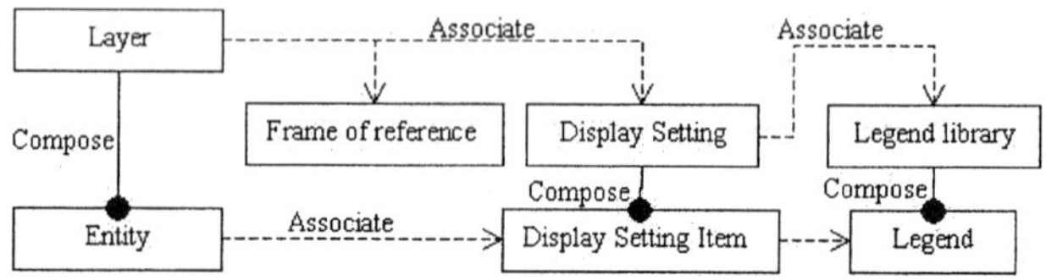

Fig. 2. E-R map in Geo-Union Storage Layer

2.2 Geo-Union Service Layer

Service Layer is responsible for management and access of GIS data, and composed of Geo-Union Client and Server. Geo-Union Client, as a server, provides data accessing and processing services for Geo-Union Component Layer, while Geo-Union Server manages and retrieves spatial data from Geo-Union Storage Layer, as a client. Geo-Union Client and Server are two independent but highly related parts: the Server provides functions such as data accessing service, spatial data index, basic spatial relation query, transaction and data sharing; through those services provided by the Server, the Client provides basic GIS tools and re-development functions. Cache, part of Geo-Union Client, can reduce network load thus improve system response rate. Cache is also in Geo-Union Server, with similar but different implementation to the Client. Geo-Union Client can be used in the development of server simulator, whose Cache enables less network load and quick system response rate.

- Geo-Union Server

Geo-Union Server is the only interface to GIS data for Geo-Union client. Through interaction with Storage Layer, the Server provides Geo-Union client with following services: connection service, data accessing service (cache service included), transaction and data sharing service. Those services are implemented mainly by a series of object component, including connection object, data object and cache object. These objects have corresponding structure and function with those in Geo-Union Client.

In Geo-Union Server, a data source table is used to manage multiple distributed spatial databases. Therefore one Geo-Union Server can serve multiple Geo-Union Clients simultaneously. Geo-Union Clients can also access multiple spatial databases through Geo-Union Server (shown in figure 3).

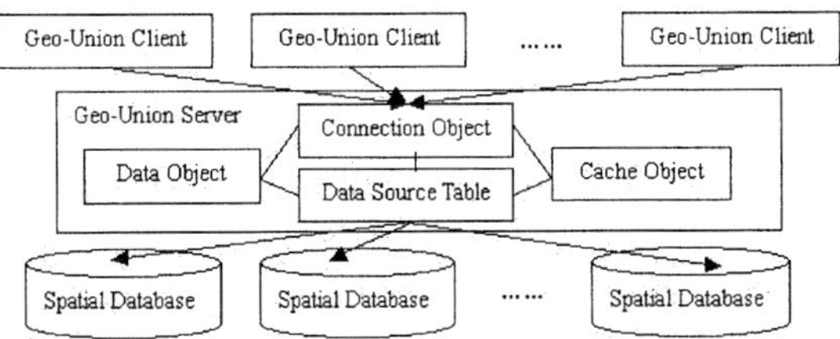

Fig. 3. Geo-Union Server and Geo-Union Client

- Geo-Union Client

Geo-Union Client is deployed on the client machine and connected with Geo-Union Server, providing GIS data accessing service and data processing service. Geo-Union client include the following functions: GIS data network access and management, GIS data object management, general spatial analysis arithmetic and spatial data cache management. Structure of Geo-Union Client is showed in figure 4. Below are the main objects in Geo-Union Client:

(1) Connection Object. Similar to Connection Object in Geo-Union Server, it manages communicating connections, GIS data access and transactions between Client and Server. It is also responsible to store and release connection relevant data objects in memory, maintain caches in Client and reduce network data flow.

(2) Data Object. Include Layer, Entity, Geometry Object, Record Collection, Map, Reference System, Display Setting, Display Setting Item, Legend Library and Legend. Considering the storage and access in Client's memory, the Client only supports two statues of Data Object: binding and dissociation. In other words, the Client can not only bind Data Object with data in Storage Layer through Geo-Union Server, but also store Data Object in its own memory. Below are status details of every kind of Data Object: (a) Layer is a collection of Entity. Layer also implements some basic spatial search operations such as K-near search, search for entities nearest to a specified entity in a layer. (b) Entity is the atomic access unit in GIS data, including Entity

Identifier, geometry object attribute, user attribute and annotation. (c) Geometry Object includes point, multipoint, line, multi-line, polygon and bitmap. Basic spatial arithmetic of relation between Geometry Objects is also provided. (d) Record Collection. As another interface to access Entities besides Layer, Record Collection is the uniform interface to access spatial and user attributes. It can store the results of searching for entities in a layer in the form of snapshot, support cursor operation, and support both immediate and batch update modes. (e) Map manages and accesses layers' structure information in Client. Through a Map Object, the system can organize layers into a practical map.

(3) Arithmetic Object. Arithmetic Object implements general spatial analysis arithmetic such as overlay analysis, network analysis and etc.

(4) Spatial Data Cache. In Spatial Data Cache, historical records of Data Object are stored to avoid retrieving repeatedly the same data from servers, thus reducing network load and user waiting time.

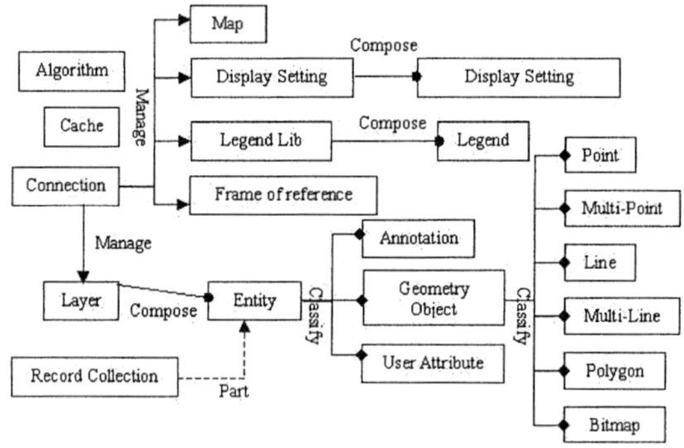

Fig. 4. Design Architecture of Geo-Union Client

2.3 Geo-Union Component Layer

Geo-Union Component Layer can provide Geo-Union Application Layer with many services. Geo-Union Component Layer is the encapsulation of Geo-Union Client and offers outer users GIS service interfaces to compose a complete GIS component library. The interfaces include Data Access Object, Map Display Object, Geometry Object, Function Object, Legend Edit and Display Object, Tool Object and etc. Composition and relation of these objects can be shown as figure 5.

- Data Access Object

Data Access Object provides GIS data manipulation functions as below: (1) GxConnection, connection object for GIS data access on the server. Operations such as connection or disconnection, layer object management, reference system management, display setting object management, legend library object management and transaction are allowed in GxConnection. (2) GxLayer, layer access object. Operations such as receiving and changing layer basic information, search and analysis based on layer, managing and searching entities in a layer, importing and exporting data are allowed. (3) GxEntity, entity object, through which users can access geometry and property data of an entity. (4) GxLegendLib, legend library. Each legend has a number greater than 0. (5) GxLegend, used for describing a legend. (6) GxReferenceSystem, used for describing a reference system. (7) GxDisplaySetting, display setting, which describes the visualization method of a layer. (8) GxDisplaySettingItem, one item in a display setting, which describes the visualization method of a kind of entities. (9) GxRecordset, search record set, composed of searched results. (10) GxFields, field connection of an entity. (11) GxField, one field of an entity. (12) GxSelection, selected entity number connection.

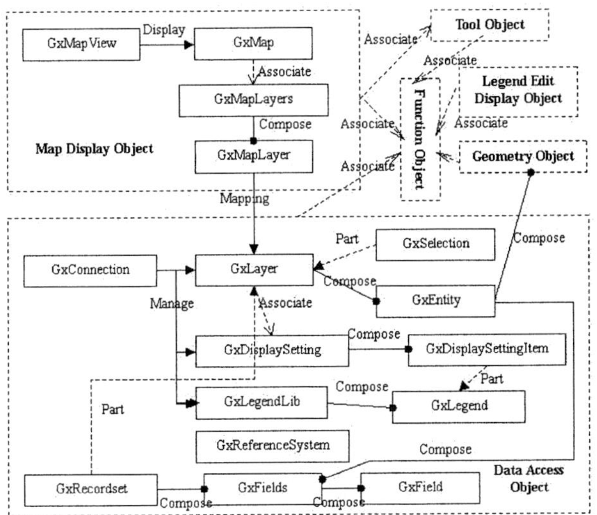

Fig. 5. Object Diagram in Component Layer of Geo-Union

- Map Display Object

Map display object can control and display GIS data, it includes: (1) GxMapView, map display control, is used for displaying map. In order to facilitate development on this control, we provide a tool management object and a group of input events. Through the tool management object GxTools, different tool objects (see also Tool Object) can be added in GxMapView to implement many special functions such as map edit, entity select, map measure and so on. The input events are all mouse or keyboard relevant and can be used in secondary development to implement customized functions. (2) GxMap is an abstract concept, can be displayed in GxMapView. One GxMap is corresponding to a collection of GxMapLayers. (3) GxMapLayers, a collection of map layers, can manage its owned map layers. (4) GxMapLayer, map layer in abstract concept, is corresponding to the map layer in storage. GxMapLayer is the map layer provided for users and includes legend configuration, reference system and annotation.

- Geometry Object

Geometry Object includes GxPoint (point entity), GxPoints (point collection), GxLine (line entity), GxLines (line collection), GxPolygon (polygon entity), GxRect (rectangle entity), GxBitmap (bitmap) and so on.

- Function Object

Function Object has two functions: one is that it can help users utilize other objects conveniently in the secondary development process; the other function is enable users to carry on special analysis in a more convenient way. (1) GxError, error handling objec. (2) GxEnumeration, enumeration object to contain the results of enumeration methods in GxConnection such as EnumLayer (enumerate all layers), EnumLegendLib (enumerate all legend libraries), EnumReferenceSystem (enumerate all reference systems), EnumDisplaySetting (enumerate all display settings) and so on. (3) GxObjectFactory, object factory. In IIS ASP programming, only ActiveX control object s can be created directly and other programmable objects must be created through controls in a indirect way. Therefore, GxObjectFactory is designed to support the creation of GIS object in all circumstances. (4) GxArray, is a kind of array, whose element can be any type or object. (5) GxSet is a kind of set, whose element can be any type or object. No element is the set can be equal to any other element. (6) GxUtitlity, a special function utility object, contains some functions which are difficult to be implemented in some visualized programming languages such as VisualBasic, Delphi, PowerBuilder. The functions includes object creating (i.e. creating a GxPoint object by x and y coordinates), special statistics and coordinate transforming. (7) GxAnalysis, special analysis object, including overlay analysis, clipping analysis and connectivity analysis.

- Legend Edit and Display Object

Legend Edit and Display Object can create, edit, manage and explore legends. It includes GxLegendEditor (legend edit control), GxLengendLibView (legend library explore control) and GxLengendView (legend display control).

- Tool Object

Tool Object implements a set of basic mouse interfaces, including GxTools (tool manage object to manage a set of other tool objects), GxPick (pick entities in a map layer), GxInputLine (input line entities in a map layer), GxInputPolygon (input polygon entities in a map layer), GxInputRect (input rectangle entities in a map layer), GxZoomIn (map zoom in), GxZoomOut (map zoom out) and GxPalm (map roam).

2.4 Geo-Union Application Layer

In Geo-Union Application Layer, users can develop customized GIS applications. This layer's main job is to obtain customized GIS application by pruning and integrating services provided by Component Layer.

3 Web Application Model of Geo-Union

Geo-Union has complete functions and flexible adaptabilities. A series of techniques to construct GIS application system are provided by Geo-Union, including server construction and application development. Geo-Union not only supports stand-alone environment, but is quite suitable for developing GIS application system, especially Web-based GIS application in network environment.

Geo-Union component layer contains a group of ActiveX controls and relative programmable objects. ActiveX controls can be embedded in Web pages directly. Web application model in Geo-Union is shown in figure 6.

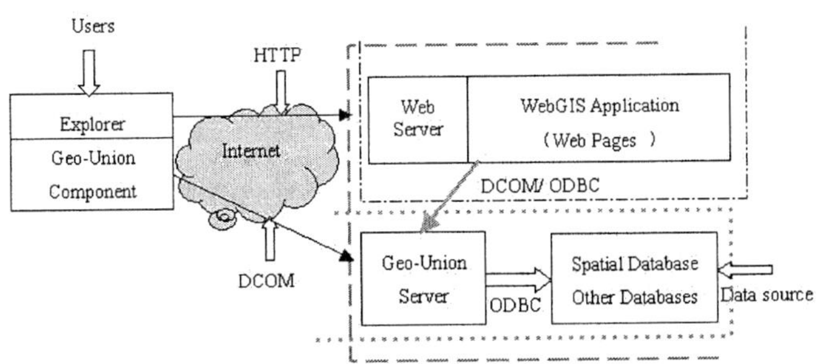

Fig. 6. Web Application Model in Geo-Union

Explorer communicates with Web server through HTTP and then gets WebGIS application module. Explorer interprets and executes the application module after receiving it. WebGIS applications are developed with Geo-Union components in specific application domain. WebGIS application and Web server compose WebGIS application server. WebGIS application can also access Geo-Union server to handle requests from Explorer.

Geo-Union server provides outer users interfaces to search and access spatial and non-spatial interfaces. As a client of Geo-Union server, Web server accesses geometry data through Geo-Union server, using ODBC or COM/DCOM. Explorer also needs accessing spatial and non-spatial data when interpreting and executing WebGIS application module. Here, as a client of Geo-Union server, explorer can search and access data using DCOM. When accessing spatial data which is transferred in vector form, explorer does not need downloading all data of the layer, but requests dynamically required entity data from Geo-Union server using entity data miss request algorithm. At the same time, as a client, Geo-Union server accesses spatial database to get spatial and non-spatial data through ODBC.

Based on commercial ORDBs, spatial database manages spatial and non-spatial data tightly relative with WebGIS application. Geo-Union server accesses database through ODBC, so the database needn't be confined in a certain specific databases.

4 Conclusion

Geo-Union system has been applied in many fields in China, such as pipe network integration information system, electric information system, water environment information system and fire emergency information system. At present, these systems are under stable and correct condition. What is more, the system keeps unfailing performance after repetitious visits.

Geo-Union system is also the result of research and application in some critical techniques of WebGIS. However, to reach higher practicability, many works left as follows: (1) As the development of Internet, more and more spatial data become available. But these data have different formats. It is still hard to share the data and services between various GIS. In order to solve this problem, people have studied standardization of GIS, but the actual open GIS has not been formed. (2) Along with the further open of Internet, the security of visiting spatial data is an unavoidable problem which WebGIS has to face. It is hoped that the secret spatial data in Web can't be achieved unlawfully. (3) WebGIS will meet the needs of thousands upon thousands users who visit Internet at the same time. How to guarantee the exactitude of concurrency and how to use the system ability farthest to meet the users' visit needs are the keys to making WebGIS worthy of its name. (4) Nowadays, intelligent agent technique is a main research direction of software domain. It provides an effective solution for establishment of complicated distributed software system. Of course, agent technique also provides a fire-new method for the establishment of WebGIS [6]. Researches on how to exert agent technique on distributed GIS construction and combine it tightly with geometry spatial metadata are not only for GIS data sharing and service sharing, but for deep GIS application cooperation and intelligent GIS information services. Furthermore, the researches provide a simple and convenient agent-based system development method for users, thus having abroad application future and important practicality [7].

Acknowledgement

This work is supported by the National Research Foundation for the Doctoral Program of Higher Education of China under Grant No. 20020001015; the 973 Program of China under Grant No.2002CB312000; the National Science Foundation of China under Grant No.60203002; the 863 Program under Grant No. 2002AA135330 and No. 2002AA134030; the Beijing Science Foundation under Grant No.4012007.

References

1. Zhang, Li, *et al*: Geographic Information System in the Internet Age (in Chinese), ACTA GEODAETICA et CARTOGRAPHICA SINICA, 27(1): 9-15 (1998).
2. Luo, Yingwei, *et al*: The Components Design for WebGIS (in Chinese), Chinese Journal of Image and Graphics, 4(A): 79-84 (1999).
3. Li, Muhua: Research and Implementation of the Componentization of Distributed WebGIS (in Chinese), [Master Dissertation]. Beijing: Peking University (2000).
4. Wu, Jian: A Study on Spatial Data Management in Component-based Distributed WebGIS (in Chinese), [Master Dissertation], Beijing: Peking University (2000).
5. Cong, Shengri: Key Issues on ORDB-Based Component GIS (in Chinese), [PhD Dissertation], Beijing: Peking University (1999).
6. M. Wooldridge and N. R. Jennings: Intelligent Agents: Theory and Practice, Knowledge Engineering Review, 10(2): 115-152 (1994).
7. Luo, Yingwei, *et al*: The Research on Geo-Agents (in Chinese), Journal of Computer Research and Development, 37(12): 1504-1512 (2000).

Adaptive Smoothing Neural Networks in Foreign Exchange Rate Forecasting

Lean Yu [1,2], Shouyang Wang [1,2], and Kin Keung Lai [3,4]

[1] Institute of Systems Science, Academy of Mathematics and Systems Sciences, Chinese Academy of Sciences, Beijing 100080, China
[2] School of Management, Graduate School of Chinese Academy of Sciences, Chinese Academy of Sciences, Beijing 100039, China
{yulean,sywang}@amss.ac.cn
[3] College of Business Administration, Hunan University, Changsha 410082, China
[4] Department of Management Sciences, City University of Hong Kong, Tat Chee Avenue, Kowloon, Hong Kong
mskklai@cityu.edu.hk

Abstract. This study proposes a novel forecasting approach – an adaptive smoothing neural network (ASNN) – to predict foreign exchange rates. In this new model, adaptive smoothing techniques are used to adjust the neural network learning parameters automatically by tracking signals under dynamic varying environments. The ASNN model can make the network training process and convergence speed faster, and make network's generalization stronger than the traditional multi-layer feed-forward network (MLFN) model does. To verify the effectiveness of the proposed model, three major international currencies (British pounds, euros and Japanese yen) are chosen as the forecasting targets. Empirical analyses reveal that the proposed novel forecasting model outperforms the other comparable models. Furthermore, experimental results also show that the proposed model is an effective alternative approach for foreign exchange rate forecasting.

1 Introduction

The difficulty in predicting foreign exchange rates, due to their high volatility and complexity, has long been an imperative concern in international financial markets as many econometric methods are unable to produce significantly better forecasts than the random walk (RW) model [1]. Recent studies provide some evidence that nonlinear models are able to produce better predictive results, ameliorating the performance of the simple RW model. Of the various nonlinear models, the artificial neural network (ANN) model has emerged as a strong alternative for predicting exchange rates. As claimed by Grudnitski and Osburn [2], neural networks are particularly well suited for finding accurate solutions in an environment characterized by complex, noisy, irrelevant or partial information. Literature documenting this research effort is quite diverse and involves different architectural designs. Some examples are presented. Early applications of neural networks in forecasting chaotic time series have been performed by Lapedes and Farker [3]. Weigend et al. [4] and Refenes et al. [5] ap-

V.S. Sunderam et al. (Eds.): ICCS 2005, LNCS 3516, pp. 523 – 530, 2005.

plied multilayer forward network (MLFN) models in their forecasts of foreign exchange prices. Weigend's model performance was tested in terms of accuracy, giving support to nonrandom behavior. Refenes' work extended Weigend's research by adding a validity test to the model's performance and compared the results with those of the forward rate, thereby providing added support to the forecasting ability of neural networks in the foreign exchange market. Tenti [6] applied recurrent neural network (RNN) models to forecast exchange rates. Hsu et al. [7] developed a clustering neural network (CNN) model to predict the direction of movements in the USD/DEM exchange rate. Their experimental results suggested that their proposed model achieved better performance relative to other indicators. De Matos [8] compared the strength of a MLFN with that of a RNN based on the forecasting of Japanese yen futures. Likewise, Kuan and Liu [9] provided a comparative evaluation of MLFN's performance and an RNN for the prediction of an array of commonly traded exchange rates. In a more recent study by Leung et al. [10], MLFN's forecasting accuracy was compared with the general regression neural network (GRNN). The study showed that the GRNN possessed a greater forecasting strength relative to MLFN with respect to a variety of currency rates. Zhang and Berardi [11] adopted an ensemble method for exchange rate forecasting and obtained better results than those under a single network model. Chen and Leung [1] used an error correction neural network (ECNN) model to predict exchange rates and good forecasting results can be obtained with their model.

Although a handful of studies exist on neural network applications in foreign exchange markets, most of the literature focuses on the MLFN [1-5, 8-10, 12-15]. However, there are several limitations to the MLFN. For example, convergence speed of the MLFN algorithm is often slow, thus making the network learning time long. Furthermore, it is easy for the optimal solution to be trapped into local minima thus making generalization capability weak. Therefore, we propose an adaptive smoothing technique to overcome these limitations to predict the daily exchange rates for three major internationally traded currencies: British pounds, euros and Japanese yen. In order to provide a fair and robust evaluation of the ASNN model relative to performance, the forecasting performance of the proposed ASNN model is compared with those of the MLFN model, which is used as the benchmark model. The rest of this article is organized as follows. Section 2 describes the ASNN model in detail. Section 3 gives an experiment and reports the results. And Section 4 concludes the article.

2 Adaptive Smoothing Neural Network for Forecasting

2.1 The Introduction of Neural Networks

Artificial neural networks (ANNs) – originally developed to mimic neural networks, in particular the human brain – are composed of a number of interconnected simple processing elements called neurons or nodes. Each node receives an input signal from other nodes or external inputs; after processing the signals locally through a transfer function, a transformed signal is output to other nodes or final outputs. ANNs are characterized by the network architecture; that is, the number of layers, the number of nodes in each layer and how the nodes are connected. In a popular form, the multi-layer feed-forward network (MLFN), all nodes and layers are arranged in a feed-forward manner. The first or the lowest layer is an input layer where external information is received. The last or the highest layer is an output layer where the network produces the model solution. In between, there are one or more hidden layers which are critical for ANNs to identify the complex patterns in the data. All nodes in adjacent layers are connected by acyclic arcs from a lower layer to a higher layer. ANNs are already one of the types of models that are able to approximate various nonlinearities in the data, and this makes them popular with academics and practitioners.

However, there are several drawbacks to the popular MLFN. First of all, the convergence speed of the MLFN algorithm is often slow, thus making network learning time long. Second, it is easy for the optimal solution obtained to be trapped into local minima, thus making generalization capability weak. Finally, the question of how to select reasonable network architecture is still an intractable problem.

In view of the above problems, in the following subsection we propose a novel algorithm to improve the MLFN by introducing adaptive smoothing techniques.

2.2 The Adaptive Smoothing Neural Network Model

In this study, adaptive smoothing techniques are used to adjust the neural network learning parameters automatically in terms of tracking signals under dynamic varying environments. This yields a new weight adjustment algorithm in virtue of quality control (QC) concept. In MLFN, model errors are usually the squared error or mean squared error (MSE). But using these error metrics makes it difficult to capture deviations between actual values and network output values (or expected values). In the process of neural network learning, adaptive smoothing algorithms can utilize ordinary error and mean absolute deviation (MAD) as a supplement of error measure to adjust the network's parameters (i.e., learning weights). With the aid of cumulative ordinary error (COE), MAD, and derivative tracking signal (TS), an adaptive smoothing neural network model can be formulated.

Assume that a network with m layers has n nodes, the transfer function of every node is usually a sigmoid function (i.e., $f(x)=\frac{1}{1+e^{-x}}$), y is an output from the output layer, O_i is an output of any unit i in a hidden layer, W_{ij} is the weight on connection from the jth to the ith unit. Suppose that there are N sample pairs (x_k, y_k) $(k = 1, 2, \ldots, N)$, the output of unit i connected with the kth sample is O_{ik}, the input of unit j connected with the kth sample is

$$net_{jk}=\sum_i W_{ij}O_{ik} \tag{1}$$

And the output of unit j connected with the kth sample is

$$O_{jk}=f(net_{jk}) \tag{2}$$

Here the error function is the squared error, i.e., $E=\frac{1}{2}\sum_{k=1}^{N}(y_k-\hat{y}_k)^2$, the cumulative ordinary error (COE) is $COE(N)=\sum_{k=1}^{N}(y_k-\hat{y}_k)$, where y_k is the actual value and $\hat{y}_k$ is the network output value. Let E_k and COE_k be a squared error and an ordinary error connected with the *kth* sample, then $E_k = (y_k - \hat{y}_k)^2$ and $COE_k = (y_k - \hat{y}_k)$. Clearly, $COE(N) = COE(N-1) + COE_N$. Meanwhile, the mean absolute deviation (*MAD*) and tracking signal (*TS*) are defined as

$$MAD\ (N) = \frac{\sum_{k=1}^{N} |y_k - \hat{y}_k|}{N} \tag{3}$$

$$TS = \frac{COE(N)}{MAD(N)} \tag{4}$$

If *TS* is "large", this means that *COE(N)* is large relative to the mean absolute deviation *MAD(N)*. This in turn says that the network output is producing errors that are either consistently positive or consistently negative. That is, a large value of *TS* implies that the network output is producing forecasts that are either consistently smaller or consistently larger than the actual values that are being forecast. Since an "accurate" forecasting system should be producing roughly one half positive errors and one half negative errors, a large value of *TS* indicates that the forecast output is not reliable. In practice, if *TS* exceeds a control limit, denoted by θ, for two or more consecutive periods, this is taken as a strong indication that the forecast errors have been larger than an accurate forecasting system can reasonably be expected to produce. In our study, the control limit θ is generally taken to be 3σ for a neural network model with the aid of the '3σ limits theory' proposed by Shewhart [16].

If the error signal indicates that adjustment action is needed, there are several possibilities. One possibility is that the model needs to be changed. To do this, input variables may be added or deleted to obtain a better representation of the time series. Another possibility is that the model being used does not need to be changed, but the estimates of the model's parameters need to be changed. When using a neural network model, this is accomplished by changing parameters (i.e., model weights and bias).

Now we present the parameter adjustment process. Define the error gradient $\delta_{jk} = \frac{\partial E_k}{\partial net_{jk}}$, then

$$\frac{\partial E_k}{\partial W_{ij}} = \frac{\partial E_k}{\partial net_{jk}} \cdot \frac{\partial net_{jk}}{\partial W_{ij}} = \frac{\partial E_k}{\partial net_{jk}} \cdot O_{ik} = \delta_{jk} \cdot O_{ik} \tag{5}$$

(i) If j is the output node, $O_{jk} = \hat{y}_k$, then

$$\delta_{jk} = \frac{\partial E_k}{\partial net_{jk}} = \frac{\partial E_k}{\partial \hat{y}_k} \cdot \frac{\partial \hat{y}_k}{\partial net_{jk}} = -(y_k - \hat{y}_k) \cdot f'(net_{jk}) \tag{6}$$

(ii) If j is not the output node, then

$$\delta_{jk} = \frac{\partial E_k}{\partial net_{jk}} = \frac{\partial E_k}{\partial O_{jk}} \cdot \frac{\partial O_{jk}}{\partial net_{jk}} = \frac{\partial E_k}{\partial O_{jk}} \cdot f'(net_{jk}) \tag{7}$$

$$\frac{\partial E_k}{\partial O_{jk}} = \sum_m \frac{\partial E_k}{\partial net_{mk}} \cdot \frac{\partial net_{mk}}{\partial O_{jk}} = \sum_m \frac{\partial E_k}{\partial net_{mk}} \cdot \frac{\partial}{\partial O_{jk}} \sum_i W_{mi} \cdot O_{ik} = \sum_m \frac{\partial E_k}{\partial net_{mk}} W_{mj} = \sum_m \delta_{mk} \cdot W_{mj} \tag{8}$$

Thus

$$\begin{cases} \delta_{jk} = f'(net_{jk})\sum_m \delta_{mk} W_{mj} \\ \dfrac{\partial E_k}{\partial W_{ij}} = \delta_{jk} \cdot O_{ik} \end{cases} \tag{9}$$

The error δ_{jk} is propagated back to the lower layers in terms of Equations (6) and (9).

In order for the network to learn, the value of each weight has to be adjusted in proportion to each unit's contribution to the total error in Equations (6) and (9). The incremental change in each weight for each learning iteration is computed using Equations (10) and (11) in the following:

$$\Delta W_{ij} = c_1 \cdot \delta_{jk} \cdot O_{ik} + c_2 \cdot \varphi_{jk} \tag{10}$$

where c_1 is a learning rate ($0 \leqslant c_1 < 1$), c_2 is a positive constant that, being less than 1.0, is the smoothing rate to smooth out the weigh changes; and

$$\varphi_{jk} = \begin{cases} 0, & |TS| \le \theta,\ (\theta = 3\sigma[16] \text{ or } \theta = 4 \cdot MAD\,[17]); \\ -COE(N), & |TS| > \theta \text{ and } TS \le 0; \\ COE(N), & |TS| > \theta \text{ and } TS > 0. \end{cases} \tag{11}$$

It should be noted that there is a difference between our weight adjustment and the traditional momentum term. The traditional momentum term is only used to accelerate the neural network learning speed, while our weight adjustment cannot only increase learning speed but can also adjust the network search path and speed network convergence and improve neural network learning performance.

For convenience, we give the detailed algorithm for ASNN in the sequel:

(1) Initialize random weights to avoid saturation in the learning process.
(2) Iterate the following procedures, until error goals are satisfactory
  a. For k=1 to N
    (i) Compute O_{ik}, net_{jk}, $COE(N)$, $MAD(N)$ and $\hat{y}_k$ (forward process)
    (ii) Compute δ_{jk} from the output layer to the preceding layer inversely (backward process)
  b. For any nodes in the same layer, compute δ_{jk} according to Equations (6) and (9)
(3) Adjust weights with Equations (10) and (11) in terms of error gradient and tracking signals.

This completes the introduction of the ASNN algorithm. Usually, we can obtain the following benefits relative to traditional MLFN algorithms. First of all, learning error limits can be controlled via the corresponding program, making the search space smaller and learning accuracy higher. Second, model parameters can be adjusted adaptively in term of tracking signals, thus making network learning more efficient. Third, the search path can be adjusted by a smoothing factor and making it easier to obtain the network optimal solution than by using the MLFN algorithm.

To summarize, adaptive smoothing neural networks can adjust the model parameters adaptively and automatically via tracking signals, thus making the network search and convergence speed faster and avoiding local minima as far as possible.

2.3 ASNN for Time Series Forecasting

An adaptive smoothing neural network can be trained by the historical data of a time series in order to capture the nonlinear characteristics of the specific time series. The model parameters (such as connection weights and nodes biases) will be adjusted iteratively by a process of minimizing the forecasting errors (e.g., MSE). For time series forecasting, the computational form of the ASNN model with three-layer network connection is expressed as

$$x_t = a_0 + \sum_{j=1}^{q} w_j f(a_j + \sum_{i=1}^{p} w_{ij} x_{t-i}) + \xi_t \tag{12}$$

where a_j (j = 0, 1, 2, …, q) is a bias on the jth unit, and w_{ij} (i = 1, 2, …, p; j = 1, 2, …, q) is the connection weight between layers of the model, $f(\bullet)$ is the transfer function of the hidden layer, p is the number of input nodes and q is the number of hidden nodes. Actually, the ASNN model in (12) performs a nonlinear functional mapping from the past observation (x_{t-1}, x_{t-2}, …, x_{t-p}) to the future values x_t, i.e.,

$$x_t = g(x_{t-1}, x_{t-2}, \cdots, x_{t-p}, v) + \xi_t \tag{13}$$

where v is a vector of all parameters and g is a nonlinear function determined by the network structure and connection weights. Thus, in some senses, the ASNN model is equivalent to a nonlinear autoregressive (NAR) model [15]. To verify the effectiveness of the ASNN model, a simulation study is presented in the following section.

3 Experiment Study

3.1 Data Sources

We use three different datasets in our forecast performance analysis. The data used are daily and are obtained from Pacific Exchange Rate Service (http://fx.sauder.ubc.ca/), provided by Professor Werner Antweiler, University of British Columbia, Vancouver, Canada. They consist of the US dollar exchange rate against each of the three currencies (EUR, GBP and JPY) with which it has been studied in this research. We take the daily data from 1 January 2000 to 31 October 2002 as in-sample data sets, and we take the data from 1 November 2002 to 31 December 2002 as evaluation test sets or out-of-sample datasets (partial data sets excluding holidays), which are used to evaluate the good or bad performance of the predictions, based on evaluation measurements. In order to save space, the original data are not listed in the paper, detailed data can be obtained from the website. In addition, to examine the forecasting performance, the normalized mean squared error (*NMSE*) [15] and directional change statistics of exchange rate movement (D_{stat}) [14, 15] are employed here.

3.2 Experimental Results

When the data are prepared, we begin the ASNN model's training and learning process. In these experiments, we prepare 752 data (two years' data excluding public holidays). We use the first 22 months' data to train and validate the network, and use the last two months' data for prediction testing. For convenience, the three-day-ahead forecasting results of three major international currencies using the proposed ASNN model are shown in Table 1.

Table 1. Forecast performance evaluation for the three exchange rates

Exchange rates	British pounds		Euros		Japanese yen	
	MLFN	ASNN	MLFN	ASNN	MLFN	ASNN
NMSE	0.5534	0.1254	0.2137	0.0896	0.2737	0.1328
D_{stat}(%)	55.00	77.50	57.50	72.50	52.50	67.50

As can be seen from Table 1, we can conclude that: (i) from the viewpoint of *NMSE* indicator, the ASNN model performs consistently better than the MLFN model; (ii) furthermore, the *NMSE* of the MLFN model is much larger than that of the ASNN model, indicating that adaptive smoothing techniques can effectively control error changes and significantly improve network performance; and (iii) from the D_{stat} point of view, the correct number of direction of exchange rate movements increases when using the ASNN model. Among these, the increase in the British pound rate is the largest, while the increase in the Japanese yen rate is the smallest. This suggests that there may be some additional factors that need to be studied in relation to the Japanese yen. One possible reason is that the Japanese yen exchange rate is more volatile than that of the British pound; another might be that the market for yen is bigger and more efficient than the market for British pounds. However, we also find that it is feasible to predict exchange rates using the ASNN model, and that the results are promising.

4 Concluding Remarks and Future Research

This exploratory research examines the potential of using an ASNN model to predict main international currency exchange rates. Our empirical results suggest that the ASNN forecasting model may provide better forecasts than the traditional MLFN forecasting model. The comparative evaluation is based on a variety of statistics such as *NMSE* and D_{stat}. For all currencies included in our empirical investigation, the ASNN model outperforms the traditional MLFN model in terms of *NMSE* and D_{stat}. Furthermore, our experimental analyses reveal that the *NMSE* and D_{stat} for three currencies using the ASNN model are significantly better than those using the MLFN model. This implies that the proposed ASNN forecasting model can be used as a feasible solution for exchange rate prediction.

However, our work also highlights some problems that need to be addressed further. For example, the accuracy of rolling forecasting is still unsatisfactory for certain currencies, such as the Japanese yen. Of course, the above problems show possible directions for future work in formulating a generic adaptive smoothing neural network prediction model for exchange rate prediction as follows:

(i) As foreign exchange markets constitute a very complex system, more factors that influence the exchange rate movement should be considered in future research.

(ii) A new adaptive smoothing algorithm that improves the traditional MLFN model should be added to neural network software packages so that users working in other domains can more easily utilize new neural network models in their work.

References

[1] Chen, A.S., Leung, M.T.: Regression neural network for error correction in foreign exchange forecasting and trading. *Computers and Operations Research*, 31, (2004) 1049-1068.

[2] Grudnitski, G., Osburn, L.: Forecasting S&P and gold futures prices: an application of neural networks. *Journal of Futures Market*, 13, (1993) 631-643.

[3] Lapedes, A., Farber, R.: Nonlinear signal processing using neural network prediction and system modeling. Theoretical Division, Los Alamos National Laboratory, *NM Report*, (1987) No. LA-UR-87-2662.

[4] Weigend, A.S., Huberman, B.A., Rumelhart, D.E.: Generalization by weight-elimination with application to forecasting. In: Lippman, R.P., Moody, J.E. and Touretzky, D.S. (Eds), *Advances in Neural Information Processing Systems*, 3, Morgan Kaufman, San Mateo, CA, (1991) 875-882.

[5] Refenes, A.N., Azema-Barac, M., Chen, L., Karoussos, S.A.: Currency exchange rate prediction and neural network design strategies. *Neural Computing and Applications*, 1, (1993) 46-58.

[6] Tenti, P.: Forecasting foreign exchange rates using recurrent neural networks. *Applied Artificial Intelligence*, 10, (1996) 567-581.

[7] Hsu, W., Hsu, L.S., Tenorio, M.F.: A neural network procedure for selecting predictive indicators in currency trading. In: Refenes, A.N. (Ed), *Neural Networks in the Capital Markets*, New York: John Wiley and Sons, (1995) 245-257.

[8] De Matos, G.: Neural networks for forecasting exchange rate. *M. Sc. Thesis*. The University of Manitoba, Canada (1994).

[9] Kuan, C.M., Liu, T.: Forecasting exchange rates using feedforward and recurrent neural networks. *Journal of Applied Econometrics*, 10 (4), (1995) 347-364.

[10] Leung, M.T., Chen, A.S., Daouk, H.: Forecasting exchange rates using general regression neural networks. *Computers and Operations Research*, 27, (2000) 1093-1110.

[11] Zhang, G.P., Berardi, V.L.: Time series forecasting with neural network ensembles: an application for exchange rate prediction. *Journal of the Operational Research Society*, 52, (2001) 652-664.

[12] Brooks, C.: Linear and nonlinear (non-) forecastability of high frequency exchange rates. *Journal of Forecasting* 16, (1997) 125-145.

[13] Gencay, R.: Linear, nonlinear and essential foreign exchange rate prediction with simple technical trading rules. *Journal of International Economics*, 47, (1999) 91-107.

[14] Yao, J.T., Tan, C.L.: A case study on using neural networks to perform technical forecasting of forex. *Neurocomputing*, 34, (2000) 79-98.

[15] Yu, L.A., Wang, S.Y., Lai, K.K.: A novel nonlinear ensemble forecasting model incorporating GLAR and ANN for foreign exchange rates. *Computers and Operations Research*, (2004) In Press.

[16] Shewhart, W. A.: *Economic Control of Quality of Manufactured Product*, New York, (1931).

[17] Chase, R.B., Aquilano, N.J., Jacobs, R.F.: *Production and Operations Management: Manufacturing and Services*, McGraw-Hill, (1998).

Credit Scoring via PCALWM*

Jianping Li [1,Ψ], Weixuan Xu[1], and Yong Shi[2]

[1] Institute of Policy and Management,
Chinese Academy of Sciences, Beijing 100080, P.R.China
ljp@mail.casipm.ac.cn, jianpingli@yahoo.com
[2] Chinese Academy of Sciences Research Center on Data Technology
and Knowledge Economy, Beijing 100039, P.R.China
yshi@gscas.edu.cn

Abstract. We have presented a principal component analysis linear-weighted model (PCALWM) for credit scoring in [5,6], this article is a further study on this model. We revised the application procedure in the credit scoring, and tested it by a larger real-life credit card dataset. In comparison with some well-known scores, the empirical results of the PCALWM can achieve a favorable KS distance. The study on some application features of this model in the credit decision-making shows that the model can help the credit issuers to select the best trade-off among the enterprise stratagem, marketing and credit risk management.

Keywords: Credit scoring, Classification, Principal component analysis.

1 Introduction

The growth in consumer credit outstanding over the last fifty years is truly spectacular in the major countries, for example, in US, from the $9.8 billions in the year 1946 to $20,386 in 2003[2] .Not all of this growth is due to the borrowing on credit lines, Credit card (and debit cards) has become increasingly important as a method of money transmission.

However, from the perspective of history, the increase of total outstanding of consumer credit is always accompanying with more much consumers' bankruptcy and default, which are causing plenty of headaches for banks and other credit issuers. From 1980 to 2003, the number of individual bankruptcy filings in the US had increased by 560%[11]. How to predict bankruptcy and avoid huge charge-off loss becomes a critical issue of credit issuers.

In the consumer credit assessment, credit scoring is widely used. Credit scoring provides a suite of decision-support tools to aid the credit analyst. In credit scoring

* This research has been partially supported by a grant from National Natural Science Foundation of China (#70472074) and the President Fund of Chinese Academy of Sciences (CAS) (yzjj946).

Ψ Corresponding author.

V.S. Sunderam et al. (Eds.): ICCS 2005, LNCS 3516, pp. 531 – 538, 2005.

system, the classification tools, which are derived from analyzing similar data for previous customers, are used to categorize new applicants or existing customers as good or bad. There are lots of classification tools available now, such as statistics methods like discriminant analysis, regression, classification tree, nearest-neighbor approach, nonstatistical method like linear programming, neural networks, genetic algorithms, expert systems, etc [12,13].

The purpose of this paper is to further analyze the principle components analysis linear-weighted model (PCALWM) which we had presented to assess the consumer credit [5,6]. We will apply this model to a larger real-life credit card database to evaluate its performance and study the characteristics in credit decision. This study will make the model more applicable and help the credit issuers to make right decision.

2 The Basic Principals and Application Stages of PCALWM in Credit Scoring

Principle components analysis (PCA) is a well-known, widely-used statistical technique, and it is a purely data-driven way of exploring the data. The central idea is to reduce the dimensionality of a data set consisting of a large number of interrelated variables, while retaining as much as possible of the variation present in the data set. The PCA approach uses all of the original variable to obtain a smaller set of new variables (principal components--PCs)that they can used to approximate the original variables. The greater the degree of the correlation between the original variables, the fewer the number of new variables required. PCs are uncorrelated and ordered so that the first few retain most of the variation present in the original set.

PCA approach was applied to the credit risk management cooperated with other methods, Alici[1] used this method to predict the corporation bankruptcy with neural networks. Recently PCA has attracted much attention in the risk analysis in larger scale data and consumer credit data [3,8,14].

The PCALWM based credit scoring system includes 4 stages, data preparation, training, testing, and predicting. The main stages show by figure 1.

2.1 Data Preparation

This stage consist of collecting relevant data, merging and cleaning the data, defining the classes, selecting the variables, etc. we will get a dataset with suitable format which is fit for the PCA computing after this stage.

2.2 Training Stage

We classify the data set into two parts, one for training to construct the predicting model and the other for testing to validate the model. The different rate of the training data in the total sample will impact the model's efficiency, and a good classification of the training data rate maybe get a good classification model.

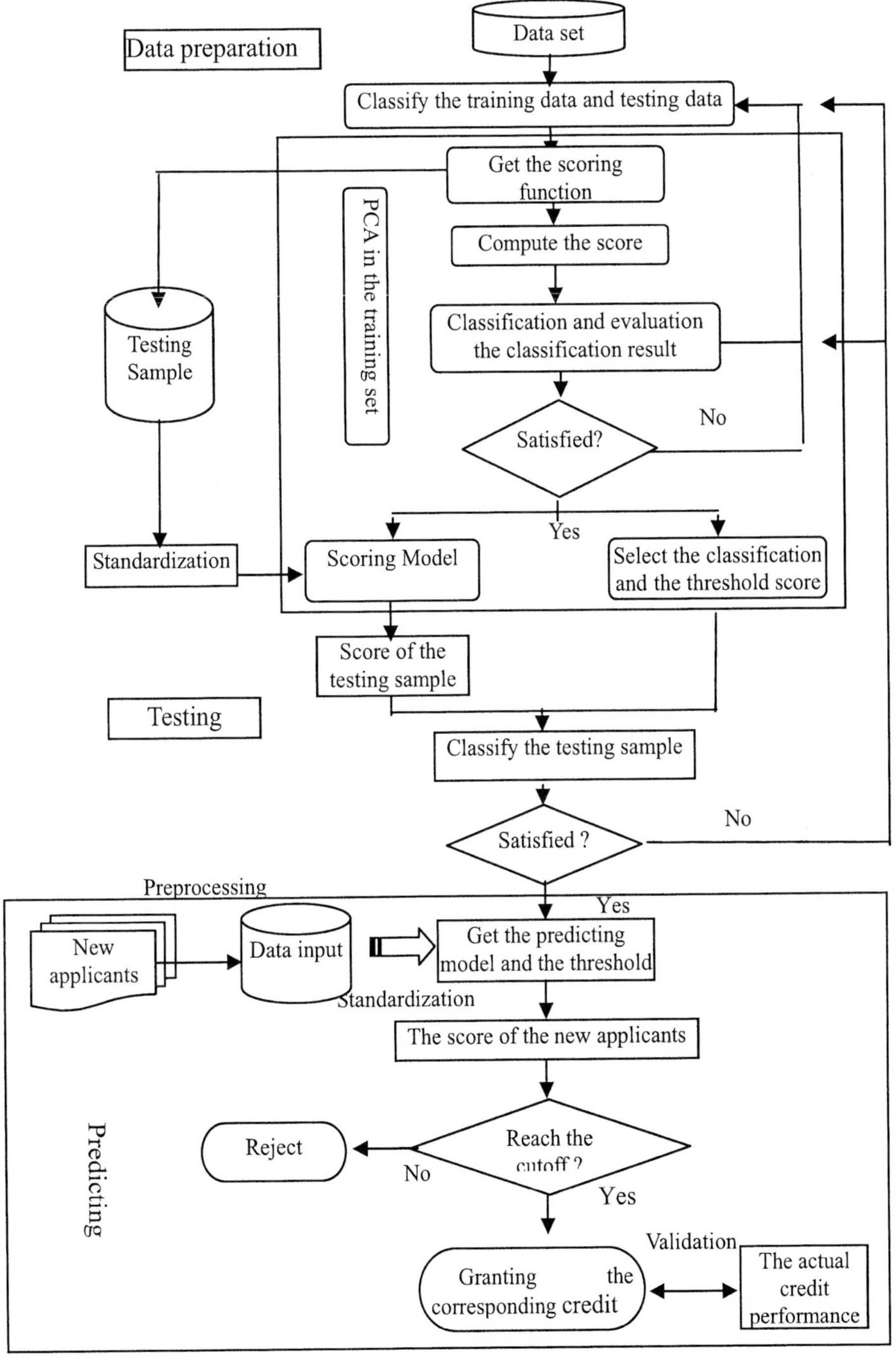

Fig. 1. The application procedure of PCALWM in credit scoring

The training stage is to achieve predicting model and set up the different credit grade. Supposed we have selected principal factors, the scoring function of the principal factors in the training sample could be expressed as:

$$F = a_1 z_1 + a_2 z_2 + \ldots + a_s z_s \tag{2.1}$$

z_i is the ith principal factor, a_i is its contribution rate :

$$a_i = \frac{\lambda_i}{\sum_{m=1}^{k} \lambda_m} \qquad a_{tl} = \frac{\sum_{m=1}^{l} \lambda_m}{\sum_{m=1}^{k} \lambda_m} \qquad (i, l = 1, 2, \ldots, k)$$

(λ_i is the ith eigenvalue sorted by descending, a_{tl} is the cumulative variance contributions of principal factor z_1 to z_l). Then,

$$Z_i = b_{i1} x_1 + b_{i2} x_2 + \ldots + b_{in} X_n \tag{2.2}$$

Take the equation (2) into equation (1), we get:

$$F = c_1 x_1 + c_2 x_2 + \ldots + c_n X_n \tag{2.3}$$

$$c_j = \sum_{i=1}^{s} a_i b_{ij} \qquad (i = 1, 2, \ldots, s;\ j = 1, 2, \ldots, n)$$

According to the score and the distribution of the bad credit in the training sample, we can set up the different grade by a certain method and standard. Then one can classify the different class based on the score. If the classification result is not ideal, we will classify and compute the data again.

2.3 Testing Stage

This stage is to validate the model and the credit grade which get from the training stage. We will choose the right score model and credit grade in this stage.

We use the same procedures in the training stage. But in data standardization here we will take the average and standard deviation of the training sample, so we get the data matrix $X_i = (x^*_{i1}, x^*_{i2}, \ldots, x^*_{ip})^T \quad i = 1, 2, \ldots, m$.

We use the equation 2.3 to compute the credit score in the testing sample, and classify the credit grade with the set grade in training stage. Similarly, if the classification result is not satisfied, we would classify the originate data and compute again.

2.4 Predicting the New Applicants

In this stage, we use the score model to predict the new applicants credit score and make the credit decision. After computing the new applicants credit score with the certain procedure as in the testing stage, we compare it with the credit standard which get from the training and testing stage. If the score reaches the threshold, we grant the corresponding credit, otherwise reject.

3 Empirical Study

3.1 Data Set

We use real life credit card data from a major US bank to conduct the experiment. This data set is composed by 5000 records, and two classes have been defined: good credit and bad credit. 65 variables are selected for personal credit assessment. There are including history payment, balance, transaction, account open, etc. we chose the SPSS soft to perform all of the computations.

3.2 Computation Result and Analysis

In fact, the practitioners have tried a number of quantitative techniques to conduct the credit card portfolio management. Some examples of known scores are (1) Behavior Score developed by Fair Isaac Corporation (FICO) [15]; (2) Credit Bureau Score also developed by FICO [15]; (3) First Data Resource (FDR)'s Proprietary Bankruptcy Score [4]; (4) Multiple-criteria Score [9,10] and (5) Dual-model Score [7]. These methods can be generally regarded as two-group classification models, and use either statistical methods or neural networks to compute the Kolmogorov-Smirnov (KS) value that often used to measure how far apart the distribution functions of the scores of the goods and bads are[13]. The resulting KS values from the learning process are applied to the real-life credit data warehouse to predict the percentage of bankrupt accounts in the future.

Formally, if $n_G(s)$ and $n_B(s)$ are the numbers of goods and bads with scores in a sample of n, where there are n_G goods and n_B bads, the $p_G(s)=\frac{n_G(s)}{n_G}$ and $p_B(s)=\frac{n_B(s)}{n_B}$ are the probabilities of a good and bad having a score s, then the KS distance is

$$KS=\max_s\left|p_G(s)-p_B(s)\right| \tag{3.1}$$

We chose 2000 records as a training set as random to compute the score model, and set three testing samples with 1000, 2000, 3000 records in the other 3000 records as random. The KS distances show as the table 1, the biggest is 0.6222 and lowest is 0.524.

Table 1. KS values of the different samples

	Traning2000	Testing1000	Testing2000	Testing3000
KS value	0.6223	0.5993	0.563	0.524

The KS values of some well-known score models in our 1000 sample are as the table 2.

Table 2. KS values of the different scores

	FICO	Credit bureau score	FDC proprietary bankruptcy score	Decision trees score	Multiple-criteria Linear Program score	PCALWM score
KS value	0.553	0.456	0.592	0.602	0.595	0.599

Among these scores, the PCALWM score is better than any other scores, except the decision tree score, so the PCALWM can achieve a good KS distance, and have a good classification.

4 Two Characteristics in Credit Decision-Making of PCALWM

In the practice, credit policy is the trade-off of the enterprise's stratagem, marketing and risk management. Loose credit policy is helpful to the market expanding, while strict credit policy for the risk management. General speaking, the loose credit policy will increase the type I error, and decrease the type II error to the credit issuers, while the strict credit policy makes the opposite effect. So, if the model can give the movement of the two type error with different credit granting rate efficiently, the model will greatly help to the credit issuers to make the trade-off in the actual credit decision.

For the encouragement of the primary empirical result, we will further study the application of PCALWM in the credit decision, mainly on:

- ✧ The classification error with different training sample rate, that to say how the training sample rate affects the classification accuracy.
- ✧ the classification error under the different granting rate with certain training sample rate. This will help the credit issuers to determine the granting rate under certain risk level.

4.1 The Training Sample Rate to the Classification Error

We will study the different classifying training and testing sample how to affect the classification error. In other words , under what training sample rate, the model can get the better classification. We chose different training sample rate at random to construct the predict model, and in the granting rate of 75%, compared the two type error and the average error.

We chose five group training sample at random, including 500, 1000, 2000, 3000, 4000 records respectively, corresponding to the 10%, 20%, 40%, 60%, 80% of the total sample, and the five testing sample are the corresponding rest records, which including 4500, 4000, 3000, 2000, 1000 records. Table 3 shows the classification error of the five testing samples.

The results showed that with the incensement of the training rate, type I error and average error decreased, but the type II error no obviously change. The type I error

Table 3. The classification error of testing sample in different training rate

Training rate	10%	20%	40%	60%	80%
Type I error	0.3952	0.3481	0.3003	0.2652	0.2393
Type II error	0.1787	0.1749	0.1740	0.1763	0.1758
Average error	0.2113	0.2058	0.1937	0.1880	0.1832

deceased rapidly, but the speed tends to slow. This results show that a bigger training rate will help to PCALWM the get a better classification, especially to decrease the type I error.

4.2 The Classification Error Under Different Granting Rate with a Certain Training Rate

We chose seven granting rate groups with 80% training rate, figure 2 displays the error movements.

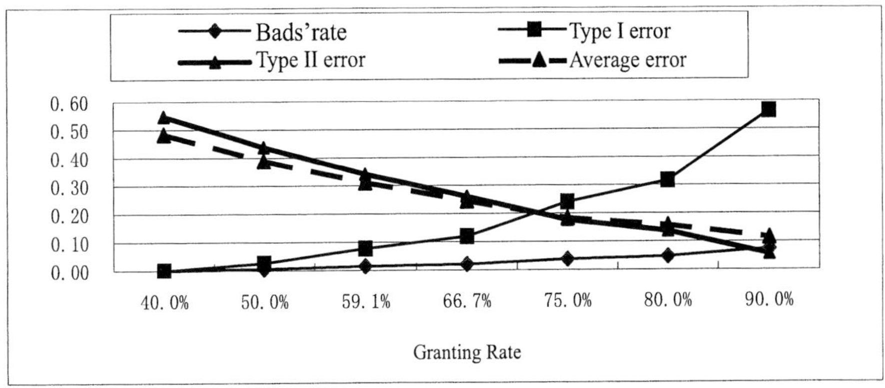

Fig. 2. The classification errors of testing sample in different granting rate

The results shows that as the granting rate increases, type I error increases, while type II error decreases sharply, and the bads' rate in granted group increases slowly. This shows the model has a good classification.

In practice, based on the tradeoff of marketing, enterprise stratagem and risk management, the credit issuers can make the choice of the two type error and average error to determine the best grant standard.

5 Summary and Conclusion

This paper has a further study on the principal component analysis linear-weighted model for credit scoring. It presents the application procedures in the credit scoring, tested by a larger real-life credit card data, and some application features in the credit decision-making. The empirical results and application characteristics of PCALWM show the modle will help the credit issuers to make the best trade-off among the enterprise stratagem, marketing and credit risk management.

Compared with some well-known credit scores, such as FICO behavior score, credit bureau score, first data resources' proprietary bankruptcy score and set enumeration decision tree score, Multiple criteria linear programming score, our model has a favorable classification.

Because of the great potential value in the consumer credit assessment, we will continue to perfect this model, we hope to get better results through more logical data processing and more suitable sample scale selecting, etc.

References

1. Alici,Y. (1995). Neural networks in corporate failure prediction: The UK experience. In: Proceedings of 3rd International Conference Neural Networks in the Capital Markets, A. N. Refenes, Y. Abu-Mostafa, J. Moody, and A. Weigend, Eds. London, UK, 393-406.
2. Board of Governors of the Federal Reserve Syetem http://www.federalreserve.gov/releases/g19/hist/cc_hist_mt.txt
3. Guo, Q, Wu, W., Massart, D.L., Boucon, C., Jong, S. de.(2002). Feature selection in principal component analysis of analytical data. Chemometrics and Intelligent Laboratory Systems, 61:123-132
4. http://www.firstdatacorp.com
5. Li, J.P, Liu, J.L, Xu, W.X, Shi, Y. (2004). An Improved Credit Scoring Method for Chinese Commercial Banks. In: S.Y. Wang and X.G. Yang eds., Advance in Risk Management, Lecture Notes in Decision Sciences, Global–Link Publisher, Hong Kong, Vol.4:63-73
6. Li, J.P, Xu, W.X. (2004). Consumer Credit Assessment via Principal Component Analysis Linear-Weighted Model. Chinese Journal of Management, Vol.12, (2): 63-73
7. Lin, Y.(2002). Improvement On Behavior Scores by Dual-Model Scoring System. International. Journal of Information Technology and Decision Making. (1): 153-164.
8. Sebzalli, Y.M., Wang, X.Z. (2001). Knowledge discovery from process operational data using PCA and fuzzy clustering. Engineering Applications of Artificial Intelligence, (14): 607–616
9. Shi, Y., Wise, M., Luo, M., Lin, Y(2001). Data Mining in Credit Card Portfolio Management: A Multiple Criteria Decision Making Approach. In: Koksalan, M., Zionts, S. (eds.): Multiple Criteria Decision Making in the New Millennium. Springer, Berlin 427-436.
10. Shi, Y., Peng, Y., Xu, W., Tang, X. (2002). Data Mining Via Multiple Criteria Linear Programming: Applications in Credit Card Portfolio Management. International Journal of Information Technology and Decision Making. 1 :131-151.
11. The American Banker Association. http://www.aba.com/default.htm
12. Thomas, L.C.(2000). A survey of credit and behavioural scoring: forecasting financial risk of lending to consumers. International Journal of Forecasting, 16:149-172
13. Thomas, L.C., Edelman, D.B., Crook, J.N.(2002). Credit Scoring and Its Application. Society for Industrial and Applied Mathematics, Philadelphia
14. Westad, F., Hersleth, M., Lea, P.r, Martens, H. (2003). Variable Selection in PCA in sensory descriptive and consumer data. Food Quality and Preference, (14):463–472
15. www.fairisaac.com

Optimization of Bandwidth Allocation in Communication Networks with Penalty Cost*

Jun Wu[1,3], Wuyi Yue[2], and Shouyang Wang[3]

[1] Institute of Intelligent Information and Communications Technology,
Konan University, Kobe 658-8501 Japan
wujun@iss.ac.cn

[2] Department of Information Science and Systems Engineering,
Konan University, Kobe 658-8501 Japan
yue@konan-u.ac.jp

[3] Academy of Mathematics and Systems Science,
Chinese Academy of Sciences, Beijing 100080 China
sywang@amss.ac.cn

Abstract. In this paper, we present an analysis on optimization and risk management in Communication Networks (CNs). The model is proposed for offline traffic engineering optimization, which takes a centralized view of bandwidth allocation, performance control, and risk control of network profit shortfall. First, we introduce a linear penalty cost in the CN optimization model and derive the optimal bandwidth capacity with the penalty cost. Then, we use the mean-variance approach to analyze the profit shortfall risk in CNs. Finally, numerical results are shown to reveal the impact of the penalty cost on the CNs performance.

1 Introduction

Traffic engineering is a process to optimize resource utilization and network performance [1], [2]. It has greatly improved network performance by using the emerging technologies, such as Multi-Path Label Switching and Optical Channel Trails [3], [4].

There are two forms of traffic engineering: online planning and offline planning. In past works, the offline optimization problem was formulated as a deterministic Multi-Commodity Flow (MCF) model, where demand of each channel was given as a fixed quantity [5], [6].

Recently, there were some works concerning with the stochastic traffic engineering. Mitra and Wang presented a stochastic traffic engineering framework for optimizing bandwidth provisioning and path selection in CNs [7]. Mitra and Wang also developed an optimization framework for the network service provider

* This work was supported in part by GRANT-IN-AID FOR SCIENTIFIC RESEARCH (No. 16560350) and by MEXT.ORC (2004-2008), Japan and in part by NSFC and MADIS, China.

V.S. Sunderam et al. (Eds.): ICCS 2005, LNCS 3516, pp. 539–547, 2005.

to manage profit in a two-tier market [8]. Mitra and Wang furthered their studies in [7], [8] and developed the efficient frontier of mean revenue and revenue risk [9]. Wu, Yue and Wang presented a stochastic model for optimizing bandwidth allocation in [10], which took a centralized view of bandwidth allocation, performance control, and risk of network profit shortfall. They analyzed the loss rate constraint and risk averseness in the CN optimization model and showed the impact of the loss rate constraint and the risk averseness on the network performance.

In this paper, based on the model presented in [10], we introduce a linear penalty cost in the optimization model for network bandwidth allocation. Whenever there is unsatisfied traffic demand with the limitation of network bandwidth, a linear penalty function will be added in the objective function. Next, we study the risk of the network profit shortfall by using the mean-variance approach [11], [12]. Finally, numerical results are shown to reveal the impact of the penalty cost on the network performance.

The rest of this paper is organized as follows. In Section 2, we present the system model that we consider in this paper and present the notations and preliminaries. In Section 3, we formulate the optimization model and derive the optimal bandwidth capacity with the linear penalty cost. In Section 4, we analyze the network profit shortfall risk by using the mean-variance approach. In Section 5, we give some numerical results to show the impact of the penalty cost on the network performance. Conclusions are given in Section 6.

2 System Model

A Communication Network (CN) is supposed to have users and a service provider. The CN should derive its unit revenue by serving demand including voice, packet data, image and full-motion video. For unit bandwidth capacity allocated to the network, a unit cost will be charged. For unsatisfied traffic demand with the limitation of network bandwidth, a linear penalty cost will be added in the objective function. The objective of this system is to maximize the CN mean profit of the network.

Let (N, L) denote a CN composed of nodes v_i ($v_i \in N$, $1 \leq i \leq N$) and links l ($l \in L$), where N is the set of all nodes and L is the set of all links. Let V denote the set of all node pairs, $v \in V$ denote an arbitrary node pair where $v = (v_i, v_j)$ and $v_i, v_j \in N$, C_l denote the maximal bandwidth capacity of link l, $R(v)$ denote an admissible route set for $v \in V$, ξ_s ($s \in R(v)$) denote the amount of capacity provisioned on route s, D_v ($v \in V$) denote the traffic load between node pair $v \in V$, b_v ($v \in V$) denote the amount of bandwidth capacity provisioned between node pair v, which can be routed on one or more routes, then $b_v = \sum_{s \in R(v)} (\xi_s)$.

In this paper, we consider the CN to be a whole system. We let $b > 0$ denote the amount of bandwidth capacity provisioned in the CN, then we have $b = \sum_{v \in V} (b_v)$. Let $D > 0$ denote the traffic demand in the whole CN, then we have $D = \sum_{v \in V} (D_v)$, which is characterized by a random distribution with its probability density function $f(x)$ and cumulative distribution function $F(x)$. $b \wedge D$ is the actual traffic load transmitted in the CN, where $\wedge$ represents the choice of the smaller between b and D. Let r denote the unit revenue by serving the traffic demand, so the total revenue of the CN is $r \times (b \wedge D)$. Let c denote the unit cost for unit bandwidth capacity allocated in the CN, so the total cost is $c \times b$. Let q denote the linear penalty cost for each unsatisfied traffic demand, so the total penalty cost is $q \times (D-b)^+$, where "+" represents the choice of the positive part of $(D-b)$. To avoid unrealistic and trivial cases, we assume that $r > q > 0$, $r > c > 0$.

3 Optimal Bandwidth Capacity in the CN with Penalty Cost

Based on the model presented in [10], in this paper, we add a linear penalty cost in the optimization model for network bandwidth allocation to evaluate the system performance. We let q denote the unit penalty cost for the unsatisfied traffic demand.

Let $\pi(b, D)$ denote the random profit function by serving traffic demand in the CN with the linear penalty cost, namely,

$$\pi(b, D) = r(b \wedge D) - q(D-b)^+ - cb. \tag{1}$$

Let $\Pi(b, D)$ denote the mean profit function with the linear penalty cost as follows:

$$\begin{aligned} \Pi(b, D) = r \int_0^b x f(x)dx + rb \int_b^{+\infty} f(x)dx \\ -q \int_b^{+\infty} (x-b) f(x)dx - cb. \end{aligned} \tag{2}$$

The objective function of the system is

$$\Pi^* = \max_{b>0}\{\Pi(b, D)\} \tag{3}$$

subject to

$$P(b \geq \delta D) \geq 1 - \epsilon \tag{4}$$

and

$$b \leq C_{max} \tag{5}$$

where Π^* is the optimal profit function. $P(b \geq \delta D) \geq 1 - \epsilon$ is the loss rate constraint with δ $(0 \leq \delta \leq 1)$ and ϵ $(0 \leq \epsilon \leq 1)$ as the parameters defined in [10]. $C_{max} > 0$ is the maximal capacity that can be allocated in the CN.

With the above formulation, we can derive the optimal capacity of bandwidth allocation. First, we analyze the property of the CN mean profit function $\Pi(b, D)$.

The first order derivative of $\Pi(b, D)$ with respect to b is given as follows:

$$\frac{d\Pi(b,D)}{db} = (r+q-c) - (r+q)F(b). \tag{6}$$

The second order derivative of $\Pi(b, D)$ with respect to b is given as follows:

$$\frac{d^2\Pi(b,D)}{db^2} = -(r+q)f(b). \tag{7}$$

With the assumptions in Section 2, we know that $f(b) \geq 0$, $r+q > 0$, hence,

$$\frac{d^2\Pi(b,D)}{db^2} \leq 0. \tag{8}$$

Therefore, we can say that $\Pi(b, D)$ is a concave function of b. So, the optimal bandwidth capacity without constraints is

$$F^{-1}\left(\frac{r+q-c}{r+q}\right) \tag{9}$$

where $F^{-1}(\cdot)$ is the inverse function of $F(\cdot)$.

Finally, if we consider the loss rate constraint and the maximal capacity constraint as in [10], the optimal bandwidth capacity for the CN is

$$\left[F^{-1}\left(\frac{r+q-c}{r+q}\right) \vee \delta F^{-1}(1-\epsilon)\right] \wedge C_{max}. \tag{10}$$

where $\vee$ represents the choice of the larger value between the two components.

4 Risk Analysis in the CN with Penalty Cost

The mean-variance analysis, which was first introduced by Markowitz [11], had been a standard tool in financial risk management. It uses a parameter α ($0 \leq \alpha \leq 1$) to characterize the risk averseness, which is a quantitative balance between the mean profit and the risk of its shortfall [12]. When α increases from 0 to 1, it indicates the willingness to sacrifice the mean profit to avoid risk of its variance.

Due to the random arrival users, the profit is also uncertain and is dependent on the distribution of the demand. So, in many cases, the optimal profit can not been obtained as desired. Based on this, we define the risk as the deviation from the optimal profit in this paper.

The random profit function and the mean profit function of the CN are given by Eq. (1) and Eq. (2) presented in Section 3, respectively. By using the method of integral by parts, Eq. (2) becomes

$$\begin{aligned}\Pi(b,D) = &-(r+q)\int_0^b F(x)dx - q\int_0^{+\infty} xf(x)dx \\ &+(r+q-c)b.\end{aligned} \tag{11}$$

The variance profit function can be obtained by the formula given as follows:

$$Var[\pi(b,D)] = E[(\pi(b,D))^2] - (\Pi(b,D))^2 . \tag{12}$$

By using the mean-variance approach to investigate the risk of profit shortfall, the objective function, which is denoted by Φ^*, is given as follows:

$$\Phi^* = \max_{b>0} \left\{ \Pi(b,D) - \alpha Var[\pi(b,D)] \right\} \tag{13}$$

where α is the risk averseness parameter, $\pi(b,D)$ is the random profit function given by Eq. (1), $\Pi(b,D)$ is the mean profit function given by Eq. (11), and $Var[\pi(b,D)]$ is the variance function given by Eq. (12).

We consider a fully distributed communication network, where the traffic demand offered to the whole CN forms a Poisson process with arrival rate $\lambda > 0$. The interarrival times are exponentially distributed with rate λ. Let X be a random variable representing the time between successive demand arrivals in the Poisson process, then we have the probability distribution function $F_X(x)$ and the probability density function $f_X(x)$ of X as follows:

$$F_X(x) = \begin{cases} 1 - e^{-\lambda x}, & x > 0 \\ 0, & x \le 0, \end{cases} \tag{14}$$

$$f_X(x) = \begin{cases} \lambda e^{-\lambda x}, & x > 0 \\ 0, & x \le 0. \end{cases} \tag{15}$$

The mean and varianceof the exponential distribution are $1/\lambda$ and $1/\lambda^2$, respectively.

Based on the assumption of the traffic demand, Eq. (11) can be obtained as follows:

$$\Pi(b,D) = -\frac{r+q}{\lambda} e^{-\lambda b} + \frac{r}{\lambda} - cb. \tag{16}$$

By using the definition of expectation and method of integral by parts, we can obtain the first component in Eq. (12) as follows:

$$\begin{aligned} E\left[(\pi(b,D))^2\right] &= e^{-\lambda b}\left(-\frac{2b}{\lambda}r^2 - \frac{2}{\lambda^2}r^2 + \frac{2}{\lambda^2}q^2 + \frac{2qcb}{\lambda} + \frac{2rcb}{\lambda}\right) \\ &\quad - \frac{2rqb}{\lambda}e^{-2\lambda b} + \frac{2r^2}{\lambda^2} - \frac{2rcb}{\lambda} + c^2b^2. \end{aligned} \tag{17}$$

With the similar method, we can obtain the second component in Eq. (12) as follows:

$$\begin{aligned} (\Pi(b,D))^2 &= e^{-\lambda b}\left(-\frac{2r^2 - 2rq}{\lambda^2} + \frac{2rcb + 2qcb}{\lambda}\right) + \frac{r^2}{\lambda^2} \\ &\quad - \frac{r^2 + 2rq + q^2}{\lambda^2} e^{-2\lambda b} - \frac{2rcb}{\lambda} + c^2b^2. \end{aligned} \tag{18}$$

Substituting Eqs. (17) and (18) into Eq. (12), we can obtain that

$$Var[\pi(b,D)] = e^{-2\lambda b}\left(-\frac{r^2+2rq+q^2}{\lambda^2} - \frac{2qcb}{\lambda}\right) + \frac{r^2}{\lambda^2} + e^{-\lambda b}\left(\frac{2rq}{\lambda^2} + \frac{2q^2}{\lambda^2} - \frac{2br^2}{\lambda}\right). \quad (19)$$

5 Numerical Results

In this section, based on the assumption of traffic demand in a CN presented in Section 4, we give some numerical results to show the impact of the penalty cost on the network performance.

According to the engineering experience, we choose several different arrival rates to represent the different cases of traffic load in the CN as follows: $\lambda = 0.01, 0.1, 0.5, 0.9$. Where $\lambda = 0.01$ represents the case that the traffic load in the CN is low, $\lambda = 0.1$ and $\lambda = 0.5$ represent the cases that the traffic load in the CN is normal, and $\lambda = 0.9$ represents the case that the traffic load in the CN is heavy.

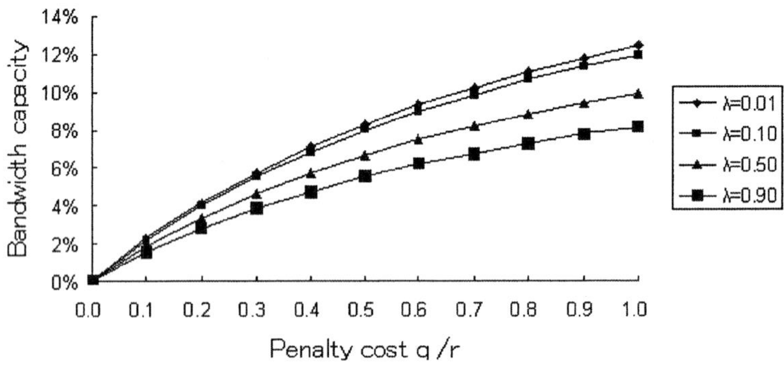

Fig. 1. Impact of penalty cost on bandwidth capacity of the CN

5.1 Impact on the Bandwidth Capacity

In this subsection, we study the impact of the penalty cost on the optimal bandwidth capacity of the CN. Note that the optimal bandwidth capacity without penalty cost presented in [10], is $F^{-1}\left(\frac{r-c}{r}\right)$. However, in this paper the optimal bandwidth capacity with the penalty cost is given by Eq. (9).

Based on the above preparation, we show the numerical results. We choose the unit revenue r as the benchmark of the linear penalty cost q. The horizontal axis (q/r) of Fig. 1 corresponds to the increase of the linear penalty cost. The ordinate axis (b/b^*) of Fig. 1 corresponds to the percentage difference of the optimal bandwidth capacity from the benchmark b^*, where b^* is the optimal

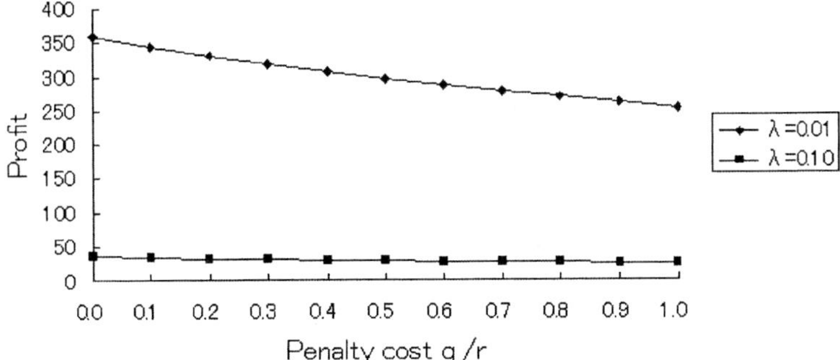

Fig. 2. Impact of penalty cost on mean profit function of the CN

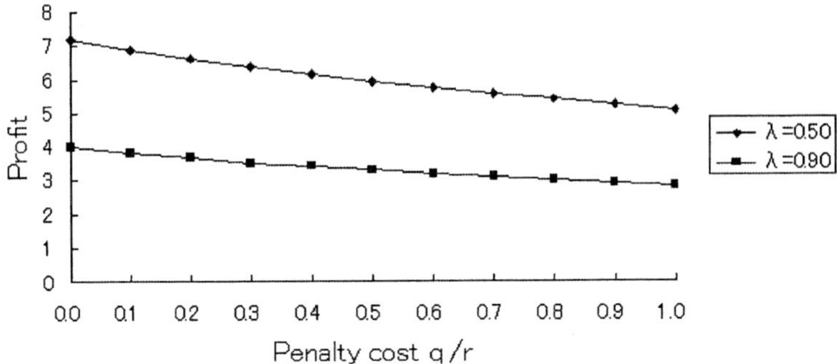

Fig. 3. Impact of penalty cost on mean profit function of the CN

bandwidth capacity without the penalty cost, and b is the optimal bandwidth capacity with the linear penalty cost. Our numerical results include the optimal bandwidth capacity obtained without penalty cost presented in [10], which is one point in the ordinate axis corresponding to $q/r = 0.0$ in Fig. 1.

For comparing with the model presented in [10], we choose the CN system parameters as follows: the unit revenue $r = 7.5$, the unit cost $c = 1.5$. Let the percentage difference of the penalty cost increase from 0.0 to 1.0 by 0.1 each step with all other parameters unchanged.

From the numerical results shown in Fig. 1, we can conclude that:

(1) In all curves, the impact of the penalty cost on the bandwidth capacity increases as the penalty cost increases.
(2) The curve with a smaller arrival rate has a quicker increasing speed than the curve with a larger arrival rate.
(3) With the same penalty cost, the heavier the traffic load in the CN is, the less the impact of the penalty cost on the bandwidth capacity will be.

Comparing with the results presented in [10] without linear penalty cost, the numerical results in our paper reveal a distinct impact of linear penalty cost on the network bandwidth capacity. It implies that if we consider the penalty cost, the CN needs to be allocated more bandwidth capacity to guarantee the network performance.

5.2 Impact on the Mean Profit Function

In this subsection, we study the impact of the penalty cost on the mean profit function. Note that the mean profit function $\Pi(b, D)$ without penalty cost, which is presented in [10], is given as follows:

$$\Pi(b, D) = r \int_0^b xf(x)dx + rb \int_b^{+\infty} f(x)dx - cb.$$

However, in this paper the mean profit function with the penalty cost is given by Eq. (2).

We choose the same system parameters as those given in Subsection 5.1. Let the percentage difference of the penalty cost increase from 0.0 to 1.0 by 0.1 each step with all other parameters unchanged.

We choose the unit revenue r as the benchmark of the penalty cost q. The horizontal axes of Figs. 2 and 3 correspond to the increase of the penalty cost q/r. The ordinate axes of Figs. 2 and 3 correspond to the mean profit $\Pi(b, D)$ presented in Eq. (16). The unit of the ordinate axes of Figs. 2 and 3 corresponds to a unit price of the mean profit. Our numerical results include the mean profit obtained without penalty cost presented in [10], which are the points in the ordinate axes corresponding to $q/r = 0.0$ in Figs. 2 and 3.

From the numerical results shown in Figs. 2 and 3, we can conclude that:

(1) In all curves, the impact of the penalty cost on the mean profit function increases as the penalty cost increases.
(2) The curve with a smaller arrival rate has a quicker decreasing speed than the curve with a larger arrival rate.
(3) With the same penalty cost, the heavier the traffic load in the CN is, the less the mean profit will be.

Comparing with the model without the penalty cost presented in [10], the numerical results in our paper reveal a distinct impact of the penalty cost on the network optimal profit. Moreover, the numerical results with different arrival rates almost have the same increasing speed and impact on the mean profit function.

6 Conclusions

In this paper, we presented a stochastic model for optimizing bandwidth allocation in Communication Networks with the linear penalty cost. The model is proposed for offline traffic engineering optimization taking a centralized view of bandwidth allocation, performance control, and risk of profit shortfall. We have derived the optimal bandwidth allocation capacity with the linear penalty cost. We have analyzed the risk averseness in the CNs in the mean-variance framework. We have given numerical results to compare our model with the previous model presented in [10] and shown the impact of the linear penalty cost on the network performance. We can conclude that the linear penalty cost has distinct impact on the network performance. The implications presented in this paper have good insights for traffic engineering design and planning.

References

[1] D. Awduche, A. Chiu, A. Elwalid, I. Widjaja and X. Xiao: Overview and Principles of Internet Traffic Engineering, RFC 3272, IETF (2002)

[2] X. Xiao, A. Hannan, B. Bailey and L. M. Ni: Traffic Engineering with MPLS in the Internet. IEEE Network **14** (2000) 28-33

[3] P. Aukia et al.: RATES: A Server for MPLS Traffic Engineering. IEEE Network **14** (2000) 34-41

[4] A. Elwalid, C. Jin, S. Low, and I. Widjaja: Mate: MPLS Adaptive Traffic Engineering. Proc. of IEEE INFOCOM (2001) 1300-1309

[5] D. Mitra and K. G. Ramakrishnan: A Case Study of Multiservice Multipriority Traffic Engineering Design for Data Networks. Proc. of IEEE GLOBECOM (1999) 1077-1083

[6] S. Suri, M. Waldvogel, D. Bauer and P. R. Warkhede: Profile-based Routing and Traffic Engineering. J. Computer Communications **26** (2003) 351-365

[7] D. Mitra and Q. Wang: Stochastic Traffic Engineering, with Applications to Network Revenue Management. Proc. of IEEE INFOCOM (2003)

[8] D. Mitra and Q. Wang: Risk-aware Network Profit Management in A Two-tier Market. Proc. of 18th International Telegraffic Congress (2003)

[9] D. Mitra and Q. Wang: Stochastic Traffic Engineering for Demand Uncertainty and Risk-aware Network Revenue Management. ACM SIGMETRICS Performance Evaluation Review **32** (2004) 1-1

[10] J. Wu, W. Yue and S. Wang: Traffic Engineering Design and Optimization for Multimedia Communication Networks. IEICE Technical Report **104** (2005) 19-24

[11] H. M. Markowitz: Portfolio Selection, Efficient Diversification of Investments. Yale University Press, New Haven (1959)

[12] S. Wang and Y. Xia: Portfolio Selection and Asset Pricing. Springer-Verlag, Berlin (2002)

Improving Clustering Analysis for Credit Card Accounts Classification

Yi Peng[1], Gang Kou[1], Yong Shi[1,2,3,*], and Zhengxin Chen[1]

[1] College of Information Science & Technology, University of Nebraska at Omaha, Omaha, NE 68182, USA
{ypeng, gkou, yshi, zchen}@mail.unomaha.edu
[2] Graduate School of Chinese Academy of Sciences, Beijing 100039, China
[3] The corresponding author

Abstract. In credit card portfolio management, predicting the cardholders' behavior is a key to reduce the charge off risk of credit card issuers. The most commonly used methods in predicting credit card defaulters are credit scoring models. Most of these credit scoring models use supervised classification methods. Although these methods have made considerable progress in bankruptcy prediction, they are unsuitable for data records without predefined class labels. Therefore, it is worthwhile to investigate the applicability of unsupervised learning methods in credit card accounts classification. The objectives of this paper are: (1) to explore an unsupervised learning method: cluster analysis, for credit card accounts classification, (2) to improve clustering classification results using ensemble and supervised learning methods. In particular, a general purpose clustering toolkit, CLUTO, from university of Minnesota, was used to classify a real-life credit card dataset and two supervised classification methods, decision tree and multiple-criteria linear programming (MCLP), were used to improve the clustering results. The classification results indicate that clustering can be used to either as a stand-alone classification method or as a preprocess step for supervised classification methods.

Keywords: Credit Card Accounts Classification, Clustering analysis, CLUTO, unsupervised learning method.

1 Introduction

One of the major tasks in credit card portfolio management is to reliably predict credit cardholders' behaviors. This task has two impacts in credit management: (1) identify potential bankrupt accounts and (2) develop appropriate policies for different categories of credit card accounts. To appreciate the importance of bankrupt accounts prediction, some statistics are helpful: There are about 1.2 billion credit cards in circulation in US. The total credit card holders declared bankruptcy in 2003 are 1,625,208 which are almost twice as many as the number of 812,898 in 1993 [4]. The total credit

* This research has been partially supported by a grant from National Natural Science Foundation of China (#70472074).

V.S. Sunderam et al. (Eds.): ICCS 2005, LNCS 3516, pp. 548 – 553, 2005.

card debt at the end of the first quarter 2002 is about $660 billion [1]. Bankrupt accounts caused creditors millions of dollars lost each year. In response, credit card lenders have made great effort to improve traditional statistical methods and recognized that more sophisticated analytical tools are needed in this area. Development of appropriate policies for various groups of credit card accounts also has a great impact on credit card issuers' profits. From the creditor's standpoint, the desirable policies should help to keep the profitable customers and minimize the defaults.

Under this strong business motivation, various techniques have been developed and applied to credit card portfolio management. The survey conducted by Rosenberg and Gleit [5] about quantitative methods in credit management provides a comprehensive review of major analytical techniques. According to Rosenberg and Gleit, discriminant analysis is the most popular quantitative tool in credit analysis at that time. In addition, integer programming, decision trees, expert systems, neural networks, and dynamic models are also available. Although these methods have made considerable progress in bankruptcy prediction, most of them are supervised learning methods. That is, they require previous knowledge about credit card accounts to make prediction. Supervised learning methods have both advantage and disadvantage. Supervised learning can normally achieve high prediction accuracy if the historical data and the data need to be analyzed have similar characteristics. The disadvantage is that supervised learning methods can not be used if there are no data records with known class labels. Therefore, it is worthwhile to investigate the usefulness of unsupervised learning in credit card accounts classification. Compare with supervised learning methods, unsupervised classification methods require no previous class labels of records. Rather, they classify data records based on their underlying structures or characteristics.

The objectives of this paper are: (1) to explore an unsupervised learning method: cluster analysis, for credit card accounts classification, (2) to improve clustering classification results using ensemble and supervised learning methods. In other words, this paper is trying to investigate the applicability of clustering both as a stand-alone classification method and as a preprocess step for supervised classification methods. In particular, a general purpose clustering toolkit, CLUTO, from university of Minnesota, was used to classify a real-life credit card dataset. Due to the fact that unsupervised classification methods normally have lower prediction accuracies than supervised method, two supervised classification method, decision tree and multiple-criteria linear programming (MCLP) [6], were used to improve the clustering results based on the clustering analysis results.

This paper is organized as follows: section 2 briefly introduces cluster analysis and CLUTO, section 3 outlines the features of credit card dataset used in the paper, section 4 illustrates how the proposed process can be applied to credit card accounts classification and reports the results, and section 5 summarizes the paper.

2 Cluster Analysis and CLUTO

Cluster analysis refers to group data records into clusters based on some criterion functions. Cluster can be defined as a collection of data records that are similar to one another within the same cluster and dissimilar to the records in other clusters.

Clustering is categorized as an unsupervised classification method because it does not require predefined classes. Instead, clustering algorithms automatically group data records by recognizing their underlying structures and characteristics. Clustering can be applied to different data types, including interval-scaled variables, binary variables, nominal variables, ordinal variables, and ratio-scaled variables, and various application areas, including pattern recognition, image processing, life science, and economic science [3].

According to different criteria, clustering methods can be classified into different categories. Han and Kamber [3] divided clustering methods into five approaches: partitioning, hierarchical, density-based, grid-based, and model-based. Partitioning methods construct a partition of the original dataset into *k* clusters that optimizes the partitioning criterion and *k* is predefined. Hierarchical methods can be either agglomerative or divisive. Agglomerative algorithms assign each data object to its own cluster and then repeatedly merging clusters until a stopping criterion is met [7]. Divisive algorithms treat all data records as one cluster and then repeatedly splitting clusters until a stopping criterion is met. Density-based algorithms cluster data records according to their density. Grid-based algorithms use multi-resolution data structure and represent datasets in n-dimensional feature space. Model-based algorithms try to optimize the fit between the data and some mathematical models.

Various commercial and free softwares have been designed to implement different clustering algorithms. CLUTO is a general purpose clustering toolkit developed by Department of Computer Science & Engineering, University of Minnesota [2]. CLUTO was chose in this research for three major reasons. First, CLUTO provides multiple classes of clustering algorithms and uses multiple similarity/distance functions. Second, CLUTO has been successfully used in application areas like information retrieval, customer purchasing transactions, web, GIS, science, and biology (CLUTO 2003). Third, CLUTO is free software.

3 Credit Card Dataset Description

The raw data came originally from a major US bank. It contains 6000 records and 102 variables (38 original variables and 64 derived variables) describing cardholders' behaviors. The 6000 credit card records were randomly selected from 25,000 real-life credit card records. The data were collected from June 1995 to December 1995 (seven months) and the cardholders were from twenty-eight States in USA. This dataset has been used as a classic working dataset by credit card issuers for various data analyses to support the bank's business intelligence. Each record has a class label to indicate its' credit status: either Good or Bad. Bad indicates a bankrupt credit card account and Good indicates a good status account. Within the 6000 records, 960 accounts are bankrupt accounts and 5040 are good status accounts. The 38 original variables can be divided into four categories: balance, purchase, payment, cash advance, in addition to related variables. The category variables represent raw data of previous six or seven consecutive months. The related variables include interest charges, date of last payment, times of cash advance, and account open date. The 64 derived variables are created from the original 38 variables to reinforce the comprehension of cardholder's behaviors, such as times overlimit in last two years, calculated interest rate, cash as percentage of balance, purchase as percentage to balance, payment as percentage to balance, and purchase as percentage to payment. These variables are not static; rather, they are evolving. New variables which are considered important can be added and variables which are proved to be trivia or irrelative in separating can be removed. Among these variables, 8 variables were selected for clustering computation following expert advices. The 8 variables are: Interest charge Dec. 95, Interest charge Dec. 95 as percent of credit line, Number of months since last payment, Credit line, Number of times delinquency in last two years, Average payment of revolving accounts, Last balance to payment ratio, and Average OBT revolving accounts. Due to the space limit, the selection criteria for classification variables will not be discussed here.

4 Empirical Studies of Cluster Analysis

As stated in the introduction, this section described two empirical studies of cluster analysis. The first study applied CLUTO to the credit card dataset to generate a set of classification results. The second study tested whether decision tree and MCLP can improve clustering classification results. Decision tree was chosen for study because it is a well-known supervised classification method. MCLP was chosen because it has been demonstrated superior performance in our previous study [6]. The research procedures for the first and second study were summarized in Method 1 and 2.

Method 1
Input: The data set A = $\{A_1, A_2 A_3, \ldots, A_{6000}\}$
Output: The clustering analysis results.

Step 1. Select 8 attributes from $A_i = (a_1, a_2, a_3, \ldots, a_{64})$, Generate the 6000 credit card data set.

Step 2. Apply the various clustering and analysis algorithms implemented in CLUTO to do a 2 to 20 way clustering analysis.

Step 3. Choose 3 clustering results according to certain criteria.

Step 4. A majority vote committee of 3 results will generate a final analysis result. The performance measures of the classification will be decided by majorities of the committee. If more than 2 of the committee members give out right classification result, then the clustering analysis C_i for this observation are successful, otherwise, the analysis is failed.

END

Method 2
Input: The data set A = $\{A_1, A_2 A_3, \ldots, A_{6000}\}$, The clustering analysis results from Method 1.

Output: The classification results; The optimal solution from MCLP, $X^* = (x_1^*, x_2^*, x_3^*, \ldots, x_8^*)$; The Decision Tree from See5.

Step 1. Manually selecting several clusters that have highest separation between Good and Bad accounts to form a training dataset T.

Step 2. Apply the two-group MCLP model to the training dataset T to compute the compromise solution $X^* = (x_1^*, x_2^*, \ldots, x_8^*)$ as the best weights of all 8 variables with given values of control parameters (b, α^*, β^*).

Step 3. Apply See5 to the training dataset T to compute the Decision Tree and its classification result.

END

The results of applying Method 1 and 2 were summarized in Table 1 and 2. The average predictive accuracies for Bad and Good groups using CLUTO are 81.17% and 66.87% and the average of Type I and Type II error rate of using CLUTO is 36.26%. Compared with these results, ensemble technique improves the classification accuracies (Bad: 91.25%, Good: 69.84%) and reduces the average of Type I and Type II error rates (32.88%). Both supervised learning methods, decision tree and MCLP, achieved better classification results for Bad records. MCLP generated better average Type I and Type II error rate (34.69%).

Table 1. Clustering analysis results by CLUTO

CLUTO							
	Bad		Good				AVG of
	Correctly Identified	Accuracy	Correctly Identified	Accuracy	Type I Error	Type II Error	Type I and II Error
1	874	91.04%	3002	59.56%	2.78%	69.99%	36.39%
2	759	79.06%	3674	72.90%	5.19%	64.28%	34.73%
3	893	93.02%	2754	54.64%	2.38%	71.91%	37.14%
4	784	81.67%	3595	71.33%	4.67%	64.83%	34.75%
5	760	79.17%	3603	71.49%	5.26%	65.41%	35.33%
6	687	71.56%	3595	71.33%	7.06%	67.78%	37.42%
7	856	89.17%	2871	56.96%	3.50%	71.70%	37.60%
8	621	64.69%	3869	76.77%	8.06%	65.35%	36.70%

5 Conclusions

Classification of credit cardholders' behavior is an important data mining application in banking industry. Situations in which there are no predefined class labels call for unsupervised classification methods. Based on this observation, this paper investigated two roles of an unsupervised classification method, cluster analysis, using a real-life credit card dataset. Cluster analysis can be used as a stand-alone classification method or as a preprocess step for supervised classification methods. The empirical results of this paper indicated that as a stand-alone classification method, cluster analysis generated lower classification rates than supervised methods. However, when combined with supervised methods, the classification results can be improved considerably.

Table 2. Classification results via Ensemble analysis, MCLP and See5

Clustering Ensemble Analysis							
	Bad		Good		Type I Error	Type II Error	AVG of Type I and II Error
	Correctly Identified	Accuracy	Correctly Identified	Accuracy			
Overall	876	91.25%	3520	69.84%	2.33%	63.44%	32.88%
Multi-Criteria Linear Programming							
	Bad		Good		Type I Error	Type II Error	AVG of Type I and II Error
	Correctly Identified	Accuracy	Correctly Identified	Accuracy			
Training	839	100.00%	719	90.67%	0.00%	8.11%	4.05%
Overall	882	91.88%	3246	64.40%	2.35%	67.04%	34.69%
Decision Tree							
	Bad		Good		Type I Error	Type II Error	AVG of Type I and II Error
	Correctly Identified	Accuracy	Correctly Identified	Accuracy			
Training	839	100.00%	792	99.87%	0.00%	0.12%	0.06%
Verifying	867	90.31%	3030	60.12%	2.98%	69.86%	36.42%

References

1. Cardweb.com, The U.S. Payment Card Information Network, as April 23, 2004, available online at: http://www.cardweb.com/cardlearn/stat.html.
2. CLUTO 2003, available online at: http://www-users.cs.umn.edu/~karypis/cluto/index.html.
3. Han, J. W. and M. Kamber. "Data Mining: Concepts and Techniques", Morgan Kaufmann Publishers, 2001.
4. New Generation Research, Inc., April 2004, available online at: http://www.bankruptcydata.com/default.asp.
5. Rosenberg, E., and A. Gleit. 1994. Quantitative methods in credit management: a survey. *Operations Research.* 42(4) 589-613.
6. Kou, G., Y. Peng, Y. Shi, M. Wise and W. Xu, 2004 Discovering Credit Cardholders' Behavior by Multiple Criteria Linear Programming *Annals of Operations Research* (forthcoming).
7. Zhao, Y. and G. Karypis. 2002. Clustering in life sciences, technical reports from Computer Science and Engineering, University of Minnesota, available online at: https://wwws.cs.umn.edu/tech_reports/index.cgi?selectedyear=2002&mode=printreport&report_id=02-016.

A Fuzzy Index Tracking Portfolio Selection Model*

Yong Fang and Shou-Yang Wang**

Institute of Systems Science,
Academy of Mathematics and Systems Science,
Chinese Academy of Sciences, Beijing 100080, China
yfang@amss.ac.cn
swang@iss.ac.cn

Abstract. The investment strategies can be divided into two classes: passive investment strategies and active investment strategies. An index tracking investment strategy belongs to the class of passive investment strategies. The index tracking error and the excess return are considered as two objective functions, a bi-objective programming model is proposed for the index tracking portfolio selection problem. Furthermore, based on fuzzy decision theory, a fuzzy index tracking portfolio selection model is also proposed. A numerical example is given to illustrate the behavior of the proposed fuzzy index tracking portfolio selection model.

1 Introduction

In financial markets, the investment strategies can be divided into two classes: passive investment strategies and active investment strategies. Investors who are adopting active investment strategies carry out securities exchange actively so that they can find profit opportunity constantly. Active investors take it for granted that they can beat markets continuously. Investors who are adopting passive investment strategies consider that the securities market is efficient. Therefore they cannot go beyond the average level of market continuously. Index tracking investment is a kind of passive investment strategy, i.e., investors purchase all or some securities which are contained in a securities market index and construct an index tracking portfolio. The securities market index is considered as a benchmark. The investors want to obtain a similar return as that of the benchmark through the index tracking investment.

In 1952, Markowitz [6, 7] proposed the mean variance methodology for portfolio selection. It has served as a basis for the development of modern financial theory over the past five decades. Konno and Yamazaki [5] used the absolute deviation risk function to replace the risk function in Markowitz's model to formulate a mean absolute deviation portfolio optimization model. Roll [8] used

* Supported by the National Natural Science Foundation of China under Grant No. 70221001.

** Corresponding author.

V.S. Sunderam et al. (Eds.): ICCS 2005, LNCS 3516, pp. 554–561, 2005.

the sum of the squared deviations of returns on a replicating portfolio from benchmark as the tracking error and proposed a mean variance index tracking portfolio selection model. Clarke, Krase and Statman [2] defined a linear tracking error which is the absolute deviation between the managed portfolio return and the benchmark portfolio return. Based on the linear objective function in which absolute deviations between portfolio and benchmark returns are used, Rudolf, Wolter and Zimmermann [9] proposed four alternative definitions of a tracking error. Furthermore, they gave four linear optimization models for index tracking portfolio selection problem. Consiglio and Zenios [3] and Worzel, Vassiadou-Zeniou and Zenios [11] studied the tracking indexes of fixed-income securities problem. In this paper, we will use the excess return and the linear tracking error as objective functions and propose a bi-objective programming model for the index tracking portfolio selection problem. Furthermore, we use fuzzy numbers to describe investors' vague aspiration levels for the excess return and the tracking error and propose a fuzzy index tracking portfolio selection model.

The paper is organized as follows. In Section 2, we present a bi-objective programming model for the index tracking portfolio selection problem. In Section 3, regarding investors' vague aspiration levels for the excess return and linear tracking error as fuzzy numbers, we propose a fuzzy index tracking portfolio selection model. In Section 4, a numerical example is given to illustrate the behavior of the proposed fuzzy index tracking portfolio selection model. Some concluding remarks are given in Section 5.

2 Bi-objective Programming Model for Index Tracking Portfolio Selection

In this paper, we assume that an investor wants to construct a portfolio which is required to track a securities market index. The investor allocates his/her wealth among n risky securities which are component stocks contained in the securities market index. We introduce some notations as follows.

r_{it}: the observed return of security i $(i = 1, 2, \cdots, n)$ at time t $(t = 1, 2, \cdots, T)$;

x_i: the proportion of the total amount of money devoted to security i $(i = 1, 2, \cdots, n)$;

I_t: the observed securities market index return at time t $(t = 1, 2, \cdots, T)$.

Let $x = (x_1, x_2, \cdots, x_n)$. Then the return of portfolio x at time t $(t = 1, 2, \cdots, T)$ is given by

$$R_t(x) = \sum_{i=1}^{n} r_{it} x_i.$$

An excess return is the return of index tracking portfolio x above the return on the index. The excess return of portfolio x at time t $(t = 1, 2, \cdots, T)$ is given by

$$E_t(x) = R_t(x) - I_t.$$

The expected excess return of index tracking portfolio x is given by

$$E(x) = \sum_{t=1}^{T} \frac{1}{T} \left(R_t(x) - I_t\right).$$

Roll [8] used the sum of squared deviations between the portfolio and benchmark returns to measure the tracking error of index tracking problem. Rudolf, Wolter and Zimmermann [9] used linear deviations instead of squared deviations to give four definitions of the linear tracking errors. We adopt the tracking error based on the mean absolute downside deviations to formulate the index tracking portfolio selection model in this paper. The tracking error based on the mean absolute downside deviations can be expressed as

$$T_{DMAD}(x) = \sum_{t=1}^{T} \frac{1}{T} | \min\{0, R_t(x) - I_t\}|.$$

Generally, in the index tracking portfolio selection problem, the track error and the excess return are two important factors which are considered by investors. An investor tries to maximize the expected excess return. At the same time, the investor hopes that the return of portfolio equals the return of the index approximatively to some extent in the investment horizon. Hence, the expected excess return and the tracking error can be considered as two objective functions of the index tracking portfolio selection problem.

In many financial markets, the securities are no short selling. So we add the following constraints:

$$x_1, x_2, \cdots, x_n \geq 0, \; i = 1, 2, \cdots, n.$$

We assume that the investor pursues to maximize the excess return of portfolio and to minimize the tracking error under the no short selling constraint. The index tracking portfolio selection problem can be formally stated as the following bi-objective programming problem:

$$\begin{aligned}
\text{(BP)} \; & \max \; E(x) \\
& \min \; T_{DMAD}(x) \\
& \text{s.t.} \; \sum_{i=1}^{n} x_i = 1, \\
& \qquad x_1, x_2, \cdots, x_n \geq 0, \; i = 1, 2, \cdots, n.
\end{aligned}$$

The problem (BP) can be reformulated as a bi-objective linear programming problem by using the following technique. Note that

$$\left| \min\{0, a\} \right| = \frac{1}{2}|a| - \frac{1}{2}a$$

for any real number a. Thus, by introducing auxiliary variables $b_t^+, b_t^-, t = 1, 2, \cdots, T$ such that

$$b_t^+ + b_t^- = \frac{|R_t(x) - I_t|}{2},$$

$$b_t^+ - b_t^- = \frac{R_t(x) - I_t}{2}, \tag{1}$$

$$b_t^+ \geq 0, \ \ b_t^- \geq 0, \ t = 1, 2, \cdots, T, \tag{2}$$

we may write

$$T_{DMAD}(x) = \sum_{t=1}^{T} \frac{2b_t^-}{T}.$$

Hence, we may rewrite problem (BP) as the following bi-objective linear programming problem:

$$\begin{aligned}(\text{BLP}) \ \max\ & E(x) \\ \min\ & \sum_{t=1}^{T} \frac{2b_t^-}{T} \\ \text{s.t.}\ & (1), (2) \ \text{ and all constraints of (BP).}\end{aligned}$$

Thus the investor may get the index tracking investment strategies by computing efficient solutions of (BLP). One can use one of the existing algorithms of multiple objective linear programming to solve it efficiently.

3 Fuzzy Index Tracking Portfolio Selection Model

In an investment, the knowledge and experience of experts are very important in an investor's decision-making. Based on experts' knowledge, the investor may decide his/her levels of aspiration for the expected excess return and the tracking error of index tracking portfolio. In [10], Watada employed a non-linear S shape membership function, to express aspiration levels of expected return and of risk which the investor would expect and proposed a fuzzy active portfolio selection model. The S shape membership function is given by:

$$f(x) = \frac{1}{1 + \exp(-\alpha x)}.$$

In the bi-objective programming model of index tracking portfolio selection proposed in Section 2, the two objectives, the expected excess return and the tracking error, are considered. Since the expected excess return and the tracking error are vague and uncertain, we use the non-linear S shape membership functions proposed by Watada to express the aspiration levels of the expected excess return and the tracking error.

The membership function of the expected excess return is given by

$$\mu_E(x) = \frac{1}{1 + \exp\left(-\alpha_E \left(E(x) - E_M\right)\right)},$$

where E_M is the mid-point where the membership function value is 0.5 and α_E can be given by the investor based on his/her own degree of satisfaction for the

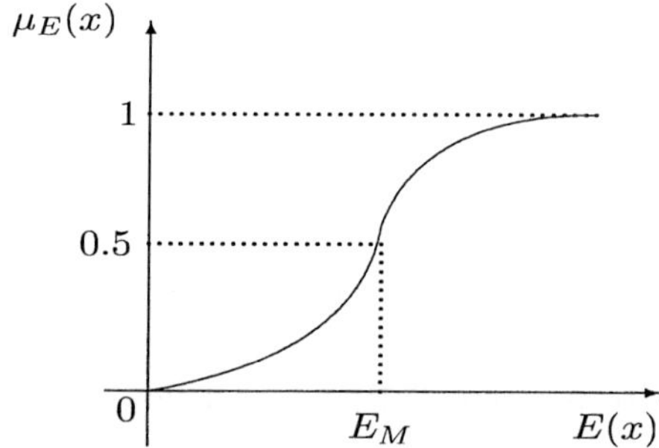

Fig. 1. Membership function of the goal for expected excess return

expected excess return. Figure 1 shows the membership function of the goal for the expected excess return.

The membership function of the tracking error is given by

$$\mu_T(x) = \frac{1}{1 + \exp(\alpha_T(T_{DMAD}(x) - T_M))},$$

where T_M is the mid-point where the membership function value is 0.5 and α_T can be given by the investor based on his/her own degree of satisfaction regarding the level of tracking error. Figure 2 shows the membership function of the goal for the tracing error.

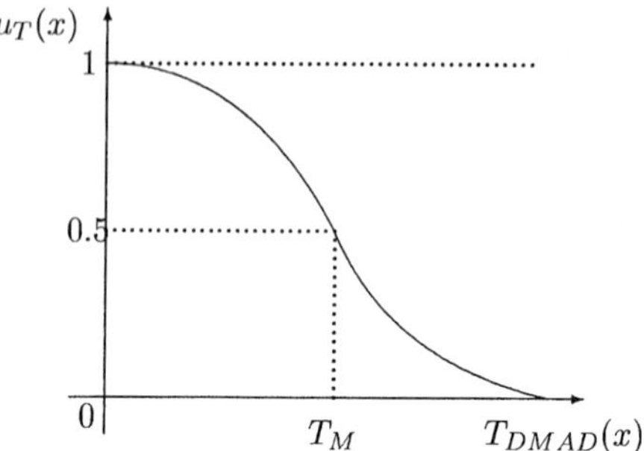

Fig. 2. Membership function of the goal for tracking error

Remark1: α_E and α_T determine the shapes of membership functions $\mu_E(x)$ and $\mu_T(x)$ respectively, where $\alpha_E > 0$ and $\alpha_T > 0$. The larger parameters α_E and α_T get, the less their vagueness becomes.

According to Bellman and Zadeh's maximization principle [1], we can define

$$\lambda = \min\{\mu_E(x), \mu_T(x)\}.$$

The fuzzy index tracking portfolio selection problem can be formulated as follows:

$$\begin{aligned}(\text{FP})\ \max\ & \lambda \\ \text{s.t.}\ & \mu_E(x) \geq \lambda, \\ & \mu_T(x) \geq \lambda, \\ & \text{and all constraints of (BLP).}\end{aligned}$$

Let $\eta = \log\frac{1}{1-\lambda}$, then $\lambda = \frac{1}{1+\exp(-\eta)}$. The logistic function is monotonously increasing, so maximizing λ makes η maximize. Therefore, the above problem can be transformed to an equivalent problem as follows:

$$\begin{aligned}(\text{FLP})\ \max\ & \eta \\ \text{s.t.}\ & \alpha_E\left(E(x) - E_M\right) - \eta \geq 0, \\ & \alpha_T\left(T_{DMAD}(x) - T_M\right) + \eta \leq 0, \\ & \text{and all constraints of (BLP),}\end{aligned}$$

where α_E and α_T are parameters which can be given by the investor based on his/her own degree of satisfaction regarding the expected excess return and the tracking error.

(FLP) is a standard linear programming problem. One can use one of several algorithms of linear programming to solve it efficiently, for example, the simplex method.

Remark2: The non-linear S shape membership functions of the two factors may change their shape according to the parameters α_E and α_T. Through selecting the values of these parameters, the aspiration levels of the two factors may be described accurately. On the other hand, different parameter values may reflect different investors' aspiration levels. Therefore, it is convenient for different investors to formulate investment strategies by using the proposed fuzzy index tracking portfolio selection model.

4 Numerical Example

In this section, we will give a numerical example to illustrate the proposed fuzzy index tracking portfolio selection model. We suppose that the investor considers Shanghai 180 index as the tracking goal. We choose thirty component stocks form Shanghai 180 index as the risky securities. We collect historical data of the thirty stocks and Shanghai 180 index from January, 1999 to December, 2002. The data are downloaded from the web-site www.stockstar.com. We use one month as a period to get the historical rates of returns of forty eight periods.

The values of the parameters α_E, α_T, E_M and T_M can be given by the investor according his/her aspiration levels for the expected excess return and the tracking error. In the example, we assume that $\alpha_E = 500$, $\alpha_T = 1000$, $E_M = 0.010$ and $T_M = 0.009$. Using the historical data, we get an index tracking portfolio selection strategy by solving (FLP). All computations were carried out on a WINDOWS PC using the LINDO solver. Table 1 shows the obtained expected excess return and tracking error of portfolio by solving (FLP). Table 2 shows the investment ratio of the obtained fuzzy index tracking portfolio.

Table 1. Membership grade λ, obtained expected excess return and obtained tracking error

λ	η	excess return	tracking error
0.9431	2.8095	0.0152	0.0062

Table 2. Investment ratio of the obtained fuzzy index tracking portfolio

Stock	1	2	3	4	5	6	7	8	9	10
Ratio	0.0000	0.0000	0.0620	0.0254	0.0000	0.0408	0.0180	0.1389	0.0324	0.0082
Stock	11	12	13	14	15	16	17	18	19	20
Ratio	0.1440	0.1488	0.0130	0.0000	0.0000	0.0000	0.1889	0.0000	0.0000	0.0000
Stock	21	22	23	24	25	26	27	28	29	30
Ratio	0.0276	0.0000	0.0000	0.0124	0.1001	0.0000	0.0395	0.0000	0.0000	0.0000

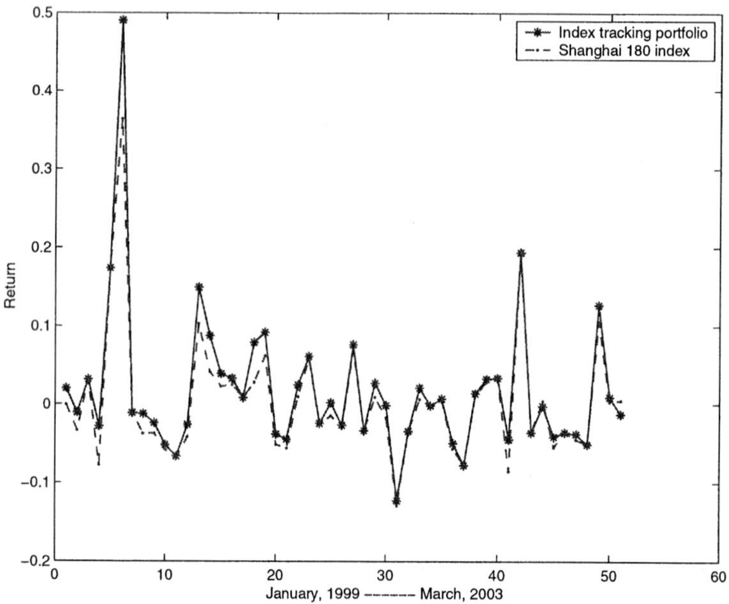

Fig. 3. The deviations between the returns of the obtained index tracking portfolio and the returns on the benchmark Shanghai 180 index

Figure 3 shows the deviations between the returns of the obtained index tracking portfolio and the returns on the benchmark Shanghai 180 index for each month from January, 1999 to March, 2003. From Figure 3, we can find that the obtained fuzzy index portfolio by solving (FLP) tracks Shanghai 180 index efficiently.

5 Conclusion

Regarding the expected excess return and the tracking error as two objective functions, we have proposed a bi-objective programming model for the index tracking portfolio selection problem. Furthermore, investors' vague aspiration levels for the excess return and the tracking error are considered as fuzzy numbers. Based on fuzzy decision theory, we have proposed a fuzzy index tracking portfolio selection model. An example is given to illustrate that the proposed fuzzy index tracking portfolio selection model. The computation results show that the proposed model can generate a favorite index tracking portfolio strategy according to the investor's satisfactory degree.

References

1. Bellman, R., Zadeh, L.A.: Decision Making in a Fuzzy Environment. Management Science 17 (1970) 141–164.
2. Clarke, R.G., Krase, S., Statman, M.: Tracking Errors, Regret, and Tactical Asset Allocation. Journal of Portfolio Management 20 (1994) 16–24.
3. Consiglio, A., Zenios, S.A.: Integrated Simulation and Optimization Models for Tracking International Fixed Income Indices. Mathematical Programming 89 (2001) 311–339.
4. Fang, Y., Wang, S.Y.: Fuzzy Portfolio Optimization: Theory and Methods. Tsinghua University Press, Beijing, 2005.
5. Konno, H., Yamazaki, H.: Mean Absolute Portfolio Optimization Model and Its Application to Tokyo Stock Market. Management Science 37(5) (1991) 519–531.
6. Markowitz, H.M.: Portfolio Selection. Journal of Finance 7 (1952) 77–91.
7. Markowitz, H.M.: Portfolio Selection: Efficient Diversification of Investment. John Wiley & Sons, New York, 1959.
8. Roll, R.: A Mean Variance Analysis of Tracking Error - Minimizing the volatility of Tracking Error will not Produce a More Efficient Managed Portfolio. Journal of Portfolio Management 18 (1992) 13–22.
9. Rudolf, M., Wolter, H.J., Zimmermann, H.: A Linear Model for Tracking Error Minimization. Journal of Banking and Finance 23 (1999) 85–103.
10. Watada, J.: Fuzzy Portfolio Model for Decision Making in Investment. In: Yoshida, Y. (eds.): Dynamical Asspects in Fuzzy Decision Making. Physica-Verlag, Heidelberg (2001) 141–162.
11. Worzel, K.J., Vassiadou-Zeniou, C., Zenios, S.A.: Integrated Simulation and Optimization Models for Tracking Indices of Fixed-income Securities. Opreations Research 42 (1994) 223–233.

Application of Activity-Based Costing in a Manufacturing Company: A Comparison with Traditional Costing

Gonca Tuncel, Derya Eren Akyol, Gunhan Mirac Bayhan, and Utku Koker

Department of Industrial Engineering, University of Dokuz Eylul,
35100 Bornova-Izmir, Turkey
{gonca.tuncel, derya.eren, mirac.bayhan}@deu.edu.tr

Abstract. Activity-Based Costing (ABC) represents an alternative paradigm to traditional cost accounting system and has received extensive attention during the past decade. Rather than distorting the cost information by using traditional overhead allocation methods, it traces the cost via activities performed on the cost objects (production or service activities) giving more accurate and traceable cost information. In this paper, the implementation of ABC in a manufacturing system is presented, and a comparison with the traditional cost based system in terms of the effects on the product costs is carried out to highlight the difference between two costing methodologies. The results of the application reveal the weak points of traditional costing methods and an S-Curve which exposes the undercosted and overcosted products is used to improve the product pricing policy of the firm.

1 Introduction

The customer driven environment of today's manufacturing systems and the competitive pressure of the global economy force manufacturing services and organizations to become more flexible, integrated and highly automated in order to increase their productivity at reduced costs. But it is impossible to sustain competitiveness without an accurate cost calculation mechanism [1]. Proposed by [2], as an alternative method to traditional cost accounting methods, ABC assigns costs to activities using multiple cost drivers, then allocates costs to products based on each product's use of these activities [3], [4]. Using multiple activities as cost drivers, it reduces the risk of distortion and provides accurate cost information [3].

In an ABC system, the total cost of a product equals the cost of the raw materials plus the sum of the cost of all value adding activities to produce it [4]. In other words, the ABC method models the usage of the organization resources by the activities performed and links the cost of these activities to outputs, such as products, customers, and services [5]. Each product requires a number of activities such as design, engineering, purchasing, production and quality control. Each activity consumes resources of different categories such as

V.S. Sunderam et al. (Eds.): ICCS 2005, LNCS 3516, pp. 562–569, 2005.

the working time of the manager. Cost drivers are often measures of the activities performed such as number of units produced, labor hours, hours of equipment time, number of orders received.

In traditional cost accounting systems, direct materials and labor are the only costs that can be traced directly to the product. By using the ABC system, activities can be classified as value-added and non-value-added activities. In order to improve the performance of the system, non-value-added can be eliminated.

Despite the advantages of providing accurate costs, it requires additional effort and expense in obtaining the information needed for the analysis [6]. However, a proper design tool can help to reduce time used for modeling and overcome the difficulties present in designing a cost model.

The primary objective of this paper is to develop an ABC system for a sanitary-ware company and to compare the results of ABC with traditional costing methods. In other words, the aim of ABC analysis is to guide improvement efforts of management in the right direction by providing accurate information about activities.

The organization of the paper is as follows: In section 2, the methodology of ABC is explained. A case study is presented in section 3 to illustrate the application of ABC in a company. Finally, in section 4, the conclusions and the future research directions are given. Some suggestions are offered to improve the performance of the company.

2 Activity-Based Costing (ABC)

ABC is an economic model that identifies the cost pools or activity centers in an organization and assigns costs to cost drivers based on the number of each activity used. Since the cost drivers are related to the activities, they occur on several levels:

1. Unit level drivers which assume the increase of the inputs for every unit that is being produced.
2. Batch level drivers which assume the variation of the inputs for every batch that is being produced.
3. Product level drivers which assume the necessity of the inputs to support the production of each different type of product.
4. Facility level drivers are the drivers which are related to the facility's manufacturing process. Users of the ABC system will need to identify the activities which generate cost and then match the activities to the level bases used to assign costs to the products.

While using the ABC system, the activities which generate cost must be determined and then should be matched to the level drivers used to assign costs to the products.

The implementation of the ABC system has the following steps:

1. Identifying the activities such as engineering, machining, inspectionetc.
2. Determining the activity costs

3. Determining the cost drivers such as machining hours, number of setups and engineering hours.
4. Collecting the activity data
5. Computing the product cost

3 Implementation of the ABC Method: A Case Study

The implementation study presented here took place in one of the leading sanitaryware companies in Turkey [7]. It has a production capacity of 6.8 thousand tons. Like many sanitaryware manufacturers, the company includes some common processes. First stage of the processes is done in the bulk-preparement department which is in charge of preparing the essential quantities of bulk bearing. Recipes are prepared for each production run and according to these ingredient recipes, the bulks are prepared by the electronically controlled tanks. After the bulk is ready, it can be sent to two different departments, pressured and classical casting departments.

In pressured casting department, the bulk is given shape by the pressured casting process. The process uses automated machines and has shorter cycle times when compared to classical casting department. However, the bulk used in this department must have the characteristics of strength and endurance. The output of this department is sent to glazing department.

Classical casting is the second alternative to produce sanitaryware products. In this department, most of the operations are performed by direct labor. The cycle times of the products are longer than the pressured casting department. The output of this department is "shaped and casted" bulk and sent to the glazing department, too.

In glazing department, shaped bulk is glazed. This stage can be defined as polishing the products with a protective material from external effects. The output of this department is sent to tunnel oven.

In tunnel oven, the products are heated over 1200 degrees Celsius. After staying in theses ovens for a period of time, the products are inspected by the workers. However, some of the output may have some undesired characteristics like the scratches, etc. In this case, these products are sent to the second heat treatment where they are reworked and heated again. The proper output is sent to packaging while the defected ones are sent to waste.

Finally, in packaging department, products are packaged and shrunken.

Company has been using a process costing logic for obtaining the proper costing of its products. Process costing is also a widely used costing tool for many companies. This method recognizes the following cost pools:

- Direct Labor: All workers taking place in the production are recognized as direct labor and this pool is the most common pool used in every stage.
- LPG-Electricity hot water: These items are important costing element in casting departments.
- Packaging: This cost is observed in the final stage of the firm called final packaging. It includes packaging and shrinking of the products.

– Overheads: This is also common cost pool for all the stages in the firm. It includes depreciation, rents, indirect labor, materials, and miscellaneous costs.

In order to perform ABC calculations, process costing sheets are used. Because process costing aggregates the cost in the case of more than one product, it is difficult to obtain the correct results. The correct results with process costing can only be achieved when there is a single homogenous product. After analyzing the output of bulk preparement department, it is seen that bulk prepared for the two departments are the same which brings an opportunity for the analysis.

Firstly, time sheets are prepared to learn how much of the labor time is used to perform the relevant activities. Workers filled the time sheet and as a result, the activity knowledge of the processes and related percentage times are obtained. The time sheets are edited and necessary corrections are made. While some of the activities are merged, some of them are discarded. After this stage, the wage data is requested from the accounting department and labor cost of the activities are determined.

If the amount of the activities that each product consumes can be learnt, the product costing can be calculated by taking the sum of the costs of the activities consumed by these products. Observations and necessary examinations are done to reach this knowledge and the conversion costs are loaded onto the activities. Cost of rent is loaded according to the space required for each activity and similarly, the power, depreciation and other conversion costs are loaded on the activities according to the activities' consumption of the resources. If activity a uses 2/5 and activity b covers 1/5, while activity c covers 2/5 of the department space, then activity a is given 2/5, b is given 1/5 and c is given 2/5 of rent. All the conversion costs are distributed by using the same logic.

The activities performed during classical casting are: molding, drilling assembly holes, rework, equipment clean-up, drilling function holes and carrying. The activity driver for molding is the processing time, for drilling assembly holes is the number of holes, for rework is the number of outputs, for equipment clean-up is the total processing time, for drilling function holes is the total number of function holes and for carrying are the number of processed units in this department, unit area the product allocates on the carrier and the total area of carried goods.

The activities performed during pressured casting are: molding, drilling lateral holes, setup, cutting excess materials from the products, general rework, carrying WIP, drilling assembly holes, ring preparation, drilling some subcomponents washbasins, washbasin drilling, rework of back of the products, X-patch control, bringing the WIP, helping to other departments, filling forms, mold check-up, WIP control, equipment and personnel control, drilling water holes. The activity driver for molding is the number molds used, for drilling lateral holes is the number of output, for setup are the incoming units, setup hour, setup for other products, setup for basins, setup for other products, for cutting excess materials from the products is the incoming units, for general rework is also the incoming units, for carrying WIP is the number of output, for drilling assembly hole and for ring preparation is the number of output, for drilling some sub-components washbasins is the number of subcomponents, for washbasin drilling is the number of washbasins, rework of back of the products and X-patch control is the number of output, for bringing the WIP is the incoming WIP, for helping to other departments is the number of reservoirs, for filling forms, mold check-up, WIP control, equipment and personnel control, drilling water holes is the incoming units.

The activities performed during glazing are maintenance, glazing with hand, WIP transfer, rework, closet shaking, planning, routine controls and prism data entry. The activity driver for maintenance is the number of reworked parts, for glazing with hand is the weight of parts to be glazed, for WIP transfer and rework is the number of reworked parts, for closet shaking is the number of closets, for planning, routine controls and prism data entry is the planned units.

The activities performed in the tunnel oven department are tunnel oven activity, transfer to oven, transfer from oven, heat control, oven security control, planning, routine controls, prism data entry, taking records, second heat treatment activity, WIP feeding to the oven, rework of heat processed units and special treatment. The activity driver for tunnel oven activity, transfer to oven, transfer from oven, heat control and oven security control is the total area covered by the heated units, for planning, routine controls, prism data entry and taking records is the total sum of outputs, for second heat treatment activity and WIP feeding to the oven is total area covered by all units, for rework of heat processed units and special treatment is the total area covered by outputs.

The activities performed in the packaging department are packaging, shrinking, clean-up, organizing the finished products, maligning, and product function control and transferring the outputs. The activity driver for packaging, and shrinking is the packaged volume, for clean-up is the total number of units cleaned, for organizing the finished products, maligning, product function control and transferring the outputs is the total number of outputs.

After finding the product costs of the firm, the costs obtained by process costing methodology and the ABC results of the product costs are compared. From the results, it is seen that there are significant differences between some of the product costs obtained by the two methods.

4 Conclusions

ABC utilizes the activity concept and by using the activities, ABC can successfully link the product costs to production knowledge. How a product is produced, how much time is needed to perform an activity and finally how much money is absorbed by performing this task are answered by the help of ABC studies.

An S-Curve will exist after the comparison of the traditional and ABC costs [8]. The following notation is used for the comparison of the two product cost values. % Bias of ABC from traditional costing is found for each product.
% bias = (ABC cost/traditional cost)*100.

Product costs under ABC and traditional costing views can be used to design a table (Table 1) which illustrates %bias of ABC and traditional costs.

Table 1. Comparison of the two product cost values

Products	% Bias	Products	% Bias
Product 70	10.415	Product 28	102.938
Product 23	54.796	Product 56	103.417
Product 64	60.439	Product 48	105.092
Product 12	65.043	Product 4	108.269
Product 59	69.485	Product 63	108.905
Product 47	71.071	Product 53	109.488
Product 62	75.279	Product 46	112.454
Product 31	77.736	Product 11	114.009
Product 1	78.621	Product 54	117.228
Product 37	78.953	Product 50	124.241
Product 3	79.965	Product 8	124.84
Product 22	80.527	Product 6	133.134
Product 30	80.731	Product 44	134.926
Product 42	83.265	Product 7	162.302
Product 19	84.533	Product 14	164.557
Product 58	89.282	Product 29	169.465
Product 36	89.531	Product 5	219.781
Product 20	96.504	Product 15	317.231
Product 26	96.688	Product 39	584.387
Product 24	100.077	Product 32	1198.251
Product 55	100.521		

Note that the table includes only the products that are sold. Products that are not sold but left as semi products are not involved.

The results show 3 significant regions:

1. Products undercosted by traditional costing:
The costs of these products are higher than the "process-costing" and when plotted are greater than the value 100 at y axis. Before the ABC study was implemented, firm was not aware of this hidden loss of money for each of these products. By the help of the ABC study, people noticed this "hidden loss" and pricing decisions are examined.
2. Products overcosted by the traditional methods:
These products are the ones which have smaller values than the value 100 of y-axis. By the help of the ABC, firm realized that, the costs of these products are lower than the costs obtained by process costing. A "hidden profit zone" occurs for these products. Without any intention, a hidden benefit is gained from these products by traditional costing. But noticing the real ABC costs of these products, the firm can re-price these outputs and can gain a competitive advantage.
3. Products which are costed almost the same:
These are the products whose costs result nearly the same by using the two methods. A narrow band of 20 percentage (%80-%120) can be accepted for this region.

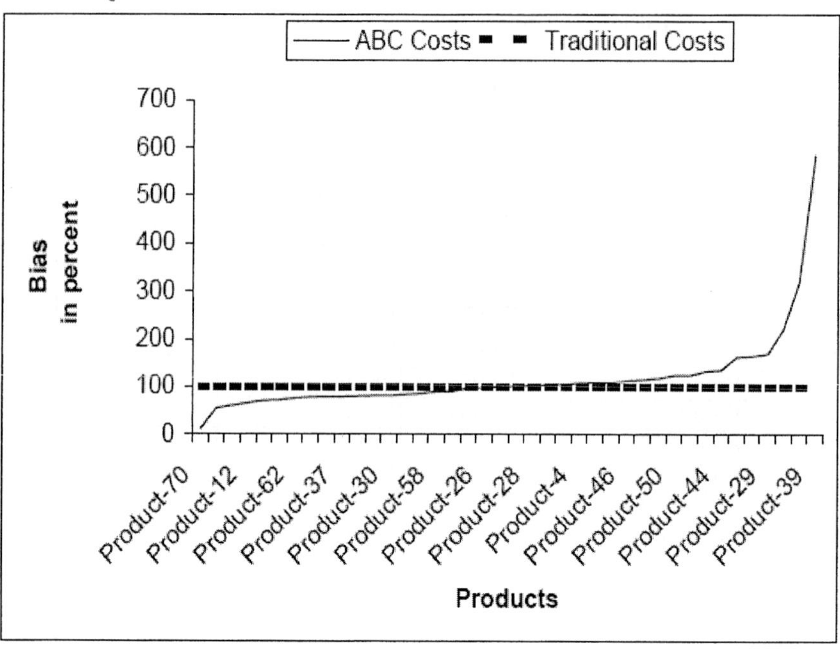

Fig. 1. S-Curve

From the ABC analysis, it is seen that while some of the products are undercosted, and overcosted, some of them give nearly the same results.

In the analysis, the cost calculations of the bulk preparement department are performed using traditional costing method. When the outputs of the processes are identical or nearly identical, then ABC must be avoided to implement. Because ABC consumes lots of time, data and efforts, implementation of it has also a cost. For identical outputs, both ABC and traditional give the same results, so it is not logical to implement ABC.

By the help of this analysis, product-pricing decisions gain importance. Pricing decisions can be done under the scope of these hidden losses and profits. According to the ABC results, the company will probably increase or decrease some of its product costs to gain more competitive advantage.

The costing process of the company highlighted some weaknesses of the information system, too. These problems are also expected to be fixed in the future.

As it is seen in this application, ABC is capable of monitoring the hidden losses and profits of the traditional costing methods. The existence of S-Curve shows which ones of the products are under or overcosted. As a further work, in the company, performance analysis are expected to be done and especially Balanced Score Card (BSC) implementations can be performed. The existence of the ABC database is an advantage for BSC applications since its financial phase recommends an ABC implementation. Kaizen applications and BSC can give the firm great advantages in the short and long run under the scope of ABC.

The studies in literature promote that ABC is a promising method to support pricing, product mix, and make or buy decisions in manufacturing companies. ABC is a fruitful area for researchers and has high potential for novel applications such as the hybridization of ABC with meta-heuristics methods as Artificial Neural Networks, Genetic Algorithms, and Simulated Annealing.

References

1. Ozbayrak, M., Akgun, M., and Turker, A.K.: Activity-based cost estimation in a push/pull advanced manufacturing system. Int. J. of Production Economics, vol. 87 (2004) 49-65.
2. Cooper, R. and Kaplan, R. S.: How cost accounting distorts product costs. Management Accounting, vol.69 (1988) 20-27.
3. Kim, G., Park, C. S., and Kaiser, M. J.: Pricing investment and production activities for an advanced manufacturing system, Engineering Economist, vol. 42, no. 4 (1997) 303-324.
4. Gunasekaran, A., and Sarhadi, M.: Implementation of activity-based costing in manufacturing. Int. J. of Production Economics, vol.56-57 (1998) 231-242.
5. Ben-Arieh, D. and Qian L.: Activity-based cost management for design and development stage. Int. J. of Production Economics, vol.83 (2003) 169-183.
6. Lewis, R. J.: Activity-based models for cost management systems. Quorum Books, West-port, CT (1995).
7. Koker, U.: Activity-based costing: Implementation in a sanitaryware company. M.Sc. Thesis, Department of Industrial Engineering, University of Dokuz Eylul (2003).
8. Cokins, G.: Activity-based cost management making it works. McGraw-Hill. Inc. (1997).

Welfare for Economy Under Awareness

Ken Horie[1,*] and Takashi Matsuhisa[2,**]

[1] Advanced Course of Electronic and Computer Engineering, Ibaraki National College of Technology
abc9872000jp@yahoo.co.jp

[2] Department of Natural Sciences, Ibaraki National College of Technology, Nakane 866, Hitachinaka-shi, Ibaraki 312-8508, Japan
mathisa@ge.ibaraki-ct.ac.jp

Abstract. We present the extended notion of pure exchange economy under uncertainty, called an *economy with awareness structure*, where each trader having a strictly monotone preference makes decision under his/her awareness and belief, and we introduce a generalized notion of equilibrium for the economy, called an *expectations equilibrium in awareness.* We show the existence theorem of the equilibrium and the fundamental welfare theorem for the economy, i.e., an allocation in the economy is ex-ante Pareto optimal if and only if it is an expectations equilibrium allocation in awareness.

1 Introduction

This article relates economies and distributed belief. We shall present a generalized notion of economy under uncertainty, called an *economy with awareness structure*, where each trader makes decision in his/her awareness and belief under incomplete information. The purposes are: First, to introduce an extended notion of expectations equilibrium for the economy, called an *expectations equilibrium in awareness.* Secondly, to show the fundamental welfare theorem for the extended economy under expectations equilibrium in awareness.

Main Theorem. *In a pure exchange economy under uncertainty, the traders are assumed to have an awareness structure and they are risk averse. Then an allocation in the economy is ex-ante Pareto optimal if and only if it is an expectations equilibrium allocation in awareness for some initial endowment with respect to some price system.*

In Economic theory and its related fields, many authors have investigated several notions of equilibrium in an economy under asymmetric information.[1] They have studied the relationships between these equilibrium concepts (e.g.:

---

* Lecture presenter.

** Corresponding author. Partially supported by the Grant-in-Aid for Scientific Research(C)(2)(No.14540145) in the Japan Society for the Promotion of Sciences.

[1] See the literatures cited in F. Forges, E. Minelli, and R. Vohla, *Incentive and the core of exchange economy - Survey*, Journal of Mathematical Economics 38 (2002), 1–41.

V.S. Sunderam et al. (Eds.): ICCS 2005, LNCS 3516, pp. 570–577, 2005.

The existence theorem of equilibrium, the core equivalence theorem and the no trade theorem etc.) One of the serious limitations of their analysis is to assume 'partition' structure as information the traders receive. From the epistemic point of view, the partition structure represents the traders' knowledge satisfying the postulates: 'Truth' **T** (what is known is true), the 'positive introspection' **4** (that we know what we do) and the 'negative introspection' **5** (that we know what we do not know). The postulate **5** is indeed so strong that describes the hyper-rationality of traders, and thus it is particularly objectionable.

This raises the question to what extent results on the information partition structure (or the equivalent postulates of knowledge). The answer is to strengthen the results: We shall weaken the conditions in the partition. This relaxation can potentially yield important results in a world with imperfectly Bayesian agents.

The idea has been performed in different settings. Geanakoplos [5] showed the no speculation theorem in the extended rational expectations equilibrium under the assumption that the information structure is reflexive, transitive and *nested.* The condition 'nestedness' is interpreted as a requisite on the 'memory' of the trader. Einy et al [4] extended the core equivalence theorem of Aumann [1] to the equivalence theorem between the ex-post core and the rational expectations equilibria for an economy under asymmetric information. Recently, Matsuhisa [6] gives an extension of the theorem into an economy under awareness structure. In his line we establish the fundamental theorem for welfare into the generalized economy.

This article is organized as follows: In Section 2 we propose the model: An economy with awareness structure and an expectations equilibrium in awareness. In Section 3 we state explicitly the fundamental theorem for welfare economics and sketch the proof. Finally we conclude by remarks.

2 The Model

Let Ω be a non-empty *finite* set called a *state space* and 2^{Ω} the field 2^{Ω} consisting of all subsets of Ω. Each member of 2^{Ω} is called an *event* and each element of Ω called a *state.* We denote by T the set of the *traders.* We shall present a model of awareness according to E. Dekel et al [3].[2]

2.1 Awareness, Belief and Information

A *belief structure* is a tuple $\langle \Omega, (B_t)_{t\in T}\rangle$ in which $B_t : 2^{\Omega} \to 2^{\Omega}$ is trader t's *belief* operator. The interpretation of the event B_tE is that 't believes E.' An *awareness structure* is a tuple $\langle \Omega, (A_t)_{t\in T}, (B_t)_{t\in T}\rangle$ in which $\langle \Omega, (B_t)_{t\in T}\rangle$ is a belief structure and A_t is t's *awareness* operator on 2^{Ω} defined by

$$\textbf{PL} \qquad A_tE = B_tE \cup B_t(\Omega \setminus B_tE) \quad \text{for every } E \text{ in } 2^{\Omega}.$$

[2] A different approach of awareness models is discussed in R. Fagin, J.Y. Halpern, Y. Moses and M.Y. Vardi, *Reasoning about Knowledge.* The MIT Press, Cambridge, Massachusetts, London, England, 1995.

The interpretation of $A_t E$ is that 't is aware of E.' The property **PL** says that t is aware of E if he believes it or if he believes that he does not believe it.

We shall give the generalized notion of information partition in the line of Bacharach [2] as follows.

Definition 1. The *associated information structure* $(P_t)_{t\in T}$ with awareness structure $\langle \Omega, (A_t), (B_t)\rangle$ is the class of t's *associated information functions* $P_t : \Omega \to 2^\Omega$ defined by $P_t(\omega) = \bigcap_{E\in 2^\Omega}\{E \,|\, \omega \in B_t E\}$. (If there is no event E for which $\omega \in B_t E$ then we take $P_t(\omega)$ to be undefined.) We denote by $\mathrm{Dom}(P_t)$ the set $\{\omega \in \Omega \mid P_t(\omega) \neq \emptyset\}$, called the *domain of* P_t.

The mapping P_t is called *reflexive* if:

Ref $\quad \omega \in P_t(\omega) \quad$ for every $\omega \in \mathrm{Dom}(P_t)$,

and it is said to be *transitive* if:

Trn $\quad \xi \in P_t(\omega)$ implies $P_t(\xi) \subseteq P_t(\omega)$ for any $\xi, \omega \in \mathrm{Dom}(P_t)$.

Furthermore P_t is called *symmetric* if:

Sym $\quad \xi \in P_t(\omega)$ implies $P_t(\xi) \ni \omega$ for any ω and $\xi \in \mathrm{Dom}(P_t)$.

Remark 1. M. Bacharach [2] introduces the *strong epistemic model* equivalent to the Kripke semantics of the modal logic **S5**. The strong epistemic model is a tuple $\langle \Omega, (K_t)_{t\in T}\rangle$ in which t's *knowledge operator* $K_t : 2^\Omega \to 2^\Omega$ satisfies the five postulates: For every E, F of 2^Ω,

N $K_t\Omega = \Omega$, **K** $K_t(E \cap F) = K_t E \cap K_t F$, **T** $K_t F \subseteq F$;
4 $K_t F \subseteq K_t K_t F$, **5** $\Omega \setminus K_t F \subseteq K_t(\Omega \setminus K_t F)$.

t's associated information function P_t induced by K_t makes a partition of Ω, called t's *information partition*, which satisfies the postulates **Ref**, **Trn** and **Sym**. This is just the Kripke semantics corresponding to the logic **S5**; the postulates **Ref**, **Trn** and **Sym** are respectively equivalent to the postulates **T**, **4** and **5**. The strong epistemic model can be interpreted as the awareness structure $\langle \Omega, (A_t), (B_t)\rangle$ such that B_t is the knowledge operator. In this situation it is easily verified that A_t must be the *trivial* operator, [3] and that $\mathrm{Dom}(P_t) = \Omega$.

2.2 Economy with Awareness Structure

A pure exchange economy *under uncertainty* is a structure

$$\mathcal{E} = \langle T, \Omega, \mathbf{e}, (U_t)_{t\in T}, (\pi_t)_{t\in T}\rangle$$

[3] I.e. $A_t(F) = \Omega$ for every $F \in 2^\Omega$.

consisting of the following structure and interpretations: There are l commodities in each state of the state space Ω; the consumption set of each trader t is $\mathbf{R}^l_+$; an *initial endowment* is a mapping $\mathbf{e} : T \times \Omega \to \mathbf{R}^l_+$ with which $\mathbf{e}(t, \cdot) : \Omega \to \mathbf{R}^l_+$ is called t's *initial endowment*; $U_t : \mathbf{R}^l_+ \times \Omega \to \mathbf{R}$ is t's von-Neumann and Morgenstern utility function; π_t is a subjective prior on Ω for $t \in T$. For simplicity, π_t is assumed to be *full support* for all $t \in T$. That is, $\pi_t(\omega) \neq 0$ for every $\omega \in \Omega$.

Definition 2. A pure exchange economy *with awareness structure* is a structure $\mathcal{E}^A = \langle \mathcal{E}, (A_t)_{t\in T}, (B_t)_{t\in T}, (P_t)_{t\in T} \rangle$, in which $\mathcal{E}$ is a pure exchange economy under uncertainty, and $\langle \Omega, (A_t)_{t\in T}, (B_t)_{t\in T}, (P_t)_{t\in T} \rangle$ is an awareness structure with $(P_t)_{t\in T}$ the associated information structure. By the *domain* of the economy $\mathcal{E}^A$ we mean $\mathrm{Dom}(\mathcal{E}^A) = \cap_{t\in T}\mathrm{Dom}(P_t)$. We always assume that $\mathrm{Dom}(\mathcal{E}^A) \neq \emptyset$.

Remark 2. An economy under asymmetric information is an economy $\mathcal{E}^A$ with the awareness structure $\langle \Omega, (A_t)_{t\in T}, (B_t)_{t\in T} \rangle$ given by the strong epistemic model, and that $\mathrm{Dom}(\mathcal{E}^A) = \Omega$.

We denote by $\mathcal{F}_t$ the field of $\mathrm{Dom}(P_t)$ generated by $\{P_t(\omega) | \; \omega \in \Omega\}$ and denote by $\Pi_t(\omega)$ the atom containing $\omega \in \mathrm{Dom}(P_t)$. We denote by $\mathcal{F}$ the join of all $\mathcal{F}_t (t \in T)$ on $\mathrm{Dom}(\mathcal{E}^A)$; i.e. $\mathcal{F} = \vee_{t\in T}\mathcal{F}_t$, and denote by $\{\Pi(\omega) | \; \omega \in \mathrm{Dom}(\mathcal{E}^A)\}$ the set of all atoms $\Pi(\omega)$ containing ω of the field $\mathcal{F} = \vee_{t\in T}\mathcal{F}_t$. We shall often refer to the following conditions: For every $t \in T$,

A-1 $\sum_{t\in T} \mathbf{e}(t, \omega) > 0$ for each $\omega \in \Omega$.
A-2 $\mathbf{e}(t, \cdot)$ is $\mathcal{F}$-measurable on $\mathrm{Dom}(P_t)$.
A-3 For each $x \in \mathbf{R}^l_+$, the function $U_t(x, \cdot)$ is at least $\mathcal{F}$-measurable on $\mathrm{Dom}(\mathcal{E}^A)$, and the function: $T \times \mathbf{R}^l_+ \to \mathbf{R}$, $(t, x) \mapsto U_t(x, \omega)$ is $\Sigma \times \mathcal{B}$-measurable where $\mathcal{B}$ is the σ-field of all Borel subsets of $\mathbf{R}^l_+$.
A-4 For each $\omega \in \Omega$, the function $U_t(\cdot, \omega)$ is strictly increasing on $\mathbf{R}^l_+$, continuous, strictly quasi-concave and *non-satiated* on $\mathbf{R}^l_+$. [4]

2.3 Expectations Equilibrium in Awareness

An *assignment* is a mapping $\mathbf{x} : T \times \Omega \to \mathbf{R}^l_+$ such that for each $t \in T$, the function $\mathbf{x}(t, \cdot)$ is at least $\mathcal{F}$-measurable on $\mathrm{Dom}(\mathcal{E}^A)$. We denote by $\mathcal{A}ss(\mathcal{E}^A)$ the set of all assignments for the economy $\mathcal{E}^A$. By an *allocation* we mean an assignment $\mathbf{a}$ such that $\mathbf{a}(t, \cdot)$ is $\mathcal{F}$-measurable on $\mathrm{Dom}(\mathcal{E}^A)$ for all $t \in T$ and $\sum_{t\in T} \mathbf{a}(t, \omega) \leq \sum_{t\in T} \mathbf{e}(t, \omega)$ for every $\omega \in \Omega$. We denote by $\mathcal{A}lc(\mathcal{E}^A)$ the set of all allocations.

We introduce the revised notion of trader's expectation of utility in $\mathcal{E}^A$. By t's *ex-ante* expectation we mean $\mathbf{E}_t[U_t(\mathbf{x}(t, \cdot)] := \sum_{\omega\in\mathrm{Dom}(P_t)} U_t(\mathbf{x}(t, \omega), \omega)\pi_t(\omega)$ for each $\mathbf{x} \in \mathcal{A}ss(\mathcal{E}^A)$. The *interim* expectation $\mathbf{E}_t[U_t(\mathbf{x}(t, \cdot)|P_t]$ is defined by $\mathbf{E}_t[U_t(\mathbf{x}(t, \cdot)|P_t](\omega) := \sum_{\xi\in\mathrm{Dom}(P_t)} U_t(\mathbf{x}(t, \xi), \xi)\pi_t(\{\xi\} \cap A_t(\{\xi\})|P_t(\omega))$ on

[4] That is, for any $x \in \mathbf{R}^l_+$ there exists an $x' \in \mathbf{R}^l_+$ such that $U_i(x', \omega) > U_i(x, \omega)$.

$\mathrm{Dom}(P_t)$. It should be noted that we use not the usual notion of posterior $\pi_t(\{\xi\}|P_t(\omega))$ but the revised one $\pi_t(\{\xi\} \cap A_t(\{\xi\})|P_t(\omega))$.[5]

A *price system* is a non-zero function $p : \Omega \to \mathbf{R}^l_+$ which is $\mathcal{F}$-measurable on $\mathrm{Dom}(\mathcal{E}^A)$. We denote by $\Delta(p)$ the partition on Ω induced by p, and denote by $\sigma(p)$ the field of Ω generated by $\Delta(p)$. The *budget set* of a trader t at a state ω for a price system p is defined by $B_t(\omega, p) := \{ x \in \mathbf{R}^l_+ \mid p(\omega) \cdot x \leq p(\omega) \cdot \mathbf{e}(t, \omega) \}$. Define the mapping $\Delta(p) \cap P_t : \mathrm{Dom}(P_t) \to 2^\Omega$ by $(\Delta(p) \cap P_t)(\omega) := \Delta(p)(\omega) \cap P_t(\omega)$. We denote by $\mathrm{Dom}(\Delta(p) \cap P_t)$ the set of all states ω in which $\Delta(p)(\omega) \cap P_t(\omega) \neq \emptyset$. Let $\sigma(p) \vee \mathcal{F}_t$ be the smallest σ-field containing both the fields $\sigma(p)$ and $\mathcal{F}_t$.

Definition 3. An *expectations equilibrium in awareness* for an economy $\mathcal{E}^A$ with awareness structure is a pair $(p, \mathbf{x})$, in which p is a price system and $\mathbf{x}$ is an assignment satisfying the following conditions:

EA1. $\mathbf{x}$ is an allocation;
EA2. For all $t \in T$ and for every $\omega \in \Omega$, $\mathbf{x}(t, \omega) \in B_t(\omega, p)$;
EA3. For all $t \in T$, if $\mathbf{y}(t, \cdot) : \Omega \to \mathbf{R}^l_+$ is $\mathcal{F}$-measurable on $\mathrm{Dom}(\mathcal{E}^A)$ with $\mathbf{y}(t, \omega) \in B_t(\omega, p)$ for all $\omega \in \Omega$, then

$$\mathbf{E}_t[U_t(\mathbf{x}(t, \cdot))|\Delta(p) \cap P_t](\omega) \geq \mathbf{E}_t[U_t(\mathbf{y}(t, \cdot))|\Delta(p) \cap P_t](\omega)$$

pointwise on $\mathrm{Dom}(\Delta(p) \cap P_t)$;
EA4. For every $\omega \in \mathrm{Dom}(\mathcal{E}^A)$, $\sum_{t \in T} \mathbf{x}(t, \omega) = \sum_{t \in T} \mathbf{e}(t, \omega)$.

The allocation $\mathbf{x}$ in $\mathcal{E}^A$ is called an expectations equilibrium *allocation* in awareness for $\mathcal{E}^A$.

We denote by $EA(\mathcal{E}^A)$ the set of all the expectations equilibria of a pure exchange economy $\mathcal{E}^A$, and denote by $\mathcal{A}(\mathcal{E}^A)$ the set of all the expectations equilibrium allocations in awareness for the economy.

3 The Results

Let $\mathcal{E}^A$ be the economy with awareness structure and $\mathcal{E}^A(\omega)$ the economy with complete information $\langle T, (\mathbf{e}(t, \omega))_{t \in T}, (U_t(\cdot, \omega))_{t \in T} \rangle$ for each $\omega \in \Omega$. We denote by $\mathcal{W}(\mathcal{E}^A(\omega))$ the set of all competitive equilibria for $\mathcal{E}^A(\omega)$.

3.1 Existence of Equilibrium in Awareness

Theorem 1. *Let $\mathcal{E}^A$ be a pure exchange economy with awareness structure satisfying the conditions* **A-1**, **A-2**, **A-3** *and* **A-4**. *Then there exists an expectations equilibrium in awareness for the economy; i.e.,* $EA(\mathcal{E}^A) \neq \emptyset$.

[5] A discussion why this improvement of the notion of posterior is needed is given in T. Matsuhisa and S.-S. Usami, *Awareness, belief and agreeing to disagree*, Far East Journal of Mathematical Sciences 2(6) (2000) 833–844.

Before proceeding with the proof we shall note that:

Lemma 1. *The event* $(\Delta(p)\cap P_t)(\omega)$ *can be decomposed into the disjoint union* $(\Delta(p)\cap P_t)(\omega)=\cup_{k=1}^{p}\Pi(\xi_k)$. *Furthermore, for* $\mathbf{x}\in \mathcal{A}ss(\mathcal{E}^A)$, $\mathbf{E}_t[U_t(\mathbf{x}(t,\cdot))|\Delta(p)\cap P_t](\omega) = \sum_{k=1}^{p} \frac{\pi_t(\Pi(\xi_k))}{\pi_t((\Delta(p)\cap P_t)(\omega))} U_t(\mathbf{x}(t,\xi_k),\xi_k)$. □

Proof of Theorem 1. In view of the existence theorem of a competitive equilibrium for an economy with complete information,[6] it follows that there exists a $(p^*(\omega),\mathbf{x}^*(\cdot,\omega)) \in \mathcal{W}(\mathcal{E}^A(\omega))$ for each $\omega\in\Omega$ by the conditions **A-1**, **A-2**, **A-3** and **A-4**. Define the pair $(p,\mathbf{x})$ as follows: For each $\omega\in\Omega$, denote $\mathbf{x}(t,\xi) := \mathbf{x}^*(t,\omega)$ for all $\xi\in\Pi(\omega)$ and $\omega\in\mathrm{Dom}(\mathcal{E}^A)$, and set $p(\xi) := p^*(\omega)$ for all $\xi\in\Pi(\omega)$ and $\omega\in\mathrm{Dom}(\mathcal{E}^A)$, $p(\xi) := p^*(\omega)$ for $\omega\notin\mathrm{Dom}(\mathcal{E}^A)$. Then we can verify that $(p,\mathbf{x})$ is an expectations equilibrium in awareness for $\mathcal{E}^A$: For **EA3**. On noting that $\mathcal{E}^A(\xi)=\mathcal{E}^A(\omega)$ for any $\xi\in\Pi(\omega)$, it follows that $(p(\xi),\mathbf{x}(t,\xi))\in\mathcal{W}(\mathcal{E}^A(\omega))$ for every $\omega\in\Omega$, and thus we can observe **EA3** by Lemma 1. The other conditions in Definition 3 are easily verified. □

3.2 Fundamental Theorem for Welfare Economics

An allocation $\mathbf{x}$ in $\mathcal{E}^A$ is said to be *ex-ante Pareto-optimal* if there is no allocation $\mathbf{a}$ such that $\mathbf{E}_t[U_t(\mathbf{a}(t,\cdot))] \geq \mathbf{E}_t[U_t(\mathbf{x}(t,\cdot))]$ for all $t\in T$ with at least one inequality strict. We can now state our main theorem.

Theorem 2. *Let* $\mathcal{E}^A$ *be an economy with awareness structure satisfying the conditions* **A-1**, **A-2**, **A-3** *and* **A-4**. *An allocation is ex-ante Pareto optimal if and only if it is an expectations equilibrium allocation in awareness for some initial endowment* $\mathbf{w}$ *with respect to some price system such that* $\sum_{t\in T}\mathbf{w}(t,\omega) = \sum_{t\in T}\mathbf{e}(t,\omega)$ *for each* $\omega\in\mathrm{Dom}(\mathcal{E}^A)$.

Proof. Follows immediately from Propositions 1 and 2 as below. □

Proposition 1. *Let* $\mathcal{E}^A$ *be an economy with awareness structure satisfying the conditions* **A-1**, **A-2**, **A-3** *and* **A-4**. *Then an allocation* $\mathbf{x}$ *is ex-ante Pareto optimal if it is an expectations equilibrium allocation in awareness with respect to some price system.*

Proposition 2. *Let* $\mathcal{E}^A$ *be an economy with awareness structure satisfying the conditions* **A-1**, **A-2**, **A-3** *and* **A-4**. *If an allocation* $\mathbf{x}$ *is ex-ante Pareto optimal in* $\mathcal{E}^A$ *then there are a price system and an initial endowment* $\mathbf{e}'$ *such that* $\mathbf{x}$ *is an expectations equilibrium allocation in awareness with* $\sum_{t\in T}\mathbf{e}'(t,\omega) = \sum_{t\in T}\mathbf{e}(t,\omega)$ *for each* $\omega\in\mathrm{Dom}(\mathcal{E}^A)$.

[6] C.f.: Theorem 5 in G. Debreu, *Existence of competitive equilibrium*, in: Handbook of Mathematical Economics, Volume 2, K.J.Arrow and M.D.Intriligator (eds), North-Holland Publishing Company, Amsterdam, 1982, 697–744.

3.3 Proof of Propositions 1 and 2

Before proving the propositions we first establish

Proposition 3. *Let $\mathcal{E}^A$ be an economy with awareness structure satisfying the conditions* **A-1**, **A-2**, **A-3** *and* **A-4**. *Then $\mathcal{A}(\mathcal{E}^A) = \{\mathbf{x} \in \mathcal{A}lc(\mathcal{E}^A) \mid$ There is a price system p such that $(p(\omega), \mathbf{x}(\cdot, \omega)) \in \mathcal{W}(\mathcal{E}^A(\omega))$ for all $\omega \in \mathrm{Dom}(\mathcal{E}^A)\}$.*

Proof. Let $\mathbf{x} \in \mathcal{A}(\mathcal{E}^A)$ and $(p, \mathbf{x}) \in EA(\mathcal{E}^A)$. We shall show that $(p(\omega), \mathbf{x}(\cdot, \omega)) \in \mathcal{W}(\mathcal{E}^A(\omega))$ for any $\omega \in \mathrm{Dom}(\mathcal{E}^A)$. Suppose to the contrary that there exist a trader $s \in T$ and states $\omega' \in \mathrm{Dom}(\mathcal{E}^A), \omega_0 \in (\Delta(p) \cap P_s)(\omega')$ with the property: There is an $\mathbf{a}(s, \omega_0) \in B_s(\omega_0, p)$ such that $U_s(\mathbf{a}(s, \omega_0), \omega_0) > U_s(\mathbf{x}(s, \omega_0), \omega_0)$. Define the $\mathcal{F}$-measurable function $\mathbf{y} : T \times \Omega \to \mathbf{R}^l_+$ by $\mathbf{y}(t, \xi) := \mathbf{a}(t, \omega_0)$ for $\xi \in \Pi(\omega_0)$, and $\mathbf{y}(t, \xi) := \mathbf{x}(t, \xi)$ otherwise. It follows immediately by Lemma 1 that $\mathbf{E}_s[U_s(\mathbf{x}(s, \cdot))|\Delta(p) \cap P_s](\omega') < \mathbf{E}_s[U_s(\mathbf{y}(s, \cdot))|\Delta(p) \cap P_s](\omega')$, in contradiction.

The converse will be shown: Let $\mathbf{x} \in \mathcal{A}ss(\mathcal{E}^A)$ with $(p(\omega), \mathbf{x}(\cdot, \omega)) \in \mathcal{W}(\mathcal{E}^A(\omega))$ for any $\omega \in \mathrm{Dom}(\mathcal{E}^A)$. Set the price system $p^* : \Omega \to \mathbf{R}^l_+$ by $p^*(\xi) := p(\omega)$ for all $\xi \in \Pi(\omega)$ and *omega* $\in \mathrm{Dom}(\mathcal{E}^A)$, and $p^*(\xi) := p(\omega)$ for $\omega \notin \mathrm{Dom}(\mathcal{E}^A)$. We shall show that $(p^*, \mathbf{x}) \in EA(\mathcal{E}^A)$: $\mathbf{x}(t, \cdot)$ is $\mathcal{F}$-measurable and $\mathbf{x}(t, \omega) \in B_t(\omega, p^*)$ on $\mathrm{Dom}(\mathcal{E}^A)$ for all $t \in T$. It can be plainly observed that **EA1**, **EA2** and **EA4** are all valid. For **EA3**: Let $\mathbf{y}(t, \cdot) : \Omega \to \mathbf{R}^l_+$ be an $\mathcal{F}$-measurable function with $\mathbf{y}(t, \omega) \in B_t(\omega, p^*)$ for all $\omega \in \mathrm{Dom}(\mathcal{E}^A)$. Since $(p^*(\omega), \mathbf{x}(\cdot, \omega)) \in \mathcal{W}(\mathcal{E}^A(\omega))$ it follows that $U_t(\mathbf{x}(t, \omega), \omega) \geq U_t(\mathbf{y}(t, \omega), \omega)$ for all $t \in T$ and for each $\omega \in \mathrm{Dom}(\mathcal{E}^A)$. By Lemma 1, $\mathbf{E}_t[U_t(\mathbf{x}(t, \cdot))|\Delta(p^*) \cap P_t](\omega) \geq \mathbf{E}_t[U_t(\mathbf{y}(t, \cdot))|\Delta(p^*) \cap P_t](\omega)$ for all $\omega \in \mathrm{Dom}(\Delta(p^*) \cap P_t)$, and so $(p^*, \mathbf{x}) \in EA(\mathcal{E}^A)$, in completing the proof. □

Proof of Proposition 1. It follows from Proposition 3 that $(p(\omega), \mathbf{x}(\cdot, \omega)) \in \mathcal{W}(\mathcal{E}^A(\omega))$ at each $\omega \in \mathrm{Dom}(\mathcal{E}^A)$. By the fundamental theorem of welfare in the economy $\mathcal{E}^A(\omega)$, we can plainly observe that for all $\omega \in \mathrm{Dom}(\mathcal{E}^A)$, $\mathbf{x}(\cdot, \omega)$ is Pareto optimal in $\mathcal{E}^A(\omega)$, and thus $\mathbf{x}$ is ex-ante Pareto optimal. □

Proof of Proposition 2. It can be shown that *for each $\omega \in \Omega$ there exists $p^*(\omega) \in \mathbf{R}^l_+$ such that $(p^*(\omega), \mathbf{x}(\cdot, \omega)) \in \mathcal{W}(\mathcal{E}^A(\omega))$ for some initial endowment $\mathbf{e}'(\cdot, \omega)$ with $\sum_{t \in T} \mathbf{e}'(t, \omega) = \sum_{t \in T} \mathbf{e}(t, \omega)$. Proof:* First it can be observed that for each $\omega \in \Omega$ there exists $p^*(\omega) \in \mathbf{R}^l_+$ such that $p^*(\omega) \cdot v \leq 0$ for all $v \in G(\omega) = \{ \sum_{t \in T} \mathbf{x}(t, \omega) - \sum_{t \in T} \mathbf{y}(t, \omega) \in \mathbf{R}^l \mid \mathbf{y} \in \mathcal{A}ss(\mathcal{E}^A)$ and $U_t(\mathbf{y}(t, \omega), \omega) \geq U_t(\mathbf{x}(t, \omega), \omega)$ for all $t \in T\}$ for each $\omega \in \mathrm{Dom}(\mathcal{E}^A)$: In fact, on noting that that $G(\omega)$ is convex and closed in $\mathbf{R}^l_+$ by the conditions **A-1**, **A-2**, **A-3** and **A-4**, the assertion immediately follows from the fact that $v \leq 0$ for all $v \in G(\omega)$ by the separation theorem[7]: Suppose to the contrary. Let $\omega_0 \in \Omega$ and $v_0 \in G(\omega_0)$ with $v_0 > 0$. Take $\mathbf{y}^0 \in \mathcal{A}ss(\mathcal{E}^A)$ such that for all t, $U_t(\mathbf{y}^0(t, \omega), \omega_0) \geq U_t(\mathbf{x}(t, \omega_0), \omega_0)$ and $v_0 = \sum_{t \in T} \mathbf{x}(t, \omega_0) - \sum_{t \in T} \mathbf{y}^0(t, \omega_0)$. Let $\mathbf{z} \in \mathcal{A}lc(\mathcal{E}^A)$ be defined by $\mathbf{z}(t, \xi) := \mathbf{y}^0(t, \omega_0) + \frac{v_0}{|T|}$ if $\xi \in \Pi(\omega_0)$, $\mathbf{z}(t, \xi) := \mathbf{x}(t, \xi)$ if not. By **A-4**

[7] C.f.: Lemma 8, Chapter 4 in K. J. Arrow and F. H. Hahn, *General competitive analysis*, North-Holland Publishing Company, Amsterdam, 1971. p.92.

it follows that for all $t \in T$, $\mathbf{E}_t[U_t(\mathbf{z})] \geq \mathbf{E}_t[U_t(\mathbf{x})]$, in contradiction to which $\mathbf{x}$ is ex-ante Pareto optimal. By a similar argument in the proof of the second fundamental theorem of welfare economics,[8] we can verify that $(p^*(\omega), \mathbf{x}(\cdot, \omega)) \in \mathcal{W}(\mathcal{E}^A(\omega))$ for some initial endowment $\mathbf{e}'$ with $\sum_{t \in T} \mathbf{e}'(t, \omega) = \sum_{t \in T} \mathbf{e}(t, \omega)$.

Now, let p be the price system defined by: $p(\xi) := p^*(\omega)$ for all $\xi \in \Pi(\omega)$ and $\omega \in \mathrm{Dom}(\mathcal{E}^A)$, $p(\xi) := p^*(\omega)$ for $\omega \notin \mathrm{Dom}(\mathcal{E}^A)$. Further we extend $\mathbf{e}'$ to the initial endowment $\mathbf{w}$ for $\mathcal{E}^A$ by $\mathbf{w}(t, \xi) := \mathbf{e}'(t, \omega)$ for all $\xi \in \Pi(\omega)$ and $\omega \in \mathrm{Dom}(\mathcal{E}^A)$. It can be observed that $\mathbf{w}(t, \cdot)$ is $\mathcal{F}$-measurable with $\sum_{t \in T} \mathbf{w}(t, \omega) = \sum_{t \in T} \mathbf{e}'(t, \omega)$.

To conclude the proof we shall show that $(p, \mathbf{x}) \in EA(\mathcal{E}^A)$. *Proof:* For each $\omega \in \mathrm{Dom}(\mathcal{E}^A)$, there exists ξ such that $\xi \in (\Delta(p) \cap P_t)(\omega) = \Delta(p)(\xi) = \Pi(\xi)$, and so we can observe by **A-3** that for each $\mathbf{x} \in \mathcal{A}lc(\mathcal{E}^A)$, $\mathbf{E}_t[U_t(\mathbf{x}(t, \cdot))|(\Delta(p) \cap P_t)](\omega) = U_t(\mathbf{x}(t, \xi), \xi)$. We shall verify **EA3** only: Suppose to the contrary that there exists $s \in T$ with the two properties: (i) there is an $\mathcal{F}$-measurable function $\mathbf{y}(s, \cdot) : \Omega \to \mathbf{R}^l_+$ such that $\mathbf{y}(s, \omega) \in B_s(\omega, p)$ for all $\omega \in \Omega$; and (ii) $\mathbf{E}_s[U_s(\mathbf{y}(s, \cdot))|(\Delta(p) \cap P_s)](\omega_0) > \mathbf{E}_s[U_s(\mathbf{x}(s, \cdot)|(\Delta(p) \cap P_s)](\omega_0)$ for some $\omega_0 \in \mathrm{Dom}(\Delta(p) \cap P_s)$. In view of the above equation it follows from (ii) that there exists $\xi \in (\Delta(p) \cap P_t)(\omega_0)$ with $U_s(\mathbf{y}(s, \xi), \xi) > U_s(\mathbf{x}(s, \xi), \xi)$, and thus $\mathbf{y}(s, \xi) > \mathbf{x}(s, \xi)$ by **A-4**. Thus $p(\xi) \cdot \mathbf{y}(s, \xi) > p(\xi) \cdot \mathbf{x}(s, \xi)$, in contradiction. □

4 Concluding Remarks

Our real concern in this article is about relationship between players' beliefs and their decision making, especially when and how the players take corporate actions under their decisions. We focus on extending the fundamental theorem of welfare economics into an economy with traders having 'awareness and belief' model. We have shown that the nature of the theorem is dependent not on common-belief nor on the partition structure of traders' information, but on the structure of awareness and belief when each player receives information.

References

1. Aumann, R. J.: Markets with a continuum of traders. Econometrica 32 (1964) 39–50
2. Bacharach, M. O.: Some extensions of a claim of Aumann in an axiomatic model of knowledge. Journal of Economic Theory 37 (1985) 167–190.
3. Dekel, E., Lipman, B.L., Rustichini, A.: Standard state-space models preclude unawareness. Econometrica 66 (1998) 159–173
4. Einy, E., Moreno, D., and Shitovitz, B.: Rational expectations equilibria and the ex-post core of an economy with asymmetric information. Journal of Mathematical Economics 34 (2000) 527–535
5. Geanakoplos, J.: Game theory without partitions, and applications to speculation and consensus, Cowles Foundation Discussion Paper No.914 (1989)
6. Matsuhisa, T.: Core equivalence in economy under awareness. In the Proceedings of Game Theory and Mathematical Economics, Warsaw, GTME 2004 (To appear).

[8] C.f.: Proposition 16.D.1 in A. Mas-Colell, M. Whinston, and J. Green, *Microeconomics Theory*. Oxford University Press, 1995, pp. 552–554.

On-line Multi-attributes Procurement Combinatorial Auctions Bidding Strategies

Jian Chen and He Huang

School of Economics and Management, Tsinghua University, Beijing, 100084, China
jchen@mail.tsinghua.edu.cn
huangh02@mails.tsinghua.edu.cn

Abstract. Based on the work of Krishna and Rosenthal (1996) about combinatorial auctions bidding equilibrium analysis and Che's (1993) research about one-unit multi-attributes auctions, we construct a multi-attributes procurement combinatorial auction (MAPCA) for 2 objects, through a first-score, sealed-bid format. There are two kinds of bidders: n simple bidders and m diverse bidders considered in this model. With some assumptions, we finally obtain the equilibrium bidding strategies for the both two kinds of bidders.

1 Introduction

Combinatorial auctions have been applied in a variety of environments involving economic transactions, and they have the potential to play an important role in electronic procurement transactions for supply chain management. Examples are Net Exchange (www.nex.com) who procures transportation services for Sears Logistics (Ledyard et al. 2002), procurement combinatorial auctions for truck services at Home Depot (Elmaghraby and Keskinocak 2003), and IBM's procurement combinatorial auction for Mars Incorporated (Hohner et al. 2003). The characteristics of combinatorial auctions that we will treat in the procurement context are items involving multiple attributes. Comparing with normal combinatorial auctions, the property of multiple attributes is somewhat unique in procurement auctions. When an auctioneer procures items by auctions, she may allow the potential suppliers to bid for various products bundles, which is called Procurement Combinatorial Auctions (CAs).

The topics on one-unit auctions' bidding strategies and winning probability are abundantly addressed. However, discussions about CAs' bidding strategies are inadequate. Krishna and Rosenthal (1996) considered situations where multiple objects are auctioned simultaneously by means of a second-price sealed-bid auction, and derived the bidding equilibrium strategies for the synergy bidder. Rosenthal and Wang (1996) studied a simultaneous-auction model with synergies, common values, and constructed strategies. Che (1993) studies design competition in one item procurement by developing a model of two-dimensional (price and quality) auctions. The purpose of this paper is to discuss bidding strategies in a specific multi-attributes procurement CA setting.

V.S. Sunderam et al. (Eds.): ICCS 2005, LNCS 3516, pp. 578 – 585, 2005.

2 The Model

In this model, there are 2 identical objects to be procured by a first-score sealed-bid auction. And two kinds of bidders, i.e. simple bidder and diverse bidder are considered. Each simple bidder is interested in only one object, while each diverse bidder, say diverse bidder *j*, is interested in these two objects with synergy effect. For each object, there are *n* interested simple bidders. We can also regard the situation as two identical auctions started in two separated rooms; each room has *n* simple bidders, and diverse bidders are apt to bid in both rooms. Each bidder has a privately known supply cost function for the object. Other assumptions are as follows.

Bidding vector is $\{p,q\}\in R_+^2$, *p*, *q* denote bidding price and quality-bid separately. Especially, quality-bid, in our model, is a unified number to measure all non-price factors of the objects. The scoring function of the bidding vectors has a quasi-linear form

$$S(q,p)=s(q)-p\ . \tag{1}$$

where $s(.)$ is strictly increasing with respect to *q*.

Number of simple bidders, for each object, is *n*; number of diverse bidders is *m*.

The cost function of simple bidder *i* is $c(q,\theta_i)$. θ_i denotes type of simple bidder i, which is private information of bidder i, and θ_i is drawn from interval $[\underline{\theta},\overline{\theta}]$ with identical distribution $F(.)$, which is common knowledge. $c(.,.)$ is increasing with respect to q and θ_i. Type of diverse bidder *j* is denoted as θ_j which is drawn from interval $[\underline{\theta}_g,\overline{\theta}_g]$ with the identical distribution $F_g(.)$; θ_j is also private information. The cost function of the diverse bidder is $c_g(q,\theta_j)$ for each object. $[\underline{\theta}_g,\overline{\theta}_g]$ and $F_g(.)$ are common knowledge; $c_g(.,.)$ is increasing with respect to q and θ_j.

If winning, profit of simple bidder i, with bid $\{p,q\}\in R_+^2$, is

$$\pi_i(q,p)=p-c(q,\theta_i)\ . \tag{2}$$

Profit of the diverse bidder *j*, with bid $\{p,q\}\in R_+^2$, is

$$\begin{cases} p-c_g(q,\theta_j), & if\ \ win\ \ one \\ 2p-2c_g(q,\theta_j)+\alpha_j, & if\ \ win\ \ both \end{cases}\ . \tag{3}$$

where α_j, i.e., synergy effect, is a non-negative constant which is diverse bidder *j*'s private information.

With type θ_i, simple bidder *i*'s bidding strategy is formula (4): [Riley and Samuelson, 1981; Che, 1993; Chen and Huang 2004]

$$\begin{cases} q(\theta_i) = \arg\max[s(q) - c(q,\theta_i)], \\ p(\theta_i) = c(q,\theta_i) + \int_{\theta_i}^{\bar{\theta}} c(q(t),t)[\frac{1-F(t)}{1-F(\theta_i)}]^n dt \end{cases} . \quad (4)$$

Diverse bidders know each other's profit function form, but do not know other diverse bidders' synergy effect value, say α_j, which are drawn from $[0, \overline{\alpha}]$, and their identical distribution $F_\alpha(.)$ is common knowledge. The model constructed above is called MAPCA in this paper.

3 Bidding Strategy Analysis

For simple bidders, Let $S_0(\theta_i) \equiv \max[s(q) - c(q,\theta_i)], \forall \theta_i \in [\underline{\theta}, \overline{\theta}]$. By using envelope theorem, $S_0(.)$ is a strictly decreasing function, therefore its inverse function exits. Let $v_i \equiv S_0(\theta_i)$, v_i can be regarded as scoring capability of simple bidder i. Let $H(v_i) \equiv 1 - F(S_0^{-1}(v_i))$, then distribution $F(.)$ of θ_i is transformed into distribution $H(.)$ of v_i. Let $b \equiv S(q_s(\theta_i), p)$, where $q_s(\theta_i) = \arg\max[s(q) - c(q,\theta_i)]$. Obviously, b is the score evaluated by the bidding of simple bidder with type θ_i. Let $B(.)$ denote the scoring function with the independent variable v_i, and $B(.)$ is strictly increasing with v_i. Define $G(.) \equiv H(B^{-1}(.))$ and $L(b) = \{H(B^{-1}(b))\}^n$ for convenience. Intuitively, $L(b)$ is the probability of all n simple bidders who are interested in one of the two items get score less than b.

Similarly, for diverse bidders, let $S_g(\theta_j) \equiv \max[s(q) - c_g(q,\theta_j)], \forall \theta_j \in [\underline{\theta}_g, \overline{\theta}_g]$. By envelope theorem, $S_g(.)$ is a strictly decreasing function, therefore its inverse function exits. Let $v \equiv S_g(\theta_j)$, v is scoring capability of diverse bidder j. Let $H_g(v) \equiv 1 - F_g(S_0^{-1}(v))$, then distribution $F_g(.)$ of θ_j is transformed into distribution $H_g(.)$ of v, which also implies that the one-to-one mapping relationship from v to θ . Let $b \equiv S(q_s(\theta_j), p)$, where $q_s(\theta_j) = \arg\max[s(q) - c_g(q,\theta_j)]$, and b is the score of the bidding by diverse bidder with type parameter θ_j. Let $B_g(v)$ denote the scoring function with v, and $B_g(.)$ is strictly increasing with v. Define $G_g(.) \equiv H_g(B_g^{-1}(.))$ and $K(b) \equiv \{H_g(B_g^{-1}(b))\}^{m-1}$. Actually, $K(b)$ is the probability of all other m-1 diverse bidders get score less than b. Note that $p - c(q_s(\theta_j), \theta_j) = v - b$, and then we have the expected revenue of any diverse bidder with type θ_j who bids quality-bid q_s and price p as follows,

$$\begin{aligned}&\pi((q_s, p)|\theta_j)\\ &= L^2(s(q_s)-p)\times K(s(q_s)-p)\times(2(p-c)+\alpha_j)\\ &\quad+2L(s(q_s)-p)\times(1-L(s(q_s)-p))\times K(s(q_s)-p)\times(p-c)\end{aligned} \quad . \tag{5}$$

where, q_s represents quality bidding, p is price bidding, c denotes the corresponding cost with quality bidding q_s. Now, we would show an important proposition (lemma 1) to illustrate any diverse bidder with any specific type, say θ_j , will report quality-bid to maximize $s(q)-c_g(q,\theta_j)$.

LEMMA 1. The diverse bidder with type θ_j follows the bidding strategy in which quality-bid $q_s(\theta_j)$ is

$$q_s(\theta_j) = \arg\max[s(q)-c_g(q,\theta_j)] \ . \tag{6}$$

Proof: Assume there exists another equilibrium strategy (q, p) , and $q \neq q_s(\theta_j)$, then construct a bid (q', p') ,

$$\text{where} \quad q' = q_s(\theta_j), \quad p' = p + s(q_s(\theta_j)) - s(q) \ . \tag{7}$$

$$\text{Note} \qquad S(q', p') = S(q, p) = S \ . \tag{8}$$

Then, with bid (q', p') , the expected revenue of diverse bidder is below,

$$\begin{aligned}&\pi((q', p')|\theta_j)\\ &= L^2(S)K(S)[2(p+s(q_s(\theta_j))-s(q)-c_g(q'))+\alpha_j]\\ &\quad+2L(S)(1-L(S))K(S)(p+s(q_s(\theta_j))-s(q)-c_g(q'))\\ &= L^2(S)K(S)\{2\times[(p-c_g(q)+(s(q_s(\theta_j))-c_g(q_s(\theta_j))-(s(q)-c_g(q)))]+\alpha_j\}\\ &\quad+2L(S)(1-L(S))K(S)[(p-c_g(q)+(s(q_s)-c_g(q_s(\theta_j))-(s(q_1)-c_g(q)))]\end{aligned} \quad . \tag{9}$$

Note $q_s(\theta_j) = \arg\max[s(q)-c_g(q,\theta_j)]$, therefore,

$$\pi((q', p')|\theta_j) \geq L(S)K(S)\cdot\{L(S)[2(p-c_g(q))+\alpha_j]+2(1-L(S))\cdot(p-c_g(q))\} \tag{10}$$

Thus, we have

$$\pi((q', p')|\theta_j) \geq \pi((q, p)|\theta_j) \ . \tag{11}$$

This completes the proof. □

Without loss of generality, let $s(q_s(\theta_j)) = 1$; from LEMMA 1 and (5), we know the optimization problem of the diverse bidder with type θ_j can be expressed as follows:

$$\max_p \pi((q_s, p)|\theta_j)$$
$$= \max_p \{L(1-p)\cdot K(1-p)\cdot[(L(1-p)\cdot(2(p-c)+\alpha_j)+2(1-L(1-p))\cdot(p-c)]\}\,. \tag{12}$$
$$= \max_p \{\alpha_j L^2(1-p)K(1-p)+2L(1-p)K(1-p)(p-c)\}$$

where c denotes the corresponding cost of the diverse bidder with her equilibrium quality-bid $q_s(\theta_j)$; c is also the optimal cost of the bidder. Actually, $q_s(\theta_j)$ and c are determined by type θ_j uniquely. Let $l(.)=L'(.), k(.)=K'(.)$, by first derivative of (12) with respect to p, we have

$$\frac{d\pi}{dp} = c-p-L(1-p)\cdot\frac{\alpha_j\cdot l(1-p)\cdot K(1-p)+\alpha_j\cdot L(1-p)\cdot k(1-p)/2-K(1-p)}{l(1-p)\cdot K(1-p)+L(1-p)\cdot k(1-p)}\,. \tag{13}$$
$$=0$$

Now considering the relationship between the optimal cost c of diverse bidder with θ_j and her price bid p, from (13) we have

$$c = p+L(1-p)\cdot\frac{\alpha_j\cdot l(1-p)\cdot K(1-p)+\alpha_j\cdot L(1-p)\cdot k(1-p)/2-K(1-p)}{l(1-p)\cdot K(1-p)+L(1-p)\cdot k(1-p)}\,. \tag{14}$$

Actually, (14) is the correspondence from optimal cost c to possible optimal price bid p, and we can abstractly express (14) as $c=c_p(p)$. The only special feature used in the argument is that $c_p(.)$ is convex. Convexity of $c_p(.)$ is not easy to characterize in terms of the primitive assumptions of our model. However, if $G_g(.)$ and $G(.)$ are uniform distribution, then the function $c_p(.)$ is indeed convex. If $c_p(.)$ is not convex, the situation becomes more complicated and worth pursuing in the future correlated works. Therefore, if $c_p(.)$ is convex, then there exit at most 2 roots of equation (14).

Corresponding to different optimal cost c, denote $p_g(c)$ as the solutions of (14). Define $p_g^*(c)$ as the unique root or greater one of the two roots, i.e.,

$$p_g^*(c) = \begin{cases} p_g(c), & \textit{for unique solution} \\ \max\{p_g(c)\}, & \textit{for other case} \end{cases}\,. \tag{15}$$

By convexity of $c_p(.)$, the derivative of $d\pi/dp$ is non-positive at the point of the greater root. That means the second derivative of expected revenue with respect to p, at the greater root, is non-positive. It is the second-order sufficient condition for local maximizer. Consequently, we are sure that $p_g^*(c)$ is the local maximizer, while the smaller root is not.

More specific, let $\overline{\alpha}=2$, that means $\alpha_j \in [0,2]$. And consider the excessive case in which the diverse bidder *j*'s cost is 0 and $\alpha_j = 2$, for individual rationality, the price bidding of her will not less than -1. Therefore we assume the domain of *p* is $[-1,1]$. Assume $G_g(.)$ and $G(.)$ are all uniform distributions, then $G_g(.)=(1-p)/2$, $G(.)=(1-p)$. Correspondingly, we have

$$K(1-p)=((1-p)/2)^{m-1}, p\in[-1,1],\ L(1-p)=(1-p)^n, p\in[0,1],$$
$$k(1-p)=[(1-m)/2]\cdot[(1-p)/2]^{m-2},\ l(1-p)=-n\cdot(1-p)^{n-1}.$$

Note that, when $p\in[-1,0]$, the diverse bidder will definitely surmounts all simple bidders with probability 1. Then (12) can be rewritten as

$$\max_p \pi((q_s,p)|\theta_j)=\max_p[\alpha_j\cdot K(1-p)+2\cdot K(1-p)\cdot(p-c)]\ . \tag{16}$$

Therefore, formula (14) can be expressed as follows

$$c=\begin{cases} p+\dfrac{(n\alpha_j+\alpha_j\cdot(m-1)/2)\cdot(1-p)^n+(p-1)}{n+m-1}, & 0<p\le 1 \\ p+\dfrac{p+\alpha_j\cdot(m-1)/2-1}{m-1}, & \dfrac{-\alpha_j}{2}<p\le 0 \end{cases}\ . \tag{17}$$

The convexity of (17) is obvious. Define $p_g^-(c)$ as the solution (price bidding) of (17) in the domain $p\in[-1,0]$.

Since the domain of expected revenue (16) is a closed set, in order to find the global maximizer, what we should do is to compare the boundary solutions with interior solution. Next, define c^* as follows,

$$c^*=\max\{c:\pi(0|c)-\pi(p_g^*(c)|c)\ge 0\}\ . \tag{18}$$

Because

$$\frac{d\{\pi(0|c)-\pi(p_g^*(c)|c)\}}{dc}=-2K(1)+2L(1-p_g^*(c))\cdot K(1-p_g^*(c))<0\ . \tag{19}$$

It follows that if there exits c^*, such that

$$\pi(0|c^*)-\pi(p_g^*(c^*)|c^*)=0\ . \tag{20}$$

then for all *c*: $0<c<c^*$, we have

$$\pi(0|c)>\pi(p_g^*(c)|c)\ . \tag{21}$$

Define another special cost point c^- as follows

$$c^-=\max\{c:\pi(p_g^-(c)|c)-\pi(0|c)\ge 0\}\ . \tag{22}$$

Because

$$\frac{d\{\pi(p_g^-(c)|c)-\pi(0|c)\}}{dc}=-2K(1-p_g^-(c))+2K(1)<0\ . \tag{23}$$

It follows that if there exits c^-, such that

$$\pi(0|c^-)-\pi(p_g^-(c^-)|c^-)=0\ . \tag{24}$$

Then for all *c*: $0<c<c^-$, we have

$$\pi(0|c)<\pi(p_g^-(c)|c)\ . \tag{25}$$

Basically, in order to compare boundary solutions with interior solutions, we defined c^* and c^- respectively. In other words, c^* is the separating cost point below which zero price bidding is better than positive interior solution, while c^- is the separating cost point below which negative price bidding is better than zero price bidding. The following result is immediate.

Theorem 1. The following constitutes an equilibrium of MAPCA.

a) The simple bidder with type θ_i follows the strategy as (4);

b) The diverse bidder with type θ_j follows the strategies:

$$\begin{cases} q_s(\theta_j)=\arg\max[s(q)-c_g(q,\theta_j)] \\ p(\theta_j)=\begin{cases} p_g^*(c_g(q_s(\theta_j))) & if\ \ c^*<c_g(q_s(\theta_j),\theta_j) \\ 0 & if\ \ c^-<c_g(q_s(\theta_j),\theta_j)\le c^* \\ p_g^-(c_g(q_s(\theta_j))) & if\ \ 0<c_g(q_s(\theta_j),\theta_j)<c^- \end{cases} \end{cases}\ . \tag{26}$$

Intuitively, theorem 1 illustrates that one bidder either with synergy effect or not will submit same flavor quality-bid, i.e., optimal quality to maximize the difference of quality score minus the corresponding cost. As to price bidding, two kinds of bidders are apt to bid lower price when they have lower optimal cost. However, there exists two special cost "jump" points to make diverse bidders bid relative lower price: zero price or negative price. It is not so strange if we consider the situation in which synergy effect is very strong for any specific diverse bidder. The synergy effect can be regarded as business reputation, advertisement considerations and market share competition, which depends on different business surroundings.

4 Conclusions

In this paper, we construct a multi-attributes procurement combinatorial auction (MAPCA) for 2 identical objects, through a first-score, sealed-bid, simultaneous format. There are two kinds of bidders, simple bidder and diverse bidder, are considered in our model. With some assumptions, we obtain the equilibrium bidding strategies for the two kinds of bidders. Basically, the difference between our work and past correlated research (Krishna and Rosenthal (1996)) are: multi-attributes vs. single attribute, first-score vs. second-price, and procurement auction vs. selling auction. Also, this piece of research can not be realized without Che's work (Che 1993) about one-unit multi-attributes procurement auction. Compared with previous work about multi-attributes procurement combinatorial auctions' bidding analysis (Chen and Huang (2004)), this model extends to multiple diverse bidders who have synergy effect in the auction. Future work includes extensions to heterogeneous items assumption of our model.

Acknowledgements

This paper was partly supported by National Science Foundation of China under grant 70231010 and 70321001.

References

1. Che, Y. K.: Design competition through multidimensional auctions. The Rand Journal of Economics.Vol.24 (1993) 668–681
2. Chen J., H. Huang: Bidding Strategies Analysis for Procurement Combinatorial Auctions. Proceedings of The Fourth International Conference on Electronic Business. (2004) 41–45
3. Elmaghraby, W., P. Keskinocak: Technology for transportation bidding at the Home Depot. C. Billington, T. Harrison, H. Lee, J. Neale, eds. The Practice of Supply Chain Management: Where Theory and Applications Converge. Kluwer. (2003)
4. Hohner, G., J. Rich, E. Ng, G. Reid, A. J. Davenport, J. R. Kalagnanam, H. S. Lee, C.: An. Combinatorial and quality-discount procurement auctions with mutual benefits at Mars Incorporated. Interface. 33(1), (2003) 23–35
5. Krishna, V., Rosenthal, R.W.: Simultaneous Auctions with Synergies. Games and Economic Behavior, Vol. 17, (1996) 1–31
6. Ledyard, J. O., M. Olson, D. Porter, J. A. Swanson, D. P. Torma: The first use of a combined value auction for transportation services. Interfaces, 32(5), (2002) 4–12
7. Parkes, D.: Iterative Combinatorial Auctions: Achieving Economic and Computational Efficiency. Doctoral thesis, University of Pennsylvania (2001)
8. Rosenthal, R.W., Wang, R. Q.: Simultaneous Auctions with Synergies and Common Values. Games and Economic Behavior, Vol.17 (1996) 32–55

An Algebraic Method for Analyzing Open-Loop Dynamic Systems

W. Zhou, D.J. Jeffrey, and G.J. Reid

Department of Applied Mathematics, The University of Western Ontario, London, Ontario, Canada N6A 5B7

Abstract. This paper reports on the results of combining the MAPLE packages DYNAFLEX and RIFSIMP. The DYNAFLEX package has been developed to generate the governing dynamical equations for mechanical systems; the RIFSIMP package has been developed for the symbolic analysis of differential equations. We show that the output equations from DYNAFLEX can be converted into a form which can be analyzed by RIFSIMP. Of particular interest is the ability of RIFSIMP to split a set of differential equations into different cases; each case corresponds to a different set of assumptions, and under some sets of assumptions there are significant simplifications. In order to allow RIFSIMP to conduct its analysis, the governing equations must be converted from trigonometric form into a polynomial form. After this is done, RIFSIMP can analyze the system and present its results either graphically, or in list form. The mechanical systems considered are restricted to open-loop systems, because at present, closed-loop systems require too much computation by RIFSIMP to permit analysis.

Keywords: DYNAFLEX, RIFSIMP, Case Splitting, Symbolic Simplification, Graph Theory, Computer Algebra.

1 Introduction

A principal goal of multibody dynamics is the automatic generation of the equations of motion for complex mechanical systems [1]. Numerically based programs exist, for example ADAMS, JAMES and WORKING MODEL, that can generate the governing equations [2]; they are commercial programs and are in widespread use in the automotive, aerospace and robotics industries [3]. However, being numerically based programs they have several drawbacks.

- It is difficult to check the equations of motion, because they are represented by large volumes of numerical data.
- One cannot develop any physical insight.
- Closed-form solutions for the numerical equations are not possible.
- When used for simulations, they are inefficient because the equations are effectively re-assembled at each time step, and this may include many multiplications by 0 and 1.

V.S. Sunderam et al. (Eds.): ICCS 2005, LNCS 3516, pp. 586–593, 2005.

An alternative approach to equation generation uses symbolic processing. DYNAFLEX (`http://real.uwaterloo.ca/~dynaflex/`) is a Maple package that generates the equations of motion for a mechanical system from an input description based on a graph-theoretic method [4]. The equations are in a completely analytical form, and this offers several advantages [5].

- The structure of the equations is easily obtained and manipulated, giving the user a physical insight into the system.
- The equations can be easily exchanged with other groups of engineers or design groups.
- Real-time simulations are facilitated.

The symbolic models generated by DYNAFLEX are usually too complex to be solved symbolically, but there is still the possibility of symbolically pre-processing the output of DYNAFLEX before attempting a solution. The purpose of this paper is to apply the package RIFSIMP to this task. The RIFSIMP package analyzes ODE- and PDE-systems and returns canonical differential forms. The basic form is Reduced Involutive Form (RIF), and the package has the following features [6]:

- Computation with polynomial nonlinearities.
- Advanced case splitting capabilities for the discovery of particular solution branches with desired properties.
- A visualization tool for the examination of the binary tree that results from multiple cases.
- Algorithms for working with formal power series solutions of the system.

When RIFSIMP is applied to an equation system, it identifies special cases during the reduction. The analysis of the system then splits according to the special cases. In a full case analysis, some cases can be very complicated while others are simple enough to be analytically solvable. The canonical form generated by RIFSIMP is of low (0 or 1) differential index, which is suitable for the application of numerical solvers (whereas the output of DYNAFLEX may not have been suitable). An important option with RIFSIMP is the possibility of excluding special cases that are known to be not of interest. Thus if RIFSIMP detects a special case, say $m = 0$, but we know that this is of no physical interest, then we can pass this information to RIFSIMP as an inequation $m \neq 0$ appended to the input system.

1.1 Open-Loop and Closed-Loop Systems

The problems addressed by DYNAFLEX can be separated into two classes: open-loop systems and closed-loop systems. The terminology comes from the graph-theoretic method which is the basis of DYNAFLEX. The example in the next section will show the basic graph-theory method. For the present, we can note that an open-loop system corresponds to a system such as a robot arm, in which several components, such as joints and arm segments, are joined together in a configuration that terminates at a free end. This is in contrast to a framework, such as a four-bar mechanism, in which the components connect back to the structure being considered.

When a structure forms a closed loop, then DYNAFLEX will generate constraint equations that describe the drop in the degrees of freedom that accompanies the closing of a loop. From the point of view of this paper, we have discovered that the RIFSIMP package takes a great deal of time and memory to analyze closed-loop systems, but can make good progress with open-loop ones. This is what is reported here.

2 Example System Analysed Using Dynaflex

In order to keep the examples within printable limits, we shall use a simple spinning top as an example. A top is an axisymmetric body that spins about its body-fixed symmetry axis. It can precess about a vertical (Z) axis, and nutate about the rotated X axis. Figure 1 shows gravity acting in the $-Z$ direction. The center of mass is located at C, and the spinning top is assumed to rotate without slipping on the ground; this connection is modelled by a spherical (ball-and-socket) joint at O. The joint coordinates at O are represented by Euler angles (ζ, η, ξ), in the form of 3-1-3 Euler angles, meaning that they correspond to precession, nutation, and spin, respectively.

2.1 The System Graph

The system graph for the top is shown in Figure 2. The graph consists of nodes and edges. The nodes correspond to centres of coordinates, while the edges describe the system. Thus in the figure, we see nodes labelled 1,2,3 which are connected by edges e1, e2, e3,e4.

Node (1) can be thought of as the datum, or ground, on which the system is resting. Nodes (2) and (3) denote the top. The geometric fact that the top is connected to the ground is described by the edge e2, connecting node 1 to node 2. The fact that the top is free to spin, but not free to slide is described by e2 being specified as a spherical joint (one can think of a ball in a socket). The fact

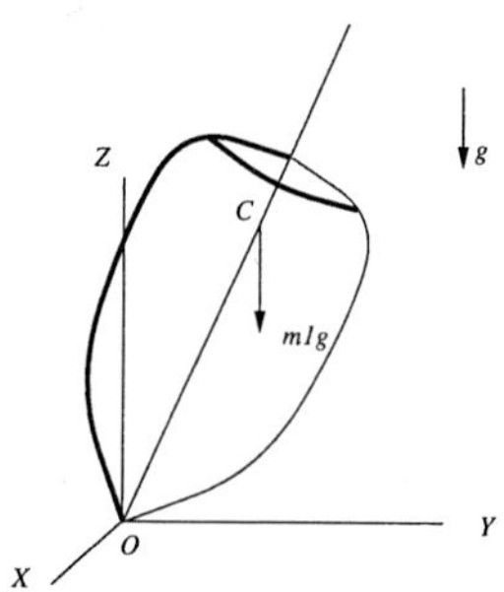

Fig. 1. The three-dimensional top. The centre of mass is at C and OC$= l$. The mass is here denoted $m1$, giving a gravitational force equal to $m1g$ acting in the $-Z$ direction

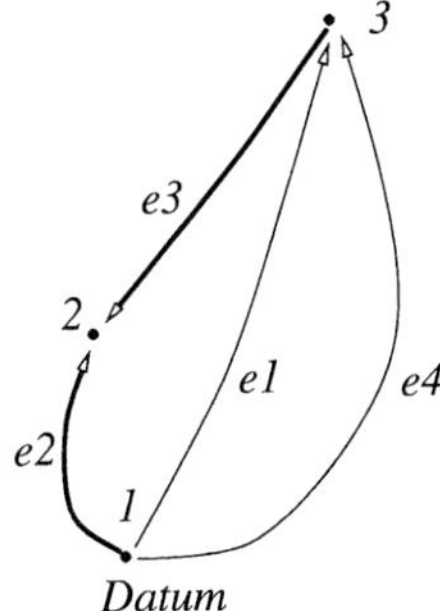

Fig. 2. The graph-theoretic method for the top

that the body is a top is described by edge e1 being specified as a rigid body with a diagonal moment of inertia. The position of the centre of mass is specified by e3, which formally is a rigid-arm element. Finally, gravity is defined by e4, being a force element.

It is important to note that the graph consisting of {e2, e3} is denoted by heavy lines, whereas elements e1 and e4 are drawn in light lines. The elements {e2,e3} correspond to the open loop referred to above.

2.2 Maple Input File

The Maple input file corresponding to the above system is as follows.

```
NOofedges:=4; NOofnodes:=3; Datum:=1;
#Note that Datum stands for the ground node.

edge[1]:=table([(1)=N, (2)=[1,3], (3)=BE_R,
     (4)=table([inert=[[x,0,0], [0,y,0], [0,0,z]], mass=c ])]);
edge[2]:=table([(1)=Y, (2)=[1,2], (3)=JE, (4)=SPH ]);
edge[3]:=table([(1)=Y, (2)=[3,2], (3)=AE_R,
     (4)=table([coords=[0,0,c] ])]);
edge[4]:=table([(1)=N, (2)=[1,3], (3)=FDE,
     (4)=table([type=PD, fz=-m1*g, force=gl ])]);
Iedge:=[];
```

2.3 The Symbolic Equations

The equations produced by DYNAFLEX from the above input file are presented below. By default, DYNAFLEX assigns its own notation for quantities such as Euler angles, moments of inertia, etc. Although DYNAFLEX notation is convenient of its internal programming, it results in equations which are difficult, and even ugly, to read when printed out for human use. Therefore we have edited the raw DYNAFLEX output format to simplify the notation to bring it in line with what human readers are used to seeing. The centre of mass is a distance l from the point of contact, the mass is m, the moment of inertia about the symmetry axis is C, and about a perpendicular axis is A. The Euler angles are given above.

$$C\ddot{\zeta}(t)\cos^2\eta(t) + C\ddot{\xi}(t)\cos\eta(t) - A\ddot{\zeta}(t)\cos^2\eta(t) - C\dot{\eta}(t)\dot{\xi}(t)\sin\eta(t)$$
$$-2C\dot{\eta}(t)\dot{\zeta}(t)\sin\eta(t)\cos\eta(t) + 2A\dot{\eta}(t)\dot{\zeta}(t)\sin\eta(t)\cos\eta(t) + A\ddot{\zeta}(t) = 0\ , \quad (1)$$
$$-mg\sin\eta(t) - A\dot{\zeta}^2(t)\sin\eta(t)\cos\eta(t) + A\ddot{\eta}(t) + C\dot{\xi}(t)\dot{\zeta}(t)\sin\eta(t)$$
$$+C\dot{\zeta}^2(t)\sin\eta(t)\cos\eta(t) = 0\ , \quad (2)$$
$$-C(-\ddot{\xi}(t) - \ddot{\zeta}(t)\cos\eta(t) + \dot{\eta}(t)\dot{\zeta}(t)\sin\eta(t)) = 0\ . \quad (3)$$

3 Automatic Symbolic Simplification Using RifSimp

The package RIFSIMP can process systems of polynomially nonlinear PDEs with dependent variables $u_1, u_2, \ldots u_n$, which can be functions of several independent variables. For the present application the only independent variable is time. The variables u_i can obey differential equations of varying order. RIFSIMP takes as its input a system of differential equations and a ranking of dependent variables and derivatives. RIFSIMP orders the dependent variables lexicographically[1] and the derivatives primarily by total derivative order:

$$u_1 \prec u_2 \prec \ldots \prec u_n \prec u_1' \prec u_2' \prec \ldots \prec u_n' \prec u_1'' \prec \ldots \quad (4)$$

Then equations are classified as being either leading linear (i.e. linear in their highest derivative with respect to the ordering $\prec$) or leading nonlinear (i.e. nonlinear in their highest derivative).

RIFSIMP proceeds by solving the leading linear equations for their highest derivatives until it can no longer find any such equations. Leading nonlinear equations (the so-called constraints), are treated by methods involving a combination of Gröbner Bases and Triangular Decomposition. It differentiates the leading nonlinear equations and then reduce them with respect to the leading linear equations. If zero is obtained, it means the equation is included in the ideal generated by the leading linear equations. If not, it means that this equation is a new constraint to the system. This is repeated until no new constraints are found.

Theorem: (Output RIFSIMP form)
If v is a list of derivatives and w is a list of all derivatives (including dependent variables) lower in ranking than v, then the output of RIFSIMP has the structure

$$v = f(t, w) \quad (5)$$

subject to a list of constraint equations and inequations

$$g(t, w) = 0, \quad h(t, w) \neq 0 \quad (6)$$

Proof. See [6] and references therein.

[1] Readers who are not familiar with the ideas of term ordering in polynomial systems can read this as meaning that the variables are placed in alphabetical order.

Theorem: (Existence and Uniqueness)
Given an initial condition $g(t^0, w^0) = 0, h(t^0, w^0) \neq 0$, there is a local analytic solution with this initial condition.

Proof. See [6].

4 Application of RifSimp to Example Problem

In order to apply RIFSIMP to the example above, we proceed as follows.

1. Change coordinates using $\cos\eta = \frac{1-u(t)^2}{1+u(t)^2}, \sin\eta = \frac{2u(t)}{1+u(t)^2}$ to get a rational polynomial differential system instead of trigonometric nonlinear differential system. (RIFSIMP does not allow trigonometric functions.)
2. Give this polynomial differential equation system to RIFSIMP, specifying the case split option to get normalized differential equations.
3. Use RIFSIMP case tree to analyze the different cases.

With no assumptions on the parameters, we obtain 24 cases, each split corresponds to some quantity being zero or not. The Maple output lists, for each split, exactly what the quantity is.

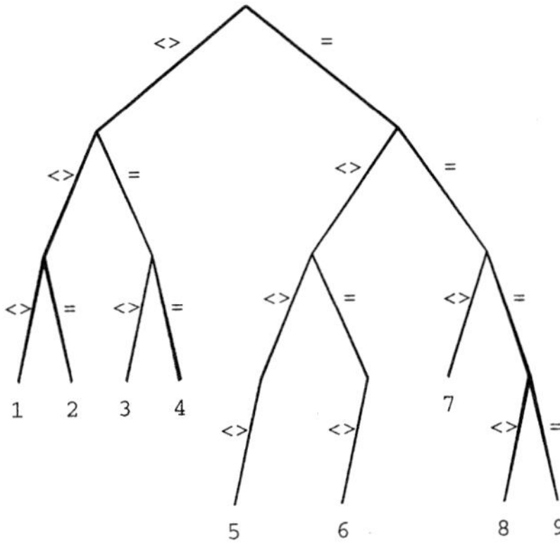

Fig. 3. The case-split tree after the exclusion of cases that are not of physical interest

An examination of the output case-split list shows that many of the cases can be removed on physical grounds. Thus RIFSIMP identifies that there is division by g, by m, and by l and therefore flags these cases as points of case splitting. Since we know in this application that these cases are not of physical interest, we can exclude them automatically from the case analysis by including a list of inequations in the RIFSIMP parameters. After including $g \neq 0$, $m \neq 0$, and $l \neq 0$, RIFSIMP returns a tree containing 9 cases, as shown in Figure 3.

5 Some Special Cases

The importance of the RIFSIMP analysis is the identification of special cases. The general ("generic") case cannot be simplified further. However, any analysis of a mechanical system should look for special cases, either to exploit these particularly simple cases or to avoid them. The example shows that RIFSIMP can be used to find these cases automatically.

5.1 First Group of Special Cases

We group two cases together that differ in that RIFSIMP identifies one case as being $A \neq 0$ and one as $A = 0$. However, the equation containing A is identically zero because of the condition $u = 0$, and hence the value of A is irrelevant. The equations simplify to

$$\ddot{v}(t) = -\ddot{w}(t), \quad u(t) = 0 \tag{7}$$

These equations can be integrated in closed form to

$$u(t) = 0, \quad v(t) = v_1 t + v_0, \quad w(t) = -v_1 t + w_0 \tag{8}$$

where v_1, v_0, w_0 are given by the initial conditions. From the definition of $\eta(t) = \arccos((1 - u(t)^2)/(1 + u(t)^2))$, we have $\eta = 0$ which means that the top is spinning vertically without any inclination with respect to axis Z.

5.2 Second Group of Special Cases

Again RIFSIMP identifies $A = 0$ and $A \neq 0$ separately. The common equations are

$$\ddot{w}(t) = 0, \quad \dot{u}(t) = 0, \quad \dot{v}(t) = \frac{gml}{C\dot{w}(t)} \tag{9}$$

and with constraint: $u(t)^2 - 1 = 0$. The analytic solution is:

$$u(t) = u_0, \quad v(t) = \frac{gmlt + v_0 w_1 C}{w_1 C}, \quad w(t) = w_1 t + w_0 \tag{10}$$

and with the constraints: $u_0^2 - 1 = 0$ and v_0, w_0, w_1 are given by the initial conditions. Also from the definition of $\eta(t) = arccos((1 - u(t)^2)/(1 + u(t)^2))$, we have $\eta = \pi/2$, which means the top is moving horizontally in the $x - y$ plane, i.e. it is precessing without nutation.

6 Conclusion and Future Work

Because of space limitations, we have restricted the example problem to being a simple one. Further experiments with other mechanical systems, such as double and triple pendulums, were made but not included here. However, from the example, we can identify successes and limitations of RIFSIMP. We first point to the success of RIFSIMP in identifying the special cases of top motion. In the simple example shown here, these might seem well known, but the important point is that these were identified *automatically* rather than by human inspection.

One of the limitations of the combination of RIFSIMP and DYNAFLEX is the fact that DYNAFLEX generates a plethora of parameters. Too many parameters can cause a serious degradation of the performance of a computer algebra system, as well as leading to a large number of special cases that can prevent a human user from seeing the patterns of interest in the output. This is reflected in the 24 special cases initially identified by RIFSIMP. Controlling the consequences of a large number of parameters will be vitally important to further applications.

This paper has concentrated on open loop systems, because RIFSIMP has been most successful in these cases. Closed loop systems generate further constraint equations that can cause RIFSIMP to exhaust computing resources before completing its analysis. One reason for this is the symbolic inversion of matrices in order to obtain RIF form. Computing techniques for handling large expressions have been developed in other contexts, see for example [7]. Combining these with RIFSIMP will increase the complexity of problems that can be solved.

Acknowledgments

We gratefully acknowledge John McPhee, Chad Schmidt and Pengfei Shi for their development of DYNAFLEX and for making their system available to us.

References

1. Schiehlen, W. *Multibody Systems Handbook*; Springer-Verlag: Berlin, 1990.
2. P. Rideau. *Computer Algegbra and Mechanics, The James Software*; Computer Algebra in Industry I, 1993 John Wiley & Sons Ltd.
3. Pengfei Shi, John McPhee. *Symbolic Programming of a Graph-Theoretic Approach to Flexible Multibody Dynamics*; Mechanics of Structures and Machines, 30(1), 123-154(2002).
4. P. Shi, J. McPhee. *Dynamics of flexible multibody systems using virtual work and linear graph theory*; Multibody System Dynamics, 4(4), 355-381(2000).
5. Christian Rudolf. *Road Vehicle Modeling Using Symbolic Multibody System Dynamics, Diploma Thesis*; University of Waterloo in cooperation with University of Karlsruhe(2003).
6. Reid, G.J., Wittkopf, A.D., and Boulton, A. *Reduction of Systems of Nonlinear Partial Differential Equations to Simplified Involutive Forms.* Eur. J. Appl. Math. **7** (1996): 604-635.
7. R. M. Corless, D. J. Jeffrey. *Two Perturbation Calculation in Fluid Mechanics Using Large-Expression Management*; J. Symbolic Computation(1996) 11, 1-17.

Pointwise and Uniform Power Series Convergence

C. D'Apice, G. Gargiulo, and R. Manzo

University of Salerno, Department of Information Engineering and Applied Mathematics, via Ponte don Melillo,84084 Fisciano (SA), Italy
{dapice,gargiulo,manzo}@diima.unisa.it

Abstract. Since the introduction of CAS (Computer Algebra Systems), educators are experimenting new ways of teaching with the aim to enhance many aspects of the learning mathematics. In particular, visualization can aid the understanding of concepts. The graphical potentialities of many CAS, in fact, allow students to discover concepts, relationships, rules, so as to construct their knowledge for themselves. The aim of this work is to present a Mathematica notebook that enables students to use visualization skills to better grasp the mathematical concepts of the pointwise and uniform convergence of power series.

1 Visualization in Mathematics Teaching

Visualization, as a technique for teaching mathematics, has become importante specially in those countries where graphics calculators are widely used. Mathematical concepts, ideas and methods have a great wealth of visual relationships, and their use and manipulation is clearly very beneficial from the point of view of their presentation.

Visualization means illustration of an object, fact, process, concept, and its result can be graphic, numeric, or algebraic. Today the term is mostly used for graphic illustrations. Visualization tools can be used in an interpretative and expressive way (Gordin, Edelson and Gomez, 1996).

Interpretative tools can support learners in extracting meaning from the information being visualized, in clarifying text and abstract concepts making them more comprehensible. Expressive visualization can be used to visually convey meaning in order to communicate a set of beliefs.

When using a calculator as a visualization tool the immediate feedback is of central importance. The illustration of concepts helps students to understand them, to associate them with something that is not abstract and the visual association aids to memorize them. Student become able to "see" the meaning of abstract patterns, interacting visually with mathematics.

Mathematica, being a powerful tool for computing, programming, data analysis and visualization of information, is able to catch students' interests, to challenge them and to make learning more interesting.

The use of Mathematica notebooks, as a support to traditional lessons, can help students in understanding some concepts, since they are personally involved in experimentation and discovery. We have introduced them in the traditional curriculum for some years, within a research addressed to the development of innovative ways of mathematics teaching using Mathematica.

V.S. Sunderam et al. (Eds.): ICCS 2005, LNCS 3516, pp. 594–601, 2005.

Here we describe a notebook that can offer useful visualization of the concepts of pointwise and uniform convergence of power series.

2 Pointwise and Uniform Convergence of Power Series

The notebook, we are going to illustrate, has been realized with the aim to aid students to reach a deeper knowledge of the concepts of pointwise and uniform convergence of power series. Students used it after traditional lessons on the above contents in order to explore the meaning of convergence.

To introduce the convergence of power series, we recall the following definitions.

Definition 1. *Let I be a subset of R,* $\{f_n\}$, $f_n : I \to \mathbb{R}$, *be a sequence of functions, and let* $f : I \to \mathbb{R}$ *be another function. We say that* $f_n \to f$

- *pointwise on I if* $\forall x \in I$ and $\forall \varepsilon > 0 \; \exists N(x, \epsilon)$ *such that* $n > N \Rightarrow |f_n(x) - f(x)| < \epsilon$
- *uniformly on I if* $\forall \epsilon > 0 \; \exists N(\epsilon)$ *such that* $n > N \Rightarrow |f_n(x) - f(x)| < \epsilon, \forall x \in I$.

The graphics in Figure 1 illustrates how, for sufficiently large n, the n-th function $f_n(x) = xe^{-nx}$ lies completely below any given level (in the form $\frac{1}{en}$). In this case, for simplicity, the (uniform) limit is identically zero, so that close to the limit simply means small. In the case of (pointwise but) non uniform convergence, the n-th function stays below any given level, but not globally: the maximum is essentially the same for all functions (here $f_n(x) = nxe^{-nx}$).

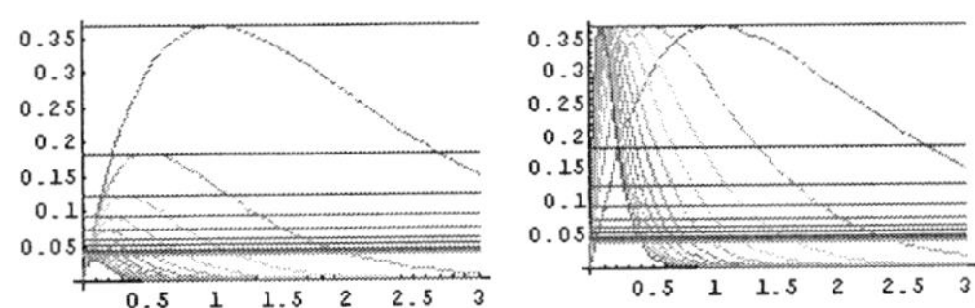

Fig. 1. On the left $f_n(x) = xe^{-nx}$, on the right $f_n(x) = nxe^{-nx}$

Definition 2. *The series* $s(x) = \sum_{k=1}^{\infty} u_k(x)$ *converges pointwise to the function s if the sequence* $s_n = \sum_{k=1}^{n} u_k(x)$ *converges pointwise to the function s.*

Definition 3. *The series* $s(x) = \sum_{k=1}^{n} u_k(x)$ *converges uniformly to the function s if the sequence* $s_n(x)$ *converges uniformly to the function s.*

Let us go to visually study the convergence of some power series.

2.1 Series $\sum_{k=0}^{\infty} z^k$

Let us consider the geometric series $\sum_{k=0}^{\infty} z^k$ which converges within the circle of convergence $|z|<1$ and diverges outside $\sum_{k=0}^{\infty} z^k = \frac{1}{1-z}$. Let us analyze the behaviour of the partial sums of the geometric series in the complex plane using the following function:

```
s[z_, n_ : 20, opts___] := ListPlot[({Re[#1], Im[#1]} &) /@
Drop[FoldList[Plus, 0, Table[z^k, {k, 0, n}]], 1],
PlotJoined → True, PlotStyle → RGBColor[1, 0, 0],
AspectRatio → Automatic, Background → GrayLevel[1], opts]
```

Visualize the behavior of the first 40 partial sums of the series in a point, $\frac{3}{5}+\frac{4i}{5}$, on the boundary of the convergence circle.

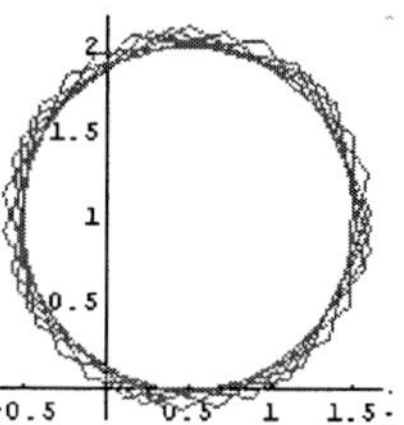

Fig. 2. $s\left[\frac{3}{5}+\frac{4i}{5},40\right]$

As we can see the geometric series does not converge, since the points representing the partial sums do not "head" towards a definite point. Let us consider (Fig. 3) the behaviour of the first 200 partial sums in a point, $0.99\left(\frac{3}{5}+\frac{4i}{5}\right)$, inside the circle of convergence. The geometric series converges but since the point is near the circle the

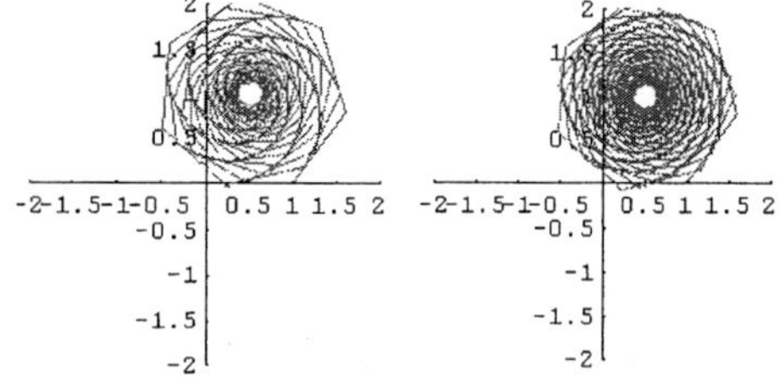

Fig. 3. On the left $s\left[0.99\left(\frac{3}{5}+\frac{4i}{5}\right),200\right]$, on the right $s\left[0.995\left(\frac{3}{5}+\frac{4i}{5}\right),20\right]$

convergence is slow: the sums slowly spiral towards their limit. Let us consider (Fig. 3) a point, $0.995\left(\frac{3}{5}+\frac{4i}{5}\right)$,nearer the circle: the series converges but the convergence is even slower than before.

In a point, $0.9\left(\frac{3}{5}+\frac{4i}{5}\right)$, far from the boundary of the convergence circle, the convergence is faster as we can see in the following graphics.

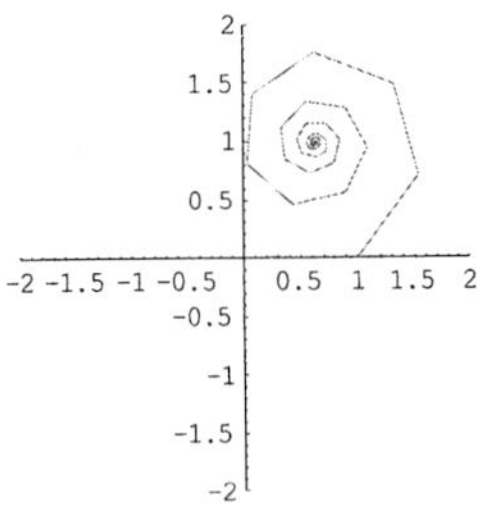

Fig. 4. $s\left[0.9\left(\frac{3}{5}+\frac{4i}{5}\right),200\right]$

2.2 Series $\sum_{k=1}^{\infty}\frac{z^k}{k}$

Let us analyze the convergence of the series $\sum_{k=1}^{\infty}\frac{z^k}{k}=-Log[1-z]$. First, we consider the same point, $\frac{3}{5}+\frac{4i}{5}$, as before, on the boundary of the circle of convergence (Fig. 5). This time, the series converges (slowly) also in this point. Then let us analyze the behaviour in a point, $0.99\left(\frac{3}{5}+\frac{4i}{5}\right)$, inside the circle (Fig. 5).

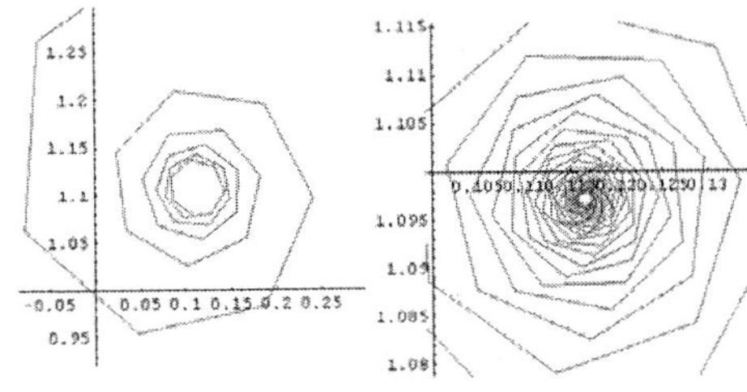

Fig. 5. On the left $s\left[\left(\frac{3}{5}+\frac{4i}{5}\right),40\right]$, on the right $s\left[0.99\left(\frac{3}{5}+\frac{4i}{5}\right),20\right]$

We "zoom out" the graphics and we consider a point still closer to the boundary and a point farther from the boundary (Fig. 6).

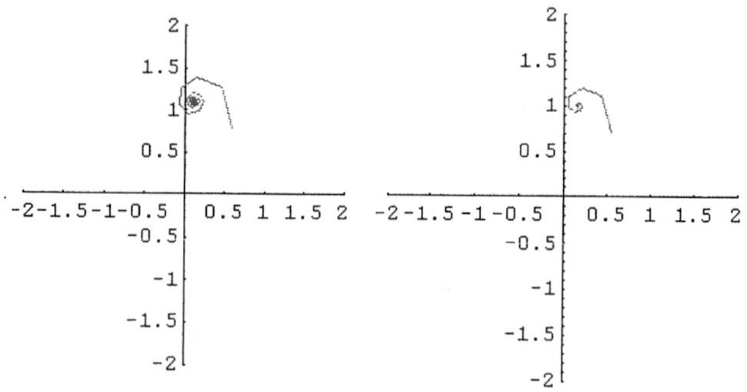

Fig. 6. On the left $s\left[0.995\left(\frac{3}{5}+\frac{4i}{5}\right),400\right]$, on the right $s\left[0.9\left(\frac{3}{5}+\frac{4i}{5}\right),400\right]$

3 Uniform Convergence

Let us define a general function that allows to study the behaviour of the partial sums of a power series in a chosen point.

```
sgeneral[termgen_, value_, numterms_: 20, colours_: 5, opts_] :=
Module[{listsumspoints},
listsumspoints = ({Re[#1], Im[#1]} &) /@ Drop[
    FoldList[Plus, 0, Table[termgen /. z → value, {k, numterms}]], 1];
Show[Graphics[Table[{Hue[i/colours],
    Line[{listsumspoints[[i]], listsumspoints[[i + 1]]}]},
   {i, Length[listsumspoints] - 1}]], opts]]
```

Let us try to understand the meaning of uniform convergence, observing the behaviour of the partial sums in different points, in polar coordinates (i.e. we pick up points in given directions and on several circles around the origin).

```
Show[Table[sgeneral[z^k/k, r ⅇ^(ⅈ θ), 20, 5, DisplayFunction → Identity],
  {r, .5, 1, .02}, {θ, 0, 2π, π/8}],
DisplayFunction → $DisplayFunction, PlotRange → {{-2, 2}, {-3, 3}},
Axes → True, AspectRatio → 1, Background → GrayLevel[1]];
```

As we can see the convergence is faster in the points far from the boundary of the convergence circle, and becomes slow near the boundary.

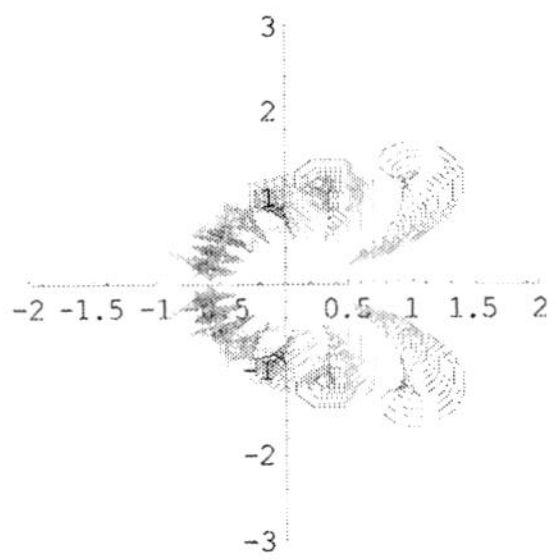

Fig. 7. The series with general term $\frac{z^k}{k}$

The series does not converge uniformly, in fact the convergence "rates" depends in an essential manner on the point.

With different general terms, $\frac{z^k}{kLog[k+1]}$ and $\frac{z^k}{\sqrt{k}}$ (Fig. 7), we see the same general pattern as before, but the convergence rate is a trifle higher (spirals are someway smoother).

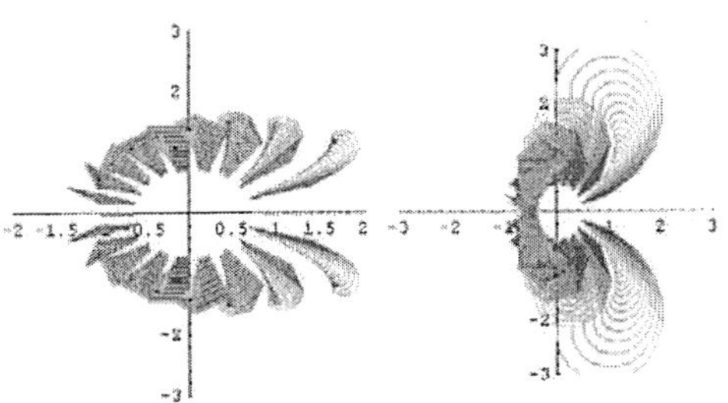

Fig. 8. The series with general term $\frac{z^k}{kLog[k+1]}$ on the left and $\frac{z^k}{\sqrt{k}}$ on the right

The geometric series again around the point 0.95 and still closer to the boundary around the point 0.95 (Fig. 8).

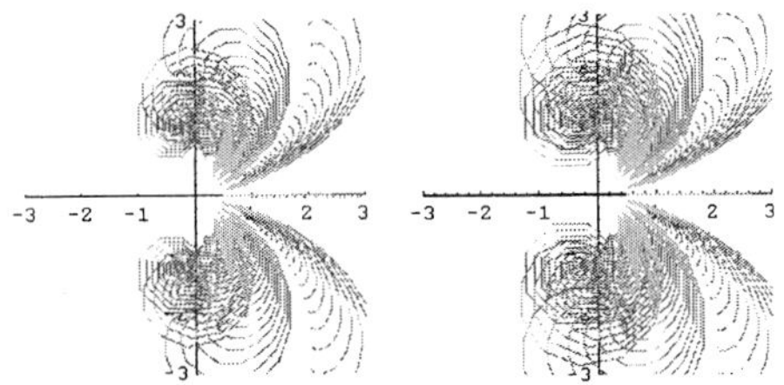

Fig. 9. The geometric series

Below are shown two examples of uniform convergence, the series with general term $\frac{z^k}{k^2}$ and $\frac{z^k}{k^4}$ (Fig. 9).

We "see" that the spirals have the same general ("uniform") structure. The second series is still "more uniform" example: as Euler would have said, an "augmented convergence".

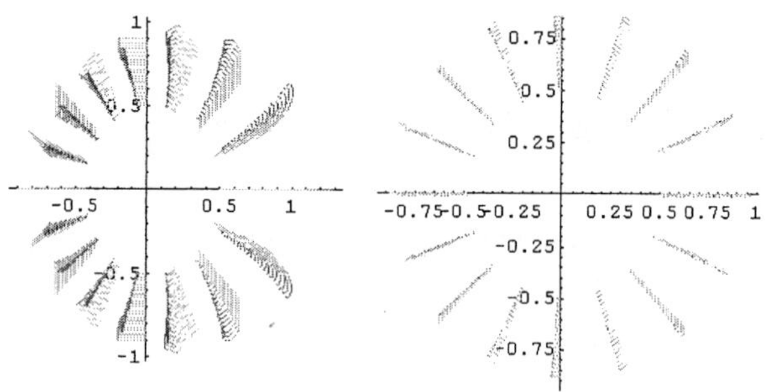

Fig. 10. The partial sums of series with general term $\frac{z^k}{k^2}$ on the left and $\frac{z^k}{k^4}$ on the right

The logarithmic series around its singular point and "very close" to its boundary singular point $z = 1$.

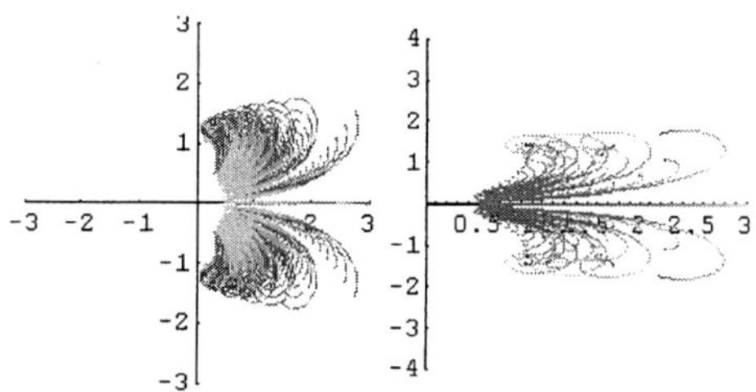

Fig. 11. The logarithmic series

4 Conclusions

The realization of this notebook encourages the use of computer graphical capabilities in teaching mathematics contents in University courses. The results of their use can be seen as a step towards the introduction of innovative modes in the process of teaching and learning. As a result of the experimentation students have enhanced their understanding of the illustrated concepts, learning in a more enjoyable way.

References

1. G. Albano, C. D'Apice, G. Gargiulo: Visualisation of four dimensional curves, Electronic Proceedings ICTMT4 (International Conference on Technology in Mathematics Teaching), Plymouth, 1999;
2. Albano, G., Cavallone, A., D'Apice, C., Gargiulo, G.: Mathematica and didactical innovation: a quadric use case, Electronic Proceedings of IMACS-ACA Conferences on Applications of Computer Algebra, Madrid, (1999);
3. Cavallone, A., D'Apice, C., Marsella, M., Salerno, S.: A didactical laboratory: image filtering, Electronic Proceedings of IMACS-ACA Conferences on Applications of Computer Algebra, Madrid, (1999);
4. D'Apice, C., De Simone, T., Manzo, R., Tibullo, V.: MOSFET: a Virtual Laboratory with Mathematica, Proceedings of ICTMT6 Volos, Greece, (2003), 252-257;
5. D'Apice, C., Manzo R., Tibullo, V.: Enhancing Mathematical Teaching-Learning Process by Mathematica, Proceedings of the International Mathematica Symposium 2003, London, (2003), 137-144.
6. Giusti, E. : Analisi Matematica II, Bollati Boringhieri, Torino, 1994.
7. Miranda, C. ; Lezioni di Analisi Matematica, parte seconda, Liguori Editore, Napoli, 1985.

Development of SyNRAC

Hitoshi Yanami[1,2] and Hirokazu Anai[1,2]

[1] Information Technology Core Laboratories, Fujitsu Laboratories Ltd,
Kamikodanaka 4-1-1, Nakahara-ku, Kawasaki 211-8588, Japan
yanami@flab.fujitsu.co.jp, anai@jp.fujitsu.com

[2] CREST, Japan Science and Technology Agency,
Kawaguchi Center Building, 4-1-8, Honcho, Kawaguchi 332-0012, Japan

Abstract. We present newly implemented procedures in SyNRAC, which is a Maple package for solving real algebraic constraints derived from various engineering problems. The current version of SyNRAC has added quantifier elimination (QE) by cylindrical algebraic decomposition (CAD) as well as QE by virtual substitution for weakly parametric linear formulas. We also show an application of CAD-based QE to the common Lyapunov function problem.

1 Introduction

Recently symbolic computation methods have been gradually applied to solving engineering problems, which has been caused by the efficient symbolic algorithms introduced and improved for these few decades and by the advancement of computer technology that has hugely increased the CPU power and memory capacity.

We have been developing a Maple toolbox, called SyNRAC, for solving real algebraic constraints. SyNRAC stands for a Symbolic-Numeric toolbox for Real Algebraic Constraints and is aimed at being a comprehensive toolbox including a collection of symbolic, numerical, and symbolic-numeric solvers for real algebraic constraints derived from various engineering problems. When we say a real algebraic constraint, what we have in mind is a first-order formula over the reals. Our main method is quantifier elimination (QE), which removes the quantified variables in a given formula to return a quantifier-free equivalent.

In this paper we present the newly implemented procedures in SyNRAC. In [1] two types of special QE methods as well as some simplification procedures of quantifier-free formulas had been implemented in SyNRAC. Besides, the current version of SyNRAC provides the following:

- general QE by cylindrical algebraic decomposition (CAD)
- special QE by virtual substitution for weakly parametric linear formulas.

CAD-based QE is called general in the sense that it can deal with any type of formula, without thought of the efficiency. Historically, the CAD-based approach

V.S. Sunderam et al. (Eds.): ICCS 2005, LNCS 3516, pp. 602–610, 2005.

preceded the special QE methods we had already implemented in SyNRAC. We implemented special QE first because there was a good class of formulas to which many practical problems could be reduced and a much more efficient special QE method was applicable [2, 3].

The latter procedure is an improved special QE procedure based on [4] for a subclass of the linear formulas—the *weakly parametric* linear formulas. We have implemented more efficient QE specialized for the class, which can reduce the size of an elimination set roughly by a factor of two at each stage.

This paper is organized as follows. We briefly describe CAD and show the commands on CAD in SyNRAC in Section 2. In Section 3, we present special QE by virtual substitution for weakly parametric linear formulas and compare the procedure in SyNRAC with a previous one. In Section 4, we show an example problem to which SyNRAC's CAD command can apply. We end with a conclusion in Section 5.

2 Cylindrical Algebraic Decomposition

Cylindrical algebraic decomposition (CAD) was discovered by Collins in 1973; see [5] for his monumental work. Collins also proposed a general QE algorithm based on CAD, which provided a powerful method for solving real algebraic constraints.

Let A be a finite subset of $\mathbb{Z}[x_1, \ldots, x_n]$. An *algebraic decomposition* for A is a collection of mutually disjoint, semi-algebraic, A-invariant sets that partitions the Euclidean n-space E^n. To define the term *cylindrical*, we explain three parts of a CAD procedure—the projection phase, the base phase, and the lifting phase.

In the projection phase of a CAD, the PROJ function plays a central role. Let r be an integer greater than 1. PROJ maps a finite set of integral polynomials in r variables to a finite set of integral polynomials in $r-1$ variables: for $A_r \subset \mathbb{Z}[x_1, \ldots, x_r]$, $\mathrm{PROJ}(A_r) \subset \mathbb{Z}[x_1, \ldots, x_{r-1}]$. For a given $A \subset \mathbb{Z}[x_1, \ldots, x_n]$, we obtain a list

$$A = A_0 \overset{\mathrm{PROJ}}{\mapsto} A_1 \overset{\mathrm{PROJ}}{\mapsto} A_2 \overset{\mathrm{PROJ}}{\mapsto} \cdots \overset{\mathrm{PROJ}}{\mapsto} A_{n-1},$$

where $A_i \subset \mathbb{Z}[x_1, \ldots, x_{n-i}]$.

In the base phase we partition E^1 by using a set of univariate polynomials $A_{n-1} \subset \mathbb{Z}[x_1]$; we find all the real zeros of A_{n-1} and partition E^1 into A_{n-1}-invariant regions that consist of the zeros of A_{n-1} and the remaining open intervals. These points and intervals are called sections and sectors, respectively.

The lifting phase inductively constructs a decomposition of E^{i+1} from the decomposition of E^i, $i = 1, 2, \ldots, n-1$. Suppose D is a decomposition of E^i. A lifting of D is a decomposition $\bar{D}$ of E^{i+1} obtained by decomposing the space $R \times E^1$ by using A_{n-i-1} for each region $R \in D$ and putting all of them together. Let R be a region of a decomposition D of E^i. $R \times E^1$ is decomposed by the following; Take a point $(p_1, \ldots, p_i)$ in R and substitute it for $(x_1, \ldots, x_i)$ in each polynomial in A_{n-i-1} to obtain a set of univariate polynomials in x_{i+1}; Partition E^1 into, say, L_0, L_1, …, L_{2k+1} by using the roots of the polynomials in x_{i+1}; Regard $R \times L_0$, $R \times L_1$, ..., $R \times L_{2k+1}$ as the resulting decomposition. The condition for this process to work is that A_{n-i-1} is *delineable* on R, in other words, every pair of polynomials in A_{n-i-1} has no intersections on R. In such a case the decomposition is independent of the choice of a sample point. A decomposition D of E^r is *cylindrical* if it is constructed by iterating the above lifting method, i.e., $r = 1$ and E^1 is decomposed as in the base phase, or $r > 1$ and D is a lifting of some cylindrical decomposition D' of E^{r-1}.

Given a formula φ one can construct a CAD for the polynomials of the atomic formulas in φ. The point for CAD-based QE is that the truth/falsehood of φ is determined regionwise because each region in the CAD is A-invariant. See [5] for details.

It is the PROJ function that is crucial in a CAD procedure. The fewer polynomials PROJ produces, the more efficient the CAD program becomes. But PROJ must be constructed to maintain the delineability and make the lifting phase possible. Some improvements in the projection phase of CAD are found in [6, 7, 8].

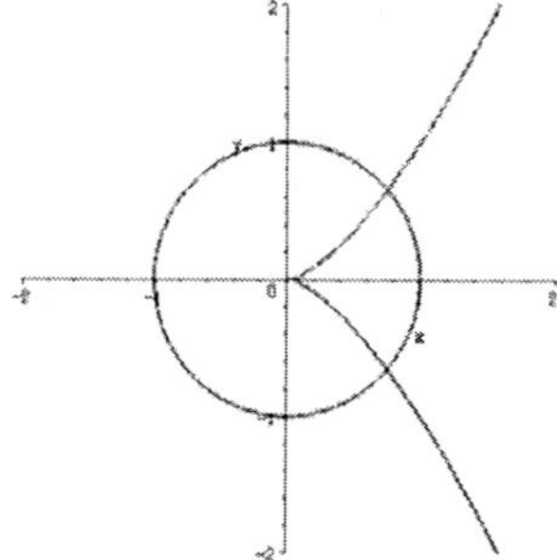

Fig. 1. The graph of A

Here we show some examples of CAD commands in SyNRAC. We construct a CAD for $A := \{x^2 + y^2 - 1, x^3 - y^2\} \subset \mathbb{Z}[x, y]$. The graph of the two polynomials in A is shown in Fig. 1. The `Projection` command repeats PROJ and returns `P[1]`$= \mathrm{PROJ}^0(A) = A$ and `P[2]`$= \mathrm{PROJ}(A)$.

```
> read "synrac";
> A:=[ x^2 + y^2 - 1, x^3 - y^2 ]:
> P:=Projection(A, [y,x]):
> P[1]; P[2];

                    [x^2 + y^2 - 1, x^3 - y^2]

                   [x^2 - 1, x^2 - 1 + x^3, x]
```

Next the `Base` command partitions E^1 by using `P[2]` and returns a list of points that represent respective sections or sectors. A rational point is taken as a sample point for a sector, and a vanishing polynomial and an isolated interval are taken for a section. There are four real roots (sections) in `P[2]` and they make five open intervals (sectors).

```
> Base(P[2], x);

  [-2, [x + 1, [-1, -1]], -1/2, [x, [0, 0]], 3/8,
      [x^2 - 1 + x^3, [3/4, 7/8]], 15/16, [x - 1, [1, 1]], 2]
```

Lastly the `Lifting` command makes a stack for each section or sector. Out of nine regions, we have the third and the fifth ones displayed. The third region is a sector with a rational sample point `[-1/2]` and the stack on it is represented in a list of five sample points of sections/sectors. The data for the fifth sector are shown in a similar way. The data for It is similar for the fifth sector, which has `[3/8]` as a sample point.

```
> L:=Lifting(P, [y,x]):
> op(L[3]); op(L[5]);

  [3], [-1/2], [-2, [-3 + 4 y^2, [-1, 0]], 0, [-3 + 4 y^2, [0, 1]], 2]

  [5], [3/8], [-2, [-55 + 64 y^2, [-1, -1/2]], -1/2,
      [512 y^2 - 27, [-1/2, 0]], 0, [512 y^2 - 27, [0, 1/2]], 1/2,
      [-55 + 64 y^2, [1/2, 1]], 2]
```

3 Special QE for Weakly Parametric Linear Formulas

We gave a description of a special QE algorithm by virtual substitution for linear formulas [9]. We firstly review special QE for linear formula and then explain the weakly parametric case. We will show some experimental results to compare the efficiency of the two QE procedures.

3.1 Special QE by Virtual Substitution for Linear Formulas

In 1988 Weispfenning [10] proposed a QE algorithm for linear formulas by virtual substitution. A formula is called *linear* if its atomic subformulas are all linear with respect to its quantified variables. In other words, every equality or inequality in the formula can be represented in

$$a_0 + a_1x_1 + \cdots + a_nx_n \;\rho\; 0,$$

where x_i are quantified variables, a_i are free, and ρ is a relational operator, i.e., one of $\{=, \neq, \leq, <\}$.

Let $\psi(p_1, \ldots, p_m) := Q_1x_1 \cdots Q_nx_n\varphi(p_1, \ldots, p_m, x_1, \ldots, x_n)$ be a linear formula, where $Q_i \in \{\forall, \exists\}$ and φ is quantifier-free. By using the equivalence $\forall x\varphi(x) \Longleftrightarrow \neg(\exists x \neg\varphi(x))$, we can change the formula into its equivalent of the form $(\neg)\exists x_1 \cdots (\neg)\exists x_n \; (\neg)\varphi$. The possible negation $(\neg)$ attached to φ can be easily eliminated (use De Morgan's law and rewrite the atomic subformulas), which is not essential part of QE. Therefore all we are required to do is eliminate $\exists x$ in $\exists x\varphi$. In general case, all the quantifiers in the formula can be eliminated inside out by changing the innermost one to $\exists x_i$ and then removing it.

To eliminate the quantified variable $\exists x$ in $\exists x\varphi$, we use a theorem in [10] saying that there exists a finite set S of x-free terms such that

$$\exists x\varphi \Longleftrightarrow \bigvee_{t\in S} \varphi(x//t)$$

holds. We note that there is a procedure transforming the expression $\varphi(x/t)$ obtained from φ by substituting t for x into an equivalent formula [11]. We denote the resulting formula by $\varphi(x//t)$. Such a set S is called an *elimination set* for $\exists x\varphi$. We obtain a quantifier-free formula equivalent to $\exists x\varphi$ by substituting all candidates in S into φ *disjunctively*.

This theorem indicates that smaller elimination sets would increase the efficiency of a QE procedure. Loos and Weispfenning [12] and Weispfenning [11] have presented a smaller elimination set to improve the algorithm. All of these algorithms had been implemented in SyNRAC.

3.2 Special QE for Weakly Parametric Linear Formulas

Weispfenning [4] has further improved the algorithm for a subclass of linear formulas. A linear formula is called *weakly parametric* if the coefficients of the quantified variables in every atomic formula $a_0 + a_1x_1 + \cdots + a_nx_n \ \rho\ 0$ are all constant. Therefore free variables occur only in the constant term a_0 in a weakly parametric linear formula.

In such a case, one can tell beforehand which half-interval should be used for each inequality and can exploit the information to dispense with as much as half an elimination set per quantified variable. See [4] for details.

3.3 Comparison of the Two Procedures

We compare a linear QE procedure previously implemented in SyNRAC and a specialized one for weakly parametric linear formulas. We have used nine example problems that were taken from the design problem of a PI controller. Problem 1 has five quantified variables. The number of variables increases one by one. So Problem 9 has 13 variables to be eliminated. We show Problems 1 and 2 here, leaving the others out because of a lack of space.

Problem1
&Ex($[k_1, k_2, c_0, c_1, c_2]$, &and($[0 \le c_2 + (1/4\,p_1 - 1)p_2{}^3,\ 0 \le -c_2 + (1/4\,p_1 + 1)p_2{}^3,$
$0 \le c_1 + (1/2\,p_1 - 4)p_2{}^2,\ 0 \le -c_1 + (1/2\,p_1 + 4)p_2{}^2,\ 0 \le k_1 + c_0 + (p_1 - 6)p_2,$
$0 \le -k_1 - c_0 + (p_1 + 6)p_2,\ 0 \le k_2 + 2\,p_1 - 4,\ 0 \le -k_2 + 2\,p_1 + 4,\ -1 \le c_0,\ c_0 \le 1,$
$1 \le c_1,\ c_1 \le 3/2,\ 1/2 \le c_2,\ c_2 \le 3/2]$))

Problem2
&Ex($[k_1, k_2, c_0, c_1, c_2, c_3]$, &and($[0 \le c_3 + (1/4\,p_1 - 1)p_2{}^4,\ 0 \le -c_3 + (1/4\,p_1 + 1)p_2{}^4,$
$0 \le c_2 + (1/2\,p_1 - 6)p_2{}^3,\ 0 \le -c_2 + (1/2\,p_1 + 6)p_2{}^3,\ 0 \le c_1 + (p_1 - 15)p_2{}^2,$
$0 \le -c_1 + (p_1 + 15)p_2{}^2,\ 0 \le k_1 + c_0 + (2\,p_1 - 18)p_2,\ 0 \le -k_1 - c_0 + (2\,p_1 + 18)p_2,$
$0 \le k_2 + 4\,p_1 - 10,\ 0 \le -k_2 + 4\,p_1 + 10,\ -1 \le c_0,\ c_0 \le 1,\ 1 \le c_1,\ c_1 \le 3/2,$
$1/2 \le c_2,\ c_2 \le 3/2,\ 1/2 \le c_3,\ c_3 \le 3/2]$))

Table 1. Linear QE and weakly parametric linear QE

Problem	Linear (general)		Weakly parametric linear	
	time (sec)	# atoms	time (sec)	# atoms
1	10.406	112	0.093	36
2	2.708	608	0.156	92
3	12.281	3072	0.320	232
4	513.406	69632	0.626	560
5	—	—	1.360	1312
6	—	—	2.968	3008
7	—	—	6.733	6784
8	—	—	16.375	14848
9	—	—	39.828	32768

On a Pentium IV 2.4 GHz processor with 1 GB memory

The results are shown in Table 1. The previous procedure could not solve Problem 5 and beyond in an hour, while the algorithm specialized for weakly parametric formula solved all of the example problems. Let us compare the computing time per elimination. At Problem 1 the procedure for weakly parametric formulas has computed $(0.406/0.093)^{1/5} \approx 1.34$ times faster than the previous one. The corresponding ratios grow gradually and reach 2.31 at Problem 4. We suppose that simplification procedures have a more effect on the computing time when a formula gets larger.

4 Application of CAD-Based QE

In this section we show an application of CAD to a practical problem. The common Lyapunov function problem is a problem that studies the existence of a common Lyapunov function for a set of linear time-invariant systems. The problem often arises in stability analysis and control design of various types of control systems such as uncertain systems, fuzzy systems, switched systems, etc. We focus on one of the main types of common Lyapunov functions, a *common quadratic Lyapunov function* (CQLF). See [13, 14, 15] for details.

One important issue of a CQLF problem is to find an existence condition of CQLF. For a given set of stable constant linear systems, we can verify whether the systems share a CQLF and construct the CQLF if they do with some numerical semidefinite programming (SDP) package. We consider here the problem of finding symbolic existence conditions on system matrices such that these systems share a common Lyapunov function. An existence condition provides us stability regions of parameters for control systems. Although there are some attempts to resolve this so far, only partial results are obtained. The CQLF problems to compute an existence condition can be solved by using QE systematically; see [16].

Common Lyapunov Function Problem: We consider a set of continuous-time linear time-invariant systems

$$\dot{\mathbf{x}} = \mathbf{A}_{ci}\mathbf{x},\ \mathbf{x} \in \mathbb{R}^n,\ \mathbf{A}_{ci} \in \mathbb{R}^{n\times n},\ i = 1,\dots,q. \tag{1}$$

The set of systems (1) is said to have a CQLF if there exists a symmetrical positive definite matrix $\mathbf{P} = \mathbf{P}^T > 0$ such that the following Lyapunov inequalities

$$\mathbf{P}\mathbf{A}_{ci} + \mathbf{A}_{ci}^T\mathbf{P} < 0,\ \forall i = 1,\dots,q \tag{2}$$

are satisfied. Then the CQLF is $\mathbf{V}(\mathbf{x}) = \mathbf{x}^T\mathbf{P}\mathbf{x}$.

Solving Common Lyapunov Function Problem by QE: We consider two Hurwitz stable continuous-time linear time-invariant systems, i.e., the case $n = 2$ and $q = 2$ in (1):

$$\dot{\mathbf{x}} = \mathbf{A}_{ci}\mathbf{x},\ \mathbf{x} \in \mathbb{R}^2,\ \mathbf{A}_{ci} \in \mathbb{R}^{2\times 2},\ i = 1, 2. \tag{3}$$

Moreover let $\mathbf{A}_{c1}, \mathbf{A}_{c2}$ be

$$\mathbf{A}_{c1} = \begin{bmatrix} x & 0 \\ y & -1 \end{bmatrix},\ \mathbf{A}_{c2} = \begin{bmatrix} 0 & 1 \\ -1 & -2 \end{bmatrix}, \tag{4}$$

respectively, where $x, y \in \mathbb{R}$ are system parameters. We have the following theorem.

Theorem 1. *(R.N.Shorten et.al. [14]) A necessary and sufficient condition for the two second-order systems (3) to have a CQLF is*

$$Re(\lambda(co(\mathbf{A}_{c1}, \mathbf{A}_{c2}))) < 0, \tag{5}$$

$$Re(\lambda(co(\mathbf{A}_{c1}, \mathbf{A}_{c2}^{-1}))) < 0, \tag{6}$$

where co(.) denotes the convex hull (polytope) of matrices: $co(\mathbf{X}, \mathbf{Y}) = \{\alpha\mathbf{X} + (1-\alpha)\mathbf{Y} : \alpha \in [0, 1]\}$, $\lambda(\mathbf{X})$ denotes the eigenvalues of matrix $\mathbf{X}$ and Re(.) denotes the real part of a complex number.

This implies that our desired condition is that all roots of characteristic polynomials of $\mathrm{co}(\mathbf{A}_{c1}, \mathbf{A}_{c2})$ and $\mathrm{co}(\mathbf{A}_{c1}, \mathbf{A}_{c2}^{-1})$ locate within a left half of the Gaussian plane for $\alpha \in [0, 1]$. The conditions can be reduced to a set of polynomial inequalities by using the well-known Liènard-Chipart criterion. Then we can apply QE for the polynomial inequalities.

Now we compute feasible regions of x, y so that the systems (3) have a CQLF by QE. As mentioned above, by applying the Liènard-Chipart criterion to characteristic polynomials of $\mathrm{co}(\mathbf{A}_{c1}, \mathbf{A}_{c2})$ and $\mathrm{co}(\mathbf{A}_{c1}, \mathbf{A}_{c2}^{-1})$, the CQLF existence condition can be described by the following formula:

$$\begin{aligned} \forall\alpha\ \ ((0 \le \alpha \le 1) \Rightarrow (&2 - \alpha - x\alpha > 0\ \wedge\ 1 + y\alpha - \alpha^2 - x\alpha^2 - y\alpha^2 > 0\ \wedge \\ &1 - 2\alpha - 2x\alpha - y\alpha + \alpha^2 + x\alpha^2 + y\alpha^2 > 0)). \end{aligned} \tag{7}$$

Applying CAD-based QE in SyNRAC to (7), the output turns out to be

$$(x < 0\ \wedge\ -2 - 2\sqrt{-x} < y < 2\sqrt{-x} - 2x). \tag{8}$$

5 Conclusion

We have presented a newly developed functions in Maple-package SyNRAC. The current version of SyNRAC provides quantifier elimination by virtual substitution up to quadratic formulas and CAD-based QE, as well as some standard simplifiers. The new features greatly extend the applicability and tractability of SyNRAC for solving real algebraic constraints in engineering. As an application of SyNRAC, we have treated a common Lyapunov function problem.

We proceed to implement other known QE algorithms and improve them, and are setting about developing symbolic-numeric algorithms. We also plan to develop a toolbox for parametric robust control design on MATLAB using SyNRAC as a core engine.

References

1. Anai, H., Yanami, H.: SyNRAC: A Maple-package for solving real algebraic constraints. In: Proceedings of International Workshop on Computer Algebra Systems and their Applications (CASA) 2003 (Saint Petersburg, Russian Federation), P.M.A. Sloot et al. (Eds.): ICCS 2003, LNCS 2657, Springer (2003) 828–837
2. Anai, H., Hara, S.: A parameter space approach for fixed-order robust controller synthesis by symbolic computation. In: Proceedings of IFAC World Congress on Automatic Control b'02. (2002)
3. Anai, H., Yanami, H., Hara, S.: SyNRAC: a maple-package for solving real algebraic constraints toward a robust parametric control toolbox. In: Proceedings of SICE Annual Conference 2003 (Fukui, Japan). (2003) 1716–1721
4. Weispfenning, V.: Simulation and optimization by quantifier elimination. Journal of Symbolic Computation **24** (1997) 189–208 Special issue on applications of quantifier elimination.
5. Collins, G.E.: Quantifier elimination for real closed fields by cylindrical algebraic decomposition. In Caviness, B., Johnson, J., eds.: Quantifier Elimination and Cylindrical Algebraic Decomposition. Texts and Monographs in Symbolic Computation. Springer, Wien, New York (1998) 85–121
6. Hong, H.: An improvement of the projection operator in cylindrical algebraic decomposition. In Caviness, B., Johnson, J., eds.: Quantifier Elimination and Cylindrical Algebraic Decomposition. Texts and Monographs in Symbolic Computation. Springer, Wien, New York (1998) 166–173
7. McCallum, S.: An improved projection operation for cylindrical algebraic decomposition. In Caviness, B., Johnson, J., eds.: Quantifier Elimination and Cylindrical Algebraic Decomposition. Texts and Monographs in Symbolic Computation. Springer, Wien, New York (1998) 242–268
8. Brown, C.W.: Improved projection for cylindrical algebraic decomposition. Journal of Symbolic Computation **32** (2001) 447–465
9. Anai, H., Yanami, H.: SyNRAC: a Maple-package for solving real algebraic constraints. In: Proceedings of International Conferences on Computational Science. Volume 2657 of LNCS., Springer (2003) 828–837
10. Weispfenning, V.: The complexity of linear problems in fields. Journal of Symbolic Computation **5** (1988) 3–27

11. Weispfenning, V.: Quantifier elimination for real algebra—the quadratic case and beyond. Applicable Algebra in Engineering Communication and Computing **8** (1997) 85–101
12. Loos, R., Weispfenning, V.: Applying linear quantifier elimination. The Computer Journal **36** (1993) 450–462 Special issue on computational quantifier elimination.
13. Shorten, R.N., Narendra, K.S.: Necessary and sufficient conditions for the existence of a common quadratic lyapunov function for m stable second order linear time-invariant systems. Proc. of ACC (2000) 359–363
14. Shorten, R.N., Narendra, K.S.: Necessary and sufficient conditions for the existence of a common quadratic lyapunov function for two stable second order linear time-invariant systems. Proc. of ACC (1999) 1410–1414
15. Mori, Y.: Investigation on common lyapunov function: Toward complete analysis. Kyoto Institute of Technology, Doctoral Thesis, (in Japanese) (2002)
16. Nguyen, T.V.: Common lyapunov function problem: Quadratic and infinity-norm functions. Kyoto Institute of Technology, Master Thesis (2003)

A LiE Subroutine for Computing Prehomogeneous Spaces Associated with Complex Nilpotent Orbits

Steven Glenn Jackson and Alfred G. Noël

Department of Mathematics,
University of Massachusetts,
Boston, MA 02125-3393, USA
{jackson, anoel}@math.umb.edu

Abstract. We develop a LiE subroutine to compute the irreducible components of certain prehomogeneous spaces that are associated with complex nilpotent orbits. An understanding of these spaces is necessary for solving more general problems in the theory of nilpotent orbits and the representation theory of Lie groups. The output is a set of LaTeX statements that can be compiled in a LaTeX environment in order to produce tables. Although the algorithm is used to solve the problem in the case of exceptional complex reductive Lie groups [2], it does describe these prehomogeneous spaces for the classical cases also. Complete tables for the exceptional groups can be found at
http://www.math.umb.edu/~anoel/publications/tables/.

1 Introduction

It has been long established that the representation theory of Lie groups is of fundamental importance to physics, chemistry and mathematics. However it is less known to scientists that this theory is currently being applied in many engineering applications, in computer science and in the development of financial mathematics. The theory is still being developed by mathematicians who are more and more participating in interdisciplinary projects. Consequently, progress in the representation theory of groups should be of interest to scientists and engineers working at the cutting edges of their fields.

The representation theory of semisimple Lie groups (and general reductive Lie groups) is especially important and challenging. Mathematicians have developed a large number of techniques to create different representations of a given Lie group. An understanding of the nilpotent orbits of such groups would provide important information about their possible representations. In the study and classification of such orbits the theory of prehomogeneous spaces plays an important role.

Definition 1 (M. Sato [5]). *Let G be a connected semisimple complex Lie group, V a finite dimensional vector space over $\mathbb{C}$ and ρ a rational representation of G in V. Then the triple (G, ρ, V) is called a* prehomogeneous vector space *if*

V.S. Sunderam et al. (Eds.): ICCS 2005, LNCS 3516, pp. 611–618, 2005.

G admits a Zariski open dense orbit Ω in V. If ρ is irreducible then (G, ρ, V) is said to be irreducible.

We often write (G, V) instead of (G, ρ, V) and gx instead of $(\rho(g))(x)$.

Let $\mathfrak{g}$ be a semisimple complex Lie algebra and G its adjoint group. It is a fact that the number of nilpotent orbits of G in $\mathfrak{g}$ is finite. Let X be a representative of a nilpotent orbit $\mathcal{O}$ of G in $\mathfrak{g}$. Then from the Jacobson-Morozov theorem X can be embedded into a triple (H, X, Y) with H semisimple and Y nilpotent such that

$$[H, X] = 2X, \quad [H, Y] = -2Y, \quad [X, Y] = H.$$

Such an (H, X, Y) is called an $\mathfrak{sl}_2$*-triple*. Given such a triple, the action of ad_H determines a grading $\mathfrak{g} = \bigoplus_{i \in \mathbb{Z}} \mathfrak{g}_i$, where $\mathfrak{g}_i = \{Z \in \mathfrak{g} : [H, Z] = iZ\}$.

It is a fact that $\mathfrak{g}_0$ is a reductive Lie subalgebra of $\mathfrak{g}$. Let G_0 be the connected subgroup of G such that $\mathrm{Lie}(G_0) = \mathfrak{g}_0$. In 1959, Kostant [4] showed that $(G_0, \mathfrak{g}_2)$ is a prehomogeneous vector space. Later, Vinberg [7] generalized Kostant's result by showing that all $(G_0, \mathfrak{g}_i)$ pairs are prehomogeneous vector spaces.

Our goal is to describe the irreducible components of the G_0-modules $\mathfrak{g}_i$ for all nilpotent orbits of the Lie group G in $\mathfrak{g}$.

It is enough to consider the simple complex Lie algebras. We have developed a general theory for classical complex simple Lie groups by exploiting the fact that their nilpotent orbits are parametrized by certain partitions of an integer related to their rank. Such parametrization is not available for exceptional groups. As a result our subroutine is needed in order to solve the problem in the cases where G is of exceptional type. As a byproduct we obtain another method of computing these prehomogeneous spaces for any fixed simple complex classical Lie group. All these results are found in [2].

In order to give the reader the flavor of the general result we provide a complete description of the result in the case of the general linear group.

2 Type GL_n

Let GL_n be the set of $n \times n$ complex invertible matrices. Then $\mathfrak{g} = \mathfrak{gl}_n$, the set of all $n \times n$ complex matrices. It is known that the nilpotent orbits of the group $G = GL_n$ on $\mathfrak{g}$ are parametrized by partitions of n. Therefore, if X is a nilpotent element of $\mathfrak{g}$ then up to conjugation we can take X as sum of Jordan blocks of size d_i where the d_i's are parts of the partition $\mathbf{d} = [d_1, \ldots, d_k]$ representing the class of X. For example if $n = 8$ and $\mathbf{d} = [3, 3, 2]$ then

$$X = \left(\begin{array}{ccc|ccc|cc} 0 & 1 & 0 & 0 & 0 & 0 & 0 & 0 \\ 0 & 0 & 1 & 0 & 0 & 0 & 0 & 0 \\ 0 & 0 & 0 & 0 & 0 & 0 & 0 & 0 \\ \hline 0 & 0 & 0 & 0 & 1 & 0 & 0 & 0 \\ 0 & 0 & 0 & 0 & 0 & 1 & 0 & 0 \\ 0 & 0 & 0 & 0 & 0 & 0 & 0 & 0 \\ \hline 0 & 0 & 0 & 0 & 0 & 0 & 0 & 1 \\ 0 & 0 & 0 & 0 & 0 & 0 & 0 & 0 \end{array}\right).$$

The Jacobson-Morozov theorem tells us that we can embed X in a triple (H, X, Y). In this case we can choose H to be the following diagonal matrix of trace zero:

$$H = \mathrm{diag}(d_1 - 1, d_1 - 3, \ldots, -d_1 + 3, -d_1 + 1, \ldots, d_k - 1, \ldots, -d_k + 1)$$

Let $V = \mathbb{C}^n$ be the standard representation of $\mathfrak{g}$. Since H is semisimple we must have $V = \bigoplus_{i \in \mathbb{Z}} V_i$, where V_i is the H-eigenspace of eigenvalue i.

Theorem 1. *Maintaining the above notation,* $G_0 \simeq \prod GL(V_i)$ *and each* G_0-*module* $\mathfrak{g}_l \simeq \bigoplus_{i-j=l} V_i \otimes V_j^*$.

Proof. Let G^H be the centralizer of H in G. Then G^H preserves each eigenspace V_i of H. Hence $G^H \simeq \prod GL(V_i)$. But under the adjoint action $G_0 = G^H$. Let $\mathfrak{g}_0 \simeq \bigoplus \mathfrak{gl}(V_i)$ be the Lie algebra of G_0; then as a $\mathfrak{g}_0$-module we can identify $\mathfrak{g}$ with $V \otimes V^*$, which is equivalent to $\bigoplus_{i,j} V_i \otimes V_j^*$. Moreover, as $\mathfrak{g}_0$-module, $\mathfrak{g} \simeq \bigoplus \mathfrak{g}_l$. It follows that each $\mathfrak{g}_0$-module $\mathfrak{g}_l$ is identified exactly with $\bigoplus_{i-j=l} V_i \otimes V_j^*$ since ad_H is a derivation. □

Here is an example. Let X be the representative of $\mathbf{d} = [3, 3, 2]$ given above. Then

$$H = \mathrm{diag}(2, 0, -2, 2, 0, -2, 1, -1).$$

Since $\dim V_2 = \dim V_0 = \dim V_{-2} = 2$ and $\dim V_1 = \dim V_{-1} = 1$ we have

$$G_0 = GL_2 \times GL_2 \times GL_2 \times GL_1 \times GL_1$$

while

$$\begin{aligned}
\mathfrak{g}_1 &\simeq (V_2 \otimes V_1^*) \oplus (V_1 \otimes V_0^*) \oplus (V_0 \otimes V_{-1}^*) \oplus (V_{-1} \otimes V_{-2}^*), \\
\mathfrak{g}_2 &\simeq (V_2 \otimes V_0^*) \oplus (V_1 \otimes V_{-1}^*) \oplus (V_0 \otimes V_{-2}^*), \\
\mathfrak{g}_3 &\simeq (V_2 \otimes V_1^*) \oplus (V_1 \otimes V_{-2}^*) \\
\mathfrak{g}_4 &\simeq (V_2 \otimes V_{-2}^*).
\end{aligned}$$

3 Description of the Algorithm

Maintaining the above notations, let $\Delta = \{\alpha_1, \ldots, \alpha_l\}$ be the Bourbaki set of simple roots of $\mathfrak{g}$ and R^+ the positive root system generated by Δ. Then there is a one-to-one correspondence between nilpotent orbits of G in $\mathfrak{g}$ and a subset of Dynkin diagrams whose nodes are labeled with elements of the set $\{0, 1, 2\}$. This is the *Dynkin-Kostant classification* [4]. Algorithms for labeling the Dynkin diagrams are found in [1]. For example, if $\mathfrak{g} = \mathfrak{sl}_8$ (the set of 8×8 matrices of trace 0) and $\mathcal{O}$ is the nilpotent orbit corresponding to the partition $\mathbf{d} = [3, 3, 2]$ then $\mathcal{O}$ is associated with the following diagram:

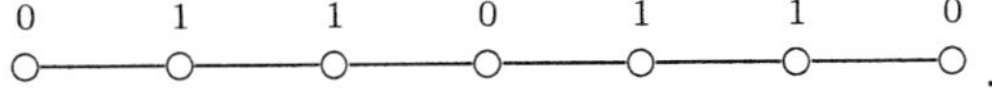

The labels represent the evaluations of the simple roots α_i on the diagonal matrix:

$$\hat{H} = \text{diag}(2, 2, 1, 0, 0, -1, -2, -2).$$

with $\alpha_i(\hat{H}) = \hat{h}_{i,i} - \hat{h}_{i+1,i+1}$. Moreover H is conjugate to $\hat{H}$ under the Weyl group S_8 (the symmetric group on eight letters). See [1] for more details. It follows that $\mathfrak{g}_i \simeq \bigoplus_{\alpha(\hat{H})=i} \mathfrak{g}^\alpha$ where $\mathfrak{g}^\alpha$ is the root space associated with the root α.

It is useful to observe that the type of the semisimple part of the reductive Lie group G_0 can be read from the diagram. It corresponds to the group generated by the roots labeled with zero. In the above example the semisimple part of G_0 is of type $A_1 \oplus A_1 \oplus A_1$.

Definition 2. *Let v be a non-zero element of the root space $\mathfrak{g}^\lambda \subseteq \mathfrak{g}_i$ and X_α a non-zero vector in any $\mathfrak{g}^\alpha \subseteq [\mathfrak{g}_0, \mathfrak{g}_0]$ with $\alpha > 0$. If $[X_\alpha, v] = 0$ for all such X_α's then we say that v is a* highest weight vector *of G_0 on $\mathfrak{g}_i$ and we call λ a* highest weight *of $\mathfrak{g}_i$.*

We shall describe an algorithm which computes the highest weights of $\mathfrak{g}_i$ given a labeling of the Dynkin diagram corresponding to the nilpotent orbit under consideration. The subroutine is written in the language LiE. Readers who are not familiar with LiE may consult [6]. The highest weights will be expressed in terms of the fundamental weights. For the reader's convenience all LiE generic functions will be written in boldface characters. The subroutine name is *irrdmodule()*. Its input consist of a labeling string (usually the Bala-Carter label attached to the nilpotent orbit), the Dynkin-Kostant labels and the type of the simple group. The subroutine returns a set of LaTeX statements that generate a table containing all the highest weights of all non-zero $\mathfrak{g}_i$ for $i > 0$.

3.1 Algorithm

```
irrdmodule(tex bala;vec label; grp g ) =
{
# Formatting statements to be used for exceptional groups
# bala: Bala-Carter notation of the nilpotent orbit
# label: Dynkin-Kostant label of the nilpotent orbit
# g: complex simple Lie algebra
grptype = " "; if g == G2 then grptype = "\gtwo";fi; if g == F4 then
grptype = "\ffour";fi; if g == E6 then grptype = "\esix";fi; if g == E7
then grptype = "\eseven";fi; if g == E8 then grptype = "\eeight";fi;
printflag = 0;
setdefault (g); n =n_pos_roots; l = Lie_rank; alpha = pos_roots;
if size(label) == l then
lookfordegree = 0; modcounter = n;
while modcounter > 0 do
degree = null(n); spaceofinterest = null(n,l);
```

```
# Compute degree positive roots
counter = 0; for i = 1 to n do for j = 1 to l do
degree[i] = degree[i] + label[j] * alpha[i][j];od;

# Build the g_i vector space specified by lookfordegree
if degree[i] == lookfordegree then counter = counter+1;
spaceofinterest[counter] = alpha[i]; fi;od;
space = null( counter, l);
for i = 1 to counter do space[i] = spaceofinterest[i]; od;
dimspace = counter;
modcounter = modcounter - dimspace;
if modcounter < 0  then print(" Negative modcounter"); break;fi;
if dimspace > 0 && lookfordegree > 0 then

# Set up the output LaTeX statement
latexline = " ";
if printflag == 1 then latexline = " \cline3-5 &"; fi;
if printflag == 0 then printflag = 1;
latexline = "\hline " + bala + " & " + grptype;
for i = 1 to l do
latexline = latexline+"{"+label[i] + "}" ; od; fi;
latexline = latexline + " & " + lookfordegree + " & " + dimspace +
" & \ba";

# Find all the highest weights of g_i
hw = all_one ( dimspace );
for i = 1 to l do
if label[i] == 0 then
for;j = 1 to dimspace do candidate = alpha[i] + space[j];

# Check if candidate is in space. If yes then it is not a highest weight
for k = 1 to dimspace do
if candidate == space [k] then hw[j] = 0; break;fi; od; od;fi; od;

# Compute highest weights
numhw = 0;
for i = 1 to dimspace do numhw = numhw + hw[i]; od;
rhighestweights = null(numhw,l);
counter = 0;
for i = 1 to dimspace do
if hw[i] == 1 then
counter = counter+1; rhighestweights[counter] = space[i]; fi; od;

# Convert to fundamental weights
if numhw> 0 then
highestweights =rhighestweights * Cartan;
for i = 1 to numhw do latexline = latexline + "(";
for j = 1 to l do
```

```
latexline = latexline + highestweights[i][j];
if j < l then latexline = latexline + "," fi; od;
latexline = latexline + ") \\ "; od;
latexline = latexline + "\ea \\"; fi;
print (latexline); fi;

# Process the next g_i vector space
lookfordegree = lookfordegree+1;
od;
else print ("label is incompatible with group type");
fi;
}
```

Example 1. We compute the highest weights of each of the four $\mathfrak{g}_i$ modules associated with the nilpotent orbit $\mathbf{d} = [3, 3, 2]$ when $\mathfrak{g} = \mathfrak{sl}_8$. Since $\mathfrak{g}_0$ is reductive we only need to consider the action of its semisimple part $[\mathfrak{g}_0, \mathfrak{g}_0]$. The subroutine is placed in a file called *phmod* which has to be loaded in the LiE environment. Here is the LiE session that produces Table 1:

```
LiE version 2.2 created on Nov 22 1997 at 16:50:29 Authors: Arjeh
M. Cohen, Marc van Leeuwen, Bert Lisser. Mac port by S. Grimm
Public distribution version

type '?help' for help information
type '?' for a list of help entries.

> read phmod
> irrdmodule("2,2,1,0,0,-1,-2,-2",[0,1,1,0,1,1,0],A7)
```

Observe that since the number of positive roots of $\mathfrak{sl}_8$ is 28 and the semisimple part of $\mathfrak{g}_0$ is of type $A_1 \oplus A_1 \oplus A_1$ the sum of the dimensions of the $\mathfrak{g}_i$'s is correct. By the "Theorem of the Highest Weight" [3–p. 279], $\mathfrak{g}_i$ is completely determined as $\mathfrak{g}_0$-module by its highest weights. Moreover, when interpreting the results given in the table, one should be aware that the action of the semisimple part of $\mathfrak{g}_0$ on $\mathfrak{g}_i$ is completely determined by those coefficients associated with the nodes of Dynkin-Kostant label 0; the other coefficients affect only the action of the center of $\mathfrak{g}_0$, which in any case must act by scalars on each irreducible component of $\mathfrak{g}_i$. Let us denote the seven fundamental weights of $\mathfrak{sl}_8$ by ω_i for $1 \leq i \leq 7$. Then (disregarding the action of the center) we see that

$$\mathfrak{g}_1 \simeq V^{\omega_1} \oplus V^{\omega_4} \oplus V^{\omega_4} \oplus V^{\omega_7}$$
$$\mathfrak{g}_2 \simeq V^0 \oplus V^{\omega_1+\omega_4} \oplus V^{\omega_4+\omega_7}$$
$$\mathfrak{g}_3 \simeq V^{\omega_1} \oplus V^{\omega_7}$$
$$\mathfrak{g}_4 \simeq V^{\omega_1+\omega_7}.$$

Table 1.

<table>
<tr><th colspan="5">Example in $\mathfrak{sl}_8$ revisited</th></tr>
<tr><th>Elementary divisors</th><th>Label</th><th>i</th><th>dim $\mathfrak{g}_i$</th><th>Highest weights of $\mathfrak{g}_i$</th></tr>
<tr><td rowspan="4">2, 2, 1, 0, 0, −1, −2, −2</td><td rowspan="4">0110110</td><td>1</td><td>8</td><td>(1, 1, −1, 0, 0, 0, 0)
(0, −1, 1, 1, −1, 0, 0)
(0, 0, −1, 1, 1, −1, 0)
(0, 0, 0, 0, −1, 1, 1)</td></tr>
<tr><td>2</td><td>9</td><td>(0, −1, 1, 0, 1, −1, 0)
(1, 0, 0, 1, −1, 0, 0)
(0, 0, −1, 1, 0, 0, 1)</td></tr>
<tr><td>3</td><td>4</td><td>(1, 0, 0, 0, 1, −1, 0)
(0, −1, 1, 0, 0, 0, 1)</td></tr>
<tr><td>4</td><td>4</td><td>(1, 0, 0, 0, 0, 0, 1)</td></tr>
</table>

where V^λ denotes the irreducible $[\mathfrak{g}_0, \mathfrak{g}_0]$-module of highest weight λ. We note that since $\mathfrak{g}_{-i}$ is dual to $\mathfrak{g}_i$, the module structure of $\mathfrak{g}_{-i}$ is easily determined from that of $\mathfrak{g}_i$.

The correctness of the algorithm should be evident to anyone familiar with the representation theory of complex reductive Lie groups. The computation of highest weights is accomplished by finding the set of vectors in each $\mathfrak{g}_i$ which are annihilated by all positive simple root spaces in $[\mathfrak{g}_0, \mathfrak{g}_0]$. Such root spaces correspond to the simple roots of Dynkin-Kostant label 0. Since $\mathfrak{g}_0$ acts by the adjoint action, the subroutine returns only the roots β such that $\mathfrak{g}^\beta$ lies in $\mathfrak{g}_i$ and $[X_\alpha, X_\beta] = 0$ for all positive α with $\mathfrak{g}^\alpha \subset [\mathfrak{g}_0, \mathfrak{g}_0]$. This is exactly the set of highest weights of $\mathfrak{g}_i$, which is recorded in the array *rhighestweights*. Finally, we express the highest weights in terms of the fundamental weights using the Cartan matrix of $\mathfrak{g}$.

A worst case analysis of the code reveals the algorithm is $\mathcal{O}(\text{rank}(\mathfrak{g}) \times (n\_pos\_roots)^3)$. This is due to the fact that the control variable of the outer **while** loop, *modcounter*, is bounded by the number of positive roots and that the more intensive internal **for** loop is at worst executed $(\text{rank}(\mathfrak{g}) \times (n\_pos\_roots)^2)$ times. Of course we are assuming that the LiE internal functions are very fast. From our experience we believe that they are optimal. On average the subroutine performs very well. We use it to compute the irreducible modules of the prehomogeneous spaces associated to all the nilpotent orbits of the exceptional complex simple groups. Complete tables for the exceptional groups can be found at http://www.math.umb.edu/~anoel/publications/tables/. For more information see [2]. The computations were carried on an IMac G4 with speed 1GHz and 1Gb SDRAM of memory.

4 Conclusion and Future Work

We presented a LiE implementation of a simple algorithm for computing the module structures of a large class of prehomogeneous spaces, namely those associated with nilpotent orbits in the adjoint representations of complex reductive Lie groups. In [2] we have developed general methods for the classical types, taking advantage of the parametrization of the nilpotent orbits by partitions. We developed the algorithm presented here to solve the problem for groups of exceptional type. The algorithm as implemented in LiE is stable and of a low complexity. Our implementation is clear, which makes analysis very easy. The correctness of the algorithm is a consequence of the well-known "Theorem of the Highest Weight." Several other authors have treated special cases of this problem; our results generalize theirs to a complete solution for complex reductive groups.

We are currently working on extending these results to non-compact real reductive Lie groups. In light of the Kostant-Sekiguchi correspondence, it suffices to consider the action of the subgroup $K_{\mathbb{C}} \subset G_{\mathbb{C}}$ whose real form is maximal compact in G on the complex symmetric space $G_{\mathbb{C}}/K_{\mathbb{C}}$. This work has already started and we have obtained preliminary results.

References

1. Collingwood, D. H. and McGovern, W. M. *Nilpotent orbits in semisimple Lie algebras.* Van Nostrand Reinhold Mathematics Series, New York (1992).
2. Jackson, S. G. and Noël, A. G. *Prehomogeneous spaces associated with complex nilpotent orbits.* To appear in Journal of Algebra.
3. Knapp, A. W. *Lie groups beyond an introduction.* Second edition, Birkhaüser Progress in Mathematics **140** (2002)
4. Kostant, B. *The principal three dimensional subgroup and the Betti numbers of a complex Lie group.* Amer. J. Math., Vol. 81, (1959) 973-1032.
5. Sato, M. and Kimura, T. *A classification of irreducible prehomogeneous vector spaces and their relative invariants.* Nagoya Math. J. 65 (1977), 1-155.
6. Van Leeuwen, M. A. A., Cohen, A. M., and Lisser, B. *LiE: A package for Lie group computations*, Computer Algebra Nederland, Amsterdam, Netherlands (1992)
7. Vinberg, E. B. *On the classification of the nilpotent elements of graded Lie algebras*, Doklady Academii Nauk SSSR 225 (1975b), 745-748 (Russian). English translation: Soviet Math. Doklady **16** (1975), 1517 - 1520.

Computing Valuation Popov Forms

Mark Giesbrecht[1], George Labahn[1], and Yang Zhang[2]

[1] School of Computer Science, University of Waterloo,
Waterloo, ON, N2L 3G1, Canada
{mwg, glabahn}@uwaterloo.ca

[2] Dept. of Mathematics and Computer Science, Brandon University,
Brandon, MB, R7A 6A9, Canada
zhangy@brandonu.ca

Abstract. Popov forms and weak Popov forms of matrices over non-commutative valuation domains are defined and discussed. Two new algorithms to construct these Popov forms are given, along with a description of some of their applications.

1 Introduction

Over the last decade, there have been significant advances in computing with symbolic systems of linear recurrence and differential equations. At the core of this new technology are computations with Ore polynomials, which provide a unified algebraic structure for these different domains. The Ore polynomials form a noncommutative polynomial ring which acts on a space of functions as a differential or difference (recurrence) operator. Matrices of such polynomials describe linear differential and differential systems of equations, which are clearly an important topic in computer algebra systems. Casting differential and recurrence operators as polynomial-like objects allows us to potentially employ the now very well developed and very efficient algorithms for matrices of polynomials in this new setting. This paper is a step towards efficient algorithms for such systems.

It is well known that the Hermite normal form of a polynomial matrix has degree bounds that can result in large row or column degree values. While this is not a disadvantage for some applications (such as linear solving) it can be a significant problem in such computations as matrix-polynomial GCDs. A second normal form, now known as the Popov normal form, was introduced by Popov [11] and has the property that row degrees are controlled. These normal forms have been successfully applied in such applications as control theory and linear systems (see for example, Kailath[8]).

Recently the Popov normal form and some variations (the shifted Popov form and the weak Popov form) have attracted considerable interest in the computer algebra community, for example, in Villard [13], Beckermann, Labahn and Villard [4], and Mulders and Storjohann [9]. While the normal forms previously mentioned all deal with matrices of commutative polynomials, Popov

V.S. Sunderam et al. (Eds.): ICCS 2005, LNCS 3516, pp. 619–626, 2005.

normal forms and their variations also exist for some matrices of noncommutative polynomials, particularly Ore polynomials. Ore domains were introduced in the 1930's as a unified way of capturing differential and recurrence operators (c.f., Ore [10], Jacobson [7]). In 1943 Jacobson [7] considered the Hermite and Smith forms of matrices over skew polynomial rings. In the computer algebra area, Abramov and Bronstein [1] gave a method to calculate the ranks of matrices over skew polynomial rings using a type of normal form reduction of these matrices. Recently Beckermann, Cheng and Labahn [3] used fraction-free methods to compute weak Popov form of matrices over skew polynomial rings.

In this paper, we consider the more general problem of defining and computing Popov and weak Popov forms for matrices over *valuation rings*. The motivation for this question comes from several directions: Mulders and Storjohann [9] extended the notion of weak Popov forms to the discrete valuation rings, but failed to define Popov forms. Meanwhile, as they point out, their methods have the problem that they do not necessarily terminate. We give a definition of Popov forms of matrices over valuation Ore domains, and provide an algorithm for such forms which is guaranteed to terminate. Our methods lead to a unified treatment of many polynomial rings currently studied in Computer Algebra, for example, commutative polynomial rings ([13], [4] and [9]), skew polynomial rings ([3] and [1]) and multivariate polynomials ([6] and example 4). Finally, we remark that our method can be used to extend the work on reduced bases for matrices over commutative valuations rings by von zur Gathen [14] to some noncommutative cases ([6]).

It is worth noting that in many cases valuation methods are quite different from term ordering methods, which are the usual methods to deal with polynomials. When using term ordering methods, one only considers the leading terms at each reduction step. Valuation methods usually consider the whole polynomial. In fact, any term ordering induces a valuation but the converse is not true.

2 Valuations on Ore Domains

Noncommutative valuation theory dates back to the 1930's with the work of Artin, Noether and others. Presentations can be found in Schilling [12] and later in Artin [2], with a more recent treatment given in Cohn [5]. In this paper, we restrict such noncommutative valuations to the more practically motivated Ore domain case.

Let us first recall some definitions and discuss some properties of valuations on skew fields. Some results have already been discussed in different settings for commutative cases.

Definition 1. *An ordered group Γ is a (not necessarily commutative) group with a total ordering $\alpha \geq \beta$, which is preserved by the group operation:*

$$\alpha \geq \beta,\ \alpha' \geq \beta' \Rightarrow \alpha + \alpha' \geq \beta + \beta' \quad \textit{for all } \alpha, \alpha', \beta, \beta' \in \Gamma.$$

Usually, Γ will be augmented by a symbol ∞ to form a monoid with the operation

$$\alpha + \infty = \infty + \alpha = \infty + \infty = \infty \quad \text{for all } \alpha \in \Gamma,$$

and the ordering $\infty > \alpha$ for all $\alpha \in \Gamma$.

Definition 2. *Let R be a ring. Then a valuation on R with values in an ordered group Γ (the value group) is a function ν on R with values in $\Gamma \cup \{\infty\}$ which satisfies:*

(V.1) $\nu(a) \in \Gamma \cup \{\infty\}$ and ν assumes at least two values,
(V.2) $\nu(ab) = \nu(a) + \nu(b)$, and
(V.3) $\nu(a + b) \geq min\{\nu(a), \nu(b)\}$, for every pair of elements $a, b \in R$.

Lemma 1. *([5]) A valuation ν on a ring R has the following properties:*

(a) $\nu(0) = \infty$ so that $\nu(0) > \gamma$ for all $\gamma \in \Gamma$ (by (V.2)).
(b) ker $\nu = \{a \in R \mid \nu(a) = \infty\}$ is a proper ideal of R (by (V.1)). Therefore, ker $\nu = 0$ if R is a skew field or a simple ring. Furthermore, $R/$ker ν is an integral domain.
(c) $\nu(1) = 0$ and $\nu(a) = \nu(-a)$.
(d) If $\nu(a) \neq \nu(b)$, then $\nu(a + b) = min\{\nu(a), \nu(b)\}$. □

Remark 1. Note that ker ν is different from the set $\{a \in R \mid \nu(a) = 0\}$.

In the case of a skew field K with valuation ν, there are two important properties of the valuation ring $V := \{a \in K \mid \nu(a) \geq 0\}$. First, the valuation ring is *total*: for every $k \in K$, either $k \in V$, or $k \neq 0$ and $k^{-1} \in V$. Second, the valuation ring is *invariant*: for every $a \in V$ and $k \in K \setminus \{0\}, k^{-1}ak \in V$. Usually a total invariant subring of a skew field is called a *valuation ring.* Conversely, given any skew field K and a valuation ring V in K, one can form a valuation which gives rise to V. Moreover we also have the following properties:

Lemma 2. *With our previous notation we have:*

(a) Given $a, b \in V$, then a is a left (and right) multiple of b if and only if $\nu(a) \geq \nu(b)$.
(b) The units set of V is $U = \{a \in V \mid \nu(a) = 0\}$. Furthermore, the non-units of V is $M = \{a \in V \mid \nu(a) > 0\}$, a maximal ideal of V.
  (i) $\nu(a) \geq \nu(b) \Leftrightarrow a = rb = br'$ for some $r, r' \in V$.
  (ii) $\nu(a) = \nu(b) \Leftrightarrow a = rb = br'$ for some $r, r' \in U$.
  (iii) $\nu(a) > \nu(b) \Leftrightarrow a = rb = br'$ for some $r, r' \in M$.

Proof.

(a) Assume that for some $r, s \in V$, we have $a = br = sb$. Then $\nu(a) = \nu(b) + \nu(r) = \nu(s) + \nu(b) \geq \nu(b)$ follows from the definition, $\nu(s) \geq 0$ and $\nu(r) \geq 0$. Conversely, if $b = 0$, then $\nu(b) = \infty$ implies $\nu(a) = \infty$. Therefore $a = 0$ so that a is both a left and right multiple of b. Now we can assume that $\nu(a) \geq \nu(b), b \neq 0$. This implies that $\nu(ab^{-1}) \geq 0, \nu(b^{-1}a) \geq 0$, that is, both $ab^{-1} \in V$ and $b^{-1}a \in V$. Therefore $a = (ab^{-1})b = b(b^{-1}a)$.

(b) Part (i) follows by (a). For (ii) it is easy to see that $\nu(r) = 0$ by the same proof as in (a). Therefore $r \in U$ and part (iii) follows from the definition of M. □

Example 1. (Cohn [5]) Let K be a skew field with a valuation ν and an automorphism σ of K such that $\nu(a^\sigma) = \nu(a)$ for all $a \in K$. Let γ be in the value group Γ (or in an ordered extension of Γ) and define a valuation on the skew polynomial $K[x;\sigma]$ by the rule

$$w(\sum a_i x^i) = min_i\{i\gamma + \nu(a_i)\}.$$

Then w is a valuation on $K[x;\sigma]$ and can be uniquely extended to a valuation of the function field $K(x;\sigma)$. If the residue-class skew field of K under ν is $\bar{K}$ and the automorphism induced on $\bar{K}$ by σ is $\bar{\sigma}$, then the residue-class field of $K(x;\sigma)$ is either $\bar{K}(x;\sigma^{-j})$ (when $j\gamma$ is the least multiple of γ which lies in Γ, that is, $j = 1$ if $\gamma \in \Gamma$ or there may exist a j such that $j\gamma \in \Gamma$ if $\gamma \notin \Gamma$) or $\bar{K}$ (when no multiple of γ lies in Γ).

The following lemma is well-known in the commutative case and gives the division rule in the sense of valuations. Now we extend it to the noncommutative case.

Lemma 3. *Let ν be a valuation on a skew field K with the valuation ring V, S be a subskew field of K that lies in V and $M = \{a \in V \mid \nu(a) > 0\}$. Assume that S maps isomorphically to the residue class skew field of ν, or equivalently, that $V = S \oplus M$ as abelian group. If $0 \neq a, b \in K$ such that $\nu(a) \geq \nu(b)$, then there exists a unique $s \in S$ such that either $a = sb$ or $\nu(a - sb) > \nu(b)$. Equivalently, there exists a unique $s \in S$ such that $a - sb = mb$ for some $m \in M$.*

Proof. From Lemma 2, $\nu(a) \geq \nu(b)$ implies that $a = rb$ for some $r \in V$. Since $V = S \oplus M$ there exists a unique $s \in S$ such that $r = s + m$ for some $m \in M$. Thus $a = rb = (s + m)b = sb + mb$, and so $a - sb = mb$. Clearly, $m = 0$ if and only if $a = sb$. If $m \neq 0$, then by Lemma 2 we have $\nu(mb) > \nu(b)$ and so $\nu(a - sb) > \nu(b)$. □

Definition 3. *Let ν be a valuation on a skew field K with the valuation ring V. For $a, b \in V$, we say that b right divides a with respect to ν, denoted $b \mid_\nu a$, if there exists $d \in V$ such that $\nu(a) = \nu(db)$. The element d is called a right valuation quotient of a by b with respect to ν if $a = db$, or if $a \neq db$ and $\nu(a - db) > \nu(a)$.*

It is easy to prove the following lemma.

Lemma 4. *$b \mid_\nu a$ if and only if there exists a right valuation quotient of a by b.*

In the remainder of this paper we will assume that ν is a valuation on a skew field K with the valuation ring V.

3 Valuation Popov Forms

In this section we define Popov and weak Popov forms for matrices over valuation domains. Unlike some papers discussing (weak) Popov forms, we consider the row vectors as a left module over a base ring and define the Popov form by two kinds of reductions. In this way the termination properties of reductions can be easily obtained, and one can use previously existing algorithms to compute these forms.

Definition 4. *Let* $\boldsymbol{v} = (v_1, \cdots, v_m) \in V^m$. *We define the pivot element* $piv(\boldsymbol{v}) \in \{v_1, \cdots, v_m\}$ *as the rightmost element with minimum valuation in* $\boldsymbol{v}$.

Next we define two kinds of reductions which are used for constructing the weak Popov form and the Popov form.

Definition 5. *Given* $\boldsymbol{a}, \boldsymbol{b}, \mathbf{c} \in V^m$, *we say that*

(a) $\boldsymbol{a}$ *weakly reduces to* $\mathbf{c}$ *modulo* $\boldsymbol{b}$ *in one step with respect to* ν *if and only if* $piv(\boldsymbol{b})$ *and* $piv(\boldsymbol{a})$ *have the same index,* $piv(\boldsymbol{b}) \mid_\nu piv(\boldsymbol{a})$ *and* $\mathbf{c} = \boldsymbol{a} - q_1\boldsymbol{b}$, *where* $q_1 \in V$ *is a valuation quotient of* $piv(\boldsymbol{a})$ *by* $piv(\boldsymbol{b})$.
(b) $\boldsymbol{a}$ *reduces to* $\mathbf{c}$ *modulo* $\boldsymbol{b}$ *in one step with respect to* ν *if and only if* $piv(\boldsymbol{b}) \mid_\nu d$, *where* d *is an entry that appears in* $\boldsymbol{a}$ *with the same index of* $piv(\boldsymbol{b})$ *and* $\mathbf{c} = \boldsymbol{a} - q_2\boldsymbol{b}$, *where* $q_2 \in V$ *is a valuation quotient of* d *by* $piv(\boldsymbol{b})$.

Definition 6. *A nonzero vector* $\boldsymbol{a}$ *in* V^m *is (resp. weakly) reduced with respect to a set* $S = \{\boldsymbol{s}_1, \ldots, \boldsymbol{s}_l\}$ *of nonzero vectors in* V^m *if* $\boldsymbol{a}$ *cannot be (resp. weakly) reduced modulo any one of* $piv(\boldsymbol{s}_i)$ *with respect to* ν, $i = 1, \ldots, l$.

Furthermore, a set $S = \{\boldsymbol{s}_1, \ldots, \boldsymbol{s}_l\}$ *of vectors is called a (resp. weakly) reduced set if any vector* $\boldsymbol{s}_i$ *is (weakly) reduced with respect to* $S \setminus \{\boldsymbol{s}_i\}$.

Definition 7. *Let* $\gamma = \{\boldsymbol{r}_1, \ldots, \boldsymbol{r}_n\}$ *be the set of row vectors of matrix* $V^{n\times m}$.

(a) $V^{n\times m}$ *is called in weak Popov form if* γ *is a weakly reduced set.*
(b) $V^{n\times m}$ *is called in Popov form if*

(i) γ *is a reduced set.*
(ii) Rows are in an ascending chain with respect to ν, *that is,* $\nu(piv(\boldsymbol{r}_1)) \leq \nu(piv(\boldsymbol{r}_2)) \leq \cdots \leq \nu(piv(\boldsymbol{r}_n))$.

Note that our definition is the same as the usual one if we choose V as a commutative polynomial ring with degree valuation. A Popov form is a weak Popov form.

The following algorithm gives a method to construct a weak Popov form and also a Popov form.

Algorithm: `(resp. Weak) Popov form for` $V^{n\times m}$

Input: ▸ Row vectors $\boldsymbol{r}_1, \ldots, \boldsymbol{r}_n$ of matrix $A \in V^{n\times m}$;

Output: ▸ Row vectors $\boldsymbol{p}_1, \ldots, \boldsymbol{p}_n$ of a (resp. weak) Popov form of $A \in V^{n\times m}$;

Initialization: $\boldsymbol{p}_1 := 0, \ldots, \boldsymbol{p}_n := 0$;

(1) Check if $\{\boldsymbol{r}_1, \ldots, \boldsymbol{r}_n\}$ is in a (resp. weak) Popov form;

(2) If not

(3) Swap rows of A to make them into a ascending chain w.r.t. ν;

(4) For $i = 2$ to $i = n$ do

(5) If $\boldsymbol{r}_i$ is (resp. weakly) reducible modulo $\boldsymbol{r}_1$,

(6) $\boldsymbol{r}_i := \boldsymbol{r}_i - q_i \boldsymbol{r}_1$, where q_i is the valuation quotient;

(8) end if

(9) goto (1);

(10) else return: $\boldsymbol{p}_1 := \boldsymbol{r}_1, \ldots, \boldsymbol{p}_n := \boldsymbol{r}_n$;

Theorem 1. *If $\nu(V)$ is well-defined, then the preceding two algorithms are correct.*

Proof. We only need to prove the algorithms terminate. Note that every reduction increases the valuation, and forms an ascending chain. Therefore the algorithms will terminate after a finite number of steps since $\nu(V)$ is well-defined. □

The following gives a second method to construct (resp. weak) Popov forms:

Algorithm: `Constructing (resp. weak) Popov form for` $V^{n\times m}$

Input: ▸ row vectors $\gamma := \{\boldsymbol{r}_1, \ldots, \boldsymbol{r}_n\}$ of a matrix $A \in V^{n\times m}$;

Output: ▸ row vectors $\rho := \{\boldsymbol{p}_1, \ldots, \boldsymbol{p}_n\}$ of (resp. weak) Popov form of $A \in V^{n\times m}$;

(1) Initialization: $\boldsymbol{p}_1 := 0, \ldots, \boldsymbol{p}_n := 0$;

(2) For $i = 1$ to $i = n$ do

(3) $\boldsymbol{p}_i := \boldsymbol{r}_i$ (resp. weakly) reduced modulo $\gamma := \{\gamma \setminus \{\boldsymbol{r}_i\}\} \cup \{\boldsymbol{p}_i\}$;
end do;

(4) Swap rows such that they are in an ascending chain with respect to ν if Popov form required;

Theorem 2. *If $\nu(V)$ is well-defined, then the above algorithm is correct.*

Proof. The above algorithm terminates after a finite number of steps since $\nu(V)$ being well-defined implies that the steps in (3) terminate.

We only need to check that ρ is a (weakly) reduced set with respect to ν. The first two steps produce $\rho = \{\boldsymbol{p}_1, \boldsymbol{r}_2, \cdots, \boldsymbol{r}_n\}$ and $\rho = \{\boldsymbol{p}_1, \boldsymbol{p}_2, \boldsymbol{r}_3, \cdots, \boldsymbol{r}_n\}$. Therefore we have to prove $\boldsymbol{p}_1$ cannot be (weakly) reduced modulo $\boldsymbol{p}_2$. If this was the case, then $\boldsymbol{p}_1$ could be (weakly) reduced modulo $\boldsymbol{r}_2$, a contradiction. By induction, it is easy to prove that ρ is a (weakly) reduced set. □

4 Applications and Examples of Valuation Popov Forms

In the rest of this paper we mention some applications of valuation Popov forms. First, recall that the definition of the rank of a matrix over an Ore domain is different from the usual definition of rank (for example, see [5]). Given a matrix $A \in V^{n\times m}$, let M be the row space of A and M' be the reduced form with respect to (weak) reduction. From the previous algorithms, M and M' generate the same row space. Therefore we have the following.

Proposition 1. *The rank of M generated by the row space of $A \in V^{n\times m}$ is invariant under (weak) reduction with respect to any valuation.*

Proposition 2. *The rank of a (weak) Popov form equals its number of nonzero rows.*

Proof. This follows from the row set of (weak) Popov form being a (weakly) reduced set. □

Example 2. The following example given in [9] is used to show that their procedures do not necessarily terminate. Using our algorithms one can easily obtain a Popov form as follows.

Let A be a matrix over a power series $F[[x]]$, where F is a field, that is,

$$A = \begin{bmatrix} x + x^2 + x^3 + \cdots \\ 1 \end{bmatrix} \in F[[x]]^{2\times 1}.$$

Assume that ν is the usual valuation on the power series domain $F[[x]]$. Then $\nu(x + x^2 + x^3 + \cdots) = 1$ and $\nu(1) = 0$. Therefore the valuation quotient is $x + x^2 + x^3 + \cdots$. Note that $\nu(0) = \infty$. Our algorithm then computes the Popov form as $\begin{bmatrix} 1 \\ 0 \end{bmatrix}$.

Example 3. Different valuations on polynomial rings may induce different Popov forms. For example, let R be a ring with a discrete valuation function $\nu(x)$. To extend the valuation $\nu(x)$ from R to the polynomial ring $R[t]$ we define, for an element $f[t] = a_0 + a_1 t + \cdots + a_n t^n$,

$$\nu(f[t]) = \min_i\{\nu(a_i) + i\} \tag{1}$$

It is well-known that ν defines a valuation function on $R[t]$ which extends $\nu(x)$. The algorithm for valuation Popov forms produces a Popov form which is different from the ones discussed in [4], [8] and [9].

Example 4. In some cases our algorithms can be used for multivariate domains. For example, let L be a finite dimensional Lie algebra over a commutative field K and let $U(L)$ be its universal envelope. It is well-known that $U(L)$ has a filtration

$$U^{-1} = 0;\ U^i = K + L + L^2 + \cdots + L^i$$

whose associated graded algebra is a polynomial algebra. This filtration then defines a valuation function $\nu(x)$ by the following rule:

$$\nu(0) = \infty, \quad \nu(x) = -i \text{ if } x \in U^i \setminus U^{i-1}. \tag{2}$$

Therefore, given any matrix over $U(L)$, we can use our algorithms to get Popov and weak Popov forms, and hence also easily obtain ranks.

5 Conclusion

In this paper we discuss some properties of valuations in noncommutative domains and define reductions for vectors in terms of valuations. For matrices over Ore domains we define and describe Popov and weak Popov forms in terms of valuations. Algorithms to construct these Popov forms are given and are shown to terminate. Further discussion and properties of these forms will appear in a forthcoming paper [6].

Acknowledgements

The authors are grateful to the anonymous referees for their helpful comments.

References

1. Abramov, S., Bronstein, M.: Linear Algebra for skew-polynomial matrices. INRIA (preprint)
2. Artin, E.: Geometric Algebra. Interscience Publishers, Inc. 1957
3. Beckermann, B., Cheng, H., Labahn, G.: Fraction-free row reduction of matrices of Ore polynomials. Proceedings of ISSAC'02, ACM Press, New York, 8-15
4. Beckermann, B., Labahn, G., Villard, G.: Shifted Normal Forms of Polynomial Matrices. Proceedings of ISSAC'99, ACM Press, ACM Press, New York, 189-196
5. Cohn, P. M.: Skew Fields. Cambridge University Press, 1995
6. Giesbrecht, M., Labahn, G., Zhang, Y.: Popov forms for multivariate polynomial matrices. (Preprint)
7. Jacobson, N.: The Theory of Rings. American Math. Soc., New York, 1943
8. Kailath, T.: Linear Systems. Prentice Hall, 1980
9. Mulders, T., Storjohann, A.: On lattice reduction for polynomial matrices. Journal of Symbolic Computation 35(2003), 377-401
10. Ore, O.: Theory of non-commutative polynomials. Annals of Mathematics 34(22): 480-508, 1933
11. Popov, V.: Invariant description of linear, time-invariant controllable systems. SIAM J. Control. Vol.10(2), 252-264, 1972
12. Schilling, O.: The Theory of Valuations. Math. Surveys No. IV, Amer. Math. Soc., New York 1950
13. Villard, G.: Computing Popov and Hermite forms of polynomial matrices. Proceeding of ISSAC'96, ACM Press, New York, 250-258
14. von zur Gathen, J.: Hensel and Newton methods in valuation rings. Mathematics of Computation vol.42(1984), 637-661

Modeling and Simulation of High-Speed Machining Processes Based on Matlab/Simulink

Rodolfo E. Haber[1,2,*], J.R. Alique[1], S. Ros[1], and R.H. Haber[3]

[1] Instituto de Automática Industrial (CSIC),
km. 22,800 N-III, La Poveda, 28500
Madrid, Spain
rhaber@iai.csic.es
[2] Escuela Politécnica Superior,
Ciudad Universitaria de Cantoblanco,
Calle Francisco Tomás y Valiente, 11
28049 – Madrid, Spain
Rodolfo.Haber@ii.uam.es
[3] Departamento de Control Automático,
Universidad de Oriente, 90400, Cuba

Abstract. This paper shows the mathematical development to derive the integral-differential equations and the algorithms implemented in MATLAB to predict the cutting force in real time in high speed machining processes. This paper presents a cutting-force-based model able to describe a high-speed machining process. The model considers the cutting force as an essential output variable in the physical processes taking place in high-speed machining. For the sake of simplicity, only one type of end mill shapes is considered (i.e., cylindrical mill) for real-time implementation of the developed algorithms. The developed model is validated in slot-milling operations. The results corroborate the importance of the cutting-force variable for predicting tool wear in high-speed machining operations.

Keywords: modeling; complex systems; high-speed machining.

1 Introduction

One of the basic tasks manufacturing systems have to perform today is machining, especially high-speed machining (HSM), a real case of a complex electromechanical system [1]. The idea of characterizing the machining process using mathematical models to yield an approximate description of the physical phenomenon aroused the interest of many researchers [2]. That work has been carried on, and it has enabled computational tools to be developed for modeling and simulating the conventional machining process, relying on classic modeling and identification strategies.

* Corresponding author.

V.S. Sunderam et al. (Eds.): ICCS 2005, LNCS 3516, pp. 627 – 634, 2005.

At the present time, modeling high-speed machining processes as a complex electromechanical process, especially high-speed milling, is a very active area of investigation that is peppering the scientific community with challenges [3,4]. High-speed machining has now been adopted and put into regular use at many companies, and yet certain points of how to monitor cutting-tool condition have been worked out only partially as yet, largely for lack of a mathematical model of the process that can feasibly be used in real-time applications. This paper is aimed at deriving a mathematical model and characterizing the high-speed cutting process in high-speed milling operations through the study of the dynamic behavior of cutting force. Previous work concerning conventional machining has been borne in mind as the essential starting point for this work [5].

A mathematical model is essential for understanding the dynamic behavior of metalworking processes and improving their operation on the basis of time or frequency responses. Increasingly, both experimental models and analytical models are being used in metalworking for the obtaining, manufacturing and processing (e.g., shaping and cutting) of metal materials. When a new product or metal part (e.g., mold) is being designed, the item must pass simulation testing on the process involved (e.g., the cutting process) before work ever begins on the item's physical manufacture. How useful simulations ultimately are depends largely on how faithfully the mathematical models they use describe the real behavior of the cutting process. Moreover, computational methods are essential to yield an adequate accuracy in the prediction and real-time implementation of the model.

The main goal of this paper is to derive a mathematical model from the characterization of the physical processes taking place during high-speed machining and to implement the model in MATLAB/SIMULINK. The paper is organized into five sections. Section 1 describes the state of the art, analyzes some previous work on the subject, gives a quick description of high-speed machining and finally outlines some questions related with the implementation of HSM models in MATLAB. Section 2 presents the kinematics of the model, considering the geometrical form of the helix for the cylindrical mill. Section 3 addresses the mathematical formulation of the proposed model and sets up the integral-differential equations for calculating cutting force in the time domain. Section 4 gives the results yielded by the simulations and tests in real time. Lastly, the paper frames some conclusions, looks ahead to some possible future work and discusses the lines opened to investigation.

2 Geometrical Model of the Tool

Geometrical modeling of the helical cutting edge includes the kinematic and dynamic analysis of the cutting process. Predicting cutting forces requires a system of coordinates, the helix angle and the angular distance of a point along the cutting edge [6]. The mathematical expressions that define this geometry in a global coordinate system are presented below in the geometric model, using classic vector notation.

Vector $\vec{r}(z)$ (figure 1) drawn from point O to a point P in cylindrical coordinates is expressed mathematically in equation 1.

$$\vec{r}_j = x_j\vec{i} + y_j\vec{j} + z_j\vec{k} = r(\phi_j)(\sin\phi_j\vec{i} + \cos\phi_j\vec{j}) + z(\phi_j)\vec{k} \tag{1}$$

where ϕ_j is the radial rake angle of a point *P* at tooth *j*. Point *P* lies at an axial depth of cut a_p in the direction of axis *Z*, at a radial distance $r(z)$ on the *XY* plane, with an axial rake angle $\kappa(z)$ and a radial lag angle of $\psi(z)$.

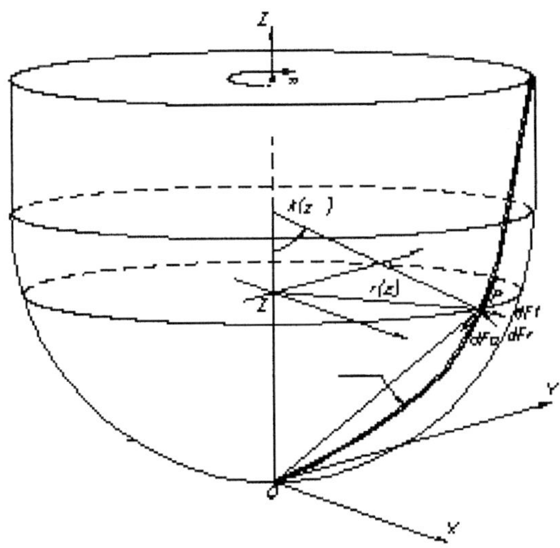

Fig. 1. Tool geometry

The geometry of the tool is represented mathematically, considering that the helical cutting edge wraps parametrically around a cylinder. The mathematical model dictated for the cutting edge considers that the edge is divided into small increments, where the cutting coefficients can vary for each location. The initial point of reference to the cutting edge of the tool (j = 1) is considered to be the angle of rotation when $z = 0$ is ϕ. The radial rake angle for the cutting edge *j* in a certain axial position z is expressed as:

$$\phi_j(z) = \phi + \sum_{n=1}^{j} \phi_p - \psi(z) \tag{2}$$

The lag angle $\psi(z)$ appears due to the helix angle θ. This angle is constant in the case of a cylindrical mill. In the generalized model for the geometry of a mill with helical teeth, the tool diameter may differ along the length of the tool, depending on the shape of the tool. An infinitesimal length of this cutting edge may be expressed as

$$dS = |dr| = \sqrt{r^2(\phi) + (r'(\phi))^2 + (z'(\phi))^2}\, d\phi,\ r'(\phi) = \frac{dr(\phi)}{d\phi},\ z' = \frac{dz(\phi)}{d\phi} \tag{3}$$

Chip thickness changes as a function of the radial rake (ϕ) and axial rake (κ):

$$h_j(\phi_j) = s_{tj} \sin\phi_j \cdot \sin\kappa \tag{4}$$

For a cylindrical mill, the following conditions are defined for finding the general solution:

$$r(z) = \frac{D}{2} \qquad \kappa = 90^{\circ} \quad \psi = k_\theta z \quad k_\theta = (2\tan\theta)/D \tag{5}$$

3 Dynamic Model of Cutting Forces

The force differentials (dF_t), (dF_r), (dF_a) act on an infinitesimal element of the cutting edge of the tool [7]:

$$\begin{aligned} dF_t &= K_{te}dS + K_{tc}h_j(\phi,\kappa)db \\ dF_r &= K_{re}dS + K_{rc}h_j(\phi,\kappa)db \\ dF_a &= K_{ae}dS + K_{ac}h_j(\phi,\kappa)db \end{aligned} \tag{6}$$

It is also considered that $db = \dfrac{dz}{\sin\kappa}$. In order to facilitate finding the mathematical relations inherent in this set-up, very small time increments are used. The positions of the points along the cutting edge are evaluated with the geometrical model presented herein above.

Furthermore, the characteristics of a point on the cutting surface are identified using the properties of kinematic rigidity and the displacements between the tool and the workpiece. The constants or cutting coefficients (K_{tc}, K_{rc}, K_{ac}, K_{te}, K_{re}, K_{ae}) can be found experimentally using cutting forces per tooth averaged for a specific type of tool and material [8,9]. We might point out that these coefficients are highly dependent on the location (axial depth) of the cutting edge. How these coefficients are found shall not be addressed in this paper.

Cutting forces can be evaluated employing a system of Cartesian coordinates. After transforming and total cutting forces as a function of ϕ are:

$$\begin{aligned} F_x(\phi) &= \sum_{j=1}^{Nf}(F_{xj}(\phi_j(z))) = \sum_{j=1}^{Nf}\int_{z1}^{z2}\left[-dF_{rj}\sin\phi_j\sin\kappa_j \quad -dF_{tj}\cos\phi_j \quad -dF_{aj}\sin\phi_j\cos\kappa_j\right]dz \\ F_y(\phi) &= \sum_{j=1}^{Nf}(F_{yj}(\phi_j(z))) = \sum_{j=1}^{Nf}\int_{z1}^{z2}\left[-dF_{rj}\cos\phi_j\sin\kappa_j \quad dF_{tj}\sin\phi_j \quad -dF_{aj}\cos\phi_j\cos\kappa_j\right]dz \\ F_z(\phi) &= \sum_{j=1}^{Nf}(F_{zj}(\phi_j(z))) = \sum_{j=1}^{Nf}\int_{z1}^{z2}\left[-dF_{rj}\cos\kappa_j \quad 0 \quad -dF_{aj}\sin\kappa_j\right]dz \end{aligned} \tag{7}$$

where z_1 and z_2 are the integration limits of the contact zone at each moment of cutting and can be calculated from the geometrical model described herein above. For the numerical calculation, the axial depth of cut is divided into disks having an infinitesimal height dz. The differentials of the cutting forces are calculated along the length of the cutting edge in contact, and they are summed to find the resulting forces for each axis $F_x(\phi)$, $F_y(\phi)$, $F_z(\phi)$ in an angle of rotation.

The exact solution can be found by substituting, making $\kappa = 90°$, and integrating, we obtain the exact solution for cylindrical mill:

$$F_{x,j}(\phi_j(z)) = \left\{ \begin{array}{l} \dfrac{s_{tj}}{4k_\beta}\left[-K_{tc}\cos 2\phi_j(z) + K_{rc}\left[2\phi_j(z) - \sin 2\phi_j(z)\right]\right] \\ + \dfrac{1}{k_\beta}\left[K_{te}\sin\phi_j(z) - K_{re}\cos\phi_j(z)\right] \end{array} \right\}_{z_{j,1}(\phi_j(z))}^{z_{j,2}(\phi_j(z))}$$

$$F_{y,j}(\phi_j(z)) = \left\{ \begin{array}{l} \dfrac{-s_{tj}}{4k_\beta}\left[K_{tc}(2\phi_j(z) - \sin 2\phi_j(z) + K_{rc}\cos 2\phi_j(z)\right] \\ + \dfrac{1}{k_\beta}\left[K_{te}\cos\phi_j(z) - K_{re}\sin\phi_j(z)\right] \end{array} \right\}_{z_{j,1}(\phi_j(z))}^{z_{j,2}(\phi_j(z))} \quad (8)$$

$$F_{z,j}(\phi_j(z)) = \frac{1}{k_\beta}\left[K_{ac}s_{tj}\cos\phi_j(z) - K_{ae}\phi_j(z)\right]_{z_{j,1}(\phi_j(z))}^{z_{j,2}(\phi_j(z))}$$

where $z_{j,1}(\phi_j(z))$ and $z_{j,2}(\phi_j(z))$ are the lower and upper limits, respectively, that establish the axial depth of cut at lip j of the mill.

4 Simulations and Experimental Validation

The algorithms were implemented in MATLAB, drawing upon the mathematical models [10]. Despite MATLAB is not a computer algebra system, it is very useful as software tool for doing numerical computations with matrices and vectors. It can also display information graphically and includes many toolboxes for several research and applications areas. MATLAB was chosen for the advanced numerical methods that are available, the possibility it affords of running simulations and applications in real time and the portability of the programs that are developed with its use (i.e., it is possible to generate C/C++ programs from MATLAB files). Neither Matlab toolbox was used for programming the model.

The main difficulties are choosing the cutting coefficients and the properties of the materials, which were taken from earlier work on the subject [11]. The workpiece-material properties that were used for the simulation in the MATLAB environment are the properties of GGG-70 cast iron with nodular graphite. In this study, in the simulation and in the real tests, two cutting conditions for high-speed milling operations were regarded: Vc=546 m/min, sp=14500 rpm, f=1740 mm/min, a_p = 0.5 mm, a_e = 0, a_s = 12 mm, θ=30°, H = 25.0 mm, D = 12.0 mm where D is the tool diameter [mm], H is the total tool cutting-edge height [mm], θ is the helix angle [degrees], a_p is the axial depth of cut along the Z axis [mm], a_e is the radial depth of cut at the starting point (ϕ_{st}) [mm], a_s is the radial depth of cut at the stopping point (ϕ_{ex}) [mm], f is the feedrate [mm/min], Vc is the cutting speed [m/min], sp is the spindle speed in [rpm].

The constants used in the simulation and in the experimental validation were K_{tc} = 2172 N/mm^2, K_{rc} = 850 N/mm^2, K_{te}=17.3 N/mm, K_{re} = 7.8 N/mm, K_{ac} = 726 N/mm^2, K_{ae} = 6.7 N/mm. These constants or cutting coefficients referring to the material and the tool were drawn from the available literature, due to their similarity to the characteristics of the tool/material set-up used in the study in question.

The real-time model validation tests were conducted at a KONDIA HS1000 high-speed machining center equipped with a Siemens 840D open CNC. Actual cutting force signal was measured using a Kistler 9257 dynamometric platform installed on the testbed. Measurement was done by means of a DAQBOARD-2005 data-acquisition card at a sampling frequency of 50 kHz. A Karnasch 30.6472 cylindrical mill 12 mm in diameter was selected to validate the model developed under the procedure described herein. The chosen test piece, measuring 200x185x50 mm, was made of GGG-70 iron and was machined in a spiral pattern. The real cutting conditions chosen were the same as considered above for the simulation. A view of the cutting tool used in the tests (a), the workpiece and its profile to be mechanized (b) and the machine-tool laboratory (c) is shown in figure 2.

Fig. 2. Cutting tool for experiments, b) experimental piece, c) partial view of the Laboratory for machine tool research

Figure 3 shows the real behavior of cutting forces F_x, F_y and F_z and the resulting force F_{qT} for the case analyzed. The model's response is shown as a solid line.

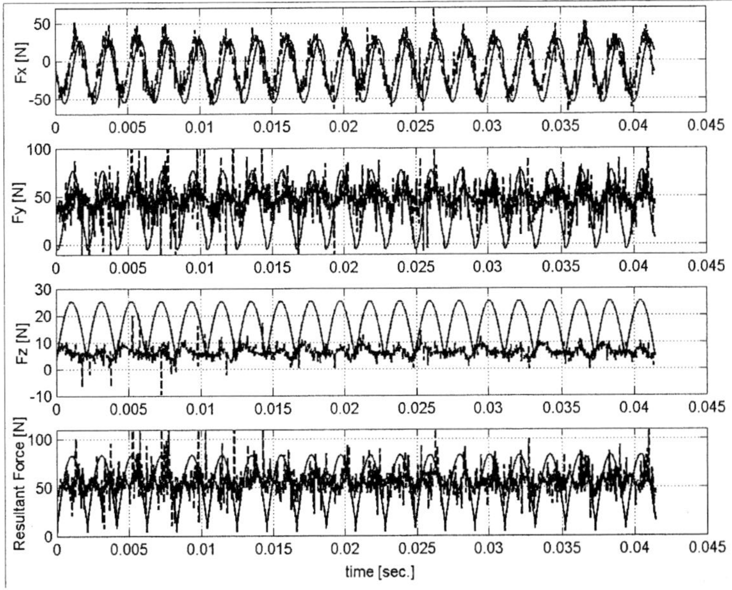

Fig. 3. Measured (straight line) and predicted by model (dashed line) cutting force for a new tool in high speed slot cutting

The average resulting cutting force $\overline{F}_e$ estimated by the model is 56.7N and the average resulting cutting force $\overline{F}_{qT}$ measured in a real high-speed cutting operation is 55.6N. The error criterion $\overline{E} = \left(\overline{F}_{qT} - \overline{F}_e\right) \cdot 100 \Big/ \overline{F}_{qT}$, is 4.5%.

5 Conclusions

This paper reports some results in modeling the high-speed cutting process and the validation of the model in question. The mathematical model implemented and assessed in MATLAB has two fundamental, co-dependent parts. The first is a multiple-input system that defines the model's kinematics, where constants and variables describe the tool geometry, the material type and the cutting parameters. The second addresses the dynamics, represented by integral-differential equations. In the case of cylindrical mill the exact analytical solution is found, inasmuch as the limits of integration and the boundary conditions along the tool geometry can be pre-established.

On the basis of the literature to which we have had access, this paper constitutes the first successful attempt in the implementation and application of computationally efficient algorithms for predicting cutting force in high-speed machining processes. With such models available, there is now a short-term way of dealing with essential questions concerning the surface finish and stability of high-speed cutting operations.

Despite the success in the application of Matlab, computer algebra in the formulation and solution of more complex cutting tools, as well as the use of the interface to communicate Mathematica and Matlab will be explored during future research.

References

1. Haber R. E., J.E. Jiménez, J.L. Coronado, A. Jiménez: Modelo matemático para la predicción del esfuerzo de corte en el mecanizado a alta velocidad, Revista Metalurgia Madrid 40(4) (2004) 247-258.
2. Bukkapatnam S.T.S, A. Lakhtakia, S.R.T. Kumara: Analysis of sensor signals shows turning on a lathe exhibits low-dimensional chaos, Physics Review E 52 (3) (1995) 2375-2387.
3. Haber R. E., J.E. Jiménez, C.R. Peres, J.R. Alique: An investigation of tool wear monitoring in a high-speed machining process, Sensors and Actuators A: Physics 116 (2004) 539-545.
4. Cantillo K. A., R. E. Haber Guerra, A. Alique, R. Galán, CORBA-based open platform for processes monitoring. An application to a complex electromechanical process. In: Bubak M., Albada G.D., Sloot P.M.A., Dongarra J. (eds.): Computational Science ICCS2004, Lecture Notes in Computer Science 3036 (2004) 531-535.
5. Altintas Y.: Manufacturing Automation: metal cutting mechanics, machine tool vibrations, and CNC design, Cambridge University Press, USA, 2000.
6. Yucesan G., Y. Altintas: Prediction of ball end milling forces, J. Eng. Ind. Trans. ASME 1 (1) (1996) 95-103.
7. Altintas Y., P. Lee: Mechanics and dynamics of ball end milling forces, J. Manuf. Sci. Eng.-Trans. ASME 120 (1998) 684-692.
8. Fu H. J., R.E. Devor, S.G. Kapoor: A mechanistic model for the prediction of the force system in face milling operation, J. Eng. Ind. Trans. ASME 106 (1) (1984) 81-88.
9. 9. Budak E., Y. Altintas, E.J.A. Armarego: Prediction of milling force coefficients from orthogonal cutting data, J. Eng. Ind. Trans. ASME 118 (1996) 216-224.
10. The Math Works Inc. Users manual Matlab 6.5 Release 13, Natick, Massachusetts, 2003.
11. Engin S., Y. Altintas: Mechanics and dynamics of general milling cutters Part I: helical end mills Int. J. Mach. Tools Manuf. 41 (2001) 2195-2212.

Remote Access to a Symbolic Computation System for Algebraic Topology: A Client-Server Approach*

Mirian Andrés, Vico Pascual, Ana Romero, and Julio Rubio

Departamento de Matemáticas y Computación, Universidad de La Rioja
{miriam.andres, vico.pascual, ana.romero, julio.rubio}@dmc.unirioja.es

Abstract. Kenzo is a Symbolic Computation system created by Sergeraert for computing in Algebraic Topology. It is programmed in Common Lisp and this programming language also acts as user interface. In this paper, a prototype to provide remote access for Kenzo is presented. This has been accomplished by using Corba technology: clients have been developed both in Java and Common Lisp (the server is always in Common Lisp, being a wrapper of the original Kenzo program). Instead of using one CORBA IDL to encode each data structure, our approach incorporates a generic way of transfering every data structure through XML strings; specifically, by means of an XML extension of MathML. This research should be understood as a first step towards building a distributed computation system for Algebraic Topology.

Introduction

Nowadays, Internet appears as a suitable tool for performing scientific computations in a distributed collaborative way. One of the fields where this idea can be applied is that of Symbolic Computation. Computer Algebra packages are being extended to interconnect them. In fact, some Computer Algebra systems, such as Distributed Maple, are already capable of performing distributed computing.

It is clear that this trend could be explored in any other Symbolic Computation system, in particular in systems devoted to algorithmic Algebraic Topology. The leader systems in this field are EAT [5] and Kenzo [6]. Both systems are written in the Common Lisp programming language and have obtained some results (specifically, homology groups) which had never been determined before using either theoretical or computational methods. Since the programs are very time (and space) consuming, it would be interesting to begin a study about the task of making them distributed. As a first step, we have considered the possibility of providing a remote access for Kenzo.

The organization of this paper is as follows. The next section presents some preliminary ideas which are needed for our task: some basic concepts in Algebraic

* Partially supported by SEUI-MEC, project TIC2002-01626.

V.S. Sunderam et al. (Eds.): ICCS 2005, LNCS 3516, pp. 635–642, 2005.

Topology, some aspects of Symbolic Computation systems (especially Kenzo and EAT), and an explanation of the previous work. In Section 2 we describe the prototype of remote access to Kenzo that we have developed, which makes use of CORBA and XML. Section 3 presents some examples of the use of our programs. The paper ends with a conclusions section and the bibliography.

1 Preliminaries

1.1 Notions of Algebraic Topology

In this section, some basic concepts of Algebraic Topology are introduced.

Definition 1. *A* chain complex $C = (C_p, d_p)$ *is a family of R-modules* $\{C_p\}_{p\in \mathbb{Z}}$ *(R is a ring), with R-modules homomorphisms* $\{d_p\}_{p\in \mathbb{Z}}$ *(the* differential application*)* $d_p : C_p \to C_{p-1}$ *such that* $d_{p-1} \circ d_p = 0$.

We consider chain complexes formed by free $\mathbb{Z}$-modules. The fact that C_p is a free $\mathbb{Z}$-module implies that it is generated, and therefore each element of C_p can be expressed as a linear combination of the set of its generators: $\sum \lambda_i \sigma_i,\quad \lambda_i \in \mathbb{Z}$. An element σ_i of a basis is called a *generator*. A product $\lambda_i \sigma_i$ is called a *term* or *monomial*, and a sum of these terms is called a *combination*.

1.2 Symbolic Computation in Algebraic Topology

There are some Symbolic Computation systems for Algebraic Topology, EAT and Kenzo being the most important ones. Created by Sergeraert and coworkers, and written in Common Lisp, they can be used to compute homology groups of infinite topological spaces, namely loop spaces.

We present here a simple example of computation with Kenzo, which consists on computing the differential of the chain complex (delta 3) applied to the combination $3 * 4 + 2 * 6$ of degree 2. The solution obtained is the combination (of degree 1) $3 * 0 - 2 * 2 + 2 * 4$.

```
> (dffr (delta 3) (cmbn 2 3 4 2 6))
(:cmbn 1 (3 . 0) (-2 . 2) (2 . 4))
```

The system allows the computation of more complicated results, for instance homology groups of chain complexes, one of the fundamental problems of Algebraic Topology. It is possible to obtain easily calculations that an expert on topology would need a lot of time and effort to get.

Some important features of these systems are:

- Intensive use of functional programming to deal with infinite data.
- Organization in two layers of data structures. The first one corresponds to the algebraic structures they work with, such as a chain complex. The second one corresponds to the elements of these structures, as for example a combination inside a chain complex.

- They have obtained some results which had never been determined before using either theoretical or computational methods.
- However, the usability of Kenzo and EAT presents some barriers. Since their user interface is Common Lisp itself (a not very extended programming language), they do not provide a friendly front end.

This last aspect of Kenzo was one of the reasons why we tried to provide a remote access for it, allowing in this way the use of other (better known) languages to construct the user interface. Since our goal is distributed computing, we try to do this in a scalable way, that is, in a way that would allow us in the future to extend our remote access system in order to perform collaborative computations among different computers.

1.3 Previous Work

In a previous work, we considered reconstructing (part of) the system in Java, a programming language very extended nowadays, especially in Internet. There are big differences between this language and Common Lisp, so we considered an intermediate step: we rebuilt (some fragments of) EAT in the programming language ML, that is not so far from Common Lisp as Java. So, we have implemented two prototypes to perform computations in Algebraic Topology in ML and Java respectively.

The next step was to interoperate between them and the Common Lisp version. To build our exchange language, we used XML [9], a metalanguage that allows a flexible representation of the information. We considered MathML [7] and OpenMath [8], two standards that give XML representations for mathematical objects (the second one also used to exchange mathematical objects between symbolic computation systems). We represented simple data with MathML, and for more complicated structures we defined our own DTD as an extension of that of MathML. OpenMath was not used because content dictionaries are strictly organised and we found it difficult to apply them in our case. With this exchange format the three systems were able to interchange XML documents among them, but only in a local non-distributed way (see [2]).

This was the starting point of this work, consisting on developing a remote access for Kenzo, with a client-server architecture. We would have the Kenzo program in a (server) machine, and several clients (with their own languages) could access it from different computers. A client would invoke some calculations in Algebraic Topology, the server would then process the computation and finally send the result to the client.

2 Development of the Prototype

As mentioned before, our goal was to construct a prototype that allows an access to the Kenzo program from different machines with other programming languages. Our first idea was to try to reuse the system developed previously and make it work with different computers. We had the XML exchange language but this was an incomplete picture because an infrastructure that would allow the interchange between the client and the server was necessary. It could be CORBA [4], a middleware that allows pieces of programs, called objects, to communicate with one another regardless of programming languages and operating systems, especially used in a local net. The objects of CORBA have an interface written in IDL (an interface description language described in the CORBA specification). For each object, this interface is the same for client and server but the implementations in both sides can be written in different languages.

In a previous work (included as a poster in ISSAC 04 and presented in [3]) we have examined some of our first trials towards this remote access, and some problems found. The main problem was to find a way to exchange the data in Kenzo between client and server. In a first attempt, we tried to define an IDL file for each of the structures we wanted to work with. For instance, an interface for a momomial (one of the simplest data that appear in the program) could be as presented below.

```
module kenzo
interface Monomial
//Attributes
readonly attribute long cffc;
readonly attribute string gnrt;
//Methods
void setMnm(in long cffc, in string gnrt);
long getCffc();
string getGnrt(); ; ;
```

It would be clearly a tedious task to construct an IDL for each type in Kenzo (the program defines 15 classes and about 40 types; see [6]). If we worked in this way, we would have an IDL explosion and the running of CORBA would be complicated. Our prototype solved this problem with the combination of CORBA and XML in a natural way, inserting two string attributes in the object interface which will contain the XML representation of the call to the Kenzo function and the result. We obtain in this way a generic IDL for every Kenzo elements.

The IDL file we use to interoperate between client and server is showed below. The two attributes `obj` and `result` will contain the XML representation of the call to the Kenzo function we want to compute and its result. With `setObject` the client is able to set the operation and its arguments (in XML). When the client calls `computeXML`, the server obtains the XML representation of the operation with `getObject`, computes it and inserts in `result` the XML representation of the solution. Finally, `getResult` allows the client to recapture the result of the operation.

```
module kenzo
interface XMLObject
//Attributes
readonly attribute string obj;
readonly attribute string result;
//Methods
void setObject(in string obj);
string getObject();
string getResult();
void computeXML(); ; ;
```

The use of this IDL file presents some advantages. It can be easily implemented in both client and server, and the same interface can be used to exchange all types of data (XML plays the leading-role to represent each structure). The use of only one interface makes easier the running of CORBA, because for each object the IDL file must be in both client and server sides. Besides, if we work with Java or ML in the client, we can reuse the code we had written in the previous work for the local system.

The exchange format used is an XML dialect, a proper extension of MathML [7]. Some labels of MathML are used, such as <cn> for integers. However, the data of MathML are simpler than those of Kenzo, so some other labels have been created for other algebraic structures (for combinations we use <cmb>, for monomials <mnm>...). For instance, the combination (of degree 2) $4 * 5 + 2 * 6$ is represented as follows:

```
<cmb>
<dgr>
<cn> 2 </cn>
</dgr>
<list>
<mnm>
<coef>
<cn> 4 </cn>
</coef>
<gnr>
<cn> 5 </cn>
</gnr>
</mnm>
<mnm>
<coef>
<cn> 2 </cn>
</coef>
<gnr>
<cn> 6 </cn>
</gnr>
</mnm>
</list>
</cmb>
```

With the showed interface, it is not difficult to implement the IDL methods in the server. The main task consists on writting some "parsers", functions that translate XML strings to Kenzo objects and vice versa (some of them written previously for the local system). With the IDL implementation built, the server program is an easy sequence of orders to create the CORBA object and set it ready to be required by the clients.

For the clients' side, we also need to implement the parsers. It has been done for Common Lisp and Java, reusing again our code. The client programs use CORBA to connect with the server and invoke its operations. First, the client builds the XML format of the operation wanted and assigns it to `obj`. Next, it invokes `computeXML` of the server and the server gets the representation, translates it to Lisp, computes it and assigns the XML of the solution to `result`. Finally, with `getResult` the client gets the result, and then translates it to a Kenzo object.

3 Examples

We consider now some examples with a Java client. It can be called in the command line, and to facilitate the communication with the user we have also written a Java applet.

As a first example, we consider the multiplication of a combination by an integer, which is invoked as it is showed below. Once computed, the wanted combination is showed on the screen.

```
C:\>java JavaClient "n-cmbn" "nil" 3 "cmbn" 2 4 5 2 6
Result: (Cmbn 2 (12 5) (6 6))
```

More complicated operations are those that require the ambient space where the computation is carried out (first layer of data structures). In the previous example, the operation is independent of the chain complex, so we define the ambient as "nil". However, the differential of a combination is defined in the chain complex, so we must specify the ambient. To encode this space, Kenzo uses functional programming, so this was a difficulty we found when representing this ambient in XML. To solve it, we encoded the ambient space with the tree of

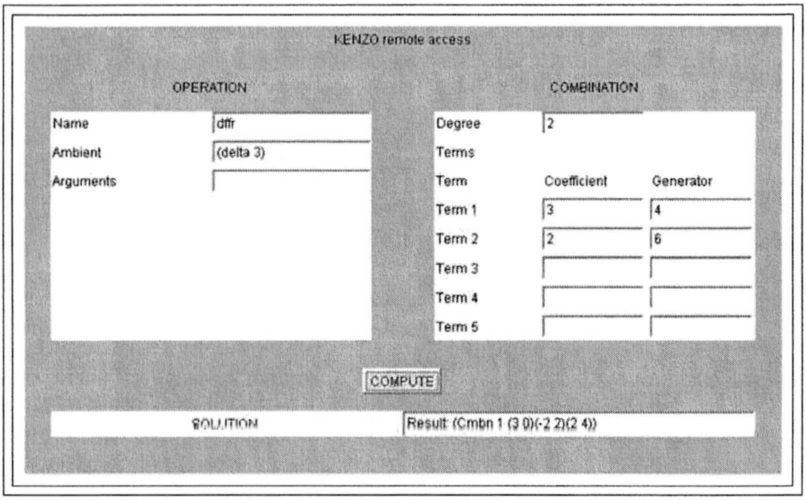

Fig. 1. Example: differential of a combination

calls that generates it. In this case, we compute the differential of a combination in the chain complex (delta 3).

```
C:\>java JavaClient "dffr" "(delta 3)" "cmbn" 2 3 4 2 6
Result: (Cmbn 1 (3 0) (-2 2) (2 4))
```

We can also use the Java Applet to introduce the parameters in several text boxes. In Figure 1 we include an image of the applet that corresponds to the differential of a combination presented before.

4 Conclusions and Further Work

In this paper we have presented a prototype based on CORBA that provides a remote access for Kenzo. The leading-role in the data exchange is played by an XML format which is a proper extension of MathML. The inclusion of XML inside a CORBA object allows the use of only one IDL for every structure in Kenzo, which makes easier the implementation and the running of CORBA. Moreover, with our IDL interface we can use the server program without any change, with only a little work for the wrapper of the system that allows making use of it from a CORBA object.

It is a small prototype that works only with few of all the algebraic structures that appear in Kenzo. Besides, the programs are totally based in CORBA, a technology designed for computers connected by a local net, so our system can not be used directly to interoperate in Internet.

Obviously, one of our short-term goals is to complete the prototype, to offer the total functionality of Kenzo. For this task, we only will have to design the XML representation for each structure and the parsers to translate them, but in the part of CORBA and IDL interfaces the work is already done, our XML format is generic. Another short-term goal is to try other technologies that allow us to work outside a local net.

The prototype built can be understood as a first step towards a distributed computation system, where some other technologies such as Web Services [1] (using for instance WSDL for the description of the object instead of IDL, and SOAP for the exchange of messages) or Grid Computing could be considered. The infrastructure could consist for example on several services cooperating in the same calculation, coordinated by a specific Web Service. Even if this work shows a first step in this direction, many difficult research questions should be addressed to reach this goal.

References

1. Alonso, G., Casati, F., Kuno, H., Machiraju, V.: Web Services. Concepts, Architectures and Applications. Springer, 2004.
2. Andrés, M., García, F. J., Pascual, V., Rubio, J.: XML-Based interoperability among symbolic computation systems. In Proceedings WWW/Internet 2003, Vol. II, Iadis Press (2003) 925-929.

3. Andrés, M., Pascual, V., Romero, A., Rubio, J.: Distributed computing in Algebraic Topology: first trials and errors, first programs. In e-proceedings IAMC 2004. http://www.orcca.on.ca/conferences/iamc2004/abstracts/04002.html
4. Object Management Group: Common Object Request Broker Architecture (CORBA). http://www.omg.org.
5. Rubio, J., Sergeraert, F., Siret, Y.: EAT: Symbolic Software for Effective Homology Computation. Institut Fourier, Grenoble, 1997. ftp://ftp-fourier.ujf-grenoble.fr/pub/EAT.
6. Dousson, X., Sergeraert, F., Siret, Y.: The Kenzo program. Institut Fourier, Grenoble, 1999. http://www-fourier.ujf-grenoble.fr/~sergerar/Kenzo/.
7. Ausbrooks, R. et al.: Mathematical Markup Language (MathML) Version 2.0. 2003. http://www.w3.org/TR/2001/REC-MathML2-20010221/.
8. Caprotti,O., et al (eds): The OpenMath standard. 2000. http://www.openmath.org/standard
9. Bray, T. et al. (eds): Extensible Markup Language (XML) 1.0. 2003. http://www.w3.org/TR/REC-xml/.

Symbolic Calculation of the Generalized Inertia Matrix of Robots with a Large Number of Joints

Ramutis Bansevičius[1], Algimantas Čepulkauskas[2], Regina Kulvietienė[2], and Genadijus Kulvietis[2]

[1] Kaunas University of Technology, Donelaičio 73, Kaunas 3006, Lithuania
bansevicius@cr.ktu.lt
[2] Vilnius Gediminas Technical University, Saulėtekio 11, Vilnius 2040, Lithuania
{algimantas_cepulkauskas, regina_kulvietiene, genadijus_kulvietis}@gama.vtu.lt

Abstract. The aim of this paper is to simplify numerical simulation of robots with a large number of joints. Many numerical methods based on different principles of mechanics are developed to obtain the equations that model the dynamic behavior of robots. In this paper, the efficiency of computer algebra application was compared with the most popular methods of forming the generalized inertia matrix of robots. To this end, the computer algebra system was used. Expressions for the generalized inertia matrix of the robots with a large number of joints have been derived, using the computer algebra technique with the following automatic program code generation. As shown in the paper, such an application could drastically reduce the number of floating point product operations, needed for efficient numerical simulation of robots.

1 Introduction

Manipulator and robot systems possess several specific qualities in both a mechanical and a control sense. In the mechanical sense, a feature specific to manipulation robots is that all the degrees of freedom are "active", i.e., powered by their own actuators, in contrast to conventional mechanisms in which motion is produced primarily by the so-called kinematics degrees of freedom. Another specific quality of such a mechanism is their variable structure, ranging from open to closed configurations, from one to another kind of boundary conditions. A further feature typical of spatial mechanisms is redundancy reflected in the excess of the degrees of freedom for producing certain functional movements of robots and manipulators. From the control standpoint, robot and manipulator systems represent redundant, multivariable, essentially nonlinear automatic control systems [13]. A manipulation robot is also an example of a dynamically coupled system, and the control task itself is a dynamic task [14].

The methods that model the dynamic behavior of manipulators are divided into two types: methods that solve the inverse dynamic problem and those that

V.S. Sunderam et al. (Eds.): ICCS 2005, LNCS 3516, pp. 643–650, 2005.

give the solution to the direct dynamic problem. In the former, the forces exerted by the actuators are obtained algebraically for certain configurations of the manipulator (position, velocity, and acceleration). On the other hand, the direct dynamic problem computes the acceleration of joints of the manipulator once the forces exerted by the actuators are given. This problem is part of the process that must be followed to perform the simulation of the dynamic behavior of the manipulator. This process is completed once are calculated the velocity and position of the joints by means of the process of numerical integration in which the acceleration of the joints and the initial configuration are data input to the problem. So, the methods may be divided with respect to the laws of mechanics on the basis of which motion equations are formed. Taking this as a criterion, one may distinguish methods based on Lagrange–Euler's (L-E), Newton–Euler's (N-E), Gibbs–Appell's (G-A) and other equations. The property of whether the method permits the solution of the direct or inverse problem of dynamics may represent another criterion. The direct problem of dynamics refers to determining the motion of the robot for known driving forces (torques), and the inverse problem of dynamics to determining driving forces for the known motion. Clearly, the methods allowing both problems of dynamics to be solved are of particular importance. The number of floating-point multiplications (divisions) / additions (subtractions) required to form a model is the most important criterion to compare the methods. This criterion is also important from the point of view of their on-line applicability.

The algorithms developed to solve the direct dynamic problem use, regardless of the dynamics principle from which they are derived, one of the following approaches [2], [9]:

- calculation of the acceleration of the joints by means of the method proposed and solution of a system of simultaneous equations;
- recursive calculation of the acceleration of the joints, propagating motion and constraint forces throughout the mechanism.

The algorithms derived from the methods that use the first approach require the calculation of the generalized inertia matrix and the bias vector [2]. The generalized inertia matrix is also used in advanced control schemes, as well as in parameter estimation procedures. For this reason its calculation, by means of simple and efficient procedures, is also beneficial to other fields, not only to motion simulation of mechanical systems. The generalized inertia matrix can be obtained through the Hessian of kinetic energy of the mechanical system with respect to generalized velocities; however, the most computationally efficient algorithms are not based on this procedure. The best known method that follows this first approach was proposed by Walker and Orin [15] who have developed (using N-E equations) the method of a composed rigid body, in which the generalized inertia matrix is obtained recursively with a complexity $O(n^2)$. Angeles and Ma [1] have proposed another method that follows this approach, based on the calculation of the natural orthogonal complement of the manipulator kinematics constraint equations with a complexity $O(n^3)$, using Kane's equations to obtain the bias vector.

On the other hand, algorithms derived from the methods that use the second approach usually have a complexity $O(n)$. These algorithms do not obtain the generalized inertia matrix, and for this reasons their application is limited to system motion simulations. The best known method among those that use the second approach is the articulated body method developed by Featherstone [6]. The number of required algebraic operations is lower to those needed in the composed rigid body method, but only for the systems that contain nine or more bodies. In [12], Saha has symbolically performed the Gaussian elimination to obtain a decomposition of the generalized inertia matrix. As an application of this decomposition, he proposed an $O(n)$ direct dynamic algorithm with a computational complexity very similar to that of [6].

The complexity of the numerical algorithms mentioned above for forming the generalized inertia matrix will be compared with computer algebra realization. The computer algebra technique application in the formation of the generalized inertia matrix of robots is very attractive, because it allows the analytic work to be pushed before the numerical integration of the system of nonlinear differential equations starts. This approach has been successfully applied to the inverse dynamic problem of the robot [3].

2 Algorithm for Calculating the Generalized Inertia Matrix

The algorithm for calculating the generalized inertia matrix has been constructed using the Uicker–Kahn method [12], based on the L-E equations, that is very convenient for computer algebra implementation [5]. The same approach was used to solve the inverse dynamic problem [3], but formation of the generalized inertia matrix must be considered more carefully, because the matrix must be recalculated at every step of numerical integration time of the robot dynamic model [14], [15]. The equations of the direct dynamic problem formulated by Vucobratovic [14] are contained in the following matrix expression:

$$H(\vec{q})\ddot{\vec{q}} = \vec{P} - \vec{C}(\vec{q}, \dot{\vec{q}}) + \vec{G}(\vec{q}), \tag{1}$$

where $H(\vec{q})$ is the generalized inertia matrix; $\vec{q}, \dot{\vec{q}}, \ddot{\vec{q}}$ are generalized coordinates, velocity, and acceleration of the robot, respectively; $\vec{P}$ is the generalized force vector; $\vec{C}(\vec{q}, \dot{\vec{q}})$ is the vector of Coriolis and centrifugal effects; $\vec{G}(\vec{q})$ is the vector of gravity effect. The bias vector $(\vec{C}(\vec{q}, \dot{\vec{q}}) + \vec{G}(\vec{q}))$ could be calculated separately, using the computer algebra approach for the inverse dynamic problem, presented in the previous work [13].

The elements of the generalized inertia matrix, according to the Uicker–Kahn method, could be expressed in the following form [2], [11]:

$$H_{ij} = \sum_{k=1}^{j} \left[trace \left(\frac{\partial W_j}{\partial q_i} J_j \frac{\partial W_k}{\partial q_k} \right) \right], \tag{2}$$

where H_{ij} are the elements of the generalized inertia matrix; q_i is a generalized coordinate of the i-th joint; J_j is the inertia matrix of the j-th link with respect to the local coordinate system.

The transformation matrix W_i between the i-th local coordinate system and the reference system can be expressed as

$$W_i = A_0^1 A_1^2 ... A_{i-1}^i, \tag{3}$$

where A_{k-1}^k is a (4×4) – homogenous transformation matrix between two local coordinate systems, and it is of the form:

$$A_{k-1}^k = \begin{bmatrix} \tilde{A}_{k-1}^k & \vec{b}_{k,k-1} \\ O & I \end{bmatrix}, \tag{4}$$

where $\tilde{A}_{k-1}^k$, $\vec{b}_{k,k-1}$ are rotation and transition transformations between two local coordinates; O and I mean zero and unit matrices, respectively. Transformation matrices are of the shape [15]:

$$\tilde{A}_{k-1}^k = \begin{bmatrix} \cos q_k & -\cos \alpha_k \sin q_k & \sin \alpha_k \sin q_k \\ \sin q_k & \cos \alpha_k \cos q_k & -\sin \alpha_k \cos q_k \\ 0 & \sin \alpha_k & \cos \alpha_k \end{bmatrix}, \tag{5}$$

$$\vec{b}_{k,k-1} = \begin{bmatrix} a_k \\ d_k \sin \alpha_k \\ d_k \cos \alpha_k \end{bmatrix}, \tag{6}$$

where α_k, a_k, d_k are kinematic parameters of the joint k.

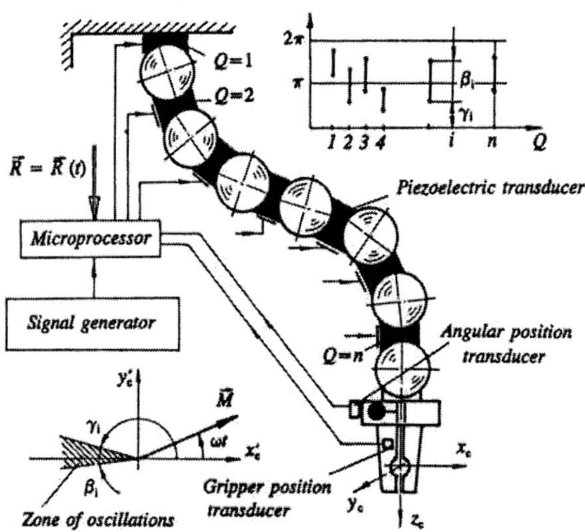

Fig. 1. The scheme of the flexible robot with a large number of joints

The flexible robot with a large number of joints is shown schematically in Figure 1. The robot is composed of cylindrical piezoceramic transducers and spheres, made from passive material, in this case, from steel [4], [5]. The contact force between the spheres and cylindrical piezoceramic transducers is maintained with the aid of permanent magnets. Here the resonant oscillations of each piezoelectric transducer are controlled by a microprocessor that switches on and off the high frequency and high voltage signal from the signal generator. The phase and duration of every pulse, applied to the electrodes of transducers, are synchronized with the rotation of an unbalanced rotor, mounted in the gripper of the robot. High-frequency resonant mechanical oscillations of ultrasonic frequency cause motions (rotations) in all directions and, at the contact zone, they turn to continuous motion.

The external torque vector, appearing in the gripper and rotating on the plane perpendicular to the gripper direction, is calculated by the computer algebra approach described in [3]. Dynamic simulation of this kind of flexible robots is a very complicated problem, because there are two types of motions – continuous and vibration [7], [13].

3 Computer Algebra Implementation

The methods using N-E or G-A equations are in principle complex because of the complexity to eliminate the constraints by forces and moments. Moreover, they do not directly show the algebraic values of the forces and moments due to the action of actuators. The L-E equations provide an opportunity of direct regarding the equations as functions of the system control inputs [13]. However, the inherent unsuitability of applying L-E equations lies in the need to calculate the partial derivatives (see formula (1), which is not a trivial numerical procedure, but very convenient for computer algebra techniques.

In the algorithm for automatic generation of the analytical model, it will be assumed that the parameters of a robot (length, mass, inertia, etc.) are known and will be treated as constants. Joint coordinates, as well as their derivatives, will be treated as independent variables, i.e., as symbols. Using the computer algebra technique, the Uicker–Kahn method is very convenient, because it enables us to obtain equations of motion in closed form and may be applied to solving either the direct or the inverse problem of dynamics.

The Uicker–Kahn method was implemented using VIBRAN [8]. The sparse matrix technology was used in this program to achieve the best performance. In order to compare the various results and algorithms, only two joints (6-degrees-of-freedom) of the proposed robot are considered.

All the elements of the generalized inertia matrix H_{ij} are calculated in the program making use of formula (2). The elements of the generalized inertia matrix have been computed for the discussed flexible robot with 6-degrees-of-freedom. Table 1 contains the kinematic parameters of this robot in Denavit–Hartenberg's notation [5], [10], [14].

Table 1.

N	q_i	α_i	A_i	d_i
1	q_1	0	0	0
2	q_2	90°	0	0
3	q_3	0	0.04	0
4	q_4	−90°	0	0
5	q_5	−90°	0	0
6	q_6	0	0	0.04

In order to avoid the numerical computation of the trigonometric function a substitution was applied $S_i = \sin q_i$, $C_i = \cos q_i$.

The fragment of analytical calculation performed for the generalized inertia matrix of the flexibile robot by the VIBRAN program, illustrated in Figure 2 (the expression of the coefficients should be generated automatically).

Twenty one element (only the symmetric part of the generalized inertia matrix) have been calculated in total and some of them are equal to zero. A special VIBRAN procedure generates two FORTRAN subroutines from the obtained analytical expressions of the generalized inertia matrix [3], [8]. The code of the first generated subroutine contains a dictionary of monomials included into the

```
SUBROUTINE robo01(A,O)
IMPLICIT REAL(A-Z)
DIMENSION A(1),O(1)
S1 =A(1)
C1 =A(2)
.....
C6 =A(12)
O(61)=S5*C4
O(63)=S6*S4
O(66)=C5*S4
O(228)=C6*C5
O(62)=O(228)*C4
END
SUBROUTINE robo(A,B)
DIMENSION A(1),B(1),O( 230)
CALL robo01(A,O)
Y1=+.1296E-3*O(1)
Y2=-.9964E-4*O(2)
.....
B(1)=+.8326E-4+Y1+Y2+Y3+Y4+Y5+Y6+Y7+Y8+Y9+Y10+Y11
*+Y12+Y13+Y14+Y15+Y16+Y17+Y18+Y19+Y20+Y21+Y22+Y23
*+Y24+Y25+Y26+Y27+Y28+Y29+Y30+Y31+Y32+Y33+Y34+Y35
.....
END
```

Fig. 2. Fragment of the code of two subroutines for numerical simulation

expressions of robot's matrices. This dictionary of monomials is sorted in ascending order of the monomial multiindices to reduce the number of floating point multiplications. The second code of the generated subroutine contains calculations of common members included in all the expressions and all the elements of robot's matrices. The generated subroutines can be immediately compiled and directly used for numerical simulation.

The number of floating point product operations, required to form the generalized inertia matrix of the robot by the Uicker–Kahn method, numerically depends on n^4 (n – number of degrees-of-freedom) and, vice versa, the recursive methods based on N-E or G-A equations mainly depend on the number of degrees-of-freedom. When using the computer algebra technique, there emerge some differences. By virtue of the Uicker–Kahn method the expressions for the elements of the generalized inertia matrix are found in closed form, meanwhile, other well-known algorithms yield only recursive equations. This fact indicates that only the numerical implementation is possible and therefore this method is suitable for the direct dynamics problem only. The code presented in Figure 2 contains only 144 floating point products and 186 sums. The computational complexity of the proposed approach is comparable with that of the most efficient algorithms known so far, as shown in Table 2.

Table 2.

Authors	Principle	Products ($n = 6$)	Sums ($n = 6$)
Walker and Orin [9]	N-E	$12n^2$+!56n!–!27(741)	$7n^2$!+!67n!–!56(598)
Angeles and Ma [1]	N-E	n^3!+!17n^2!–!21n!+!8(710)	n^3!+!14n^2!–!16n!–!+!5(629)
Mata et al. [11]	G-A	11.5n^2!+!19.5n!–!49(482)	8.5n^2!+!31.5n!–!69(426)
This work	L-E	144	186

Some remarks could be made to explain these results. First of all, computer algebra systems work very efficiently with a large number of short expressions, which enables an effective simplification of these expressions during analytical computation. It appears that a lot of numerical methods are developed especially to avoid numerical differentiation and most of them are recursive, which is inconvenient for analytical computation. However, the calculation of derivatives is a very simple procedure for computer algebra systems.

4 Conclusions

The expressions for the generalized inertia matrix of the robots with a large number of joints have been obtained using the Uicker–Kahn method, based on Lagrange–Euler's equations, and realized by the computer algebra technique. The computational complexity of the proposed approach is comparable with that of the most efficient algorithms known so far.

The proposed analytical implementation of Uicker–Kahn's method drastically reduces the number of floating point operations, particularly for the robots with a large number of joints. This approach performs the efficient simulation of dynamic behavior of the robot.

References

1. Angeles, J., Ma, O.: Dynamic simulation of n-axis serial robotic manipulators using a natural orthogonal complement, Int. J. Rob. Res. 7 (5) (1988) 32–37
2. Balafoutis, C.A., Patel, R.V.: Dynamic Analysis of Robot Manipulators: A Cartesian Tensor Approach, Kluwer Academic Press, Boston (1991)
3. Bansevicius, R., Cepulkauskas, A., Kulvietiene, R., Kulvietis, G.: Computer Algebra for Real-Time Dynamics of Robots with Large Number of Joints. Lecture Notes in Computer Science, Vol. 3039. Springer-Verlag, Berlin Heidelberg New York (2004) 278–285
4. Bansevicius R., Parkin R., Jebb, A., Knight, J.: Piezomechanics as a Sub-System of Mechatronics: Present State of the Art, Problems, Future Developments. IEEE Transactions on Industrial Electronics, vol. 43, (1) (1996) 23–30
5. Barauskas, R., Bansevicius, R., Kulvietis, G., Ragulskis, K.: Vibromotors for Precision Microrobots. Hemisphere Publishing Corp., USA (1988).
6. Featherstone, R., Orin, D. E.: Robot dynamics: equations and algorithms. Proceedings of the 2000 IEEE International Conference on Robotics and Automation, San Francisco (2000) 826–834
7. Knani J.: Dynamic modelling of flexible robotic mechanisms and adaptive robust control of trajectory computer simulation. Applied Mathematical Modelling , Vol. 26. (12) (2002) 1113–1124
8. Kulvietiene, R., Kulvietis, G.: Analytical Computation Using Microcomputers. LUSTI, Vilnius (1989)
9. Mata, V., Provenzano, S., Valero, F., Cuadrado, J., I.: An $O(n)$ algorithm for solving the inverse dynamic problem in robots by using the Gibbs–Appell formulation, Proceedings of Tenth World Congress on Theory of Machines and Mechanisms, Oulu, Finland, (1999), 1208–1215
10. Mata, V., Provenzano, S., Valero, F., Cuadrado, J., I.: Serial-robot dynamics algorithms for moderately large numbers of joints. Mechanism and Machine Theory, 37 (2002) 739–755
11. Rovetta, A., Kulvietis, G.: Lo sviluppo di software per il controllo dinamico di robot industriali. Dipartimento di Meccanica, Politecnico di Milano, Milano (1986)
12. Saha, S.K.: A decomposition of the manipulator inertia matrix, IEEE Trans. Rob. Autom. 13 (2) (1997) 301–304
13. Surdhar, J., S., White, A., S.: A parallel fuzzy-controlled flexible manipulator using optical tip feedback. Robotics and Computer-Integrated Manufacturing, Vol. 19 (3) (2003) 273–282
14. Vucobratovic, K., M., Kircanski M., N.: Real-time Dynamics of Manipulation Robots, Springer-Verlag, Berlin Heidelberg New York (1985)
15. Walker, M.W., Orin, D.E.: Efficient dynamic computer simulation of robotic mechanisms, J. Dyn. Syst. Meas. Control 104 (1982) 205–211

Revisiting Some Control Schemes for Chaotic Synchronization with Mathematica

Andrés Iglesias[1,2,⋆] and Akemi Galvez[2]

[1] Department of Computer Science, University of Tsukuba, Laboratory of Advanced Research, Building B, Room # 1025, Kaede Dori, 305-8573, Tsukuba, Japan
[2] Department of Applied Mathematics and Computational Sciences, University of Cantabria, Avda. de los Castros, s/n E-39005, Santander, Spain
iglesias@unican.es
uc8031@alumnos.unican.es
http://personales.unican.es/iglesias

Abstract. An interesting topic in dynamical systems is *chaotic synchronization*, that is, the possibility to synchronize the behavoir of several chaotic systems. Among the variety of available schemes, those dealing with some sort of control mechanisms have received increasing attention in recent years. In this work, we applied the program `Mathematica` to study the control strategies of chaotic synchronization. In our opinion, the powerful symbolic, numeric and graphic capabilities of Mathematica make this software an ideal tool to analyze the problem on hand.

1 Introduction

A very interesting issue in dynamical systems is the so-called *chaotic synchronization*, that is, the possibility to synchronize the behavior of several chaotic systems. Among the different schemes for chaotic synchronization, those based on the application of some kind of control have received increasing attention during the last few years. This paper analyzes some control strategies for chaotic synchronization by using the program Mathematica [16], one of the most popular and widely used computer algebra systems. One of the most remarkable Mathematica features is the integration of very powerful symbolic, numerical and graphical capabilities within a uniform framework. This feature is especially useful here because the equations describing the dynamics of the chaotic systems are nonlinear, and hence, an adequate combination of numerical and symbolic procedures is usually required in order to analyze their behavior. The reader is referred to [8, 11, 12] for previous works about the application of Mathematica to the analysis of chaotic systems.

⋆ Corresponding author.

V.S. Sunderam et al. (Eds.): ICCS 2005, LNCS 3516, pp. 651–658, 2005.

The structure of the paper is as follows: in Section 2 we describe some control schemes for chaotic synchronization. They have been organized for clarity into two subsections, devoted to those methods based on designing a controller and those based on the stabilization of the error dynamics. Finally, the paper closes with the conclusions and some further remarks.

2 Control Schemes for Chaotic Synchronization

Recently, it has been shown that chaotic systems can be synchronized by applying some kind of control [1, 2, 3, 4, 5, 7]. For example, [2] presents the synchronization of a couple of Lorenz systems by using active control, while in [3] and [4] controllers for the Duffing and logistic systems respectively have been described. In particular, in [3] the authors tried to answer the following question: *given a chaotic system and an arbitrary reference signal, is it possible to design a controller based on this reference signal so that the output of the chaotic system follows this signal asymptotically?* As remarked by the authors, the synchronization of chaotic systems belongs to this class of problems: it is enough to consider the output of one of the chaotic systems as the reference signal. In fact, the approach in [3] is more general, since the reference signal could be not only the output of a chaotic system but also any other (linear or nonlinear) signal.

2.1 Designing a Controller for Chaotic Synchronization

The basic idea of this method is to consider a chaotic system in the form:

$$\begin{cases} x'(t) = y(t) \\ y'(t) = f(x(t), y(t), t) + u(t) \end{cases} \tag{1}$$

where $f(x(t), y(t), t)$ is a nonlinear function of x, y and t and $u(t)$ is the controller to be designed. If $z(t)$ is the reference signal to be followed by the system, the proposed control function $u(t)$ takes the form [3]:

$$u(t) = -f(x(t), y(t), t) - \alpha_1 x(t) - \alpha_2 y(t) + [z''(t) + \alpha_2 z'(t) + \alpha_1 z(t)] \tag{2}$$

It can proved (see [3] for details) that the error between the output of the chaotic system and the reference signal converges to zero as long as the constants α_1 and α_2 are both greater than zero for any initial condition. It is important to remark that this control function $u(t)$ does not contain any information about the system structure which generates the signal $z(t)$. As a consequence, this scheme can be applied to a very wide variety of situations, such as, for instance, inhomogeneous driving.

To illustrate this method, let us consider the Duffing system given by the couple of ordinary differential equations:

```
In[1]:= duf1={x'[t]==y[t],y'[t]==1.8 x[t]-0.1 y[t]-
              x[t]^3+1.1 Cos[0.4 t],x[0]==0,y[0]==2}
```

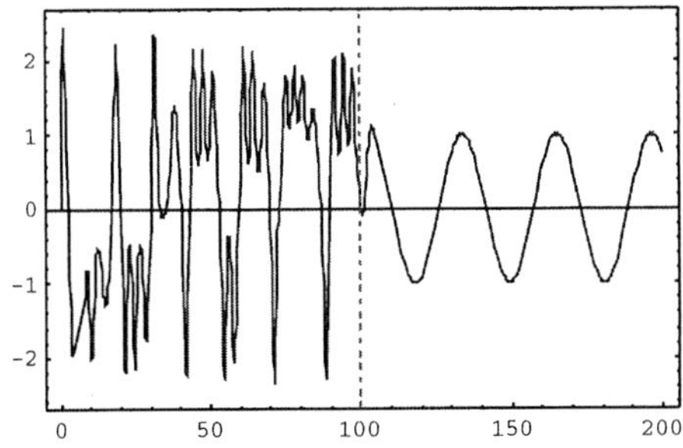

Fig. 1. Synchronization of a Duffing system to a driving signal given by $z(t) = sin(0.2\,t)$. The figure shows the temporal series of the x variable of such a system before and after the application of the synchronization procedure (indicated by the vertical dashed line)

The chaotic behavior of this system in the interval (0,100) is depicted in Figure 1 (to the left of the dashed line):

```
In[2] := NDSolve[duf1,{x,y},{t,0,100},MaxSteps->20000]//Flatten
In[3] := x1[t_]=x[t] /. %;
```

Now, we would like to force the system to follow the sinusoidal signal $z(t) = sin(0.2\,t)$. To accomplish this, we apply the control function, $u(t)$, as follows:

```
In[4] := z[t_]:=Sin[0.2t];
In[5] := u[t_]=-1.8 x[t]+0.1 y[t]+x[t]^3-1.1 Cos[0.4 t]-x[t]-y[t]+
             z''[t]+z'[t]+z[t];
```

According to (1), for this particular choice of the reference signal, we consider the system:

```
In[6] := duf2={x'[t]==y[t],y'[t]==1.8 x[t]-0.1 y[t]-x[t]^3+
            1.1 Cos[0.4 t]+u[t],x[100]==0.044674,y[100]==-0.61367}
```

where the initial conditions were obtained from the previous integration of the system duf1. Now, we integrate duf2 and display the resulting temporal series for the x variable:

```
In[7] := NDSolve[duf2,{x,y},{t,100,200},MaxSteps->20000]//Flatten;
In[8] := sx[t_]:=If[0≤t≤100,x1[t],x[t] /. %];
In[9] := Show[{Plot[sx[t],{t,0,200},PlotStyle->RGBColor[0,0,1],
             PlotRange->{-2.7, 2.7},Frame->True],
             Graphics[{RGBColor[1,0,0],Dashing[{0.01}],
             Line[{{100,-2.7},{100,2.7}}]}]}]
```

$Out[9] := See\,Figure\,1$

Figure 1 shows the evolution of the variable x of the Duffing system before and after the application of the control function (indicated by the vertical dashed line). From this figure, the good performance of this synchronization method between the chaotic system and the driving signal becomes clear.

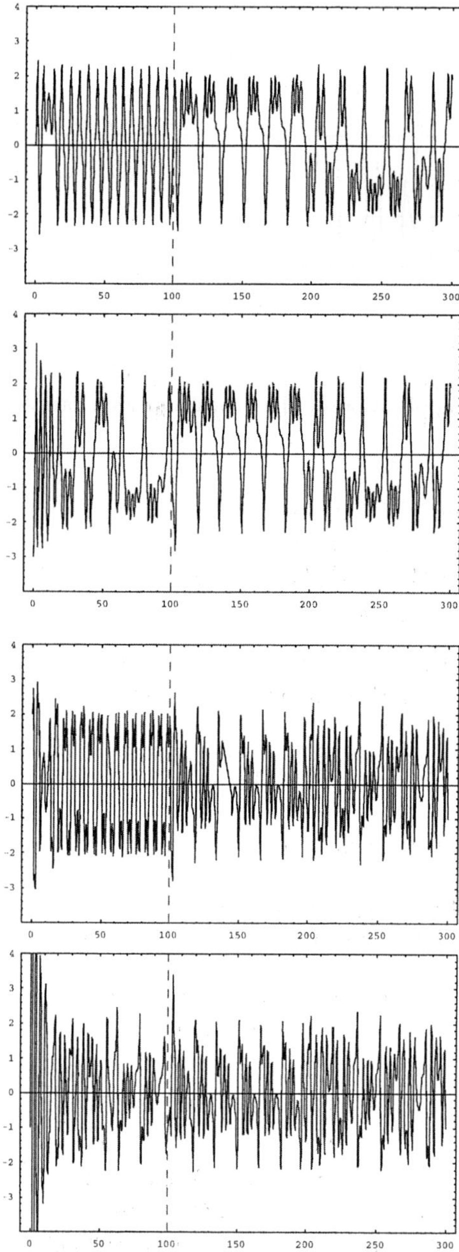

Fig. 2. Synchronization of two Duffing systems using active control in which the signal of the drive acts as the driving signal for the response: temporal series of the drive and response systems for the variables x (top) and y (bottom) respectively

As remarked above, this method is useful to achieve chaos synchronization by simply considering the output of a system (the drive) as the reference signal

$z(t)$ to be injected into the response. In the next example we consider two non-identical (since they are given by different parameter values) chaotic Duffing systems:

```
In[10]:= duf3={x3'[t]==y3[t],y3'[t]==1.8 x3[t]-0.1 y3[t]-
           x3[t]^3+1.1 Cos[t],x3[0]==0,y3[0]==2};
In[11]:= duf4={x4'[t]==y4[t],y4'[t]==1.8 x4[t]-0.1 y4[t]-
           x4[t]^3+1.1 Cos[0.4 t],x4[0]==-3,y4[0]==-1};
```

Integrating these equations in the interval (0,100), we can observe a chaotic behavior for both variable x and y as shown in Figure 2 (to the left of the dashed line):

```
In[12]:= rul1=NDSolve[Union[duf3,duf4],{x3,y3,x4,y4},{t,0,100},
           MaxSteps->20000]//Flatten;
```

Now, we force the system to follow the output of the drive system $z(t) = x_3(t)$. Hence, after $t = 100$ the method is applied with $u(t)$ given by:

```
In[13]:= u[t_]=-1.8 x4[t]+0.1 y4[t]+x4[t]^3-1.1 Cos[0.4 t]-
           x4[t]-y4[t]+x3''[t]+x3'[t]+x3[t];
```

so that the second system becomes:

```
In[14]:= duf4={x4'[t]==y4[t],y4'[t]==1.8 x4[t]-0.1 y4[t]-x4[t]^3+
        1.1 Cos[0.4 t]+u[t],x4[100]==-0.429856,y4[100]==-1.73792};
```

Finally, we integrate the resulting system and display the temporal series for the variables x and y of the drive and response systems, as shown in Figure 2.

```
In[15]:= rul2=NDSolve[Union[duf3,duf4],{x3,y3,x4,y4},{t,100,300},
           MaxSteps->30000]//Flatten;
In[16]:= {sx3[t_],sx4[t_],sy3[t_],sy4[t_]}=If[0≤t≤100,
           # /. rul1,# /. rul2]& /@ {x3[t],x4[t],y3[t],y4[t]};
In[17]:= Show[Plot[#,{t,0,300},PlotStyle->RGBColor[1,0,0],
           PlotRange->{-4,4},Frame->True],
           Graphics[{RGBColor[1,0,0],Dashing[{0.02}],
           Line[{{100,-4},{100,4}}]}]}]& /@
           {sx3[t],sx4[t],sy3[t],sy4[t]}
```

Out[17] := *See Figure* 2

2.2 Chaotic Synchronization via Stabilization of the Error Dynamics

In [9] a method that allows synchronizing drive-response connections that do not synchronize in the Pecora-Carroll [15] sense (that is, connections with positive conditional Lyapunov exponents) was presented. The key idea of the method is considering a convex combination of the drive and the response subsystems as the new driving signal. In this combination, the component associated with the response system acts as a chaos suppression method stabilizing the dynamics of this system. Then, the component associated with the drive imposes the behavior of the drive into the stabilized response synchronizing both systems.

Also in [7] a strategy to synchronize two strictly different chaotic oscillators (a Duffing and a Van der Pol oscillators as the drive and the response systems, respectively) was designed by following the next steps: firstly, the problem of chaos synchronization was seen as a stabilization one, where the goal is to stabilize the dynamical system of the synchronization error. Then, the chaos suppression problem was solved through a robust asymptotic controller, which is based on geometrical control theory. To estimate the differences between the models, they used a nonlinear function, which can be interpreted as a state variable in a equivalent extended system. Thus, the controller allows the energy "excess" from the slave system to be absorbed and that the oscillators behave in a synchronous way.

A similar idea was proposed in [1] for a couple of Rossler systems and in [5] to synchronize two nonlinear chaotic electronic circuits. For example, in [1] the authors considered an active control to be applied to the second system of a couple drive-response in order to get synchronization. The aim of this feedback control is to achieve the asympotic stability of the zero solution of the error system (i.e., the differences between the drive and the response). The performance of this method can be illustrated by considering the Rossler model, described by a set of three differential equations. The drive system is given by:

```
In[18]:= ros1={x1'[t]==-y1[t]-z1[t],y1'[t]==x1[t]+0.2 y1[t],
     z1'[t]==0.2+z1[t](x1[t]-5.7),x1[0]==0.5,y1[0]==1,z1[0]==1.5};
```

For the second system we introduce three control functions $u_1(t)$, $u_2(t)$ and $u_3(t)$ as:

```
In[19]:= ros2={x2'[t]==-y2[t]-z2[t]+u1[t],y2'[t]==x2[t]+0.2 y2[t]+
              u2[t],z2'[t]==0.2+z2[t](x2[t]-5.7)+u3[t],
              x1[0]==2.5,y1[0]==2,z1[0]==2.5};
```

To determine those functions, the authors considered the error system as the difference between both Rossler systems and defined the control functions as:

```
In[20]:={u1[t],u2[t],u3[t]}={V1[t],V2[t],x1[t]z1[t]-x2[t]z2[t]+V3[t]};
```

There are many possible choices for these control functions $V_1(t)$, $V_2(t)$ and $V_3(t)$. For example, we can choose:

In[21]:= $$\begin{pmatrix} \texttt{V1[t]} \\ \texttt{V2[t]} \\ \texttt{V3[t]} \end{pmatrix} = \begin{pmatrix} -1 & 1 & 1 \\ -1 & -1.2 & 0 \\ 0 & 0 & 4.7 \end{pmatrix} \begin{pmatrix} \texttt{x2[t]} - \texttt{x1[t]} \\ \texttt{y2[t]} - \texttt{y1[t]} \\ \texttt{z2[t]} - \texttt{z1[t]} \end{pmatrix};$$

For this choice, the difference system becomes:

```
In[22]:= rhs[Equal[a_,b_]]:=b;
In[23]:= {sys2,sys1}=rhs/@ #& /@ (Take[#,3]& /@{ros2,ros1});
```

```
In[24]:= sys2-sys1 //Simplify
```

$$Out[24] := -\begin{pmatrix} x2[t]-x1[t] \\ y2[t]-y1[t] \\ z2[t]-z1[t] \end{pmatrix}$$

whose characteristic matrix is $-I$, where I is the diagonal matrix. Therefore, the difference system has all its eigenvalues negative implying that the error converges to zero as time t goes to infinity and hence the synchronization of the two Rossler systems is achieved (see [1] for details).

3 Conclusions and Further Remarks

These previous proposals can be better understood as particular cases of a more general theory, for which controlled synchronization is seen as the procedure to find, given both the transmitter (or drive) and the receiver (or response) as well as the corresponding output function, a suitable mechanism to control the response system such that the drive and response will asymptotically synchronize [14]. In general, this problem can be describe as follows: given a transmitter $x' = f(x)$ with output $y = g(x)$ and a receiver $\overline{x}' = h(\overline{x}, u)$ systems, where we assume that both x and $\overline{x}$ are n-dimensional vectors, and u is the unknown control function, the goal is to obtain this control function such that x and $\overline{x}$ asymptotically synchronize.

Of course, there are many different ways to solve this problem, mostly based on the use of a feedback control of the form $u = \theta(\overline{x}, y)$ where θ is a smooth function depending on the receiver $\overline{x}$ and the output of the transmitter y. However, it must be said that this problem of finding a suitable output feedback controller achieving synchronization between the transmitter and the receiver is, in general, a very difficult (or even impossible) task and only particular examples (as those shown above) have been described in the literature. Other interesting references on controlled synchronization can be found in [6, 10, 17].

All the commands have been implemented in Mathematica version 4 [16]. The symbolic capabilities and the powerful Mathematica functional programming [13] have been extensively used to make the programs shorter and more efficient. In our opinion Mathematica is a very useful tool to analyze the interesting problem of chaotic synchronization through control schemes in a user-friendly and easy way.

References

1. Agiza, H.N., Yassen, M.T.: Synchronization of Rossler and Chen chaotic dynamical systems using active control, Phys. Lett. A **278** (2001) 191-197
2. Bai, E.W., Lonngren, K.E.: Synchronization of two Lorenz systems using active control, Chaos, Solitons and Fractals **8** (1997) 51-58
3. Bai, E.W., Lonngren, K.E.: Synchronization and control of chaotic systems, Chaos, Solitons and Fractals **10** (1999) 1571-1575

4. Bai, E.W., Lonngren, K.E.: A controller for the logistic equations, Chaos, Solitons and Fractals **12** (2001) 609-611
5. Bai, E.W., Lonngren, K.E., Sprott, J.C.: On the synchronization of a class of electronic circuits that exhibit chaos, Chaos, Solitons and Fractals **13** (1999) 1515-1521
6. Blekhman, I.I., Fradkov, A.L., Nijmeijer, H., Yu, A.: On self-synchronization and controlled synchronization, Syst. Contr. Lett. **31** (1997) 299-305
7. Femat, R., Solis, G.: On the chaos synchronization phenomena, Phys. Lett. A **262** (1999) 50-60
8. Gutiérrez, J.M., Iglesias, A.: A Mathematica package for the analysis and control of chaos in nonlinear systems, Computers in Physics **12**(6) (1998) 608-619
9. Gutiérrez, J.M., Iglesias, A.: Synchronizing chaotic systems with positive conditional Lyapunov exponents by using convex combinations of the drive and response systems, Phys. Lett. A **239**(3) (1998) 174-180
10. Huijberts, H.J.C., Nijmeijer, H., Willems, R.: Regulation and controlled synchronization for complex dynamical systems, Int. J. Robust Nonlinear Contr. **10** (2000) 363-377
11. Iglesias, A., Gutiérrez, J.M., Ansótegui, D., Carnicero, M.A.: Transmission of digital signals by chaotic synchronization. Application to secure communications, In: Keranen, V., Mitic, P., Hietamaki, A. (Eds.) Innovation in Mathematics. Proceedings of the Second International Mathematica Symposium-IMS'97, Computational Mechanics Publications, Southampton, England (1997) 239-246
12. Iglesias, A., Gálvez, A.: Analyzing the synchronization of chaotic dynamical systems with Mathematica: Parts I-II. Computational Science and its Applications-ICCSA'2005. Lecture Notes in Computer Science (2005) (*in press*)
13. Maeder, R.: Programming in Mathematica, Second Edition, Addison-Wesley, Redwood City, CA (1991)
14. Nijmeijer, H.: A dynamical control view on synchronization, Physica D **154** (2001) 219-228
15. Pecora, L.M., Carroll, T.L.: Synchronization in chaotic systems, Phys. Rev. Lett. **64** (1990) 821-823
16. Wolfram, S.: The Mathematica Book, Fourth Edition, Wolfram Media, Champaign, IL & Cambridge University Press, Cambridge (1999)
17. Yu, A., Pogromsky, Nijmeijer, H.: Observer based robust synchronization of dynamical systems, Int. J. of Bifurc. and Chaos **8** (1998) 2243-2254

Three Brick Method of the Partial Fraction Decomposition of Some Type of Rational Expression

Damian Słota and Roman Wituła

Institute of Mathematics, Silesian University of Technology, Kaszubska 23, 44-100 Gliwice, Poland
{d.slota, rwitula}@polsl.pl

Abstract. The purpose of this paper is to present a new method of the partial fraction decomposition, the so called "three brick method". The method is of algebraic nature and it is based on simple reduction tasks. The discussed algorithm has been successfully implemented in Mathematica software package.

1 Introduction

The partial fraction decomposition is a very important topic in symbolic and computational mathematics, for example in inverse Laplace transform. "The residuum method" which is commonly applied to find the inverse Laplace transform, although very legible from the theoretical point of view, turns out as very tedious from the computational point of view (especially if the calculations are finally meant to get rid of the complex notation). The latter often tends to be "the bottleneck" in the whole process of calculating the inverse Laplace transform. Thus, an alternative method of the decomposition of rational functions into partial fractions has evolved considerable interest [2].

In the paper a new method of the partial fraction decomposition, called here "three brick method", is discussed. It is a simple and very efficient method of the decomposition of rational functions:

$$W(s) = \frac{as+b}{\left(s^2+\alpha s+\beta\right)\left(s^2+\gamma s+\delta\right)} \tag{1}$$

into partial fractions. It is a decomposition method of pure algebraic nature and is based on simple reduction tasks.

2 Three Brick Method

Let α, β, γ, $\delta \in \mathbb{C}$, $\alpha \neq \gamma$. The case of $\alpha = \gamma$ is expressed in formula (14). Let us put:

$$M_1 := s^2 + \alpha s + \beta \ , \qquad M_2 := s^2 + \gamma s + \delta \tag{2}$$

V.S. Sunderam et al. (Eds.): ICCS 2005, LNCS 3516, pp. 659–662, 2005.

and $M = M_1 M_2$. The method of decomposition of rational function $W(s)$ in the form of $(a, b \in \mathbb{C})$:

$$\frac{as+b}{M} = \frac{As+B}{M_1} + \frac{Cs+D}{M_2} \,, \tag{3}$$

is described below.

First, the following system of equations $(I_{eq}, II_{eq}, III_{eq})$ is created:

$$\begin{cases} I_{eq}: & \frac{1}{M_1} - \frac{1}{M_2} = \frac{M_2 - M_1}{M} = \frac{(\gamma-\alpha)s + \delta - \beta}{M} \,, \\ II_{eq}: & \frac{s}{M_1} - \frac{s}{M_2} = \frac{s(M_2 - M_1)}{M} = \frac{(\gamma-\alpha)s^2 + (\delta-\beta)s}{M} \,, \\ III_{eq}: & \frac{2}{M_1} - \frac{1}{M_2} = \frac{2M_2 - M_1}{M} = \frac{s^2 + (2\gamma-\alpha)s + 2\delta - \beta}{M} \,. \end{cases} \tag{4}$$

The next step is to eliminate the summands containing s^2 from the numerators of the fraction at the right side of equations II_{eq} and III_{eq}, thus the fourth equation is generated $(IV_{eq} := II_{eq} - (\gamma - \alpha)\, III_{eq})$:

$$\frac{s - 2\gamma + 2\alpha}{M_1} - \frac{s - \gamma + \alpha}{M_2} = \frac{[\delta - \beta - (\gamma-\alpha)(2\gamma-\alpha)]s - (\gamma-\alpha)(2\delta-\beta)}{M} \,. \tag{5}$$

The next two steps of the algorithm involve the elimination of the summands containing s and the free term from the numerators of the fractions at the right side of equations I_{eq} and IV_{eq}; $V_{eq} := \left(2\gamma - \alpha - \frac{\delta-\beta}{\gamma-\alpha}\right) I_{eq} + IV_{eq}$:

$$\frac{s + \alpha - \frac{\delta-\beta}{\gamma-\alpha}}{M_1} - \frac{s + \gamma - \frac{\delta-\beta}{\gamma-\alpha}}{M_2} = \frac{\alpha\delta - \beta\gamma - \frac{(\delta-\beta)^2}{\gamma-\alpha}}{M} \tag{6}$$

and $VI_{eq} := (2\delta - \beta)\, I_{eq} + \frac{\delta-\beta}{\gamma-\alpha}\, IV_{eq}$:

$$\frac{\frac{\delta-\beta}{\gamma-\alpha}s + \beta}{M_1} - \frac{\frac{\delta-\beta}{\gamma-\alpha}s + \delta}{M_2} = \frac{\left(-\alpha\delta + \beta\gamma + \frac{(\delta-\beta)^2}{\gamma-\alpha}\right)s}{M} \,. \tag{7}$$

Thanks to the next operation: $\frac{1}{k}(-aVI_{eq} + bV_{eq})$, where

$$k := \begin{vmatrix} \alpha & \beta \\ \gamma & \delta \end{vmatrix} - \frac{(\delta-\beta)^2}{\gamma-\alpha} \,, \tag{8}$$

the following decomposition is derived:

$$\frac{as+b}{M} = \frac{1}{k}\left(\frac{\left(b - a\frac{\delta-\beta}{\gamma-\alpha}\right)s + b\alpha - a\beta - b\frac{\delta-\beta}{\gamma-\alpha}}{M_1} + \right.$$
$$\left. - \frac{\left(b - a\frac{\delta-\beta}{\gamma-\alpha}\right)s + b\gamma - a\delta - b\frac{\delta-\beta}{\gamma-\alpha}}{M_2} \right) \,. \tag{9}$$

Remark. The following equivalent equations hold:

$$k = 0 \quad \Leftrightarrow \quad \delta\alpha^2 - \gamma(\delta+\beta)\alpha + \gamma^2\beta + (\delta-\beta)^2 = 0 \ , \tag{10}$$

concurrently, the discriminant of the last polynomial of the second order, in consideration of variable α (assuming that $\delta \neq 0$) has the following form:

$$\Delta_\alpha = (\gamma^2 - 4\delta)(\delta-\beta)^2 \ . \tag{11}$$

Accordingly $\Delta_\alpha < 0$, if, and only if, the discriminant of the trinomial M_2 is negative and if $\beta \neq \delta$. The case of $\beta = \delta$ is expressed in formulas:

$$\frac{1}{(s^2+\alpha s+\beta)(s^2+\gamma s+\beta)} = \frac{1}{\beta(\alpha-\gamma)}\left(\frac{s+\alpha}{s^2+\alpha s+\beta} - \frac{s+\gamma}{s^2+\gamma s+\beta}\right) , \tag{12}$$

$$\frac{s}{(s^2+\alpha s+\beta)(s^2+\gamma s+\beta)} = \frac{1}{\gamma-\alpha}\left(\frac{1}{s^2+\alpha s+\beta} - \frac{1}{s^2+\gamma s+\beta}\right) . \tag{13}$$

If $\alpha = \gamma$ (assuming that $M_1 \neq M_2$) the following decomposition is derived:

$$\frac{as+b}{(s^2+\alpha s+\beta)(s^2+\alpha s+\delta)} = \frac{1}{\delta-\beta}\left(\frac{as+b}{s^2+\alpha s+\beta} - \frac{as+b}{s^2+\alpha s+\delta}\right) . \tag{14}$$

3 Algorithm in Mathematica

The algorithm discussed in the previous section will next be implemented in Mathematica [1, 3]. The summands connected with trinomials M_1 and M_2 shall be expressed in separate notations, so that the transformations are performed separately on both fractions. We start with defining the trinomials:

In[1]:= $\mathrm{m1} = \mathrm{s}^2 + \alpha\,\mathrm{s} + \beta;\ \ \mathrm{m2} = \mathrm{s}^2 + \gamma\,\mathrm{s} + \delta;$

Let us form three equations (I_{eq}, II_{eq}, III_{eq}):

In[2]:= $\mathrm{eq1} = \{1/\mathrm{m1}, -1/\mathrm{m2}\};$
$\mathrm{eq2} = \{\mathrm{s}/\mathrm{m1}, -\mathrm{s}/\mathrm{m2}\};$
$\mathrm{eq3} = \{2/\mathrm{m1}, -1/\mathrm{m2}\};$

The summands containing s^2 are eliminated (IV_{eq}):

In[3]:= $\mathrm{eq4} = \mathrm{Simplify}[\mathrm{eq2} - (\gamma-\alpha)\mathrm{eq3}]$

Out[3]= $\left\{\frac{\mathrm{s}+2\,\alpha-2\,\gamma}{\mathrm{s}^2+\mathrm{s}\,\alpha+\beta}, -\frac{\mathrm{s}+\alpha-\gamma}{\mathrm{s}^2+\mathrm{s}\,\gamma+\delta}\right\}$

The next summands containing s are eliminated (V_{eq}):

In[4]:= $\mathrm{eq5} = \mathrm{Simplify}\left[\left(2\gamma - \alpha - \frac{\delta-\beta}{\gamma-\alpha}\right)\mathrm{eq1} + \mathrm{eq4}\right]$

Out[4]= $\left\{\frac{\alpha^2-\beta+\mathrm{s}\,(\alpha-\gamma)-\alpha\,\gamma+\delta}{(\mathrm{s}^2+\mathrm{s}\,\alpha+\beta)\,(\alpha-\gamma)}, \frac{\beta-\alpha\,\gamma+\gamma^2+\mathrm{s}\,(-\alpha+\gamma)-\delta}{(\alpha-\gamma)\,(\mathrm{s}^2+\mathrm{s}\,\gamma+\delta)}\right\}$

The elimination of the free term (VI_{eq}):

In[5]:= eq6 = Simplify$\left[(2\delta-\beta)\text{eq1}+\left(\frac{\delta-\beta}{\gamma-\alpha}\right)\text{eq4}\right]$

Out[5]= $\left\{\frac{\beta(\alpha-\gamma)+s(\beta-\delta)}{(s^2+s\alpha+\beta)(\alpha-\gamma)},\frac{(-\alpha+\gamma)\delta+s(-\beta+\delta)}{(\alpha-\gamma)(s^2+s\gamma+\delta)}\right\}$

The definition of parameter k:

In[6]:= $k=\text{Det}\left[\begin{pmatrix}\alpha & \beta\\ \gamma & \delta\end{pmatrix}\right]-\frac{(\delta-\beta)^2}{\gamma-\alpha};$

Final decomposition:

In[7]:= up = Total$\left[\frac{1}{k}(-a\,\text{eq6}+b\,\text{eq5})\right]$

Out[7]=
$$\frac{-\frac{a(\beta(\alpha-\gamma)+s(\beta-\delta))}{(s^2+s\alpha+\beta)(\alpha-\gamma)}+\frac{b(\alpha^2-\beta+s(\alpha-\gamma)-\alpha\gamma+\delta)}{(s^2+s\alpha+\beta)(\alpha-\gamma)}}{-\beta\gamma+\alpha\delta-\frac{(-\beta+\delta)^2}{-\alpha+\gamma}}+\frac{\frac{b(\beta-\alpha\gamma+\gamma^2+s(-\alpha+\gamma)-\delta)}{(\alpha-\gamma)(s^2+s\gamma+\delta)}-\frac{a((-\alpha+\gamma)\delta+s(-\beta+\delta))}{(\alpha-\gamma)(s^2+s\gamma+\delta)}}{-\beta\gamma+\alpha\delta-\frac{(-\beta+\delta)^2}{-\alpha+\gamma}}$$

The Apart instruction cannot be used for the derivation of above decomposition:

In[8]:= Apart$[(a\,s+b)/(\text{m1 m2})]$

Out[8]= $\frac{b+a\,s}{(s^2+s\alpha+\beta)(s^2+s\gamma+\delta)}$

4 Conclusion

In this paper a new method of the partial fraction decomposition, the three brick method, is presented. It is a simple and effective method of an algebraic nature, based on simple reduction procedures. The discussed algorithm has successfully been used in Mathematica software package. The advantage of the method is the possibility to perform computations for symbolic data, which is not always possible by using the standard Mathematica commands (Apart).

References

1. Drwal, G., Grzymkowski, R., Kapusta, A., Słota, D.: Mathematica 5. WPKJS, Gliwice (2004) (in Polish)
2. Grzymkowski, R., Wituła, R.: Computational Methods in Algebra, part I. WPKJS, Gliwice (2000) (in Polish)
3. Wolfram, S.: The Mathematica Book, 5th ed. Wolfram Media, Champaign (2003)

Non Binary Codes and "Mathematica" Calculations: Reed-Solomon Codes Over GF (2^n)*

Igor Gashkov

Karlstad University, Department of Engineering Sciences, Physics and Mathematics 65188 Karlstad Sweden
Igor.Gachkov@kau.se

Abstract. The effect of changing the basis for representation of Reed-Solomon codes in the binary form may change the weight distribution and even the minimum weight of codes. Using "Mathematica" and package "Coding Theory" we give some examples of effective changing of basis which gives the binary code with greatest minimum distance [1] (Ch.10. §5. p 300) and codes which have the same distance independently of basis changing.

1 Reed-Solomon Code

A Reed-Solomon (RS) code over the Galois Field GF(q) is a special BCH - code having the length of the code words equal to the number of nonzero elements in the ground field. The RS - codes are cyclic and have as generator polynomial

$$g(x) = (x - a^b)(x - a^{b+1})...(x - a^{d-1}) \tag{1}$$

where a is a primitive element of the field GF(q) and d is the code distance.
The elements of GF(q) can be represented as m-vector of elements from GF(p). Choosing p = 2 we get the binary codes by substituting for each symbol in GF(2^m) the corresponding binary m - vector. We will use the package" Coding Theory" [2], [5].

```
In[1]:=<<CodingTheory.m;BIP=BinaryIrrPolynomials[3,x]
```

Out[1] = $\{1+x^2+x^3, 1+x+x^3\}$

```
In[2]:=r=BIP[[2]];ShowBinaryGaloisField[r, x, a, a]
```

GF(8) is received by extending the field Z_2 by the irreducible polynomial $r(x) = 1 + x + x^3$ and observe that a is a primitive element in this extension field.

Out[2] =

Log	Vector	Polynomial		Min. polynomial
0	(1, 0, 0)	1	1	$1+x$
1	(0, 1, 0)	a	a	$1+x+x^3$
...		...		...
6	(1, 0, 1)	a^6	$1+a^2$	$1+x^2+x^3$

* The research was support by The Royal Swedish academy of Sciences

V.S. Sunderam et al. (Eds.): ICCS 2005, LNCS 3516, pp. 663 – 666, 2005.

2 The Effect of Changing the Basis

Example 1. We construct the generator polynomial for the RS-code of length 7 over the field GF(8) with code distance 4.

```
In[4] := b = 1; d = 4;
Rt = Table[GF[[2+Mod[b+i,Length[GF]-1],3]],{i,0,d-2}];
g = Product[x - Rt[[i]], {i, 1, Length[Rt]}];g =
Collect[PolynomialMod[PolynomialMod[Expand[g],r/.x ->
a], 2], x];g//.CD
```

Out[4]=

$$(x-a)(x-a^2)(x-a^3) \quad x^3+(a^2+1)x^2+ax+a^2+1$$
$$x^2a^6+a^6+xa+x^3$$

We have got the RS - code with the generator polynomial g with parameters [Length=7, Dimension=4, Distance=4] over the field GF(8). First of all we control that binary subspace (over GF(2)) give the code with parameters [7,1,7] with generator polynomial f(x) (Out[5]).

```
In[5] := CmRt = Complement[a^Union[Flatten[
Table[CyclotomicCoset[Exponent[Rt, a][[i]], 2, n], {i,
1, Length[Rt]}]]]], Rt];
g1=g*Product[x - CmRt[[i]], {i, 1, Length[CmRt]}]
f = Collect[PolynomialMod[PolynomialMod[
Expand[g1],r/.x -> a], 2], x]
```

Out[5]=

$$x^6+x^5+x^4+x^3+x^2+x+1$$

With 4 information bits our RS – code (over GF(8)) has 4096 code words, and we can construct all these code words and we calculate all code vectors over GF(8)

$$\{\{0,0,0,a^2,a+a^2,a^2,1+a\},\{0,0,0,1,a^2,1,a\},...\langle 4091\rangle...,$$
$$\{1+a+a^2,a+a^2,1,1,1,1+a+a^2,a^2\},\{1+a+a^2,a+a^2,1,a^2,1+a,1+a^2\}\}$$

and overwrite as

$$\{\{0,0,0,a^2,a^4,a^2,a^3\},\{0,0,0,1,a^2,1,a\},\{0,0,0,a^4,a^6,a^4,a^5\},...\langle 4089\rangle...,$$
$$\{a^5,a^4,1,0,a^6,a^4,a^4\},\{a^5,a^4,1,1,1,a^5,a^2\},\{a^5,a^4,1,a^2,a^3,a,a^6\}\}$$

All code vectors with minimum weight can we calculate using Mathematica

$$\{\{0,0,0,a^2,a^4,a^2,a^3\},\{0,0,0,1,a^2,1,a\},\{0,0,0,a^4,a^6,a^4,a^5\},...\langle 239\rangle...,$$
$$\{a^5,a^3,a^4,0,0,0,a^3\},\{a^5,a^4,0,0,a,0,a^6\},\{a^5,a^4,a^3,0,0,1,0\}\}$$

Its means that we have all code vectors with weight 4 and first coordinate is 1.

$$\begin{pmatrix}
1 & a^2 & 1 & a & 0 & 0 & 0 \\
1 & 0 & a^5 & a^4 & a^3 & 0 & 0 \\
1 & 0 & a^3 & 0 & a^5 & a^3 & 0 \\
1 & 0 & a^3 & a^3 & 0 & 0 & a^4 \\
1 & 0 & a^4 & 0 & 1 & 0 & a^2 \\
1 & 0 & a & 0 & 0 & a^6 & a^3 \\
1 & 0 & a^2 & 1 & 0 & a^4 & 0 \\
1 & 0 & 0 & a^5 & a^2 & a^6 & 0 \\
1 & 0 & 0 & a^4 & 0 & a^2 & a \\
1 & 0 & 0 & a & a^4 & 0 & 1 \\
1 & 0 & 0 & 0 & a^6 & a & a^6 \\
1 & a^6 & 0 & 0 & a^3 & 0 & a \\
1 & a^6 & a^5 & 0 & 0 & a^2 & 0 \\
1 & a & 0 & 0 & 0 & 1 & a^2 \\
1 & 1 & 0 & 0 & a & a^4 & 0 \\
1 & a^5 & 0 & a^2 & 0 & a^5 & 0 \\
1 & a^5 & a^6 & 0 & 0 & 0 & a^5 \\
1 & a^4 & a & 0 & a^2 & 0 & 0 \\
1 & a^4 & 0 & a^5 & 0 & 0 & a^3 \\
1 & a^3 & 0 & a^6 & a^6 & 0 & 0
\end{pmatrix}$$

We can see that a change of the basis (the representation of the elements of the field GF(8) as binary vector) may change the minimum weight of the code. If we take the standard basis: $1 \to (0,0,1); a \to (0,1,0); a^2 \to (1,0,0)$ we get the binary code with parameters [3*7 = 21, 3*4 = 12, 4] with the same code distance , we have vector (1, a, 0, 0, 0, 1, a^2) , but if we change basis as :

$$1 \to (0,0,1); a \to (0,1,0); a^2 \to (1,1,1) \tag{2}$$

we obtained a code with parameters [21, 12, 5] and weight polynomial

$$w(x) = x^{21} + 21x^{16} + 168x^{15} + ... + 168x^6 + 21x^5 + 1$$

We finally consider the dual to the found code. As usual we find the weight polynomial of the dual code by using the Mc Williams Identity:

```
In[6] :=McWilliamsIdentity[W+1,21,x]
```

Out[6]= $1+210x^8+280x^{12}+21x^{16}$

We see that the dual code has the parameters [21, 9, 8] (Golay code).

3 Conclusion

With the introduction of computers and computer algebra and some application program as "Mathematica "and special packages [3] [4] gives possibility to solve some problems which solving without computers is very difficult. Using above construction we can find some Reed-Solomon codes which have the same distance independently of basis changing.

Example 2. We construct the generator polynomial for the RS-code of length 7 over the field GF (8) with code distance 3.
The same procedure as (**Example 1**) gives the generator polynomial of RS code

$$(x-a^4)(x-a^5)$$

which have the same code distance as independently of basis changing.

```
In[7] :=  CyclotomicCoset[5, 2,Length[GF]- 1]
CyclotomicCoset[2, 2, Length[GF]  - 1]
```

Out[7]=

{5, 3, 6} {2, 4, 1}

References

1. MacWilliams, F. J., and Sloane, N. J. A. (1977) The Theory of Error - Correcting Codes. North - Holland, Amsterdam.
2. I Gachkov (2003) *Error Correcting codes with Mathematica,* Lecture note in Computer science LNCS 2657 737-746
3. I Gachkov (2004) *Computation of weight enumerators of binary linear codes using the package " Coding Theory "* 6th International Mathematica Symposium Banff Canada eProceedings 16 p.
4. I. Gashkov (2004) *Constant Weight Codes with Package CodingTheory.m in Mathematica*. Lecture note in Computer science LNCS 3039 370-375
5. I. Gashkov (2004) *Package CodingTheory.m in Mathematica.* Download the package Mathematica web page http://library.wolfram.com/infocenter/MathSource/5085/

Stokes-Flow Problem Solved Using Maple

Pratibha[1] and D.J. Jeffrey

Department of Applied Mathematics,
The University of Western Ontario, London, Ontario, Canada
[1] Present address: Information Technology Development Agency (ITDA),
Government of Uttaranchal,272-B, Phase II, Vasant Vihar,
Dehradun, India 248 006

Abstract. An unusual boundary-value problem that arises in a fluid-mechanical application is solved to high precision, as a challenge problem in scientific computation. A second-order differential equation must be solved on $(0,\infty)$, subject to boundary conditions that specify only the asymptotic behaviour of the solution at the two ends of the solution domain. In addition, the solution is required to high accuracy to settle a conjecture made by previous authors. The solution is obtained by computing multiple series solutions using Maple.

1 Introduction

Computations of fluid flow sometimes lead to unusual problems in the solution of ordinary differential equations. The present problem comes from a paper by O'Neill and Stewartson [1]. The problem posed is an excellent test case for the application of computer algebra systems, such as MAPLE, to scientific computation. A function $A(s)$ satisfies the differential equation

$$s^3 K' A'' + sA'[s^2 K'' + 3\,s\,K' + 2\,K] - A[s^2 K'' + 4sK' + 2K] = -s^2 X''\ , \quad (1)$$

where $K = s^{-1} - \coth s$ and $X = \coth s - 1$, subject to the boundary conditions that $A(s)$ is no more singular than s^{-2} at the origin and that $A(s)$ decays at infinity. The equation has a regular singular point at $s = 0$ and an irregular point at infinity. In terms of this function, constants k_1 and k_2 must be calculated according to the formulae

$$\begin{aligned} k_1 = \frac{4}{5} + \frac{1}{2}\int_0^\infty \Bigg\{ & \left(A + \frac{3}{5s^2}\right)\left[2s\,\mathrm{csch}^2\,s - (\coth s - 1)(1 + 2s + s^2\,\mathrm{csch}^2\,s)\right] \\ & -\frac{3}{5}\,e^{-2s}\left[\frac{1}{s^3} + \frac{1}{s^2} + \frac{2}{3s} + \frac{5}{3}(2s-1)\,\coth s\right] \\ & -\frac{3}{5}\left[\frac{2}{s^3} - \left(\frac{3}{s^2} + \frac{2}{s}\right)(\coth s - 1)\right]\Bigg\}\ ds, \end{aligned} \quad (2)$$

$$k_2 = \frac{1}{5} + \frac{1}{4}\int_0^\infty \left[4\,s\,A + s^2\,A''\,K + 2\,(\coth s - 1) - 4\,e^{-2s}/5s\right]\ ds. \quad (3)$$

V.S. Sunderam et al. (Eds.): ICCS 2005, LNCS 3516, pp. 667–670, 2005.

The challenge is to verify (by computation) that the two constants are equal.

The problem is an ideal one for exploring ways in which MAPLE can contribute to computational problems. Because of the singular points in the equation, the standard way to solve them numerically is to generate the first steps of a numerical solution using a series expansion about the origin, and about infinity. The truncation of the series and the change from one integration method to another generate some inaccuracies that might be acceptable in other contexts. However, since the question concerns the difference between the constants at the fifth and sixth significant figures, we need a highly accurate solution. Here we show that such a solution can be obtained by MAPLE with relatively little effort on the part of the user, the work being done by routines that we developed in [2].

The use of series expansions in the solution of differential equations has always been hampered by several difficulties. The first difficulty is the laborious nature of their derivation. This is compounded by the second difficulty, which is the slow rate of convergence of the series. In order to obtain acceptable accuracy, many terms are needed, but the calculation of these is tedious. The final difficulty is their radius of convergence. Often a singularity in the complex plane will prevent the series converging at all points of interest, although the function is well behaved on the real line.

All of these difficulties can be overcome in this case using MAPLE. The new routines developed in [2] allow us to obtain large numbers of terms in the series with little effort and quickly. Using this fact we can re-expand the solution at different points along the axis, thus working around the convergence problem. We shall thus obtain a highly accurate solution and prove that the two constants k_1 and k_2 are indeed equal.

2 Solutions About $s = 0$

The differential equation (1) has a regular singular point at the origin and another in the complex plane at $s = i\pi$. We denote the homogeneous solutions about $s = 0$ as $A_h^{(0)}$ and the particular integral as $A_p^{(0)}$. Using MAPLE we found each as a series expression correct to 75 terms.

$$A_h^{(0)} = s^{(-2+\sqrt{10})}[1 + \sum_{\substack{n=2 \\ n \text{ even}}}^{74} a_n\, s^n], \tag{4}$$

where the first few constants in the series are

$$a_2 = \frac{13}{270} - \frac{2\sqrt{10}}{135}, \qquad a_4 = -\frac{31}{2268} + \frac{967\sqrt{10}}{226800}, \qquad a_6 = \frac{5}{6804} - \frac{313\sqrt{10}}{1360800}.$$

The particular integral $A_p^{(0)}$, correct to 75 terms is found as

$$A_p^{(0)} = \sum_{\substack{n=-2 \\ n \text{ even}}}^{74} b_n\, s^n, \tag{5}$$

where

$$b_{-2} = -\frac{3}{5}, \qquad b_0 = -\frac{6}{25} \tag{6}$$

$$b_2 = -\frac{107}{1575}, \qquad b_4 = \frac{394}{307125} \tag{7}$$

We reject the other homogeneous solution because asymptotically it behaves as $O(s^{-2-\sqrt{10}})$, for small values of s. Hence for small s, the general solution of differential equation (1) is

$$A(s) = A_p^{(0)} + c\,A_h^{(0)}, \tag{8}$$

where the constant c has to be determined.

3 Series Solutions About Other Expansion Points

The solutions $A_h^{(0)}$ and $A_p^{(0)}$ converge only for $s < \pi$. This is unavoidable because of the singularity on the imaginary axis. We expand the functions $A_p^{(0)}$ and $A_h^{(0)}$ along the real axis by analytic or numerical continuation. We use the solutions $A_h^{(0)}$ and $A_p^{(0)}$ to give us boundary conditions at a new expansion point. We begin with the series solutions about $s = 5/2$, an ordinary point of the differential equation (1). We find the homogeneous solution $A_h^{(5/2)}$ of the homogeneous version of the differential equation (1) (i.e. right-hand side is zero) subject to the boundary conditions

$$A_h^{(5/2)} = A_h^{(0)}\bigg|_{s=5/2}, \qquad \frac{d}{ds}A_h^{(5/2)} = \frac{d}{ds}A_h^{(0)}\bigg|_{s=5/2}. \tag{9}$$

Similarly, we find the particular integral $A_p^{(5/2)}$. The general solution of the differential equation (1) about $s = 5/2$ is

$$A(s) = A_p^{(5/2)} + c\,A_h^{(5/2)}. \tag{10}$$

Continuing in this way, we find the solutions at expansion points $s = 4, 6, 8$ etc. To achieve high accuracy, all the solutions were found correct to 75 terms.

4 Asymptotic Solution

About $s = \infty$, we obtain an asymptotic expansion as follows.

For large s, $\coth(s) \sim 1$. We therefore replace $\coth s$ with 1 and find the particular integral a_p and the homogeneous solution a_h for the resulting equation to be

$$a_h = e^{-2s}, \qquad a_p = -2\,s^2\,e^{-2s}. \tag{11}$$

We neglect the other homogeneous solution which is $1 - 2s$ for large s.

Higher order approximations are found in the same manner. Hence for large s, the differential equation (1) has the asymptotic solution

$$A(s) \sim a_p + C\,a_h \tag{12}$$

where the constant C has to be determined.

5 The Constants C and c

We find the constants C and c by matching the two solutions, for s small and for s large. The two solutions and their derivatives are equal for those values of s where these solutions converge. By solving the two equations, we get the constants c and C as given in Table (1).

Table 1. Values of constants C and c for different N

N	C	c
30	3.8687549168	0.218368701355
50	3.8797383823	0.218369454539
74	3.8797393835	0.218369454587

6 Definite Integrals k_1 and k_2

Since we have series expansions at all points, the evaluation of the integral is equally straightforward. The integrals k_1 and k_2 are found as

$$k_1 = -0.2622674637, \tag{13}$$
$$k_2 = -0.2622674637. \tag{14}$$

The programs used for the series solutions of ODEs and the programs used in this particular computation can be obtained from the authors.

References

1. M.E. O'Neill and K. Stewartson 1967 "On the slow motion of a sphere parallel to a nearby plane wall" *J. Fluid Mech.* **27**, 705–724.
2. Pratibha 1995 "Maple tools for hydrodynamic interaction problems", PhD Thesis, The University of Western Ontario, London, Canada.

Grounding a Descriptive Language in Cognitive Agents Using Consensus Methods

Agnieszka Pieczynska-Kuchtiak

Institute of Information Science and Engineering, Wroclaw University of Technology, Wybrzeze Wyspianskiego 27, 50-370 Wroclaw, Poland
agnieszka.pieczynska-kuchtiak@pwr.wroc.pl

Abstract. In biological systems the process of symbol grounding is realised at a neural level. But in case of artificial subjects it must be done by certain organisations of private data structures and data collections, and by internal data processing and recognition algorithms. In this work a possible way of implementation of the well - known problem of symbol grounding is presented. The idea of an algorithm of grounding formulas in a private agent's data structure is given. It is assumed that from the agent's point of view stored perceptions are the original source of any meaning accessible for cognitive processes of symbol grounding. This algorithm is used when the state of a particular object from external world is not known. As a results the logic formulas with modal operators and descriptive connectives are generated.

1 Introduction

The problem of symbol grounding has been widely discussed in artificial intelligence, robotics, cognitive science and linguistics [2], but it still fails to practical solution applied by artificial systems [11]. Complete study of the grounding problem is very difficult task and practical solutions of this problem seem to be a big challenge. In biological systems the process of grounding is realised at a neural level. But in case of artificial subjects it must be done by certain organisations of private data structures and data collections, and by internal data processing and recognition algorithms [1].

The symbol grounding is strictly related to the problem of generating statements about the external world. According to the cognitive linguistics and the phenomenology of knowledge each external formula generated by the cognitive agent needs to be grounded in relevant knowledge structures. These relevant structures are called the grounding experience of the related formula [9]. In this case a descriptive language consisted of logic formulas with modal operator of belief or possibility one is taken into account. External formulas are the reflection of agent grounded opinion about the current state of external world. They describe the states of object from external world. Such an approach to define the epistemic concept of belief is an alternative way to definition of formulas' semantics, because each formula is treated by the cognitive agent as true, if and only if this formula is satisfied by the overall state of agent-encapsulated knowledge [4], [5]. In this work a possible way of

V.S. Sunderam et al. (Eds.): ICCS 2005, LNCS 3516, pp. 671 – 678, 2005.

grounding descriptive language, represented by the algorithm for grounding formulas in the agent's knowledge structures, is proposed. This algorithm uses consensus in the agent's knowledge structures, is proposed. This algorithm uses consensus methods [3]. The process of grounding is carried out when the state of particular object from external world is not known for an agent and the empirical verification is not possible.

2 Assumptions

It is assumed that the agent is independent entity equipped with sensors and situated in some world. This world is consisted of the atomic objects $O=\{o_1,o_2,...,o_Z\}$. Each object $o_z \in O$ at the particular time point $t_n \in T$, $T=\{t_0,t_1,t_2,..\}$ possesses or doesn't posses the property $P_i \in P$, $P=\{P_1,P_2,...,P_K\}$. The states of the objects from O are a target of agent's cognitive processes.

2.1 Agent's Internal Organization

The states of the objects from external world at the particular time point t_k are remembered by the agent in the structure called base profile $BP(t_n)$.

Definition. Let the base profile be given as the following system of mathematical relations:

$$BP(t_n)= <O,P^+_1(t_n),P^-_1(t_n),P^\pm_1(t_n),...,P^+_K(t_n),P^-_K(t_n),P^\pm_K(t_n)>, \quad (1)$$

It is assumed that:

1. $O=\{o_1,o_2,...,o_Z\}$, where each o denotes a unique cognitive representation of a particular object of the world W.
2 For each i=1,2,...,K, relations $P^+_i(t_n)\subseteq O$, $P^-_i(t_n)\subseteq O$, $P^+_i(t_n)\subseteq O$ hold. For each $o_z \in O$ the relation $o_z \in P^+_i(t_n)$ holds if and only if the agent has perceived that this object o_z possesses atomic property P_i. For each $o_z \in O$ the relation $o_z \in P^-_i(t_n)$ holds if and only if the agent has perceived that this object o_z does not posses the atomic property P_i. For each $o_z \in O$ the relation $o_z \in P^\pm_i(t_n)$ holds if and only if the agent doesn't known the state of an object o_z in relation to the property P_i.

Obviously, for each i=1,2,...,K, the conditions $P^+_i(t_n)\cap P^-_i(t_n)=\emptyset$, $P^+_i(t_n)\cap P^\pm_i(t_n)=\emptyset$, $P^-_i(t_n)\cap P^\pm_i(t_n)=\emptyset$ hold.

All base profiles are ordered and stored in agent's private temporal database $KS(t_c)$.

Definition. The overall state of stored perceptions is given as a temporal data base:

$$KS(t_c) =\{BP(t_n): t_n \in T \text{ and } t_n \le t_c\}. \quad (2)$$

2.2 Agent's Language of Communication

The agent is provided with the descriptive language of communication. This language makes it possible for an agent to generate the external messages (logic formulas).

Definition. The agent's language is defined as a pair L:

$$L = \{A, F\}, \tag{3}$$

where A denotes the alphabet consisted of terms $O=\{o_1,o_2,...,o_Z\}$, descriptive connectives: negation ($\neg$), conjunction ($\wedge$), exclusive alternative ($\underline{\vee}$) and alternative ($\vee$); modal operators of belief (*Bel*) and possibility (*Pos*). F denotes the set of logic formulas:

$$F = \Phi \cup \Pi, \tag{4}$$

where Φ is a set of formulas with belief (Bel) operator:

$$\Phi=\{Bel(P_i(o_z)\wedge P_j(o_z)),\ Bel(P_i(o_z)\wedge\neg P_j(o_z)), Bel(\neg P_i(o_z)\wedge P_j(o_z)),\ Bel(\neg P_i(o_z)\wedge \neg P_j(o_z)),\ Bel(P_i(o_z)\underline{\vee} P_j(o_z)),\ Bel(P_i(o_z)\vee P_j(o_z))\} \tag{5}$$

For example if the agent at the time point t_c generates the message $\varphi_c=Bel(P_i(o_z)\wedge P_j(o_z))$ it is equivalent to the spoken language: *I belief that the object o_z at the time point t_c possesses the property P_i and property P_j.*
Π is a set of logic formulas with possibility operators:

$$\Pi=\{Pos(P_i(o_z)\wedge P_j(o_z)),\ Pos(P_i(o_z)\wedge\neg P_j(o_z)),\ Pos(\neg P_i(o_z)\wedge P_j(o_z)),\ Pos(\neg P_i(o_z)\wedge\neg P_j(o_z))\} \tag{6}$$

For example if the agent at the time point t_c generates the message $\varphi_c=Pos(P_i(o_z)\wedge P_j(o_z))$ it is equivalent to the spoken language: *It is possible that object o_z at the time point t_c possesses both the property P_i and P_j.* Each formula is interpreted from the agent's point of view as external, logical representation of beliefs on the current state of the object $o_z\in O$ [4], [5]. It is assumed that at the same time t_c it is possible to generate the only one message with belief operator and few with the possibility one.

3 The Idea of an Algorithm for Grounding Formulas

In this section the idea of the algorithm for grounding formulas will be given. If the state of particular object $o_z\in O$ is not known, then the grounding algorithm is used.

This algorithm is consisted of six steps [6], [7], [8]:

Step 1. Elimination of perceptions
All the base profiles $BP(t_n)$, in which the state of an object o_z in relation to the at least one property P_i or P_j is not known, are rejected. In this way the set $KS'(t_c)$ is created.

Definition. The set $SW'(t_c)$ of base profile is defined as:

$$SW'(t_c) = \{BP(t_k) |\ ((o_z\in P^+_i(t_k) \text{ and } o_z\in P^+_j(t_k)) \text{ or } (o_z\in P^+_i(t_k) \text{ and } o_z\in P^-_j(t_k)) \text{ or } (o_z\in P^-_i(t_k) \text{ and } o_z\in P^+_j(t_k)) \text{ or } (o_z\in P^-_i(t_k) \text{ and } o_z\in P^-_j(t_k)) \text{ and } k\le c,\ i,j\in K,\ i\ne j\}. \tag{7}$$

According to (7) $SW'(t_c)$ is the set of all base profiles in which the state of an object o_z was known for an agent in relation to the both properties P_i and P_j.

Step 2. Classification of perceptions
All the base profiles from $KS'(t_c)$ are grouped into four classes $C^m(t_c)$, m=1,2,3,4 of observations. The criterion of the membership of particular base profile $BP(t_n)$ to the proper class $C^m(t_c)$ is the state of the object o_z in relation to the properties P_i and P_j.

In particular, the following classes of base profiles are taken into account:

a) $C^1(t_c) = \{BP(t_n)$:
$t_n \le t_c$ and $BP(t_n) \in KS'(t_c)$ and both $o_z \in P^+_i(t_n)$ and $o_z \in P^+_j(t_n)$ hold for $BP(t_n)\}$
Class $C^1(t_c)$of data consists of all base profiles $BP(t_n)$ stored up to the time point t_c, in which the object o_z has been found as having both properties P_i and P_j.
b) $C^2(t_c) = \{BP(t_n)$:
$t_n \le t_k$ and $BP(t_n) \in KS'(t_c)$ and both $o_z \in P^+_i(t_n)$ and $o_z \in P^-_j(t_n)$ hold for $BP(t_n)\}$
c) $C^3(t_c) = \{BP(t_n)$:
$t_n \le t_c$ and $BP(t_n) \in KS'(t_c)$ and both $o_z \in P^-_i(t_n)$ and $o_z \in P^+_j(t_n)$ hold for $BP(t_n)\}$
d) $C^4(t_c) = \{BP(t_n)$:
$t_n \le t_c$ and $BP(t_n) \in KS'(t_c)$ and both $o_z \in P^-_i(t_n)$ and $o_z \in P^-_j(t_n)$ hold for $BP(t_n)\}$
Interpretations for $C^2(t_c)$, $C^3(t_c)$ and $C^4(t_c)$ are similar to $C^1(t_c)$.
Each class $C^m(t_c)$ is semantically correlated with one logic formula without modal operator in the following way:

$$
\begin{aligned}
&P_i(o_z) \wedge P_j(o_z) \text{ is related to the content of } C^1(t_c) \\
&P_i(o_z) \wedge \neg P_j(o_z) \text{ is related to the content of } C^2(t_c) \\
&P_i(o_z) \wedge \neg P_j(o_z) \text{ is related to the content of } C^2(t_c) \\
&\neg P_i(o_z) \wedge P_j(o_z) \text{ is related to the content of } C^3(t_c) \\
&\neg P_i(o_z) \wedge \neg P_j(o_z) \text{ is related to the content of } C^4(t_c) \\
&P_i(o_z) \underline{\vee} P_j(o_z) \text{ is related to the content of } C^5(t_c)=C^2(t_c) \cup C^3(t_c) \\
&P_i(o_z) \vee P_j(o_z) \text{ is related to the content of } C^6(t_c)=C^1(t_c) \cup C^2(t_c) \cup C^3(t_c)
\end{aligned} \tag{8}
$$

Step 3. Consensus profiles computing
For each class $C^m(t_c)$, m=1,2,3,4 an agreement of the knowledge called consensus profile $CBP^m(t_c)$, m=1,2,3,4 is computed. In order to determine consensus profiles some postulates need to be applied. See Section 4.

Step 4. Distance between each consensus profile and current base profile computing
The agent computes the distance d_m, m=1,2,3,4 between each consensus profile $CBP^m(t_c)$ and current base profile $BP(t_c)$. This distance reflects numerical similarity between each consensus profile and current base profile. One of the possible ways for determining such distance is computing of the total cost of transformation one object into another. In this case transformation $CBP^m(t_c)$ into $BP(t_c)$ is based on the objects movements between the sets $cP^+_{im}(t_c)$, $cP^-_{im}(t_c)$ and $cP^{\pm}_{im}(t_c)$ in order to obtain the sets $P^+_i(t_c)$, $P^-_i(t_c)$ and $P^{\pm}_i(t_c)$.

Step 5. Choice function value computing
The agent computes the values of the choice function $V_w(X)$, w=1,2,...,6. Each of these value is derived from a subset of $\{d_1,d_2,d_3,d_4\}$:

$$
\begin{array}{l}
V_1(X) \text{ is derived for } P_i(o_z)\wedge P_j(o_z) \text{ from the set of distances } X=\{d_1\} \\
V_2(X) \text{ is derived for } P_i(o_z)\wedge\neg P_j(o_z) \text{ from the set of distances } X=\{d_2\} \\
V_3(X) \text{ is derived for } \neg P_i(o_z)\wedge P_j(o_z) \text{ from the set of distances } X=\{d_3\} \\
V_4(X) \text{ is derived for } \neg P_i(o_z)\wedge\neg P_j(o_z) \text{ from the set of distances } X=\{d_4\} \\
V_5(X) \text{ is derived for } P_i(o_z)\underline{\vee} P_j(o_z) \text{ from the set of distances } X=\{d_2,d_3\} \\
V_6(X) \text{ is derived for } P_i(o_z)\vee P_j(o_z) \text{ from the set of distances } X=\{d_1,d_2,d_3\}
\end{array} \tag{9}
$$

The aim of this step is to create the set of logic formulas without modal operators that are considered as candidates for external messages. Smaller the average value of distances of the set X and the variance of distance values of the set X bigger choice function value.

Step 6. Decision procedure
The formula with belief operator is chosen as external, for which the correlated choice function value is maximal. If there are more than one maximal choice function, value the cardinality of the related class of perceptions $C^m(t_c)$ are taken into account. If these cardinalities are mutually closed then all formulas with possibility operator are chosen as external.

4 Consensus Profiles Computing

In the algorithm for grounding formulas some conceptions from the theory of consensus have been applied [3], [10]. In the third step of this algorithm for each class of perceptions the consensus profile $CBP^m(t_c)$, m=1,2,3,4 is computed. Consensus profile is an agreement of the knowledge stored in each class $C^m(t_c)$. On this stage of an algorithm the knowledge that is represented by the set of base profiles into single profile is transformed.

Definition. Let the consensus profile be given as:

$$
CBP^m(t_n) = \langle O, cP^+_{1m}(t_n), cP^-_{1m}(t_n), cP^\pm_{1m}(t_n), \ldots, cP^+_{Km}(t_n), cP^-_{Km}(t_n), cP^\pm_{Km}(t_n)\rangle, \tag{10}
$$

It is assumed that for $y\in\{1,2,\ldots,7\}$:

1. $o_z\in cP^+_{im}(t_c)$ if and only if the relation is consistent with at least one postulate PS_y
2. $o_z\in cP^-_{im}(t_c)$ if and only if the relation is consistent with at least one postulate PS_y
3. $o_z\in cP^\pm_{im}(t_c)$ if and only if the relation is consistent with at least one postulate PS_y
4. for each i=1,2,...,K, the conditions $cP^+_{im}(t_n)\cap cP^-_{im}(t_n)=\varnothing$, $cP^+_{im}(t_n)\cap cP^\pm_{im}(t_n)=\varnothing$, $cP^-_{im}(t_n)\cap cP^\pm_{im}(t_n)=\varnothing$ hold.

Each postulate PS_y is divided into two parts: in the first parts the set of premises is specified and in the second – the conclusions. The premises represent some conditions that should be fulfilled in order to execute the action from conclusions.

Definition. (postulate's fulfillment)

The postulate PS_y, $y \in \{1,2,...,7\}$ is fulfilled by the structure S if and only if at least one of the following condition holds:

1. All the premises $\delta = \{\delta_1,...,\delta_x\}$ and all the conclusions $\phi = \{\phi_1, ..., \phi_z\}$ of the postulate PS_y: $\delta = \{\delta_1,...,\delta_x\}$ are true:

$$S \models PS_y \text{ if and only if } (\neg\exists\, \delta_i \in \delta \mid \delta_i =^{bool} \text{false}) \text{ and } (\neg\exists \phi_i \in \phi \mid \phi_i =^{bool} \text{false}), \tag{11}$$

2. At least one premise of the postulate PS_y, $y=\{1,2,\ldots,7\}$ is false.

$$S \models PS_y \text{ if and only if } (\exists\, \delta_i \in \delta \mid \delta_i =^{bool} \text{false})$$

where:
$i, x, z \in N$
$S \in U_S = U_C x U_{CBP}$
U_C is universe of the all class of perceptions $C^m(t_c)$
U_{CBP} is universe of all consensus profiles
δ - set of all premises of the postulate PS_y
ϕ - set of all conclusions of the postulate PS_y

4.1 Postulates for Consensus

In order to determine consensus profile $CBP^m(t_c)$ some rational requirements (postulates) need to be applied. Following postulates are taken into account:

Postulate PS_1 *(unanimity of knowledge)*
If object o_z possesses (doesn't posses) the property P_i in each base profile $BP(t_k) \in C^m(t_c)$, $k \leq c$, then in a consensus profile $CBP(t_c)$ possesses (doesn't posses) the property P_i.

Postulate PS_2 *(superior of knowledge)*
If object o_z possesses (doesn't posses) the property P_i in at least one base profile $BP(t_k) \in C^m(t_c)$, $k \leq c$ and there doesn't exist base profile $BP(t_n) \in C^m(t_c)$, $k \leq c$, $k \neq n$, in which the object o_z doesn't posses (possesses) the property P_i, then in a consensus profile $CBP(t_c)$ the object o_z possesses (doesn't posses) the property P_i.

Postulate PS_3 *(the rule of majority)*
If the number of base profiles $BP(t_k) \in C^m(t_c)$, $k \leq c$ in which the object o_z possesses (doesn't posses) the property P_i is bigger then the number of base profiles $BP(t_n) \in C^m(t_c)$, $k \leq c$, $k \neq n$, in which the objects o_z doesn't posses (possesses) the property P_i, then in a consensus profile $CBP(t_c)$ the object o_z possesses (doesn't posses) the property P_i.

Postulate PS_4 *(the rule of majority of knowledge)*
If the number of base profiles $BP(t_k) \in C^m(t_c)$, $k \leq c$ in which the object o_z possesses (doesn't posses) the property P_i is bigger then the number of base profiles $BP(t_n) \in C^m(t_c)$, $k \leq c$, $k \neq n$, in which the objects o_z doesn't posses (possesses) the

property P_i the state of an object o_z is unknown, then in a consensus profile $CBP(t_c)$ the object o_z possesses (doesn't posses) the property P_i.

Postulate PS_5 *(restrictive rule of majority of knowledge)*
If the number of base profiles $BP(t_k) \in C^m(t_c)$, $k \le c$ in which the object o_z possesses (doesn't posses) the property P_i is bigger then the sum of the number of base profiles $BP(t_n) \in C^m(t_c)$, $k \le c$, $k \ne n$, in which the objects o_z doesn't posses (possesses) the property P_i and the base profiles $BP(t_n) \in C^m(t_c)$, $k \le c$, $k \ne n$, in which the state of an objects o_z is unknown then in a consensus profile $CBP(t_c)$ the object o_z possesses (doesn't posses) the property P_i.

4.2 Algorithm for Determining Consensus Profile

Taking the postulates PS_1-PS_5 into account the algorithm for the consensus profile computing is given. In consequence the structure S is obtained that fulfills the postulates PS_1-PS_5.

Algorithm (consensus profile computing)
Input: The set $C^m(t_c) \subset SW(t_c)$, $m \in \{1,2,3,4\}$
Output: The structure $S=\{C^m(t_c), CBP^m(t_c)\}$, $m \in \{1,2,3,4\}$, $S \in U_S$, U_s is universe of all structures S.

Start.
Let $cP^+_{im}(t_c)=cP^-_{im}(t_c)=cP^{\pm}_{im}(t_c)=\phi$, $P^+_i(t_n) \in BP(t_n)$, $P^-_i(t_n) \in BP(t_n)$, $P^{\pm}_i(t_n) \in BP(t_n)$, $n \le c$.
For each property $P_i \in P$
For each object $o_z \in O$

Step 1.
If in each base profile $BP(t_n) \in C^m(t_c)$ $o_z \in P^+_i(t_n)$ then $cP^+_{im}(t_c) := cP^+_{im}(t_c) \cup \{o_z\}$.

Step 2.
If in each base profile $BP(t_n) \in C^m(t_c)$, $o_z \in P^-_i(t_n)$ then $cP^-_{im}(t_c) := cP^-_{im}(t_c) \cup \{o_z\}$.

Step 3.
If there is at least one base profile $BP(t_n) \in C^m(t_c)$, in which $o_z \in P^+_i(t_n)$ and there doesn't exist any profile $BP(t_n) \in C^m(t_c)$, in which $o_z \in P^-_i(t_n)$, then $cP^+_{im}(t_c) := cP^+_{im}(t_c) \cup \{o_z\}$.

Step 4.
If there is at least one base profile $BP(t_n) \in C^m(t_c)$, in which $o_z \in P^-_i(t_n)$ and there doesn't exist base profile $BP(t_n) \in C^m(t_c)$, in which $o_z \in P^+_i(t_n)$, then $cP^-_{im}(t_c) := cP^-_{im}(t_c) \cup \{o_z\}$.

Step 5.
If the number of base profiles $BP(t_n) \in C^m(t_c)$, in which $o_z \in P^+_i(t_n)$ is bigger then the number of base profiles $BP(t_k)$ in which $o_z \in P^-_i(t_k)$, then $cP^+_{im}(t_c) := cP^+_{im}(t_c) \cup \{o_z\}$.

Step 6.
If the number of base profiles $BP(t_n) \in C^m(t_c)$, in which $o_z \in P^-_i(t_k)$ is bigger then the number of base profiles $BP(t_k)$ in which $o_z \in P^+_i(t_n)$, then $cP^-_{im}(t_c) := cP^-_{im}(t_c) \cup \{o_z\}$.

Step 7.
For each base profile $BP(t_n) \in C^m(t_c)$: $cP^{\pm}_{im}(t_c) := cP^{\pm}_{im}(t_c) \cup P^{\pm}_i(t_n) \cup (P^+_i(t_n) \backslash cP^+_{im}) \cup (P^-_i(t_n) \backslash cP^-_{im}(t_c))$.

Stop.

5 Conclusions

Although the problem of symbol grounding has been studied for quite a long time, there is still a lack of mathematical models for simplest languages' grounding.
The aim of this paper was to investigate the problem of grounding modal formulas generated individually by the agents after applying dedicated algorithm. Presented algorithm uses consensus methods and represents the process of grounding logic formulas in agent's knowledge structures. These formulas describe the states of objects from external world in relation to some properties. Proposed solution realises also an original way of defining semantics of modal formulas, because logic formulas are related to the internal representations of an object rather than to ontologically existing entity.

References

1. Coradeschi S., Saffiotti A., An Introduction to the Anchoring Problem, Robotics and Autonomous Systems 43, (2003), 85-96.
2. Harnad, S.:The Symbol Grounding Problem. Physica, 42, 335-236
3. Katarzyniak, R., Nguyen, N.T.: Reconciling Inconsistent Profiles of Agents' Knowledge States in Distributed Multiagent Systems Using Consensus Methods. Systems Science, Vol. 26 No. 4, (2000) 93-119.
4. Katarzyniak, R., Pieczynska-Kuchtiak, A.: Formal Modelling of the Semantics for Communication Languages in Systems of Believable Agents. In: Proc. of ISAT'2001, Szklarska Poreba, (2001), 174-181
5. Katarzyniak, R., Pieczynska-Kuchtiak, A.: Intentional Semantics for Logic Disjunctions, Alternatives and Cognitive agent's Belief. In: Proc. of the 14th International Conference on System Science, Wroclaw, Poland, (2001), 370-382
6. Katarzyniak, R., Pieczynska-Kuchtiak, A.: A Consensus Based Algorithm for Grounding Belief formulas in Internally Stored Perceptions. Neural Network World, 5, (2002) 671-682
7. Katarzyniak, R., Pieczynska-Kuchtiak, A.: Distance Measure Between Cognitive Agent's Stored Perceptions. In: Proc. of IASTED International Conference on Modelling, Identyfication and Control, MIC'2002, Innsbruck, Austria (2002) 517-522
8. Katarzyniak, R., Pieczynska-Kuchtiak, A.: Grounding Languages in Cognitive Agents and Robots. In: Proc. of Sixteenth International Conference on System Engineering, Coventry (2003) 332-337
9. Katarzyniak, R., Pieczynska-Kuchtiak, A.: Grounding and extracting modal responses in cognitive agents: AND query and states of incomplete knowledge. International Journal of Applied Mathematics and Computer Science, 14(2), (2004), 249-263.
10. Nguyen, N.T.: Consensus System for Solving Conflicts in Distributed Systems. Information Sciences, 147, (2002) 91-122
11. Vogt P., Anchoring of semiotics symbols, Robotics and Autonomous Systems, 43, (2003) 109-120.

Fault-Tolerant and Scalable Protocols for Replicated Services in Mobile Agent Systems*

JinHo Ahn[1] and Sung-Gi Min[2,**]

[1] Dept. of Computer Science, Kyonggi University, Republic of Korea,
jhahn@kyonggi.ac.kr

[2] Dept. of Computer Science & Engineering, Korea University, Republic of Korea,
sgmin@korea.ac.kr

Abstract. To enhance scalability of replicated services a large number of mobile agents attempt to access in mobile agent systems, we present a new strategy to apply an appropriate passive replication protocol to each replicated service according to its execution behavior because deterministic services require weaker constraints to ensure their consistency than non-deterministic ones. For this goal, two passive replication protocols are introduced for non-deterministic services and for deterministic services respectively. They both allow visiting mobile agents to be forwarded to and execute on any node performing a service agent, not necessarily the primary agent. Especially, in case of the protocol for deterministic services, after a backup service agent has received a mobile agent request and obtained the delivery sequence number of the request from the primary service agent, the backup agent, not the primary one, is responsible for processing the request and coordinating with the other replica service agents.

1 Introduction

Mobile agent paradigm is considered as a promising vehicle for developing distributed computing systems such as grid computing, e-commerce, active networks and embedded systems because it provides a number of advantages such as reduction of network traffic and asynchronous interaction and so on unlike the traditional client server paradigm [1, 5, 7]. As the mobile agent system is gaining popularity and the number of mobile agent's users rapidly increases, a large number of mobile agents may concurrently be transferred to a node supporting a particular service. In this case, the service agent on the node can be a performance bottleneck and if the agent fails, the execution of all the transferred mobile agents be blocked. In order to solve these problems, the service agent function should be replicated at multiple nodes. This approach may balance the load caused by the mobile agents and even if some service agents crash, continuously allow the other service agents to provide the mobile agents with the

* This work was supported by Korea Research Foundation Grant.(KRF-2002-003-D00248).

** Corresponding author. Tel.:+82-2-3290-3201; fax:+82-2-953-0771.

V.S. Sunderam et al. (Eds.): ICCS 2005, LNCS 3516, pp. 679–686, 2005.

service. There are two approaches used in distributed systems to potentially be applied for satisfying the goal: active and passive replication [8]. Among the two approaches, the passive replication approach [3, 4] has three desirable features we focus on. First, the approach enables its consistency to be guaranteed even if replicated service agents are performed in a non-deterministic manner. Thus, it can be applied to every replicated service regardless of the execution behavior of the service. Second, it needs lower processing power during failure-free execution than the active replication one. Third, mobile agents has only to use a unicast primitive, not a multicast one because they send service requests only to the primary service agent. But, the traditional passive replication approach may result in some scalability and performance problems when being applied to the mobile agent system as a fault-tolerant technique for replicated services. In other words, to the best of our knowledge, previous works [3, 4] uniformly applied the traditional passive replication approach to each replicated service regardless of whether it is deterministic or non-deterministic. But, in this approach, every mobile agent request should be sent only to the primary service agent, which processes the request and coordinates with the other live replicas and then returns a response of the request to the mobile agent. This special role of the primary is necessarily required to ensure the consistency for non-deterministic services. Moreover, the traditional passive replication approach forces all visiting mobile agents to be transferred to and execute their works in order only on the node running the primary service agent of each domain. These inherent features may cause the extreme load condition to occur on the primary service agent when a large number of mobile agents are forwarded to the service domain and access its resources. Thus, this previous strategy may not achieve high scalability and performance. This paper presents a scalable strategy to apply an appropriate passive replication protocol to each service according to its execution behavior because deterministic services require weaker constraints to ensure their consistency than non-deterministic ones. For this goal, two passive replication protocols are designed in this paper. The first protocol for non-deterministic services is named $PRPNS$ and the second protocol for deterministic services, $PRPDS$. They both allow visiting mobile agents to be forwarded to and execute on any node performing a service agent, not necessarily the primary agent. Especially, in case of the second protocol $PRPDS$, after a backup service agent has received a mobile agent request and obtained the delivery sequence number of the request from the primary service agent, the backup agent, not the primary one, is responsible for processing the request and coordinating with the other replica service agents. Due to this feature, $PRPDS$ is more lightweight than $PRPNS$ tolerating non-deterministic servers such as multi-threaded servers.

2 The Proposed Passive Replication Protocols

In here, we attempt to improve scalability of mobile agent systems by using the appropriate passive replication protocol for each replicated service domain according to whether the service is deterministic or non-deterministic. For this purpose, two passive replication protocols $PRPNS$ and $PRPDS$ are introduced.

When mobile agents are concurrently transferred to a non-deterministic replicated service, the consistency is ensured by applying the traditional passive replication protocol. When mobile agent a_j attempts to use the service of a non-deterministic replicated service domain u_i, a_j should be transferred from its current node to the node where the primary service agent of u_i, u_i^{prim}, executes. Then, the following phases are performed:

(Phase 1). When u_i^{prim} receives a request message from mobile agent a_j, u_i^{prim} processes the message. After the execution, it generates $(response, psn, next\_state)$, where $response$ is the response of the message, psn identifies the processed request message and $next\_state$ is the state of u_i^{prim} updated by processing the request.

(Phase 2). u_i^{prim} sends all backup service agents the update message $(response, psn, next\_state, j, reqid)$ using **View Synchronous Multicast**($VSCAST$) [8] respectively, where j identifies a_j and $reqid$ is the send sequence number of the request message. In here, $VSCAST$ is the multicast communication primitive used to ensure correctness of the passive replication approach. When each backup service agent receives the update message, it updates its state using $next\_state$, maintains $(response, j, reqid)$ in its buffer and then sends an acknowledgement message to u_i^{prim}. In this case, $(response, j, reqid)$ is needed to ensure the exactly once semantics despite u_i^{prim}'s failure.

(Phase 3). After receiving an acknowledgement message from every live backup service agent, u_i^{prim} sends $response$ to a_j.

Therefore, this protocol forces all visiting mobile agents to be transferred to and execute their works in order only on the node running the primary service agent of each domain. This behavior may cause the extreme load condition to occur when a large number of mobile agents are forwarded to the service domain to use the non-deterministic service. For example, in figure 1, three mobile agents a_t^l, a_s^m and a_r^n are concurrently transferred to the node $node_i^2$ where the primary service agent of u_i, u_i^2, is running. The mobile agents perform the service of u_i in order via only u_i^2 using the above mentioned protocol respectively. Thus, we introduce a passive replication protocol for non-deterministic service, $PRPNS$, to solve the problem by allowing mobile agents to be forwarded to and execute on each a node performing a service agent, not necessarily the primary agent. The protocol forces each service agent p to forward every request received from mobile agents to the primary service agent. Afterwards, the primary agent performs the phases 1 through 2 previously mentioned to satisfy the consistency condition for non-deterministic service. After receiving an acknowledgement from every normal backup service agent, the primary agent sends the response of the request to the service agent p, which forwards it to the corresponding mobile agent.

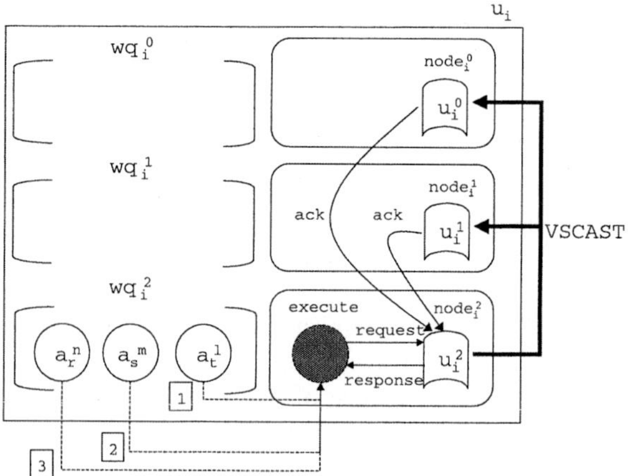

Fig. 1. an execution of three mobile agents accessing a replicated service u_i in the traditional passive replication protocol

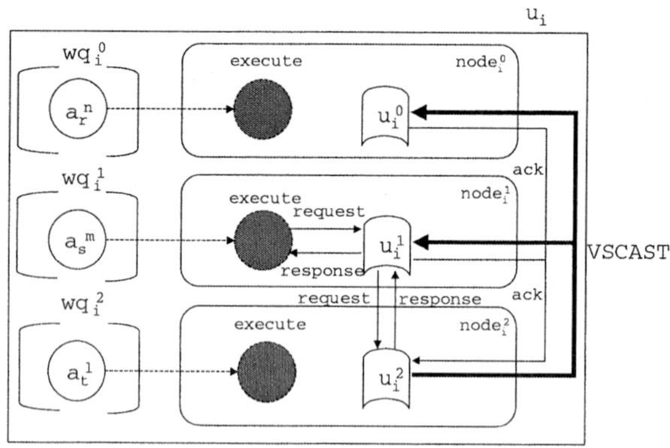

Fig. 2. an execution of three mobile agents accessing a replicated service u_i in the protocol *PRPNS*

Figure 2 illustrates how the protocol *PRPNS* executes. In this figure, mobile agents a_t^l, a_s^m and a_r^n are transferred to and execute on the nodes $node_i^2$, $node_i^1$ and $node_i^0$ for accessing service u_i. Then, the mobile agents perform their tasks via u_i^2, u_i^1 and u_i^0 respectively. For example, if u_i^1 receives a request from a_s^m, it forwards the request to the primary service agent u_i^2. Afterwards, u_i^2 processes the request and coordinates with the other service agents u_i^1 and u_i^0 using *VSCAST*. When receiving acknowledgements from the two backup service agents, it sends the response of the request to u_i^1, which forwards the response to a_s^m. From this example, we can see that *PRPNS* improves scalability of replicated services in case a large number of mobile agents attempt to use a particular service simultaneously. If a mobile agent accesses a service via a backup service agent in this protocol, two more messages are required per request compared with the traditional protocol. However, the additional message cost is not significant because each service domain is generally configured on a local area network like Ethernet.

A deterministic replicated service requires weaker constraints to ensure the consistency than a non-deterministic one. In other words, after the primary service agent has determined the processing order of every request from mobile agents, it is not necessary that only the primary agent handles all requests, and coordinates with the other replica service agents like in the protocol $PRPNS$. With this observation, we attempt to use a lightweight passive replication protocol, $PRPDS$, for each deterministic service domain. The proposed protocol has the following features.

- Each mobile agent can use a service via the primary service agent or a backup one.
- Only the primary service agent determines the processing order of every request from mobile agents.
- After a backup service agent has received a mobile agent request and obtained the order of the request from the primary service agent, the backup agent, not the primary one, processes the request and coordinates with the other replica service agents including the primary agent.

Due to these desirable features, this protocol enables each visiting mobile agent to be forwarded to and execute on a node running any among replicated service agents in the service domain. If mobile agent a_j is transferred to the node where a backup service agent u_i^{backup} executes, $PRPDS$ executes the following phases. Otherwise, the three phases of $PRPNS$ are performed.

(Phase 1). When u_i^{backup} receives a request message from a_j, u_i^{backup} asks the primary service agent u_i^{prim} the *psn* of the request message. In this case, after u_i^{prim} determines the *psn* of the message, it notifies u_i^{backup} of the *psn*. Then, u_i^{prim} processes the request message and saves $(response, psn,\ next\_state,\ j, reqid)$ of the message in its buffer. When receiving the *psn*, u_i^{backup} processes the corresponding request and generates $(response, psn,\ next\_state)$ of the request.

(Phase 2). u_i^{backup} sends the other service agents the update message $(response,\ psn, next\_state,\ j, reqid)$ using $VSCAST$ respectively. When each service agent except for u_i^{prim} receives the update message, it updates its state using $next\_state$, maintains $(response, j, reqid)$ in its buffer and then sends an acknowledgement message to u_i^{backup}. If u_i^{prim} receives the update message from u_i^{backup}, it just removes the element $(response, psn, next\_state,\ j, reqid)$ for the message from its buffer, saves $(response, j, reqid)$ in the buffer and sends an acknowledgement message to u_i^{backup}.

(Phase 3). Once u_i^{backup} receives an acknowledgement message from every other live service agent, it sends *response* to a_j.

Figure 3 shows an execution of three mobile agents a_t^l, a_s^m and a_r^n attempting to use service u_i via u_i^2, u_i^1 and u_i^0 in the protocol $PRPDS$. In this figure, only the

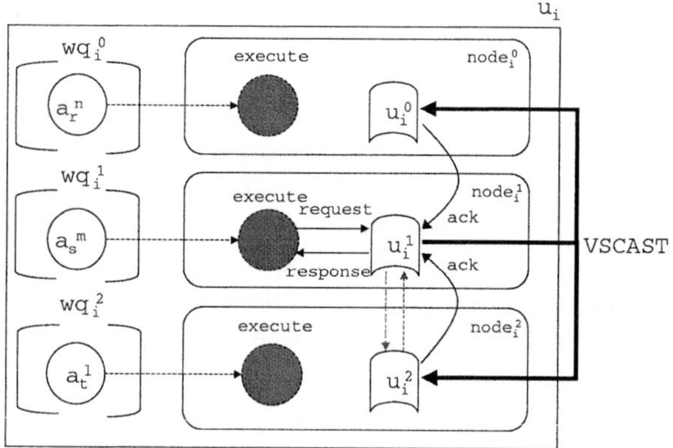

Fig. 3. an execution of three mobile agents accessing a replicated service u_i in the protocol $PRPDS$

procedure is pictured to execute in $PRPDS$ when a_s^m sends a request to u_i^1. In this case, u_i^1 asks the primary agent u_i^2 the psn of the request. u_i^2 determines the psn and sends it to u_i^1. Then, u_i^2 processes the request and saves $(response, psn, next\_state, j+1, reqid)$ of the request in its buffer. Meanwhile, after obtaining the psn of the request from the primary agent, u_i^1 processes the request and coordinates with the other service agents u_i^2 and u_i^0 using $VSCAST$ to satisfy the consistency condition for deterministic service. When u_i^1 receives an acknowledgement from every live agent, it sends the response to a_s^m. From this figure, we can see that the protocol $PRPDS$ may significantly improve scalability of replicated services by enabling each visiting mobile agent to use a particular service via any among replicated service agents on the service domain and the request processing and coordination load to be distributed between a set of service agents.

2.1 Recovery

If service agents fail on a service domain, the proposed two protocols, $PRPNS$ and $PRPDS$, should perform their recovery procedures to satisfy each consistency condition for nondeterministic service or deterministic service despite the failures respectively. Firstly, in case of the crash of the primary service agent u_i^{prim}, the two protocols execute both each appropriate recovery procedure in the following cases.

(Case 1). The primary service agent u_i^{prim} fails before finishing the phase 1.

In this case, a new primary service agent $u_i^{prim'}$ is elected among all backup service ones. Since no response of each request can be received from u_i^{prim}, the request will be resent to $u_i^{prim'}$, which has only to perform the phases 1 through 3.

(Case 2). The primary service agent u_i^{prim} fails after completing the phase 1, but before sending the response to the corresponding mobile agent.

Like in case 1, a new primary agent $u_i^{prim'}$ is selected. To ensure linearizability in this case, either all the backup service agents receive the update message, or none of them receive it. $VSCAST$ is used to satisfy the atomicity condition of linearizability in case of an asynchronous mobile agent system with an unreliable failure detector. If no backup service agent receives the update message, this case is similar to case 1. Otherwise, $(response, j, reqid)$ in the phase 2 is used to ensure the exactly-once semantics. In other words, when the request from the corresponding mobile agent is sent to the new primary service agent $u_i^{prim'}$ again, the latter immediately sends the response of the request to the first without handling the request.

(Case 3). The primary service agent u_i^{prim} fails after finishing the phase 3. Like in case 1, a new primary service agent $u_i^{prim'}$ is selected and identified.

Secondly, when a backup service agent u_i^{backup} crashes, $PRPNS$ and $PRPDS$ perform their recovery procedures as follows. In $PRPNS$, u_i^{backup} is just removed from its service group. On the other hand, $PRPDS$ executes a corresponding recovery procedure in each following case.

(Case 1). Backup service agent u_i^{backup} fails before asking the primary service agent u_i^{prim} the *psn* of each request in the phase 1.

In this case, u_i^{backup} is just removed from its group. Afterwards, u_i^{backup}'s failure is detected because of no response of the request from it.

(Case 2). Backup service agent u_i^{backup} fails before finishing the phase 1.

(Case 2.1). The primary u_i^{prim} fails.

The other backup service agents select a new primary service agent, $u_i^{prim'}$, among them.

(Case 2.2). The primary u_i^{prim} is alive.

Detecting u_i^{backup}'s crash, u_i^{prim} retrieves every update information, which is form of $(response, psn,\ next\_state,\ j, reqid)$, from its buffer and sends it to the other live backup service agents respectively. In this case, $VSCAST$ is used to ensure linearizability.

(Case 3). Backup service agent u_i^{backup} fails after sending the update message to the other service agents in the phase 2, but before sending the response to the corresponding mobile agent.

As u_i^{backup} sent the update message to the other service agents by using $VSCAST$, the entire consistency is ensured. Therefore, the service group of u_i^{backup} has only to remove u_i^{backup}.

(Case 4). Backup service agent u_i^{backup} fails after completing the phase 3. u_i^{backup} is removed from its group and mobile agents can detect that u_i^{backup} fails.

3 Conclusion

This paper proposed a new strategy to improve scalability of mobile agent systems by applying an appropriate passive replication protocol to each replicated service according to its execution behavior, deterministic or non-deterministic. For this purpose, we presented the two passive replication protocols, $PRPNS$ and $PRPDS$, for non-deterministic and deterministic services respectively. While ensuring linearizability [6], they both allow visiting mobile agents to be forwarded to and execute their tasks on any node performing a service agent, not necessarily the primary agent. Especially, the more lightweight protocol $PRPDS$ allows any service agent to process each mobile agent request and coordinate with the other replica service agents after receiving the request and obtaining its delivery sequence number from the primary agent. Thus, if $PRPDS$ is well-combined with existing load balancing schemes [2], the request processing and coordination load can be evenly distributed among a set of deterministic and replicated service agents based on the workload of each service agent.

References

1. P. Bellavista, A. Corradi and C. Stefanelli. The Ubiquitous Provisioning of Internet Services to Portable Devices. *IEEE Pervasive Computing*, Vol. 1, No. 3, pp. 81-87, 2002.
2. H. Bryhni, E. Klovning and O. Kure. A Comparison of Load Balancing Techniques for Scalable Web Servers. *IEEE Network*, 14:58-64, 2000.
3. N. Budhiraja, K. Marzullo, F. B. Schneider and S. Toueg. The primary-backup approach. *Distributed Systems(S. Mullender ed.*, ch. 8, 199-216, Addison-Wesley, second ed., 1993.
4. X. Defago, A. Schiper and N. Sergent. Semi-Passive Replication. *In Proc. of the 17th IEEE Symposium on Reliable Distributed Systems*, pp. 43-50, 1998.
5. M. Fukuda, Y. Tanaka, N. Suzuki, L.F. Bic and S. Kobayashi. A Mobile-Agent-Based PC Grid. *In Proc. of the Fifth Annual International Workshop on Active Middleware Services*, pp. 142-150, 2003.
6. M. Herlihy and J. Wing. Linearizability: a correctness condition for concurrent objects. *ACM Transactions on Progr. Languages and Syst.*, 12(3):463-492, 1990.
7. K. Rothermel and M. Schwehm. Mobile Agents. In A.Kent and J.G.Williams(Eds.):*Encyclopedia for Computer Science and Technology*, 40(25):155-176, 1999.
8. M. Wiesmann, F. Pedone, A. Schiper, B. Kemme and G. Alonso. Understanding Replication in Databases and Distributed Systems. *In Proc. of the 21st International Conference on Distributed Computing Systems*, pp. 464-474, 2000.

Multi-agent System Architectures for Wireless Sensor Networks

Richard Tynan, G.M.P. O'Hare, David Marsh, and Donal O'Kane

Adaptive Information Cluster (AIC),
University College Dublin
Belfield, Dublin 4, Ireland

{richard.tynan, gregory.ohare, david.marsh, donal.okane}@ucd.ie

Abstract. Traditionally Multi-Agent Systems have been thought of in terms of devices that possess a relatively rich set of resources e.g. power, memory, computational ability and communication bandwidth. This is usually necessary due to the complex deliberation and negotiation processes they require to fulfil their goals. Recently, networked devices have become available on the millimeter scale called Wireless Sensor Networks (WSNs), which pose new challenges because of their constrained resources. However what these devices lack in resources they make up for in numbers due to their small inexpensive nature. In this paper we identify some of the existing MAS architectures for WSNs, and we propose some novel architectures of our own.

1 Introduction

Modern computing networks are comprised of a wide range of heterogeneous devices that may possess various levels of resources and are interconnected using differing networking technologies. This gives rise to a device continuum with high-powered servers connected using high-speed GB/s LANs at one extreme and Mobile Phones connected using GPRS at the other. Multi-Agent Systems for subsets of this spectrum of devices are numerous. Diao et al [1] use autonomic management agents for Web server optimization. At the other end of the scale, Gulliver's Genie uses agents for context sensitive tourist information delivery in the mobile computing arena [2].

Contemporary advances in microprocessor fabrication has lead to a dramatic reduction in the size and power consumed by computational devices. Sensing technology, batteries and radio hardware have also followed a similar minaturisation trend. These factors combined have enabled the production of match-box scale, battery powered devices capable of complex processing tasks called Wireless Sensor Networks, thus heralding a new era for ubiquitous sensing technology. Previous deployments of these networks have been used in many diverse fields such as wildlife habitat monitoring [3], traffic monitoring [4] and lighting control [5].

The device spectrum for computer networks now has a new resource challenged device on which an agent could potentially be deployed - a WSN node. Agents have been deployed previously on WSNs primarily to process the raw data in an intelligent fashion with a view to reducing transmissions - the single biggest factor in determining

V.S. Sunderam et al. (Eds.): ICCS 2005, LNCS 3516, pp. 687–694, 2005.

the longevity of a network [6], [7]. Agents have also been used in a lighting control application for energy conservation [5]. The research to date has mainly focussed on the application and not on the potential architectures for the MASs. In this paper we identify some existing MAS architectures and propose some new architectures.

2 Device Hierarchy

When considering the diverse range of devices that can comprise a modern computer network where an agent may be deployed it is useful to take into account of the resources available at each device. For our purposes we have identified a five tier hierarchy of devices. Figure 1 depicts this hierarchy and a device typical of each layer.

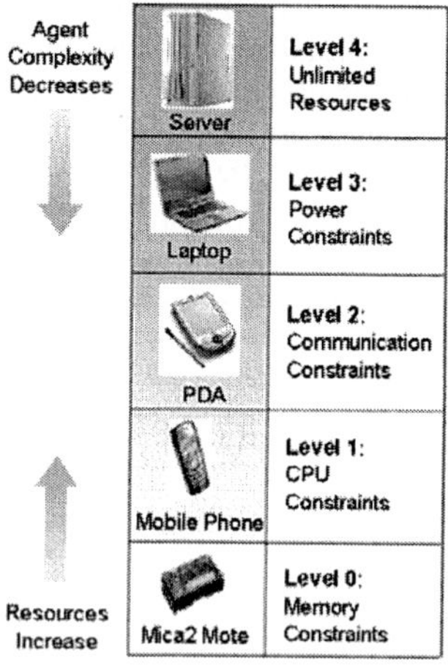

Fig. 1. Hierarchy of devices according to resources available

A more precise description of this hierarchy can be seen in table 1. A Level 4 device we consider to have unbounded resources in terms of the 4 criteria. Devices of this nature are connected to a mains supply and have vast memory and computational resources. Also their communication occurs over high-speed networks so network bandwidth is very high. Below this level the first distinction we make is power. We now enter the mobile computing arena based on battery powered devices. The next distinction that occurs is based on communication capabilities i.e. a laptop is usually connected to other nodes through a wireless LAN whereas PDAs and mobile phones use 3G technology and GPRS. The criteria separating level 2 and level 1 devices are CPU constraints. Due to the more complex operation of a PDA, compared to the mobile phone, it requires a faster CPU which would allow deeper agent deliberation. The final distinction between devices is that of memory. Modern mobile phone memory is of the order of megabytes whereas, a WSN node typically only has a few kilobytes.

Table 1. Table representing the various constraints of a device at a given level. An x indicates that this constraint is not a concern for an agent on a device in this layer. An o implies that the agent must be aware of this constraint

	Power	Communication	CPU	Memory
Level 4	x	x	x	x
Level 3	o	x	x	x
Level 2	o	o	x	x
Level 1	o	o	o	x
Level 0	o	o	o	o

So what implication does this have for MASs? An agent must be aware of its resident devices abilities and factor these into its deliberation about future actions. However this hierarchy becomes more interesting when we look at mobile agents. When an agent migrates between devices of different levels in the hierarchy the agent may opportunistically take advantage of the differing resources on a device. An agent migrating to a lower level device must be aware of this and tailor its reasoning process accordingly. For instance, when migrating from a level 4 to a level 3 device, an agent must be aware that, should the device's power become critically low, it should have a route available to it to leave the device. In the case of moving from a level 2 to a level 1 device, the agent may have to adapt to a simplified version of the agent platform that does not provide services common to larger platforms. When moving to a level 0 device, it is likely that the encoding of the agent's mental state will have to be radically reduced. Of course, when moving up through the hierarchy, an agent should become aware, or be made aware, that its operation is no longer as restricted as before, and tailor its behavior accordingly.

3 Wireless Sensor Networks

Wireless Sensor Networks (WSNs) are made possible by the minaturisation of their components, namely microprocessors, sensors and radio communication hardware. The main components for a WSN are a base-station and one or more sensor nodes. The sensor nodes relay their sensed data either directly to the base station or through each other (termed multi-hopping) depending on the scale of the network. In turn the base station sends commands to the nodes to, for example, increase their sampling frequency.

Multi-hopping, while useful in extending the reach or scale of a WSN, does have its pitfalls. The cost of transmitting a packet can be greatly increased depending on the distance a node is away from the base station. Secondly, since nodes nearest the base station, i.e. 1 hop away, will not only have to send their data but also that of all other nodes greater than a single hop, there will be a greater demand placed on the power supply of these nodes. It means that, in general, a nodes lifespan is inversely proportional to the number of hops it is away from the base station. To alleviate this problem, multiple base stations can be used, with the nodes only transmitting data to their local station. A second solution creates a hierarchy of nodes with varying power and transmission capabilities. Higher power nodes act as a networked backbone for the lower powered sensors. One final alternative is to populate the area closest to the base station more densely than in outlying areas.

In some WSNs the nodes of the network can perform considerable processing tasks. This facilitates the examination of raw sensed data in an intelligent fashion with transmissions only containing summarized information rather than the raw data i.e. a single packet containing the average temperature over say ten seconds could be transmitted instead of ten individual packets containing the temperature as sampled at each second. Processing capabilities can vary from platform to platform. We have chosen the Mica2 Mote [8] for our experimentation. It has 4KB of RAM, 128KB of instruction memory and 512KB of flash memory for data storage purposes. They are also equipped with a suite of sensory modalities, heat, light, sound, barometric pressure and humidity.

4 Agents for Wireless Sensor Networks

Having seen that current research endeavors have shown not only the feasibility but also the utility for agents on WSNs we now turn our attention to candidate architectures that may prove appropriate.

4.1 WSN Nodes as Perceptors

The first architecture model considers individual nodes as perceptors for a single agent, as depicted in figure 2. On the perceive part of the perceive-deliberative-act cycle, the agent requests a sensed value from each node, which it then uses as input to the deliberative part of that cycle. This has a number of benefits and restrictions. Chief among the benefits is ease of programming, since there is a single control point in the system. The system will also be more synchronized with regard to the timing of transmissions, and hence more efficient. This is because any collisions during transmission necessitates a costly retransmission, so synchronization brings direct energy benefits.

The restrictions of this architecture are typical of centralized systems. The primary disadvantage is that the system is not scalable. There are two main reasons for this. Firstly, since each node must be queried to generate a perceptor value, as the number of nodes increases so does the length of the deliberative cycle. With current node radio speeds, the length of the cycle could quickly grow past one second. This is compounded by the second scalability problem, which is that with larger numbers of nodes, multi-hop communication is increasingly likely to be needed. This invariably leads to even greater time delays for the agent while it gathers its data, while energy consumption also rises. An additional problem is that if two or more agents wish to use the network as a set of perceptors, they will have to coordinate with each other to synchronize access, and of course they will be subjected to additional delays while waiting for the other agents to finish with the network. Larger networks also increase the likelihood of having unconnected parts of the network. In this architecture, these parts would remain inactive as long as they are unable to receive messages from the agent. This sort of architecture is probably most suited to small-scale networks which are not likely to have more than

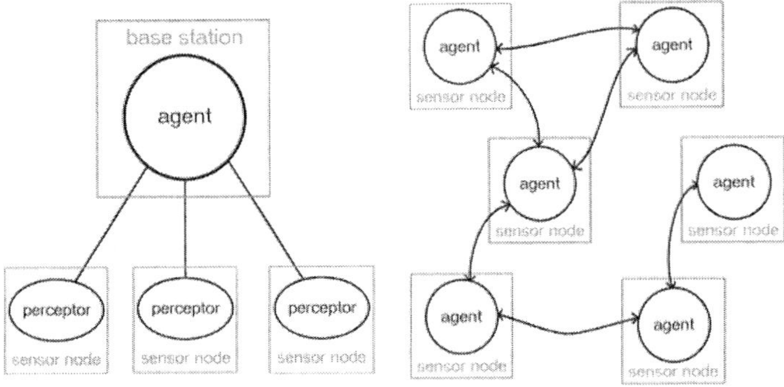

Fig. 2. WSN as perceptor to central base station agent (left), collaborative agent approach where agents are considered peers (right)

a single data sink (i.e. agent that wishes to use the network). As such it rarely exists in practice.

4.2 One Agent per Sensor Node

In this architecture, each node in the network hosts an agent which is responsible for the behavior of that node (figure 2). This approach represents a distributed method of network control. The agents can co-ordinate their behavior to achieve both local and global goals, and so the network may be employed in more useful ways than in the simple perceptor case. For example, network services such as packet routing can be built within the framework of collaborating agents, whereas this is not possible in the centralized case where a layer independent of the controlling agent would be required.

One of the primary benefits of this architecture is scalability. Local tasks can be performed without multi-hopping messages across the network to a central controller, which leads to immediate energy and latency gains. Concurrency is another gain associated with decentralizing the responsibility for processing. This allows the network to perform more work in a given time period than could be achieved by other means. The effects of network partitioning are also alleviated somewhat by using this architecture, since partitioned regions can still perform local tasks.

Disadvantages of this method include the difficulty of creating systems of agents that can realize global goals using only local information. The cost of negotiation between agents is an overhead avoided by the centralized architecture. In a network of heterogeneous nodes, i.e. nodes with varying degrees of sensing and in particular computational resources, an agent on a faster node may experience little performance gain because of the limitations of the surrounding agents that must run on slower nodes. This architecture is most suitable when the sensors have some form of local actuation, as seen in [5], and [9].

4.3 Mobile Agents

This scheme uses an agent with migratory capabilities which visits a set of nodes, accumulating state as it does so. This is a typical method for data aggregation/fusion applications and is shown in figure 3. Benefits include low overall computational costs, as no more than one node is active at a time. Since a single transmission and reception event is all that is required of an individual node in order to facilitate the agent's migration, communication is also generally quite low. Simplicity is another benefit, since the agent manages both the task of collecting data and its visitation itinerary. The negative aspects of network partitioning have less impact on this method since an agent can simply alter its visitation itinerary to go to the nodes it can reach until connectivity is restored.

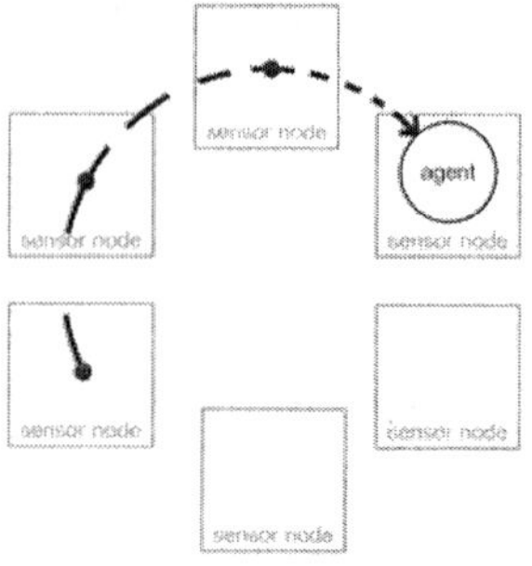

Fig. 3. Agent navigates the network for its control and data harvesting requirements

One disadvantage of this architecture is that if the node the agent is currently on fails (for example the battery depletes before the agent can migrate elsewhere), all data that has been accumulated by it so far would be lost. Methods to counter this include leaving copies of the state of the agent on previously visited nodes, which could "revive" the agent using this information in the event that the original was lost. Secondly, if latency is an important concern then this method is likely to be less efficient compared to the nodes-as-agents architecture because of the lack of concurrency. If a network is very large, it may take too much time for a single agent to gather all the data. This scheme is likely to be useful where nodes collect a lot of data before processing it [7].

Our initial experience with implementing MASs for WSNs has highlighted the advantages and disadvantages of the previous architectures, and motivated the development of the following 2 combined approaches. These attempt to harness the advantages of a number of the previous architectures while minimizing their disadvantages.

4.4 Hybrid Approach

If we roughly describe the advantages of the three methods above as being simplicity, robustness and efficiency respectively, and their weaknesses as being scalability, complexity and latency, then we can attempt to combine them in such a way as to reduce the impact of the weaknesses while retaining the advantages. This leads to the following hybrid approaches as illustrated in figure 4.

Collaborative Agents: Each agent is a perceptor type group. As was mentioned above, the first architecture is not suitable for larger networks of sensors. However, if the network is divided into numerous small groups of nodes, each one could contain a single agent that is responsible for several nodes. The system can be viewed at a high level as a collection of collaborating agents, and at a low level as a collection of nodes being managed by one agent. At the level of the individual agent, most of the problems associated with the lack of scalability will no longer apply once the number of nodes the agent is responsible for is low enough. While network partitions may still cause problems, the chance of them occurring in a small group of nodes is much reduced, and the groups of nodes could be organized around such disturbances. The disadvantages of the collaborative agent architecture are lessened by combining it with the first approach. Since there are less agents, co-ordination cost between agents are reduced. Heterogeneous networks could be organized in such a way that every agent operates on an approximately equivalent set of nodes, which should mask the inequalities present when each node has its own agent.

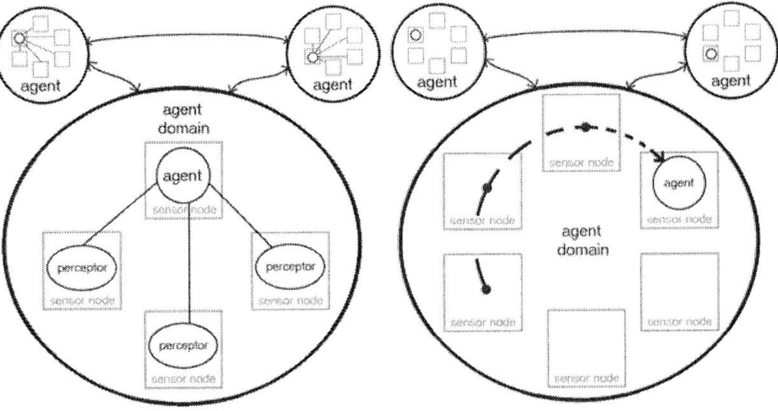

Fig. 4. Agents are allotted a subset of the network for control based on the perceptor model. Agent cooperation then performs high level control and harvesting (left). The agent controls and senses its environment by migrating to the sensor nodes under its control (right)

Collaborative Agents: Each agent is the domain for a mobile agent. Combining the second and third architectures also reduces some of the problems that they have on their own. Once again, the network can be divided into small groups of nodes, but in this instance each group has a single mobile agent that migrates between the constituent nodes. The latency issue for the migrating agent architecture is reduced by reducing the number of nodes it has to traverse. The probability of an agent being trapped on a node that is going to fail is also reduced as the number of nodes it travels among lessens. For the same reasons as in the first combined architecture, the disadvantages of the collaborative approach are reduced when combined with the mobile agent case.

5 Conclusions

Recent work has shown that multi-agent systems can be deployed on the new wave of computationally limited devices that have emerged in the last few years. One of the latest classes of these devices is the wireless sensor node. In line with the general trend of adapting agents to work on decreasingly powerful equipment, the use of agents on such simple platforms brings a new set of challenges. We detailed the restrictions that exist at the different levels of sophistication in order to illuminate the problems faced by an agent when executing at the various levels. We then provided a brief introduction to wireless sensors networks, specifying their components, the capabilities of these components, the pressing research issues and an example of some WSN hardware. Finally we outline MAS architectures that we believe are appropriate to use on WSNs. We give details of their respective advantages and disadvantages, and with these points as a guideline we posit the situations in which they are likely to be useful. Included is a pair of hybrid architectures, with the elements combined in order to minimize their negatives while maintaining their positive qualities.

Acknowledgements

This material is based on works supported by Science Foundation Ireland under Grant No. 03/IN.3/I361. We would also like to thank the Irish Research Council for Science, Engineering and Technology (IRCSET) for their support.

References

1. Diao, Y., Hellerstein, J.L., Parekh, S., Bigus, J.P.: Managing Web Server Performance with AutoTune Agents. IBM Systems Journal **42** (2003) 136–149
2. O'Grady, M.J., O'Hare, G.M.: Gullivers's genie: Agency, mobility and adaptivity. Computers and Graphics Journal, Special Issue on Pervasive Computing and Ambient Intelligence - Mobility, Ubiquity and Wearables Get Together. **28** (2004)
3. Mainwaring, A., Polastre, J., Szewczyk, R., Culler, D., Anderson, J.: Wireless Sensor Networks for Habitat Monitoring. In: International Workshop on Wireless Sensor Networks and Applications. (2002)
4. Coleri, S., Cheung, S.Y., Varaiya, P.: Sensor Networks for Monitoring Traffic. In: Allerton Conference on Communication, Control and Computing. (2004)
5. Sandhu, J., Agogino, A., Agogino, A.: Wireless Sensor Networks for Commercial Lighting Control: Decision Making with Multi-agent Systems. In: AAAI Workshop on Sensor Networks. (2004)
6. Marsh, D., Tynan, R., O'Kane, D., O'Hare, G.: Autonomic wireless sensor networks. Engineering Applications of Artificial Intelligence (2004)
7. Qi, H., Wang, X., Iyengar, S.S., Chakrabarty, K.: Multisensor Data Fusion in Distributed SensorNetworks Using Mobile Agents. In: 4th International Conference Information Fusion. (2001)
8. Hill, J.: System Architecture for Wireless Sensor Networks. PhD thesis, UC Berkeley (2003)
9. Braginsky, D., Estrin, D.: Rumor routing algorithm for sensor networks. In: WSNA. (2002)

ACCESS: An Agent Based Architecture for the Rapid Prototyping of Location Aware Services

Robin Strahan, Gregory O'Hare, Conor Muldoon, Donnacha Phelan, and Rem Collier

Department of Computer Science, University College Dublin, Dublin 4, Ireland
{Robin.Strahan, Gregory.OHare, Conor.Muldoon, Donnacha.Phelan, Rem.Collier}@ucd.ie
http://emc2.ucd.ie

Abstract. We describe the Agents Channelling ContExt Sensitive Services (ACCESS) architecture, an open agent-based architecture that supports the development and deployment of multiple heterogeneous context sensitive services. We detail the ACCESS architecture and describe the scenario of an individual arriving in a city and using his ACCESS enabled PDA to secure lodgings.

1 Introduction

This paper describes the Agents Channelling ContExt Sensitive Services (ACCESS) architecture, an open agent-based architecture that supports the development and deployment of multiple heterogeneous context sensitive services. A key motivation for the development of ACCESS was the recognition that most location-aware context sensitive applications exhibit a common core of functionality. That is, such systems commonly employ: location-sensing technologies, dynamic generation of maps that are customised to the user, support for the management of the users context, and service brokerage.

In this paper we review some related systems, detail the ACCESS architecture paying particular attention to three facets: map caching, user profiling, and agent tuning and finally, we demonstrate the interaction between ACCESS and a mobile user by presenting the scenario of an individual arriving in a city and seeking a hotel room.

2 Related Work

There are a number of existing systems that aim to provide a basis for the development of agent-based context aware systems.

MyCampus is an agent-based environment for context-aware mobile services developed at Carnegie Mellon University [1]. The system aims to aid a PDA equipped user in carrying out different tasks (planning events, send messages, find other users, etc) by accessing Intranet and Internet services. The information from these services is filtered by use of context, such as the user's location, their class schedule, the location of their friends and the weather. Personal preferences are also used to tailor the information provided to the user.

V.S. Sunderam et al. (Eds.): ICCS 2005, LNCS 3516, pp. 695 – 702, 2005.

The SALSA framework [2] allows developers to implement autonomous agents for ubiquitous computing systems. It uses WiFi to communicate and estimates the user's position by triangulation of at least three 802.11 access points. It uses an open source Instant Messaging Server (Jabber) to notify the state of people and agents, and to handle the interaction between people, agents, and devices through XML messages. Agents abstract the complexities associated with the collaboration of users and the opportunistic interaction with services of ubiquitous computing environments.

The COBRA (COntext BRoker Architecture) [3] is a broker-centric agent architecture that aims to reduce the cost and difficulty of building context-aware systems. At its centre is an autonomous agent called the Domain Context Broker that controls a context model of a specific domain or location. It updates this model, using data from heterogeneous sources (physical sensors, Web, pattern behaviours, etc) and attempts to resolve any detected context conflicts. It shares this information with independent agents within its domain. The use of broker-centric allows computational intensive activities such as gathering and interpreting data to be preformed by a suitable device rather than on resource limited devices.

3 The ACCESS Architecture

As an agent-based architecture intended to support the rapid development and deployment of multiple heterogeneous context sensitive mobile services, the Agents Channelling ContExt Sensitive Services (ACCESS) architecture seeks to provide support for multi-user environments by offering personalization of content, through user profiling and context, as well as support for mobile lightweight intentional agents and intelligent prediction of user service needs.

The ACCESS architecture has been designed to extend Agent Factory (AF), a well-established framework [6][7] for the development and deployment of multi-agent systems. The ACCESS architecture enhances the basic infrastructure delivered by the AF run-time environment through the implementation of additional agents, which deliver functionality that is common to many context-sensitive applications. These ACCESS Management Agents form a cohesive layer into which multiple heterogeneous context-sensitive services may be plugged. In addition, the existing AF development methodology has been augmented to aid the developer in producing ACCESS-compliant services. The resulting ACCESS toolkit facilitates the rapid prototyping of context-sensitive services by allowing service developers to focus on the service business logic, and not the infrastructure required to deliver it.

4 ACCESS Management Agents

The ACCESS Management agents implement the core functionality of the ACCESS architecture, which includes context management, user profiling, map generation, content delivery, account management, and location sensing. The User Profiling Agent, acting in tandem with the Activity Analyzer Agent and the Context Agent, undertakes personalization. The User Profiling Agent provides a mechanism to allow agents to request a user's preferences, adopting a neighbourhood-based approach to identify this set of similar users. Implicit profile information is obtained using data mining techniques on recorded user activity, for instance when examining service usage, what was used, when it was used, and where it was used is noted.

The notion of user context is the concept that ACCESS employs to determine not only what services the user may require, but also where and when to offer them. A user's context is considered a combination of their location, preferences and previous activities. It is the responsibility of the Context Agent to create and manage user specific hotspots. A hotspot is a region of space-time that is limited by specified bounds (e.g. an area of 100 square meters centred around a particular shop, during business hours). The Context Agent informs the Profiling Agent when a user encounters a hotspot, which then decides whether the hotspot breach is relevant to the user.

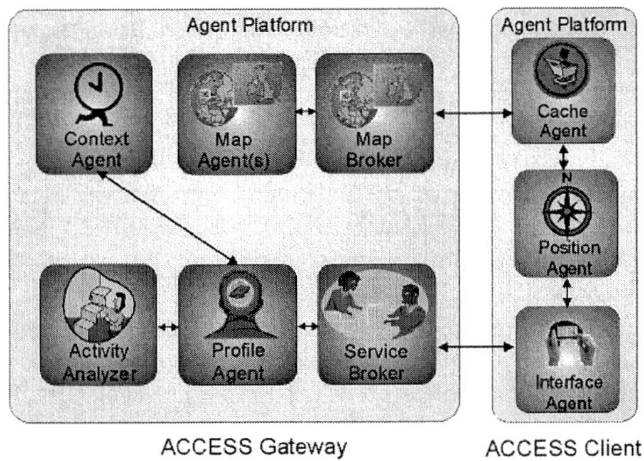

Fig. 1. Access Management Agents

The Map Server element is made up of the Map Broker and one or more Map Agents. The Map Broker manages the dynamic generation of maps for a specified location. When the Map Broker is requested to generate a map, it communicates with a number of Map Agents to determine which of them can produce the most suitable map. The selected Map Agent then generates the map segment which can subsequently be merged with service specific content overlay. This peer community of Map Agents allows for distribution of load and for a degree of failure protection.

The Position Agent is responsible for periodically informing subscribed agents of updates in the user or device's position. The Position Agent currently determines position by use of a GPS receiver.

To provide service specific functionality, abstract Service Manager and Service Delivery roles are provided that a service developer may extend. These roles encapsulate the requisite agent-based functionality for interacting with the core infrastructure. The Service Manager role is responsible for advertisement of the service by registering with a Service Broker Agent and for registering the service with other ACCESS compliant services that it may wish to use. The Service Delivery role acts as the users representative within a service and collaborates with the Service Manager Role and an

Interface Agent. The service developer must extend this role with additional behaviour patterns to interact with service specific agents.

The Interface Agent manages the user's interaction with the Graphical User Interface within the system. Additionally it provides a means by which it may dynamically extend its functionality at run-time through the use of Java introspection for the provision of service specific functionality.

4.1 Map Caching

A key feature of any application or service built using the ACCESS architecture is the ability to provide dynamically generated maps. But with ACCESS being primarily aimed at resource-limited devices such as PDAs or smart phones, it is desirable to minimise the detriment in making use of this feature.

There are two methods are used to minimise the cost of using the map feature; the first focuses on minimising the delay between a device requesting a map and rendering it, the second attempts to minimise both the time delay and the communication or bandwidth cost in displaying a map.

The first method consists of pre-emptive or active caching. This is accomplished by attempting to determine the user's future direction of travel. The Cache Agent tracks the user's heading and calculates the average heading. When the user approaches the edge of the current map, this average heading is used to establish which map segments are requested by the Cache Agent from the Map Server. So, for example, if the average heading was 90°, then the Cache Agent would request three map segments covering the area to the east of the user's current location. In addition, knowledge communicated to the Cache Agent from other agents, such as user location, history from the Activity Analyzer Agent etc, can influence which map segments are requested. This pre-fetching of the segments results in the map segment being immediately available for rendering, thus minimising the delay in changing the currently displayed map segment to the next segment.

The second method attempts to remove the bandwidth or communication cost in sending and receiving data between the Cache and Map Agents. This is done through the use of passive caching. Each map segment that is received from the Map Server, including segments used in active caching, is stored locally (cached) on the device by the Cache Agent. When a new map segment is required by the user, the Cache Agent will attempt to retrieve a suitable map segment from this cache and will only request a map segment from the Map Server if there are no segments in the cache that meets the user's requirements. If several suitable segments from the cache are found, the Cache Agent selects the segment whose centre is closest to the user's current location, thereby attempting to maximise the amount of time before another segment is required.

4.2 User Profiling

The fact that ACCESS is a tool kit for the deployment of context sensitive services places some unique constraints on the design of the User Profiling Agent, such as the fact that anticipating the profiling requirements of individual service developers is difficult, if not impossible. In an attempt to overcome this obstacle ACCESS users are provided with a 'generic' ACCESS profile containing personal data such as gender, address, age, job description, education etc., which service developers can augment with a service specific profile. Using this generic profile, the User Profiling Agent can identify a set of similar users to use as recommendation partners for the active user. Users *must* explicitly complete their generic profile when registering with ACCESS, whether a user is required to supply a set of service specific preferences is a matter for the individual service developer. Should a user not enter a service profile, some personalisation is possible from the ACCESS profile. Items being recommended to the user may also have their own profiles listing their distinguishing features (for instance in a hotel finder application each hotel might have a profile listing its amenities). Again this is a matter for the service developer.

The ACCESS User Profiler Agent adopts a neighbourhood-based approach in identifying a set of similar users to use as recommendation partners for the active user. The measure of similarity adopted to identify this set is neither a correlation measure nor a vector-based cosine similarity metric. Instead, a simple overlap measure, where, for each user, the number of preferences rated identically to the active user, constitutes the overlap similarity value.

$$\mathrm{sim}(x,y) = \frac{|\mathrm{preferences}(x) \cap \mathrm{preferences}\ (y)|}{|\mathrm{preferences}(x) \cup \mathrm{preferences}\ (y)|}$$

The number of overlapping items in the active user (x)'s profile and a test user (y)'s profile is simply counted. The union of all their preferences, as a normalising factor, divides this value. This method, known as the Jaccard similarity metric, is quite a simple way of measuring user similarity, but we can take the similarity value obtained from this method as indicative of the ACCESS user profiles true similarity, because the ratings in each user's profile are obtained by explicit means (that is, users consciously elected to give a certain value to a given question), rather than implicitly-required ratings, which can sometimes be misconstrued or inaccurate reflections of users profiles.

4.3 Agent Tuning

Collaborative Performance Tuning is an autonomic procedure by which ACCESS management agents collectively alter their response times so as to adapt to the dynamic utilities and requirements of the software systems that they represent. The tuning framework enables inactive agents to go into a temporary hibernation state without affecting their ability to react to their environment in a timely manner. The hibernation subsystem monitors an agent's commitments and reduces the number of redundant CPU cycles wasted on agents whose mental state does not change between iterations of their control algorithm. Once an agent goes into hibernation, other agents on the platform may opportunistically reduce their response times to take advantage of the additional computational resources made available.

Other systems [9][10] have been developed for the collective negotiation of CPU resources. Collaborative Performance Tuning differs from these systems in that a BDI approach is used that utilises the notions of joint intensions and mutual beliefs thus coercing agents into forming coalitions and to act as a team. This prevents agents getting into prisoner dilemma type scenarios in which agents making locally optimal decisions create socially unacceptable behaviour patterns in which all agents concerned are worse off as an inevitable consequence of rational choice.

As the inference engine executes, it examines the number of commitments an agent has on its current deliberative cycle. if the agent does not have any commitments it is said to be inactive on that cycle. The rationale for this is that given an agent's current mental state, if an agent is inactive on a particular cycle and its mental state will not change on the next cycle, the agent need not deliberate on the next cycle as it knows that its mental state will not be altered.

At various stages throughout execution ACCESS agents may collectively alter their response times to adapt to the evolving system requirements. Joint intentions are utilized whereby rather than unilaterally reducing its response time an agent must adopt a commitment to make it intentions mutually believable to all active agents on the platform. This implies that an agent must communicate with and receive a response from all other active agents before taking action. Thus ensuring that agents act in a manner that is beneficial to the team as a whole. The problem with collective decision-making however is that decisions can only be made as fast as the slowest decision maker. Within the ACCESS architecture the Position Agent often operates with quite a large response time, agents cannot self-tune or pre-empt the Position Agent because to do so they would have to act unilaterally. This prevents Collaborative Performance Tuning being used for tasks with a low level of granularity such as rendering maps, which with modern just-in-time compilers would be completed prior to the termination of the negotiation process. Therefore Collaborative Performance Tuning is primarily used to improve efficiency in medium to high-level quality of service problems or to improve the general performance of the system over the entire course of its execution as the system adapts to dynamic environment conditions.

5 A User Scenario

To illustrate the usefulness and appropriateness of ACCESS we consider a fictitious character Joe arriving at an airport equipped with a PDA hosting ACCESS. Initially when Joe loads ACCESS at the airport he receives an advertisement for Sos, an accommodation finding application, which his Interface Agent received from a Service Broker. Joe clicks the advertisement to download Sos and to start the Service. The Interface Agent contacts an appropriate Service Manager whose name it received from the Service Broker to begin using Sos. When the Interface Agent registers with the Sos Service Manager it is delegated to a Service Delivery Agent. The Sos Service Delivery Agent then checks to see if Joe has been registered with an additional Meeting Service. The Meeting Service is used to keep track of the user's diary. If Joe has been registered the Service Delivery Agent of Sos requests the Service Delivery Agent of the Meeting Service to inform it of meetings relevant to Joe's current spatial context. The Meeting Service is aware of the locations of meetings that Joe will be attending and thus returns an appropriate list.

Joe is now presented with a screen (Fig. 3) allowing him to select his room preferences and a drop down list containing points of interest that has been populated with relevant meetings and popular tourist attractions. Selecting an option from this list indicates that Joe would like his hotel to be located within the vicinity of the chosen item. Once the room details have been selected, the Profile Agent is contacted to obtain an appropriate list of hotels. The Profile Agent factors in implicit and explicit preferences, tariffs charged for the advertisement of hotels in addition to proximity of user's location and points of interest when generating this list. Once Joe receives the list and selects a hotel, the Sos Service Delivery Agent pushes the advertisement for a Bus Service. When Joe loads the Bus Service it receives the destination hotel location, which it uses to work out an appropriate route. The Caching Agent is informed of this route, which it uses to pre-emptively request the generation of maps for the impending journey. The Bus Service Delivery Agent collaborates with an additional Location Service, which operates in the background without the Joe's knowledge. The Location Service works with the Caching Agent in obtaining maps from the map server centred on Joe's position for tracking purposes.

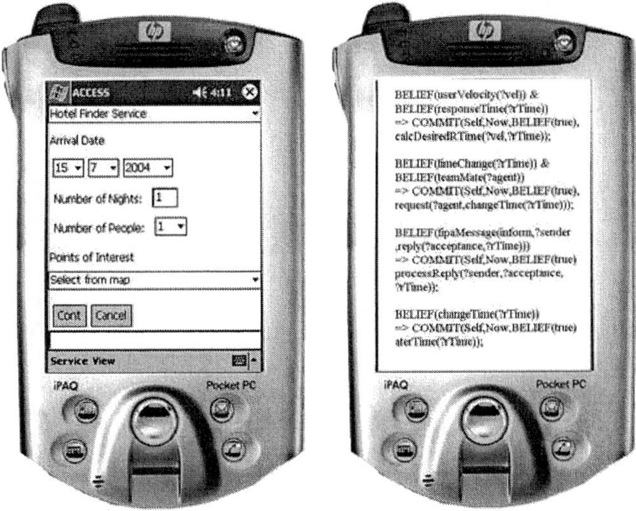

Fig. 3. Room Select Screen and Position Agent tuning Mental State

When Joe gets onto the bus and starts travelling the Position Agent realizes that Joe's average velocity has increased and that it needs to reduce its response time. Fig. 3 illustrates the Position Agent's mental state. The Position adopts a commitment to reduce its response time and thus informs the other team members of the situation. On receiving this information the Interface Agent adopts a commitment to have its response time increased whereas the Cache Agent adopts a commitment to reduce its response time. Once all agents receive replies from their teammates their response times are altered. The Agents will maintain these altered response times so long as Joe's average velocity remains within a certain range.

6 Conclusions

This paper has described ACCESS, a generic agent based architecture for the rapid development and role out of location aware services. The key characteristics and differentiators of this architecture are the provision of lightweight intentional mobile agents, which offer an agent tuning ability, support dynamic profile updates, dynamic map generation and a rich concept of context. ACCESS context awareness enables and underpins degradation or enhancement of content to suit the device context and user needs.

Acknowledgements

We gratefully acknowledge the support of Enterprise Ireland (grant ATRP/01/209) and Science Foundation Ireland through Modelling Collaborative Reasoners (SFI Investigator Award).

References

1. Sadeh N., Chan T., Van L., Kwon O. and Takizawa K., Creating an Open Agent Environment for Context-aware M-Commerce, in Agentcities: Challenges in Open AgentEnvironments, LNAI, Springer Verlag (2003).
2. Rodríguez M. and Favela J., A Framework for Supporting Autonomous Agents in Ubiquitous Computing Environments, Proceedings of System Support for Ubiquitous Computing Workshop at the Fifth Annual Conference on Ubiquitous Computing (UbiComp 2003), Seattle, Washington, (2003).
3. Chen M, An Intelligent Broker Architecture for Context-Aware Systems, PhD Dissertation Proposal, UMBC (2002).
4. Pearce J., IBM: Pervasive Computing is the future, ZD Net article, Jan 30 2003 (2003).
5. Creese S., Future Challenges in Pervasive Computing Environments, SC Infosec article, Mar 5 2003 (2003).
6. Collier, R., Agent Factory: A Framework for the Engineering of Agent-Oriented Applications, *Ph.D. Thesis*, Computer Science Dept., University College Dublin, Ireland (2001).
7. Collier, R.W., O'Hare G.M.P., Lowen, T., Rooney, C.F.B., Beyond Prototyping in the Factory of the Agents, 3rd Central and Eastern European Conference on Multi-Agent Systems (CEEMAS'03), Prague, Czech Republic (2003).
8. Dey, and G. Abowd, Towards a Better Understanding of Context and Context-Awareness, Proceedings of the CHI 2000 Workshop on The What, Who, Where, When, and How of Context-Awareness (2000).
9. Soh, L.-K., Tsatsoulis, C., Agent-Based Argumentative Ne-gotiations with Case-Based Reasoning. AAAI Fall Symposium Series on Negotiation Methods for Autonomous Cooperative Systems, North Falmouth, Mass. (2001)
10. Walsh, W.E. and Wellman, M.P., A market protocol for decentralized task allocation: Extended version. In The Proceedings of the Third International Conference on Multi-Agent Systems (ICMAS-98) (1998).

Immune-Based Optimization of Predicting Neural Networks

Aleksander Byrski and Marek Kisiel-Dorohinicki

Institute of Computer Science,
AGH University of Science and Technology,
Mickiewicz Avn. 30, 30-059 Cracow, Poland
{olekb,doroh}@agh.edu.pl

Abstract. Artificial immune systems turned out to be an interesting technique introduced into the area of *soft computing*. In the paper the idea of an immunological selection mechanism in the agent-based optimization of a neural network architecture is presented. General considerations are illustrated by the particular system dedicated to time-series prediction. Selected experimental results conclude the work.

1 Introduction

Looking for solutions for many difficult problems, researchers very often turned to the processes observed in nature. Such biologically and socially inspired computational models were succesfully applied to various problems and they still belong to a very important research area of *soft computing*. Artificial neural networks, which principles of operation originate from phenomena occuring in the brain, evolutionary algorithms based on the rules of organic evolution, multi-agent systems originating from the observation of social processes are the examples of such approaches. Also artificial immune systems, inspired by the human immunity, recently began to be the subject of increased researchers' interest. Such techniques are often combined together to create hybrid systems, that by the effect of synergy exhibit some kind of intelligent behaviour, which is sometimes called *computational intelligence* as opposed to rather symbolic artificial intelligence [1].

Following this idea a hybrid optimization technique of evolutionary multi-agent systems have been applied to a wide range of different problems [8]. In this paper immunological mechanisms are proposed as a more effective alternative to classical energetic ones used in EMAS. Particularly the introduction of *negative selection* may lead to early removing of improper solutions from the evolving population, and thus skip the process of energetic evaluation [7] and speed up the computation. Such approach may be justified in specific cases of the problems, that require relatively time-consuming fitness evaluation. As an example of such a problem the optimization of neural network architecture is considered in the paper.

V.S. Sunderam et al. (Eds.): ICCS 2005, LNCS 3516, pp. 703–710, 2005.

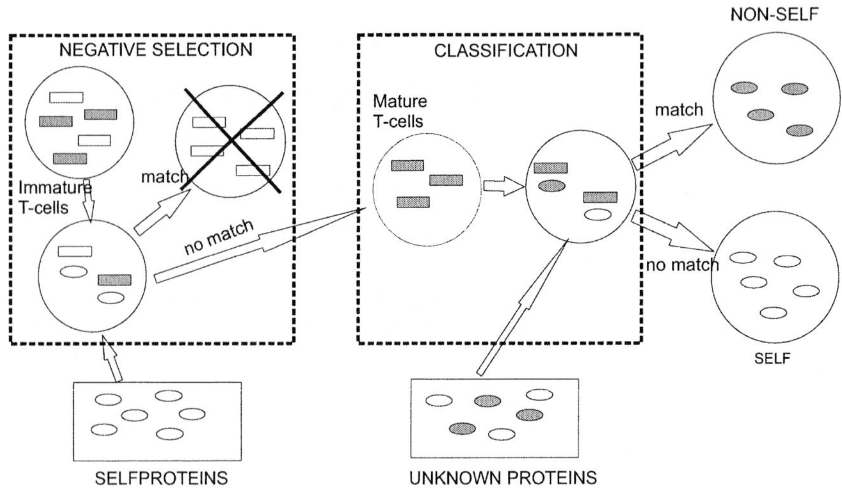

Fig. 1. Negative selection mechanism in artificial immune systems

Below, after a short presentation of the basics of human immunity and artificial immune systems, the details of the proposed approach are given. First the idea of neural multi-agent system for time-series prediction is described. Then the application of immune-based selection mechanism in agent-based neural network architecture optimization system is proposed. Some preliminary results obtained for a typical simple prediction problem allow for the first conclusions to be drawn.

2 Artificial Immune Systems

An immune system plays a key role in maintaining the stable functioning of the human body. At the cellular layer lymphocytes are responsible for detection and elimination of disfunctional endogenous cells, termed infectious cells and exogenous microorganisms, infectious non-self cells such as bacteria and viruses. One of the most important adaptation mechanisms in the human immunity is based on the process of negative selection, which allows for removal of lymphocytes that recognize self cells [5].

Different approaches inspired by the human immune system were constructed and applied to enhance algorithms solving many problems, such as classification or optimization [12]. Basic principles of immune-based algorithms utilising negative selection may be clearly illustrated on a classical problem of machine learning – concept learning – as described below.

Concept learning can be framed as the problem of acquiring the definition of a general category given a sample of positive and negative training examples of these categories' elements [10]. Thus the solutions can be divided into

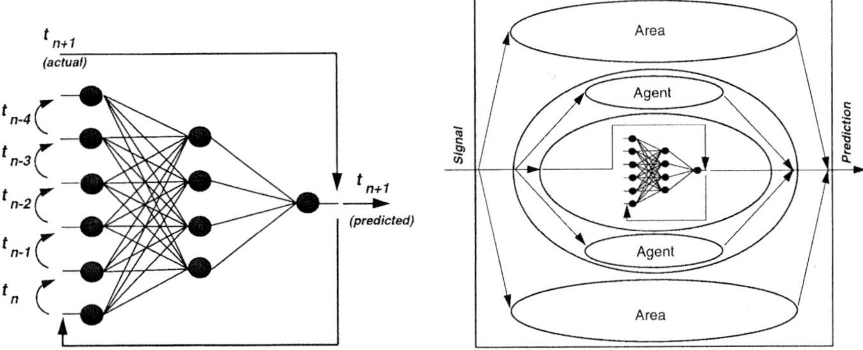

Fig. 2. A single predicting neural network (a) and the whole multi-agent system (b)

two groups: self (positive) and non-self (negative). An artificial immune system, which consists of a large number of lymphocytes can be trained to recognize these solutions in the process of negative selection. In the beginning the lymphocytes are created, but at this point they are considered immature. Next, the affinity binding of these cells to present self-cells (good solutions of some problem) is evaluated. Then the lymphocytes that bind themselves to "good" cells are eliminated. Finally, lymphocyte-cells that survive entire process are considered mature (see fig. 1). Mature lymphocytes are presented with the cells that have unknown origin (they may be self, or non-self cells), and they are believed to have possibility of classifying them [11].

3 Neural Agents for Time-Series Prediction

Prediction (or *forecasting*) is a generation of information about the possible future development of some process from data about its past and present behaviour. Time-series prediction consists in searching for some trends in the sequence of values of some variable [6].

Many examples from the literature show that neural networks may be successfully used as a mechanism to model the characteristics of a signal in a system for a time-series prediction [9]. The choice of a particular architecture of the network is to a large extent determined by a particular problem. Usually the next value of the series is predicted on the basis of a fixed number of the previous ones. Thus the number of input neurons correspond to the number of values the prediction is based on, and the output neuron(s) give prediction(s) of the next-to-come value(s) of the series. The multi-layer perceptron (MLP) in (see fig. 2a) should predict t_{n+1} value of the series, basing on some previous values, which are given on the inputs of the first layer. When t_{n+1} value is predicted, the inputs are shifted, and the value t_{n+1} is given as the input to the last neuron of the first layer. A network may be supervisory trained, using the comparison between values predicted and received as an error measure.

A multi-agent predicting system is a population of intelligent agents performing independent analysis of incoming data and generating predictions [4, 3]. In this case subsequent elements of the input sequence(s) are supplied to the environment, where they become available for all agents (see fig. 2b). Each agent may propose its own predictions of (a subset of) the next-to-come elements of input obtained from the posessed neural network. Of course the network is trained by the agent using incoming data. On the basis of predictions of all agents, prediction of the whole system may be generated e.g. by the means of PREMONN algorithm [2]. In such a system introduction of evolutionary processes allows for searching for a neural network architecture and learning parameters most suitable for the current problem and system configuration, just like in evolutionary multi-agent systems.

4 Immune-Based Selection in Evolutionary Multi-agent Systems

The idea of *EMAS* was already described and succesfully applied to several difficult problems including function optimization and prediction [3, 2].

The key idea of *EMAS* is the incorporation of evolutionary processes into a multi-agent system at a population level. This means that besides interaction mechanisms typical for agent-based systems (such as communication) agents are able to *reproduce* (generate new agents) and may *die* (be eliminated from the system). Inheritance is to be accomplished by an appropriate definition of reproduction, which is similar to classical evolutionary algorithms. A set of parameters describing basic behaviour of an agent is encoded in its genotype, and is inherited from its parent(s) – with the use of mutation and recombination. The proposed principle of selection corresponds to its natural prototype and is based on the existence of non-renewable resource called *life energy.* Energy is gained and lost when agents execute actions. Increase in energy is a reward for "good" behaviour of an agent, decrease – a penalty for "bad" behaviour (which behaviour is considered "good" or "bad" depends on the particular problem to be solved). At the same time the level of energy determines actions an agent is able to execute. In particular, low energy level should increase possibility of death and high energy level should increase possibility of reproduction [7].

In the simplest case the evaluation of an agent (its phenotype) is based on the idea of agent rendezvous. Assuming some neighbourhood structure in the environment, agents evaluate their neighbours, and exchange energy. Worse agents (considering their fitness) are forced to give a fixed amount of their energy to their better neighbours. This flow of energy causes that in successive generations, survived agents should represent better approximations of the solution.

In order to speed up the process of selection, based on the assumption that "bad" phenotypes come from the "bad" genotypes, a new group of agents (acting as lymphocyte T-cells) may be introduced. They are responsible for recognizing and removing agents with genotypes similar to the genotype pattern posessed by these lymphocytes. Of course there must exist some predefined affinity function, e.g. using real-value genotype encoding, it may be based on the percentage difference between corresponding genes. These agents-lymphocytes may be created in the system in two ways:

1. vaccination – during system initialisation lymphocytes are created with random genotype patterns, or with the patterns generated by some other tecnnique designed to solve a similar problem,
2. causality – after the action of death, the late agent genotype is transformed into lymphocyte patterns by means of mutation operator, and the newly created group of lymphocytes is introduced into the system.

In both cases, the new lymphocytes must undergo the process of negative selection. In a specific period of time, the affinity of the immature lymphocytes patterns to the "good" agents (posessing relative high amount of energy) is tested. If it is high (lymphocytes recognize "good" agents as "non-self") they are removed from the system. If the affinity is low – probably they will be able to recognize "non-self" individuals ("bad" agents) leaving agents with high energy intact.

5 Immunological Multi-agent System for Neural Network Optimization

The immunologically-enhanced selection is driven by two types of agents (computational agents and lymphocytes) and in case of neural network optimization may be described as follows.

First, the population of computational agents is initialized with random problem solutions (neural networks and learning parameters) and the initial energy level. Every agent is able to perform predictions of subsequent time-series values, after it acquires its actual value and modifies the weights of its neural network in one step of the learning process. During the described earlier *agent rendezvous* action it compares its fitness with neighbours and exchanges a portion of energy with them. When the level of energy reaches certain level, an agent looks for the neighbour that is able to reproduce, and they create a new agent, and then a fixed portion of the parents' energy is passed to their child. An agent dies when its energy falls below certain level – it is removed from the system and a lymphocyte is created based on its genotype.

The specific task for a lymphocyte agent is affinity testing – existing lymphocytes search the population for agents similar to the genotype pattern contained in a lymphocyte. If the lymphocyte survived the process of negative selection – it causes the removal of similar agent. Otherwise lymphocyte is removed from the system. In the proposed approach, the affinity is determined using the relation of the corresponding genes in the following way. Let $p = [p_1 \dots p_n]$ be a

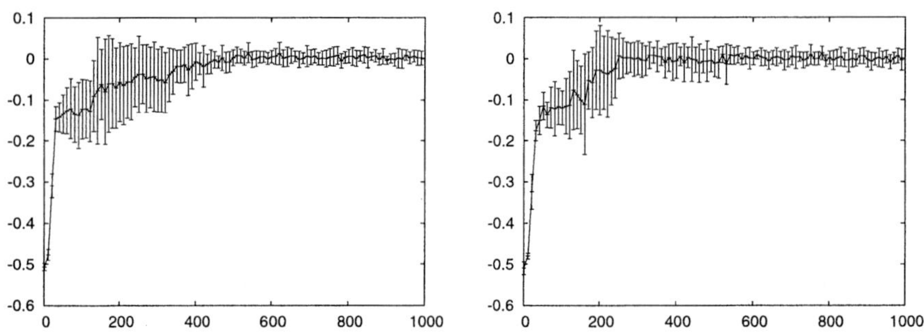

Fig. 3. Prediction error obtained for EMAS (a) and iEMAS (b)

paratope (genotype pattern owned by the lymphocyte) and $e = [e_1 \ldots e_n]$ be an epitope (genotype owned by the antigen – in the described system it is simply the genotype of the tested agent), and n is the length of the genotype. If d_{max} is the maximum deviation and c_{min} is the minimum count of corresponding genes, the construction of the set of corresponding genes may be considered:

$$G = \{p_i : \left|\frac{p_i}{e_i}\right| \leq d_{max}\}$$

The lymphocyte is considered as stimulated (its affinity reached minimal level) when $\overline{\overline{G}} \geq c_{min}$, and considered as non-stimulated otherwise.

6 Preliminary Experimental Studies

The system was designed and implemented using distributed evolutionary multi-agent platform Ant.NET developed at AGH-UST (*http://antnet.sourceforge.net*).

The experiments were performed in order to show whether the introduction of the immunological-based selection mechanism into EMAS will make the classical energetic selection more effective. A distorted sinus function was used as a benchmark for prediction.

Evolutionary multi-agent system consisted of one evolutionary island, with initial population of 100 agents. The genotype of the agent cotained the description of a feed-forward neural network (multi-layered perceptron) – count of neurons in hidden and input layers, learning coefficients. The networks were taught by the means of backpropagation algorithm with momentum. Before maturing, every lymphocyte undergone the negative selection process. During 100 system steps, if immature lymphocyte was stimulated by an agent with energy of 150 or higher, the cell was removed from the system.

In figure 3 the prediction error is presented in terms of fitness of the best solution in consecutive steps of the system activity. Each plot shows an average (and the value of standard deviation) of the prediction error obtained from the 10 runs of optimization with the same parameters. Comparing the plots, it seems that introduction of the immunological selection allows for a bit quicker obtaining of the desired prediction accuracy (prediction error close to zero). Yet this result is not convincing and must be verified by further experiments.

7 Conclusion

In the paper an agent-based approach to optimization of the architecture of a predicting neural network was considered. An immune-inspired selection mechanism was proposed as an alternative to energetic selection used in evolutionary multi-agent systems. As the preliminary experimental results show, it allows for improper individuals to be removed from the system faster than using classical energetic selection, and thus allows for obtaining slightly better solutions in comparable time.

The approach seems to be especially adequate for solving problems in which fitness evaluation takes relatively long time, just like in the considered case of neural network architecture optimization, which requires training of the neural network each time it is evaluated [3]. By means of negative selection mechanism, agents that probably will not be able to construct an efficient neural network, may be removed before they complete the process of training.

Of course up till now it is too early to compare the method with various other optimization heuristics known from the literature. The work in progress should allow for such a comparison.

References

1. P. Bonissone. Soft computing: the convergence of emerging reasoning technologies. *Soft Computing*, 1(1):6–18, 1997.
2. A. Byrski and J. Bałamut. Evolutionary neural networks in collective intelligent predicting system. In L. Rutkowski, editor, *Seventh International Conference on Artificial Intelligence and Soft Computing*. Springer Verlag, 2004.
3. A. Byrski, M. Kisiel-Dorohinicki, and E. Nawarecki. Agent-based evolution of neural network architecture. In M. Hamza, editor, *Proc. of the IASTED Int. Symp.: Applied Informatics*. IASTED/ACTA Press, 2002.
4. K. Cetnarowicz, M. Kisiel-Dorohinicki, and E. Nawarecki. The application of evolution process in multi-agent world (MAW) to the prediction system. In M. Tokoro, editor, *Proc. of the 2nd Int. Conf. on Multi-Agent Systems (ICMAS'96)*. AAAI Press, 1996.
5. W.H. Johnson, L.E. DeLanney, and T.A. Cole. *Essentials of Biology*. New York, Holt, Rinehart and Winston, 1969.
6. Nikola K. Kasabov. *Foundations of Neural Networks, Fuzzy Systems, and Knowledge Engineering*. The MIT Press, 1996.
7. M. Kisiel-Dorohinicki. Agent-oriented model of simulated evolution. In William I. Grosky and Frantisek Plasil, editors, *SofSem 2002: Theory and Practice of Informatics*, volume 2540 of *LNCS*. Springer-Verlag, 2002.

8. M. Kisiel-Dorohinicki, G. Dobrowolski, and E. Nawarecki. Agent populations as computational intelligence. In Leszek Rutkowski and Janusz Kacprzyk, editors, *Neural Networks and Soft Computing*, Advances in Soft Computing, pages 608–613. Physica-Verlag, 2003.
9. Timothy Masters. *Neural, Novel and Hybrid Algorithms for Time Series Prediction.* John Wiley and Sons, 1995.
10. T. M. Mitchell. *Machine learning.* McGraw-Hill, 1997.
11. S. Wierzchoń. *Artificial Immune Systems [in polish].* Akademicka oficyna wydawnicza EXIT, 2001.
12. S. Wierzchoń. Function optimization by the immune metaphor. *Task Quaterly*, 6(3):1–16, 2002.

Algorithm of Behavior Evaluation in Multi-agent System

Gabriel Rojek[1], Renata Cięciwa[2], and Krzysztof Cetnarowicz[3]

[1] Department of Computer Science in Industry,
AGH University of Science and Technology,
Al. Mickiewicza 30, 30-059 Kraków, Poland
rojek@agh.edu.pl

[2] Department of Computer Networks,
Nowy Sącz School of Business – National-Louis University,
ul. Zielona 27, 33-300 Nowy Sącz, Poland
rcieciwa@wsb-nlu.edu.pl

[3] Institute of Computer Science,
AGH University of Science and Technology,
Al. Mickiewicza 30, 30-059 Kraków, Poland
cetnar@agh.edu.pl

Abstract. Behavior based detection of unfavorable activities in multi-agent systems (presented in [3, 4, 5]) is an approach to the problem of detection of intruders. This approach refers to evaluation of behavior of every agent which exists in a multi-agent system. Process of behavior evaluation is distributed – every agent makes autonomous behavior evaluation of other agents. This means that an agent is evaluated separately by all agents in the environment of the secured system. That separate results of behavior evaluations have to be collected and an algorithm should be used in order to elect the *worst* agents which should be eliminated. Collecting and data processing of results of distributed behavior evaluations is the main topic of this article.

1 Introduction

Behavior based detection of unfavorable activities in multi-agent systems is inspired by ethically-social mechanisms that act in human societies. An individual in a society seems trustworthy if its behavior could be observed by others and this behavior is evaluated by majority as good and secure. The decision about trustworthy of an individual takes place in society in the decentralized and distributed way – all individuals in a society make own decisions which form one decision of the society.

Inspiration by ethically-social mechanisms in computer security systems induce decentralization of security mechanisms which should be based on observation and evaluation of behavior of agent functioning in a secured system. Actions undertaken by agents are perceived as objects, which create a sequence registered by all agents in the environment. Registered objects-actions could be processed

V.S. Sunderam et al. (Eds.): ICCS 2005, LNCS 3516, pp. 711–718, 2005.

in order to qualify whether it is a *good* or a *bad* acting agent in this particular system, in which evaluation takes place. A *bad* agent also could be named *intruder.*

2 Division Profile

Decentralization of security mechanisms is realized in multi-agent systems by means of equipping all agents with some additional goals, tasks and mechanisms. Those goals, tasks and mechanisms are named *division profile.* The name *division profile* is inspired by M-agent architecture which could be used to describe an agent (M-agent architecture was introduced among others in [1, 2]).

Description of division profile was presented in [3, 4, 5]. This article contain only some information that is crucial to the problem of collecting and data processing of the results of distributed behavior evaluations.

Each agent in a multi-agent system has his own autonomous calculated division profile. Division profile of an agent has three stages of functioning: creation of collection of *good* (*self*) sequences of actions, generation of detector set, behavior evaluation. An agent a, which division profile is at his behavior evaluation stage, has division state m_a represented as a vector:

$$m_a = (m_a^1, m_a^2, ..., m_a^{j-1}, m_a^j) \tag{1}$$

where j is the number of neighboring agents (neighboring agents are agents which are visible for agent a) and m_a^k is the coefficient assigned to neighboring agent number k. Coefficient m_a^k indicates whether the agent number k is evaluated by agent a as *good* or *bad.* Coefficient m_a^k is a number of counted matches between:

- detectors of agent a which evaluates behavior and possesses division state m_a,
- sequence of actions undertaken by agent number k.

Marking the length of a detector as l and the length of sequence of actions as h, the coefficient m_a^k is a number from a range $\langle 0, h - l + 1\rangle$. The maximum of counted matches is equal $h - l + 1$, because every fragment of sequence of actions, which has a length equal to the length of a detector, can match only one detector.

3 Problem of Distributed Behavior Evaluation

In order to choose an agent which should be removed from the system, division states of all agents should be collected and the algorithm of data processing should be used. This problem was solved in our earlier work, but presented solutions seems to be not sufficient in order to obtain a flexible and self-adopting method.

Simulations that in every constant time period Δt each agent in the system executes his full life cycle are presented in [3, 5]. An exemplary agent a calculates his division state m_a in every life cycle and chooses agent (or agents) number k, that $m_a^k = \max(m_a^1, m_a^2, ..., m_a^{j-1}, m_a^j)$. Then agent a send a demand of deleting agent (or agents) number k to the environment. To this demand there is coefficient o_a^k equal to the m_a^k attributed.

The environment calculates the sum of coefficients and liquidates an agent (or agents) number n which fulfills two requirements:

- $o_*^n = \max(o_*^1, o_*^2, ..., o_*^{j-1}, o_*^j)$, where $o_*^p = o_1^p + o_2^p + ... + o_{j-1}^p + o_j^p$ $(1 \leq p \leq j)$;
- $o_*^n > OU$, where OU is constant.

Periodically, after a constant time period Δt, the calculated sums of coefficients are set to 0. Constant coefficient OU is introduced in order to get tolerance for behavior that is evaluated as *bad* in a short time, or is evaluated as *bad* by a small amount of agents.

Presented in [4] simulations are more adequate to a real-world multi-agent system. In that simulations all actions of agents are asynchronous. In each constant time period Δt number of activated agents can be different. Used asynchronous model of agent systems forced some modification in behavior based detections of intruders, particular in algorithm of data processing of behavior evaluations. If an agent number k tries to undertake any action, the environment asks neighboring agents for their "opinion" about him.

Exemplary agent a in a case of receiving a request (from the environment) of evaluation of an agent number k calculates his division state m_a and sends:

- coefficient o_a^k equal to coefficient m_a^k,
- additional information:
  - `true` if agent a evaluates agent number k as *the worst* ($m_a^k = \max(m_a^1, m_a^2, ..., m_a^{j-1}, m_a^j)$),
  - `false` in other cases.

The environment sums these coefficients and eliminates an agent (or agents) number n which fulfills two requirements:

- $o_*^n > OU$, where OU is constant and $o_*^p = o_1^p + o_2^p + ... + o_{p-1}^p + o_{p+1}^p + ... + o_{j-1}^p + o_j^p$ $(1 \leq p \leq j)$,
- more than 50 per cent of agents evaluate agent number n as *the worst.*

Liquidation of an agent only on the base of the sum of coefficients seemed to be not possible in that research.

It can be noticed that in mentioned solutions there is problem of setting the constant OU. The constant OU was set empirically after series of tests. The constant OU have to be reset in every new type of test. Another inconvenience of presented solution in asynchronous model of simulation is the need to calculate the whole division state m_a (all coefficients assigned to all agents), when evaluated is only one agent number k. In order to reduce computational complexity of behavior based detection of unfavorable activities and make this security solutions more flexible some ideas are presented and tested in the next section.

4 Algorithms of Distributed Behavior Evaluation – Propositions and Experiments

To reduce number of operations undertaken by each agent, which are connected with behavior evaluation, a modification of algorithm of evaluation process was proposed. An agent a in a case of receiving a request of evaluation of an agent number k sends back only the coefficient o_a^k, where $o_a^k = f_{eval}(m_a^k)$. An agent a do not have to calculate the whole division state m_a, but only the coefficient m_a^k. The environment sums gained coefficients. If final sum of all received coefficients o_*^k is larger than $1/2 * j$ agent k is eliminated (j is the number of agents). The function $f_{eval}()$ can be called function of evaluation which should be selected in order to assure removing intruders and leaving good agents.

In order to confirm effectiveness of proposed solutions and to select proper function of evaluation, a multi-agent system with asynchronously acting agents was implemented. In the simulated environment there exist two types of resources: resources of type A and resources of type B. This situation reflect these operations in computer system which should be executed in couples e.g. opening / closing a file. Resources are used by agents, but refilling all resources is only possible when each type of resources reach the established low level. The simulated system has three types of agents:

- *type g=0* – agents which take one unit of randomly selected (A-50%, B-50%) resource in every full life cycle;
- *type g=1* – agents which take one unit of randomly selected (A-75%, B-25%) resource in every full life cycle; type g=1 agents can be treated as intruders, because increased probability of undertaking only actions of one type can cause blocking the system (what is presented in [3, 4]);
- *type g=2* – agents which take one unit of A resource in every full life cycle; type g=2 agents are also called intruders.

We have simulated the case in which initially there are 64 agents of type g=0, 8 agents of type g=1 and 8 agents of type g=2. All agents in the system are equipped with the division profile mechanisms with parameters $h = 18$ and $l = 5$. The simulations are run to 2000 constant time periods Δt and 10 simulations were performed.

4.1 Linear Function of Evaluation

First a case was simulated in which all agents use linear function for behavior evaluation. An agent a sends back to the environment the coefficient o_a^k in the range $0 \leq o_a^k \leq 1$. This coefficient is given by

$$o_a^k = \frac{m_a^k}{h - l + 1} \tag{2}$$

where $h - l + 1$ is the maximum of counted matches of agent a.

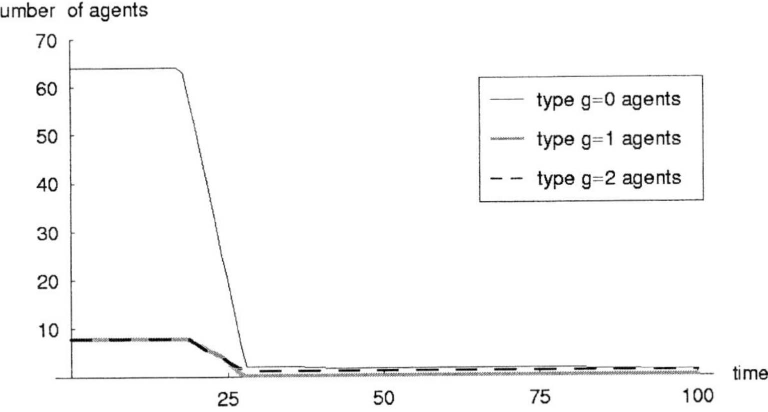

Fig. 1. Number of agents in separate time periods, agents using linear function of evaluation

Diagram in Fig. 1 shows the average numbers of agents in separate time periods. This diagram shows only 100 time periods Δt, during next time periods number of agents remains at the same level.

After series of tests it turned out that intruders were eliminated as well as good agents. Analysis of obtained results indicates that most of agents were deleting successively from 19 constant time period Δt to 28 constant time period Δt. The results of presented simulation enforced a reconfiguration of agents' behavior evaluation algorithm.

Example of three agents: to find a reason for rapidly deleting all agents we have chosen randomly:

- agent number 0 – an agent of type g = 0,
- agent number 1 – an agent of type g = 1,
- agent number 2 – an agent of type g = 2.

The final sum of coefficient gained during their evaluation process was analyzed. Average amount of coefficients returned to the environment is shown on diagram in Fig. 2. The agent number 2 obtains only minimal or maximal values of coefficient. The most of coefficients returned to the environment during evaluation process of agents number 0 and 1 are larger than 0.5. The final sum of coefficients is $o_*^0 = 48.6$, $o_*^1 = 49.7$ and $o_*^2 = 51.0$. The final sum of coefficients for all three exemplary agents is above condition $1/2 * j$ (j is equal to 80 at the time of this evaluation). As a result, agent number 0 is eliminated as well as agent 1 and 2 – intruders.

4.2 Discrete Function of Evaluation

On diagram in Fig. 2 we observe that amount of returned coefficient for both agents number 1 and number 2 reaches maximal value for $o_a^k = 1$. Thus we have

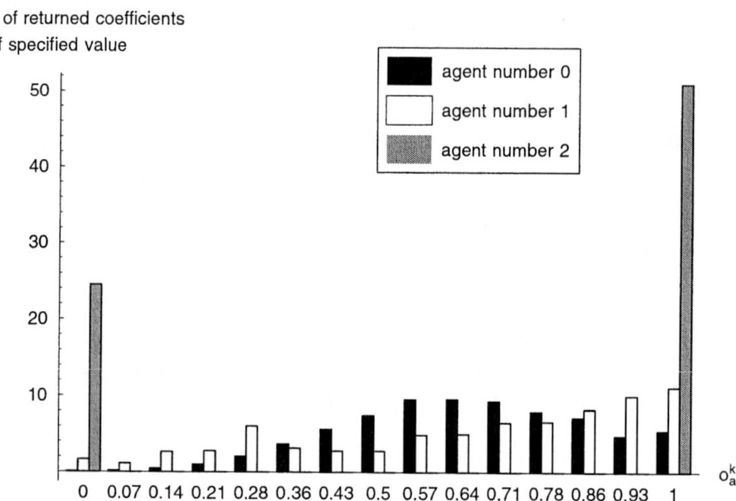

Fig. 2. Average amount of coefficient returned during evaluation process of exemplary three agents

simulated a case in which all coefficients o_a^k, sent to the environment, have value from the set {0,1}. This coefficient is given by

$$o_a^k = \begin{cases} 1, & \text{if } m_a^k = h - l + 1 \\ 0, & \text{if } m_a^k \neq h - l + 1 \end{cases} \tag{3}$$

Diagram in Fig. 3 shows the average number of agents in separate time periods. All good agents have remained in the system, bad agents were deleting successively from 19 constant time period Δt to 28 constant time period Δt, agents of type g=1 were eliminated only in 71%. Obtained results indicate that discrete function of evaluation can be inefficient for agents of type g=1, because, as we can see in Fig.2, they are evaluated also in a way that does not indicate the maximal value of coefficients in division states.

4.3 Power Function of Evaluation

In order to increase a weight of high coefficients we have simulated also a case in which all agents use the power function for behavior evaluation. An agent in a case of receiving a request (from the environment) of evaluation of an agent k sends back to the environment the coefficient o_a^k in the range $0 \leq o_a^k \leq 1$. The coefficient o_a^k is given by

$$o_a^k = \left(\frac{m_a^k}{h - l + 1} \right)^4 \tag{4}$$

Diagram in Fig. 4 shows the average number of agents in separate time periods. All bad agents were deleting successively from 19 constant time period Δt

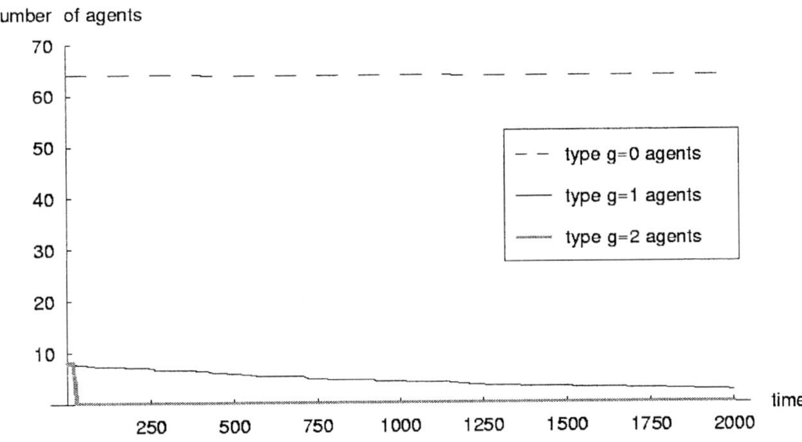

Fig. 3. Number of agents in separate time periods, agents using discrete function of evaluation

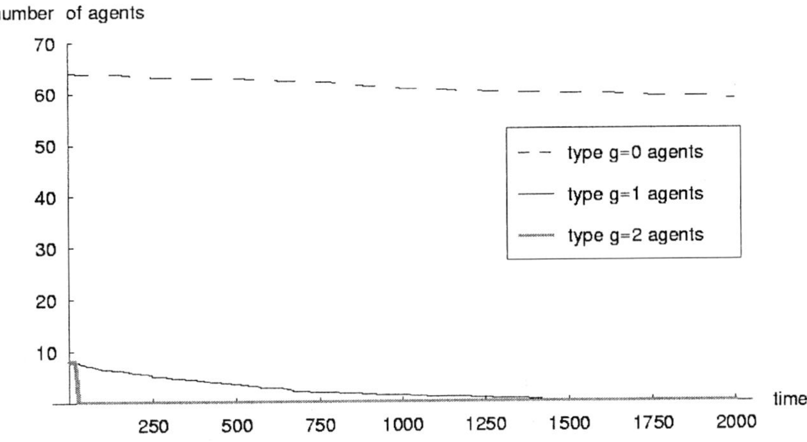

Fig. 4. Number of agents in separate time periods, agents using power function of evaluation

to 28 constant time period Δt, the agents of type g=1 were eliminated in 99% and good agents were also eliminated in 8%.

Randomly selection of resources cause that in a particular situation actions undertaken by a good agent could be considered as undesirable for a system. In the environment, with agents using power function of evaluation, such agent is immediately eliminated.

5 Conclusion and Further Research

In this paper the problem of collecting and data processing of distributed behavior evaluations in the multi-agent system was discussed. Three functions of evaluation: linear, discrete and power were proposed and researched. Proposed mechanisms in algorithm of behavior evaluation allow to reduce computational complexity of earlier algorithms presented in [3, 4, 5].

In order to confirm the effectiveness of proposed conceptions a multiagent system with agents acting asynchronously was implemented. The results obtained in Sect. 4.1 demonstrate that using the linear function for behavior evaluation causes elimination of intruders as well as good agents. However results obtained in Sect. 4.2 and Sect. 4.3 indicate that the proposed methods detect an abnormal behavior of agents in the system. Discrete function for behavior evaluation can be useful in case of application in which it should be sure that any good agent is not removed, but elimination of all intruders is not crucial. Power function of evaluation should be used in application in which elimination (or only indication) of all agents that can be considered as undesirable is crucial.

Further research in behavior based detection of unfavorable activities in multiagent systems should include methods for precise recognizing and eliminating all intruders from the system. One of the goals is to extend algorithms of distributed behavior evaluation with mechanisms, which makes it possible to take into account also the final sum of coefficients obtained during earlier live cycles of an agent. Such algorithms could let us avoid situations, in which good agents are eliminated because of short-term deterioration of their actions.

References

1. Cetnarowicz, K.: M-agent architecture based method of development of multiagent systems. Proc. of the 8th Joint EPS-APS International Conference on Physics Computing, ACC Cyfronet, Kraków (1996)
2. Cetnarowicz, K., Nawarecki, E., Żabińska, M.: M-agent Architecture and its Application to the Agent Oriented Technology. Proc. of the DAIMAS'97, St. Petersburg (1997)
3. Cetnarowicz, K., Rojek, G.: Unfavourable Behviour Detection with the Immunological Approach. Proceedings of the XXVth International Autumn Colloquium ASIS 2003, MARQ, Ostrava (2003) 41–46
4. Cetnarowicz, K., Cięciwa, R., Rojek, G.: Behavior Based Detection of Unfavorable Activities in Multi-Agent Systems. In MCPL, Conference on Management and Control of Production and Logistics, Santiago - Chile, (2004) 325–330
5. Cetnarowicz, K., Rojek, G.: Behavior Based Detection of Unfavorable Resources, in: Lecture Notes in Computer Science, Proceedings of Computational Science - ICCS 2004: 4th International Conference, Springer-Verlag, Heidelberg (2004) 607–614

Formal Specification of Holonic Multi-agent Systems Framework

Sebastian Rodriguez, Vincent Hilaire, and Abder Koukam

UTBM,
Systems and Transports Laboratory,
90010 Belfort Cedex,
France,
Tel: +33 384 583 009, Fax +33 384 583 342
vincent.hilaire@utbm.fr

Abstract. Even if software agents and multi-agent systems (MAS) are recognized as both useful abstractions and effective technologies for modeling and building complex distributed applications, they are still difficult to engineer. When massive number of autonomous components interact it is very difficult to predict the behavior of the system and guarantee that the desired functionalities will be fulfilled. Moreover, it seems improbable that a rigid unscalable organization could handle a real world problem. This paper presents a holonic framework where agents exhibit self-organization according to the tasks at hand. We specify formally this framework and prove some properties on the possible evolutions of these systems.

Keywords: Holonic Multi-Agent Systems, self-organised system, formal specification, model checking.

1 Introduction

Even if software agents and multi-agent systems (MAS) are recognized as both useful abstractions and effective technologies for modeling and building complex distributed applications, they are still difficult to engineer. When massive number of autonomous components interact it is very difficult to predict the behavior of the system and guarantee that the desired functionalities will be fulfilled. Moreover, it seems improbable that a rigid unscalable organization could handle a real world problem. The aim of this paper is to present a formally specified framework for holonic MAS which allows agent to self-organise. We prove some pertinent properties concerning the self-organising capabilities of this framework.

The term holon was originally introduced in 1967 by the Hungarian Philosopher Arthur Koestler[7] to refer to natural or artificial structures that are neither wholes nor parts in an absolute sense. According to Koestler, a holon must respect three conditions: (1) being stable, (2) having the capability of autonomy

V.S. Sunderam et al. (Eds.): ICCS 2005, LNCS 3516, pp. 719–726, 2005.

and, (3) being capable of cooperation. Holonic organizations have proven to be an effective solution to several problems associated with hierarchical self organized structures (e.g. [10], [1], [12]). In many MAS applications, an agent that appears as a single entity to the outside world may in fact be composed of several agents. This hierarchical structure corresponds to the one we find in Holonic Organizations. Frameworks have been proposed to model specific problem domains, mainly in Flexible Manufacturing Systems (FMS) and Holonic Manufacturing Systems (HMS), such as PROSA (PROSA stands for Product-Resource-Order-Staff Architecture) , [14] and MetaMorph[9]. However, the Holonic paradigm has also been applied in other fields such as e-health applications [11].

Our framework isn't application domain dependent so it can be easily reused. This framework is based upon organizational concepts which have been successfully used in the MAS domain [6,13]. We base our approach on the Role-Interaction-Organization (RIO) Methodology. RIO uses a specific process and a formal notation OZS that is described in [6]. OZS is a multi-formalisms notation that integrates in Object-Z classes [3] a statechart [5]. OZS classes have then constructs for specifying functional and reactive aspects. We have defined a formal semantics for OZS [4]. This semantic is based upon the translation of Object-Z and statecharts into transition systems and allows the use of theorem proover and model checker.

The paper is organized as follows : section 2 presents RIO and specifies the holonic framework, the section 3 describes proven properties and eventually section 4 concludes.

2 RIO and Holonic Framework Overview

In this section we present the RIO framework and its extension for Holonic MAS. We use the OZS formalism which is based upon the integration of Object-Z and statecharts. Object-Z is an object oriented extension of Z and thus uses the set theory and first order predicate logic. Statecharts add hierarchy of state, parallelism and broadcasted communication to finite state automata. Each concept of the RIO Framework is specified by an OZS class.

2.1 RIO Classes

The RIO Methodology is based on three main concepts. A *Role* is an abstraction of the behaviour of an acting entity. For example, we can see an university as an organization with several roles such as *Researcher*, *Professor*, etc.

We have chosen to specify it by the *Role* class. This class represents the characteristic set of attributes whose elements are of [*Attribute*] type. These elements belong to the *attributes* set. A role is also defined by stimulus it can react to and actions it can execute. They are specified by *stimulus* set and *actions* set respectively. The [*Attribute*], [*Event*] and [*Action*] types are defined as given types which are not defined further.

The reactive aspect of a role is specified by the sub-schema *behaviour* which includes a statechart. It is to say that the *behaviour* schema specifies the different states of the role and transitions among these states. The *obtainConditions* and *leaveConditions* attributes specify conditions required to obtain and leave the role. These conditions require specific capabilities or features to be present in order to play or leave the role. Stimuli, which trigger reactions in the role behaviour, must appear in one transition at least. The action belonging to the statechart transitions must belong to the *actions* set.

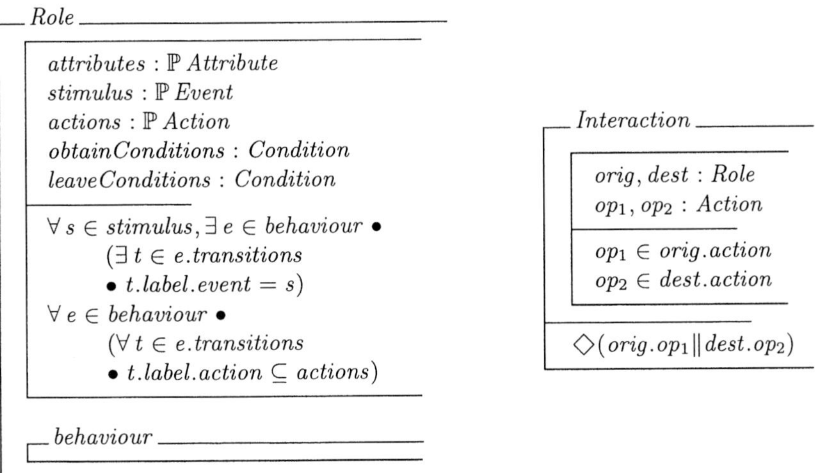

An *interaction* is specified by a couple of role which are the origin and the destination of the interaction. The role *orig* and *dest* interacts by the way of operations op_1 and op_2. These operations are combined by the $\|$ operator which equates output of op_1 and input of op_2. The $\diamond$ symbol is a temporal logic operator which states that eventually the predicate is true. In order to extend interaction to take into account more than two roles or more complex interactions involving plan exchange one has to inherit from the *Interaction* class. For example, the two roles *Proferssor* and *CoursePlanner* interact at the beginning of the year. The *CoursePlanner* sends the schedule to the *Professor* role.

2.2 HMAS Framework

In this section we present a set of roles which contitutes the kernel of the HMAS Framework. They describe the behaviour and interactions of the components of a holonic organization: holons. The holons inside a holonic organization, may join or create other holons to colaborate towards a shared goal. Inside a Holon there is one that acts as the *representative* (Head) and others as members (Part) of the Holon.

In order to enable holons to dynamically change their roles, we define a satisfaction based on the progress of his current task. This satisfaction, called *instant satisfaction*, depends on the played role and is calculated using the following definition, where R_i is the role played by the holon i.:

Self Satisfaction (SS_i). Satisfaction for the agent i produced by his own work.
Collaborative Satisfaction (CS_i^H). Satisfaction produced for the Agent i by his collaboration with other member agents of the Holon H,
Instant Satisfaction (IS_i). Satisfaction produced by the work done up to the moment

$$\forall i \in HMAS\ IS_i = \begin{cases} CS_i + SS_i & if\ R_i = Part \vee R_i = Head \\ SS_i & if\ R_i = Stand - Alone \end{cases} \quad (1)$$

2.3 Framework Specification

The inheritance relationships between the different roles is presented in the figure 1.

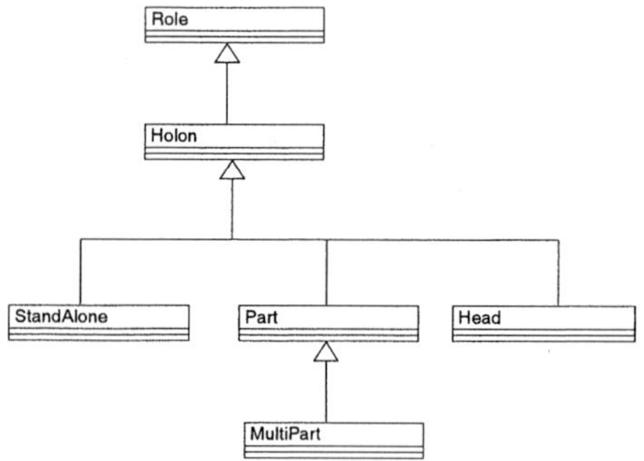

Fig. 1. Inheritance relationships between HMAS roles

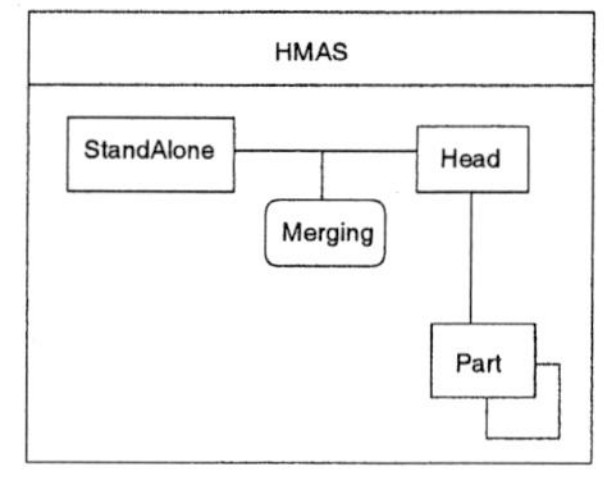

Fig. 2. RIO diagram of the HMAS framework

The RIO diagram of the figure 2 presents the possible interactions between the different roles. A *StandAlone* role player may interact with *Heads* in order to enter a specific holon. This interaction is specified by the *Merging* class which inherits from *Interaction*. *Part* role players interact with their *Head* during the holon's life. These interactions are commands or requests. The *Holon* class inherits from *Role* class and defines the generic elements for all Holonic role players. These elements are the different satisfaction criterions defined in the section 2.2. *SS* stands for self-satisfaction and *IS* stands for instant satisfaction. The current task of the holon is specified by *current* an element of a given type [*Task*]. For each task, the function *NS* associate a threshold. It is the minimum value for the self-satisfaction of the holon in order to pursue the current task. The available services for other holons are specified by the function *availableServices*. All following roles inherit from *Holon* and add specific attributes, operations and behaviours.

Holon
Role

$satisfied : \mathbb{B}$
$SS, IS : \mathbb{R}$
$current : Task$
$NS : Task \mapsto \mathbb{R}$
$availableServices : Service \rightarrow \mathbb{B}$

A *StandAlone* role is the entry point of an holonic organization. Each holon which is satisfied by its progress and with no engagement with other holons, plays this role. As soon as its satisfaction is less than the threshold defined by the *NS* function, the *StandAlone* holon searches a holon to merge with.

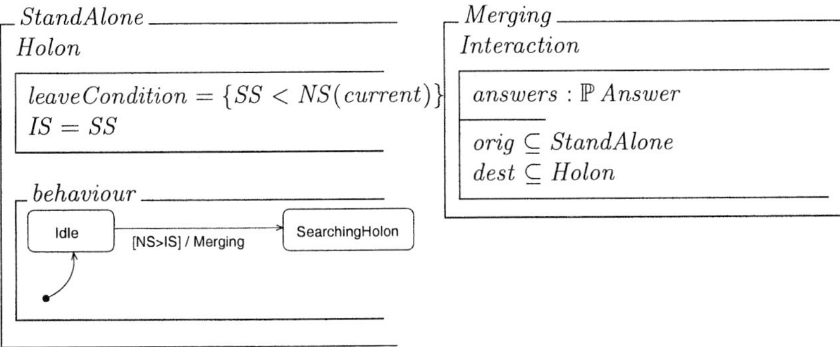

The [Answer] given type specifies the answers given by heads in response to a merge request. This interaction requires that the origin of the interaction is a holon playing the StandAlone role and the destination can be any Holon (a Head or another StandAlone holon).

The *Part* class specifies a role which is part of a holon . A holon part of a bigger holon knows the other members of the holon which are the elements of the *others* set. It also knows the head of its holon, *myHead*. The *Part* role has one more satisfaction criterion than a Holon, *CollaborativeSatisfaction*. It may also be engaged on some of its available services. These engagements are specified by the *engagements* function

Part
Holon

$others : \mathbb{P}\, Holon, myHead : Head$
$CS : \mathbb{R}$
$engagements : Service \mapsto \mathbb{B}$

$IS = CS + SS$
$\mathrm{dom}\ engagements \subseteq \mathrm{dom}\ availableServices$
$leaveCondition = \{CS < 0\}$

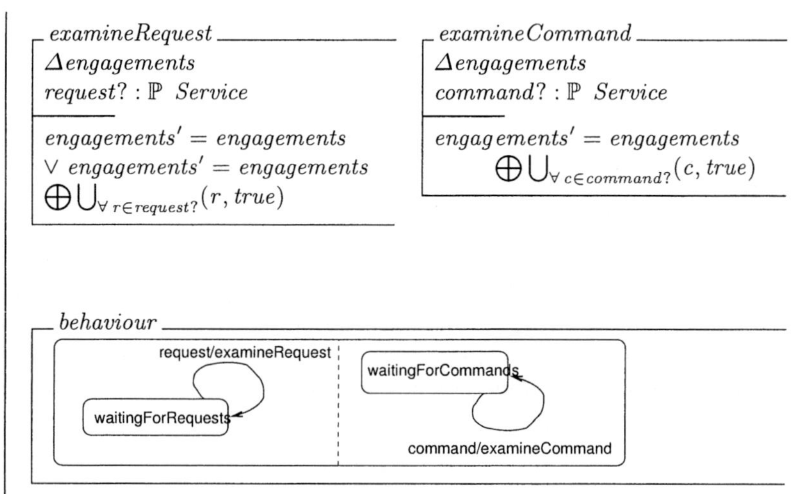

Finally, a holon may play the *Head* role. In this case this holon is the representative of the members of the Holon. Thus, it will examine the request of other holons to join his Holon.

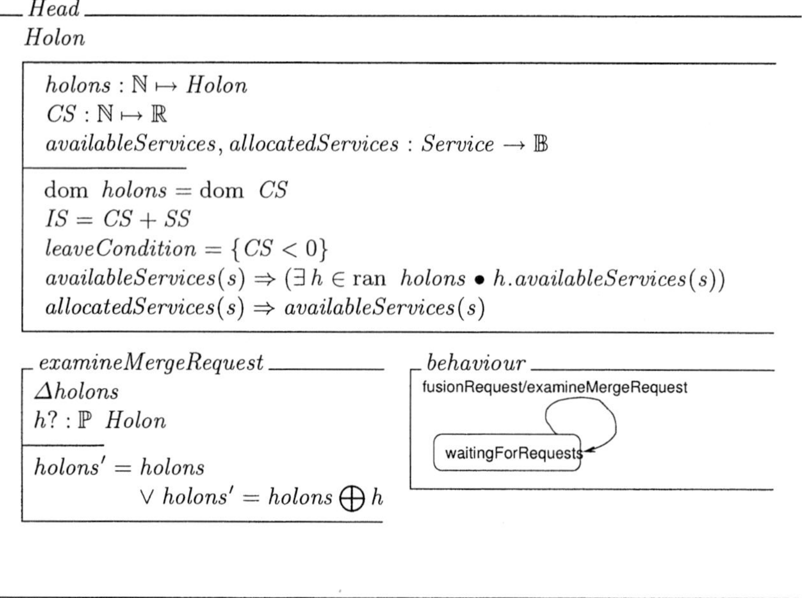

3 Proofs

OZS semantics [4] is based upon transition systems as defined in [8]. It means that for each OZS specification there is an associated transition system. This transition system represents the set of possible computations the specification can produce. With such transition systems and software tools like SAL [2] one can verify specification properties.

Among the tools proposed by SAL we have chosen the SAL model checker which enables the verification of the satisfiability of a property. The SAL model-checker proves or refutes validity of Linear Temporal Logic (LTL) formulas relatively to a transition system. To establish the satisfiability of history invariant H one must actually establish that $\neg H$ is not valid. This technique is the simplest to use but is limited by the specification state space.

The first property we have proven may be interpreted as "if the holon's satisfaction is not enough then it will try to merge". This property is specified as follows :

$$\forall\, a : HMASAgent \bullet a.is < a.ns \Rightarrow \Diamond(StandAlone \notin a.playing)$$

It states that for all agent of an holonic MAS if its instant satisfaction becomes less than the necessary satisfaction eventually this agent will not play the StandAlone role. It will try to be engaged in a holon and thus play either the Part or the Head role. In other words, if the agent can not accomplish its task alone, it will create a holon to cooperate with other having a share objective.

The second property we have proven may be interpreted as "if the holon's satisfaction evolves and becomes less than necessary satisfaction the system will try to reorganise". This property is specified as follows :

$$\forall\, a : HMASAgent \bullet instant = a.playing \wedge a.is < a.ns \Rightarrow \Diamond(a.playing \neq instant)$$

It states that for all agent of an holonic MAS whatever role it plays if its instant satisfaction becomes less than the necessary satisfaction eventually this agent will change the role it is playing.

4 Conclusion

In this paper we have presented a framework for the design of Holonic MAS. This framework is based upon roles the agent can play and satisfactions which characterise the progression of the agent towards achievement of its goals. We have presented this framework through its formal specification using the OZS formalism. The semantics of this formalism enables the verification of properties. We have proven two pertinent properties for this framework. The first property we have proven may be interpreted as "if the holon's satisfaction is not enough then it will try to merge". It's an important property of such self-organised systems. Indeed, it means that if one holon is unsatisfied by its current achievements it will try to merge to find complementary capabilities or services.

The second property may be interpreted as "if the holon's satisfaction evolves and becomes less than necessary satisfaction the system will try to reorganise". This property ensures that if the current holarchy doesn't correspond to the current context it will evolve in order to find a better one.

Other frameworks and methodologies have been proposed [9,14] and, although they have shown to be effective inside specific domains, a more generic framework is needed. Indeed, it is difficult to design a Holonic MAS without clear and specific definitions that can lead from the analysis in terms of holon to the design of the system. Moreover, a framework with predictable properties, such as those we have proven, constitutes a solid foundation for the development of Holonic MAS.

References

1. Hans-Jörgen Bürckert, Klaus Fischer, and Gero Vierke. Teletruck: A holonic fleet management system.
2. Leonardo de Moura, Sam Owre, Harald Rueß, John Rushby, N. Shankar, Maria Sorea, and Ashish Tiwari. SAL 2. In Rajeev Alur and Doron Peled, editors, *Computer-Aided Verification, CAV 2004*, volume 3114 of *Lecture Notes in Computer Science*, pages 496–500, Boston, MA, July 2004. Springer-Verlag.
3. Roger Duke, Paul King, Gordon Rose, and Graeme Smith. The Object-Z specification language. Technical report, Software Verification Research Center, Departement of Computer Science, University of Queensland, AUSTRALIA, 1991.
4. Pablo Gruer, Vincent Hilaire, Abder Koukam, and P. Rovarini. Heterogeneous formal specification based on object-z and statecharts: semantics and verification. *Journal of Systems and Software*, 70(1-2):95–105, 2004.
5. David Harel. Statecharts: A visual formalism for complex systems. *Science of Computer Programming*, 8(3):231–274, June 1987.
6. Vincent Hilaire, Abder Koukam, Pablo Gruer, and Jean-Pierre Müller. Formal specification and prototyping of multi-agent systems. In Andrea Omicini, Robert Tolksdorf, and Franco Zambonelli, editors, *Engineering Societies in the Agents' World*, number 1972 in Lecture Notes in Artificial Intelligence. Springer Verlag, 2000.
7. Arthur Koestler. *The Ghost in the Machine*. Hutchinson, 1967.
8. Zohar Manna and Amir Pnueli. *Temporal Verification of Reactive Systems: Safety*. Springer, 1995.
9. Francisco Maturana. *MetaMorph: an adaptive multi-agent architecture for advanced manufacturing systems*. PhD thesis, The University of Calgary, 1997.
10. Sebastian Rodriguez, Vincent Hilaire, and Abderrafia Koukam. Towards a methodological framework for holonic multi-agent systems. In *Fourth International Workshop of Engineering Societies in the Agents World*, Imperial College London, UK (EU), 29-31 Octubre 2003.
11. M. Ulieru and A. Geras. Emergent holarchies for e-health applications: a case in glaucoma diagnosis. In *IECON 02 [Industrial Electronics Society, IEEE 2002 28th Annual Conference of the]*, volume 4, pages 2957– 2961, 2002.
12. Gero Vierke and Christian Russ. Agent-based configuration of virtual entreprises.
13. Michael Wooldridge, Nicholas R. Jennings, and David Kinny. A methodology for agent-oriented analysis and design. In Oren Etzioni, Jörg P. Müller, and Jeffrey M. Bradshaw, editors, *Proceedings of the Third Annual Conference on Autonomous Agents (AGENTS-99)*, pages 69–76, New York, May 1-5 1999. ACM Press.
14. J. Wyns. *Reference architecture for Holonic Manufacturing Systems - the key to support evolution and reconfiguration*. PhD thesis, Katholieke Universiteit Leuven, 1999.

The Dynamics of Computing Agent Systems

M. Smołka*, P. Uhruski, R. Schaefer, and M. Grochowski

Institute of Computer Science, Jagiellonian University, Kraków, Poland
{smolka, uhruski, schaefer, grochows}@ii.uj.edu.pl

Abstract. The paper presents the Multi Agent System (MAS) designed for the large scale parallel computations. The special kind of diffusion-based scheduling enables to decompose and allocate the migrable computing agents basing only of the local information. The paper introduces the formal model of the MAS under consideration in order to depict the roles of agent behavior and the whole system dynamics. The optimal scheduling problem for MAS as well as the way of its verification was presented in terms of such model. The brief report of the test results is stressed in the section 6.

1 Motivation

The Multi-Agent System (MAS) governed by the local diffusion scheduling is a reasonable alternative for the centrally governed distributed computing systems. Although the idea of mobile task diffusion is well known since more then ten years (see. e.g. [6]) we propose the policy that consists in on-demand task partitioning and task remaping obtained by the dynamic agent creation and migration. It was implemented and initially tested for the regularly and irregularly concurrent computations (see [2, 1, 5, 7]). The brief report of the test results is stressed in the section 6. The paper introduces the formal model of the MAS under consideration in order to depict the roles of agent behavior and the whole system dynamics. The optimal scheduling problem for MAS as well as the way of its verification was presented in terms of such model.

2 Formal Description of the Architecture

The MAS under consideration that allows the diffusion governed scheduling is a collection of: *a computational environment* (MAS platform) and *a computing application* composed of mobile agents called *Smart Solid Agents* (SSA). The computational environment is a triple $(\mathbf{N}, B_H, perf)$, where:

$\mathbf{N} = \{P_1, \ldots, P_n\}$, where P_i is the Virtual Computation Node (VCN). Each VCN can maintain more than one agent (the number of hardware processors usage is not relevant in our assumptions).

* This author has been supported by the State Committee for Scientific Research of the Republic of Poland under research grants 2 P03A 003 25 and 7 T07A 027 26.

V.S. Sunderam et al. (Eds.): ICCS 2005, LNCS 3516, pp. 727–734, 2005.

B_H is the connection topology $B_H = \{N_1, \ldots, N_n\}, N_i \subset \mathbf{N}$ is an immediate neighborhood of P_i (including P_i as well).

$perf = \{perf_1, \ldots, perf_n\}, perf_i : \mathbb{R}_+ \to \mathbb{R}_+$ is a family of functions, which describes relative performance of all VCN with respect to the total memory request M^i_{total} of all allocated agents. If M^i_{total} on P_i is small, $perf_i$ turns back the constant value, which depends only on the CPU architecture. If M^i_{total} is larger, the $perf_i$ decreases due to the intensive swap utilization.

Each SSA is represented by the pair $A_i = (T_i, S_i)$ where: T_i is the computational task executed by agent, including all data required for computation, and S_i stands for the shell responsible for the agent's logic. The index i stands for an unambiguous agent identifier.

Each task T_i has to denominate the current requirement for computational power (E_i, M_i) where: E_i is the task remaining time measured in units common for all application tasks, and M_i is the RAM requirement in bytes. Another important condition we imposed for the task is that it must allow pausing and continuation of it's computation. Pausing is needed for the hibernating task in case of agent migration or partitioning, and continuation is needed to restore the paused job. In particular it can be designed in such a way that it can work from one checkpoint to the next one, and during this checkpoint operation, it saves its present state. Moreover each task T_i can be partitioned into two subtasks $T_i \to \{T_{i_1}, T_{i_2}\}$ such that $E_i > E_{i_j}$, $M_i > M_{i_j}$, $j = 1, 2$. The task partitioning rule depends strongly on the computational problem to be solved (see [3]).
The state of *the computing application* is the triple $(\mathbf{A}_t, G_t, Sch_t)$, $t \in [0, +\infty)$ where:

$\mathbf{A}_t$ is the set of application agents, $\mathbf{A}_t = \{A_{\xi_j}\}_{\xi_j \in I_t}$, I_t is the set of indices of agents active at the time t,

G_t is the tree representing agents partitioning at the time t. All agents constitute the set of nodes $\bigcup_{\xi \in \Theta} A_\xi$, $\Theta = \bigcup_{j=0}^{t} I_j$, while G_t edges show the partitioning history. All information on how to rebuilt G_t is spread among all agents such that each of them knows only its neighbors in the tree.

$\{Sch_t\}_{t \in [0,+\infty)}$ is the family of functions such that $Sch_t : \mathbf{A}_t \to \mathbf{N}$ is the current schedule of application agents among the MAS platform servers. The function is represented by the sets ω_j of agents' indices allocated on each $P_j \in \mathbf{N}$. Each of ω_j is locally stored and managed by P_j.

The shell S_i communicates with both T_i and the local server $P_j = Sch(A_i)$. It supports inter–task communication and queries task requirements for resources as well as implements the necessary logic to perform scheduling. Each server $P_j \in \mathbf{N}$ periodically asks all local agents (allocated on P_j) for their requirements and computes the local load concentration

$$L_j = \frac{E^j_{total}}{perf_j(M^j_{total})} \text{ where } E^j_{total} = \sum_{i \in \omega_j} E_i \text{ and } M^j_{total} = \sum_{i \in \omega_j} M_i \qquad (1)$$

Then P_j communicates with neighboring servers and establishes

$$\mathbf{L}_j = \{(L_\zeta, E^\zeta_{total}, M^\zeta_{total}, perf_\zeta)\} \text{ where } \zeta \text{ is such that } P_\zeta \in N_j \tag{2}$$

as well as the set of node indices Q_j such that

$$k \in Q_j \iff k \neq j,\ P_k \in N_j,\ L_j - L_k > 0 \tag{3}$$

The current values of both $\mathbf{L}_j$ and Q_j are available to the local agents.

3 Diffusion of the Smart Solid Agent

We introduce the *binding energy* $\mathbf{E}_{i,j}$ of the agent A_i allocated on VCN P_j characterized by the following conditions: $\mathbf{E}_{i,j}$ is a descending function of E_i and a nonascending function of L_j. One of the simplest form of the binding energy utilized in computational tests (see section 6) is $\mathbf{E}_{ij} = max\{E_{min}, (\alpha_1 E_i + \alpha_2 L_j)\}$ where α_1, α_2 are the proper scaling parameters and E_{min} stands for the minimum binding energy assigned to each agent.

We assume that the agent A_i may dynamically evaluate its binding energy for other nodes from the neighborhood N_j using the information contained in $\mathbf{L}_j$. The current value of the binding energy gradient is a vector defined by:

$$\begin{array}{c} \nabla^t_{i,j} = ((j,l), \mathbf{E}_{i,l} - \mathbf{E}_{i,j}) \text{ where } P_j = Sch(A_i) \text{ and } l \in Q_j \text{ is such that} \\ \mathbf{E}_{i,l} - \mathbf{E}_{i,j} = \max_{\zeta \in Q_j}\{\mathbf{E}_{i,\zeta} - \mathbf{E}_{i,j}\} \end{array} \tag{4}$$

An agent A_i allocated on P_j migrates to P_l indicated by $\nabla^t_{i,j}$ if the binding energy $\mathbf{E}_{i,l}$ on the destination VCN exceeds the current $\mathbf{E}_{i,j}$ more than ϵ. The threshold ϵ stands for the migration parameter.
In general Smart Solid Agent $A_i = (T_i, S_i)$ currently allocated on $P_j \in \mathbf{N}$ can perform the following actions:

- **(a-1)** Execute task T_i (solve and communicate with other agents).
- **(a-2)** Pause T_i .
- **(a-3)** Continue T_i .
- **(a-4)** Denominate own load requirements (E_i, M_i).
- **(a-5)** Compute $\nabla^t_{i,j}$ and check the condition $\mathbf{E}_{i,l} - \mathbf{E}_{i,j} > \epsilon$.
- **(a-6)** Partition $T_i \to \{T_{i_1}, T_{i_2}\}$, create child agents $\{A_{i_j} = (T_{i_j}, S_{i_j})\}, j = 1, 2$.
- **(a-7)** Migrate to $P_l \in \mathbf{N}, l \neq j$.
- **(a-8)** Disappear.

These actions allow A_i to accomplish two goals:

- **(G-1)** Perform computation of carried task by executing action (a-1) and then perform action (a-8) when the task is done.
- **(G-2)** Find a better execution environment. We suggest following the algorithm utilizing actions (a-2) - (a-8).

If $Q_j = \emptyset$ **then** continue T_i
  else { compute $\nabla^t_{i,j}$;
    If $\mathbf{E}_{i,l} - \mathbf{E}_{i,j} > \epsilon$
      then { pause T_i; migrate along the gradient $\nabla^t_{i,j}$; continue T_i }
      else { Partition $T_i \to \{T_{i_1}, T_{i_2}\}$;
        create $\{A_{i_j} = (T_{i_j}, S_{i_j})\}, j = 1, 2$;   // G_t gets modified
        disappear }
  }.

The overall SSA intention is to accomplish the goal (G-1) in the shortest possible time. If the agent recognizes the local VCN resources as insufficient, it tries to accomplish the goal (G-2). On the other hand, P_j may force $\{A_i\}, i \in \omega_j$ to realize goal (G-2) when its performance is endangered.

4 The MAS Dynamics

The diffusion-based scheduling strategy has proven to be simple and efficient (cf. section 6) but it lacks a theoretical background allowing to determine if it is optimal or quasi-optimal in any sense. In this section we shall try to state a formal mathematic model for multi-agent computations. Consider for a while that we are given a space of all possible agents and denote it by $\mathbf{A}$. We shall consider discrete-time evolution of a given MAS. Let us introduce the notion of the *vector weight of an agent* which is the mapping $w : \mathbb{N} \times \mathbf{A} \longrightarrow \mathbb{R}^2_+$ whose components are E_i and M_i as introduced earlier. Assume that we know how the total weight of child agents after partition depends on their parent's weight before partition and that this dependency is componentwise, i.e. we know the functions $f^1, f^2 : \mathbb{R}_+ \to \mathbb{R}_+$ such that in the case of partition $A \to \{A_1, A_2\}$ we have

$$w^i_{t+1}(A_1) + w^i_{t+1}(A_2) = f^i(w^i_t(A))$$

for $i = 1, 2$. Such an assumption seems realistic, in simple cases f^i can be even the identity.

Next denote by $W : \mathbb{N} \times \mathbf{N} \longrightarrow \mathbb{R}_+$ the *total weight of all agents allocated on a virtual node at any time*, i.e.

$$W_t(P) = \sum_{Sch_t(A)=P} w_t(A)$$

(obviously we put 0 if no agent is maintained by P). The main idea of introducing such a notion is the need to find a global quantity describing the state of the system appropriately and allowing us to avoid considering the dynamics of a single agent.

In the sequel we shall assume that the number of virtual nodes $\sharp\mathbf{N} = N$ is *fixed*. Thus we can consider W_t as a nonnegative vector in $\mathbb{R}^{2N}$ whose j-th component corresponds to E^j_{total} and $(N + j)$-th component corresponds to

M^j_{total}. In fact we shall treat W_t as a *stochastic (vector-valued) process*. Now we shall state the equations of evolution of W_t (i.e. *state equations* of our system). Let F_t be a time-dependent stochastic nonnegative vector field on $\mathbb{R}^{2N}_+$ describing the dynamics of our system in 'established' state, i.e. when there are neither migrations nor partitions. Let $u^E_{ij,t}(W_t), u^M_{ij,t}(W_t) \in [0,1]$ denote the proportions of the weight components of agents migrating from node i to node j to the corresponding components of the total weight of all agents at node i and let $u^E_{ii,t}(W_t), u^M_{ii,t}(W_t)$ denote the proportions of the weight components of splitting agents to the corresponding components of the total weight of all agents at node i. Then the state equations have the form:

$$\begin{cases} W^i_{t+1} = F^i_t\left(\left[(1-\sum_{k=1}^N u^E_{ik,t}(W_t))W^i_t\right]_{i=1}^N\right) \\ \qquad +f^1\left(u^E_{ii,t}(W_t)W^i_t\right) - (\sum_{k=1}^N u^E_{ik,t}(W_t))W^i_t \\ \qquad +\sum_{j\neq i} u^E_{ji,t}(W_t)W^j_t \\ W^{N+i}_{t+1} = F^{N+i}_t\left(\left[(1-\sum_{k=1}^N u^M_{ik,t}(W_t))W^{N+i}_t\right]_{i=1}^N\right) \\ \qquad +f^2\left(u^M_{ii,t}(W_t)W^{N+i}_t\right) - (\sum_{k=1}^N u^M_{ik,t}(W_t))W^{N+i}_t \\ \qquad +\sum_{j\neq i} u^M_{ji,t}(W_t)W^{N+j}_t \end{cases} \tag{5}$$

for $i = 1,\dots,N$. Note that $u = (u^E, u^M)$ where $u^E, u^M : \mathbb{N}\times\mathbb{R}^N_+ \longrightarrow [0,1]^{N\times N}$ is in fact a *control* of our system. From the nature of our problem it is easy to see that any control strategy must satisfy the following conditions for any $t \in \mathbb{N}$ and $x \in \mathbb{R}^N_+$

$$\begin{cases} u^E_{ij,t}(x)\cdot u^E_{ji,t}(x) = 0, u^M_{ij,t}(x)\cdot u^M_{ji,t}(x) = 0 \text{ for } i \neq j, \\ \sum_{k=1}^N u^E_{ik,t}(x) \le 1, \sum_{k=1}^N u^M_{ik,t}(x) \le 1 \qquad \text{for } i = 1,\dots,N. \end{cases} \tag{6}$$

The first pair of equalities means that at a given time migrations between two nodes may happen in only one direction. The remaining conditions mean that the number of agents leaving a node must not exceed the number of agents present at the node just before the migration.
The initial state of equation (5)

$$W_0 = \hat{W} \tag{7}$$

usually has all except one components equal to 0 which means that a system starts with one agent or a batch of agents placed on a single VCN but of course in general we need not make such an assumption.

5 The Optimal Scheduling Problem

Now let us propose two examples of cost functionals which seem appropriate for multi-agent computations. The first is the expected total time of computations

$$V(u;\hat{W}) = E(\min\{t \ge 0 : \sum_{i=1}^N W^i_t = 0\}), \tag{8}$$

The second takes into account the mean load balancing over time. It has the following form

$$V(u;\hat{W}) = E(\sum_{t=0}^{\infty}\sum_{i=1}^{N}(L_t^i - \overline{L}_t)^2) \tag{9}$$

where $L_t^i = \frac{W_t^i}{perf_i(W_t^{N+i})}$ is the load concentration and $\overline{L}_t = \frac{1}{N}\sum_{i=1}^{N} L_t^i$ is its mean over all nodes. Let us introduce the following notation for the set of admissible controls

$$\mathbf{U} = \{u = (u^E, u^M) : \mathbb{N} \times \mathbb{R}_+^N \to [0,1]^{2(N\times N)} | \ u \text{ satisfies conditions (6)}\}.$$

Assume that V has either the form (8) or (9). Given an initial configuration $\hat{W}$ our *optimal scheduling problem* is now to find such control strategy $u^* \in \mathbf{U}$ that

$$V(u^*;\hat{W}) = \min\{V(u;\hat{W}) : \ u \in \mathbf{U}, \ W \text{ is a solution of } (5),(7)\}. \tag{10}$$

6 Numerical Tests

This section presents performed experiments and discusses how they fit presented analytical model. The results cover two major problem domains examined in course of our experiments: Mesh Generator (MG) and Hierarchic Genetic Strategy (HGS). The presented tests are broadly described in [5, 7] while papers [1, 2] contain the MAS implementation details. Computations were run in the same physical environment with LAN of 30 to 50 PCs connected with 100MBit network.

6.1 Mesh Generator

The MG is a CAD/CAE task implementation. Each agent is equipped with a single part of the partitioned solid, for which the mesh has to be generated. The application had the following execution path:

1. Create all the agents, equip each with the part of the solid to generate the mesh for. The agent's size varies from very small to large depending on the solid partitioning and each piece shape - the more complex the shape is, the bigger mesh is generated.
2. Agents are put on the MAS network and distribute freely using the encoded diffusion rule. After a single agent finds satisfying executing environment it starts computations.
3. Single solid piece meshes are send to the master application and joined together.

The MG implements evolution and migration cases of the system dynamics, so the main points of results analysis was to check if the local, diffusion based scheduling policy does not put too much overhead on the total execution time and if the available resources were fully utilized from the very beginning - any utilization 'holes' would signal synchronization points or diffusion rule not performing local scheduling properly or being not efficient.

In short, we have formulated the following conclusions concerning the results:

- Diffusion based scheduling allows all agents to spread quickly among computers on the network.The time needed to allocate tasks is small in comparison to the whole computation time.
- Available resources were utilized at 96%, the application used up to 32 agents.
- The load is perfectly balanced according to the given local diffusion law.

It is clear from these conclusions that the diffusion based, local scheduling policy is well suited for such regular problems. The results also confirm that the communication factor may be omitted for certain class of applications, since the total communication overhead is minimal. Also the communication required for the diffusion (examine a VCN node neighborhood and reevaluate the binding energy for every agent) did not degrade the system efficiency.

6.2 HGS

The HGS [4] is a stochastic, hierarchical genetic algorithm optimizing a given function on a defined domain. The application produces agents dynamically in the course of the runtime. A single HGS agent is a container, which starts executing its internal populations as soon as it founds suitable computation environment. The total amount of agents at the execution time may be different even for a particular input data. The algorithm execution path is the following:

1. Create single agent with the initial populations, agent migrates to find suitable execution environment.
2. Agent starts computing and the amount of internal populations changes due to the genetic evolution.
3. If during the computation agent's internal populations amount grows beyond a fixed number, the agent is split and new one is created out of part of the populations set.

The HGS agents perform all three elements of MAS dynamics: evolution, migration and partitioning. The communication overhead (performing agent's migration requires agent's to pass the initial populations) and dynamic scheduling efficiency were tested within this experiment. On a basis of this experiment the following conclusions concerning the diffusion scheduling were drawn:

- Diffusion based scheduling deals properly and effectively with the dynamic rescheduling of the computing units.
- The communication overhead is around 5% of total execution time with application using up to 300 actively computing agents.

Therefore the final conclusion may be stated that the diffusion based local scheduling performs well in case of irregular stochastic problem with dynamic amount of agents. It is very adaptive and local scheduling evaluation does not significantly decrease the solution's effectiveness.

In addition we have also tested the effectiveness of the dynamic, diffusion based scheduling versus centralized, greedy scheduling policy (a Round Robin solution) utilizing low level network message passing mechanisms (Java RMI). Please see [7] for detailed results, which showed that the dynamic scheduling shows moderate speedup loses (up to 30% of total computation time), which disappear when agent's amount grows - in case of 300 agents, diffusion based scheduling performs 10% better that Round Robin.

7 Conclusions and Further Research

The diffusion analogy as well as the MAS technology give way to an effective design of a local diffusion-based scheduling strategy for a distributed environment. Its effectiveness is achieved by the low complexity of local scheduling rules and the lack of intensive communication required by centralized schedulers. The formal description introduced in sections 4 and 5 provides the discrete equation of evolution and the characterization of admissible controls as well as the cost functional for computing MAS. Such considerations put our problem into the solid framework of the stochastic optimal control theory which provides us with tools such as Bellman-type principles allowing us to study the optimality of control strategies as well as the MAS asymptotics.

References

1. Grochowski M., Schaefer R., Uhruski P.: An Agent-based Approach To a Hard Computing System - Smart Solid. *Proc. of the International Conference on Parallel Computing in Electrical Engineering (PARELEC 2002)*, 22-25 September 2002, Warsaw, Poland. IEEE Computer Society Press 2002, pp. 253-258.
2. Uhruski P., Grochowski M., Schaefer R.: Multi-agent Computing System in a Heterogeneous Network. *Proc. of the International Conference on Parallel Computing in Electrical Engineering (PARELEC 2002)*, 22-25 September 2002, Warsaw, Poland. IEEE Computer Society Press 2002, pp. 233-238.
3. Schaefer R., Flasiński M., Toporkiewicz W.: Optimal Stochastic Scaling of CAE Parallel Computations. *Lecture Notes in Computer Intelligence*, Vol. 1424, Springer 1998, pp.557-564.
4. Wierzba B., Semczuk A., Kołodziej J., Schaefer R.: Hierarchical Genetic Strategy with real number encoding. *Proc. of the 6th Conf. on Evolutionary Algorithms and Global Optimization* Łagów Lubuski 2003, Wydawnictwa Politechniki Warszawskiej 2003, pp. 231-237.
5. Grochowski M., Schaefer R., Uhruski P.: Diffusion Based Scheduling in the Agent-Oriented Computing Systems. *Lecture Notes in Computer Science*, Vol. 3019, Springer 2004, pp. 97-104.
6. Luque E., Ripoll A., Corts A., Margalef T.: A distributed diffusion method for dynamic load balancing on parallel computers. *Proc. of EUROMICRO Workshop on Parallel and Distributed Processing*, San Remo, Italy, January 1995. IEEE CS Press.
7. Momot J., Kosacki K., Grochowski M., Uhruski P., Schaefer R.; Multi-Agent System for Irregular Parallel Genetic Computations. *Lecture Notes in Computer Science*, Vol. 3038, Springer 2004, pp. 623-630.

A Superconvergent Monte Carlo Method for Multiple Integrals on the Grid

Sofiya Ivanovska, Emanouil Atanassov, and Aneta Karaivanova

Institute for Parallel Processing - Bulgarian Academy of Sciences,
Acad. G. Bonchev St., Bl.25A, 1113 Sofia, Bulgaria
{sofia, emanouil, anet}@parallel.bas.bg

Abstract. In this paper we present error and performance analysis of a Monte Carlo variance reduction method for solving multidimensional integrals and integral equations. This method combines the idea of separation of the domain into small subdomains with the approach of importance sampling. The importance separation method is originally described in our previous works [7, 9]. Here we present a new variant of this method adding polynomial interpolation in subdomains. We also discuss the performance of the algorithms in comparison with crude Monte Carlo. We propose efficient parallel implementation of the importance separation method for a grid environment and we demonstrate numerical experiments on a heterogeneous grid. Two versions of the algorithm are compared - a Monte Carlo version using pseudorandom numbers and a quasi-Monte Carlo version using the Sobol and Halton low-discrepancy sequences [13, 8].

1 Introduction

Consider the problem of approximate calculation of the multiple integral

$$I[f] = \int_G f(x)p(x)\,dx, \quad p(x) \geq 0, \quad \int_G p(x)\,dx = 1,$$

where $p(x)$ is a probability density function. The crude Monte Carlo quadrature is based on the probability interpretation of the integral:

$$I_N[f] = \frac{1}{N}\sum_{n=1}^{N} f(x_n), \tag{1}$$

where $\{x_n\}$ is distributed according to $p(x)$. The error is proportional to $\sigma[f]N^{-\frac{1}{2}}$:

$$\epsilon_N[f] \sim \sigma[f]N^{-1/2},$$

where

$$\sigma[f] = \left(\int_G (f(x)p(x) - I[f])^2\,dx\right)^{1/2}.$$

V.S. Sunderam et al. (Eds.): ICCS 2005, LNCS 3516, pp. 735–742, 2005.

There are various ways to improve the convergence rate of Monte Carlo integration. The following theorem, due to Bachvalov, establishes a lower bound for both Monte Carlo and deterministic integration formulae for smooth functions:

Theorem 1. *(Bachvalov [4, 5]) There exist constants $c(s,k)$, $c'(s,k)$, such that for every quadrature formula $I_N[f]$ which is fully deterministic and uses the function values at N points there exists a function $f \in \mathbf{C}_k^s$ such that*

$$\left| \int_{\mathrm{E}^s} f(x)\, dx - I_N[f] \right| \geq c(s,k) \, \|f\| \, N^{-\frac{k}{s}}$$

and for every quadrature formula $I_N[f]$, which involves random variables and uses the function values at N points, there exists a function $f \in \mathbf{C}_k^s$, such that

$$\left\{ \mathrm{E} \left[\int_{\mathrm{E}^s} f(x)\, dx - I_N[f] \right]^2 \right\}^{1/2} \geq c'(s,k) \, \|f\| \, N^{-\frac{1}{2}-\frac{k}{s}}.$$

From this theorem it follows that Monte Carlo methods have advantage over deterministic methods, especially in high dimensions. In order to obtain the optimal convergence rate for functions with bounded k-th order derivatives, one widely used technique is *stratification.* Split the integration region G into M subdomains:

$$G = \bigcup_{j=1}^{M} D_j, \quad D_i \cap D_j = \emptyset, \quad i \neq j, \quad \sum_{j=1}^{M} N_j = N,$$

$$p^{(j)}(x) = p(x)/\overline{p}_j, \quad \overline{p}_j = \int_{D_j} p(x)\, dx.$$

In the subdomain D_j use a Monte Carlo integration formula $I_{N_j}^{(j)}[f]$, which utilizes N_j random points $\xi_1^{(j)}, \ldots, \xi_{N_j}^{(j)}$. The stratified Monte Carlo formula is

$$I_N[f] = \sum_{j=1}^{M} \frac{\overline{p}_j}{N_j} I_{N_j}^{(j)}[f].$$

The error of the stratified method is given by

$$\epsilon_N \sim N^{-1/2}\sigma_s, \quad \sigma_s^2 = \sum_{j=1}^{M} {\sigma^{(j)}}^2, \quad \sigma^{(j)} = \left(\int_{D_j} \left(f(x)\, p_j(x) - I_{N_j}[f]\right)^2 dx \right)^{\frac{1}{2}}.$$

In our method we achieve $\sigma^{(j)} = \mathcal{O}(N^{-\frac{k}{s}})$, if the function f is in $\mathbf{C}_k^s$, and therefore we attain the optimal convergence rate (see Sec. 2).

Another possible idea for improving the convergence of Monte Carlo methods is to replace the pseudorandom numbers with terms of a low-discrepancy sequence [6, 12]. The most widely used measure for estimating the quality of the distribution of a low-discrepancy sequence $\tau = \{x_i\}_{i=1}^{N}$ in E^s is its discrepancy

$$D_N^* = \sup_{I \subset \mathrm{E}^s} \left| \frac{A_N(\tau, I)}{N} - \mu(\mathrm{E}) \right|.$$

Low-discrepancy sequences have order of convergence of the discrepancy $\mathcal{O}\left(N^{-1} \log^s N\right)$, which becomes order of convergence of the corresponding quasi-Monte Carlo method for integrating a function over E^s:

$$Q_N[f] = \frac{1}{N} \sum_{j=1}^{N} f(x_i). \tag{2}$$

Therefore quasi-Monte Carlo methods offer higher conversion rate (at the expense of more stringent smoothness requirements). Formula 2 can be considered as a quasi-Monte Carlo version of formula 1. Designing a quasi-Monte Carlo version of a Monte Carlo method is not always straight-forward. We studied the possibility of using the Halton and Sobol low-discrepancy sequences in our algorithm in Sec. 2.

Monte Carlo methods usually show good parallel efficiency. In Sec. 3 we describe our parallelization approach, which is suitable for grid environments. We obtained good parallel efficiency in both Monte Carlo and quasi-Monte Carlo versions.

2 Description of Our Algorithm

We introduce adaptive division of the domain into smaller subdomains, so that smaller subdomains are used where the function values are larger. In every subdomain we approximate the function by a polynomial, using function values at some fixed points. The input parameters of our algorithm are the dimension s, the smoothness order k, the number of base points M, the number of steps N, the number of random points in each cube m, the number of points used for one application of crude Monte Carlo method R. First we select M points $a_1, \ldots, a_M$ in the unit cube E^s, following the procedure, discussed in [2], and compute the respective coefficients $c_1, \ldots, c_M$. For every coordinate x_j, we estimate the values of

$$g(x_j) = \int_{\mathrm{E}^{s-1}} |f(x_1, \ldots, x_{j-1}, x_j, x_{j+1}, \ldots, x_s)| \, dx_1, \ldots, dx_{j-1} dx_{j+1}, \ldots, dx_s$$

at points $x_j = \frac{r}{N}$, using crude Monte Carlo algorithm with R points. Then we approximate the one-dimensional function g by a piece-wise linear function $\tilde{g}$, using these values. For every coordinate we choose $N+1$ points ξ_r^i, so that $\xi_0^i = 0$, $\xi_N^i = 1$, and

$$\int_{\xi_{r-1}^i}^{\xi_r^i} \tilde{g}(t)\, dt = \int_{\xi_r^i}^{\xi_{r+1}^i} \tilde{g}(t)\, dt, \quad r = 1, \ldots, N-1.$$

These steps can be considered as pre-processing, and can be done in parallel, if we use more than one CPU. Using the points ξ_r^i, we partition the cube E^s into N^s subdomains D_j. Approximate $\int_{D_j} f(x)\,dx$ by the formula:

$$\mu\left(D_j\right)\frac{1}{m}\sum_{i=1}^{m}\left(f(\eta_i)-L_j(f,\eta_i)\right)+\mu\left(D_j\right)\sum_{i=1}^{M}c_i f(T_j(a_i)),$$

where T_j is the canonical linear transformation that maps E^s onto D_j, the points $\eta_i \in D_j$ are uniformly distributed random points, and $L_j(f,\eta_i)$ is the polynomial approximation to f, obtained using the values $f(T_j(a_i))$ (see [2]). Summing these unbiased estimates, we obtain an estimate of the integral over E^s. The variance of this estimate is a sum of the variances σ_j^2 in every subdomain D_j. The order of convergence is $\mathcal{O}\left(N^{-k-s/2}\right)$, when using $\mathcal{O}(N^s)$ points and so it is optimal in the sense of Theorem 1. An aposteriory estimate of the error is obtained by using the empirical variances $\tilde{\sigma_j}$.

The number of function evaluations is $N^s(m+M)+Rs(N-1)$. In most practical situations the time for performing them dominates in the whole computation.

We studied also quasi-Monte Carlo variants of the algorithm. In Monte Carlo methods one uses pseudorandom number generators in order to sample uniformly distributed random variables. The idea of quasi-Monte Carlo methods is to replace the pseudorandom numbers with terms of a low-discrepancy sequence. When integrating a function in the s-dimensional unit cube, the dimension of the low-discrepancy sequence is frequently higher than s, if a more complex algorithm is used. We note that when used in parallel computations, low-discrepancy sequences can offer exact reproducibility of results, unlike Monte Carlo. In our algorithm the coordinates of the r-th term of the sequence can be used to produce the coordinates of the points η_i. Since we have s coordinates and m points, the constructive dimension of the algorithm becomes sm. We tested two of the most popular families of low-discrepancy sequences - the sequences of Sobol and Halton. Since the rate of convergence of the quasi-Monte Carlo algorithm has a factor of $\log^{sm} N$, we did not observe improvement with respect of the Monte Carlo version (see Sec. 4).

3 Parallel Implementation of the Algorithm Suitable for Computational Grids

Monte Carlo methods are inherently parallelizable, and both coarse-grain and fine-grain parallelism can be exploited. We take into account the fact that the Grid is a potentially heterogeneous computing environment, where the user does not know the specifics of the target architecture. Therefore parallel algorithms should be able to adapt to this heterogeneity, providing automated load-balancing. Monte Carlo algorithms can be tailored to such environments, provided parallel pseudo-random number generators are available. The use of quasi-Monte Carlo algorithms poses more difficulties. In both cases the efficient implementation of the algorithms depends on the functionality of the corresponding packages for generating pseudorandom or quasirandom numbers.

As a package of parallel pseudo-random generators we used the SPRNG ([10]). For generating the scrambled Halton and Sobol sequences, we used our ultra-fast generators ([1, 3]), which provide the necessary functionality:

- portable
- use assembly language for best performance on various Intel Pentium and AMD processors
- provide a fast-forward operation, important for our parallelization approach.

Our parallelization is based on the master-slave paradigm, with some ramifications. We divide the work into chunks, corresponding to the subdomains, which are requested from the master process by the slave processes. In order to increase the efficiency, the master also performs computations, while waiting for communication requests. Thus we achieve overlap of computations and communications, and we do not lose the possible output of the master process. When using low-discrepancy sequences, we take care in both master and slave processes to fast-forward the generator exactly to the point that is needed. The scrambling that is provided by the generators enables aposteriory estimation of the error in the quasi-Monte Carlo case.

4 Numerical Experiments

Our numerical tests are based on the next two examples, which are taken from paper of Moskowitz and Caflisch [11].

Example 1. The first example is Monte Carlo integration over $\mathrm{E}^5 = [0,1]^5$ of the function

$$f_1(x) = \exp\left(\sum_{i=1}^{5} a_i\, x_i^2 \,\frac{2+\sin(\sum_{j=1,j\neq i}^{5} x_j)}{2}\right),$$

where $\mathbf{a} = (1, \frac{1}{2}, \frac{1}{5}, \frac{1}{5}, \frac{1}{5})$.

Example 2. The second example is Monte Carlo integration over $\mathrm{E}^7 = [0,1]^7$ of the function

$$f_2(x) = e^{1-(\sin^2(\frac{\pi}{2}x_1)+\sin^2(\frac{\pi}{2}x_2)+\sin^2(\frac{\pi}{2}x_3))} \times \arcsin\left(\sin(1) + \frac{x_1+\cdots+x_7}{200}\right).$$

Table 1 shows the results for 5-dimensional integral. The smoothness that is used is 4 or 6. And just for comparison we show results with crude Monte Carlo method that uses the same number of functional values as our algorithm. In

Table 1. Results for **Example 1** with number of cubes N and number of points per cube 10

N	k	SPRNG	Sobol	Halton	crude MC
2	4	5.15e-04	3.16e-04	3.72e-04	1.52e-02
	6	7.06e-05	1.43e-05	2.34e-05	7.40e-03
3	4	8.01e-05	4.38e-05	4.77e-05	5.51e-03
	6	2.15e-07	7.41e-07	3.19e-07	2.68e-03
4	4	1.21e-05	1.01e-05	2.09e-05	2.68e-03
	6	2.95e-07	3.45e-07	1.35e-07	1.31e-03
5	4	1.78e-06	3.03e-06	2.87e-06	1.54e-03
	6	2.79e-08	3.61e-08	4.09e-08	7.48e-04

Table 2. Results for **Example 2** with number of cubes N and number of points per cube 10

N	k	SPRNG	Sobol	Halton
2	4	1.25e-05	3.12e-05	3.29e-05
	6	9.49e-07	3.61e-07	7.34e-07
3	4	1.31e-06	5.30e-06	4.67e-06
	6	7.57e-08	7.61e-08	8.61e-08
4	4	2.94e-08	1.82e-06	4.73e-07
	6	5.99e-09	1.07e-08	1.44e-08
5	4	1.23e-07	3.21e-07	5.43e-07
	6	8.43e-10	1.61e-09	1.20e-08

both examples the errors obey the theoretical laws, i.e., the convergence rate is $\mathcal{O}\left(N^{-\frac{1}{2}}\right)$ for the crude Monte Carlo method, and $\mathcal{O}\left(N^{-\frac{k}{s}-\frac{1}{2}}\right)$ for our adaptive method, when smoothness k is used. Table 3 shows CPU time for our algorithm when N is equal to 14 and the number of random points per cube m is equal to 40. We used different computers for these tests - three of the computers have Pentium processors running at different clock speeds and one has a Motorola G4 processor. All of the computers had the Globus toolkit installed. In Table 4 we compare the estimated time for running our algorithm on all those computers in parallel, in case of perfect parallel efficiency, with the measured execution time. We used the MPICH-G2 implementation of MPI, which is the most general approach for running parallel jobs on computational grids. In this way we successfully utilized machines with different endianness in the same computation. We obtained roughly the same accuracy with low-discrepancy sequences as with pseudo-random numbers, due to the high effective dimension. The CPU time of our implementations of the quasi-Monte Carlo versions was frequently smaller, due to the efficient generation algorithms. Quasi-Monte Carlo algorithms are more difficult to parallelize under such constraints, but this can be done if the generators of the low-discrepancy sequences have the necessary functionality.

Table 3. Time and efficiency for **Example 1** with number of points 14 and number of points per cube 40

	P4/2.8GHz	P4/2GHz	P4/2GHz	G4/450MHz
SPRNG	102	117	118	413
Sobol	91	106	96	393
Halton	91	106	106	393

Table 4. Parallel efficiency measurements

	Estimated Time	Measured Time	Efficiency
SPRNG	34	41	83%
Sobol	30	34	88%
Halton	31	34	91%

5 Conclusions

An adaptive Monte Carlo method for solving multiple integrals of smooth functions has been proposed and tested. This method is an improved version of the importance separation method. The importance separation method is combined with polynomial interpolation in the subdomains. A quasi-Monte Carlo version of the algorithm was also studied. The obtained results are not an efficient parallel implementation of the algorithm has been achieved using a version of the master-slave computing paradigm, enabling the execution of the programs in heterogeneous Grid environments. The ideas of this implementation can be extended to other Monte Carlo and quasi-Monte methods.

Acknowledgments. This work is supported by the Ministry of Education and Science of Bulgaria under Grant # I-1405/04.

References

1. E. Atanassov. Measuring the Performance of a Power PC Cluster, Computational Science - ICCS 2002 (P. Sloot, C. Kenneth Tan, J. Dongarra, A. Hoekstra - Eds.), LNCS 2330, 628–634, Springer, 2002.
2. E. Atanassov, I. Dimov, M. Durchova. A New Quasi-Monte Carlo Algorithm for Numerical Integration of Smooth Functions, Large-Scale Scientific Computing (I. Lirkov, S. Margenov, J. Wasniewski, P. Yalamov - Eds.), LNCS 2907, 128–135, Springer, 2004.
3. E. Atanassov, M. Durchova. Generating and Testing the Modified Halton Sequences, Numerical Methods and Applications (I. Dimov, I. Lirkov, S. Margenov, Z. Zlatev - Eds.), LNCS 2542, 91–98, Springer, 2003.
4. N.S. Bachvalov. On the approximate computation of multiple integrals, Vestnik Moscow State University, Ser. Mat., Mech., Vol. 4, 3–18, 1959.

5. N.S. Bachvalov. Average Estimation of the Remainder Term of Quadrature Formulas, USSR Comput. Math. and Math. Phys., Vol. 1(1), 64–77, 1961.
6. R.E. Caflisch. Monte Carlo and quasi-Monte Carlo methods, Acta Numerica, Vol. 7, 1–49, 1998.
7. I. Dimov, A. Karaivanova, R. Georgieva and S. Ivanovska. Parallel Importance Separation and Adaptive Monte Carlo Algorithms for Multiple Integrals, Numerical Methods and Applications (I. Dimov, I. Lirkov, S. Margenov, Z. Zlatev - Eds.), LNCS 2542, 99–107, Springer, 2003.
8. J. H. Halton. On the efficiency of certain quasi-random sequences of points in evaluating multi-dimensional integrals, Numer. math., 2, 84–90, 1960.
9. A. Karaivanova. Adaptive Monte Carlo methods for numerical integration, Mathematica Balkanica, Vol. 11, 391–406, 1997.
10. M. Mascagni. SPRNG: A Scalable Library for Pseudorandom Number Generation. Recent Advances in Numerical Methods and Applications II (O. Iliev, M. Kaschiev, Bl. Sendov, P.S. Vassilevski eds.), Proceeding of NMA 1998, World Scientific, Singapore, 284–295, 1999.
11. B. Moskowitz and R.E. Caflisch. Smoothness and dimension reduction in quasi-Monte Carlo methods, J. Math. Comput. Modeling, 23, 37–54, 1996.
12. H. Niederreiter. Random number generation and quasi-Monte Carlo methods, SIAM, Philadelphia, 1992.
13. I.M. Sobol. On the distribution of point in a cube and the approximate evaluation of integrals, USSR Computational Mathematics and Mathematical Physics, 7, 86–112, 1967.

A Sparse Parallel Hybrid Monte Carlo Algorithm for Matrix Computations

Simon Branford, Christian Weihrauch, and Vassil Alexandrov

Advanced Computing and Emerging Technologies Centre,
School of System Engineering,
The University of Reading
Whiteknights, P.O. Box 225
Reading, RG6 6AY, UK
{s.j.branford, c.weihrauch, v.n.alexandrov}@reading.ac.uk

Abstract. In this paper we introduce a new algorithm, based on the successful work of Fathi and Alexandrov, on hybrid Monte Carlo algorithms for matrix inversion and solving systems of linear algebraic equations. This algorithm consists of two parts, approximate inversion by Monte Carlo and iterative refinement using a deterministic method.

Here we present a parallel hybrid Monte Carlo algorithm, which uses Monte Carlo to generate an approximate inverse and that improves the accuracy of the inverse with an iterative refinement. The new algorithm is applied efficiently to sparse non-singular matrices. When we are solving a system of linear algebraic equations, $Bx = b$, the inverse matrix is used to compute the solution vector $x = B^{-1}b$.

We present results that show the efficiency of the parallel hybrid Monte Carlo algorithm in the case of sparse matrices.

Keywords: Monte Carlo Method, Matrix Inversion, Sparse Matrices.

1 Introduction

The problem of inverting a real $n \times n$ matrix (MI) or solving a system of linear algebraic equations (SLAE) is of unquestionable importance in many scientific and engineering applications: e.g real-time speech coding, digital signal processing, communications, stochastic modelling, and many physical problems involving partial differential equations. The direct methods of solution require $O(n^3)$ sequential steps when using the usual elimination or annihilation schemes (e.g. non-pivoting Gaussian elimination, Gauss-Jordan methods) [10]. Consequently the computation time for very large problems, or for real-time solution problems, can be prohibitive and this prevents the use of many established algorithms.

It is known that Monte Carlo methods give statistical estimates for elements of the inverse matrix, or for components of the solution vector of SLAE, by performing random sampling of a certain random variable, whose mathematical expectation is the desired solution [11, 13]. We concentrate on Monte Carlo methods for MI and/or solving SLAE, since, *first*, only $O(NT)$ steps are required to find an element of the inverse matrix or component of the solution

V.S. Sunderam et al. (Eds.): ICCS 2005, LNCS 3516, pp. 743–751, 2005.

vector of SLAE (where N is the number of chains and T is an estimate of the chain length in the stochastic process, both of which are independent of n, the size of the matrix) and, *second*, the sampling process for stochastic methods is inherently parallel.

Coarse grained Monte Carlo algorithms for MI and SLAE have been proposed by several authors [2, 3, 6, 7, 8, 9]. In this paper we will explore a parallel sparse algorithm for inverting diagonally dominant matrices. We will show that this algorithm works effectively and efficiently when deployed over multiple processors.

The idea of Monte Carlo for matrix computations is presented in Section 2; the parallel hybrid algorithm is described in Section 3; in Section 4 we detail how we implemented the algorithm for sparse matrices; the results of the experimental runs of this implementation are presented in Section 5; and we conclude the work in Section 6.

2 Monte Carlo Matrix Computations

Assume that the system of linear algebraic equations (SLAE) is presented in the form:

$$Bx = b \tag{1}$$

where B is a real square $n \times n$ matrix, $x = (x_1, x_2, ..., x_n)^t$ is a $1 \times n$ solution vector and $b = (b_1, b_2, ..., b_n)^t$.

Assume the general case $\|B\| > 1$. We consider the splitting

$$B = \hat{B} - C \tag{2}$$

where off-diagonal elements of $\hat{B}$ are the same as those of B, and the diagonal elements of $\hat{B}$ are defined as $\hat{b}_{ii} = b_{ii} + \gamma_i||B||$, choosing in most cases $\gamma_i > 1$ for $i = 1, 2, ..., n$. We further consider $\hat{B} = B_1 - B_2$ where B_1 is the diagonal matrix of $\hat{B}$, e.g. $(b_1)_{ii} = \hat{b}_{ii}$ for $i = 1, 2, ..., n$. As shown in [9] we could transform the system (1) to

$$x = Tx + f \tag{3}$$

where $T = \hat{B}^{-1}C$ and $f = \hat{B}^{-1}b$. The multipliers γ_i are chosen so that, if it is possible, they reduce the norm of T to be less than 1 and thus reducing the number of Markov chains required to reach a given precision. We consider two possibilities, first, finding the solution of $x = Tx + f$ using Monte Carlo (MC) method if $\|T\| < 1$ or finding $\hat{B}^{-1}$ using MC and after that finding B^{-1}. Then, if required, obtaining the solution vector is $x = B^{-1}b$.

Consider first the stochastic approach. Assume that $\|T\| < 1$ and that the system is transformed to its iterative form (3). Consider the Markov chain given by:

$$s_0 \rightarrow s_1 \rightarrow \cdots \rightarrow s_k, \tag{4}$$

where the $s_i, i = 1, 2, \cdots, k$, belongs to the state space $S = \{1, 2, \cdots, n\}$. Then for $\alpha, \beta \in S, p_0(\alpha) = p(s_0 = \alpha)$ is the probability that the Markov chain starts at state α and $p(s_{j+1} = \beta | s_j = \alpha) = p_{\alpha\beta}$ is the transition probability from state α to state β.The set of all probabilities $p_{\alpha\beta}$ defines a transition probability matrix $P = \{p_{\alpha\beta}\}_{\alpha,\beta=1}^{n}$ [1, 2, 3].

We say that the distribution $(p_1, \cdots, p_n)^t$ is acceptable for a given vector g, and that the distribution $p_{\alpha\beta}$ is acceptable for matrix T, if $p_\alpha > 0$ when $g_\alpha \neq 0$, and $p_\alpha \geq 0$, when $g_\alpha = 0$, and $p_{\alpha\beta} > 0$ when $T_{\alpha\beta} \neq 0$, and $p_{\alpha\beta} \geq 0$ when $T_{\alpha\beta} = 0$ respectively. We assume $\sum_{\beta=1}^{n} p_{\alpha\beta} = 1$, for all $\alpha = 1, 2, \cdots, n$. Generally, we define

$$W_0 = 1, W_j = W_{j-1} \frac{T_{s_{j-1}s_j}}{p_{s_{j-1}s_j}} \tag{5}$$

for $j = 1, 2, \cdots, n$.

Consider now the random variable $\theta[g] = \frac{g_{s_0}}{p_{s_0}} \sum_{i=1}^{\infty} W_i f_{s_i}$. We use the following notation for the partial sum:

$$\theta_i[g] = \frac{g_{s_0}}{p_{s_0}} \sum_{j=0}^{i} W_j f_{s_j}. \tag{6}$$

Under condition $\|T\| < 1$, the corresponding Neumann series converges for any given f, and $E\theta_i[g]$ tends to (g, x) as $i \to \infty$. Thus, $\theta_i[g]$ can be considered as an estimate of (g, x) for i sufficiently large. To find an arbitrary component of the solution, for example, the rth component of x, we should choose, $g = e(r) = (\underbrace{0, ..., 1}_{r}, 0, ..., 0)$ such that

$$e(r)_\alpha = \delta_{r\alpha} = \begin{cases} 1 & if \quad r = \alpha \\ 0 & otherwise \end{cases} \tag{7}$$

It follows that

$$(g, x) = \sum_{\alpha=1}^{n} e(r)_\alpha x_\alpha = x_r. \tag{8}$$

The corresponding Monte Carlo method is given by:

$$x_r = \hat{\Theta} = \frac{1}{N} \sum_{s=1}^{N} \theta_i[e(r)]_s \tag{9}$$

where N is the number of chains and $\theta_i[e(r)]_s$ is the approximate value of x_r in the sth chain. It means that using Monte Carlo method, we can estimate only one, few or all elements of the solution vector. We consider Monte Carlo with uniform transition probability (UM) $p_{\alpha\beta} = \frac{1}{n}$ and almost optimal Monte Carlo method (MAO) with $p_{\alpha\beta} = \frac{|T_{\alpha\beta}|}{\sum_{\beta=1}^{n} |T_{\alpha\beta}|}$, where $\alpha, \beta = 1, 2, \ldots, n$. Monte Carlo MI is obtained in a similar way [1].

To find the inverse $M^{-1} = \{m_{rr'}^{(-1)}\}_{r,r'=1}^{n}$ of some matrix M, we must first compute the elements of matrix $A = I - M$, where I is the identity matrix.

Clearly, the inverse matrix is given by

$$M^{-1} = \sum_{i=0}^{\infty} A^i \tag{10}$$

which converges if $\|A\| < 1$.

To estimate the element $m^{(-1)}_{rr'}$ of the inverse matrix M^{-1}, we let the vector f be the following unit vector

$$f_{r'} = e(r') \tag{11}$$

We then can use the following Monte Carlo method for calculating elements of the inverse matrix M^{-1}:

$$m^{(-1)}_{rr'} \approx \frac{1}{N} \sum_{s=1}^{N} \left[\sum_{(j|s_j=r')} W_j \right] \tag{12}$$

where $(j|s_j = r')$ means that only

$$W_j = \frac{A_{rs_1} A_{s_1 s_2} \ldots A_{s_{j-1} s_j}}{p_{rs_1} p_{s_1 s_2} \ldots p_{s_{j-1} p_j}} \tag{13}$$

for which $s_j = r'$ are included in the sum (12).

Since W_j is included only into the corresponding sum for $r' = 1, 2, \ldots, n$, then the same set of N chains can be used to compute a single row of the inverse matrix, which is one of the inherent properties of Monte Carlo making them suitable for parallelisation.

The *probable error* of the method, is defined as $r_N = 0.6745\sqrt{D\theta/N}$, where $P\{|\bar{\theta} - E(\theta)| < r_N\} \approx 1/2 \approx P\{|\bar{\theta} - E(\theta)| > r_N\}$, if we have N independent realizations of random variable (r.v.) θ with mathematical expectation $E\theta$ and average $\bar{\theta}$ [11].

In the general case, $\|B\| > 1$, we make the initial split $B = \hat{B} - C$. From this we compute $A = B_1^{-1} B_2$, which satisfies $\|A\| < 1$ (by careful choice, of $\hat{B}$, we make $\|A\| < 0.5$, which gives faster convergence). Then we generate the inverse of $\hat{B}$ by (12) and from this we recover B^{-1}, using an iterative process.

3 Hybrid Algorithm for Matrix Inversion

In this paper we consider diagonally dominant matrices, which means that the first split (2) of the input matrix and the recovery process are not required. Further we consider a hybrid algorithm [8], which uses an iterative refinement to improve the accuracy of the inverse generated by the Monte Carlo method.

The iterative refinement works on the Monte Carlo generated inverse to improve the accuracy. We set $D_0 = B^{-1}$, where B^{-1} is the Monte Carlo generated inverse. The iterative process is $R_{i-1} = I - D_{i-1}B$ and $D_i = (I + R_{i-1})D_{i-1}$ (for $i = 1, 2, \ldots$), which continues until $\|R_{i-1}\| < \gamma$.

3.1 Hybrid Monte Carlo Algorithm

1: Read in matrix B and broadcast to all processes

1 Input matrix B, parameters ϵ, δ and γ

2 Broadcast matrix B, and parameters ϵ, δ and γ, to all processes

2: Calculate intermediate matrices (B_1, B_2)

1 Split $B = B_1 - B_2$, where $B_1 = diag(B)$ and $B_2 = B_1 - B$

3: Calculate matrix A and norm

1 Compute the matrix $A = B_1^{-1} B_2$

2 Compute $\|A\|$ and the number of Markov Chains $N = \left(\frac{0.6745}{\epsilon(1-\|A\|)}\right)^2$

4: Calculate matrix P

1 Compute the probability matrix, P

5: Calculate matrix M, by MC on A and P

1 Broadcast matrices A and P, and parameters ϵ, δ and γ

2 For i = $proc_{start}$ to $proc_{stop}$ (where $proc_{start}$ and $proc_{stop}$ are calculated to evenly distribute the work between the available processors)

2.1 For j = 1 to N

Markov Chain Monte Carlo Computation

2.1.1 Set $W_0 = 1$, $SUM[k] = 0$ for $k = 1, 2, \ldots, n$ and $point = i$

2.1.2 Generate a random $nextpoint$, selected based on the probabilities in row $point$

2.1.3 If $A[point][nextpoint] = 0$ then goto *2.1.2*, else continue

2.1.4 Compute $W_j = W_{j-1} \frac{A[point][nextpoint]}{P[point][nextpoint]}$

2.1.5 Set $SUM[nextpoint] = SUM[nexpoint] + W_j$

2.1.6 If $|W_j| > \delta$ set $point = nextpoint$ and goto *2.1.2*

2.2 End of loop j

2.3 Then $m_{ik} = \frac{SUM[k]}{N}$ for $k = 1, 2, \ldots, n$

3 End of loop i

6: Calculate $\hat{B}^{-1}$

1 Compute the Monte Carlo inverse $B^{-1} = MB_1^{-1}$

2 The master process collects the parts of B^{-1} from the slaves

7: Iterative Refinement

1 Broadcast matrix B

2 Set $D_0 = B^{-1}$ and $i = 1$

2.1 Broadcast matrix D_{i-1}

2.2 Calculate $proc_{start}$ to $proc_{stop}$ rows of R_{i-1} and D_i using the iterative refinement $R_{i-1} = I - D_{i-1}B$ and $D_i = (I + R_{i-1})D_{i-1}$

2.3 The master process collects the parts of D_i from the slaves

2.4 If $\|R_{i-1}\| \geq \gamma$ set $i = i + 1$ and goto *2.1*

8: Save Inverse

1 The inverse is $B^{-1} = D_i$

4 Implementation

Sparse matrices are common to many scientific problems. In these a typical MI or SLAE problem will have a matrix with a handful of non-zero elements in a row of size in the tens of thousands (or greater). In this work, while considering sparse matrices, we have used the compressed sparse row (CSR) matrix format.

We implemented the Monte Carlo (*Step 5*) and the iterative refinement (*Step 7*) of the hybrid Monte Carlo method in parallel, since these are the most computationally intensive steps in the algorithm (Section 3.1). The master starts by computing *Step 1* to *Step 4*, which is the serial part of the algorithm. For matrix P only the values array is generated, since the row and the column array are identical to those for matrix A.

Each processing element (PE), including the master, computes an evenly distributed part of rows of the solution matrix B^{-1}, during the Monte Carlo, *Step 5*. Therefore the matrices A, P and B_1 (which is stored as a vector, since it is a diagonal matrix) are broadcast to all PE's. These are used by the PE's to compute part of B^{-1} and then the master collects these matrix parts together.

The master starts the iterative refinement, *Step 7*, by sending matrices B and D_{i-1} to the slaves. Each PE computes an evenly distributed number of rows of R_{i-1}, and then uses this to compute the corresponding rows of D_i. The PE's also calculate norm of the part of R_{i-1} it has calculated. The master then collects the parts of D_i together and calculates $\|R_{i-1}\|$, which it uses to decide if the result of the refinement stage is accurate enough. If not then another iteration is performed.

5 Results

For the experiments we generated diagonally dominant banded matrices, with 5 non-zero elements defined on either side of the leading diagonal. The off-diagonal elements were generated randomly, with the leading diagonal elements selected to give $\|A\| \approx 0.5$.

We used eight Dual Intel® Xeon^{TM} 2.8 GHz nodes, each equipped with 512 KB second level cache and 1 GB main memory, and connected via a switched 1 Gbit network. For the experimental runs only one processor per node was used. LAM-MPI [4, 12] was used for the interprocess communication.

For the calculations floating point accuracy has been used, since this provides enough significant figures in the workings to generate an inverse with the required (10^{-2}) accuracy (measured by $\|I - BD_i\|$, where D_i is the final inverse generated by the parallel hybrid algorithm).

The value for δ, one of the input parameters, was picked to keep T, the average length of the Markov Chains, about equal to $\sqrt{N}$. This produces good results from the Monte Carlo calculations [5], without generating non-necessary elements for the Markov Chain.

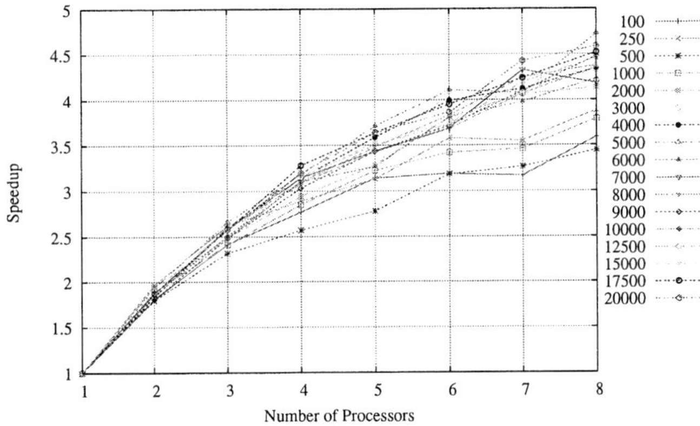

Fig. 1. Speedup of Parallel Hybrid Monte Carlo

As can be seen in Figure 1, for the majority of the matrix sizes we tested there is a speedup of 4 to 4.5 when using eight processors. The smaller matrix sizes ($n \leq 1000$) produced worse performance, but in these cases the total running time was less than a second and this makes the results unreliable.

The efficiency of the parallelisation tails off when using more than four processors. Since the computation time for these problems is small ($n = 20000$ on one processor took 31 seconds) the benefits of splitting the work to more than eight processors will be outweighed by the communication overhead.

Figure 2 shows the percentage of time spent in communication, serial operation and parallel operation when using eight processors for experimental runs on varying matrix sizes. The results show that the communication overhead decreases as the matrix size increases, with a greater amount of time being spent in both serial and parallel computations.

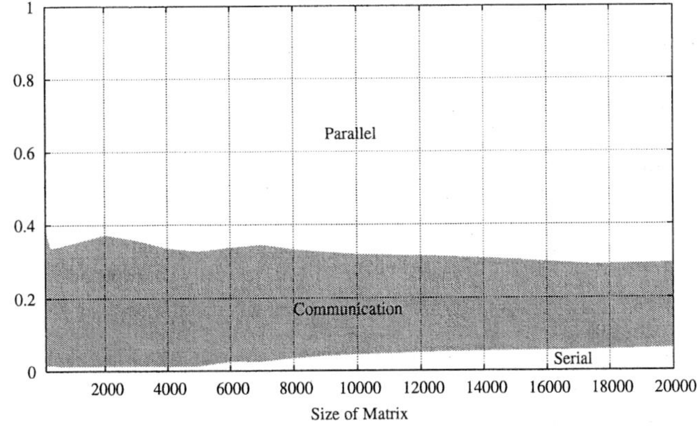

Fig. 2. Communication and Computation Time on Eight Processors

The communication overhead does not decrease as quickly as could be expected since the amount of calculation, in relation to the communication overhead, does not increase by that much as the matrix size grows. This is because the matrices used for these experiments contain the same number of non-zero elements in each row, which means that the sparsity increases linearly with the dimension of the matrix.

6 Conclusion

We found that our implementation achieves a reasonably good speedup. The sparse parallel algorithm, described in section 3.1, makes effective use of the available processors to invert the matrix. The communication overhead gets smaller as the matrix size increases, with a greater percentage of the time being spent in parallel and serial.

We could improve the efficiency by calculating a more accurate inverse, which would mean more time required for the Monte Carlo or iterative refinement steps with little increase in the communication needed. Also to be considered is the effectiveness of the algorithm on larger dimensions and matrices of varying sparsity.

References

1. V.N. Alexandrov. Efficient Parallel Monte Carlo Methods for Matrix Computation. In *Mathematics and Computers in Simulation*, number 47, pages 113–122, Netherlands, 1998. Elsevier.
2. V.N. Alexandrov and A. Karaivanova. Parallel Monte Carlo Algorithms for Sparse SLAE Using MPI. In *LNCS 1697*, pages 283–290. Springer, 1999.
3. V.N. Alexandrov, A. Rau-Chaplin, F. Dehne, and K. Taft. Efficient Coarse Grain Monte Carlo Algorithms for Matrix Computation Using PVM. In *LNCS 1497*, pages 323–330. Springer, August 1998.
4. G. Burns, R. Daoud, and J. Vaigl. LAM: An Open Cluster Environment for MPI. In *Proceedings of Supercomputing Symposium*, pages 379–386, 1994.
5. I.T. Dimov and V.N. Alexandrov. A New Highly Convergent Monte Carlo Method for Matrix Computations. In *Mathematics and Computers in Simulation*, number 47, pages 165–181, Netherlands, 1998. Elsevier.
6. I.T. Dimov, V.N. Alexandrov, and A. Karaivanova. Resolvent Monte Carlo Methods for Linear Algebra Problems. In *Mathematics and Computers in Simulation*, number 155, pages 25–36, Netherlands, 2001. Elsevier.
7. I.T. Dimov, T.T. Dimov, and T.V. Gurov. A New Iterative Monte Carlo Approach for Inverse Matrix Problem, 1998.
8. B. Fathi Vajargah and V.N. Alexandrov. Coarse Grained Parallel Monte Carlo Algorithms for Solving Systems of Linear Equations with Minimum Communication. In *in Proc. of PDPTA*, pages 2240–2245, Las Vegas, 2001.
9. B. Fathi Vajargah, B. Liu, and V.N. Alexandrov. Mixed Monte Carlo Parallel Algorithms for Matrix Computation. In *Lecture Notes in Computational Science - ICCS 2002*, number 2330, pages 609–618, Berlin Heidelberg, 2002. Springer Verlag.

10. G.H. Golub and C.F. Van Loan. *Matrix Computations.* The Johns Hopkins University Press, Baltimore and London, third edition, 1996.
11. I.M. Sobol. *Monte Carlo Numerical Methods.* Nauka, Moscow, 1973. (in Russian).
12. J.M. Squyres and A. Lumsdaine. A Component Architecture for LAM/MPI. In *Proceedings, 10th European PVM/MPI Users' Group Meeting*, number 2840 in Lecture Notes in Computer Science, pages 379–387, Venice, Italy, September / October 2003. Springer-Verlag.
13. J.R. Westlake. *A Handbook of Numerical Matrix Inversion and Solution of Linear Equations.* Wiley, New York, 1968.

Parallel Hybrid Monte Carlo Algorithms for Matrix Computations

V. Alexandrov[1], E. Atanassov[2], I. Dimov[2], S. Branford[1], A. Thandavan[1], and C. Weihrauch[1]

[1] Department of Computer Science, University of Reading
[2] IPP, Bulgarian Academy of Sciences

Abstract. In this paper we consider hybrid (fast stochastic approximation and deterministic refinement) algorithms for Matrix Inversion (MI) and Solving Systems of Linear Equations (SLAE). Monte Carlo methods are used for the stochastic approximation, since it is known that they are very efficient in finding a quick rough approximation of the element or a row of the inverse matrix or finding a component of the solution vector. We show how the stochastic approximation of the MI can be combined with a deterministic refinement procedure to obtain MI with the required precision and further solve the SLAE using MI. We employ a splitting $A = D - C$ of a given non-singular matrix A, where D is a diagonal dominant matrix and matrix C is a diagonal matrix. In our algorithm for solving SLAE and MI different choices of D can be considered in order to control the norm of matrix $T = D^{-1}C$, of the resulting SLAE and to minimize the number of the Markov Chains required to reach given precision. Further we run the algorithms on a mini-Grid and investigate their efficiency depending on the granularity. Corresponding experimental results are presented.

Keywords: Monte Carlo Method, Markov Chain, Matrix Inversion, Solution of System of Linear Equations, Grid Computing.

1 Introduction

The problem of inverting a real $n \times n$ matrix (MI) and solving system of linear algebraic equations (SLAE) is of an unquestionable importance in many scientific and engineering applications: e.g. communication, stochastic modelling, and many physical problems involving partial differential equations. For example, the direct parallel methods of solution for systems with dense matrices require $O(n^3/p)$ steps when the usual elimination schemes (e.g. non-pivoting Gaussian elimination, Gauss-Jordan methods) are employed [4]. Consequently the computation time for very large problems or real time problems can be prohibitive and prevents the use of many established algorithms.

It is known that Monte Carlo methods give statistical estimation of the components of the inverse matrix or elements of the solution vector by performing random sampling of a certain random variable, whose mathematical expectation

V.S. Sunderam et al. (Eds.): ICCS 2005, LNCS 3516, pp. 752–759, 2005.

is the desired solution. We concentrate on Monte Carlo methods for MI and solving SLAEs, since, firstly, only $O(NL)$ steps are required to find an element of the inverse matrix where N is the number of chains and T is an estimate of the chain length in the stochastic process, which are independent of matrix size n and secondly, the process for stochastic methods is inherently parallel.

Several authors have proposed different coarse grained Monte Carlo parallel algorithms for MI and SLAE [7, 8, 9, 10, 11]. In this paper, we investigate how Monte Carlo can be used for diagonally dominant and some general matrices via a general splitting and how efficient mixed (stochastic/deterministic) parallel algorithms can be derived for obtaining an accurate inversion of a given non-singular matrix A. We employ either uniform Monte Carlo (UM) or almost optimal Monte Carlo (MAO) methods [7, 8, 9, 10, 11]. The relevant experiments with dense and sparse matrices are carried out.

Note that the algorithms are built under the requirement $\|T\| < 1$. Therefore to develop efficient methods we need to be able to solve problems with matrix norms greater than one. Thus we developed a spectrum of algorithms for MI and solving SLAEs ranging from special cases to the general case. Parallel MC methods for SLAEs based on Monte Carlo Jacobi iteration have been presented by Dimov [11]. Parallel Monte Carlo methods using minimum Makrov Chains and minimum communications are presented in [5, 1]. Most of the above approaches are based on the idea of balancing the stochastic and systematic errors [11]. In this paper we go a step further and have designed hybrid algorithms for MI and solving SLAEs by combining two ideas: iterative Monte Carlo methods based on the Jacobi iteration and deterministic procedures for improving the accuracy of the MI or the solution vector of SLAEs.

The generic Monte Carlo ideas are presented in Section 2, the main algorithms are described in Section 3 and the parallel approach and some numerical experiments are presented in Section 4 and 5 respectively.

2 Monte Carlo and Matrix Computation

Assume that the system of linear algebraic equations (SLAE) is presented in the form:

$$Ax = b \tag{1}$$

where A is a real square $n \times n$ matrix, $x = (x_1, x_2, ..., x_n)^t$ is a $1 \times n$ solution vector and $b = (b_1, b_2, ..., b_n)^t$.

Assume the general case $\|A\| > 1$. We consider the splitting $A = D - C$, where off-diagonal elements of D are the same as those of A, and the diagonal elements of D are defined as $d_{ii} = a_{ii} + \gamma_i \|A\|$, choosing in most cases $\gamma_i > 1, i = 1, 2, ..., n$. We further consider $D = B - B_1$ where B is the diagonal matrix of D, e.g. $b_{ii} = d_{ii} i = 1, 2, ..., n$. As shown in [1] we could transform the system (1) to

$$x = Tx + f, \tag{2}$$

where $T = D^{-1}C$ and $f = D^{-1}b$. The multipliers γ_i are chosen so that, if it is possible, they reduce the norm of T to be less than 1 and reduce the number of Markov chains required to reach a given precision. We consider two possibilities, first, finding the solution of $x = Tx + f$ using Monte Carlo (MC) method if $\|T\| < 1$ or finding D^{-1} using MC and after that finding A^{-1}. Then, if required, obtaining the solution vector is found by $x = A^{-1}b$. Following the Monte Carlo method described in [7, 11] we define a Markov chain and weights $W_0 = 1, W_j = W_{j-1}\frac{T_{s_{j-1}s_j}}{p_{s_{j-1}s_{j-1}}}$ for $j = 1, 2, \cdots, n$. Consider now the random variable $\theta[g] = \frac{g_{s_0}}{p_{s_0}}\sum_{i=1}^{\infty} W_i f_{s_i}$. We use the following notation for the partial sum:

$$\theta_i[g] = \frac{g_{s_0}}{p_{s_0}}\sum_{j=0}^{i} W_j f_{s_j}. \tag{3}$$

Under condition $\|T\| < 1$, the corresponding Neumann series converges for any given f, and $E\theta_i[g]$ tends to (g, x) as $i \to \infty$. Thus, $\theta_i[g]$ can be considered as an estimate of (g, x) for i sufficiently large. To find an arbitrary component of the solution, for example, the rth component of x, we should choose, $g = e(r) = (\underbrace{0, ..., 1}_{r}, 0, ..., 0)$ such that

$$e(r)_\alpha = \delta_{r\alpha} = \begin{cases} 1 & if \quad r = \alpha \\ 0 & otherwise \end{cases} \tag{4}$$

It follows that $(g, x) = \sum_{\alpha=1}^{n} e(r)_\alpha x_\alpha = x_r$.

The corresponding Monte Carlo method is given by:

$$x_r = \hat{\Theta} = \frac{1}{N}\sum_{s=1}^{N} \theta_i[e(r)]_s,$$

where N is the number of chains and $\theta_i[e(r)]_s$ is the approximate value of x_r in the sth chain. It means that using Monte Carlo method, we can estimate only one, few or all elements of the solution vector. We consider Monte Carlo with uniform transition probability (UM) $p_{\alpha\beta} = \frac{1}{n}$ and Almost optimal Monte Carlo method (MAO) with $p_{\alpha\beta} = \frac{|T_{\alpha\beta}|}{\sum_{\beta=1}^{n}|T_{\alpha\beta}|}$, where $\alpha, \beta = 1, 2, \ldots, n$. Monte Carlo MI is obtained in a similar way [3].

To find the inverse $A^{-1} = C = \{c_{rr'}\}_{r,r'=1}^{n}$ of some matrix A, we must first compute the elements of matrix $M = I - A$, where I is the identity matrix. Clearly, the inverse matrix is given by $C = \sum_{i=0}^{\infty} M^i$, which converges if $\|M\| < 1$.

To estimate the element $c_{rr'}$ of the inverse matrix C, we let the vector f be the following unit vector $f_{r'} = e(r')$.

We then can use the following Monte Carlo method for calculating elements of the inverse matrix C:

$$c_{rr'} \approx \frac{1}{N}\sum_{s=1}^{N}\left[\sum_{(j|s_j=r')} W_j\right], \tag{5}$$

where $(j|s_j = r')$ means that only

$$W_j = \frac{M_{rs_1}M_{s_1s_2}\dots M_{s_{j-1}s_j}}{p_{rs_1}p_{s_1s_2}\dots p_{s_{j-1}p_j}} \tag{6}$$

for which $s_j = r'$ are included in the sum (5).

Since W_j is included only into the corresponding sum for $r' = 1, 2, \dots, n$, then the same set of N chains can be used to compute a single row of the inverse matrix, which is one of the inherent properties of MC making them suitable for parallelization.

3 The Hybrid MC Algorithm

The basic idea is to use MC to find the approximate inverse of matrix D, refine the inverse (filter) and find A^{-1}. In general, we follow the approach described in [7, 11] We can then find the solution vector through A^{-1}. According to the general definition of a regular splitting [2], if A, M and N are three given matrices satisfying $A = M - N$, then the pair of matrices M, N are called regular splitting of A, if M is nonsingular and M^{-1} and N are non-negative.

Therefore, let A be a nonsingular diagonal dominant matrix. If we find a regular splitting of A such that $A = D - C$, the SLAE $x^{(k+1)} = Tx^{(k)} + f$, where $T = D^{-1}C$, and $f = D^{-1}b$ converges to the unique solution x^* if and only if $\|T\| < 1$ [2].

The efficiency of inverting diagonally dominant matrices is an important part of the process enabling MC to be applied to diagonally dominant and some general matrices. Consider now the algorithm which can be used for the inversion of a general non-singular matrix A. Note that in some cases to obtain a very accurate inversion of matrix D some filter procedures can be applied.

Algorithm1: Finding A^{-1}.

1. **Initial data:** Input matrix A, parameters γ and ϵ.
2. **Preprocessing:**
   - 2.1 **Split** $A = D - (D - A)$, where D is a diagonally dominant matrix.
   - 2.2 **Set** $D = B - B_1$ where B is a diagonal matrix $b_{ii} = d_{ii}$ $i = 1, 2, ..., n$.
   - 2.3 **Compute** the matrix $T = B^{-1}B_1$.
   - 2.4 **Compute** $||T||$, the Number of Markov Chains $N = (\frac{0.6745}{\epsilon} \cdot \frac{1}{(1-||T||)})^2$.
3. **For** i=1 to n;
   - 3.1 **For** j=1 to j=N;
     - **Markov Chain Monte Carlo Computation:**
     - 3.1.1 **Set** $t_k = 0$(stopping rule), $W_0 = 1$, $SUM[i] = 0$ and $Point = i$.
     - 3.1.2 **Generate** an uniformly distributed random number *nextpoint*.
     - 3.1.3 **If** $T[point][netxpoint]! = 0$.
     - **LOOP**
       - 3.1.3.1 **Compute** $W_j = W_{j-1} \frac{T[point][netxpoint]}{P[point][netxpoint]}$.
       - 3.1.3.2 **Set** $Point = nextpoint$ and $SUM[i] = SUM[i] + W_j$.
       - 3.1.3.3 **If** $|W_j| < \gamma$, $t_k = t_k + 1$

3.1.3.4 **If** $t_k \geq n$, end LOOP.

3.1.4 **End If**

3.1.5 **Else** go to step 3.1.2.

3.2 **End of loop j.**

3.3 **Compute** the average of results.

4. **End of loop i.**
5. **Obtain** The matrix $V = (I - T)^{-1}$.
6. **Therefore** $D^{-1} = VB^{-1}$.
7. **Compute** the MC inversion $D^{-1} = B(I - T)^{-1}$.
8. **Set** $D_0 = D^{-1}$ (approximate inversion) and $R_0 = I - DD_0$.
9. **use filter procedure** $R_i = I - DD_i$, $D_i = D_{i-1}(I + R_{i-1})$, $i = 1, 2, ..., m$, where $m \leq k$.
10. **Consider the accurate inversion of D** by step 9 given by $D_0 = D_k$.
11. **Compute** $S = D - A$ where S can be any matrix with all non-zero elements in diagonal and all of its off-diagonal elements are zero.
12. **Main function** for obtaining the inversion of A based on D^{-1} step 9:

    12.1 **Compute** the matrices $S_i, i = 1, 2, ..., k$, where each S_i has just one element of matrix S.

    12.2 **Set** $A_0 = D_0$ and $A_k = A + S$

    12.3 **Apply** $A_k^{-1} = A_{k+1}^{-1} + \frac{A_{k+1}^{-1} S_{i+1} A_{k+1}^{-1}}{1 - trace(A_{k+1}^{-1} S_{i+1})}$, $i = k-1, k-2, ..., 1, 0$.
13. **Print**the inversion of matrix A.
14. **End** of algorithm.

4 Parallel Implementation

We have implemented the algorithm proposed on a cluster of PCs and an IBM® SP machine under MPI. We have applied master/slave approach and we have run also on a miniGrid incorporating both the cluster and the SP machine.

Inherently, Monte Carlo methods for solving SLAE allow us to have minimal communication, i.e. to partition the matrix A, pass the non-zero elements of the dense (sparse) matrix to every processor, to run the algorithm in parallel on each processor computing $\lceil n/p \rceil$ rows (components) of MI or the solution vector and to collect the results from slaves at the end without any communication between sending non-zero elements of A and receiving partitions of A^{-1} or x. The splitting procedure and refinement are also parallelised and integrated in the parallel implementation. Even in the case, when we compute only k components $(1 \leq k \leq n)$ of the MI (solution vector) we can divide evenly the number of chains among the processors, e.g. distributing $\lceil kN/p \rceil$ chains on each processor. The only communication is at the beginning and at the end of the algorithm execution which allows us to obtain very high efficiency of parallel implementation.

In addition an iterative filter process is used to improve the accuracy of the Markov Chain Monte Carlo calculated inverse. The iterative filter process is initialised by setting $D_0 = D^{-1}$ (where D^{-1} is the inverse from the Monte Carlo calculations). Then iteratively $R_i = I - DD_i$ and $D_{i+1} = D_i(I + R_i)$. These iterations continue $(i = 1, 2, \ldots)$ until $\|R_i\| < \gamma$.

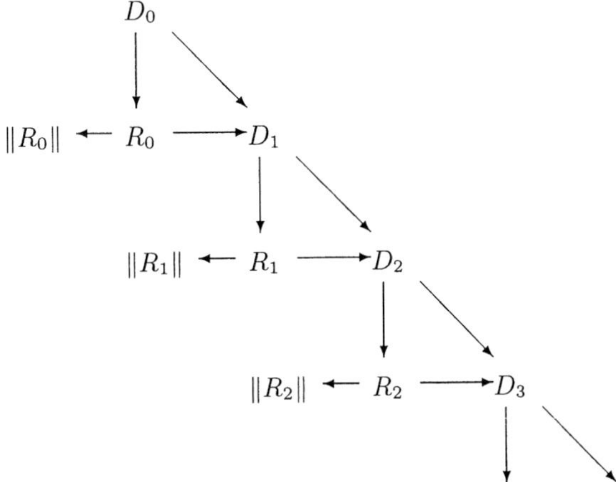

Fig. 1. Data Dependency Graph

The data dependency graph for the iterative filter (Figure 1) directs us to the method for parallelising the iterative filter. Each iteration is calculated separately, with a master process coordinating the calculations slaves and deciding on whether further iterations are required.

The master starts an iteration by sending D_i to each of the slaves. Each of the slaves calculates: n/p columns of R_i; the partial row sums of R_i (required for calculating $\|R_i\|$); and n/p columns of D_{i+1}. The slaves then send the columns of D_{i+1} and the partial row sums of R_i to the master, which calculates $\|R_i\|$ and so decides if the inverse, D_{i+1} is of the required accuracy.

In this way we can obtain for the parallel time complexity of the MC procedure of the algorithm $O(nNL/p)$ where N denotes the number of Markov Chains. According to central limit theorem for the given error ϵ we have $N \geq \left(\frac{0.6745}{\epsilon \times (1-\|T\|)}\right)^2$, L denotes the length of the Markov chains and L$\leq \left(\frac{\log(\gamma)}{\log\|T\|}\right)$, where ϵ, γ show the accuracy of Monte Carlo approximation [3]. Parameters ϵ, γ are used for the stochastic and systematic error. Note that if rough approximations of the MI or solution of the SLAE is required we need to run only the Monte Carlo component of the algorithm. If a higher precision is required we need to also run the filter procedures, which will add complexity of $O(n^3/p)$ in case of sparce matrices for example. The absolute error of the solution for matrix inversion is $\left\|I - \hat{A}^{-1}A\right\|$, where A is the matrix whose inversion has to be found, and $\hat{A}^{-1}$ is the approximate MI. The computational time is shown in seconds.

5 Experimental Results

The algorithms run on partition of a 32 processor IBM® SP machine and a workstation cluster with a 100 Mbps Ethernet network. Each workstation had an Intel® Pentium III processor with 256 MB RAM and a 30 GB harddisk. Each workstation was running SUSE Linux 8.1. The MPI environment used was LAM MPI 7.0.

We have carried test with low precision $10^{-1} - 10^{-2}$ and higher precision $10^{-5} - 10^{-6}$ in order to investigate the balance between stochastic and deterministic components of the algorithms based on the principle of balancing of errors (e.g. keeping the stochastic and systematic error of the same order) [7]. We have also compared the efficiency of parallel Monte Carlo and Quasi-Monte Carlo methods for solving SLAEs.

Our results show that all the algorithms scale very well. The results show that if we need to refine the results using filter procedure the proportion of the filter procedure time grows with the growth of the matrix size, so we need to limit these procedures if possible. The last table shows that Quasi-Monte Carlo is faster in finding rough approximation of the solution of SLAE. The second table shows that is important to balance computations in a Grid environment and communicate with larger chunks of data. For example in this case this can lead to more than twice reducing the computational time on the same number of processors.

Table 1. MC with filter procedures on the cluster

Matrix Size	Time (Dense Case) in seconds			
	4 proc.	8 proc.	12 proc.	16 proc.
250	59.269	24.795	16.750	14.179
500	329.072	177.016	146.795	122.622
1000	1840.751	989.423	724.819	623.087

Table 2. MC with filter procedures on the miniGrid

Matrix Size	Time (MC, Dense Case) in seconds	
	16 proc. (4 SP and 12 cluster)	16 proc. (8 SP and 8 cluster)
250	729.208	333.418
500	4189.225	1945.454

Table 3. MC vs QMC without filtering on the cluster (matrix size 250 by 250)

Matrix Size	Time (Dense Case) in seconds			
	4 proc.	8 proc.	12 proc.	16 proc.
MC	48.819	20.909	13.339	9.691
QMC	0.744	0.372	0.248	0.186

6 Conclusion

In this paper we have introduced a hybrid Monte Carlo/deterministic algorithms for Matrix Computation for any non-singular matrix. We have compared the efficiency of the algorithm on a cluster of workstations and in a Grid environment. The results show that the algorithms scale very well in such setting, but a careful balance of computation should be maintained. Further experiments are required to determine the optimal number of chains required for Monte Carlo procedures and how best to tailor together Monte Carlo and deterministic refinement procedures.

References

1. B. Fathi, B.Liu and V. Alexandrov, *Mixed Monte Carlo Parallel Algorithms for Matrix Computation* , Lecture Notes in Computer Science, No 2330, Springer-Verlag, 2002, pp 609-618
2. Ortega, J., *Numerical Analysis*, SIAM edition, USA, 1990.
3. Alexandrov V.N., *Efficient parallel Monte Carlo Methods for Matrix Computation,* Mathematics and computers in Simulation, Elsevier **47** pp. 113-122, Netherlands, (1998).
4. Golub, G.H., Ch., F., Van Loan, *Matrix Computations,* The Johns Hopkins Univ. Press, Baltimore and London, (1996)
5. Taft K. and Fathi Vajargah B., *Monte Carlo Method for Solving Systems of Linear Algebraic Equations with Minimum Markov Chains.* International Conference PDPTA'2000 Las Vegas, (2000).
6. Sobol I.M. *Monte Carlo Numerical Methods.* Moscow, Nauka, 1973 (in Russian).
7. Dimov I., Alexandrov V.N. and Karaivanova A., *Resolvent Monte Carlo Methods for Linear Algebra Problems,* Mathematics and Computers in Simulation, Vo155, pp. 25-36, 2001.
8. Fathi Vajargah B. and Alexandrov V.N., *Coarse Grained Parallel Monte Carlo Algorithms for Solving Systems of Linear Equations with Minimum Communication,* in Proc. of PDPTA, June 2001, Las Vegas, 2001, pp. 2240-2245.
9. Alexandrov V.N. and Karaivanova A., *Parallel Monte Carlo Algorithms for Sparse SLAE using MPI,* LNCS 1697, Springer 1999, pp. 283-290.
10. Alexandrov V.N., Rau-Chaplin A., Dehne F. and Taft K., *Efficient Coarse Grain Monte Carlo Algorithms for matrix computation using PVM,* LNCS 1497, pp. 323-330, Springer, August 1998.
11. Dimov I.T., Dimov T.T., et all, *A new iterative Monte Carlo Approach for Inverse Matrix Problem,* J. of Computational and Applied Mathematics **92** pp 15-35 (1998).

An Efficient Monte Carlo Approach for Solving Linear Problems in Biomolecular Electrostatics

Charles Fleming[1], Michael Mascagni[1,2], and Nikolai Simonov[2,3]

[1] Department of Computer Science and
[2] School of Computational Science,
Florida State University, Tallahassee, FL, 32306-4530, USA
[3] ICM&MG, Lavrentjeva 6, Novosibirsk 630090, Russia

Abstract. A linear (elliptic) problem in molecular electrostatics is considered. To solve it, we propose an efficient Monte Carlo algorithm. The method utilizes parallel computing of point potential values. It is based on the walk-in-subdomains technique, walk-on-spheres algorithm, and an exact treatment of boundary conditions.

1 Introduction

Recently [1, 2], we proposed an advanced Monte Carlo technique for calculating different properties of large organic molecules. To use this approach, the problems under consideration have to be treated in the framework of continuous media models. In particular, to investigate electrostatic properties of molecules, we adopted one of the most popular and commonly used models, namely, the implicit solvent model. This means that solvent and ions dissolved within are treated as a continuous medium, whose properties are characterized by dielectric permittivity, ϵ_e, whereas the molecule under investigation is described explicitly. The solute (molecule) is thought of as a cavity with dielectric constant, ϵ_i, which is much less than that of the exterior environment.

Mathematically, this classical electrostatic approach leads to a boundary value problem for the Poisson equation satisfied by the electrostatic potential, $u(x)$:

$$-\nabla\epsilon\nabla u(x) = \rho(x)\ ,\ x \in \mathbb{R}^3\ , \tag{1}$$

where ϵ is the position-dependent permittivity and $\rho(x)$ is the charge distribution [3]. Usually, the molecule under consideration is described geometrically as a union of spherical atoms with partial point charges at the sphere centers. This means that $\rho(x) = \sum_{m=1}^{M} q_m \delta(x - x^{(m)})$, where $q^{(m)}$ are the charges and $x^{(m)}$ are the atomic centers and hence the charge locations.

The charge distribution in the solvent is determined by the dissolved mobile ions. In the framework of a continuous medium approach, their positions in physical space are described by the Boltzmann statistical distribution; this leads to the non-linear Poisson-Boltzmann equation for $u(x)$ in the solvent [3].

V.S. Sunderam et al. (Eds.): ICCS 2005, LNCS 3516, pp. 760–765, 2005.

For small potential values, this equation may be linearized, thus leading to the relation

$$\Delta u(x) - \kappa^2 u(x) = 0 \ , \ x \in G_1 \ , \tag{2}$$

where κ^2 is a positive constant called the Debye-Hückel screening parameter.

All the difficulties that one encounters in solving such boundary-value problems arise from the geometrical features of the model of the molecule. The geometry comes through the boundary conditions that have to be satisfied by the limiting values of functions and their normal derivatives on different sides of the boundary, Γ:

$$u_i(y) = u_e(y) \ , \ \epsilon_i \frac{\partial u_i}{\partial n(y)} = \epsilon_e \frac{\partial u_e}{\partial n(y)} \ , \ y \in \Gamma \ . \tag{3}$$

Here, Γ is either the surface of the molecule or the boundary of the so-called ion-exclusion layer. One of the possible variants is when these surfaces coincide.

One of the distinctive and attractive features of the Monte Carlo approach is that there is no need to perform cumbersome and otherwise labor-consuming preparatory approximations of the molecular surfaces and boundary conditions. In fact, a primitive analytic description based on a list of atomic centers and radii is absolutely adequate for type of random-walk simulations we carry out.

Another favorable aspect of the Monte Carlo algorithms we employ is their natural, immanent parallelizability. For linear electrostatic problems, the desired computational entity can often be represented as a linear functional of point potential values. This makes possible further distribution of the calculations and therefore computing these values in parallel.

2 Energy: Linear Case

We consider the problem of calculating the electrostatic free energy of a large biomolecule. In the linear case, this energy is equal to

$$E = \frac{1}{2} \sum_{m=1}^{M} u^{(m)} q^{(m)} \ , \tag{4}$$

where $u^{(m)}$ is the regular part of the electrostatic potential at the center of the mth atom. To estimate $u^{(m)}, m = 1, \ldots, M$ we make use of the walk-on-spheres algorithm. The specific form of the molecule' geometrical model makes it possible to simulate the required Markov chains in such a way that the chain converges to the boundary geometrically, and the last point lies exactly on Γ.

Let $G_i = \bigcup_{m=1}^{M} B(x^{(m)}, r^{(m)})$ be the inside of the molecule. Here, $B(x, r)$ denotes the ball centered at the point x, and r is its radius (see Figure 1). For a given distribution of point charges in G_i, the electrostatic potential can be

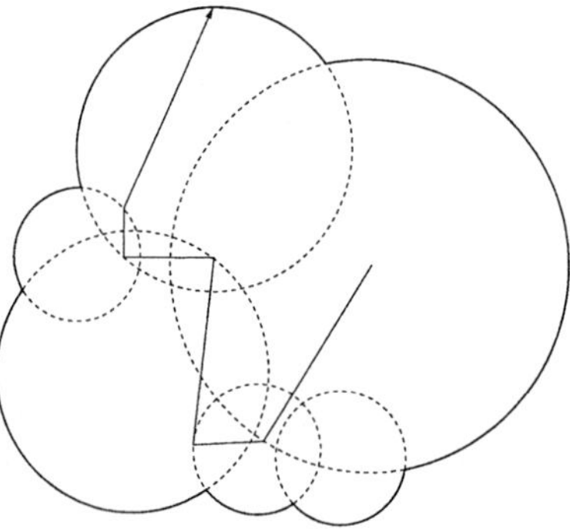

Fig. 1. An example of a computational domain in the energy calculation. The domain is a union of atomic spheres, and the line follows a walk-on-spheres/walk-in-subdomains trajectory that begins at the center of the largest sphere and ends as it leaves the interior of the computational domain

represented as $u_i(x) = u^0(x) + g(x)$, where $g(x) = \sum_{m=1}^{M} \frac{q^{(m)}}{4\pi\epsilon_i} \frac{1}{|x - x^{(m)}|}$. From this it follows that the regular part of the electrostatic potential, u^0, satisfies the Laplace equation in G_i. To find $u^{(m)} = u^0(x^{(m)})$, we construct a Monte Carlo estimate based on combination of the walk-on-spheres and walk-in-subdomains algorithms [4]. Let $\xi[u^0](x_0)$ be a Monte Carlo estimate for $u^0(x_0)$. Using the probabilistic representation, we set $\xi[u^0](x_0) = u^0(x^*)$, where x^* is the exit point of Brownian motion from G. The most efficient way to simulate these exit points is to use the natural representation of G as a union of intersecting spherical subdomains. For every ball, the exit point can be sampled directly. By the Poisson integral formula for a function, u, that satisfies the Laplace equation, at every point $x \in B(x^{(m)}, r^{(m)})$ we have $u(x) = \int_{S(x^{(m)},r^{(m)})} p_p(x \to y)\, u(y) d\sigma(y)$, where $p_p(x \to y) = \frac{1}{4\pi r^{(m)}} \frac{(r^{(m)})^2 - |x - x^{(m)}|^2}{|x - y|^3}$ is the Poisson kernel. The Markov chain $\{x_i, i = 0, 1, \ldots, N\}$ is defined recursively. Given $x_i \in B(x_i^{(m)}, r_i^{(m)})$, the next point is chosen randomly on $S(x_i^{(m)}, r_i^{(m)})$ in accordance with density $p_p(x_i \to x_{i+1})$. By definition, x_{i+1} either hits the boundary or lies in some other ball. With probability one, the path of this Markov chain terminates at the boundary after a finite number of steps, N. With some natural assumptions about G, which are always valid for molecular geometries, $\mathbb{E}N < \infty$. We can then set $x^* = x_N$.

3 Boundary Conditions

Next, we have to construct estimates for boundary values of the potential, u^0. By definition $\xi[u^0](x_N) = \xi[\psi](x_N) - g(x_N)$, where ψ is the unknown boundary value of u. One approach for constructing a Monte Carlo estimate for ψ is to randomize a finite-difference approximation of the boundary condition with step-size h [1]. Hence, we have $\psi(x_N) = \mathbb{E}(u(x_{N+1})|x_N) + O(h^2)$, where $x_{N+1} = x_N - hn$ with probability $p_i = \dfrac{\epsilon_i}{\epsilon_i + \epsilon_e}$, $x_{N+1} = x_N + hn$ with the complementary probability, p_e, and n is the external normal vector. This introduces an $O(h)$ bias into resulting Monte Carlo estimate.

Here, we propose another approach that speeds up the computations and eliminates the bias. This approach is new and is specifically adapted to the boundary flux-type conditions that include normal derivatives. Consider an auxiliary sphere $S(x_N, a)$, which does not contain point charges. Then, application of Green's formula to functions u_i, u_e and $\Phi_\kappa(x) = -\dfrac{1}{4\pi}\dfrac{\sinh(\kappa(a-r))}{r\sinh(\kappa a)}$, where $r = |x - x_N|$, leads to the following mean-value relation:

$$\begin{aligned} u(x_N) = & \frac{\epsilon_e}{\epsilon_e + \epsilon_i}\int_{S_e(x_N,a)} \frac{1}{2\pi a^2}\frac{\kappa a}{\sinh(\kappa a)} u_e \\ & + \frac{\epsilon_i}{\epsilon_e + \epsilon_i}\int_{S_i(x_N,a)} \frac{1}{2\pi a^2}\frac{\kappa a}{\sinh(\kappa a)} u_i \\ & - \frac{(\epsilon_e - \epsilon_i)}{\epsilon_e + \epsilon_i}\int_{\Gamma\bigcap B(x_N,a)\setminus\{x_N\}} 2\frac{\partial \Phi_0}{\partial n} Q_{\kappa,a} u \\ & + \frac{\epsilon_i}{\epsilon_e + \epsilon_i}\int_{B_i(x_N,a)} [-2\kappa^2 \Phi_\kappa] u_i \ , \end{aligned} \tag{5}$$

where $Q_{\kappa,a}(r) = \dfrac{\sinh(\kappa(a-r)) + \kappa r\cosh(\kappa(a-r))}{\sinh(\kappa a)} < 1$, and Γ is the surface of the molecule.

Randomization of this formula provides a procedure to extend the Markov chain $\{x_i\}$ after it hits the boundary without introducing any bias. The algorithm works as follows. With probability p_e the next point is chosen in the exterior of the molecule, $\overline{G}_e$, and with the complementary probability, p_i, $x_{N+1} \in \overline{G}_i$. In the exterior, the next point of the Markov chain is chosen isotropically on the boundary of $B(x_N, a)\bigcap G_e$. The random walker in this case has non-zero probability of terminating. On $S(x_N, a)$ the termination probability is equal to $1 - \dfrac{\kappa a}{\sinh(\kappa a)}$. Note that $B(x_N, a)\bigcap G_e$ is not convex. This means that there can be two intersections with its boundary. In this case we can choose randomly one of them. In the interior, we choose an isotropic direction first. Then, with the corresponding probabilities, the next point is sampled on the chosen straight line either on the boundary of $B(x_N, a)\bigcap G_i$, or inside the domain. From (5) it follows that $\psi(x_N) = \mathbb{E}u(x_{N+1})$.

4 Random Walks Inside and Outside

The further construction is common to both methods of treating these boundary conditions.

For internal points, we construct a Markov chain of the walk-in-subdomains starting at x_{N+1}, and set $u(x_{N+1}) = \mathbb{E}(u^0(x_2^*) + g(x_{N+1})|x_{N+1})$ where $x_2^* \in \Gamma$ is the next exit point.

For $u(x_{N+1}) \in G_e$, we can use the walk-on-spheres algorithm. A random walker of this Markov chain survives at every step with probability $\frac{\kappa d_i}{\sinh(\kappa d_i)}$, where d_i is the distance to the boundary. With probability one this chain terminates. This happens either inside G_e or when the distance d_i from the point to the boundary becomes less than a prescribed number, ε. In the second case we return to estimating ψ on the boundary. Since the probability of terminating is bounded away from zero, the mean number of steps in the Markov chain is finite.

The same approach works also for the case when the ion-exclusion layer is introduced into the molecule model. This means that the whole space is divided into three domains: the internal part of the molecule, G_i, the external, G_e, and the layer between them, G_0. In G_0 we have $\epsilon = \epsilon_e$ and $\kappa = 0$. On ∂G_i the mean-value relationship (5) then reads:

$$
\begin{aligned}
u(x_N) = {} & \frac{\epsilon_e}{\epsilon_e + \epsilon_i} \int_{S_e(x_N,a)} \frac{1}{2\pi a^2} u_e \\
& + \frac{\epsilon_i}{\epsilon_e + \epsilon_i} \int_{S_i(x_N,a)} \frac{1}{2\pi a^2} u_i \\
& - \frac{\epsilon_e - \epsilon_i}{\epsilon_e + \epsilon_i} \int_{\Gamma \bigcap B(x_N,a) \setminus \{x_N\}} 2\frac{\partial \Phi_0}{\partial n} u \ ,
\end{aligned} \tag{6}
$$

where $\Phi_0(x) = -\frac{1}{4\pi}\left(\frac{1}{|x - x_N|} - \frac{1}{a}\right)$. On ∂G_e the dielectric constants are equal on both sides. This means that the third term in (5) equals zero.

5 Experiments, Conclusions, and Future Work

To test the proposed algorithms we applied our Monte Carlo estimates to a simple model problem with a known analytic solution. For a spherical (one-atom) molecule, the exact value of E scaled by $q^2/(4\pi\epsilon_i R)$ equals $\epsilon_i/(\epsilon_e(1+\kappa R)) - 1$. For the values $\epsilon_i = 4.0$, $\epsilon_e = 78.5$, $\kappa = 0$ one obtains the exact solution of -0.9490. Our calculation based on the walk-on-spheres method provided the result -0.9489 with a 0.1% statistical error in 0.7 seconds for $a = 0.1$ and in 1.4 seconds for $a = 0.03$. The algorithm based on the finite-difference approximation of the boundary condition with $h = 0.001$ provided the result with the same accuracy in 5.6 seconds. All calculations were carried out on an ordinary desktop computer with a 1.3 GHz P4 processor running Windows 2000, which is already a fairly slow computer.

Thus we see roughly a speed increase of one order of magnitude when progressing from the finite-difference approach to the boundary condition to this new method. In addition, this comes with the elimination of the $O(h)$ bias of the finite-difference technique. While we carried out these very simple computations on an ordinary PC, it is important to note that the specific form of the functional (4) we computed makes it possible to simulate in parallel M independent random walks for estimating every $u^{(m)}$ very much more rapidly in a parallel setting.

We believe these techniques will be very important in the eventual creation of a suite of Monte Carlo approaches to problems and computations in biomolecular electrostatics. Thus, we plan to continue this line of research towards this long-term goal. In the short term we plan to investigate other molecular surfaces and their implications for these methods, the direct solution of the nonlinear Poisson-Boltzmann equation via Monte Carlo, and the calculation of forces and other types of functionals of the potential and its partial derivatives.

References

1. Mascagni, M. and Simonov, N.A.: Monte Carlo method for calculating the electrostatic energy of a molecule. In: Lecture Notes in Computer Science, Vol. 2657. Springer-Verlag, Springer-Verlag, Berlin Heidelberg New York (2003) 63–74.
2. Mascagni, M. and Simonov, N.A.: Monte Carlo methods for calculating some physical properties of large molecules. SIAM Journal on Scientific Computing. **26** (2004) 339–357.
3. Davis, M.E. and McCammon, J.A.: Electrostatics in biomolecular structure and dynamics. Chem. Rev. **90** (1990) 509–521
4. Simonov, N.A.: A random walk algorithm for the solution of boundary value problems with partition into subdomains. in Metody i algoritmy statisticheskogo modelirovanija, Akad. Nauk SSSR Sibirsk. Otdel., Vychisl. Tsentr, Novosibirsk (1983) 48–58.

Finding the Smallest Eigenvalue by the Inverse Monte Carlo Method with Refinement*

Vassil Alexandrov[1] and Aneta Karaivanova[2]

[1] School of Systems Engineering, University of Reading, Reading RG6 6AY, UK
v.n.alexandrov@rdg.ac.uk
http://www.cs.reading.ac.uk/people/V.Alexandrov.htm
[2] IPP - BAS, Acad. G. Bonchev St., Bl.25A, 1113 Sofia, Bulgaria
anet@parallel.bas.bg
http://parallel.bas.bg/~{}anet/

Abstract. Finding the smallest eigenvalue of a given square matrix A of order n is computationally very intensive problem. The most popular method for this problem is the Inverse Power Method which uses LU-decomposition and forward and backward solving of the factored system at every iteration step. An alternative to this method is the Resolvent Monte Carlo method which uses representation of the resolvent matrix $[I-qA]^{-m}$ as a series and then performs Monte Carlo iterations (random walks) on the elements of the matrix. This leads to great savings in computations, but the method has many restrictions and a very slow convergence.

In this paper we propose a method that includes fast Monte Carlo procedure for finding the inverse matrix, refinement procedure to improve approximation of the inverse if necessary, and Monte Carlo power iterations to compute the smallest eigenvalue. We provide not only theoretical estimations about accuracy and convergence but also results from numerical tests performed on a number of test matrices.

Keywords: Monte Carlo methods, eigenvalues, Markov chains, parallel computing, parallel efficiency.

1 Introduction

Let A be a real symmetric matrix. Consider the problem of evaluating the eigenvalues of A, i.e. the values of λ for which

$$Au = \lambda u \tag{1}$$

holds. Suppose, the n eigenvalues of A are ordered as follows $|\lambda_1| > |\lambda_2| \geq \ldots \geq |\lambda_{n-1}| > |\lambda_n|$.

* Supported by the Ministry of Education and Science of Bulgaria under Grant No. I1405/04.

V.S. Sunderam et al. (Eds.): ICCS 2005, LNCS 3516, pp. 766–774, 2005.

It is known that the problem of calculating the smallest eigenvalue of A is more difficult from numerical point of view than the problem of evaluating the largest eigenvalue. Nevertheless, for many applications in physics and engineering it is important to estimate the smallest one, because it usually defines the most stable state of the system which is described by the considered matrix.

2 Background

2.1 Inverse Power Method

One of the most popular methods for finding extremal eigenvalues is the power method ([6], [8]) which for a given matrix A is defined by the iteration

$$x^{new} = Ax^{old}.$$

Except for special starting points, the iterations converge to an eigenvector corresponding to the eigenvalue of A with largest magnitude (**dominant eigenvalue**). The least squares solution μ to the overdetermined linear system

$$\mu x_k = x_{k+1}$$

is an estimate for λ_1 which is called the **Raleigh quotient**.

$$\mu = \frac{x_k^T x_{k+1}}{x_k^T x_k}.$$

Suppose that we want to compute the eigenvector corresponding to the eigenvalue of A of smallest magnitude. Letting (λ_1, e_1) through (λ_n, e_n) denote the eigenpairs of A, the corresponding eigenpairs of $C = A^{-1}$ are $(1/\lambda_1, e_1)$ through $(1/\lambda_n, e_n)$. If λ_n is the eigenvalue of A of smallest magnitude, then $1/\lambda_n$ is C's eigenvalue of largest magnitude and the power iteration $x^{new} = A^{-1}x^{old}$ converges to the vector e_n corresponding to the eigenvalue $1/\lambda_n$ of $C = A^{-1}$. When implementing the inverse power method, instead of computing the inverse matrix A^{-1} we multiply by A to express the iteration $x^{new} = A^{-1}x^{old}$ in the form $Ax^{new} = x^{old}$. Replacing A by its LU factorization yields

$$(LU)x^{new} = x^{old}. \tag{2}$$

In each iteration of the inverse power method, the new x is obtained from the old x by forward and back solving the factored system (2). This scheme is computationally more efficient. With k iterations, the number of arithmetic operations is $O(kn^2)$ for Power method and $O(n^3 + kn^2)$ for Inverse Power Method.

2.2 Resolvent Monte Carlo Method

Given a matrix A, consider an algorithm based on Monte Carlo iterations by the resolvent matrix $R_q = [I - qA]^{-1} \in \mathbb{R}^{n\times n}$. The following representation

$$[I - qA]^{-m} = \sum_{i=0}^{\infty} q^i C^i_{m+i-1} A^i, \quad |q\lambda| < 1 \tag{3}$$

is valid because of behaviors of binomial expansion and the spectral theory of linear operators. Let remind that the eigenvalues of the matrices R_q and A are connected by the equality $\mu = \frac{1}{1-q\lambda}$, and the eigenvectors coincide. The following expression (see [4])

$$\mu^{(m)} = \frac{([I - qA]^{-m} f, h)}{([I - qA]^{-(m-1)} f, h)} \longrightarrow_{m\to\infty} \mu = \frac{1}{1 - q\lambda}, \quad f \in \mathbb{R}^n, h \in \mathbb{R}^n, \tag{4}$$

is valid (f and h are vectors in $\mathbb{R}^n$). For negative values of q, the largest eigenvalue μ_{max} of R_q corresponds to the smallest eigenvalue λ_{min} of the matrix A.

Moreover, if $|\lambda'_{max} < 1|$ where λ'_{max} is the largest eigenvalue of the matrix $A' = \{|a_{ij}|\}^n_{i,j=1}$, the following statement holds, [4], [2]:

$$([I - qA]^{-m} f, h) = E\left\{\sum_{i=0}^{\infty} q^i C^i_{m+i-1} (A^i f, h)\right\}.$$

Now we can construct Monte Carlo method. Define Markov chain $k_0 \to k_1 \to \ldots \to k_i$ ($1 \le k_j \le n$ are natural numbers) with initial and transition densities, $Pr(k_0 = \alpha) = p_\alpha$, $Pr(k_j = \beta | k_{j-1} = \alpha) = p_{\alpha\beta}$, (see [4],[2]). Now define the random variables W_j using the following recursion formula:

$$W_0 = \frac{h_{k_0}}{p_{k_0}}, \quad W_j = W_{j-1} \frac{a_{k_{j-1}k_j}}{p_{k_{j-1}k_j}}, \quad j = 1, \ldots, i. \tag{5}$$

Than it can be proven that (see [4], [2]):

$$\lambda_{min} \approx \frac{1}{q}\left(1 - \frac{1}{\mu^{(m)}}\right) = \frac{(A[I - qA]^{-m} f, h)}{([I - qA]^{-m} f, h)} =$$

$$\frac{E\sum_{i=1}^{\infty} q^{i-1} C^{i-1}_{i+m-2} W_i f_{x_i}}{E\sum_{i=0}^{\infty} q^i C^i_{i+m-1} W_i f_{x_i}} = \frac{E\sum_{i=0}^{l} q^i C^i_{i+m-1} W_{i+1} f_{x_i}}{E\sum_{i=0}^{l} q^i C^i_{i+m-1} W_i f_{x_i}}, \tag{6}$$

where $W_0 = \frac{h_{k_0}}{p_{k_0}}$ and W_i are defined by (5). The coefficients C^n_{i+m} are calculated using the presentation $C^i_{i+m} = C^i_{i+m-1} + C^{i-1}_{i+m-1}$.

This method has strong requirements about matrices for which it can be applied. The systematic error consists of two parts:

- an error from the Power method applied on the resolvent matrix $[I - qA]^{-1}$ which determines the value of the parameter m in the following way: The rate of convergence is

$$O\left(\left(\frac{1+|q|\lambda_n}{1+|q|\lambda_{n-1}}\right)^m\right).$$

The parameter m has to be chosen such that $(\frac{\mu_2}{\mu_1})^m < \varepsilon$.
- an error which comes from the series expansion of the resolvent matrix - it determines the value of the parameter l (length of each random walk). Considering the presentation

$$([I - qA]^{-m} f, g) \approx \sum_{i=0}^{l} q^i C^i_{m+i-1}(A^i f, g) = \sum_{i=1}^{l} u_i,$$

the parameter l has to be chosen such that $|u_{l+1}| < \varepsilon_2$.
- rounding errors - to maintain them at a low level the condition $(\max_{1\leq i\leq l} |u_i|)\,\varepsilon < \alpha\mu$ must be satisfied, where $\alpha < 1$ represents the requested precision and ε is the machine precision parameter.

Unfortunately, the parameter l can not be relatively large because the binomial coefficients C^l_{m+l-1} grow exponentially with l. This is serious restriction for using the resolvent Monte Carlo.

The Resolvent Monte Carlo method computes the smallest eigenvalue with only $O(lN)$ operations, where l is the average length of the Markov chains and N is the number of walks. But this method has a lot of restrictions, and, also, gives very rough approximation.

3 Inverse Monte Carlo with Refinement

Here we propose a method that includes fast Monte Carlo scheme for matrix inversion, refinement of the inverse matrix (if necessary) and Monte Carlo power iterations for computing the largest eigenvalue of the inverse (the smallest eigenvalue of the given matrix).

3.1 Monte Carlo for Computing the Inverse Matrix

To find the inverse $A^{-1} = C = \{c_{rr'}\}^n_{r,r'=1}$ of some matrix A, we must first compute the elements of matrix $M = I - A$, where I is the identity matrix. Clearly, the inverse matrix is given by

$$C = \sum_{i=0}^{\infty} M^i, \tag{7}$$

which converges if $\|M\| < 1$ (sufficient condition).

To estimate the element $c_{rr'}$ of the inverse matrix C, we define Markov chain $s_0 \rightarrow s_1 \rightarrow \cdots \rightarrow s_k$, where the $s_i, i = 1, 2, \cdots, k$, belongs to the state space $S = \{1, 2, \cdots, n\}$. Then for $\alpha, \beta \in S, p_0(\alpha) = p(s_0 = \alpha)$ is the probability that the Markov chain starts at state α and $p(s_{j+1} = \beta | s_j = \alpha) = p_{\alpha\beta}$ is the transition probability from state α to state β. The set of all probabilities $p_{\alpha\beta}$ defines a transition probability matrix $P = \{p_{\alpha\beta}\}_{\alpha,\beta=1}^{n}$.

We then can use the following Monte Carlo method (see, for example, [11], [3]) for calculating elements of the inverse matrix C:

$$c_{rr'} \approx \frac{1}{N} \sum_{s=1}^{N} \left[\sum_{(j|s_j=r')} W_j \right], \tag{8}$$

where $(j|s_j = r')$ means that only

$$W_j = \frac{M_{rs_1} M_{s_1 s_2} \ldots M_{s_{j-1} s_j}}{p_{rs_1} p_{s_1 s_2} \cdots p_{s_{j-1} p_j}} \tag{9}$$

for which $s_j = r'$ are included in the sum (8).

Since W_j is included only into the corresponding sum for $r' = 1, 2, \ldots, n$, then the same set of N chains can be used to compute a single row of the inverse matrix, which is one of the inherent properties of MC making them suitable for parallelization.

Furthermore, if necessary we transform matrix A to matrix D where D is diagonally dominant matrix, we find its inverse and after that apply a filter procedure in order to find A^{-1} [5].

The *probable error* of the method, is defined as $r_N = 0.6745\sqrt{D\theta/N}$, where $P\{|\bar{\theta} - E(\theta)| < r_N\} \approx 1/2 \approx P\{|\bar{\theta} - E(\theta)| > r_N\}$, if we have N independent realizations of random variable (r.v.) θ with mathematical expectation $E\theta$ and average $\bar{\theta}$ [11].

3.2 Refinement

Given a square nonsingular matrix A, Monte Carlo gives fast but very rough estimation of the elements of the inverse matrix C_0. Under the condition $||R_0|| \leq \delta < 1$ where $R_0 = I - AC_0$, the elements of A^{-1} may be computed to whatever degree of accuracy is convenient using the iterative process: $C_m = C_{m-1}(I + R_{m-1})$, $R_m = I - AC_m$ for $m = 1, 2, \ldots$. It can be shown that

$$||C_m - A^{-1}|| \leq ||C_0|| \frac{\delta^{2^m}}{1 - \delta}.$$

It follows that once we have an initial approximation C_0 of the inverse such that $||I - AC_0|| \leq k < 1$ we can use the above iterative scheme and than the number of correct digits increases in geometric progression.

If we have inverted matrix D then we need to apply the filter procedure to obtain the A^{-1} using the formula

$$A_k^{-1} = A_{k+1}^{-1} + \frac{A_{k+1}^{-1} S_{i+1} A_{k+1}^{-1}}{1 - trace(A_{k+1}^{-1} S_{i+1})}, i = k-1, k-2, ..., 1, 0.$$

3.3 Evaluation of the Smallest Eigenvalue

Once we have the matrix $C = A^{-1}$ we apply the Monte Carlo power iterations to find its largest eigenvalue performing random walks on the elements of the matrix C:

$$\lambda_{max}(C) \approx \frac{E\{W_m f_{k_i}\}}{E\{W_{m-1} f_{k_{i-1}}\}}.$$

Here $W_j = \frac{C_{k_0 k_1} C_{k_1 k_2} \dots C_{k_{j-1} k_j}}{p_{k_0 k_1} p_{k_1 k_2} \dots p_{k_{j-1} p_j}}$, and m is sufficiently large to ensure the convergence of the power method which is $O(\frac{\lambda_n}{\lambda_{n-1}})^m$.

The computational cost of power Monte Carlo is mN, where m is determined by the convergence of the power iterations and N is the number of walks.

3.4 Restrictions of the Method and How to Avoid Them

Suppose that $\lambda_n < \lambda_{n-1}$; in this case the power method is convergent. The only possible restriction in the above scheme is the requirement $||M|| < 1$ for the convergence of the Monte Carlo inversion. This can easily be avoided using the presented bellow algorithm for inverting general type matrices using Monte Carlo:

Algorithm: Finding A^{-1}.

1. **Initial data:** Input matrix A, parameters γ and ϵ.
2. **Preprocessing:**
   2.1 **Split** $A = D - (D - A)$, where D is a diagonally dominant matrix.
   2.2 **Set** $D = B - B_1$ where B is a diagonal matrix $b_{ii} = d_{ii}$ $i = 1, 2, ..., n$.
   2.3 **Compute** the matrix $T = B^{-1} B_1$.
   2.4 **Compute** $||T||$, the Number of Markov Chains $N = (\frac{0.6745}{\epsilon} . \frac{1}{(1-||T||}) ^2$.
3. **For** i=1 to n;
   3.1 **For** j=1 to j=N;
   Markov Chain Monte Carlo Computation:
   3.1.1 **Set** $t_k = 0$(stopping rule), $W_0 = 1$, $SUM[i] = 0$ and $Point = i$.
   3.1.2 **Generate** an uniformly distributed random number $nextpoint$.
   3.1.3 **If** $T[point][nextpoint]! = 0$.
   LOOP
   3.1.3.1 **Compute** $W_j = W_{j-1} \frac{T[point][nextpoint]}{P[point][nextpoint]}$.
   3.1.3.2 **Set** $Point = nextpoint$ and $SUM[i] = SUM[i] + W_j$.
   3.1.3.3 **If** $|W_j| < \gamma$, $t_k = t_k + 1$
   3.1.3.4 **If** $t_k \geq n$, end LOOP.
   3.1.4 **End If**
   3.1.5 **Else** go to step 3.1.2.
   3.2 **End of loop j.**
   3.3 **Compute** the average of results.
4. **End of loop i.**

5. **Obtain** The matrix $V = (I - T)^{-1}$.
6. **Therefore** $D^{-1} = VB^{-1}$.
7. **Compute** the MC inversion $D^{-1} = B(I - T)^{-1}$.
8. **Set** $D_0 = D^{-1}$ (approximate inversion) and $R_0 = I - DD_0$.
9. **use filter procedure** $R_i = I - DD_i$, $D_i = D_{i-1}(I + R_{i-1})$, $i = 1, 2, ..., m$, where $m \leq k$.
10. **Consider the accurate inversion of D** by step 9 given by $D_0 = D_k$.
11. **Compute** $S = D - A$ where S can be any matrix with all non-zero elements in diagonal and all of its off-diagonal elements are zero.
12. **Main function** for obtaining the inversion of A based on D^{-1} step 9:
    12.1 **Compute** the matrices $S_i, i = 1, 2, ..., k$, where each S_i has just one element of matrix S.
    12.2 **Set** $A_0 = D_0$ and $A_k = A + S$
    12.3 **Apply** $A_k^{-1} = A_{k+1}^{-1} + \frac{A_{k+1}^{-1} S_{i+1} A_{k+1}^{-1}}{1 - trace(A_{k+1}^{-1} S_{i+1})}$, $i = k - 1, k - 2, ..., 1, 0$.
13. **Print**the inversion of matrix A.
14. **End** of algorithm.

4 Parallel Implementation and Numerical Tests

The method presented in this paper consists of 3 procedures:

- Monte Carlo inversion
- Refinement
- Power Monte Carlo iterations

We study the convergence and the parallel behaviour of each of the above procedures separately in order to have the overall picture.

It is well known that Monte Carlo algorithms have high parallel efficiency. In fact, in the case where a copy of the non-zero matrix elements of A is sent to each processor, the execution time for computing an extremal eigenvalue on p processors is bounded by $O(lN/p)$ where N is the number of performed random walks and l is the mean value of the steps in a single walk. This result assumes that the initial communication cost of distributing the matrix, and the final communication cost of collecting and averaging the distributed statistics is negligible compared to the cost of generating the Markov chains and forming the chosen statistic.

The numerical tests presented include example with 128×128 matrix, where we have applied rough Monte Carlo estimate and no refinement. In this case, the procedure for finding the largest eigenvalue of the inverse matrix using power Monte Carlo iterations, has no convergence. After that we applied a refinement procedure to find more accurate inverses with $\epsilon = 10^{-6}$ and 10^{-8}. Applying power Monte Carlo iterations to find the largest eigenvalue of the refined inverse matrix gives the value $\lambda Inv_{max} = 3.3330$ which is closed enough to the value given by the MATLAB eigenvalue procedure which is 3.3062.

In addition to convergence test, we also performed parallel computations to empirically confirm the parallel efficiency of this Monte Carlo method. A lot of parallel tests concerning Monte Carlo inversion and refinement can be found in [1], here we test the Monte Carlo part for computing the desired eigenvalue. The parallel numerical test was performed on a Compaq Alpha parallel cluster with 8 DS10 processors each running at 466 megahertz using MPI to provide the parallel calls. Each processor executes the same program for N/p trajectories (here p is the number of processors), and at the end of the trajectory computations, a designated host processor collects the results of all realizations and computes the desired average values. The results shown in *Figure* that the high parallel efficiency of Monte Carlo methods is preserved for this problem as well.

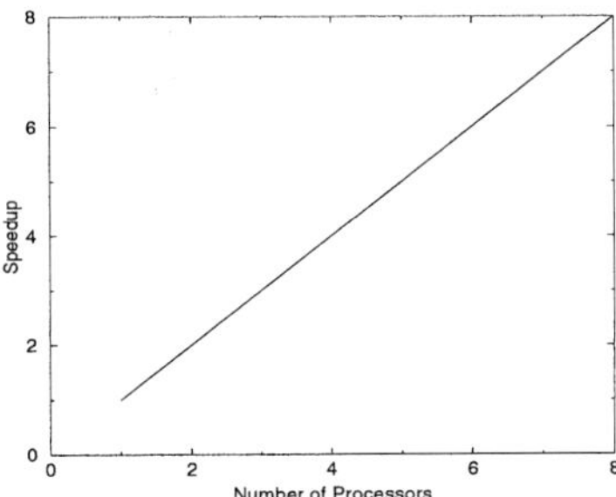

Fig. 1. Computing the largest eigenvalue of the inverse matrix - Speedup vs number of processors

5 Conclusion

The Monte Carlo methods with refinement have been proposed and extensively testes for solving systems of linear algebraic equations (see, for example, [1, 5]). In this paper we propose a method that includes Monte Carlo procedure for finding the inverse matrix, refinement procedure to improve approximation of the inverse if necessary, and Monte Carlo power iterations to compute the smallest eigenvalue. This method gives better results than the Resolvent Monte Carlo.

References

1. V. N. Alexandrov, S.Branford, and C. Weihrauch, A New Parallel Hybrid Monte Carlo Algorithm for Matrix Computations', *Proc. Of PMAA 2004*, September 2004.
2. I. Dimov, V. Alexandrov, A. Karaivanova, Resolvent Monte Carlo Methods for Linear Algebra Problems, *Mathematics and Computers in Simulations*, Vol. . **55**, 2001, pp. 25-36.

3. I. Dimov, T. Dimov, T. Gurov, A new iterative Monte Carlo Approach for Inverse Matrix Problem, *J. of Comp. and Appl. Mathematics*, **92**, 1998, pp. 15-35.
4. I. Dimov, A. Karaivanova, Parallel computations of eigenvalues based on a Monte Carlo approach, *Journal of MC Methods and Appl.*, Vol.**4**,Num.1, 1998 pp.33-52.
5. B. Fathi, B. Liu and V. Alexandrov, Mixed Monte Carlo Parallel Algorithms for Matrix Computations, *Computational Science - ICCS2002*, LNCS **2330**, pp. 609-618, 2002.
6. G. H. Golub, C.F. Van Loon, *Matrix computations*, **The Johns Hopkins Univ. Press**, Baltimore, 1996.
7. J.H. Halton, Sequential Monte Carlo Techniques for the Solution of Linear Systems, *SIAM Journal of Scientific Computing*, Vol.**9**, pp. 213-257, 1994.
8. William W. Hager, *Applied Numerical Linear Algebra*, Prentice Hall International (UK) Limited, London, 1989.
9. J.M. Hammersley, D.C. Handscomb, *Monte Carlo methods*, **John Wiley & Sons, inc.**, New York, London, Sydney, Methuen, 1964.
10. G.A. Mikhailov, *Optimization of the "weight" Monte Carlo methods*, Nauka, Moscow, 1987.
11. I.M. Sobol *Monte Carlo numerical methods*, Nauka, Moscow, 1973.

On the Scrambled Sobol´ Sequence

Hongmei Chi[1], Peter Beerli[2], Deidre W. Evans[1], and Micheal Mascagni[2]

[1] Department of Computer and Information Sciences,
Florida A&M University, Tallahassee, FL 32307-5100
hchi@cis.famu.edu

[2] School of Computational Science and Information Technology,
Florida State University, Tallahassee, FL 32306-4120

Abstract. The Sobol´ sequence is the most popular quasirandom sequence because of its simplicity and efficiency in implementation. We summarize aspects of the scrambling technique applied to Sobol´ sequences and propose a new simpler modified scrambling algorithm, called the multi-digit scrambling scheme. Most proposed scrambling methods randomize a single digit at each iteration. In contrast, our multi-digit scrambling scheme randomizes one point at each iteration, and therefore is more efficient. After the scrambled Sobol´ sequence is produced, we use this sequence to evaluate a particular derivative security, and found that when this sequence is numerically tested, it is shown empirically to be far superior to the original unscrambled sequence.

1 Introduction

The use of quasirandom, rather than random, numbers in Monte Carlo methods, is called quasi–Monte Carlo methods, which converge much faster than normal Monte Carlo. Quasi–Monte Carlo methods are now widely used in scientific computation, especially in estimating integrals over multidimensional domains and in many different financial computations.

The Sobol´ sequence [21, 22] is one of the standard quasirandom sequences and is widely used in quasi–Monte Carlo applications. The efficient implementation of Sobol´ sequence uses Gray codes. We summarize aspects of this technique applied to Sobol´ sequences and propose a new scrambling algorithm, called a multiple digit scrambling scheme. Most proposed scrambling methods [1, 8, 16, 19] randomized a single digit at each iteration. In contrast, our multi-digit scrambling scheme, which randomizes one point at each iteration, is efficient and fast because the popular modular power-of-two pseudorandom number generators are used to speed it up.

The construction of the Sobol´ sequence uses linear recurrence relations over the finite field, $\mathbb{F}_2$, where $\mathbb{F}_2 = \{0, 1\}$. Let the binary expansion of the nonnegative integer n be given by $n = n_1 2^0 + n_2 2^1 + ... + n_w 2^{w-1}$. Then the nth element of the jth dimension of the Sobol´ sequence, $x_n^{(j)}$, can be generated by

$$x_n^{(j)} = n_1 \nu_1^{(j)} \oplus n_2 \nu_2^{(j)} \oplus ... \oplus n_w \nu_w^{(j)} \tag{1}$$

V.S. Sunderam et al. (Eds.): ICCS 2005, LNCS 3516, pp. 775–782, 2005.

where $\nu_i^{(j)}$ is a binary fraction called the ith direction number in the jth dimension. These direction numbers are generated by the following q-term recurrence relation

$$\nu_i^{(j)} = a_1\nu_{i-1}^{(j)} \oplus a_2\nu_{i-2}^{(j)} \oplus ... a_q\nu_{i-q+1}^{(j)} \oplus \nu_{i-q}^{(j)} \oplus (\nu_{i-q}^{(j)}/2^q). \tag{2}$$

We have $i > q$, and the bit, a_i, comes from the coefficients of a degree-q primitive polynomial over $\mathbb{F}_2$. Note that one should use a different primitive polynomial to generate the Soboĺ direction numbers in each different dimension. Another representation of $\nu_i^{(j)}$ is to use the integer $m_i^{(j)} = \nu_i^{(j)} * 2^i$. Thus, the choice of q initial direction numbers $\nu_i^{(j)}$ becomes the problem of choosing q odd integers $m_i^{(j)} < 2^i$. The initial direction numbers, $\nu_i^{(j)} = \frac{m_i^{(j)}}{2^i}$, in the recurrence, where $i \leq q$, can be decided by the $m_i^{(j)}$'s, which can be arbitrary odd integers less than 2^i. The Gray code is widely used in implementations [4, 11] of the Soboĺ sequence.

The direction numbers in Soboĺ sequences come recursively from a degree-q primitive polynomial; however, the first q direction numbers can be arbitrarily assigned for the above recursion (equation (2)). Selecting them is crucial for obtaining high-quality Soboĺ sequences. The top pictures in both Fig. 1 and Fig. 2 show that different choices of initial direction numbers can make the Soboĺ sequence quite different. The initial direction numbers for the top picture in figure (1) is from Bratley and Fox's paper [4]; while top picture in figure (2) results when the initial direction numbers are all ones.

Soboĺ [22] realized the importance of initial direction numbers, and published an additional property (called Property A) for direction numbers to produce more uniform Soboĺ sequences; but implementations [11] of Soboĺ sequences showed that Property A is not really that useful in practice. Cheng and Druzdzel [5, 20] developed an empirical method to search for initial direction numbers, $m_i^{(j)}$, in a restricted space. Their search space was limited because they had to know the total number of quasirandom numbers, N, in advance to use their method. Jackel [10] used a random sampling method to choose the initial $m_i^{(j)}$ with a uniform random number u_{ij}, so that $m_i^{(j)} = \lfloor u_{ij} \times 2^{i-1} \rfloor$ for $0 < i < q$ with the condition that $m_i^{(j)}$ is odd.

Owing to the arbitrary nature of initial direction numbers of the sequence, poor two-dimensional projections frequently appear in the Soboĺ sequence. Morokoff and Caflisch [18] noted that poor two-dimensional projections for the Soboĺ sequence can occur anytime because of the improper choices of initial direction numbers. The bad news is that we do not know in advance which initial direction numbers cause poor two-dimensional projections. In other words, poor two-dimensional projections are difficult to prevent by trying to effectively choose initial direction numbers. Fortunately, scrambling Soboĺ sequences [8, 19] can help us improve the quality of the Soboĺ sequence having to pay attention to the proper choice of the initial direction numbers.

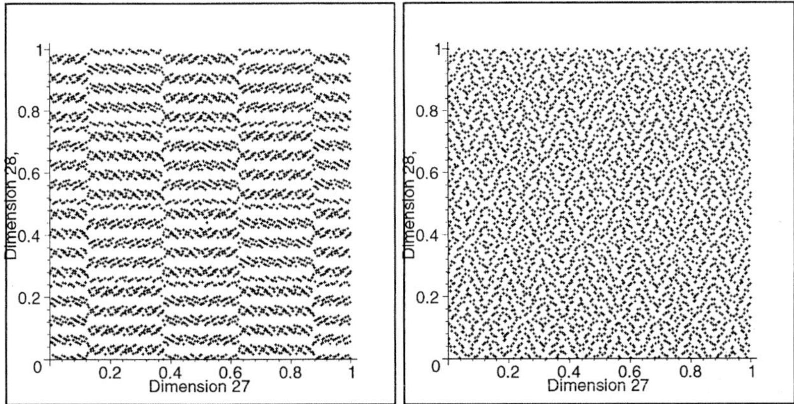

Fig. 1. Left: 4096 points of the original Sobol sequence and the initial direction numbers are from Bratley and Fox's paper [4]; right: 4096 points of the scrambled version of the Sobol sequence

2 Scrambling Methods

Recall that Sobol sequence is defined over the finite field, $\mathbb{F}_2$ [13]. Digit permutation is commonly thought effective in the finite field, $\mathbb{F}_p$. When digit permutation is used to scramble a quasirandom point over $\mathbb{F}_p$, the zero is commonly left out. The reason is that permuting zero (assuming an infinite string of trailing zeros) leads to a biased sequence in the sense that zero can be added to the end of any sequence while no other digit can. So this strategy for pure digital permutation, where zero is not changed, is not suitable for the Sobol sequence because the Sobol sequence is over $\mathbb{F}_2$. For example, we could write 0.0101 as 0.01010000 if we want to scramble 8 digits. If zero is left out, the scrambled results for 0.0101 and 0.01010000 are same. Otherwise, the bias may be introduced.

The linear permutation [8] is also not a proper method for scrambling the Sobol sequence. Let $x_n = (x_n^{(1)}, x_n^{(2)}, \ldots, x_n^{(s)})$ be any quasirandom number in $[0,1)^s$, and $z_n = (z_n^{(1)}, z_n^{(2)}, \ldots, z_n^{(s)})$ be the scrambled version of the point x_n. Suppose that each $x_n^{(j)}$ has a b-ary representation as $x_n^{(j)} = 0.x_{n1}^{(j)} x_{n2}^{(j)} \ldots x_{nK}^{(j)} \ldots$, where K defines the number of digits to be scrambled in each point. Then we define

$$z_n^{(j)} = c_1 x_n^{(j)} + c_2, \text{for } j = 1, 2, .., s, \tag{3}$$

where $c_1 \in \{1, 2, ..., b-1\}$ and $c_2 \in \{0, 1, 2, ..., b-1\}$. Since the Sobol sequence is built over $\mathbb{F}_2$, one must assign 1 to c_1 and 0 or 1 to c_2. Since the choice of c_1 is crucial to the quality of the scrambled Sobol sequence, this linear scrambling method is not suitable for the Sobol sequence or any sequence over $\mathbb{F}_2$.

As stated previously, the quality of the Sobol sequence depends heavily on the choices of initial direction numbers. The correlations between different dimensions are due to improper choices of initial direction numbers [5]. Many methods [5, 10] to improve the Sobol´ sequence focus on placing more uniformity into the initial direction numbers; but this approach is difficult to judge by any measure. We concentrate on improving the Sobol´ sequence independent of the initial direction numbers. This idea motivates us to find another approach to obtain high-quality Sobol´ sequences by means of scrambling each point.

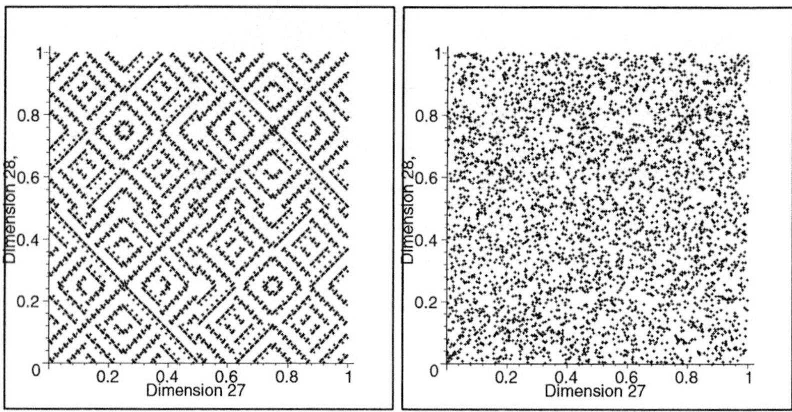

Fig. 2. Left: 4096 points of the original Sobol´ sequence with all initial direction numbers ones [23], right: 4096 points of the scrambled version of the Sobol´ sequence

3 An Algorithm for Scrambling the Sobol´ Sequence

We provide a new approach for scrambling the Sobol´ sequence, and measure the effectiveness of this approach with the number theoretic criterion that we have used in [6]. Using this new approach, we can now scramble the Sobol´ sequence in any number of dimensions.

The idea of our algorithm is to scramble k bits of the Sobol´ sequence instead of scrambling one digit at a time. The value of k could be any positive integer as long as we could find a suitable Linear Congruential Generators (LCG) for it. Assume x_n is nth Sobol´ point, and we want to scramble first k bits of x_n. Let z_n be the scrambled version of x_n. Our procedure is described as follows:

1. $y_n = \lfloor x_n * 2^k \rfloor$, is the k most-significant bits of x_n, to be scrambled.
2. $y_n^* = a y_n \pmod{m}$ and $m \geq 2^k - 1$, is the linear scrambling, applied to this integer.
3. $z_n = \frac{y_n^*}{2^k} + (x_n - \frac{y_n}{2^k})$, is the reinsertion of these scrambled bits into the Sobol´ point.

The key step of this approach is based on using LCGs as scramblers. LCGs with both power-of-two and prime moduli are common pseudorandom number generators. When the modulus of an LCG is a power-of-two, the implementation is cheap and fast due to the fact that modular addition and multiplication are just ordinary computer arithmetic when the modulus corresponds to a computer word size. The disadvantage, in terms of quality, is hard to obtain the desired quality of pseudorandom numbers when using a power-of-two as modulus. More details are given in [14, 15]. So LCGs with prime moduli are chosen in this paper.

The rest of our job is to search for a suitable and reliable LCG as our scrambler. When the modulus of a LCG is prime, implementation is more expensive. A special form of prime, such as a Merssene[1] or a Sophie-Germain prime[2], can be chosen so that the costliest part of the generation, the modular multiplication, can be minimized [15].

To simplify the scrambling process, we look to LCGs for guidance. Consider the following LCG:

$$y_n^* = ay_n \pmod{m}, \tag{4}$$

where m is chosen to be a Merssene, $2^k - 1$, or Sophie-Germain prime in the form of $2^{k+1} - k_0$, k is the number of bits needed to "scramble", and a is a primitive root modulo m [12, 7]. We choose the modulus to be a Merssene or Sophie-Germain [15] because of the existence of a fast modular multiplication algorithms for these primes. The optimal a should generate the optimal Sobol sequence, and the optimal a's for modulus $2^{31} - 1$ are tabulated in [7]. A proposed algorithm for finding such optimal primitive root modulus m, a prime, is described [6].

Primarily, our algorithm provides a practical method to obtain a family of scrambled Sobol sequences. Secondarily, it gives us a simple and unified way to generate an optimal Sobol sequence from this family. According to Owen's proof [19], after scrambling, the Sobol sequence is still a (t, s)-net with base 2. However, using our algorithm, we can begin with the worse choices for initial direction numbers in the Sobol sequence: all initial direction numbers are ones. The results are showed in Fig.2. The only unscrambled portion is a straight line in both pictures. The reason is that the new scrambling algorithm cannot change the point with the same elements into a point with different elements.

4 Geometric Asian Options

Here, we present the valuation of a complex option, which has a simple analytical solution. The popular example for such problems is a European call option on the geometric mean of several assets, sometimes called a geometric Asian option. Let K be the strike price at the maturity date, T. Then the geometric mean of N assets is defined as

$$G = (\prod_{i=1}^{N} S_i)^{\frac{1}{N}},$$

where S_i is the ith asset price. Therefore the payoff of this call option at maturity can be expressed as $\max(0,\ G - K)$. Boyle [3] proposed an analytical solution

[1]If $2^q - 1$ and q are primes, then $2^q - 1$ is a Merssene prime.

[2]If $2q + 1$ and q are primes, then $2q + 1$ is a Sophie-Germain prime.

for the price of a geometric Asian option. The basic idea is that the product of lognormally distributed variables is also lognormally distributed. This property results because the behavior of an asset price, S_i, follows geometric Brownian motion [2]. The formula for using the Black-Scholes equation [2, 9] to evaluate a European call option can be represented as

$$
\begin{aligned}
C_T &= S * Norm(d_1) - K * e^{-r(T-t)} * Norm(d_2), \qquad (5)\\
\text{with } d_1 &= \frac{ln(S/K) + (r + \sigma^2)(T - t)}{\sigma\sqrt{T-t}},\\
d_2 &= d_1 - \sigma\sqrt{T-t},
\end{aligned}
$$

where t is current time, r is a risk-free rate of interest, which is constant in the Black-Scholes world, and Norm(d_2) is the cumulative normal distribution. Because the geometric Asian option has an analytical solution, we have a benchmark to compare our simulation results with analytical solutions. The parameters used for our numerical studies are as follows:

Number of assets	N
Initial asset prices, $S_i(0)$	100, for $i = 1, 2, ..., N$
Volatilities, σ_i	0.3
Correlations, ρ_{ij}	0.5, for $i < j$
Strike price, K	100
Risk-free rate, r	10%
Time to maturity, T	1 year

The formula for computing the analytic solution for a geometric Asian option is computed by a modified Black-Scholes formula. Using the Black-Scholes formula, we can compute the call price by using equation (5) with modified parameters, S and σ^2, as follows:

$$
\begin{aligned}
S &= Ge^{(-A/2+\sigma^2/2)T}\\
A &= \frac{1}{N}\sum_{i=1}^{N}\sigma_i^2 \qquad (6)\\
\sigma^2 &= \frac{1}{N^2}\sum_{i=1}^{N}\sum_{i=j}^{N}\rho_{ij}\sigma_i\sigma_j.
\end{aligned}
$$

We followed the above formula in equation (5) and (6), computed the prices for different values of $N = 10$ and $N = 30$, with $K = 100$, and computed $p = 12.292$ and $p = 12.631$ respectively. For each simulation, we had an analytical solution, so we computed the relative difference between that and our simulated solution with the formula $\frac{|p_{qmc}-p|}{p}$, where p is the analytical solution and p_{qmc} is the price obtained by simulation. For different N, we computed p_{qmc} by simulating the asset price fluctuations using geometric Brownian motion. The results are shown in Fig.3.

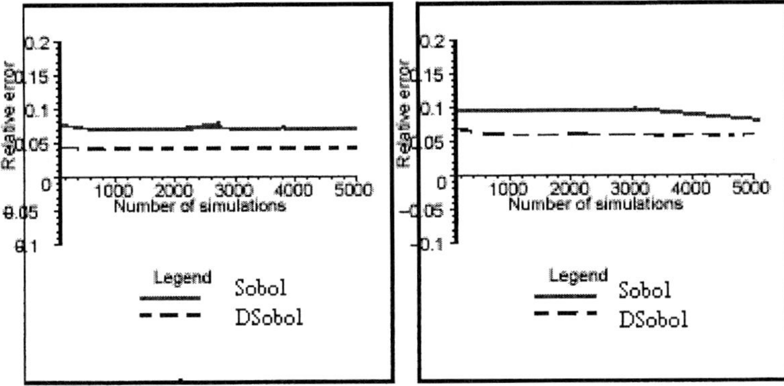

Fig. 3. Left: geometric mean of 10 stock prices; right: geometric mean of 30 stock prices. Here the label "Sobol" refers to the original Sobol´ sequence [4], while "DSobol" refers to our optimal Sobol´ sequence

From equation (5), we can see that random variables are drawn from a normal distribution. Each Sobol´ point must be transformed into a normal variable. The favored transformation method for quasirandom numbers is the inverse of the cumulative normal distribution function. The inverse normal function provided by Moro [17] was used in our numerical studies. From Fig. 3, it is easily seen that the optimal Sobol´ sequence performs much better than the original Sobol´ sequence.

5 Conclusions

A new algorithm for scrambling the Sobol´ sequence is proposed. This approach can avoid the consequences of improper choices of initial direction numbers that negatively impact the quality of this sequence. Therefore, our approach can enhance the quality of the Sobol´ sequence without worrying about the choices of initial direction numbers. In addition, we proposed an algorithm and found an optimal Sobol´ sequence within the scrambled family. We applied this sequence to evaluate a complex security and found promising results even for high dimensions. We have shown the performance of the Sobol´ sequence generated by our new algorithm empirically to be far superior to the original sequence. The promising results prompt us to use more applications to test the sequences, and to reach for more general scrambling techniques for the Sobol´ sequence.

References

1. E. Atanassov. A new efficient algorithm for generating the scrambled sobol´ sequence. In *Numerical Methods and Applications (LNCS 2542)*, pages 81–90, New York, 2003. Springer-Verlag.

2. F. Black and M. Scholes. The pricing of options and corporate liabilities. *Journal of Political Economy*, **81**:637–659, 1973.
3. P. Boyle. New life forms on the option landscape. *Journal of Financial Engineering*, **2**(3):217–252, 1992.
4. P. Bratley and B. Fox. Algorithm 659: Implementing soboĺ's quasirandom sequnence generator. *ACM Trans. on Mathematical Software*, **14**(1):88–100, 1988.
5. J. Cheng and M.J. Druzdzel. Computational investigation of low-discrepancy sequences in simulation algorithms for bayesian networks. In *Uncertainty in Artificial Intelligence: Proceedings of the Sixteenth Conference (UAI-2000)*, pages 72–81, San Francisco, CA, 2000. Morgan Kaufmann Publishers.
6. H. Chi, M. Mascagni, and T. Warnock. On the optimal Halton sequences. *Mathematics and Computers in Simulation*, To appear, 2005.
7. G. A. Fishman and L. R. Moore. An exhaustive analysis of multiplicative congruential random number generators with modulus $2^{31} - 1$. *SIAM J. Sci. Stat. Comput.*, **7**:24–45, 1986.
8. H. S. Hong and F. J. Hickernell. Algorithm 823: Implementing scrambled digital sequences. *ACM Transactions on Mathematical Software*, **29**(2):95–109, june 2003.
9. J. Hull. *Options, Future and Other Derivative Secutrities.* Prentice-Hall, New York, 2000.
10. P. Jackel. *Monte Carlo Methods in Finance.* John Wiley and Sons, New York, 2002.
11. S. Joe and F. Y. Kuo. Remark on Algorithm 659: Implementing Soboĺ's quasirandom sequence generator. *ACM Transactions on Mathematical Software*, **29**(1):49–57, March 2003.
12. D. E. Knuth. *The Art of Computer Programming, vol. 2: Seminumerical Algorithms.* Addison-Wesley, Reading, Massachusetts, 1997.
13. R. Lidl and H.Niederreiter. *Introduction to Finite Fields and Their Applications.* Cambridge University Press, Cambridge, 1994.
14. M. Mascagni. Parallel linear congruential generators with prime moduli. *Parallel Computing*, **24**:923–936, 1998.
15. M. Mascagni and H. Chi. Parallel linear congruential generators with Sophie-Germain moduli. *Parallel Computing*, **30**:1217–1231, 2004.
16. J. Matousek. On the l2-discrepancy for anchored boxes. *Journal of Complexity*, **14**:527–556, 1998.
17. B. Moro. The full monte. *Risk*, **8**(2) (February):57–58, 1995.
18. W.J. Morokoff and R.E. Caflish. Quasirandom sequences and their discrepancy. *SIAM Journal on Scientific Computing*, **15**:1251–1279, 1994.
19. A.B. Owen. Randomly permuted(t,m,s)-netsand (t,s)-sequences. *Monte Carlo and Quasi-Monte Carlo Methods in Scientific Computing*, **106** in Lecture Notes in Statistics:299–317, 1995.
20. S. H. Paskov and J. F. Traub. Faster valuation of financial derivatives. *J. Portfolio Management*, **22**(1):113–120, Fall 1995.
21. I.M. Soboĺ. On the distribution of points in a cube and the approximate evaluation of integrals. *USSR Comput. Math. and Math. Phy.*, **7**(4):86–112, 1967.
22. I.M. Soboĺ. Uniformly distributed sequences with additional uniformity properties. *USSR Comput. Math. and Math. Phy.*, **16**:236–242, 1976.
23. S. Tezuka. *Uniform Random Numbers, Theory and Practice.* Kluwer Academic Publishers, IBM Japan, 1995.

Reconstruction Algorithm of Signals from Special Samples in Spline Spaces*

Jun Xian[1] and Degao Li[2]

[1] Department of Mathematics, Zhejiang University, Hangzhou, 310027, P.R. China
mathxj@163.com
[2]Information Engineering School, Jiaxing University, Jiaxing, 314001, P.R. China

Abstract. In this paper, we introduce integral sampling and study the reconstruction of signals based on non-uniform average samples in spline subspace. By using a new method, we obtain a new reconstruction formula.

1 Introduction

In digital signal and image processing, digital communication, etc., a continuous signal is usually represented and processed by using its discrete samples. For a bandlimited signal of finite energy, it is completely characterized by its samples, and is described by the famous classical Shannon sampling theorem. However, in many real applications sampling points are not always regular and sampled value may not be values of a signal f precisely at times x_k for the inertia of the measurement aparatus. As for the signal spaces, they are not always bandlimited signal of finite energy. The problem arose initially in the design of an interferometer in which the interferogram is obtained using a continuously moving mirror, but may also have bearing in other problems in which the data are smoothed by an integrating sensor, such as a CCD array with slow response time compared to the sample integral. So we need to give the reconstruction of signals from samples of its integral. As special shift-invariant spaces and non-bandlimited spaces, spline subspaces yield many advantages in their generation and numerical treatment so that there are many practical applications for signal or image processing[1-15]. In this paper, we discuss the reconstruction of signal from samples of its integral in spline subspaces.

The outline of this paper is in the following. In Section 2, we introduce the concept of integral sampling and give the reconstruction formula from regular incremental integral samples in spline subspaces. In section 3, numerical results are included. The conclusion is given in Section 4.

* This work is supported in part by the Mathematical Tanyuan Foundation and China Postdoctoral Science Foundation.

V.S. Sunderam et al. (Eds.): ICCS 2005, LNCS 3516, pp. 783–787, 2005.

2 Reconstruction of Signal from Regular Incremental Integral Samples in Spline Subspaces

Suppose a and b are given constants that satisfy $b - a = 1$. Let

$$y_k = \int_{-\infty}^{k+b} f(t)dt, \ y_{k-1} = \int_{-\infty}^{k+a} f(t)dt, \ z_k = y_k - y_{k-1}.$$

We refer to $\{z_k\}$ as the set of regular integral samples. The problem is how to reconstruct the signal f from $\{z_k\}$ (regular incremental integral samples). We now introduce some notations and lemma that will be used in Section 2. In this paper, the Fourier transform of f is defined by $\hat{f}(\omega) = \int_{\mathbb{R}} f(x)e^{-ix\omega}dx$. The space $V_N = \{\sum\limits_{k\in Z} c_k\varphi_N(\cdot - k) : \{c_k\} \in \ell^2\}$ is spline subspace generated by $\varphi_N = \chi_{[0,1]} * \cdots * \chi_{[0,1]}$ (N convolutions), $N \geq 1$. It is well-known that the space V_N is a special shift-invariant space.

Lemma 2.1. *Let $y(t) = \int_{-\infty}^{t} f(x)dx$ and $f \in V_N$, then $y \in V_{N+1}$.*

Theorem 2.1[8]**.** *For arbitrary $f \in V_N$, we have*

$$f(x) = \sum_{k\in Z} f(k + \frac{N+1}{2})S(x-k),$$

where $\hat{S}(\omega) = \frac{\hat{\varphi}_N(\omega)}{\sum \varphi_N(k+\frac{N+1}{2})e^{-ik\omega}}$ and V_N is spline subspace generated by $\varphi_N = \chi_{[0,1]} * \cdots * \chi_{[0,1]} (N \geq 1$ convolutions).

By Theorem 2.1 and Lemma 2.1, we have the following result.

Theorem 2.2. *Let*

$$y_k = \int_{-\infty}^{k+\frac{1}{2}+1} f(t)dt, y_{k-1} = \int_{-\infty}^{k+\frac{1}{2}} f(t)dt, z_k = y_k - y_{k-1}.$$

Then for any $f \in V_N$, we have reconstruction formula

$$f(t) = \sum_{k\in Z} z_k h(t-k),$$

where h is defined by

$$h_k = h_{1k} + h_{k+1}, \hat{h}_1(\omega) = i\omega\frac{\hat{\varphi}_{N+1}(\omega)}{\sum \varphi_{N+1}(k+\frac{N}{2}+1)e^{-ik\omega}},$$

$h_k(\cdot) = h(\cdot - k)$ and $h_{1k}(\cdot) = h_1(\cdot - k)$.

Actually, Theorem 2.2 shows the reconstruction formula from regular incremental integral samples in V_N. The incremental integral samples can be regarded as special weighted samples. The interpolation function h is implementary. This will be shown in Section 3.

3 Numerical Examples

In the section, we will give some numerical tests. So we need a brief and applicable formation of sampling function h in Theorem 2.2.

In Theorem 2.2, we have relation $h_k = h_{1k} + h_{k+1}$. Taking the Fourier transform of both sides of the above equality, we have the following equality:

$$\hat{h}(\omega) = \frac{\hat{h}_1(\omega)}{1 - e^{-i\omega}}.$$

For precision, we give the following equality

$$\hat{h}(\omega) = \frac{i\omega}{1 - e^{-i\omega}} \frac{\hat{\varphi}_{N+1}(\omega)}{\sum \varphi_{N+1}(k + \frac{N}{2} + 1)e^{-ik\omega}}.$$

By the inverse Fourier transform, we can find interpolation function h.

Let $N = 3$ in our numerical tests. Then

$$\hat{h}(\omega) = \frac{i\omega}{1 - e^{-i\omega}} \frac{e^{-\frac{4}{2}i\omega}(\frac{\sin\frac{\omega}{2}}{\frac{\omega}{2}})^4}{\frac{1}{48}e^{2i\omega} + \frac{23}{48}e^{i\omega} + \frac{1}{24}e^{-i\omega} + \frac{23}{48}}$$

We will reconstruct the signal $f(x) = \varphi_3(x) + 2\varphi_3(x-1) + 3\varphi_3(x+1) \in V_3$ from its integral samples$\{z_k = \int_{k+\frac{1}{2}}^{k+\frac{3}{2}} f(t)dt\}$.

It is obvious that $suppf \subseteq [-1, 4]$. So we only think about finite sampling set, that is, $k = -1, 0, 1, 2, 3, 4$ in reconstruction formula. The sample values of the function f in integer point $\{f(k)\}_{k=-1,0,1,2,3,4}$ are marked by $*$ in Figure 1. In fact, $\{z_k = \langle f, u(\cdot - k)\rangle\}$ is the result of $\{f(k)\}_{k=-1,0,1,2,3,4}$ perturbed by noise. The noisy signal sampling points $\{z_k\}$ marked by □ and are connected by "- -" in Figure 1.

We will give the reconstruction from the weighted sampling point $\{z_k\}$. In Figure 2, the original signal is represented by continuous line. Reconstruction signal using the above mentioned algorithm is represented by "-.-".

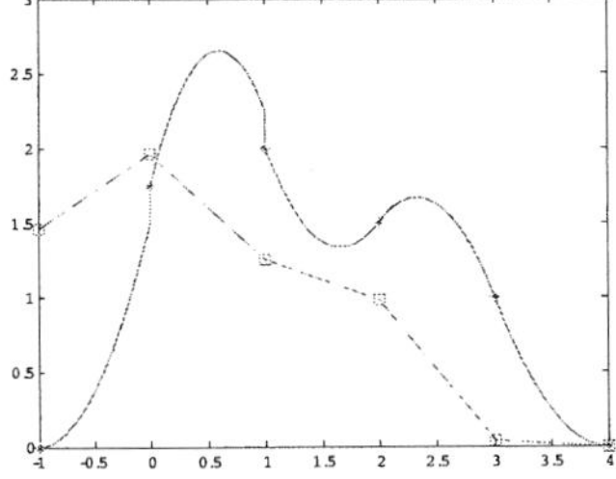

Fig. 1. The noisy signal sampling point $\{z_k\}$ marked by □ and connected by "- -".$\{f(k)\}_{k=-1,0,1,2,3,4}$ marked by "$*$"

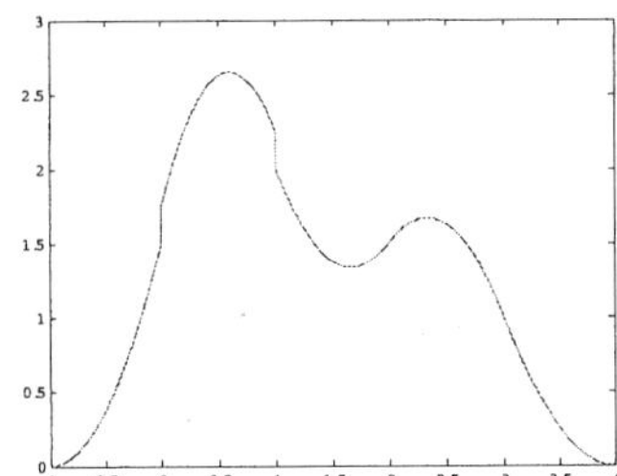

Fig. 2. The original signal is represented by continuous line. The reconstruction signal is represented by "-.-"

To measure the accuracy of the reconstruction, we compute the mean square error(MSE) by the formula $MSE = \sum_{t \in D} (f(t) - \sum_{k=-1}^{4} z_k \varphi_3(t-k))^2$, where $D = \{-1, -0.5, 0, 0.5, 1, 1.5, 2, 2.5, 3, 3.5, 4\}$. It is easy to know MSE of the function $f(t)$ is 1.5466e-15.

The graph of the reconstruction signal is almost identical to the graph of the original signal in Figure 2. MSE reflects the observation.

4 Conclusion

In this paper, we introduce the new concept of integral sampling where a signal is perfectly reconstructed from the samples of the integral of the signal. The explicit reconstruction formula is obtained in the case of regular sampling in the spline shift-invariant spaces. Finally we successfully give a numerical example to demonstrate out results. Due to the limitation of four-page papers, we omit all proofs.

References

1. Aldroubi, A., Gröchenig, K.: Beuling-Landau-type theorems for non-uniform sampling in shift invariant spline spaces. J. Fourier. Anal. Appl. **6** (2000) 93-103
2. Aldroubi, A., Gröchenig, K.: Non-uniform sampling and reconstruction in shift-invariant spaces, SIAM Rev. **43** (2001) 585-620, .
3. Aldroubi, A., Unser, M., Eden, M.: Cardinal spline filters: Stability and convergence to the ideal sinc interpolator. Singal. Processing. **28** (1992) 127-138
4. Chui, C. K.: An introduction to Wavelet, Academic Press, New York,1992
5. Jerri, A. J.: The Gibbs phenomenon in Fourier analysis, splines and wavelet approximations. Mathematics and its Applications, 446. Kluwer Academic Publishers, Dordrecht, 1998.
6. Gröchenig, K., Janssen, A., Kaiblinger, N., Norbert, P.: Note on B-splines, wavelet scaling functions, and Gabor frames. IEEE Trans. Inform. Theory. **49(12)** (2003) 3318-3320
7. Liu, Y.: Irregular sampling for spline wavelet subspaces. IEEE Trans. Inform. Theory. **42** (1996) 623-627
8. Sun, W. C., Zhou, X. W.: Average sampling in spline subspaces. Appl. Math. Letter. **15** (2002) 233-237
9. Sun, W. C., Zhou, X. W.: Reconstruction of bandlimited signals from local averages. IEEE Trans. Inform. Theory. **48** (2002) 2955-2963
10. Unser, M., Blu, T.: Fractional splines and wavelets. *SIAM Rev.* **42(1)** (2000) 43-67
11. Van De Ville, D., Blu, T., Unser, M., etc.: Hex-splines: a novel spline family for hexagonal lattices. IEEE Trans. Image Process. **13(6)** (2004) 758–772
12. Walter, G. G.: Negative spline wavelets. J. Math. Anal. Appl. **177(1)** (1993) 239-253
13. Wang, J.: Spline wavelets in numerical resolution of partial differential equations, International Conference on Wavelet Analysis and its application, AMS/IP Studies in Advanced Mathematics. **25** (2002) 257-276

14. Xian, J., Lin, W.: Sampling and reconstruction in time-warped spaces and their applications. Appl. Math. Comput. **157(1)** (2004) 153-173
15. Xian, J., Luo, S. P., Lin, W.: Improved A-P iterative algorithm in spline subspaces. Lecture Notes in Computer Science. **3037** (2004) 60-67

Fast In-place Integer Radix Sorting

Fouad El-Aker

Computer Science Department, University of Petra,
P.O. Box 940650, Amman 11194, Jordan
elaker_fouad@maktoob.com
elaker_fouad@yahoo.ca

Abstract. This paper presents two very efficient sorting algorithms. MSL is an O(*N***B*) in-place radix sorting algorithm, where *N* is the input size and *B* is the keys length in bits. This paper presents an implementation of MSL that is sub-linear in practice, for uniform data, on Pentium 4 machines. We also present an O(*N***logN*) hybrid quicksort that has a non-quadratic worst case.

1 Introduction

Right to left LSD and left to right MSD are O(*N***B*) radix sorting algorithms. *N* is the input size and *B* is the length of keys in bits. LSD and MSD use an extra space of size *N*. ALR [4] and MSL [1] process bits left to right, however unlike MSD, ALR and MSL are in-place and cache friendly. MSD, ALR and MSL execute recursively, for every partition. LSD body code is executed only *B/D* times, where *B* is the length of keys in bits and *D* is the used digit size. This makes the design of algorithms faster than LSD quite difficult. LSD is faster than MSD in [6]. We present a sub-linear run time MSL implementation suitable for sorting 31 bits and 63 bits integers in Java in this paper. MSL implementation in Section 2 uses small digit sizes increasing data cache friendliness. MSL loops were implemented reducing the number of instructions, and therefore increasing instruction cache friendliness. In addition, section 3 presents a non-quadratic implementation of quicksort, called switch sort. Hybridizing switch sort and MSL does not improve over MSL. Section 4 presents the test results. Section 5 gives the conclusions and future work.

2 MSL and Smaller Digit Sizes

MSL and ALR use a permutation loop in order to avoid reserving an extra array of size *N*, which is performed in MSD. The main steps of MSL are presented in [1]. MSL permutation loop, shuffles keys into their target groups. In the circular list of keys, $K = <K_1, K_2, .., K_L>$, assume the Target Address (K_J) = Array Location (K_{J+1}), where J is not equal to L, and Target Address (K_L) = Array Location (K_1). Digit extraction and group end address lookup are used in computing a key's target address. MSL permutation loop moves keys in a circular list K to their target addresses. K_1 is named the initial key in K, and is computed prior to the permutation loop.

V.S. Sunderam et al. (Eds.): ICCS 2005, LNCS 3516, pp. 788–791, 2005.

Many permutation loops are required in order to shuffle all keys in a group to their target addresses. ALR searches for K_1 sequentially, and preserves the property that all keys to the left of the current K_1 key are in their correct target addresses. MSL searches groups' information sequentially for the left most group, G_{Left}, which has at least one key possibly not in its target address. MSL uses the top key in G_{Left}, as K_1.

In [5], sections 4 and 5, digit size 6 was determined as appropriate for radix sorting algorithms. This is because of data cache friendliness. The importance of cache friendliness in radix sorting algorithms is emphasized in [5] and [4]. MSL also cuts to insertion sort for group sizes 20 or less, same as [4].

3 Switch Sort

Hybridized quicksort [3] implementation in this paper selects one from many pivot computations and is described in this section. Assume that we are interleaving the execution of a constant number, K, of divide and conquer algorithms whose worst cases are $f_1, f_2, .. f_K$. The list of algorithms is denoted $AL = (A_1, A_2, .. A_K)$. When A_J in AL performance is degenerate, we interrupt A_J and switch execution to the next algorithm in the circular list AL. AL worst case is equal to $K * f_W$ provided that the following conditions hold. (1) $f_W = Min\ (f_1, f_2, .. f_K)$. (2) We can determine that the current call to A_J is futile in constant time. (3) We can switch execution to the next algorithm in the circular list AL without loosing the processing done so far. If each A_J in AL executes a futile call, execution returns to A_W after circular calls to other algorithms in AL. A quicksort example is shown and described next.

```
SS(int A[], int A_Pivot, int l, int r, int min, int
max){
if ( r - l + 1 <= 10 ) insertionSort (Array, l, r );
} else {  Step 1: switch (Apply_Pivot) {
  case 0 : Pivot = max/2 + min/2 ; break;
  case 1: Pivot = Median_3 (A) ; break; }
  Step 2: Pos = partition (A, Pivot, l, r) ;
  Step 3.1:  Compute R%;
  Step 3.2: if (R < 0.05) A_Pivot = A_Pivot ^1;  // xor
  Step 4: Quicksort(A, A_Pivot, l, Pos, Pivot,
    max);
    Quicksort(A, A_Pivot, Pos+1, r,  min, Pivot);
 }
}
```

We measure balanced partitioning in quicksort to determine that the current call is futile. The partitioning ratio is defined as the size of the smaller partition divided by the size of the input group in quicksort. Let P% be the minimum acceptable partitioning ration, over all the algorithms in *AL*, equals 5% in Step 3.2. R% is the partitioning ration for the current quicksort call. When R% < P%, Step 3.2, partitioning is named degenerate or a failure. *AL* code above has only quicksort implementations, and a switch statement is used to decide which pivot computation to use, see Step 1 above. We call the algorithm switch sort (SS). Step 3.2 selects an alternative pivot computation for recursive calls. Max-Min average pivot computation in the first line in Step 1 is an adaptive implementation of taking the middle value of the input range in radix exchange [7]. Median of three quicksort passes down the actual lower partition max and the actual upper partition min. Radix exchange always divides the input range by half on recursive calls, independent of data. *AL* worst case is *O(2 * NlgN)*, where the worst case of radix exchange is *O(2 * NlgN)*.

4 Experimental Results

In Table 1, MSL run time is non-significantly sub-linear in experiments. The test data is uniform. The machine used for the displayed results is 3 GHz Pentium 4, 1GB RAM, 1MB level 2 cache, and 16 KB level 1 cache, with 400 MHz RAM speed. MSL sub-linear run time was confirmed on other Pentium 4 machines. In Table 1, add the sizes at columns headings to the sizes at each row to get the array size at a cell. Row 30M+ (30 millions+) and column +5M refer to the cell for the array size 35 millions.

In Table 1, MSL running time for array size 35 millions is 4000 milliseconds, and for array size 70 millions is 7875 milliseconds, 31 bits integers. In Table 1, the running time for array size 25 millions, is 4032 milliseconds, and for array size 50 millions, is 7735 milliseconds, for 63 bits integers.

Cutting to insertion sort is an important factor in MSL. On the other hand, we could not improve the running time of MSL by hybridizing MSL with switch sort. MSL and switch sort are compared against other algorithms in Table 2.

Table 1. MSL running times in milliseconds. Sizes are multiple of $M=10^6$

31Bits	+1M	+2M	+4M	+5M	+6M	+8M	+10M
N=0+	93	188	437	657	890	1234	1469
N=10M+	1563	1671	1875	1984	2094	2266	2500
N=20M+	2578	2678	2891	2984	3078	3281	3484
N=30M+	3594	3703	3906	4000	4125	4344	4547
N=40M+	4671	4766	5000	5079	5218	5390	5625
N=50M+	5782	5875	6062	6172	6313	6532	6734
N=60M+	6890	6953	7172	7313	7453	7656	7875
63Bits	+1M	+2M	+4M	+5M	+6M	+8M	+10M
N=0+	141	250	594	843	1125	1531	1859
N=10M+	2016	2157	2485	2594	2765	3016	3281
N=20M+	3469	3578	3859	4032	4141	4438	4735
N=30M+	4860	5016	5328	5469	5625	5891	6203
N=40M+	6375	6500	6797	6953	7125	7406	7735

LSD, digit size 8 (LSD8) is faster than LSD with digit size 16, LSD16, and other digit sizes, on the test machine. LSD processes the total keys bits. MSL processes only the distinguishing prefixes, but is recursive (section 1). In Table 2, MSL has half the run time of LSD8 for 63 bits data. In addition, MSL is better than LSD8 for larger 31 bits arrays. See size 16 and 32 millions as well as MSL sub-linear run time in Table 1. Switch sort (SS), is faster than LSD16, 63 bits longs data. Switch sort is also faster than the two algorithms, which Switch sort alternates, quicksort and Max-Min Average (MMA). Java built in tuned quicksort (JS), which is a tuned implementation of [2], is used in Table 2, instead of our own slower median of three quicksort.

Table 2. MSL running times in milliseconds. Sizes are multiple of $M=10^6$

31Bits	1/2M	1M	2M	4M	8M	16M	32M
MSL	47	93	188	437	1234	2094	3703
LSD8	47	109	234	454	938	1875	3859
LSD16	94	234	500	1031	2047	4250	8656
JS	109	234	516	1062	2219	4640	9672
SS	109	250	500	1031	2141	4500	9546
MMA	109	234	500	1031	2172	4516	9438
63Bits	1/2M	1M	2M	4M	8M	16M	32M
MSL	62	141	250	594	1531	2765	5016
LSD8	172	344	672	1328	2719	5563	10953
LSD16	250	516	1015	2031	4563	8609	18891
JS	156	329	672	1422	2969	6203	12922
SS	140	313	640	1344	2781	5860	12203
MMA	141	312	641	1360	2829	5906	12359

5 Conclusion and Future Work

MSL is a sub-linear in-place radix-sorting algorithm, for uniform data. Switch sort is a non-quadratic implementation of quicksort. Future work includes low run time algorithms and models for sorting as well as for other problems.

References

1. Al-Badarneh Amer, El-Aker Fouad: Efficient In-Place Radix Sorting, Informatica, 15 (3), 2004, pp. 295-302.
2. J. L. Bentley, and M. D. McIlroy: Engineering a Sort Function, Software-Practice and Experience, 23 (1), 1993, pp. 1249-1265.
3. F. El-Aker, and A. Al-Badarneh: MSL: An Efficient Adaptive In-place Radix Sorting Algorithm, ICCS, Part II, 2004, pp. 606-609.
4. Maus, A.: ARL: A Faster In-place, Cache Friendly Sorting Algorithm, Norsk Informatikkonferranse, NIK'2002, 2002, pp. 85-95.
5. N. Rahman and R. Raman: Adapting radix sort to the memory hierarchy, Proc. 2nd Workshop on Algorithm Engineering and Experiments, ALENEX, 2000.
6. Sedgewick, R.: Algorithms in Java, Parts 1-4, 3rd Ed., Addison-Wesley, 2003.

Dimension Reduction for Clustering Time Series Using Global Characteristics

Xiaozhe Wang[1], Kate A. Smith[1], and Rob J. Hyndman[2]

[1] Faculty of Information Technology, [2] Department of Econometrics and Business Statistics, Monash University, Clayton, Victoria 3800, Australia
{catherine.wang, kate.smith}@infotech.monash.edu.au
rob.hyndman@buseco.monash.edu.au

Abstract. Existing methods for time series clustering rely on the actual data values can become impractical since the methods do not easily handle dataset with high dimensionality, missing value, or different lengths. In this paper, a dimension reduction method is proposed that replaces the raw data with some global measures of time series characteristics. These measures are then clustered using a self-organizing map. The proposed approach has been tested using benchmark time series previously reported for time series clustering, and is shown to yield useful and robust clustering.

1 Introduction

Clustering time series has become an important topic in data mining [1,3], motivated by several research challenges including similarity searching of bioinformatics sequences. This paper focuses on "whole clustering" [2] using a variety of statistical measures to capture the time series global characteristics, which departs from the common methods of clustering time series based on distance measures applied to the actual values [1,2]. The proposed method seeks to provide a novel method for clustering time series with high dimensionality, varying lengths, and missing value. The dimension reduction is performed by a feature extraction process. These features are: trend, seasonality, serial correlation, non-linearity, skewness, kurtosis, self-similarity, and chaos. For additional dimension reduction and visualization, a self-organizing map (SOM) is used to cluster the features. A total of 15 measures are calculated (9 on the 'raw' data and 6 on the 'decomposed' data) and fed into the clustering process.

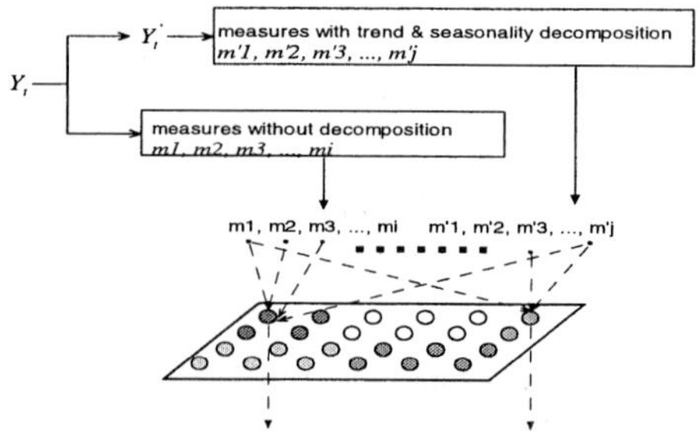

Fig. 1. Method framework with neural network SOM architecture for clustering

V.S. Sunderam et al. (Eds.): ICCS 2005, LNCS 3516, pp 792–795, 2005

2 Measuring Characteristics of Time Series

A time series is the simplest form of temporal data and is a sequence of real numbers collected regularly in time, denoted as $Y_1,\ldots,Y_n$. For each of the features, we attempted to find the most appropriate way to measure the presence of the feature, ultimately scaling the metric to (0,1) to indicate the degree of presence of the feature.

- **Trend and seasonality** are common features of time series, and it is natural to characterize a time series by its degree of trend and seasonality. Once the trend and seasonality of a time series have been measured, we can de-trend and de-seasonalize the time series to enable additional features such as noise or chaos to be more easily detected. We have used the basic decomposition model in [4]:

> - $Y^*_t = T_t + S_t + e_t$, where $Y^*_t = f_\lambda(Y_t)$, $f_\lambda(u) = (u^\lambda - 1)/\lambda$ denotes a Box-Cox transformation, T_t denotes the trend at time t and S_t denotes the seasonal component at time t.;The parameter λ is chosen to make e_t as normal as possible. Where the minimum of $\{Y_t\}$ is non-negative, we choose $\lambda \in (-1,1)$ to minimize the Shapiro-Wilk statistic.
> - For seasonal data, the decomposition uses the STL procedure with fixed seasonality. If the data are nonseasonal, $S_t = 0$ and T_t is estimated using a penalized regression spline with smoothing parameter chosen using crossvalidation; The detrended data is $Y_t' = f_\lambda^{-1}(Y_t^* - \hat{T}_t)$, the deseasonalised data is $Y_t'' = f_\lambda^{-1}(Y_t^* - \hat{S}_t)$ and the detrended and deseasonalised data is $Y_t''' = f_\lambda^{-1}(Y_t^* - \hat{T}_t - \hat{S}_t)$ where $\hat{T}_t$ and $\hat{S}_t$ denote the trend and seasonal estimates.
> - The measures of trend and seasonality are: $1 - Var(Y_t')/Var(Y_t)$ and $1 - Var(Y_t'')/Var(Y_t)$

- The **periodicity** of the seasonal pattern is used as an additional measure. We measure the period using the following algorithm:

> - Detrend time series using a regression spline with 3 knots.
> - Find $r_k = Corr(Y_t, Y_{t-k})$ (autocorrelation function) for all lags k up to 1/3 of series length. And look for peaks and troughs in autocorrelation function.
> - Frequency is first peak provided: a) there is also a trough before it, b) the difference between peak and trough is at least 0.1, c) the peak corresponds to positive correlation.
> - If no such peak is found, frequency is set to 1.

- To measure the degree of **serial correlation** of the data set, we use Q_h the Box-Pierce statistic [4] where $Q_h = n\sum_{k=1}^{h} r_k^2$.
- To measure **non-linear autoregressive structure**, nonlinear time series models have been used extensively in recent years to model complex dynamics not adequately represented using linear models [5]. We have used Teräsvirta's neural network statistic for nonlinearity [6].
- **Skewness** is a measure of lack of symmetry and **kurtosis** is a measure of a distribution's peakedness. For the standard normal distribution, the skewness and kurtosis are both zero. For univariate data Y_t, the skewness coefficient is

$S = n^{-1}s^{-3}\sum_{t=1}^{n}(Y_t-\bar{Y})^3$ and the kurtosis coefficient is $K = n^{-1}s^{-4}\sum_{t=1}^{n}(Y_t-\bar{Y})^4 - 3$, where $\bar{Y}$ is the sample mean and s is the sample standard deviation.

- Processes with **long-rang dependence** have attracted a good deal of attention from probabilists and theoretical physicists. The definition of **self-similarity** most related to the properties of time series is the self-similarity parameter or Hurst exponent (*H*). *H* is estimated using *H=d+0.5* by a class of fractional autoregressive integrated moving-average (FARIMA) processes of FARIMA(0,*d*,0) by maximum likelihood [7].
- Nonlinear dynamical systems often exhibit **chaos**, which is characterized by a Lyapunov Exponent (LE). For a one-dimensional discrete time series, we used the method by Hilborn [8] to calculate LE:

- We consider the rate of divergence of nearby points in the series by looking at the trajectories h periods ahead. Suppose Y_j and Y_i are two points such that $|Y_j-Y_i|$ is small. Then we define $\lambda(Y_i,Y_j) = h^{-1}\log|Y_{j+h}-Y_{i+h}|/|Y_j-Y_i|$.
- We estimate the Lyapunov exponent of the series by averaging these values over all i: $\lambda = n^{-1}\sum_{i=1}^{n}\lambda(Y_i,Y_j)$, where Y_j is the closest point to Y_i such that $i \neq j$ .Scaling transformations

In order to present the clustering algorithm with scaled data in the (0,1), we perform a statistical transformation of the data. For example, to map the correlation measure Q in the range $(0,\infty)$ to a scaled value q in the range (0,1) we use the transformation: $q = (e^{aQ}-1)/(b+e^{aQ})$, where a and b are constants chosen so that q satisfies the following conditions: q has 90th percentile of 0.10 when Y_t is standard normal white noise and q has value of 0.9 for a well-known benchmark data set.

3 Clustering and Experimental Evaluation

The central property of SOM is that it forms a nonlinear projection of a high-dimensional data manifold on a regular, low-dimensional (usually 2D) grid [9]. The clustered results can show the data clustering and metric-topological relations of the data items. It has a very powerful visualization output and is useful to understand the mutual dependencies between the variables and data set structure.

To determine the effectiveness of the measures in the proposed approach, we used the time series clustering benchmark datasets, "Reality check Dataset" (www.cs.ucr.edu/~eamonn/TSDMA). The data contains 14 time series of 1000 points in each which normalized into (0,1). Then 15 measures extracted from the data are used as inputs to the SOM. To compare with the benchmarking clusters generated using hierarchical clustering of the raw data points, we re-interpreted the clusters generated by the SOM into a hierarchical structure in (Fig. 2 right). Compared to the hierarchical clustering result (Fig. 2 left) [2], similar clusters have been obtained from our approach. But our clustering results are arguable better, or at least more intuitive. For example, series 1&4 and 9&10 have been grouped far from each other based on the hierarchical clustering using actual points (Fig. 2 left), but a visual inspection of these series shows that they are actually quite similar in character. Using the global measures as inputs, the clustering algorithm is aware of the "whole picture" and recognizes the similarity of these four time series (Fig. 2 right).

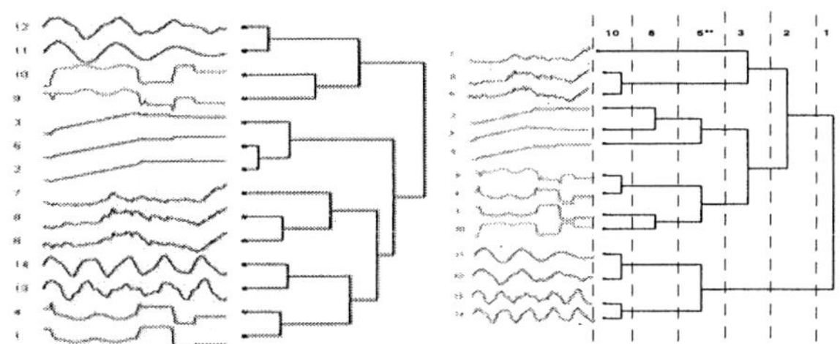

Fig. 2. Clustering results using hierarchical representation (left) and SOM (right)

4 Conclusions and Future Research

In this paper, we have proposed a new method for time series clustering and compared the results to a benchmarked time series clustering data. The empirical results demonstrate that our proposed clustering approach is able to cluster time series with high dimensionality, varying lengths, and missing value. Using only a finite set of global measures as input for clustering, we can still achieve useful clustering. In fact, the knowledge provided to the clustering algorithm by the global measures appears to benefit the quality of the clustering results. In future, a more comprehensive metrics to summarize the time series features will be explored and applied to more datasets.

References

1. Bradley, P. S., Fayyad, U. M.: *Refining Initial Points for K-means Clustering.* In: the 15th international conference on machine learning, Madison, (1998).
2. Keogh, E., Lin, J., Truppel, W.: *Clustering of Time Series Subsequences is Meaningless: Implications for Past and Future Research.* In: the 3rd IEEE International Conference on Data Mining, Melbourne.
3. Wang, C., & Wang, X. S.:*Supporting Content-based Searches on Time Series via Approximation.* In: the 12th international conference on scientific and statistical database management, Berlin, (2000).
4. Makridakis, S., Wheelwright, S. C., Hyndman, R. J.: Forecasting Methods and Applications (3rd Ed.): John Wiley & Sons, Inc., (1998).
5. Harvill, J. L., Ray, B. K., Harvill, J. L.: *Testing for Nonlinearity in a Vector Time Series.* Biometrika, 86, (1999), 728-734.
6. Blake, A. P., Kapetanios, G.: A Radial Basis Function Artificial Neural Network Test for Neglected Nonlinearity. The Econometrics Journal, 6(2), (2003), 357-373.
7. Haslett, J., & Raftery, A. E.: *Space-Time Modelling with Long-Memory Dependence: Assessing Ireland's Wind Power Resource* (With Discussion). Applied Statistics, 38, (1989), 1-50.
8. Hilborn, R. C.: *Chaos and Nonlinear Dynamics: an Introduction for Scientists and Engineers.* New York: Oxford University Press, (1994).
9. Kohonen, T.: Self-Organizing Maps (Vol. 30): Springer Verlag, (1995).

On Algorithm for Estimation of Selecting Core*

Youngjin Ahn[1], Moonseong Kim[1],
Young-Cheol Bang[2], and Hyunseung Choo[1]

[1] School of Information and Communication Engineering,
Sungkyunkwan University, 440-746, Suwon, Korea +82-31-290-7145
watchman@skku.edu
{moonseong, choo}@ece.skku.ac.kr
[2] Department of Computer Engineering,
Korea Polytechnic University, 429-793, Gyeonggi-Do, Korea +82-31-496-8292
ybang@kpu.ac.kr

Abstract. With the development of the multicast technology, the real-time strategy among the group applications using the multicast routing is getting more important. An essential factor of these real-time applications is to optimize the Delay- and delay Variation-Bounded Multicast Tree (DVBMT) problem. In this paper, we propose Estimation of Selecting Core (ESC) algorithm for DVBMT solution. Furthermore, it is expected that ESC algorithm results in the better any other DVBMT solutions.

1 Introduction

For the QoS transmission, not only the tree cost as a measure of bandwidth efficiency is one of the important factors, but also networks supporting real-time traffic are required to receive messages from source node in a limited amount of time. Therefore, we should consider the multicast end-to-end delay and delay variation problem [2]. In this paper, we study the delay variation problem under the upper bound on the multicast end-to-end delay. We propose an efficient algorithm in comparison with the Delay and Delay Variation Constraint Algorithm (DDVCA [3]), known as the best algorithm so far. Even if the time complexity of our algorithm is same as the one of DDVCA, the proposed algorithm would have the better performance than DDVCA has in terms of the multicast delay variation. The rest of the paper is organized as follows. In Section 2, we state the network model for the multicast routing, and the problem formulation. Section 3 presents the details of the proposed algorithm. Finally, section 4 concludes this paper.

2 Network Model for Multicasting and DVBMT

We consider that a computer network is represented by a directed graph $G = (V, E)$ with n nodes and l links or arcs, where V is a set of nodes and E is a set

* This work was supported in parts by Brain Korea 21 and the Ministry of Information and Communication in Republic of Korea. Dr. H. Choo is the corresponding author.

V.S. Sunderam et al. (Eds.): ICCS 2005, LNCS 3516, pp. 796–800, 2005.

of links, respectively. Each link $e = (i,j) \in E$ is associated with delay $d(e) \geq 0$. The delay of a link, $d(e)$, is the sum of the perceived queueing delay, transmission delay, and propagation delay. We define a path as sequence of links such that (u,i), (i,j), $\ldots$, (k,v), belongs to E.

Let $P(u,v) = \{(u,i),\ (i,j),\ \ldots,\ (k,v)\}$ denote the path from node u to node v. If all $u,\ i,\ j,\ \ldots,\ k,\ v$ are distinct, then we say that it is a simple directed path. For a given source node $s \in V$ and a destination node $d \in V$, $(2^{s \rightarrow d}, \infty)$ is the set of all possible paths from s to d.

$$(2^{s \rightarrow d}, \infty) = \{\ P_k(s,d) \mid \text{all possible paths from } s \text{ to } d,\ {}^{\forall}s,\ d \in V,\ {}^{\forall}k \in \Lambda\ \}$$

where Λ is an index set. The path-delay of P_k is given by $\phi_D(P_k) = \sum_{e \in P_k} d(e)$, ${}^{\forall}P_k \in (2^{s \rightarrow d}, \infty)$. $(2^{s \rightarrow d}, \Delta)$ is the set of paths from s to d for which the end-to-end delay is bounded by Δ. Therefore $(2^{s \rightarrow d}, \Delta) \subseteq (2^{s \rightarrow d}, \infty)$.

For the multicast communications, messages need to be delivered to all receivers in the set $M \subseteq V \setminus \{s\}$ which is called the multicast group, where $|M| = m$. The path traversed by messages from the source s to a multicast receiver, m_i, is given by $P(s, m_i)$. Thus multicast routing tree can be defined as $T(s,M) = \bigcup_{m_i \in M} P(s, m_i)$ and the messages are sent from s to M through $T(s,M)$.

The multicast end-to-end delay constraint, Δ, represents an upper bound on the acceptable end-to-end delay along any path from source node to a destination node. The multicast delay variation, δ, is the maximum difference between the end-to-end delays along the paths from the source to any two destination nodes.

$$\delta \ = \ max\{\ |\phi_D(P(s,m_i)) - \phi_D(P(s,m_j))|,\ {}^{\forall}m_i, m_j \in M,\ i \neq j\ \}$$

The issue defined and discussed in [2], initially, is to minimize multicast delay variation under multicast end-to-end delay constraint. The authors referred to this problem as Delay- and delay Variation-Bounded Multicast Tree (DVBMT) problem. The DVBMT problem is to find the tree that satisfies

$$min\{\ \delta_\alpha \mid {}^{\forall}P(s,m_i) \in (2^{s \rightarrow m_i}, \Delta),\ {}^{\forall}P(s,m_i) \subseteq T_\alpha,\ {}^{\forall}m_i \in M\ {}^{\forall}\alpha \in \Lambda\ \},$$

where T_α denotes any multicast tree spanning $M \cup \{s\}$.

3 The Proposed ESC Algorithm

Description of ESC Algorithm: We designate the proposed algorithm as Estimation of Selecting Core (ESC). ESC algorithm gives an explicit solution about the DVBMT problem. We define the $MODE$ function, since the location of the core node influences the multicast delay variation. In addition to the $MODE$ function, we measure the delay variation for each mode, using the CMP function. The $MODE$ and CMP functions are deployed as follows.

$$MODE(c) = \begin{cases} \text{I} & if \;\; c = s \\ \text{II} & if \;\; ^{\exists}m \text{ in } P(s,c), \text{ where } ^{\forall}m \in M \\ \text{III} & if \;\; ^{\exists}s \text{ in } P(m,c), \text{ where } ^{\forall}m \in M \\ \text{IV} & if \;\; \text{II and III} \\ \text{V} & otherwise \end{cases},$$

where s is source node.

$$CMP(x) = \begin{cases} |(ds_{core} + max\_delay) - ds_{m_k}| & if \;\; x \; = \; \text{II or III or IV} \\ dv_{core} & otherwise \end{cases},$$

where $core = MODE^{-1}(x)$, $max\_delay = max\{min\{\phi_D(P(core, m))\} \mid {}^{\forall}m \in M \backslash \{m^* \in M \mid {}^{\exists}m^* \text{ in } P(s, core) \bigvee {}^{\exists}s \text{ in } P(m^*, core)\}\}$. ds_{v_i} and dv_{v_i} are defined in Fig. 1.

```
Proposed Algorithm (G(V,E), M, s, Δ)
Input:  A directed graph G(V,E), M is the multicast group with m = |M|,
        a source node s, a end-to-end delay bound Δ.
Output: The multicast tree T such that φD(P(s,mi)) ≤ Δ,
        ∀P(s,mi) ⊆ T, ∀mi ∈ M, and has a small multicast delay variation.
01. Begin
02. dvmin = ∞; core = ∅; candidate = ∅; compare = ∞; T = ∅;
    /* candidate : candidates of a core node */
    /* compare   : measure for the 5 modes   */
03. For ∀vi ∈ V Do
04.     dsvi = min{φD(P) | ∀P ∈ (2^(s→vi), ∞)};
05.     ds_predvi = v s.t. (v,vi) ∈ P(s,vi) with dsvi;
06. For ∀mk ∈ M Do
07.     ppd(mk,vi) = min{φD(P) | ∀P ∈ (2^(mk→vi), ∞)}, ∀vi ∈ V;
08.     dest_pred(mk,vi) = v s.t. (v,vi) ∈ P(mk,vi) with ppd(mk,vi), ∀vi ∈ V;
09. For ∀vi ∈ V Do
10.     maxi = max{ppd(mk,vi) | ∀mk ∈ M};
11.     mini = max{ppd(mk,vi) | ∀mk ∈ M};
12.     dvvi = maxi − mini;
13.     If dvvi < dvmin and dsvi + maxi ≤ Δ
14.         then dvmin = dvvi; core = vi
15. If core = ∅ then ∄T
16. For ∀vi ∈ V Do
17.     If dvvi = dvcore and dsi + maxi ≤ Δ
18.         then candidate = candidate ∪ vi;
19. For ∀c ∈ candidate Do
20.     If CMP(MODE(c)) < compare
21.         then compare = CMP(MODE(c)); core = c;
22. For ∀mk ∈ M Do
23.     T = T ∪ P(core,mk);
24. T = T ∪ P(s,core);
25. Return T
26. End Algorithm.
```

Fig. 1. Pseudo Code of ESC Algorithm

In Fig. 1, Lines 9-21 gives an account of the process to search the suitable core node. Giving a specification, Line 9-15 picks out the minimum dv_{v_i}, comparing the difference between the maximum delay variation and the minimum delay variation for each node, and verifying the upper bound Δ. If there isn't such a core node, the algorithm will be terminated because the multicast tree cannot be constructed within Δ. In Lines 16-18, ESC algorithm stores the candidate cores, after looking for the node which has the same value as minimum dv_{v_i}. Lines 19-21 finds out the best core that has the minimum delay variation, using the measure CMP for the 5 modes. These cases are described as follows. The total time complexity of the proposed algorithm is $O(mn^2)$, being equal to DDVCA.

Five-Mode Core Selection: As above functions $MODE$ and CMP, $MODE$ I is that the core node corresponds to the source node. In this case, CMP keeps dv_{core}. Otherwise, $MODE$s II~V are executed.

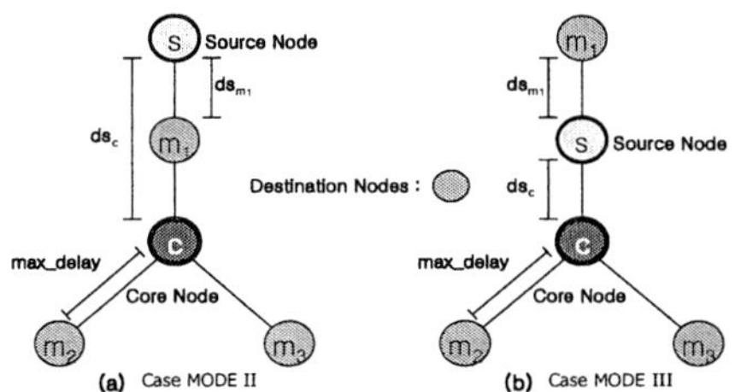

Fig. 2. Main Idea of $Mode$ Function

In the case of $MODE$ II [1], Fig. 2 (a) shows that there exists the destination node from the source node to the core node. CMP which can be computed stores the sum of the delay from the core node to the destination node and max_delay. The third case just satisfies $MODE$ III. In this case, ESC algorithm initially adds the delay from the source node to the core node to max_delay and subtracts the delay from the source node to the adjacent destination node, and stores the value with CMP. If the value is negative, CMP accepts the absolute value of it. The fourth case, MODE which satisfies II and III should find the factor to determine CMP, having situation that the destination nodes exist around the source node. The measure of the comparison is determined with the minimum delay from the source to the associated destinations. We can decide this situation because the shortest delay value from the source node affects the delay variation of the created tree. Therefore, ESC algorithm selects either $MODE$ II or III. In $MODE$ V, if ESC algorithm doesn't satisfy $MODE$s I~IV, CMP stores dv_{core}.

4 Conclusion

In this paper, we consider the transmission of a message that guarantees certain bounds of the end-to-end delays as well as the multicast delay variations computer network. The time complexity of ESC algorithm is $O(mn^2)$, which is the same as that of DDVCA. Furthermore, it is expected that ESC algorithm results in the better minimum multicast delay variation than DDVCA.

References

1. M. Kim, Y.-C. Bang, and H. Choo, "Efficient Algorithm for Reducing Delay Variation on Bounded Multicast Trees," Springer-Verlag Lecture Notes in Computer Science, vol. 3090, pp. 440-450, September 2004.

2. G. N. Rouskas and I. Baldine, "Multicast routing with end-to-end delay and delay variation constraints," IEEE JSAC, vol. 15, no. 3, pp. 346-356, April 1997.
3. P.-R. Sheu and S.-T. Chen, "A fast and efficient heuristic algorithm for the delay- and delay variation bound multicast tree problem," Information Networking, Proc. ICOIN-15, pp. 611-618, January 2001.

A Hybrid Mining Model Based on Neural Network and Kernel Smoothing Technique

Defu Zhang, Qingshan Jiang, and Xin Li

Department of Computer Science, Xiamen University, 361005, China
dfzhang@xmu.edu.cn

Abstract. Neural networks as data mining tools are becoming increasingly popular in business. In this paper, a hybrid mining model based on neural network and kernel smoothing technique is developed. The kernel smoothing technique is used to preprocess data and help decision-making. Neural network is employed to predict the long trends of stock price. In addition, some trading rules involving trading decision-making are considered. The China Shanghai Composite Index is as case study. The return achieved by the hybrid mining model is four times as large as that achieved by the buy and hold strategy, so the proposed model is promising and certainly warrants further research.

1 Introduction

In this era of information blast, people meet huge amount of information everyday and are drowning in information. Information will provide huge profits for people if information can be utilized and processed, otherwise, it will become the burden of people. The goal of data mining is to bring the practice of information processing closer to providing the real answer and decision-making for the different investors and organizations. However, traditional data mining tools have been very difficult to meet the need of people. People have to find new computational models to get useful information from huge amount of data.

Neural networks (NNs) are a class of generalized nonlinear nonparametric models inspired by studies of the human brain. They are robust and have good learning and generalization capabilities [1], and are appropriate for clustering and prediction problems. For a more detailed description of NNs, the interested readers are referred to the papers in [2-4]. Due to NNs can mine valuable information from a mass of history information, so they have become one of the most efficient and useful mining tools [1, 5-8]. Especially, some researches [8-10] have shown that NNs performed better than conventional statistic approaches in financial forecasting. Despite NNs as data mining tool have many advantages, however, they still have some drawbacks, for example, overfitting and poor explanation capability and so on, which significantly affect the performance of NNs. In order to enhance the forecasting capability of NNs, many researchers have improved NNs by combining other techniques [11]. In this paper, the kernel smoothing technique, which is use to filter 'noise' and help decision-making, is combined with NNs to develop a mining model. The actual results show that this model is efficient, and some interesting results are obtained.

V.S. Sunderam et al. (Eds.): ICCS 2005, LNCS 3516, pp. 801 – 805, 2005.

2 The Hybrid Mining Model

Many researches [12] have shown that the closing price and the trading volume are the most important factors that affects stock market, so the two data are selected to predict the future trends of stock market. Due to the original financial data are very complex, and generally contain noise. In order to make model more effective, these data are usually preprocessed before training. In this paper, a kernel smoothing technique [12] is used to filter noise. In detail, for a time series $Y: y_1, \cdots, y_n$, a smoothed time series $\hat{Y}: \hat{y}_1, \cdots, \hat{y}_n$ is calculated as following:

$$\hat{y}_i = \frac{\sum_{t=1}^{n} K_h(i-t) y_t}{\sum_{t=1}^{n} K_h(i-t)}, i = 1,2,\cdots,n, \; K_h(x) = \frac{1}{h\sqrt{2\pi}} e^{-\frac{x^2}{2h^2}}.$$

In addition, due to the activation function is a sigmoid function that squashes input data to [0,1], so the two data are scaled in [0.1,0.9] respectively.

The network architecture is shown in Fig.1.

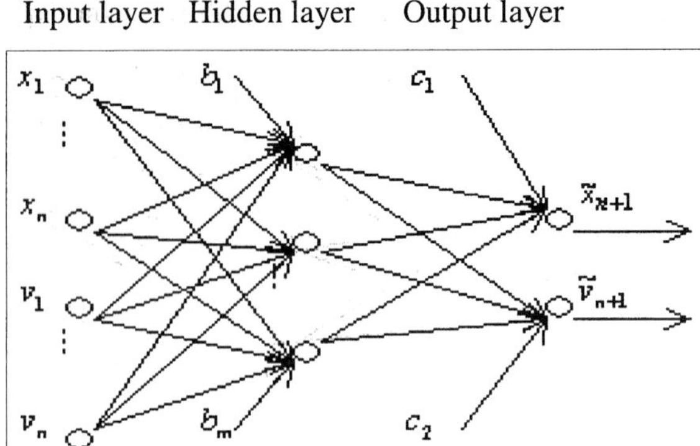

Fig. 1. A network architecture used by this paper

In order to make mode more effective, the following mean squared error is used.

$$MSE = \frac{1}{2} \sum_{i=1}^{N} \alpha (\tilde{x}_{i+n} - x_{i+n})^2, \tag{1}$$

where $\alpha = \begin{cases} \beta & if\, (\tilde{x}_{i+n} - x_{i+n-1})(x_{i+n} - x_{i+n-1}) > 0 \\ 2-\beta & else \end{cases}$, $\beta \in (0,1)$, N is the total number of training pattern pairs, and $\tilde{x}_{i+n}$ is the computed output, x_{i+n} is the target output. (1) denotes that the incorrect predicted directions are penalized more heavily than the correct predicted directions.

In order to avoid the poor explanation capability of NNs and give valuable suggestions for our investment decision-making, some trading rules are considered. The trading rules 1 is as follows:

$$\text{The trading signs}=\begin{cases}\text{Buy}, & \text{if } R_{BP} > R_{BT}, \\ \text{Sell}, & \text{if } R_{SP} < R_{ST}, \\ \text{Do nothing, otherwise.} & \end{cases}$$

where $R_{BP} = \dfrac{\max\limits_{1\le i\le 30} (\tilde{x}_{n+i}) - x_n}{x_n}$, $R_{SP} = \dfrac{\min\limits_{1\le i\le 30} (\tilde{x}_{n+i}) - x_n}{x_n}$, R_{SP} and R_{BP} are different constants and denote the return of predict selling and buying respectively.

All the seasoned investors know the stock trend is very important to make money, and the past trend will affect the future trend. Therefore, we obtain the trading rules 2 by combining the past trends and the trading rule 1. Namely, the trading signs are sell if $R_{SP} < R_{ST}$ and the trend in the past two days is down. The trend can be obtained by the kernel smoothing techniques in the process of data preprocessing. We compare the return of the trading rule 1 with the return of the trading rule 2 (see Fig.2), the computational results have shown that the trading rule 2 is more effective. The mining model can tell us when to buy or sell the stock according to these rules.

3 Computational Results and Conclusions

The hybrid mining model was implemented in C++ for Dos on a PC. The performance of this model was evaluated by trading the China Shanghai Composite Index from May 1996 to 15 Dec. 2003. The training patterns and test patterns are the data of 300 and 200 trading days respectively. Other parameters are designed as follows: $n = 20, m = 5, R_{BT} = 0.06, R_{ST} = 0.041, \beta = 0.4$. The initial weights and thresholds are in [-0.5, 0.5]. All the training patterns are selected randomly to train network about 3000 times, then all the testing patterns are tested. The process is repeated until the stop criteria are met.

The comparisons of returns between the mining model and the buy and hold strategy can be seen in Fig.2. All returns were calculated after taking the actual transaction costs for each transaction into consideration, namely the transaction costs for buying and selling is 1% respectively. For Fig.2, the return of the mining model with the trading rule 1 and 2 is 3.12 and 4.33 respectively, and the return of the buy and hold strategy is 1.31. The return of the trading rule 2 is about four times as large as that of the buy and hold strategy. So the mining model is superior to the buy and hold strategy. In addition, from Fig.2, it can know that the presented mining model performs better in bear market than in bull market, it denotes that the mining model has better capability of controlling risk. What is more, we find an interesting result, namely, the performance by training neural network every 150 days is better than that by training neural network every day. It directly improves the computational speed of the hybrid model.

The results of trading about seven years to Shanghai Composite Index have shown that the mining model was encouraging. This mining model can have actual application

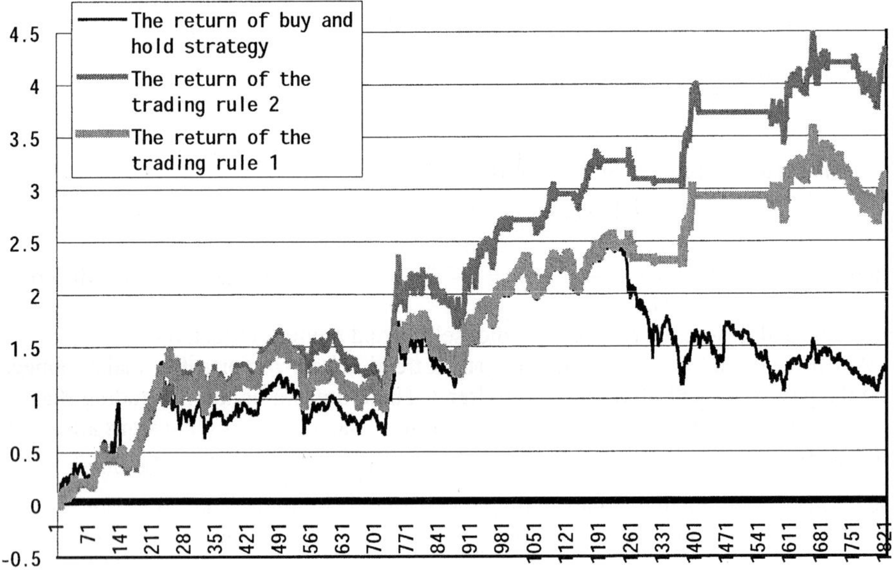

Fig. 2. Comparisons of returns between mining model and buy and hold strategy

and is very efficient for combinational investment. The future work is to strive for making the model more adaptive to the application and applying this model to other market and individual stocks.

References

1. Sushmita Mitra, Sankar K. Pal, and Pabitra Mitra: Data Mining in Soft Computing Framework: A Survey. IEEE Transactions on Neural Networks (2002) 13(1): 3–14
2. Hertz, J. Krogh, A., and Palmer R.G.: Introduction to the theory of neurocomputing. Addison-Wesley, Reading, MA (1991)
3. Widrow, B.; Rumelhart, D.E.; and Lehr, M.A.: Neural networks: applications in industry, business and science. Communications of the ACM (1994) 37(3): 93–105
4. Kate A.Smith, Jatinder N.D. Gupta: neural networks in business: techniques and applications for the operations researcher. Computers & Operations Research (2000) 27:1023-1044
5. A. J. Chapman: Stock market trading systems through neural networks: Developing a model. International Journal of Applied Expert Systems (1994) 2(2): 88-100
6. Hoptroff R.: The principles and practice of time series forecasting and business modelling using neural nets. Neural Computing & Applications (1993) 1: 59–66
7. Zhang G, Patuwo B, Hu M: Forecasting with artificial neural networks: the state of the art. International Journal of Forecasting (1998) 14: 35–62
8. Refenes,A.N., A.Zapranis, and G. Francis: Stock performance modeling using neural network: A comparative study with regression models. Neural Network (1994) 5: 961-970
9. Y.S.Abu-Mostafa, A.F. Atiya, M. Magdon-Ismail, and H. White: Neural networks in financial engineering. IEEE Transactions on Neural Networks (2001) 12(4): 653-656

10. Saad E, Prokhorov E, Wunsch D.: Comparative study of stock trend prediction using time delay, Recurrent and probabilistic neural networks. IEEE Transactions on Neural Networks, (1998) 9:1456–70
11. Paul G. Harrald and Mark Kamstra: Evolving Artificial Neural Networks to Combine Financial Forecasts. IEEE Transactions on Evolutionary Computation 1997 1(1): 40-52
12. 12.Andrew W. Lo, Harry Mamaysky, Jiang Wang: Foundations of technical Analysis: Computational algorithms, statistical inference, and empirical implementation. http://www.nber.org/papers/w7613, National Bureau of Economic Research (2000)

An Efficient User-Oriented Clustering of Web Search Results

Keke Cai, Jiajun Bu, and Chun Chen

Deptartment of Computer Science, Zhejiang University,
Hangzhou, China, 310027
{caikeke, bjj, chenc}@zju.edu.cn

Abstract. As a featured function of search engine, clustering display of search results has been proved an efficient way to organize the web resource. However, for a given query, clustering results reached by any user are totally identical. In this paper, we explored a user-friendly clustering scheme that automatically learns users' interests and accordingly generates interest-centric clustering. The basis of this personal clustering is a keyword based topic identifier. Trained by users' individual search histories, the identifier provides most of personal topics. Each topic will be the clustering center of the retrieved pages. The scheme proposed distinguishes the functionality of clustering from that of topic identification, which makes the clustering more personal and flexible. To evaluate the proposed scheme, we experimented with sets of synthetic data. The experimental results prove it an effective scheme for search results clustering.

1 Introduction

For the overwhelming information on Internet, conventional keyword-based Internet search usually return hundreds of thousands of retrieved results, most of which are irrelevant to what enquired. To solve this problem, automatic clustering technique has been proposed. Its main idea is to group retrieved pages by common topics, which is usually evaluated by the relevance among pages returned. Some schemes [1], [2] compare the direct or indirect hyperlinks of pages to gauge their relevance, whereas others, such as [3], utilize common keywords contained in different pages. These methods enable effective identification of retrieved results, however most of them rely particularly on retrieved web pages themselves to cluster. Consequently, for a given query the clustering results are always static.

Recently, user-oriented personal service has been widely deployed in the domain of search engine. A clustering scheme called content ignorant was invented [4], [5], which performs the clustering of queries beforehand for pages clustering to build upon. Though it provides a dynamic and rapid response for web surfing, its expansibility restrains because it can only establish mappings between queries and URLs that have already been preprocessed. In this paper, we explore a user-friendly clustering scheme that automatically learns user' interests and simultaneously generates interest-centric clustering. Differed from conventional pages clustering schemes, our clustering is designed to make the clustering more individual but not to provide user the most detailed clustering. The basis of this personal clustering is a keyword-based topic identifier, which is fully trained by user's individual search histories. When a

V.S. Sunderam et al. (Eds.): ICCS 2005, LNCS 3516, pp. 806–809, 2005.

user submits a query to search engine, it will automatically generate a set of topics with individual characteristics, which will be utilized in later page clustering.

2 Personal Clustering Scheme

The proposed clustering model consists of three basic components: Personal Topic Identifier, Query Executor and Personal Clustering Generator. Personal topic identifier recognizes personal topics of each user; Query executor processes user's current query and returns retrieved web pages; Personal clustering generator constructs the final clustering display of the retrieved web pages. Among these, personal topic identifier is the kernel. It monitors and records user's behaviors, and then denotes each visited page as a weighted keyword vector. It is believed that users' interests may lead to frequent visit to web pages of specific topics, which correspondingly causes the frequent co-occurrence of specific keywords. In this paper, such frequent keyword sets are regarded as the personal topics.

Equation 1 represents weight computation of term t_j to page p_i. It is a modification of TF-IDF weighting scheme [6]. Terms with weight more than system-predefined threshold will be selected as page-related keywords. Influence factor $log(N/n_{i,j})$ reflects the frequency of t_j in the condition of all web pages. $f_{i,j}$ is the normalized frequency of term t_j in web page p_i. Parameter α reflect the position of term t_j in page p_i, and it is more than 1 if *tj* appears in the title of p_i, otherwise it is 1.

$$W_{i,j} = \alpha * f_{i,j} * log(N/n_{i,j}) . \quad (1)$$

Considering the number restriction of clusters as well as the requirement of the minimal overlap among clusters, we modify the CLOSET+ [7] frequency mining algorithm to finish the process of personal topics identification. To avoid serious overlap between closed keyword sets, we refine the results set so that if two keyword sets are similar enough, they are emerged. Equation 2 is the similarity definition between two keywords set ks_1 and ks_2 by cosine distance.

$$sim(ks_1, ks_2) = ks_1 \cdot ks_2 \,/\, \| ks_1 \| \, \| ks_2 \| . \quad (2)$$

Based on the identified clustering topics as well as the retrieved query results, the finial clustering can be easily realized through our defined keywords-based cluster identifier. For a flat set of clustering $C = \{C_1 \ldots C_n\}$, it consists of a set of rules $f_i \rightarrow C_i$, i = 1...n. f_i is a frequent keyword set, which represents the corresponding topic information of C_i. A web page p is considered belonging to the C_i, if p matches the rule of $f_i \rightarrow C_i$ Function *ovp*() in equation 3 measures the match degree between pages and clusters, and the domain of its value ranges between 0 and 1. *kp* represents the keyword vector of page p. We cluster p to C_i if their overlap is no less than the predefined threshold *minSim*. It is possible that some retrieved cannot be mapped to any cluster. In our paper, all these pages will be assigned to a separated cluster.

$$ovp(p, C_i) = |kp \cap f_i| \,/\, |f_i| . \quad (3)$$

Table 1 shows the personal precision of three kinds of queries under personal clustering (*PC*) and normal clustering (*NC*). For clustering *PC* and *NC*, their personal precision *Personal* is defined as a *cosine* distance against *MC*, which is another clustering results obtained through manual definition. As show in Table 1, our personal clustering method presents satisfactory adaptability in various queries, and its personal precision averagely reach about 83 percent. It shows that this clustering method can perfectly capture any topic information mostly concerned by user, and accordingly form a set of user-topic related clusters.

$$Personal(AC) = \frac{1}{|MC|} \sum_{AC_i \in AC} max_{AC_i}(sim(MC_j, AC_i)). \tag{4}$$

Table 1. Personal precision comparisons

Data Source	Pages Number	Clusters Number			Personal Precision (percent)	
		MC	*PC*	*NC*	*PC*	*NC*
Q1	99	4	3	8	89.2	57.8
Q 2	465	6	8	16	82.1	34.8
Q 3	1220	9	7	27	79.8	10.5

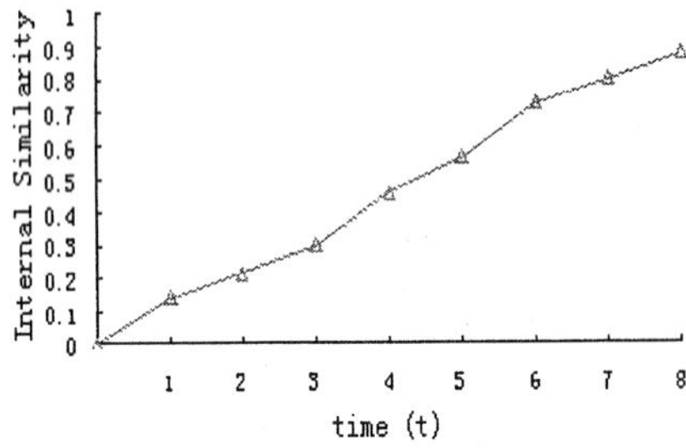

Fig. 2. Internal similarity of clusters

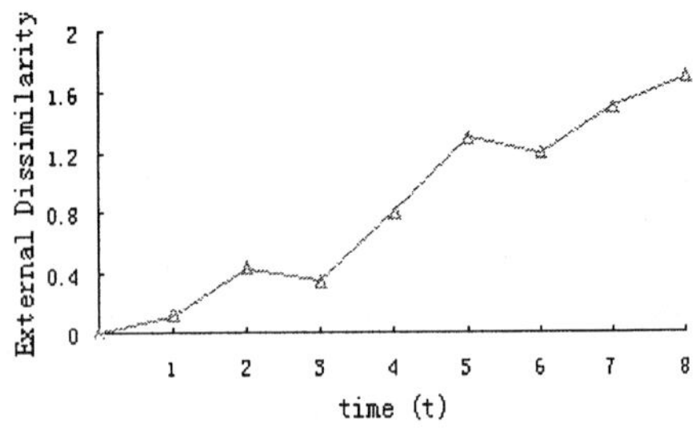

Fig. 3. External dissimilarity of clusters

$$internalSim = \frac{1}{|C|} \sum_{C_i \in C} \frac{1}{|C_i|} \sum_{\substack{x \in C_i \\ y \in C_i}} sim(x, y). \tag{5}$$

Fig. 2 and 3 show the quality of our clustering as client' query become more and more concentrated on some special topics. Internal similarity *internalSim* represents inner cohesiveness of each cluster, and external dissimilarity *dep* describes the independence of different clusters. The definition of *dep* is built upon the overlap of different clusters [8]. Given a page *p* clustered to C_i, we define O_p as the number of its occurrence in other clusters.

As shown in Fig. 2, the form of user query topics directly decides the clustering accuracy. With the gradual enrichment of user query topics information, the clustering becomes more and more accurate. Fig. 3 shows that this clustering can dynamically adapt user query, and capture any personal topics well and truly.

$$dep(C) = \frac{1}{|C|} \sum_{C_i \in C} \frac{1}{|C_i|} \sum_{p \in C_i} (O_p - 1). \qquad (6)$$

3 Conclusions

This paper presents an efficient means for the personal clustering of retrieved web pages by learning users' interests with respect to their query histories, and further using this information to realize a meaningful clustering. With the enrichment of user history records, the clustering will pick and return even more accurate results.

References

1. Hou, J. and Zhang, Y., Utilizing Hyperlink Transitivity to Improve Web Page Clustering. Proceedings of the Fourteenth Australasian Database Conference. Adelaide, Australia (2003) 49-57
2. D. S. Modha and W. S. Spangler, Clustering hypertext with applications to web searching. In Proceedings of the 11th ACM Conference on Hypertext and Hypermedia. San Antonio, TX (2000) 143-152
3. Zhang Dong, Towards Web Information Clustering. PhD thesis. Southeast University, Nanjing, China (2002)
4. D. Beeferman and A. Berger, Agglomerative clustering of a search engine query log. In Proceedings of the Sixth ACM SIGKDD International Conference on Knowledge Discovery and Data Mining. Boston, MA (2000) 407-416
5. Mark Hansen, Elizabeth Shriver, Using navigation Data to Improve IR functions in the Context of Web Search. CIKM. (2001) 135-142
6. Ricardo Baeza-Yates, Berthier Ribeiro-Neto, Modern information retrieval. ACM Press. (1999)
7. J. Wang, J. Han, and J. Pei, CLOSET+:Searching for the Best Strategies for Mining Frequent Closed Itemsets. In Proc. 2003 ACM SIGKDD. Int. Conf. on Knowledge Discovery and Data Mining (KDD'03). Washington, D.C (2003)
8. F. Beil, M. Ester, and X. Xu., Frequent term-based text clustering. In Proc 8th Int. Conf. on Knowledge Discovery and Data Mining KDD. (2002)

Artificial Immune System for Medical Data Classification

Wiesław Wajs[1], Piotr Wais[2], Mariusz Święcicki[3], and Hubert Wojtowicz[2]

[1] Institute of Automatics, University of Mining and Metallurgy, Kraków, Poland
wwa@ia.agh.edu.pl
[2] National Technical University in Krosno
waisp@poczta.onet.pl
hubwoj@pwsz.krosno.pl
[3] Institute of Computer Modelling, Cracow University of Technology, Kraków, Poland
mswiecic@uck.pk.edu.pl

Abstract. The article presents application of artificial immune algorithms in classification of vectorized medical data sets. Artificial immune network was created and trained for the purpose of arterial blood gasometry parameters (pH, pCO2, pO2, HCO3) classification. Training data originates from the Infant Intensive Care Unit of the Polish – American Institute of Pediatry, Collegium Medicum, Jagiellonian University in Cracow.

1 Introduction

Biological phenomenons, and particularly organic processes are of a dynamic character. Diagnosis and therapy are of the same character. In approaching diagnosis and making therapeutic decision phenomenons happening in time are investigated. Every doctor knows that it is rare to make a decision about diagnosis and treatment on the basis of only one clinical observation. Usually it is based on several patient's examinations, regular analysis of many biophysical and biochemical parameters, or imaging examination. On a particular level of these examinations preliminary diagnosis is determined, and later a final one. Basing on determined diagnosis and on the evaluation of the sickness process dynamics decision about the treatment is taken.

Initial stabilization of the infants state is a difficult task. To achieve it the doctor analyses repeatedly many parameters related to the patient's health condition. These parameters are: birth anamnesis, physiological examinations (body's weight, dieresis), results of additional examinations (biochemical, micro biological, imaging) and readings from the monitoring instruments (pulseximeter, cardio monitor, invasive measurement of the arterial pressure, respiratory mechanics monitor). It isn't rare when doctor appraises simultaneously over fifty variables. Analysis of this amount of data is hard and requires experience, all the more if the decision about the treatment should be made quickly. In result of carried out analysis doctor makes decision about the treatment expecting positive results, which are expressed by the desired changes

V.S. Sunderam et al. (Eds.): ICCS 2005, LNCS 3516, pp. 810–812, 2005.

in the results of additional examinations, readings of the monitoring instruments and physical examination. The whole process can be verified by comparing it to the model of respiratory insufficiency progress carried out by the doctor. Creation of this model is based on theoretical and empirical knowledge found in scientific literature and also on the doctor's experience. Intensification of gas exchange disorders are best reflected by arterial blood gasometry parameters examined in the context of currently used methods of ventilation support. For that reason as an output values four directly measured parameters of arterial blood gasometry: pH, pCO_2, pO_2, HCO_3 and oxygen parameter (pO_2/FiO_2) were chosen. Alkalis deficiency and hemoglobin oxygen saturation were omitted as derivatives of the four parameters mentioned above.

2 An Artificial Immune Network for Signal Analysis

Main goal of the system is a classification of signals and to achieve it several problems had to be solved. First of them is connected with an algorithm of learning the immune network. The next problem is related to the structures of data, which are responsible for representation of signals.

The input signal for the system, is interpreted as an antibody (Ab) so the task of immune network is to find an antigen Ag that will be suitable for Ab. The Ag-Ab representation will partially determine which distance measure shall be used to calculate their degree of interaction. Mathematically, the generalized shape of a molecule (*m*), either an antibody (Ab) or an antigen (Ag), can be represented by a set of real-valued coordinates $m = <m1, m2, ..., mL>$, which can be regarded as a point in an *L*-dimensional real-valued space.

$$D = \sqrt{\sum_{i=1}^{L} (ab_i - ag_i)^2} \quad (1)$$

The affinity between an antigen and an antibody is related to their distance that can be estimated via any distance measure between two vectors, for example the Euclidean or the Manhattan distance. If the coordinates of an antibody are given by $<ab1, ab2, ..., abL>$ and the coordinates of an antigen are given by $<ag1, ag2, ..., agL>$, then the distance (*D*) between them is presented in equation (1), which uses real-valued coordinates. The measure distances are called Euclidean shape-spaces.

Algorithm used in the research is described in papers [1] [2] by de Castro and Von Zuben. Modified version of immune net algorithm was adapted to signals classification. The learning algorithm lets building of a set that recognizes and represents the data structural organization. The more specific the antibodies, the less parsimonious the network (low compression rate), whilst the more generalist the antibodies, the more parsimonious the network with relation to the number of antibodies. The suppression threshold controls the specificity level of the antibodies, the clustering accuracy and network plasticity.

2.1 Overview and Analysis of the Research

Currently immune network performs unsupervised pattern recognition of the gathered medical data. In the process of further research description of the data classes will be added, which is prepared by an expert medicine doctor. The idea is to categorize patients according to the hospitalization group they are initially assigned to by the doctor. Next goal after creating a reliable classification system is to develop a time series predicting system to help doctors in forecasting patient's condition ahead in time.

A major issue in the research is the scarcity of available data from hospitalization. Efficient system of new data acquisition is one of the main concerns and to achieve it in the future a creation of an agent system [3] is planned for gathering information through the Internet from several hospitals and adding it to the database.

3 Multidimensional Medical Data Classification

In the given example use of the AIS is presented in the unsupervised pattern recognition of the medical data. Using a database, functioning for the few years on the Infant Intensive Care Ward of the Polish – American Institute of Pediatrics Collegium Medicum Jagiellonian University of Cracow, a network was created, which final task is going to be the classification of arterial blood gasometry parameters. In the process of training previous values of gasometry, respirator settings and surfactant administration were used as an input data. Training of the network consists of two phases. First phase is a learning of the network. Training data set is comprised of blood parameters values starting from the time (t) – time series are of the same length. Second phase tests the network's generalization abilities by presenting it input vectors from the test dataset. Elements of the training dataset – time series are interpreted by the network as antigens, which are stimulating particular antibody creating an artificial immune network. Training dataset for the artificial immune network comprises of pattern series, which are represented by the vectors. As a result of this time series included in the dataset can be interpreted as a point coordinates in n-dimensional space, which n-value is dependant on the pattern series length.

4 Summary

In the paper functioning of the artificial immune system and example of its application in unsupervised pattern recognition of medical multidimensional data was presented. Perspectives for creating a reliable classification and time series modeling systems based on the artificial immune algorithm were discussed.

References

[1] De Castro, L. N., Von Zuben, F. J. (2000a), An Evolutionary Immune Network for Data Clustering, Proc. of the IEEE SBRN, pp. 84-89.

[2] De Castro, L. N., Von Zuben, F. J. (2000b), The Clonal Selection Algorithm with Engineering Applications, GECCO'00 – Workshop Proceedings, pp. 36-37.

[3] Wajs Wiesław, Wais Piotr, Autonomous Agent for Computer System and Computerized System of Automatics, Some Analogy and Difference, International Workshop Control and Information Technology, IWCIT 2001.

EFoX: A Scalable Method for Extracting Frequent Subtrees ⋆

Juryon Paik, Dong Ryeol Shin, and Ungmo Kim

Dept. of Computer Engineering, Sungkyunkwan University,
300 Chunchun-dong, Jangan-gu, Suwon-si, Gyeonggi-do 440-746, Republic of Korea
quasa277@gmail.com, {drshin, umkim}@ece.skku.ac.kr

Abstract. The more web data sources provide XML data, the greater information flood problem has been caused. Hence, there have been increasing demands for efficient methods of discovering desirable patterns from a large collection of XML data. In this paper, we propose a new and scalable algorithm, EFoX, to mine frequently occurring tree patterns from a set of labeled trees. The main contribution made by our algorithm is that there is no need to perform any tree join operation to generate candidate sets.

1 Introduction

The first step toward mining information from XML documents is to find *frequent subtrees* occurring in a large collection of XML trees. However, the discovery of frequent subtrees appearing in a large-scaled tree-structured dataset is not an easy task to do. The main difficulties arise in candidate subtrees enumerations and in pattern tree matching. The problem here is that the enumeration of candidates is typically made by join operations; as the number of XML documents increases, the efficiency of previously developed algorithms deteriorates rapidly. Hence, it is required to find more scalable and less burdensome methods for extracting frequent subtrees. In this paper, we propose a novel algorithm *EFoX (Extract Frequent subtrees of Xml trees)* for efficiently finding frequent subtrees from a set of XML documents. The proposed algorithm not only reduces significantly the number of tree prunings, but also simplifies greatly each round by avoiding time-consuming join operations.

2 Related Work

Wang and Liu [4] considered mining of collections of paths in ordered trees by using Apriori [1] technique. Asai et al. [2] proposed FREQT for mining labeled

⋆ This research was supported in part by University IT Research Center Project funded by the Korean Ministry of Information and Communication and by Korea Science & Engineering Foundation (R01-2004-000-10755-0).

V.S. Sunderam et al. (Eds.): ICCS 2005, LNCS 3516, pp. 813–817, 2005.

ordered trees. FREQT uses rightmost expansion notion to generate candidate trees by attaching new nodes to the rightmost edge of a tree. Zaki [5] proposes two algorithms, TreeMiner and PatternMatcher, for mining embedded subtrees from ordered labeled trees. The common problem of the previous approaches is identified as follows: Each tree pruning round during generating candidate sets requires to perform expensive join operations. Therefore, as the number of XML documents increases, the efficiency for extracting frequent subtrees deteriorates rapidly since both the cost of join operations and the number of pruning rounds add up.

3 Algorithm EFoX

In this section, we present a new algorithm EFoX for efficiently extracting frequent subtrees from a given set of trees. To this end, EFoX consists of two steps; The first one is to create and maintain KidSet to avoid join operations entirely and reduce the number of candidates. The second step is to extract frequent subtrees incrementally based on data stored in KidSet.

3.1 Generating KidSet

Let $D = \{T_1, T_2, \ldots, T_i\}$ be a set of trees and let $|D|$ be the number of trees in D.

Definition 1 (key). *Let K_d be a collection of node labels assigned on the nodes at depth d in every tree in D. (Assume that depth of root node is 0.) We call each member in K_d by a key.*

At this point, note that there may exist some nodes labeled with the same names in D. Thus, for each key we need to identify the list of trees in which the key belongs.

Definition 2 (KidSet). *A KidSet, $[K]^d$, is defined as a set of pairs (k_d, t_{id}) where k_d is a key in K_d and t_{id} is a list of tree indexes in which k_d belongs.*

According to some minimum support, a collection of KidSets can be classified as two classes.

Definition 3 (Frequent). *Given some minimum support σ and a pair (k_d, t_{id}), the key k_d is called **frequent** if $|t_{id}| \geq \sigma \times |D|$.*

Definition 4 (Frequent Set). *Given a KidSet, $[K]^d$, a pair in $[K]^d$ is called **Frequent Set** if its key is frequent. Otherwise, it is called **Candidate Set**. We denote Frequent Set and Candidate Set by $[F]^d$ and $[C]^d$, respectively.*

It is required to consider the characteristic of tree structure. This consideration stems from the fact that same labels can be placed several times throughout a XML tree.

Cross-Reference operation. Let $\mathcal{FS} = \{[F]^0, [F]^1, \ldots, [F]^d\}$ be an initial set of Frequent Sets and $\mathcal{CS} = \{[C]^0, [C]^1, \ldots, [C]^d\}$ be an initial set of Candidate Sets. Let i be one of integers $0 < i \leq d$. One round of `cross-reference` operation consists of following two phases:

- *Phase 1.* Compute ($[F]^i$ *vs.* $[C]^{i-1}$) and ($[C]^i$ *vs.* $[F]^{i-1}$ *to* $[F]^0$).
  The purpose of this step is to eliminate pairs from Candidate Sets, which are actually the pairs included in any previous Frequent Sets. We reduce a number of pairs in Candidate Sets significantly due to this phase.
- *Phase 2.* Compute ($[C]^i$ *vs.* $[C]^{i-1}$).
  This step adopts Apriori-style. However, the efficiency is dramatically improved since there exists no join operation between labels to generate candidate sets. Through this phase, we **derives** (not generate) Candidate Set with union operation. Because of the **hierarchical** KidSet structure, it is not required to generate candidate paths and additional candidate pairs by using join operations. At every end of a round, $[C]^i$ unions all pairs not belonged to any Frequent Sets between $[F]^0$ and $[F]^{i-1}$. Consequently, two consecutive Candidate Sets are always processed to find additional frequent elements.

3.2 Mining Frequent Subtrees

The final set of Frequent Sets contains all keys which are frequent. However, not every Frequent Set has elements, and not every key has connection each other. We don't consider the empty sets because an empty Frequent Set implies that any label can be placed. The firstly appeared nonempty Frequent Set stores the keys which are root nodes of frequent subtrees. Based on those nodes, frequent subtrees are incrementally obtained by forming paths with keys in the rest Frequent Sets. During construction of frequent subtrees, they are extracted from the trees as shown by Theorem 1.

Theorem 1 Correctness of EFoX. *The bigger frequent subtree is always expanded from one of the smaller frequent subtrees.*

Proof. omitted due to lack of space.

4 Evaluation

To evaluate the effectiveness and the scalability of our algorithm, we carried out a couple of experiments on artificially generated medium-size data sets by tree generator described in [3]. We evaluated EFoX over the generic algorithm FSM. The basic idea of behind FSM is to identify frequent subtrees through a repeated process of enumerating and pruning candidate subtrees. Two synthetic data set, S15 ($\sigma = 0.15$) and T10K ($T = 10000$), were tested. The generator mimics three target frequent trees reflecting predefined parameters. We refer the reader to the paper [3] for a detailed explanation of using parameters.

The execution time for data set S15 with varying number of input trees is shown in Fig. 1(a). It can be seen that EFoX demonstrates a dramatic improvement over the generic frame FSM. Fig. 1(b) shows the execution time for data set T10K with changing minimum support. The execution time decreases as the minimum support increases in both algorithms, but there is significant time difference between them. We stopped to run FSM since the running time exceeds several hours.

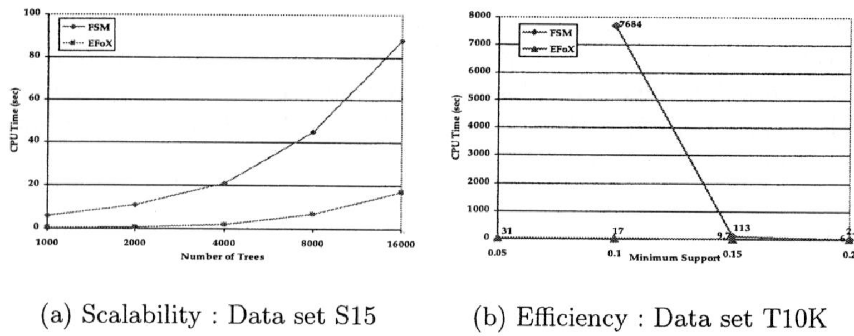

(a) Scalability : Data set S15 (b) Efficiency : Data set T10K

Fig. 1. Execution Time

5 Conclusion

A new type of tree mining has been defined in the paper. A new algorithm, EFoX, is proposed to discover frequent subtrees. The algorithm uses a special data structure, KidSet, to manipulate frequent node labels and tree indexes. EFoX is evaluated with artificial medium-size data sets. Our experimental results show that EFoX performs much better than generic frame FPM, both in terms of scalability and efficiency. We need further research to adapt the current proposed methodology to both synonym and polysemy of each element of XML trees, for use in other XML mining operations, namely, clustering or classification.

References

1. R. Agrawal and R. Srikant. Fast algorithms for mining association rules in large databases. *In Proc. of the 12th International Conference on Very Large Databases*, pp487–499, 1994.
2. T. Asai, K. Abe, S. Kawasoe, H. Arimura, H. Sakamoto, and S. Arikawa. Efficient substructure discovery from large semi-structured data. *In Proc. of the 2nd SIAM International Conference on Data Mining (ICDM'02)*, 2002.

3. A. Termier, M-C. Rousset, and M. Sebag. TreeFinder : a First step towards XML data mining. *In Proc. of IEEE International Conference on Data Mining (ICDM'02)*, pp.450–457, 2002.
4. K. Wang, and H. Liu. Schema discovery for semistructured data. *In Proc. of the 3rd International Conference on Knowledge Discovery and Data Mining (KDD'97)*, pp.271–274, 1997.
5. M. J. Zaki. Efficiently mining frequent trees in a forest. *In Proc. of the 8th ACM SIGKDD International Conference on Knowledge Discovery and Data mining (KDD'02)*, pp.71–80, 2002.

An Efficient Real-Time Frequent Pattern Mining Technique Using Diff-Sets

Rajanish Dass and Ambuj Mahanti

Indian Institute of Management Calcutta
{rajanish, am}@iimcal.ac.in

Abstract. Frequent pattern mining in real-time is of increasing thrust in many business applications such as e-commerce, recommender systems, and supply-chain management and group decision support systems, to name a few. A plethora of efficient algorithms have been proposed till date. However, with dense datasets, the performances of these algorithms significantly degrade. Moreover, these algorithms are not suited to respond to the real-time need. In this paper, we describe BDFS(b)-diff-sets, an algorithm to perform real-time frequent pattern mining using diff-sets. Empirical evaluations show that our algorithm can make a fair estimation of the probable frequent patterns and reaches some of the longest frequent patterns much faster than the existing algorithms.

1 Introduction

In recent years, business intelligence systems are playing pivotal roles in fine-tuning business goals such as improving customer retention, market penetration, profitability and efficiency. In most cases, these insights are driven by analyses of historic data. Now the issue is, if the historic data can help us make better decisions, how real-time data can improve the decision making process [1].

Frequent pattern mining for large databases of business data, such as transaction records, is of great interest in data mining and knowledge discovery [2], since its inception in 1993, by Agrawal et al. Researchers have generally focused on the frequent pattern mining, as it is complex and the search space needed for finding all frequent itemsets is huge [2]. A number of efficient algorithms have been proposed in the last few years to make this search fast and accurate. Among these, a number of effective vertical mining algorithms have been recently proposed, that usually outperforms horizontal approaches. Despite many advantages of the vertical format, the methods tend to suffer, when the tid-list cardinality gets very large as in the case of dense datasets [3]. Again, these algorithms have limited themselves to either breadth first or depth first search techniques. Hence, most of the algorithms stop only after finding the exhaustive (optimal) set of frequent itemsets and do not promise to run under user defined real-time constraints and produce some satisficing (interesting sub-optimal) solutions due to their limiting characteristics[4, 5].

V.S. Sunderam et al. (Eds.): ICCS 2005, LNCS 3516, pp. 818 – 821, 2005.

2 Business Issues of Real-Time Frequent Pattern Mining

Using up-to-date information, getting rid of delays, and using speed for competitive advantage is what the real-time enterprise is about. There are numerous areas where real-time decision making plays a crucial role. These include areas like real-time customer relationship management, ,real-time supply chain management systems real-time enterprise risk and vulnerability management, real-time stock management and vendor inventory, real-time recommender systems, real-time operational management with special applications in mission critical real-time information as is used in the airlines industry, real-time intrusion and real-time fraud detection, real-time negotiations and other areas like real-time dynamic pricing and discount offering to customers in real-time. More than that, real-time data mining will have tremendous importance in areas where a real-time decision can make the difference between life and death – mining patterns in medical systems.

3 BDFS(b)-Diff-Sets: An Efficient Technique of Frequent Pattern Mining in Real-Time Using Diff-Sets

Algorithm BDFS(b)-diff-sets:

Initialize the allowable execution time τ.

Let the initial search frontier contain all 3-length candidate patterns. Let this search frontier be stored as a global pool of candidate patterns. Initialize a set called Border Set to null.

Order the candidate patterns of the global pool according to their decreasing length (resolve ties arbitrarily). Take a group of most promising candidate patterns and put them in a block b of predefined size.

- *Expand (b)*

```
    Expand (b: block of candidate patterns)
    If not last_level
      then
      begin
            Expand1(b)
          end.
```

Expand$_1$(b):

1. *Count support for each candidate pattern in the block b by intersecting the diff-set list of the items in the database.*
2. *When a pattern becomes frequent, remove it from the block b and put it in the list of frequent patterns along with its support value. If the pattern is present in the Border Set increase its subitemset counter. If the subitemset counter of the pattern in Border Set is equal to its length move it to the global pool of candidate patterns.*
3. *Prune all patterns whose support values < given minimum support. Remove all supersets of these patterns from Border Set.*
4. *Generate all patterns of next higher length from the newly obtained frequent patterns at step 3. If all immediate subsets of the newly generated pattern are frequent then put the pattern in the global pool of candidate patterns else put it in the Border Set if the pattern length is > 3.*
5. *Take a block of most promising b candidate patterns from the global pool.*
6. *If block b is empty and no more candidate patterns left, output frequent patterns and exit.*
7. *Call Expand (b) if enough time is left in τ to expand a new block of patterns, else output frequent patterns and exit.*

Fig. 1. Algorithm BDFS(b)-diff-sets

4 Empirical Evaluation

The following figures shows the empirical evaluation of BDFS(b)-diff-sets. We have found that BDFS(b)-diff-sets compares well with the existing best performing algorithms in time of completion and scalability. Real-time performance of BDFS(b)-diff-sets (in Fig. 9 and 10) show that it is always ahead of time while providing outputs. We have made detailed performance evaluation based on empirical analysis using commonly used synthetic and real-life dense datasets. Thus, we have demonstrated that real-time frequent pattern mining can be done successfully using BDFS(b)-diff-sets. We believe this study will encourage use of AI heuristic search techniques in real-time frequent pattern mining.

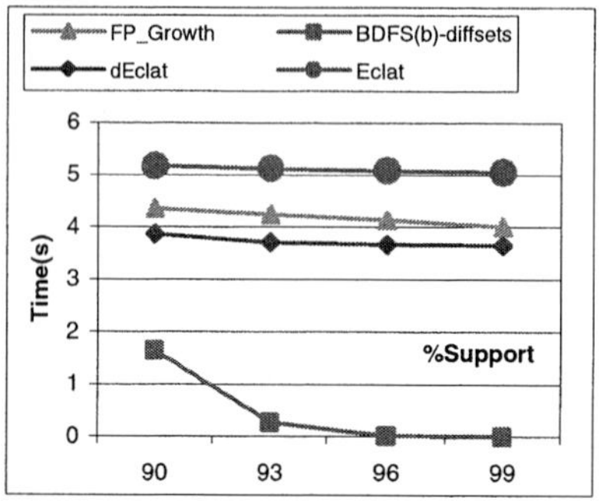

Fig. 2. Time comparison of FP-Growth, Eclat and dEclat with BDFS(b)-diffsets (b= 20880) on PUMSB, N=2113, T=74, D=49046

Fig. 3. Time comparison of FP-Growth with BDFS(b)-diffsets for T10I8D100K, b=100K

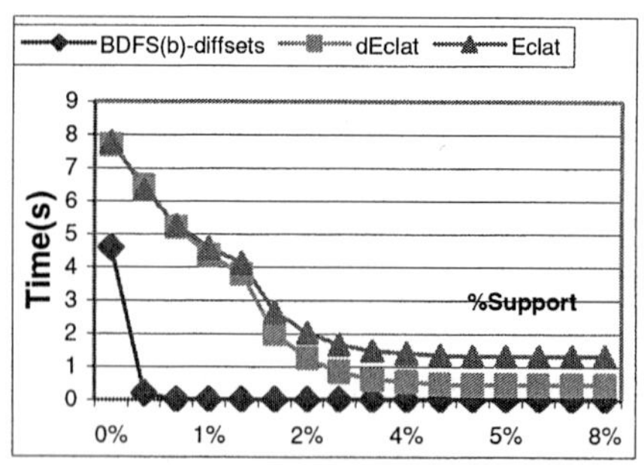

Fig. 4. Time comparison of Eclat and dEclat with BDFS(b)-diffsets for T10I8D100K, b=100K

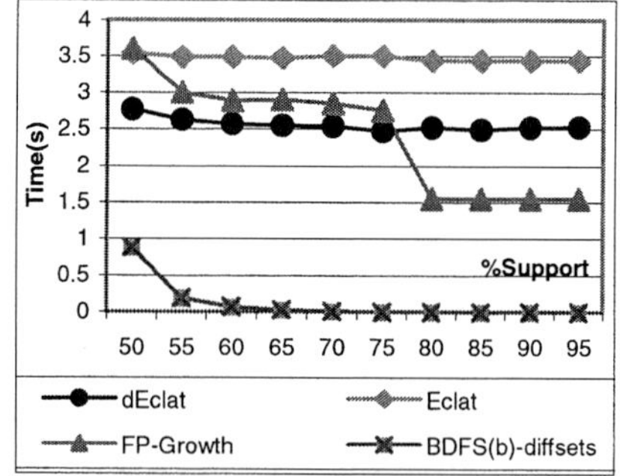

Fig. 5. Time comparison of FP-Growth, Eclat and dEclat with BDFS(b)-diffsets (b=2088K) for PUMSB*, N=2088 T= 50.5, D = 49046

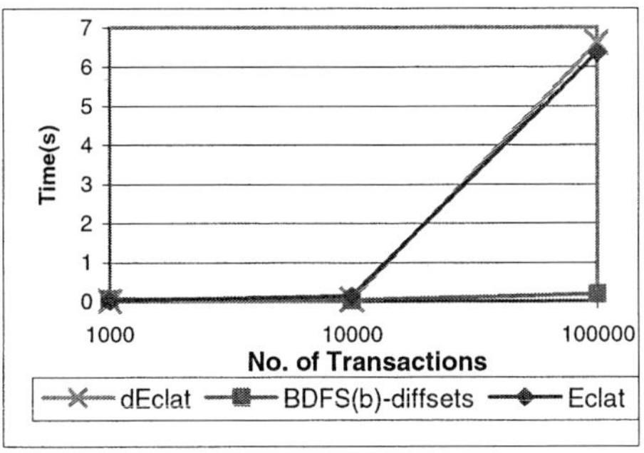

Fig. 6. Scalability evaluation of BDFS(b)-diffsets with Eclat and dEclat supp=0.5%, b = 100K for T10I8D1K,10K and 100K

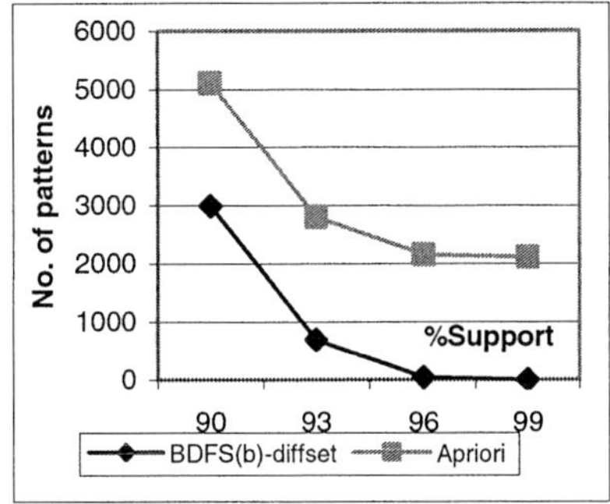

Fig. 7. Number of patterns checked by Apriori and BDFS(b)-diffsets (b=208800) for Pumsb, N=2113,T=74, D=49046, with varying support

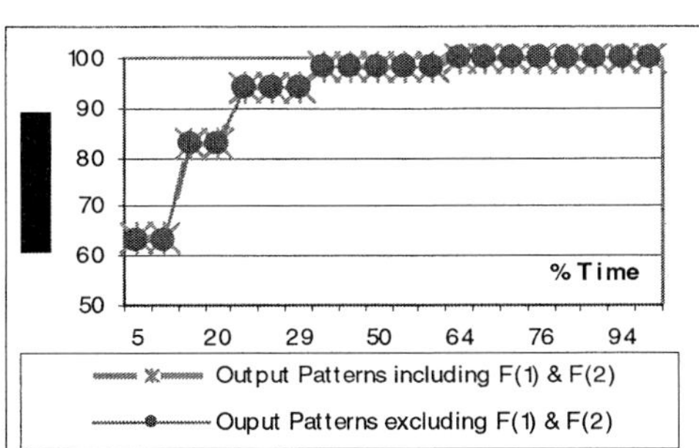

Fig. 8. Time-Patterns % of BDFS(b) for b=75K and 65% supp for Chess (N=75, T=37, D=3196)

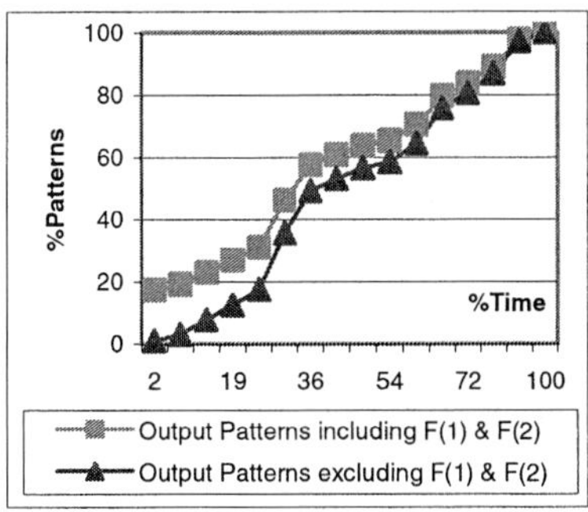

Fig. 9. Time-Patterns % for b=75K and 65% supp for T10I8D100K

References

1. Gonzales, M.L., *Unearth BI in Real-time*. 2004, Teradata.
2. Goethals, B., *Memory Issues in Frequent Pattern Mining*, in *Proceedings of SAC'04*. 2004, ACM: Nicosia, Cyprus.
3. Zaki, M.J. and K. Gouda. *Fast Vertical Mining Using Diffsets*. in *9th International Conference on Knowledge Discovery and Data Mining*. 2003. Washington, DC.
4. Dass, R. and A. Mahanti. *Frequent Pattern Mining in Real-Time – First Results*. in *TDM2004/ACM SIGKDD 2004*. 2004. Seattle, Washington USA.
5. Dass, R. and A. Mahanti. *An Efficient Technique for Frequent Pattern Mininig in Real-Time Business Applications*. in *38th IEEE Hawaii International Conference on System Sciences (HICSS 38)*. 2005. Big Island: IEEE.
6. Lee, W., et al. *Real time data mining-based intrusion detection*. in *DARPA Information Survivability Conference & Exposition II*. 2001. Anaheim, CA , USA: IEEE Xplore.

Improved Fully Automatic Liver Segmentation Using Histogram Tail Threshold Algorithms

Kyung-Sik Seo

Dept. of Electrical & Computer Engineering,
New Mexico State University, Las Cruces, NM, USA
nmsu2@hanmail.net

Abstract. In order to remove neighboring abdominal organs of the liver, we propose an improved fully automatic liver segmentation using histogram tail threshold (HTT) algorithms. A region of interest of the liver is first segmented. A left HTT (LHTT) algorithm is performed to eliminate the pancreas, spleen, and left kidney. After the right kidney is eliminated by the right HTT (RHTT) algorithm, the robust liver structure is segmented. From the results of experiments, the improved automatic liver segmentation using HTT algorithms has strong similarity performance as manual segmentation by medical doctor.

1 Introduction

A computed tomography (CT) is currently a conventional and excellent tool for diagnosis of the liver in medical imaging technology. Liver segmentation using the CT has been performed often [1, 2]. As previous researches have depended on semi-automatic liver segmentation such as a seed point, a rand mark, a reference image, and training data, fully automatic liver segmentation based on the spine was proposed [3]. However, proposed liver segmentation has problems because of pixel similarity of neighboring abdominal organs such as the spleen, pancreas, and kidneys.

In order to segment neighboring organs, we propose improved fully automatic liver segmentation using histogram tail threshold (HTT) algorithms. The region of interest (ROI) of the liver structure is first segmented. Then liver segmentation using HTT algorithms are processed to eliminate neighboring abdominal organs.

2 Improved Liver Segmentation

As pre-processing, the ROI of the liver is extracted using multi-modal threshold [3-5], C-class maximum a posteriori [6], and binary morphological (BM) filtering [7-9]. Figure 1 shows the ROI extraction of the liver.

The left HTT (LHTT) is presented to remove neighboring abdominal organs such as the pancreas, spleen, and left kidney from the ROI. Let $I_{ROI}(m,n)$ be the gray-level ROI and $h_{ROI}(k_1,k_2)$ be the histogram of I_{ROI} with the range, $[k_1,k_2]$. Let I_{LHTT} be the LHTT image. Then the LHTT algorithm is proposed:

V.S. Sunderam et al. (Eds.): ICCS 2005, LNCS 3516, pp. 822–825, 2005.

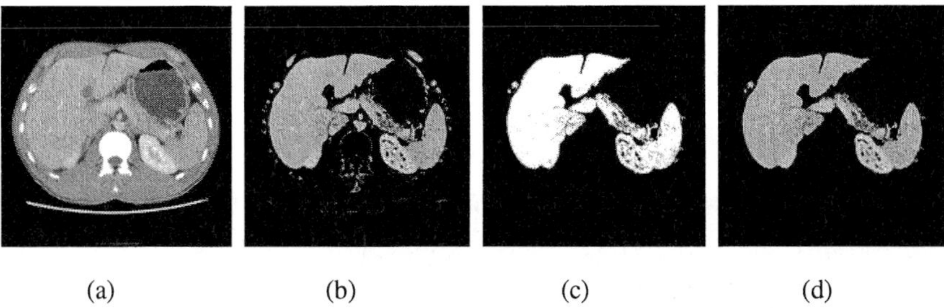

(a) (b) (c) (d)

Fig. 1. ROI extraction of the liver: (a) CT image, (b) ROI after MMT, (c) ROI after C-class MAP decision, (d) Gray-level ROI

- Find $k_{\max}$ where $k_{\max}$ is the gray-level value when $h_{ROI}(k)$ is the maximum value.
- Calculate the left histogram tail interval $k_{LI} = (k_{\max} - k_1)$.
- Find the left histogram tail threshold value.
- Create the LHTT image $I_{LHTT} = \{(m,n) \mid k_1 \le I_{ROI}(m,n) \le k_{LHTT}\}$.

Figure 2 shows an example of the LHTT algorithm.

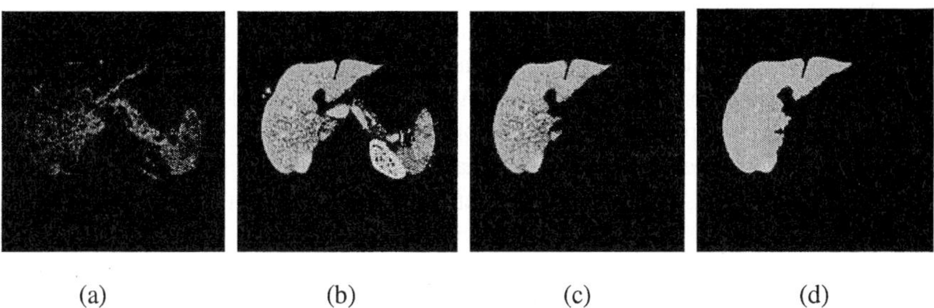

(a) (b) (c) (d)

Fig. 2. Left histogram tail threshold: (a) LHTT image, (b) Difference image between I_{ROI} and I_{LHTT}, (c) Segmented liver structure by area estimation and C-class MAP decision, (d) Robust liver segmentation by BM filtering and gray-level transformation

As the right kidney is adjacent to the lower liver part and has the same gray-level values as liver vessels, this kidney creates problems in segmenting a liver structure. The RHTT is presented for extracting and removing the right kidney from the liver image. Let I_{RHTT} be the RHTT image. Then the RHTT algorithm is proposed:

- Find $k_{\max}$ where $k_{\max}$ is the gray-level value when $h_{ROI}(k)$ is the maximum value.
- Calculate the right partial histogram interval $k_{RI} = (255 - k_{\max})$.

- Find the right partial threshold value $k_{RHTT} = (k_{\max} + k_{RI} / 8)$.
- Create the RHTT image $I_{RHTT} = \{(m,n) \mid k_{RHTT} \leq I_{ROI}(m,n) \leq 255\}$.

Figure 3 shows an example of the RHTT algorithm.

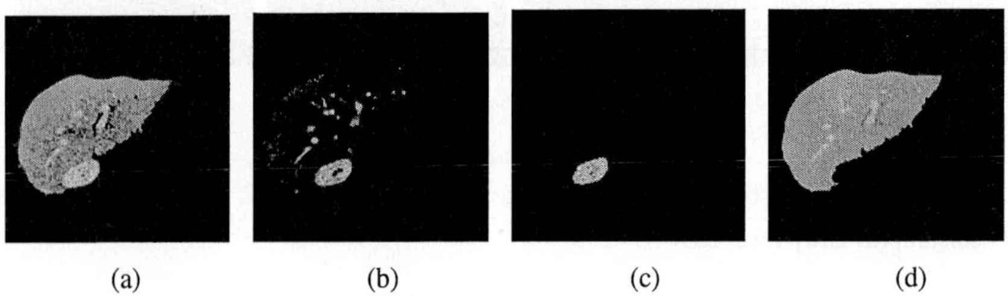

(a) (b) (c) (d)

Fig. 3. Example of right histogram tail threshold: (a) Sample image, (b) RHTT image, (c) Extraction of right kidney after the area and angle estimation, (d) Elimination of the right kidney

3 Experiments and Analysis

40 slices of eight patients were selected for testing the improved segmentation method and one medical doctor in Chonnam National University Hospital segmented the liver structure by the manual method. In order to evaluate performance of the improved algorithm, three different segmentation methods were compared by using normalized average area (NAA) with the image size and area error rate (AER) [3]. Figure 4(a) shows NAA using error bars. Average NAAs of each method were 0.1972, 0.1593, and 0.1644. Also, Figure 4(b) shows the comparison of average AER per each patient based on manual segmentation to segmentation before and after HTT. The former is 4~40 % and the latter is 5~10 %. Total average AER for all patients is 23.9198 % for the former and 7.4223 % for the latter. From the results of this comparison, automatic segmentation using HTT algorithms matches the results of manual segmentation more closely than segmentation without HTT algorithms.

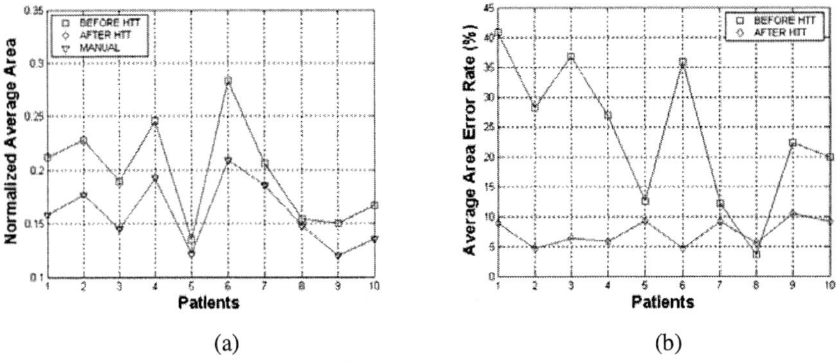

(a) (b)

Fig. 4. Results: (a) Normalized average area, (b) Average area error rate

4 Conclusions

In this paper, we proposed the improved fully automatic liver segmentation using histogram tail threshold (HTT) algorithms. 40 slices of eight patients were selected to evaluate performance of the improved algorithm. From the results of experiments, the improved automatic liver segmentation has strong similarity performance as manual segmentation by medical doctor.

References

1. Bae, K. T., Giger, M. L., Chen, C. T., Kahn, Jr. C. E.: Automatic segmentation of liver structure in CT images. Med. Phys.,Vol. 20. (1993) 71-78
2. Gao, L., Heath, D. G., Kuszyk, B. S., Fishman, E. K.: Automatic liver segmentation technique for three-dimensional visualization of CT data. Radiology, Vol. 201. (1996) 359-364
3. 3.Seo, K., Ludeman, L. C., Park S., Park, J.: Efficient liver segmentation based on the spine. LNCS, Vol. 3261. (2004) 400-409.
4. 4.Orfanidis, S. J.: Introduction to signal processing. Prentice Hall, Upper Saddle River NJ (1996)
5. 5.Schilling, R. J., Harris, S. L.: Applied numerical methods for engineers. Brooks/Cole Publishing Com., Pacific Grove CA (2000)
6. Ludeman, L. C.: Random processes: filtering, estimation, and detection. Wiley & Sons Inc., Hoboken NJ (2003)
7. Gonzalez, R. C., Woods, R. E.: Digital image processing. Prentice Hall, Upper Saddle River NJ (2002)
8. Shapiro, L. G., Stockman, G. C.: Computer vision. Prentice-Hall, Upper Saddle River NJ (2001)
9. Parker, J.R.: Algorithms for image processing and computer vision. Wiley Computer Publishing, New York (1997)

Directly Rasterizing Straight Line by Calculating the Intersection Point

Hua Zhang[1,2], Changqian Zhu[1], Qiang Zhao[2], and Hao Shen[2]

[1] Department of Computer and Communication Engineering, Southwest Jiaotong University, Chengdu, Sichuan, 610031 P.R.China
cqzhu@home.swjtu.edu.cn
[2] Institute of Computer Applications, China Academy of Physics Engineering, P.O. Box 919 Ext. 1201, Mianyang, Sichuan, 621900 P.R.China
{hzhang, zhaoq, shh}@caep.ac.cn

Abstract. In this paper, a method of line scan-conversion is presented by calculating the intersection point of straight line with the middle scan line of screen. With this method, the pixel's screen coordinate of spatial line could also be obtained. Moreover, raster precision is exactly the same as that of Bresenham's middle algorithm; and more than or at least one pixel could be obtained at a computational effort of one inner loop. However, to Bresenham's middle algorithm, only one pixel is obtained in one inner loop's calculating.

1 Introduction

Line scan-conversion is to raster a spatial straight line onto a 2D raster display. There is a famous and classical algorithm, named as Bresenham's middle algorithm [1]. In Bresenham's algorithm, the next point is obtained by checking the sign of error, and only integer computation is used in inner loop calculating. With regard to Bresenham's middle algorithm, there are many fast algorithms and implementations, such as in [2], [3], [4], [5], [6], [7], and [8]. Some of these methods emphasize on the distribution of line's direction; others require that the line drawn is not to be very short. Chen [9] et al. presented statistical results of actual applications for line scan-conversion. In those statistical results, about 38.58 percent of lines are not horizontal, vertical and diagonal; and about 87.79 percent of lines are not larger than 17 pixels. In this paper, a method independent of line direction and line length is presented. We also assume that the line's slope has the rang of 0 to 1, which is an assumption often made when discussing line scan-conversion. Other slope's line can be handled by reflections about the principal axes.

2 Line Rasterization

In fig. 1, for a line's slope with the range of 0 to 1(when the slope is equal to zero, this is a special case that will be discussed in following section), two regions could be obtained by subdividing the first octant with line 2, which has the slope of half one. Our algorithm of line rasterization in each region could be represented as following steps.

V.S. Sunderam et al. (Eds.): ICCS 2005, LNCS 3516, pp. 826 – 829, 2005.

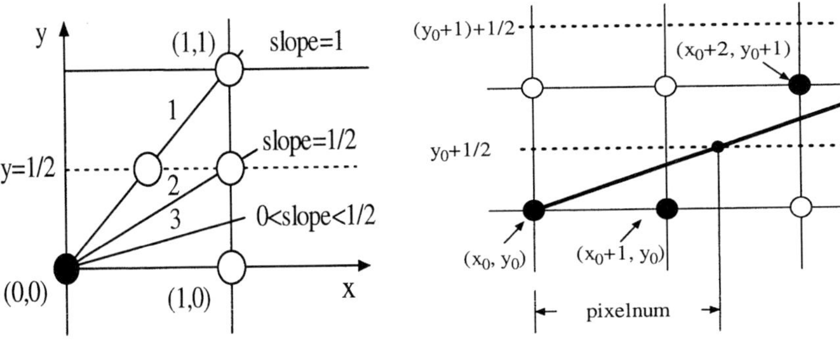

Fig. 1. Basis of line rasterization **Fig. 2.** An example of line rasterization

For a fixed two different end points with coordinates of ($x0$, $y0$) and (xn, yn), the line equation could be obtained, as in:

$$y = \frac{yn - y0}{xn - x0} x + (y0 - \frac{yn - y0}{xn - x0} x0) \tag{1}$$

Coordinates of intersection point between the line and ***y=y+1/2*** could be obtained by submitting ***y=y+1/2*** to the upper line equation. Now, submitting ***y=y+1/2*** into (1) and rearranging terms, the x component of intersection point could be represented as following (see fig. 2):

$$x = \frac{(xn - x0)(2(y - y0) + 1) + 2(yn - y0)x0}{2(yn - y0)} \tag{2}$$

Assume x coordinates of current start pixel obtained is xc. Let the integer variable ***pixelnum*** be the distance of xc and x (see fig. 2). Now, the distance of x and xc is as following:

$$pixelnum = x - xc$$

Now, ***pixelnum*** pixel(s) could be written by increasing in x direction. For example, in fig. 1, line 1, 2, and 3 has ***pixelnum*** of 0, 1, and 2, respectively. In line 1, there is no pixel is written in x direction from the start pixel. In line 2, there is one pixel (1, 0) is written in x direction from the start pixel. In line 3, there are two pixels, (1, 0) and (2, 0), written in x direction from start pixel. Note when x coordinates exceed xn, writing pixels is completed.

When the first loop is completed, point (1, 1) is selected as a new start point for line 1, because vertical distance from (1, 1) to line 1 is not greater than vertical distance from (1, 0) to line 1. Similarly, point (2, 1) and (3, 1) is selected as the new start point for line 2 and line3, respectively.

Repeating the upward processes until the line reach the end of line, the process of line rasterization is completed. The **C** code could be represented as following:

//The line end points are (x0,y0) and (xn,yn), assumed not equal. The slope of this line has //the rangeof 0 to 1, except 0.

```
LineRasterization (X0, y0, xn, yn){
  x=x0;y=y0;deltax=xn-x0;deltay=yn-y0;
  Pixelnum=0;pixelseq=0;
  While(x<=deltax){
     pixelnum=( deltax*(2*(y-y0)+1)+
              2*x0*deltay ) /(2*deltay) - x;
     For(pixelseq=0;pixelseq<=pixenum;x++){
        if(x>deltax) return ;
        writepixel(x,y);
        pixelseq++;
     }
     y++;
  }
}
```

2.1 Example

Consider line from (0, 1) to (6, 4). Rastering this line yields initial calculations: x=0, y=1, *deltax*=6, *deltay*=3, *pixelnum*=0, and *pixelseq*=0.

After running through the main loop and inner loop, coordinates could be shown in table 1.

Table 1. Coordinates of Rasterization

x	*pixelnum*	*writepixel*	Bresenham's results
0	1	(0, 1)	(0, 1)
		(1, 1)	(1, 1)
2	1	(2, 2)	(2, 2)
		(3, 2)	(3, 2)
4	1	(4, 3)	(4, 3)
		(5, 3)	(5, 3)
6	1	(6, 4)	(6, 4)

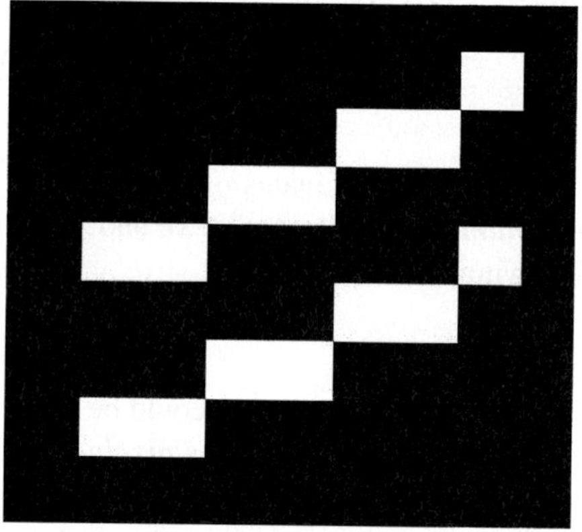

Fig. 3. Result of Rasterization

From these results, ***x*** is increased in each main loop; two pixels are written in each inner loop; and the coordinates of our algorithm for rasterization are the same as those of Bresenham's middle algorithm. Fig. 3 is a rasterized result for both Bresenham's algorithm and our algorithm. Note that the upper line has coordinates of (0, 4) and (6, 7), which are translated 3 pixels in ***y*** direction, and every pixel is represented by

20×20 pixels on screen. The upper line is rasterized by Bresenham's middle algorithm. The lower line is rasterized by our method. From fig. 3, same rasterized result could be noticed.

2.2 Slope Equal to Zero

When the slope of line is equal to zero, that is to say, $yn - y0 = 0$, the upper algorithm could not be used to raster the line. In this case, the line is parallel to horizontal axis. Therefore, the line could be rasterized from $x0$ to xn only with the increment of **1** in x direction(y is constant in the rasterization process).

3 Conclusions

A method of line scan-conversion is presented by directly calculating the intersection point of straight line with the middle scan line of screen. From this method, the coordinates of actual pixels on screen could be obtained. Moreover, the rasterized result is exactly the same as that of Bresenham's algorithm, and at least one pixel could be obtained in one inner loop's running. How fast about our method in actual applications is a further work, which we propose for future work.

References

1. Bresenham, J.E.: Algorithm for Computer Control and Digital Plotter. IBM Syst. J. Vol. 4, No.1, April (1965) 25-30
2. Dorst, L., Semulders, A.W.M.: Discrete Representation of Straight Lines. IEEE Trans. on Patter Analysis and Machine Intelligence, Vol. PAMI-6, No. 4, July (1984) 450-463
3. Foley, J.D., van Dam, A., Feiner, S.K., Hughes, J.F.: Computer Graphics: Principles and Practice. 2nd edn. In C, Addition-Wesley (1996) 72-80
4. Rokne, J.G., Wu, X.: Fast Line Scan-conversion. ACM Trans. on Graphics, Vol. 9, No. 4, Oct. (1990) 370-388.
5. Wu, X., Ronkne, J.G.: Double-step Incremental Generation of Lines and Circles. Computer Vision, Graphics and Image Processing, Vol. 37, No. 3, March (1987) 331-344
6. Suenaga, Y., Kamae, T., Kabayashi, T.: A High-speed Algorithm for Generation of Straight Lines and Circular Arcs. IEEE Trans. Computers, Vol. C-28, No. 10, Oct. (1979) 728-736
7. Gill, G.W.: N-step Incremental Straight-Line Algorithm. IEEE Computer Graphics and Applications, Vol. 5, No. 3, May (1994) 66-72
8. Brons, R.: Linguistic Methods for the Description of a Straight Line on a Grid. Computer Graphics and Image Process, Vol. 9 (1979) 183-195
9. Chen, J.X., Wang, X., Bresenham J.E.: The Analysis and Statistic of Line Distribution. IEEE Computer Graphics and Applications, Vol. 22, No. 6 (2002) 100-107

PrefixUnion: Mining Traversal Patterns Efficiently in Virtual Environments

Shao-Shin Hung, Ting-Chia Kuo, and Damon Shing-Min Liu

Department of Computer Science and Information Engineering,
National Chung Cheng University,
Chiayi, Taiwan 621, Republic of China
{hss, ktc91,damon}@cs.ccu.edu.tw

Abstract. Sequential pattern mining is an important data mining problem with broad applications. Especially, it is also an interesting problem in virtual environments. In this paper, we propose a projection-based, sequential pattern-growth approach, called *PrefixUnion*. Meanwhile, we also introduce the relationships among transactions, views and objects. According to these relationships, we suggest two mining criteria — *inter-pattern growth* and *intra-pattern growth*, which utilize these characteristics to offer ordered growth and reduced projected database. As a result, the large-scale VRML models could be accessed more efficiently, allowing for a real-time walk-through in the scene.

1 Introduction

The interactive walkthrough system provides a virtual environment with complex 3D models [1, 3, 4]. On the other side, sequential pattern mining is one of the main topics in data mining methods [2, 6, 7, 8]. In this paper, we propose a mining mechanism based on *inter-pattern growth* and *intra-pattern growth*. These two pattern growth criteria are used to minimize useless pattern growth by finding these patterns, whose projected-patterns are the same, and letting them union beforehand. This results in less access times and much better performance.

The rest of this paper is organized as follows. In Section 2, the related works are given. In Section 3, the mining problem of virtual environment sequential patterns is introduced along with the notation that is used throughout the paper. The *PrefixUnion* mining algorithm is suggested in Section 4. To evaluate the efficiency of the *PrefixUnion* algorithm, our experimental results are presented in Section 5. Finally, we conclude our study in Section 6.

2 Related Works

2.1 Sequential Patterns Mining

Sequential pattern mining problem was first introduced in [6]. With the motivation of avoiding or substantially reducing the expensive candidate generation and pruning, the *FreeSpan* [9] and *PrefixSpan* [10] were proposed. On the other side, they still

V.S. Sunderam et al. (Eds.): ICCS 2005, LNCS 3516, pp. 830 – 833, 2005.

have some non-trivial costs. One is that the full length original sequence must be retained in each projected database because a pattern can be generated by any subsequence combination. This leads to many duplicated sequences involved. The other is that the growth of a subsequence is explored at any split point in a candidate sequence resulting in several possible new subsequences.

3 Problem Formulation

3.1 Notations

In this section, we introduce the terms used in our problem and mining algorithm. Let $\Sigma = \{l_1, l_2, ..., l_m\}$ be a set of *m* literals called *objects* (also called *items*) [1, 6]. A *view v* is denoted by $v = <\chi_1, \chi_2, ..., \chi_k>$, is an unordered list of objects such that each objects $\chi_i \in \Sigma$. The *view v* is defined as whatever the user stays and observes during the processing of virtual environments. A *sequence S*, denoted by $\{v_1, v_2, ..., v_n\}$, is an ordered list of *n* views. Let the database *D* be a set of sequences (also called transactions). Each sequence records each user's traversal path in walkthrough system. A sequence $\beta = <\beta_1, \beta_2, ..., \beta_\kappa>$ is a *subsequence* (or is called *contained*) of sequence $\alpha = <\alpha_1, \alpha_2, ..., \alpha_n>$ if there exists $1 \le i_1 < i_2 < ... < i_\kappa \le n$ such that $\beta_1 \subseteq \alpha_1, \beta_2 \subseteq \alpha_2, ..., \beta_k \subseteq \alpha_k$ holds. The *support* of a sequence *p* in the sequence database *D* is defined as the number of the sequences which contain this pattern *p*. A *frequent sequence* is a sequence whose support is equal to or more than the user defined threshold (also called *min_support*). A *frequent pattern* is a *maximal sequence* that is *frequent*. Finally, let *P* be a set of all frequent patterns in *D*.

4 *PrefixUnion Mining Algorithm*

In order to realize our pattern-growth approach, we will define two kinds of pattern-growth types. One is the *intra-view growth* – the growth of pattern is bounded by the

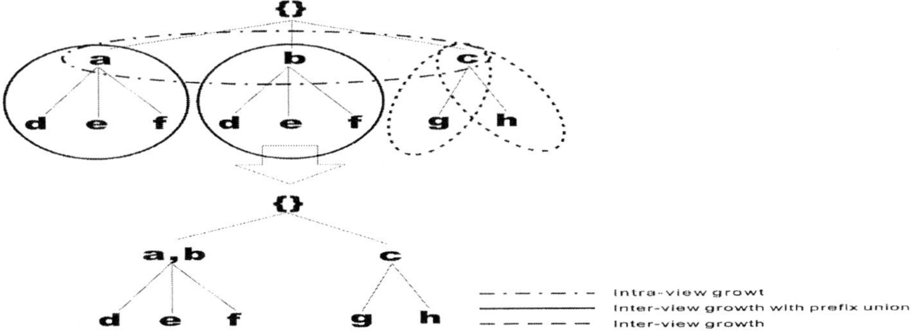

Fig. 1. Scenario of differences among intra-view growth, inter-view growth and inter-view growth with *PrefixUnion*

view boundary. The other is the *inter-view growth* – the growth of pattern is to select an object in next view. The pattern-growth algorithm is based on recursively constructing the patterns. The *α-projected* sub-database is defined as the set of subsequences in the database which the *suffixes* of the sequences have the common *prefix* *α*. In order to demonstrate this concept, these relationships are shown in the Figure 1.

In Figure 1, objects *a*, *b*, and *c* are contained in the same view (i.e., the intra-view growth). Object *a* and its siblings belong to the inter-view growth. Object *b* also has the same case. Object c and its siblings is another different example of inter-view growth (i.e. *c---g* and *c---h*). Only the object *a* and object *b* can be applied by *PrefixUnion* approach. After the processing, both objects *a* and *b* are merged. By this *PrefixUnion* property, the search space is reduced sharply in each step. This property improves much better performance, especially in the presence of small minimum support threshold.

5 Experimental Results

In this section, the effectiveness of the proposed mining algorithm is investigated. The data set is about 437 MB of 1,256 objects in the system. The traversal path consist of approximately 10 ~ 15 views from one end to the other end.

In the Figure 2, our algorithm outperforms the pattern growth method. If the locations of any two objects are too far in the virtual environments, it is naturally assumed that there are no relationships between the objects. In other words, the relationships are limited and restricted compared to the IBM dataset.

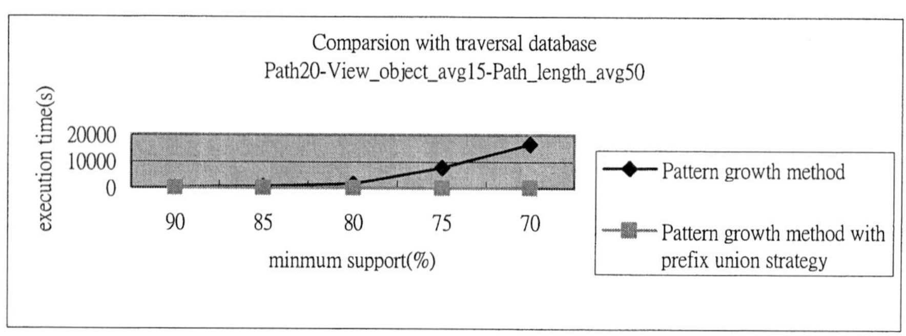

Fig. 2. Comparison of traditional pattern growth mining algorithm and our *PrefixUnion* mining algorithm in our virtual environments

6 Conclusions and Future Work

We have extended the applications of mining. With properties of intra-view pattern growth and inter-view pattern growth added, it is more precise and useful for us to discover the frequent traversal patterns. Besides, we also consider how to efficiently mining the necessary patterns in order to speed up the computations.

References

[1] Daniel G. Aliaga, and Anselmo Lastra, "Automatic Image Placement to Provide a Guaranteed Frame Rate", *Proceedings of the 26th Annual Conference on Computer Graphics and Interactive Techniques*, pp. 307-316, 1999.

[2] Ming-Syan Chen, Jong Soo Park, and Philip S. Yu, "Efficient Data Mining for Path Traversal Patterns", *IEEE Transactions on Knowledge and Data Engineering*, Vol. 10, Issue 2, pp. 209-221, 1998.

[3] Y. Nakamura, S. ABE, Y. Ohsawa, and M. Sakauchi, "A Balanced Hierarchical Data Structure Multidimensional Dada with Efficient Dynamic Characteristic", *IEEE Transactions on Knowledge and Data Engineering*, Vol. 5, No. 4, pp. 682-694, 1993.

[4] Y. Nakamura and T. Tamada, "An Efficient 3D Object Management and Interactive Walkthrough for the 3D Facility Management System", *Proc. IECON'94*, Vol. 2, pp. 1937-1941, 1994.

[5] T. Morzy, M. Wojciechowski, and M. Zakrzewicz, "Pattern-Oriented Hierarchical Mining", *Proc. of the 3rd East European Conference on Advances in Databases and Information Systems (ADBIS'99)*, Maribor, Slovenia, LNCS 1691, Springer-Verlag, 1999.

[6] Rakesh Agrawal, Tomasz Imielinski and Arun N. Swami, "Mining Association Rules between Sets of Items in Large Databases", *Proceedings of the 1993 ACM SIGMOD International Conference on Management of Data*, pp.207-216, 1993.

[7] Rakesh Agrawal, and R. Srikant, "Mining Sequential Patterns", *Proceedings of the 1995 International Conference on Data Engineering (ICDE'95)*, pp.3-14, 1995.

[8] Rakesh Agrawal, and R. Srikant, "Mining Sequential Patterns: Generalizations and Performance Improvements", *Proceeding Fifth International Conference on Extending Database Technology (EDBT'96)*, pp.3 -17, Mar 1996.

[9] J.pei, J.han, B. M-Asl, J. Wang, H. pinto, Q. Chen, U. Dayal, and M.-C. Hsu, "FreeSpan: Frequent Pattern-Projected Sequential Pattern Mining", *Proceedings of the 2000 ACM SIGKDD International Conference Knowledge in Database (KDD'90)*, pp. 355-359, August 2000.

[10] J.pei, J.han, B. M-Asl, J. Wang, H. pinto, Q. Chen, U. Dayal, and M.-C. Hsu, "PrefixSpan: Mining Sequential Patterns Efficiently by Prefix-Projected Pattern Growth", Proceedings of the 2001 International Conference Data Engineering (ICDE'01), pp. 215-224, April 2001.

Efficient Interactive Pre-integrated Volume Rendering

Heewon Kye[1,3], Helen Hong[2], and Yeong Gil Shin[1,3]

[1] School of Computer Science and Engineering, Seoul National University
{kuei, yshin}@cglab.snu.ac.kr
[2] School of Computer Science and Engineering, BK21: Information Technology, Seoul National University, San 56-1 Shinlim-dong Kwanak-gu, Seoul 151-742, Korea
hlhong@cse.snu.ac.kr
[3] INFINITT Co., Ltd., Taesuk Bld., 275-5 Yangjae-dong Seocho-gu, Seoul 137-934, Korea

Abstract. Pre-integrated volume rendering has become one of the most efficient and important techniques in three dimensional medical visualization. It can produce high-quality images with less sampling. However, two important issues have received little attention throughout the ongoing discussion of pre-integration: Skipping over empty-space and the size of lookup table for a transfer function. In this paper, we present a novel approach for empty-space skipping using the *overlapped-min-max block*. Additionally, we propose a new approximation technique to reduce the dependent texture size so that it decreases the size of texture memory and the update time. We demonstrate performance gain and decreasing memory consumption for typical renditions of volumetric data sets.

1 Introduction

Pre-integrated volume rendering is a technique for reconstructing the continuous volume rendering integral. Utilizing a pre-processed look-up table (called pre-integration table), this method not only eliminates a lot of artifacts but also reduces the sampling rate for rendering. However, since this method uses two consecutive sample values as an index for the pre-integration table which is constructed before rendering for a given classified function, conventional acceleration techniques such as empty space skipping or interactive classification methods are not applied as it is.

Skipping empty space has been extensively exploited to accelerate volume rendering. However, pre-integrated volume rendering samples two consecutive points as a line segment, previous empty-space skipping methods could not be directly applied. The pre-integration table is indexed by three integration parameters: two consecutive sample values and the distance between those samples. To accelerate the pre-integration step, Engel *et al.* reduced the dimensionality of the table from three to two by summing a constant sampling distance [1]. Even though they used a two-dimensional pre-integration table, it is still bulky when rendering high-precision data such as 12 bits-per-voxel data which is common in medical applications. A 12-bit image requires 256 times more memory and updating time than an 8-bit image.

In this paper, we present a novel data structure, called the *overlapped-min-max block* for applying empty-space scheme to the pre-integrated volume rendering, and a new approximation technique for reducing the dimensionality of the table from two to

V.S. Sunderam et al. (Eds.): ICCS 2005, LNCS 3516, pp. 834 – 837, 2005.

one. We implement them on recent consumer graphics hardware and on software-only shear-warp rendering [2] and ray-casting [3]. With our accelerations, the rendering and classification speed is much faster for medical datasets while maintaining the image quality.

2 Overlapped Min-max Block for Empty-Space Skipping

Traditional rendering methods sample a value at a point in three-dimensional space to get a color and opacity. If a block is entirely transparent, additional samplings in the block can be skipped. Pre-integrated volume rendering samples two points to get their color and opacity. Since two sampling points form a line segment, or a slab, all the blocks that intersect the line segment should be transparent for skipping the sampling process. As shown in Fig. 1, a line segment may intersect at most three blocks in two-dimensional representation. Retrieving information three-times from the lookup table degrades the rendering performance. In addition, there is an overhead to determine which blocks are transparent (there are two cases such as Fig. 1a and Fig. 1b).

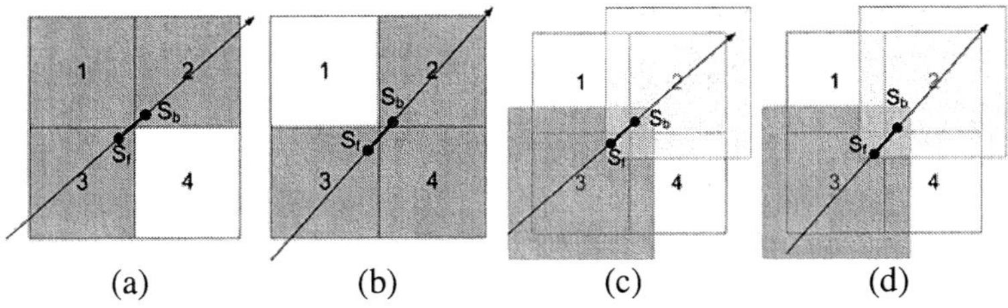

Fig. 1. The overlapped min-max table for pre-integrated volume rendering. There are two sampling point S_f and S_b, and four blocks from *block1* to *block4*. S_f is in *block3* and S_b is in *block2*. To skip the line segment S_fS_b, (a) *block1*, *block2*, and *block3* have to be transparent, (b) *block2*, *block3*, and *block4* have to be transparent in the previous block structure. Moreover, there is overhead to determine the current situation is either (a) or (b). In our method, *the overlapped-min-max block*, each block holds some region jointly with its neighbors such as (c) and (d). To skip the line segment, we can test only one block, *block2* (or *block3*)

Making each line segment belonging to only single block, we can efficiently decide whether we skip or not by testing one block. For this, we modify the region covered by each block. Each block covers some region of which thickness is at least the sampling distance as shown in Fig. 1c and Fig. 1d. By overlapping the region of each block, we can easily test whether the block of a line segment is transparent using only that block. This scheme is especially efficient on graphics hardware for its simplicity.

3 Efficient Pre-integration Table

In accelerated pre-integrated rendering, opacity of i-th sample (α_i) is written as:

$$1-\alpha_i = 1-\alpha(s_f, s_b) \approx \exp\left(-\frac{1}{s_b - s_f}\left(T(s_b)-T(s_f)\right)\right) \qquad (1)$$

$$where,\quad T(s) := \int_0^s \tau(s)ds.$$

Because equation (1) comprehends the ray segment integral of a transfer function, we do not need to consider the maximum of the Nyquiest frequencies of the transfer functions $\tau(s)$ with the scalar field s. Therefore, it is sufficient to sample a volume with relatively low frequency. The 2D lookup table to obtain α_i, requires a texture of which size is N^2, where N is density range. In order to generate color images, each texture entry requires 4 bytes. When an image has 12-bit pixel depth, the required texture size becomes 4096^2 and the required memory is 4096^2x4byte=64MB. The bigger lookup table brings the longer generation time and lower cache-efficiency. In this paper, we propose an 1D-texture lookup method that needs only 4096 entries. Because scalar values s are usually quantized equation (1) can be rewritten as:

$$1-\alpha_i \approx \sqrt[s_b-s_f]{\exp\left(-\sum_{s=s_f}^{s_b}\tau(s)\right)} = \sqrt[s_b-s_f]{\prod_{s=s_f}^{s_b}\exp(-\tau(s))} \approx \frac{1}{s_b-s_f}\sum_{s=s_f}^{s_b}\exp(-\tau(s)) = \left(S(s_b)-S(s_{f-1})\right)/(s_b-s_f) \tag{2}$$

$$\textit{where,}\quad S(x)=\sum_{s=0}^{x}\exp(-\tau(s)),\ S(-1)=0.$$

We modify a geometric average of transparency, into an arithmetic average as shown in equation (2). Therefore, only 1D-texture $S(x)$ needs to be stored to get the α_i. The color can be formulized in a similar fashion.

4 Experimental Results

Experiments have been performed on a PC equipped with Pentirum4 2.8GHz processor, 1GB main memory and an ATI 9800. Table 1 summarizes the performance using empty-space skipping (ESP). Obviously, skipping empty-space gains more performance. In hardware rendering, rendering time is reduced two times or more by using ESP. In software rendering, the gain of ESP is much bigger (10-30 times faster than w/o ESP). Since the angio dataset contains more empty space than the head dataset, the performance improvement factor of the angio dataset is bigger than that of the head dataset.

Table 1. A performance comparison in case of using empty-space skipping (msec)

Dataset	3D Hardware		Ray-Casting		Shear-Warp	
	with ESP	w/o ESP	with ESP	w/o ESP	with ESP	w/o ESP
BigHead	83.8	302.1	897.7	12409	747.6	7904
Angio	95.1	392.1	737.6	16183	339.6	9984

If a transfer function is fixed using 2D-texture (88.5ms) slightly faster than our 1D-texture (95.1ms) since 1D-texture requires two texture loads for classification while 2D-texture requires one. However, when the transfer function is changed, using our 1D-texture (101.2ms) is much faster than 2D-texture (824.4ms) because of lookup table creation time. Fig. 2 show rendered images of BigHead and Angio volume data, respectively. There is no noticeable difference between using 1D-texture and 2D-texture.

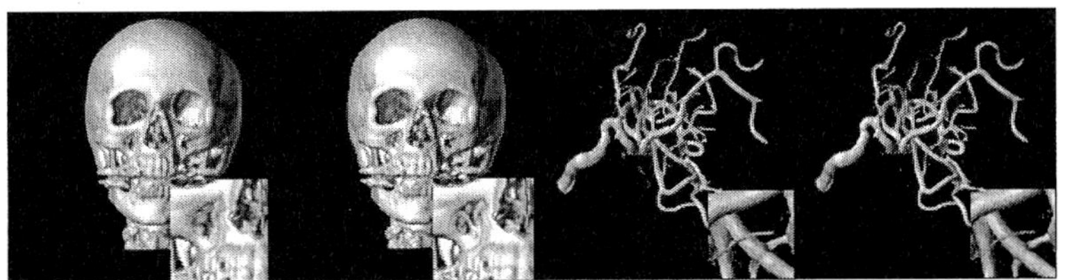

Fig. 2. The comparison of image quality for BigHead and Angio volume using (left) 1D-texture, (right) and 2D-texture

5 Conclusions

In this paper, we have proposed a new method to accelerate traversal and classification of both hardware and software based pre-integrated volume rendering. Using the overlapped-min-max block, empty-space skipping can be accomplished more efficiently and can be easily implemented in a hardware-based method. To reduce the classification time and memory consumption, a new approximation method of a lookup table is also proposed. With regard to image quality, we have presented the minimum bound of error theoretically. Experimental results show that our method produces the same quality of rendered images as the previous classification method of pre-integration.

References

1. K. Engel, M. Kraus, and T. Ertl. High-Quality Pre-Integrated Volume Rendering Using Hardware-Accelerated Pixel Shading. *Eurographics / SIGGRAPH Workshop on Graphics Hardware,* Los Angeles, CA, USA, August 2001.
2. P. Lacrout and M. Levoy, Fast Volume Rendering Using a Shear-Warp Factorization of the Viewing Transformation. *Proceedings of SIGGRAPH 94*, 451-458, 1994.
3. M. Levoy, Display of Surfaces from Volume Data, *IEEE Computer Graphics & Application*, 8: 29-37, 1988.

Ncvtk: A Program for Visualizing Planetary Data

Alexander Pletzer[1,4], Remik Ziemlinski[2,4], and Jared Cohen[3,4]

[1] RS Information Systems
[2] Raytheon
[3] The College of New Jersey
[4] Geophysical Fluid Dynamics Laboratory, Princeton NJ 08542, USA

Abstract. Ncvtk is a visualization tool offering a high degree of interactivity to scientists who need to explore scalar and vector data on a longitude-latitude based grid. Ncvtk was born out of four recent trends: open source programming, the availability of a high quality visualization toolkit, the emergence of scripting, and the wide availability of fast and inexpensive hardware.

1 Why Ncvtk?

In geophysical sciences as in many other areas, the volume of data generated by simulation codes has steadily increased over time. Present climate modeling codes can generate tens of Gigabytes (Gbytes) of data for a single run. Plowing through such amounts of data requires visualization tools that go well beyond simple plotting utilities. The challenge then is to provide a simple visualization tool that can be used by non-experts.

We have developed Ncvtk [1] to offer climate scientists a means to explore data interactively and intuitively. Ncvtk was designed from the ground up to be straightforward to use. No prior programming or other experience is required. We believe users should be no more than a few mouse clicks away from seeing their data. Moreover, in contrast to a number of proprietary programs, Ncvtk is compiled from source and thus can accommodate rapid advances in software libraries, compiler, and hardware. This distinguishes Ncvtk from most commercial visualization tools.

Ncvtk focuses on planetary science data, which lie on a spherical domain and are elevation- and/or time-dependent. Typically, the maximum elevation/depth is three orders of magnitudes smaller than the horizontal length scale. For most purposes, the data are heavily stratified in elevation. Therefore, it is convenient to regard the data as lying flat on the surface of a sphere with the most interesting phenomena taking place on the longitude-latitude plane.

2 Building Blocks

We have chosen to build Ncvtk by using the Visualization Toolkit (VTK) [2] as our foundation. VTK is a powerful object oriented, 3D visualization library,

V.S. Sunderam et al. (Eds.): ICCS 2005, LNCS 3516, pp. 838–841, 2005.

which is open-source and is being actively developed. VTK is known to run on many Unix flavors (Linux x86, ia64, Mac OS X) and Windows. Contrary to other visualization packages, VTK does not provide a visual programming interface or a custom scripting language. Instead, VTK is written in C++ while offering bindings to the Java, Tcl and Python programming languages.

We settled our choice on Python, an object oriented scripting language that offers a vast collection of modules (graphical user interface, regular expression engine, calendar arithmetic, etc.). Such capability is not readily available, out-of-the-box in C/C++. Numerically demanding tasks are carried out using the Python Numeric module [3], whose performance we found to be on par with compiled code. By some fortuitous coincidence, it was realized that VTK methods expecting a raw pointer array can also take a Python Numeric array. This allows communication between Python and C++ without the need for data copying.

The climate simulation codes of our users comply to the Flexible Modeling System (FMS) [4] guidelines, which require data to be written in the self-described NetCDF file format. Attributes of fields such as name, physical units, axes dependency, etc. are stored within the file. Ncvtk assumes that the simulation codes loosely follow the CDC conventions [5] for describing valid data range, missing values, whether the data are packed or not, and whether the grid is rectilinear or warped. NetCDF data may consist of floats, doubles or shorts (when packed). Ncvtk will automatically cast the data to working precision (float or double) without the need for user intervention.

3 Example

Some of Ncvtk's capabilities are now illustrated. Figure 1 was obtained from a run performed by the non-hydrostatic, atmospheric code ZETAC, which solves the compressible fluid equations for velocity and pressure using a terrain following vertical coordinate. This makes ZETAC ideal for studying cyclogenesis in the vicinity of mountain ridges [6].

A color pipeline reading a color map from an external file was used to represent the topology and a piecewise linear opacity function was applied to indicate the presence of clouds. The last layer consists of labeled temperature contours and colored pressure contours, all at surface level. The horizontal wind velocity field at surface level is shown as white arrow glyphs. The opacity of the glyphs increases with wind magnitude. To emphasize the hour of the day, an external light simulating the sun was added.

Each visual pipeline (color plot, opacity, contour, vector, etc.) comes with its own graphical user interface containing a field selector, an elevation slider and specific widgets for user interactivity. Any number of pipelines can be instantiated. A pipeline can be removed simply by destroying the graphical user interface window.

In this onion skin approach, an arbitrary number of fields can be displayed simultaneously, layered on top of each other. Field values may also be probed by moving the mouse over regions of interest. Since only two-dimensional fields

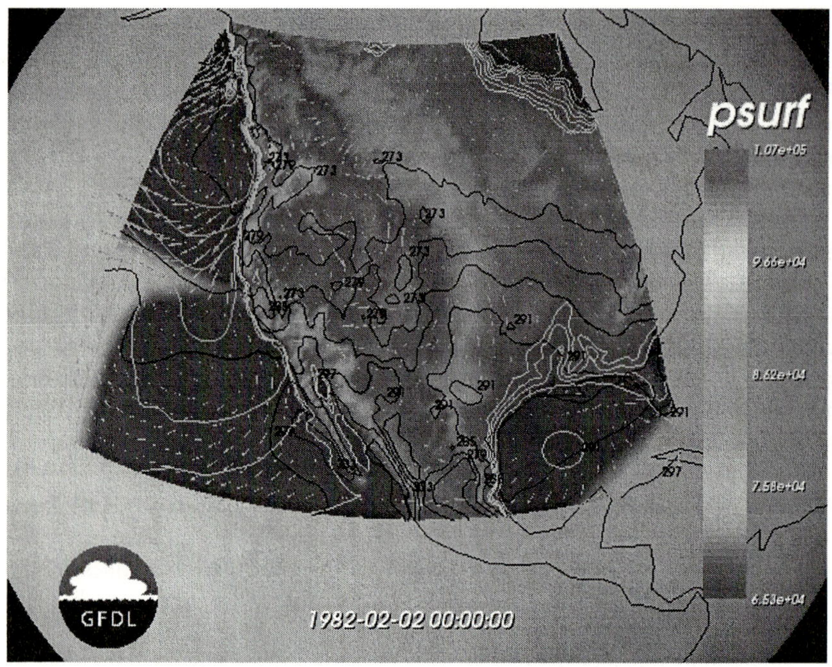

Fig. 1. Non-hydrostatic atmospheric simulation obtained using the ZETAC code

need to be stored at any given time and elevation, the memory footprint remains modest even in the case of Gbyte-sized data sets. This feature allows Ncvtk to run on personal computers with typically less than 1 Gbyte of memory.

Note that most of the heavy number crunching involves computing structured grid coordinates and evaluating minimum/maximum data ranges across space and time. Recall that the data of a single field can amount to Gbytes. To preserve responsiveness in Ncvtk, the data range calculation is performed by subsampling in time and elevation.

4 Summary

Ncvtk was released in October 2004 and an executable version has since been made available to users at the Geophysical Fluid Dynamics Laboratory. Before Ncvtk, the tool of choice was Ncview [7], a program using color texture to represent scalar fields on a plane. Ncview is straightforward to use and provided our initial inspiration. Like Ncvtk, Ncview reads data from a NetCDF file and is targeted at planetary sciences. Ncview's main advantage over Ncvtk is that it is lightweight and fast. We have attempted to emulate Ncview's ease of use while adding many more features including support for multiple fields, multiple staggered grids, warped coordinates, opacity and contour plot, and vector fields.

A workshop on Ncvtk has recently been organized [8] attracting about 30 participants. This provided constructive feedback to Ncvtk's developers. It appears that scientists especially value accuracy, robustness and ease-of-use.

By implementing Ncvtk in Python, we were able to leverage on a vast body of existing Python modules and so cover significant ground in a few months of development time. Few application specific programming languages that ship with many commercial visualization packages can claim to offer comparable scope to that of a generic programming language. Among the many Python features that played a prominent role in Ncvtk one can cite lists and hash tables to represent object collections, module unit tests to allow classes to be tested in isolation, and on demand, event-driven programming. Thanks to Python's dynamic nature, a scripting programming interface that maps the GUI functionality on a one-to-one basis could be written with little extra effort. In retrospect, Python's biggest strength may have been, however, the error handling that comes built into the language. Python failures do not lead to segmentation violation or other forms of hard crashes, conferring to Ncvtk a high degree of robustness. We believe that many projects could benefit from the shift from compiled code to scripting.

Ncvtk's design emphasizes scalability in performance and functionality. So far we have only scratched the surface of VTK' capability; there remains ample room to add more functionality to Ncvtk. Extensions that are presently being considered include vertical slices of scalar fields, texture based visualization of vector fields, the ability to read files from several NetCDF files simultaneously, support for several scene renderers, a mechanism for adding user extensions (add-ons), and the ability to plot vertical profiles. The challenge will be to allow Ncvtk to evolve while remaining simple to use, robust and predictable.

Acknowledgments

We would like to thank Drs C. L. Kerr and S. T. Garner for providing the impetus for this work and for providing initial test data. Ncvtk is graciously hosted by SourceForge [9], which includes a CVS repository, a bug tracking tool, a feature request tool, and mailing lists. Ncvtk would not exist without the freely available VTK library [2].

References

1. http://ncvtk.sourceforge.net
2. http://public.kitware.com/VTK/
3. http://www.pfdubois.com/numpy/
4. http://www.gfdl.noaa.gov/~fms/
5. http://www.cdc.noaa.gov/cdc/conventions/cdc_netcdf_standard.shtml
6. Orlanski, I., Gross, B. D.: Orographic modification of cyclode development. J. Atmos. Sci., **51**, 589–611.
7. http://meteora.ucsd.edu/~pierce/ncview_home_page.html
8. http://ncvtk.sourceforge.net/ncvtk_tutorial.pdf
9. http://sourceforge.net

Efficient Multimodality Volume Fusion Using Graphics Hardware

Helen Hong[1], Juhee Bae[2], Heewon Kye[2], and Yeong Gil Shin[2]

[1] School of Computer Science and Engineering,
BK21: Information Technology, Seoul National University
hlhong@cse.snu.ac.kr
[2] School of Computer Science and Engineering, Seoul National University,
San 56-1 Shinlim 9-dong, Kwanak-gu, Seoul 151-742, Korea
{jhbay, kuei, yshin}@cglab.snu.ac.kr

Abstract. We propose a novel technique of multimodality volume fusion using graphics hardware that solves the depth cueing problem with less time consumption. Our method consists of three steps. First, it takes two volumes and generates sample planes orthogonal to the viewing direction following 3D texture mapping volume rendering. Second, it composites textured slices each from different modalities with several compositing operations. Third, alpha blending for all the slices is performed. For the efficient volume fusion, a pixel program is written in HLSL(High Level Shader Language). Experimental results show that our hardware-accelerated method distinguishes the depth of overlapping region of the volume and renders them much faster than conventional ones on software.

1 Introduction

In clinical medicine, several different modalities such as Positron Emission Tomography (PET), Computed Tomography (CT), and Magnetic Resonance Imaging (MRI) are useful for radiologists and surgeons to help diagnosis support and treatment planning. These modalities give complementary information when they are shown simultaneously. Thus many medical applications require visual output generated from multimodality volumes rather than only one volume.

Several methods have been proposed to combine multiple volumes obtained from multimodality imaging. Cai and Sakas [1] proposed three different intermixing approaches. The shortcoming of the method was a lack of precise depth cueing among the multiple volumes, even though they tried to use Z-buffer value. Jacq and Roux [2] presented a multi-volume rendering focused on material classification which applied different material percentages and merging rules at each sampling point. Zuiderveld et al. [3] described how to cache calculated voxel properties using sized hash table that contribute to the final result. Previous researches have performed multimodality volume fusion on software. However, the software-based multimodality fusion has a limitation in depth presentation and processing time. In this paper, we propose an efficient hardware-accelerated solution to combine multimodality volumes at interactive rates.

V.S. Sunderam et al. (Eds.): ICCS 2005, LNCS 3516, pp. 842 – 845, 2005.

2 Multimodality Volume Fusion

Fig. 1 shows the pipeline of our method for multimodality volume fusion. We assume reference and float volume datasets have the same orientation by setting rigid transformation that would be sufficient for the registration of the datasets [4].

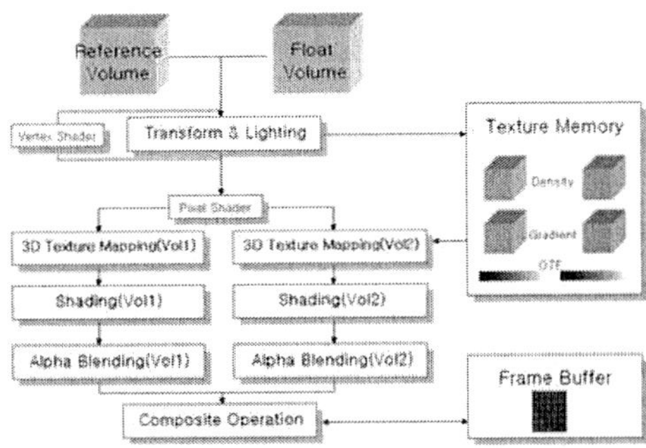

Fig. 1. The pipeline of multimodality volume fusion

2.1 Depth Compositing of the Volume

Since multiple volumes have different size and orientation, it needs correct depth composition and orientation. Fig. 2 shows different ways to combine two slices from each volume. In Fig. 2(a), it is required to set the orders of each slice of the volumes which is a troublesome work. We combine each texture with same texture coordinates from different volumes before blending as shown in Fig. 2(b). Two voxels (ultimately texels) referred to the same texture coordinates are combined by the compositing operations and then they are mapped to the same slice.

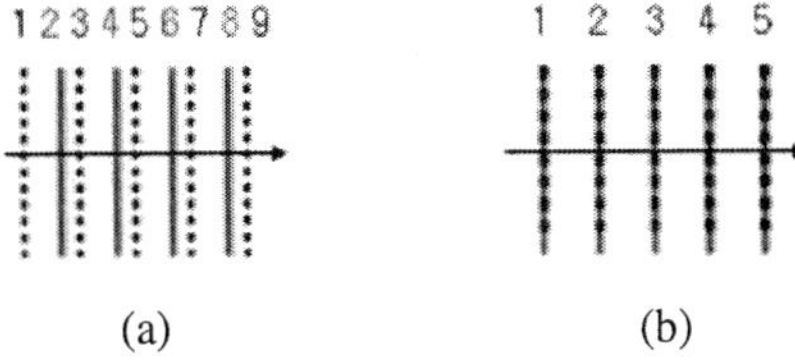

Fig. 2. A comparison of blending textures (black dotted line: slice of reference volume, red solid line: slice of float volume) (a): alpha blending in slice orders (b): combine textures before alpha blending

2.2 Compositing Operations for Volume Fusion

Several compositing operations are experimented to find an appropriate combining method; some are referred from the previous compositing researches [7][8].

The results of combined color and opacity of over, XOR, plus and weighted addition operation are shown from Eq. (1) to Eq. (4), respectively.

$$C = \frac{\alpha_A \cdot C_A + (1-\alpha_A)\alpha_B \cdot C_B}{\alpha_A + (1-\alpha_A)\cdot \alpha_B} \qquad (1)$$
$$\alpha = \alpha_A + (1-\alpha_A)\cdot \alpha_B$$

$$C = \frac{(1-\alpha_B)\cdot \alpha_A \cdot C_A + (1-\alpha_A)\alpha_B \cdot C_B}{(1-\alpha_B)\cdot \alpha_A + (1-\alpha_A)\cdot \alpha_B} \qquad (2)$$
$$\alpha = (1-\alpha_B)\cdot \alpha_A + (1-\alpha_A)\cdot \alpha_B$$

$$C = C_A + C_B \qquad (3)$$
$$\alpha = \alpha_A + \alpha_B$$

$$C = \frac{weight \cdot \alpha_A \cdot C_A + (1-weight)\cdot \alpha_B \cdot C_B}{weight \cdot \alpha_A + (1-weight)\cdot \alpha_B} \qquad (4)$$
$$\alpha = weight \cdot \alpha_A + (1-weight)\cdot \alpha_B$$
$$(0 \le weight \le 1)$$

where C indicates color, α indicates alpha, A refers to the first volume, and B to the second volume.

2.3 Weighted Opacity for Surface Extraction

To see a float volume through a reference volume, we consider weight to the opacity as shown in Eq. (5). The weight value tells whether the pixel is a part of the surface or a part of the homogeneous region by gradient magnitude. In the preprocessing step, gradient magnitude is used as normal for shading. Here, we use gradient magnitude as a numerical divergence which implies that the surface has higher value while homogeneous part has a lower value. During fusion process, we let the homogeneous area to be transparent so that we could see through the reference volume by eliminating the opacity value of the homogeneous area between reference and float volume of the overlapping area.

$$\text{weight} = \begin{cases} 0 & if(|\nabla| < c) \quad (c: constant) \\ 1 & else \end{cases} \qquad (5)$$

3 Experimental Results

All our implementation and tests have been performed on an Intel Pentium IV PC containing 2.53 GHz CPU with ATI Radeon X800 256 MB RAM. Our method

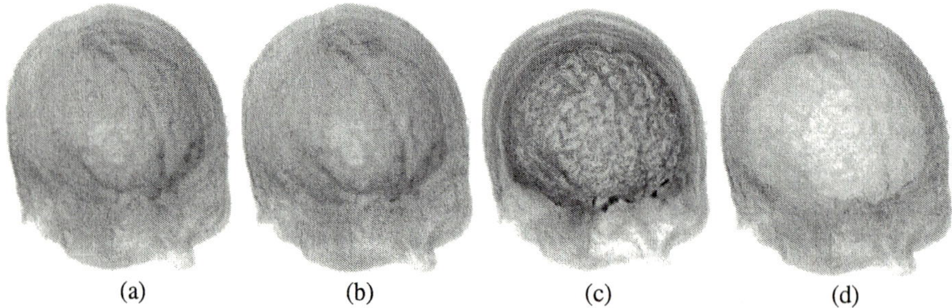

Fig. 3. The results of compositing operations at opacity 8. (a) over (b) XOR(exclusive OR) (c)weight addition (d) plus operation

has been applied to the brain of 256 x 256 x 96 MR images and 128 x 128 x 40 PET images with 2 bytes-intensity. The performance of our method is evaluated with the aspects of visual inspection and processing time.

Fig. 3 displays the results using several compositing operations. Fig. 3(a) and Fig. 3(b) show the result of over operation and XOR(exclusive OR). Fig. 3(c) using weight addition displays rather a dark but manifest result compared to others. Fig. 3(d) represents a result of plus operation to be mainly white because the colors are simply added which induces overflow.

The average of the processing time of the fusion volume supported by software is 3703 msec while fusion on hardware is 625 msec, respectively. We gained about 5 to 6 times of speed enhancement on rendering time.

4 Conclusion

We have developed an efficient multimodality volume fusion method using graphics hardware. Our method presents several approaches of combining two different volume datasets into a singular image representation. MR and PET images of the brain have been used for the performance evaluation with the aspects of visual inspection and processing time. Our method shows the exact depth of each volume and the realistic views with interactive rate in comparison with the software-based multimodality volume fusion. Distinguishable and fast result of the proposed method can be successfully utilized in medical diagnosis.

References

1. Cai W., Sakas G., Data Intermixing and Multi-volume Rendering, Computer Graphics Forum (1999), 18(3): 359-368.
2. Jaeq, J., Roux, C., A Direct Multi-volumes Rendering Methods Aiming at Comparison of 3D Images and Methods, IEEE Trans. On Information Technology in Biomedicine (1997) 1(1):30-43.
3. Zuiderveld, K.J., Viergever, M.A., Multi-modal Volume Visualization using Object-oriented methods, in Proc. IEEE/ACM Volume Visualization '94 Symposium (1994) 59-66.
4. Hong, H., Shin, Y., Intensity-based registration and combined visualization of multimodal brain images for noninvasive epilepsy surgery planning, Proc. of SPIE Medical Imaging (2003).

G^1 Continuity Triangular Patches Interpolation Based on PN Triangles

Zhihong Mao, Lizhuang Ma, and Mingxi Zhao

Dept. of Computer Science and Engineering, Shanghai Jiao Tong University, PR China
{mzh_yu, ma-lz, zhaomx}@sjtu.edu.cn

Abstract. There are currently many methods for triangular local interpolation: Given triangular meshes P in three dimension space, the given flat triangles are based only on the three vertices and three normal vectors, or PN triangles, construct a smooth surface that interpolates the vertices of P. In this paper a completely local interpolation scheme is presented to guarantee to join patches G^1 continuously around boundary curves of each PN triangle.

1 Introduction

It is well-known that the interpolation of curved triangular patches over PN triangular meshes, each triangle is based on the three vertices and the corresponding normal vectors, is an important tool in computer aided geometric design. The common approach [1,2,3] is to firstly create a cross boundary tangent vector field for each boundary and then to construct patches that agree with these cross boundary fields. Steven [4] gave a unifying survey of the published methods. Stephen Mann [5] discusses a method for increasing the continuity between two polynomial patches by adjusting their control points. But all these methods must know the information of two adjacent patches and they are not completely local method.

Stefan Karbacher [6] present a non-linear local subdivision scheme for the refinement of triangle meshes. Overveld [7] gave an algorithm for polygon subdivision based on point-normals. Alex Vlachos [8] introduced curved point-normal (PN) triangles to replace the flat triangle. In these methods the authors only consider the point-normals of the input triangle and they are completely local. But they only construct C^0 continuity meshes. In this paper our objective is to construct a smooth G^1 continuity surface.

2 Triangular G^1 Local Interpolation Based on PN Triangles

In this paper we shall employ the quartic Bézier polynomial to define a triangular patch with linear normal patch:

$$\rho(\lambda_1,\lambda_2,\lambda_3)=\sum_{i=0}^{4}\sum_{j=0}^{4-i}\mathrm{b}_{i,j,k}\mathrm{B}^4{}_{i,j,k}(\lambda_1,\lambda_2,\lambda_3);\quad B^4{}_{i,j,k}(\lambda_1,\lambda_2,\lambda_3)=\frac{4!}{i!\,j!k!}\lambda_1{}^i\lambda_2{}^j\lambda_3{}^k \tag{2.1}$$

$$Q_1(\lambda_1,\lambda_2,\lambda_3)=\lambda_1 q_{1,0,0}+\lambda_2 q_{0,1,0}+\lambda_3 q_{0,0,1} \tag{2.2}$$

V.S. Sunderam et al. (Eds.): ICCS 2005, LNCS 3516, pp. 846–849, 2005.

$b_{i,j,k}$ and $q_{i,j,k}$ are respectively the control points of curved patches of the surface ρ_i and corresponding linear normal patches Q_l .

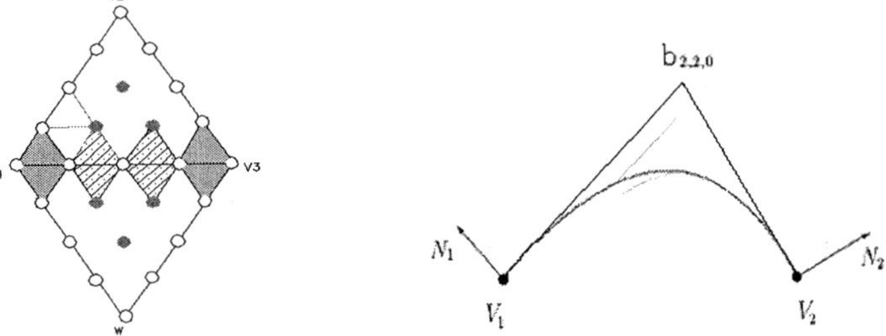

Fig. 1. Two patches meeting along a common boundary

Fig. 2. Determination of $b_{2,2,0}$

As illustrated schematically in Fig.1, these two patches will share boundary edge V_1V_3. To meet with G^1 continuity, each of the four panels of four control points must be coplanar. We get G^1 continuity by adjusting the inner control points [5]. As illustrated by red lines, an inner point has four constrains, but only three freedoms. In order to solve the problem, we make the gray triangle degenerate to become a point. So for quartic Bézier polynomial control points we have:

$$b_{4,0,0} = b_{3,1,0} = b_{3,0,1}\,;\; b_{0,4,0} = b_{1,3,0} = b_{0,3,1}\,;\; b_{0,0,4} = b_{0,1,3} = b_{1,0,3} \quad (2.3)$$

2.1 Determination of Boundary Curve

In the following parts of this section we only consider the edge $e_3\,(V_1V_2, \lambda_3 = 0)$ (see Fig. 2). We have $b_{4,0,0} = b_{3,1,0}$ and $b_{0,4,0} = b_{1,3,0}$, the next is to determine $b_{2,2,0}$. Here we decide the point $b_{2,2,0}$ by the intersection of three planes: 1) The first plane is decided by the point V_1 and the corresponding normal N_1; 2) The second plane is decided by the point V_2 and the corresponding normal N_2; 3) The third plane is decided by two lines, the one line is V_1V_2, another line is $((V_1+V_2)/2+(N_1+N_2)/2)$.

2.2 Construct the Boundary Normal Curves

From equation (2.2) we can get $Q_1(\lambda_1, \lambda_2, 0) = \lambda_1 q_{1,0,0} + \lambda_2 q_{0,1,0}$ (on edge e_3), $Q_l(\lambda_1, \lambda_2, 0)$ must interpolate N_1 and N_2: $q_{1,0,0} = \omega_1 N_1$; $q_{0,1,0} = \omega_2 N_2$, Where ω_1, ω_2 are positive constants. If the boundary curve is G^1 continuity, we must have

$$Q_1 \bullet \left(\frac{\partial \rho}{\partial \lambda_1} - \frac{\partial \rho}{\partial \lambda_2}\right) = 0\,, \text{ or } \; [\lambda_1 q_{1,0,0} + \lambda_2 q_{0,1,0}] \bullet [(b_{4,0,0} - b_{3,1,0})\lambda_1^{\,3}$$

$$+ 3(b_{3,1,0} - b_{2,2,0})\lambda_1^{\,2}\lambda_2 + 3(b_{2,2,0} - b_{1,3,0})\lambda_1\lambda_2^{\,2} + (b_{1,3,0} - b_{0,4,0})\lambda_2^{\,3}] = 0 \quad (2.4)$$

The equation equals zero means that the coefficients of each term must be zero and we have $b_{4,0,0} = b_{3,1,0}$ and $b_{0,4,0} = b_{1,3,0}$, so equation (2.4) can be simplified to the following equation:

$$(q_{1,0,0} + q_{0,1,0}) \bullet (V_1 - V_2) = 0 \tag{2.5}$$

Now the problem become to finding reasonable values for ω_1, then use equation (2.5) to compute ω_2. Here we set $\omega_1 = 1$.

2.3 Decide the Inner Control Points

The next is to decide the still unknown inner control points $b_{1,2,1}, b_{2,1,1}$ and $b_{1,1,2}$. In order to construct cross boundary G^1 continuity, we must hold (the edge e_3):

$$Q_l(\lambda_1, \lambda_2, 0) \bullet (\frac{\partial p}{\partial \lambda_2} - \frac{\partial p}{\partial \lambda_3}) = (\lambda_1 q_{1,0,0} + \lambda_2 q_{0,1,0}) \bullet (\frac{\partial p}{\partial \lambda_2} - \frac{\partial p}{\partial \lambda_3}) = 0 \tag{2.6}$$

The equation equals zero means that the coefficients of each term must be zero and associate with equation (2.5), we can get following equations:

$$N_1 \bullet (V_1 - b_{2,1,1}) = 0;\ N_2 \bullet (V_2 - b_{1,2,1}) = 0;\ \omega_2 N_2 \bullet (V_2 - b_{2,1,1}) + \omega_1 N_1 \bullet (V_2 - b_{1,2,1}) = 0 \tag{2.7}$$

Using the same computation for e_1 and e_2, we have:

$$N_3 \bullet (V_3 - b_{1,1,2}) = 0;\ N_2 \bullet (V_2 - b_{1,2,1}) = 0;\ \omega_3 N_3 \bullet (V_3 - b_{1,2,1}) + \omega_4 N_2 \bullet (V_3 - b_{1,1,2}) = 0$$

$$N_3 \bullet (V_3 - b_{1,1,2}) = 0;\ N_1 \bullet (V_1 - b_{2,1,1}) = 0;\ \omega_5 N_1 \bullet (V_1 - b_{1,1,2}) + \omega_6 N_3 \bullet (V_1 - b_{2,1,1}) = 0 \tag{2.8}$$

Unite equation (2.7) and (2.8), we get six equations, but we have 9 unknown values. Here we optimized the fairness of the given patch by reducing its curvature [9]. A standard measure for the surface quality in geometric modeling is the thin plate energy: $E(s) \approx \int_{\Delta} F^2_{uu} + 2F^2_{uv} + F^2_{vv}$. where Δ denotes the domain triangle $V_1V_2V_3$. Now we can determine the free parameters by minimizing E(s).

3 Conclusion

Methods for local interpolation of triangulated, parametric data have existed for many years and received more and more attention. However it is difficult to construct cross boundary continuity surface. So the presented local methods are C^0 continuity. In this paper we present a completely local method to construct G^1 boundary continuity patches based on the point-normals of the inputted flat triangles. The examples used and shown (see Fig.3,4) demonstrate that this algorithm can produce a very smooth mesh from an initial coarse mesh model. However in order to construct G^1 boundary

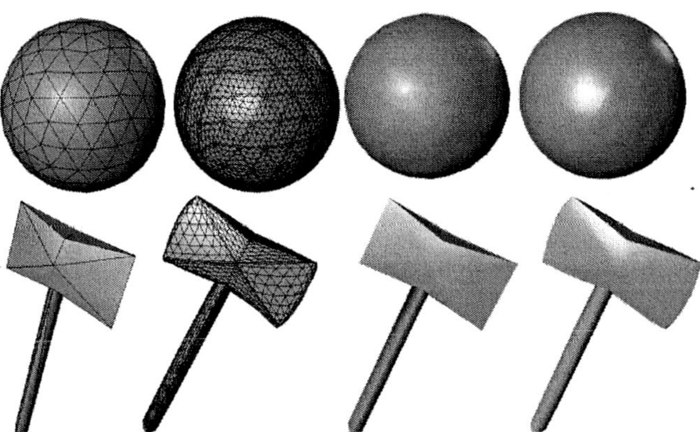

Figs. 3,4. (from left to right) (a) initial mesh (b) mesh with our method (c) shading model (initial mesh) (d) shading for our scheme

continuity patches, we make the triangle attached to each vertex degenerate to become a point and introduce non-regular points. The investigation about non-regular points is not involved in this paper.

Acknowledgments

The work was partially supported by national natural science foundation of China (Grand No. 60373070 and No. 60173035) and 863 High Tech Project of China (Grant No. 2003AA411310).

References

[1] H. chiyokura and F. Kimura. Design of solids with free-form surfaces. Computer Graphics, 17(3):289-298, 1983.

[2] G. Herron. Smooth closed surfaces with discrete triangular interpolants. CAGD, 2(4):297-306, December 1985.

[3] T. Jensen. Assembling triangular and rectangular patches and multi-variate splines. In G. Farin, editor, Geometric Modeling: Algorithms and New Trends, pages 203-220. SIAM.

[4] Steve Mann, Charles Loop, Michael Lounsbery, etc. A survey of parametric scattered data fitting using triangular interpolants. In Curve and surface Modeling. SIAM.

[5] S. Mann. Continuity adjustments to triangular bezier patches that retain polynomial precision. Research Report CS-2000-01,2000.

[6] Stefan Karbacher, Stefan Seeger and Gerd Häusler. A Non-linear Subdivision scheme for Triangle Meshes. Proc. Of Vision, Modeling and Visualization: 163-170, 2000.

[7] C.W.A.M. van Overveld and b.Wyvill. An algorithm for polygon subdivision based on vertex normals. In Computer Graphics International : 3–12, 1997.

[8] Vlachos Alex, Jörg Peters, Chas Boyd and Jason L. Mitchell. Curved PN Triangles. ACM Symposium on Interactive 3D Graphics :159-166, 2001.

[9] Moreton H. and C. Sequin, Functional optimization for fair surface design, SIGGRAPH 92 proceeding, 167-176.

Estimating 3D Object Coordinates from Markerless Scenes

Ki Woon Kwon[1], Sung Wook Baik[2], and Seong-Whan Lee[1,*]

[1] Korea University, Seoul 136-713, Korea
{kwkwon, swlee}@image.korea.ac.kr
[2] Sejong University, Seoul 143-747, Korea
sbaik@sejong.ac.kr

Abstract. This paper presents a novel method for estimating the coordinates of a 3D object using the four vertices of a quadrangle and the camera motion parameters. Estimation of 3D object coordinates from 2D images of video is a studied problem in augmented reality. However, most solutions are dependent on fiducial markers in video or known coordinate systems which are required with superimposition of virtual object on frames. In this paper, we begin with the fact that the rectangular objects in 3D real world are projected the perspective quadrangle onto image planes. We can estimate 3D object coordinates from 4 vertices of quadrangular objects through transformation of image coordinates. The camera motion parameters between pairs of successive frames in a sequence are calculated using epipolar geometry.

1 Introduction

An AR system should be able to [1] 1) combine real environments and computer-generated virtual objects, 2) operate virtual objects interactively with the change in the real world, and 3) align virtual graphic objects onto real environments. When a virtual object is superimposed in a reference frame, the frame should contain one plane with which a 3×3 planar homography can be found [2-5].

The homography is the transformation modeling the 2D movement of coplanar points under perspective projection. To obtain another planar homography between two consecutive frames in a sequence, different methods calculating camera motion to use multiple planes have been considered [4].

Kutulakos and Vallino proposed a system that can represent 3D graphic objects using four pairs of prior affine basis points that correspond to a sequence of images extracted from two uncalibrated affine cameras [6]. Another system involves a perspective camera model. This is more difficult to estimate the projective reconstruction from perspective views than using affine reconstruction from orthographic views [7].

In this paper, we estimate the direction of the Z-axis and the vertices of a quadrangle that is defined in a reference frame. The consecutive frames are computed for the essential stereoscopic matrix using epipolar geometry, and the estimated coordinates of the 3D object are determined from the camera motion parameters [8].

* Author for Correspondence.

V.S. Sunderam et al. (Eds.): ICCS 2005, LNCS 3516, pp. 850 – 853, 2005.

2 Estimating the Coordinates of a 3D Object

The rectangle is deemed to be one side of a rectangular parallelepiped. Consequently, its X-, Y-and Z-axes are at right angles to each other. Based on this fact, the Z-axis of the rotation angle is determined via complex rotations of the X-and Y-axes in the real world. The estimated direction of the Z-axis can be used to calculate the angles among the X-, Y-and Z-axes, and these angles are used when overlaying a 3D graphic object on the frame image.

To estimate the direction of the Z-axis, the image coordinates are rotated by applying the Euler-angle to each axis. The vertices and center point of a quadrilateral which the user designates from a reference frame, are applied to Equation 1.

$$\begin{bmatrix} 1+D\cdot E & -C+A\cdot H & B+A\cdot J \\ C+D\cdot H & 1+D\cdot F & -A+D\cdot I \\ -B+D\cdot J & A+D\cdot I & 1+A\cdot G \end{bmatrix} \tag{1}$$

where

$$A = a\cdot\sin\theta\,,\ B = b\cdot\sin\theta\,,\ C = c\cdot\sin\theta\,,\ D = (1-\cos\theta)$$

$$E = a^2-1\,,\ F = b^2-1\,,\ G = c^2-1\,,\ H = ab\,,\ I = bc\,,\ J = ac$$

Then, the unit vector in the direction of the Z-axis can be calculated. Equation 1 is a transformation matrix used for rotating an object about an arbitrary axis.

3 Calculating Camera Motion Parameters

We develop another method for extracting camera motion parameters using a monoscopic system. Most of the video sequences used are pictures in which the intrinsic parameters of the camera are unknown. Therefore, the intrinsic parameters of the camera are set to a fixed skew of 0, an aspect ratio of 1, and the principle point is in the center of the quadrangle. The extrinsic parameters are calculated as an essential matrix. A pair of successive frames has the similar property with images of left and right camera. To get the cross product with vector and matrix, $\boldsymbol{S}$ is a skew symmetric matrix. And, the $\boldsymbol{R}\cdot\boldsymbol{S}$ matrix, an essential matrix is computed by Equation 2.

$$\boldsymbol{q}_r^T\cdot(\boldsymbol{R}\cdot\boldsymbol{S})\cdot\boldsymbol{q}_l = \boldsymbol{0} \tag{2}$$

The essential matrix $R\cdot S$ is computed using Equation 2, using an 8-point algorithm. In addition, E in a 3 × 3 matrix is computed using Equation 3 and 4, and the property of the epipolar constraint. In order to calculate the essential matrix, we expand the equation,

$$\boldsymbol{x}'^T\,\boldsymbol{E}\boldsymbol{x} = 0 \tag{3}$$

For 8 point correspondences, Equation 6 becomes

$$Ae = 0 \tag{4}$$

where

$$A = \begin{bmatrix} u'_1 u_1 & u'_1 v_1 & u'_1 & v'_1 u'_1 & v'_1 v_1 & v'_1 & u_1 & v_1 & 1 \\ u'_2 u_2 & u'_2 v_2 & u'_2 & v'_2 u'_2 & v'_2 v_2 & v'_2 & u_2 & v_2 & 1 \\ \vdots & \vdots & \vdots & \vdots & \vdots & \vdots & \vdots & \vdots & \vdots \\ u'_8 u_8 & u'_8 v_8 & u'_8 & v'_8 u'_8 & v'_8 v_8 & v'_8 & u_8 & v_8 & 1 \end{bmatrix}$$

and $e = (e_{11}, e_{12}, e_{13}, e_{21}, e_{22}, e_{23}, e_{31}, e_{32}, e_{33})$.

In Equation 4, $(u_{1\cdots8}, v_{1\cdots8})$ and $(u'_{1\cdots8}, v'_{1\cdots8})$ are obtained points from the images of the left and right cameras, where $e_{1..8}$ are the components of essential matrix E. Since Equation 4 is too time-consuming to process, $e_{1..8}$ are computed using singular values decomposition (SVD).

4 Experimental Results and Analysis

The proposed method, which estimates the coordinates of a 3D object using a planar structure for video-based AR, has been tested for the camera motion information to the coordinates of a 3D object created on an image in a sequence of frames. Our method applies the camera motion parameters to the coordinates of a 3D object in the frames of a video, and compares the estimated direction of the Z-axis with the direction of the real Z-axis. Fig. 1 shows the measured difference between the estimated and real directions of the Z-axis. The accumulated error increases towards the end of sequences. Fig. 2 shows examples of superimposition of an object located at different backgrounds in video sequence.

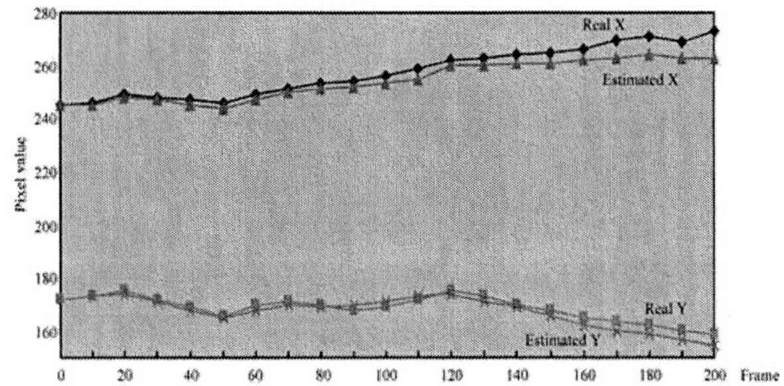

Fig. 1. Comparison for the registration between the estimated and real directions of Z-axes

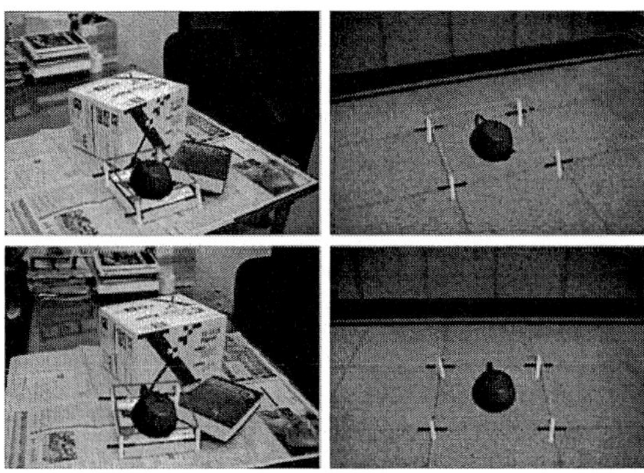

Fig. 2. Superimposition of a teapot in video sequence

References

1. Ronald T. Azuma: A Survey of Augmented Reality, Teleoperators and Virtual Environments, Vol. 6, No. 4, pp.355-385, 1997
2. Gilles Simon and Marie-Odile Berger: Estimation for Planar Structures, IEEE Computer Graphics and Applications, Vol. 22, pp.46-53, 2002
3. Simon J.D. Prince, Ke Xu and Adrian David Cheok: Augmented Reality Camera Tracking with Homographies, IEEE Computer Graphics and Applications, Vol. 22, pp.39-45, 2002
4. Gilles Simon, Andrew W. Fitzgibbon and Andrew Zisserman: Markerless Tracking using Planar Structures in the Scene, Proc. International Symp. Augmented Reality, pp.137-146, 2000
5. Peter Sturm: Algorithms for Plane-Based Pose Estimation, Proc. of the Conference on Computer Vision and Pattern Recognition, pp.1010-1017, 2000
6. Kiriakos N. Kutulakos and James R. Vallino: Calibration-Free Augmented Reality, IEEE Trans. on Visualization and Computer Graphics, Vol. 4, pp.1-20, 1998
7. Yong duek Seo and Ki Sang Hong: Calibration-Free Augmented Reality in Perspective, IEEE Trans. on Visualization and Computer Graphics, Vol. 4, No. 6, pp.346-359, 2000
8. Kumar Rakesh, Sawhney, H. Sawhney and Allen R. Hanson: 3D model acquisition from monocular image sequences, Proc. of the Conference on Computer Vision and Pattern Recognition, pp.209-215, 1992

Stochastic Fluid Model Analysis for Campus Grid Storage Service

Xiaofeng Shi, Huifeng Xue, and Zhiqun Deng

College of Automation, Northwestern Polytechnical University,
P.O.Box 183#, Xi'an 710072, China
{Xiaofengshi2002, zhiqundeng}@tom.com

Abstract. Campus grid storage service is to aggregate the storage resources in the servers of Campus Grid Center and colleges (or institutes, departments), and the storage resources of personal computers in the campus network. It provides storage resources registration, allocation, scheduling, and release services for users by three levels storage architecture. Due to the storage nodes' dynamites, the total storage space that nodes contribute will dynamically change with time. To study the performance of the storage service, the stochastic fluid model is adopted. By this analytical model, we got the mathematical results as follows: the function between the storage allocation probability and the number of nodes is got; if more nodes join the campus grid, the aggregated storage space will be larger, and then the available storage resources will be more; if the storage resources allocation rate is larger than the storage resources release rate, then the available storage resources will decrease.

1 Introduction

The campus grid is mainly to aggregate all kinds of resources in the campus networks, and provide the services that the campus network currently cannot afford. The Campus grid is not just to provide the High Performance Computing, but also provide other services [1] at a low cost, such as storage service, computing service, and etc.

This paper mainly focuses on the campus grid storage service, which aggregates the storage resources that nodes contribute. Due to nodes' dynamical join and departure, the stochastic fluid model [2,3, and references therein] is adopted to analyze the performance the storage service.

Stochastic fluid model, an important analytic model, has drawn considerable attention in such applications as the performance of P2P cache in [4,5]. In these papers, by this model they got good results from analysis of the cache clusters and P2P cache.

Our work based on [5] is to apply the stochastic fluid model to study the performance of the campus grid storage service. We model the aggregated storage space as a single infinite storage space. Nodes' dynamical join and departure result in the storage space dynamical change, as can be modeled by the stochastic fluid model. Finally, we get the performance of the storage service.

2 Campus Grid Storage Service Structure

The campus grid storage service is implemented by three levels architecture (see Fig.1):

V.S. Sunderam et al. (Eds.): ICCS 2005, LNCS 3516, pp. 854 – 857, 2005.

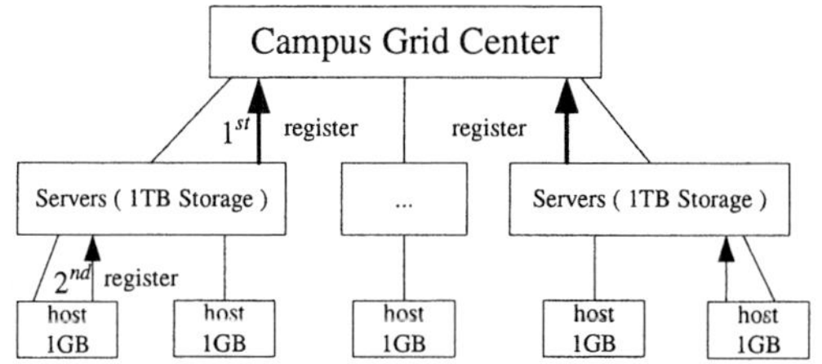

Fig. 1. Nodes' organization architecture of the Campus Grid

1. The first level, Campus Grid Center servers. They manage the registration information of servers in the colleges. The registration information includes the whole storage space, colleges' identifiers, servers' addresses, and etc. Each of these servers contributes at least 1TB storage space.
2. The second level, servers distributed in each college or institute, and department. They manage the registration information of PCs or workstations. Each of these servers contributes at least 1TB storage space.
3. The third level, Hosts (PCs or workstations). They contribute and request the storage resources. They can store data in the campus grid storage space. Each host can contribute 1GB or more storage space.

Due to storage nodes dynamical join and departure, Campus Grid Center servers will actively detect the state (work well, temporary stop, stop) of servers in the colleges. Similarly, servers in colleges need to detect the hosts' working state.

3 General Stochastic Fluid Model for Campus Grid Storage Space

The whole campus grid system can be viewed as a virtual single storage pool. The total storage space of the universe campus grid system is c, which is considered as the total storage space contributed by all the nodes at a given time. $x(t)$ is the total available storage space number (GB). Here, one GB storage space as the minimum unit to contribute and allocate storage resources. If a node joins the system, the storage space number increased is $J(t)$. And if a node leaves the system, the storage space number (GB) reduced is $L(t)$. We define

$$\theta = \textit{Storage resource allocation rate - Storage resource release rate} \quad (1)$$

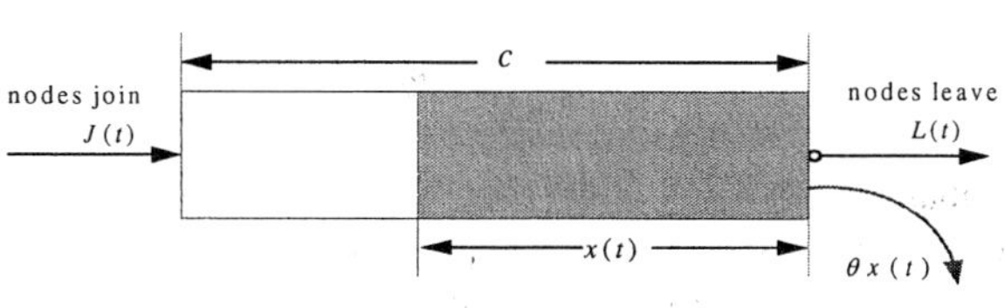

Fig. 2. The general stochastic fluid model for campus grid storage space

From the above definition, it can be deduced that the storage space number in the system at time t is: $x(t) = J(t) - L(t) - \theta x(t)$. The general stochastic fluid model for campus grid storage space can be seen in Fig.2.

4 Storage Dynamical Model

N is the total number of nodes. Nodes go up and down independently of each other, the time until a given up (or down) node goes down (or up) is exponentially distributed. The birth rate (up) of each node is λ. The death rate (down) is μ. In general, the request of each node randomly arrives based on Poisson processes at a rate σ.

If there are i nodes that are up, there are x storage spaces that are available. Here $i = 0,1,\dots N$ and $x \in [0,c]$. For each node, the get probability to allocate the storage resources requested is G, which can be defined by $G = x / c$.

When a node goes down, the storage space contributed by that node will go down. The remaining number of the storage space in the system is $\Delta_d(i) * x$. Here, $\Delta_d(i)$ is the down reduction coefficient. Once a node goes up, the storage space will join the system. There is no storage space reduced in the system and the remaining number of the storage space is $\Delta_u(i+1) * x$. The up reduction coefficient is $\Delta_u(i+1)$. In fact, it can be seen that $0 \le \Delta_d(i) \le 1$ and $\Delta_u(.) = 1$.

Duo to the nodes' change, the resulting storage space change is $J(t) - L(t)$, which can be induced by

$$J(t) - L(t) = i\sigma(1 - x/c) \tag{2}$$

We adopted the simple stochastic fluid model proposed by [5]. Then

$$G = \frac{E[X]}{c} = \frac{1}{(1+\rho)^N} \sum_{i=1}^{N} \binom{N}{i} \rho^i v_i \tag{3}$$

Here, the vector v can be seen in [5].

5 Simulation Analysis

If we do not consider the effect of storage space allocation and release, then: $\theta = 0$. The simulation result can be seen in Fig.3 as data1 curve shows. Else, we consider this, and we set $\theta = 0.3$ and $\mu = 0.02$. The simulation result can be seen in Fig.3 as data2 curve shows.

In Fig.3, we can see that the storage allocation probability will be higher if the number of nodes is larger. Then, once users request the storage resources in the campus grid, the probability that the resources are allocated is high. The data2 curve shows that if the storage resources allocation rate is larger than the storage resources release rate, the probability that the space is allocated will be low.

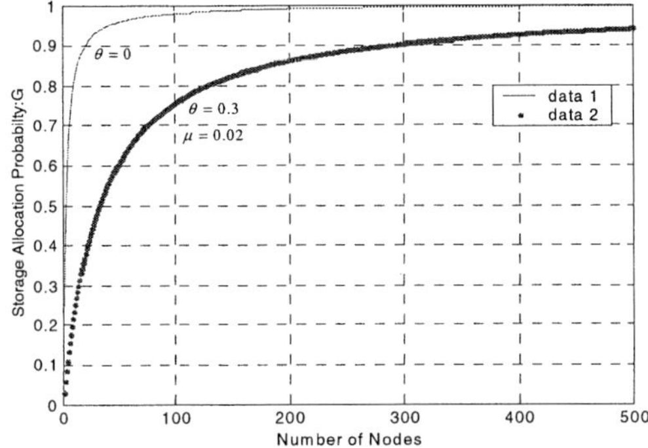

Fig. 3. The relation between number of nodes and the storage resources allocation probability

6 Conclusions

In this paper, we have built a general simple virtual storage pool model for the dynamical storage space due to nodes' join and departure. We explore the stochastic fluid analytical models for the purpose of campus grid storage service performance. We have got the results as follows: if more nodes join the campus grid, the aggregated storage space will be larger, and then the available storage resources will be more; if the storage resources allocation rate is larger than the storage resources release rate, then the available storage resources will decrease.

References

1. Zhiqun Deng, Zhicong Liu, Guanzhong Dai, Xinjia Zhang and Dejun Mu. Nodes' Organization Mechanisms on Campus Grid Services Environment. Lecture Notes in Computer Science, Vol. 3251. Springer-Verlag, Berlin Heidelberg New York (2004) 247-250
2. Nelly Barbot and Bruno Sericola. Distribution of busy period in stochastic fluid models. Communications in Statistics-Stochastic Models, (2001) 17(4)
3. Vidyadhar G. Kulkarni. Fluid Models for Single Buffer Systems, Frontiers in Queuing: Models and Applications in Science and Engineering. Ed. J. H. Dshalalow. CRC Press, (1997) 321-338
4. Florence Clévenot, Philippe Nain. A Simple Fluid Model for the Analysis of the Squirrel P2P Caching System. Proceedings of the IEEE INFOCOM 2004
5. Florence Clévenot, Philippe Nain, Keith W. Ross. Stochastic Fluid Models for Cache Clusters. Performance Evaluation, Vol.59 (1), (2005) 1-18

Grid Computing Environment Using Ontology Based Service

Ana Marilza Pernas[1] and Mario Dantas[2]

[1] Department of Informatics, Federal University of Pelotas,
99010-900 Pelotas, Brazil
ampernas@ufpel.edu.br

[2] Department of Informatics and Statistics, Federal University of Santa Catarina,
88040-900 Florianopolis, Brazil
mario@inf.ufsc.br

Abstract. Grid computing environments can share resources and services in a large-scale. These environments are being considered as an effective solution for many organizations to execute distributed applications to obtain high level of performance and availability. However, the use of a grid environment can be a complex task for an ordinary user, demanding a previous knowledge of the access requirements from a virtual organization. In order to improve the search of resources and its selection, in this paper we propose the use of ontology as alternative approach to help the use of grid services. This paradigm can provide a description of available resources, leading users to desire operations and describing resources syntax and semantics that can form a common domain vocabulary. Our experimental result indicates that our proposal allows a better comprehension about available resources.

1 Introduction

A grid environment can be understood as a set of services, provided by institutions to be used from another. Its architecture may be viewed as service-oriented [1], where two entities have a special importance: the producer and the consumer of the service. In this vision, owners offer services to be used, according to some restrictions that must be satisfied before given access permission to consumer. Because of these restrictions and rules of sharing, the service oriented grid architecture requires a scheme to support the interoperability between applications from the users and a high level of access transparency for resources.

Considering a service oriented approach, in this article we propose the use of ontology for the description of available resources in a grid computing environment. The motivation of this research is based on the clear advantages in using ontology, as some projects have shown (e.g. [2, 3, 4]), to have a common domain of concepts shared among ordinary users. The use of ontology for the semantics description of a vocabulary can provide a clear understanding of characteristics and properties of classes and relations.

V.S. Sunderam et al. (Eds.): ICCS 2005, LNCS 3516, pp. 858–861, 2005.

The paper is organized as follows. The development of the proposed ontology based service is described in section 2. In section 3, we present some related works. In section 4 we present our conclusions and future work.

2 Ontology for Grid Resources Description

In this section we present the ontology created. In addition to the axioms, other two structures help the ontology in the description of resources:

- **Metadata** reflects the information related to a data. In this research, metadata stores information about computational resources;
- **Semantics View** - stores information related to the present state of a resource. Thus, when a request comes, these structure returns information about that moment.

Even not commonly used in other ontologies, these structures can improve the ontology action, returning answers more quickly. The ontology was implemented using the OWL Full language [5], and was edited using Protégé-2000 editor [6].

The ontology works in the directory of a grid configuration. Fig. 1 (modified from [7]) shows that consumers' queries come to the ontology. To obtain information about computational resources, metadata receives information from catalogs and data files, while semantic views communicates with the Metacomputing Directory Service (MDS), which provides a distributed access to the grid structure and information related to system components.

2.1 Ontology Development

To probe how to describe the concepts related to grid computational resources, we search for the vocabulary utilized by the following projects: NPACI [8], ESG [9]; NASA's Information Power Grid [10] and the Distributed ASCI Supercomputer Project 2 [11]. After our search we create the documentation required to

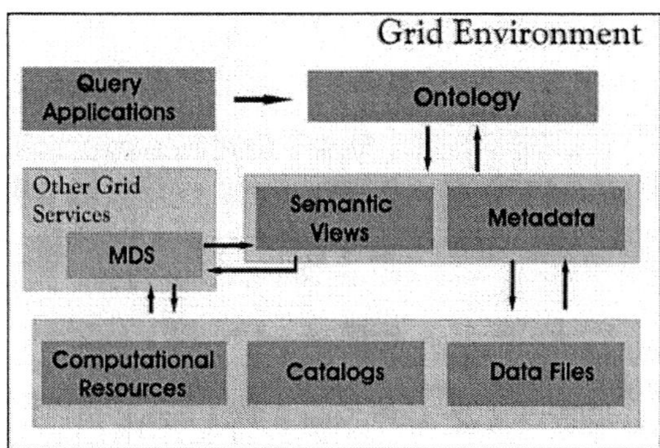

Fig. 1. Grid architecture using the ontology approach

```
(findall ? typeofmachine
(exists ?unix
(and ('typeOS'? unix ?typeofmachine)
  (=('name'?AIX)('typeOS'?typeofmachine))
  (exists ?typeofmachine
   (and
    (>('diskSpaceGB' ?typeofmachine) 40)
    (>('maxMemGB' ?typeofmachine)128))))))
```

Fig. 2. Axiom to access restriction

build the ontology, witch consists of all concepts, described in a formally way and mapped in classes and instances, that will form the ontology.

Our next task was to reproduce this documentation to the OWL language, using the Protégé editor. In the editor, we also created axioms, employing the PAL (Protégé Axiom-Language). Fig. 2 shows an axiom to access restriction, where is only possible to occur if the operating system from the resource is AIX, disk space greater then 40 GB and memory greater then 128 GB. Metadata were also created using Protégé and OWL language.

2.2 Ontology Based Service

To develop the grid service we created an application, using Java Language, to allow the service to interact with the ontology. The application was designed with three modules. The first module provides a list of all classes and instances defined in the ontology. In the second module, consumers can probe for metadata from any class listed by the first module. The third module allows a search of any computational resource, where a consumer can visualize the entire configuration. The service was defined using the Globus toolkit [12] and it is characterized by the application executing in a grid configuration. The application was configured to accept incoming connections from consumers of a grid. In this new environment, computational resources are presented using a more clear description, as we verified with users from our Federal University configuration.

3 Related Work

Concerning to the resources selection in a grid environment, is possible to find some research works (e.g. [3, 13]), where authors use the ontology to help the use of a grid environment. Some research works are related to apply the ontology in existing grid environments. One of theses efforts is the Semantic Grid [1], a grid infrastructure which has the goal to support applications related to e-Science.

4 Conclusion and Future Work

In this paper we presented a research work to provide access transparency to users of grid configurations. Our approach was based on ontology. We first presented some concepts of ontology and service oriented in grid configurations.

The environment of our prototype was described starting with the methodology used, followed by some characteristics of the development and finally how the ontology base service works. The system has proved to be an efficient and friendly approach to provide grid resources to consumers. As a future research work we are planning to enhance the system to allow some dynamic changes, such as metadata or inclusion, on the application. Other work is to create an ontology to agriculture field and use the application on more wide and complex grid environment.

References

1. De Roure, D., Jennings, N.R., Shadbolt N.R.: The Semantic Grid: A Future e-Science Infrastructure. In: Grid Computing: Making The Global Infrastructure a Reality, F. Berman, A.J.G. Hey and G. Fox (eds), Southern Gate, Chinchester, England: John Wiley Sons, 1080p. (2003) 437-470.
2. Pernas, A. M.: Ontologias Aplicadas a Descrio de Recursos em Grids Computacionais. Dissertation (Master Degree), Federal University of Santa Catarina, Florianpolis, Brazil (2004).
3. Tangmunarunkit, H., Decker, S., Kesselman, C.: Ontology-based Resource Matching - The Grid meets the Semantic Web. 1th Workshop on Semantics in Peer-to-Peer and Grid Computing (SemPGrid) at the 12th International World Wide Web Conference, Budapest, May (2003).
4. Heine, F., Hovestadt, M., Kao, O.: Towards Ontology-Driven P2P Grid Resource Discovery. 5th IEEE/ACM International Workshop on Grid Computing, Pittsburgh, USA,November(2004).
5. McGuinness, D., Van Harmelen, F.: Web Ontology Language Overview. (2004). Available online http://www.w3.org/TR/owl-features/.
6. Noy, N., Fergerson, R., Musen, M.: The knowledge model of Protege-2000: Combining interoperability and flexibility. 12th Int. Conference on Knowledge Engineering and Knowledge Management(EKAW), French Riviera, October (2000) 2-6.
7. Goble, C., De Roure, D.: Semantic Web and Grid Computing. September (2002). Available online: http://www.semanticgrid.org/documents/swgc/.
8. NPACI-National Partnership for Advanced Computational Infrastructure: Partnership Report. (2000). Available online: http://www.npaci.edu/.
9. Foster, I., Middleton, D., Williams, D.: The Earth System Grid II: Turning Climate Model Datasets into Community Resources. January (2003). Available online: https://www.earthsystemgrid.org/about/docs/ESGOverviewSciDACPINapa_v8.doc.
10. IPG, Information Power Grid - Nasa's Computing and Data Grid: What is the IPG? October (2002). Available online: http://www.ipg.nasa.gov/aboutipg/what.html.
11. Verstoep, K.: The Distributed ASCI Supercomputer 2 (DAS-2). May (2000). Available online: http://www.cs.vu.nl/das2/.
12. Foster I., Kesselman, C.: The Globus Project: a Status Report. In: Proc. of 7th Heterogeneous Computing Workshop (HCW 98), March (1998) 4-18.
13. Pouchard, L. et.al.: An Ontology for Scientific Information in a Grid Environment: the Earth System Grid. In: Proc. of the 3th IEEE/ACM International Symposium on Cluster Computing and the Grid, Japan, Tokyo, May (2003) p. 626.

Distributed Object-Oriented Wargame Simulation on Access Grid*

Joong-Ho Lim, Tae-Dong Lee, and Chang-Sung Jeong**

School of Electrical Engineering in Korea University,
1-5ka, Anam-Dong, Sungbuk-Ku, Seoul 136-701, Korea
{jhlim, lyadlove}@snoopy.korea.ac.kr
csjeong@charlie.korea.ac.kr

Abstract. This paper presents the design and implementation of Distributed Object-oriented Wargame Simulation(DOWS) on Access Grid(AG). DOWS is an object-oriented distributed simulation system based on a director-actor model. DOWS on AG supports a collaborative environment by providing a virtual venue with high quality audio and real-time video interactive interface for remote users, and allows a groups of users in remote sites to easily participate in the whole simulation. We design an efficient communication scheme between application and AG so that DOWS can be incorporated on AG for collaboration purpose.

1 Introduction

The simulation can be adapted in many fields such as construction, traffic, measurement and so on. One of them is a war game simulation in military field. Although there are few battles in the world, the military should continue to train their army to improve the operation skill of battle. However, the government must spend expensive budgets to check their strategies and maintain their battle ability, because the real training operation of the army must require high cost and time. Therefore, war-game simulation is regarded as one of the most effective ways to test new strategies and the operation capability of army's force.

However, it requires a large volume of multimedia data as well as many complex processes, and there exists some limit in number of operators and observers due to the space limitation. Also, it is difficult to have video/audio services in collaborative environment. In this paper, we present a Distributed Object-oriented Wargame Simulation(DOWS) system which provides not only distributed simulation environment but also collaborative environment using Access Grid(AG). We shall describe an efficient communication scheme which integrates application into AG so that DOWS can be easily incorporated onto AG for collaborative purpose.

* This work has been supported by KIPA-Information Technology Research Center, University research program by Ministry of Information & Communication, and Brain Korea 21 projects in 2005.

** Corresponding author.

V.S. Sunderam et al. (Eds.): ICCS 2005, LNCS 3516, pp. 862–865, 2005.

In section 2, we briefly describe the Access Grid, and in section 3, briefly explain DOWS. In section 4, we present an architecture for DOWS on AG. In section 5, we conclude with future works.

2 Access Grid

Grid computing is a specialized form of distributing computing where the application is distributed over a wide-area network and computing nodes are geographically far-off from each other[2]. It comprises Computational, Data and Access Grid. Computational Grid can share computing resources at remote site for high performance computing. Data Grid provides an integrated view for the distributed data. Access Grid(AG), developed by Argonne national laboratory, provides various services such as video/audio conferencing, data sharing, shared application, text-based communication and certificates management for collaboration[3]. It enables remote users to interact each other, and exchange information through video and audio services.

AGTk(Access Grid Toolkit) has been developed using Grid technology[1], and provides all the services offered in Grid. It is composed of AG Server, AG Client and virtual venues. An AG client joins into a virtual venue to collaborate with other members in the venue, and interact through AG server.

3 Dows System

DOWS is an object-oriented simulation system based on a director-actor model which can be mapped efficiently on object-oriented and distributed simulation. Director-actor model consists of actors and directors which interact with others. Actor represents a simulation entity or a sub-model in the simulation, and director is a participant in the simulation which controls the actors. Each director generates commands or events to its associated actors which in turn activate the actual simulations by interacting with other actors in the same or different subgroups. A set of actors may be designated as actor group so that each director can issue commands to each member in the actor group simultaneously. In director-actor model, simulation is carried out by actors interacting with each other. Each actor can represent a simulation entity participating in one simulation model or a sub-model in the whole simulation. DOWS consists of four major components which interact with each other on a distributed environment: director, actor, agent and coordinator. First, directors instantiate the simulation model and participate in the simulation concurrently through the interactive communication with the agents which are in charge of the parallel simulation of its associated sub-models using several actors. Second, each actor corresponds to a simulation entity of the simulation model or a sub-model which is a part of the whole simulation model. Several actors can be executed either in one processor or several processors interconnected through a network on a distributed environment. Each actor sends an event message to the channel of the other actor, and maintains the synchronization. Third, the agent provides an efficient virtual real time simulation environment which integrates the coordination among directors and actors by supporting time synchronization, simulation message transfer, and network fault detection. each agent can play a role of a message router, and coordinates each other to forward messages between actors and directors. Last, the coordinator runs the whole simulation by sending start message to each actor, or suspends it by sending a block signal to the agents whose input thread in turn notifies its corresponding actors to block.

4 DOWS on Access Grid

AG supports collaborative applications through the channel sharing. We classify users into operator and observer: The former can control simulation by issuing commands to the actors directly, and the latter only view the simulation controlled by the former. After login, DOWS operator and observer applications can be started by operators and observers respectively. Figure 1(a) show Venue client and server. Venue server supports various services such as venue creation, synchronization, authorization, registration, and data broadcasting. As in figure 1(b)(c), Venue client creates a relay station for each DOWS operator application, which interacts with other observer applications through its own relay station. Relay stations has a role which relays messages between DOWS application and

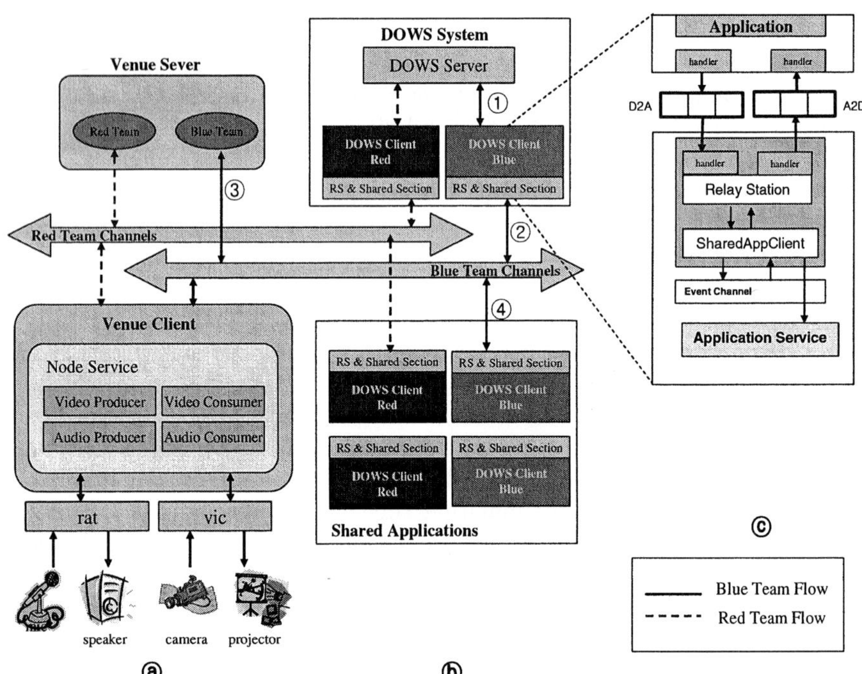

Fig. 1. An architecture of DOWS on AG

AG, and use two message queues between them. Each relay station receives the updated data in the message queue sent from its corresponding DOWS operator application using its message handler, and propagates them to other relay stations in the same venue using the event channel provided by application service in AGTk. Then, all the observer application in the same venue receives the updated data from the message queue sent by the relay station. Application service APIs provided by AGTk support a mechanism for event channel, discovery, coherence, and synchronization among relay stations. Relay station is implemented using python script language, and makes use of the functions in the SharedAppClient class which exploits application service APIs provided by AGTk.

For example, as shown in figure 1(a), two venues called "Blue Team Venue" and "Red Team Venue" are created in the Venue server. Suppose DOWs server, Venue clients and server exist already. Each venue client creates a relay station, which in turn generates its corresponding DOWS client. Entering the name of team venue, we can view the real time simulation updates in our team using the relay stations which relays messages between DOWS application and AG. Note that DOWS application cannot communicate directly with AG.

5 Conclusion

This paper have presented the design and implementation of DOWS on AG which supports a collaborative environment for distributed applications. We have shown that the distributed application can be efficiently incorporated into AG environment by proposing a relay station which makes use of event channel and application service APIs in AGTk. Therefore, it provides a way which enables users to integrate applications on a collaborative environment with high quality audio and real-time video interactive interfaces in distributed systems. As a future work, we are working on the integration of computational grid into our system for the performance improvement of computation intensive simulation arising in DOWS.

References

1. Access Grid http://www.accessgrid.org/
2. Stevens, R., Papka, M.E. and Disz, T.,"Prototyping the workspaces of the future," Internet Computing, IEEE, pp 51-58, July-Aug. 2003.
3. Ho, H. C. and Yang, C. T. and Chang C. C., "Building an E-learning Platform by Access Grid and Data Grid Technologies,"Proceedings of the 2004 IEEE International Conference on e-Technology, pp 121-126 2004.
4. Ernest H. Page and Roger Smith,"Introduction to Military Training Simulation: A Guide for Discrete Event Siulationists," Proceedings of the 1998 Winter Simulation Conference, pp 33-40, 1998.

RTI Execution Environment Using Open Grid Service Architecture*

Ki-Young Choi, Tae-Dong Lee, and Chang-Sung Jeong**

School of Electrical Engineering in Korea University,
1-5ka, Anam-Dong, Sungbuk-Ku, Seoul 136-713, Korea
{2xx195, lyadlove}@snoopy.korea.ac.kr
csjeong@charlie.korea.ac.kr

Abstract. HLA(High Level Architecture) has been developed to promote interoperability and reusability within the modeling and simulation community. RTI(Run Time Infrastructure) is a software implementation of HLA which is composed of three components: libRTI, FedExec, and RTIExec. However, the previous RTI has an unfriendly execution environment with static and manual resource allocation and execution. In the paper, we present a RTI execution environment using OGSA(Open Grid Service Architecture) which addresses the problems, so called RTI-G. It supports easy-to-useness, transparency and performance by providing users with dynamic resource allocation and automatic execution. Moreover, it provides an unified view of RTI and Grid services by integrating them as web services on OGSA.

1 Introduction

Running a large-scale distributed simulation may need a large amount of computing resource residing in geographically different locations. HLA(High Level Architecture)[1] provides application developers with a powerful framework for reuse and interoperability of distributed simulation. However, RTI(Run Time Infrastructure), the implementation of HLA, was not designed to support software applications that need to integrate instruments, displays, computational and information resources managed by diverse organizations. Moreover, the existing RTI(Run Time Infrastructure) has an unfriendly execution environment with static and manual resource allocation and execution. Therefore, the execution of RTI was a painful manual process. Recently, Grid technology has been introduced to address the resource allocation and execution problems in distributed virtual organizations. In the paper, we present a RTI execution environment using OGSA(Open Grid Service Architecture)[2], so called RTI-G, which supports an easy-to-use, transparent execution RTI environment while achieving

* This work has been supported by KIPA-Information Technology Research Center, University research program by Ministry of Information & Communication, and Brain Korea 21 projects in 2005.

** Corresponding author.

V.S. Sunderam et al. (Eds.): ICCS 2005, LNCS 3516, pp. 866–869, 2005.

high performance by providing users with dynamic resource allocation and automatic execution using various services such as MDS, GRAM, and GridFTP in Globus toolkit 3(GT3)[3] implemented based on OGSA. MDS(Metacomputing Directory Service) provides a standard mechanism for publishing and discovering resource status and configuration information, and GRAM(Globus Resource Allocation Manager) executes the executables in the allocated resource. GridFTP is used to transfer I/O files and executables into the remote resources.

In section 2, we describe the design of RTI-G based on OGSA, and in section 3, the experimental result. In section 4, we give a conclusion.

2 Design of RTI-G

RTI-G is a grid-enabled implementation of RTI. That is, using services in GT3 based on OGSA, each service of RTI-G is designed to provide automatic and dynamic execution environment. The architecture of RTI-G consists of two layers: RTI service layer and Grid layer. In this section, we describe about each layers respectively more in detail.

2.1 RTI Service Layer

RTI layer is composed of three services: libRTI-G, FedExec-G, RTIExec-G. RTI software can be executed on a stand alone workstation or executed over an arbitrarily complex network. RTIExec-G manages the creation and destruction of federation. Each federation is characterized by a single and global FedExec-G, which manages federates joining and resigning the federation. libRTI-G provides RTI service interface to federate developers. The whole simulation in HLA is accomplished through the interaction among libRTI-G, RTIExec-G, and FedExec-G.

OGSA integrates key Grid technologies with Web services mechanisms to create a distributed system framework based on OGSI(Open Grid Services Infrastructure). Each RTI service is defined as a grid service using WSDL(Web Service Description Language) in OGSA to be efficiently incorporated in the Grid environment, and hence to be accessible by standard Grid clients.

Using WSDL, a grid service is defined about how a client interacts with its service instance. That is, the Grid service description is embodied in the serviceType of the instance, along with its associated serviceTypes, portTypes, serviceDataDescriptions, messages, and types definitions. Therefore, in order to be compatible with the grid services, each service in RTI layer should be specified in WSDL. There are two methods for describing the service of RTI layer in WSDL. The first is to write the service in WSDL directly, which is the most versatile method, providing the total control over the description of service PortType. However, it is not user-friendly because WSDL is a rather verbose language. The second is to generate WSDL from an interface language. For example, we can generate WSDL automatically from a Java interface or an IDL interface, which is the easiest method, but not versatile, because very complicated interfaces are not always converted correctly to WSDL. Since our

RTI is implemented using C++ language, we choose the first method to easily convert our RTI in C++ into WSDL.

We deploy the services in RTI service layer onto GT3 container using GWSDL and WSDD(Web services Deployment descriptor) so that they can be used identically as various other grid services defined on OGSA, thus providing an unified environment of RTI and Grid services. Using the service browser, we can invoke RTI service, and create a service instance as follows: First, user request a service by sending its GSH(Grid Service Handle) to the RTI service factory, which creates the instance. A GSH is a stable name for a Grid service, but does not allow client to actually communicate with the Grid service. RTI service factory maps GSH to GSR(Grid service Reference), a WSDL document which describes how a client communicate with the Grid service.

2.2 Grid Layer

Grid layer comprises several services such as MDS, GRAM, and GridFTP implemented in GT3. MDS is used to allocate the resource with high performance by monitoring resource status and configuration information, and GRAM to execute the RTI service in the allocated resource. GridFTP is used to transfer the RTI service executables into the remote resources.

3 Experiments

For the evaluation of RTI-G, we execute Nexio, a networked shooting game on RTI-G and RTI respectively, and compare their performance. In Nexio, two players can play shooting game, and communicate through services in RTI. For each case, we measure the data transfer rate between players. The data size of each object in the game is 10Kbytes, and with the advancement of each stage, the number of objects is increased by 100. In the first case, player 1 requests the join into a federation, and then RTIExec and FedExec are created subsequently on the server specified by player 1. Then, player 2 in different host joins into the federation as in figure 1(a). In the second case, player 1 requests the join into a federation, and then RTIExec and FedExec are created in the server allocated by RTI service factory after accessing MDS and selecting the one with the best performance as in figure 1(b). RtiExec service is created as RTIservice instance, and makes FedExec service is generated through fork operation on the allocated server. After that, player 2 joins into the federation, and they exchange information about objects in the game.

We used each client machines with same specification to reduce side effects according to different computing power of each machine. All client host has 850MHZ CPU, 256MB memory, 512MB of swap, running under the Linux OS. There are three servers with different specs as shown in figure 1(c). In the second case, the best server is selected using MDS by RTI service factory. As we see in figure 1(d) which shows the data transfer rate vs total object size, RTI-G has the better performance than RTI.

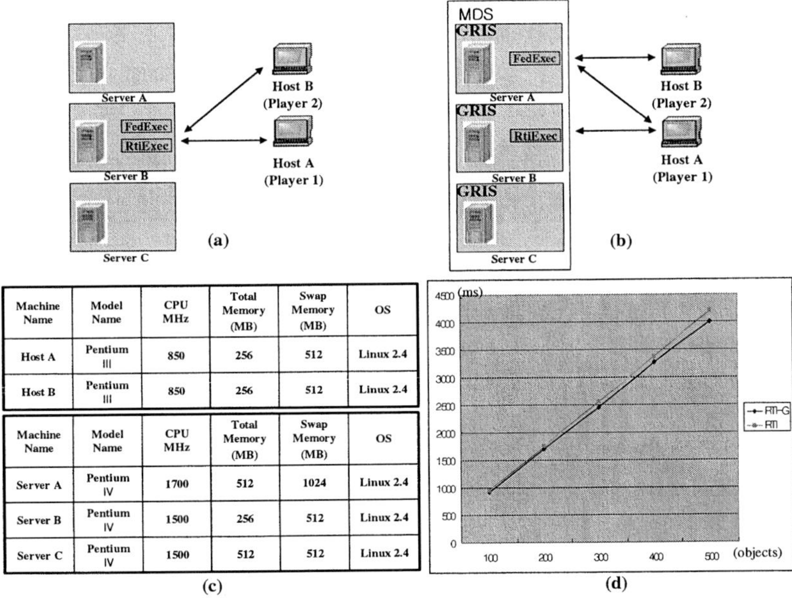

Machine Name	Model Name	CPU MHz	Total Memory (MB)	Swap Memory (MB)	OS
Host A	Pentium III	850	256	512	Linux 2.4
Host B	Pentium III	850	256	512	Linux 2.4

Machine Name	Model Name	CPU MHz	Total Memory (MB)	Swap Memory (MB)	OS
Server A	Pentium IV	1700	512	1024	Linux 2.4
Server B	Pentium IV	1500	256	512	Linux 2.4
Server C	Pentium IV	1500	512	512	Linux 2.4

Fig. 1. (a)The first case using RTI, (b)The second case using RTI-G, (c) Resource performance, (d) Comparison of data transfer rate on RTI and RTI-G

4 Conclusion

In the paper, we have presented RTI-G which supports an easy-to-use, transparent execution RTI environment while achieving high performance by providing users with dynamic resource allocation and automatic execution using various services in GT3. The specification of RTI-G in GWSDL and the deployment of RTI-G services in RTI service layer onto GT3 container provide an integrated unified environment of RTI and Grid services, thus supporting easy-to-use, transparency, and performance. Also, we have described the experimental result for RTI-G by comparing it with RTI. As a future work, we are designing a RTI portal for the efficient execution and performance monitoring of RTI with application to distributed war game simulation.

References

1. IEEE Standard for Modeling and Simulation,"High Level Architecture (HLA) Federate Interface Specification," IEEE Std 1516.1-2000 54.
2. D.Talia, "The Open Grid Services Architecture: where the grid meets the Web," Internet Computing, IEEE , Volume: 6 , Issue: 6, Nov.-Dec., pp.67-71, 2002.
3. I. Foster, C. Kesselman, "Globus: A Metacomputing Infrastructure Toolkit, Intl J. Supercomputer Applications," 11(2), pp. 115-128, 1997.

Heterogeneous Grid Computing: Issues and Early Benchmarks

Eamonn Kenny[1], Brian Coghlan[1], George Tsouloupas[2], Marios Dikaiakos[2], John Walsh[1], Stephen Childs[1], David O'Callaghan[1], and Geoff Quigley[1]

[1] Department of Computer Science, Trinity College Dublin, Ireland
{ekenny, coghlan, walshj1, childss, ocallwd, gquigle}@cs.tcd.ie
[2] Dept. of Computer Science, University of Cyprus, 1678, Nicosia, Cyprus
{georget, mdd}@ucy.ac.cy

Abstract. A heterogeneous implementation of the current LCG2/EGEE grid computing software is supported in the Grid-Ireland infrastructure. The porting and testing of the current software version of LCG2 is presented for different flavours of Linux, namely Red Hat 7.3, Red Hat 9 and Fedora Core 2. The GridBench micro-benchmarks developed in CrossGrid are used to compare the different platforms.

1 Introduction

Grid-Ireland uses the LCG[1] and EGEE[2] grid middleware, which originally assumed reference ports to Red Hat 7.3 and Microsoft Windows, but subsequently this has been revised to include Scientific Linux 3 (SL3). Unfortunately this is a very restrictive situation, counter to the original heterogeneous ethos of grid computing. As a result of our interest in heterogeneity, we at Trinity College Dublin began porting to non-reference platforms in October 2003. Subsequently EGEE have almost finished porting the current LCG2 grid implementation to Scientific Linux on 32-bit and 64-bit architectures.

2 Porting for Heterogeneity

The LCG2/EGEE software components are shown in the form of a dependency graph in Figure 1. Grid-Ireland wished, in the first instance, that the porting of the LCG2 software to other platforms would focus on the ability to execute Globus and EDG jobs on worker nodes, and that replica management, R-GMA and VOMS would be supported.

To avail of the base functionality requires Globus and various EDG support packages. Since Globus 2.4.3 is known to have many bugs, the University of Wisconsin-Madison corrects these and packages all the necessary components as part of the Virtual Data Toolkit (VDT) [3]. We have assisted Maarten Litmaath

V.S. Sunderam et al. (Eds.): ICCS 2005, LNCS 3516, pp. 870–874, 2005.

in CERN to port VDT-1.1.14 to IRIX and Fedora Core 2. A Red Hat 9 port is already provided by VDT. A port exists for Globus to Mac OS X and AIX but the VDT version must be ported to both of these platforms.

Grid-Ireland also wished that MPI, replica management and the OpenPBS client be provided on each worker node. In some cases Torque might be required since newer versions of operating systems are not always provided for in OpenPBS. Also the R-GMA information system producer and consumer APIs and the VOMS client were required.

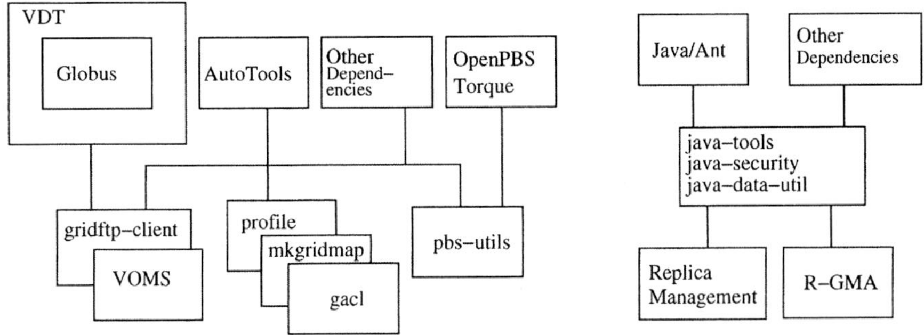

Fig. 1. LCG2/EGEE software components

At the moment there is no requirement by Grid-Ireland for the workload management system (WMS) but it appears that there is logging of WMS events from worker nodes to the resource broker. This logging activity can be disabled but we would prefer to retain this desirable feature. WMS consists of many modules, but it might be able to be refactored to delineate those specific to the event logging, so that just this functionality needs to be ported. It should be noted that if the whole of WMS were ported successfully then almost everything will have been ported because it depends on so many other packages.

There are a number of on-going issues, but we have successfully ported the functionality for job submission to Fedora Core 2, IRIX 6.5.14 and 6.5.17m, AIX 5.2L and Red Hat 9. We also plan to do this for Mac OS X v10.3 very soon, and a number of other platforms if the need arises within Grid-Ireland.

3 Micro-benchmarking Results

Presently EDG job submission is possible for the following non-reference platforms: SGI IRIX 6.5, Fedora Core 2, Red Hat 7.3 and 9 as described in [4], where preliminary results using a fast fourier transform (FFT) were used to show differences in computational speed between different architectures. A routine such as a FFT cannot be independently used as a benchmark since it gives no explicit information about I/O, CPU, caching, floating point or disk write speed.

The CrossGrid GridBench benchmarks[5] developed at the University of Cyprus provides more precise benchmarking.

GridBench is a tool for benchmarking Grids. It consists of *i)* the GridBench Framework and *ii)* the GridBench Suite. The GridBench framework provides mechanisms for defining and executing benchmarks as well as archiving and

Table 1. Micro-Benchmark Results

OS Type	Version	Bonnie (kB/s)	HPL (GFlop/s)	b_eff_io (MB/s)	epdhrystones (dhrystones)	Whetstone (MIPS)	epflops-4 (MFlops)
Fedora Core	2	193536	0.648350	46.712	3858611	707.86	825
Red Hat	7.3	159431	0.696549	52.515	4201680	742.71	898
Red Hat	9	154663	0.729079	45.318	4255319	750.35	899

Table 2. Machine Specifications

OS Type	Version	CPU Speed (GHz)	memsize/Total (MB)	epstream (add)	epstream (triad)
Fedora Core	2	2.8	951/2044	1600	960
Red Hat	7.3	2.8	979/2044	1600	960
Red Hat	9	2.8	872/2044	1600	960

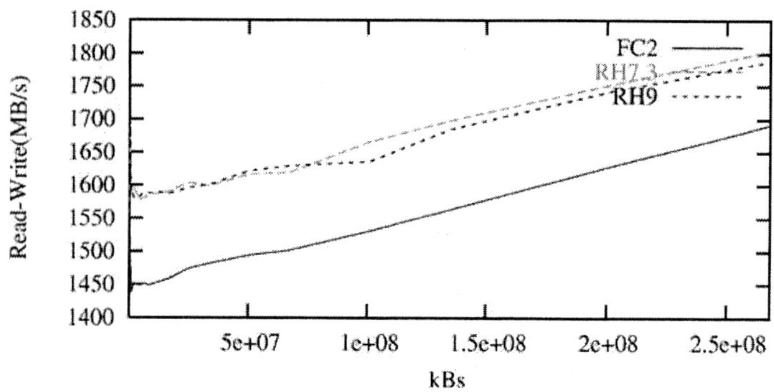

Fig. 2. Cachebench for Fedora Core 2, Red Hat 7.3 and Red Hat 9

managing the results. The Gridbench Framework also provides a user-friendly graphic user interface serving as a "virtual workbench" for conducting Grid benchmarks and tests. The GridBench Suite provides a set of benchmarks aiming to characterize Grid resources at different levels (e.g. the worker-node level or the cluster level). The Gridbench Framework and Suite are described in [5, 6]. Gridbench is currently being enhanced for fully automating the process of benchmarking heterogeneous resources.

A collaboration between TCD and UCY has resulted in the use of 11 of the GridBench benchmarks from the GridBench suite. Each of these modules, along with a suitable job description language (JDL) file, are required for the job submission to a single workernode to obtain the benchmark results. Initially all 11 modules were sent simultaneously as a batch job, but it was found that the results obtained were inconsistent, with fluctuating wall clock speeds; this was due to the load tailing off as the next submitted job arrived on the worker node. To achieve consistent load between submissions, each job submission must be staggered in time. The results in Table 1 and Figure 2 were computed on 3 worker nodes with identical hardware specifications (see Table 2).

4 Conclusions

The base worker node port of the LCG2/EGEE grid software for Globus and EDG job submission is now completed for Fedora Core 2, Red Hat 9, IRIX and AIX. The results presented in Table 1 and 2 show conclusively that Fedora Core 2 is fastest for write access to the disk (see bonnie in Table 1), while being slower for computational performance (see cachebench in Fig. 2, epflops in Table 1 and HPL in Table 1). In the case of integer computations it is markedly slower (see epdhrystones in Table 1). Results were also obtained for blasbench and the gridbench file transfer benchmark (gbftb) with no noticable differences.

At first the Red Hat 7.3 RPMs were used to perform all the benchmarks on each Linux platform. To ensure that we could be confident of the validity of our results it was necessary to build mpich-1.2.6 and GridBench under Fedora Core 2 using the gcc-3.3.3 compiler. Compiling Gridbench under Fedora Core 2 with identical optimization options to that of Red Hat 7.3 gave no performance gains.

Acknowledgements

We would like to thank IBM, Dell and DIAS for sponsoring us with machines to perform the software ports, and Science Foundation Ireland for funding this effort. Most of all we would like to thank the deployment group in CERN and Vincenzo Ciaschini at INFN for all their help in porting to each platform.

References

1. LHC: Large hadron collider computing grid project. http://lcg.web.cern.ch/LCG/ (2004)
2. EGEE: Enabling grids for e-science in europe. http://www.eu-egee.org/ (2004)
3. VDT: Virtual data toolkit. http://www.cs.wisc.edu/vdt/ (2004)
4. Kenny, E., Coghlan, B., Walsh, J., Childs, S., O'Callaghan, D., Quigley, G.: Heterogeneity of Computing Nodes for Grid Computing. Submitted to EGC 2005 (2004)

5. Tsouloupas, G., Dikaiakos, M.D.: Gridbench: A tool for benchmarking grids. In: Proceedings of the 4th International Workshop on GridComputing (GRID2003), Phoenix, AZ, IEEE (2003) 60–67
6. Tsouloupas, G., Dikaiakos, M.D.: Characterization of computational grid resources using low-level benchmarks. Technical Report TR-2004-5, Dept. of Computer Science, University of Cyprus (2004)

GRAMS: Grid Resource Analysis and Monitoring System[1]

Hongning Dai, Minglu Li, Linpeng Huang, Yi Wang, and Feng Hong

Department of Computer Science & Engineering, Shanghai Jiao Tong University,
1954 Hua Shan Road, 200030 Shanghai, P.R. China
{hndai, li-ml, huang-lp, wangsuper, hongfeng}@cs.sjtu.edu.cn

Abstract. In this paper we propose GRAMS, a resource monitoring and analysis system in Grid environment. GRAMS provides an infrastructure for conducting online monitoring and performance analysis of a variety of Grid resources including computational and network devices. Based on analysis on real-time event data as well as historical performance data, steering strategies are given for users or resource scheduler to control the resources. Besides, GRAMS also provides a set of management tools and services portals for user not only to access performance data but also to handle these resources.

1 Introduction

Grids offer us a new vision, infrastructure and trend for the coordinated resources sharing, problem solving and services orchestration in dynamic, multi-institutional virtual organizations [1]. A recent trend in government and academic research is the development and deployment of computational grids [2]. Computational grids are large-scale distributed systems that typically consist of high-performance compute, storage, and networking resources. Several computational grids in China are NHPCE (1999-2001) [7], CNGrid (2002-2006) [8], ChinaGrid (2002-2005) [9], E-Science Grid (2002-2005) [10], Spatial Information Grid (2001-2005) [11] and Shanghai-Grid (a city grid project to enhance the digitalizing of Shanghai)[3].

The primary goal of Shanghai-Grid is to develop a set of system software for the information grid and establish an infrastructure for the grid-based applications. This project will build an information grid tailored for the characteristics of Shanghai and support the typical application of grid-based traffic jam control and guidance. Currently we are trying to build up a system to monitor and control the resources, services and applications that make up the Grid. We have found it difficult to ensure that the resources in the Grid are working correctly to support grid application. It is also cumbersome to predict system performance and to control the resources in the Grid. These difficulties have led to our development of a system for monitoring and analysis of grid resources.

[1] Supported by the National Grand Fundamental Research 973 Program of China under Grant No.2002CB312002, the Grand Project (No.03dz15027) and Key Project (No.025115033) of the Science and Technology Commission of Shanghai Municipality.

V.S. Sunderam et al. (Eds.): ICCS 2005, LNCS 3516, pp. 875 – 878, 2005.

In this paper, we present a system namely GRAMS (Grid Resource Analysis and Monitoring System) which can observe performance events on each grid node, collect these event data, and analyze performance data to determine what actions should be taken.

2 GRAMS Architecture

GRAMS is based on OGSA [5] and GMA [4]. The GRAMS architecture is shown in Fig.1. The GRAMS includes several main components: Monitor Managers, Directory Service, Consumers (Customer Service, Archival Service, and Performance Analysis Service), Steering Service and User Interface. All the services can be deployed to different distributed environments.

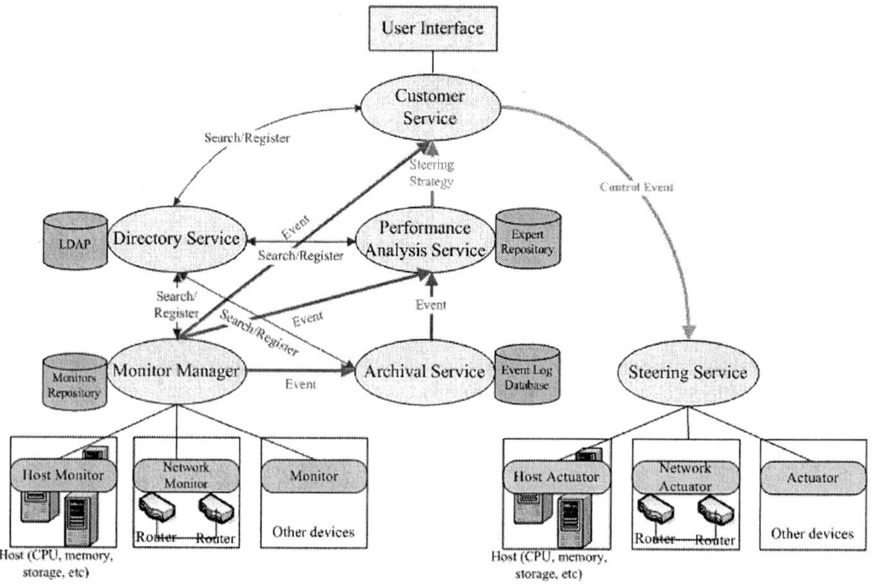

Fig. 1. Architecture of GRAMS framework

Directory Service stores information about producers, consumers and performance data. With the directory service, producers and consumers can publish and discover information that they request. This published information includes registration of producers and consumers, the types of event data, the characteristics of the data, and the ways to gain access to that data. The directory service contains only information about which event instances can be provided or accepted, not the event data itself.

Monitor Managers are used to manage monitors that measure and gather a variety of kinds of information for monitoring and analyzing. The monitors include: host monitor that is used to measure system property, network monitor to catch network info, and other monitor to capture other useful data from other devices of the Grid systems. The Monitor Manager publishes the monitors' existences in directory service entries, which tell consumers how to locate them. When the subscription between the monitors and customer service is established, the monitors will send events to customer service through the data channel of Monitor Manager until the subscription is terminated. Monitor repository in Monitor Manager will store the information about monitors (including their locations, their structures and kinds of performance metrics).

Archival Service is a database system which is used to store detailed monitoring information as well as performance results collected. It gained large amounts of long-term event data from Monitor Manager service for later use.

Performance Analysis Service is an expert system which acquires huge amounts of data from Monitor Manager as well as Archival Service, and produces resource management rules and scheduling strategies.

By using **User interface** provided by **Customer Service**, users can conduct online monitoring and performance analysis as well as making resource management decisions.

Steering Service receives requests from Customer Service to perform actions on the given resources or devices.

Interactions among GRAMS components and services are divided into controlling streams and data streams. Controlling streams are used to perform tasks including conducting activities of monitors and services, subscribing and querying requests for performance data, registering, querying and receiving information of Directory Service. In data streams delivery, a stream channel is used to transfer data (monitor data, performance data and results) between producers and consumers. Each monitor manager has only one connection to a consumer for delivering all kinds of data. All the connections are secure sockets based on Grid Security Infrastructure (GSI) [6] and Public Key Infrastructure (PKI).

3 Conclusions

In this paper we proposed the architecture of GRAMS which is a flexible, scalable monitoring and performance analysis system in Grid environments. We have implemented an early prototype of GRAMS, which already proves it feasible and effect.

There are still a lot works that should be done to improve the system. Fuzzy rules in expert system will be extended to cater for more complex and richer semantic knowledge. Performance prediction and instrumental tracing for Grid-based applications are the centers that our next step work will focus on. We will try to integrate current monitoring tools into GRAMS to serve more types of Grid monitor users and provide more functionalities and features, such as sampling, tracing and profiling [12].

References

1. Foster, C. Kesselman, and S, Tuecke, "The Anatomy of the Grid: Enable Scalable Virtual Organizations", *Int. J. of H. Performance Computing Applications*, 15 (2001) pp.200-222.
2. I. Foster and C. Kesselman, *The Grid: Blueprint for a New Computing Infrastructure*, Morgan Kauffmann, 1999.

3. M. Li, H. Liu, C. Jiang, and W. Tong, "Shanghai-Grid in Action: The First Stage Projects towards Digital City and City Grid", *LECT NOTES COMPUT SC 3032*: 2004, pp.616-623.
4. B. Tierney, R. Aydt, and W. Smith, "A Grid Monitoring Architecture", *Technical report*, Performance Working Group, Grid Forum, January 2002. http://www-didc.lbl.gov/GGF-PERF/GMA-WG/papers/GWD-GP-16-3.pdf
5. I. Foster, C. Kesselman, J. Nick, and S. Tuecke, "Grid Services for Distributed System Integration", *IEEE Computer*, June 2002, pp.37–46.
6. I. Foster, C. Kesselman, G. Tsudik, and S. Tuecke, "A security architecture for computational grids", In *Proceedings of the 5th ACM Conference on Computer and Communications Security (CCS-98)*, ACM Press, New York, Nov. 3–5 1998, pp.83–92.
7. National High Performance Computing Environ-ment (NHPCE): http://vega.ict.ac.cn
8. China National Grid Project (CNGrid): http://www.grid.org.cn/
9. ChinaGrid (China Education and Research Grid): http://grid.hust.edu.cn/platform/chinagrid.jsp
10. China E-science Grid Project: http://www.most.gov.cn/English/index.htm
11. Spatial Information Grid (SIG): http://www.nudt.edu.cn/newweb/intercommunion/whatissig.htm
12. Michael Gerndt, et al. "Performance Tools for the Grid: State of the Art and Future", APART-2 White Paper on Grid Performance Analysis, http://www.lpds.sztaki.hu/~zsnemeth/apart/repository/gridtools.pdf.

Transaction Oriented Computing (Hive Computing) Using GRAM-Soft

Kaviraju Ramanna Dyapur and Kiran Kumar Patnaik

Department of Computer Science,
Birla Institute of Technology (Deemed University), Mesra, India
{kaviraju15, kkpatnaik}@bitmesra.ac.in

Abstract. The promise of Hive Computing – Organizations being able to acquire all the power they need for only as long as is necessary – is incredibly compelling. Hive Computing has experienced significant success in bringing productivity gains, high performance and cost savings to business applications such as; Enterprise Resource Planning (ERP), Customer Relationship Management (CRM), Supply Chain Management (SCM) and B2B portals (E-commerce sites). Grid Computing has brought productivity gains, high performance and cost savings, but in some places it is largely incomplete, i.e. when it comes to the questions of mission-critical computing in general and transaction oriented in particular. This paper discusses a new approach to the development, deployment and management of mission-critical applications – called Hive Computing – that is designed to complement and extend the vision of Grid Computing.

1 Introduction

Hive computing assumes that failure is the rule, not the exception. Computers and other hardware are vulnerable to numerous technical and human-induced flaws and catastrophic failures. Hive Computing takes the lead from the designers of TCP in assuming that computers are unreliable, and designs systems that can deal with rather than fear system failure. Hive Computing is a comprehensive and integrated approach to the development, deployment and management of transaction-oriented applications. The goal of Hive Computing is to enable business and other organizations to leverage the rising performance and falling process of commodity computers to construct mission-critical computing solutions that are both reliable and affordable. Hive Computing is derived from Grid Computing, but especially meant for transaction-oriented tasks.

This paper presents a functionality of GRAMSoft that takes the Hive computing approach to Transaction oriented applications in business applications resulting in productivity gains, high performance and cost savings.

2 Hive Computing Based Infrastructure

The key functions of Hive are; receiving and processing *requests* that are sent to it by a client and an administrator, *b*roadcasting request to the Hive and execution of job by

V.S. Sunderam et al. (Eds.): ICCS 2005, LNCS 3516, pp. 879 – 882, 2005.

a *worker* which is an individual computer that has three characteristics; 1) all workers must be located on the same logical network, 2) regardless of their exact hardware configuration, the same software is deployed on all workers and 3) all workers are dedicated to the Hive. Also administers a worker and maintaining database of status of sub-tasks and each worker status (idle or busy).

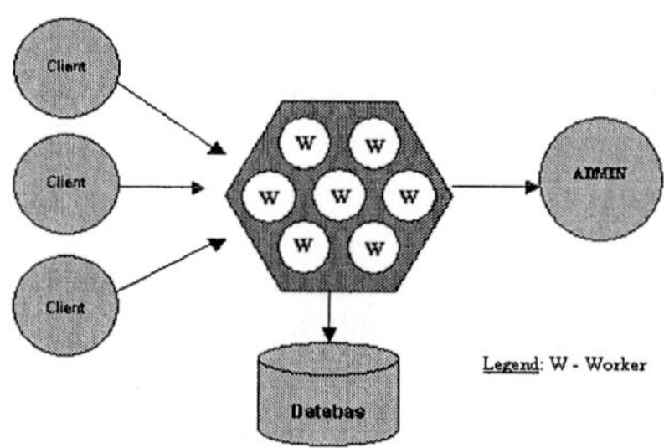

Fig. 1. Key components of Hive

3 Software to Build Hive Computing Infrastructure

The tool "GRAM-Soft" that forms a Hive Computing infrastructure containing transaction-oriented resources. This software involves: Mobile Agents (MAs), Meta-Computing (MC), Resource Co-Allocation.

3.1 Mobile Agents in GRAM-Soft

In our GRAM-Soft, MAs are the carriers of tasks so as to execute them and return with the result to the system from where user has submitted his job. Also, they are meant for job migrations from one resource to another resource (upon its failure on one resource) and resumption of its execution from the point where it has been suspended. When the sub-task (sub-task may contain several files) is under execution on one computer, status of each file (Under Progress, Completed or Suspended) will be recorded in a "Task_Status" file. This file information decides whether or not sub-task execution is to be resumed in another capable computer. The sub-task, containing only those files that are failed to execute (can be known by looking at "Task_Status" file) is migrated to first available capable computer and its execution is resumed.

3.2 Meta Computing in GRAM-Soft

MC allows applications to use collection of computational resources on an as needed basis, without regard to physical location of resources. Existing MC features in Grid-based softwares are; locating and allocating computational resources, Authentication of resources and Process creation. Additional MC features in GRAM-Soft are; execution ordering on processes, grouping of jobs and rescheduling (if necessary). This introduces five challenging resource management problems such as, site autonomy problem, heterogeneous substrate problem, policy extensibility problem, co-allocation problem and online control problem. *"No existing Meta-Computing Resource Man-*

agement Systems address all the problems". GRAM-Soft also addresses the issues prevalent in resource co-allocation.

4 Architecture of GRAM-Soft

Principal components of GRAM-Soft are:

- □ GRAM Client Library: Used by application or co-allocator acting on behalf of an application. It interacts with the GRAM gatekeeper at a remote site to perform mutual authentication and transfer a request containing resource specification.
- □ MAs: Responsible for migrating sub-tasks between different capable computers.
- □ GRAM Gatekeeper: Performs Mutual authentication of user and resource, determines local user name for the remote user and starts a job manager, which executes as that local user which actually handles the request. Here gatekeeper must run privileged programs all the time.

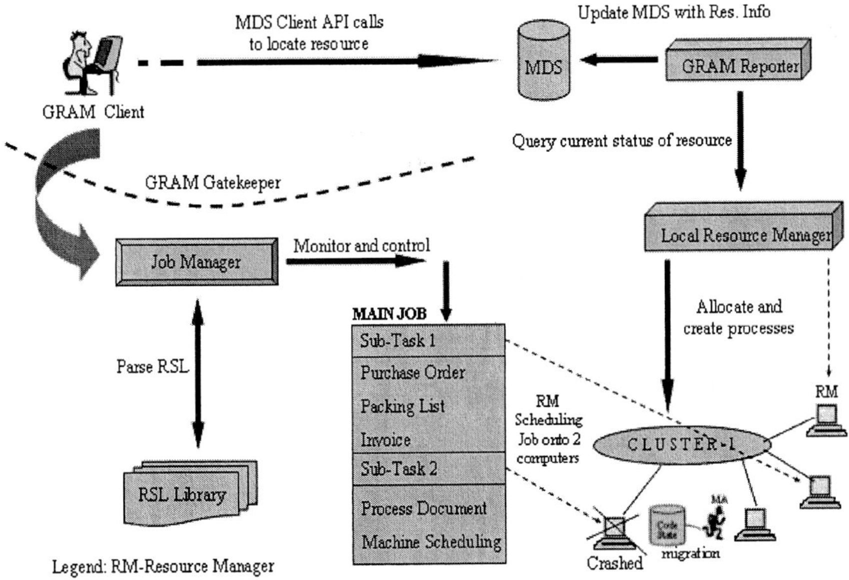

Fig. 2. GRAM-Soft Scheduling Model

- □ RSL Parsing Library: It contains resource characteristics like, main memory available, CPU processing rate, secondary memory available, and resource type.
- □ Job Manager: Initiates execution of sub-task's files and monitors their status, performs callback of sub-task termination. Job manager is destroyed once the sub-task for which it was responsible is terminated.
- □ GRAM Reporter: It is responsible for storing into MDS various information like job schedules, status of sub-tasks' files and computer status (idle or busy).
- □ MDS: Provides information about current availability and capabilities of the resources.

5 Conclusions

Here, we have focused on creating a heterogeneous Hive that contains set of transaction oriented high performance resources required for time-critical transactions. Such infrastructure should be developed as a decentralized job management system for which we believe mobile agents, meta-computing and resource co-allocation could serve as a Hive infrastructure. Regardless of precisely how businesses choose to leverage its power, there is no doubt that – by building on and extending the idea of Grid Computing – Hive Computing will provide businesses and other organizations with tremendous levels of flexibility and drive down the cost of reliability. In this work, we will develop a heterogeneous agent-based hierarchical model, resource scheduling (meta-computing environment) and co-allocation to meet the requirements of the scalability. Our future work will focus on the implementation of a resource allocation management in a Hive infrastructure using GRAM-Soft tool that contains methodology of the mobile agents, meta-computing and resource co-allocation.

References

[1] Chris O'Leary -Hive Computing for Transaction Processing Grids, Tsunami Research

[2] J. Gehring and A. Reinefeld. MARS - A Framework for Minimizing the Job Execution Time in a Metacomputing Environment. *Future Generation Computer Systems*, 12(1): 1996.

[3] J. Weissman. The Interference Paradigm for Network Job Scheduling. *Proceedings of the Heterogeneous Computing*

[4] K. Czajkowski, I. Foster – Resource Management Architecture for Metacomputing Systems.

[5] K. Czajkowski, I. Foster and C. Kesselman - Resource Co-Allocation in Computational Grids

[6] K. Czajkowski, I. Foster, C. Kesselman, S. Martin, W. Smith, S. Tuecke. Resource Management Architecture for Meta-Computing Systems.

[7] F. Berman, R. Wolski, J. Schopf and G. Shao. Application level scheduling on distributed heterogeneous networks. In proceedings of Supercomputing'96. ACM Press, 1995.

[8] S. Zhou. LSF: Load sharing in large-scale heterogeneous distributed systems. In proc. Workshop on Cluster Computing, 1992.

Data-Parallel Method for Georeferencing of MODIS Level 1B Data Using Grid Computing

Yincui Hu[1], Yong Xue[1,2,*], Jiakui Tang[1], Shaobo Zhong[1], and Guoyin Cai[1]

[1] State Key Laboratory of Remote Sensing Science, Jointly Sponsored by the Institute of Remote Sensing Applications of Chinese Academy of Sciences and Beijing Normal University, Institute of Remote Sensing Applications, Chinese Academy of Sciences, P. O. Box 9718, Beijing 100101, China

[2] Department of Computing, London Metropolitan University, 166-220 Holloway Road, London N7 8DB, UK

{huyincui@163.com, y.xue@londonmet.ac.uk}

Abstract. Georeference is a basic function of remote sensing data processing. Geo-corrected remote sensing data is an important source data for Geographic Information Systems (GIS) and other location services. Large quantity remote sensing data were produced daily by satellites and other sensors. Georeferenceing of these data is time consumable and computationally intensive. To improve efficiency of processing, Grid technologies are applied. This paper focuses on the parallelization of the remote sensing data on a grid platform. According to the features of the algorithm, backwards-decomposition technique is applied to partition MODIS level 1B data. Firstly, partition the output array into evenly sized blocks using regular domain decomposition. Secondly, compute the geographical range of every block. Thirdly, find the GCPs triangulations contained in or intersect with the geographic range. Then extract block from original data in accordance with these triangulations. The extracted block is the data distributed to producer on Grid pool.

1 Introduction

Large quantity imagery data of remote sensing produced daily by variable satellites and other sensors. The processing of remote sensing data is computationally intensive. It requires parallel and high-performance computing techniques to achieve good performance. G eo-corrected remote sensing data are source data for geographic information systems and other location services. When the correction formulation is complicate, the large image georeference will require a mass of time. Over the past decade, Grid has become a powerful computing environment for data intensive and computing intensive applications. Cannataro (2000) proposed to develop data mining services within a Grid infrastructure as the deployment platform for high performance distributed data mining and knowledge discovery.

Researchers have aimed to develop Grid platforms for remote sensing data processing. Aloisio *et al.* (2004) proposed Grid architecture for remote sensing data

* Corresponding author.

V.S. Sunderam et al. (Eds.): ICCS 2005, LNCS 3516, pp. 883 – 886, 2005.

processing and developed a Grid-enabled platform, SARA/Digital Puglia with his research group. SARA/Digital Puglia (Aloisio *et al.* 2003) is a remote sensing environment developed in a joint research project. Our research group has developed a grid-based remote sensing environment, which is the High-Throughput Spatial Information Processing Prototype System in Institute of Remote Sensing Applications, Chinese Academy of Sciences (Cai *et al.* 2004, Hu *et al.* 2004, Wang et al. 2003).

This paper focuses on the parallelization of the remote sensing data on a grid platform. First, we discussed the algorithm of rectifying remote sensing image. Second, data partition for georeference on grid is introduced. Finally, we analyzed the result of georeference on Grid platform.

2 Georeference Implementation on the Grid

2.1 MODIS Level 1B Data

MODIS level 1B products are obtained from **Mod**erate Resolution **I**maging **S**pectroradiometer (MODIS) carried on EOS satellite and have been calibrated. MODIS includes 36 spectral bands extending from the visible to the thermal infrared wavelengths (Running *et al.,* 1994). The MODIS Level 1B products contain longitude and latitude coordinates. These coordinates can be used as Ground Control Points (GCP) to rectify the image.

2.2 Parallel Rectification on Grid

There are three important components to rectify an image, i.e. transformation model selection, coordinate transformation and resample. In this paper, triangle warping model is selected to transform the coordinates. Moreover, cubic resample method is used. The triangular is build from the longitude and latitude coordinates.

The image rectification steps for interpolating, transforming and resampling can be integrated into one routine. This rectify routine can be repeated to correct every part of the large image. Now that the rectify routine can be data parallel processing in Grid, how to partition the data and how to merge the results are the main questions confront us.

2.3 Data Partition Strategy

The partition strategy influences the process efficiency and determines the merge strategy. So to select an efficient partition method is very important. The value of the output pixel is interpolated by value of pixels around its location in original image. The original location is obtained by triangle transformation from GCPs triangulations. According to the features of the algorithm, backwards-decomposition technique is applied to partition MODIS level 1B data. It comprises four steps as follows:

Firstly, partition the output array into evenly sized blocks using regular domain decomposition. Secondly, compute the geographical range of every block. Thirdly, find the GCPs triangulations contained in or intersect with the geographical range.

Then extract block from original data in accordance with these triangulations. The extracted block is the data that will be distributed to producer on Grid pool.

3 Experiments and Analysis

Our experiments were performed on a Grid-computing environment, which is in the High-Throughput Spatial Information Processing Prototype System (HIT-SIP) based on Grid platform in Institute of Remote Sensing Applications, Chinese Academy of Sciences. Test data are MODIS level 1B products. The data format is HDF. The data is partitioned into many parts by backwards-decomposition techniques. The experiment result is shown in table1.

Table 1. Experiment results

Test data size	Sequential Time (sec)	No. of parts	Execution time (sec)
768 MB	Out of memory	10	1208
		80	856
83 MB	404	8	358
		80	281
20 MB	282	4	130
		8	110

The experiment shows that data-parallel georeference is efficient especially for those large-size data. The computer shows errors "out of memory, unable to allocate memory" when rectifying the large data (768 MB) in a computer with 512MB memory. This situation could be solved with the help of the resources in Grid pool. The large data is decomposed into small parts and distributed to the Grid.

4 Conclusions

As an important new field in the distributed computing arena, Grid computing focuses on intensive resource sharing, innovative applications, and, in some cases, high-performance orientation. We implemented data-parallel georeference in HIT-SIP platform. The experiments indicate that Grid is efficient for data-parallel georeference. The efficiency could be improved especially for those large data. For those processing that need large memory, Grid can also provide enough resources to solve the problem. Ongoing work on HIT-SIP includes developing middleware of remote sensing processing and providing remote sensing processing services for Internet consumers.

Acknowledgement

This publication is an output from the research projects "CAS Hundred Talents Program" and "Monitoring of Beijing Olympic Environment" (2002BA904B07-2) and “Remote Sensing Information Processing and Service Node” funded by the MOST, China and "Aerosol fast monitoring modeling using MODIS data and middlewares development" (40471091) funded by NSFC, China.

References

1. Aloisio, G., Cafaro, M.: A Dynamic Earth Observation System. Parallel Computing. 29 (2003), 1357-1362
2. Aloisio, G., Cafaro, M., Epicoco, I., Quarta, G.: A Problem Solving Environment for Remote Sensing Data Processing. Proceeding of ITCC 2004: International Conference on Information Technology: Coding and Computing. Vol.2. IEEE Computer Society, 56-61.
3. Cai GY, Xue Y, Tang JK, Wang JQ, Wang YG, Luo Y, Hu YC, Zhong SB, Sun XS, 2004. Experience of Remote Sensing Information Modelling with Grid Computing. Lecture Notes in Computer Science, Vol. 3039. Springer-Verlag, 989-996.
4. Cannataro, M: Clusters and Grids for Distributed and Parallel Knowledge Discovery. Lecture Notes in Computer Science. 1823(2000), 708-716
5. Hu YC, Xue Y, Wang JQ, Sun XS, Cai GY, Tang JK, Luo Y, Zhong SB, Wang YG, Zhang AJ.: Feasibility Study of Geo-spatial Analysis Using Grid Computing. Lecture Notes in Computer Science, Vol. 3039. Springer-Verlag (2004) 956-963.
6. Lanthier, M., and Nussbaurm, D: Parallel Implementation of Geometric Shortest Path Algorithms. Parallel Computing. 29(2003), 1445-1479.
7. Pouchard, L; Cinquini, L; Drach, B; Middleton, D; Bernholdt, D; Chanchio, K; Foster, I; Nefedova, V; Brown, D; Fox, P; Garcia, J; Strand, G; Williams, D; Chervenak, A; Kesselman, C; Shoshani, A; Sim, A.: An Ontology for Scientific Information in a Grid Environment: The Earth System grid. 3rd IEEE/ACM International Symposium on Cluster Computing and the GRID. IEEE Computer Society (2003) 626-632.
8. Roros, D. -K. D., Armstrong, M. P: Using Linda to Compute Spatial Autocorrelation in Parallel. Computers & Geosciences. 22(1996), 425-432
9. Roros, D. -K. D., Armstrong, M. P.: Experiments in the Identification of Terrain Features Using a PC-Based Parallel Computer. Photogrammetric Engineering & Remote Sensing. 64(1998), 135-142
10. Wang, S. and Armstrong, M. P.: A Quadtree Approach to Domain Decomposition for Spatial Interpolation in Grid Computing Environments. Parallel Computing. 29(2003) 1481-1504.
11. Jianqin Wang, Yong Xue, and Huadong Guo, 2003, A Spatial Information Grid Supported Prototype Telegeoprocessing System. In Proceedings of 2003 IEEE International Geoscience and Remote Sensing Symposium (IGARSS’2003) held in Toulouse, France on 21-25 July 2003, v 1, p 345-347.

An Engineering Computation Oriented Grid Project: Design and Implementation[1]

Xianqing Wang[1,2], Qinhuai Zeng[2], Dingwu Feng[2], and Changqin Huang[2,3]

[1] Guangdong Institute of Science and Technology, Zhuhai, 519090, P.R. China
[2] Hunan University of Arts and Science, Changde, 415000, P.R. China
[3] College of Computer Science, Zhejiang University, Hangzhou, 310027, P.R. China
xqwang_kgy@126.com, djzqh@163.com, cqhuang@zju.edu.cn

Abstract. This paper describes a Service-bAsed Grid project for Engineering computation, named SAGE. Based on the Globus toolkit, a grid-service-based architecture oriented to engineering computation is presented. To give grid users good usability, task definition and resource discovery are visually conducted. Whilst, a Quality of Service (QoS) driven user-centric scheduling strategy is proposed, two scheduling methods and steering-enabled visual interfaces are applied for different types of grid users. Result processing can be visualized in the aid of a PC-cluster and a stereopticon. The practices suggest that these mechanisms improve the convenience and QoS.

1 Introduction

Grid computing has a promising future in large-scale scientific computation. The Globus toolkit [1] is the most popular grid environment and de facto standard, and it has adopted OGSA [2] architecture to provide applications with grid service level at present. As a special type of large-scale computation, engineering computation is very hardy-solved because of requirements for many valuable computing resources or specialist instrumentations, grid computing is a suited computing infrastructure for the high capability of distributed and heterogeneous resources sharing. However, grid-based engineering computation needs to solve some important issues, such as narrowing the gap between currently deployed grid services and the would-be user community.

2 Overview of Proposed Architecture

To meet requirements for scientists in engineering computation field in grid environments, we propose the SAGE project (Service-bAsed Grid project for Engineering computation) that aims to use Grid technology to establish an enabling environment for large-scale scientific and engineering research. The overview of architecture is shown in Fig.1, it consists of four components: a) The interface component is a portal

[1] A Project Supported by Scientific Research Fund of Hunan Provincial Education Department, China (Grant No. 04A037).

V.S. Sunderam et al. (Eds.): ICCS 2005, LNCS 3516, pp. 887 – 890, 2005.

of the system, It not only provides a visual interface for development and steering of grid-based applications, but also gives these system administrator a friendly management portal. It comprises a task submission window and two types of grid scheduling interfaces. b) Task-level logic component is responsible for task management, community policy and user administration. It includes virtual organization register, user manager, task definition, task import and export, task submission, task scheduling, task monitoring, task rescheduling, file transfer, and result processing. c) Grid service logic part consists of all services and service management mechanisms. All services are divided into three types: advanced general service, engineering computation modeling and post-processing service, basic grid service. The former two types are constructed on the basis of the latter. Service management contains: service deployment, service location, service composition, service scheduling and service-level performance steering. d) Grid Security Infrastructure is the last hierarchy in our architecture. It is based on the Globus Toolkit's security mechanism, main characteristics exist: parallelized subtask-based authorization, community policy and task delegation management [3].

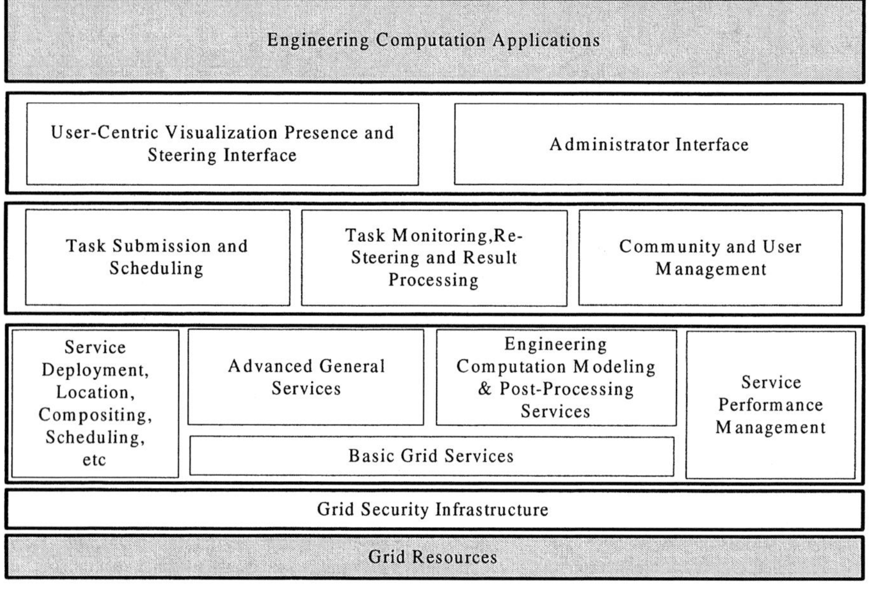

Fig. 1. Engineering computation architecture

3 Definitions of Grid Task and Resource Discovery

A session for each user is defined in the Grid. The session is created when one logs in, and destroyed after one logs out. In the session, a user defines computation task, possesses resources, monitors task execution and does post-process for the results. The Grid provides a graphical user interface to input a new task or to update an existing task. The data include executable program, resources and schedule requirements.

When the Grid accepts the data, an instance of task class is initialized and added to logical resources allocation queue. The task's state is initialized to "WAITING".

Resources information services provide two interfaces, a hierarchical resources tree, and a resources list. The content of the tree is the detailed information regarding hosts in virtual organizations, where local hosts are registered. It also shows the architecture of those virtual organizations. In the list, each item contains information of an individual host. The information on the list can be customized for display. Generally it consists of host name, node count, CPU count, speed and free percentage of CPU, free memory size, free secondary storage, network bandwidth and latency. In this project, resources are divided into two classes, computing power and visualization devices.

4 Task Scheduling and QoS Management

This architecture adopts aggregated QoS driven visual scheduling, the scheduling could be performed through two types of visual windows, scheduling and QoS sessions. These visual methods provide users with a direct awareness and a friendly interaction. A visual scheduling framework is shown in Fig. 2.

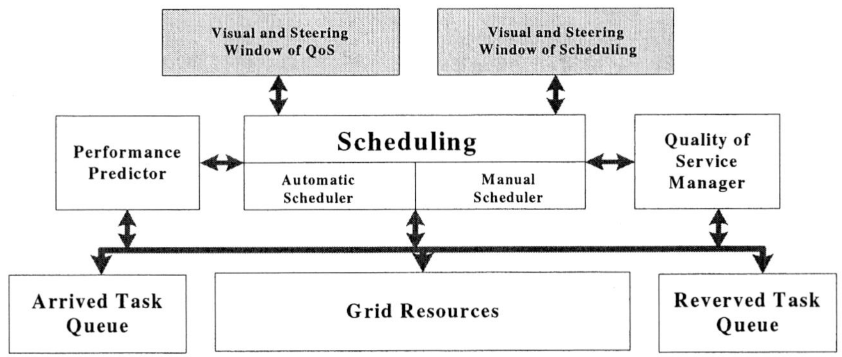

Fig. 2. A visual scheduling framework

According to different types of users, the visual scheduling is performed in both manual and automatic manner with a monitoring mechanism. In addition to the arrived task queue, a reserved task queue is used for arranging pre-scheduled tasks. Hence reservation-based scheduling can perform well. The scheduling algorithms during the automatic scheduling can be selected by users, as various conventional algorithms can be integrated into the system. The idea is based on the fact that different algorithms are suitable for respective environments and objectives.

In our architecture, a self-designed simple performance predictor serves for the scheduler by predicting and computing the previously mentioned performance metrics. It analyzes the performances of the tasks to be scheduled in advance; furthermore, it evaluates QoS parameters of scheduled tasks. QoS manager is responsible for accepting these required performance values from the input, for setting performance threshold values that are conditions of triggering adjustment mechanisms and warning user, and for managing the events of post-scheduling and the interactions for a better performance. In the project, aggregated QoS is modeled based on four performance metrics. The relevant scheduling algorithms are based on Dual-Component and Dual-Queue Distributed Schedule Model (D^3SM), which is described in the literature [4].

5 Visualization for Results

The project supports the collaborative visualization of meshes and visual steering of CFD/CSM simulations. The visualization will be implemented by our on-going project named ViG (Visualization in Grid). ViG makes use of the DAWNING PC-cluster. The cluster has 9 rendering nodes, and each node has a dual-head FX5200 display card connected to a projector. The projectors are deployed to generate stereoscopic 3 x 3 images on the large display wall. ViG will provide input to the development of this service and will develop a fully Grid-enabled extension of standard rendering APIs to allow scientists to run Grid applications through the ViG user interface.

6 Conclusions

The Grid project for engineering computation (SAGE) supplies the users with a visual application development and deployment environment for engineering computation. The whole is characterized by grid service. Whilst, definition of grid task & resource discovery are visually conducted, QoS driven user-centric scheduling, are designed and implemented. Result processing can be visualized in the aid of a PC-cluster and a stereopticon. In practice, the convenience and QoS are significantly improved, however, there exist a great shortage in the whole system, which are yet the important aspects of our future work.

References

1. I. Foster, and C. Kesselman (eds.), The Grid: Blueprint for a New Computing Infrastructure, Morgan Kaufmann, 1999.
2. J. Nick, I. Foster, et al, The Physiology of the Grid: An Open Grid Services Architecture for Distributed Systems Integration, Open Grid Service Infrastructure, Global Grid Forum, 2002.
3. CQ Huang, GH Song et al, An Authorization Architecture Oriented to Engineering and Scientific Computation in Grid Environments, ACSAC 2004, LNCS 3189, pp. 461–472, 2004.
4. CQ Huang, DR Chen et al, Performance-Driven Task and Data Co-scheduling Algorithms for Data-Intensive Applications, APWeb 2004, LNCS 3007, pp. 331–340, 2004.

Iterative and Parallel Algorithm Design from High Level Language Traces

Daniel E. Cooke and J. Nelson Rushton

Computer Science Department, Texas Tech University, Lubbock, Texas, U.S.A. 79409
{dcooke, nrushton}@coe.ttu.edu

Abstract. We present a high level language called SequenceL. The language allows a programmer to describe functions in terms of abstract relationships between their inputs and outputs, and the semantics of the language are capable of automatically discovering and implementing the required algorithms, including iterative and parallel control structures in many cases. Current implementations do not produce code of comparable efficiency to that of a good human programmer. Current implementations can, however, be used as a tool to guide human programmers in discovering and comparing options for parallelizing their solutions. This paper describes the language and approach, and illustrates this kind of guidance with a simple example.

1 Introduction and Related Work

SequenceL is a high-level language in which algorithms for implementing a solution are automatically derived from a high level description of that solution. In this paper we introduce how the language can be exploited in the design of parallel algorithms.

The beginning point of this effort can be traced to [1], which overlaps the iterative morphisms found in [6]. SequenceL's semantic work-horse is the *Normalize-Transpose-Distribute* (NTD), which is used to simplify and decompose structures, based upon a dataflow like execution strategy. The NTD semantic achieves a goal similar to that of the Lämmel and Peyton-Jones, boilerplate elimination. [4,5]

2 Normalize-Transpose-Distribute (NTD)

SequenceL functions have typed arguments, where the type of an argument is simply its level of nesting (scalar, sequence of scalars (i.e. vector), sequence of sequences of scalars, (i.e., matrix) etc.). When a SequenceL function receives arguments, which are *overtyped* (i.e., nesting level higher than expected), the remaining arguments are *normalized* -- duplicated to match the length of the overtyped arguments.

For example, consider the definition of dot product in SequenceL:

```
dotprod: vector * vector -> scalar
dotprod(A,B) = sum(A*B)
```

V.S. Sunderam et al. (Eds.): ICCS 2005, LNCS 3516, pp. 891 – 894, 2005.

Given the call dotprod((2,3,4) (100,10,1)), instantiating the function results in sum(((2,3,4) * (100,10,1)). Since * expects scalars for both arguments, NTD is performed, resulting in sum(200, 30, 4), which evaluates to 234, the desired result.

3 The Environment and Matrix Multiply

From the standpoint of specification, the *i,jth* element of the product matrix of *a* and *b* is the dot product of the *i'th* row of a *a* with the *j'th* column of *b*. Here is SequenceL code for the algorithm:

```
mmrow: vector * matrix → matrix
mmrow(A,B)  ::= dotprod(A,transpose(B))
```

This function appears to compute only a single row of the matrix product *AB*, but will compute all rows if a matrix is passed as an argument, due to NTD. Here is the trace of a sample execution:

(mmrow, (((1, 2, 4), (10, 20, 40), (11, 12, 14)), ((1, 2, 4), (10, 20, 40), (11, 12, 14)))) (1)

Since the first argument is a matrix rather than a vector as expected, normalize and transpose are performed yielding

((mmrow,((1,2,4),((1,2,4),(10,20,40),(11,12,14)))), (2)
(mmrow,((10,20,40),((1,2,4),(10,20,40),(11,12,14)))),
(mmrow,((11,12,14),((1,2,4),(10,20,40),(11,12,14)))))

Now the language interpreter instantiates the body of the *mmrow* function and the transposes are performed (we write dp for dotprod to save space):

((dp,((1,2,4),(transpose,((1,2,4),(10,20,40),(11,12,14))))), (3)
(dp,((10,20,40),(transpose,((1,2,4),(10,20,40),(11,12,14))))),
(dp,((11,12,14),(transpose,((1,2,4),(10,20,40),(11,12,14))))))

After the transposes, the SequenceL *dp* function as defined in section 2 is invoked:

((dp,((1,2,4),((1,10,11),(2,20,12),(4,40,14)))), (4)
(dp,((10,20,40),((1,10,11),(2,20,12),(4,40,14)))),
(dp,((11,12,14),((1,10,11),(2,20,12),(4,40,14)))))

Note that *dp* takes two vectors as input. From where we left off in step *4* of the trace, we know that the second argument of each *dp* reference is a two-dimensional structure, so another *NTD* is performed on each *dp* resulting in *9* dp references:

(((dp,((1,2,4),(1,10,11))), (dp,((1,2,4),(2,20,12))), (dp,((1,2,4),(4,40,14)))), (5)
((dp,((10,20,40),(1,10,11))), (dp,((10,20,40),(2,20,12))), (dp,((10,20,40),(4,40,14)))),
((dp,((11,12,14),(1,10,11))), (dp,((11,12,14),(2,20,12))), (dp,((11,12,14),(4,40,14)))))

which are evaluated (potentially in parallel) for the final result:

(((65, 90, 140), (650, 900, 1400), (285, 430, 720)))

The parallelisms discovered between step *5* and the final result are micro-parallelisms, and therefore, should be carried out iteratively. As it turns out, steps *4* and *5* of the trace are of keen interest in designing parallel or concurrent algorithms. Either point could be used as the basis for the algorithms. In the next two subsections we evaluate these options and how they can be realized in concurrent JAVA codes.

3.1 Parallelizing Based on Step 5 of the Trace

Step *5* indicates that we are combining every element of each row of the first matrix with every element of each column of the second, forming a cartesian product of sorts from the two matrices. To accomplish this in concurrent JAVA we would have the following code segments, extracted from a total of about 60 lines of code.

```
... public void run()     {
      s = 0;
      for (k=0;k<=m.length-1;k++)
            { s += m[rs][k] * m[k][cs]; }}
public static void main (String args()) { ...
      for(i=0;i<=(r*c)-1;i++)
            {mat[i].start();                ... }
```

Notice that the $\sim O(n^2)$ algorithm that forks the *run* threads is in the main program and that we are indeed computing each element of the result concurrently. This loop follows a nested $\sim O(n^2)$ algorithm that initialized each of the threads with the proper subscript values for *rs* and *cs*.

3.2 Parallelizing Based on Step 4 of the Trace

Step *4* of the SequenceL trace indicates another option for the parallelization of the desired computation. In this approach, each row of the first matrix is combined with all columns of the second. This suggests moving the $\sim O(n^2)$ algorithm into the concurrently executed *run* threads, reducing the overhead involved in separating threads and spawning concurrent tasks from $\sim O(n^2)$ to $O(n)$.

```
... public void run()     {
      s = new int(m.length);
      for (rs=0;rs<=m.length-1;rs++)
            for (k=0;k<=m.length-1;k++)
                  {s[rs]+=m[rs][k]*m[k][cs];}}
public static void main (String args()) {...
      for(i=0;i<=r-1;i++)
            {mat[i].start();}               ... }
```

This approach concurrently computes the result of each row of the resultant matrix and follows a rule of thumb in developing concurrent codes: parallelize outer loops.

It is interesting to note that though this implementation is substantially more efficient than the previous one, it did not occur to us until after we examined the trace of matrix multiplication in SequenceL.

4 Discussion

Since completing the changes to the language last year, we have conducted additional experiments to discover parallel algorithms for problems including Gaussian Elimination, Quicksort, Discrete Wavelet Transforms, Newton-Raphson solutions to linear equations, Fast Fourier Transform, among others. We have also investigated applications to remote sensing problems and to the prototyping of Guidance, Navigation, and Control problems for processing onboard the Space Shuttle. These experiments have shown SequenceL to be an easy-to-use and promising tool for exploring ideas and problem solutions in a variety of domains.

References

1. Daniel E. Cooke and A. Gates, "On the Development of a Method to Synthesize Programs from Requirement Specifications," *International Journal on Software Engineering and Knowledge Engineering*, Vol. 1 No. 1 (March, 1991), pp. 21-38.
2. Daniel E. Cooke, "An Introduction to SEQUENCEL: A Language to Experiment with Nonscalar Constructs," *Software Practice and Experience,* Vol. 26 (11) (November, 1996), pp. 1205-1246.
3. Daniel E. Cooke and Per Andersen, "Automatic Parallel Control Structures in SequenceL," *Software Practice and Experience*, Volume 30, Issue 14 (November 2000), pp. 1541-1570.
4. R. Lämmel and S. Peyton-Jones, *"Scrap your boilerplate: a practical design pattern for generic programming,"* in Proceedings of TLDI 2003, ACM Press.
5. R. Lämmel and S. Peyton-Jones, *"Scrap more boilerplate: reflection, zips, and generalised casts,"* to appear in Proceedings of ICFP 2004, ACM Press.
6. Meijer, E. and Fokkinga, M.M. and Paterson, R., "Functional Programming with Bananas, Lenses, Envelopes and Barbed Wire" *FPCA* (Springer-Verlag, 1991) LNCS Series Vol. 523, pp. 124—144.

An Application of the Adomian Decomposition Method for Inverse Stefan Problem with Neumann's Boundary Condition

Radosław Grzymkowski and Damian Słota

Institute of Mathematics, Silesian University of Technology, Kaszubska 23, 44-100 Gliwice, Poland
{r.grzymkowski, d.slota}@polsl.pl

Abstract. In this paper the solution of one-phase inverse Stefan problem with Neumann's boundary condition is presented. This problem consists of the reconstruction of the function which describes the heat flux on the boundary, when the position of the moving interface is well-known. The proposed solution is based on the Adomian decomposition method and the least square method.

1 Introduction

In this paper we solve the one-phase inverse Stefan problem, which consists of the reconstruction of the function which describes the heat flux on the boundary, when the position of the moving interfaces is well-known. This kind of problem becomes an inverse design problem. The conditions for the existence and uniqueness of the solution of this problem are given in the literature [3]. The solution is based on the Adomian decomposition method and the least square method.

The Adomian decomposition method was developed by G. Adomian [1, 2]. This method is useful for solving a wide class of problems [2, 4]. Using this method we are able to solve non-linear operator equation:

$$F(u) = f \tag{1}$$

where $F : H \to G$ is a non-linear operator, f is a known element from Hilbert space G and u is the sought element from Hilbert space H. Operator $F(u)$ can be written as:

$$F(u) = L(u) + R(u) + N(u) \ , \tag{2}$$

where L is the invertible linear operator, R is the remaining linear operator and N is a non-linear operator. The solution of the equation (1) is sought in the form of a functional series:

$$u = \sum_{i=0}^{\infty} g_i \ . \tag{3}$$

After some manipulations we obtain the following recurrent formula:

$$\begin{aligned} g_0 &= g^* + L^{-1}(f) \ , \\ g_n &= -L^{-1}R(g_{n-1}) - L^{-1}(A_{n-1}) \ , \qquad n \geq 1 \ , \end{aligned} \tag{4}$$

V.S. Sunderam et al. (Eds.): ICCS 2005, LNCS 3516, pp. 895–898, 2005.

where g^* is a function dependent on the initial and boundary conditions and L^{-1} is the inverse operator.

2 Inverse Stefan Problem

Let $D \subset \mathbb{R}^2$ be a domain (Figure 1). We seek an approximate solution of the following problem:

For the given position of freezing front Γ_g, the distribution of temperature u in domain D is calculated as well as function $q(t)$ on boundary Γ_1, which satisfies the following equations:

$$\frac{\partial^2 u(x,t)}{\partial x^2} = \frac{1}{a}\frac{\partial u}{\partial t}(x,t) , \qquad \text{in } D , \tag{5}$$

$$u(x,0) = \varphi(x) , \qquad \text{on } \Gamma_0 , \tag{6}$$

$$-\lambda \frac{\partial u(x,t)}{\partial x} = q(t) , \qquad \text{on } \Gamma_1 , \tag{7}$$

$$u(\xi(t),t) = u^* , \qquad \text{on } \Gamma_g , \tag{8}$$

$$-\lambda \frac{\partial u(x,t)}{\partial x} = \kappa \frac{d\xi(t)}{dt} , \qquad \text{on } \Gamma_g , \tag{9}$$

where a is the thermal diffusivity, λ is the thermal conductivity, κ is the latent heat of fusion per unit volume, and u, t and x refer to temperature, time and spatial location, respectively.

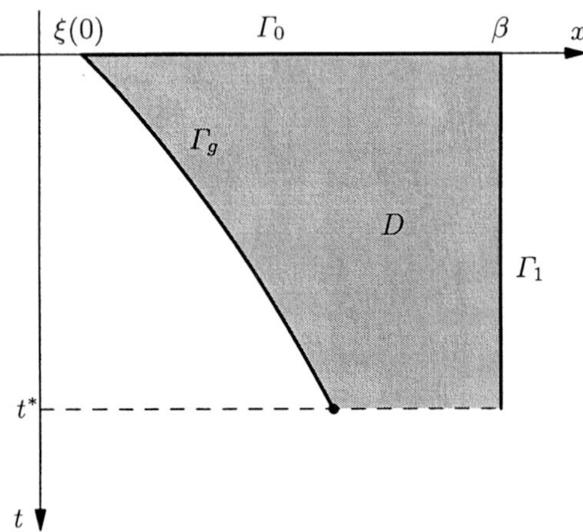

Fig. 1. Domain formulation of the problem

3 Solution of the Problem

In the problem under consideration we have operator equations (1), where:

$$L(u) = \frac{\partial^2 u}{\partial x^2} \ , \quad R(u) = -\frac{1}{a}\frac{\partial u}{\partial t} \ , \quad N(u) = 0 \ , \quad f = 0 \ .$$

The inverse operator L^{-1} is given by:

$$L^{-1}(u) = \int\limits_{\xi(t)}^{x} \int\limits_{x}^{\beta} u(x,t)\, dx\, dx \ . \tag{10}$$

Hence, using the boundary condition (7) and (8) we obtain:

$$\begin{aligned} g_0(x,t) &= \frac{1}{\lambda}\, q(t)\, \big(\xi(t) - x\big) + u^* \ , \\ g_n(x,t) &= -\frac{1}{a} \int\limits_{\xi(t)}^{x} \int\limits_{x}^{\beta} \frac{\partial g_{n-1}(x,t)}{\partial t}\, dx\, dx \ , \qquad n \geq 1 \ . \end{aligned} \tag{11}$$

We seek an approximation solution in the form:

$$u_n(x,t) = \sum_{i=0}^{n} g_i(x,t) \ , \qquad n \in \mathbb{N} \ . \tag{12}$$

Because the g_i (11) are dependent on the unknown function $q(t)$, we derived this function in the form:

$$q(t) = \sum_{i=1}^{m} p_i\, \psi_i(t) \ , \tag{13}$$

where $p_i \in \mathbb{R}$ and the basis functions $\psi_i(t)$ are linearly independent. The coefficients p_i are selected to show minimal deviation of function (12) from the conditions (6) and (9) (considering the assumed measure). To construct the measure of error the least square method is applied. Thus we seek the minimum of the following functional:

$$J(p_1, \ldots, p_m) = \int\limits_{\xi(0)}^{\beta} \big(u_n(x,0) - \varphi(x)\big)^2\, dx + \\ + \int\limits_{0}^{t^*} \Big(\lambda\, \frac{\partial u_n(x,t)}{\partial x}\Big|_{x=\xi(t)} + \kappa\, \frac{d\xi(t)}{dt}\Big)^2\, dt \ . \tag{14}$$

Substituting equations (12), (11) and (13) to functional J and differentiating it with respect to the coefficients p_i $(i = 1, \ldots, m)$ and equating the obtained derivatives to zero, the system of the linear algebraic equations is obtained. In the course of solving this system, coefficients p_i are determined, and thereby the approximated distributions of temperature in the domain D and the heat flux on the boundary Γ_1 are obtained.

4 Numerical Example

The theoretical considerations introduced in the previous sections will be illustrated with an example, in which: $\beta = 1$, $a = 1$, $\lambda = 2$, $\xi(t) = a\,t$, $\varphi(x) = e^{-x}$, $\kappa = \lambda/a$, $u^* = 1$, $t^* = 1$. Then the exact solution of the inverse Stefan problem can be found from the following functions: $u(x,t) = e^{a\,t-x}$ for $(x,t) \in D$, and $q(t) = \lambda\, e^{a\,t-\beta}$ for $t \in [0,t^*]$. As basis functions we take $\psi_i(t) = t^{i-1}$, $i = 1,\ldots,m$. For the calculations, we assume $m \in \{2,3,\ldots,6\}$ and $n \in \{2,3,\ldots,7\}$.

Figure 2 shows the exact and the reconstructed heat flux on the boundary Γ_1 for a different number of elements of the sum (12), $n \in \{4,7\}$ and for six basis functions $\psi_i(t)$. The results obtained show that functions $q(t)$ and $u(x,t)$ are reconstructed very well.

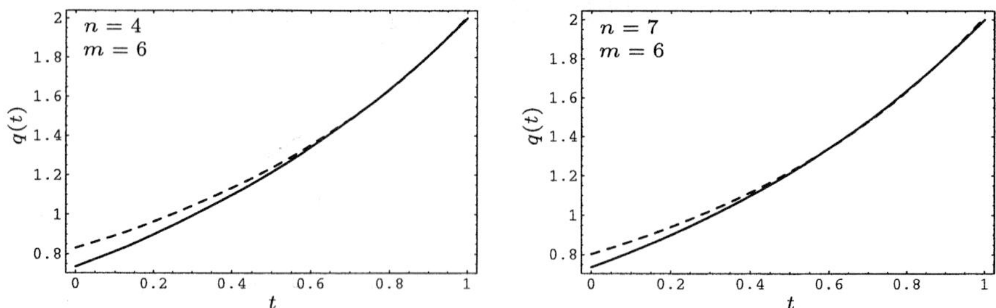

Fig. 2. The heat flux on the boundary Γ_1 (solid line – exact value, dash line – reconstructed value)

5 Conclusion

We have presented a method for solving the one-phase inverse Stefan problem which consists of the reconstruction of the function describing the heat flux on the boundary, when the position of the moving interface is well-known. The method makes use of the Adomian decomposition method and the least square method. The calculations presented show that this method is effective for solving this problem.

References

1. Adomian, G.: Stochastic Systems. Academic Press, New York (1983)
2. Adomian, G.: A review of the decomposition method in applied mathematics. J. Math. Anal. Appl. **135** (1988) 501–544
3. Goldman, N.L.: Inverse Stefan problem. Kluwer, Dordrecht (1997)
4. Lesnic, D.: Convergence of Adomian's decomposition method: periodic temperatures. Computers Math. Applic. **44** (2002) 13–24

Group Homotopy Algorithm with a Parameterized Newton Iteration for Symmetric Eigen Problems

Ran Baik[1], Karabi Datta[2] and Yoopyo Hong[2]

[1] Department of Computer Engineering, Honam University, Gwangju 506-090, Korea

[2] Department of Mathematical Sciences, Northern Illinois University, DeKalb, IL 60115, USA

baik@honam.ac.kr

{dattak, hong}@math.niu.edu

Abstract. In this paper we develop a general Homotopy method called the Group Homotopy method to solve the symmetric eigenproblem. The Group Homotopy method overcomes notable drawbacks of the existing Homotopy method, namely, (i) the possibility of breakdown or having a slow rate of convergence in the presence of clustering of the eigenvalues and (ii) the absence of any definite criterion to choose a step size that guarantees the convergence of the method. On the other hand, the Group Homotopy method maintains attractive features of the ordinary Homotopy method such as the natural parallelism and the structure preserving properties.

1 Introduction

Let S and A be two real $n \times n$ symmetric matrices. Define a mapping $H : \mathbb{R} \longrightarrow M_n$ such that $H(t) = A(t) \equiv (1-t)S + tA, \quad t \in [0,1]$. Note that $A(t) = S + t(A - S)$ so that the matrix $A(t)$ can be made close to S by choosing t small enough. The objective is to obtain the set of all the eigenpairs $\left\{ \begin{pmatrix} u_k \\ \lambda_k \end{pmatrix} \right\}$, $k = 1, \ldots, n$ of $A(1) = A$ by successively obtaining the set of eigenpairs $\left\{ \begin{pmatrix} u_k(i) \\ \lambda_k(i) \end{pmatrix} \right\}, k = 1, \ldots, n$ of $A(t_i), 0 = t_0 < t_1 < \cdots < t_n = 1$, starting from the set of eigenpairs $\left\{ \begin{pmatrix} x_k \\ \alpha_k \end{pmatrix} \right\}, k = 1, \ldots, n$ of $A(0) = S$[1,2,3,6,7,8]. The procedure is called the Homotopy method. A major problem that arises in implementing a Homotopy method for the eigenproblem is: how to choose $\Delta t_i \equiv t_{i+1} - t_i$ properly so that the all eigenpairs of $A(t_i + \Delta t_i)$ can be obtained from the eigenpairs (possibly clustered) of $A(t_i)$ using Newton's method with a reasonable number of iterations. In this paper we provide an answer to the question. It is well known that there exists a small neighborhood of each eigenpair of $A(t_{i+1})$ such that Newton's method converges in the small neighborhood.

V.S. Sunderam et al. (Eds.): ICCS 2005, LNCS 3516, pp. 899–907, 2005.

Thus the question is: how close should $A(t_i)$ be to $A(t_{i+1})$ to have a guarantee that each eigenpair of $A(t_i)$ is in that convergence neighborhood of the corresponding eigenpair of $A(t_{i+1})$ under Newton's method. [4] provides an answer to above question whenever the eigenvalues of $A(t_i)$ are not clustered. We define $gap^*(A(t_i)) \equiv \min_k\{gap_k(A(t_i))\}$ where $gap_k(A(t_i)) = |\lambda_k(i) - \lambda_{k+1}(i)|$, and $\lambda_k(i)$ are the eigenvalues of $A(t_i)$ in decreasing order for $k = 1, \ldots, n-1$. We show that if Δt_i is chosen such that $\Delta t_i \leq \dfrac{gap^*(A(t_i))}{q}$ for some $q \geq 4$, then all eigenpairs of $A(t_i)$ will converge to the corresponding eigenpairs of $A(t_i + \Delta t_i)$ under Newton's method. The method will fail when $gap^*(A(t_i))$ becomes extremely small for some $t_i \in [0, 1)$, that is, the matrix $A(t_i)$ has the clustered eigenvalues. To overcome the problem we introduce the group *gap*, *Gap** in section 2. It is noted that under the group gap we always obtain a sufficiently large step size Δt_i so that the Homotopy method terminates in a finite number of steps. We also discuss problems that may occur due to the large step size Δt_i when the Newton's method is applied to obtain the successive eigenpairs. We show in section 3 that this problem can be easily resolved by a simple modification of the Newton's method. The new Homotopy method with group gap, Gap* is called the Group Homotopy method. The Group Homotopy method with the parameterized Newton's iteration never fails and is guaranteed to give all eigenpairs of A in a finite number of steps. Furthermore, the choice of the initial matrix S can be arbitrary even though a good choice of the initial matrix will facilitate the convergence. An algorithm of the Group Homotopy method is presented in section 3.

2 Properties of the Group Homotopy Method

In this section we develop a method called the Group Homotopy method which is a generalization of the Individual Homotopy Method. This is done to overcome the difficulties that arise in the Individual Homotopy method. Assume that an arbitrary real symmetric matrix $S \in M_n$ is chosen as the initial matrix. As we have observed previously, the Individual Homotopy method fails whenever the gap between some adjacent eigenvalues of the matrix $A(t_i)$ becomes quite small. This situation may occur when some eigenvalues of $A(t_i)$ are clustered, i.e., there are groups of a consecutive eigenvalues in which the distance between any two groups is much larger than the gaps among the eigenvalues that belong to the same group. We resolve the difficulty by using a concept, called the **group gap.** For that purpose, we define the clustering of eigenvalues as follows:

Definition 1. *Let* $\sigma(A) = \{\lambda_i\}_{i=1,\ldots,n}$ *be the set of eigenvalues of a symmetric matrix* A *in decreasing order. We say that two adjacent eigenvalues* λ_k *and* λ_{k+1} *are clustered if* $|\lambda_k - \lambda_{k+1}| < Gap^*, 1 \leq k \leq n-1$ *where* Gap^* *is a given positive number.*

Definition 2. *We say that a subset* $G = \{\lambda_k, \lambda_{k+1}, \ldots, \lambda_{k+m}\} \subseteq \sigma(A)$ *is a group of clustering of eigenvalues if* $|\lambda_{k+i} - \lambda_{k+i+1}| < Gap^*$ *for all* $i = 0, \ldots, m-1$ *where* $|\lambda_{k-1} - \lambda_k| \geq Gap^*$ *and* $|\lambda_{k+m} - \lambda_{k+m+1}| \geq Gap^*$.

Now we define the Group Homotopy Method. Let S and A be given real $n \times n$ symmetric matrices such that S is the initial matrix and A is the target matrix. Then $A(t_{i+1}) = A(t_i) + \Delta t_i(A - S)$ where $0 \leq t_0 < t_1 < \ldots < t_n = 1$, $\Delta t_i = t_{i+1} - t_i$, $A(t_0) = S$ and $A(t_n) = A$.

Definition 3. *We say a Homotopy method is the Group Homotopy method if the step size* Δt_i *is determined by* Gap^*.

Before we develop the Group Homotopy method we make the following brief remarks.

Remarks

(A) Advantages of using Gap^***:** The stepsize $\Delta t_i = \dfrac{gap^*(A(t_i))}{q}$ in the Individual Homotopy are determined completely by the eigenvalues of $A(t_i)$ where the eigenpath of the Homotopy is determined a'priori by the initial matrix S. On the other hand Gap^* in the Group Homotopy is a constant that we are free to choose to group the eigenvalues of $A(t_i)$ so that the resulting stepsize Δt_i is large enough.

(B) How to choose Gap^***:** For the choice of Gap^*, one can consider the following: (1) the dimension of matrices involved, (2) magnitude of the eigenvalues of matrices involved, and (3) the nearness of the initial matrix to the target matrix. Furthermore, a good choice of Gap^* should be the one that can be computed easily while providing a large enough step size for the Homotopy method to terminate in a reasonable amount of time. The following choice of Gap^* is used in this paper: $Gap^* = \delta\|A - S\| = \delta\|A - S\|$, $\delta > 0$ where $\|\cdot\|$ is a matrix norm.

(C) Justification of the choice $Gap^* = \delta\|A - S\|$**:** We will justify our choice of Gap^* and show how it leads to a simple determination of Δt_i. For some positive scalars c_1 and c_2, we have $\|A - S\|^2 = c_1\|A - S\|_F^2 = c_1 c_2 \sum_{k=1}^{n} (\alpha_k - \lambda_k)^2$ where α_k's and λ_k's are the eigenvalues of the matrices S and A, respectively. Thus, $\delta\|A - S\|$ measures overall nearness of the initial eigenvalues to the target eigenvalues. It also implicitly considers the eigenvectors of both matrices A and S. Set $\Delta t_i = \dfrac{Gap^*}{q\|A - S\|}$, $q > 0$. Note that $\Delta t_i = \dfrac{Gap^*}{q\|A - S\|} = \dfrac{\delta\|A - S\|}{q\|A - S\|} = \dfrac{\delta}{q}$ for some $\delta > 0$. Thus if we set $\delta = gap^*(A(t_i))$ then $\Delta t_i = \dfrac{gap^*(A(t_i))}{q}$, or the Group Homotopy method simply reduces to the previous Individual Homotopy method. Furthermore, note that if δ is chosen to be a positive number then $\Delta t_i = \dfrac{\delta}{q}$ is a fixed constant, regardless of i. Thus under the Group Homotopy the total number of steps is not more than $\left[\dfrac{q}{\delta}\right]$, a nearest integer larger than

$\frac{q}{\delta}$. When we have a good choice of initial matrix S, we might set $\Delta t = 1$. It is precisely the case when $q\|A - S\| < \frac{gap^*(S)}{q}$, for some $q > 0$. In this case if we set $Gap^* = q\|A - S\|$ then $\Delta t = 1$ and $Gap^* < \frac{gap^*(S)}{q}$, $q > 0$ implies each group of clustering eigenvalues of $S = A(0)$ consists of exactly one eigenvalue of S. It is also easy to construct an example of the case where a group of the clustering eigenvalues contains all eigenvalues of $A(t_i)$ in the Group Homotopy.

2.1 Properties of the Gap*

Suppose symmetric matrices A and S are given, $Gap^* = \delta\|A - S\|$, $\delta > 0$. Then $\Delta t = \frac{\delta}{q}$, $q > 0$. Since $\|A(t_{i+1}) - A(t_i)\|_2 \leq \|A(t_{i+1}) - A(t_i)\|$ for any matrix norm $\|\cdot\|$ where $\|\cdot\|_2$ is the spectral matrix norm, $\|A(t_{i+1}) - A(t_i)\|_2 = \Delta t\|A - S\|_2 = \frac{\delta}{q}\|A - S\|_2 \leq \frac{\delta}{q}\|A - S\| = \frac{Gap^*}{q}$, $q \geq 0$. Since the inequality holds for any matrix norm $\|\cdot\|$, we use the Frobenius matrix norm, $\|\cdot\|_F$ to compute Gap^* in the examples provided in section 3, i.e.,

$$\|A(t_{i+1}) - A(t_i)\|_2 < \frac{Gap^*}{q}$$

where $Gap^* = \delta\|A - S\|_F$, for some $\delta > 0$.

Suppose the eigenvalues of $A(t_i)$ are partitioned as groups of clustering eigenvalues; $\sigma(A(t_i)) = \cup_{k=1}^{g} G_k$, where $G_k = \{\lambda_{k_1}(i) \geq \cdots \geq \lambda_{k_{s_k}}(i)\}$ is a group of clustering eigenvalues of $A(t_i)$, $s_1 + \cdots + s_g = n$. We partition the eigenvalues of $A(t_{i+1})$ conformally to form the groups correspond to the grouping in $\sigma(A(t_i))$, $\sigma(A(t_{i+1})) = \cup_{k=1}^{g} G'_k$, where $G'_k = \{\lambda_{k_1}(i+1) \geq \cdots \geq \lambda_{k_{s_k}}(i+1)\}$, $s_1 + \cdots + s_g = n$. We call G'_k the **counterpart** of G_k.

Lemma 4. *Suppose B, $C \in M_n$ are symmetric matrices and let $\lambda_1 \geq \cdots \geq \lambda_n$ and $\alpha_1 \geq \cdots \geq \alpha_n$ be the eigenvalues of B and C, respectively. Then $\|B - C\|_2 = \rho(B - C) \geq \max_k\{|\lambda_k - \alpha_k|\}$, where $\rho(B - C) = \max_k\{|\lambda_k(B - C)|/\lambda_k(B - C)$ are the eigenvalues of $(B - C)\}$. Consequently, if $\|B - C\|_2 \leq \frac{Gap^*}{q}$, for $q > 0$, then $|\lambda_k - \alpha_k| \leq \frac{Gap^*}{q}$ for all $k = 1, \cdots, n$.*

From Lemma 4 we have $\max_k\{|\lambda_k(i+1) - \lambda_k(i)|\} \leq \frac{Gap^*}{q}$ where $\lambda_k(i+1)$ and $\lambda_k(i)$ are the eigenvalues of $A(t_{i+1})$ and $A(t_i)$, respectively. Therefore, the kth eigenvalue of $A(t_{i+1})$ must be within $\frac{Gap^*}{q}$ distance from the kth eigenvalue of $A(t_i)$ for all $k = 1, \ldots, n$.

Theorem 5. *Let $G \subseteq \sigma(A(t_i))$ be a group of clustering eigenvalues and G' be the counterpart in $\sigma(A(t_{i+1}))$. Then*

$$(i)\ |\lambda_j(i) - \lambda_j(i+1)| \leq \frac{Gap^*}{q} \ \ for\ \lambda_j(i) \in G\ \ and\ \ \lambda_j(i+1) \in G',$$

$$(ii)\ |\lambda_s(i) - \lambda_t(i+1)| \geq (1 - \frac{1}{q})Gap^*\ \ for\ \lambda_s(i) \in G\ \ and\ \lambda_t(i+1) \notin G',$$

$$(iii)\ |\lambda_s(i+1) - \lambda_t(i+1)| \geq (1 - \frac{2}{q})Gap^*\ for\ \lambda_s(i+1) \in G'\ and\ \lambda_t(i+1) \notin G'$$

Implications of Theorem 5: Suppose that two successive matrices $A(t_i)$ and $A(t_{i+1})$ are obtained via group gap Gap^* in the process of the Group Homotopy. Then the statement (i) asserts that the jth eigenvalue in a group $G' \subseteq \sigma(A(t_{i+1}))$ must be within $\frac{Gap^*}{q}$ distance from the jth eigenvalue in $G \subseteq \sigma(A(t_i))$. That is for a large enough $q > 0$, the eigenvalues can change only a small amount between two successive steps of the Group Homotopy method.

Statement (ii) shows that there is a big enough gap between the eigenvalues in G and the eigenvalues that are not in the counterpart G'. Thus we see that the eigenvalues in G are more likely to converge to the eigenvalues in G' under the Newton iteration. From the statement (iii) we see that for a large enough $q > 0$ there is a gap between any two G''s in $\sigma(A(t_{i+1}))$ that are counterparts of groups in $\sigma(A(t_i))$. Thus a group of the clustered eigenvalues tends to remain as a group of the clustered eigenvalues during the Group Homotopy.

3 Group Homotopy Algorithm

Now we prove the main result of the Group Homotopy with the parameterized Newton's iteration. In the following we let $\sigma(A(t_i)) = \{\alpha_1, \ldots, \alpha_n\}$, $\alpha_n \leq \cdots \leq \alpha_1$ and $\sigma(A(t_{i+1})) = \{\lambda_1, \ldots, \lambda_n\}$, $\lambda_n \leq \cdots \leq \lambda_1$ for the notational convenience.

A Parametrized Newton Iteration :[5] Suppose $\begin{pmatrix} x \\ \alpha \end{pmatrix} \in \mathbb{R}^{n+1}$ is given and let $A \in M_n$ be a symmetric matrix. Assume $\alpha \notin \sigma(A)$.

Then for $l = 0, 1, \ldots,$ define $x^{(l+1)} = \frac{1}{\hat{\beta}^{(l)}}(\alpha^{(l)}I - A)^{-1}x^{(l)}, \alpha^{(l+1)} = \alpha^{(l)} - \frac{\beta^{(l)}}{\left(\hat{\beta}^{(l)}\right)^2}$

where $\beta^{(l)} = x^{(l)T}(\alpha^{(l)}I - A)^{-1}x^{(l)}$ and $\hat{\beta}^{(l)} = \|(\alpha^{(l)}I - A)^{-1}x^{(l)}\|_2$.

We describe two algorithms for simultaneously finding the eigenpairs of an $n \times n$ real symmetric matrix A. This method is a good choice if $\|S - A\|_F < gap^*(S)$, where S is the initial matrix.

3.1 Parameterized Newton's Method with Orthogonalization

Let $A = (a_{ij}) \in M_n$ be a given symmetric matrix with eigenpairs $\begin{pmatrix} u_k \\ \lambda_k \end{pmatrix}$ for $k = 1, \ldots, n$.The following algorithm simultaneously computes $\begin{pmatrix} u_k \\ \lambda_k \end{pmatrix}$, $k = 1, \ldots, n$, using the parameterized Newton's Method with orthogonalization.

Choose $S = diag(\alpha_1^{(0)}, \ldots, \alpha_n^{(0)})$ be the initial matrix where the diagonal elements of S are arranged in decreasing order. Let $\epsilon > 0$ be the given tolerance.

Step 1: Set the initial eigenpairs as $\mathcal{D} = \left\{ \begin{pmatrix} x_k^{(0)} \\ \alpha_k^{(0)} \end{pmatrix} \right\}$ for $k = 1, \ldots, n$ where $x_k^{(0)}$ is the kth column of identity matrix.

Step 2: For $k = 1, \cdots, n$ do (in parallel) (i) Compute $\|(\alpha_k^{(0)} I - A) x_k^{(0)}\|_2$ (ii) If $\|(\alpha_k^{(0)} I - A) x_k^{(0)}\|_2 < \epsilon$, then go to Step 5, otherwise go to Step 3.
End

Step 3: Parameterized Newton's Iteration:
For $k = 1, \cdots, n$ do (in parallel),
3(a) For $i = 0, 1, \cdots$ do until convergence
(i) Solve $(\alpha_k^{(i)} I - A) y_k^{(i)} = x_k^{(i)}$. (ii) Compute $\beta_k^{(i)} = (x_k^{(i)})^T y_k^{(i)}$.
(iii) Compute $\overset{\wedge}{\beta}{}_k^{(i)} = \|y_k^{(i)}\|_2$. (iv) $x_k^{(i+1)} = \dfrac{1}{\overset{\wedge}{\beta}{}_k^{(i)}} y_k^{(i)}$.
(v) Compute $p_i = \dfrac{\beta_k^{(i)}}{\hat{\beta}_k^{(i)}}$. (vi) If $(1 - |p_i|) < \epsilon$ go to 3(b).
If $|p_i| < \epsilon$ go to 3(c) (vii) Compute $\alpha_k^{(i+1)} = \alpha_k^{(i)} - \dfrac{1}{\hat{\beta}_k^{(i)}} p_i$.
3(b) Set $\lambda_k = \alpha_k^{(i)} - \dfrac{1}{\hat{\beta}_k^{(i)}}$, $u_k = x_k^{(i+1)}$ 3(c) Set $\lambda_k = \alpha_k^{(i)} - \dfrac{1}{\hat{\beta}_k^{(i)}}$
and then compute the corresponding normalized eigenvector u_k of A.
End
End.

Step 4: Let m be the number of eigenpairs obtained from Step 3. If $m = n$, then all n distinct eigenpairs have been obtained, go to Step 5. If $m < n$, then obtain $(n - m)$ remaining eigenpairs as follows: **4-(i)** Denote by $\{\hat{G}_k\}_{k=1,\ldots,m}$ the m groups of eigenpairs as defined below.

$$\hat{G}_k = \left\{ \begin{pmatrix} x_{kj}^{(0)} \\ \alpha_{kj}^{(0)} \end{pmatrix} \Big/ \begin{pmatrix} x_{kj}^{(0)} \\ \alpha_{kj}^{(0)} \end{pmatrix} \text{ converges to } \begin{pmatrix} u_k \\ \lambda_k \end{pmatrix} \right\} \text{ for } j = 1, \ldots s_k$$

where $s_1 + s_2 + \cdots + s_m = n$ and s_k is the number of eigenpairs in a Group $\hat{G}_k$. Set p = number of Groups of $\hat{G}_k$ having number of eigenpairs $s_k > 1$. If $p = 0$ then go to Step 5.

4-(ii) For each of the above p groups do the following steps, in parallel.

Compute:$\min_{1 \le j \le s_k} \|(\alpha_{k_j}^{(0)} I - A) X_{k_j}^{(0)}\|_2$. Discard one eigenpair which achieves the minimum (Note that there is one eigenpair in each group that achieves the minimum).

4 -(iii) Orthogonalize the other vectors in each of the above p groups (in parallel): Set $l = m$. For $j = 1, \ldots, (s_k - 1)$ do
Orthogonalize $\{u_1, u_2, \ldots, u_l, x_{k_j}^{(0)}\} \longrightarrow \{u_1, u_2, \ldots, u_l, \hat{x}_{k_j}^{(0)}\}$ using the modified Gram-Schmidt process.

(a) Using Step 3 with eigenpair $\begin{pmatrix} \hat{x}_{k_j}^{(0)} \\ \alpha_{k_j}^{(0)} \end{pmatrix}$, obtain a new eigenpair $\begin{pmatrix} \hat{x}_{k_j} \\ \hat{\alpha}_{k_j} \end{pmatrix}$.

(b) Set $\begin{pmatrix} \hat{x}_{k_j} \\ \alpha_{k_j} \end{pmatrix} \to \begin{pmatrix} u_{l+1} \\ \lambda_{l+1} \end{pmatrix}$. (c) $l \leftarrow l+1$. End

4-(iv) Let p' = number of distinct eigenpairs obtained in 4(iii) where $p' \leq (n-m)$.Set $m \leftarrow m+p'$ If $m < n$ go to 4(i) otherwise go to step 5. Step 5. End.

Table 1. ($q = 4$, $n = 50$)

eigenvalues	t_0	t_1	t_2	t_3	t_4	Eigenvalues
of $A(t_i)$	0	0.25	0.5	0.75	1.0	from MATLAB
1st	150.00	457.8705	879.5562	1308.3704	1739.0537	1739.0537
⋮	⋮	⋮	⋮	⋮	⋮	⋮
20th	36.00	32.0262	32.6110	27.9100	23.4161	23.4161
⋮	⋮	⋮	⋮	⋮	⋮	⋮
40th	-87.00	94.4645	-96.4081	-104.6880	-112.8265	-112.8265
⋮	⋮	⋮	⋮	⋮	⋮	⋮
50th	-147.00	-164.1900	-215.4465	-285.9817	-363.2445	-363.2445
M		46	44	45	45	
B		4	6	5	5	
C(D)		8(6)	8(5)	10(4)	8(4)	

3.2 The Group Homotopy Algorithm with Parameterized Newton's Iteration

Given an $n \times n$ symmetric matrix A, the following algorithm simultaneously computes the eigenpairs $\begin{pmatrix} u_k \\ \lambda_k \end{pmatrix}$, for $k = 1, \ldots, n$ of A, using the Group Homotopy Method.

Step 0: Set $t_0 = 0$. Set $S = A(0) = diag(\alpha_1^{(0)}, \ldots, \alpha_n^{(0)})$, where $\alpha_k^{(0)}$, $k = 1, \ldots, n$ are arbitrarily chosen real numbers and the corresponding initial eigenvectors are the columns of the identity matrix.

Step 1: Choose $q \geq 4$. Set $\Delta t = \dfrac{1}{q}$.

Step 2: For $i = 1, \ldots, q$ do (i)Compute $t_i = t_{i-1} + \Delta t$ (ii) If $t_i \geq 1$ then set $t_i = 1$ (iii) Apply Algorithm 3.1 to obtain all the eigenpairs of $A(t_i) = A(t_{i-1}) + \Delta t(A - A(0))$ (iv)If $t_i < 1$ go to step 2(i) otherwise stop. End.

Numerical Results. Since $\|\cdot\|_2 \leq \|\cdot\|_F$ where $\|\cdot\|_F$ is the Frobenius matrix norm and $\|\cdot\|_F$ is easy to compute, we use $\|A - S\|_F$ to compute Δt_i in the following examples. Let M = the number of eigenpairs obtained using the modified Newton iteration. B = the number of eigenpairs recovered with modified Gram-Schmidt process using the modified Newton iteration. C = the maximum number of iterations at each step. D = the maximum number of iterations needed for the orthogonalization.

Example 1: Consider the matrix
$A = [a_{ij}], \in M_n$, $a_{ij} = i$ if $i > j$, $a_{ij} = (-1)^i 3 * i$ if i = j, $a_{ij} = j$ if $j > i$.
Choose the initial matrix S such that $A(0) = S = diag(a_{11}, a_{22}, \ldots, a_{nn})$.
$\|A - A(0)\|_F = 1779.2$. The result of Table 1 uses $q = 4$. From the results of numerical experiments below, it is clear that if q is large then the number of iterations in each step is significantly lower (see last row of each table) and also the number of eigenpairs recovered using algorithm 3.1 (Step 4) is also less (row representing B).

Since $Gap^* < gap^*(S)$, we choose $\Delta t = 1$. It shows that the choice of S is very important to accomplish the experiment in an efficient way.

4 Conclusions

We have presented a new Homotopy method, called the Group Homotopy method for computing the eigenvalues and eigenvectors of a real symmetric matrix. The important features of this method are:

(i) Unlike the other methods for symmetric eigenvalue problem, the new method does not transform the initial matrix to a tridiagonal matrix, and is thus suitable for large and sparse problems.
(ii) The method can take advantage of the structure preserving linear systems solvers, especially designed for structure matrices, such as Toeplitz. Thus the method is expected to be much faster than the existing methods in most cases.
(iii) Unlike the other existing Homotopy methods, the step-size is predetermined at the outset according to a criterion presented in the paper and is never altered at any iteration during the execution of the algorithm.
(iv) The algorithm is parallel in nature.

References

1. M.T. Chu : On a Numerical Treatment for the Curve-Tracing Of the Homotopy Method. Numerische Mathematik, **42**.(1983) 323–329
2. M.T. Chu : A Simple Application of the Homotopy Method to Symmetric Eigenvalue Problems. Linear Algebra Appl., **59**.(1985) 85–90
3. M.T. Chu : A Note on The Homotopy Method for Linear Algebraic Eigenvalue Problems. Linear Algebra Appl., **108**.(1988) 225–236
4. Karabi Datta and Yoopyo Hong and Ran Baik Lee : An Application of Homotopy Method For Eigenvalue Problem of A Symmetric Matrix. IMACS, Series in Computational and Applied Mathematics, **3**.(1996) 367–376
5. Karabi Datta and Yoopyo Hong and Ran Baik Lee : Parameterized Newton's iteration for computing an Ejgenpairs of a real Symmetric Matrix in an Interval, Computational Methods in Applied Mathemaics, **3**.(2003) 517–535

6. Roger A. Horn and Charles R. Johnson: Matrix Analysis. Cambridge University Press, (1985)
7. K. Li and T.Y. Li: An Algorithm for Symmetric Tridiagonal Eigen Problems: Divide and Conquer with Homotopy Continuation. SIAM J. Sci. Comput., **14**.(1993) 735–751
8. Beresford N. Parlett: The Symmetric Eigenvalue Problem. Prentice-Hall, Inc.Englewood Cliffs,(1980)

Numerical Simulation of Three-Dimensional Vertically Aligned Quantum Dot Array

Weichung Wang[1] and Tsung-Min Hwang[2]

[1] Department of Applied Mathematics, National University of Kaohsiung, Kaohsiung 811, Taiwan
wwang@nuk.edu.tw

[2] Department of Mathematics, National Taiwan Normal University, Taipei 116, Taiwan
min@math.ntnu.edu.tw

Abstract. We study the electronic properties of quantum dot arrays formed by 2 to 12 vertically aligned quantum dots numerically. Numerical schemes in grid points choosing, finite differences, matrix reduction, and large-scale eigenvalue problem solver are discussed. The schemes allow us to compute all the desired energy states and the wave functions efficiently. Numerical experiment results are presented.

Keywords: Semiconductor quantum dot array, the Schrödinger equation, energy levels, wave function, numerical simulation.

1 Vertically Aligned Quantum Dot Array Model

Recent advances in fabrication and varied applications of semiconductor quantum dot array (QDA) have attracted intensive studies in theoretical, experimental, and numerical. The energy state spectrum and the corresponding wave functions of a QDA system is of basic physical interest and is crucial for designing applications like photoelectric devices.

A main challenge for simulating a three-dimensional QDA is to solve very large scale eigenvalue problems for only several interior eigenvalues that are of interest. Aiming at the QDA that disk-shaped co-axial InAs QDs are vertically aligned and embedded in a cylindrical GaAs matrix (see left part of Figure 1), we develop efficient numerical schemes overcome the difficulties.

The QDA is modelled by the Schrödinger equation, in the cylindrical coordinate, as

$$\frac{\hbar^2}{2m_\ell(\lambda)}\left[\frac{\partial^2 F}{\partial r^2}+\frac{1}{r}\frac{\partial F}{\partial r}+\frac{1}{r^2}\frac{\partial^2 F}{\partial \theta^2}+\frac{\partial^2 F}{\partial z^2}\right]+c_\ell F=\lambda F, \tag{1}$$

where $\hbar$ is the reduced Plank constant, λ is the total electron energy, $F = F(r,\theta,z)$ is the wave function, $m_\ell(\lambda)$ and c_ℓ are the electron effective mass and confinement potential in the ℓth region. The index ℓ is used to distinguish the

V.S. Sunderam et al. (Eds.): ICCS 2005, LNCS 3516, pp. 908–911, 2005.

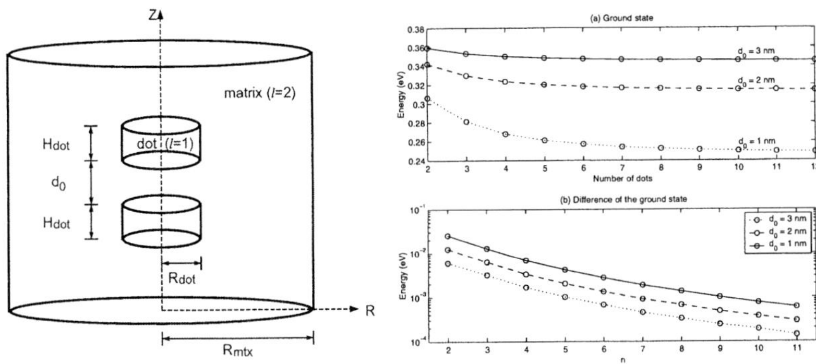

Fig. 1. Left part: Structure schema of a cylindrical vertically aligned quantum dot array and the heterostructure matrix. Right part: Ground state energies for various spacer layer distances d_0 and number of quantum dots

region of the QDs (for $\ell = 1$) from that of the matrix (for $\ell = 2$). The non-parabolic effective mass approximation [1] is given as

$$\frac{1}{m_\ell(\lambda)} = \frac{P_\ell^2}{\hbar^2}\left(\frac{2}{\lambda + g_\ell - c_\ell} + \frac{1}{\lambda + g_\ell - c_\ell + \delta_\ell}\right), \tag{2}$$

where P_ℓ, g_ℓ, and δ_ℓ are the momentum, main energy gap, and spin-orbit splitting in the ℓth region, respectively. For Eq. (1), the Ben Daniel-Duke interface conditions [2] are imposed on the interface of the two different materials as

$$\frac{\hbar^2}{2m_1(\lambda)}\frac{\partial F(r_I,\theta_I,z_I)}{\partial n_-} = \frac{\hbar^2}{2m_2(\lambda)}\frac{\partial F(r_I,\theta_I,z_I)}{\partial n_+} \tag{3}$$

where (r_I, θ_I, z_I) denotes the position on the interface and the n_+ and n_- denote the corresponding outward normal derivatives on the interface. Finally, Dirichlet boundary conditions are prescribed on the boundary (top, bottom, and wall) of the matrix.

2 Numerical Schemes

This section discusses the numerical schemes used for solving the three-dimensional QDA Schrödinger equation (1) to compute the electron energy levels and the associated wave function in the system.

We first discretize the domain by choosing mesh points. Regular uniform mesh points are chosen in the azimuthal angle θ coordinate. Non-uniform mesh points are used in the radial coordinate r and the natural axial coordinate z with the following two special treatments. First, in the heterojunction area, fine meshes are used to capture the rapid change of the wave functions. Secondly, a half of the mesh length is shifted in the radial coordinate to avoid incorporating the pole condition [3].

Based on the grid points, Eq. (1) is discretized by the 3D centered seven-point finite difference method

$$\frac{-\hbar^2}{2m_\ell(\lambda)}\left(\frac{F_{i+1,j,k}-2F_{i,j,k}+F_{i-1,j,k}}{(\Delta r)^2}+\frac{1}{r_i}\frac{F_{i+1,j,k}-F_{i-1,j,k}}{2\Delta r}\right.$$
$$\left.+\frac{1}{r_i^2}\frac{F_{i,j+1,k}-2F_{i,j,k}+F_{i,j-1,k}}{(\Delta\theta)^2}+\frac{F_{i,j,k+1}-2F_{i,j,k}+F_{i,j,k-1}}{(\Delta z)^2}\right)+c_\ell F_{i,j,k}=\lambda F_{i,j,k},$$

where $F_{i,j,k}$ is the approximated value of wave function F at the grid point (r_i, θ_j, z_k) for $\ell = 1, 2$, $i = 1, \ldots, \rho$, $j = 1, \ldots, \mu$, and $k = 1, \ldots, \zeta$. In the heterojunctions, two-point finite differences are applied on the interface conditions of the QDs. The numerical boundary values for the matrix in the z- and r-direction are zeros according to the Dirichlet boundary conditions.

Assembling the finite difference discretizations results in a $\rho\mu\zeta$-by-$\rho\mu\zeta$ 3D eigenvalue problem. By reordering the unknown vector and using the fast Fourier transformation to tridiagonalize matrices $T_k(\lambda)$ (for $k = 1, \ldots, \zeta$), we obtain μ independent $\rho\zeta$-by-$\rho\zeta$ eigenvalue problems with the form

$$\widetilde{T}_j(\lambda)\widetilde{F}_j = \widetilde{D}_j(\lambda)\widetilde{F}_j, \tag{4}$$

for $j = 1, \ldots, \mu$, where $\widetilde{T}_j(\lambda)$ and $\widetilde{D}_j(\lambda)$ are $\rho\zeta$-by-$\rho\zeta$ matrices. Each of the eigenvalue problems in the form of (4) is called a 2D eigenvalue problem, since the grid points of the unknowns in $\widetilde{F}_j$ have the same θ value. By multiplying the common denominator of (4), we can then form the cubic λ-matrix polynomial

$$\mathbf{A}(\lambda)\mathbf{F} = (\lambda^3 A_3 + \lambda^2 A_2 + \lambda A_1 + A_0)\mathbf{F} = 0, \tag{5}$$

where A_0, A_1, A_2, and A_3 are independent to λ. The cubic eigenvalue problem can then be solved efficiently by (i) the cubic Jacobi–Davidson method to compute the smallest positive eigenvalue representing the ground state energy, and (ii) the explicit deflation scheme to estimate the successive smallest positive eigenvalues (i.e. the excited energy states). See [4,5] for detail.

3 Results and Discussions

In our numerical experiments, we assume that H_{dot} and R_{dot} of the QDs are 3 and 7.5 nm, respectively. For the matrix, $R_{mtx} = 37.5$ nm and 6 nm matrix layer are assumed above the top and below the bottom of the QDA. The material parameters used in the experiments are $c_1 = 0.0000$, $g_1 = 0.4200$, $\delta_1 = 0.4800$, $P_1 = 0.7730$, $c_2 = 0.7700$, $g_2 = 1.5200$, $\delta_2 = 0.3400$, and $P_2 = 0.8071$. Numerical simulation findings are summarized as follows.

The ground state energy of the QDA are affected by the number of QDs and the spacer layer distances d_0. Right top of Figure 1 (a) shows the computed ground state energies versus the number of QDs. It is clear that more QDs in the QDA results in lower ground state energy for a fixed d_0. Furthermore, for a fixed number of QDs, smaller spacer layer distances lead to lower ground state energies. Right bottom part of Figure 1 shows the differences (in logarithm)

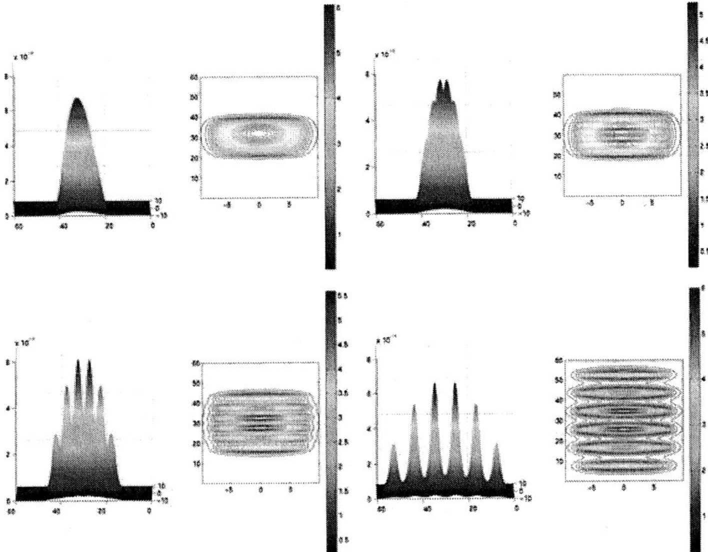

Fig. 2. Wave functions corresponding to the ground state energy. The quantum dot array contains six quantum dots and $d_0 = 0,\ 0.5,\ 2,\ 6$ nm, respectively

of the energies for the QDAs containing n and $n+1$ QDs for $n = 1, \cdots, 11$. For various $d_0 = 1, 2, 3$ nm, the ground state energies decrease exponentially in a similar manner. To be specific, the energy differences can be nicely fitted by the linearly least-squares lines with slope -0.407. Figure 2 demonstrates wave functions corresponding to the ground state energy for QDA formed by six quantum dots and $d_0 = 0,\ 0.5,\ 2,\ 6$ nm. The results suggest how the wave functions change as the d_0 is decreased.

References

1. O. Voskoboynikov, C. P. Lee, and O. Tretyak. Spin-orbit splitting in semiconductor quantum dots with a parabolic confinement potential. *Phys. Rev. B*, 63:165306, 2001.
2. D. J. BenDaniel and C. B. Duke. Space-charge effects on electron tunnelling. *Phys. Rev.*, 152(683), 1966.
3. M.-C. Lai. A note on finite difference discretizations for Poisson equation on a disk. *Numer. Methods Partial Differ. Equ.*, 17(3):199–203, 2001.
4. W. Wang and T.-M. Hwang and W.-W. Lin and J.-L. Liu. Numerical Methods for Semiconductor Heterostructures with Band Nonparabolicity *Journal of Computational Physics*, 190(1):141–158, 2003.
5. T.-M. Hwang, W.-W. Lin, J.-L. Liu, and W. Wang. Jacobi–davidson methods for cubic eigenvalue problems. *Numerical Linear Algebra with Applications* To appear.

Semi-systolic Architecture for Modular Multiplication over GF(2^m)

Hyun-Sung Kim[1] and Il-Soo Jeon[2]

[1] Kyungil University, School of Computer Engineering, 712-701, Kyungsansi, Kyungpook Province, Korea
[2] Kumho Nat'l Inst. of Tech., School of Electronic Eng., 730-703, Gumi, Kyungpook Province, Korea

Abstract. This paper proposes a new algorithm and an architecture for it to compute the modular multiplication over GF(2^m). They are based on the standard basis representation and use the property of irreducible all one polynomial as a modulus. The architecture, named SSM(Semi-Systolic Multiplier) has the critical path with 1-D_{AND}+1-D_{XOR} per cell and the latency of m+1. It has a lower latency and a smaller hardware complexity than previous architectures. Since the proposed architecture has regularity, modularity and concurrency, they are suitable for VLSI implementation.

1 Introduction

The arithmetic operations in the finite field have several applications in error-correcting codes, cryptography, digital signal processing, and so on [1]. Information processing in such areas usually requires performing multiplication, inverse/division, and exponentiation. Among these operations, the modular multiplication is known as the basic operation for public key cryptosystems over GF(2^m) [2-3].

Numerous architectures for modular multiplication in GF(2^m) have been proposed in [2-8] over the standard basis. Wang et al. in [5] proposed two systolic architectures with the MSB-first fashion with less control problems as compared to [4]. Jain et al. proposed semi-systolic multipliers [6]. Its latency is smaller than those of the other standard-basis multipliers. Kim in [7] proposed a bit-level systolic array with a simple hardware complexity with the LSB-first modular multiplication. Thus, further research for efficient circuit for cryptographic applications is needed. To reduced the system complexity, Itoh and Tsujii designed two low-complexity multipliers for the class of GF(2^m), based on the irreducible AOP (All One Polynomial) and the irreducible equally spaced polynomial [8]. Later, Kim in [2] proposed various AOP architectures based on LFSR(Linear Feedback Shift Register) architecture.

This paper proposes a new algorithm and a parallel-in parallel-out semi-systolic array architecture to compute the modular multiplication over finite field GF(2^m). They are based on the standard basis representation and use the property of irreducible AOP as a modulus. Let D_{AND} and D_{XOR} be the latency of AND and XOR gate, respectively. The architecture has the critical path with 1-D_{AND}+1-D_{XOR} per cell and the latency of m+1. It could be used to secure cryptosystem application.

V.S. Sunderam et al. (Eds.): ICCS 2005, LNCS 3516, pp. 912 – 915, 2005.

2 Semi-systolic Architecture

GF(2^m) is the finite extension field of finite field GF(2) [2]. An arbitrary element A over GF(2^m) can be represented with $\{1, \alpha, \alpha^2, \cdots, \alpha^{m-1}\}$, which is based on the standard basis, i.e., $A=A_{m-1}\alpha^{m-1}+A_{m-2}\alpha^{m-2}+\cdots+A_1\alpha+A_0$. A polynomial of the form $f(x)= f_m x^m+f_{m-1}x^{m-1}+\cdots + f_1 x+ f_0$ is called an irreducible polynomial if and only if a divisor of $f(x)$ is 1 or $f(x)$. Assume that a polynomial of the form $f(x)= f_m x^m+f_{m-1}x^{m-1}+\cdots + f_1 x+ f_0$ over GF(2) is called an AOP (All One Polynomial) with degree m if f_i=1 for i=0,1, ⋯ ,m. It has been shown that an AOP is irreducible if and only if m+1 is the prime and 2 is the primitive modulo m+1. Let a set $\{1, \alpha, \alpha^2, \cdots ,\alpha^{m-1}\}$ be generated by α which is a root of AOP $f(x)$ and be the standard basis. In the standard basis, an element A over GF(2^m) is presented by $A=A_{m-1}\alpha^{m-1}+A_{m-2}\alpha^{m-2}+\cdots+A_1\alpha+A_0$. A set with $\{1, \alpha, \alpha^2, \cdots, \alpha^{m-1}, \alpha^m\}$ is called an extended basis of $\{1, \alpha, \alpha^2, \cdots,\alpha^{m-1}\}$. In the extended basis, an element a over GF(2^m) is represented by $a=a_m\alpha^m+a_{m-1}\alpha^{m-1}+\cdots+a_1\alpha+a_0$ with $A_i=a_m+a_i$ ($0\le i \le m$-1). Thus, an element over GF(2^m) has two different representations. Let $F(x)=x^m+x^{m-1}+\cdots +x+1$ be an irreducible AOP of degree m: and let α be a root of $F(x)$. i.e., $F(\alpha)=\alpha^m+\alpha^{m-1}+\cdots +\alpha+1$. Then, we have $\alpha^m=\alpha^{m-1}+\cdots +\alpha+1$, $\alpha^{m+1}=1$.

The multiplication of elements a and b over GF(2^4) in the extended basis can be performed by ab mod p with $p=\alpha^{m+1}+1$ which applied the property of AOP as a modulus. Let the result of this multiplication, ab mod p, be $r=r_m\alpha^m+r_{m-1}\alpha^{m-1}+\cdots+r_1\alpha+r_0$. The recurrence equation for the MSB first algorithm with the property of AOP is as follows: $r=ab \bmod p=\{\cdots[[ab_m]\alpha \bmod p+ab_{m-1}]\alpha \bmod p+\cdots+ab_1\}\alpha \bmod p+ab_0$. From the equation, a new algorithm to compute ab mod p can be derived as following Algorithm 1.

[Algorithm 1] Modular Multiplication
Input : $a=(a_m,a_{m-1},\cdots,a_1,a_0)$, $b=(b_m,b_{m-1},\cdots,b_1,b_0)$
Output : $r=ab$ mod p
Initial value : $r^{m+1}=(r_m,r_{m-1}, \cdots,r_1,r_0) =(0,0, \cdots,0,0)$
Step 1 for $i=m$ to 0
Step 2 $r^i=Circular\_Left(r^{i+1})+ ab_i$

where $Circular\_Left(x)$ is the 1-bit-left-circular shift of x and r^i is used to represent the i-th intermediate result for the final result r. In the above algorithm, the modular reduction is performed just by using 1-bit-left-circular-shift operation. Specially, all the operations in the for loop can be performed bit by bit in parallel.

Let a, b, and b^2 be an elements in GF(2^4). Then a and b with an extended basis $\{1, \alpha,\alpha^2,\alpha^3,\alpha^4\}$ can be represented as follows: $a=a_4\alpha^4+a_3\alpha^3+a_2\alpha^2+a_1\alpha+a_0$, $b=b_4\alpha^4+b_3\alpha^3+b_2\alpha^2+b_1\alpha+b_0$.

When $p=\alpha^5+1$ is used as a modular in the extended basis, we have

$$\begin{aligned} r &= ab \bmod p \\ &= a(b_4\alpha^4+b_3\alpha^3+b_2\alpha^2+b_1\alpha+b_0) \bmod p \\ &= \{\cdots[[ab_4]\alpha \bmod p+ab_3]\alpha \bmod p+\cdots+ab_1\}\alpha \bmod p+ab_0 \\ &= r_4\alpha^4+r_3\alpha^3+r_2\alpha^2+r_1\alpha+r_0. \end{aligned}$$

Fig.1 shows a multiplier named SSM based on Algorithm 1 over GF(2^4). SSM is composed of $(m+1)(m+1)$ basic cells. It is the parallel architecture which a_i and b_i ($0\leq i \leq m$) are inputted at the same time. The $(m+1)$-bits of data a are inputted from the top row and transmitted to the adjacent cells following each row. But the data b_i in each row is broadcasted to all cells in the same row at the same time. Let D_{AND} and D_{XOR} be the latency of AND and XOR gate, respectively. SSM has a critical path with 1-D_{AND}+1-D_{XOR} per cell. SSM in Fig. 1 can be generalized for arbitrary m as well as m=4. Fig. 2 shows the basic cells for SSM.

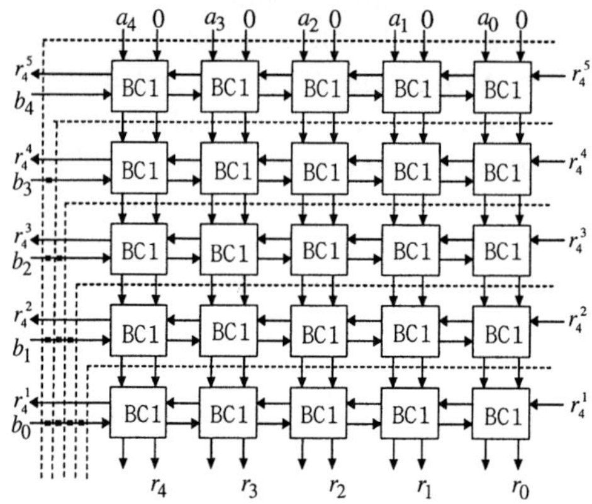

Fig. 1. SSM over GF(2^4)

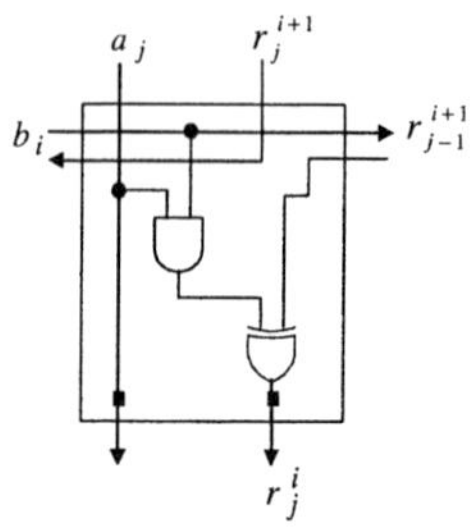

Fig. 2. Basic cell of SSM

3 Comparison and Analysis

Table 1 shows the comparison between the proposed architecture and pervious two architectures.

Table 1. Comparisons

Properties	Jain [6]	Kim [2]	SSM
Basic architecture	Semi-systolic	LFSR	Semi-systolic
Irreducible polynomial	Generalized	AOP	AOP
Number of cell	m^2	m+1	$(m+1)^2$
Cell complexity	2 AND, 2 XOR 3 latches	1 AND, 1 XOR 2 REG, 1 MUX	1 AND, 1 XOR 2 latches
Latency	m	2m+1	m+1
Critical path	1-D_{AND} +1-D_{XOR}	1-D_{AND}+ log m(1-D_{XOR})	1-D_{AND} +1-D_{XOR}

It is assumed that AND and XOR represent 2-input AND and XOR gates, respectively, and latch for 1-bit latch. Let D_{AND} and D_{XOR} be the latency of AND and XOR gate, respectively. As a result, the proposed architecture, SSM, has lower latency and smaller complexity than previous architectures in [6] and [2].

4 Conclusions

This paper proposed a new algorithm and a parallel-in parallel-out semi-systolic array architecture to compute the modular multiplication over finite field GF(2^m). The property of irreducible AOP was used as a modulus to get a better hardware and time complexity. Proposed architecture has lower latency and smaller hardware complexity than previous architectures as shown in Table 1. Since SSM has regularity, modularity and concurrency, they are suitable for VLSI implementation.

References

[1] D.E.R.Denning, Cryptography and data security, Addison-Wesley, MA, 1983.

[2] H.S.Kim, Bit-Serial AOP Arithmetic Architecture for Modular Exponentiation, PhD. Thesis, Kyungpook National University, 2002.

[3] S.W.Wei, "VLSI architectures for computing exponentiations, multiplicative inverses, and divisions in GF(2^m)," IEEE Trans. Circuits and Systems, 44, 1997, pp.847-855.

[4] C.S.Yeh, S.Reed, and T.K.Truong, "Systolic multipliers for finite fields GF(2^m)," IEEE Trans. Comput., vol.C-33, Apr. 1984, pp.357-360.

[5] C.L.Wang and J.L.Lin, "Systolic Array Implementation of Multipliers for Finite Fields GF(2^m)," IEEE Trans. Circuits and Systems, vol.38, July 1991, pp796-800.

[6] S. K. Jain and L. Song, "Efficient Semisystolic Architectures for finite field Arithmetic," IEEE Trans. on VLSI Systems, vol. 6, no. 1, Mar. 1998.

[7] H.S.Kim, "Efficient Systolic Architecture for Modular Multiplication over GF(2^m)," will be appeared in PARA 2004 proceeding, 2005.

[8] T. Itoh and S. Tsujii, "Structure of parallel multipliers for a class of finite fields GF(2^m)," *Info. Comp.*, vol. 83, pp.21-40, 1989.

[9] S.Y.Kung, VLSI Array Processors, Prentice-Hall, 1987.

Meta Services: Abstract a Workflow in Computational Grid Environments

Sangkeon Lee and Jaeyoung Choi

School of Computing, Soongsil University,
1-1 Sangdo-dong, Dongjak-gu, Seoul 156-743, Korea
sglee@ss.ssu.ac.kr
choi@ssu.ac.kr

Abstract. We define a concept of Meta services which describes a way of abstracting and mapping a workflow to a service in computational Grid environments. By using Meta services, a workflow in a Grid environment could adapt various service concepts such as Grid services, Web services, and portal services without modification to the workflow. And the converted Meta services could be shared and reused by users. Furthermore, historical performance data could be included in the Meta service, making effective execution of the workflow possible.

1 Introduction

Grid computing is a new distributed computing infrastructure which supports the sharing of large-scale resources that are geographically distributed. GRID systems provide higher performance and a wider range of resource sharing than traditional distributed systems and they allow advanced application functionalities [1]. Nowadays, research for applying workflows to a Grid environment is actively being studied.

GridFlow [2] is a workflow management system, which uses agent-based resource management and a local resource scheduling system, Titan. It focuses on the scheduling of time-critical grid applications in a cross-domain and highly dynamic grid environment by using a fuzzy timing technique and performance prediction. MyGrid [3] provides ontology-based services such as resource discovery, workflow enactment, and distributed query processing for integration. It is a middleware research to support biological environments on Grids. The GridLab project [4] is developing application tools and middleware services for Grid environments. Services provided by GridLab include resource brokering, data management, and adaptive services. And Condor [5] provides a workload management system for computation-intensive jobs and a scheduling of dependencies using DAGMan. Pegasus [6] is a workflow mapping and planning system that can map abstract application specific workflows to an executable form for DAGMan. MSF (Meta Scheduling Framework) [7] provides a graphical workflow composition service and workflow management service for computational Grid.

From our experience of constructing a Grid portal with MSF (Meta Scheduling Framework) [7], we need to define a new concept of services to integrate the

V.S. Sunderam et al. (Eds.): ICCS 2005, LNCS 3516, pp. 916–919, 2005.

workflow described with JCML (Job Control Markup Language) to the portal without modification of the workflow. The JCML is a workflow description language used in MSF. We also found that sharing workflows is essential for the Grid portal users. Because most particular Grid portals are designed to support specific scientific research, sharing workflows among the Grid users is a very useful feature. In this paper, we present Meta services in order to map existing Grid workflows to Meta services without modification of description of the workflows. And the Meta services provide service specific information that allows the sharing of workflows among users and resource information for effective scheduling of Grid resources. And historical performance data is required for effective scheduling of Grid resources. However all prior research does not provide such functionalities. Other Grid research such as Smith's work [8], Pegasus, and Gridflow shows that historical performance data can be used for effective scheduling of Grid resources. Therefore we define a concept of Meta services to meet these requirements.

2 Meta Services

Meta services map an existing Grid workflow to a service by overriding attributes of the Grid workflow. Most of workflows are described by XML or they can be converted to XML documents. And Meta services describe how to override attributes in XML documents with service parameter and service specific information. The mapping is shown in Figure 1.

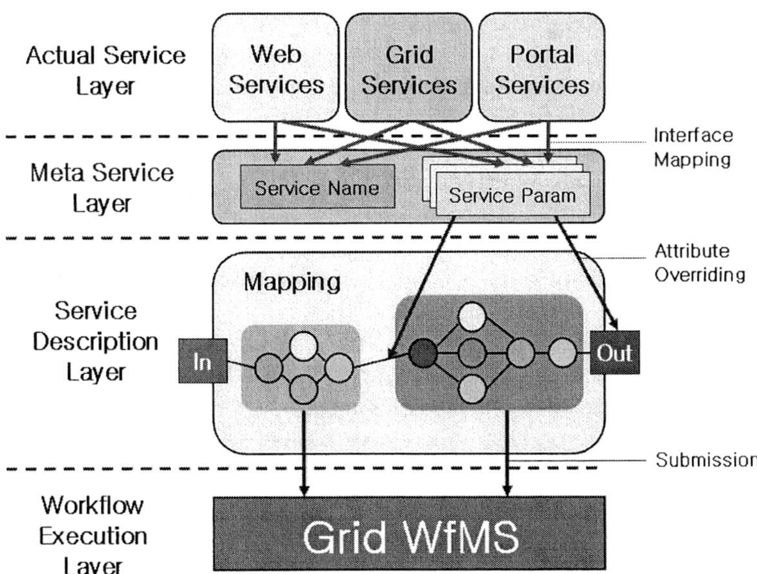

Fig. 1. Meta Services: Mapping workflows to services

Because a Meta service is accessed by it's name and parameters, the interface of the Meta service is easily mapped to the corresponding Web service or Grid service, which contains one operation within one porttype. Moreover it is also easily mapped to a Portal service written in java servlet or CGI. When a Meta service is called, the service name is used as an identifier of the service, the service parameters override attributes of the workflow in the service, and the service description includes the service logic composed by the workflow. Service description of a Meta service contains the followings:

- Workflow description
  - Declaration of workflow units
  - A service logic to connect the workflow units
- Information for sharing workflows
  - User's permission
- Resource information for effective scheduling
  - Resource restriction
  - Historical performance data

The declaration of workflow units in Meta services includes one or more workflows which a user wants to use, and declares them as workflow units. The units are linked to each other, and then executed within a loop or a condition of the declaration. In an extreme case, a Meta service may include only one service. When Meta services are called, attributes of the workflows are overridden with service parameters or values defined by the user. In executing the services, instances of the workflows are generated with parameters by overriding attributes of the workflows. Because workflows are generated by calling the corresponding services with parameters, users could easily generate and reuse workflows.

A portal provides common services to its users, and the users may define frequently used services as Meta services in the portal. For this purpose, Meta services might be declared as public or other users might be permitted to use them. As a result, permitted users could read and execute the common Meta services, but their instances are generated in the users' repositories. Therefore, if a user wants to modify the service, he could modify it with a copy of the service.

And a user may want to avoid executing the workflow on faulty resources. If historical performance data is collected and recorded on the Meta services, a scheduler could save the time required to find appropriate resources. Meta services maintains lists of MER (most effective resources) and MFR (most faulty resources). Moreover, though the same services are executed on several resources with the same performance and specifications, it may be difficult to expect the same performance on Grid environments. And after a user has executed the same Meta service many times, he may find that the service is executed efficiently on specific resources or it is executed unsuccessfully on other resources. That is, Meta services are designed so that the users may execute workflows on their preferred resources by using resource profile information.

3 Conclusion

In this paper, we presented a concept of Meta services. The concept of Meta services is combined with the MSF system in order to provide more reusable and adaptable workflow management environments. Users of the combined system could browse, search, and call the Meta services with the name and parameters of a service. A Meta service is mapped to the corresponding JCML workflows, and successfully executed in MSF.

The aim of this paper is to introduce a concept of Meta services which enables the followings: Meta services are ported to various service environments such as Grid services, Web services, and portal services. Meta Services could provide user's permission information to share workflows among portal users. Meta services also provide a resource profiling information, which could be used for efficient scheduling of resources. We applied the concept of Meta Services to extend our previous work, MSF, but the concept can be applied to other WfMS. In the near future, we will integrate heterogeneous Grid workflows with Meta services.

Acknowledgement

This work was supported by Korea Research Foundation under contract KRF-2004-005-D00172.

References

1. I. Foster and C. Kesselman, ed., The Grid: Blueprint for a New Computing Infrastructure, Morgan Kaufmann, (1998)
2. J. Cao, S. A. Jarvis, S. Saini and G. R. Nudd, GridFlow: Workflow Management for Grid Computing, 3rd International Symposium on Cluster Computing and the Grid, (2003) 12-15
3. R. Stevens, A. Robinson and C. Goble, myGrid: Personalized bioinformatics on the information grid, Bioinformatics, 19(1), (2003) 302-304
4. G. Allen et al., Enabling applications on the Grid - a GridLab overview. Intl. Journal on High Performance Computing Applications, 17(4), (2003) 449-466
5. M. Litzkow, M. Livny and M. Mutka, Condor - A Hunter of Idle Workstations, 8th International Conference of Distributed Computing Systems, (1998) 13-17
6. Y. Gil, C. Kesselman, G. Mehta, S. Patil, M. Su, K. Cahi, M. Livny, Pegasus Mapping Scientific Workflows onto the Grid. E. Deelman, J. Blythe, Across Grids Conference (2004)
7. Seogchan Hwang, Jaeyoung Choi: MSF: A Workflow Service Infrastructure for Computational Grid Environments, LNCS 3292, 445-448 (2004)
8. W. Smith, I. Foster, V. Taylor, Predicting Application Run Times Using Historical Information, Proc. IPPS/SPDP '98 Workshop on Job Scheduling Strategies for Parallel Processing (1998)

CEGA: A Workflow PSE for Computational Applications[1]

Yoonhee Kim

Dept. of Computer Science, Sookmyung Women's University, Korea
yulan@sookmyung.ac.kr

Abstract. The importance of problem solving environment (PSE)s over Grid has been emphasized owing to the heterogeneity and the large volume of resources involved and the complexity of computational applications. This paper proposes a Grid-enabled PSE called as Computing Environment for Grid Applications (CEGA) and discusses how it is evolving to develop a computational application in a style of workflow model and incorporates Grid computing services to extend its range of services and handle information for development, deployment, execution and maintenance for an application as well as an application requirement itself. The paper describes the architecture of CEGA and its implementation for development, execution and visualization of an application.

1 Introduction

The concept of Grid computing has been investigated and developed to enlarge the concept of distributed computing environment to create infrastructure that enables integrated services for resource scheduling, data delivery, authentication, delegation, information service, management and other related issues [1]. As the Grid provides integrated infrastructure for solving problems, interfacing services such as web portal to access Grid services, PSEs (Problem Solving Environments) have been developed to improve the collaboration among Grid services and reduce significantly the time and effort required to develop, run, and experiment with large scale Grid applications.

However, most PSEs to support parallel and distributed computing focus on providing environments for successful execution of applications and providing reasonable resource scheduling schemes. There have been several application-specific tools and PSEs to utilize Grid environment efficiently. ASC Grid Portal [2] is a PSE for large-scale simulation in astrophysics. Hotpage [3] is another PSE targeted toward high performance computing applications. Cactus [4] provides a problem solving environment for developing large-scale distributed scientific applications. GrADS [5] is a toolkit to help users to build applications over heterogeneous resources with ease of use. Similarly, UNICORE [6] provides graphical user interface to access heterogeneous resources uniformly. However, an effort on generalizing a PSE to support development and execution of applications (i.e. applications in workflow management), has been not fully investigated.

[1] This Research was supported by the Sookmyung Women's University Researc Grants 2003.

V.S. Sunderam et al. (Eds.): ICCS 2005, LNCS 3516, pp. 920 – 923, 2005.

A Computing Environment for a Grid Application (CEGA) has been developed to provide a computing environment for a computational application in a workflow model. The CEGA provides an efficient Graphical User Interface (GUI) approach for developing, running, evaluating and visualizing large-scale parallel and distributed applications that utilize computing resources connected by local and/or wide area network. To support Grid services through CEGA, the server creates application configuration using Resource Specification Language (RSL), which runs over the Globus toolkit [7].

The organization of the remaining sections of the paper is as follows. I present an overview of the CEGA architecture in Section 2. In Section 3 I describe the major functionality in workflow editor for development and visualization in detail. The conclusion is followed in Section 5.

2 The Architecture of CEGA

The CEGA is a workflow based problem solving environment for grid computational applications. It provides transparent computing and communication services for large scale parallel and distributed applications. The architecture of CEGA consists of CEGA portal services, CEGA system serves and Globus (see Fig.1) Globus is the Grid middleware toolkit to launch jobs and collects their results over Grid resources. CEGA Portal Service help users to generate a workflow of an application, based on the application logic, specify the input/output and runtime requirements to provide various execution environments without having big efforts for changing their experiment options. In addition, the status of job executions is reported back to a user using visualization tools, which are attached to the editor. To interface between the Portal and Globus, System Services collect and handle data from the Editor and Globus services, steer the workflow over Globus. In addition, it monitors the status of job execution. That is, CEGA provides two important services over Grid services: 1) Problem Solving Environment: to assist in the development of large scale parallel and distributed applications; and 2) Evaluation Tool: to analyze the performance of parallel and distributed applications under different resource environments.

3 CEGA Services

3.1 CEGA Editing Service

Graphical user interface helps users to develop, modify and execute an application easily over Grid resources. CEGA editor provides workflow patterns which includes various job patterns and linkages. User can use customized workflow patterns which a user can directly modify on; as well as built-in patterns. As Object List Tree is added in the left side of the window, new object is easily added to the graph with just drag & drop scheme. For editing service, open, save, new, execute and monitor interface is developed on the top menu. Once an application is developed within the editor, the information is passed to the Workflow Engine for workflow management to proceed application execution.

3.2 CEGA Execution Service

An application is defined as a set of jobs; which are connected to one another based on their job dependency. The execution of jobs can be executed either in parallel or sequence based on the dependency. When Job Control service collects application workflow information from the editor, it controls the order of execution of the jobs based on the dependency over Grid environment as the results of analysis of the workflow information. The service generates a set of activities in Resource Specification language (RSL), one of which can run over Globus in parallel. The deployment of the data in appropriate locations among the execution of activities is done by the control of the service based on the workflow. The system service checks the job dependency on the graph and divides multiple phases, which means making execution groups of jobs in parallel. Each group is mapped to grid resources by means of job scheduling service in Globus. Beside of job steering, it provides cooperative administration for dynamic Globus environment such as administrating Grid nodes and coordinating other Grid services, GRAM, MDS, and FTP as examples. The monitoring service also provides runtime monitoring and collects runtime log data in XML including starting and ending time, and total execution time. Whenever a user wants to monitor the runtime data, the Engine provides them to a visualization tool.

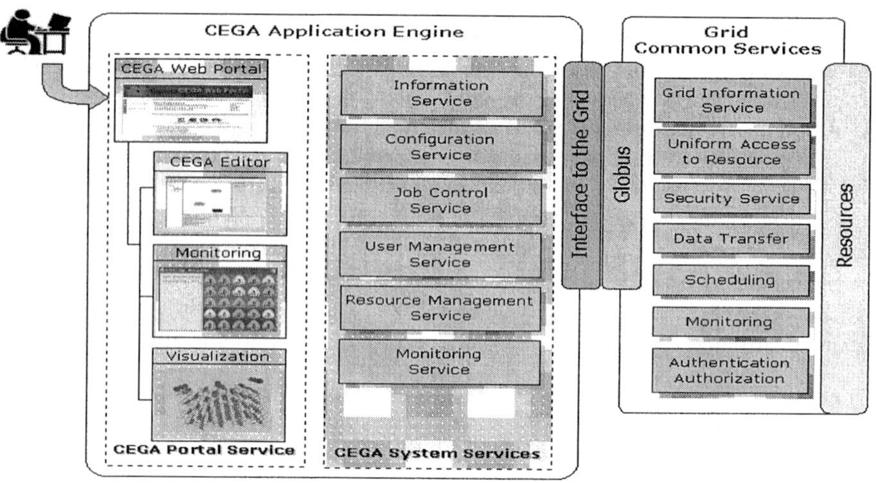

Fig. 1. The Architecture of CEGA

3.3 CEGA Monitoring and Visualization Service

The Workflow Engine controls and monitors the application execution based on the application workflow. A user is informed as the Engine provides the status data to the Editor in real time. The monitoring information is classified as two types: one is parsing status caused by syntax error before execution and the other is execution status. The information of monitoring syntax error and job execution result is displayed on the bottom window in text. At the same time, the color of each job model shows a current status of job execution in 6 states: ready, active, done, failed, suspended, and pending. And when the execution is successfully done, the result of the application is stored in the location requested at the beginning of the application development. The visualization of the result of the application is dependent of the characteristics of the application. CEGA can basically provide the result in text mode and choices of some plugged in graphic tools (See Fig.1).

Currently, CEGA has been developed for a potential energy surface calculation in molecular dynamics simulation, which requires high performance parallel computing. It also provide interface to customize experimental options easily for diverse execution conditions. It reduces deployment cost for experimentation.

4 Conclusions

The paper describes the architecture of CEGA to provide a parallel and distributed programming environments; it provides an easy-to-use graphical user interface that allows users to develop, run and visualize parallel and distributed applications running on heterogeneous computing resources connected by networks. To support Grid services through CEGA the server creates application configurations using Resource Specification Language (RSL), which runs over the Globus toolkit. This paper shows the functionality of the workflow editor to develop an application and generate application configuration over Grid environment. As it is adapt a workflow model, it can be easily extended to other application domain when diverse workflow patterns are added. Currently, the extended workflow models are under development.

References

1. Foster, C. Kesselman, "The Grid:Blueprint for a New Computing Infrasteructure," Morgan-Kaufmann, 1998.
2. Astrophysics Simulation Collaboratory: ASC Grid Portal, http://www.ascportal.org
3. HotPage, http://hotpage,npaci.edu
4. Cactus Code, http://www.cactuscode.org
5. GrADS Project, http://nhse2.cs.rice.edu/grads/index.html
6. Romberg,M., "The UNICORE Architecture Seamless Access to Distributed Resources," High Performance Distributed Computing, 1999
7. Globus Project, http://www.globus.org
8. J. Novotny, "The Grid Portal Development Kit," Concurrency: Practice and Experience, Vol.00, pp1-7, 2000

A Meta-heuristic Applied for a Topologic Pickup and Delivery Problem with Time Windows Constraints

Jesús Fabián López Pérez

Post-Graduate Program of Management Science, FACPYA UANL, Monterrey, México
fabian.lopez@e-arca.com.mx

Abstract. Our problem is about a routing of a vehicle with product pickup and delivery and with time window constraints. This problem requires to be attended with instances of medium scale $(\text{nodes} \geq 100)$. A strong active time window exists $(\geq 90\%)$ with a large factor of amplitude $(\geq 75\%)$. This problem is NP-hard and for such motive the application of an exact method is limited by the computational time. This paper proposes a specialized genetic algorithm. We report good solutions in computational times below 5 minutes.

Keywords: Logistics, Genetic Algorithms, NP-Hard, Time Windows.

1 Problem Definition and Bibliographical Review

The objective is to determine the optimal route for a distribution vehicle. The vehicle departs from a distribution center and returns to the same point at the end of the route. An optimal route is defined as that which visits all the clients in such a way that we incur a minimal cost. We define a cost matrix which identifies the time or distance required to go from each client to all others. The problem constraints are as follows.

a. Each client visited has a requirement of product to be delivered and a load to be collected. We have to observe a finite load capacity for the vehicle all the time.
b. The time window identified for each client is defined by an opening hour and a closing hour. The time window width is equal to the difference between the closing hour and the opening hour. The visit to each client must be within the time window. It is not permitted to arrive before the opening hour nor after the closing hour.

Our bibliographical summary of previous investigations include Applegate et al. 1998 [1]; Dumas & Solomon 1995 [2]; Eijl Van 1995 [3]. The outlined problem is combinatoric in nature and is catalogued as NP-Hard, Tsitsiklis 1992 [4]. Regarding routing application area, the less investigated variant is the one which has to do with the physical product distribution, Mitrovic 1998 [5]. The instances that have been typically tested are characterized by time windows with a low percentage of overlapping, Ascheuer et al. 2001 [6]. The computational complexity for the solution of the SPDP-TW depends strongly on the structure of the time windows that are defined for each customer. The experimental results obtained by Ascheuer et al. proved that the TSP-TW is particularly difficult to be solved for instances with

V.S. Sunderam et al. (Eds.): ICCS 2005, LNCS 3516, pp. 924–928, 2005.

more than 50% of active nodes with time window constraints. Ascheuer, Jünger & Reinelt 2000 [7] worked with instances up to 233 nodes. They reported 5.95 minutes of computational time for an instance of 69 nodes. All the greater instances required more than 5 hours of computational time. They conclude that the instances on the limit up to 70 nodes can be solved to optimality by exact methods like the Branch & Cut algorithms (B&C).

2 Methodology Proposed

Our methodology proposes 6 routines. We have 4 preprocessing routines, the Genetic Algorithm and finally one another routine for post-processing. We expose the 6 phases:

1. **Network topology decomposition phase based on a *"shortest path algorithm (SPP)"*.** We consider here the topology corners (street corners) that are required to model the traffic constraints that we have to observe in order to arrive to each customer in the network. This pre-processing strategy contributes to reduce the computational complexity during all the subsequent phases. In order to simplify our formulation we define an empirical assumption. We are going to use a constant "4" as the quantity of network arcs (north, south, east & west) that we require to model a typical city street corner. With the previous assumption we can establish that, if we setup a network with N1 nodes, we would obtain only N2 nodes, where $N1 \approx 4N2$. We use an SPP algorithm to pre-calculate the optimal sub-tour required to move from each customer to each one of the rest. All these preprocessed sub-tours fill the *N2* cost matrix to be used in the next phases.
2. **Compressing & clustering phase through a *"neighborhood heuristic"*.** The *N2* nodes are grouped to setup a reduced quantity of *N3* meta-nodes *(where: $N3 < N2$).* Our assumption here is that we have driving times near to zero between the nodes that are going to be grouped. This assumption is going to be considered valid through all the phases where we work with the reduced version of the network. Taking in mind this, we can inference that the obtained solution for the reduced network can be derived as an optimal for the original network as well. The heuristic that we use here to group the nodes is by geographically neighborhood and also by time windows structure similarity. Starting from a group of nodes to be grouped in a meta-node, the time window structure of this meta-node is defined by the latest opening time and by the earliest closing time. We use in the algorithm a 50% compression factor for the grouping phase which means that $N2=2*N3$.
3. **Discriminate compressing phase through a *"k nearest nodes heuristic"*.** The network arcs with greater cost are eliminated from the matrix. The logic of the previous assumption is because of those arcs have a smaller probability to appear in the optimal solution. For each *N3* node in the network, we maintain only the *"k"* arcs with the smallest cost, *where* $k<<N3$. We use a conservative 20% discriminate factor in order to reduce the probability that the optimal solution go out from the search space. This empirical assumption means that the matrix that will be transferred to the next phase will be reduced and defined by $N3 \times N4$, *where* $N4 = 20\% * N3$ in an "incidence sense". This means that although the dimensionality of the matrix is still the same $(N3 \times N3)$, the quantity of non-zero elements in the matrix is reduced to an equivalent matrix size of $N3 \times N4$.
4. **Aggressive Branch & Cut phase.** the initial math formulation we use here is quite similar as we may found in a basic TSP problem [6]. We have that X_{ij} formulation means the existence of an arc from *"i" to "j"* in the route. The procedure to add sub-tour elimination constraints are also included on this phase. Starting with *N3 meta-nodes*, the objective is to find as quickly as possible, the first feasible solution that cover the time window and

vehicle capacity constraints. The logic that we apply here is to iteratively generate cuts within a Branch & Cut scheme. For that purpose we identify in the incumbent solution, the node with the greater deviation in relation to the time window and/or the vehicle capacity constraint. This node is named *"pivot node"*. Then we verify the nodes of the tour that can be identified as *"related"* in order to re-sequence the position of the *pivot node* within the tour. This relation of the *pivot node* is exploded in the generation of the cut. The logic that we apply here to generate the cut assures that the *pivot node "k"* use at least one of the arcs that connect it to one of the *related nodes "j"*. This procedure continues until is found the first feasible solution. At this stage we use Xpress Ver 15.10 ©

$$\ni I = \{1..N_3\}\,(\textit{network nodes})$$

$$k \in I \ \ (\textit{pivot node})$$

$$j \subseteq I \ \ \{1..m\}(\textit{related nodes to k})$$

$$\sum_{j=1}^{m} (x_{jk} + x_{kj}) \geq 1 \quad \forall k \subseteq I$$

5. **Evolutionary phase.** our objective here is to approximate the optimal solution for the compact version of the network. Maintain in the pool of constraints a cut unnecessarily, means to take out the optimal solution or at least a better solution, from the search space. Our computational experience indicates that the quantity of cuts that get to be accumulated in the pool is meaningful (15-40 cuts). The goal is to identify which cuts of the pool are necessary to be eliminated. The cut elimination procedure can not be seen as an individual process for each cut, since the presence and/or the elimination of any cut can commit simultaneously the presence and/or the elimination of other(s). Identify which cuts must be eliminated, can be seen as a combinatoric sub-problem. We then propose an evolutionary strategy to attend this sub-problem. A binary codification permits to represent the elimination (0) and the presence (1) of a cut in the pool. Our Genetic Algorithm applies a tournament selection operator with a 50% crossing factor. The reproduction method applied was by means of a two random crossing points throughout the chromosome length. The mutation factor is initialized with a 5% value and it is auto-adjusted in each generation depending on the percentage of individuals in the population with identical genetic material content. Upon increasing the degeneracy level in the population, is applied an exponential growth curve in the mutation factor with 50% as an asymptotic limit. The elitism factor is limited to 15% of the population. The fitness function we calculate in this evolutionary stage is related with two objectives. The first is the route cost minimization. The second is the infeasibility level we calculate in the route as we may delete some cut(s) in the chromosome. The infeasibility level is only related with the capacity and time windows constraints for the SPDP-TW formulation. Our objective on this stage is to find the subset of cuts that can be deleted from the pool and at the same time we obtain a feasible solution. The final solution at this stage is when we found a minimal cost route which is still feasible as well. We remark here that our methodology is different from the conventional approach in the evolutionary bibliography. Our review indicates that most of the evolutionary approaches to attend routing problems take in mind the genes as the nodes or sequence nodes in the route. We treat with a modified problem. The genes represent the presence or elimination of a cut in the math formulation of the problem that is actually modeling the route.
6. **Uncompressing phase to generate a route for the original network.** The post-processing phase has the objective of translating the solution obtained in the compact version of the network to one another that will be topology sense equivalent to the original network. We have here 2 routines. The first routine is focused in determining the optimal sequence on which the *N3 meta-nodes* should be disaggregated to return to the *N2 nodes* obtained in phase 2. Starting from a selected meta-node, we construct only the valid arcs to the previous and to the next meta-node. This procedure is propagated to the rest of the

meta-nodes in the network. The second conversion routine makes use of the topology information generated in the first phase of our general methodology. Its objective is to substitute the sequence of the tour defined by the N2 nodes according to the cardinal movements that are required to obtain the N1 nodes that are present on the original network.

3 Experimental Development and Results

We applied an *"Experimental Design"* through the use of 4 experimental instruments: (1) B&C Algorithm (Xpress© Ver 15.10); (2) Steady Sate Genetic Algorithm (Evolver© Ver 6.0); (3) Generational Genetic Algorithm (Solver© Ver 4.0); (4) Proposed Genetic Algorithm. We will calculate a "percentage of optimality" for the statistic test of the hypothesis.

$$optimality\ \% = 1 - \frac{GA\ solution - Lower\ Bound}{Lower\ Bound}$$

where Lower Bound = Best solution reached by B & C within 5 hours limit

We gave treatment to instances with more than 70% of active time windows and with a minimal width of 75%. The dimension of the tested instances are defined by *w*, where $(100 \le w \le 120)$. The genetic parameters applied for the implementation of each GA's # 1, 2 & 3 were adapted empirically and separately to different values accordingly to the best case scenario. That means that the parameters were tuned for each GA. The experimental design was applied for a sample of 40 instances. All these instances were randomly generated. Only the B&C instrument was limited up to 5 hours of computational time. We remark here that although 5 hours of computational time is not evidence of optimality, we can report that we obtain the optimal solution in 38 of 40 instances. For the previous described GA instruments, the "% of optimality" was applied in 4 successive moments of time (minute #3, #5, #8 and #10). We define the following statistic parameters: *(1) Mean (m) & (2) Standard deviation (s).* The T Student test *"P(x> 90%)"* applied for each element (m_{ij}, s_{ij}) calculates the probability that the *algorithmic instrument "j"* in the *time interval "i"* obtains at least a 90% of optimality. Table 1 shows the values calculated for the "T" statistic. Table 2 shows the probability coefficients "P" Value.

Table 1. "T" Statistic Values

	Algorithmic instruments to be compared					
	B&C Algorithm (Control Group)	Basic Genetic Algorithm (Evolver)	Basic Genetic Algorithm (Frontline)	Proposed Genetic Algorithm		
				P(x>90%)	P(x>92.5%)	P(x>95%)
3th Minute	NA	-0.404	-0.558	2.426	1.069	-0.091
5th Minute	-2.426	-0.116	-0.307	3.313	1.539	0.111
8th Minute	-1.280	0.162	0.317	4.851	4.105	0.830
10th Minute	-0.700	0.400	0.903	6.298	5.577	1.328

Table 2. Probability "P Values"

	Algorithmic instruments to be compared					
	B&C Algorithm (Control Group)	Basic Genetic Algorithm (Evolver)	Basic Genetic Algorithm (Frontline)	Proposed Genetic Algorithm		
				P(x>90%)	P(x>92.5%)	P(x>95%)
3th Minute	NA	34%	29%	99%	85%	46%
5th Minute	<1%	45%	38%	100%	93%	54%
8th Minute	10%	56%	62%	100%	100%	79%
10th Minute	24%	65%	81%	100%	100%	90%

4 Discussion and Conclusions

Our B&C implementation obtains the optimal solution for 38 of 40 instances that are particularly difficult to be solved and where the investigation is focused. In addition we tested some "toy" instances with lees than 70 nodes and with less than 60% of active time windows. The computational times were very favorable since we report times below 3 minutes. Our proposed GA #3 obtains satisfactory solutions (***optimality*** $\geq 90\%$) and in reasonable computational times ***(3*** $\leq t \leq$ ***5 minutes)***. Both "out of the self" GA's (1 & 2), are significantly inferiors since these never surpass 90% of optimality before the fifth minute. We conclude:

1. *We can establish that the proposed methodology reaches a percentage of optimality* $\geq$ *90% in a computational time* $\geq$ *5 minutes (0.001 significance level).*
2. Table 1 & 2 shows that the proposed GA offers solutions within an acceptable optimality range and with computational times that make feasible its implementation. However, we should establish that our methodology can assure only 54% of confidence when is required to reach an optimality $\geq$ 95% in a computational time $\geq$ 5 minutes.

References

[1] Applegate, D; Bixby, R; Chvátal, V; (1998), On the solution of traveling salesman problems. "Documenta Mathematica Extra Volume ICM III", USA.
[2] Dumas, Y; Desrosiers, J; Solomon, M. (1995), An algorithm for the traveling salesman problem with time windows, "Operations Research 43(2)", USA, pp. 23-25.
[3] Eijl Van, C. (1995), A polyhedral approach to the delivery man problem, "Tech Report 95–19", Dep of Maths and Computer Science, Eindhoven Univ of Technology, the Netherlands.
[4] Tsitsiklis, J. (1992), Special cases of traveling salesman and repairman problems with time windows, "Networks No. 22", USA.
[5] Mitrovic, Snezana. (1998), Pickup and Delivery Problem with Time Windows, "Technical Report SFU CMPT TR 1998-12", Canada, pp 38-39.
[6] Ascheuer, N; Fischetti, M; Grotschel, M. (2001), Solving ATSP with time windows by branch-and-cut, Springer-Verlag, Germany.
[7] Ascheuer, N; Jünger, M; Reinelt, G. (2000), A branch & cut algorithm for the ATSP with precedence constraints, "Comput. Optimization and Applications 17(1)", USA, pp. 2-7.

Three Classifiers for Acute Abdominal Pain Diagnosis – Comparative Study

Michal Wozniak

Chair of Systems and Computer Networks, Wroclaw University of Technology,
Wybrzeze Wyspianskiego 27, 50-370 Wroclaw, Poland
michal.wozniak@pwr.wroc.pl

Abstract. The inductive learning algorithms are the very attractive methods generating hypothesis of the target concept on the base of the set of labeled examples. This paper presents some of the rules generation methods, their usefulness for the rule-base classifier and their quality of classification for the medical decision problem.

1 Introduction

Machine learning [1,5] is the attractive approach for building decision support systems. In the paper we compare the heuristic classifier (given by experts) and three another generated by the chosen inductive learning methods.

The content of the work is as follows. Section 2 introduces idea of the inductive decision tree algorithms and learning sets of rules method. In Section 3 we describe mathematical model of the acute abdominal pain decision problem. Next section presents results of the experimental investigations of the algorithms. Section 4 concludes the paper.

2 Algorithms

We chose three of the inductive learning algorithm: (1) C4.5 algorithm given by R. J. Quinlan [2], (2) Fuzzy Decision Tree Algorithm FID 3.0 given by C. Janikow [3] and (3) Rule generation algorithm - AQ given by R. Michalski [4].

2.1 Inductive Decision Tree

Algorithms C4.5 and FID are based on "Top Down Induction of Decision Tree" (TDIST) procedure[2]. The central choice in the TDIDT algorithm is selecting which attribute to test at each node in the tree. The chosen algorithms use the information gain that measures how well the given attribute separates the training examples according to the target classification. This measure based on the Shanon's entropy of set *S*. The information gain of an attribute *A* relative to the collection of examples *S*, is defined as

V.S. Sunderam et al. (Eds.): ICCS 2005, LNCS 3516, pp. 929–932, 2005.

$$Gain(S,A) = Entropy(S) - \sum_{c \in values(A)} \frac{|S_v|}{|S|} Entropy(S_v), \quad (1)$$

where $values(A)$ is the set of all possible values for attribute A and S_v is the subset of S for which $A = v$.

2.2 Learning Set of Rules

The algorithms like CN2 [1] or AQ [4] based on the learning one rule (LOR) strategy, removing data it covers, then iterating the process. The LOR method is similar to the ID3. The LOR algorithms follow only the most promising branch in the tree at the each step – returns only one rule, which covers at least some of the examples.

3 Model of Acute Abdominal Pain Diagnosis

The mathematical model of the diagnosis of acute abdominal pain (AAP) was simplified by experts from the Clinic of Surgery, Wroclaw Medical Academy. It leads to the following classification of the AAP: appendicitis, (2)divercitulitis, (3) small-bowel obstruction, (4) perforated peptic ulcer, (5) cholecystitis, (6) pancreatitis, (7) non-specyfic abdominal pain, (8)rare disorders of "acute abdominal".

Although the set of symptoms necessary to correctly assess the existing APP is pretty wide, in practice for the diagnosis, results of 36 (non-continuos) examinations are used, whose are presented in table I.

Table 1. Clinical features considered

no	feature	no	feature	no	Feature
1	sex	13	appetite	25	systolic blood pressure
2	age	14	bowels	26	diastolic blood pressure
3	site on onset	15	micturition	27	movement
4	site on present	16	previous indigestion	28	distension
5	intensity	17	jaundice	29	tenderness
6	aggravating factors	18	previous similar pain	30	Blumberg's sign
7	relieving factors	19	previous surgery (abdominal)	31	guarding
8	progress	20	drugs	32	rigidity
9	duration	21	mood	33	swellings
10	character on onset	22	color	34	Murphy's sign
11	Character on present	23	temperature	35	abdominal auscultation (bowel sounds)
12	nausea and vomiting	24	pulse	36	rectal examinations

3.1 Heuristic Decision Tree

The experts-physicians gave the decision tree depicted on Fig.1. Numbers of leafs are the numbers of diagnosis presented above. The numbers in the nodes are corresponded with the following diagnosis: (9) acute enteropathy, (10) acute disorders of the digestive system, (11). others.

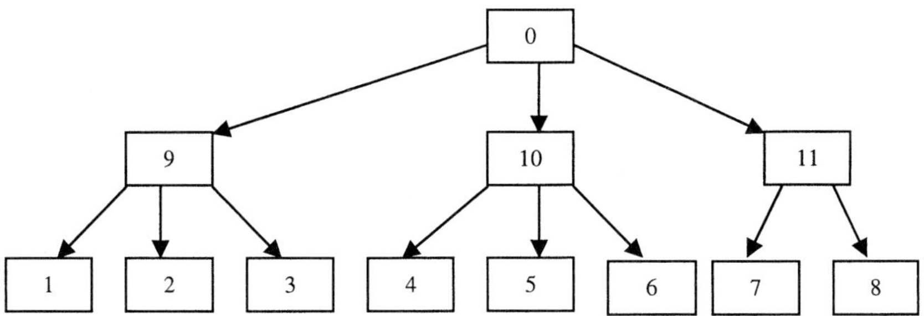

Fig. 1. Heuristic classifier for the APP diagnosis problem

4 Experimental Investigation

The presented algorithms C4.5, FID and AQ were used for creating rules for APP decision problem. Their frequencies of correct classification were compared with quality of heuristic classifier. The set of data has been gathered in the Surgery Clinic Wroclaw Medical Academy. It contains 476 learning examples.

For each learning method the following experiment was made:

- from the learning set 40 examples was chosen (according with frequency of the class appearance); this set was use for test,
- the rest of examples (436) were training ones.

This procedure was repeated 20 times for each of the algorithms. The results of the experiments are presented in Table 2 .

Table 2. Frequency of correct classification

Class number	**Heuristic**	**AQ**	**C4.5**	**FID**
1	79,1	90,5	95,8	86,7
2	88,2	55,0	92,3	100
3	93,1	95,0	95,8	100
4	67,1	90,0	95,6	66,7
5	82,5	98,8	86,9	83,3
6	84,4	85,0	96,2	75,0
7	84,7	97,5	91.5	75,0
8	88,2	75,0	92,3	50,0
average	**83,0**	**85,8**	**93,3**	**80,4**

The results of test are clear. The classifier given by C4.5 algorithm is always better than heuristic one. The AQ and FID algorithm gives the better results for some of class, but for another the frequency of correct classification is very low.
Experts revised the structures of classifiers given by inductive learning algorithms. They confirmed that all of rules were correct and maybe the heuristic classifier was incomplete.

5 Conclusion

The methods of inductive learning were presented. The classifiers generated by those algorithms were applied to the medical decision problem (recognition of Acute Abdominal Pain). The results of test were compared with recognition quality of heuristic algorithm.

It must be emphasised that we have not proposed a method of "computer diagnosis". What we have proposed are the algorithms whose can be used to help the clinician to make his own diagnosis. The superiority of the presented empirical results for the inductive learning classifiers over heuristic one demonstrates the effectiveness of the proposed concept in such computer-aided medical diagnosis problems. Advantages of the proposed methods make it attractive for a wide range of applications in medicine, which might significantly improve the quality of the care that the clinician can give to his patient.

This work is supported be The Polish State Committee for Scientific Research under the grant which is realizing in years 2005-2007.

References

1. Mitchell T.M., *Machine Learning*, McGraw-Hill Comp., Inc, New York 1997
2. Quinlan J.R., *C4.5: Programs for Machine Learning,* Morgan Kaufman, San Mateo, CA 1993
3. Janikow C.Z., Fuzzy Decision Tree: Issues and Methods, *IEEE Trans on Man and Cybernetics,* vol 28, Issue 1, 1998
4. Michalski R.S., A Theory and Methodology of Inductive Learning, *Artificial Intelligence,* no 20, 1983, pp.111-116
5. Puchala E., A Bayes Algorithm for Multitask Pattern Recognition Problem – Direct Approach, Lecture Notes in Computer Science, no 2659, 2003.
6. Walkowiak K., A Branch and Bound Algorithm for Primary Routes Assignment in Survivable Connection Oriented Networks, *Computational Optimization and Applications*, Kluwer Academic Publishers, February 2004, Vol. 27.

Grid-Technology for Chemical Reactions Calculation

Gabriel Balint-Kurti[1], Alexander Bogdanov[2], Ashot Gevorkyan[3], Yuriy Gorbachev[4], Tigran Hakobyan[5], Gunnar Nyman[6], Irina Shoshmina[2], and Elena Stankova[2]

[1] University of Bristol, School of Chemistry,
Bristol BS8 1TS, UK
Gabriel.Balint-Kurti@bristol.ac.uk
[2] Institute for High Performance Computing and Information Systems,
199397 St.-Petersburg, Russia
(bogdanov, irena, lena)@csa.ru
[3] Institute for Informatics and Automation Problems NAS of Armenia,
375014 Yerevan, Armenia
g_ashot@sci.am
[4] St.-Petersburg State Polytechnical University,
195251 St.-Petersburg, Russia
gorbachev@csa.ru
[5] Yerevan Physical Institute,
375036 Yerevan, Armenia
hakob@mail.yerphi.am
[6] Goteborg University,
412 96 Goteborg, Sweden
nyman@chem.gu.se

Abstract. We discuss a possible strategy for implementing a grid-based approach to realizing the immense computational resources required to compute reactive molecular scattering cross sections and rate constants.

1 Introduction

Evaluation of chemical reaction cross-sections and rate-constants as well as more detailed scattering characteristics is one of the great challenges for computational technologies. The main difficulties are connected with the multi-dimension and multi-scale character of the problem. Both the accuracy and the computations of detailed state-to-state scattering cross sections are limited by the availability of computational resources and by the efficiency of the computational algorithm used in their calculation. We examine here a possible solution to these problems through the use of Grid technology. The main concept of Grid technology is the solution of a computational problem through *coordinated dynamic resource sharing within a multi-institutional virtual organizations* [1]. Some attempts at such Grid based solutions have already been implemented [2 - 4]. In these cases the Grid infrastructure supports the usage of standard quantum chemistry programs (Gaussian [5] and GAMESS [6]). This paper discusses a possible solution of the reactive scattering problem using a Grid based technology.

V.S. Sunderam et al. (Eds.): ICCS 2005, LNCS 3516, pp. 933 – 936, 2005.

2 Description of Reactive Scattering Problem

We consider principally the process $A + (BC)_m \rightarrow (AB)_n + C$ which may take place with or without involving the formation of a quasi-bound or an activated complex $(ABC)^*$. When the process occurs via the formation of a quasi-bound complex, both the classical and quantum mechanical treatment of the process show chaotic behavior. The computation of the cross sections and rate constants require large, often unattainably large, amounts of computational resources on a conventional computer. Here we discuss an alternative approach that effectively utilizes Grid technologies.

The problem may be divided into two parts that may be solved separately. The first part being that of calculating the potential energy surface (PES) $V(q_1,q_2,\gamma)$ and the second part being the calculation of reaction probabilities and rate constants (or other properties). PESs are usually obtained from *ab initio* quantum chemical calculations (we consider these results as initial input to our problem), but they are represented in Cartesian co-ordinates that are usually not convenient for further calculations. For the scattering calculations the so-called curvilinear co-ordinates are the most convenient ones to use[7].

One of the computational problems consists in the transformation of the potential energy surface to the new co-ordinates $U(x^0,x^1,x^2) \equiv V(q_1,q_2,\gamma)$. It is also convenient to use certain analytic models for the PESs and to just determine their parameters. The PESs we develop depend on the fairly small number of parameters and these may be optimized to fit different sets of *ab initio* data [8]. In any case this problem may be solved separately. Since this problem is closely related to the inversion of a large dimensional matrix the corresponding methods may be used for its solution and these algorithms can also be parallelized. This procedure may easily be transferred to a Grid technology, implemented on a Grid of loosely connected computational systems. The calculation of scattering cross sections may be reduced to four sequential problems. The first one is the generation of the classical geodesic equations on the Lagrange surface of body system (at first this method of investigation classical dynamical systems was introduced by Krilov [9]) for the evolution of the system along the reaction co-ordinate and calculation of trajectory tube distribution; the second one is the problem of the solution Shroedinger equation on geodesic trajectory tubes; the third one is the calculation of the transition S-matrix elements and the fourth one is the calculation of transition probabilities [10, 11].

3 Grid for Chemical Reactions Calculation

The basic principles of the Grid for chemical reaction calculation can be stated as follows: 1) Present computer and data resources as a single virtual environment by developing a web portal on the Grid technology. 2) Build an easy-to-use user web interface for providing access to these resources. 3) Facilitate the sharing of results of research. 4) Organize archiving of input, output, and intermediate data.

As the basic software for Chemical Grid computing the Globus code will be used because of the following distinguishing advantages [12]. As the scheduling system weplan to use the Nimrod/G tool [13, 14], which is designed to manage the computational process including the transfer of input data and of the results of the calculations.

Returning to the problems arising in solving the reactive scattering problem that has been partly discussed in section 3, one must complete the following list of steps:

1) Computation of PES. 2) Parameterization of PES. 3) Solution of geodesic trajectory problems on Lagrange surfaces of the three-body system and the construction of the trajectory tubes distribution. 4) Calculation of the quantum system evolution on the trajectory tubes. 5) Computation of S-matrix elements and reaction probabilities for a set of initial phases φ_s and collision energies E_k^i. 6) Averaging of S-matrix element amplitudes over distribution of trajectory tubes, i.e. calculation of probabilities of elementary reactive quantum transitions. 7) Visualization of the results.

A Grid implementation of the solution of the reactive scattering problem implies the establishment of some infrastructure that should include algorithms for the *ab initio* calculation of PESs and the corresponding databases for different scattering partners. The next step then is the parameterization of the PESs using for instance the nonlinear optimization algorithm described in [9]. The Chemical Grid should contain a number of models that will permit the parameterizations of different aspects of PESs to varying accuracies. The classic geodesic trajectory problem calculation on the Lagrange surfaces for the reactive collision should be implemented for use in a distributed computational environment. Also the quantum reactive scattering part for the different trajectory tubes can be parallelized.

Our experience in complex problem solving using the Grid [15] will be used to achieve these goals. We wish to create a Chemical Grid that will allow the incorporation of both the data banks of different PESs and other properties accumulated by other scientists and our own computational codes.

Acknowledgments

This work was partly supported by the INTAS grant 03-51-4000 and ISTC A-823.

References

1. Foster, I, Kesselman, C. and Tuecke, S.: "The Anatomy of the Grid: Enabling Virtual Organisations", International Journal of Supercomputer Applications, 15 (3) 2001, pp. 200-222.
2. Nishikava, T., Nagashima, U. and Sekiguchi, S. "Design and Implementation of Intelligent Scheduler for Gaussian Portal on Quantum Chemistry Grid" In: P.M.A.Sloot, D.Abramson,, A.V.Bogdanov, J.J.Dongarra, A.Y.Zomaya, Yu.E.Gorbachev, eds, Proceedings, Part 3, Computational Science – ICCS 2003, in series Lecture Notes in Computer Science, v. 2659, pp. 244-253, Springer Verlag, ISBN 3-540-40194-6.

3. Baldrige, K.K., Greenberg, J.P. "Management of Web and Associated Grid Technologires for Quantum Chemistry Computation" In: P.M.A.Sloot, D.Abramson,, A.V.Bogdanov, J.J.Dongarra, A.Y.Zomaya, Yu.E.Gorbachev, eds, Proceedings, Part 4, Computational Science – ICCS 2003, in series Lecture Notes in Computer Science, v. 2658, pp. 111-121, Springer Verlag, ISBN 3-540-40194-6.
4. Sudholt, W., Baldridge, K., Abramson, D., Enticott, C. and Garic, S. "Parameter Scan of an Effective Group Difference Pseudopotential Using Grid Computing", New Generation Computing 22 (2004) 125-135.
5. http://www.gaussian.com
6. https://gridport.npaci.edu/GAMESS
7. Pack, R. T. and Parker, G.A., Quantum reactive scattering in three dimensions using hyperspherical (APH) coordinates. Theory J. Chem. Phys. 87, (1987) 3888.
8. Gevorkyan, A.S., Ghulian, A.V. and Barseghyan, A.R. "Modeling of the Potential Energy Surface of Regrouping Reaction in Collinear Three-Atom Collision System Using Nonlinear Optimization" In: P.M.A.Sloot, D.Abramson, A.V.Bogdanov, J.J.Dongarra, A.Y.Zomaya, Yu.E.Gorbachev, eds, Proceedings, Part 2, Computational Science – ICCS 2003, in series Lecture Notes in Computer Science, v. 2658, pp. 545-554, Springer Verlag, ISBN 3-540-40194.
9. Krylov, N.S., Works by abroad of statistical physics, publishing company Academy of Scince SSSR [in Russian] ed A. Fok, Moscow (1950), p.205
10. Bogdanov, A.V., Gevorkyan, A.S. and Grigoryan, A.G., Bifurcations in trajectory problem as a cause of internal-time singularities and the onset of quantum (wave) chaos, Tech. Phys. Lett. 25, (1999) 637
11. Bogdanov, A.V., Gevorkyan, A.S., Grigoryan, A.G., Internal Time Peculiarities as a Cause of Bifurcations Arising in Classical Trajectory Problem and Quantum Chaos Creation in Three body System, AMS/IP Studies in Advanced Mathematics, V.13. p. 69-80, (1999)
12. Johnston, W. E., The NASA IPG Engineering Team, and The DOE Science Grid Team Implementing production Grid
13. Abramson, D., Giddy, J., Kotler, L.: High Performance Parametric Modeling with Nimrod/G: Killer Application for the Global Grid? International Parallel and Distributed Processing Symposium (IPDPS), Cancun, Mexico (May 2000) 520- 528; http://www.csse.monash.edu.au/~davida/nimrod/
14. Shoshmina, I., Bogdanov, A.V. and Abramson D. "Whither the Grid?" Proceedings of the International Conference "Distributed Computing and Grid Technologies in Science and Education". July 2004, Dubna, Russia. (accepted for publication)
15. Krzhizhanovskaya, V.V., Gorbachev, Yu.E., Sloot, P.M.A. A Grid-based Problem-solving Environment for Simulation of Plasma Enhanced Chemical Vapor Deposition. In: Book of abstracts of the International Conference "Distributed Computing and Grid Technologies in Science and Education". 29 June - 2 July 2004, Dubna, Russia. Publ: JINR, Dubna, 2004. pp.89-90.

A Fair Bulk Data Transmission Protocol in Grid Environments

Fanjun Su[1], Xuezeng Pan[1], Yong lv[2], and Lingdi Ping[1]

[1] College of Computer Science, Zhejiang University,
Hangzhou 310027, China
suwang@zju.edu.cn
[2] College of Electrical Engineering, Zhejiang University,
Hangzhou 310027, China

Abstract. In this paper, we propose FHSTCP (Fair High-Speed TCP) as an improvement of HSTCP, which adds a fair factor to eliminate the difference of congestion window caused by different RTT and adopts block-pacing to reduce the burstiness. Simulation results show that FHSTCP can alleviate the RTT unfairness meanwhile keeping advantages of HSTCP.

1 Introduction

Recently there appear many high-speed networks with bandwidth larger than 1Gbps, even than 10Gbps. Through high-speed networks, data intensive grid application can transfer high-bandwidth real time data, images, and video. TCP [1] performs badly in high-speed networks [2]. Some improvements have been made, such as HighSpeed TCP (HSTCP) [2], Scalable TCP (STCP) [3], BIC [4]. However, in [4] the author points out that HSTCP has very severe RTT (Round Trip Time) unfairness. We define the RTT unfairness of two competing flows to be the throughput ratio. In this paper, a fair protocol named FHSTCP (Fair High-Speed TCP) is proposed, and we give a relative fair criterion to evaluate the protocol. FHSTCP adds a fair factor to eliminate the difference of congestion window between flows with different RTT. Block-pacing scheme is adopted to reduce the burstiness caused by fair factor of long RTT flows. The performance of FHSTCP is evaluated using ns2 [5].

2 FHSTCP

TCP and HSTCP use the following algorithm to adjust their congestion window:

$$\text{ACK: } w \leftarrow w + a(w)/w$$

$$\text{Drop: } w \leftarrow w - w \times b(w)$$

Where w denotes congestion window size. For standard TCP, a(w)=1, b(w)=0.5, which is not sufficient for high-speed networks, so HSTCP makes a(w) and b(w) become the function of current congestion window size [2].

V.S. Sunderam et al. (Eds.): ICCS 2005, LNCS 3516, pp. 937–940, 2005.

The main idea of FHSTCP is adding a factor to compensate the congestion window increment difference caused by different RTT. Let $a(w)' = \eta \times a(w)$. The fair factor is based on the value of RTT, calculating as $\eta = c \times RTT$. For example, when c = 10 and RTT = 200ms = 0.2s, $\eta = 2$. After adding fair factor, different flows can get the same $w(t)$. Because throughput $V(t) = w(t)/RTT$, FHSTCP can keep RTT unfairness to be inversely proportional to RTT ratio. After fair factor is added, long RTT flows will have a sharp congestion window increment after an RTT. To solve this problem, a block-pacing method is adopted. The congestion window will be divided into several "blocks". After the packets in one block have been sent out, other packets in another block will be sent after a time interval. This can counteract the negative effects caused by adding fair factor. We set the number of the block based on fair factor η. The details of the algorithm are shown in figure 1.

```
Setting initial value:
   k=1; // the number of blocks
   block=0;  // the packets number in a block
   number=0; // number of packets that has been sent
On receiving a new ack in congestion avoidance state:
   increment=c*RTT*a(w)/cwnd;
   // cwnd is the size of congestion window
   if(increment>1)
      increment=1;
      // avoid the increment larger than slow start
   cwnd=cwnd+increment;
   k=(int)c*RTT+1;
   block=cwnd/k;
On congestion happening:
   cwnd=cwnd*(1-b(w));
   number=0;
   time_0=now; // now is current time
On Sending date:
   if(number>block)
      time_1=now; // now is current time
      if(k>cwnd)
        k=cwnd;
      delay=RTT/k-(time_1-time_0);
      // delay is the time interval between two blocks
      // (time_1-time_0)is sending time of one block
      output(delay); // send data after a time of delay
      number=0; // begin a new count
      time_0=time_1+delay;
   else
      output( ); // send data directly
      number++;
```

Fig. 1. Pseudo-codes of FHSTCP

3 Performance Evaluation

We present a relative fairness criterion: (1) The bandwidth requirement of standard TCP should be met. (2) The fairness between different high-speed TCP flows should be guaranteed. Satisfying condition (1) also means the protocol has the TCP friendliness, because standard TCP only works well in high loss rate environment. The loss rate we choose is 10^{-3}, which is corresponding to the value of Low_P in HSTCP. To evaluate the fairness of high-speed TCP flows, fair index [6] is used as follows:

$$f(x) = \left(\sum_{i=1}^{n} x_i\right)^2 \Big/ \left(n \sum_{i=1}^{n} x_i^2\right)$$

where x_i is the link utilization of the flow i. We adopt ns2 simulator (version 2.26) [5]. The topology and configuration are shown in figure 2. The buffer size of the router is set to be the product of bandwidth and the delay of bottleneck link. We use TCP SACK for the simulation, and packet size is set to 1000 byte. The maximal congestion window is set to 1000000. FTP is the application used to transmit data through the TCP connections. To avoid phase effect [7], some web flows and short-lived TCP flows are used, together with 3~5 standard long-lived TCP flows. They act as background traffics for the simulation. DT (Drop Tail) queue management policy is used.

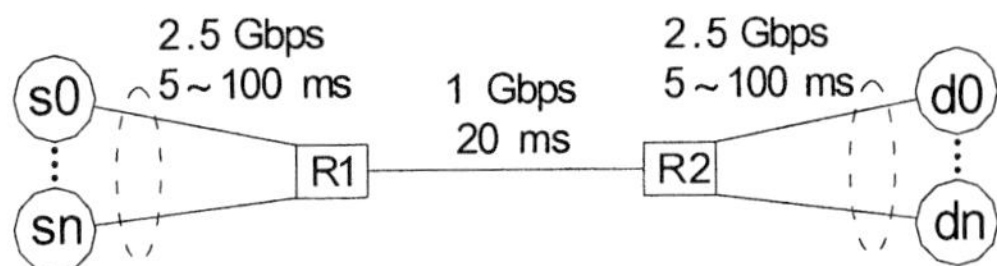

Fig. 2. Simulation topology and configuration

Some simulation results show effect of parameter c. we list our simulation results in table 1(F1, F2, F3 are FHSTCP flows whose RTT are 80, 140, 200ms, and F4 denotes background flows). We can find when c=5, the per-flow bandwidth utilization and total bandwidth utilization of FHSTCP flows decrease together. This means that a too small value of c will limit the scalability of FHSTCP. However, when c=20, short RTT FHSTCP flows will grasp more bandwidth and fairness are decreased too.

Table 1. The effect of *c* and the TCP friendliness

c	Bandwidth Utilization (%)				The fairness index of FHSTCP	Loss rate of background flows
	F1	F2	F3	F4		
5	29.79	16.80	14.38	26.96	0.90	7×10^{-5}~1×10^{-4}
10	32.22	26.61	19.36	16.35	0.96	7×10^{-5}~8×10^{-4}
20	44.85	24.69	19.92	5.55	0.88	5×10^{-4}~1×10^{-3}

What's more, background flows can get less bandwidth. In other words, the friendliness of FHSTCP decreases. Therefore, c=10 is optimal. The packet loss rate of background flows are below 10^{-3} in 3 cases, so we say FHSTCP is TCP friendly.

We also give fairness comparison of FHSTCP, HSTCP, STCP and standard TCP. Three flows with RTT being 80ms, 140ms, and 200ms respectively are used, and different algorithms, such as FHSTCP (c=10), HSTCP, STCP, and standard TCP are adopted respectively. The fairness index is calculated at different time scales. As shown in figure 3, we can find FHSTCP has a better fairness property.

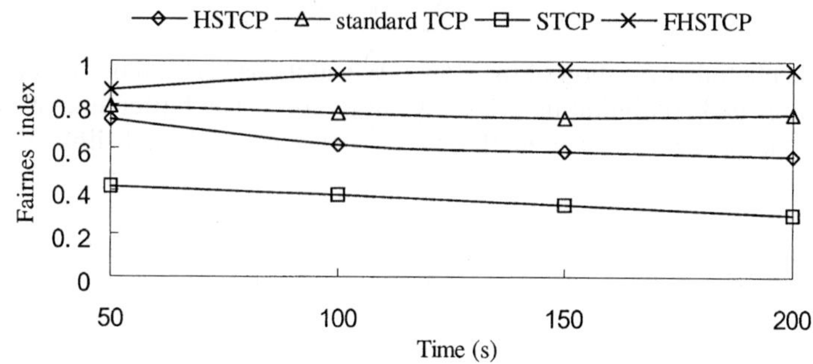

Fig. 3. The fairness of different algorithms

4 Conclusion

As an improvement of HSTCP, FHSTCP adds a fair factor to eliminate the difference of congestion window caused by different RTT and adopts block-pacing to counteract the negative effects caused by fair factor. Simulation results show that FHSTCP has better fairness, and is TCP friendliness while keeping the advantages of HSTCP.

References

1. Stevens, W.R.: TCP/IP Illustrated: The Protocols. Volume 1. Addison-Wesley (1994)
2. Floyd S.: HighSpeed TCP for large congestion windows. RFC 3649 (2003)
3. Kelly T.: Scalable TCP: Improving Performance in High-Speed Wide Area Networks. ACM Computer Communications Review 33 (2003) 83-91
4. Xu L., Harfoush K., Rhee I.: Binary Increase Congestion Control (BIC) for Fast Long-Distance Networks. Proceedings of INFOCOM (2004) 2514 -2524
5. Network simulation-ns2. http://www.isi.edu/nsnam/ns
6. Chiu D., Jain R.: Analysis of the increase and decrease algorithms for congestion avoidance in computer networks. Journal of Computer Networks and ISDN 17 (1989) 1-14
7. Floyd S., Jacobson V.: On traffic phase effects in packet-switched gateways. Internetworking: Research and Experience 3 (1992) 115-156

A Neural Network Model for Classification of Facial Expressions Based on Dimension Model

Young-Suk Shin

Department of Information Communication Engineering, Chosun University, #375 Seosuk-dong, Dong-gu, Gwangju, 501-759, Korea
ysshin@mail.chosun.ac.kr

Abstract. We present a new neural network model for classification of facial expressions based on dimension model that is illumination-invariant and without detectable cues such as a neutral expression. The neural network model on the two-dimensional structure of emotion have improved the limitation of expression recognition based on a small number of discrete categories of emotional expressions, lighting sensitivity, and dependence on cues such as a neutral expression.

1 Introduction

The work in facial expressions for human-computer intelligent interaction did not start until the 1990s. Models for recognizing facial expressions have traditionally operated on a short digital video sequence of the facial expression being made, such as neutral, then happy, then neutral[1,2,3]. All require the person's head to be easily found in the video. Therefore, continuous expression recognition such as a sequence of "happy, surprise, frown" was not handled well. And the expressions must either be manually separated, or interleaved with some reliably detectable cues such as a neutral expression, which has essentially zero motion energy.

In this paper, we present a new neural network model on the two dimensional structure of emotion for classification of facial expressions that is illumination-invariant and without detectable cues such as a neutral expression.

2 Facial Expression Representations for Invariant-Illumination and Neutral Expression

The face images used for this research were centered the face images with coordinates for eye and mouth locations, and then cropped and scaled to 20x20 pixels. The luminance was normalized in two steps. First, a "sphering" step prior to principal component analysis is performed. The rows of the images were concatenated to produce 1×400 dimensional vectors. The row means are subtracted from the dataset, X. Then X is passed through the zero-phase whitening filter , V, which is the inverse square root of the covariance matrix:

V.S. Sunderam et al. (Eds.): ICCS 2005, LNCS 3516, pp. 941 – 944, 2005.

$$V = E\{XX^T\}^{-\frac{1}{2}},\ W = XV \tag{1}$$

This indicates that the mean is set to zero and the variances are equalized as unit variances. Secondly, we subtract the local mean gray-scale value from the sphered each patch. From this process, W removes much of the variability due to lightening. Figure 1(a) shows the cropped images before normalizing. Figure 1(b) shows the cropped images after normalizing.

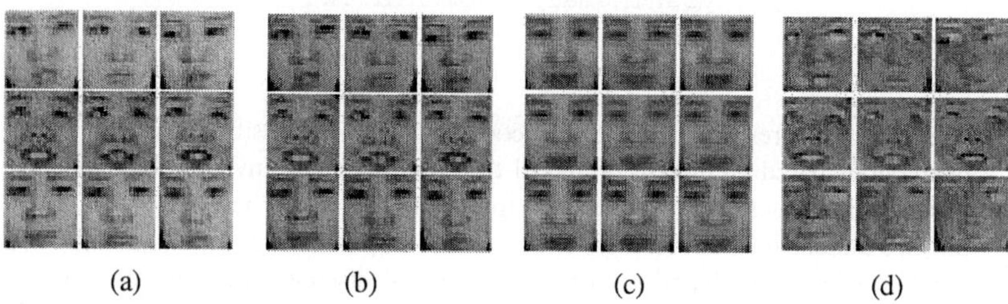

Fig. 1. (a) Images before normalizing. (b) Images after normalizing. (c) PCA representation only included the first 1 principle component. (d) PCA representation excluded the first 1 principle component

In a task such as facial expression recognition, the first 1 or 2 principal components of PCA do not address the high-order dependencies of the facial expression images, that is to say, it just displays the neutral face. Figure 1(c) shows PCA representation that included the first 1 principle component. But selecting intermediate ranges of components that excluded the first 1 or 2 principle components of PCA did address well the changes in facial expression (Figure 1(d)).

Therefore, to extract information of facial expression regardless of neutral expression, we employed the 200 PCA coefficients, P_n, excluded the first 1 principle component of PCA of the face images. The principal component representation of the set of images in W in Equation(1) based on P_n is defined as $Y_n = W * P_n$. The approximation of W is obtained as:

$$\overline{W} = Y_n * P_n^T . \tag{2}$$

The columns of $\overline{W}$ contains the representational codes for the training images (Figure 1(d)). The representational code for the test images was found by $\overline{W}_{test} = Y_{test} * P_n^T$. Best performance for facial expression recognition was obtained using 200 principal components excluded the first 1 principle component.

3 Recognition

The face images used for this research were a subset of the Korean facial expression database[4]. The data set contained 500 images, 3 females and 3 males, each image

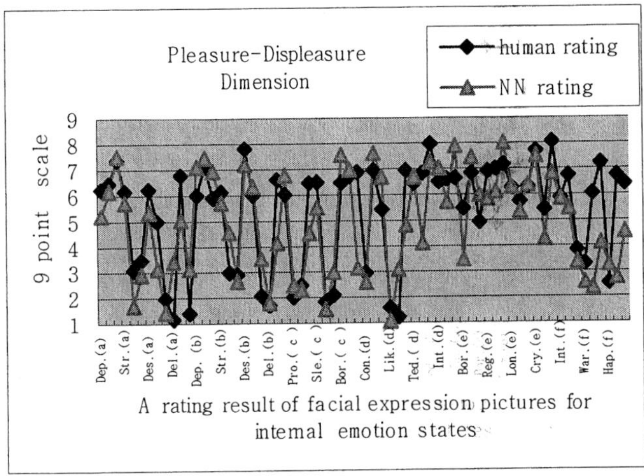

Fig. 2. A rating result of facial expression recognition in Pleasure-Displeasure dimension (Str.,strangeness;Des.,despair;Del.,delight;Pro.,proud;Sle.,sleepiness;Bor.,boredom; Con.,confusion;Lik.,likable;Ted.,tedious;Int.,intricacy;Reg.,regret;Lon.,loneliness; Cry.,crying;War.,warmness;Hap.,happiness; Dep.,depression)

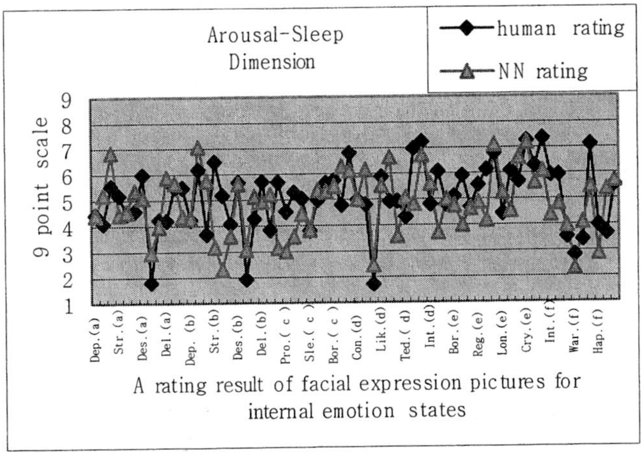

Fig. 3. A rating result of facial expression recognition in Arousal-Sleep dimension

using 640 by 480 pixels. Expressions were divided into two dimensions(Pleasure-Displeasure and Arousal-Sleep dimension) according to the study of internal states through the semantic analysis of words related with emotion by Younga et al. [5] using 83 expressive words. The system for facial expression recognition uses a three-layer neural network. The first layer contained the representational codes derived in Equation (2). The second layer was 30 hidden units and the third layer was two output nodes to recognize the two dimensions: Pleasure-Displeasure and Arousal-Sleep.

Training applies an error back propagation algorithm. The activation function of hidden units uses the sigmoid function. 500 images for training and 66 images excluded from the training set for testing are used. The 66 images for test include 11 expression images of each six people. The first test verifies with the 500 images trained already. Recognition result produced by 500 images trained previously showed 100% recognition rates. The rating result of facial expressions derived from 9 point scale on two dimension for degrees of expression by subjects was compared with experimental results of a neural network(NN).

Figure 2 and 3 show the correlation of the expression recognition between human and NN in each of the two dimensions. The statistical significance of the similarity for expression recognition between human and NN on each of the two dimensions was tested by Person correlation analysis. The correlation in the Pleasure-Displeasure dimension between human and NN showed 0.77 at the 0.01 level and 0.51 at the 0.01 level in the Arousal-Sleep dimension.

4 Conclusion

Our results allowed us to extend the range of emotion recognition and to recognize on dimension model of emotion with illumination-invariant without detectable cues such as a neutral expression. We propose that the inference of emotional states within a subject from facial expressions may depends more on the Pleasure-Displeasure dimension than Arousal-Sleep dimension. It may be analyzed that the perception of Pleasure-Displeasure dimension may be needed for the survival of the species and the immediate and appropriate response to emotionally salient, while the Arousal-Sleep dimension may be needed for relatively detailed cognitive ability for the personal internal states.

Acknowledgements. This study was supported by research funds from Chosun University, 2004.

References

1. Oliver, N. Pentland, A., Berard, F.: LAFTER:a real-time face and lips tracker with facial expression recognition. Pattern Recognition **33** (2000) 1369-1382
2. Cohen, I., Sebe, N., Garg, A., Chen, L. S., Huang, T. S.: Facial expression recognition from video sequence. Proc. Int'l Conf. Multimedia and Exp(ICME) (2002) 121-124
3. Cohen, I. :Semisupervised learning of classifiers with application to human-computer interaction. PhD thesis, Univ. of Illinois at Urbana-Champaign (2003)
4. Saebum, B., Jaehyun, H., Chansub, C.: Facial expression database for mapping facial expression onto internal state. '97 Emotion Conference of Korea, (1997) 215-219
5. Younga, K., Jinkwan, K., Sukyung, P., Kyungja, O., Chansub, C.: The study of dimension of internal states through word analysis about emotion. Korean Journal of the Science of Emotion and Sensibility, **1** (1998) 145-152

A Method for Local Tuning of Fuzzy Membership Functions

Ahmet Çinar

Firat University, Faculty of Eng. Computer Engineering , 23119, Elazig, Turkey
acinar@firat.edu.tr

Abstract. In this paper, a new method based on genetic algorithms is proposed for local tuning of fuzzy membership functions. For this purpose, the local adjustment is employed on the initial membership functions. Genetic algorithm is used to investigate discrete points that will be modified on the membership functions Hence, global adjustment does not require and the processing time required for tuning of membership functions is decreased.

1 Introduction

In this paper, the initial base values of membership functions are not changed, only local modifications are made. The basic of the study is as following:

Step 1: Discretization of initial membership functions. Step 2: Finding the membership functions that will be modified by genetic algorithms and tuning. Step 3: Reconstructing the membership functions and applying to system.

2 Discretization of Initial Membership Functions

For this process, the geometric support construction method is used. The set of polygonal segments of boundary curve is located in the any list. And then, the construction of the polygonal support is achieved such that the distance any segment and portion of the curve it represent must be bounded. Hence, refinement of curve will be more important and correct in the any highly curved regions on the boundary curve.

3 Tuning Membership Functions by Using Genetic Algorithms

Figure 1 depicts diagram of proposed method for local tuning of fuzzy membership functions.

The following sections describe each block seen above.

Block 1: Dedicating control points to membership functions.
Block 2: Fitness Function.

In this block, how to make changes on the population is described. For this goal, initial membership function is passed through certain fitness function and the obtained

V.S. Sunderam et al. (Eds.): ICCS 2005, LNCS 3516, pp. 945 – 949, 2005.

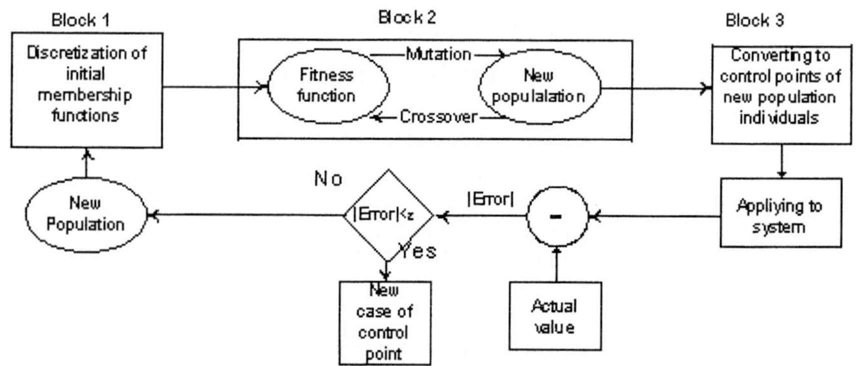

Fig. 1. The diagram of the proposed method

value is operated to genetic operator. Therefore, some individuals are founded. In this study, a function dependent on system input output is used as fitness function. It is known that, a fitness function is generated a certain integer value. Whereas, this method is not appropriate for this study due the fact that the direct fitness function did not explained. The following section describes the proposed fitness function. Figure 5 depicts genetic individuals and control points. These control points are produced by means of aforementioned method in block 1. This process is the population dedicating process on the control points and making up control points on the initial membership function.

For example; let control point value be 11. First, log value of control point is obtained, such as *m=log2(11)=3.3* . If *m* is not integer number, it is rounded for example 3.3 to 4. Therefore, population values are follows: *0001,0010,0011,0100,0101, 0110,0111,1001,1010,1011,1100*. The *m* value is used as bit number.

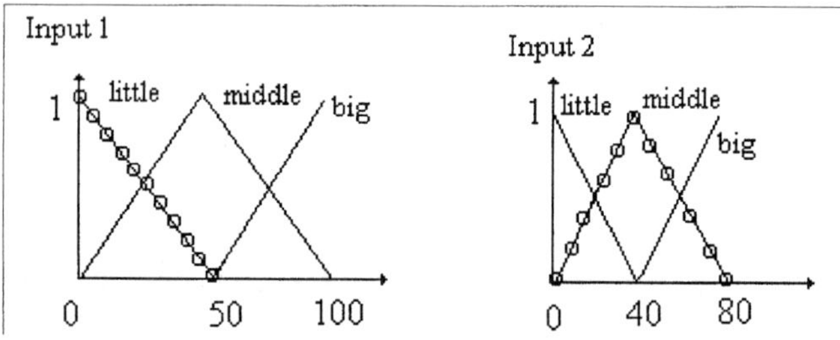

Fig. 2. Control points on the initial membership function

If the rule base table and system input output response are known, fitness function is used as following: Let us know the input output response for any system. *a[ ]*,b[ ] and *c[ ]* are three sets. Let *c[ ]* be response with respect to *a[ ]* and *b[ ]*. Control points at distance ε_l is changeable. But, we have to know which points be modified.

For that reason, *a[ ]* and *b[ ]* sets are used. Figure 3 shows control points that will change with respect to ε_l value.

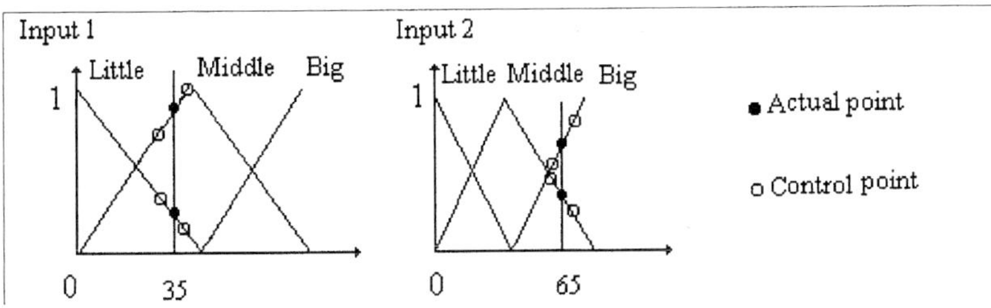

Fig. 3. The values with respect to input=35 and input2=65 and relationship control points

After finding these control points, the output value is obtained by means of any defuzzyfication method such as middle of maximum.

u_i: The i_{th} value of u membership functions

$\mu(u_i)$: The value of u_i membership functions.

$$u^* = \frac{\sum_{i=1}^{l} .\mu(u_i).u_i}{\sum_{i=1}^{l} \mu(u_i)} \tag{1}$$

If the obtained result is similar to output *c[ ]* with respect to ε_l, then any changes does not require on this population. In the proposed method, value ε_l is as half of distance between of two points as shown in figures 4. If the error is bigger than ε_l, then generic operators are used on population. Figure 4 depicts simple geometric operations for value ε_1.

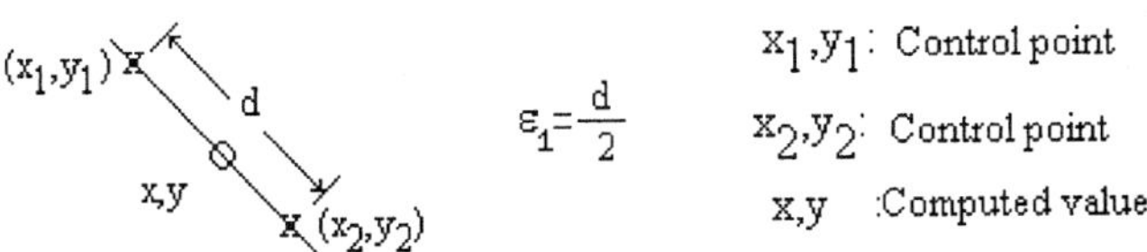

Fig. 4. The using method for value εl

The mutation is applied to discrete points changed. It is known that, only a bit is changed in the simplest manner for the mutation process. A new population after this operation will be different. This new population has to convert in the appropriate for system. This operation is achieved in block 3.

Block 3: Converting to control points of new population value.

In this block, new population values are converted to control points. Let original point be genetic individual with respect to 0010 value. Figure 5 shows this position.

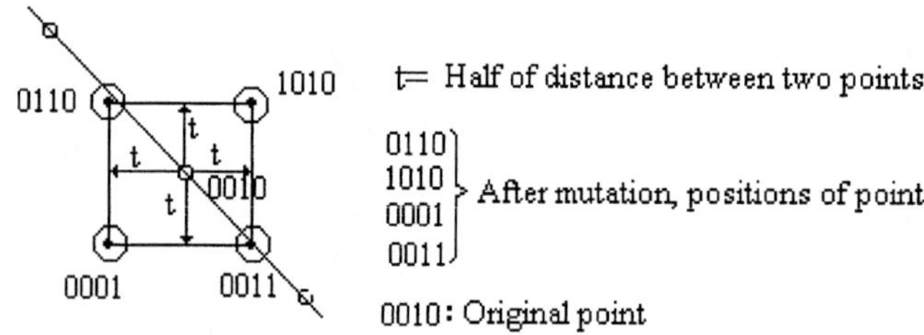

Fig. 5. Original point and its position after modification

According to using bit numbers, the actual location of point is as follows:
x,y value is 0010, x+t,y+t value is 0011, x-t,y+t value is 0001,x-t,y-t value is 0110,x+t,y-t value is 1010. If bit number is 4, there is no problem. But the bit number is bigger than 4, then only right 4 bits are used. Also, let bit number be 5 and 00100. Then only using 0100 is appropriate. If bit number is less than 4, then, a bit is added on the left side. If bit value is 010, then it is used as 0010.

In this way, after computing actual coordinate values and making up memberships functions, according to system response, modifications of membership function is considered. If it needs, new population is passed through fitness function.

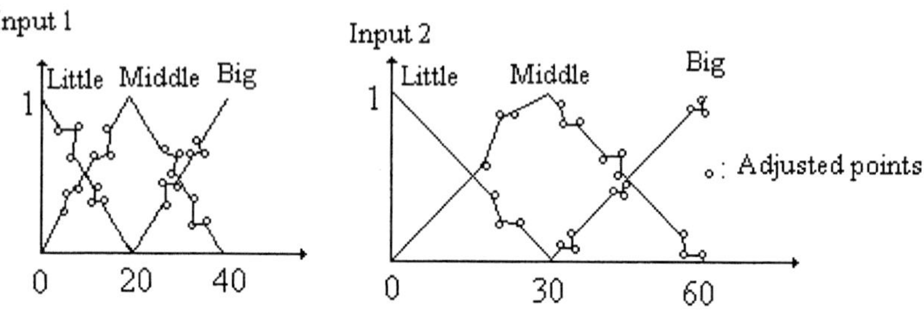

Fig. 6. The adjusted membership functions

4 Conclusion

In this study, a distinct method based on genetic algorithm for local adjustment of fuzzy membership functions is presented. Instead of global adjustment, local adjustment is achieved. Thus, the fast and accurate process is achieved with respect to system input-output response. By means of proposed method, initial fuzzy membership functions are selected, and then very much little changes on the membership function are done. Consequently the global changes such as gradient-descent method are avoided.

References

1. D.E. Goldberg, Genetic Algorithms in Search, Optimization and Machine Learning, Addison Wesley , 1989
2. H. Nomura, I. Hayashi, N.Wakai, A self-tuning method of fuzzy control by descent method, in Proc 4th IFSA congress, Engineering, pp.49-52, 1991
3. H.J.Zimmermann ,Fuzzy programming and linear programming with several objective functions, Fuzzy Sets and Systems, vol.1, pp.44-45,1978
4. Kuo Ting, Hwang Shu-Yuen, A Genetic Algorithm With Disruptive Selection, IEEE Transaction On Systems, Man, And Cybernetics-Part B: Cybernetics, Vol. 26, No. 2, April 1996
5. Wang Chi-Hsu,Wang Wei-Yen, Fuzzy B-Spline Membership Function (BMF) And Its Application In Fuzzy-Neural Control, IEEE Transactions On System Man And Cybernetics, Vol. 25, No.5, May 1995
6. Zadeh L. , Fuzzy sets , Inf. Control, Vol:8, pp:338-353, 1965.

QoS-Enabled Service Discovery Using Agent Platform*

Kee-Hyun Choi, Ho-Jin Shin, and Dong-Ryeol Shin

School of Information & Communication Engineering Sungkyunkwan University, Korea
{gyunee, hjshin, drshin}@ece.skku.ac.kr
http://nova.skku.ac.kr

Abstract. As the number of Internet services is growing rapidly, the network users need a framework to locate and utilize those services that are available in the Internet. As a result, dynamic service discovery plays an important role in certain networks, such as P2P networks. Recently many P2P mechanisms have been proposed which focus on the naming and discovery protocols. However, because these mechanisms do not provide congestion management and priority-based scheduling in P2P networks, they cannot provide an efficient communication among the nodes. Based on these observations, in this paper, we propose a QoS-based agent framework, designed to resolve such problems, and demonstrate its effectiveness.

1 Introduction

Service discovery is increasing in importance as its usage becomes more and more widespread throughout the Internet. The entire premise of the Internet is centered on the sharing of information and services. The focus of P2P networks is the mechanism by which a peer provides other peers with services and discovers the services availabe from other peers. In such an environment, a peer can be any network-aware device such as cell phones, PDA, PC or any device you can imagine that passes information in and out. In some cases, a peer might be an application distributed over several machines.

A large number of papers dealing with service discovery have recently been published. Although the schemes proposed provide efficient and fast searching mechanisms for service discovery, they do not consider QoS. With these observations in mind, we developed a new agent platform designed to provide QoS and evaluated its effectiveness by simulation, which is a main contribution of this paper. The remainder of this paper is organized as follows. Section 2 describes the proposed agent platform. In section 3, we present the simulation results for the proposed system. Finally, we conclude this paper in Section 4.

2 Proposed Agent Platforms

In this section, we describe our proposed Agent Platform (AP) architecture. The overall design of the architecture is shown in Figure 1. We adopt the FIPA[1] reference

* This research was partially supported by a grant from the CUCN, Korea and Korea Science & Engineering Foundation (R01-2004-000-10755-0).

V.S. Sunderam et al. (Eds.): ICCS 2005, LNCS 3516, pp. 950–953, 2005.

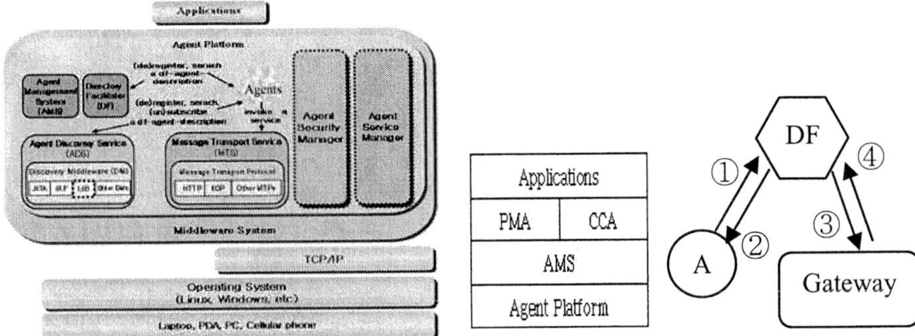

Fig. 1. Agent Platform

Fig. 2. a) Agent platform b) Setup sequence

model and add new components. The Agent Service Manager is a QoS solution for ad-hoc networks and P2P networks. By using the ASM, we allow our agent platform to provide QoS in P2P networks. The ASM uses two separated but not independent management agents. The PMA provides dynamic priority information according to the network status. When an application wants to transmit its data, it contacts the DF, in order to obtain priority information from the PMA. In this way, each application can control its traffic according to the priority assigned by the PMA. The DF updates the priority information whenever the network status changes. Figure 2-b) shows the setup procedure. Table 1 shows the priority level of each type of service. The DF classifies the traffic according to the priority information. When the DF receives a service request message, it returns the current priority information. We can control congestion by using the second special agent, which is called the CCA. The CCA also uses priority information, just as the PMA does. However, the CCA verifies the priority information by periodically sending a check messages. Thus, applications (e.g., ftp, telnet, multimedia, etc.) can adjust their sending rate accordingly. By means of these two agents, we can adjust the data rate at the local system (e.g., PC, PDA, and Notebook, etc) whenever the network traffic is changed. Moreover, the network traffic can be classified using the information in the gateway. Thus, the priority is not fixed but adjustable according to the network environment.

3 Simulations

This section presents the performance results of the priority-based service discovery (PSD) mechanism. We simulate our mechanism with/without an infrastructure (e.g., AP, gateway).

3.1 Simulation with Infrastructure

We simulate these mechanisms using a simple dumbbell network model in a wired network. The gateway has ten traffic sources and the nodes, which are attached to the gateway, are wireless nodes. These wireless nodes do not move in our simulation. In Link2, we test both the droptail queuing mechanism and the proposed priority-based

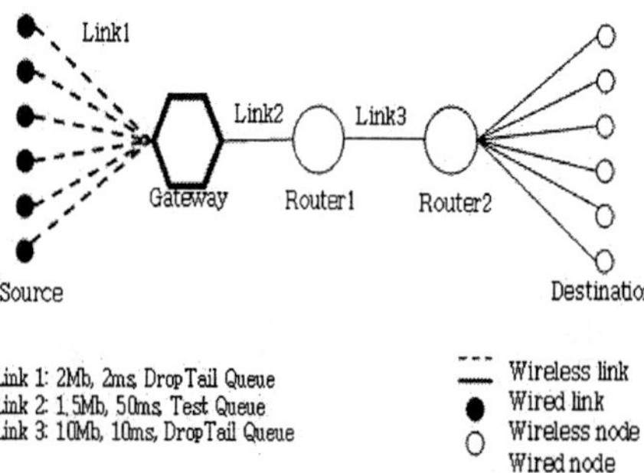

Fig. 3. Network Topology

Table 1. Priority Level

Services	priority
Service advertisements	1
Service requests	2
Event messages	3
Voice traffic	4
Multimedia	5
E-commerce	6
Light data traffic	7
Bulk data traffic	8

queuing mechanism. Figure 3 depicts the network topology. Using the NS simulator [2], a simulation was conducted in order to determine to what extent the proposed scheme improves the time-limited traffic as compared to the conventional mechanisms (droptail queue in this paper). We use TCP traffics for both the light and bulk data traffic. Each TCP flow uses an FTP application. We change the maximum congestion window size of each TCP flow according to its priority. High priority traffic is either CBR or EXP traffic. Figure 4 shows the simulation results. Even if the number of traffic sources increases, the CBR traffic using PSD does not slow down, whereas the CBR traffic with the droptail queuing mechanism does. Moreover, the TCP traffic with PSD shows the almost the same result as that with the droptail mechanism, which implies that the proposed scheme improves the performance of time-limited traffic without affecting the TCP traffic.

3.2 Simulation in Ad-Hoc Network

We modify the DFS[3] mechanism using the priority information. The DFS protocol for WLANs allocates bandwidth to flows in proportion to their weights and accounts for variable packet sizes. The DFS schedules packets for transmission based on their eligibility. Because of the distributed nature of the DFS protocol, collisions may occur, which causes priority reversal and undermine the fairness. Although DFS provides fairness and better throughput, the backoff interval is directly proportional to the *Packet-length/Weight* ratio in the linear scheme of DFS. For this reason, we use the priority-based backoff algorithm in enhanced DFS[4]. The scenario and parameters used in this paper are identical to those used in DFS. We perform simulations based on the implementation described in [3]. The channel bandwidth is set to 2Mbps. The simulation environment consists of n nodes. All of the nodes are stationary and are in transmission range of each other, in order to simulate a broadcast LAN. The number of nodes is always even. In this paper, identical flows refer to flows that are always backlogged and have equal packet size. Figure 5 shows the throughput results for the PSD, DFS and DCF mechanisms. We note that PSD not only guarantees fairness but also improves the throughput of time-limited traffics.

4 Conclusions

With the increasing number of people using mobile devices such as PDAs, cell-phones, notebooks, etc., the need to connect them to the Internet for ubiquitous sharing and data access is increasing. An efficient service discovery infrastructure plays an important role in such a dynamic environment. We reviewed a number of existing service discovery frameworks and found that they all suffer from one common problem, namely, a lack of QoS-based service discovery. In this paper, we introduced an agent platform designed to support QoS management. The simulation results show that the proposed mechanism improves the performance of time-limited traffic, without affecting that of other traffic, and guarantees fairness in an ad-hoc network.

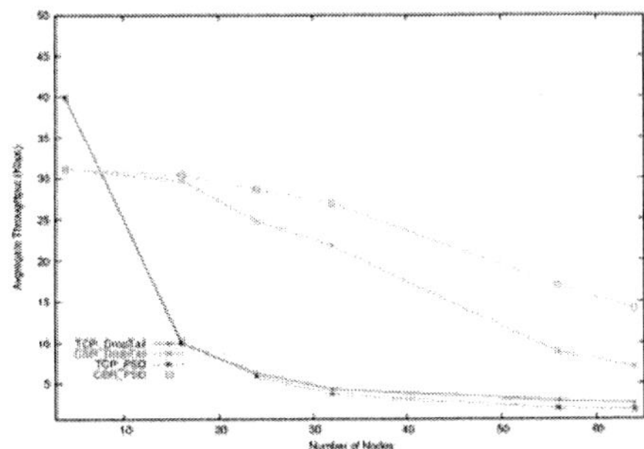

Fig. 4. Aggregated throughput in wired network

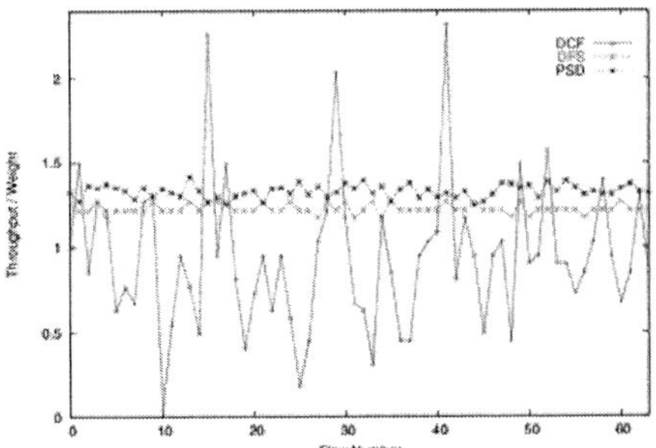

Fig. 5. Throughput of 64 flows in wireless network

References

1. FIPA: Agent Management Specification, Dec. 2002.
2. K. Fall and K. Varadhan, "ns Notes and documentation", tech. rep., VINT Project, UC-Berkeley and LBNL(2003)
3. N. H. Vaidya, P. Bahl, and S. Gupta, "Distributed Fair Scheduling in a Wireless LAN", In Proc. Of ACM MOBICOM2000, Boston, MA USA(200) 167–178
4. K.H Choi, H.J. Shin and Dong-Ryeol Shin, "Delay and Collision Reduction Mechanism for Distributed Fair Scheduling in WIreless LANs", ICCSA, May, 2004.

A Quick Generation Method of Sequence Pair for Block Placement

Mingxu Huo and Koubao Ding

Dept. of Information and Electronic Engineering, Zhejiang University, Hangzhou 310027, P.R. China
{huomingxu,dingkb}@zju.edu.cn

Abstract. Sequence Pair (SP) is an elegant representation for the block placement of IC Design, and it is usually imperative to generate the SP from an existing placement. A quick generation method and one concise algorithm are proposed instead of the original unfeasible one. It is also shown that if the relations of any two blocks are either vertical or horizontal, the solution space size of a representation is $(n!)^2$ if it is P*-admissible. The analytical and experimental results of the algorithm both show its superiority in running time.

1 Introduction

With the rapid increase of IC complexity, floorplanning and block placement become more important, where floorplanning can be regarded as placement with soft module blocks. Most of such complex combinatorial optimization problems are NP-Hard [1], and therefore heuristic approaches such as Simulated Annealing algorithms are widely used to generate good layouts. The representation of geometrical topological relations of blocks is one of the crucial factors in evaluating values of the cost functions. Without some additional procedures, most representations proposed in the literature usually cannot include optimal solution, while Sequence Pair (SP) [1], BSG [2], TCG [3] and TCG-S [4] etc are P*-admissible representations [1, 4] that can represent the most general floorplans and contain a complete structure for searching an optimal solution. The original method to generate SP is named Gridding [1], which is too complicated to be implemented. However, it is imperative to find an effective and efficient approach to generate SP from arbitrary existing placement.

2 Generation Method

The topological relations of any two non-overlap module blocks are horizontal and vertical, i.e., left to, right to, above and below [5]. Diagonal relations also exist, but they can be simply degenerated by preferring horizontal relations to vertical ones, or by adopting the relation definitions in [3]. An HV-Relation-Set (HVRS) for a set of blocks is a set of horizontal or vertical relations for all block pairs [6], and a Feasible-HVRS involves all the block relations excluding the non-realizable ones.

A Sequence Pair (SP) is an ordered pair of sequences Γ_+ and Γ_-, each of which is a sequence of n block names [1], e.g., in Fig. 1 is one of the possible placements of the

V.S. Sunderam et al. (Eds.): ICCS 2005, LNCS 3516, pp. 954 – 957, 2005.

apte circuit from MCNC benchmark[1] circuits, the SP from which is (Γ_+, Γ_-) = (670185423, 120634785). A sequence pair corresponds to a Feasible-HVRS.

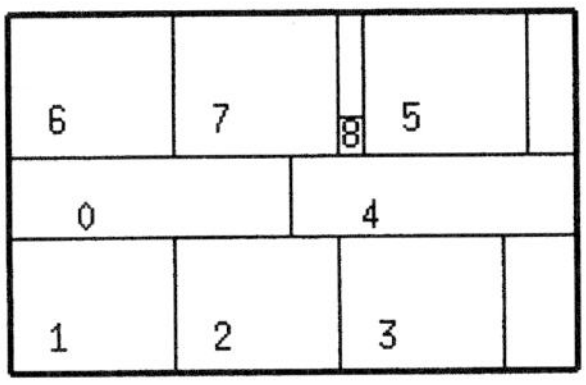

Fig. 1. A placement of the apte MCNC benchmark circuit. Its SP is (670185423, 120634785)

A generation plane is an infinite plane with numerous different positions. All blocks are treated as symbol points, each of which will be embedded in one unique position in the generation plane according to the Feasible-HVRS, and each embedded symbol point introduces two slope lines crossing at the symbol point, as in Fig. 2(a). The totality of all the slope lines introduced by embedded symbol points is an oblique grid, which divide the generation plane into different positions (Fig. 2(b)).

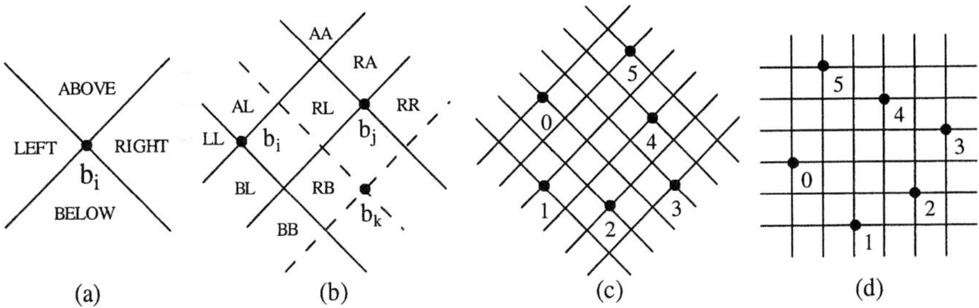

Fig. 2. The generation plane. (a) divided into 4 positions by b_i (b) 9 positions by b_i and b_j, 16 positions after b_k embedded (c) oblique grid after 6 symbols embedded (d) after rotated by –45 degree. The SP of these six blocks is (051423,120345)

Take an arbitrary order of the n blocks, and then embed them orderly. For blocks b_i and b_j, b_j must be embedded in one of the 4 unique positions according to b_i, named: LEFT, BELOW, RIGHT and ABOVE (Fig. 2(a)), to satisfy their relation. Blocks must satisfy their relations with all embedded ones by locating at a proper position.

When all blocks are embedded, there will be 2n lines and n^2 intersection points. The n symbol points locate on n of n^2 intersection points, and no two symbols locate on the same line simultaneously. Fig. 2(c) shows the generation plane after blocks 0 to 5 of Fig. 1 are embedded. As can be easily seen from Fig. 2(d), the sequences of the block names in column (on the vertical lines) and in row (on horizontal lines) are the Γ_+ and Γ_- of the Sequence Pair to be generated, respectively.

[1] See: http://www.cse.ucsc.edu/research/surf/GSRC/MCNCbench.html

3 A Generation Algorithm

Make use of two linked list spx and spy to represent the Γ_+ and Γ_- respectively, and generate spx and spy separately by inserting the block indices into the lists. The relations according to their Feasible-HVRS define the orders of two blocks in the lists respectively, and in either list, b_i is before b_k if b_i is before b_j and b_j is before b_k. The following program section illustrates the construction of spx.

Construction of list spx in C++ (the block indices are 0,1,2,...,n-1)

```
spx.push_back(0);                      //insert the first block
for(i = 1; i < n; ++ i ) {             //insert others orderly
for(posx = spx.begin(); posx != spx.end();++posx) {
j = *posx; r = relation(i,j); //check relations orderly
if(r == LEFT || r == ABOVE) { //found proper position
spx.insert(posx, i); break;     // insert and stop check
} } }
spx.push_back(i);                      // append to list end
}
```

4 Analytical and Experimental Results

The relation of every two given blocks has four possibilities. Without loss of generality, the block names to be embedded in sequence are assumed as 1, 2, 3, ... n. Before embedding the kth block, every position introduced by the k–1 embedded block symbols is different from others, providing the proper number of unique positions.

Let p_k denote the different ways for the kth block to be embedded, which is evidently k^2. This generation process comprises the sequence to embed blocks one by one, so the total ways for the embedment of n blocks can be calculated by multiplying all p_k, and thus the solution space size of P*-admissible representations is $(n!)^2$.

In general, it needs n(n-1)=O(n^2) time complexity to determine the Feasible-HVRS of n blocks. All the positions in the list can be assumed equiprobable, and thus the average comparison times for inserting the block b_{k+1} into a list with k blocks is:

$$C_{k+1} = \frac{1}{k+1}\sum_{i=0}^{k} i = \frac{k}{2} \quad (k = 0,1,2,...,n-1) \tag{1}$$

Replacing k + 1 with k, let C_k denote the average comparison times for the kth block, the total average comparison times C to sum up is in Eq. (2). Although its complexity is still O(n^2), C in Eq. (2) is only a quarter of the former n(n-1) running time.

$$C = \sum_{k=1}^{n} C_k = \sum_{k=1}^{n} \frac{k-1}{2} = \frac{1}{4}n(n-1) \tag{2}$$

Table 1 shows the running process of the algorithm to generate the spx from the placement in Fig. 1. The count of comparisons is 20, approximating to a quarter of n(n-1)=9*8=72. If complicated data structures employed, it will be much faster.

Table 1. Run of the algorithm after block 0 inserted in list spx. A, B, L and R stand for relations between two blocks of Above, Below, Left to and Right to respectively

#	Comparisons	List	#	Comparisons	List
1	B0	0 *1*	**5**	R0, R1, A4	0 1 *5* 4 2 3
2	B0, R1,	0 1 *2*	**6**	A0	*6* 0 1 5 4 2 3
3	R0, R1, R2	0 1 2 *3*	**7**	R6, A0	6 *7* 0 1 5 4 2 3
4	R0, R1, A2	0 1 *4* 2 3	**8**	R6, R7, R0, R1, L5	6 7 0 1 *8* 5 4 2 3

Experiments on all the MCNC benchmarks are run repeatedly in comparison with the generation routine from Parquet [7] on running time. Our results save average 91.5% time. And the more complex the circuit is, the more time is saved.

5 Conclusion

In this paper, we introduce a new method to generate Sequence Pair, and show on the basis of the method that for any P*-admissible placement representation if it takes for granted that every two blocks only have vertical and horizontal relations, the size of its solution space should be as large as $(n!)^2$. A quick generation algorithm is also proposed according to the method to reduce the running time. Analytical and experimental results of the algorithm both show its high performance.

References

1. Murata, H., et al., Rectangle-packing-based module placement. 1995 IEEE/ACM ICCAD. Digest of Technical Papers, (1995):. 472-479.
2. Nakatake, S., et al., Module packing based on the BSG-structure and IC layout applications. IEEE TCAD Of Integrated Circuits And Systems, (1998). 17(6): 519-530.
3. Jai-Ming, L. and C. Yao-Wen, TCG: a transitive closure graph-based representation for non-slicing floorplans. Proceedings of the 38th DAC, (2001): 764-769.
4. Lin, J.M. and Y.W. Chang, TCG-S: Orthogonal coupling of P*-admissible representations for general floorplans, in 39th DAC, Proceedings. (2002). 842-847.
5. Onodera, H., Y. Taniguchi, and K. Tamaru, Branch-and-bound placement for building block layout. 28th ACM/IEEE DAC. Proceedings 1991, (1991): 433-439.
6. Murata, H., et al., A mapping from sequence-pair to rectangular dissection. Proceedings of the ASP-DAC '97. 1997, (1997): 625-633.
7. Adya, S.N. and I.L. Markov, Fixed-outline floorplanning through better local search. Proceedings 2001 IEEE ICCD: VLSI in Computers and Processors. 2001, (2001): 328-334.

A Space-Efficient Algorithm for Pre-distributing Pairwise Keys in Sensor Networks*

Taekyun Kim[1], Sangjin Kim[2], and Heekuck Oh[1]

[1] Department of Computer Science and Engineering, Hanyang University, Korea
{tkkim, hkoh}@cse.hanyang.ac.kr
[2] School of Internet Media Engineering, Korea University of Technology and Education, Korea
sangjin@kut.ac.kr

Abstract. We propose a space-efficient key pre-distribution scheme based on quasi-orthogonal finite projective plane. This approach, compared to the previous approaches, guarantees full connectivity and the uniqueness of pairwise keys. Moreover, the size of the key ring depends not on the size of the key pool, but on the size of the network. The actual order of the key ring size is only $O(\sqrt{N})$, where N is the size of the network. As a result, our approach provides better scalability than previous approaches.

1 Introduction

We are concerned with establishing pairwise symmetric keys for sensor nodes. These keys are required to provide authenticity of sensor nodes and secrecy of information exchanged between them. Current approach is to initialize each sensor node with some secret information before deployment [1, 2, 3]. After deployment, these sensor nodes perform a protocol with neighboring nodes to establish the required pairwise keys. Some systems [2] need boostrapping to create the new pairwise keys, whereas the other systems [1, 3] pre-distribute randomly selected pairwise keys. Our method is different from the previous probabilistic methods [1, 3] in that we store keys in sensor nodes in a deterministic way.

2 The Proposed Approach

Definition 1. *An **FPP (Finite Projective Plane)** is a geometrical system* $\mathscr{G} = (\mathscr{P}, \mathscr{L}, \mathscr{I})$ *satisfying the following conditions, where* $\mathscr{P}$ *is a finite set of points,* $\mathscr{L}$ *is a finite set of lines, and* $\mathscr{I}$ *is an incidence relation between* $\mathscr{P}$ *and* $\mathscr{L}$.

- *For all points* $P, Q \in \mathscr{P}$ *with* $P \neq Q$*, there exists a unique line* $l \in \mathscr{L}$ *that passes through* P *and* Q.
- *There exist at least 3 points.*
- *For every line* $l \in \mathscr{L}$*, there exists a point not incident with* l.
- *Every line passes through at least 3 points.*
- *Every pair of distinct lines intersect.*

* This work was supported by the Ministry of Information and Communication, Korea, under the university HNRC-ITRC program supervised by the IITA.

V.S. Sunderam et al. (Eds.): ICCS 2005, LNCS 3516, pp. 958–962, 2005.

$\{1,2,3\}\{1,4,5\}\{1,6,7\}\{2,4,6\}$
$\{2,5,7\}\{3,5,6\}\{3,4,7\}$

(a) order $n = 2$

$\{1,2,3,4\}\{1,5,6,7\}\{1,8,9,10\}\{1,11,12,13\}\{2,5,10,12\}$
$\{2,6,8,13\}\{2,7,9,11\}\{3,5,9,13\}\{3,6,10,11\}\{3,7,8,12\}$
$\{4,5,8,11\}\{4,6,9,12\}\{4,7,10,13\}$

(b) order $n = 3$

Fig. 1. Example of FPPs

$N_1 : \{K_1,K_2,K_3\}$ $N_4 : \{K_1,K_4,K_5\}$
$N_6 : \{K_1,K_6,K_7\}$ $N_2 : \{K_2,K_4,K_6\}$
$N_5 : \{K_2,K_5,K_7\}$ $N_3 : \{K_3,K_5,K_6\}$
$N_7 : \{K_3,K_4,K_7\}$

(a) Using single FPP

$N_1 : \{K_1,K_2,K_3\},\{K'_3,K'_4,K'_7\}$ $N_4 : \{K_1,K_4,K_5\},\{K'_1,K'_2,K'_3\}$
$N_6 : \{K_1,K_6,K_7\},\{K'_1,K'_4,K'_5\}$ $N_2 : \{K_2,K_4,K_6\},\{K'_1,K'_6,K'_7\}$
$N_5 : \{K_2,K_5,K_7\},\{K'_2,K'_4,K'_6\}$ $N_3 : \{K_3,K_5,K_6\},\{K'_2,K'_5,K'_7\}$
$N_7 : \{K_3,K_4,K_7\},\{K'_3,K'_5,K'_6\}$

(b) Using two FPPs

Fig. 2. Construction of key rings for each node using an FPP of order $n = 2$

Definition 2. *For any $n \geq 2$, if each line of an FPP is incident with exactly $n+1$ points, we call this FPP a* ***finite projective plane of order*** *n.*

If $\mathcal{G}$ is an FPP of order n, then there are n^2+n+1 points and lines in total. An FPP of order n does not exists for all $n \geq 2$. However, it is known that FPPs of order $n = p^k$ exists, where p is a prime number and k is a positive integer. If we map points to numbers and lines to sets, a finite projective plane of order n can be shown as Fig. 1. There are many FPPs for a given order. Each FPP in Fig. 1 is just one of them.

In our method, the lines of an FPP is mapped to sensor nodes and points on a line to the key ring assigned to the corresponding node. For example, Fig 2.(a) shows the key ring assigned to each node when we use a single FPP of order $n = 2$. We denote N_i as a sensor node ID, and K_i as a symmetric key. In this case, the maximum number of nodes we can deploy is $N = n^2+n+1$ and the size of key ring in each node is $K = n+1$. The relationship between N and K is $N = K(K-1)+1$. Since every pair of distinct lines intersect in an FPP, each pair of node is guaranteed to have a single common key in their respective key rings. However, if we use this common key as the pairwise key, all pairwise keys are not unique. For example, the pairwise key of pairs (N_1,N_4), (N_1,N_6), (N_4,N_6) are all the same.

To make pairwise keys unique, we use an another FPP as shown in Fig 2.(b). Since we used two FPPs, each pair of nodes have a pair of common keys. For example, N_1 and N_4 shares K_1 and K'_3. We use $h(K_1||K'_3)$ as the pairwise key between N_1 and N_4, where h is collision-resistant hash function from $\{0,1\}^*$ to $\{0,1\}^k$, where k is the key length in bits. However, if we randomly assign the second FPP without any constraint, we may still get pairs that uses the same pairwise keys. For example, in Fig 2.(b), N_3, N_4, and N_5 all have K_5 and K'_2 in common. This problem occurs if and only if a triple of nodes that share a common key in the first FPP also share a common key in the second FPP. This problem is independent of the order of FPP used. That is, the problem occurs if this condition holds without regard to the order of FPP used.

To remedy this problem, we interchange the nodes and keys in the FPP. Since FPP satisfies the principle of duality, the resulting plane is also an FPP. We interchange them because we are not concerned with what specific keys are in each node's key ring but with the combinations of nodes that share a common key. Fig 3.(a) shows an interchanged FPP of Fig 2.(a). To the right of the FPP, we show the combinations of nodes that share the same key. To make each pairwise keys unique, no two combinations appearing on the same row in the first FPP must not also appear on any row in the second FPP. The FPP of Fig 3.(b) satisfies this condition. We say that FPP of Fig 3.(a) and Fig 3.(b) are quasi-orthogonal to each other. More formally, quasi-orthogonality of FPP is defined as below.

(a)	(b)
$\{N_1,N_4,N_6\}:(1,4),(1,6),(4,6)$	$\{N_1,N_2,N_4\}:(1,2),(1,4),(2,4)$
$\{N_1,N_2,N_5\}:(1,2),(1,5),(2,5)$	$\{N_1,N_3,N_6\}:(1,3),(1,6),(3,6)$
$\{N_1,N_3,N_7\}:(1,3),(1,7),(3,7)$	$\{N_3,N_4,N_7\}:(3,4),(3,7),(4,7)$
$\{N_2,N_4,N_7\}:(2,4),(2,7),(4,7)$	$\{N_2,N_3,N_5\}:(2,3),(2,5),(3,5)$
$\{N_3,N_4,N_5\}:(3,4),(3,5),(4,5)$	$\{N_4,N_5,N_6\}:(4,5),(4,6),(5,6)$
$\{N_2,N_3,N_6\}:(2,3),(2,6),(3,6)$	$\{N_1,N_5,N_7\}:(1,5),(1,7),(5,7)$
$\{N_5,N_6,N_7\}:(5,6),(5,7),(6,7)$	$\{N_2,N_6,N_7\}:(2,7),(2,6),(6,7)$

Fig. 3. Quasi-orthogonal FPPs of order $n = 2$

Definition 3. *Two FPPs $\mathscr{G}_1 = (\mathscr{P}_1, \mathscr{L}_1, \mathscr{I}_1)$ and $\mathscr{G}_2 = (\mathscr{P}_2, \mathscr{L}_2, \mathscr{I}_2)$ with the same order are* ***quasi-orthogonal*** *if the following conditions hold:*

1. *$\mathscr{P}_1 = \mathscr{P}_2$,*
2. *For every line $l_1 \in \mathscr{L}_1$ and $l_2 \in \mathscr{L}_2$, there exist at most two points incident with both l_1 and l_2.*

Fig 4 shows two quasi-orthogonal FPPs of order $n = 3$.

$\{N_1,N_5,N_8,N_{11}\},\{N_1,N_2,N_6,N_7\}$	$\{N_1,N_3,N_5,N_{12}\},\{N_1,N_4,N_7,N_8\}$
$\{N_1,N_3,N_9,N_{10}\},\{N_1,N_4,N_{12},N_{13}\}$	$\{N_1,N_6,N_{10},N_{11}\},\{N_1,N_2,N_9,N_{13}\}$
$\{N_2,N_4,N_5,N_9\},\{N_5,N_6,N_{10},N_{12}\}$	$\{N_2,N_3,N_4,N_{10}\},\{N_2,N_5,N_7,N_{11}\}$
$\{N_3,N_5,N_7,N_{13}\},\{N_3,N_4,N_6,N_8\}$	$\{N_2,N_6,N_8,N_{12}\},\{N_3,N_6,N_7,N_9\}$
$\{N_7,N_8,N_9,N_{12}\},\{N_2,N_8,N_{10},N_{13}\}$	$\{N_3,N_8,N_{11},N_{13}\},\{N_4,N_5,N_6,N_{13}\}$
$\{N_4,N_7,N_{10},N_{11}\},\{N_2,N_3,N_{11},N_{12}\}$	$\{N_4,N_9,N_{11},N_{12}\},\{N_5,N_8,N_9,N_{10}\}$
$\{N_6,N_9,N_{11},N_{13}\}$	$\{N_7,N_{10},N_{12},N_{13}\}$

Fig. 4. Quasi-orthogonal FPPs of order $n = 3$

Theorem 1. *Given two quasi-orthogonal FPPs of order n, there is a way to assign key rings to nodes in a network of size n^2+n+1 or smaller, so that all of the pairwise keys are unique. Furthermore, the size of key ring in each node is $2n+2$.*

Proof. If we use two FPPs, uniqueness of pairwise keys are violated if and only if the followings are satisfied.

1. 3 nodes share a common key in the first FPP.
2. These 3 nodes also share a common key in the second FPP.

Since the quasi-orthogonality of FPPs guarantees that no three nodes share a common key in both FPPs, the argument is correct.

The followings are the steps used to construct key rings for sensor nodes of size N.

i) **Step 1:** Construct an FPP of order $n = p^k \geq \sqrt{N}$, where p is prime and k is a positive integer. Given an FPP of order n, the maximum supportable network size is n^2+n+1. Therefore, if $n \geq \sqrt{N}$, the maximum supportable network size is larger than N. If we consider incremental addition, the number of nodes that will be deployed after initial deployment must be taken into consideration.
ii) **Step 2:** Find a quasi-orthogonal FPP of the FPP constructed in step 1.
iii) **Step 3:** Assign a key to each row of the two quasi-orthogonal FPPs. Since the assigned keys must be different from each other, the total number of keys used is $2N = 2n^2+2n+2$.
iv) **Step 4:** Find the dual FPPs by interchanging keys and nodes of the two quasi-orthogonal FPPs constructed in step 3.
v) **Step 5:** Assign a pair of key rings to each node. Without loss of generality, this procedure can be done at random, i.e., the node ID is meaningless at this point. Furthermore, there is no restriction for making a pair of key rings as long as each ring is from the different dual FPPs. This allows us $N!$ choices of assigning key ring pairs to nodes. From the principle of duality, the key ring size of each node is $2K = 2n+2$.

This completes the proof. □

$N_1: \{K_1,K_2,K_3,K_4\},\{K'_1,K'_2,K'_3,K'_4\}$ $N_2: \{K_1,K_5,K_6,K_7\},\{K'_1,K'_6,K'_{10},K'_{12}\}$
$N_8: \{K_1,K_8,K_9,K_{10}\},\{K'_2,K'_7,K'_9,K'_{12}\}$ $N_{11}: \{K_1,K_{11},K_{12},K_{13}\},\{K'_3,K'_6,K'_9,K'_{11}\}$
$N_2: \{K_2,K_5,K_{10},K_{12}\},\{K'_4,K'_5,K'_6,K'_7\}$ $N_6: \{K_2,K_6,K_8,K_{13}\},\{K'_3,K'_7,K'_8,K'_{10}\}$
$N_7: \{K_2,K_7,K_9,K_{11}\},\{K'_2,K'_6,K'_8,K'_{13}\}$ $N_9: \{K_3,K_5,K_9,K_{13}\},\{K'_4,K'_8,K'_{11},K'_{12}\}$
$N_{10}: \{K_3,K_6,K_{10},K_{11}\},\{K'_3,K'_5,K'_{12},K'_{13}\}$ $N_3: \{K_3,K_7,K_8,K_{12}\},\{K'_1,K'_5,K'_8,K'_9\}$
$N_4: \{K_4,K_5,K_8,K_{11}\},\{K'_2,K'_5,K'_{10},K'_{11}\}$ $N_{12}: \{K_4,K_6,K_9,K_{12}\},\{K'_1,K'_7,K'_{11},K'_{13}\}$
$N_{13}: \{K_4,K_7,K_{10},K_{13}\},\{K'_4,K'_9,K'_{10},K'_{13}\}$

Fig. 5. Key rings allocated to each node when using two FPPs of order $n = 3$

Fig 5 shows key rings assigned to each node when using two quasi-orthogonal FPPs depicted in Fig 4. Here, we assumed that first set of nodes $\{N_1,N_3,N_5,N_{12}\}$ share K'_1, the second set of nodes $\{N_1,N_4,N_7,N_8\}$ share K'_2 and so on. However, the order of assigning keys to second FPP is irrelevant and does affect the uniqueness of pairwise keys. The size of key ring assigned to each node is $2K \approx 2\sqrt{N}$.

3 Conclusion

In this paper, we have shown that it is possible to deterministically construct a key ring for sensor nodes having following characteristics: (i) it guarantees fully direct connectivity, (ii) it guarantees uniqueness of pairwise keys, (iii) the size of the key ring stored at each node depends on the size of the network, whereas previous approach depends on the size of the key pool, (iv) the order of the key ring stored at each node is $O(\sqrt{N})$, where N is the size of the network.

References

1. Chan, H., Perrig, A., Song, D.: Random Key Predistribution Schemes for Sensor Networks. Proc. of the IEEE Symp. on Security and Privacy, IEEE Press (2003) 197–215
2. Deng, J., Han, R., Mishra, S.: Security Support For In-Network Processing in Wireless Sensor Networks. Proc. of the 1st ACM Workshop on the Security of Ad Hoc and Sensor Networks, ACM Press (2003), 83–93
3. Eschenauer, L., Gligor, V.D.: A Key-Management Scheme for Distributed Sensor Networks. Proc. of the 9th ACM Conference on Computer and Communications Security, ACM Press (2002), 41–47

An Architecture for Lightweight Service Discovery Protocol in MANET*

Byong-In Lim, Kee-Hyun Choi, and Dong-Ryeol Shin

School of Information and Communication Engineering,
Sungkyunkwan University
440-746, Suwon, Korea, +82-31-290-7125
{lbi77, gyunee, drshin}@ece.skku.ac.kr

Abstract. As the number of Internet services grows, it is becoming more important for the network users to be able to locate and utilize those services that are of interest to them. As a result, a service discovery protocols are coming to play an increasingly important role in highly dynamic networks. In this paper, we describe a new architecture for service discovery called LSD (Lightweight Service Discovery). The LSD is a peer-to-peer cache-based service discovery protocol for ad hoc environments. Through the LSD protocol, small handheld mobile devices with wireless connectivity can detect each other in their neighboring devices, and perceive whether particular shared services are available or not. In addition, the LSD solves the overhead problems which arise when traditional discovery protocols are applied to mobile ad hoc networks.

1 Introduction

As the number of Internet services grows, service discovery protocols are coming to play an increasingly important role in highly dynamic networks. Consequently, a number of service discovery protocols have been proposed for wired networks. Prominent among these [1] are the SLP, Salutation, UPnP and UDDI. These conventional approaches to service discovery protocols can be classified into two main categories. These systems were mostly designed for an administrated network context and are either based on mediator based discovery approaches (e.g. UDDI, Jini) or on maintaining a network-wide multicast tree based on direct or peer-to-peer discovery approaches (e.g. UPnP, Salutation, SLP, when operated without directory agents). In ad hoc networks, the provision for service discovery is mandatory since they are designed for irregular and unexpected changes in network topology. In a service discovery environment, basic services that are readily available like discovery and delivery in wired networks have to be re-designed in order to take into account the wireless network's properties. Furthermore, whenever the consumer rearranges the components in such a system, the system must automatically adapt its configuration as necessary. In order to satisfy these diverse requirements, we designed a modified framework based on the existing service discovery algorithms [2, 3] and

* This research was partially supported by a grant from the CUCN, Korea and Korea Science & Engineering Foundation (R01-2004-000-10755-0).

V.S. Sunderam et al. (Eds.): ICCS 2005, LNCS 3516, pp. 963–966, 2005.

implemented a lightweight service discovery protocol suitable for ad-hoc environments.

The remainder of the paper is organized as follows. Section 2 discusses the LSD Architecture. Service Propagation Algorithm is described in section 3. Finally, this paper is concluded in section 4.

2 Design of LSD Architecture

Figure 1 shows the LSD (Lightweight Service Discovery) architecture and corresponding components in a single mobile device.

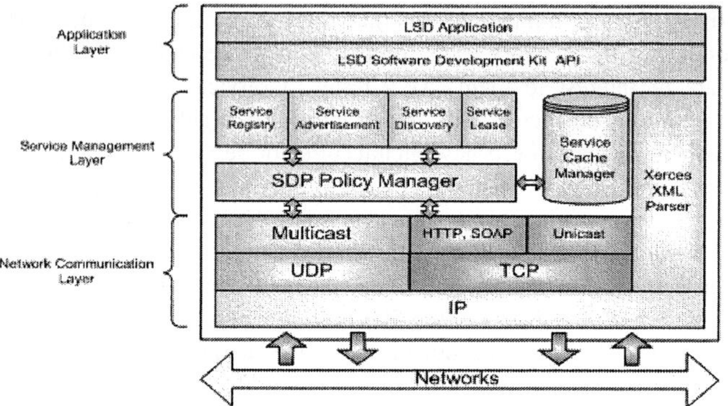

Fig. 1. LSD System Architecture

In Figure 1, the LSD consists of three layers: the application layer, the service management layer, and the network communication layer. The application layer provides the user with applications such as audio-video players and printing applications. The service management layer provides services associated with discovery. In the service management layer, we use a cache manager that registers information about the services offered by peers in the network. All of the devices in the network should listen to all advertisement messages, known as delta messages and save them in their local caches for a given period of time. When a user wants to find a particular service, they begin by searching for the service in their local caches at first. If no information is available locally, the user sends a request message through multicasting. The SDP Policy Manager in the layer is responsible for enforcing policies designed to control platform behavior. These policies are registered with the SDP Policy Manager which is responsible for ensuring that all components of the platform are in compliance with the specified policies. These policies can specify caching preferences such as the refresh rate or replacement strategy or describe advertisement preferences such as the frequency or time-to-live, etc. The network communication layer is associated with two protocols. The first is UDP which is used for multicasting; the second is TCP which is used unicasting. We send advertise/request messages by using UDP, and retrieve the data using http or soap based on TCP.

3 Service Propagation Algorithm of LSD

The LSD pays attention to an efficient service discovery to exploit the nature of highly dynamic ad-hoc networks. Based on the Konark [3], it has developed a new algorithm which attempts to balance the convergence time and network bandwidth. The key difference between Gossip-Konark and LSD is service data structure and the convergence algorithm involved in joining a network. The Gossip-Konark has a structure of tree-based hierarchal services. On the other hand, the LSD has a structure of flat services that have a difference in patterns and message procedures of Query, Advertisement. Moreover, the LSD uses a dummy advertisement messages. When a node joins to new network, if it has no local service, it sends the dummy advertisement message. When a node receives a service message, it sends the difference between its relevant services and service information in the received service messages, called a delta message. Thus network overload is reduced. It also avoids the storm of concurrent multicasts by randomly assigning the multicast time interval. When a node receives the advertisement message, the LSD prevents multicast storm problem using RNDWAIT time similar to the Konark. Moreover, if the node which receives the service message verifies that its service registry has the same information as that in the received service message, it then remains silent and no multicasts occur. Figure 2 shows the message transmission procedure for the LSD. The incremental discovery procedure is able to reduce network traffic for service discovery due to delta messages.

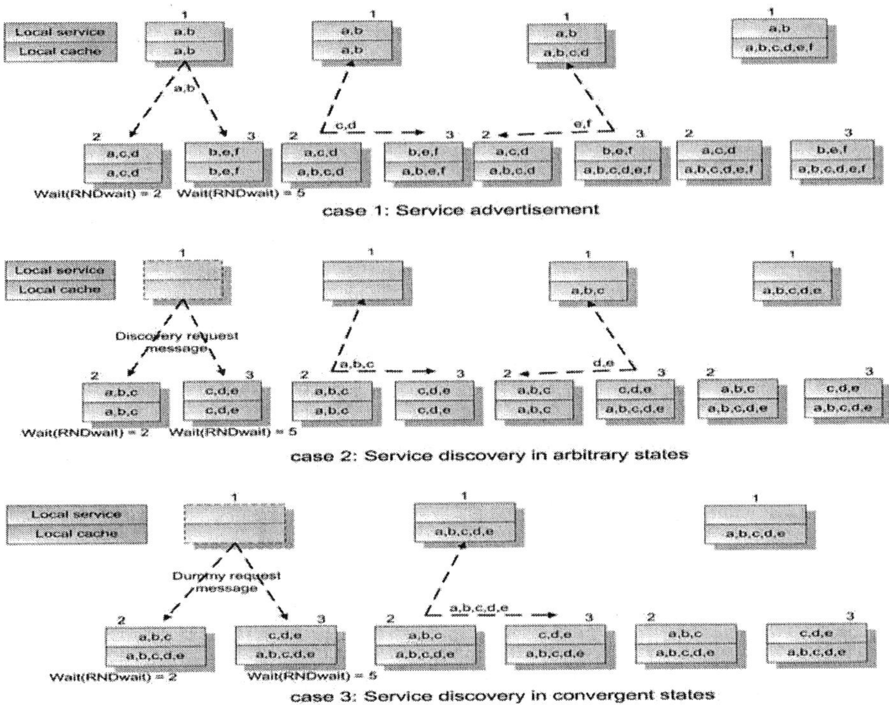

Fig. 2. Service Propagation Algorithm in LSD

Suppose that an ad-hoc network of interest consists of three nodes: node 1, node 2, and node 3. Assume that node 2 and node 3 have random waiting times (RNDWAIT) of 2 and 5 seconds, respectively. We have three cases according to the different situations as illustrated in Figure 2.

Case 1: service advertisement

(a) Node 1 sends its own service advertisement message to other nodes using the multicast. At this time, node 2 can send delta message based on the advertisement earlier than node 3, because node 2 has RNDWAIT=2, whereas node 3 has RNDWAIT=5.

(b) Listening to advertisement and delta messages from other nodes, node 3 sends service e, f to other nodes, because node 3 already receives services a, b, c, d.

Case 2: service discovery in arbitrary states, i.e., non-convergent states

(a) When node 1 joins to a new network, node 1 sends out a dummy request messages to other nodes. From these messages, node 2 multicasts a delta message ahead of other nodes, for the same reason as the case 1.

(b) Node 3 makes a delta message based on the request and then sends it to other nodes, which results in fast discovery.

Case 3: service discovery in convergent states (where all nodes know the service information within the network)

(a) When node 1 joins to a new network in convergent state, node 1 sends out a dummy request message to other nodes. Then node 2 can send delta message earlier than other nodes as illustrated as in case 1 and case 2.

(b) Since the network already reaches the convergence of service information, the number of multicasts and resulting traffic road is reduced. One dummy request message is enough to reach the convergent state.

4 Conclusion

Unfortunately existing approaches to service trading are not well suited for these highly dynamic topologies since they rely on centralized servers and have high communication costs due to periodic query flooding. In this paper, we proposed a novel scheme to correct and optimize discovery protocol in case of ad-hoc environments. Currently we are performing a simulation and doing a comprehensive performance test for our prototyped system.

References

1. F. Zhu, M. Mutka, and L. Ni, "Classification of Service Discovery in Pervasive Computing Environments," Michigan State University, East Lansing, available at http://www.cse.msu.edu/~zhufeng/ServiceDiscoverySurvey.pdf MSU-CSE-02-24.
2. M. Nidd, "Service discovery in DEAPspace," IEEE Pers. Commun., vol.8, pp. 39–45, Aug. 2001.
3. C. Lee, A. Helal, N. Desai, V. Verma, B. Arslan, "Konark: A System and Protocols for Device Independent, Peer-to-Peer Discovery and Delivery of Mobile Services," IEEE System. Man. Cybernetic., vol. 33, no. 6, pp. 682-696, Nov. 2003

An Enhanced Location Management Scheme for Hierarchical Mobile IPv6 Networks

Myung-Kyu Yi

Dept. of Computer Science & Engineering Korea University,
1,5-Ga, Anam-Dong, SungBuk-Gu, Seoul 136-701, South Korea
kainos@disys.korea.ac.kr

Abstract. In this paper, we propose an enhanced location management scheme for minimizing signaling costs in hierarchical Mobile IPv6 (HMIPv6) networks. If the mobile node's mobility is not local, in our proposal, the mobile node sends location update messages to correspondent nodes in the same way as Mobile IPv6 (MIPv6). After the creation of a spatial locality of the mobile node's movement, the mobile node sends location update messages to the correspondent nodes in same way as HMIPv6. Therefore, our proposal can reduce signaling costs in terms of packet transmission delays in HMIPv6 networks. Therefore, our proposal offers considerable performance advantages to MIPv6 and HMIPv6.

1 Introduction

Mobile IPv6 (MIPv6) provides an efficient and scalable mechanism for host mobility within the Internet [1]. It allows an IPv6 node to be mobile and arbitrarily change its location on the IPv6 Internet while still maintaining existing connections. However, MIPv6 causes in a high signaling cost when it updates the location of an Mobile Node (MN) if it moves frequently [2]. Thus, Hierarchical Mobile IPv6 (HMIPv6) is proposed by IETF to reduce signaling costs [2]. It uses a new MIPv6 node called the Mobility Anchor Point (MAP) to handle Mobile IP registration locally. It is well known that the performance of HMIPv6 is better than that of MIPv6 [2]. This is especially true when the basic assumption is that 69% of a user's mobility is local. If the user's mobility is not local, performance of HMIPv6, in terms of delays for packet delivery, is worse than that of MIPv6, due to the encapsulation processing by the MAP. Since all packets from a Correspondent Node (CN) to an MN are first delivered through the MAP, it is possible that the MAP can become bottlenecked. Therefore, the load of the search and tunnelling processes increase on the MAP as the number of MNs increase in the foreign or home networks. It is a critical problem for the performance of HMIPv6 networks. To overcome these drawbacks, we will introduce an enhanced location management scheme, called AHMIPv6, for minimizing signaling costs in HMIPv6 networks. In our proposal, location registration with the Home Agent (HA) and MAP is exactly the same as that in HMIPv6. However, the MN sends a Binding Update (BU) message to the CN with either on-Link

V.S. Sunderam et al. (Eds.): ICCS 2005, LNCS 3516, pp. 967–970, 2005.

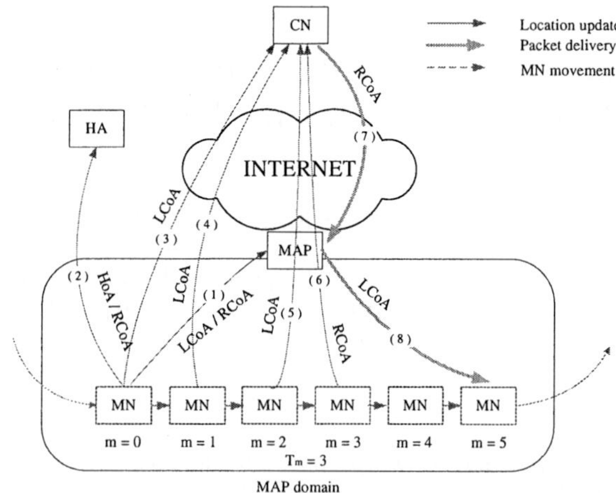

Fig. 1. The System Model for the Proposed Scheme in HMIPv6 Networks

Care-of-Address (LCoA) or Regional CoA (RCoA), based on the geographical locality properties of the MN's movements. To do so, we define m and T_m as the number of subnet crossings and the value of the movement threshold to decide whether the MN's mobility is local or not, respectively. If m is less than T_m (i.e., $m < 3$ in Fig. 1), the MN sends a BU message to the CN with the LCoA. Otherwise, if m is greater than or equal to T_m, the MN sends a BU message to the CN with the RCoA. Therefore, the CN can directly send packets to the MN without the intervention of the MAP, and before the creation of geographical locality properties of the MN's movements.

The rest of the paper is organized as follows. Section 2 describes the proposed procedures of location update and packet delivery using the enhanced mobility management scheme called AHMIPv6. Finally, our conclusions are presented in Section 3.

2 Protocol Description

This section describes the location update and packet delivery procedure.

2.1 Location Update Procedure

Similar to HMIPv6, each MN has two addresses, an RCoA and an LCoA. Each MN has a value of m and T_m. While m is the number of subnet crossings within the MAP domain, T_m is the threshold value which decides whether the MN sends a BU message to the CN with the LCoA or the RCoA. Whenever an MN enters a new MAP domain, it sets the value of m to zero. T_m can be adjusted based on the user's mobility pattern and current traffic load.

Fig. 2 presents the procedure for a location update by sending the BU message. The procedures for a location update are as follows:

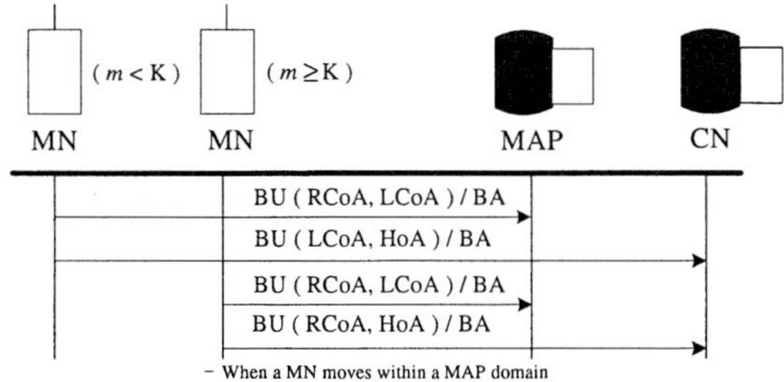

Fig. 2. The Proposed Location Update Procedure

- If an MN moves to a different MAP domain, then:

1) the MN obtains two CoAs: an LCoA and an RCoA.
2) Then, it registers with its MAP and HA by sending a BU message, and it sets the value of m to zero.

- Otherwise, if an MN moves within the same MAP domain, then:

1) the MN gets a new LCoA.
2) The MN registers with its MAP by sending a BU message.

After registration with the MAP, the MN compares the value of m with T_m.

Case 1. If the value of m is less than T_m, then:

3-1) the MN sends a $BU_{[HoA,LCoA]}$[1] message to the CN.

Case 2. Otherwise, if the value of m is greater than or equal to T_m, then:

3-2) the MN sends a $BU_{[HoA,RCoA]}$[2] message to the CN. After the sending of the $BU_{[HoA,RCoA]}$ message, the MN does not send any other BU messages to the CN before it moves out of the MAP domain.

As a result, the MN performs registration with the CNs using an RCoA or an LCoA, depending on its mobility pattern.

2.2 Packet Delivery Procedure

Fig. 3 presents the procedure for packet delivery in our proposal. In our proposal, the packet delivery procedure is exactly the same to that in MIPv6 or HMIPv6. The CN does not need to consider whether it has an RCoA or an LCoA of the MN in its binding cache entry. If a CN has an LCoA of the MN in its binding cache entry, it sends packets directly to an MN. Thus, the proposed scheme can achieve optimal routing, the same as MIPv6. Otherwise, if a CN has an RCoA

[1] BU with the binding between the MN's HoA and LCoA.
[2] BU with the binding between the MN's HoA and RCoA.

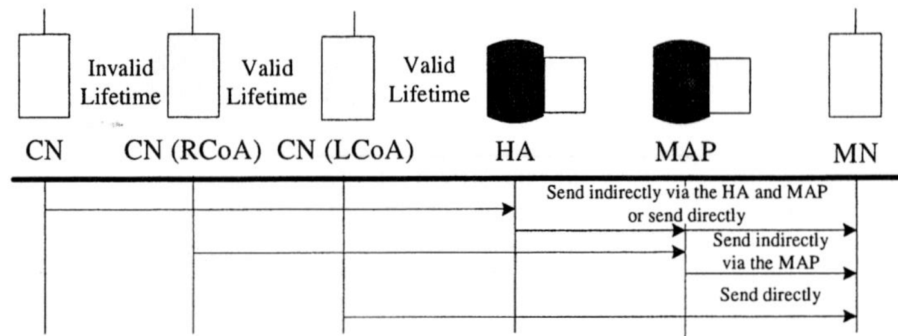

Fig. 3. The Proposed Packet Delivery Procedure

of the MN in its binding cache entry, it must send the packets to the MAP first using an RCoA. Then, the MAP encapsulates and forwards them directly to the MN. If a CN has no binding cache for the MN, it sends the packets to the home link, then the HA intercepts and tunnels them to the MN using the LCoA or RCoA.

3 Conclusions

In this paper, we proposed an enhanced location management scheme for minimizing signaling costs in HMIPv6 networks. In our proposal, location registration with the HA and MAP is exactly the same as that in HMIPv6. However, the MN sends a BU message to the CN with either an LCoA or an RCoA, based on the geographical locality properties of the MN's movements. To do so, we define m and T_m as the number of subnet crossings and the value of the movement threshold, to decide whether the MN's mobility is local or not, respectively. If m is less than T_m, the MN sends a BU message to the CN with an LCoA. Otherwise, if m is greater than or equal to T_m, the MN sends a BU message to the CN with an RCoA. Therefore, the CN can directly send packets to the MN without the intervention of the MAP, before the creation of geographical locality properties of the MN's movements. Therefore, our proposal achieves significant performance improvements by using the MN's selection to send a BU message to the CN, either with an LCoA or an RCoA.

References

1. D. B. Johnson, C. E. Perkins, and J. Arkko,"Mobility support in IPv6," IETF Request for Comments 3775, June 2004.
2. H. Soliman, C. Castelluccia, K. El-Malki, and L. Bellier, "Hierarchical Mobile IPv6 mobility management (HMIPv6)", IETF Internet draft, draft-ietf-mipshop-hmipv6-02.txt (work in progress), June 2004.

A Genetic Machine Learning Algorithm for Load Balancing in Cluster Configurations

M.A.R. Dantas and A.R. Pinto

Department of Informatics and Statistics,
Federal University of Santa Catarina,
88040-900 Florianopolis Brazil
{mario, arpinto}@inf.ufsc.br

Abstract. Cluster configurations are a cost effective scenarios which are becoming common options to enhance several classes of applications in many organizations. In this article, we present a research work to enhance the load balancing, on dedicated and non-dedicated cluster configurations, based on a genetic machine learning algorithm. Classifier systems are learning machine algorithms, based on high adaptable genetic algorithms. We developed a software package which was designed to test the proposed scheme in a master-slave Cow and Now environment. Experimental results, from two different operating systems, indicate the enhanced capability of our load balancing approach to adapt in cluster configurations.

1 Introduction

Dedicated and non-dedicated cluster configurations offer a cost/effective high-performance computing if idle resources from these clusters can be successful harnessed. Several package environments (such as OSCAR (Open Source Cluster Application Resource)[1] and Linux Virtual Server [2]) were developed to provide functions of a single system image. These environments improve the use of cluster architectures, providing many important mechanisms to help application programmers to achieve a high performance. The load balancing issues in distributed systems (such as cluster environments) have been studied by a number of researches and many points have already been addressed [3, 4, 5].

However, a well known problem in implementing software environments for single system image is the use of an effective technique to distribute processes to processors in a fair based policy [6]. As many research works have already demonstrated (e.g. [5, 6]), in a cluster of computers it is high the possibility of a specific node become heavily loaded and others lightweight loaded.

The load balancing problem is recognized as a NP-complete problem [7]. Therefore, application developers commonly use heuristic or stochastic methods because those two paradigms can obtain sub-optional results in a reasonable period of time. In the literature we can observe an expressive utilization of genetic algorithm.

V.S. Sunderam et al. (Eds.): ICCS 2005, LNCS 3516, pp. 971–974, 2005.

In this article, we present an approach to schedule processes using a classifier system. This method has an interesting feature of machine learning [8] based on a genetic algorithm. The results were obtained considering two configurations. The first configuration was a cluster of tightly machines (Cow) using Linux with the OSCAR single system image. The other environment was an ordinary network of workstations (Now) with loosely coupled machines executing the Windows operating system. The proposed system was implemented in Java.

The paper is organized as follows. Section 2 presents some concepts of scheduling and load balancing approaches. Concepts of genetic algorithms and classifier systems are shown in section 3. Our proposal is described in section 4. Experimental results of a parallel application are presented in section 5. Finally, in section 6 we draw some conclusions about the present research and our contribution, indicating some future directions for this research work.

2 Scheduling and Load Balancing in Parallel and Distributed Systems

The taxonomy presented in [4], considers the global scheduling as static and dynamic. The static approach requires previous knowledge of processes behavior and its dependence. Therefore, the static technique is more suitable for homogeneous environments, because this method knows processes behavior at the compilation time. On the other hand, the dynamic scheduling is used when we do not know any previous requirements of processes and environments. Thus, in this paradigm we should implement queries which interact frequently with the configuration to gather information of the load of the processors. In addition, a strict dynamic method considers the migration of processes when a processor became overload for some reason. However, this technique is a complex task and has a high computational cost (e.g. [9]). Therefore, it is also possible to implement a more relaxed approach of dynamic scheduling migration called *on time assignment scheduling* (OTS).

3 Genetic Algorithms and Classifier Systems

Genetic algorithms are designed to perform a search, based on mechanisms of natural selection. In nature the most adapted individuals have more chances to survive. In addition, they provide their genetic code to their descendants. An individual in a genetic algorithm is represented by its chromosome, which is usually a string. In a new generation, new artificial creatures (strings) are created using fragments of the most adapted individuals from the old generation.

A classifier system is a genetic based machine learning algorithm, that can learn syntactically simple rules (called classifiers) to guide its performance in an arbitrary environment [8]. A classifier system has 3 main modules: rule and message system, a special class of production system; apportionment of credit algorithm; genetic algorithm. A production system is a computational scheme that employs rules as the main algorithmic goal. Rules usually present the following form:

If <condition> then <action>

If the classifier system receives the message "101011" the first classifier will be activated and the action "100" will be executed. During the creation of a classifier system, all classifiers have the same fitness. When a classifier is chosen, it has to pay a certain amount of its fitness to the apportionment of credit algorithm. The amount of fitness is determined by a pre-defined tax. The paid amount is the product between the pre-defined tax and the classifier fitness. This amount will be paid in the next classifier system consultation, if the previous classifier sent an action that improves the system performance. Otherwise, the amount will be accumulated until a classifier gives a positive action. After a pre-defined number of consults, genetic algorithm is activated. The population of classifiers is replaced.

4 Proposed Environment

A large number of process scheduling researches that using genetic algorithm are based on the simulation of the environment. In contrast, we decided to design and implement a solution using real cluster configurations.

The first step in developing a global solution to schedule process, based on classifier systems, is to choose a distributed paradigm. In our case we decided to implement a master-slave approach. Our classifier system proposal is to develop a modified threshold method. This algorithm provides a process threshold indicating the number of processes which each can execute. We consider the average response time variation of processes as entry condition to the classifier. The configuration of classifiers will be: *<if a grow occurs or a decrement>* + *<variation rate of average response time>*:*<if the threshold should be incremented or decremented>*+*<threshold rate variation>*

5 Experimental Results

The performance of the classifier system was compared to random and threshold methods. In the random method a destination node is randomly chosen to receive processes. On the other hand, the threshold method pre-defines a maximum number of processes.

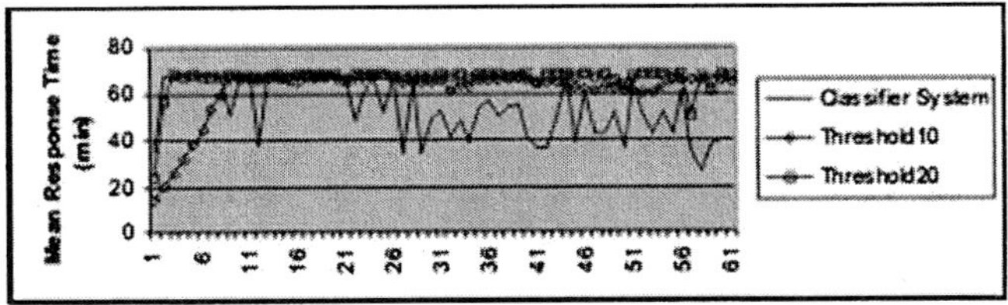

Fig. 1. Second experimental results

The first experiment was executed in the non-dedicated environment. In the last experiment we consider an OSCAR cluster configuration. The processes submitted to system in frits experiment was a for loop to calculate a sum. On the other hand, the processes submitted to OSCAR configuration calculated the prime numbers by Eratostenes ciev. We consider the average response time metric, therefore a small average response time mean a better performance. The second experiment was idealized for test the robustness of our proposal. Therefore, we execute in the dedicated cluster for 20 hours, with interval of 20 minutes. As figure 1 shows the system had an interesting behavior. The proposed classifier system shows a high adaptability that was our primarily goal.

6 Conclusion and Future Work

The experimental results present in this article show the high adaptive characteristic of our proposed classifier system implementation. In addition, it demonstrates the capability of learning about cluster configurations (non-dedicated and dedicated environments). These facts are confirmed by intervals where we had low performance compared to the following enhanced performance. As future work we are planning a scheme to store classifiers, therefore when the system begins the execution classifiers would be adapted to the reality which they will execute.

References

1. Benoit Ligneris, Stephen Scott, Thomas Naughton, Neil Gorsuch, Open Source Cluster Application Resource (OSCAR): Design, Implementation and interest for the [Computer] Scientific Community, First OSCAR Symposium, Sherbrooke May 11-14, Canada, 2003.
2. Zhang, W. (April, 2004) "Linux Virtual Server for Scalable Network Services", Available online: http://www.linuxvirtualserver.org/docs/scheduling.html.
3. Dantas, M.A.R. and Zaluska, E.J., Efficient Scheduling of MPI Applications on Networks of Workstations, FGCS Journal, V 13, pp. 489-499, 1998.
4. Casavant T.H. and Khul J.G., "A Taxonomy of Scheduling in General-Purpose Distributed Computing Systems", IEEE Trans. On Software Eng., vol. l4, no. 2, pp. 141-154, Feb. 1988.
5. Zhou S., "A Trace-driven Simulation Study of Dynamic Load Balancing", IEEE Trans. On Software Eng., vol. l4, no. 9, pp. 1327-1341, sep. 1988.
6. Dantas, M.A.R., Queiroz W.J. and Pfitscher G.H., "An Efficient Threshold Approach on Distributed Workstation Clusters", in HPC in Simulation, pp. 313-317, Washington, USA, 2000.
7. Papadimitriou, C. and Steilglitz, K., Combinatorial Optimization: Algoritms and Complexity, Dover Publications, 1998.
8. Goldberg, D.E., "Genetic Algorithm in Search, Optimization, and Machine Learning", Addison-Wesley, 1989.
9. Powell, M.L. and Miller, B.P., "Process migration in DEMOS/MP", in Proc. 9th ACM Symp. Operat. Syst. Principles, 1983, pp.110-119.

A Parallel Algorithm for Computing Shortest Paths in Large-Scale Networks*

Guozhen Tan and Xiaohui Ping

Department of Computer Science, Dalian University of Technology,
Dalian, 116023, P.R. China
gztan@dlut.edu.cn

Abstract. This paper presents the Optimality Theorem in distributed parallel environment. Based on this theorem, a parallel algorithm using network-tree model is presented to compute shortest paths in large-scale networks. The correctness of this algorithm is proved theoretically and a series of computational test problems are performed on PC cluster. Factors such as network size and level of the network, which take effect on the performance, are discussed in detail. Results of the experiments show that the proposed parallel algorithm is efficient in computing shortest paths in large-scale networks, especially when the network size is great.

1 Introduction

The recent research interest in large-scale networks has been focused on partitioning the network and constructing new network model. The previous contributions are the Level graph model[2][3], the Hierarchical Encoded Path View (HEPV) model[4] and the Network-Tree Model (NTM)[5]. These models all have some defects. The Network-Tree Model can be constructed from arbitrary network, and the route optimization can be constrained within a few sub-networks, which greatly reduce the searching scope and significantly improve the computing efficiency.

In this paper, we present a parallel algorithm based on network-tree model to solve the problem and test it in a series of networks. The experiment results validate that the parallel algorithm achieve high performance in speedup and efficiency.

2 Network-Tree Model

Network-Tree Model is the definitional combination of network and tree.Any network can be divided into a number of sub-networks, and each sub-network will be constructed to be one node of network-tree. We use macro-node to denote the node of network-tree to distinguish from the node of original network.

* This work is supported by the National Science Foundation of China(60373094).

V.S. Sunderam et al. (Eds.): ICCS 2005, LNCS 3516, pp. 975–978, 2005.

Network-Tree Model(NTM): In a given network N,let the network-tree be $N\_Tree = (T, H)$, where T is the macro-node set of N. If only one macro-node exists in T, $H = \phi$, otherwise, H is a binary relation defined on T. There is only one macro-node m_r called the root of N_Tree. If $T - \{m_r\} \neq \phi$,there exists a collection of disjoint trees $T_1, T_2, \cdots, T_n (n > 0)$. For arbitrary $i (1 \leq i \leq n)$, there is one unique macro-node $m_i \in T_i$ satisfies $< m_r, m_i > \in H$. For $H - \{< m_r, m_1 >, \cdots, < m_r, m_n >\}$, there exists a collection of disjoint binary relations $H_1, H_2, \cdots, H_n$,and $\forall i, 1 \leq i \leq n$, H_i is a binary relation defined on T_i, and (T_i, H_i) is also a network-tree defined above, and we call it the sub-tree of the root macro-node m_r.

Definition 1: In network-tree model $N\_Tree = (T, H)$,for any macro-node $m_c \in T$ and its parent macro-node $m_p \in T$, the node set of $m_c \cap m_p$ is called the connecting node set of m_c, denoted by R_{m_c}.

Let $sp_m(u, v)$ denote the shortest path from u to v in network m, $sp_{T'}(u, v)$ denote the shortest path from u to v in network-tree T', and $l(\pi)$ denote the distance of path π.

Definition 2: Network-tree model $N\_Tree = (T, H)$, if for each sub-tree T' of N_Tree, with macro-node m_r being the root of T', $\forall u, v \in V(m_r)$, $l(sp_{m_r}(u, v)) = l(sp_{T'}(u, v))$, thus N_Tree is called Expanded Network-Tree Model.

Definition 3: Network-tree model $N\_Tree = (T, H)$, let the transitive closure of macro-node m be $B_m = \{m_p | (m_p \in T) \wedge (m_p = m \vee (\exists m_q \in B_m)(< m_p, m_q > \in H))\}$.

Definition 4: In network model $N = (V, A, W)$, $\forall u, v \in V$, let the virtual arc from u to v be $\overline{a_{uv}} = (u, v, sp_N(u, v))$, with $w(\overline{a_{uv}}) = l(sp_N(u, v))$.

3 A Parallel Algorithm Based on Network-Tree Model

3.1 Optimality Theorem

Theorem 1: In the Expanded Network-Tree Model, for given nodes s and t, the optimal path π from s to t can be represented as following equation: $\pi = sp_{m_1}(s = v_0, v_1) + sp_{m_2}(v_1, v_2) + \cdots + sp_{m_h}(v_{h-1}, v_h) + sp_{m_{h+1}}(v_h, v_{h+1}) + sp_{m_{h+2}}(v_{h+1}, v_{h+2}) + \cdots + sp_{m_{n-1}}(v_{n-2}, v_{n-1}) + sp_{m_n}(v_{n-1}, v_n = t)$ where $n \geq 1, 1 \leq h \leq n, < m_h, m_{h-1} > \in H, < m_{h-1}, m_{h-2} > \in H, \cdots, < m_2, m_1 > \in H$, and $< m_{h+1}, m_{h+2} > \in H, \cdots, < m_{n-1}, m_n > \in H. v_{i+1} \subseteq R_{m_{i+1}}$ with $0 \leq i < h$, and $v_i \subseteq R_{m_{i+1}}$ with $h < i < n$.

3.2 Parallel Dividing Method and Distributed and Parallel Network Tree Shortest Path Algorithm (*DP_NTSP* Algorithm)

First we get the expanded network-tree through preprocessing . According to theorem 1, the optimal path is consisted of a series of sub-paths in each macro-node of the transitive closures. Then we pre-compute the distances from arbitrary node to connecting nodes in each macro-node,and write the distance data to a file,so the master processor can compute an optimal shortest path consisting of virtual arcs according to the distances. Then the master distributes the virtual arcs to the slaves, and each slave computes the required actual path.

Description of *DP_NTSP* Algorithm. We use the multiple-program-multiple-data (MPMD) model to implement the parallel algorithm.Suppose the source node is s, and the destination node is t.

Master program:
*Step*1 : Creats p copies of the slave program.
*Step*2 : According to s and t,computes the transitive closures $B_{\sigma(s)}$ and $B_{\sigma(t)}$,let $B = B_{\sigma(s)} \bigcup B_{\sigma(t)}$.
*Step*3 : Reads the distance data contained by B from the file and computes an optimal path consisting of virtual arcs.
*Step*4 :While there is an active slave program,
4[a] receives the results from the slave program.
4[b] If all the virtual arcs have not been allocated, sends a new virtual arc to the slave program i; Otherwise, stop the execution of slave i.
*Step*5 :Creates a shortest path according to the received results and terminates.

Slave program:
*Step*1 : Creates the code for the computation of shortest paths.
*Step*2 : While there is a virtual arc to handle,
2[a] receives from the master program a virtual arc or the instruction to stop execution.
2[b] gets the network data that related to the virtual arc and computes the actual path that corresponds to the virtual arc.
2[c] sends the results to the master program.

4 Computational Results

In our computational work, we choose six networks. The detailed parameters of each network are shown in table 1.We test the execution time of *DP_NTSP* algorithm in each network; the computational results are presented in Fig.1.

Table 1. parameters of the testing networks

Network	k	$l_1 = b_1$	$l_2 = b_2$	$l_3 = b_3$	w_1	w_2	w_3	Nodes	Arcs
$g1$	2	6	10	/	5	9	/	3,721	14,640
$g2$	2	8	8	/	9	5	/	4,225	16,640
$g3$	3	2	5	8	6	6	6	6,561	25,920
$g4$	3	4	4	6	2	3	6	9,409	37,248
$g5$	3	6	6	6	5	6	8	47,089	187,488
$g6$	3	6	7	8	5	7	9	113,569	452,928

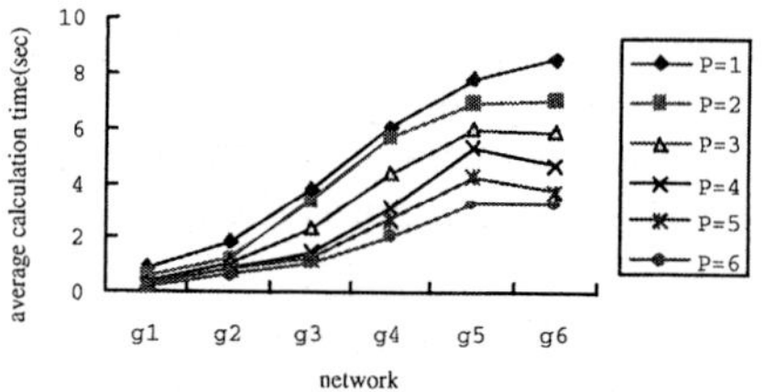

Fig. 1. Calculation time in different networks

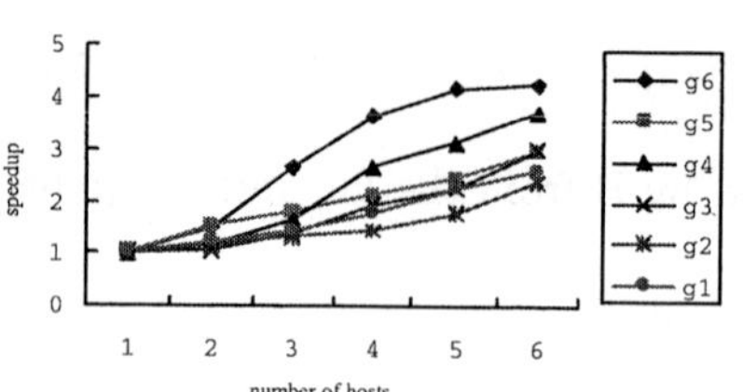

Fig. 2. speedup curves

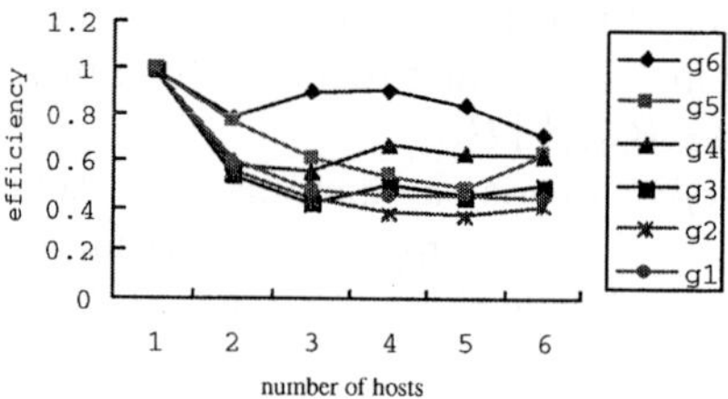

Fig. 3. efficiency curves

From Fig.1 we can see that with the number of hosts increasing, the execution time of algorithm is relatively decreased. In Fig.2 and Fig.3, it is apparent that with the network scale increasing, the speedup and efficiency of the algorithm are improved notably. The results show that the proposed parallel algorithm can well solve the optimal path problem in large-scale networks.

References

1. Koutsopoulos. Commentaries on Fastest Paths in Time-dependent Networks for IVHS Application. IVHS Journal, 1993, 1(1): 89-90
2. J.Shapiro, J.Waxman D.Nir. Level Graphs and Approximate Shortest Path Algorithms, Networks, 1992, 22: 691-717
3. Chan-Kyoo Park, Kiseok Sung, Seungyong Doh, Soondal Park. Finding a path in the hierarchical road networks. 2001 IEEE Intelligent Transportation Systems Conference 1998, 10 (2): 163-179
4. Ning Jing, Yun-Wu Wang, and Elke A. Rundensteiner. Hierarchical Encoded Path Views for Path Query Processing: An Optimal Model and Its Performance Evaluation. IEEE Transactions on Knowledge and Data Engineering, 1998, 10(3): 409 432
5. Guozhen Tan, Xiaojun Han, and Wen Gao. Network-Tree Model and Shortest Path Algorithm. The 3rd International Conference on Computational Science. Melbourne, Australia: Springer, Lecture Notes on Computer Science (LNCS 2659), 2003: 537-547

Exploiting Parallelization for RNA Secondary Structure Prediction in Cluster

Guangming Tan, Shengzhong Feng, and Ninghui Sun

Institute of Computing Technology,Chinese Academy of Sciences
{tgm, fsz, snh}@ncic.ac.cn

Abstract. RNA structure prediction remains one of the most compelling, yet elusive areas of computational biology. Many computational methods have been proposed in an attempt to predict RNA secondary structures. A popular dynamic programming (DP) algorithm uses a stochastic context-free grammar to model RNA secondary structures, its time complexity is $O(N^4)$ and spatial complexity is $O(N^3)$. In this paper, a parallel algorithm, which is time-wise and space-wise optimal with respect to the usual sequential DP algorithm, can be implemented using $O(N^4/P)$ time and $O(N^3/P)$ space in cluster. High efficient utilization of processors and good load balancing are achieved through dynamic mapping of DP matrix to processors. As experiments shown, dynamic mapping algorithm has good speedup.

1 Introduction

Although computational methods such as phylogenetic comparative methods [2] to predict RNA secondary structure only provide an approximate RNA structural model, they have been widely used in the research of RNA secondary structures. In phyloge netic comparative methods, a stochastic context-free grammar (SCFG)[2] is used to model RNA secondary structure. The core of this method is dynamic programming (DP) algorithm [1][2][3]. Applications of SCFG-based RNA models require the alignment of an RNA sequence to a model. The computational complexity of polynomial-time dynamic programming algorithms for a sequence of length L and a model with K states is $O(K*L^2+B*L^3)$, where B is the number of bifurcation states in model [3] or $O(L^4)$ when B~M. The alignment runtime on sequential computers is too high for many RNAs of interest such as ribosomal RNA (rRNA).

Cluster systems [4] have become very popular because of their cost/performance and scalability advantages over other parallel computing systems, such as supercomputers. Load balancing is very important to the performance of application programs in cluster systems. Good load balancing strategy improves the efficiency of processors, thus reduces the overall time of application programs. In this paper, we present an efficient parallel algorithm for aligning an RNA sequence to an SCFG model. The parallel algorithm consumes $O(N^4/P)$ time and $O(N^3/P)$ space, it is time-wise and space-wise optimal with respect to sequential algorithm.

The rest of paper is structured as follows. In section 2 we briefly review the SCFG model of RNA secondary structure and the DP algorithm. In section 3, we propose load balancing parallel algorithm through dynamic partitions of DP matrix. In section 4 we show experiment results and evaluate the performance of the parallel algorithm. Section 5 concludes this paper.

V.S. Sunderam et al. (Eds.): ICCS 2005, LNCS 3516, pp. 979 – 982, 2005.

2 SCFG-Based Algorithm in RNA Secondary Structure Prediction

Eddy [3] described sequence-structure alignment algorithm in detail (The notations and definitions in this paper are referred to [2][3], we will not give their descriptions in detail). An RNA sequence can be aligned to a CM to determine the probability that the sequence belongs to the modeled family. The most likely parse tree generating the sequence determines the similarity score. Given an input sequence $x=x_1 \ldots x_N$ of length N and a CM G with K states numbered in preorder traversal. The DP algorithm iteratively calculates a 3-dimensional DP matrix M(i,j,v) for all i = 1,…j+1, j = 0,…N, v = 1…K. M(i,j,v) is the log-odds score of the most likely CM parse tree rooted at state k that generates the subsequence $x_i \ldots x_j$. The matrix is initialized for the smallest subtrees and subsequence, i.e. subtrees rooted at E-states and subsequence of length 0. The iteration then proceeds outwards to progressively longer subsequences and larger CM subtrees.

$$M(i,j,v)=\begin{cases} \max\limits_{y \in c_v}\{M(i,j,y)+\log t_v(y) & \text{if } S(v)\in\{DEL,BEG\} \\ e_v(x_i,x_j)+\max\limits_{y\in c_v}\{M(i+1,j-1,y)+t_v(y)\} & \text{if } S(v)\in MP \text{ and } d\geq 2 \\ e_v(x_i)+\max\limits_{y\in c_v}\{M(i+1,j,y)+t_v(y)\} & \text{if } S(v)\in\{ML,IL\} \text{ and } d\geq 1 \\ e_v(xj)+\max\limits_{y\in c_v}\{M(i,j-1,y)+t_v(y) & \text{if } S(v)\in\{MR,IR\} \text{ and } d\geq 1 \\ \max\limits_{i-1\leq mid\leq j}\{M(i,mid,kleft)+M(mid+1,j,kright)\} & \text{if } S(v)\in BIF \\ 0 & \text{if } S(v)\in E \\ -\infty & \text{otherwise} \end{cases} \quad (1)$$

At the end of the iteration M(1,N,1) contains the score of the best parse of the complete sequence with the complete model.

3 Parallel RNA Secondary Structure Prediction

As many other DP algorithms, the DP algorithm for RNA secondary structure prediction exhibits the character of a wave-front computation [5], that is, each matrix element depends on a set of previously computed elements. However, it is the wave-front computation that increases the difficulty of the parallelization of DP algorithm. Due to the character of a wave-front computation, our parallel algorithm focuses on DP matrix mapping to processors in the process of filling the DP matrix. The performance of parallel algorithm greatly depends on the strategy of the partition of DP matrix.

A simple matrix mapping method is to statically partition the matrix along the diagonal as depicted in [4][5]. The utility efficiency of processors is very low in the static partitioning algorithm since one processor becomes idle at each stage of p stages. A solution for p<n is to repartition the matrix among the processors when the number of idle processors exceeds some threshold t. The t value directly determines the performance of algorithm. When t = 1, repartitioning occurs at every stage. The communication among processor is too frequent. When t = p, it never repartitions and corresponds to the static partitioning algorithm. In SCFG model based RNA secondary structure prediction algorithm the case where t = p/2 is discussed. This threshold can avoid the extremes of repartitioning too frequently and too infrequently.

Without loss of generality assume that p evenly divides into log(n/p)+1 stages, each of which is further divided into p/2 +1 phases, each phases is composed of two steps. The phases and steps are analogous to the stages and steps of the static partitioning algorithm, respectively. Repartitioning occurs between stages. Along the diagonal the DP matrix is partitioned into strips numberd 1,2…, log(n/p)+1 from the left of M such that strip i has $n/2^i$ diagonals for 1<=i<=log(n/p). The last strip has p diagonals. Starting with the first strip, the ith strip is filled at stage i using the static partitioning algorithm. Then previously computed entries are repartitioned among the p processors and stage i+1 begins. Processors i (i<=0<p-1) evenly partitions the entries stored in local into two halves by rows. The low half entries are sent to processor 2*i+1, except for processor 0, the high half entries are sent to processor 2*i. Accordingly, processor i receive entries from processor i/2.

Actually, the matrix is evenly partitioned by p both vertically and horizontally at each stage. For all 1<=s<=log(n/p),1<=j<=p/2,$M^{(s)}_{(i+1)j}$ denotes the diagonal block computed in processor i in j phrase of stage s. At the last stage log(n/p)+1, each processor computes one row along diagonal and processor 0 computes the last cell. Both DP matrix and entries in the first phrase of each stage are triangular. (see Fig. 1)

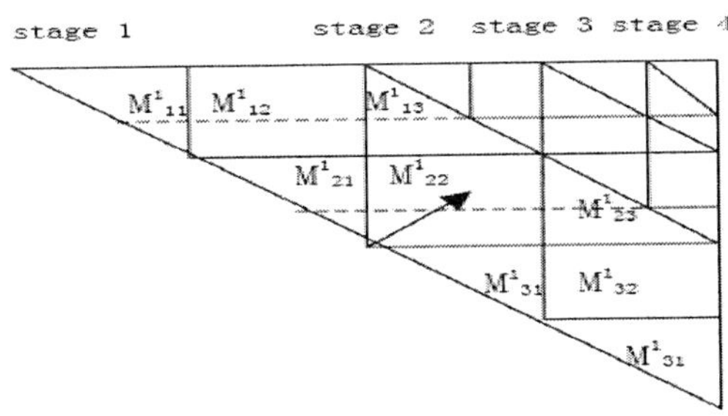

Fig. 1. The wavefront computation in dyanmic portioning algorithm. The dashed represents the repartition

The overall time can again be decomposed two parts: a computation time T_{comp}, a communication time T_{comm} , the communication time comprises a static communication time T_{stcomm} and a repartition time $T_{repartition}$. The computation time T_{comp} and the static communication time T_{stcomm} are $O(n^4/p)$ and $O(n^3)$, respectively. At each stage s, processor i sends no more than $(n/2^{s-1}p)(n-n/2^s)$ entries to processor 2*i. Hence, the total repartitioning time $T_{repartition} = \sum_{s=1}^{\log(n/p)} (n/2^{s-1}p)(n-n/2^s)\log(p) = O(n^2\log(p)/p)$. The overall time is $T_{comp} + T_{stcomm} + T_{repartition} = O(n^4/p)+ O(n^3)+ O(n^2\log(p)/p) = O(n^4/p)$. The maximum space occurred at the first stage, it is $O(n^3/p)$.

4 Performance Evaluation

We have implemented the presented two parallel algorithms using MPI and experimentally evaluated their performance on a NUMA cluster systems. Each node consists of four 2.4GHz AMD Opteron CPUs and 8GB memory. The nodes are connected by 2Gbits/s Myrinet switch.

As the experiment shown, The experimental evidence indicates that the computation time dominates the communication time for all of the runs. The dynamic partitioning algorithm greatly reduces computation time. Table.1 give the results for RNA sequences length 120(5S_ecoli), 377(RnaseP), 1542 (lsu_ecoli) and 2904 (ssu_ecoli). The maximum speedup occurred using 32 processors for two algorithms. Due to relative contribution of the communication time is great, the maximum speedup for all input length is difference. The lower speedup is caused by the more communication cost. For the dynamic partitioning algorithm, the repartitioning time contribute significantly to the communication time. However, thanks to the increasing efficiency of processors, the speedup of the dynamic partitioning algorithm is fine.

Table 1. The run time of dynamic partitioning algorithm. Time: seconds

Length	processors	overall	compute	communication	seepdup
120	2	0.734	0.703	0.031	1.47
	4	0.445	0.355	0.09	2.42
	8	0.328	0.078	0.25	3.29
	16	0.238	0.172	0.066	4.53
	32	0.219	0.051	0.168	4.93
377	2	10.953	10.758	0.195	1.51
	4	5.676	5.375	0.301	2.92
	8	2.801	2.688	0.113	5.92
	16	1.566	1.324	0.242	10.59
	32	0.98	0.676	0.304	16.92
1542	2	169.731	169.562	0.172	1.55
	4	85.172	84.609	0.563	3.09
	8	42.707	42.273	0.434	6.16
	16	22.051	21.113	0.938	11.94
	32	13.223	10.543	2.68	19.91
2904	2	2774.266	2771.047	3.219	1.64
	4	1442.59	1379.574	63.016	2.96
	8	859.289	702.219	157.07	4.97
	16	512.27	347.523	164.747	8.34
	32	218.258	77.207	141.051	19.57

5 Conclusion

We have given parallel DP algorithm for RNA secondary structure prediction in NUMA cluster systems. The parallel algorithm is cost-optimal with respect to the most efficient sequential algorithm. The experimental results shows that employing a good load balancing technique yields superior performance. The RNA secondary structure prediction algorithms mostly are DP algorithms. Our parallel DP algorithm skeleton can be used in other DP algorithms in computational biology by little modification.

References

1. E Rivas and S Eddy. A dynamic programming algorithm for RNA structure prediction including pseudoknots. J. Mol. Biol. 285:2053-2068. 1999
2. S Eddy and R Durbin. RNA sequence analysis using covariance models. Nucl. Aicds Res. 22(11):2079-2088. 1994
3. S Eddy. A memory-efficient dynamic programming algorithm for optimal alignment of a sequence to an RNA secondary structure. BMC Bioinformatics, 3:18, 2002
4. A Grama, A Gupta, G Karypis, V Kumar. Introduction to Parallel Computing. Addison Wesley, 2003
5. Parallel Dynamic Programming. PhD dissertation by Phillip Gnassi Bradford, 1994.

Improving Performance of Distributed Haskell in Mosix Clusters

Lori Collins, Murray Gross, and P.A. Whitlock

Department of Computer and Information Sciences, Brooklyn College,
2900 Bedford Avenue, Brooklyn, NY 11210-2889
magross@its.brooklyn.cuny.edu

Abstract. We present experimental results demonstrating the performance improvement obtained in our distributed Haskell implementation when we replaced scheduling according to a micromanagement paradigm with contention-driven scheduling with automatic load balancing. This performance enhancement has important implications in the area of automatic run-time optimization.

1 Introduction

We modified version 5.04 of Glasgow distributed Haskell (GdH)[1, 2] to take advantage of the special properties of a cluster of computers managed by the Mosix distributed operating system[3], which provides optimized task distribution with automatic load balancing. Standard GdH uses a PVM-based[4] micromanaging scheduler to lock a single process on each real processor. Experience with conventional parallel programs convinced us that performance would be maximized by executing multiple computations on each processor. Below we present background and test results for modifications we made to standard GdH to maximize performance by using Mosix distribution and load balancing instead of the standard GdH scheduling.

2 Parallel/Distributed Haskell and the Mosix Environment

While functional languages provide a number of advantages over procedural languages [5], four characteristics are particularly important in the area of parallel and distributed execution: (1) functions are first class values, i.e., passable from one function to another as ordinary data; (2) a value bound to a variable cannot change (referential transparency); (3) in Haskell, function evaluation is lazy and (4) side effects are forbidden except in the rare cases such as I/O where they cannot be avoided. The first-class status of functions, referential transparency and the prohibition against side effects provide potentially radical reductions in coordination delay and overhead. Lazy execution means that computation of values is delayed until they are actually required. This guarantees that once a

V.S. Sunderam et al. (Eds.): ICCS 2005, LNCS 3516, pp. 983–986, 2005.

value is computed, it may be freely used by any process in the computation without fear of change. Overall computing time is reduced since processes block on values only while they are currently being computed, a necessary delay. A value not yet available will be computed by whichever process first requires it.

The effect of these specific language characteristics causes parallel Haskell programs to serialize on critical paths rather than on the programmer's conceptual scaffolding. Since it is clearly impossible to improve performance beyond what is obtained on critical paths, optimal performance is obtained without programmer intervention.

Our distributed calculations are run on a cluster of 15 computers currently running Debian Linux with OpenMosix patches[3]. Mosix makes a cluster of separate computers run like a symmetric multiprocessor (SMP) by dynamic process migration[3]. This is ideal in heterogeneous configurations, since it automatically enables load balancing on processors by forcing process migrations to achieve load balancing across available CPU's. Requests for system services are trapped by Mosix, executed on the home machine when they cannot be performed on the worker machines, and the results are transmitted back to the worker machine. A high-efficiency file system[6] permits processors to access files on other processors with minimal overhead. Data can be distributed across the cluster and there are only a few clearly defined conditions that prevent dynamic migration of tasks from one machine to another.

3 Modification to the GdH Compiler

In the original GdH design, PVM[4] was used for all remote process creation and communication. Rather than building a virtual machine of virtual processors provided by PVM processes, our modified version of the run-time system builds a virtual machine of virtual processors created by forking local processes, which may be migrated to remote processors by Mosix. The send() and recv() calls to PVM were replaced by local system calls. We use UDP-based interprocess communication on the loop back interface. After error testing of our new GdH-BC system we proceeded to test whether the hypothesized efficiencies had actually been achieved[7].

4 Performance Results

The initial tests compared the performance of runs on a single processor of the ffactsm_seq program, complied with serial GHC[8], with the program ffactsm_par, compiled by our modified GdH-BC and run on fourteen or fifteen real processors. The tests consisted of twenty-six sets of arguments with execution times varying from seconds to hours. Each experiment was repeated three times. The serial and parallel calculations gave the same answers in all cases. The speedup, S, between the serial and parallel calculation is given by:

$$S = \frac{\text{Execution time on a single processor, best sequential algorithm}}{\text{Execution time using } p \text{ processors}}, \quad (1)$$

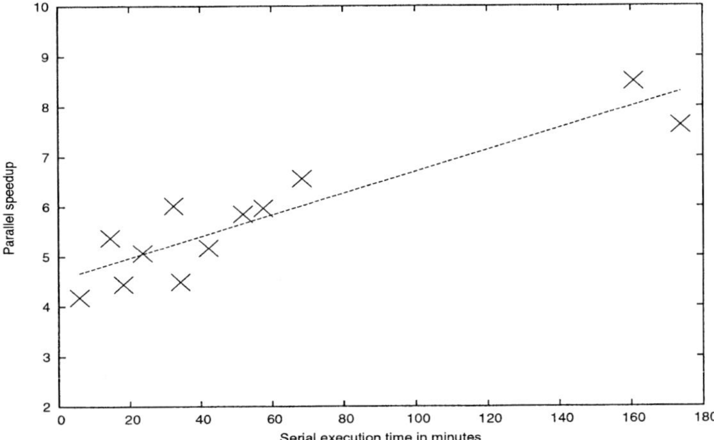

Fig. 1. Observed speedup, as defined by Eq. (1), when the test program was run in parallel using GdH-BC as compared to the serial execution times. The line is a linear fit to the data

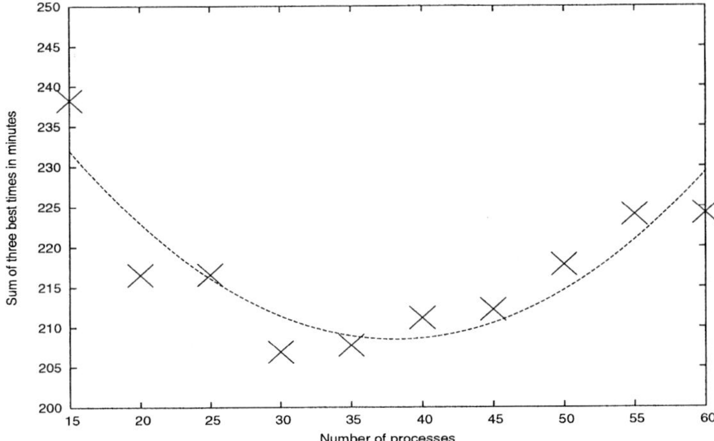

Fig. 2. Total timings from parallel runs of the test program with a fixed number of processes. The line is a fit to the data

and is shown in Figure 1. As expected we saw a decrease in execution time with our parallel calculations in comparison to our serial execution times. Mosix gathers execution time statistics before distributing a calculation, so quickly completed calculations are not distributed as extensively as longer calculations.

Our claim that best performance would be achieved by executing multiple processes on each processor is confirmed by Figure 2. In the figure, we have displayed the total time of the three best runs versus the number of processes requested per run. Once there are more processes than the fifteen real processors the execution time decreases until an equilibrium point is reached and the running time begins to increase with the number of processes. We think it safe to say that the increase in running time when the number of processes increases beyond a "balance point" is the result of increases in system overhead associated with process coordination and load balancing.

5 Conclusions

We conclude that the original design of the GdH run-time system, which precluded more than one PVM session per processor, imposes an unnecessary performance barrier. We find that at least under some circumstances, the contention driven supervisory model in our GdH-BC outperforms the micromanaged scheduling provided by the original PVM-based version of the system. While further testing is required, we believe that our version of the Haskell system provides significant performance advantages over competitors in implementation of parallel/distributed computation.

Acknowledgments

The cluster of computers was made possible by ONR Grant N00014-96-1-1-1057. The initial development of the GdH-BC compiler was done by Dino Klein and Qing Shou. Testing of the compiler was begun by Qing Shou and Kerim Simsek. One of us, L.C., is supported by NSF Grant number HRD0217542.

References

1. Trinder, P.W., Loidl, H-W., and Pointon R.F.: Parallel and distributed haskells. Journal of Functional Programming **12** (2002) 469–510.
2. Pointon, R.F., Trinder, P.W., and Loidl H-W.: The Design and Implementation of GdH. In: Markus Mohnen, Pieter W. M. Koopman (Eds.) Implementation of Functional Languages, 12th International Workshop, Lecture Notes in Computer Science **2011** (2001) 53–70
3. Barak, A., La'adan, O., and Shiloh, A.: Scalable cluster computing with MOSIX for LINUX. Proc. 5th Annual Linux Expo, Atlanta, GA (1999) 95–100
4. Sunderam, V.: PVM: A Framework for Parallel Distributed Computing. Concurrency: Practice & Experience **2** No. 4 (1990) 315–339
5. Hughes, J.: Why functional programming matters. Computer Journal **32** (1990) 1–23
6. Amar, L., Barak, A., and Shiloh A.: The MOSIX Parallel I/O System for Scalable I/O Performance. Proc. 14-th IASTED Int. Conference on Parallel and Distributed Computing and Systems, Cambridge, MA (2002) 495–500
7. http://146.245.249.159/~kerim/timing.pdf
8. Peyton Jones, S.L, Hall, C., Hammond, K., and Partain, W.: The Glasgow Haskell compiler: a technical overview. UK Joint Framework for Information Technology (JFIT) Technical Conference, Keele, (1993) 249–257

Investigation of Cache Coherence Strategies in a Mobile Client/Server Environment

C.D.M. Berkenbrock and M.A.R. Dantas

Department of Informatics and Statistics (INE),
University of Santa Catarina (UFSC),
88040-900 Florianopolis, Brazil
{diacui, mario}@inf.ufsc.br

Abstract. In this article, we present an investigation case study based on an implementation and performance analysis of three different cache coherence strategies over a real wireless environment. Our research work considers the broadcasting timestamp, the cache coherency schema with incremental update propagation and the amnesic terminals strategies. These strategies are based on periodic broadcast of invalidation reports. The performance of these strategies is analysed through an ordinary real environment. In this environment we compare the impact of invalidation report size, broadcast interval and cache size in mobile devices.

1 Introduction

Data replication in mobile devices can represent an interesting mechanism to reduce latency, data transfer from servers and collision avoidance in wireless environments. However, this approach can only be considered successful if users can find essential data in the local cache. As a result, mobile users can avoid the access to the wireless broadcast link . On the other hand, as disconnection is frequently observed in wireless configurations data stored verification in a mobile device is a complex task for several classes of applications.

In this article, we present an investigation case study based on an implementation and performance analysis research considering three different cache coherence strategies over a real wireless environment. The broadcasting timestamp (TS)[1], cache coherence schema with incremental update propagation (CCS-IUP) [2] and amnesic terminals (AT) [1] strategies are based on a periodical broadcast of cache invalidation report (IR). Our experimental results reflect the impact of the size of IR and the cache size of mobile devices when utilizing a database application.

The paper is organized as follows. In section 2 we presented the experimental environment architecture. In section 3 we discuss the results obtained in a real wireless configuration. Finally, in section 4 we present conclusions of the research work.

V.S. Sunderam et al. (Eds.): ICCS 2005, LNCS 3516, pp. 987–990, 2005.

2 Experimental Environment Architecture

The client/server paradigm for distributed computing in wired networks is a well known architecture with many proposals for cache coherence [3]. This approach is also considered in the literature for wireless environments (e.g. [4, 5]), however with special concern about disconnection and clients moving.

In this article, we consider the TS, CCS-IUP and AT techniques, because these strategies have invalidation report (IR) and stateless servers. IR is a lightweight mechanism for wireless networks that informs only the necessary items that were modified. The stateless servers are not obliged to keep information related to mobile devices in its area and it is not necessary to know about the cache of clients.

The environment architecture developed to execute our experiments is shown in figure 1. The architecture has three main components: a server, clients and a database system. In addition, we implemented the three protocols strategies, proposed in the literature.

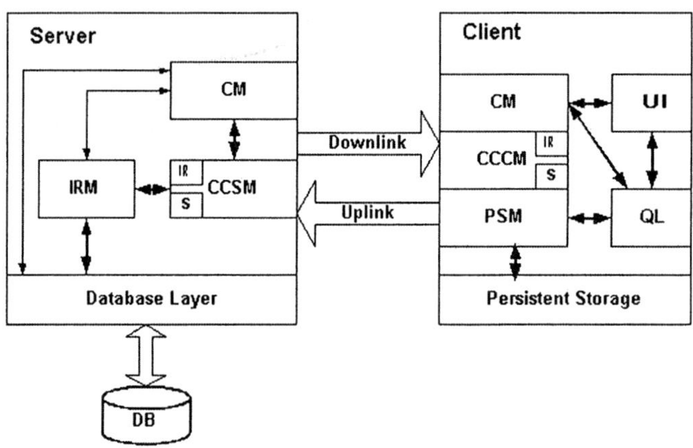

Fig. 1. The environment architecture

Inside the server the modules are communication manager(CM), cache coherence server manager (CCSM) and invalidation report manager (IRM). CM is responsible for clients link management with the server side. The CCSM verifies which strategy is being used and adds the appropriated parameters inside the IR for this strategy. The IRM is responsible for the modification of the IR when the database is modified.

On the other hand, clients are characterized by the modules communication manager (CM), cache coherence client manager (CCCM), persistent storage manager (PSM), query list (QL) and user interface (UI). The CM client manages the connection with the server. The CCCM is responsible for receiving IRs from the server and applies the cache modification following a specific strategy (S).

The PSM executes any modification necessary into the cache of the mobile device. The QL stores all the registers that are being modified and UI is a friendly interface to mobile users to access the environment.

The database system component consider for our experiments was the Mckoi SQL [6]. We employed this database, because it is open source JAVA SQL database system package and provides interesting functions to develop applications (e.g. trigger and referential integrity).

3 Experimental Results

In the server database we used 50 items, which were accessed by two mobile devices. The broadcast window used was 10 intervals, similar to others research works (e.g. [4, 5]), with 20 seconds between intervals. In addition, the experimental tests were executed using the TS, CCS-IUP and AT considering that mobile users have 5, 15 and 25 percent of item from the database locally cached.

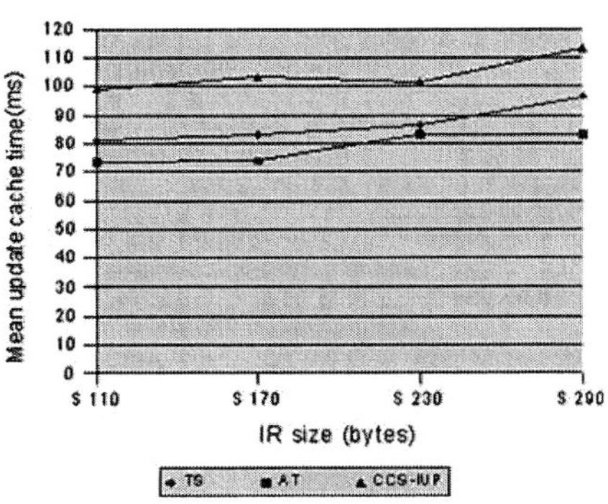

Fig. 2. Mean update cache time vs. IR size

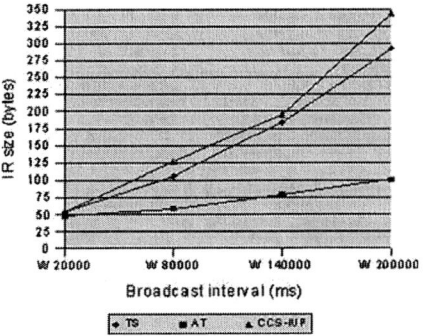

Fig. 3. IR size vs. Broadcast interval

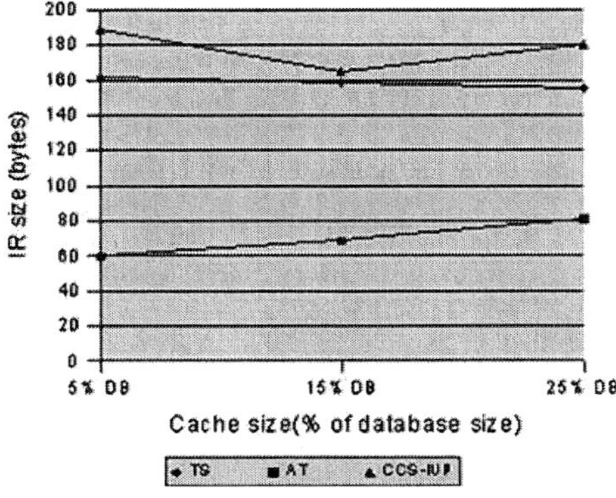

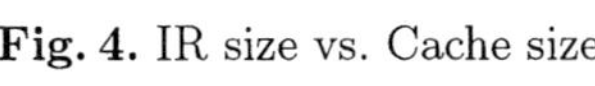

Fig. 4. IR size vs. Cache size

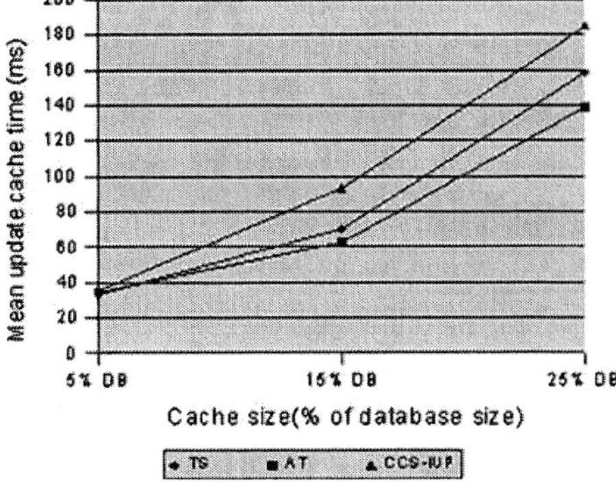

Fig. 5. Mean update cache time vs. Cache size

In figure 2 our results indicates that the TS and CCS-IUP spend more time to update the cache than the AT approach. Figure 3 shows IR sizes related to broadcast intervals. The graphic shows that the TS and CCS-IUP have a tendency to increase IR sizes when broadcast intervals increase.

The relation of the notification size against the cache size is presented in figure 4. The increase of the cache size resulted in the increase of the IR size for AT. The TS registered a small decrement of the IR size and the CCS-IUP did not get a regular behavior. In figure 5, as we expected, it is possible to verified that the increase of the cache size directly implies in the time to update the data store in mobile clients.

4 Conclusions

In this paper we presented an investigation case study based on an implementation and performance analysis of three different strategies for cache coherence in wireless networks.

The experimental results demonstrated facilities and constrains of the strategies based on effects from the invalidation report (IR) size, broadcast interval and mobile clients cache size. The results demonstrated different performance for the parameters considered. The AT strategy had a interesting performance, but with less facilities compared to the other two approaches.

Acknowledgements

This research was partially supported by CNPQ (Brazilian National Research Council) under the grant 132145/2004-9.

References

1. Daniel Barbar; and Tomasz Imielnski. Sleepers and workaholics: caching strategies in mobile environments (extended version). The VLDB Journal, (1995) 4(4):567602.
2. Hyunsik Chung and Haengrae Cho. Data caching with incremental update propagation in mobile computing environments. In Australian Computer Journal (1998).
3. Magnus E. Bjornsson and Liuba Shrira. BuddyCache: high-performance object storage for collaborative strong-consistency applications in a WAN. Proceedings of the 17th ACM SIGPLAN conference on Object-oriented programming, systems, languages, and applications,(2002) 26-39.
4. Joanne Holliday and Divyakant Agrawal and Amr El Abbadi. Disconnection modes for mobile databases. Wirel. Netw, (2002) 391-402.
5. Jin Jing and Ahmed Elmagarmid and Abdelsalam Sumi Helal and Rafael Alonso. Bit-sequences: an adaptive cache invalidation method in mobile client/server environments. Mob. Netw. Appl, (1997) 115-127.
6. Mckoi SQL Database System. Available online: http://www.mckoi.com/database/.

Parallel Files Distribution

Laurentiu Cucos[1] and Elise de Doncker[1]

Western Michigan University
{lcucos ,elise}@cs.wmich.edu
http://aegis.cs.wmich.edu/~lcucos
http://www.cs.wmich.edu/~elise

Abstract. For parallel computations requiring massive data Input/Output, one of the goals is to maximize the use of the underling storage topology, in particular exploit the benefit of using local disks. This paper presents a mechanism to distribute independent data records from multiple files to multiple processing nodes and vice-versa. An allocation problem is solved using the Max-Flow algorithm. We give timing results using various read/distribute protocols. One application is to re-distributing tasks (regions) for restarting parallel numerical integration runs.

1 Introduction and Motivation

In order to maintain a minimum transport cost between producers, consumers and warehouses, it is preferable to have all three within the same system, in close proximity, or use paths of easy access between one another. In addition, is desired to have a small number of moves or stops between producers and consumers.

As in an industrial scenario, parallel computations that produce, store and later consume large amounts of data, have to follow similar principles.

Let us consider an arbitrary topology of high speed connections between a number of organizations, each providing a number of machines with either local storage and/or access to shared partitions. The optimum distribution would be achieved when the data is generated, saved, and loaded in the same computing node (keeping all other data transfers at a minimum); however, this may not always be possible or efficient. For irregular problems, some nodes may generate more data than they can store, reload and re-process, or just more data than other nodes. Under these conditions it would be difficult to predict and control how the data should be distributed.

Addressing mainly scenarios where data need to be read only once, this paper presents an efficient distribution engine.

As an application we consider redistributing regions in between calls to numerical multivariate integration codes. ParInt (Parallel Integration project) [1, 5] implements adaptive partitioning (as well as other) strategies for multivariate integration.

Section 2 below outlines a general equal-distribution algorithm, Section 3 gives experimental results and Section 4 concludes the paper.

V.S. Sunderam et al. (Eds.): ICCS 2005, LNCS 3516, pp. 991–994, 2005.

2 Static Distribution

To avoid synchronization overheads, and since the total amount of work is known, it is more efficient to compute the data distribution in advance. While in most cases data can be easily preallocated by distributing equal shares among all machines, non-trivial interconnection networks require more sophistication.

Consider the following general allocation problem [4]:

Problem: Given a set of n machines, M $= \{m_i\}_{1\leq i\leq n}$, a collection of p files, $S = \{s_j\}_{1\leq j\leq p}$, where s_j has $|s_j|$ records, and a unidirectional bipartite graph $G = \{(M, S), E\}$, where $E = \{(v_i, u_j)\}$ with $v_i \in M$ and $u_j \in S$, which defines a mapping between machines and files, compute the item distribution so that each machine gets an equal share of records. In case that an equal distribution cannot be achieved, find a solution with records shipped between machines at a minimum communication cost.

Example: Four computing nodes m_1, m_2, m_3, m_4, participating in a computation requiring data from three file servers FS_1, FS_2, FS_3; where m_1, m_2, m_4 can access FS_1, m_1, m_3 can access FS_2, and all can access FS_3. All machines are interconnected by the same type of network.

Solution: An efficient distribution can be achieved using a Min-Cut Max-Flow algorithm [3] and the following procedure.

1. Augment the graph with two new nodes: *source* and *sink*, and edges between all set-nodes and source, and all machine-nodes and sink.
2. Assign infinite capacities to edges between sets and machines, $|S_i|$ between set S_i and source, and $\lceil \frac{\sum_{i=1}^{p}|S_i|}{n} \rceil$ between each machine and the sink.
3. Apply the Max-Flow algorithm to the newly created graph.
4. If the resulting flow is $\sum_{i=1}^{p} |S_i|$, there exists an equal distribution of the sets; the distribution is given by the flow in the inner edges. Otherwise, the sets cannot be equally distributed without additional communication and go to step 5.
5. Equally distribute the records not allocated in step 3 to machines with a smaller number of records.

3 Experiments

We implemented the system PFD which is a parallel read/write engine to distribute data records from/to files located over multiple machines. It provides a transparent and efficient way to incorporate data reading, distributing, and processing in the same program. The system is layered over the MPI message passing library [2].

PFD was mainly developed for ParInt, to save/load large numbers of integration regions; however it can be easily integrated in any application that reads/writes large numbers of independent records.

Given a collection of records distributed over a number of machines, PFD:

- reads each record exactly once (when the record is processed);
- minimizes the communication and efficiently distributes all records among all participating nodes;
- provides an easy and transparent way to use local disks;
- uses a buffer of configurable size for interprocess communication;
- provides an array of options to read and distribute the data such that:
  - each node will process the same number of records (EQUAL),
  - or allow faster nodes to process more data (COMPETE),
  - or, only nodes with direct access to data will process records (LOCAL).

The system is written in C, and can be integrated with the user application trough a small set of read/write/pack/unpack functions.

We performed a set of experiments reading 1GB of data organized in 8KB records and using different distributions, ranging from all data being stored in

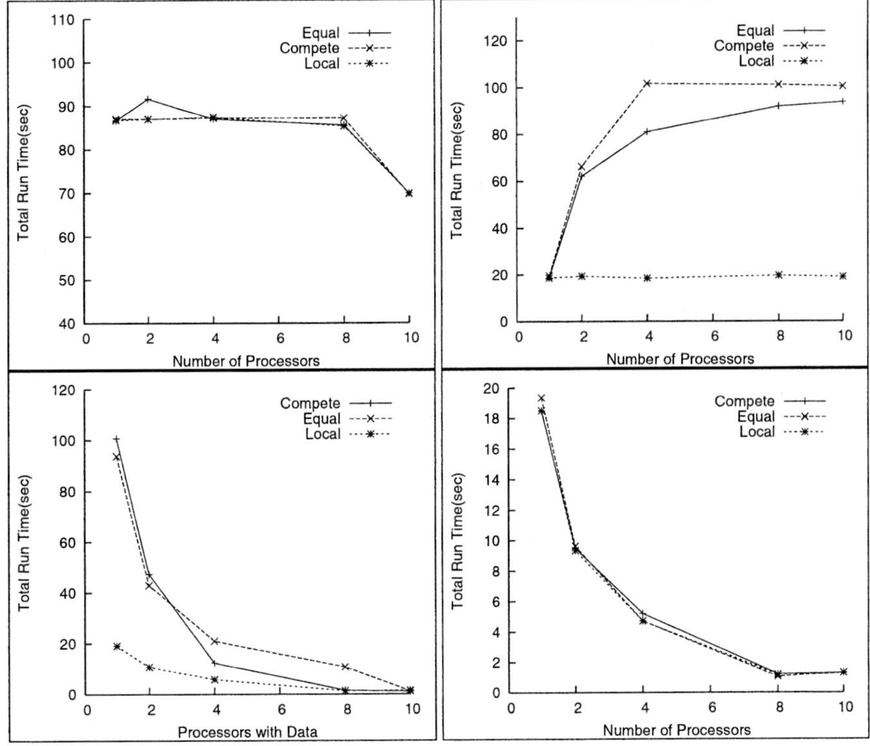

Fig. 1. *Top-Left:* Distribute records from a shared partition; *Top-Right:* Distribute records from one non-shared partition; *Bottom-Left:* Distribute records located in a subgroup of machines; *Bottom-Right:* Distribute records shared equally by the participants

one shared partition, to all data located in one local disk. Up to 10 readers are selected from a set of 1.2GHz AMD Athlon processor nodes, communicating on a Fast Ethernet network.

Figure 1 *(Top-Left)* shows results for data located in a shared partition. In this case, the runtime is independent of the read method (EQUAL, COMPETE, or LOCAL), and of the number of processors, and is mainly driven by the network transfer speed. Note that the shared partition of this experiment was located on a disk significantly faster than the local disks used in the following experiments.

Figure 1 *(Top-Right)* shows results for data located in one non-shared partition. While the total read time is around 20 sec, the performance quickly depreciates as more processors are involved in the computation. We attribute this behavior to the bottleneck at the distribution node.

Figure 1 *(Bottom-Left)* and *(Bottom-Right)* show results for data non-uniformly distributed, and uniformly distributed, respectively, over participating machines. There is a significant improvement in runtime and scalability. In the latter case, the read time was not affected by the distribution method (Equal, Compete, or Local).

4 Conclusions

This paper outlines a mechanism to distribute independent data records from multiple files to multiple processing nodes and vice-versa. We present a solution to the general allocation problem of sharing the data equally among the participating compute nodes, given an arbitrary storage topology.

Although using local disks normally binds data producers with consumers, our system maintains the benefit offered by this connection, while giving at the same time a set of options to share data with non-producers.

Acknowledgments

This work is supported in part by Western Michigan University, and by the National Science Foundation under grants ACI-0000442, ACI-0203776, EIA-0130857.

References

1. http://www.cs.wmich.edu/parint, PARINT web site.
2. http://www-unix.mcs.anl.gov/mpi/index.html, MPI web site.
3. CORMEN, T. H., LEISERSON, C. E., AND RIVEST, R. L. *Introduction to Algorithms.* The MIT press, 1994.
4. CUCOS, L. *Load Sharing Strategies in Distributed Environments.* PhD thesis, Western Michigan University, December 2003.
5. ZANNY, R., DE DONCKER, E., KAUGARS, K., AND CUCOS, L. PARINT1.2 *User's Manual.* Available at `http: //www.cs.wmich.edu/parint`.

Dynamic Dominant Index Set for Mobile Peer-to-Peer Networks*

Wei Shi, Shanping Li, Gang Peng, and Xin Lin

College of Computer Science, Zhejiang University,
Hangzhou, P.R. China 310027
shiwei@zj165.com
shan@cs.zju.edu.cn
{e_pglmary, alexlinxin}@hotmail.com

Abstract. Efficient locating mechanism is the key issue of establishing peer-to-peer (P2P) resource sharing for mobile Ad Hoc networks (MANET). In this paper, a simple concept of Dynamic Dominant Indexing Set (DDIS) is proposed. By periodically advertising information of shared resources on the backbone of current network topology in a distributed fashion, DDIS gains great improvement on user response time, fault tolerance and scalability. Simulations indicate that DDIS simplifies the searching process and greatly reduces the response delay and thus, it is more appropriate for mobile peer-to-peer source sharing especially in emergent situations.

1 Introduction

Inviting P2P model into mobile ad hoc networks (MP2P) to address problems encountered during the development of mobile ad hoc networks (MANETs) is proved to be a promising method. Like P2P computing, one of the major challenges of MP2P computing is resource locating and routing. In order to improve users' experience, mechanisms are seriously required to quickly locate the desired information.

To address resources locating problem in MP2P, we should take following factors into account: 1) in order to lookup the resources more accurately, locating mechanism should guarantee the freshness of search results; 2) many scenarios in MP2P environment, for example emergency rescue and combat, require searching response time as short as possible; 3) being aware of the entire network topology is a mission impossible for mobile devices; and 4) individual peer's fault will not cause the crash of the whole system.

Establishing a dynamic lookup set should be an efficient and feasible approach in MP2P networks. In this paper, we propose a locating mechanism in MP2P

* This paper is supported by National Natural Science Foundation of China (No. 60473052) and Zhejiang Provincial Natural Science Foundation of China (No.602032).

V.S. Sunderam et al. (Eds.): ICCS 2005, LNCS 3516, pp. 995–999, 2005.

environment called Dynamic Dominant Index Set (DDIS). DDIS dynamically constructs the index set which is based on the Minimum Connected Dominating Set (MCDS). MCDS is the minimum set of nodes such that every node in the network is either in the set or is a neighbor of a node in the set. According to the information of shared resources periodically published on MCDS, peers can locate the resources quickly by consulting a limited number of peers rather than bothering more other peers. However, finding the MCDS of a specific network has been proved to be a NP-complete problem [1]. We improved the Dominating Pruning (DP) algorithm that proximately solves the MCDS problem in a distributed fashion. With the help of the improved DP algorithm, the DDIS mechanism can build the MCDS-based index set of current MP2P network topology in a more efficient way.

The paper is organized as follows. In the next section, the related work is discussed. Section 3 introduces our searching mechanism. Simulation results and analyzes are presented in section 4. Finally we conclude our study in section 5.

2 Related Work

In [2] the author put forward a locating mechanism in MP2P network, named PDI. PDI gains considerably good performance. The Dominating Pruning (DP) algorithm [1], computes the approximation of MDCS in a distributed manner according to 2-hop neighborhood information. But DP algorithm only concentrates on constructing an efficient routing algorithm by reducing the broadcast redundancy in MANET and does not take into account the MP2P-specific factors. DDIS improves DP to make it more suitable for building an effective search index in MP2P environment.

3 The DDIS Mechanism

Here, the unit disk graph [3],$G=(V,E)$, is introduced to represent a MP2P network, where V represents a set of mobile peers and E represents a set of edges. We use $N(u)$ to represent the neighbor set of u. $N(N(u))$ represents the neighbor set of $N(u)$. And we assume that u (sender) and v (receiver) are neighbors.

In our scenario, when a peer launches or forwards an advertising packet, it would decide the forward list around itself. The forward list in the advertising packet would notify his neighbors who belong to the index set and should take the responsibility to cache the metadata in the advertising packet. Meanwhile, by analyzing the forward list in the advertising packet, the receivers can calculate their own forward lists in a more accurate fashion, which will reduce the broadcast redundancy.

Peer willing to share his resources periodically launches a “ADV” broadcast containing the peer's id, the information of his shared resource, the expiring time of shared resource, and the forward list of that packet. Due to the nature of broadcasting in MANET, all neighbors around the source peer will get the packet and only the peers in the forward list of that packet are suggested to cache and forward the packet. Others can just simply discard it. Then the receivers calculate their own forward lists and rebroadcast the packet from the source peer. This process continues until the whole network is covered. The forward list of v in DDIS can be computed in the following way, note that $F(u)$ here denotes as the forward list of u passed to v:

1. Let $F(u,v) = []$ (empty list), $Z = \phi$ (empty set) and $K = \cup S_i$ where $S_i = N(v_i) \cap (U(u,v) - N(N(v) \cap F(u)))$, $S_i = N(v_i \cap P(u,v))$ for $v_i \in B(u,v)$.
2. If there exists any peer w in P that v can notify only through v_n, $F(u,v) = F(u,v) \| v_n$, $Z = Z \cup S_n$, $K = K - S_n$, and $S_j = S_j - S_n$ for all $S_j \in K$. This step repeats until no peer in P that v can notify only through v_n.
3. Find set S_k with the maximum size in K. (In case of a tie, the one with the smallest identification k is selected.)
4. $F(u) = F(u,v) \parallel v_k$, $Z = Z \cup S_k$, $K = K - S_k$, and $S_j = S_j - S_k$ for all $S_j \in K$.
5. If no new node is added to Z, exit; otherwise, goto step 3.

Here $B(u,v) = N(v) - N(u)$. DP algorithm evaluates the number of peers in $U(u,v) = N(N(v)) - (N(u) \cup N(v))$, so called "evaluating peers", which can be covered by v_i in decision of including/excluding v_i in/from $F(u,v)$. However, the evaluating peers can be even fewer because the peers in the forward list of u will notify the ones in $N(N(v) \cap F(u))$. So these peers can be safely excluded from $U(u,v)$ in DDIS and the evaluating peers can be denoted as $P(u,v) = U(u,v) - N(N(v) \cap F(u))$, recall that $U(u,v)$ represents the uncovered peers 2 hops away from v in DP algorithm.

In selection process of building forward list, DDIS firstly picks up the neighbors that will consequentially be included in v's forward list because without them v can not inform all the peers in $P(u,v)$. In DP algorithm, these peers, what we call "key neighbors" in this paper, will also be inevitably included in forward list. But since DP algorithm is a kind of greedy algorithm and the selection sequence is random, there may exist unnecessary peers in resulting forward list whose coverage is a part of union of key peers' neighbors. By choosing key neighbors firstly, the forward list of DDIS can be smaller than the one that DP algorithm builds.

With the help of the improved DP algorithm, DDIS can deliver the metadata of shared resources on the dominant index set of current network topology more efficiently. Peers in DDIS frequently check whether the cached metadata is expired. Querying peer firstly checks his cached data to see whether there is information of the required resources or, otherwise, just asks his neighbors to find where the required resources are.

4 Performance Results

In this section, we present the simulation results of our DDIS mechanism. We used an IEEE 802.11 standard MAC layer and a standard physical layer de-

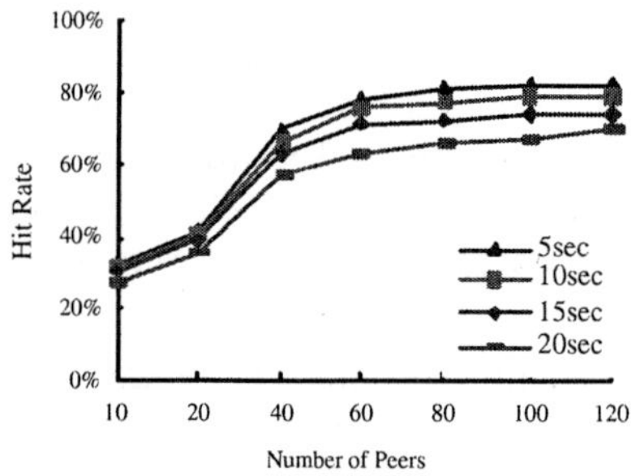

Fig. 1. Hit rate of DDIS with different advertising intervals

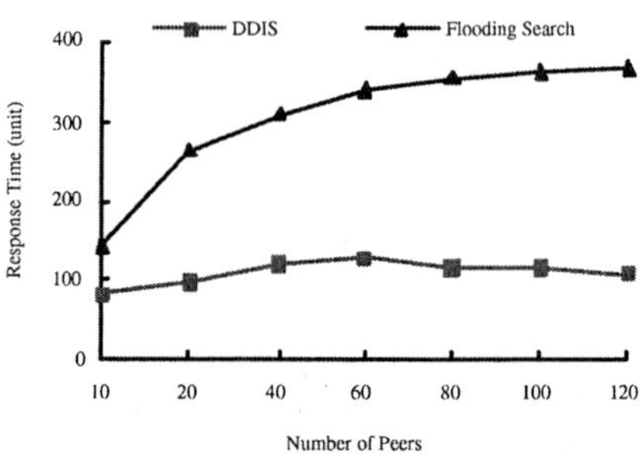

Fig. 2. Response time of the two mechanisms

ploying two-ray ground propagation as radio propagation model. N_{Peers} peers roaming in an area of $1000m \times 1000m$. Fig. 1 illustrates the hit rate of DDIS with different advertising intervals. As we can see, the shorter advertising interval we set, the higher hit rate we get. However, the performance improvement is achieved at the expense of heavier overhead. Balancing the accuracy of queries and the overhead DDIS must pay needs more efforts of study. From Fig. 2, the main advantage of DDIS can be learned. By dynamically caching the metadata of the peers on the backbone of current network, DDIS gains great improvement on user response time compared with flooding search mechanism. Our goal is to build a small index set to affect all the peers in the network. By putting the metadata around the participants, they can quickly locate their favorable resources without more efforts. Since the required information is put closely to the requiring peers, the response time of DDIS changes little when the number of participants grows. The flooding search mechanism, on the contrary, by broadcasting in the network to locate the target peer which may be far away from the querying peer takes about average 2 to 3 times longer than DDIS to response the users' queries.

5 Conclusions and Future Work

In this paper, we put forward a novel concept of DDIS. In fact, DDIS is a kind of replication of index center. By selecting peers in MDCS, the advertisements are delivered close to every peer, which is especially valuable in emergent situations where response time is much more the key issue.

References

1. H. Lim and C. Kim: Flooding in wireless ad hoc networks. Computer Communications Journal, 24(3-4): 353-363, 2001.

2. C. Lindemann and O. Waldhorst: A Distributed Search Service for Peer-to-Peer File Sharing in Mobile Applications. Proc. The 2nd International Conference on Peer-to-Peer Computing, September 2002.
3. B.N. Clark, C.J. Colbourn, and D.S. Johnson: Unit Disk Graphs. Discrete Math, vol. 86, pp. 165-177, 1990.

Task Mapping Algorithm for Heterogeneous Computing System Allowing High Throughput and Load Balancing[1]

Sung Chune Choi and Hee Yong Youn

School of Information and Communications Engineering,
Sungkyunkwan University,
440-746, Suwon, Korea +82-31-290-7952
{choisc, youn}@ece.skku.ac.kr

Abstract. The applicability and strength of heterogeneous computing systems are derived from their ability to match computing nodes to appropriate tasks since a suite of different machines are interconnected. A good mapping algorithm offers minimal expected completion time and machine idle time. In this paper we propose a new task scheduling algorithm allowing higher performance than the existing algorithms such as the Min-min, Max-min, and Sufferage algorithm. It is achieved by task swapping approach based on the expected completion time and ready time of each machine. Extensive computer simulation validates the proposed algorithm along with the earlier ones.

Keywords: Heterogeneous computing, load balancing, task mapping, throughput, scheduling.

1 Introduction

In heterogeneous computing (HC) environment, a suite of different machines are interconnected to provide a variety of computational capabilities and maximize their combined performance to execute tasks having diverse requirements. There exist a number of different types of HC systems. This paper focuses on mixed-machine HC systems, where a number of high-performance heterogeneous machines are interconnected through high-speed links [1].

In HC system the application is decomposed into tasks, where each task is computationally homogeneous. The applicability and strength of HC systems are derived from their ability to match the computing resources to appropriate tasks. Here each task is assigned to one of the machines which is best suited for its execution to minimize the execution time. Therefore, an efficient mapping scheme allocating the application tasks to the machines is needed.

The general problem of mapping tasks to the machines is a well known NP-complete problem and several mapping algorithms have been proposed to approxi-

[1] This research was supported by the Ubiquitous Autonomic Computing and Network Project, 21st Century Frontier R&D Program in Korea and the Brain Korea 21 Project in 2004. Corresponding author: Hee Yong Youn.

V.S. Sunderam et al. (Eds.): ICCS 2005, LNCS 3516, pp. 1000 – 1003, 2005.

mate its optimal solution in the literature. The representative batch mode mapping algorithms are Min-min, Max-min, and Sufferage algorithm [2,3,4]. A good mapping algorithm compromises between matching for smallest expected completion time and load balancing to minimize the machine idle time. Since the previous algorithms have some limitations, we propose a new scheduling algorithm solving them. It is achieved by task swapping approach based on the expected completion time and ready time of each machine. Computer simulation reveals that the proposed algorithm consistently outperforms the earlier algorithms for various degree of task and machine heterogeneity.

2 Related Work

At first, we define some metrics used throughout the paper, which are the expected execution time (EET), ready time (RT), expected completion time (ECT), and makespan. The EET_{ij} is the estimated execution time for task i (t_i) on machine j (m_j) if m_j has no load when t_i is assigned. If an HC system of m machines has t tasks, we can obtain a $t \times m$ EET matrix. The RT_j is the time that m_j becomes ready after completing the execution of the tasks that are currently assigned to it. The ETC_{ij} is the time at which m_j completes t_i after finishing any previously assigned tasks. From the definitions above, it is easy to get $ECT_{ij} = RT_i + EET_{ij}$. The makespan for a complete schedule is then defined as the time duration from the start to the time the entire tasks are completed. Makespan is a measure of the throughput of an algorithm. In other words, the ready time of a machine after tasks are assigned will be smaller if the makespan is decreased.

The Min-min algorithm computes each task's Minimum Completion Time (MCT) over the available hosts and the task with the minimum MCT is assigned to the best host. The motivation behind the Min-min algorithm is that assigning tasks to the hosts completing them fastest will lead to overally reduced makespan. The Max-min algorithm is similar to the Min-min algorithm except that task with the maximum earliest completion time is assigned. The Max-min algorithm might outperform the Min-min algorithm when there exist more short tasks than long tasks. The Sufferage algorithm assigns a machine to a task that would 'suffer' most in term of expected completion time if that particular machine is not assigned to it.

3 The Proposed Scheduling Algorithm

The proposed new scheduling algorithm shown in Figure 1 is divided into two parts. The initialization step of Line (1) to (4) is similar to the ones in the Min-min and Max-min algorithm. It differs from the Min-min algorithm in that a task is mapped onto a machine as soon as it arrives at the scheduler like the MCT (minimum completion time) algorithm. Each task is examined to determine the machine providing earliest completion time, and then the pre-allocation table and the temporary RT table are updated to reassign the tasks. The MCT algorithm is fast and simple, but it may not assign a task to the best matched machine since it does not consider subsequently arriving tasks those better match the machine. To remedy this sort of miss-matching, each task is reassigned by calculating the expected completion time and the ready time of each machine in Line (5) to (13). In each iteration of the for loop, a task is selected arbitrarily, and the task having earlier completion time than this one is found if it exists. If such task exists, the two tasks are swapped if at least one of the following two conditions is satisfied; i) the ECTs of the two machines are decreased, ii) even though the ECT of one machine is increased, the maximum ready time is decreased. The main objective of the proposed approach is for better matching and load balancing at the same time. When all the iterations of the inner for loop are completed, the temporary RT table of each machine is updated.

```
/* a task is mapped onto a machine as soon as it arrives at the scheduler */
(1) for all m_j (in a fixed arbitrary order)
(2)     calculate the ECT according to EET and RT, and then update
(3)     find the machine m_j with the minimum earliest completion time
(4)     update the temporary RT table and update the pre-allocation table
(5) for each t_k that assigned to each machine (in a fixed arbitrary order)
(6)     for each task t_i for m_j (in a fixed arbitrary order)
(7)         calculate the diff_value of t_i and t_k to m_j and m_l
(8)         if (Temporary RT_l + EET_kl) < (Maximum ready time in temporary RT table)
                && diff_value of task t_i assigned to m_j ≥the diff_value of m_l
(9)                 swap task between m_j and m_l
(10)    for every machine m_j
(11)        if(Temporary RT_j + EET_kj) < (Maximum ready time in temporary RT table)
(12)                deallocate t_k and allocate t_k to m_j
(13)                update the temporary RT table
(14) update the ready time table based on the tasks that were assigned to the machines
(15) update the expected completion time table
```

Fig. 1. The proposed scheduling algorithm

4 Performance Evaluation

Table 1 shows the improvement of the EET of the Max-min, Min-min, and Sufferage algorithm using the proposed algorithm for eight machine system. Here,

Table 1. Improvement of the EETs with the proposed algorithm for eight machine system

Task/ Machine heterogeneity	Max-min	Min-min	Sufferage
Low/ Low	30.1 %	3.5 %	3.4 %
Low/ High	27.6 %	3.7 %	5.3 %
High/ Low	26.7 %	1.3 %	0.5 %
High/ High	29 %	3.7 %	2.6 %

consistent EET table was obtained from the inconsistent EET table by sorting the execution times of the tasks on all machines. Note that the proposed algorithm outperforms the existing algorithms for various degree of task and machine heterogeneity.

5 Conclusion

The mapping algorithms in distributed systems aim at different measure such as makespan, load balancing, and throughput. A good mapping algorithm needs to compromise the conflicting measures. In this paper we have proposed a new task scheduling algorithm which is better than the existing algorithms in terms of throughput and load balancing. It is achieved by task swapping approach based on the expected completion time and ready time of each machine. In inconsistent heterogeneity mode, the makespan of the proposed algorithm is lower than the existing algorithms because it swaps the tasks among the machines for archiving good load balancing. Also, the proposed algorithm has lower time complexity than others.

As a future work we will develop a more optimized algorithm that considers the quality of service, and carry out comprehensive performance evaluation. We will also implement the newly proposed task scheduling algorithm in an actual heterogeneous environment for testing and refinement.

References

1. H. J. Siegel, J. K. Antonio, R. C. Metzger, M. Tan, and Y. A. Li, "Heterogeneous computing." In A. Y. Zomaya (ed.), Parallel and Distributed Computing Handbook, New York, NY: McGraw-Hill, 1996, 725-761.
2. Maheswaran, M. Ali, S. Siegal, H.J. Hensgen, D. Freund, R.F., "Dynamic matching and scheduling of a class of independent tasks onto heterogeneous computting systems," HCW'99, 1999, 30-44.
3. O.H. Ibarra and C. E. Kim, "Heuristic algorithms for scheduling independent tasks on non-identical processors," Journal of the ACM, 24(2), April 1977, 280-289.
4. R. F. Freund, M. Gherrity, S. Ambrosius, M. Campbell, M. Halderman, D. Hensgen, E. Keith, T. Kidd, M. Kussow, J. D. Lima, F. Mirabile, L. Moore, B. Rust, and H. J. Siegel, "Scheduling resources in multiuser, heterogeneous, computing environments with SmartNet," HCW'98, 1998, 184–199.
5. Ali, S. Siegel, H.J. Maheswaran, M. Hensgen, D. "Task execution time modeling for heterogeneous computing systems," HCW'2000, May 2000, 185 – 199
6. M.Y. Wu and W. Shu, "A High-Performance Mapping Algorithm for Heterogeneous Computing Systems," International Parallel and Distributed Processing Symposium (IPDPS), April 2001.
7. Arnaud Giersch, Yves Robert, Frédéric Vivien, "Scheduling Tasks Sharing Files on Heterogeneous Master-Slave Platforms", 12th Euromicro Conference on Parallel, Distributed and Network-Based Processing (PDP'04), February 2004, 11 - 13.

An Approach for Eye Detection Using Parallel Genetic Algorithm

A. Cagatay Talay

Department of Computer Engineering, Istanbul Technical University, 34469 Istanbul, Turkey
talay@cs.itu.edu.tr

Abstract. In this paper, a new reliable method for detecting human eyes in an arbitrary image is devised. The approach is based on searching the eyes with Parallel Genetic Algorithm. As the genetic algorithm is a computationally intensive process, the searching space for possible face regions is limited to possible eye regions so that the required timing is greatly reduced. The algorithm works on complex images without constraints on the background, skin color segmentation and so on. The eye detection process works predictably, fairly, reliably and regardless of the perspective.

1 Introduction

Eye detection is a crucial aspect in many useful applications ranging from face recognition/detection to human computer interface, driver behavior analysis, or compression techniques like MPEG4. A large number of works have been published in the last decade on this subject. Generally the detection of eyes consists of two steps: locating face to extract eye regions and then eye detection from eye window. The face detection problem has been faced up with different approaches: neural network, principal components, independent components, skin color based methods [1, 2]. Each of them imposes some constraints: frontal view, expressionless images, limited variations of light conditions, hairstyle dependence, uniform background, and so on. A very exhaustive review has been presented in [3]. On the other side many works for eye or iris detection assume either that eye windows have been extracted or rough face regions have been already located [4, 5-7]. No much works have been presented in literature that search directly eyes in whole images, except for active techniques: they exploit the spectral properties of pupil under near IR illumination.

The main objectives of this work is to propose an eyes detection algorithm that is applicable in real time with a standard camera, in a real context such as people driving a car (then with a complex background), and skipping the first segmentation step to extract the face region as commonly done in literature. The rest of the paper is organized as follows: Section 2 gives brief information about the Parallel Genetic Algorithms. The search process of eyes is described in Section 3. Finally, in Section 4 conclusions and future works are presented.

V.S. Sunderam et al. (Eds.): ICCS 2005, LNCS 3516, pp. 1004 – 1007, 2005.

2 Genetic Algorithms

A sequential GA proceeds in an iterative manner by generating new populations of strings from the old ones. Every string is the encoded version of a tentative solution. An evaluation function associates a fitness measure to every string indicating its suitability to the problem. The algorithm applies stochastic operators such as selection, crossover, and mutation on an initially random population in order to compute a whole generation of new strings. Unlike most other optimization techniques, GAs maintain a population of tentative solutions that are competitively manipulated by applying some variation operators to find a global optimum. For nontrivial problems this process might require high computational resources, and thus a variety of algorithmic issues are being studied to design efficient GAs. With this goal, numerous advances are continuously being achieved by designing new operators, hybrid algorithms, termination criteria, and more [8]. We adopt one such improvement consisting in using parallel GAs (PGAs) and incorporating some advanced heuristics into an overall genetic algorithm.

PGAs are not just parallel versions of sequential genetic algorithms. In fact, they reach the ideal goal of having a parallel algorithm whose behavior is better than the sum of the separate behaviors of its component sub-algorithms, and this is why we directly focus on them. Several arguments justify our work. First of all, GAs are naturally prone to parallelism since the operations on the representations are relatively independent from each other. Besides that, the whole population can be geographically structured [9] to localize competitive selection between subsets, often leading to better algorithms.

Using PGAs often leads to superior numerical performance even when the algorithms run on a single processor [10, 11]. However, the truly interesting observation is that the use of a structured population, either in the form of a set of islands [12] or a diffusion grid [9], is responsible for such numerical benefits. As a consequence, many authors do not use a parallel machine at all to run structured-population models and still get better results than with serial GAs [13, 14]. Hardware parallelization is an additional way of speeding up the execution of the algorithm, and it can be attained in many ways on a given structured-population GA. Hence, once a structured-population model is defined, it could be implemented in any uniprocessor or parallel machine. There exist many examples of this modern vision of PGAs, namely, a ring of panmictic GAs on a MIMD computer, a grid of individuals on uniprocessor/MIMD/SIMD computers, and many hybrids.

3 Process of Eye Detection

The idea behind this study is quite simple: the eyes can be easily located in the image since the iris is always darker than the sclera no matter what color it is. In this way the edge of the iris is relatively easy to detect as the set of points that are disposed on a circle. The first step in applying PGAs to the problem of feature selection for eye detection is to map the pattern space into a representation suitable for genetic search. Since the main interest is in representing the space of all possible subsets of the original feature list, the simplest form for image base representations considers each feature as a binary gene. Each individual chromosome is then represented as a fixed-length binary string standing for some subset of the original feature list. In this method, first of all the pupil and the edge of the eye are extracted, in addition the position of the eyes is detected more accurately. For the extraction of the eye area, chromosome of individual is set as the first former array composed position information and the size of eye's outlines and pupil. Moreover, fitness of individual is obtained from evaluation function, which pays attention to three features of eyes (difference between white of eye and pupil, color, shape, and size of pupil, edge of eye). The eye area is extracted by chromosome information on the individual with the maximum fitness when evolution completed.

For detection, the pupil is expressed in circle, and the outlines of eye are expressed in the oval, and the pupil is assumed to be at centers of eye. It is defined that the chromosome of individual is composed as X and Y coordinates which shows center of eyes, radius of circle that shows size of pupil and shape of oval which shows outlines of eyes. The first former array of each center coordinates is ten bits, a radius of the circle and oval major axis and minor axis is six bits, total 36 bits, makes. In PGA random initialization is used. Next, whether the defined chromosome is suitable as the eye is decided according to the evaluation function. Since this method pays attention to the features of eye, "Eyes have white of the eye and pupil", "The shape of the pupil is near circle, and the color is a black", "The outline of eye can be approximated to the oval", the following three are used as an evaluation function.

Eye has white of the eyes and pupil, it is feature that high density difference between white area and black area. In a word, if there is a big change in the density value of the pixel in a certain area, it is concluded this area is near eyes or the areas around eyes. Then, products of two high-ranking values of the density difference in the area are used as feature. Where eyes are enclosed, the density change is large in the boundary between white of the eye and pupil, the product has a large value.

It is a feature of the pupil that shape is a circular arc, because upper part of pupil is hidden by above eyelid. The color of pupil looks black in the brightness image and the turn of the pupil in enclosed with white of the eye. Therefore, the portion of black pixel in circular arc and the number of white pixels in circular arc surroundings are used as among of features. In Fig. 1, it is evaluated that area A where eyes are shown has high value. But the case of area B (skin) or area C (eyebrow) does not have proper value of white and black pixels in the circular arc, then the evaluation is bad.

Moreover, the gap of the center is evaluated by obtaining the difference of the radius of circle which is in scribe to the pupil and the circle with the string. As shown in Fig. 2, if the circular arc center shifts from the center of the circle, the difference of

Region A

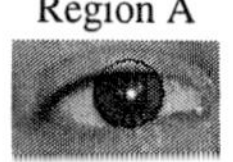

Region B

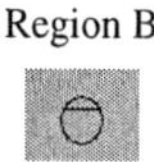

Region C

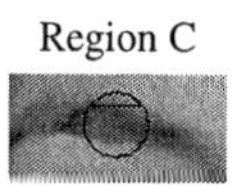

Fig. 1. Evaluation of Circle

Fig. 2. Gap of center

case, black pixels are few and the evaluation is good, but the case of eyebrow and hair, is badly evaluated because a lot of black pixel exists. The difference of shape between oval and outline of eye is obtained by using edge image. The sum of distance from 12 points is obtained, top, bottom, right and left. If the sum distance is big value, the oval and the shape of eyes are greatly different, if the sum is small, those shapes looks like.

4 Conclusions and Future Work

In this study, an effective algorithm for eyes detection in arbitrary images is presented. PGA is used to detect the eyes according to the information based on the features of eyes, like shape of the eye and pupil, white of eye and pupil. The proposed technique does not impose any constraint on the background and does not require any preprocessing step for face segmentation. Eye is detected without receiving charge of the lighting and effect of face of direction. High detection rates have been obtained. The results are surprisingly good also when the eyes are not completely open.

References

1. H. A. Rowley, S. Baluja, T. Kanade: Neural Network-Based Face Detection. IEEE Trans. on Pattern Analysis and Machine Intelligence Vol. 20,No. 1, Jan. 1998, pp 23-38.
2. R. Hsu, M. Mottleb, A. K. Jain: Face Detection in Color Images. IEEE Transaction on Pattern Anlysis and Machine Intelligence Vol24, No. 5, May 2002.
3. M. H. Yang, D. Kriegman, N. Ahuja: Detecting Faces in Images: A Survey. IEEE Transaction on Pattern Analysis and Machine Intelligence Vol. 24, No. 1 January 2002.
4. T. Kawaguchi, M. Rizon: Iris detection using intensity and edge information. Pattern Recognition 36 (2003) 549-562.
5. S. Baskan, M. Bulut, V. Atalay: Projection based method for segmentation of human face and its evaluation. Pattern Recognition Letters 23 (2002) 1623-1629.
6. S. Sirohey, A. Rosenfiled, Z. Duric: A method of detection and tracking iris and eyelids in video. Pattern Recognition 35 (2002) 1389-1401
7. M. Rizon, T. Kawaguchi: Automatic extraction of iris with reflection using intensity information. Proc. of the 3th Iasted Conf. on Vis. Imaging and image processing, 2003.
8. J.H. Holland: Adaptation in natural and artificial sys.. U. of Michigan Pr., Ann Arbor, 1975.
9. P. Spiessens and B. Manderick: A massively parallel genetic algorithm. Proceedings of the 4th Int. Conf. on Genetic Algorithms. R.K. Belew (Ed.) Morgan Kaufmann, (1991).
10. V.S. Gordon and D. Whitley: Serial and parallel genetic algorithms as function optimizers. Procs. of the 5th ICGA. S. Forrest (Ed.) Morgan Kaufmann, (1993).
11. F. Herrera and M. Lozano: Gradual distributed real-coded genetic algorithms. Technical Report #DECSAI-97-01-03, (1997). (Revised version 1998).
12. R. Tanese: Distributed genetic algorithms. Proc. of 3rd ICGA. J.D. Schaffer (Ed.) (1989).
13. V.S. Gordon and D. Whitley: Serial and parallel genetic algorithms as function optimizers. Procs. of the 5th ICGA. S. Forrest (Ed.) Morgan Kaufmann, (1993).
14. E. Alba, J.F. Aldana, and J.M. Troya: A genetic algorithm for load balancing in parallel query evaluation for deductive relational databases. Procs. of the I. C. on ANNs and GAs. D.W. Pearson, N.C. Steele, and R.F. Albrecht (Eds.) Springer-Verlag, (1995).

Graph Representation of the Nested Software Structure

Leszek Kotulski

Institute of Comp. Sci., Jagiellonian University,
Kraków, Poland
kotulski@ii.uj.ed.pl

Abstract. The use of the UML notation for software specification leads usually to lots of diagrams showing different aspects and components of the software system in a several view. Complex components are constructed by composing in parallel some elementary components and as a result the overall architecture of the system is described as a hierarchical composition of primitive components, which at the execution time may be deployed on distributed environment. The task of specifying such a system quickly becomes unmanageable without the help of some structuring tools. This paper discusses how the UML package's nested structure can be maintained and visualized with the help of a graph transformation mechanism based on the edNLC class of grammar.

1 Introduction

The ULM [12] is a recent approach to strengthen the effort of designing an object-oriented modeling language where all main issues of system analysis and design are taken into account. The UML provides deployment diagrams to show a system's network topology and software components that live on the network nodes. For the description of the objects replication and migration, remote interactions and dynamics network topologies, which are important issues in distributed object systems these techniques promise support, however, seems to be insufficient. Modeling distributed systems by distributed graph transformation, a powerful and flexible description, is obtained by applying graph transformation to network structures [10]. Distributed graphs are, in this case, structured with two abstraction levels: the network and the local level. The main drawback of above proposition is lack of support any structuralization mechanisms, such for example as the UML packages. In the paper we show that a graph transformation mechanism is also suitable for the visualization nested software structures. The formal background of this proposition are the ULM notation and the Rosenberg's edNLC class of grammars [5], developed next by Flasiński [2,3] and author [4,7,8].

2 Nested Diagrams

Deployment diagrams, used in the UML, show the configuration of run time processing elements (nodes) and software components (processes and objects) that

V.S. Sunderam et al. (Eds.): ICCS 2005, LNCS 3516, pp. 1008 – 1011, 2005.

execute on them. Nodes are physical processing resources, and are drawn as a 3D rectangle. Each node contains software components. The software components on different nodes can communicate across the physical connections between the nodes.

A component unit represents a modular, deployable, and replaceable part of a system that encapsulates an implementation and exposes a set of interfaces. Some components can be represented by the UML packages that provide a tool for organizing the products of analysis, design, and coding. Let us consider a management system based on a Data Warehouse concept, which stores of data obtained form an ERP systems (see Fig. 1). However, the ERP system consists from many cooperating subsystems (such as `finance`, `sales`, `manufactures` and `human resource` packages – as in fig. 2), because of that someone may want to see direct associations among these subsystems and the Data Warehouse repository. This simple example shows, that exist a necessity of introduction more than only two levels of abstraction (representing hardware or software components). That means that the method of the presentation allocated software components should create several abstraction levels (analogically as in the object oriented programming). We describe here the component and package concept using the graph transformation notation. Like a group [6] the package can be specified as a graph which visualize connections among its components. As a result, the overall architecture of the system is described as a hierarchical composition of primitive components.

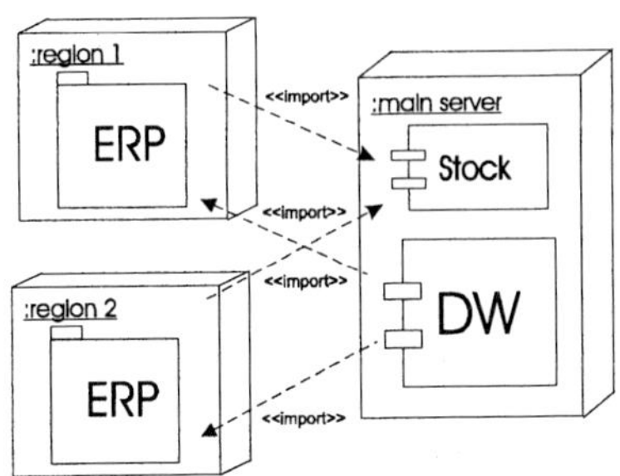

Fig. 1.

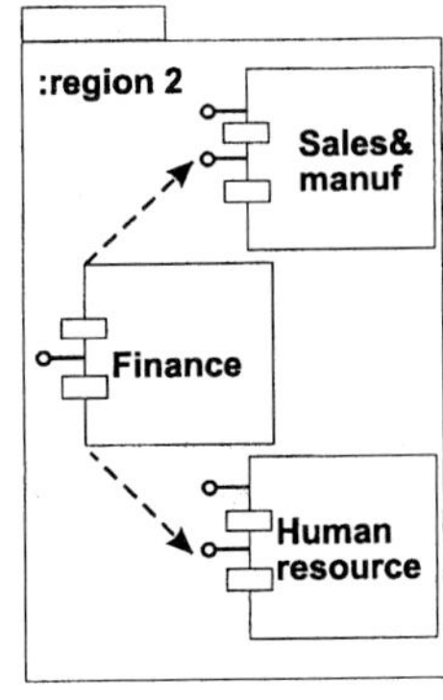

Fig. 2.

Considering visual languages both import and export interface concepts are supported [10, 11, 13]. Component diagrams in the UML knows only export interfaces, but Botch, Rumbaing and Jacobson [1] suggest that in a system consisting from a few dozen components the UML notation should directly specify imports using packages and <<import>> stereotypes. Moreover, in distributed systems the motivation of the import interface visualization is not only a possibility of formal checking the consistency of the modeled system but also the formal specification of stub instances at the implementation level.

To simplify an example we reduce components interconnection to the following: `DW` calls each of the `ERP` components to take theirs monthly summary, `Finance` calls appropriately `Human resource` and `Sales` for receiving daily reports and `Stock` (to evaluate securities). In such a case, the UML diagram presented in fig. 2 can be expressed as the attributed graph presented in fig. 3. It describes two levels of a component's configuration: the top level – representing the `ERP` as a final component and the internal level – representing the `ERP` internal structure; each component appearing inside internal structure can also has its own nesting representation; so it can create the hierarchical (tree) structure of components (where leafs are object instances).

The presented solution is based on the node label controlled graph grammars [2,3,4,5,7,8]. Graph nodes used for describing of a distributed system can be labeled as follows: P - for a package component, E – for an export interface, I – for an import interface (stub), O – for an object instance implementing component, N – for a computing node representation. All additional information about the graph node can be described by its attributes. Graphically we expose only component's name. The only correct labeling pattern[1] is included is the following set of triples (E,b,P), (E,b,O), (P,c,I), (O,b,G), (O,c,I), (N,d,G), (N,d,O), (I,l,E), (N,n,N). Edge labels b, c, d, l and n are appropriately abbreviation for **be**longing, **c**alls, **d**eploy, **l**inked and **n**ode interconnection.

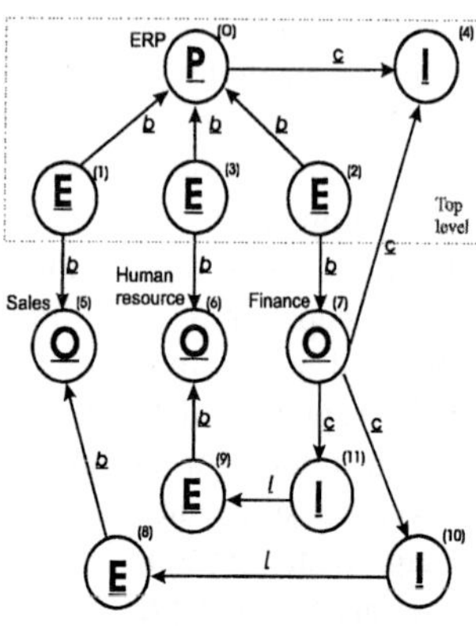

Fig. 3.

The presented transformation from the UML to the graph notation seems to be simple and intuitive. On the other hand, when size both package and allocation graphs grows to hundreds or thousands nodes we should also consider a computational complexity of the considered solution. Let us notice that the membership problem i.e the parsing of the graph in order to designate the proper sequence of productions has the critical complexity for many graph grammars (among others for graph transformation model based on graph morphism [10,11]). For this reason we suggest using of the node label controlled class of graph grammars. The introduced in [7,8] aedNLC graph grammar (as equivalent to ETPL(k) graph grammar [3]) allows one to solve parsing, membership and a new graph generation problems with $O(n^2)$ computational complexity.

The correctness of graph transformation is forced by a graph transformation rules called productions described how the left hand production graph be replaced by the right hand one with the embedding transformation defining how edges coming to/from the left hand side graph be connected with the right hand one. For example if we would like to express that some service can be offered by an object we use the following production with the embedding transformation

$$\underset{O}{\bigcirc}^{(1)} \xRightarrow{P_5} \underset{O}{\bigcirc}^{(1)} \xleftarrow{\;b\;} \underset{E}{\bigcirc}^{(2)}$$

$\mathbb{E}_5$ = {((op,out, 1), {(O,(P,true),op,out), (E,(P,true), pe, in) }), COPY_REST } interpreted as follows every edge labeled by "op" and coming from the removed node in the visualization graph ought to be replaced by the edge (labeled by op) connecting the node of the graph of the righ –hand side of the production and labeled by "O" with the node p of the rest graph and labeled by "P" and the edge (labeled by pe) connecting the node of the graph of the righ –hand side of the production and labeled by "E" with the node p of the rest graph and labeled by "P". Note that edges mentioned in the embedding transformation are the auxiliary ones and are not expressed in the graph presented in fig. 3. The full description of the set of the productions used for creation of the visualization graph is presented in the technical raport [8].

[1] Term labeling pattern for edge (v,μ,w) means triple (δ(v) ,μ,δ(w)), where δ(v) and δ(w) returns labels of v, w node indices appropriately.

3 Conclusions

In the previous chapter it is shown that the graph transformation can be useful formalism for controlling the allocation process. The deployed software component can be split onto a few abstraction levels using packages. It improve a quality of a distributed system presentation; now we can show either a nested structure of a component or an allocation graph in a flat form (where any component is either in their top or internal level of specification; two different, non nested components can be, however, described at the different abstraction level [8)]. Moreover, as the package is specified by the attributed graph, so it can formally check the correctness of its component deployment and the consistency of the generated software (i.e. assure that for each request a proper service is associated). As it shown in [7,8] the aedNLC grammars allows one to divide such an allocation graph onto a few distributed subgraphs and control their modification in a parallel way. The last property is important not only from the computational complexity point of view but also creates the practical possibility of a parallel work of several systems administrators.

References

1. G. Booch, J. Rumbaugh, I. Jacobson: The Unified Modeling Language – User Guide. *Addison Wesley Longman, Inc.* 1999.
2. Flasiński M.: Characteristic of edNLC-graph Grammars for Syntactic Pattern Recognition. *Computer Vision, Graphics and Image Processing,* 1989, vol.42, pp. 1-21.
3. Flasiński M.: Power Properties of NCL Graph Grammars with a Polynomial Membership Problem. *Theoretical Computer Science*, 1998, vol. 201, pp. 189-231.
4. Flasiński M., Kotulski L.: On the Use of Graph Grammars for the Control of a Distributed Software Allocation. *The Computer Journal*, 1992, vol. 35, pp. A167-A175.
5. Janssens D., Rozenberg G., Verraedt R.: On Sequential and Parallel Node-rewriting Graph Grammars. *Computer Graphics and Image Processing*, 1982, vol. 18, pp. 279-304.
6. Kotulski L., Jurek J., Moczurad W.: Object-Oriented Programming in the Large Using Group Concept. *Computer Systems and Software Engineering - 6th Annual European Conference,* Hague 1992, pp. 510-514.
7. Kotulski L.: Model systemu wspomagania generacji oprogramowania współbieżnego w środowisku rozproszonym za pomocą gramatyk grafowych. *Postdoctorals Lecturing Qualifications. Jagiellonian University Press,* ISBN 83-233-1391-1, 2000.
8. Kotulski L.: Parallel Allocation of the Distributed Software Using Node Label Controlled Graph Grammars. *Jagiellonian University, Inst. of Comp. Sci., Preprint no. 2003/006*
9. Kotulski L.: Nested Software Structure generated by aedNLC graph grammar – technical report. *Jagiellonian University, Inst. of Comp. Sci., Preprint no. 2005/003* *http://www.ii.uj.edu.pl/preprint/kotulski03*
10. Taentzer G., Fischer I., Koch M., Vole V.: Visual design of distributed graph transformation. *In Handbook of Graph Grammars and Computing by Graph Transformation, Volume 3: Concurrency, Parallelism, and Distribution. World Scientific,* 1999.
11. Taentzer G.: A Visual Modeling Framework for Distributed Object Computing. *In Formal methods for open object-based distributed systems, Kluwer Academic Publishers,* 2002
12. OMG-Unified Modeling Language, v1,5. *www.rational.com*
13. Zhang D., Zhang K., Cao J.: A context-sensitive Graph Grammar Formalism for the Specification of Visual Languages. *The Computer Journal*, 2001, vol 44, no. 4, pp186-200.

Transaction Routing in Real-Time Shared Disks Clusters*

Kyungoh Ohn, Sangho Lee, and Haengrae Cho

Department of Computer Engineering, Yeungnam University,
Gyungsan, Gyungbuk 712-749, Republic of Korea
hrcho@yu.ac.kr

Abstract. This paper proposes a real-time transaction routing algorithm, which allocates real-time transactions to a node in a shared disks cluster. Unlike traditional routing algorithms, which consider load balancing and transaction affinity only, our algorithm also considers transaction priorities inherent to real-time applications.

1 Introduction

Although there has been a great deal of independent research in real-time processing and shared disks (SD) cluster, their aggregation has very little attention [3]. The cluster technology enables highly available real-time database services, which are the core of many telecommunication services. The cluster can also achieve high performance real-time transaction processing by exploiting inter-node parallelism.

This paper proposes a real-time transaction routing algorithm in the SD cluster. The transaction routing is a process of the front-end router to select an execution node for an incoming transaction. The traditional transaction routing algorithms of the SD cluster have two design goals: *load balancing* and *transaction affinity* [4, 5, 6]. The load balancing means to avoid overloading individual node. The transaction affinity means to execute transactions with similar data access pattern on the same node (*affinity node*). To support real-time transactions, we have additional goal of *transaction priority*. This goal means to reduce the number of transactions missing their deadlines by considering the deadline as a priority. The contribution of this paper is to extend a well-performed traditional algorithm, named DACA (*Dynamic Affinity Cluster Allocation*) proposed by authors [4], to the real-time transaction processing.

2 Algorithm

We propose a new real-time transaction routing algorithm, named P-DACA (*Priority conscious DACA*). In this section, we first describe our underlying model

* This research was supported by University IT Research Center Project.

V.S. Sunderam et al. (Eds.): ICCS 2005, LNCS 3516, pp. 1012–1015, 2005.

and then summarize the basic idea of DACA algorithm [4]. Lastly, we present P-DACA algorithm that extends DACA for real-time transaction processing.

2.1 Assumption

A transaction router selects an execution node of each incoming transaction. The node schedules the execution of its transactions with *earliest deadline first* (EDF) policy [1]. We assume the *mixed* real-time transaction workload [2], which consists of real-time transactions and non real-time transactions. A real-time transaction is assigned with a deadline. Executing the real-time transaction after its deadline is meaningless.

We assume that the record-level locking is in effect. In real-time applications, the locking has to handle the *priority inversion* problem that lower priority transactions block the execution of higher priority transactions. To prevent the problem, we consider the locking model of [2]. Specifically, a real-time transaction aborts non real-time transactions holding locks in conflict mode. On the other hand, a non real-time transaction always waits at lock conflict. The same procedure holds between real-time transactions with different priorities. High priority transactions are always guaranteed to acquire locks without waiting.

2.2 DACA

To alleviate the routing overhead, DACA considers balancing the load of each *affinity cluster* (AC) [5]. An AC collects several transaction classes with high affinity to a given set of tables. The transaction router maintains routing parameters. Specifically, when a transaction router allocates a transaction of an affinity cluster AC_q to a node N_p, it increments both $\#\mathrm{T}(\mathrm{AC}_q)$ and $\#\mathrm{T}(N_p)$, which means the number of active transactions at AC_q and at N_p respectively. Both counters are decremented when a transaction commits or aborts.

There are two overload types: *AC overload* and a *node overload*. The AC overload implies that transactions of an AC are rushed into the system. The node overload occurs when a node N_p is allocated to several ACs and $\#\mathrm{T}(N_p)$ is over average. DACA balances the load of each node according to the overload type. If AC_q is overloaded, then DACA allocates more nodes to AC_q by *node expansion* strategy. If there is no AC overload but N_p is overloaded, then DACA distributes some ACs assigned to N_p to other node by *AC distribution* strategy.

DACA can make an optimal balance between the affinity-based routing and load balancing as follows. First, DACA tries to reduce the number of nodes allocated to the overloaded AC if the load deviation of each node is not significant. This allows DACA to reduce the frequency of inter-node cache invalidations. Next, DACA prohibits allocating both an overloaded AC and other ACs to a node. As a result, DACA can achieve high buffer hit ratio for the overloaded AC. Even though several non-overloaded ACs may be allocated to a single node, efficient handling of the overloaded AC is more important to improve the overall transaction throughput.

2.3 Priority Conscious DACA (P-DACA)

Before describing the details of P-DACA, we first illustrate the problem of DACA when transactions have priorities. Example 1 shows the problem.

Example 1: Suppose there are two ACs (AC_1, AC_2) and two nodes (N_1, N_2). N_1 is an affinity node of AC_1 and executes a transaction t_1 of priority 100. N_2 is an affinity node of AC_2 and executes a transaction t_2 of priority 60. At this time, suppose a new transaction t_3 of priority 70 arrives, and t_3 belongs to AC_1. Then DACA allocates t_3 to its affinity node N_1 if N_1 does not result in overload state. However, t_3 has lower priority than t_1 and cannot be executed until t_1 completes. So t_3 has a higher probability of missing its deadline. On the other hand, if t_3 is allocated to N_2, then it can be executed immediately. □

The goal of P-DACA is (a) to reduce the number of transactions missing their deadlines, and (b) to take advantages of affinity clustering as DACA. To achieve this goal, P-DACA performs the following three basic steps sequentially to decide where a new real-time transaction t_r will be routed to.

1. If there is an affinity node of t_r where the priority of t_r becomes the highest one, then allocate t_r to that node.
2. If there is a non-affinity node of t_r where the priority of t_r becomes the highest one, then allocate t_r to that node.
3. If there is no node that can execute t_r immediately, then allocate t_r to one of its affinity nodes.

The basic steps can be optimized if the transaction router maintains a *priority list* for each node. The priority list is a sorted list in descending order of priorities of active real-time transactions at the node. Then a new real-time transaction t_r has to be allocated to a node where t_r can be executed faster than other nodes. P-DACA checks this condition by comparing the relative position of t_r in the priority list of each node. Suppose $P(t_r)$ means the priority of t_r, and t_r is included in the affinity cluster AC_r. $\mathcal{S}(AC_r)$ is a set of affinity nodes of AC_r. The followings summarize the transaction routing algorithm of P-DACA.

1. #T(AC_r) = #T(AC_r) + 1;
2. If (t_r is a real-time transaction) then
   (a) If ($\exists N_p \in \mathcal{S}(AC_r)$, $\texttt{rank}(P(t_r), N_p) < w_1$ AND $\texttt{rank}(P(t_r), N_p) < \texttt{rank}(P(t_r), N_i)$, for all $N_i \in \mathcal{S}(AC_r)$, $i \neq p$) then goto step 4;
   (b) Else if ($\exists N_p \notin \mathcal{S}(AC_r)$, $\texttt{rank}(P(t_r), N_p) < w_2$ AND $\texttt{rank}(P(t_r), N_p) < \texttt{rank}(P(t_r), N_i)$, for all $N_i \notin \mathcal{S}(AC_r)$, $i \neq p$) then goto step 5;
   (c) Else goto step 3.
3. Select N_p, where #T(N_p) is minimum for all $N_i \in \mathcal{S}(AC_r)$;
4. If (AC_r is overloaded) then call `node_expansion`(AC_r);
5. Else if (N_p is overloaded) then call `ac_distribution`(N_p);
6. #T(N_p) = #T(N_p) + 1; Insert $P(t_r)$ into the priority list of N_p;
7. Return N_p.

At step 2, the function of $\texttt{rank}(P(t_r), N_p)$ returns the number of transactions whose priorities are higher than $P(t_r)$ at N_p. If the function returns 0, t_r will be a transaction with the highest priority at N_p. The values w_1 and w_2 are windowing constraints that limit the acceptable rank of t_r. w_1 is usually bigger than w_2 since an affinity node of t_r could complete t_r faster. Note that if w_1 and w_2 are set to 1, the algorithm works similarly to the above basic steps. A non real-time transaction is allocated to one of its affinity nodes with the lightest load (step 3). If allocating t_r would cause an AC overload or a node overload, then the resolution strategy of DACA has to be performed (step 4 and 5). Example 2 shows how P-DACA can resolve the problem of Example 1.

Example 2: Suppose that the information of ACs, nodes, and transactions are same to Example 1. Suppose also that both w_1 and w_2 are set to 1. Then the transaction router allocates t_3 to N_2 that can execute t_3 immediately. This is because (a) the affinity node of t_3 is N_1 but $\text{rank}(P(t_3), N_1) = 1 = w_1$, and (b) even though N_2 is not an affinity node of t_3 but $\text{rank}(P(t_3), N_2) = 0 < w_2$. Note that if w_1 is set to 2, t_3 is allocated to N_1.□

3 Concluding Remarks

We proposed a new transaction routing algorithm for the real-time SD cluster, named P-DACA. The notable features of P-DACA are two-fold. First, P-DACA allocates a real-time transaction to a node that guarantees higher probability of completing the transaction within its deadline. Even though the transaction could miss its deadline due to succeeding transactions with higher priority, the selection strategy is the best choice at the current state. Next, P-DACA tries to allocate a transaction to its affinity node if possible. So P-DACA can achieve high buffer hit ratio. This reduces the transaction execution time, and the probability of missing deadline can be reduced as a result.

References

1. Lam, K-Y., Kuo, T-W. (ed.): Real-Time Database Systems: Architecture and Techniques. Kluwer Academic Publishers (2000)
2. Lam, K-Y., Kuo, T-W., Lee, T.: Strategies for resolving inter-class data conflicts in mixed real-time database systems. Journal of Syst. and Soft. **61** (2002) 1-14
3. Lee, S., Ohn, K., Cho, H.: Feasibility and Performance Study of a Shared Disks Cluster for Real-Time Processing. Lecture Notes in Computer Science **3397** (2005) 518-527
4. Ohn, K., Cho, H.: Cache Conscious Dynamic Transaction Routing in a Shared Disks Cluster. Lecture Notes in Computer Science **3045** (2004) 548-557
5. Yu, P., Dan, A.: Performance Analysis Clustering on Transaction Processing Coupling Architecture. IEEE Trans. Knowledge and Data Eng. **6** (1994) 764-786
6. Yu, P., Dan, A.: Performance Evaluation of Transaction Processing Coupling Architectures for Handling System Dynamics. IEEE Trans. Parallel and Distributed Syst. **5** (1994) 139-153

Implementation of a Distributed Data Mining System

Ju Cho[1], Sung Baik[1,*], and Jerry Bala[2]

[1] Sejong University, Seoul 143-747, Korea
{jscho, sbaik}@sejong.ac.kr
[2] Datamat Systems Research, Inc.
1600 International Drive, McLean, VA 22102, USA
jbala@dsri.com

Abstract. This paper describes the implementation of a distributed data mining system. The system consists of a web server, a pre-processor for data preparation, a mediator, and agents. A distributed learning algorithm of a decision tree in an agent-mediator communication mechanism is the most important and difficult to achieve the distributed data mining in this system, in the view of implementation. The algorithm has successfully been implemented with several techniques. Its implementation is presented in a UML (Unified Modeling Language) sequence diagram.

1 Introduction

In this paper, we consider a distributed data mining (DDM) approach [1], in which the modified decision tree algorithm on an agent based framework can deal with heterogeneous data sets in the distributed environment [2,3]. The data mining based on the algorithm takes full advantage of all the available data through a mechanism for integrating data from a wide variety of data sources and is able to handle data characterized by geographic (or logical) distribution, complexity and multi-feature representations, and the vertical partitioning/distribution of feature sets.

The paper describes the implementation of an agent based distributed data mining system which consists of a web server, a pre-processor for data preparation, a mediator, and agents. The web server supports users with a web-based interface through which they can access databases located at different sites and manipulate data mining facilities. The pre-processor prepares data sets for data mining by dealing with databases in a distributed way. The implementation of data preparation for DDM is presented in section 2. The mediator coordinates the communication between several agents with security concerns such as authentication. Each agent is located at each heterogeneous data site to achieve coordinated learning through the cooperation of local learning and communication with the other agents. The implementation of the communication, between the mediator and agents for DDM, and the distributed learning algorithm of a decision tree in the communication mechanism is presented in section 3.

---

[*] Author for Correspondence.

V.S. Sunderam et al. (Eds.): ICCS 2005, LNCS 3516, pp. 1016–1019, 2005.

2 Data Preparation for Distributed Data Mining

Data preparation is a preliminary and necessary phase for data mining. In DDM, data sets located in different distributed sites should be processed and converted to appropriate data forms in advance, so that data mining can be performed in a distributed fashion. Fig. 1 shows a UML sequence diagram for data preparation. In our work, two kinds of files are generated from the database for DDM; 1) a control file (e.g. data.ctr) and 2) a learning file (e.g. data.lea). The steps of data preparation are as follows:

1. The selected database schema information and the meta-data from all of different distributed sites are exchanged via communication between the mediator and agents. Each agent deals with each database at its own location.
2. According to the exchanged information, control files are generated at the sites participating for data analysis. Each control file has all schema information obtained from all databases located at different sites.
3. The values of the class field selected by a user are transferred to other sites from the site which the class field belongs to. The class information is found in all example control files presented in Fig. 2.
4. A primary key for associating the data across different sites is selected. The only co-existing values of the primary key field in all databases are selected. And then their associated tuples are selected to build a learning file in each site.

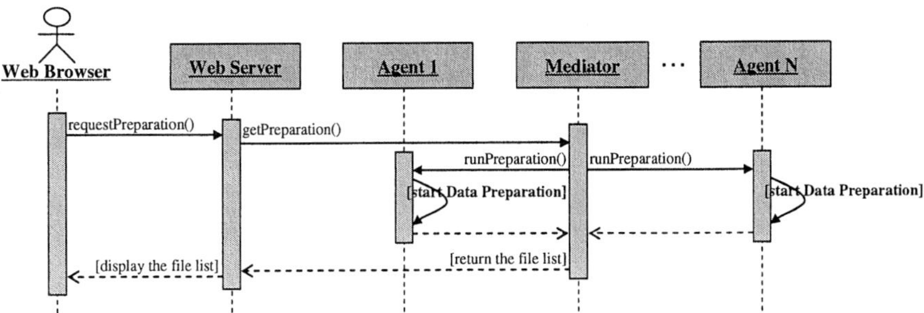

Fig. 1. UML Sequence Diagram for Data Preparation for DDM

3 Distributed Data Mining

A data mining engine was implemented in C in a previous work [4]. The communication interface within the agent was implemented in Java. The communication of the mediator and the agents was implemented in Java on RMI (Remote Method Interface). The data mining engine implemented in C is interfaced with the communication interface implemented in Java by JNI (Java Native Interface). The interfaces of core remote methods for DDM are presented in Fig. 3.

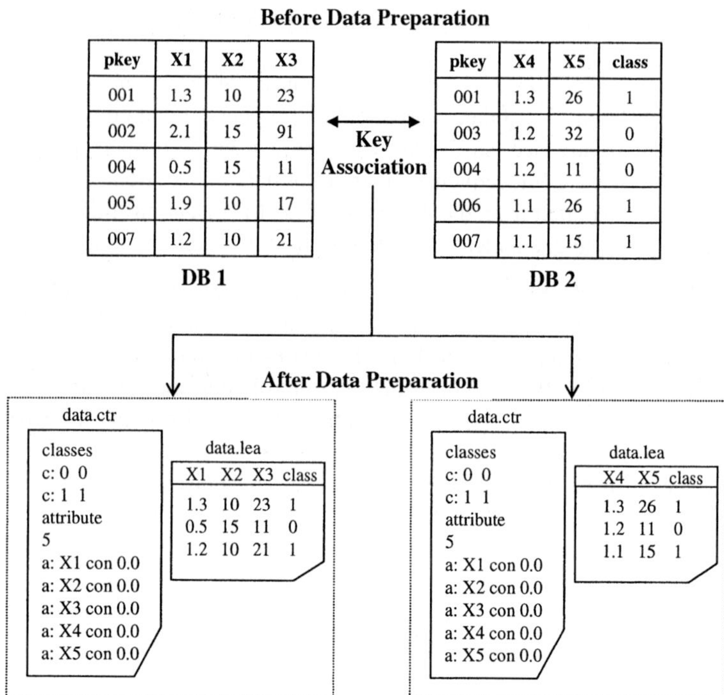

Fig. 2. Examples of Data Sets before/after Data Preparation

```
public interface Mediator extends Remote {
        ...
        public void startDataMining(String fileNm) throws RemoteException;
        public void setEntropy(String agentName, double entropy, int splitData) throws RemoteException;
        public void setDDT(String agentName, byte[] b, int len) throws RemoteException;
        ...
}
public interface Agent extends Remote {
        ...
        public void runDataMining(String fileNm) throws RemoteException;
        public void notifyRole(boolean winOrLooser) throws RemoteException;
        public void getDDT(byte[] b, int len) throws RemoteException;
        ...
}
```

Fig. 3. Interface Methods for the Communication between Mediator and Agent

Fig. 4 presents a UML sequence diagram for DDM. The distributed learning algorithm of a decision tree in an agent-mediator communication mechanism is as follows:

1. runDataMining() : Start the local data mining processes associated with local agents.
2. [find the attribute] : Find the attribute and its associated value that can best split the data into the various training classes during local mining.

3. setEntropy() : Send the best local attribute and its associated value to the mediator.
4. [select the best attribute] : Select the best attribute from the best local attributes of all the agents.
5. notifyRole() : Notify each agent of its role for the next action (splitting or waiting).
6. [split the data] : Split the data, according to the best global attribute and its associated split value, in the formation of two separate clusters of data in the selected agent.
7. setDDT() & getDDT() : Distribute the structural information in each cluster and the best attribute to the other agents through the mediator.
8. [construct the partial decision tree] : Construct the partial decision trees according to the structural information in other agents.
9. [generate rules] : Generate decision rules at each agent and notify the mediator for termination if there is no more splitting. Otherwise, go to step 2.
10. [notify the end of DDM] : Terminate.

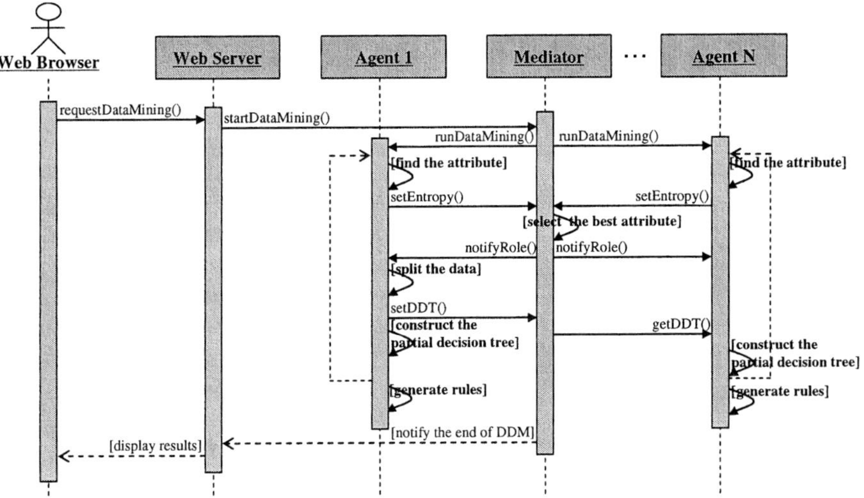

Fig. 4. UML Sequence Diagram for Distributed Data Mining

References

1. S. W. Baik, J. Bala and J. S. Cho: Agent Based Distributed Data Mining, Lecture Notes in Computer Science, Vol. 3320, pp.42-45, 2004
2. D. Caragea, A. Silvescu and V. Honavar: Decision Tree Induction from Distributed, Heterogeneous, Autonomous Data Sources, Proceedings of the Conference on Intelligent Systems Design and Applications (ISDA 03), 2003
3. C. Giannella, K. Liu, T. Olsen and H. Kargupta: Communication Efficient Construction of Decision Trees Over Heterogeneously Distributed Data, Proceedings of the Fourth IEEE International Conference on Data Mining (ICDM 04), pp.67-74, 2004
4. J. Bala, S. W. Baik, S. Gutta and P. Pachowicz: InferView: An Integrated System for Knowledge Acquisition and Visualization, Proceedings of AFCEA Federal Data Mining Symposium, 1999

Hierarchical Infrastructure for Large-Scale Distributed Privacy-Preserving Data Mining

Jinlong Wang, Congfu Xu, Huifeng Shen, and Yunhe Pan

Institute of Artificial Intelligence, Zhejiang University,
Hangzhou, 310027, China
zjupaper@yahoo.com
xucongfu@cs.zju.edu.cn
yaekee@hotmail.com
panyh@sun.zju.edu.cn

Abstract. Data Mining is often required to be performed among a number of groups of sites, where the precondition is that no privacy of any site should be leaked out to other sites. In this paper, a hierarchical infrastructure is proposed for large-scale distributed Privacy Preserving Data Mining (PPDM) utilizing a synergy between P2P and Grid. The proposed architecture is characterized with (1) its ability for preserving the privacy in data mining; (2) its ability for decentralized control; (3) its dynamic and scalable ability; (4) its global asynchrony and local communication ability. An algorithm is described to show how to process large-scale distributed PPDM based on the infrastructure. The remarks in the end show the effectiveness and advantages of the proposed infrastructure for large-scale distributed PPDM.

1 Introduction

Nowadays, data mining is one of the most important topics in large-scale distributed domains such as military, commerce and health-care etc. In the meanwhile, data privacy becomes a major concern that threatens the widespread deployment of data mining systems. This leads to a new sub-area of data mining — privacy-preserving data mining (PPDM) [1] in large-scale distributed systems.

In this paper, we propose a hierarchical infrastructure for large-scale distributed PPDM. It synthesizes both Grid and P2P, and constructs a hierarchical architecture (super computers in the top level based on Grid [2], desktop computers and other pervasive computers in the bottom level, organized with P2P [3], which could thus help to ensure Grid scalability). In this way, it can achieve higher performance and satisfy the requirements of the reality better.

The remainder of the paper is organized as follows: section 2 describes the hierarchical infrastructure. Section 3 discusses the privacy preserving technology and cryptographic primitives. Section 4 gives a large-scale distributed PPDM algorithm based on the hierarchical infrastructure. Finally, we conclude with some remarks in the end.

V.S. Sunderam et al. (Eds.): ICCS 2005, LNCS 3516, pp. 1020–1023, 2005.

2 Hierarchical Infrastructure

The main purpose of the infrastructure is to supply a platform for large-scale distributed PPDM. The overall framework is shown as Fig1 below.

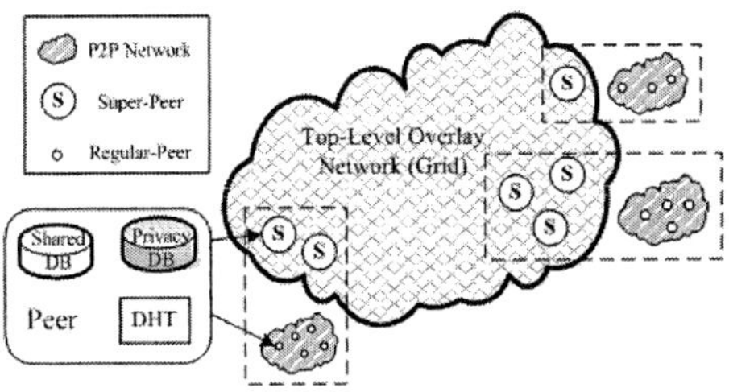

Fig. 1. Overall Framework of Architecture

Just as the scenario of Fig1, each peer (regular peer and super-peer) owns privacy database (PDB) which may not be open to other peers, shared database (SDB) which may be shared by other peers, and dynamic hash table (DHT) which is the table of route, indicating those peers the messages will be sent to. All of peers are organized into disjoint virtual groups (also called virtual organizers, VOs) by a series of protocols. Each VO maintains its own overlay network, including one or more super-peers, which are a subset of peers according to their computing resources and the data volume they store. These super-peers constitute a top-level overlay network (Grid [2]).

A VO may join and leave the system as a whole. In a VO, the numbers of peers can be dynamically changed by peers joining and leaving in any time. When a peer joins the system, it registers with one of existing peers. A peer can only have one connection to another peer higher in the hierarchy to register with, but be registered with many lower level peers. A peer communicates with its upper and lower peers through DHT. When a peer leaves the system, its lower peers must be informed to dynamically adjust the network topology to keep the hierarchical relations. The hierarchical protocol makes the larger scalable systems possible and can support the diversity of the cryptographical primitives.

Since the classification of group is according to the node distances and functions, there will be much more frequent communication among the same group than the one between groups. And, in a VO, we make use of asymmetrical bidirectional patterns to improve communication efficiency. Upward communication transfers all types of information, including system messages (peers joining or leaving) and data messages (data updating) to update up-peers, the downward communication transfers only system messages to adjust to the dynamic topology changes. In this way, the architecture can use the bandwidth more effectively, gain the advantages over Grid or P2P respectively, and thus suit a great variety of network environments better.

3 Privacy-Preserving and Cryptographic Primitives

Based on the hierarchical architecture, multiplicity of privacy-preserving methods can be used. Literature [4] provides an overview of the new and rapidly emerging research area of PPDM. Making use of these methods, we can protect sensitive raw data and knowledge.

Within the different groups, we can utilize various cryptographic methods. Even in the same group, diversity of the security methods can also be sustained based on the demands.

In intra-group, the peers receive the messages below and in the same subset with them, but not directly mine the peers below and adjacent to them, which protects the privacy of the raw data. At the same time, the peers can modulate the messages sent to protect their sensitive knowledge. Based on the peer adjacent to it, we can shield the individual data sources to the peers not adjacent to it.

In the top level, the super-peers can use the cryptograph-based techniques to encrypt their private data. And none of peers are willing to disclose their own outputs to anyone else. With the cryptographic primitives, individuals and sources are protected.

4 Large-Scale Distributed PPDM on the Architecture

In this section, we apply the architecture to PPDM in large-scale distributed systems. In the interest of clarity, we consider the system where each group comprises a super-peer.

Just as the above scenario of Fig1, within each group, the super-peers are in charge of the hierarchical information of various sharing resources, each super-peer manages a range of peers, which may be the super-peer's neighbor or in the same subnet with it. Regular peers collaboratively mine data in P2P.

In P2P, for each peer, it receives the information from other peers, and then processes these information together with its PDB. After that, it will update its own PDB and transform the message sent to the SDB according to security strategies. Next, through its own DHT, it sends relevant fresh information in its SDB to other up-peers adjacent to it for furthering processing. In this way, the peer in the infrastructure consists of the knowledge mined from those peers below and in the same subnet with it. Layer upon layer, the super-peer comprises the up-to-date knowledge mined from the same group. Finally, in the top level, on the basis of the cryptograph-based techniques, we can process data mining with the Knowledge Grid [5] among the super-peers. The large-scale distributed PPDM algorithm based on the architecture is depicted as the algorithm 1.

Within the hierarchical distributed architecture, the communication cost is evidently reduced to a large extent by never transmitting all data values but transmitting incremental and updated information. This approach does not require CPU and I/O costs significantly higher than a similar approach and its communication may be lower.

Algorithm 1: large-scale distributed PPDM

Data: Peers set $P = \{p_i | i \in N\}$(the peers constitutes the hierarchical infrastructure), and their own local databases.
Result: The knowledge mined from the hierarchical infrastructure.

```
begin
    while TRUE do
        Step 1. Peers p_i mines its local data and make use of the messages
        inputted from other peers;
        if p_i is super peer then
            break;
        Step 2. Every p_i modulates its knowledge on the basis of demand, hides
        the sensitive knowledge and sends its encrypted data and knowledge to
        up-peer p_j in DHT adjacent to p_i;
    Process PPDM among super-peers through Knowledge Grid.
end
```

5 Conclusions

In this paper, we present a dynamic, hierarchical infrastructure for PPDM in large-scale loosely connected and constantly evolving environments. The bidirectional communication method can save more bandwidth. By adopting asynchrony communication pattern, our infrastructure avoids centralized control and improves the ability in suiting dynamic changes in current commercial network environment. In particular, by virtue of the hierarchial model based on P2P-Grid, we can supply diversities of strategies for preserving both the privacy data sources and the sensitive knowledge. This property is especially important in industry, business and scientific applications.

Acknowledgements

This paper was supported by the Natural Science Foundation of China (No. 60402010) and the Advanced Research Project sponsored by China Defense Ministry (No. 413150804, 41101010207), and was partially supported by the Aerospace Research Foundation sponsored by China Aerospace Science and Industry Corporation (No. 2003-HT-ZJDX-13).

References

1. R. Agrawal, R. Srikant. Privacy-preserving data mining. SIGMOD 2000.
2. I. Foster, C. Kesselman, S. Tuecke. The Anatomy of the Grid: Enabling Scalable Virtual Organizations. International J. Supercomputer Applications, 15(3), 2001.
3. M. P. Singh. Peer-to-Peer Computing for information systems. AP2PC 2002.
4. V S. Verykios, E. Bertino, I N. Fovino, L P. Provenza, Y. Saygin, Y. Theodoridis. State-of-the-art in Privacy Preserving Data Mining. SIGMOD Record 2004, 33(1).
5. M. Cannataro, D. Talia. KNOWLEDGE GRID: An Architecture for Distributed Knowledge Discovery. Communications of the ACM, January 2003 46(1) 89-93.

Prediction of Protein Interactions by the Domain and Sub-cellular Localization Information[¶]

Jinsun Hong and Kyungsook Han[*]

School of Computer Science and Engineering, Inha University, Inchon 402-751, Korea
khan@inha.ac.kr

Abstract. There has been a recent interest in the computational methods for predicting genome-wide protein interactions due to the availability of genome sequences of several species, the difficulty of detecting whole protein interactions in higher species even with the current high-throughput experimental methods, and the common perception of the data generated by high-throughput experimental methods as noisy data. However, data predicted by computational methods as well as that detected by high-throughput experimental methods inherently contain extremely many false positives. Several methods have been developed for estimating the reliability of experimental protein interaction data, but there are few for predicted interaction data. This paper presents a prediction method of protein-protein interactions using the protein domain and sub-cellular localization information, and experimental results of the method to the protein-protein interactions in human.

1 Introduction

An intrinsic problem with high-throughput methods for detecting protein-protein interactions is that data generated by the methods are extremely noisy, even more so than is the case for gene expression data, so one cannot simply use the data blindly [1]. More than half of current high-throughput data are estimated to be spurious [2]. Although it is possible to focus on interactions with higher reliability using only those supported by two or more sources of evidence, this approach invariably throws out the majority of available data [1].

In an attempt to improve the reliability of interaction data, we have recently generated a new set of interaction data using protein domain [3-5] and sub-cellular localization. This paper presents a scoring scheme for assessing the reliability of the protein interaction data predicted by a computational method called homologous interactions, and experimental results of the scheme to the protein-protein interactions in human. The scoring method is more general than a typical method of selecting co-functional or co-localized protein pairs only since some pairs of proteins interact more frequently than others despite their different functions or locations in a cell. The protein-protein interaction data predicted by the method was compared to the

[¶] This work was supported by the Ministry of Information and Communication of Korea under grant IMT2000-C3-4 and by KOSEF through the Systems Bio-Dynamics Research Center.

[*] To whom correspondence should be addressed.

V.S. Sunderam et al. (Eds.): ICCS 2005, LNCS 3516, pp. 1024 – 1027, 2005.

experimental data in HPRD (http://www.hprd.org/), and the results show that the method predicts reliable protein-protein interactions and is useful for assessing the reliability of protein-protein interactions.

2 Prediction Methods and Experimental Results

The prediction system of protein-protein interactions is composed of two parts: predicting protein interactions and assessing the reliability of the predicted interactions (Fig. 1).

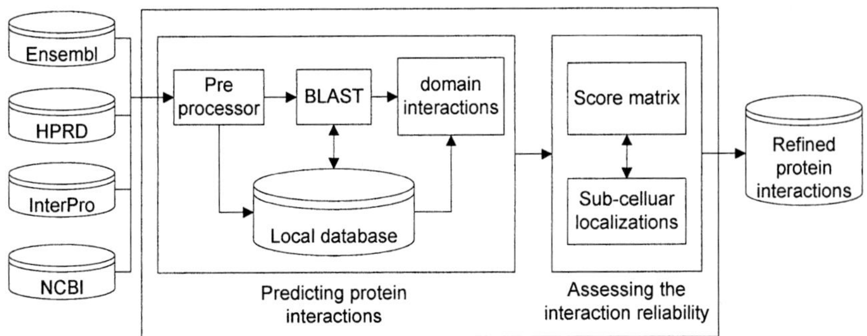

Fig. 1. Architecture of the prediction system of protein-protein interactions

The interaction prediction part extracts relevant data (for example, protein names and sequences, sub-cellular localizations, domains and their amino acid sequences, and known protein-protein interactions) from Ensembl (http://www.ensembl.org/), HPRD (http://www.hprd.org/), InterPro (http://www.ebi.ac.uk/interpro/), and NCBI (http://www.ncbi.nih.gov/), and removes redundant data. Suppose a protein X with domains a, b, and c and protein Y with domains d and e. When proteins X and Y interact each other, we can predict 6 domain-domain interactions of (a, d), (b, d), (c, d), (a, e), (b, e), (c, e), which are in turn used to predict new protein-protein interactions. PSI-BLAST [6] is used to identify domains from protein sequences.

The reliability assessing part constructs a score matrix from the interaction frequency at a specified sub-cellular localization, and the score matrix is used to evaluate the reliability of predicted protein-protein interactions. The information on protein, protein-protein interactions, and sub-cellular localization was obtained from HPRD that has human protein-protein interactions with the sub-cellular localization information. The sub-cellular localization schema of MIPS (http://www.mips.gsf.de/) was used to classify 20 sub-cellular localizations in human.

The sub-cellular localization score (S_l) was computed using equation 1, in which $I_l(p_i, p_j)$ is the number of interactions between proteins p_i and p_j in a same compartment l, and N is the number of proteins that participate in protein-protein interactions. Interactions with the sub-cellular localization score $S_l = 0$ are the cases in which source or target protein has no sub-cellular localization information.

$$S_l(p_i,p_j) = \frac{1}{N}\sum I_l(p_i,p_j) \qquad (1)$$

We constructed 2,272 domain-domain interactions between 1,712 human domains and predicted 5,188 protein-protein interactions between 4,205 human proteins. In the reliability assessing part, localization ID from 1 to 20 was assigned to 5,188 protein-protein interactions and the value of a score matrix was applied to these interactions. 3,839 protein-protein interactions have higher value than 0 and others have value of 0.

We compared the predicted data by our system to the experimental data in HPRD. Fig. 2 shows S_l scores in the range from 0 to 2 for the two data sets. A vertical axis presents an accumulated percentage at each score. In the experimental data of HPRD, there were 1,336 protein-protein interactions (37%) of 3,611 protein-protein interactions with $S_l = 0$. In the predicted data by our system, there were 1,349 protein-protein interactions (26%) of 5,188 protein-protein interactions with $S_l = 0$. This indicates that the prediction system generates more reliable interactions than the experimental data of HPRD.

Fig. 3 shows the sub-cellular localization distribution that includes cytoplasm, extra-cellular, membrane, nucleus between existing experimental 1,179 interactions obtained from BIND database and 5,188 interactions predicted by our system. In general, our system contains more protein interactions than existing experimental data in most cell compartments, and much more protein interactions in nucleus.

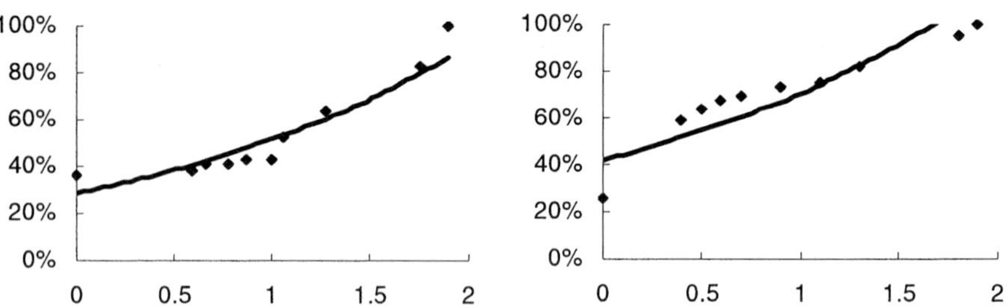

Fig. 2. S_l scores of the HPRD data (left) and those of our system (right)

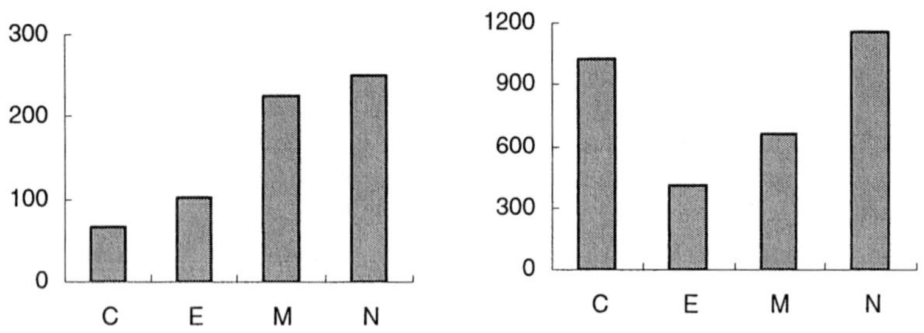

Fig. 3. The sub-cellular localization distribution of the BIND data (left) and our data (right). C: Cytoplasm, E: Extra-cellular, M: Membrane, N: Nucleus

We also computed the Interaction Generality (IG) values [7] to assess the reliability of protein-protein interactions. The IG value of an interacting protein pairs A-B is defined as the number of proteins directly interacting with protein A or B. Proteins with IG value less than 5 were 76% of proteins that participates in 3,839 predicted proteins-protein interactions.

3 Conclusions

This paper presented an intuitive, computationally feasible scoring scheme for measuring the reliability of a large-scale data set of protein-protein interactions, and for filtering potentially false positives in the data set. The number of potential artifacts can be substantially reduced be removing data with low score from the data set. Since the reliability score is computed for individual protein interactions in a data set, and the score can be directly used to filter out spurious interactions (that is, those interactions with low scores) from the data set. The overall reliability of the entire data set can also be computed by taking the average scores of the interactions in the data set. In this study we computed sub-cellular localization scores only, but the scoring scheme is general enough to incorporate any number of subscores depending on the sources of information available.

References

1. D'haeseleer, P., Church, G.: Estimating and improving protein interaction error rates. Proceedings of the Computational Systems Bioinformatics Conference (2004) 208- 215
2. von Mering, C, Krause, R., Snel, B., Cornell, M., Oliver, S.G., Fields, S., Bork, P.: Comparative assessment of large-scale data sets of protein–protein interactions. Nature 417 (2002) 399-403
3. Antonina, A., Dave, H., Steven, E.B., Tim, J. P.H., Cyrus, C., Alexey, G. M.: SCOP database in 2004: refinements integrate structure and sequence family data. Nucl. Acids. Res. 32 (2004) D226-D229
4. Einat, S., Hanah, M.: Correlated Sequence-signatures as Markers of Protein-Protein Interaction. J. Mol. Biol. 311 (2001) 681-692
5. Lappe, M., Park, J., Niggemann, O., Holm, L.: Generating protein interaction maps from incomplete data: application to fold assignment. Bioinformatics 17(2001) 149-156
6. Altschul, S.F., Madden, T.L., Schffer, A.A., Zhang, J., Zhang, Z., Miller, W., and Lipman, D.J.: Gapped BLAST and PSI-BLAST: a new generation of protein database search programs. Nucleic Acids Res. 25 (1997) 3389-3402
7. Saito, R., Suzuki, H., Hayashizaki., Y.: Interaction generality, a measurement to assess the reliability of a protein-protein interaction. Nucleic Acids Res. 30 (2002) 1163-1168

Online Prediction of Interacting Proteins with a User-Specified Protein[¶]

Byungkyu Park and Kyungsook Han[*]

Department of computer Science and Information Engineering,
Inha University, Inchon 402-751, Korea
khan@inha.ac.kr

Abstract. The protein-protein interaction data available today in databases were either determined by experimental methods or predicted by computational methods. Biologists studying a certain mechanism or function are usually interested in a small-scale network of protein-protein interactions related to the mechanism or function of their interest rather than a genome-wide network of protein-protein interactions. We have previously developed a database server that can predict interactions between the proteins submitted by a user. However, the server cannot find proteins from the databases that potentially interact with a protein given by a user. Finding potentially interacting proteins with a protein of interest is more useful than finding interactions between given proteins, but more difficult partly because it involves intensive computation of searching data in the databases and because different databases have different accession numbers and names for a same protein. This paper describes the new online system for predicting interacting proteins with a user-specified human protein and its preliminary results. We believe that this is the first online system for predicting interacting proteins with a given protein and that it is a useful resource for studying protein-protein interactions.

1 Introduction

A genome-wide network of protein-protein interactions in higher species such as human is very huge and complex, and therefore subnetworks of smaller-scale are usually preferred to a whole network for close examination. However, small-scale subnetworks of protein-protein interactions of interest are not readily available from databases partly because computing all possible subnetworks involves prohibitive computation of searching data in the databases and because different databases have different accession numbers and names for a same protein.

We previously developed a database called the human protein interaction database (HPID; http://www.hpid.org/) for predicting potential interactions between proteins submitted by users as well as for providing human protein interaction information pre-computed from existing structural and experimental data [1, 2]. However, the

¶ This work was supported by the Ministry of Information and Communication of Korea under grant IMT2000-C3-4 and by KOSEF through the Systems Bio-Dynamics Research Center.

* To whom correspondence should be addressed.

V.S. Sunderam et al. (Eds.): ICCS 2005, LNCS 3516, pp. 1028 – 1031, 2005.

previous version of HPID (HPID version 1.0) is not capable of finding potentially interacting partners for a given protein. As an extension of HPID 1.0, we now present a new online prediction system that is capable of predicting interacting proteins with a user-specified human protein. The online prediction system uses the data derived from HPID 1.0 (http://www.hpid.org/), BIND (http://bind.ca/), DIP (http://dip.doe-mbi.ucla.edu/) and HPRD (http://www.hprd.org/) for human protein interactions. It also recognizes the protein IDs of EMBL (http://www.ebi.ac.uk/embl/), Ensembl (http://www.ensembl.org/), HPRD and NCBI (http://www.ncbi.nlm.nih.gov/) to search for proteins. This paper discusses the main concepts of the work and its results.

2 Searching Interacting Proteins

The protein interactions were determined on the SCOP [3] Protein Structural Interactome MAP (PSIMAP) [4]. The structural interactions of human proteins were predicted by a homology-based assignment of the domain structures to the whole genome.

Unlike HPID 1.0, HPID 2.0 is not dependent on protein IDs to find interacting partners of proteins. Given a protein sequence, interacting partners with the protein are found by homology search of protein sequences from relevant databases. HPID 2.0 currently considers Ensembl, BIND, DIP, HPRD, and NCBI as relevant databases for homology search. To predict interacting partners of a protein, HPID 2.0 first identifies the domains of the given protein using the information from three databases of superfamily (http://supfam.mrc-lmb.cam.ac.uk/SUPERFAMILY/) [5], InterPro (http://www.ebi.ac.uk/InterProScan/) and Pfam (http://www.sanger.ac.uk/). Fig. 1 shows the overall architecture of the HPID online prediction system.

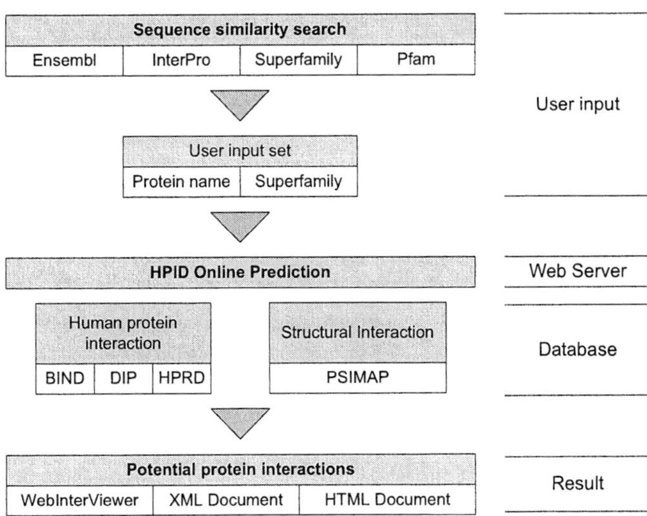

Fig. 1. The architecture of the HPID online prediction system. Potential interaction proteins are shown in three formats: protein interaction networks visualized by WebInterViewer [6-8], XML or HTML documents

3 Example of Predicting Interacting Proteins

This section shows an example of predicting interacting partners of a protein given by a user. Suppose that a user wants to know interacting proteins with human protein ENSP00000046794. HPID finds 17 interacting partners from the experimental data and 2,101 interacting partners from the prediction data of HPID. 13 out of the 17 interacting partners were also found in the prediction data; that is, the experimental and prediction data have 13 proteins in common. 4 out of the 17 proteins were not found in the prediction data. The reason that the 4 proteins were not found in the prediction data can be explained by the possibility that (1) the proteins have no superfamily assigned to them, or (2) PSIMAP does not have a pair of interacting superfamilies associated with the proteins.

Table 1 shows the superfamilies and functions of the 13 interacting partners that were found both in the experimental and prediction data. The functions of the 13 proteins can be clustered into a few functional groups, which agree with other research results [9, 10]. HPID also provides the information of superfamily, function, sub-cellular localization of potential interaction proteins, which can be further validated by experimental methods.

Table 1. Superfamilies and functions of the 13 proteins interacting with human protein ENSP0000046794

Interaction partners	Superfamily	Gene_Ontology (process, function)
ENSP00000262512	d.93.1	SH3/SH2 adaptor protein activity
ENSP00000315460	d.93.1	SH3/SH2 adaptor protein activity
ENSP00000339007	d.93.1	SH3/SH2 adaptor protein activity, epidermal growth factor receptor binding
ENSP00000339186	d.93.1	SH3/SH2 adaptor protein activity
ENSP00000244007	d.93.1	phosphoinositide phospholipase C activity, receptor signaling protein activity, calcium ion binding, hydrolase activity
ENSP00000343423	d.93.1	phosphoinositide phospholipase C activity, receptor signaling protein activity, calcium ion binding, hydrolase activity
ENSP00000307961	d.93.1	protein tyrosine phosphatase activity, hydrolase activity
ENSP00000264033	g.44.1	transcription factor activity, signal transducer activity, calcium ion binding, ligase activity
ENSP00000302269	d.93.1	transcription factor activity, guanyl-nucleotide exchange factor activity, diacylglycerol binding
ENSP00000288986	d.93.1	receptor binding, cytoskeletal adaptor activity, receptor signaling complex scaffold activity
ENSP00000263405	b.34.2	receptor binding, protein binding
ENSP00000316460	b.34.2	receptor binding, protein binding
ENSP00000221409	d.144.1	

4 Conclusion

Despite a large volume of protein-protein interaction data, small-scale networks of protein-protein interactions of interest are not readily available from databases partly because computing all possible subnetworks involves prohibitive computation of searching data in the databases and because different databases have different accession numbers and names for a same protein. We constructed an online prediction system called HPID 2.0 for searching for interaction partners of a protein given by a user. The online prediction system uses the data derived from HPID 1.0 (http://www.hpid.org/), BIND (http://bind.ca/), DIP (http://dip.doe-mbi.ucla.edu/) and HPRD (http://www.hprd.org/) for human protein interactions. It also recognizes the protein IDs of EMBL (http://www.ebi.ac.uk/embl/), Ensembl (http://www.ensembl.org/), HPRD and NCBI (http://www.ncbi.nlm.nih.gov/) to search for relevant proteins. HPID 2.0 is the first online system for predicting interacting proteins with a given protein and will be a useful resource for studying protein-protein interactions.

References

1. Han, K., Park, B., Kim, H., Hong, J., and Park, J.: HPID: The Human Protein Interaction Database. Bioinformatics 20 (2004) 2466 – 2470
2. Han, K. and Park, B.: A Database Server for Predicting Protein-Protein Interactions. Lecture Notes in Computer Science 3036 (2004) 271-278
3. Lo Conte, L. Brenner, S.E. Hubbard, T.J.P., Chothia, C., Murzin, A.G.: SCOP database in 2002: refinements accommodate structural genomics. Nucl. Acids. Res. 30 (2002) 264-267
4. Park, J., Lappe, M., Teichmann, S.: Mapping protein family interactions: intramolecular and intermolecular protein family interaction repertoires in the pdb and yeast. J. Mol. Biol. 307 (2001) 929–938
5. Madera, M., Vogel, C., Kummerfeld, S. K., Chothia, C. and Gough, J.: The SUPERFAMILY database in 2004: additions and improvements., Nucl. Acids. Res. 32 (2004) Database issue D235-D239
6. Ju, B.-H., Park, B., Park, J. and Han, K.: Visualization and analysis of protein interactions. Bioinformatics 19 (2003) 317–318
7. Ju, B.-H. and Han, K.: Complexity management in visualizing protein interaction networks. Bioinformatics 19 (2003) i177–i179
8. Han, K. and Ju, B.-H.: A fast layout algorithm for protein interaction networks. Bioinformatics 19 (2003) 1882–1887
9. Saito, R., Suzuki, H., and Hayashizaki, Y.: Interaction generality, a measurement to assess the reliability of a protein-protein interaction. Nucl. Acids. Res. 30 (2002) 1163-1168
10. Oliver, S.: Guilt-by-association goes global. Nature 403 (2000) 601-603

An Abstract Model for Service Compositions Based on Agents*

Jinkui Xie and Linpeng Huang

Dept. of Computer Science and Engineering, Shanghai Jiao Tong University,
Shanghai 200030, P.R. China
{jkxie, huang-lp}@cs.sjtu.edu.cn

Abstract. This paper presents an abstract model for service compositions based on agents. The model is based on distributed Abstract State Machine and tallies with Universal Plug and Play (UPnP) architecture standard. In the abstract model, the composition of services can be viewed as a union of agents. This model can be used to depict a number of Web Service applications, and as an instance, the paper presents a BPEL Abstract Machine.

1 Introduction

This paper presents an abstract model for service compositions. The abstract model constructs the whole service compositions system as a collection of communicating agent subsystems. Each subsystem's communication structure is analogous in the essence. The model is open and has a Universal Plug and Play (UPnP) architecture [7]. We set up the model with distributed Abstract State Machine, and give some descriptions to BPEL (Business Process Execution Language for Web Services) [6].

2 Abstract State Machines

We first define the notion of Abstract State Machines. For more details, we refer to the Lipari-Guide [4] and the ASM 1997 Guide [5]. An *Abstract State Machine* (ASM) is a tuple $(\Sigma, \Phi_{Init}, Trans)$ where Σ is a signature, Φ_{Init} is a set of Σ-formulas (the *initial conditions*), and $Trans$ is a finite set of *transition rules*. The set of *states* is the set $\mathrm{Alg}(\Sigma)$ of Σ-algebras; $[\![\cdot]\!]_q$ denotes the interpretation function of symbols of Σ in Σ-algebra q. A state q is *initial* iff q is a model of Φ_{Init} in the sense of logic, denoted as $q \models \Phi_{Init}$. In this article, we use order-sorted partial Σ-algebras.

For the service compositions' purpose, we introduce A *distributed ASM* (DASM) involves a collection of *agents* [3]. Agents are represented in global states as well. They are elements of a dynamically universe *Agent* that may

* This work is supported by "SEC E-institute: Shanghai High Institutions Grid" project 200308.

V.S. Sunderam et al. (Eds.): ICCS 2005, LNCS 3516, pp. 1032–1035, 2005.

grow and shrink. Intuitively, a run can be seen as the common part of histories of the same computation recorded by various observers.

3 Service Compositions Based on Agents

Research in the Web Service model is a hot area. In our abstract model, Web Service infrastructure can be divided into three layers: the Service Requestor Layer, the Middle Agent Layer, and the Service Layer [1]. Moreover, in this system, the Agent category has three subcategories: the *Service Requestor Agent* (SRA), the *Service Provide Agent* (SPA), and the *Coordination Agent* (COA). Among these subcategories, the agents are flexible and can transform their roles in the Web Service context, so the whole composition system is a Universal Plug and Play (UPnP) architecture [7].

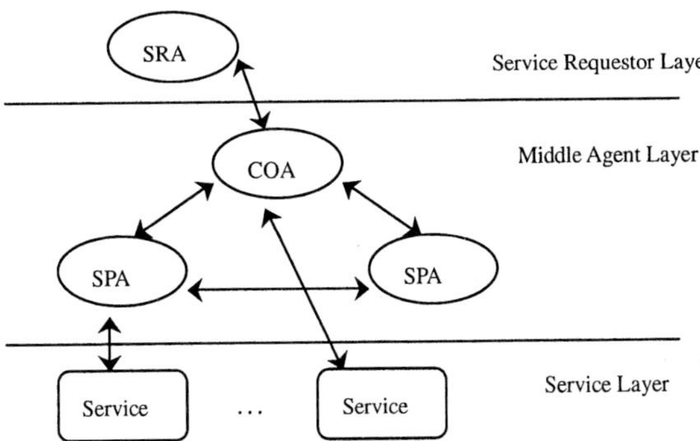

Fig. 1. Agent-based service compositions framework

The agent-based service compositions framework is presented on Fig. 1. In the model, there are three basic operations: (1) $agent_1 \circ agent_2$: it represents a composite service that performs the service $agent_1$ followed by the service $agent_2$, i.e., $\circ$ is an operator of *sequence*. (2) $agent_1 + agent_2$: it represents a composite service that behaves as either service $agent_1$ or service $agent_2$. Once one of them executes its operation the other service is discarded, i.e., $+$ is an operator of *choice*. (3) $agent_1 \parallel agent_2$: it represents a composite service that performs the services $agent_1$ and $agent_2$ independently, i.e., $\parallel$ is an operator of *concurrence*.

4 Service Compositions Abstract Machine

Based on DASM, we can present a BPEL Abstract Machine. The BPEL (Business Process Execution Language for Web Services) builds on top of WSDL (and indirectly also on SOAP) effectively introducing a stateful interaction model that allows to exchange sequences of messages between business partners (i.e. Web

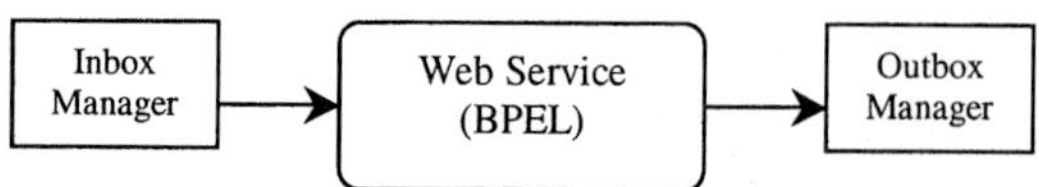

Fig. 2. High-level structure of the BPEL model

services) [6]. A BPEL document abstractly defines a Web service consisting of a collection of business *process* instances. A process instance maintains a continuous interaction with the external world (i.e., the communication network) through two interface components, called *inbox manager* and *outbox manager* [2], as shown in Fig. 2.

```
domain AGENT ≡ INBOX_MANAGER ∪ OUTBOX_MANAGER ∪ PROCESS
                ∪ ACTIVITY_AGENT ∪ HANDLER_AGENT
```

In the initial DASM state, there are only three DASM agents: the *inbox manager*, the *outbox manager* and a *dummy process*.

```
domain MESSAGE
inboxSpace: INBOX_MANAGER → MESSAGE-set
InboxManagerProgram ≡
  if inboxSpace(self) ≠ φ then
    choose p ∈ PROCESS, m ∈ inboxSpace(self)
              with match(p, m) and waiting(p)
      Assign_Message(p, m)
      if p = dummyProcess then
        new newdummy : PROCESS
          dummyProcess = newdummy
```

In general, a BPEL program combines two different types of activities: basic activities and structured activities.

```
RUNNING_AGENT ≡ PROCESS ∪ ACTIVITY_AGENT
startedExecution: PROCESS → BOOLEAN
suspended: RUNNING_AGENT → BOOLEAN
ProcessProgram ≡
  if ¬suspended(self) then
    if ¬startedExecution(self) then
      startedExecution(self) := true
      suspended(self) := true
    else
      stop self
  else
    Execute_Activity(activity(self))
ReceiveMode: RUNNING_AGENT → BOOLEAN
waitingForMessage: PROCESS → ⟨RUNNING_AGENT, ACTIVITY⟩-set
Execute_Receive(activity: RECEIVE) ≡
  let inputDescriptor = ⟨self, activity⟩
    if ¬receiveMode(self) then
```

```
      receiveMode(self) := true
      add inputDescriptor to waitingSet
    else
      if inputDescriptor ∉ waitingSet then
        receiveMode(self) := false
        suspended(self) := false
        where waitingSet = waitingForMessage(rootProcess(self))
```

A basic agent executes a single activity. Thus, its program is very similar to a process agent.

```
BasicAgentPROGRAM ≡
  if ¬suspended(self) and ¬startedExecution(self) then
    startedExecution(self) := true
    suspended(self) := true
  if suspended(self) then
    Execute_Activity(baseActivity(self))
  if ¬suspended(self) and startedExecution(self) then
    remove self from compositionAgentSet(parentAgent(self))
    stop self
```

5 Conclusions

In this paper, we introduce a distributed ASM (DASM) involves a collection of agents, then provide an abstract service architecture based on agents, which is open and universal. As an application, the paper presents a BPEL Abstract Machine. The proposed hierarchical abstract framework enhances the intelligibility and integrality of the business process. The future work of the model will focus on the complicated business interactions in the service.

References

1. V. Ermolayev, N. Keberle. Towards a Framework for Agent-Enabled Semantic Web Service Composition, International Journal of Web Service Research, Volume X, No. X, 2004.
2. R. Farahbod, U. Glässer and M. Vajihollahi. Specification and Validation of the Business Process Execution Language for Web Services, ASM 2004, LNCS 3052.
3. U. Glässer, Y. Gurevich and M. Veanes. An Abstract Communication Model, Microsoft Research Technical Report MSR-TR-2002-55, May 2002.
4. Y. Gurevich. Evolving algebras 1993: Lipari guide. In E. Börger, editor, Specification and Validation Methods, pages 9-36. Oxford University Press, 1995.
5. Y. Gurevich. May 1997 draft of the ASM guide. Technical Report CSE-TR-336-97, Univer-sity of Michigan EECS Department, 1997.
6. Specification: Business Process Execution Language for Web Services, Version 1.1, 2003. http://www-106.ibm.com/developerworks/library/ws-bpel/
7. UPnP Device Architecture V1.0. Microsoft Universal Plug and Play Summit, Seattle 2000, Microsoft Corporation, Jan. 2000.

An Approach of Nonlinear Model Multi-step-ahead Predictive Control Based on SVM

Weimin Zhong, Daoying Pi, and Youxian Sun

National Laboratory of Industrial Control Technology,
Institute of Modern Control Engineering, Zhejiang University, Hangzhou 310027, China
{wmzhong, dypi, yxsun}@iipc.zju.edu.cn

Abstract. In this paper, a support vector machine (SVM) with polynomial kernel function based nonlinear model multi-step-ahead controller is presented. A SVM based multi-step-ahead predictive model is established by black-box identification. And control law is obtained by numerical and optimization methods respectively. The effect of controller is demonstrated on a recognized benchmark problem. Simulation results show that multi-step-ahead predictive controller can be well applied to nonlinear system with good performance.

1 Introduction

Model predictive control (MPC) is a class of control algorithms in which a process model is used to predict and optimize process performance [1]. Now, attention has turned to nonlinear model predictive control (NMPC) [2] because many industrial processes are intrinsically nonlinear. Recently a new kind of learning machine called SVM [3,4], which is well used for function regression and time series prediction, can be used for nonlinear system identification and system control. This paper puts forward a new approach to NMPC based on SVM with quadratic polynomial kernel function. The paper is organized as follows: In section 2, a SVM with polynomial kernel function based multi-step-ahead predictive model and control algorithm for nonlinear systems are discussed. Simulation results on a recognized benchmark problem are given in section 3. In section 4, some conclusions are put forward. For more details about SVM regression with polynomial kernel function, see references [3,4].

2 SVM Based NMPC

2.1 Nonlinear Predictive Control Model

Assume the j-step-ahead ($j = 1 \cdots P$, P is prediction horizon) nonlinear model can be described by SVM form:

$$\begin{aligned} y_m(k+j) &= f(I_{k+j}) = f[y(k+j-1), y(k+j-2), \cdots, y(k-n+j), \\ &\quad u(k+j-1), u(k+j-2), \cdots, u(k-n+j)] \\ &= \sum_{i=1}^{nsv} a_i (I_i^{'} \cdot I_{k+j} + 1)^2 + b \end{aligned} \tag{1}$$

$$s.t. \quad u_{\min} \le u \le u_{\max}$$

V.S. Sunderam et al. (Eds.): ICCS 2005, LNCS 3516, pp. 1036 – 1039, 2005.

Where n , m are determined by approximation accuracy. Support values $a_i (i = 1,\cdots,nsv)$ and bias value b can be gotten through learning according to d pairs training data $\{I_s, y\}(s = 1,\cdots,d)$. I' is the set of support vectors from I_s. And assume when $j > M$, $u(k + j - 1) = u(k + M - 1)$, M is control horizon.

Introduce feedback correction to deal with the model error and disturbance, so the j-step-ahead closed-loop predictive output at time k is:

$$y_p(k + j) = y_m(k + j) + h_j e(k) \tag{2}$$

Where h_j is the error correcting coefficient, $e(k) = y(k) - y_m(k)$.

2.2 Nonlinear Predictive Controller

Select optimization objective function with moving horizon as below:

$$\min J(k) = \sum_{j=1}^{P} q_j [y_r(k + j) - y_p(k + j)]^2 \tag{3}$$

Where q_j are weighting coefficients, $y_r(k + j)$ is reference value from trajectory:

$$y_r(k + j) = a_r^j y(k) + (1 - a_r^j) y_{sp} \qquad y_r(k) = y(k) \tag{4}$$

Where a_r is the coefficient related with system's robustness and y_{sp} is set point value.

Case of $P = M$:

In this case, nonlinear optimization problem (3) contains M manipulate variables and M equality constraint of model outputs. Use known $y_p(k - 1 + j / k - 1)$ replace unknown $y(k - 1 + j)$, where $y_p(k - 1 + j / k - 1)$ is the closed-loop predictive output.

$$\begin{aligned} \min J(k) = {} & \underbrace{q_1[y_r(k+1) - y_p(k+1)]^2}_{J_1(k)\ \ contains\ \ u(k)} + \underbrace{q_2[y_r(k+2) - y_p(k+2)]^2}_{J_2(k)\ \ contains\ \ u(k),u(k+1)} \\ & + \cdots + \underbrace{q_M[y_r(k+M) - y_p(k+M)]^2}_{J_M(k)\ \ contains\ \ u(k),u(k+1),\cdots,u(k+M)} \end{aligned} \tag{5}$$

It's a common sense that the first-step controller output $u(k)$ outweighs the others, so we can first minimize $J_1(k)$ to get $u(k)$, and then substitute $u(k)$ into $J_2(k)$ to get $u(k+1)$. Recursively, we can get all $u(k + j - 1)$.

$$\frac{\partial J_j(k)}{\partial u(k + j - 1)} = 2q_j [y_p(k + j) - y_r(k + j)] \frac{\partial y_p(k + j)}{\partial u(k + j - 1)} = 0 \tag{6}$$

It's a cubic equation with only one unknown bounded variable, and it's easily solved.

Case of $P > M$:
Consider nonlinear optimization problem (3) with $P > M$, this is a nonlinear optimal problem with equality constraint of model output and boundary constraint of controller output. In this paper, Nelder-Mead simplex direct search method is used to optimize and obtain M unknown variables. And we will use solutions obtained by algorithm in section 2.2.1 as starting point.

3 Experimental and Simulation Results

The example is a benchmark problem used by Narendra & Parthasarathy[5], and further studied by Kambhampati[2] using a Gaussian neural network to fulfill the one-step-ahead predictive control. Set $C = 10000$, $\varepsilon = 0.001$, $h_j = 1$, $q_j = 1$.

$$y(k+1) = \frac{y(k)}{1 + y^2(k)} + u^3(k) \tag{7}$$

where y is the plant output and $u \in [-2,2]$ is the input.

150 pairs data generated by applying by a series of random numbers between [-2,2] are used to train the SVM predictive model with $m = n = 5$. And set $y_{sp} = 1.5$ and $a_r = 0.9$. Fig.2 is the system output and controller output with $P = 2, M = 2$. Fig.3 is the system output and controller output with $P = 4, M = 2$. In both cases, a disturbance rejection (d=0.1) is added when system is in steady state at time k=100, and the output response will track the set point well again quickly.

4 Conclusions

In this paper two cases of SVM with polynomial kernel function based nonlinear model multi-step-ahead predictive control strategies are investigated. The effect of both controllers is demonstrated by a benchmark problem. The analysis and experiment results show that the control strategy will provide a robust stable control of nonlinear systems with good performance in keeping reference trajectory and disturbance-rejection.

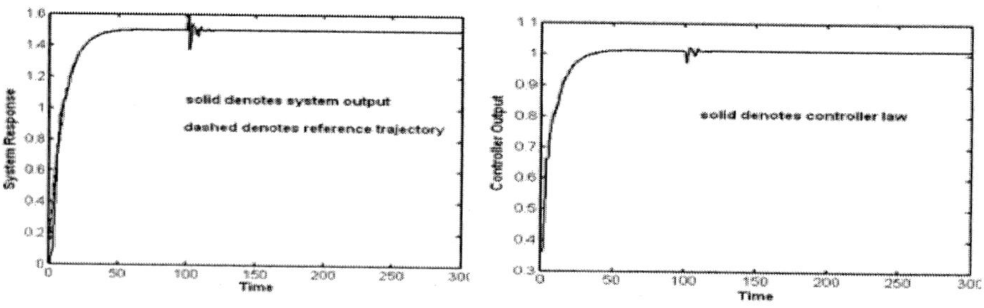

Fig. 2. System output and controller output with $P = 2, M = 2$

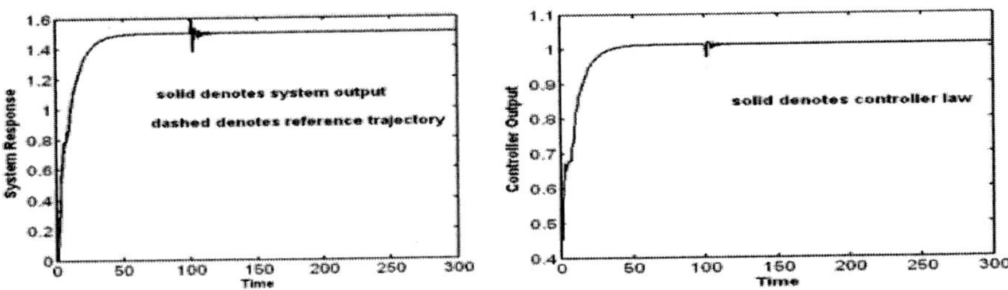

Fig. 3. System output and controller output with $P = 4, M = 2$

Acknowledgment

This work is supported by China 973 Program under grant No.2002CB312200.

References

1. Shu D.Q. : Predictive control system and its application. China Machine Press, Beijing,1996
2. Kambhampati C., Mason J.D., Warwick K.: A stable one-step-ahead predictive control of non-linear systems. Automatica, Vol.36 (2000) 485–495
3. Vapnik V.N.: The nature of statistical learning theory. Springer-Verlag, New York, USA (1995)
4. Cortes C.: Prediction of generalization ability in learning machines (PhD thesis), University of Rochester, New York, USA (1995)
5. Narendra K.S., Parthasarathy K. : Identification and control of dynamic systems using neural networks. IEEE Trans. Neural Networks, vol.1, (2000) 4-27

Simulation Embedded in Optimization – A Key for the Effective Learning Process in (About) Complex, Dynamical Systems

Elżbieta Kasperska and Elwira Mateja-Losa

Institute of Mathematics, Silesian University of Technology, Kaszubska 23, 44-100 Gliwice, Poland
{e.kasperska, e.mateja}@polsl.pl

Abstract. The purpose of this paper is to present some results of the experiments of type simulation embedded in optimization (on model type Systems Dynamics). They allow, not only, the so called, direct optimization, but extended sensitivity analysis of parameters and structures too. Authors have used languages Cosmic and Cosmos to support the learning process in (about) modelled dynamical system.

1 Introduction

We have problems understanding complex systems not only because they are rich in feedbacks [2, 3, 4, 5, 6, 7], but because these complex systems are changing their structure while we are trying to understand them using a fixed structure approach.

System Dynamics was developed in the late 1950's and early 1960's at the Massachusetts Institute of Technology's Sloan School of Management by Jay W. Forrester.

The classical concept of System Dynamics assumes, that, during the time horizon of the model run or the simulation, the structure (given a-priori) will remain constant.

During last couple of years, some ideas of structural evolution have occurred in System Dynamics modelling and simulation. Firstly Prof. Coyle took the problem of, so called, "simulation during optimization". The question was: how to "optimize" the structure in order to achieve the desired behaviour? He performed experiments, using simulation language COSMIC and COSMOS [1]. It is a software tool that automatically links dynamics simulation model to an optimization package. This facility makes it possible to apply powerful optimization techniques to:

- fine tuning of policies in the model (DIRECT OPTIMIZATION),
- sensitivity analysis of the model (BASE VECTOR ANALYSIS),
- simplification of the structure of the model (SIMPLIFICATION),
- exploring the effects of forecasting and forward planning in the model (PLANNING HORIZON).

V.S. Sunderam et al. (Eds.): ICCS 2005, LNCS 3516, pp. 1040–1043, 2005.

In this paper some of such experiments are presented and on this background the context of leaning process is considered.

2 Simulation Embedded in Optimization – Experiments on Model DYNBALANCE(3–1–III)

The model named DYNBALANCE(3–1–III), created by Kasperska, was chosen like the object of experiments type simulation embedded in optimization. The figure 1 presents the general structure of this model. The graphical convention is adapted from Forrester [4]. The main attention should be pay to the objective functions: "fob", "mmaxfo", which model the deciders preferences about of the behaviour of system, like the results of the influence of the decisions about the production, storing and selling. Below authors present the main results of some experiments, type simulation embedded in optimization.

Experiment 1

The objective function represents the profit from selling minus the total cost of production and the penalty factor connected with cost of the storage of product.

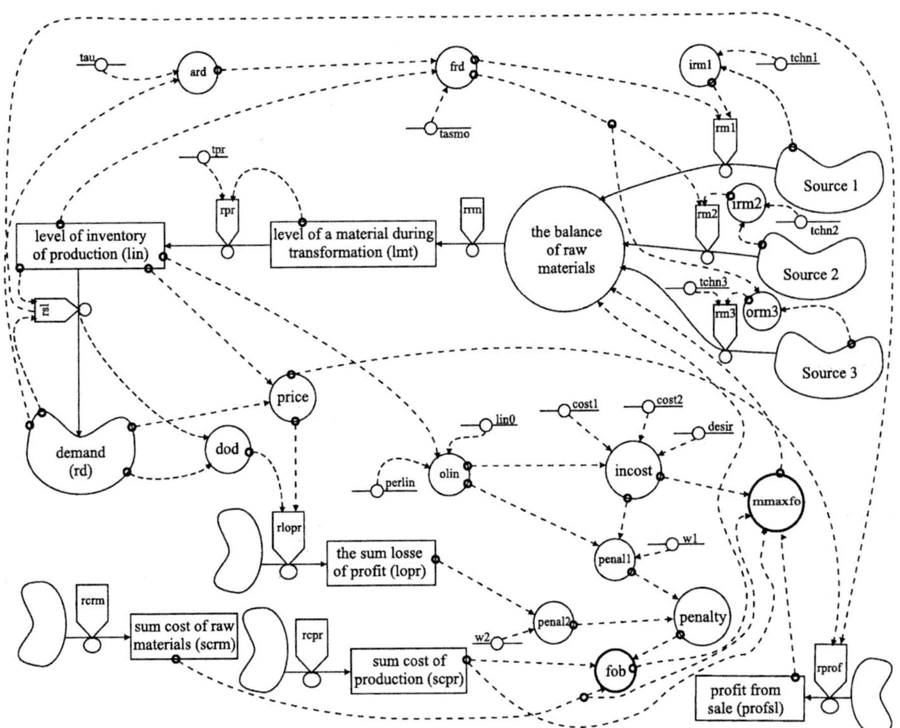

Fig. 1. Structure of model DYNBALANCE(3–1–III)

The demand for product has "ramp" characteristic. Table 1 shows the results of the experiment.

Table 1. The main results of experiment 1

Parameter	Final value	Original value	Lower limit	Upper limit
tchn1	0.000	20	0	40
tchn2	11.332	10	0	40
tchn3	39.984	20	0	40
Initial value of mmaxfo:			$0.20233E + 07$	
Final value of mmaxfo:			$0.47194E + 07$	
Final value of inventory:			221	
Final value of cost of production:			$24.295E + 05$	
Final value of price:			50	
Final value of penalty (maxpen):			$0.040792E + 07$	

The conclusion from analysing the results of experiment 1, are as follows:

- the zero value of parameter "tchn1" implies the necessity of cutting off the production from 1st raw material (it seems that first technology is not economical, under the simulation condition about the price and the unit cost and the supplying from raw materials,
- the value of parameter "tchn2" is closed to its original value and "tchn3" is nearly equal to the upper limit - as can be seen the optimization process select such values that suits the objective function in the whole horizon of simulation,
- the final value of penalty for "mmaxfo" is rather small, comparing with the final value of profit – it seems that taken politics of production was really good.

Experiment 2

In this experiment the variable "fob" was chosen, like the objective function and we are looking for a minimum of it. Like the so called "active parameters" we have chosen: tchn1, tchn2, tchn3 and like the "base-parameter candidate" – parameters: tan, tasmo, ucpr1, ucpr2, ucpr3, ucr1,ucr2, ucr3.

The initial value of objective function "fob" was: $0.29128E+08$, and the final value of "fob" was: $0.25939E + 08$. The conclusion from analysing the results of experiment 1, are as follows:

- many of parameters (for example: tan, ucr2, ucr3, ucpr1, ucpr2, ucpr3, tasmo) have kept their original value – that seems that their initial values were chosen reasonably good,
- the parameters: tchn1, tchn2, tchn3 (which were active in optimization dialog) have changed their values considerably – it means that they are sensitive parameters of model,

- this type of experiment gives more information about which parameters have more effect on improving the objective function.

3 Final Remarks and Conclusions

The purpose of the paper was to present some results of experiments of type simulation embedded in optimization on model type System Dynamics. These results show how the process of learning, in (about) complex, dynamical system, can become effective and supportive for both: deciders and us as modelers, as well. In "narrow" concept we can treat the modelling like a element of the global learning (see: three loops of learning by Radzicki, Sterman [6, 7]) but in the other side modelling is a way of learning (specially, if we consider "self – made models").

Final conclusions are as follows:

- simulation embedded in optimization can become the effective, supportive way of learning about complex, dynamical system,
- both sensitivity analysis (parameters and structures give us the key information about elements and theirs connections (which determine the dynamical behaviour of system).

References

1. Coyle, R.G. (ed.): Cosmic and Cosmos. User manuals. The Cosmic Holding CO, London (1994)
2. Coyle, R.G.: System Dynamics Modelling. A Practical Approach. Chapman & Hall, London (1996)
3. Coyle, R.G.: Simulation by repeated optimisation. J. Operat. Res. Soc. **50** (1999) 429–438
4. Forrester, J.W.: Industrial Dynamics. MIT Press, Massachusetts (1961)
5. Kasperska, E., Mateja-Losa, E., Słota, D.: Some dynamic balance of production via optimization and simulation within System Dynamics method. In: Hines, J.H., Diker, V.G. (eds.): Proc. 19th Int. Conf. of the System Dynamics Society. SDS, Atlanta (2001) 1–18
6. Radzicki, H.J.: Mr Hamilton, Mr Forrester and the foundation for evolutionary economics. In: Davidsen, P.I., Mollona, E. (eds.): Proc. 21st Int. Conf. of the System Dynamics Society. SDS, New York (2003) 1–41
7. Sterman, J.D.: Business dynamics – system thinking and modeling for a complex world. Mc Graw-Hill, Boston (2000)

Analysis of the Chaotic Phenomena in Securities Business of China

Chong Fu[1], Su-Ju Li[2], Hai Yu[1], and Wei-Yong Zhu[1]

[1] School of Information Science and Engineering, Northeastern University, 110004 Shenyang, China
{fu_chong, yu_hai, zhu_weiyong}@sohu.com
[2] College of Foreign Studies, Northeastern University, 110004 Shenyang, China

Abstract. The tendency of stock price in securities business based on the Shenzhen stock composite index was studied by using chaotic dynamics theory. The fluctuation of stock price was proved to be a kind of chaotic process of inner random. The dynamic model of Shenzhen stock composite index was established both by the restructure of phase space of the data and by the analysis of the Poincaré section and Lyapunov exponent of the data. The chaotic evolvement process of this model was analyzed in detail. This provides a new method for the investigation of modern financial system by the use of chaotic theory.

1 Introduction

In the early 1900s, French mathematician Louis Bacheiler proposed a model describing the rules of price fluctuation, which became the foundation of modern financial securities theory. This model assumes that the price fluctuation obeys normal distribution and points out that the changes of the price are independent in statistics and follows the "Random Walk" which accords with the bell-like curve. According to his theory, the price should fluctuate slightly in a small range, the violent shakeout should seldom happen, but it exceeds people's expectation greatly. This indicates that current securities theory cannot give a correct explanation to the large fluctuation of finance system.

In 1990s, Professor B.B.Mandelbrot, the academician of American Academy and founder of fractal theory, proposed a new multi fractal model to describe the fluctuation of the securities market. It accurately describes the relationship of the up and down fluctuation, revealing the essential of the frequent violent fluctuation of securities market, providing a determinative analysis method for the study of unpredictable financial system [1]. At present, most of the existing financial models are based on "Random Walk" theory, few scholars set up financial model on the basis of chaos theory. In this paper, the chaotic dynamics theory was used to analyze the chaotic property of securities market and a chaotic iteration model was established to further study the evolvement process of the complicated financial system.

V.S. Sunderam et al. (Eds.): ICCS 2005, LNCS 3516, pp. 1044 – 1047, 2005.

2 Power Spectrum Analysis of Trading Sequence

The tendency chart of composite index in Shenzhen securities between 8-17th-1992 and 8-31th-2004 is shown in Fig. 1. The sampling interval is based on day.

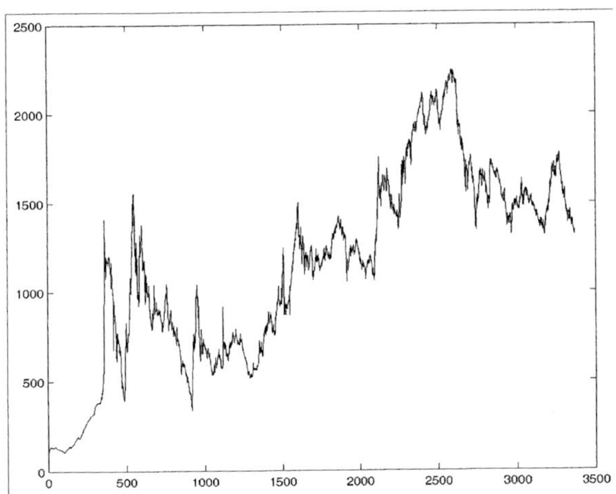

Fig. 1. Tendency chart of composite index of Shenzhen securities between 8-17th-1992 and 8-31th-2004

Power spectrum analysis is often used to identify the chaos behavior of a system. For sequence in Fig. 1, we do discrete Fourier transform, calculate its Fourier coefficient

$$F(u) = \frac{1}{N} \sum_{i=0}^{N-1} x_i \exp\{-j2\pi ux / N\}. \tag{1}$$

The result is shown in Fig. 2. The power spectrum of the chaotic system shows a peak on a broadband, which is different from periodic movements or Gaussian noise. From Fig. 2 we can see, the power changes continuously in frequency domain, it shows a downward trend in the form of negative exponent. Some irregular peaks are on the background, demonstrating period 3 phenomenon.

3 Phase Space Reconstruction and Poincaré Section

The chaotic attractor is the fundamental reason of some random movements. The first step of analyzing the chaotic property of a time sequence is reconstructing the attractor based on the experiment data. For time sequence x_0, x_1, x_2, $\ldots x_{N-1}$ (N must be large enough), we can get other $d-1$ dimension data by using the time delay method. The coordinate axes are respectively $x(t)$, $x(t+m)$, $x(t+2m)$, …, $x(t+(d-1)m)$, m is the selected delay number. All the data determine a point in a multi dimension state space, then we can use other methods to examine whether it has a chaotic attractor.

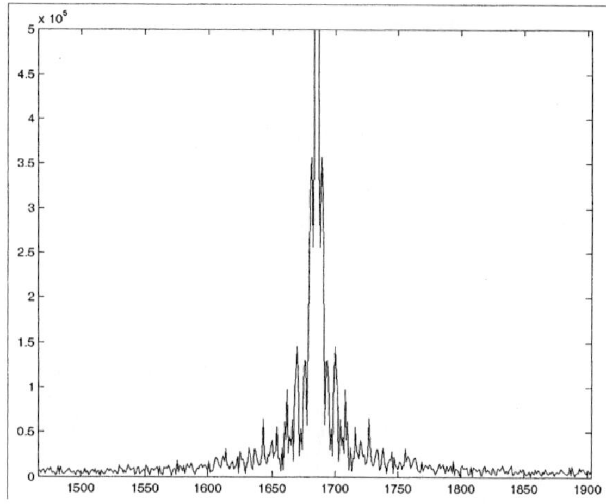

Fig. 2. Local enlarged power spectrum of Shenzhen securities composite index

The transection of attractors in phase space is Poincaré section. For a discrete time sequence in phase space, *X(t)* presents current state, while *X(t+m)* the state after delaying *m*, *X* a *d* dimension vector , the state transformations in *d* dimension space is:

$$(x(t+m), x(t),..., x(t-(d-2)m)) = F_m(x(t), x(t-m),..., x(t-(d-1)m)). \tag{2}$$

The Poincaré section of Shenzhen securities composite index is shown in Fig. 3, the distribution of the points in the figure is very regular, the dynamic model is established as [2, 3]:

$$X_{t+1} = rX_t \exp(-0.00034(1-X_t)^2). \tag{3}$$

r is a random variable, $r \sim N(0.978, 11.21)$.

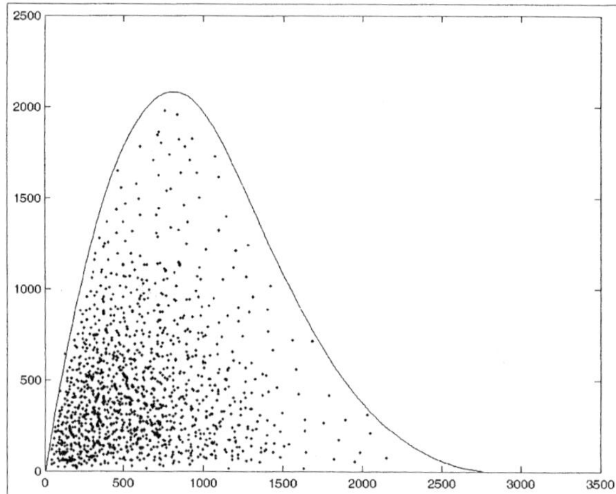

Fig. 3. Poincaré section and iterative model of Shenzhen securities composite index

4 Dynamical Characteristic Analysis of the Model

The bifurcation diagram of dynamic model defined by Eq. 3 is shown in Fig. 4 [4].

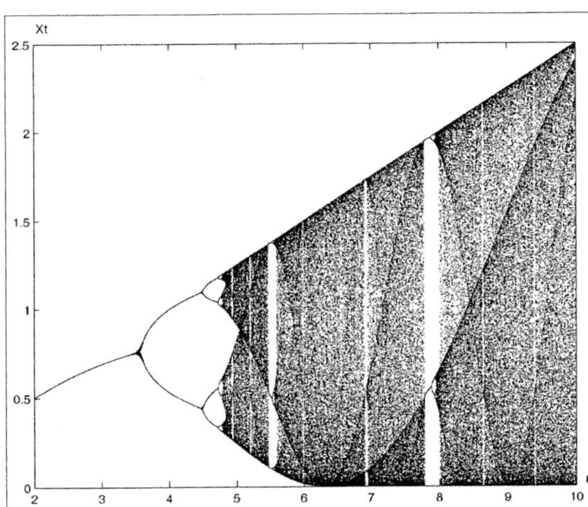

Fig. 4. Bifurcation diagram of the dynamical model defined by Eq. 3

When control parameter $r > 4.716$, the system will come into chaos process. The evolvement of this system is a chaotic process, which contains periodic orbit, semi-periodic orbit, random orbit and chaos orbit.

5 Conclusion

This paper proves that the fluctuation of stock price is a mixed process, including periodic process and a topological invariance chaotic process. It provides a new idea and method of using the chaotic theory to study the fluctuation, inner principles and tendency of securities business.

References

1. Mandelbrot, B.B.: Fractals and Scaling in Finance. Springer-Verlag, Berlin Heidelberg New York (1997)
2. Liu, X.D., Zhu W.Y., Wang G.D.: Analysis of Chaotic Behavior of the Six-roll UC Mill and its OGY Control. Journal of Northeastern University, Vol. 19. (1997) 591–594
3. Ruelle, D., Takens, F.: On the Nature of Turbulence. Common Mathematics Physics, Vol. 20 (1971) 167–192
4. Feigenbaum, M.J.: Quantitative Universality for a Class of Nonlinear Transformations. J Statistic Phys, Vol. 19 (1978) 25–52

Pulsating Flow and Platelet Aggregation

Xin-She Yang

Department of Engineering, University of Cambridge,
Trumpington Street, Cambridge CB2 1PZ, UK
xy227@eng.cam.ac.uk

Abstract. Platelet aggregation has been modelled as a nonlinear system of viscous flow and pulsating flow based on the Fogelson's continuum model. By solving nonlinear governing equations and the coupling between flow and platelet aggregates in the cylindrical coordinates using the finite element method, we can see how the platelet aggregation forms under the pulsating flow conditions. Numerical simulations show that the aggregates are stretched by the flow in the dynamic equilibrium of the link forces and background flow. In addition, significant elastic stress can be developed to maintain the shape and integrity of the platelet aggregates.

1 Introduction

Platelet aggregations are important in physiological processes such as thrombus and haemostasis. However, these processes are very complicated and many factors and activities occur at the same. For example, physiological response in haemostasis to blood vessel injury involves the cell-cell adhesive platelet aggregation and coagulation. Platelets are tiny oval structure with diameter of 2-4 micrometers. They are active in blood with a half-life of 8-12 days. Non-active platelets flow free in blood vessels in a dormant state. Activating chemicals such as ADP initiated by injury can induce platelet aggregation in the blood plasma. A platelet's surface membrane can be altered so that the platelet becomes sticky, and thus capable of adhering to other activated platelets and the vessel walls. Although platelets only consist of about 0.3% in concentration and yet occur in enormous numbers about 250 millions per millilitre [2,3,6]. Once triggered by the chemicals, the active platelets start to clot so as to restrict the blood flow at the injury site. To prevent uncontrolled overspreading of activated platelets, a chemical inhibitor, thrombin, is also involved.

There have been extensive studies on the biological and physiological effect of platelet aggregations. Some mathematical models and computational models are very instructive for the modelling of the detailed mechanism of the formation of platelet aggregates [3-6]. Most of these studies use the Navier-Stokes equations with simplified boundary conditions or the steady-state approximation. However, the real blood vessel system has complex geometries due to branching and atherosclerotic plague deposition inside the vessels. In addition, the flow velocity and pressure are even more complicated by the pulsating process from the heart pumping process. Thus, the more realistic modelling shall include the pulsating effect on the flow and

V.S. Sunderam et al. (Eds.): ICCS 2005, LNCS 3516, pp. 1048 – 1051, 2005.

platelet aggregations. In this paper, we intend to extend the existing models to include the pulsating flow and the platelet aggregations in the corresponding environment.

2 Fogelson's Model

Fogelson first formulated a continuum model for platelet aggregation process [3]. The full model consists of a system of coupled nonlinear partial differential equations. We use the simplified version of Fogelson's model with the following equations:

$$\nabla\cdot\boldsymbol{u}=\mathbf{0}, \tag{1}$$

$$\rho(\boldsymbol{u}_t+\boldsymbol{u}\cdot\nabla\boldsymbol{u})=-\nabla p+\mu\nabla^2\boldsymbol{u}+\boldsymbol{f}+\beta\nabla\sigma, \tag{2}$$

$$\sigma_t+\boldsymbol{u}\nabla\sigma=\sigma\nabla\boldsymbol{u}+(p\nabla\boldsymbol{u})^T, \tag{3}$$

$$\phi_t+\boldsymbol{u}\cdot\nabla\phi=r, \tag{4}$$

where the first two equations are the Navier-Stokes equations for incompressible fluid flow $\boldsymbol{u}=(U,V,W)$ and p pressure. The coefficient β is constant and f is the force density. The last term is due to the cohesion of platelets. The third equation is for the cohesion-stress tensor σ. The last equation is for the concentration ϕ of the active platelets and the production rate r can be considered as a function of concentrations of platelets and the activating chemicals. The governing equations are nonlinear and the flow is coupled with the formation of platelet aggregates. Thus the full solution necessitates efficient numerical methods. As the blood flow is slow and viscous, so the first two equations can be simplified for the case of viscous flow and the zero force f. We have

$$\nabla\cdot\boldsymbol{u}=0, \quad \rho\boldsymbol{u}_t=\mu\nabla^2\boldsymbol{u}-\nabla p+\beta\nabla\sigma. \tag{5}$$

In most cases, the blood flow is one-dimensional tubular flow, it is convenient to choose the local coordinates so that the z-axis is in the direction of the local blood flow. In this case, we can assume the pulsating pressure (far field) in the form

$$p_z=\partial p/\partial z=A\cos(\omega t)+\gamma, \tag{6}$$

where A is a constant. In fact, we can assume any known function form for the pressure gradient in terms of a Fourier expansion

$$p_z=g(t)=\gamma+\sum_{n=1}^{N}[A_n\cos(\omega_n t)+B_n\cos(\omega_n t)]. \tag{7}$$

3 Simulation and Results

Although the equations are coupled, they can be solved using the well-developed finite element method together with the Newton-Raphson iteration procedure for the nonlinearity. By solving the problem in the cylindrical coordinates, we can see how the platelet aggregation forms and develops under the pulsating flow conditions.

3.1 Shear Stress Distribution and Variation

To simulate the tubular flow and the stress development, we first study the flow at the bifurcation region where one large vessel is branched into two smaller vessels. For a given pulse, one can calculate the wall shear stress and stress variation with time after the peak R-wave of the given pulse. Figure 1 shows the shear stress distribution and the mean wall shear stress variation. The major feature of the computed shear stress variation is consistent with the experimental results. It is worth pointing out that there exists a peak for wall shear stress for a given pulsating pressure gradient, and thus the peak shear stress could directly affect the rupture if the stenosis or plague is presented at the region concerned. The shear stress is higher at bifurcation regions and the stress level can reach as high as 15 Pa for a blood vessel with a diameter of 1mm. As the flow is pulsating, the shear stress and pressure vary with time. The mean shear stress at wall varies significantly after the R-wave starts. It first increases quickly to a peak at t=0.065 and then decreases exponentially with time.

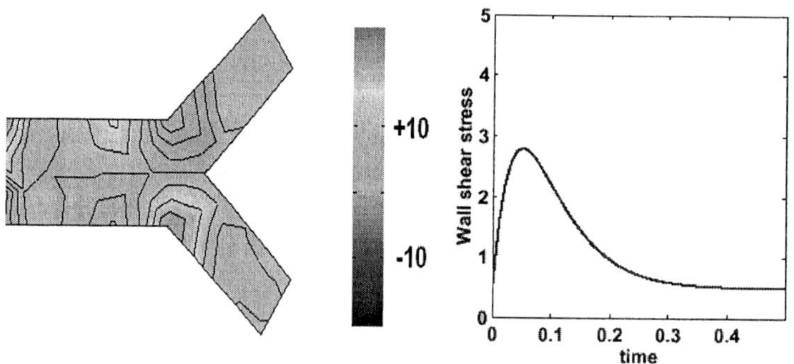

Fig. 1. Shear stress distribution at the bifurcation region and wall stress variation with time

3.2 Platelet Aggregation

In the case of pulsating flow in a cylindrical vessel with a diameter of 1mm, a small injury (point A) with a size of less than 0.1mm occurs and thus releases the ADP chemical to activate the platelet aggregation, so platelets starts to deposit at the injury site, and the size of the aggregation starts to increase. Figure 2 show the platelet aggregation at the different times t=1,10,50 seconds after the injury. The flow is from left to the right with the initial constant flow field. The boundary condition for the pressure in the far field is the pulsating function given earlier in equation (7). For simplicity, we have used the zero reaction rate $(r=0)$, $\gamma=0$, $\beta=1$, and the normalized concentration so that $\phi=0$ for no platelets and $\phi=0.5$ for solid platelet aggregates.

We can see clearly that the concentration of activated platelets is much higher at the injury site than in the blood. The aggregation rate is proportional to the

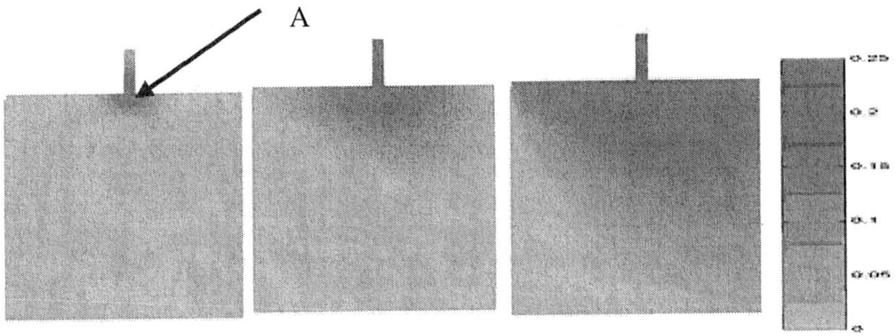

Fig. 2. Formation of platelet aggregation at different times (t=1,5,50)

concentration of the activated platelets, and thus the size and shape of the platelet aggregation can be represented in terms of the concentration as shown in the figure. The aggregates are stretched by the flow in the dynamic equilibrium of the link forces and background flow. As the platelets aggregate, the injured vessel will be blocked severely after some time, and it may take 100 seconds to reach this stage.

4 Conclusions

We have simulated the platelet aggregation by using the Fogelson's continuum model with nonlinear governing equations and the coupling between flow and platelet aggregates. By solving the problem in the cylindrical coordinates using the finite element method, we can see how the platelet aggregation forms under the pulsating flow conditions. Numerical simulations show that the aggregates are stretched by the flow in the dynamic equilibrium of the link forces and background flow. In addition, significant elastic stress can be developed to maintain the shape and integrity of the platelet aggregates.

References

1. Chakravarty, S., Mandal P. K.: Mathematical modelling of blood flow through an overlapping arterial stenosis. Math. Comput. Modelling, 19 (1994) 59-70.
2. David, T., Thomas, S., Walker, P. G.: Platelet deposition in stagnation poiint flow: an analytical and computional simulation. Medical Engineering and Physics, 23 (2001) 229-312.
3. Fogelson, A.: Continuum models of platelet aggregation: formulation and mechanical properties. SIAM J. Appl. Math., 52 (1992) 1089-1110.
4. Gurevich, K. G., Chekalina, N. D., Zakharenko, O. M.: Application of mathematical modelling to analysis of nerves growth factor interaction with platelets. Bioorganic Chemistry, 27(2000)57-61.
5. Guy, R. D., Fogelson, A. L.: Probabilistic modelling of platlet aggregation: effects of activation time and receptor occupancy. J. Theor. Biol., 219(2002) 33-53.
6. Keener, J. and Sneyd, J.: Mathematical Physiology, Springer-Verlag, Berlin Heidelberg New York (1998).
7. Zienkiewicz, O. C. and Taylor, R. L.: The Finite Element Method, Vol. I/II, McGraw-Hill, 4th Edition, (1991).

Context Adaptive Self-configuration System

Seunghwa Lee and Eunseok Lee

School of Information and Communication Engineering, Sungkyunkwan University,
300 Chunchun jangahn Suwon, 440-746, Korea
{jbmania, eslee}@selab.skku.ac.kr

Abstract. This paper suggests a *context adaptive self-configuration system*, in which the system itself is aware of its available resources, and executes a configuration suitable for each system and the characteristics of each user. A great deal of time and effort is required to manually manage computing systems, which are becoming increasingly larger and complicated. Many studies have dealt with these problems but most unfortunately have focused on 'automation', which do not reflect the system specifications that differ between systems. Therefore, this paper proposes an adaptive self-configuration system that collects diverse context information for a user and its system, reflects it to an auto-response file and executes the detail parameter setting automatically. A prototype was developed to evaluate the system and compare the results with the conventional methods of manual configuration and MS-IBM systems. The results confirmed the effectiveness of the proposed system.

1 Introduction

Recently, the rapid development of IT has contributed to various devices with increasing performances. As the devices to be administered are increasing in number and complication, they impose a heavier burden on system management [1]. For examine, a company's computing system networking hundreds of hosts may require a great deal of time, money and manual effort if they have to be manually and individually installed and updated. In order to solve it, studies such as a centralized integration management of distributed resources, and unattended installation technology to reduce the workload and periodic automatic update are being applied by various institutions [2][3]. However, these studies focused mainly on 'automation', which simply replaces the tasks manually processed. Therefore, these studies provide only a uniform service that does not reflect unique device's characteristics or the user preferences. This paper proposes a context adaptive self-configuration system, which is aware of its available resources, and executes the suitable configuration on each situation utilizing these contexts with the user's information.

The paper is presented as follows: Section 2 describes the characteristics, structure and behavior process of the proposed system; and Section 3 gives a conclusion as well as a simple description of the system evaluation through its implementation and experiments.

V.S. Sunderam et al. (Eds.): ICCS 2005, LNCS 3516, pp. 1052 – 1055, 2005.

2 Proposed System

2.1 System Overview

The word 'configuration' in this paper means newly install components, such as an OS or software, etc., that are to be administered and updated them for version management or defects healing.

For installation, the proposed system need be installed so that components to be contained reflect the existing users' preference and the system resource status. The basic configuration automatically meets these requirements through a customized auto-response file and a parameter-setting file. In addition, for updating, it monitors the frequency of component use, determines the priority, and enables works to be efficiently processed in case the remaining storage of a system or the updating time is insufficient. Currently, copying files is processed by a peer-to-peer transmission method to supplement the several disadvantages of the existing centralized distribution method and to efficiently update the files.

2.2 System Structure

Fig. 1 shows the overall structure of the proposed system, and the detail modules are as follows:

- *Component Agent*: this is installed in a managed device to monitor the changes in resources, to check the parameter setting information of the user's applications along with the frequency of use, and to periodically transmit the this information to a self-configuration system. In addition, it also receives packages or the configuration files necessary for auto installation from the self-configuration system or a host located in the same zone. Upon installation, it observes the user's behavior, and transmits the observed information to the *Context Synthesizer*.

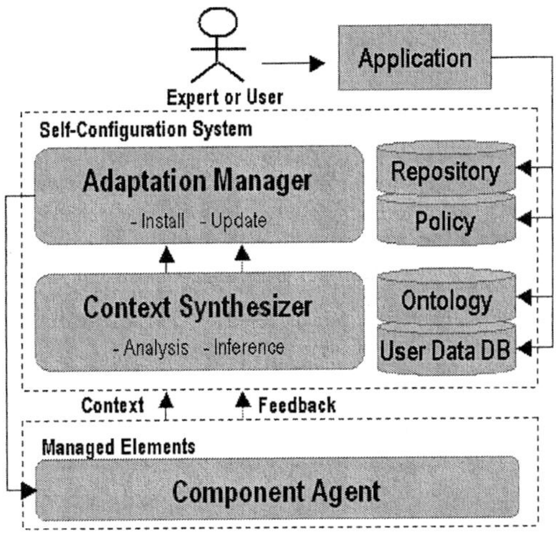

Fig. 1. Architecture of the Proposed system

- *Context Synthesizer*: Gathers the context provided by the *Component Agent* and stores the basic user data, the application preferences and usage frequencies in the *User Data DB*. The *Context Synthesizer* analyzes and deducts the current status using the stored user data and *Ontology* (various expressions on the preference data of the applications) and relays the results to the *Adaptation Manager*.

- *Adaptation Manager*: Decides and performs the appropriate tasks based on policies that were pre-defined through the data provided by the *Context Synthesizer*. The *Adaptation Manager* installs the components stored in the *Repository* or performs the various configuration tasks such as updates. An installation image on the user data and the customized 'auto-response files' for installation are also automatically created.

The system administrator or the user can adjust and manage these components, policies, Ontology, and the user data to be stored in the Repository. The proposed system can also communicate with other external Configuration Servers and share data.

2.3 System Behavior

1) Installation

The system installs the components by reflecting the existing user's preference and system resource information, and automatically executes the basic configuration through a customized auto-response file and an auto configuration file. In addition, in case a user's resource is insufficient, it automatically selects the 'minimum installation' option for the auto-response file, creating the script. Currently, such a process is notify to the user so that the final decision may be made by the user whilst an agent observes and learns the user's activities, which are reflected to determine the subsequent activities. The major workflow is as follows.

Step 1. Component Agent: monitors the system resource information and configuration information, creating a profile and transmitting it to a context synthesizer, for any changes.

Step 2. Context Synthesizer: saves the received data to the User Data DB, on which the present situation is analyzed.

Step 3. In case a user requests an installation or an administrator needs to distribute it in a lump, it retrieves the user data (preference) saved in the User Data DB, after authenticating the user.

Step 4. Adaptation Manager: creates and distributes the auto-response file and the detail parameter setting file, based on the retrieved user data. (ex. if (resource = insufficient) then checks the 'minimum installation' option of the auto-response file)

Step 5. Component Agent: installs any received file, monitors the user's activities and executes the feedback.

Step 6. Modifies and learns the User Preference, Policy.

Fig. 2. Pseudo Code for the Install Phase

2) Update

The update priority is determined by monitoring the frequency of the components, through which the system facilitates the various tasks efficiently if the remaining storage or the time required for updating is insufficient. Reflecting the *Astrolabe* [4] structure, each host organizes the components into zones, and these zones are organized hierarchically. A representative host of each zone collects a list of files on the individual hosts while a host requiring the file requests the representative host about the location of a near host holding the file, and receives it in peer-to-peer transmission method. From this, it can reduce the load of the centralized server, be resistant to various troubles and quickly copy files. The update files copied within the zone are checked for consistency through a size comparison.

Step 1. Component Agent: monitors the frequency of each application by the users, transmits it to the context synthesizer periodically and sends its own update file list to the representative host in the same zone. In addition, the representative host's collected file lists are gathered at the higher level.

Step 2. Context Synthesizer: saves the received data into the User Data DB, on which the user's preference is present.

Step 3. For any update request (update request by a user or update request at a lump by an administrator), it authenticates the user and retrieves the user's data saved in User Data DB.

Step 4. Adaptation Manager: determines the update priority according to the retrieved user data.

Step 5. The system requests the representative host in the same zone, the location of a host holding the file, catches hold of it and finally copies the file. (if the file needed does not exist at the representative host, then go to the level and submit the request to the higher level host).

Fig. 3. Pseudo Code for the Update Phase

3 Conclusion

This study proposed a prototype system for testing the basic functions and validating a system by comparing for the task time for the existing research. From the results, it could be concluded that proposed system features a shortened configuration work time than the existing systems (shortening the basic parameter setting time) and an increased the user satisfaction and usability. Meanwhile, it is expected that the proposed system might be useful in the *Ubiquitous Computing* era where a user utilizes multiple computing devices and provides them with a more convenient computing environment as a result of the reduced system management.

References

1. Paul Horn, "Autonomic Computing: IBM's Perspective on the State of Information Technology", IBM White paper, Oct.2001
2. http://www-306.ibm.com/software/tivoli
3. http://www.microsoft.com/technet/prodtechnol/winxppro/deploy/default.mspx
4. Robbert van Renesse, Kenneth Birman and Werner Vogels, "Astrolabe: A Robust and Scalable Technology for Distributed System Monitoring, Management, and Data Mining", ACM Transactions on Computer Systems, Vol.21, No.2, pp.164-206, May 2003
5. John Keeney, Vinny Cahill, "Chisel: A Policy-Driven, Context-Aware, Dynamic Adaptation Framework", In Proceedings of the Fourth IEEE International Workshop on Policies for Distributed Systems and Networks, Jun.2003
6. Anand Ranganathan, Roy H. Campbell, "A Middleware for Context-Aware Agents in Ubiquitous Computing Environments", In ACM/IFIP/USENIX International Middleware Conference 2004, Jun.2004
7. Richard S. Sutton, Andrew G. Barto, 'Reinforcement Learning: An Introduction (Adaptive Computation and Machine Learning)', The MIT Press, Mar.1998

Modeling of Communication Delays Aiming at the Design of Networked Supervisory and Control Systems. A First Approach

Karina Cantillo[1], Rodolfo E. Haber[1], Angel Alique[1], and Ramón Galán[2]

[1] Instituto de Automática Industrial – CSIC, Campo Real km 0.200, Arganda del Rey, Madrid 28500
{cantillo, rhaber, a.alique}@iai.csic.es

[2] Escuela Técnica Superior de Ingenieros Industriales. Universidad Politécnica de Madrid, Calle José Gutiérrez Abascal, nº 2, Madrid 28006
rgalan@etsii.upm.es

Abstract. This paper presents a first approach to model the communication delays with the aim of improving the design of networked supervisory and control systems (NSCS). Network delays affect the performance of NSCS. Therefore, it is necessary a complete analysis, modeling and simulation of the communication delays upon the connection technology in use (e.g., Ethernet networks). A statistical model is proposed on the basis of autocorrelation of LRD that is imposed by the bursty behavior of the network traffic.

1 Introduction

Nowadays, data networks (i.e., Ethernet networks) have become in a low-cost connection media for networked supervisory and control systems (NSCS) [1], due mainly to the improvement concerning with the speed of communication and the bandwidth. It is evident the need to perform the analysis, modeling and simulation of the communication delays in the framework of NSCS. A suitable and reliable model of communication delays allows assessing the network behavior, the overall influence of the delays on the deterioration of the NSCS performance and the sufficient conditions for closed-loop stability.

In the next section is detailed the developed framework to measure communication delays. The developed model is shown in section 3. In subsection 3.2 is presented the communication delays simulator. Finally, some conclusions and remarks are given.

2 Communication Delays Measurement over the Propose Work Environment

The case of study consists of two applications based on RT-CORBA for measuring delays samples. Applications are executed using TCP/IP protocol over an Ethernet network with 110 stations and 10 Mbs of bandwidth. *Real-Time* CORBA defines

V.S. Sunderam et al. (Eds.): ICCS 2005, LNCS 3516, pp. 1056 – 1059, 2005.

mechanisms and policies to control processor, communication and memory resources. The ACE ORB, unlike the most of CORBA implementations, shares a minimum part of ORB resources, reducing substantially the synchronization costs and the priority inversion between the process threads, offering a predictable behavior [2]. The communication interface defined for the applications is depicted below.

```
interface delaysTCPIP{
void delaymonitor(in double tclient, out double
tserv, out v1, out v2, out v3, out double proctserv);};
```

3 Data Network Delays Models: The Use of Statistical Techniques

Some network traffic models well-described in the literature are inspired in statistics method such as Poisson Processes, Markov chain, time series, heavy-tailed models, self-similar process and others. These models can be considered as black box tools able to link cause-effect and emulate network traffic behavior with a high accuracy.

The bursty behavior of the Ethernet technology in network traffic imposed the inclusion of the autocorrelation concept in the model to be defined. Poisson processes can be used for modeling the delay between continuous packets with the Exponential Distribution Function (EDF). The drawback is that these processes do not take into account the auto-correlation concept due to the independence hypothesis in data, leading erroneous simulations and analysis. In order to verify these constraints, a statistic analysis of data was performed with *Quantile-Quantile* graphic and the validation methods *Chi-Square Test*, *Anderson-Darling Test*. The results showed that EDF is not adequate to represent the samples of delay included in the measured data.

The introduction of the auto-correlation concepts in the model was then required. Initials solutions were simple chain Markov models, adjusted to a correlation of short-range dependence (SRD), characterizing just in a minimal proportion the bursty behavior. In order to reach a whole representation, auto-correlation of long-range dependence (LRD) must be considered. In order to represent the LRD autocorrelation indices the self-similar processes are used, because the LRD can be characterized only with a simple parameter, the Hurst coefficient (H).

3.1 Inclusion of Self-similar Processes in the Proposed Model

Self-similarity processes consist of repeated patterns in multiples scales. The literature shows that the network traffic is self-similar [3]. There are different methods for the Hurst coefficient estimation. The R/S method was chosen because it is easy to implement and it is very useful in other research fields. Initially, the developed program in Matlab makes use of a sample taken in a time-period with low network load. The estimated H was 0.6535. Subsequently, a sample taken in presence of higher network load was assessed, obtaining an estimated of H=0.7232. The results show how H can be used as a reliable measure parameter of the bursty behavior in the network traffic.

In order to complete the representation, a distribution function that fits the LRD characteristics must be chosen as the empiric distribution indicated by the samples. One solution is the use of processes with infinite variance called heavy-tail processes. The distribution functions Pareto, Weibull, Exponencial, Gamma, and Lognormal, were analyzed and compared using the central limit theorem and the infinite variance concept. Figure 1 shows the Log-Log plot of each one. The graphic shows that Pareto is the distribution function that better fits data behavior. A residual variance analysis was performed supporting previous result. Additionally, the Pareto distribution function is heavy-tail type that is recommended to depict the network delays.

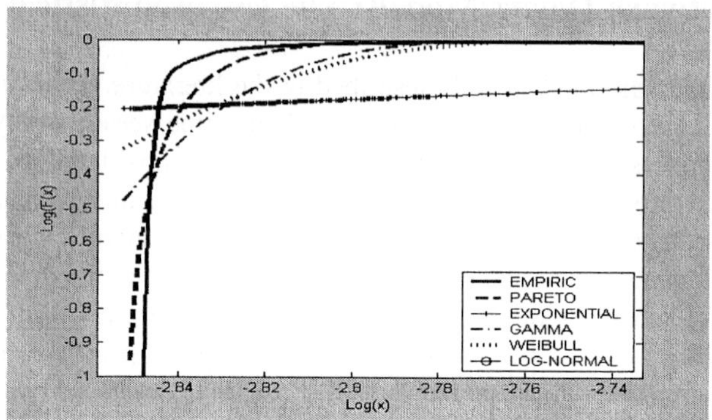

Fig. 1. Comparative Analysis: Empiric Cumulative Distribution Function (CDF) and Studied Cumulative Distribution Functions (Pareto, Exponential, Weibull, Gamma, Log-Normal)

3.2 Communication Delays Generator

The method *Fractional Gaussian Noise* (FGN) was chosen to generate pure fractal signals based on the Hurst parameter, due to its implementation easiness [4]. The developed simulator initially calculates the Hurst coefficient, based in the R/S method. This coefficient is used to generate the self-similar sample. The sample is mapped to the Pareto distribution function and then formatted to obtain a simulation close to the reality. Before the formatted procedure, the equation used is $Y_i = \left(\frac{a \times b}{Log(1 - F_n(X_i))} \right)$. F_n is the normal cumulative distribution function, a is the shape parameter, b is the minimum value of the analyzed sample and X_i is the datum obtained by FGN method. Figure 2 shows a comparison in a log-log plot between the empiric cumulative distribution of the analyzed sample and the empiric cumulative distribution of the sample obtained from the simulator.

In order to validate the results a residual variance analysis, between the sample generated by the simulator and a sample derived from a Poisson model, was accomplished. The developed simulator achieved the best representation. Although the defined model does not represent all the properties of networks traffic, as SRD behavior, this model is essential to assess the influence of the network delays in reliability and feasibility of NSCS design.

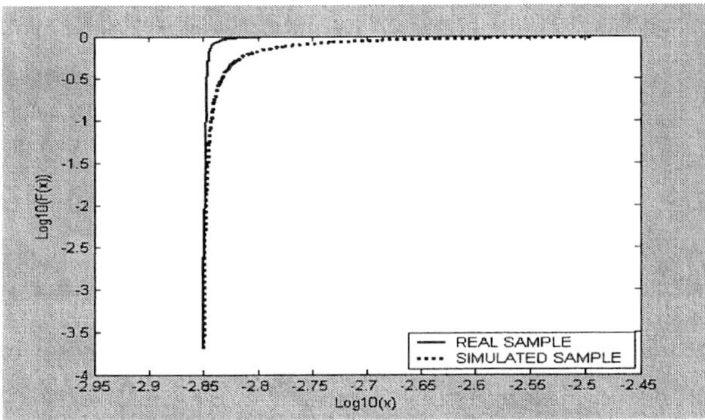

Fig. 2. Comparative Analysis: Real Sample and Simulated Sample

4 Conclusions

This paper proposes a statistical model. The bursty behavior of the network traffic imposed the use of self-similar processes using the Hurst coefficient. The results show how H can be used as a performance index of the bursty behavior in the network traffic. Likewise, Pareto distribution function shows the best performance in representing the empiric distribution function derived from samples measured in real-time. On the other hand, this paper proposes a communication delays simulator using *Fractional Gaussian Noise* method. Both model and simulator are very useful to design networked supervisory and control systems although they do not take into account the autocorrelation of SRD in the case study.

References

1. Branicky, M.S., Phillips, S.M., Zhang, W.: Stability of Networked Control System: Explicit Analysis of Delay. Proceedings of the American Control Conference (2000) 2352-2357
2. Schmidt, D.C., Mungee, S., Gaitan, S.F, Gokhale, A.: Software Architectures for Reducing Priority Inversion and Non-determinism in Real-time Object Request Brokers. Journal of Real-time Systems, Vol. 21, 1-2, (2001) 77-125
3. Leland, W.E., Taqqu, M.S., Willinger, W., Wilson, D.V.: On the self-similar nature of Ethernet traffic. IEEE/ACM Transactions on Networking, Vol. 2 (1994) 1-15
4. Paxson, V.: Fast, Approximate Synthesis of Fractional Gaussian Noise for Generating Self-Similar Network Traffic. ACM SIGCOMM CCR, Vol. 27 (1997) 5-18
5. Cantillo, K., Haber, R.E., Alique, A., Galan, R.: CORBA-Based open platform for processes monitoring. An application to a complex electromechanical process. In: Bubak M., Albada G.D., Sloot P.M.A., Dongarra J. (eds.). Computational Science ICCS2004, Lecture Notes in Computer Science 3036 (2004) 531-535

Architecture Modeling and Simulation for Supporting Multimedia Services in Broadband Wireless Networks*

Do-Hyeon Kim[1] and Beongku An[2]

[1] School of Computer & Telecommunication Engineering,
Cheju National University, Cjeju, South Korea
kimdh@cheju.ac.kr
[2] Department of Electronic, Electrical & Computer Engineering,
Hongik University, Jochiwon, South Korea
beongku@wow.hongik.ac.kr

Abstract. In this paper, we propose and evaluate two reference models for supporting wireless video broadcasting services based on ATM over LMDS in metropolitan area wireless access networks. Our proposed reference models, namely end-to-end ATM model and headend/server-to-hub ATM model, are characterized. The end-to-end ATM model supports transfer of the ATM cell between headend and set-top box located in the end of networks for the wireless video services. The headend/server-to-hub ATM model transfers the ATM cell between headend/server and hub, and transmits the LMDS frame in the MPEG-TS over LMDS wireless access network. The performance evaluation of the proposed models is performed via analysis and simulation. The results evaluate delay and jitter of two proposed models. Especially, headend/server-to-hub ATM model has better performance than end-to-end ATM model for wireless video services because the end-to-end ATM model additionally needs more delay and jitter than headend/server-to-hub ATM model to convert the ATM cell to the LMDS frame.

1 Introduction

Recently, Broadband wireless access loop technologies are represented by MMDS (Multichannel Multipoint Distribution Service) and LMDS (Local Multipoint Distribution Service). Wireless video services use the wireless access loop such as LMDS instead of the coax cable for CATV and VoD (Video on Demand) service. Especially, we need the analysis about sensitive feature about the jitter and delay caused by the transmission of the video and image data in the broadband wireless multimedia networks [3,4]. However, in our study until now there are no reference models for supporting video services in broadband wireless networks.

This paper presents two reference models of the wireless video broadcasting services based on ATM over LMDS in metropolitan area wireless access networks. The first reference model is the end-to-end ATM model, where ATM cell is transferred between the set-top box and the headend/server. The second reference model is the headend/server-to-hub ATM model. In the second model, ATM cell is transmitted be-

* This work was supported by the Korea Research Foundation Grant.(KRF-2004-002-D00376).

V.S. Sunderam et al. (Eds.): ICCS 2005, LNCS 3516, pp. 1060 – 1063, 2005.

tween the hub and the headend/server in fixed network segment based on ATM, and MPEG2-TS (Transport Stream) is transferred to the set-top box in the radio access network segment using LMDS. We evaluate the performance of the proposed models in terms of the end-to-end delay and jitter of MPEG2-TS. In conclusion, the headend/server-to-hub ATM model can reduce more the delay and the jitter than end-to-end ATM model.

2 The Proposed Architecture Models

Wireless access loop has more merits than the wired cable in the view points of the economical efficiency and network flexibility. Wireless video system based on LMDS/ATM consists of service provider, headend, hub, set-top box, etc. We propose two architecture models namely, the end-to-end ATM model and the headend/sever-to-hub ATM model, to support the wireless video services based on LMDS/ATM [5,6].

The end-to-end ATM model supports transfer of the ATM cell between headend and set-top box located in the end of the wireless CATV or VoD system. This model uses the ATM backbone network and the LMDS wireless access network. In this model, the ATM cell is translated into LMDS frame to transmit the data between hub and set-top box, and LMDS frame is conversed into the ATM cell in set-top box. Figure 1 shows the protocol stack in end-to-end ATM model which supports the ATM protocol stack in all devices such as a headend, set-top box, hub, etc.

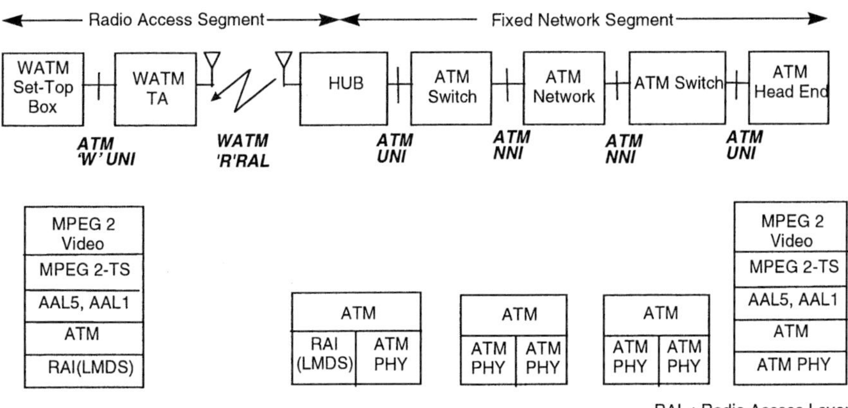

Fig. 1. End-to-end ATM model: Configuration and protocol stack

In the headend/server-to-hub ATM model, the ATM cell is transferred between headend/server and hub, and the LMDS frame included the MPEG-TS is transmitted in wireless access network. Then, the MPEG-TS frame is translated into LMDS frame to transmit the data between hub and set-top box. Especially, the main features of this proposed model is that the ATM cell is translated into MPEG2-TS frame and LMDS frame in hub, and conversed LMDS frame into the MPEG2-TS frame in set-top box.

Figure 2 describes that the protocol in headend/server-to-hub ATM model can support the ATM protocol stack in all devices such as a headend and hub. The hub removes the AAL head of the ATM cell and recombines with the MPEG2-TS stream.

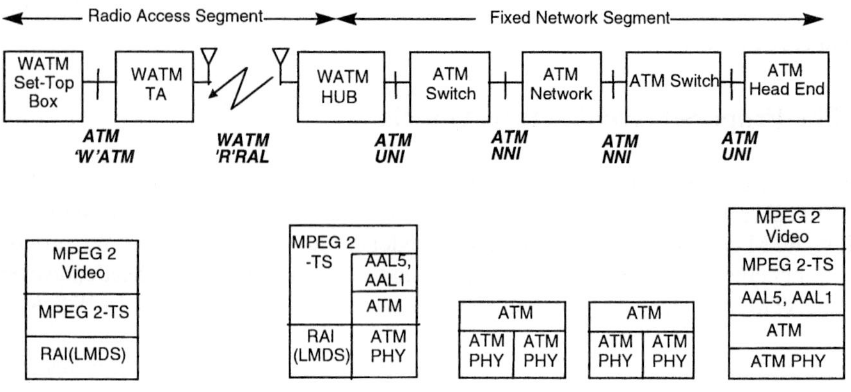

Fig. 2. Headend/server-to-hub ATM model: Configuration and protocol stack

3 Performance Evaluations

To verify the performance, we analyze jitter factors that happen in headend/server, hub and set-top box in the headend/server-to-hub model and end-to-end ATM model. We quantatively analyze the models in terms of end-to-end delay and jitter of MPEG2-TS and perform simulation using OPNET (Optimized Network Engineering Tool) by MIL3 company .

Table 1. End-to-end delay of MPEG2-TS in the end-to-end ATM model

Delay & Jitter		Useful bit rate of LMDS link					
		20MHz			40MHz		
		15.36 Mbps	23.04 Mbps	26.88 Mbps	30.72 Mbps	46.08 Mbps	53.76 Mbps
Flower Garden	Fixed Delay	1.05ms	0.41ms	0.39ms	0.38ms	0.34ms	0.33ms
	Maximum Delay	936.2ms	16.6ms	9.51ms	4.23ms	0.67ms	0.66ms
	Maximum Jitter	935.1ms	16.2ms	9.12ms	3.85ms	0.33ms	0.33ms

The simulation environment of wireless video service networks based on LMDS/ATM consists of the video source, server/headend, hub and set-top box. We assume that the distance from headend to hub sets by 50Km which is the size of a traditional CATV service area, and radio LMDS transmission link between the hub and set-top box establishes by 5Km. The transfer speed of the 155Mbps recommended by the ATM forum between server and hub is used, while the flower garden image is used.

The fixed end-to-end delay is 0.39ms and the jitter is between 0.39ms and 0.68ms at the radio channel speed 26.88Mbps used flower garden images in the end-to-end ATM model. Especially, jitter has a high probability value between 0.44ms and 0.57ms because the conversion of MPEG2-TS to AAL5 PDU and ATM cell to LMDS frame cause much jitter factor. In the table 1, we present the maximum end-to-end delay and jitter of MPEG2-TS for football and flower garden images according to radio channel speed.

Table 2. End-to-end delay of MPEG2-TS in the headend/server-to-hub ATM model

Delay & Jitter		Useful bit rate of LMDS link					
		20MHz			40MHz		
		15.36 Mbps	23.04 Mbps	26.88 Mbps	30.72 Mbps	46.08 Mbps	53.76 Mbps
Flower garden	Fixed Delay	0.84ms	0.36ms	0.35ms	0.34ms	0.33ms	0.32ms
	Maximum Delay	322.89ms	10.39ms	4.23ms	0.40ms	0.36ms	0.35ms
	Maximum Jitter	322.06ms	10.03ms	3.88ms	0.06ms	0.04ms	0.03ms

The fixed end-to-end delay is 0.35ms and the jitter is between 0.35ms and 0.42ms at the radio channel speed 26.88Mbps used flower garden images in the headend/server-to-hub ATM model. In the table 2, we present the maximum end-to-end delay and jitter of MPEG2-TS for football and flower garden images according to radio channel speed. The simulation results demonstrate the delay and jitter of two proposed models in the wireless CATV and wireless VoD service based on LMDS/ATM. Especially, headend/server-to-hub ATM model has better performance than end-to-end ATM model for wireless video services because additionally the end-to-end ATM model needs more delay and jitter than headend/server-to-hub ATM model to convert the ATM cell to the LMDS frame.

References

1. Digital Audio Video Council Technical Report, DAVIC 1.1 Specification Part 08.
2. ATM Forum Technical Committee, Audiovisual Multimedia Services : Video on Demand Specification 1.0.
3. Dipankar Raychaudhuri, "ATM-based Transport Architecture for Multi-services Wireless Personal Communication Networks," IEEE JSAC, Vol.12, No.8, pp. 1401~1414, Oct. 1994.
4. Bruno Cornaglia, Riccardo Santaniello and Enrico Scarrone, "Proposal for the protocol reference model for WATM," ATM Forum/96-1650, Dec. 1996.
5. Melbourne Barton, Daniel Pinto, et al., "Reference configuration model for wireless ATM," ATM Forum/96-1623, Dec. 1996.
6. "ISO/IEC 13818-6: MPEG-2 Digital Storage Media Command and Control," ISO/IEC JTC1/SC29/WG11, Mar. 1995.

Visualization for Genetic Evolution of Target Movement in Battle Fields

S. Baik[1], J. Bala[2], A. Hadjarian[3], P. Pachowicz[3], J. Cho[1], and S. Moon[1]

[1] College of Electronics and Information Engineering, Sejong University,
Seoul 143-747, Korea
{sbaik, jscho, sbmoon}@sejong.ac.kr
[2] School of Information Technology and Engineering,
George Mason University,
Fairfax, VA 22030, U.S.A
jbala@gmu.edu
[3] Sigma Systems Research, Inc.
Fairfax, VA 22032, U.S.A
{ahadjarian, ppach}@sigma-sys.com

Abstract. This paper focuses on the development of an interactive 3D visualization environment for displaying predicted movements by genetic evolution. The integration of interactive 3D visualization and the genetic search approach can relieve the burden of image exploitation tasks occurring due to an explosion of imagery data obtained in battle fields. The visualization component was implemented using In3D - a cross-platform Java class library designed to enable the creation of interactive 3D visualizations of complex information spaces.

1 Introduction

Recently, military missions have expanded into new and uncharted territories such as coalition operations; highly mobile operations; enhanced C4I (Command, Control, Communications, Computers and Intelligence); peacekeeping and humanitarian operations; law enforcement operations of counter-narcotics and counter-terrorism; monitoring treaties and weapons proliferation along with further "operations other than war." These increased areas of responsibility bring with them a greater need for imagery support. In this context, this explosion of available imagery data overwhelms the imagery analysts and outpaces their ability to analyze it. Consequently, the analysis tasks becoming bottlenecked in the imagery community. This situation generates an urgent need for new techniques and tools that can assist analysts in the transformation of this huge amount of data into a useful, operational, and tactical knowledge. The approach presented in this paper applies Genetic Algorithms (GAs) learning techniques to evolve new individuals in the population of movements in order to converge the evolution process toward optimal (most probable) movements. The major innovations in this approach are graphically depicted in Fig. 1.

V.S. Sunderam et al. (Eds.): ICCS 2005, LNCS 3516, pp. 1064 – 1067, 2005.

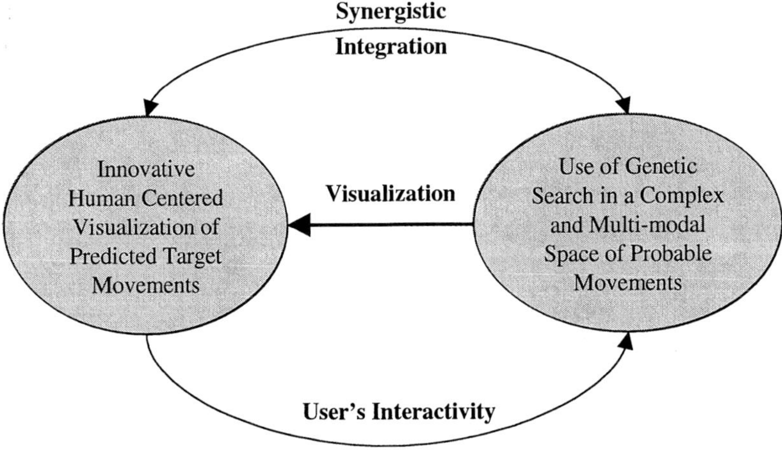

Fig. 1. Innovative concepts of Genetic Evolution Movement

2 Genetic Evolution of Target Movement

In a previous work [1], we focused on the genetic evolution approach for target movement prediction. Genetic Algorithms (GAs) [2,3] are used to generate the population of movement generation operators. The population evolves by using crossover and mutation and only the strongest elements survive, thus contributing to improved performance in terms of more probable/optimal movements. This contribution is used as an objective evaluation function to drive the generation process in its search for new and useful movement generation operators. When compared to the other researches[1], the main differentiators can be summarized in the following two points: 1) GAs use payoff (objective function) information, not derivatives or other auxiliary knowledge. Other search techniques require much auxiliary information in order to work properly; and 2) GAs use probabilistic transition rules, not deterministic rules. GAs use random choice as the tool to guide a search towards regions of the search space with likely improvement (i.e., more probable movements).

3 Interactive 3-D Visualization for Target Movement Prediction

It becomes more popular to use visualization techniques [4-6] to facilitate the development of GA systems. We focused on the development of an interactive 3D visualization environment for displaying predicted movements by genetic evolution. The system-interfacing component and visualization component was implemented using

[1] BONN Corp. has developed an enhancement to an existing movement prediction system, called the Tactical Movement Analyzer (TMA). The NATO Reference Mobility Model (NRMM) is the Army's accredited mobility performance prediction model.

In3D[2] - a cross-platform Java class library designed to enable the creation of interactive 3D visualizations of complex information spaces. The use of this library is pivotal in the implementation of the visualization environment for the genetic evolution of movements.

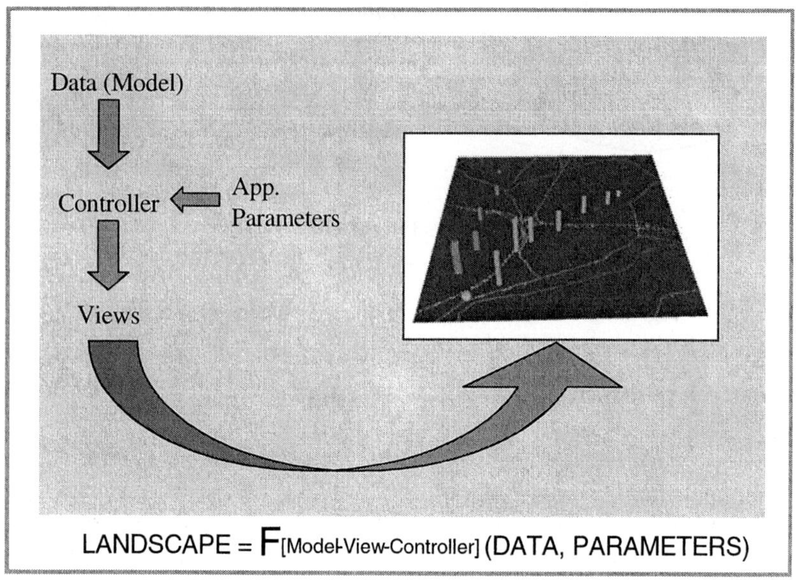

Fig. 2. A Model-View-Controller display paradigm

The visualization component implements the Model-View-Controller paradigm of separating imagery data (model) from its visual presentation (view). Interface elements (Controllers) act upon models, changing their values and effectively changing the views (Fig. 2). The Model-View-Controller paradigm supports the creation of applications which can attach multiple, simultaneous views and controllers onto the same underlying model. Thus, a single landscape (imagery and objects) can be represented in several different ways and modified by different parts of an application. The controller can achieve this transformation with a broad variety of actions, including filtering and multi-resolution, zooming, translation, and rotation. The component provides navigational aids that enhance users' explorative capabilities (e.g., the view from above provides a good overview of the information but it is not until zooming in and around and inspecting small items that the user gets a detailed understanding). Fig. 3 depicts an example of the prediction landscape.

---

[2] It was developed by Visual Insights Corp.

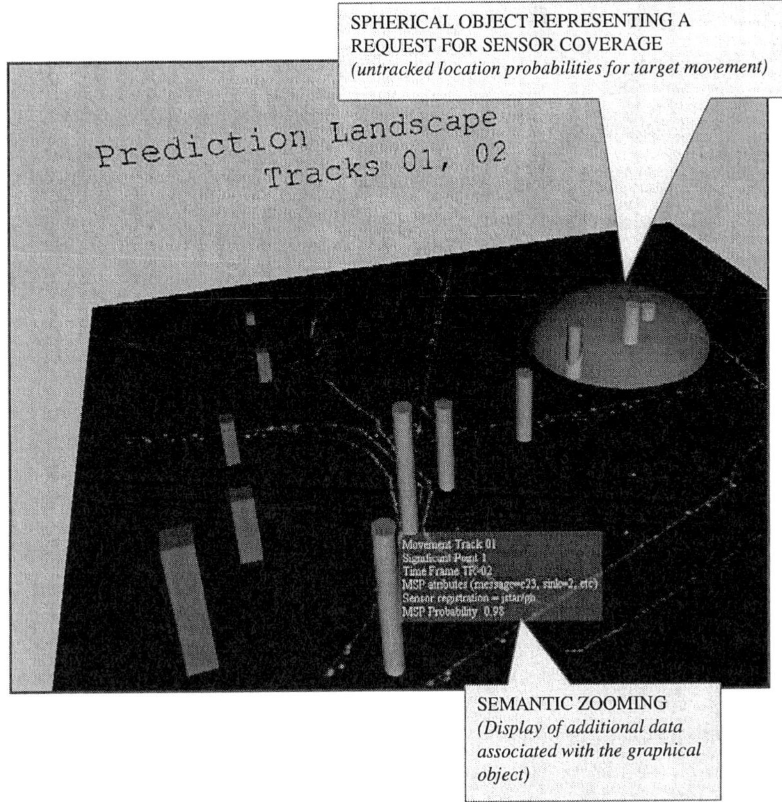

Fig. 3. Prediction landscape for Genetic Evolution of Movements

References

1. S. W. Baik, J. Bala, A. Hadjarian, and P. Pachowicz: Genetic Evolution Approach for Target Movement Prediction, Lecture Notes in Computer Science Vol. 3037, 2004
2. A. Brindle: Genetic algorithms for function optimization, Ph.D. Thesis, Computer Science Dept., Univ. of Alberta, 1981
3. K. A. DeJong: Adaptive system design: a genetic approach, IEEE Trans. Syst, Man, and Cyber, Vol. SMC-10, No. 9, pp. 566-574, Sept. 1980
4. W. B. Shine and C. F. Eick: Visualizing the evolution of genetic algorithm search processes, IEEE International Conference on Evolutionary Computation, pp. 367-372, April 1977
5. E. Hart and P. Ross: GAVEL - A New Tool for Genetic Algorithm Visualization, IEEE Trans on Evolutionary Computation, Vol. 5, No. 4, pp. 335-348, August 2001
6. A. M. Malkawi, R. S. Srinivasan, Y. K. Yi and R. Choudhary: Decision Support and design evolution: integrating genetic algorithms, CFD and visualization, Automation in Construction, Vol. 14, pp. 33-44, 2005

Comfortable Driver Behavior Modeling for Car Following of Pervasive Computing Environment[1]

Yanfei Liu and Zhaohui Wu

College of Computer Science , Zhejiang University, 38# Zheda Road, Hangzhou , Zhejiang , China, 310027
yliu@zju.edu.cn

Abstract. This paper demonstrates a novel car-following model based on driver or passengers' comfort. As we know, hasty deceleration during emergency brake will cause passengers feel uncomfortable. According to the relationship between brake acceleration and people's comfortable feeling, the comfortable model is setup. The model calculates the following car's acceleration by measuring the distance between the following car and the preceding car, the velocity of the following car, and controls the car's acceleration to make driver and passengers feel comfort. The paper combine the model with the pervasive computing concept, provoke the pervasive computing driver behavior modeling idea and turn it into reality to increase the adaptability and reliability of car's parts, when car equipped with this device, the prospect is not only the assistant driver or comfortable driver are realized in the car-following circumstance, but also the whole car's performance will be improved.

1 Introduction

The idea of pervasive computing is developing to be one of the hottest research topics at present [1]. The academic circles of all countries already have great foresight to focus on the research of the related topic [2].

The researches on driver behavior modeling have developed following mainly directions in recent years, the driver performance and capacity [3], the longitudinal driver behavior [4] and driver skill. The driver performance and capacity include mental and physical researches. There have been made a huge progress in all the directions these years [5].

Most of the early works in car-following model [6], PD-controller car following model [7], and visibility angle model [8] [9] are that drivers react immediately to the behavior of the vehicle in front of them so as to avoid imminent accidents. This paper focus on the driver's comfortable of car following.

Actually there can be as many as 50 embedded computers inside a modern car, on the other hand, a general human drivers behaviors is inherently complex. Both the

[1] This research was supported by 863 National High Technology Program under Grant No. 2003AA1Z2080, "Research of Operating System to Support Ubiquitous Computing".

V.S. Sunderam et al. (Eds.): ICCS 2005, LNCS 3516, pp. 1068 – 1071, 2005.

car's researches and driver behaviors modeling can not be separate from using as many as computing techniques, the pervasive computing must take an important roles in future research of them.

2 A Comfortable Car-Following Model

2.1 A Car Following Model Based on Space and Velocity

Car following model describes the driver longitudinal behavior shown as in Fig.1.

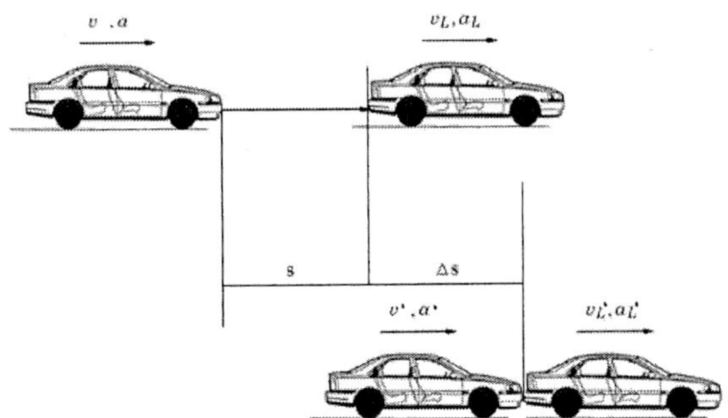

Fig. 1. Car following model based on space and velocity

At the time of collision, to the following car there is relation:

$$2a(s+\Delta s)=(v'^2 - v^2) \qquad (1)$$

For the worst condition, above equation can be simplified as following:

$$2as=v^2 \qquad (2)$$

2.2 Comfortable Car Following Model Based on Acceleration

The acceleration a can be calculated from s and v according to formula (2). Of the different a, the comfortable status are shown as table 1, the a_c represent the critical comfortable acceleration, the researches show that its value is 2 m/s^2.

Table 1. The following car's status of comfortable car following models

Area	Condition	Acceleration
Comfortable	$s > v^2/2a_c$	$a < a_c$
Uncomfortable	$v^2/2a_{max} < s < v^2/2a_c$	$a_c < a < a_{max}$
Dangerous	$s < v^2/2a_{max}$	$a > a_{max}$

2.3 The Realization of Comfortable Driver Behavior Model

Actually, the minimum brake distance $s_{\min}$ is related to the car's velocity, the car's velocity, and the friction coefficient of road surface. An experience formula of the relationship is shown as:

$$s_{min}=v^2/(2\mu g) \tag{3}$$

When considering the boundary conditions and friction coefficient, the relationships among a, s, v, μ are shown as Fig.2.:

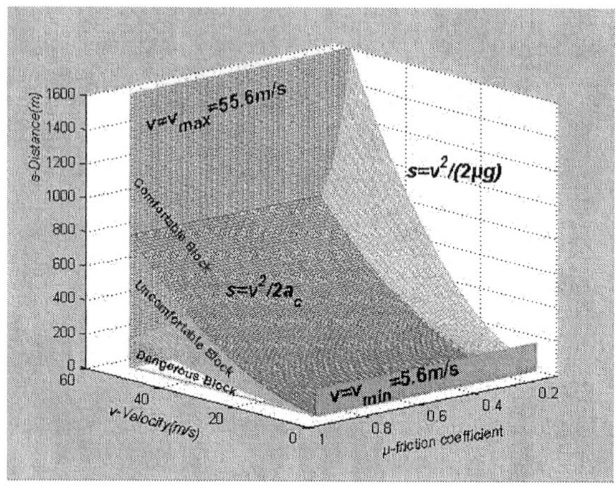

Fig. 2. The relationship among s, v, μ and a

The surfaces in Fig.2.enclosed several blocks, they are represent different physical conception, we named the relevant blocks as Table 2.

Table 2. The following car's status of comfortable car following models with COF

Block	Enclosed by	Acceleration
Comfortable block	$s > v^2/2a_c$, $s > v^2/2\mu g$, $0.2<\mu<0.9$, $v_{min}<v<v_{max}$	$a < a_c$
Uncomfortable block	$s < v^2/2a_c$, $s > v^2/2\mu g$, $0.2<\mu<0.9$, $v_{min}<v<v_{max}$	$a_c<a < a_{max}$
Dangerous Block	$s < v^2/2\mu g$, $s=0$, $0.2<\mu<0.9$, $v_{min}<v<v_{max}$	$a > a_{max}$

3 Implementation Driver Behavior Modeling of Pervasive Computing Environment

A wireless speed sensor is equipped to obtain the car's velocity, a *Bushnell Yardage Pro 800 Compact Rangefinder* displacement device is installed to detect the distance, and a dynamic friction coefficient tester is used to determine the friction coefficient.. The architecture of the model of pervasive computing environment is shown in Fig. 4.

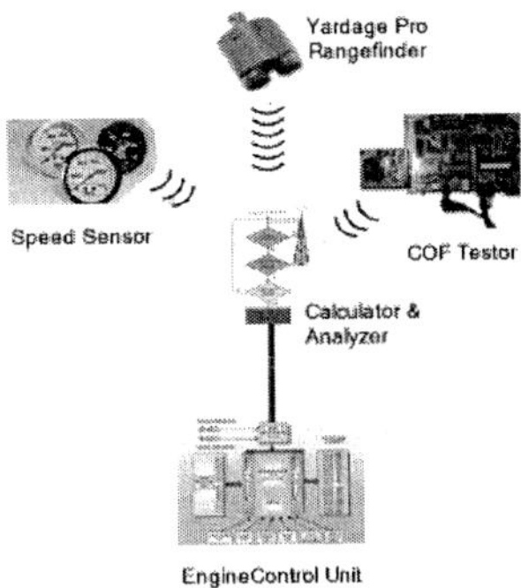

Fig. 4. The architecture of the comfortable model in pervasive computing environment

4 Experiments and Conclusions

The experiments are implemented by using a BUICK REGAL 2.5 car of GM, and the car running on a dry road surface of a highway on a sunny day.

One contribution of this paper is to demonstrate a new car-following model based on driver's comfort, while most of the former models are contribute to the safety of the car or the traffic flow throughput.

Another contribution is the introduction of pervasive computing conception to the drivers' behavior modeling researches. The paper combines the drivers' model with the pervasive computing concept, and makes the model as a pervasive computing device into reality to increase the system's adaptability and reliability.

References

1. Xu Guangyou ,The ubiquitous pervasive computing. China computer, June 9, 2003 general issue number 1222, year issue number 41.
2. Xiong Jiang, Embedded system and pervasive computing.
3. Strategies for Reducing Driver Distraction from In-Vehicle Telematics Devices: A Discussion Document (June 2003), Transport Canada TP 14133 E.
4. Driver Behavior Models for Traffic Simulation, Xiaopeng Fang, thesis of Iowa State University,2001
5. Yoshiyuki Umemura,Driver Behavior and Active Safety (Overview),Special Issue Driver Behavior and Active Safety,R&D Review of Toyota CRDL Vol. 39 No. 2
6. Johan Bengtsson, Adaptive Cruise Control and Driver Modeling, Printed in Sweden,Lund University, Lund 2001
7. Xiaopeng Fang, Hung A. Phaml and Minoru Kobayashi, PD Controller for Car-Following Models Based on Real Data
8. Pipes, L.A. and Wojcik, C.K., A contribution to theory of traffic flow, SAE Conference Proceedings – Analysis and Control of Traffic Flow Symposium,(Detroit, 1968), pp. 53-60
9. Ir. J.J. Reijmers, Traffic Guidance Systems, November 26, 2003 Et4-024 P99

A Courseware Development Methodology for Establishing Practice-Based Network Course

Jahwan Koo[1] and Seongjin Ahn[2]

[1] School of Information and Communications Engineering, Sungkyunkwan Univ.
Chunchun-dong 300, Jangan-gu, Suwon, Kyounggi-do, Korea
jhkoo@songgang.skku.ac.kr
[2] Department of Computer Education, Sungkyunkwan Univ.
Myeongnyun-dong 3-ga 53, Jongno-gu, Seoul, Korea
sjahn@comedu.skku.ac.kr*

Abstract. In this paper, we present a practice-based courseware development methodology for establishing a senior undergraduate network course for the computer-engineering department by reflecting on the rapidly changing information and communication technologies, enforcing practical education, and focusing on the existing and currently used curriculum models. Therefore, we have developed a special method, named it STM (Segmenting, Targeting, and Mapping), and applied it to the courseware development of a practice-based network course.

1 Introduction

Data communication and computer networks are major topics in the undergraduate computer-engineering curriculum. The goal of developing a courseware for network practice is to provide in-depth and meaningful networking content to teach senior undergraduate students. The first goal of this course is to teach the students how to comprehend and subsequently solve problems, specifically grasping the concepts of whole systems within the fields of network and communication systems. This is accomplished through the design and implementation, management, tuning and troubleshooting projects. The students can develop leadership skills through team projects. The second is to educate professionals in nationally competitive fields such as Internet, wired and wireless communication, home networking, and ubiquitous computing. The third is to diversify the contents of the course and the methods of instruction taking into consideration each student's ability, aptitude, and career path.

The courseware development for network practice lasted 12 months from March, 2003 to February, 2004. The objectives of this courseware are to educate students about internetworking technologies and simultaneously assist them in designing and building networks, and configure internetworking devices such as hubs, switches, routers, and wireless access points. The courseware development

* Dr. S. Ahn is the Corresponding Author.

V.S. Sunderam et al. (Eds.): ICCS 2005, LNCS 3516, pp. 1072–1075, 2005.

team was composed of a professor in charge of the development, five M.S. and Ph. D's with relevant expertise and teaching experience, and a Cisco certified internetworking professional possessing the highest level of expertise.

The paper is organized as follows. Section 2 introduces a method, called STM (Segmenting, Targeting, and Mapping), which is used to design the courseware for network practice. Finally, conclusion is drawn in Section 3.

2 STM(Segmenting, Targeting, and Mapping) Methodology

At this point, there is an increasing need to prepare systematic procedures and methods for migrating principle-based into practice-based courses. Therefore, we have developed a special method referencing in [1] [2], named it STM, and applied it to the coureware development of a practice-based network course.

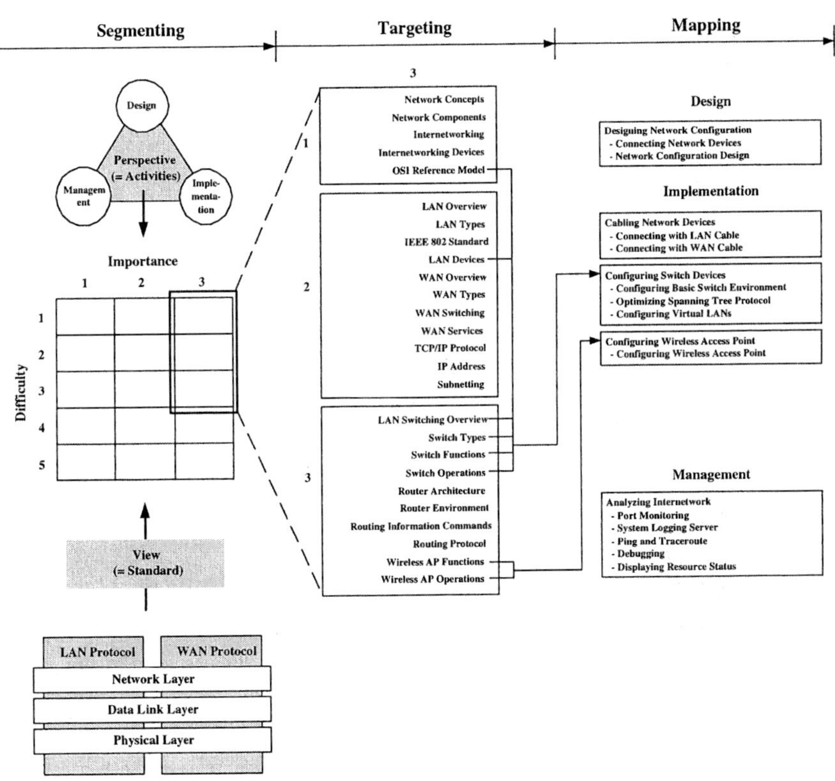

Fig. 1. STM diagram

2.1 Processes

STM Methodology is divided greatly to 3 Processes: Segmenting, Targeting, and Mapping. Segmenting is to categorize the various topics according to the degree of the difficulty and the importance. For this process, we gathered categorized approximately 30 books and publications written on the internetworking technologies according to topics. We also referred to related books published by the Cisco Press, and e-learning materials by Cisco network academy program [3] [4] [5] [6]. Notably, the categorized topics could be altered according to various perspectives and different views. Figure 1 shows the case where a series of network activities and computer communication protocols are used as perspectives and views, respectively. The categorized topics were subsequently segmented by 5 levels of difficulty and 3 levels of importance, and were targeted in consideration of the educational purpose, students' ability, aptitude, and career choice. Next, the segmented and targeted topics were mapped into the contents for network practice by integrating single or multiple topics into inclusive contents.

2.2 Output Definition

Contents to be deduced in mapping process are inverted ultimately to the contents of the final output, for example, a student workbook and a supplement CD. The student workbook is the main textbook for both the instructors and students; it is divided into 5 Parts including exercises, 3 appendixes, and a glossary. The concepts covered in this textbook enable students to develop practical experience in skills related to configuring LANs, virtual LANs, static or dynamic routing protocols, and collecting and analyzing network traffic. In addition, this textbook extends students' practical experience with WANs, ISDN, PPP, Frame Relay design, configuration, and maintenance. In addition, this textbook is complemented by a supplement CD [7], which contains presentation slides for instructors, lab activity movies presented in a multimedia format, related references, and authors' e-mail addresses for students' questions. There are 34 lab activity movies in the supplement CD to deliver accurate information and enhance understanding about network activities performed in real world.

2.3 Infrastructure Preparation: Network Lab Configuration

To define step-by-step exercises for network practice, the network lab configuration diagram must be prepared. Such physical infrastructure is defined beforehand necessarily and should be examined because it can influence considerably content of teaching material or teaching material development costs. Figure 2 shows an example of the network lab configuration diagram. The network laboratory holds about 20 students, which is modelled from the hierarchical internetworking architecture composed of one main center, two regional centers, and four branch offices. Each center and branch office are connected by T1 or 64 Kbps leased lines. Personal computers on each branch office are connected to a 10/100 Mbps workgroup switch or a 802.11b wireless access point.

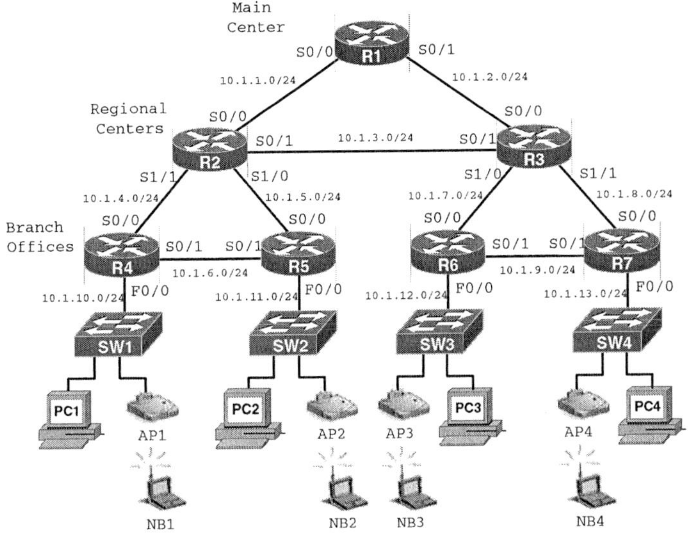

Fig. 2. Network lab configuration diagram

3 Conclusion

In this paper, we introduced the educational goals and development directions according to the change of curricula at SKKU's computer-engineering department. We presented the resources, people, time, and methods needed to develop a new courseware. We present the networking course as a proof-of-concept application of the STM methodology. In the future, we will gradually use this courseware in the network-centric courses.

References

1. J. T. Gorgone, G. B. Davis, J. S. Valacich, H. Topi, D. L. Feinstein, and H. E. Longenecker, "IS 2002 - Model Curriculum and Guidelines for Undergraduate Degree Programs in Information Systems," Association for Information Systems, 2002.
2. Computing Curricula 2001 Computer Science Final Report, December, IEEE-CS and ACM, 2001.
3. S. McQuerry, "Interconnecting cisco network devices," Cisco Systems, Inc, 2000.
4. Cisco networking academy program: first-year companion guide, Cisco Systems, Inc, 2001.
5. Cisco networking academy program: second-year companion guide, Cisco Systems, Inc, 2001.
6. Cisco networking academy program web site. [online] Available: http://cisco.netacad.net
7. Network practice course web site. [online] Available: http://songgang.skku.ac.kr/jhkoo/index1.htm

Solving Anisotropic Transport Equation on Misaligned Grids

J. Chen[1], S.C. Jardin[1], and H.R. Strauss[2]

[1] PPPL, P.O. Box 451, Princeton, NJ 08543, USA
[2] Courant Institute, NYU, 251 Mercer Street, NY 10012, USA
jchen@pppl.gov

Abstract. Triangular 3rd order Lagrange elements have been implemented previously to study the numerical error associated with grid misalignment. It has previously been found that grid misalignment strongly affects numerical accuracy in the case of linear elements. The same conclusion was obtained by higher order finite difference. Here we observe that this is also true for higher order finite elements, up to 3rd order, when the solution has a steep gradient. Three types of meshes are considered. Type t1 has one element edge fully aligned with the strong transport direction; type t2 doesn't have any alignment with that direction; type t0 is a combination of t1 and t2, i.e., partial alignment.

1 Introduction

Higher order Lagrange elements[1] have been used recently to study the numerical effect of grid misalignment[2][3] on highly anisotropic transport problems. It has been found that the numerical solution is polluted by the grid misalignment when linear elements were used. Also in [4] it showed that grid alignment is critical to obtain accurate solution numerically even for higher order finite difference methods. In [1] we found that higher order elements can reduce the numerical artifacts caused by such misalignment significantly. Here we extend the study in [1] to another 2 types of meshes: t1 and t2. The mesh applied in [1] is a combination of t1 and t2, and we refer it as t0 for convenience.

The steady-state anisotropic heat conduction equation considered here is given in the orthogonal coordinates (ξ, η) by:

$$\frac{\partial}{\partial \xi}\kappa_\xi \frac{\partial T}{\partial \xi} + \frac{\partial}{\partial \eta}\kappa_\eta \frac{\partial T}{\partial \eta} = 0. \tag{1}$$

Here T is the temperature and κ_ξ and κ_η are the conductivities along the transport axes ξ and η, respectively. Without loss of generality we assume $\kappa_\eta \equiv 1$ and $\kappa_\xi \gg 1$, so that the strong transport direction is aligned with ξ, and the anisotropy ratio is given by $\kappa_\xi/\kappa_\eta \equiv \kappa_\xi$.

The solution domain is given in the Cartesian coordinates (x, y) by the rectangular ABCD: $[0, L_x] \times [0, L_y]$ with Dirichlet boundary conditions. As shown in Fig. 1. [AB] is aligned with the x-coordinate and [AD] is aligned with the

V.S. Sunderam et al. (Eds.): ICCS 2005, LNCS 3516, pp. 1076–1079, 2005.

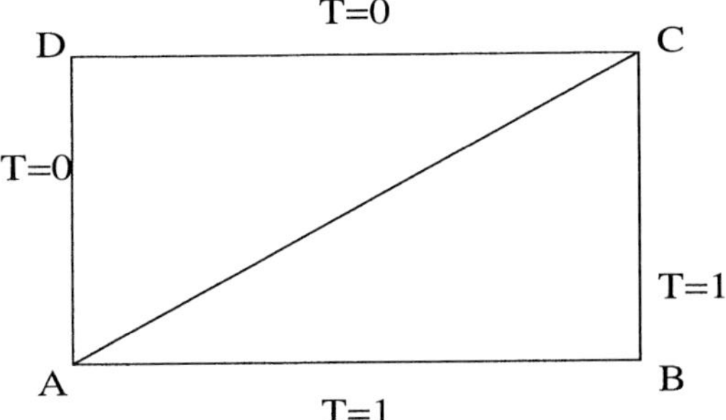

Fig. 1. Rect ABCD: $L_x = L_y = 1$. T=0 on side [AD-DC], T=1 on [AB-BC]. The strong anisotropic direction is aligned with the grid diagonal [AC]

y-coordinate. The strong transport axis ξ is aligned with the [AC] diagonal. The boundary conditions are: T=0 on side [AD-DC]; T=1 on side [AB-BC]. For infinite anisotropy, the exact solution is: $T = 0$ above the grid diagonal [AC], $T = 1$ below the grid diagonal [AC], and the width of the transition zone is zero. For a finite anisotropy, the exact solution introduces an internal layer which has non-zero transition width.

2 Mesh Setup

The unstructured triangular mesh is formed by first dividing the rectangular domain ABCD into uniform rectangular cells: $[0, N_x - 1] \times [0, N_y - 1]$. Then each of the rectangle grids is subdivided into two triangles in the following 3 ways:

Mesh t1 in Fig. 2. The strong transport direction ξ has full alignment with the element edge which is parallel to the diagonal [AC].

Mesh t2 in Fig. 2. The strong transport direction ξ has no alignment with the element edge. The elements, aligned with ξ int Mesh t1, are now aligned with the other diagonal [BD].

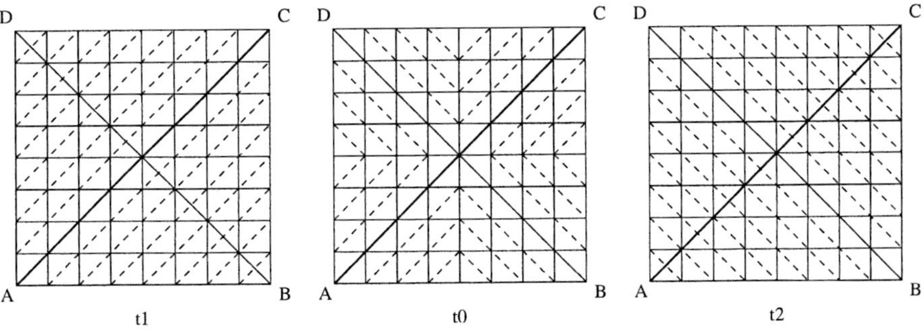

Fig. 2. Meshes. $t1$: 100% alignment; $t0$: 50% alignment; $t2$: 0% alignment

Mesh t0 in Fig. 2. This combines t1 and t2 such that the alignment is localized in the upper-right and lower-left blocks of the rectangular domain. In the upper-left and lower-right blocks, the misalignment is the same as in Mesh t2.

3 Numerical Experiments and Discussions

The numerical results for the convergence studies are presented in Table 1 and 2 for Mesh t1, t0 and t2, respectively in an ordering such that the alignment

Table 1. Profile width

	t1 mesh (100% alignment)				t0 mesh (50% alignment)				t2 mesh (0% alignment)			
N	10^2	10^3	10^4	10^5	10^2	10^3	10^4	10^5	10^2	10^3	10^4	10^5
39	0.1060	0.0353	0.0124	0.0094	0.1060	0.0354	0.0158	0.0138	0.1065	0.0441	0.0337	0.0328
49	0.1061	0.0353	0.0121	0.0077	0.1059	0.0355	0.0141	0.0113	0.1063	0.0408	0.0289	0.0275
59	0.1060	0.0350	0.0121	0.0066	0.1060	0.0351	0.0132	0.0096	0.1062	0.0386	0.0254	0.0240
69	0.1060	0.0351	0.0117	0.0059	0.1060	0.0351	0.0125	0.0084	0.1062	0.0374	0.0226	0.0213
79	0.1060	0.0350	0.0117	0.0054	0.1060	0.0350	0.0122	0.0076	0.1061	0.0365	0.0211	0.0193
89		0.0349	0.0115	0.0050		0.0350	0.0119	0.0069	0.1061	0.0361	0.0196	0.0177
99		0.0349	0.0115	0.0047		0.0350	0.0118	0.0063	0.1061	0.0357	0.0184	0.0163
109		0.0349	0.0115	0.0045			0.0117	0.0059		0.0355	0.0174	0.0152
119			0.0114	0.0043			0.0117	0.0055		0.0353	0.0166	0.0143
129			0.0114	0.0042			0.0116	0.0053		0.0352	0.0159	0.0135
139			0.0114	0.0040			0.0116	0.0051		0.0351	0.0153	0.0127
149			0.0114	0.0039			0.0115	0.0049		0.0351	0.0147	0.0121
159			0.0113	0.0039			0.0115	0.0048		0.0350	0.0143	0.0116
169			0.0113	0.0038			0.0114	0.0047		0.0350	0.0139	0.0111
179			0.0113	0.0037			0.0114	0.0046		0.0350	0.0135	0.0106
189				0.0037			0.0114	0.0045			0.0133	0.0106
199				0.0038				0.0044			0.0130	0.0099
209				0.0038				0.0043			0.0127	0.0095
219				0.0038				0.0042			0.0125	0.0092
229								0.0042				0.0090

Table 2. L_2 Norm

	t1 mesh (100% alignment)				t0 mesh (50% alignment)				t2 mesh (0% alignment)			
N	10^5	10^6	10^7	10^8	10^5	10^6	10^7	10^8	10^5	10^6	10^7	10^8
39	0.9969	0.9972	0.9972	0.9972	0.9965	0.9967	0.9967	0.9967	0.9919	0.9919	0.9920	0.9920
49	0.9974	0.9978	0.9978	0.9978	0.9971	0.9973	0.9974	0.9974	0.9933	0.9934	0.9934	0.9934
59	0.9977	0.9981	0.9982	0.9982	0.9975	0.9978	0.9978	0.9978	0.9942	0.9943	0.9943	0.9943
69	0.9979	0.9984	0.9985	0.9985	0.9977	0.9981	0.9981	0.9981	0.9949	0.9950	0.9950	0.9950
79	0.9980	0.9986	0.9987	0.9987	0.9979	0.9983	0.9983	0.9983	0.9954	0.9955	0.9956	0.9956
89	0.9982	0.9987	0.9988	0.9988	0.9980	0.9985	0.9985	0.9985	0.9958	0.9960	0.9960	0.9960
99	0.9982	0.9988	0.9989	0.9989	0.9981	0.9986	0.9987	0.9986	0.9961	0.9963	0.9963	0.9963
109	0.9983	0.9989	0.9990	0.9990	0.9982	0.9987	0.9988	0.9988	0.9964	0.9966	0.9966	0.9966
119	0.9984	0.9990	0.9991	0.9991	0.9983	0.9988	0.9989	0.9989	0.9966	0.9968	0.9968	0.9968
129	0.9984	0.9991	0.9992	0.9992	0.9983	0.9989	0.9989	0.9989	0.9968	0.9970	0.9970	0.9970
139	0.9984	0.9991	0.9992	0.9992	0.9984	0.9989	0.9990	0.9990	0.9970	0.9972	0.9972	0.9972
149	0.9985	0.9992	0.9993	0.9993	0.9984	0.9990	0.9991	0.9991	0.9971	0.9973	0.9974	0.9973
159	0.9985	0.9992	0.9993	0.9993	0.9985	0.9990	0.9991	0.9991	0.9972	0.9975	0.9975	0.9975
169	0.9985	0.9992	0.9994	0.9994	0.9985	0.9991	0.9992	0.9992	0.9973	0.9976	0.9976	0.9976
179	0.9985	0.9993	0.9994	0.9994	0.9985	0.9991	0.9992	0.9992	0.9974	0.9977	0.9977	0.9977
189	0.9985	0.9993	0.9994	0.9994	0.9985	0.9992	0.9993	0.9993	0.9975	0.9978	0.9978	0.9978
199	0.9986	0.9992	0.9995	0.9994	0.9985	0.9992	0.9993	0.9993	0.9976	0.9979	0.9979	0.9979
209	0.9986	0.9993	0.9995	0.9995	0.9986	0.9992	0.9993	0.9993	0.9977	0.9980	0.9980	0.9980
219	0.9986	0.9993	0.9995	0.9995	0.9986	0.9992	0.9994	0.9993	0.9977	0.9980	0.9981	0.9981
229		0.9994	0.9995	0.9995	0.9986	0.9993	0.9994	0.9994	0.9978	0.9981	0.9981	0.9981
239		0.9994	0.9996	0.9995		0.9993	0.9994	0.9994	0.9978	0.9982	0.9982	0.9982

is decreasing. The profile width, defined in [4], is used to detect the numerical convergence when $\kappa_\xi <= 10^5$. If the width remains the same 3 times in a row when the grid resolution is increasing, we consider the solution converged. Only 3rd order elements are applied here based on the results from [1],

When the anisotropy κ_ξ is small, $\leq 10^3$, as shown in columns 1 and 2 for each type of mesh in Table 1, the misalignment has a negligible effect. At $\kappa_\xi = 10^2$, convergence takes place at grid resolution 79 for t1 and t0 meshes, and 99 for t2 mesh. The solution on the t2 mesh is within 0.1% of the one on the t1 mesh. At $\kappa_\xi = 10^3$, both solutions on the t0 and t2 meshes have deviated from the one on the t1 mesh by about 0.3%, but convergence takes place later on the t2 mesh than on the t0 mesh.

When the anisotropy increases, the transition layer becomes thinner and the misalignment starts to come in and play a role.

At $\kappa_\xi \approx 10^3 - 10^5$, the misalignment has a medium effect. As seen from columns 3 and 4 in table 1 for the different types of mesh, the convergence takes place at a much earlier stage on the t1 mesh with full alignment. The convergence comes later on the partially aligned mesh t0, and it is seriously delayed on mesh t2 with no alignment. At $\kappa_\xi = 10^4$, solutions on the t0 mesh has a deviation from the one on the t1 mesh of about 0.8%, but about 11% from the solution on the t2 mesh with no alignment. At $\kappa_\xi = 10^5$, solutions on the t1 mesh converges at grid resolution 199, but not yet for solutions on the t0 and t2 meshes.

When the anisotropy continues to increase, $\kappa_\xi > 10^5$, numerically it becomes difficult to measure the profile width correctly since this layer becomes very thin. Therefore, the L_2 norm of the solution is used in Table 2. This norm should converges to 1 when $\kappa_\xi \to \infty$. The data in Table 2 indicate that the alignment has a strong effect. The L_2 norm approaches 1 faster on the mesh, than the t0 mesh. The L_2 norm on the t2 mesh is the slowest. Therefore, we conclude that in order to achieve good precision for problems with high anisotropy, grid alignment is an important factor to be considered even for the 3rd order elements.

Finally, it is observed that 3rd order elements only works for $\kappa_\xi <= 10^7$ since the L_2 norm remains the same at $\kappa_\xi = 10^7$ and $\kappa_\xi = 10^8$. For problem with $\kappa_\xi > 10^7$, higher order elements than 3rd order should be introduced. We will address this in the future.

References

1. J. Chen et al, Higher Order Lagrange Elements in M3D, DOE Technical Report, 2005. http://www.osti.gov/servlets/purl/836490-CNwldj/native/.
2. R. Vesey and D. Steiner, A Two-Dimensional Finite Element Model of the Edge Plasma, J. Comp. Phys. **116** 300-313 (1994).
3. R. Zanino, Advanced Finite Element Modeling of the Tokamak Plasmas Edge, J. Comp. Phys. **138** 881-906 (1997).
4. M. V. Umansky, M. S. Day, and T. D. Rognlien, On Numerical Solution of Strongly Anisotropic Diffusion Equation on Misaligned Grids. Submitted to Numerical Heat Transfer.

The Design of Fuzzy Controller by Means of Evolutionary Computing and Neurofuzzy Networks

Sung-Kwun Oh[1] and Seok-Beom Roh[2]

[1] Department of Electrical Engineering, The University of Suwon, San 2-2 Wau-ri, Bongdam-eup, Hwaseong-si, Gyeonggi-do, 445-743, South Korea
ohsk@suwon.ac.kr
[2] Department of Electrical Electronic and Information Engineering, Wonkwang University, 344-2, Shinyong-Dong, Iksan, Chon-Buk, 570-749, South Korea

Abstract. In this study, we propose a new design methodology to design fuzzy controllers. This design methodology results from the use of Computational Intelligence (CI), namely genetic algorithms and neurofuzzy networks (NFN). The crux of the design methodology is based on the selection and determination of optimal values of the scaling factors of the fuzzy controllers, which are essential to the entire optimization process. First, the tuning of the scaling factors of the fuzzy controller is carried out, and then the development of a nonlinear mapping for the scaling factors is realized by using GA based NFN.

1 Introduction

In the conventional design method to build fuzzy controllers, a control expert proposes some linguistic rules and decides upon the type and parameters of the associated membership functions. With an attempt to enhance the quality of the control knowledge conveyed by the expert, genetic algorithms (GAs) have already started playing a pivotal role. One should stress however that evolutionary computing is computationally intensive and this may be a point of concern when dealing with amount of time available to such search. For instance, when controlling a nonlinear plant such as an inverted pendulum of which initial states vary in each case, the search time required by GAs could be prohibitively high. To alleviate this shortcoming, we introduce a nonlinear mapping from the initial states of the system and the corresponding optimal values of the parameters. With anticipation of the nonlinearity residing within such transformation, in its realization we consider GA-based NFN. Bearing this mind, the development process consists of two main phases. First, using genetic optimization we determine optimal parameters of the fuzzy controller for various initial states of the dynamic system. Second, we build up a nonlinear model that captures a relationship between the initial states of the system and the corresponding genetically optimized control parameters.

2 Design Methodology of Fuzzy Controller

The block diagram of fuzzy PID controller is shown in Fig. 1. Note that the input variables to the fuzzy controller are transformed by the scaling factors (GE, GD, GH,

V.S. Sunderam et al. (Eds.): ICCS 2005, LNCS 3516, pp. 1080 – 1083, 2005.

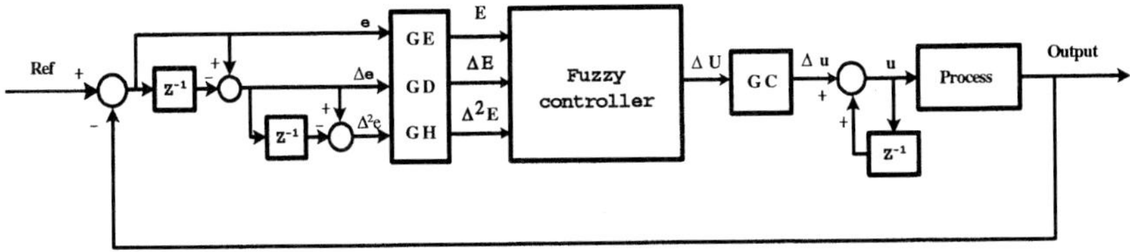

Fig. 1. An overall architecture of the fuzzy PID controller

and GC) whose role is to allow the fuzzy controller to properly "perceive" the external world to be controlled.

The above fuzzy PID controller consists of rules of the following form, cf. [3]

R_j : if E is A_{1j} and ΔE is A_{2j} and Δ^2E is A_{3j} then ΔU_j is D_j

The capital letters standing in the rule (R_j) denote fuzzy variables whereas D is a numeric value of the control action.

Genetic algorithms (GAs) are the search algorithms inspired by nature in the sense that we exploit a fundamental concept of a survival of the fittest as being encountered in selection mechanisms among species [4]. The genetic search is guided by a reproduction, mutation, and crossover. The standard ITAE expressed for the reference and the output of the system under control is treated as a fitness function [2].

It is the overall design procedure of the fuzzy PID controller realized by means of GAs that consists of the following steps

[step 1] Select the general structure of the fuzzy controller
[step 2] Define the number of fuzzy sets and set up initial control rules.
[step 3] Form a collection of initial individuals of GAs.
[step 4] All the control parameters (GE, GD, GH and GC) are tuned at the same time.

3 The Estimation Algorithm by Means of GA-Based Neurofuzzy Networks (NFN)

Let us consider an extension of the network with the fuzzy partition realized by fuzzy relations. Fig. 2 visualizes an architecture of such NFN for two-input and one-output, where each input assumes three membership functions. The circles denote processing units of the NFN. The node indicated ∏ denotes a Cartesian product, whose output is the product of all the incoming signals. And N denotes the normalization of the membership grades.

The learning of the NFN is realized by adjusting connections of the neurons and as such it follows a standard Back-Propagation (BP) algorithm

$$w(new) = w(old) + 2\cdot\eta\cdot(y_p - \hat{y}_p)\cdot\mu_i + \alpha(w_{ij}(t) - w_{ij}(t-1)) \qquad (1)$$

Where, η is a positive learning rate and α is the momentum coefficient.

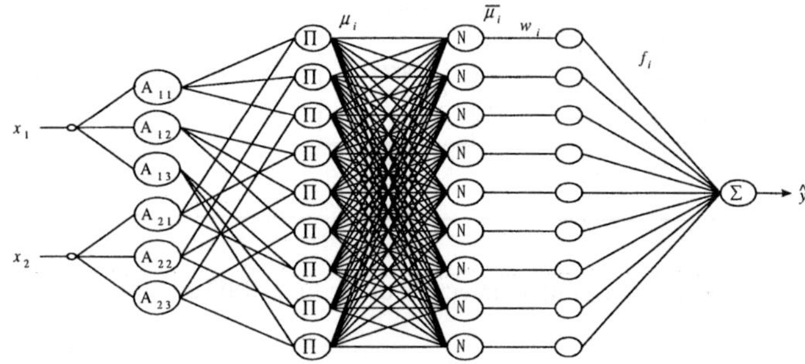

Fig. 2. NFN structure by means of the fuzzy space partition realized by fuzzy relations

4 Experimental Studies

In this section, we demonstrate the effectiveness of the fuzzy PID controller by applying it to the inverted pendulum system. Our control goal here is to balance the pole without regard to the cart's position and velocity [7]. We genetically optimize control parameters with a clear intent of achieving the best performance of the controller. GAs are powerful nonlinear optimization techniques. However, the powerful performance is obtained at the cost of expensive computational requirements and much time. To overcome this weak point, we propose the nonlinear model (GA-based NFN) which is able to estimate the control parameters quickly in the case that the initial angular positions and velocities of the inverted pendulum are selected arbitrarily within the given range. Fig. 3 demonstrates (a)pole angle (b)pole angular velocity for initial angle $\theta = 0.22$(rad) and initial angular velocity $\dot{\theta}=0.22$(rad/sec) .

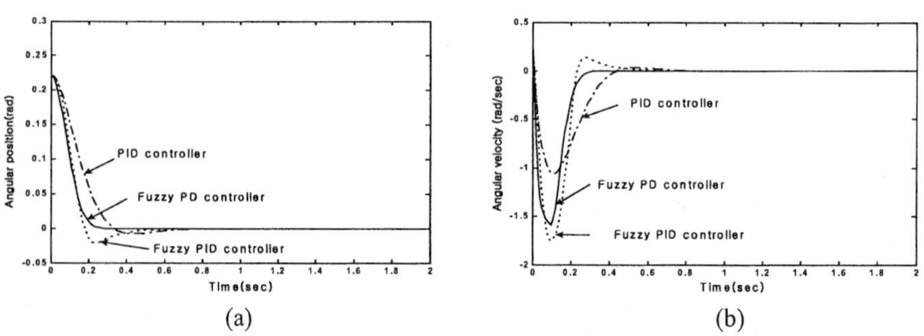

Fig. 3. (a)pole angle (b)pole angular velocity for initial angle $\theta = 0.22$(rad) and initial angular velocity $\dot{\theta}=0.22$(rad/sec)

From the above Fig. 3, we know that the fuzzy PD and fuzzy PID control effectively the inverted pendulum system. The output performance of the fuzzy controllers such as the fuzzy PD and the fuzzy PID controller including nonlinear characteristics are superior to that of the PID controller especially when using the nonlinear dynamic equation of the inverted pendulum. Moreover, the proposed estimation algorithm such as GA-based NFN generates the preferred model architectures. Especially the fuzzy PD controller describes the preferred one among the controllers.

5 Conclusions

We have proposed a two-phase optimization scheme of the fuzzy PID and PD controllers. The first phase of the design of the controller uses genetic computing that aims at the global optimization of its scaling factors where they are optimized with regard to a finite collection of initial conditions of the system under control. In the second phase, we construct a nonlinear mapping between the initial conditions of the system and the corresponding values of the scaling factors. From the simulation studies, using genetic optimization and the estimation algorithm of the GA-based neurofuzzy networks model, we showed that the fuzzy PI/PID controller controls effectively the inverted pendulum system. While the study showed the development of the controller in the experimental framework of control of a specific dynamic system (inverted pendulum), this methodology is general and can be directly utilized to any other system.

Acknowledgement. This work has been supported by KESRI(R-2004-B-274), which is funded by MOCIE(Ministry of commerce, industry and energy)

References

1. Oh, S.K., Pedrycz, W.: The Design of Hybrid Fuzzy Controllers Based on Genetic Algorithms and Estimation Techniques. Kybernetes **31** (2002) 909-917
2. Oh, S.K., Ahn, T., Hwang, H., Park, J., Woo, K.: Design of a Hybrid Fuzzy Controller with the Optimal Auto-tuning Method. Journal of Control, Automation and Systems Engineering. **1** (1995) 63-70
3. Li, H.X.: A comparative design and tuning for conventional fuzzy control. IEEE Trans. Syst., Man, Cybern. B. **27** (1997) 884-889
4. Goldberg, D.E.: Genetic algorithms in Search, Optimization, and Machine Learning. Addison-Wesley (1989)
5. Yamakawa, T.: A New Effective Learning Algorithm for a Neo Fuzzy Neuron Model. 5th IFSA World Conference. (1993) 1017-1020
6. Park, B.J., Oh, S.K., Pedrycz, W.: The Hybrid Multi-layer Inference Architecture and Algorithm of FPNN Based on FNN and PNN. 9th IFSA World Congress. (2001) 1361-1366.
7. Jang, J.R.: Self-Learning Fuzzy Controllers Based on Temporal Back Propagation. IEEE Trans. On Neural Networks. **3** (1992) 714-723
8. Oh, S.K., Rho, S.B., Kim, H.K.: Fuzzy Controller Design by eans of Genetic Optimization and NFN-Based Estimation Technique. International journal of Control, Automations, and Systems. **2**(3) (2004) 362-373

Boundary Effects in Stokes' Problem with Melting

Arup Mukherjee and John G. Stevens

Montclair State University, Montclair NJ 07043, USA
{mukherjeea, stevensj}@mail.montclair.edu

Abstract. We discuss the use of the computer algebra system Maple ® to analyze the heat transport when a heated sphere melts its way through a solid medium. The combined power of symbolic computation and numerical simulation available in Maple is used to analyze a model in which the molten layer around the moving sphere is augmented by a surrounding thermal-layer. In particular, we carefully study the effects of imposing a variety of boundary conditions.

1 Introduction

The power of large-scale simulations has had a major impact in advancing the understanding of physical processes in a multitude of applications. In many circumstances, the careful analysis of the model and an understanding of the effects of particular model parameters can create an efficient feedback mechanism which iteratively improves the model and provides insights into the underlying physical phenomena. Using different approximations for unknowns in a model, or comparing the effects of different boundary conditions are useful tools in this feedback cycle. In this paper, we use the power of Maple to advance the understanding of a model for Stokes' problem with melting. In particular, we study the effects of changing boundary conditions that affect the basic unknowns of the model. The classical Stokes' problem calculates the drag on a sphere moving at a constant velocity through a viscous, fluid medium. "With melting" assumes that the sphere maintains it surface temperature above that of the surrounding medium due to some internal heat source, the heat from the sphere is sufficient to melt a region around it, and the sphere falls through the melted region until it attains a steady state velocity. The model describes situations ranging from magma migration and core formation to that of a run-away nuclear reactor melting its way through the earth [3]. In section 2 we introduce the integral balance method and its innovative use in Stokes' flow with melting proposed by Emerman and Turcotte [1]. We study the effects of imposing improved boundary conditions for the velocity, and demonstrate that this leads to an improved estimate for the melt-layer. Section 3 highlights the differences between the melt-layer model and a more accurate two-layer approach [2] where a phenomenological thermal layer is added beyond the melt layer. The boundary condition for the temperature profile used in the melt-layer model of section 2 leads to certain anomalies which

V.S. Sunderam et al. (Eds.): ICCS 2005, LNCS 3516, pp. 1084–1087, 2005.

are resolved by the two-layer model. The thickness of the thermal-layer is obtained as a series approximation by re-applying the integral balance method on the thermal-layer. This approximation for the thermal-layer thickness is needed to impose the correct boundary condition for the temperature profile in the melt-layer. A repeat of the integral balance method in the melt-layer then yields the melt-layer thickness.

2 Integral Balance Method: Melt-Layer Model

In the integral balance method, it is assumed that heat is transported over some distance δ, and the heat conduction equation is integrated over this thermal layer thickness to obtain the integral heat balance. Appropriate boundary conditions and a trial function of a spatial variable are assumed for the temperature profile, leading to an expression involving the spatial variable and δ. Substituting this expression into the integral heat balance leads to a differential equation for δ which is solved to obtain an expression for the thermal layer thickness. Finally, the temperature profile is obtained through direct substitution. Emerman and Turcotte [1] employed the integral balance method to obtain an expression for the non-dimensional melt layer thickness, $\delta(\theta)$. The relevant geometry is displayed in Figure 1. Using the non-dimensional variables $\theta = x/R, \mathbf{y} = y/R, \delta = d/R, \mathbf{u} = u/u_0, \mathbf{v} = v/u_0$, and $\mathbf{T} = (T - T_m)/(T_0 - T_m)$, the Navier Stokes equation in lubrication form with boundary conditions $\mathbf{u}(0) = \mathbf{u}(\delta) = 0$ leads to $\mathbf{u}(\mathbf{y}) = -(3\mathbf{y}\sin\theta)(\mathbf{y} - \delta)/\delta^3$. The Stefan number is defined as $\mathtt{Ste} = \frac{c_p(T_0 - T_m)}{L + c_p(T_m - T_\infty)}$ where c_p is the specific heat capacity of melt, L is the latent heat of fusion, and T_∞ is the ambient temperature of the medium and the Peclet number is defined as $\mathtt{Pe} = u_0 R/\kappa$, where κ is the thermal diffusivity of melt. Ste and Pe form the dimensionless groups for the problem in non-dimensional form. Assuming a quadratic profile for temperature $\mathbf{T}(\mathbf{y})$ and imposing the conditions $\mathbf{T}(0) = 1$, $\mathbf{T}(\delta) = 0$ and $\partial\mathbf{T}/\partial\mathbf{y} = -\mathtt{Pe}\cos\theta/\mathtt{Ste}$, when $\mathbf{y} = \delta$ yields the temperature profile $\mathbf{T}(\mathbf{y}) = 1 + \mathbf{y}\left(\frac{\mathtt{Pe}}{\mathtt{Ste}}\cos\theta - \frac{2}{\delta}\right) + \mathbf{y}^2\left(\frac{1}{\delta^2} - \frac{\mathtt{Pe}}{\mathtt{Ste}}\frac{\cos\theta}{\delta}\right)$. Substituting the expressions for $\mathbf{u}(\mathbf{y})$ and $\mathbf{T}(\mathbf{y})$ into the integral heat balance and solving the resulting differential equation for $\delta(\theta)$ produces the melt-layer thickness as $\delta_{\mathbf{ET}}(\theta) = \delta_{\mathbf{ET}}(0)\sec\theta = f(\mathtt{Ste})\sec\theta/\mathtt{Pe}$, where $f(z) = 0.25(-3z - 20 + \sqrt{9z^2 + 280z + 400})$. Complete details of the derivation in this form can be found in [1, 2]. The energy equation for this model leads to $\partial^2\mathbf{T}/\partial\mathbf{y}^2 = 0$ when $\mathbf{y} = 0$, which yields the approximation $\delta = \delta_{\mathbf{ETO}}(\theta) = (\mathtt{Ste}/\mathtt{Pe})\sec\theta$. It is easily verified that as $\mathtt{Ste} \to 0$, $f(\mathtt{Ste}) \to \mathtt{Ste}$. Thus for small values of Ste, $\delta_{\mathbf{ETO}} \approx \delta_{\mathbf{ET}}$ while always satisfying $\delta_{\mathbf{ETO}} > \delta_{\mathbf{ET}}$.

2.1 Boundary Conditions for Velocity

The boundary condition $\mathbf{u}(\delta) = 0$ is a good approximation under the assumption $\delta << 1$, but is clear that the correct boundary condition at the boundary of the melt-layer is $\mathbf{u}(\delta) = -\sin\theta$. Solving for the velocity using this improved boundary condition results in $\mathbf{u}(\mathbf{y}) = -3\mathbf{y}\sin\theta(\mathbf{y} - \delta + \mathbf{y}\delta - \frac{2}{3}\delta^2)/\delta^3$. Using this expression for $\mathbf{u}(\mathbf{y})$ and assuming the temperature profile given above, a modified

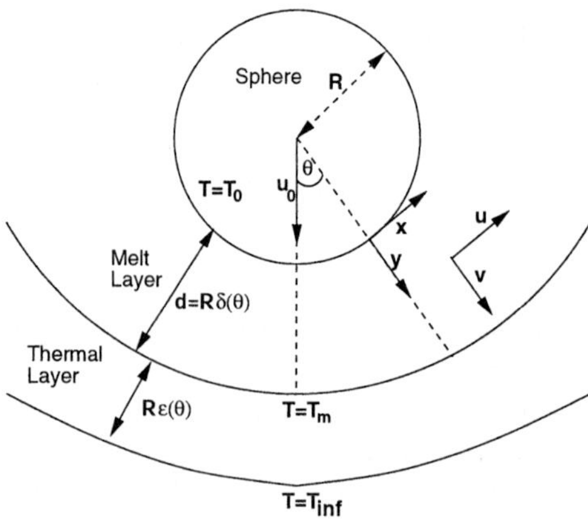

Fig. 1. Geometry for melt-layer and two–layer Stokes' Problem with Melting. The variables are discussed in detail in section 2

Table 1. Relative Percentage error $\left|\frac{\delta_{\mathbf{M}}(\theta)-\delta_{\mathbf{ET}}(\theta)}{\delta_{\mathbf{ET}}(\theta)}\right| \times 100$ for fixed $\mathtt{Ste} = 0.5$ (left) and fixed $\mathtt{Pe} = 10$ (right)

colatitude	$1 \le \mathtt{Pe} \le 50$ with $\mathtt{Ste} = 0.5$					$0.1 \le \mathtt{Ste} \le 2.0$ with $\mathtt{Pe} = 10$				
$\Downarrow$	1	5	10	15	50	0.1	0.25	0.5	1.0	2.0
$\theta = \pi/4$	2.93	0.63	0.32	0.21	0.06	0.02	0.09	0.32	0.96	2.52
$\theta = \pi/3$	5.74	1.32	0.68	0.45	0.14	0.04	0.21	0.68	1.95	4.77
$\theta = \pi/2.5$	0.13	0.65	1.94	4.98	10.67	13.26	3.67	1.94	1.32	0.41
$\theta = \pi/2.4$	17.24	5.19	2.79	1.92	0.60	0.20	0.98	2.79	6.81	13.84
$\theta = \pi/2.3$	23.58	8.00	4.45	3.09	0.98	0.37	1.67	4.45	10.06	19.02
$\theta = \pi/2.2$	34.40	13.97	8.23	5.85	1.94	0.88	3.45	8.23	16.61	28.27
$\theta = \pi/2.1$	54.71	29.72	19.64	14.75	5.43	3.28	10.04	19.64	32.82	47.32

melt layer thickness, $\delta_{\mathbf{M}}(\theta)$, is easily computed. Unlike the melt-layer expressions $\delta_{\mathbf{ETO}}(\theta)$ and $\delta_{\mathbf{ET}}(\theta)$ which become infinite as $\theta \to \pi/2$, $\delta_{\mathbf{M}}(\pi/2)$ is finite. Table 1 shows the relative percentage errors as a function of the Peclet number $\mathtt{Pe}$ and the Stefan number $\mathtt{Ste}$ as the colatitude θ changes. When $\mathtt{Ste}$ is fixed, For each value of $\mathtt{Pe}$, the errors increase as the colatitude θ approaches $\pi/2$. Also, for any fixed value of θ, the errors are a decreasing function of $\mathtt{Pe}$. Again, for any fixed value of $\mathtt{Ste}$, the errors increase monotonically as $\theta \to \pi/2$, and for any fixed value of θ, the errors increase with increasing $\mathtt{Ste}$. The third column for both tables correspond to the non-dimensional group pair $\mathtt{Ste} = 0.5$ and $\mathtt{Pe} = 10$. These values for $\mathtt{Ste}$ and $\mathtt{Pe}$ are very close to the specific example worked out in detail in [2]. In fact, if we assume that the sphere resembles a small rock penetrator of radius 3 cm melting its way through a typical rock like basalt, the relevant physical properties lead to $\mathtt{Ste} = 0.556$ and $\mathtt{Pe} = 10.33$. For values of $\mathtt{Ste}$ and $\mathtt{Pe}$ in a range where our analysis makes sense, the inequality $\delta_{\mathbf{M}} < \delta_{\mathbf{ET}} < \delta_{\mathbf{ETO}}$ always holds and the separation between $\delta_{\mathbf{M}}$ and $\delta_{\mathbf{ET}}$ increases as the equator is approached ($\theta = \pi/2$).

3 Integral Balance Method: Two-Layer Model

The boundary condition $\partial \mathbf{T}/\partial \mathbf{y} = -(\mathtt{Pe}/\mathtt{Ste})\cos\theta$ at $\mathbf{y} = \delta$ implies that the gradient of the temperature is zero at the equator. This condition is inconsistent with the temperature at this point being T_m while the surrounding solid medium has a temperate T_∞. An improved boundary condition at this melt/solid interface, based on the requirement that the latent heat required to melt the solid must be supplied by conduction, is $\partial \mathbf{T}/\partial \mathbf{y} = -(\mathtt{Pe}/\sigma)\cos\theta + \beta(\partial \mathcal{T}/\partial \mathbf{y})$, where $\beta = k_s(T_m - T_\infty)/k_l(T_0 - T_m)$, and $\sigma = c_p(T_0 - T_m)/L$ is similar to Ste but based on the actual latent heat, $\mathcal{T} = (T_s - T_\infty)/(T_m - T_\infty)$ is the non-dimensional temperature in the thermal-layer, and k_s and k_l represent the thermal conductivities for solid and liquid (melt) respectively. The geometry for the two-layer model is shown in Figure 1. If $\varepsilon(\theta)$ is the non-dimensional thermal layer thickness, the integral balance method described in section 2 can be applied to the thermal layer to obtain a series approximation for $\varepsilon(\theta)$, and $\mathcal{T}(\mathbf{y})$ is then computed by direct substitution. Finally, the resulting expression for $\mathcal{T}(\mathbf{y})$ is used in the boundary condition above and the integral balance method applied to the melt-layer to obtain its thickness, $\delta_{\mathbf{2L}}(\theta)$. The salient differences between $\delta_{\mathbf{2L}}$ and $\delta_{\mathbf{ET}}$ computed using $\mathbf{u}(\delta) = 0$ are outlined in [2]. The melt-layer thickness, $\delta_{\mathbf{ET}}$ becomes infinite, leading to a physically unrealistic unbounded melt region while $\delta_{\mathbf{2L}}(\pi/2) = 0.1325$, leading to a finite melt region.

Using the improved boundary condition $\mathbf{u}(\mathbf{y}) = -\sin(\theta)$ to determine the velocity $\mathbf{u}(\mathbf{y})$ in conjunction with the two-layer approach leads to a further improvement in the estimate of the melt-layer thickness. If $\delta_{\mathbf{2M}}(\theta)$ denotes the melt-layer thickness using this modified two-layer approach, then the relations $\delta_{\mathbf{2M}} < \delta_{\mathbf{2L}} < \delta_{\mathbf{M}} < \delta_{\mathbf{ET}} < \delta_{\mathbf{ETO}}$ hold among the different melt-layer expressions. Using the same values for σ, Ste, Pe, and β, yields $\delta_{\mathbf{2M}}(\pi/2) = 0.1306$. If the radius of the sphere is reduced, σ, Ste, and Pe increase while β decreases. The relative percentage errors, $100 \times |\delta_{\mathbf{2M}}(\theta) - \delta_{\mathbf{2L}}(\theta)|/|\delta_{\mathbf{2L}}(\theta)|$ range between 0.07 and 1.46 for $0 \le \theta \le \pi/2$.

References

1. Emerman, S. H.,Turcotte, D. L.: Stokes's problem with melting. Int. J. Heat Mass Transfer. **Vol. 26, No. 11** (1983) 1625–1630.
2. Mukherjee, A., Stevens, J., G.: Heat transport in Stokes' problem with melting: A two-layer approach. Int. J. of Heat Mass Transfer. **to appear**.
3. Rasmussen, N. C., Yellin, J., Kleitman D. J., and Stewart R. B.: Nuclear power: can we live with it? Tech. Rev. **81** (1979) 32–46.

A Software Debugging Method Based on Pairwise Testing[1]

Liang Shi[1,2], Changhai Nie[1,2], and Baowen Xu[1,2,3,4]

[1] Dept of Computer Science and Engineering, Southeast University, 210096 Nanjing, China
{Shiliang_ada, Changhainie, Bwxu}@seu.edu.cn
[2] Jiangsu Institute of Software Quality, 210096 Nanjing, China
[3] Computer School, National University of Defense Technology, 410073 Changsha, China
[4] Key Laboratory of Software Engineering, Wuhan University, 430072 Wuhan, China

Abstract. Pairwise testing is one of very practical and effective testing methods for various types of software systems. This paper proposes a novel debugging method based on pairwise testing. By analyzing the test cases and retesting with some complementary test cases, the method can narrow the factors that cause the errors in a very small range. So it can provide a very efficient and valuable guidance for the software testing and debugging.

1 Introduction

Software testing and debugging is an important but expensive part of the software development process. Much research has been aimed at reducing its cost. Among them, pairwise testing is a very important method that has been studied and applied widely for its scientificity and effectiveness in software testing with a quite small test suite, especially for the software whose faults are caused mainly by the test parameters or the interactions of a few test parameters of the system.

Pairwise testing has been used in software testing for a long time. At the beginning, through the application of design of experiments, a more efficient software testing way is obtained [4]. Much research has been done in pairwise testing. But they have been mainly aimed at reducing the size of the test suite by some methods of construction. For instance, in [1, 2] two heuristic search-based approaches are proposed, in [3] a new algebraic construction is proposed.

---

[1] This work was supported in part by the National Natural Science Foundation of China (60425206, 60373066, 60403016), National Grand Fundamental Research 973 Program of China (2002CB312000), and National Research Foundation for the Doctoral Program of Higher Education of China (20020286004).
Correspondence to: Baowen Xu, Department of Computer Science and Engineering, Southeast University, 210096 Nanjing, China. Email: bwxu@seu.edu.cn

V.S. Sunderam et al. (Eds.): ICCS 2005, LNCS 3516, pp. 1088 – 1091, 2005.

In this paper, we propose a debugging method based on the result of pairwise testing. The research about this has not been seen now. The convenient, reliable and effective debugging method can help people, who are working for testing or debugging the software, to find the errors and revise them quickly. It can improve the test efficiency and reduce test costs. Therefore, the research about the debugging method on the result of pairwise testing is quite important.

2 Model and Algorithm for Debugging of Pairwise Testing

Consider the system under test has n parameters c_i, where $c_i \in T_i$, T_i is a set with finite elements and the number of the elements is t_i, that is, $t_i = |T_i|$ ($1 \le i \le n$). We suppose the faults of the system are only caused by the parameters or the interactions of some parameters, and if an error is caused only by a combination of some k parameters where $1 \le k \le n$, every test case that includes the combination will also cause the error in testing. Let PT be a test set that covers all pair-wise combinations of parameter values. And there are m test cases in it, and each is like this: $(v_1, v_2, \ldots, v_n)$, where $v_i \in T_i$ and $1 \le i \le n$.

After testing the software with the test set PT, suppose there are l test cases in PT1 which cause errors in testing, and the other $m-l$ ones in PT2 which work well. The testing is successful for finding bugs in the software. Then we should analyze what cause the errors: a value v_i of parameter c_i; or a combination (v_i, v_j) of some two parameters c_i and c_j;…; a combination of some n-1 parameters values or a combination of all the n parameters values.

Let us consider the simplest case: the software failure is caused only by the value v_i of parameter c_i. Then the value v_i must appear in each test case of PT1 and will not be in any test case of PT2. For the general case, we can have the following conclusion:

Theorem 1. If a bug of the software under test is caused only by a combination of some k parameters where $1 \le k \le n$, the combination must appear in each test case of PT1, and must not appear in any of PT2.

Corollary 1. If the bugs of the software under test are caused by some combinations of some k parameters where $1 \le k \le n$, the combinations must only appear in PT1 and must not appear in PT2.

Corollary 2. If we can find the public composition in each test case of PT1, that is, if there are some parameter values or some parameter value combinations that appear in PT1 and don't appear PT2, then these public compositions have the maximum possibility to be the reasons that cause the errors.

Corollary 3. All the compositions that appear in PT1 and do not appear in PT2 maybe cause the errors. It is impossible for any of the compositions that appear in PT2 cause errors.

To locate the defects, we find all the compositions that are in PT1 and not in PT2, Such compositions form a set A. All the elements in the set may cause the errors. Generally the set A has more elements and need further reduction.

To reduce the set A, we design n test cases for each test case in PT1. For instance, $(v_1, v_2, ..., v_n)$ is one of the l test cases, and we can design the n test cases like these: $(*, v_2, ..., v_n)$, $(v_1, *, ..., v_n)$,..., $(v_1, v_2, ..., *)$, where "*" represents any value which is not the same as the original value in $(v_1, v_2, ..., v_n)$. Thus we need $n\times l$ additional test cases. By the result of these testing, we can reduce the two sets greatly and then obtain the conclusion.

Theorem 2. If a combination of some k values of test case $(v_1, v_2, ..., v_n)$ $(2\leq k\leq n)$ causes the system error, when testing the software with the accordingly additional n test case generated as above, there must be n-k test cases which cause the same error.

Now we describe the algorithm for debugging of pairwise testing based on the above discussion. After testing the software with the test set PT, if we find some bugs in the running of some l test cases in PT1, we can analyze the result by the following four steps:

(1) Analyze the test cases in PT1and construct the set A.
(2) Construct the additional test cases for each of the l test cases as the Theorem 2.
(3) Test the software with the additional test suite and reduce the set A by throwing of all the elements that appear in PT2.
(4) Output the set A, which contains all the possible factors that may cause the errors.

3 Case Study

In this section, we take the testing for a telephone system [1, 2] as an example to show how to use the approach proposed in this paper. Table 1 shows four parameters that define a very simple test model. The test plan shown in Table 2 has 9 test cases but it covers every pair-wise combination of parameter values.

Table 1. Parameters for Phone Call

Call	Billing	Access	Status
Local	Caller	Loop	Success
Long	Collect	Isdn	Busy
Inter	Free call	Pbx	Blocked

Table 2. Pair-Wise Test Cases for Phone Call

Call	Billing	Access	Status
Local	Collect	Pbx	Busy
Long	Free call	Loop	Busy
Inter	Caller	Isdn	Busy
Local	Free call	Isdn	Blocked
Long	Caller	Pbx	Blocked
Inter	***Collect***	***Loop***	***Blocked***
Local	Caller	Loop	Success
Long	Collect	Isdn	Success
Inter	Free call	Pbx	Success

Table 3. The Additional Test Cases

Call	Billing	Access	Status
*	Collect	Loop	Blocked
Inter	*	Loop	Blocked
Inter	Collect	*	Blocked
Inter	Collect	Loop	*

Suppose a bug is found by the sixth test case in Table 2. By our algorithm, the set A is created at the first step, where A={(Inter, Collect), (Inter, Loop), (Inter, Blocked),

(Collect, Loop), (Collect, Blocked), (Loop, Blocked), (Inter, Collect, Blocked), (Inter, Collect, Loop), (Collect, Loop, Blocked), (Inter, Loop, Blocked), (Inter, Collect, Loop, Blocked)}. At the next step, four additional test cases shown in Table 3 are generated for the one causing the error. Next we test the system again with the additional test cases. We can suppose only the second test case in the Table 3 causes error in testing. In this case, using the algorithm we can reduce the set as A ={(Inter, Loop, Blocked), (Inter, Collect, Loop, Blocked)}. That is, if the error is caused only by the interactions of some parameter values, one of the combinations (Inter, Loop, Blocked) and (Inter, Collect, Loop, Blocked) must be the cause. The other cases can be dealt with and analyzed analogously.

4 Conclusions

In general, when finding some errors in pairwise testing, we must determine which factors cause the errors and then revise them, that is, we must debug the software. At present, researches in this field are very few. There are some ways about experiment result analysis in method of orthogonal experiment, such as the method of intuitive analysis, the method of variance analysis and so on, but these methods are not practical to analyzing testing result.

The problem of software debugging based on pairwise testing is very important and complex. This paper proposes a method based on the condition that the system errors only caused by the values of some parameters or the interactions of some parameter values, and if a bug is caused only by a combination of some k parameters where $1 \leq k \leq n$, every test case that includes the combination will also cause the bug in testing. As an assistant debugging tool of pairwise testing, we have implemented it in our ETS (embedded software testing supporting system). The experiment results show that our method can improve the efficiency and decrease the cost of software testing greatly.

References

1. Cohen, D. M., et al.: The AETG System: An Approach to Testing Based on Combinatorial Design. IEEE Trans. On Software Engineering, July 1997, 23(7)
2. Tai, K. C., Lei, Y.: A Test Generation Strategy for Pairwise Testing. IEEE Trans. On Software Engineering, Jan 2002, 28(1)
3. Noritaka Kobayashi, Tatssuhio Tsuchiya, Tohru Kikuno: A New Method for Constructing Pair-wise Covering Designs for Software Testing. Information Processing Letters 81(2002) 85-91
4. Mandl, R.: Orthogonal Latin Squares: An Application of Experimental Design to Compiler Testing. *Communications of the ACM*, Vol 28, no 10, pp. 1054-1058, Oct. 1985
5. Kuhn D R and Gallo A M, Software Fault Interactions and Implications for Software Testing. IEEE Trans. On Software Engineering, June 2004, 30(6):418-421
6. Colbourn C J, Cohen M B, and Turban R C.: A Deterministic Density Algorithm for Pairwise Interaction Coverage. **IASTED Proc. of the Intl. Conf on Software Engineering (SE 2004)**, Innsbruck, Austria, February 2004:345-352

Heuristic Algorithm for Anycast Flow Assignment in Connection-Oriented Networks

Krzysztof Walkowiak

Chair of Systems and Computer Networks, Faculty of Electronics, Wroclaw University of Technology, Wybrzeze Wyspianskiego 27, 50-370 Wroclaw, Poland
Krzysztof.Walkowiak@pwr.wroc.pl

Abstract. Replication of content on geographically distributed servers can improve both performance and reliability of the Web service. Anycast is a one-to-one-of-many delivery technique that allows a client to choose a content server of a set of replicated servers. Presenting numerous mirror content servers to a client implies the difficult problem of finding the best server in terms of network performance. We formulate an optimization problem of anycast flows assignment in a connection-oriented network. This is a 0/1, NP-complete problem, which is computationally very difficult due to the size of solution space and constraints. Therefore, we propose computationally effective heuristic algorithm. To our knowledge, this is the first proposal to solve the problem of anycast flow assignment in connection-oriented network. Results of simulations are shown to evaluate performance of proposed algorithm for various scenarios.

1 Introduction

Web servers providing popular content (MP3 files, movies, electronic books, software distribution) need to scale to a large number of clients. One solution to address this problem is to augment network link capacity or processing resources of the site's server. Another, usually much more cost-effective, solution is to replicate the server or content in many locations in the Internet. This relatively simple technique provides numerous advantages, e.g. inexpensive improvement of client perceived accessibility of content, lower latency, increase of network reliability. Traffic associated with requests to replicas can be modeled as anycast flow. Anycast is a *one-to-one-of-many* technique to deliver a packet to one of many hosts. One of the most famous techniques that apply anycast traffic is Content Delivery Network (CDN) [1], [3].

Anycast paradigm is expected to be a very attractive approach to give a solution to many important issues that arise in modern computer networks. One of key elements of anycast transmission is the flow assignment problem, i.e. anycast demands should be assigned to network routes in order to optimize a selected network performance function. In this paper we address this problem and propose an efficient heuristic algorithm for static anycast flow assignment. The main novelty of our work is that we focus on connection-oriented (c-o) networks. In c-o network prior to transmitting the data, a virtual connection is established and the data is carried along this connection. Popular c-o techniques are Asynchronous Transfer Mode (ATM), MultiProtocol Label Switching (MPLS). To our knowledge, this is the first algorithm that solves the

V.S. Sunderam et al. (Eds.): ICCS 2005, LNCS 3516, pp. 1092 – 1095, 2005.

anycast flow assignment problem in c-o network. We consider an existing facility network - we don't optimize location of replica server and network topology.

The anycast non-bifurcated flow assignment (ANBFA) is as follows

Given	network topology, traffic demand pattern, location of replica servers, link capacity
Minimize (Maximize)	Objective function defined according to information on link flow and capacity, e.g. network delay, network cost, network survivability
Over	selection of replica server, routing (path assignment)
Subject to	connection-oriented flow constraints, capacity constraints

The mathematical formulation of the optimization problem can be found in [8], where the ANBFA problem is referred to as CATR (clients' assignment to replicas). Due to limited size of the paper we cannot present detailed information on anycast flow and optimization. Therefore, for more information refer to [1-9].

2 Anycast Non-bifurcated Flow Deviation Algorithm

In this section we show Anycast Non-bifurcated Flow Deviation (ANBFD) algorithm for the problem formulated above. ANBFD algorithm is based on the Flow Deviation approach proposed in [2] and widely used for network design problems [4], [7].

In c-o networks an anycast demand consists of two connections: one from the client to the server (upstream) and the second one in the opposite direction (downstream). Upstream connection is used to send user's requests. Downstream connection carries requested data. Let $\delta(i)$ denote index of the connection associated with connection i. If i is a downstream connection $\delta(i)$ must be an upstream connection and vice versa. The global anycast non-bifurcated multicommodity flow denoted by $\underline{f}^r$ is defined as a vector of flows in all arcs. We call a flow $\underline{f}^r$ feasible if for each arc a the flow of a doesn't exceed capacity of a.

Algorithm ANBFD

Let $\underline{f}^1$ denote a feasible anycast flow containing routes for all connections to be established. Let $L(\underline{g})$ denote value of the objective function for a feasible flow $\underline{g}$ and $l_j(\underline{g})$ denote a metric of arc j. The common approach is to calculate this metric as partial derivative of the objective function over arc flow [2]. We start with r:=1. Let B denote a set including connections that have not been already processed. Operator *first*(B) returns the index of first connection in set B.

<u>Step 1.</u> Find $SR(\underline{f}^r)$ defined as the set of shortest paths for each connection under metric $l(\underline{f})$. Connections i and $\delta(i)$ associated with one anycast demand are processed jointly. For each node v hosting a replica shortest paths in both directions between v and the second end node are calculated. Finally, we select such a pair of paths for which length of the downstream path is the shortest and add these paths to set $SR(\underline{f}^r)$. Set B:=P, i:=*first*(B) and go to step 2.

Step 2. Let $\underline{g} := \underline{f}^r$.

a) Find $\underline{v}$ from $\underline{g}$ by deviating flow of connections i and $\delta(i)$ to the shortest paths included in $SR\left(\underline{f}^r\right)$. Paths for other connections except i and $\delta(i)$ remain unchanged.

b) If $\underline{v}$ is a feasible flow and $L(\underline{v}) < L(\underline{g})$ set $\underline{g} = \underline{v}$.

c) If $B = \varnothing$ go to step 3. Otherwise, set B:=B-$\{i, \delta(i)\}$ and go to step 2a.

Step 3. If $\underline{g} = \underline{f}^r$ stop the algorithm, since the solution cannot be improved. Otherwise, set r:=r+1, $\underline{f}^r := \underline{g}$ and go to step 1.

In order to find feasible starting solution we can apply an algorithm based on the initial phase of the FD algorithm [2], [7]. One of the main advantages of ANBFD is that it takes into account all possible routes in the network and the calculation time of ANBFD doesn't depend on the number of candidate routes. Therefore, there is no need to reduce size of the problem given by the number of routes. Algorithm ANBFD converges in finite number of steps. However, the number of steps and consequently algorithm complexity depends on particular parameters of the considered problem (network topology, location of replicas, demand matrix).

3 Results

The test network consists of 36 nodes and 144 directed links [9]. We run experiments for various scenarios of replica location and number of replicas. Since more data is received by clients, than is sent to replicas. Therefore, we assume that upstream bandwidth is 0.1 of the downstream bandwidth. The total demand is calculated as a sum of all downstream and all upstream bandwidth requirements.

In many real life cases, a client is assigned to the closest replica in terms of the hop number. Therefore, for comparison we develop a simple heuristic, which is referred to as CR (Closest Replica). Algorithm CR assigns each anycast demand to the closest (in terms of hop number) replica. This way we obtain a unicast non-bifurcated flow assignment problem, which is solved by a non-bifurcated f FD algorithm.

We assume that the objective function represents the total flow in the network. In Table 1 we report results obtained for algorithms ANBFD and CR for the different number and locations of replica servers. Empty cells indicate that for a particular case the algorithm cannot find a feasible solution. First, quite intuitive, observation is that increasing the number of replicas decreases the network flow. Moreover, if there are more replicas in the network, demands with higher bandwidth requirements can be satisfied. If we compare ANBFD against CR, we can see that for relatively low loads both algorithms perform similarly. This can be explained by the fact that for low congestion each demand is assigned to nearest replica for both algorithms. The major advantage of ANBFD is for the case with 4 replicas – for two highest loads CR cannot find a feasible solution, while ANBFD can. This follows from the construct of ANBFD, which can reroute anycast demand to another replica, what enables selection of less congested links. Similar results were obtained for other tests.

Table 1. Performance of algorithms ANBFD and CR for various replica location and demands

Total Demand	1 replica in (30)		2 replicas (9,23)		3 replicas in (9,23,30)		4 replicas in (5,9,23,30)	
	ANBFD	CR	ANBFD	CR	ANBFD	CR	ANBFD	CR
1080	3088	3088	2125	2125	1722	1700	1362	1370
3105	9281	9425	5945	5945	4834	4805	3859	3885
5040			10502	10430	7886	7890	6331	6400
7020					11450	11515	8818	8915
9000							12225	11925
10980							14861	
11880							16146	

4 Final Remarks

The heuristic algorithm ANBFD proposed in this work is, according to our best knowledge, the first attempt to optimize anycast flows in connection-oriented networks. Numerical experiments have been conducted to evaluate performance of the algorithm for various scenarios. Results have confirmed that introducing new content servers enables substantial reduction of network flow what, consequently, decreases network delay and improves network reliability. The experiment shows that ANBFD is more efficient than the CR in solving relatively higher congested problems.

Acknowledgements. This work was supported by a research project of the Polish State Committee for Scientific Research carried out in years 2005-2007.

References

1. Awerbuch, B., Brinkmann, A., Scheideler C.: Anycasting in adversarial systems: routing and admission control. Lecture Notes in Computer Science, LNCS 2719 (2003), 1153-1168
2. Fratta, L., Gerla, M., Kleinrock, L.: The Flow Deviation Method: An Approach to Store-and-Forward Communication Network Design. Networks Vol. 3 (1973) 97–133
3. Hao, F., Zegura, E., Ammar, M.: QoS routing for anycast communications: motivation and an architecture for DiffServ networks. IEEE Communication Magazine, 6 (2002), 48-56
4. Kasprzak, A.: Designing of Wide Area Networks. Wroclaw Univ. of Tech. Press, (2001)
5. Partridge, C.,Mendez, T., Milliken, W.: Host Anycasting Service. IETF RFC 1546 (1993)
6. Peng, G.: CDN: Content Distribution Network. Technical Report (2003)
7. Pióro, M., Medhi, D.: Routing, Flow, and Capacity Design in Communication and Computer Networks. Morgan Kaufman Publishers (2004)
8. Walkowiak, K.: An exact algorithm for design of content delivery networks in MPLS environment. Journal of Telecommunications and Information Technology 2 (2004), 13-22
9. Walkowiak, K.: QoS Dynamic Routing in Content Delivery Networks, to appear in Proceedings of Networking 2005

Isotropic Vector Matrix Grid and Face-Centered Cubic Lattice Data Structures

J.F. Nystrom[1] and Carryn Bellomo[2]

[1] University of Akureyri, 602 Akureyri, Iceland
jamesn@unak.is
[2] University of Nevada, Las Vegas, Las Vegas NV 89154, USA
carryn.bellomo@ccmail.nevada.edu

Abstract. Techniques for generating data structures for isotropic vector matrix grids (or face-centered cubic lattices) are presented. Grid basics and some background mathematical foundations are also provided.

1 Introduction

An Isotropic Vector Matrix (IVM) grid, also known as the face-centered cubic (FCC) lattice, has the property that all neighbor nodes of every vertex are equally spaced therefrom[1]. There are 12 nearest neighbor nodes for each interior vertex in an IVM grid/FCC lattice arrangement. Compare now a three-dimensional Cartesian orientation of nodes where every vertex has exactly 26 neighbors; 6 of which are nearest neighbor nodes, 12 of which are next-nearest nodes, and the remaining 8 can be considered furthest neighbor nodes.

Herein we describe two techniques for constructing IVM grids. Section II gives an overview of the Vector Equilibrium (VE) cell and an algorithm[2] for generalized IVM grid construction. Section III encodes two overlapping IVM grids within a Cartesian-orientation of nodes. The Discussion includes ideas for simulations we plan to implement on IVM grids, and the Appendix describes the IVMCEM solver and the omni-directional curl operator.

2 IVM Basics and Generalized Grid Construction

The VE cell, also known as a cuboctahedron, is comprised of a central vertex surrounded by 12 nearest neighbor nodes. Figure 1 shows a VE cell and the six-element IVM vector basis. To distinguish the 12 exterior vertexes of the VE, we use `[+1]` to represent the point at the tip of the IVM basis vector $\mathbf{e}_1$ (see Fig. 1), `[0]` to designate the center of the VE cell, and `[-6]` to represent the vertex in the $-\mathbf{e}_6$ direction. The VE cell contains four hexagonal planes, each plane composed of 6 exterior vertexes and the center vertex. (The color rendering of Fig. 1 shows the *a*-plane in magenta, the *b*-plane in red, the *c*-plane in green, and the *d*-plane in aqua). The exterior vertexes of each hexagonal plane are:

[1] Herein we adopt R. Buckminster Fuller's[1] use of the term *Isotropic Vector Matrix* when referring to grids that utilize this specific geometrical arrangement of vertexes.

V.S. Sunderam et al. (Eds.): ICCS 2005, LNCS 3516, pp. 1096–1099, 2005.

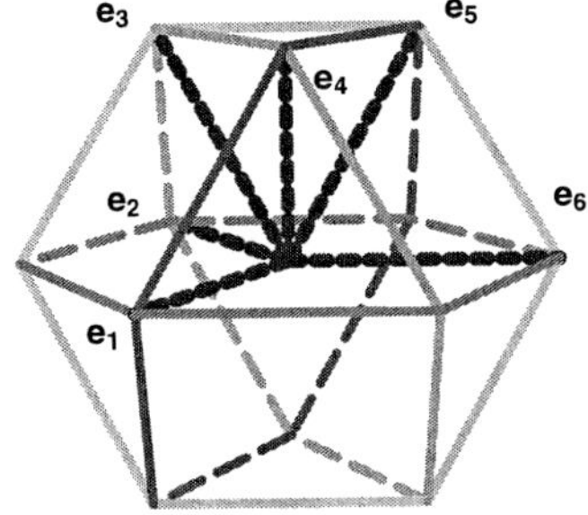

Fig. 1. VE cell and the IVM basis vectors $\mathbf{e}_1$ - $\mathbf{e}_6$

a-plane: `[+1] [-2] [+6] [-1] [+2] [-6]`; *b*-plane: `[+1] [+4] [+5] [-1] [-4] [-5]`;
c-plane: `[-2] [+4] [+3] [+2] [-4] [-3]`; *d*-plane: `[-5] [-3] [+6] [+5] [+3] [-6]`.

One way to build an IVM grid is to start with a VE and commence to make the exterior cells of the VE the center of its own VE cell. Each time we add new layer of VE to the outer shell of an IVM we create a higher *frequency* IVM grid. For each layer we add to an IVM grid, the number of nodes increases by $10F^2 + 2$; where F is the frequency of the new grid. For example, the first layer around the nuclear/center node has 12 nodes (and thus creates the VE cell). The second layer has 42 nodes, the third layer 92 nodes, and so on.

In [2] an algorithm which uses this process to build IVM grids is described in detail. The majority of the computation time for this generalized grid construction is occupied by a search into the node list for whether or not a node we are to add is in fact already in the node list. If we use a linear search technique, we obtain an $O(n^2)$ grid construction algorithm. We have implemented a binary search algorithm[3] to obtain an $O(n \log n)$ algorithm and a hashing algorithm[2] to produce an $O(n)$ grid construction algorithm.

3 The IVM Grids Within a Cartesian Grid

The algorithm described briefly in §2 creates a data structure for an IVM grid whereupon simulations can be built. Embedded in the data structure is information to access nearest neighbor vertexes; which is required, for example, to implement the *omni-directional curl operator* (see [2, 4, 5] and the Appendix) or generic finite difference stencils. We have developed another IVM grid generation technique[3] wherein nearest neighbor information is not stored, but instead is calculated using a simple index scheme. We overlay two IVM grids onto a

[2] This hashing technique was proposed and implemented by P.I. Wilson at Texas A&M University - Corpus Christi while working on a grant funded by a United States of America National Science Foundation grant # NSF MII 03-30822.

[3] R.W. Gray suggested this type of organization to one of the authors (JFN) quite a long time ago.

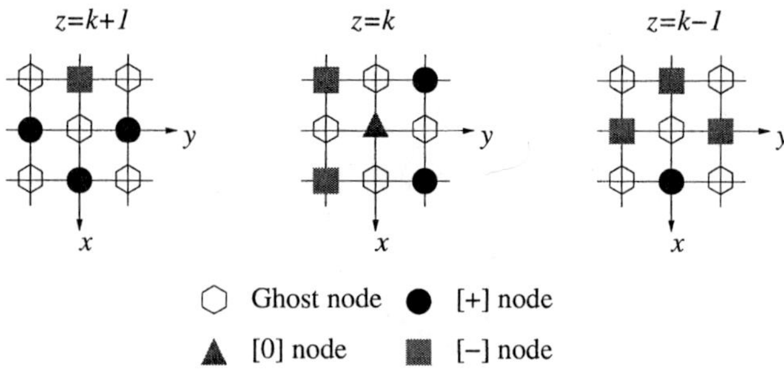

Fig. 2. Placement of VE cell vertexes and *ghosts* on Cartesian grid

three-dimensional Cartesian grid. Here we use the same orientation of lattice structure as is found in the common salt crystal, *NaCl* (see, for example, [6]). For salt, the *Na* atoms are arranged in a FCC lattice (i.e., an IVM grid) that interpenetrates a FCC lattice of *Cl* atoms. We will only be using one of the two IVM grids embedded within the Cartesian grid. The vertexes of the IVM grid not used are hereafter called *ghost* nodes.

The placement of IVM vertexes onto a Cartesian grid is shown in Fig. 2. We use a hexagon shape to indicate ghost nodes. Here we show the center and 12 exterior vertexes of a single VE cell. We maintain the same addressing notation as given in §2. The `[0]` node is shown on Fig. 2 as a solid triangle at the $z = k$ level, vertexes `[+1]` through `[+6]` are shown with a solid circle, and `[-1]` through `[-6]` are shown with a solid square.

Let the center of the VE cell, `[0]`, be located at a Cartesian index of `{i,j,k}`. Following the coordinate frame orientation in Fig. 2, the 12 exterior vertexes of the VE cell are then located with the following indexing scheme:

```
[+1] = {i+1,j,k-1}     [+2] = {i,j-1,k+1}     [+3] = {i+1,j,k+1}
[-1] = {i-1,j,k+1}     [-2] = {i,j+1,k-1}     [-3] = {i-1,j,k-1}

[+4] = {i+1,j+1,k}     [+5] = {i,j+1,k+1}     [+6] = {i-1,j+1,k}
[-4] = {i-1,j-1,k}     [-5] = {i,j-1,k-1}     [-6] = {i+1,j-1,k}
```

Using this indexing scheme it is trivial to write the IVM basis vectors $\mathbf{e}_1$ through $\mathbf{e}_6$ in terms of the Cartesian basis vectors.

4 Discussion

We have presented two disparate techniques for generating IVM grids. The first technique can produce IVM grids of generalized orientation. The second technique fixes the orientation of the IVM grid within a Cartesian framework.

There are many trade-offs associated with each choice of grid generation. The biggest difference we see relates to the parallel implementation of simulations that utilize IVM grids. The data structure created by generalized construction does not easily map to modern processor features such as caching, and can not be easily partitioned onto a multiprocessor cluster (like a *Beowulf* cluster), while

Cartesian placement method has access to all the latest parallelization research for Cartesian-based data structures.

We have used an IVM grid to construct a computational electromagnetic solver (see the Appendix). We also plan to investigate other application areas on IVM grids, including cell population dynamics, a lattice Boltzmann method (D3Q13), and cellular automata systems (including implementing ideas from the *computational cosmography*[5]).

References

1. Fuller, R.B. 1975. *Synergetics*. New York, NY: Macmillan.
2. Nystrom, J.F. 2003. Grid construction and boundary condition implementation for the isotropic vector field decomposition methodology. In Proceedings of *ACES 2003*, 745. Monterey, CA: The Applied Computational Electromagnetics Society.
3. Nystrom, J.F. and C. Bellomo. 2003. Efficient Grid Generation for the IVMCEM Solver. In Proceedings of *PIERS 2003*, 516. Cambridge, MA: The Electromagnetics Academy.
4. Nystrom, J.F. 2002. The isotropic vector field decomposition methodology. In Proceedings of *ACES 2002*, 257. Monterey, CA: The Applied Computational Electromagnetics Society.
5. Nystrom, J.F. 2004. On the Omni-directional Emergence of Form in Computation. *Lecture Notes in Computer Science* **3305**: 632.
6. Ashcroft, N.W. and N.D. Mermin. 1976. *Solid State Physics*. Philadelphia, PA: W.B. Saunders.

Appendix

The IVMCEM solver[2, 3, 4, 5] is a fully-discrete time-domain computational electromagnetic solver. This solver uses the omni-directional curl operator for the spatial part (of the Maxwell equations) and a Runge-Kutta fourth-order integrator (RK4) for the temporal advancement of a solution. In the IVMCEM solver we collocate field quantities at the grid locations using 12 doubled-valued components (6 for the electric field and 6 for the magnetic field).

The omni-directional curl operator calculates the vector curl for fields expressed in terms of the IVM basis vectors $\mathbf{e}_1$ through $\mathbf{e}_6$. For example, vector fields $\mathbf{S}$ and $\mathbf{T}$ can be written thusly:

$$\mathbf{S} = S_1\mathbf{e}_1 + S_2\mathbf{e}_2 + S_3\mathbf{e}_3 + S_4\mathbf{e}_4 + S_5\mathbf{e}_5 + S_6\mathbf{e}_6 \ ,$$
$$\mathbf{T} = T_1\mathbf{e}_1 + T_2\mathbf{e}_2 + T_3\mathbf{e}_3 + T_4\mathbf{e}_4 + T_5\mathbf{e}_5 + T_6\mathbf{e}_6 \ .$$

Let a', b', c', and d' be *plane variables*, each evaluated as a contour integrals on the hexagonal planes of the VE cell (see §2). The omni-directional curl operator calculates $\mathbf{S} = \nabla \times \mathbf{T}$ for $\mathbf{S}$ at the center of the VE cell based on values of $\mathbf{T}$ on the 12 exterior vertexes of the VE cell. Each of the six vector components of $\mathbf{S}$ depends on the contour integrals around two separate hexagonal planes.

Design of Evolutionally Optimized Rule-Based Fuzzy Neural Networks Based on Fuzzy Relation and Evolutionary Optimization

Byoung-Jun Park[1], Sung-Kwun Oh[2], Witold Pedrycz[3], and Hyun-Ki Kim[2]

[1] Department of Electrical Electronic and Information Engineering, Wonkwang University, 344-2, Shinyong-Dong, Iksan, Chon-Buk, 570-749, South Korea
[2] Department of Electrical Engineering, The University of Suwon, San 2-2 Wau-ri, Bongdam-eup, Hwaseong-si, Gyeonggi-do, 445-743, South Korea
ohsk@suwon.ac.kr
[3] Department of Electrical and Computer Engineering, University of Alberta, Edmonton, AB T6G 2G6, Canada and Systems Research Institute, Polish Academy of Sciences, Warsaw, Poland

Abstract. In this paper, new architectures and comprehensive design methodologies of Genetic Algorithms (GAs) based Evolutionally optimized Rule-based Fuzzy Neural Networks (EoRFNN) are introduced and the dynamic search-based GAs is introduced to lead to rapidly optimal convergence over a limited region or a boundary condition. The proposed EoRFNN is based on the Rule-based Fuzzy Neural Networks (RFNN) with the extended structure of fuzzy rules being formed within the networks. In the consequence part of the fuzzy rules, three different forms of the regression polynomials such as constant, linear and modified quadratic are taken into consideration. The structure and parameters of the EoRFNN are optimized by the dynamic search-based GAs.

1 Introduction

In this paper, new architectures and comprehensive design methodologies of Genetic Algorithms [1] (GAs) based Evolutionally optimized Rule-based Fuzzy Neural Networks (EoRFNN) are introduced for effective analysis and solution of nonlinear problem and complex systems. The proposed EoRFNN is based on the Rule-based Fuzzy Neural Networks (RFNN). In the consequence part of the fuzzy rules, three different forms of the regression polynomials such as constant, linear and modified quadratic are taken into consideration. The polynomial of a fuzzy rule results from that we look for a fuzzy subspace (a fuzzy rule) influencing the better output of a model, and then raise the order of polynomial of the fuzzy rule (subspace). Contrary to the former, we make a simplified form for the representation of a fuzzy subspace lowering of the performance of a model. This methodology can effectively reduce the number of parameters and improve the performance of a model. GAs being a global optimization technique determines optimal parameters in a vast search space. But it cannot effectively avoid a large amount of time-consuming iteration because GAs finds optimal parameters by using a given space (region). To alleviate the problems, the dynamic search-based GAs is introduced to lead to rapidly optimal convergence over a limited

V.S. Sunderam et al. (Eds.): ICCS 2005, LNCS 3516, pp. 1100 – 1103, 2005.

region or a boundary condition. To assess the performance of the proposed model, we exploit a well-known numerical example [2], [3].

2 Polynomial Fuzzy Inference Architecture of Rule Based FNN

The networks are classified into the two main categories according to the type of fuzzy inference, namely, the simplified and linear fuzzy inference. In this paper, we propose the polynomial fuzzy inference based RFNN (pRFNN). The proposed pRFNNs are obtained from the integration and extension of conventional RFNNs.
[Layer 1] Input layer
[Layer 2] Computing activation degrees of linguistic labels
[Layer 3] Computing firing strength of premise rules
[Layer 4] Normalization of a degree activation (firing) of the rule
[Layer 5] Multiplying a normalized activation degree of the rule by connection weight
[Layer 6] Computing output of pRFNN
The proposed pRFNN can be designed to adapt a characteristic of a given system, also, that has the faculty of making a simple structure out of a complicated model for a nonlinear system, because the pRFNN comprises consequence structure with various orders (Types) for fuzzy rules.

3 Evolutionally Optimized Rule-Based Fuzzy Neural Networks

In order to generate the proposed EoRFNN, the dynamic search based GAs is used in the optimization problems of structures and parameters. From the point of fuzzy rules, these divide into the structure and parameters of the premise part, and that of consequence part. The structure issues in the premise of fuzzy rules deal with how to use of input variables (space) influencing outputs of model. The selection of input variables and the division of space are closely related to generation of fuzzy rules that determine the structure of RFNN, and govern the performance of a model. Moreover, a number of input variable and a number of space divisions induce some fatal problems such as the increase of the number of fuzzy rules and the time required. Therefore, the relevant selection of input variables and the appropriate division of space are required. The structure of the consequence part of fuzzy rules is related to how represents a fuzzy subspace. Universally, the conventional methods offer uniform types to each subspace. However, it forms a complicated structure and debases the output quality of a model, because it does not consider the correlation of input variables and reflect a feature of fuzzy subspace. In this study, we apply the various forms in expressions of a fuzzy subspace. The form is selected according to an influence of a fuzzy subspace for an output criterion and provides users with the necessary information of a subspace for a system analysis.

4 Experimental Studies

In this experiment, we use three-input nonlinear function as in [2], [3]. This dataset was analyzed using Sugeno's method [2]. We consider 40 pairs of the original input-output data. 20 out of 40 pairs of input-output data are used as learning set and the remaining part serves as a testing set.

Fig. 1 shows topologies of EoRFNN. In order to solve a given nonlinear problem, these architectures are designed and generated in flexibility that can cope with an environment (condition).

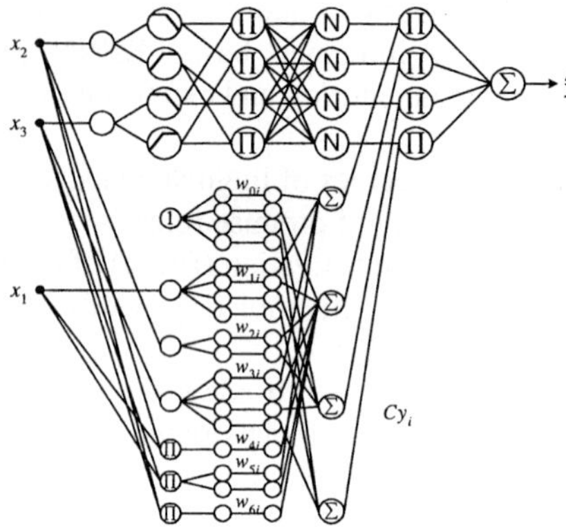

Fig. 1. Topology of EoRFNN02 with 2 inputs for the nonlinear function

Table 1 covers a comparative analysis including several previous models. Sugeno's model I and II were fuzzy models based on linear fuzzy inference method while Shinichi's models formed fuzzy rules by using learning method of neural networks. The study of literature [5] is based on fuzzy-neural networks using HCM clustering and evolutionary fuzzy granulation. Multi-FNN consists of 3 FNN structures. The proposed EoRFNNs come with higher accuracy and improved prediction capabilities.

Table 1. Comparison of performance with other modeling methods

Model		PI	E_PI	No. of rules
Linear model [3]		12.7	11.1	
GMDH [3,6]		4.7	5.7	
Sugeno's [2,3]	Fuzzy I	1.5	2.1	3
	Fuzzy II	1.1	3.6	4
Shinichi's [4]	FNN Type 1	0.84	1.22	$8(2^3)$
	FNN Type 2	0.73	1.28	$4(2^2)$
	FNN Type 3	0.63	1.25	$8(2^3)$
FNN [5]	Simplified	2.865	3.206	9(3+3+3)
	Linear	2.670	3.063	9(3+3+3)
Multi-FNN [5]	Simplified	0.865	0.956	9(3+3+3)
	Linear	0.174	0.689	9(3+3+3)
Proposed model (EoRFNN)		0.241	0.357	8(2×2×2)
		0.224	0.643	4(2×2)

5 Concluding Remarks

In this paper, new architectures and comprehensive design methodologies of Evolutionally optimized Rule-based Fuzzy Neural Networks (EoRFNN) has discussed for effective analysis and solution of nonlinear problem and complex systems. Also, the dynamic search-based GAs has introduced to lead to rapidly optimal convergence over a limited region or a boundary condition. The proposed EoRFNN is based on the Rule-based Fuzzy Neural Networks (RFNN) with the extended structure of the fuzzy rules. The structure and parameters of the proposed EoRFNN are optimized by GAs. The proposed EoRFNN can be designed to adapt a characteristic of a given system, also, that has the faculty of making a simple structure out of a complicated model for a nonlinear system. This methodology can effectively reduce the number of parameters and improve the performance of a model. The proposed EoRFNN can be efficiently carried out both at the structural as well as parametric level for overall optimization by utilizing the separate (Ⓐ) or consecutive (Ⓑ) tuning technology. From the results, Ⓑ methodologies simultaneously tuning both structure and parameters, reduce parameters of consequence part, and offer the output performance better than the Ⓐ. Namely, Ⓑ method is effective in identifying a model than Ⓐ.

Acknowledgements. This work has been supported by KESRI(I-2004-0-074-0-00), which is funded by MOCIE(Ministry of commerce, industry and energy)

References

1. Goldberg, D.E.: Genetic Algorithms in search, Optimization & Machine Learning. Addison-wesley. (1989)
2. Kang, G., Sugeno, M.: Fuzzy Modeling. Transactions of the Society of Instrument and Control Engineers. **23**(6) (1987) 106-108
3. Park, M.Y. and Choi, H.S.: Fuzzy Control System. Daeyoungsa. (1990) 143-158 (In Korean)
4. Horikawa, S.I., Furuhashi, T., Uchigawa, Y.: On Fuzzy Modeling Using Fuzzy Neural Networks with the Back Propagation Algorithm. IEEE Transactions on Neural Networks. **3**(5) (1992) 801-806
5. Park, H.S., Oh, S.K.: Multi-FNN Identification Based on HCM Clustering and Evolutionary Fuzzy Granulation. International Journal of Control, Automation and Systems. **1**(2) (2003) 194-202
6. Kondo, T.: Revised GMDH algorithm estimating degree of the complete polynomial. Transactions of the Society of Instrument and Control Engineers. **22**(9) (1986) 928-934
7. Park, H.S., Oh, S.K.: Fuzzy Relation-based Fuzzy Neural-Networks Using a Hybrid Identification Algorithm. International journal of Control, Automations, and Systems. **1**(3) (2003) 289-300
8. Park, H.S., Oh, S.K.: Rule-based Fuzzy-Neural Networks Using the Identification Algorithm of the GA Hybrid Scheme. International Journal of Control, Automations, and Systems. **1**(1) (2003) 101-110

Uniformly Convergent Computational Technique for Singularly Perturbed Self-adjoint Mixed Boundary-Value Problems

Rajesh K. Bawa[1] and S. Natesan[2]

[1] Department of Computer Science, Punjabi University, Patiala - 147 002, India
[2] Department of Mathematics, Indian Institute of Technology, Guwahati - 781 039, India

Abstract. In this paper, we propose a second–order parameter–uniform convergent hybrid scheme for self–adjoint singular perturbation problems (SPPs) subject to mixed (Robin) type conditions. The cubic spline baesd difference scheme is combined with the classical central difference scheme to obtain monotone scheme. Numerical example is provided to support the theory.

Keywords: Finite difference scheme, cubic splines, singular perturbation problems, piece-wise uniform meshes.

Subject Classification: AMS 65L10 CR G1.7.

1 Introduction

Singular perturbation problems (SPPs) arise in several branches of engineering and applied mathematics which include fluid dynamics, quantum mechanics, elasticity, chemical reactor theory, gas porous electrodes theory, etc. To solve these types of problems various methods are proposed in the literature, more details can be found in the books of Farrell et al. [2], and Roos et al. [6].

We consider the following singularly perturbed self–adjoint boundary–value problem (BVP):

$$Lu(x) \equiv -\varepsilon u''(x) + b(x)u(x) = f(x), \quad x \in D = (0,1) \tag{1}$$

$$\alpha_1 u(0) - \beta_1 u'(0) = A, \quad \alpha_2 u(1) + \beta_2 u'(1) = B, \tag{2}$$

where $\alpha_1, \beta_1, \alpha_2, \beta_2 > 0$ and $\varepsilon > 0$ is a small parameter, b and f are sufficiently smooth functions, such that $b(x) \geq \beta > 0$ on $\overline{D} = [0,1]$. Under these assumptions, the BVP (1-2) possesses a unique solution $u(x) \in C^2(D) \cap C^1(\overline{D})$. In general, the solution $u(x)$ may exhibit two boundary layers of exponential type at both end points $x = 0, 1$. Boundary-value problems of the type (1-2) arise in many applications, for instance, confinement of a plasma column by reaction pressure, theory of gas porous electrodes, performance of catalytic pellets and geophyisical fluid dynamics chemical reactions [1, 5]. In [3], the authors

V.S. Sunderam et al. (Eds.): ICCS 2005, LNCS 3516, pp. 1104–1107, 2005.

have devised HODIE schemes for singularly perturbed convection-diffusion and reaction-diffusion problems respectively.

For sufficiently small ε , classical methods on uniform meshes only work for very large number of mesh points. Nevertheless, if these methods are defined on special fitted meshes, the convergence to the exact solution is uniform in ε. Shishkin meshes are simple piecewise uniform meshes of this kind, frequently used for singularly perturbed problems. For above mentioned problem, The Shishkin mesh $\overline{\Omega}$ is constructed as follows. The domain $\overline{\Omega}$ is divided into three subintervals as $\overline{\Omega} = [0,\sigma] \cup [\sigma, 1-\sigma] \cup [1-\sigma, 1]$ for some σ such that $0 < \sigma \leq 1/4$. On the subintervals $[0,\sigma], [1-\sigma, 1]$ a uniform mesh with $N/4$ mesh–intervals is placed, where $[\sigma, 1-\sigma]$ has a uniform mesh with $N/2$ mesh intervals. It is obvious that the mesh is uniform when $\sigma = 1/4$, and it is fitted to the problem by choosing $\sigma = \min\left\{\frac{1}{4}, \sigma_0\sqrt{\varepsilon}\ln N\right\}$,where σ_0 is a constant will be fixed later. Further, we denote the mesh size in the regions$[\sigma, 1-\sigma]$ as $h^{(1)} = 2(1-2\sigma)/N$, and in $[0,\sigma], [1-\sigma, 1]$ by $h^{(2)} = 4\sigma/N$. Here, we propose an hybrid scheme which is a mixture of the cubic spline scheme with the classical central difference scheme for the BVP (1-2) on above mentioned Shishkin mesh. We apply the cubic spline difference scheme in the inner region$(0,\sigma) \cup (1-\sigma, 1)$, whereas in the outer region $(\sigma, 1-\sigma)$ we use the classical central difference scheme. This is mainly because to retain the discrete maximum principle of the difference scheme. The present method provides second–order uniform convergence throughout the domain of interest. A numerical experiment have been carried out to show the efficiency of the method.

2 ε-Uniform Hybrid Scheme

The cubic spline based scheme is analyzed for stability and convergence and it is observed that for the corresponding matrix to be a M-matrix, a very restrictive condition is needed on the mesh size, specially in the outer region where a coarse mesh is enough to reflect the behavior of the solution in that region . So, to overcome this, The following hybrid scheme is proposed in which the well known classical central difference scheme is taken in the outer region and the cubic spline scheme in boundary layer region

$$r_i^- u_{i-1} + r_i^c u_i + r_i^+ u_{i+1} = q_i^- f_{i-1} + q_i^c f_i + q_i^+ f_{i+1}, \quad i = 1, \cdots, N-1, \tag{3}$$

along with following equations for approximations at boundaries

$$\begin{cases} r_0^c u_0 + r_0^+ u_1 = q_0^- + q_0^c f_0 + q_0^+ f_1, \\ \\ r_N^- u_{N-1} + r_N^c u_0 = q_N^- + q_N^c f_{N-1} + q_N^+ f_N, \end{cases} \tag{4}$$

for $i = 1, \cdots, N/4$ and $3N/4, \cdots, N-1$

$$\begin{cases} r_i^- = \frac{-3\varepsilon}{h_{i-1}(h_i+h_{i-1})} + \frac{h_{i-1}}{2(h_i+h_{i-1})} b_{i-1}; \quad r_i^c = \frac{3\varepsilon}{h_i h_{i-1}} + b_i; \\ \\ r_i^+ = \frac{-3\varepsilon}{h_i(h_i+h_{i-1})} + \frac{h_i}{2(h_i+h_{i-1})} b_{i+1}; \end{cases} \tag{5}$$

$$\left\{ q_i^- = \frac{h_{i-1}}{2(h_i+h_{i-1})}; \quad q_i^c = 1; \quad q_i^+ = \frac{h_i}{2(h_i+h_{i-1})}, \right. \tag{6}$$

and for $i = N/4+1, \cdots, 3N/4-1$

$$\left\{ r_i^- = \frac{-2\varepsilon}{h_{i-1}(h_i+h_{i-1})}; \quad r_i^c = \frac{2\varepsilon}{h_i h_{i-1}} + b_i; \quad r_i^+ = \frac{-2\varepsilon}{h_i(h_i+h_{i-1})}, \right. \tag{7}$$

$$\left\{ q_i^- = 0; \quad q_i^c = 1; \quad q_i^+ = 0. \right. \tag{8}$$

and

$$\begin{cases} r_0^c = -\frac{3\varepsilon}{h_0}\left(\alpha_1 + \frac{\beta_1}{h_0}\right) - b_0\beta_1; \quad r_0^+ = -\frac{3\varepsilon\beta_1}{h_0^2} + \frac{b_1}{2}\beta_1; \\ q_0^- = -\frac{3\varepsilon A}{h_0}; \quad q_0^c = -\beta_1; \quad q_0^+ = -\frac{\beta_1}{2}; \\ r_N^- = -\frac{3\varepsilon\beta_2}{h_{N-1}^2} + \frac{b_{N-1}}{2}\beta_2; \quad r_N^c = -\frac{3\varepsilon}{h_{N-1}}\left(\alpha_2 + \frac{\beta_2}{h_{N-1}}\right) - \frac{b_N}{2}\beta_2; \\ q_N^- = -\frac{3\varepsilon B}{h_{N-1}}; \quad q_N^c = -\frac{\beta_2}{2}; \quad q_N^+ = -\beta_2. \end{cases} \tag{9}$$

3 Numerical Experiments

To show the accuracy of the present method, here we have implemented it to a test problem. The results are presented in the form of tables with maximum point–wise errors and rate of convergent. Table 1 display the results for the values $\varepsilon = 2^{-4}, 2^{-16}, \cdots, 2^{-40}$ and different values of N.

Table 1. Maximum pointwise errors G_ε^N, rates of convergence p and ε - uniform errors G^N corresponding to the Hybrid scheme for Example 1

ε	Number of mesh points N						
	16	32	64	128	256	512	1024
2^{-4}	2.0176e-2	4.9167e-3	1.2214e-3	3.0487e-4	7.6188e-5	1.9045e-5	4.7611e-6
	2.0369	2.0092	2.0023	2.0006	2.0001	2.0000	
2^{-16}	1.5583e-1	5.6409e-2	1.9283e-2	6.4067e-3	2.0840e-3	6.5813e-4	2.0297e-5
	1.4660	1.5486	1.5897	1.6203	1.6629	1.6971	
2^{-24}	1.5515e-1	5.6201e-2	1.9217e-2	6.3854e-3	2.0771e-3	6.5597e-4	2.0231e-5
	1.4650	1.5482	1.5895	1.6202	1.6629	1.6971	
2^{-32}	1.5512e-1	5.6192e-2	1.9214e-2	6.3844e-3	2.0768e-3	6.5588e-4	2.0228e-5
	1.4649	1.5482	1.5895	1.6202	1.6629	1.6971	
2^{-36}	1.5512e-1	5.6191e-2	1.9214e-2	6.3844e-3	2.0768e-3	6.5587e-4	2.0228e-5
	1.4649	1.5482	1.5895	1.6202	1.6629	1.6971	
2^{-40}	1.5512e-1	5.6191e-2	1.9214e-2	6.3844e-3	2.0768e-3	6.5587e-4	2.0228e-5
	1.4649	1.5482	1.5895	1.6202	1.6629	1.6971	
G^N	1.6274e-1	5.8585e-2	1.9599e-2	6.5096e-3	2.1171e-3	6.6858e-4	2.0619e-4
p_{uni}	1.4740	1.5798	1.5902	1.6205	1.6630	1.6971	

Example 1. [7] Consider the self–adjoint SPP

$$-\varepsilon u''(x) + (1+x)^2 u(x) = [4x^2 - 14x + 4](1+x)^2,\ x \in (0,1)$$
$$u(0) - u'(0) = 0, \quad u(1) = 0.$$

We use the following double mesh principle to calculate the maximum point-wise error and rate of convergence.

Let $\overline{D}_\varepsilon^N$ be a Shishkin mesh with the parameter σ altered slightly to $\overline{\sigma} = \min\left\{\frac{1}{4}, \sigma_0\sqrt{\varepsilon}\ln(N/2)\right\}$,. Then, for $i = 0, 1, \cdots N$, the ith point of the mesh $\overline{D}_\varepsilon^N$ coincides with the $(2i)$ the point of the mesh $\overline{D}_\varepsilon^{2N}$. The double mesh difference is defined as $G_\varepsilon^N = \max_{x_i \in \overline{D}_\varepsilon^N} |U^N(x_j) - U^{2N}(x_j)|$, and $G^N = \max_\varepsilon G_\varepsilon^N$, where $U^N(x_j)$ and $U^{2N}(x_j)$ respectively denote the numerical solutions obtained using N and $2N$ mesh intervals. Further, we calculate the parameter-robust orders of convergence as $p = \log_2(\frac{G_\varepsilon^N}{G_\varepsilon^{2N}})$ and $p_{uni} = \log_2(\frac{G^N}{G^{2N}})$. Here, we took $\sigma_0 = 1$. We have tabulated the results in Tables 1 which shows the maximum point–wise error and the rate of convergence for Example 1.

4 Conclusions

In this paper, We have proposed a hybrid method for the numerical solution of singularly perturbed reaction–diffusion problems. The underlying idea of the method combines both the cubic spline and classical central difference scheme. The method is of second–order convergent. One test example is studied to verify the efficiency and accuracy of the theoretical error estimates, and they reflect perfectly the same.

References

1. R.C.Y. Chin and R.Krasny. A hybrid asymptotic finite-element method for stiff two-point boundary-value problems. *SIAM J. Sci. and Stat. Comput.*, **4**:229-243,1983.
2. P.A. Farrell, A.F. Hegarty, J.J.H. Miller, E. O'Riordan, and G.I. Shishkin. *Robust Computational Techniques for Boundary Layers.* Chapman & Hall/CRC Press, 2000.
3. J.L. Gracia, F. Lisbona, and C. Clavero. High order ε-uniform methods for singularly perturbed reaction-diffusion problems. *Lecture Notes in Computer Science*, **1998**:350–358, 2001.
4. J.J.H. Miller, E. O'Riordan, and G.I. Shishkin. *Fitted Numerical Methods for Singular Perturbation Problems.* World Scientific, Singapore, 1996.
5. C.E.Pearson. On a differential equation of boundary layer type. *J. Math. Phys.*, **47**(144):134-154,1968.
6. H.-G. Roos, M. Stynes, and L. Tobiska. *Numerical Methods for Singularly Perturbed Differential Equations.* Springer, Berlin, 1996.
7. M.Stojanovic. Numerical solution of initial and singularly perturbed two-point boundary value problems using adaptive spline function approximation. *Publications de L'institut Mathematique.*, **43**(57):155-163,1988.

Fuzzy System Analysis of Beach Litter Components

Can Elmar Balas

Gazi University, Faculty of Engineering and Architecture
Civil Engineering Department, 06570 Ankara, Turkey
cbalas@gazi.edu.tr

Abstract. Tourist beaches on the southern coast of Turkey are surveyed in order to facilitate a standardised fuzzy approach to be used in litter prediction and to assess the aesthetic state of the coastal environment for monitoring programs. During these surveys the number of litter items on beaches were counted and recorded in different categories. The main source of litter on beaches was determined as "beach users". A fuzzy system was developed to predict the classification of the beaches, since uncertainty was generally inherent in beach work due to the high variability of beach characteristics and the sources of litter categories. This resulted in effective utilization of "the judgment and knowledge of beach users" in the evaluation of beach gradings.

1 Introduction

Marine litter is defined as the solid materials of human origin that are discarded at sea or reaches the sea through waterways or domestic or industrial outfall [1]. Litter in the marine environment leads to numerous problems adversely affecting coastal development sectors. Prevention at source is one of the most important strategies in enabling the reduction of litter pollution, and for this aim to be achieved strong links between measurement and management need to be realized. Five beaches on the south coast of Turkey are studied with regards to type and amount of litter on each beach. The surveys were carried out at over 100 m wide transects of the beaches.

Four site investigations were performed, first being on 20-29 October 2000, second on 10-19 November 2000, third on 15-26 March 2001 and the last on 2-13 June 2001. Field litter studies were conducted on some of the most attractive tourist beaches of the Turkish Riviera (Antalya) coast, namely Cirali, Konyaalti, Kemer, Side and Belek. For a 100m stretch of beach located on the normal access points, all litter items were enumerated and placed in their respective categories/grades [2]. Litter amounts collected ranged from 18 to 743-items/100 m stretch of beach. Litter items were graded from the best (Grade A) to worst case (Grade D) as shown in Table 1.

2 Litter Prediction by Fuzzy System

Field measurements indicated that the main beach litter item (the most abundant in terms of quantity) was the general litter category. The number of litter items in other categories was low and oil pollution was not observed. Konyaaltı beach was rated to

V.S. Sunderam et al. (Eds.): ICCS 2005, LNCS 3516, pp. 1108 – 1111, 2005.

be in bad conditions on the first three site surveys. The reasons were its nearness to the city center, the demolished refreshment kiosks due to the implementation of the municipality plan of Antalya and the difficulty in controlling the pollution on its seven-kilometer long beach. However on the last survey, the grading of Konyaaltı beach increased from "D" to "B", due to the improvement in tourism facilities implemented by the municipality of Antalya. This indicated a sharp decline in the amount of litter items on the beach with the beginning of the tourism season.

Table 1. Categories for grading of beaches

	Category	Type	A	B	C	D
1	Sewage Related Debris	General	0	1-5	6-14	>15
		Cotton Buds	0-9	10-49	50-99	>100
2	Gross Litter		0	1-5	6-14	>15
3	General Litter		0-49	50-499	500-999	>1000
4	Harmful Litter	Broken Glass	0	1-5	6-24	>25
5	Accumulations	Number	0	1-4	5-9	>10
6	Oil		Absent	Trace	Noticeable	Objectionable
7	Faeces		0	1-5	6-24	>25

The Moonlight beach and the public beach of Kemer were in good condition with a beach grading of "B". With the beginning of tourism season, these beaches were regularly cleaned. Çıralı and Belek are the selected coastal sites for 'Coastal Management and Tourism Project'. World Wide Fund is conducting the project with World Bank funding. Çıralı beach was in good condition with a beach grading of "B". There were less tourism facilities when compared to other beaches studied, hence the grade of the beach did not change with the tourism season. Side beach obtained a low degree of grading in surveys, mostly due to faeces observed. With the beginning of tourism season, the beach was regularly cleaned and the overall grade of Side beach increased to "B". The variation can be attributed to national holidays, stormy wheather conditions, demolition and construction of beach facilities and the tourism season.

A fuzzy system of artificial intelligence [3] was developed in this paper, which had input parameters of general litter and sewage related debris, and an output parameter of the grading of litter categories. In the fuzzification process of the system inputs, which were the number of general litter and sewage related debris items, the grading criteria was utilized.

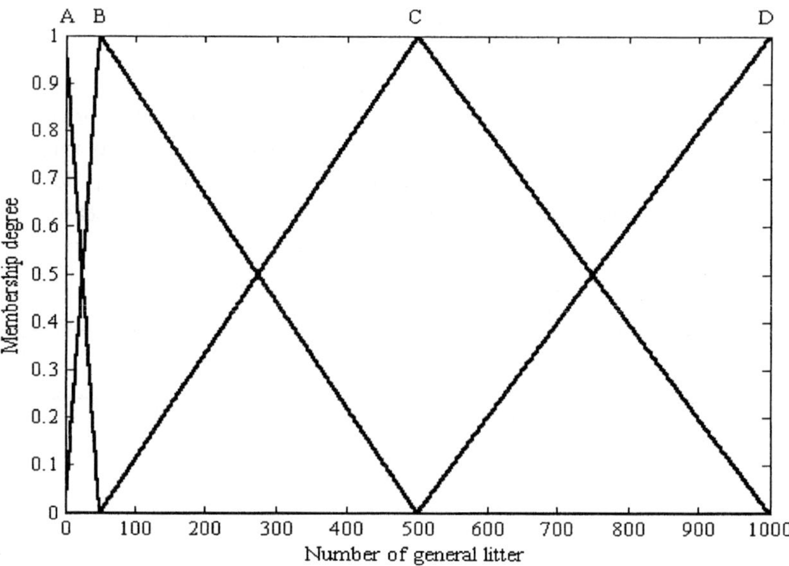

Fig. 1. Fuzzy input sets and the membership functions of the grading system for the category of general litter

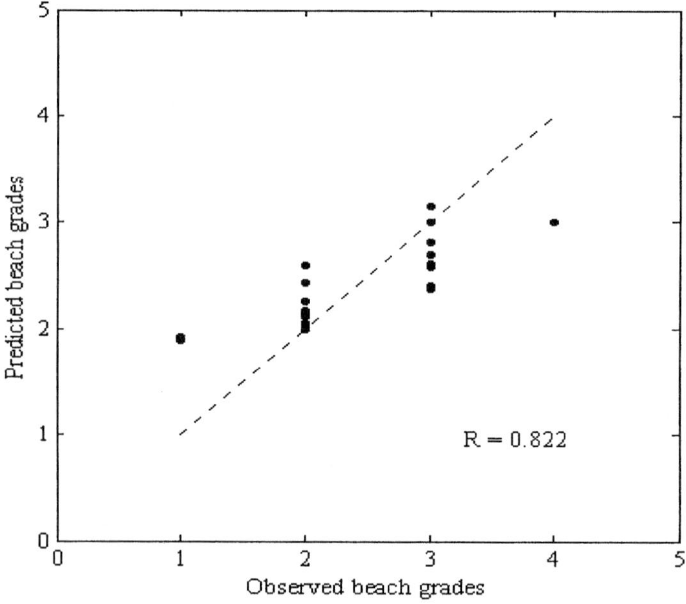

Fig. 2. Comparison of beach grades of the field study with the predicted grades of fuzzy system

The fuzzy input sets and membership functions of these variables were obtained for the grading of A, B, C, D, from the best to worst case depending on the number of litter items measured on beaches (Figure 1). Fuzzy output sets for the grading of litter categories (A: excellent, B: good, C: average and D: worst) were coded as 1,2, 3 and 4, respectively. The fuzzy rule based system was established by the Cartesian product (product space) of fuzzy input sets, which resulted the in the logical base of 16 fuzzy rules. The fuzzy sets of grades were slightly modified to assess the uncertainties inherent in other litter categories by including supplementary adjectives of "very good (B^+), to some extend good (B^-), above average (C^+), below average (C^-) and bad (D^+)", i.e. the maximum number of litter items, greater than a certain limiting value in related categories will decrease a half grade in the input fuzzy subset of rules. Therefore, potentially harmful litter, gross litter and accumulations of litter exceeding their limits given in Table 1, will decrease the grading definition in fuzzy rules for general litter, a half grade. Similar interactions for the general litter definitions of rules are available, if there is a trace of oil pollution on the beach. Likewise, the occurrence of faeces of non-human origin affects the rules in the input fuzzy set for sewage related debris. At the testing stage, the litter measurements of the field study were compared with the predictions of the fuzzy system, as illustrated in Figure 2.

3 Conclusions

Tourist beaches near Antalya region were surveyed in order to develop a standardized fuzzy system approach to be used in litter prediction. The main source of litter on Turkish Mediterranean beaches was determined as "beach users". The main advantage in using fuzzy systems was that they could consider the linguistic definitions/notes of beach users and field study teams during measurements. Therefore, they make effective use of "additional information" such as the knowledge and experience of team members. As a result, additional information inherent in the linguistic comments/refinements and judgment of study teams and beach users could be included in the grading system. Specific issues related to beach characteristics, litter assessment methodology and definition of oil pollution, which could not be included in standard procedures and/or could be easily lost in mathematical expressions/evaluations, were incorporated by using this artificial intelligence system.

References

1. Balas, C.E., Williams, A.T., Simmons, S.L and A Ergin. A Statistical Riverine Litter Propagation Model, Marine Pollution Bulletin, 42(11), 1169-1176 (2001).
2. EA/NALG, Assessment of Aesthetic Quality of Coastal and Bathing Beaches, Monitoring Protocol and Classification Scheme, Environmental Agency, Bristol, UK (2000).
3. Zadeh, L. A. Fuzzy Logic and the Calculi of Fuzzy Rules, Fuzzy Graphs, and Fuzzy Probabilities, Computer and Mathematics with Applications, Vol. 37, 35 (1999).

Exotic Option Prices Simulated by Monte Carlo Method on Market Driven by Diffusion with Poisson Jumps and Stochastic Volatility

Magdalena Broszkiewicz and Aleksander Janicki

Mathematical Institute, University of Wrocław,
pl. Grunwaldzki 2–4, 50–384 Wrocław, Poland
{janicki, msoboc}@math.uni.wroc.pl
http://www.math.uni.wroc.pl/~janicki

Abstract. We consider a broad class of stochastic models of a financial market generalizing the classical Black–Scholes model, which comprise both: stochastic volatility of Brownian type and jumps at random times. We restrict ourselves to the model, where volatility is described by the diffusion which comprises the Heston stochastic volatility defined as a diffusion of Brownian type and the Poisson jump diffusion. We provide an argument that such models perfectly match typical real–life financial phenomena comparing the so-called logarithmic returns.

Applying computer simulations methods we investigate the dependence of prices of a few selected contingent claims (specifying some different options) on the parameters of our stochastic model.

1 The Market Model

Consider a financial market where two assets S_0 and S are traded up to a fixed horizon T.

Let $(\Omega, \mathcal{F}, P)$ be a probability space and let $\{\mathcal{F}(t)\}_{0\le t\le T}$ be a P-augmented right-continuous filtration.

The riskless asset price is given by

$$S_0(t) = \exp(rt), \quad \forall t \in [0,T], \tag{1}$$

where r is non-negative deterministic risk free interest rate. The dynamics of risky asset's price is given by

$$\begin{aligned} dS(t) &= S(t_-)\{\mu dt + \sigma(V(t))[\sqrt{1-\rho^2}dW(t) + \rho dW^\sigma(t)] \\ &\quad + \alpha_1 d\widetilde{N}^{\lambda_1}(t) + \alpha_2 d\widetilde{N}^{\lambda_2}(t)\}, \\ dV(t) &= \eta dt + \gamma dW^\sigma(t), \end{aligned} \tag{2}$$

where $(W, W^\sigma)'$ is two dimensional Brownian motion, $\widetilde{N}^{\lambda_i}$ are independent compensated Poisson process (ie. $\widetilde{N}^{\lambda_i}(t) = N^{\lambda_i}(t) - \lambda_i t$), $\{V(t)\}_{0\le t\le T}$ is the

V.S. Sunderam et al. (Eds.): ICCS 2005, LNCS 3516, pp. 1112–1115, 2005.

volatility process, ρ is the correlation between asset process and its volatility, $\rho \in [-1, 1]$, μ, η, γ are constants, σ is a function that satisfies

$$\exists \underline{\sigma}, \overline{\sigma} > 0 \ \ \forall z \in \mathbb{R} \ \ \underline{\sigma} \leq \sigma(z) \leq \overline{\sigma}. \tag{3}$$

We take two independent Poisson processes to assure that both jumps up and down would be available. To achieve that it is necessary to choose parameters $|\alpha_1| < 1$ and $|\alpha_2| < 1$ of the different signs. To explain this model we suggest

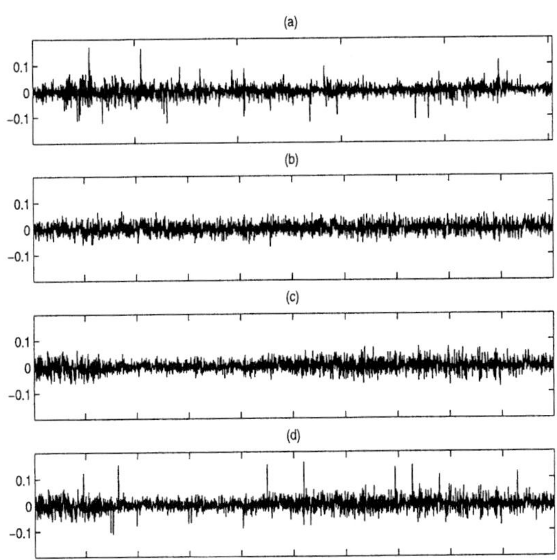

Fig. 1. Asset's price returns for various market model. On picture (a) there are real logarithmic returns of prices of IBM's stocks. Remaining pictures show logarithmic returns of prices modelled in various way: (b) simple diffusion (like Black-Scholes model), (c) stochastic volatility (like Heston model), (d) stochastic volatility and Poisson jumps

to compare the examples obtained at Fig.1. Looking at (b) and (c) we see that stochastic volatility improves the model and makes it more accurate but only (d) reflects all phenomenon which are characteristic to prices behavior. It is well known that absence of arbitrage which is equivalent to existence of martingale measure, is absolutely necessary to compute the fair price of any option. In this new measure the process of discounted price of risky asset should be a martingale. On the market defined above the Radon-Nikodym's derivative can be defined as follow

$$\begin{aligned} Z(t) := \frac{d\mathbb{Q}}{d\mathbb{P}}|_{\mathcal{F}(t)} &= \exp\left\{-\int_0^t \theta(s)dW(s) - \frac{1}{2}\int_0^t \theta(s)^2 ds\right\} \\ &\times \exp\left\{-\int_0^t \nu(s)dW(s) - \frac{1}{2}\int_0^t \nu(s)^2 ds\right\} \\ &\times \exp\left\{\ln(1-\phi_i)N^{\lambda_i}(t) + \lambda_i\phi_i\right\}, \end{aligned}$$

where θ and ν are adapted to $\{\mathcal{F}\}$ and satisfy the integrability conditions $\int_0^T (\theta(t))^2 dt < \infty$ and $\int_0^T (\nu(t))^2 dt < \infty$. From the condition $\mathbb{E}(Z(T)) = 1$ we obtain

$$(\theta(t)\sqrt{1-\rho^2} + \rho\nu(t))\sigma(t) + \lambda_1\phi_1\alpha_1 + \lambda_2\phi_2\alpha_2 = \mu - r \quad 0 \leq t \leq T \quad a.s. \ (4)$$

The processes θ and ν are interpreted as the risk premiums connected with the Brownian motions W and W^σ respectively. The premium for the risk related to Poisson processes N_i^λ are ϕ_i.

Having only one such equation we are not able to find the new measure uniquely. After Merton we can assume that the premiums coming from the Poisson processes disappear. But even than we can write the process θ only dependent on process ν.

According to Girsanov's theorem we obtain $W_0(t)$ and $W_0^\sigma(t)$ as new Wienner processes and $\widetilde{N}^{L_i}(t)$ as a new compensated Poisson processes (with intensity L_i). Since only one risky asset is traded, the premiums are not unique and nor is the martingale measure.

Applying new processes to the discounted asset price we obtain

$$d\left(\frac{S(t)}{S_0(t)}\right) = \frac{S(t_-)}{S_0(t)}\Big\{\sigma(V(t))[\sqrt{1-\rho^2}dW_0(t) + \rho dW_0^\sigma(t)] + \alpha_1 d\widetilde{N}^{L_1}(t)\alpha_2 d\widetilde{N}^{L_2}(t)\Big\}. \quad (5)$$

It is easy to see that the above process is a supermaringale and since σ satisfies the (3) condition it is also a martingale.

2 Option Prices

It is well known that the option price with payout function $\psi(S(T))$ is ruled by the equation

$$ECC(t) = S_0(t)\mathbb{E}^{\mathbb{Q}}\left[\frac{1}{S_0(T)}\psi(S(T))|\mathcal{F}(t)\right] = \frac{1}{H(T)}\mathbb{E}^{\mathbb{P}}\left[H(T)\psi(S(T))|\mathcal{F}(t)\right],$$

where $H(t) = \frac{Z(t)}{S_0(t)}$ is a state density price.

Now substituting $S_0(t)$ according to (2), and knowing that $H(0) = 0$ we obtain the option premium with payout function $\psi(S(T))$.

$$ECC(0) = e^{-rT}\mathbb{E}\left[Z(T)\psi(S(T))\right]. \quad (6)$$

Having the solution in the form of expectation like in (6), the easiest way to obtain results is using the Monte Carlo method.

In all examples containing stochastic volatility we assume that the function σ has the form $\sigma(z) = z$, $\underline{\sigma} \leq z \leq \overline{\sigma}$, And σ have the values $\underline{\sigma}$, $\overline{\sigma}$, respectively outside the above interval.

From the stochastic exponent theorem of we can compute the price of the stock in the moment of maturity ($t = T$).

$$S(T) = S(0)\exp\left\{\mu T - \frac{1}{2}\int_0^T (\sigma(V(s)))^2 ds + \sqrt{1-\rho^2}\int_0^T \sigma(V(s))dW(s)\right\}$$
$$\times \exp\left\{\rho\int_0^T \sigma(V(s))dW^\sigma(s)\right\}\exp\left\{\ln(1+\alpha_1)N^{\lambda_1} + \ln(1+\alpha_2)N^{\lambda_2}\right\}.$$

Some results of simulations are presented in the table below.

Table 1. Prices of options simulated for various models of market: BS - Black-Scholes diffusion market, SV - stochastic volatility market, SV+P - stochastic volatility with Poisson jumps market

Payout function	BS	SV	SV+P
$(S(T) - K)^+$	17.8462	17.7726	17.5570
$X\mathbb{I}_{\{S(T)>K\}}$	10.7013	9.6102	10.3223
$S(T) - \min_t(S(t))$	28.5074	22.9462	28.7744
$(q^\alpha(S) - K)^+$	0.8136	3.1839	0.2556
$(K - M(T, T_0, S))^+$	0.6422	1.1330	0.9097

In all simulations the models are parameterized in the same way. First case is standard European option and parameters are chosen so to obtain similar premium. We assume exercise price $K = S(0) = 100\$$. Second line is binary option, which guarantee to pay amount $X = 50\$$ if the price of asset in the maturity time will be greater than $K = 130\$$. Third case is an lookback option (option depend on all trajectory, not only last moment). It pays the difference between price of stock at maturity time and minimal price of asset on all time horizon. Forth case is so-called quantile option ($q_\alpha(\cdot)$ is a quantile of range α) and the last is Asian option where $M(T_0, T, S)$ is the mean price on time horizon (T_0, T), $T_0 > t_0 = 0$.

References

1. Bellamy N., Jeanblanc M.: Incompletness of markets diven by a mixed diffusion. Finance & Stochastics 4 (2000) 209–222
2. León J. A., Solé J. L., Utzet F., Vives J. On Lévy processes, Malliavin calculus and market models with jumps. Finance & Stochastics 6 (2002) 197–225
3. Protter, P.: Stochastic Integration and Differential Equations – A New Approach. Springer-Verlag, New York (1990)
4. Touzi N.: American Options Exercise Boundary When the Volatility Changes Randomly. Applied Matematics & Optimization 39 (1999) 411–422

Computational Complexity and Distributed Execution in Water Quality Management

Maria Chtepen[1], Filip Claeys[2], Bart Dhoedt[1], Peter Vanrolleghem[2], and Piet Demeester[1]

[1] Department of Information Technology (INTEC), Ghent University, Sint-Pietersnieuwstraat 41, Ghent, Belgium
{maria.chtepen, bart.dhoedt, piet.demeester}@intec.ugent.be
[2] Department of Applied Mathematics, Biometrics and Process Control (BIOMATH), Ghent University, Coupure Links 653, Ghent, Belgium
{filip.claeys, peter.vanrolleghem}@biomath.ugent.be

Abstract. Tourist beaches on the southern coast of Turkey are surveyed in order to facilitate a standardised fuzzy approach to be used in litter prediction and to assess the aesthetic state of the coastal environment for monitoring programs. During these surveys the number of litter items on beaches were counted and recorded in different categories. The main source of litter on beaches was determined as "beach users". A fuzzy system was developed to predict the classification of the beaches, since uncertainty was generally inherent in beach work due to the high variability of beach characteristics and the sources of litter categories. This resulted in effective utilization of "the judgment and knowledge of beach users" in the evaluation of beach gradings.

1 Introduction

The importance of Water Quality Management has drastically increased during the last decades as a consequence of growing environmental awareness. Compared to well-defined systems (e.g. electrical and mechanical) that can entirely be described by classical laws of physics, the behavior of ill-defined water systems is much more difficult to predict due to only partial availability of generally applicable laws. As a result of the complexity of water systems, modeling is regarded an inherent part of design, operation and optimization. The models used in practice are typically based on a combination of general physical laws (typically laws of preservation of energy and mass) and purely empirical laws. These models form an excellent tool to summarize and increase the understanding of complex interactions in biological systems.

As an example of a Water Quality Management application the WEST [1, 2] software system will be considered. WEST[1] is a versatile yet powerful tool, which thus far has mainly been applied to wastewater treatment processes. It consists

[1] World-wide Engine for Simulation and Training.

V.S. Sunderam et al. (Eds.): ICCS 2005, LNCS 3516, pp. 1116–1119, 2005.

of clearly separated Modeling and Experimentation Environments. Both environments are self-contained and consist of a graphical front-end, computational back-end and control logic. The Modeling Environment allows for the creation of executable models on the basis of high-level model descriptions, through the application of model compiler techniques. These executable models are subsequently used as a basis for the creation of Virtual Experiments (VE's) in the Experimentation Environment. The reason why the term "VE" was adopted in this case is related to the fact that WEST goes beyond plain simulation. In fact, the types of VE's that are currently supported are: Simulation, Steady-state Analysis, Optimization, Confidence Analysis, Scenario Analysis, Sensitivity Analysis and Uncertainty Analysis. VE's such as Simulation and Steady-state Analysis are said to be *atomic* since they are not hierarchically composed of other VE's. On the other hand, *compound* VE's such as Optimization, Confidence Analysis, Scenario Analysis, Sensitivity Analysis and Uncertainty Analysis all are based on the repeated execution of simulations for different parameter sets.

Modeling and Virtual Experimentation in the domain of Water Quality Management is computationally complex. Total model building times range from a couple of seconds to several hours and typical execution times for compound VE's range from a number of seconds to several weeks on a high-end PC. To reduce this complexity a new system named WDVE [3] was designed that allows for distributed execution of compound VE's. WDVE can be seen as a light-weight Grid system [4] dedicated to the Water Quality Management application area. It was built on top of technologies such as C++, STL, XML, and SOAP. WDVE was initially implemented for the WEST software system, however thanks to generic design it can be used for execution of any computationally intensive task consisting of an unordered set of jobs.

Design and implementation of the WDVE system started from a number of important requirements.

Financial. The new system shouldn't require any additional investments in terms of hardware and staffing.

Functional. The system should evidently be efficient and robust, i.e. it should deliver fast execution of jobs and should account for machine crashes and network drop-outs.

Technical. A strict demand for elegant design and coherent implementation was imposed, together with full OS portability, interoperability and limited dependence on third-party tools.

2 WDVE Architecture

WDVE consists of two major modules: Master and Slave. The Master's main functionalities are to receive VE's (Jobs) from the Application front-end, to store Jobs until they are processed, to match them against available work nodes and to initiate execution. Slaves run (possibly concurrent) Jobs on their computational resources and collect the outputs requested by the user.

Master as well as Slave were further decomposed into a number of submodules (see Fig. 1), each implementing a particular, well-defined functionality:

Dispatcher. The Dispatcher's main task is to administer the Jobs from the moment they are submitted by the Application until execution is completed and results are available to the user.

Registry. The goal of the Registry is to allow for work nodes to be statically or dynamically registered as Slaves.

Selector. The Selector implements an algorithm for selecting a Slave for the next Job to be executed.

Checker. The consistency of the state of the system is ensured by the Checker background thread, which constantly monitors the status of various entities and triggers appropriate actions.

Acceptor. This module is at the heart of each Slave. It evaluates requests from the Master for the execution of Jobs. In case a Job is accepted, a new instance of the Application back-end is created and Job execution is started.

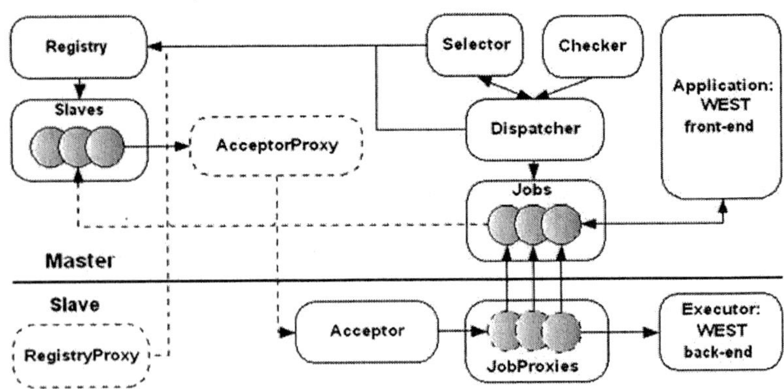

Fig. 1. WDVE: Architecture

For the transfer of input, output and status data between Master and Slaves, the gSOAP implementation of the SOAP protocol was used. Reasons for electing SOAP from the wide range of the available middleware technologies are performance, portability and the open source nature of gSOAP. In order for the specifics of the middleware not to be visible by the rest of the code, an abstraction layer was added on top of the gSOAP wrapper classes.

An important architectural question is how to represent Jobs in their various incarnations, starting from the point when a Job is submitted to the system and ending with the completed Job and its output data. It was decided to consistently use XML to this end. In the context of WDVE, Jobs descriptions consist of Executor-dependent and Executor-independent information. The Executor-independent information contains a set of Resources including various types of input and output files. The Executor-dependent information is made up of various other pieces of information that are relevant to the specific Executor that is to execute the Job.

3 Test Results and Conclusions

Tests were run comparing the efficiency of WDVE with monolithic solutions. The results that were found are in line with the typical performance behavior of other distributed systems. In Fig. 2 are some of the results that were obtained from a test using three identical Slaves that execute Jobs assigned by a Master that has a total of 1 to 100 Jobs to execute. From Fig. 2 (left) one easily concludes that in case the Jobs are short, there is nothing to be gained from distributing Jobs over multiple Slaves. However, in case of Jobs of realistic complexity, one does get a significant reduction of the total execution time (Fig. 2, right). As the number of Slaves increases, this reduction becomes more pronounced, up to a point however where no further improvement is possible. The speed-up factors obtained in practice are also always lower than those that one would theoretically expect. Evidently, this is due to the overhead inflicted by the WDVE software itself and the transmission of data over the network.

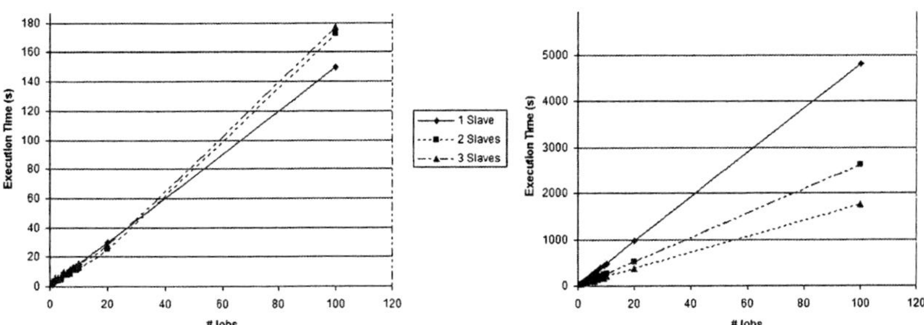

Fig. 2. WDVE: Test results for short Jobs (left) and long Jobs (right)

References

1. H. Vanhooren, J. Meirlaen, Y. Amerlinck, F. Claeys, H. Vangheluwe, and P. Vanrolleghem. WEST: Modelling Biological Wastewater Treatment. *Journal of Hydroinformatics, IWA Publishing*, 5(1), 2003.
2. HEMMIS N.V. WEST website. *http://www.hemmis.com.*
3. F. Claeys, M. Chtepen, L. Benedetti, B. Dhoedt, and P.A. Vanrolleghem. Distributed Virtual Experiments in Water Quality Management. In *6th International Symposium on Systems Analysis and Integrated Assessment in Water Management (WATERMATEX)*, pages 485–494, Beijing, China, November 3–5 2004. International Water Association.
4. EGEE. Enabling Grids for E-science in Europe (EGEE) website. *http://egee-intranet.web.cern.ch/egee-intranet/gateway.html.*

Traffic Grooming Based on Shortest Path in Optical WDM Mesh Networks*

Yeo-Ran Yoon, Tae-Jin Lee, Min Young Chung, and Hyunseung Choo

Lambda Networking Center,
School of Information and Communication Engineering,
Sungkyunkwan University, Korea
{ookuma, tjlee, mychung, choo}@ece.skku.ac.kr

Abstract. This paper investigates the static traffic grooming in WDM optical mesh networks. Our objective is to improve the network throughput and to minimize the blocking probability. As take care of this problem efficiently, we propose Shortest-path First Traffic grooming(SFT) algorithm. The comprehensive computer simulation shows that our proposed algorithm is up to about 14% superior to the existing one known to be effective.

1 Introduction

Optical wavelength-division multiplexing (WDM) is a promising technology to accommodate the explosive growth of Internet and telecommunication traffics. With each optical link capable of carrying traffic on several wavelengths, each one of which supports traffic in the Gbps range. However, the traffic requested by individual connection is still in the Mbps range. Hence, to utilize the available bandwidth efficiently, several connections have to be grouped onto the same wavelength.This requires strategic routing and wavelength assignment (RWA) of each connection [1].

The problem of RWA on sub-wavelength demands with the objectives of minimizing the network cost and optimizing network throughput is called "*traffic grooming*" problem which refers to the techniques used to combine lower-capacity components onto available wavelengths. Traffic grooming has received considerable attention recently, and there are many related work in the literature. In recent past, there have been efforts towards solving the traffic grooming problem for mesh networks. This issue has been addressed in both the static [4, 5] as well as the dynamic [3] scenarios.

In this paper, we consider the static traffic grooming for the WDM mesh networks. We propose a Shortest-path First Traffic grooming (SFT) algorithm in objective to maximize the network throughput and to minimize the blocking probability.The proposed algorithm uses effective routing method instead of previous shortest path routing, then low-capacity connection requests are efficiently

* This work was supported in parts by Brain Korea 21 and the Ministry of Information and Communication, Korea. Corresponding author: Prof. H. Choo.

V.S. Sunderam et al. (Eds.): ICCS 2005, LNCS 3516, pp. 1120–1124, 2005.

groomed together and carried. According to the computer simulation, the SFT algorithm achieves 14% improved performance in terms of the network throughput, compared with the Maximizing Single-hop traffic(MST) algorithm[5].

2 Related Works and Problem Statement

Current and future WDM networks are increasingly arranged in general mesh topologies. Indeed, much recent work has focused on grooming traffic in mesh networks [3, 4, 5]. Existing work on traffic grooming has considered two types of traffic model: static traffic model and dynamic random traffic one. In the static traffic model, all capacity demands are known in advance and do not change over time. The studies for static grooming are [4, 5]. In the dynamic random traffic model, a demand is assumed to arrive at a random time and last for a certain amount of time in random. The work in [3] studied traffic grooming issues in dynamic traffic environment.

In general when solving the traffic grooming problem, traffic requests are carried through single-hop grooming and multi-hop grooming. Single-hop grooming is a method that allows a connection to transverse a single lightpath. On the other hand, multi-hop grooming is a method that allows a connection to traverse multiple lightpaths. In multi-hop grooming, a connection can be dropped at intermediate nodes and groomed with other low-capacity connections on different lightpaths before it reaches its destination node.

The traffic grooming problem can be formulated as follows [5]. Given a network configuration (including physical topology, where each edge is a physical link, number of transceivers at each node, number of wavelengths on each fiber, and the capacity of each wavelength) and a set of connection requests with different bandwidth granularities, such as OC-12, OC-48, etc., we need to determine how to set up lightpaths to satisfy the connection requests. Because of the sub-wavelength granularity of the connection requests, one or more connections can be multiplexed on the same lightpath.

3 Shortest Path First Traffic Grooming (SFT)

Our goal is to improve the network throughput by using the efficient routing algorithm for traffic grooming. MST algorithm [5] assigns lightpaths by adaptive routing using Dijkstra's shortest-path, however it finds the shortest path under the current network status which may not be the real shortest one because the network status is changed frequently. For that reason, we apply a routing algorithm that preferentially assigns the shortest paths which are found on the original network status in the proposed SFT algorithm.

The proposed SFT algorithm uses both fixed routing based on shortest paths and adaptive routing. Firstly this algorithm only assigns lightpaths of connection requests which are available to use shortest paths in the original network status. Then connection requests which are not proper to use shortest paths go for employing adaptive routing before trying other wavelengths. If we use SFT algorithm employing this routing method, more demands are carried through shortest paths and resources are used efficiently than that of using MST. Furthermore, connection requests which are not carried through single-hop traffic grooming have more chances to be carried through multi-hop grooming.

Step 1: Constructing a virtual topology

1.1: Sort all node pairs (s_i, d_i) according to the sum of uncarried traffic requests $T(s_i, d_i)$ between s_i and d_i, and put them into a list L in a descending order.

1.2: Find a shortest path for each (s_i, d_i) and its hop count ($h(s_i, d_i)$) by using Dijkstra's shortest-path algorithm.

1.3: Setup a lightpath for each node pair (s_i, d_i) in the list L just by using the shortest path obtained in **Step 1.2** subject to transceiver constraints. If we assign the lightpath, let $T(s_i, d_i) = Max[T(s_i, d_i)\text{-}C, 0]$ and delete edges of the assigned lightpath from the physical topology. If we cannot use the shortest path obtained in **Step 1.2**, send (s_i, d_i) to list L'. And if there are not available lightpaths due to short of transceivers or deleted edges, move (s_i, d_i) to L''.

1.4: Setup lightpaths for node pairs (s_i, d_i)s in L' using the adaptive routing, subject to transceiver constraints. If we fail, move (s_i, d_i) to L'' ; Otherwise, we assign the lightpath and let T(s_i, d_i) = $Max[T(s_i, d_i)\text{-}C$, 0]. Then edges of the assigned lightpath are deleted from the physical topology.

1.5: Go to **Step 1.3** until lists L and L' become empty.

Step 2: Routing the low-speed connections on the virtual topology constructed in **Step 1**.

2.1: Route all connection requests which can be carried through single lightpath hop, and update the network status for the virtual topology.

2.2: Route the remaining connection requests using currently available spare capacity of the virtual topology based on their sum of uncarried traffic request value $T(s_i, d_i)$.

Step 3: Back to the original network status, store L'' in L and repeat **Steps 1** and **2** until entire wavelengths are fully exhausted

4 Numerical Results

The performance of SFT algorithm developed in this paper is compared against well-known MST on the NSFNET and random network topologies. NSFNET has 14 nodes and 21 links as shown in Fig. 1(a) and random networks are generated based on [2]. We compare SFT and MST algorithms in terms of network throughput. We show the results of the MST and the proposed SFT algorithm on NSFNET in Fig. 1(b). We have constructed the resulting graphs for number of wavelengths per each fiber link versus network throughput. The results show that our SFT performs better than MST algorithm with respect to network throughput at any number of wavelengths per link. As you see here, the SFT algorithm outperforms 3% $\sim$ 14% over MST.

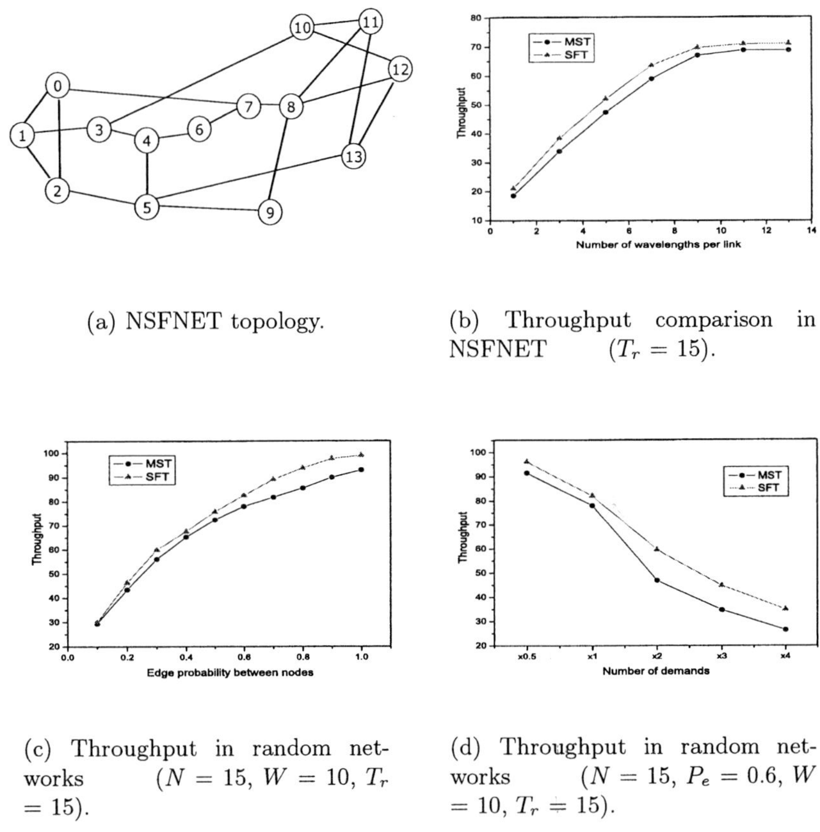

(a) NSFNET topology.

(b) Throughput comparison in NSFNET ($T_r = 15$).

(c) Throughput in random networks ($N = 15$, $W = 10$, $T_r = 15$).

(d) Throughput in random networks ($N = 15$, $P_e = 0.6$, $W = 10$, $T_r = 15$).

Fig. 1. Throughput comparison

In Fig. 1(c), we compare the network throughput of MST and SFT algorithms in random topologies with 15 nodes when the edge existence probability between node pairs(P_e) increases. We observe that the SFT algorithm demonstrates 4% $\sim$ 10% higher throughput than the MST. In Fig. 1(d), we plot the results that the number of demands versus network throughput in random topologies. We can see that the throughput of SFT algorithm acquires 5% $\sim$ 18% better performance than that of MST. And as the number of demands increases, the performance of the SFT algorithm becomes better.

5 Conclusion

We investigated the traffic grooming in WDM optical networks and proposed an improved heuristic algorithm with respect to network throughput called Shortest path First Traffic grooming (SFT). This algorithm consider the shortest path on original graph and uses as many shortest paths as possible for the demands. We showed that the SFT algorithm outperforms MST with various aspects in NSFNET and random networks.

References

1. R. Ramaswami and K. N. Sivarajan, Optical Networks: A Practical Perspective. San Francisco, CA: Morgan Kaufmann, 1998.
2. A.S. Rodionov and H. Choo, "On generating random network structures: Connected Graphs," Springer-Verlag Lecture Notes in Computer Science, vol. 3090, pp. 483 491, Sep. 2004.
3. S. Thiagarajan and A. K. Somani, "Capacity fairness of WDMnetworks with grooming capability," Opt. Networks Mag., vol. 2, no. 3, pp. 24-31, May/June 2001.
4. H. Zhu, H. Zang, K. Zhu, and B. Mukherjee, "A novel, generic graph model for traffic grooming in heterogeneous WDM mesh networks," IEEE/ACM Trans. Networking, vol. 11, pp. 285-299, Apr. 2003.
5. K. Zhu and B. Mukherjee, "Traffic grooming in an optical WDM mesh network," IEEE J. Select. Areas Commun., vol. 20, pp. 122-133, Jan. 2002.

Prompt Detection of Changepoint in the Operation of Networked Systems*

Hyunsoo Kim and Hee Yong Youn

School of Information and Communications Engineering,
Sungkyunkwan University, 440-746, Suwon, Korea
bayes1@hanmail.net
youn@ece.skku.ac.kr

Abstract. Detection of network problems is an important step in automating network management. Early detection of performance degradation can alleviate the last moment hassel of network managers. This paper focuses on a statistical method aiming at detecting changepoint as quickly as possible using Bayes factor along with the binary segmentation procedure modified for fast detection. Computer simulation verifies the effectiveness and correctness of the proposed approach.

Keywords: Bayes factor, binary segmentation, changepoint detection, network management, non-homogeneous Possion process.

1 Introduction

For detection and diagnosis of changepoint statistical analysis method has been successfully applied to a variety of networked and distributed systems. The rapid growth of networked and distributed systems throughout the workplace has given rise to the problems difficult to manage with the expertise of human operators in modern information technology field. There is an urgent need for automating the management functions to reduce the operation and management cost. Early detection of performance degradation can alleviate the last moment hassel of network mangers.

A sudden jump in the behavior of a system is a changepoint which can be estimated from the input data using the maximum likelihood estimator [2]. We need a formulation of changepoint for detecting even a slight change. In this paper the main goal is fast detection that is an important requirement for reducing potential negative impact on the network services and users. The detection problem is formulated as a changepoint problem using Bayes factor. The computation of the Bayes factor, however, gets more complex as the number of nodes increases.

* This research was supported by the Ubiquitous Autonomic Computing and Network Project, 21st Century Frontier R&D Program in Korea and the Brain Korea 21 Project in 2004. Corresponding author: Hee Yong Youn.

V.S. Sunderam et al. (Eds.): ICCS 2005, LNCS 3516, pp. 1125–1129, 2005.

To overcome the computational difficulties, we explore the binary segmentation method proposed in [3]. Our approach for detecting changepoint is mainly developing formulas by Bayesian viewpoint. The proposed detection method is designed to be sensitive to slight changes in the operation characteristics of a network, and predict the changepoint at the change point of the shape parameter values. Computer simulation verifies the effectiveness and correctness of the proposed approach.

2 Preliminaries

In the non-homogeneous Poisson process (NHPP), the interarrival time is neither independent nor identically distributed. A random variable of special interest is $N(t)$, the number of failures in the time interval $(0, t]$. The intensity function of a counting process $\{N(t), t \geq 0\}$ is defined as $\lambda(t) = m'(t)$, where $m(t)$ is the mean number of failures in the time interval $(0, t]$, often called the mean value function. The power law process for the reliability growth of repairable systems has the intensity function,

$$\lambda(t) = \alpha\beta t^{\beta-1}, \alpha > 0, \beta > 0, t > 0, \tag{1}$$

where α is a scale parameter and β is a shape parameter [1].

The likelihood function for the data $D_{(0,T]}$ with the model of $\lambda(\cdot)$ given in (1) can be written as

$$L(\alpha, \beta | D_{(0,T]}) = \left(\prod_{i=1}^{n} \lambda(t_i) \right) \exp \left(- \int_0^T \lambda(t) dt \right), \tag{2}$$

where $D_{(0,T]} = \{t_1, \cdots, t_{n+1} | 0 < t_1 \leq \cdots \leq t_n < T, T = t_{n+1}\}$. See [3].

For the data $D_{(0,T]}$ of failure times, the intensity functions are given respectively by

$$M_0 : \lambda(t) = \alpha_0 \beta_0 t^{\beta_0 - 1}, 0 < t \leq T, \text{ vs. } M_1 : \lambda(t) = \begin{cases} \alpha_1 \beta_1 t^{\beta_1 - 1}, 0 < t < \tau, \\ \alpha_1 \beta_2 t^{\beta_2 - 1}, \tau \leq t < T, \end{cases}$$

where τ is a changepoint.

3 The Proposed Changepoint Detection Method

We use the Bayes factor for comparing and testing the no-changepoint model and single changepoint model. The test is based on calculating the Bayes factor B_{10} for the single changepoint model M_1 against the no-changepoint model M_0.

The detection procedure of the changepoint consists of the following steps:

Step 1: For the complete data, compute the Bayes factor B_{10}.
Step 2: When $B_{10} > 1$, estimate a changepoint $\hat{\tau}$ for the given data.

Step 3: After a changepoint, as new data are sequentially added, compute the Bayes factor B_{10}.
Step 4: When $B_{10} > 1$, estimate a new changepoint $\hat{\tau}$ for the given data.

The Bayes factor B_{10} for the changepoint model against the no-changepoint model is

$$B_{10}(D_{(0,T]}) = B_1/B_0, \tag{3}$$

where

$$\begin{aligned}
B_1 = & \int_{[0,t_1)} \int_0^\infty \int_0^\infty \frac{2}{(1+\beta_1)^3} \cdot \frac{2\beta_2^n}{(1+\beta_2)^3} \cdot \frac{[\prod_{i=1}^n t_i)]^{\beta_2}}{K_1} d\beta_1 d\beta_2 \frac{1}{T} d\tau \\
& + \sum_{j=1}^{n-1} \int_{[t_j,t_{j+1})} \int_0^\infty \int_0^\infty \frac{2\beta_1^j}{(1+\beta_1)^3} \cdot \frac{2\beta_2^{n-j}}{(1+\beta_2)^3} \cdot \frac{q_j}{K_1} d\beta_1 d\beta_2 \frac{1}{T} d\tau \\
& + \int_{[t_n,T)} \int_0^\infty \int_0^\infty \frac{2\beta_1^n}{(1+\beta_1)^3} \cdot \frac{2}{(1+\beta_2)^3} \cdot \frac{[\prod_{i=1}^n t_i]^{\beta_1}}{K_1} d\beta_1 d\beta_2 \frac{1}{T} d\tau, \\
B_0 = & \int_0^\infty \frac{2\beta_0^n}{(1+\beta_0)^3} \cdot \frac{[\prod_{i=1}^n t_i]^{\beta_0}}{(T^{\beta_0}+\nu)^{n+\xi}} d\beta_0.
\end{aligned}$$

Here, $K_1 = (\tau^{\beta_1} + T^{\beta_2} - \tau^{\beta_2} + \nu)^{n+\xi}$ and $q_j = \left[\prod_{i=1}^j t_i\right]^{\beta_1} \cdot \left[\prod_{i=j+1}^n t_i\right]^{\beta_2}$, for $j = 1, \cdots, n-1$.

In this procedure if the Bayes factor $B_{10} > 1$, estimate a changepoint $\hat{\tau}$ by using

$$\hat{\tau} = E(\tau | D_{(0,T]}) = A/B, \tag{4}$$

where

$$\begin{aligned}
A = & \int_{[0,t_1)} \int_0^\infty \int_0^\infty \frac{2}{(1+\beta_1)^3} \cdot \frac{2\beta_2^n}{(1+\beta_2)^3} \cdot \frac{[\prod_{i=1}^n t_i]^{\beta_2}}{K_1} d\beta_1 d\beta_2 \tau d\tau \\
& + \sum_{j=1}^{n-1} \int_{[t_j,t_{j+1})} \int_0^\infty \int_0^\infty \frac{2\beta_1^j}{(1+\beta_1)^3} \cdot \frac{2\beta_2^{n-j}}{(1+\beta_2)^3} \cdot \frac{q_j}{K_1} d\beta_1 d\beta_2 \tau d\tau \\
& + \int_{[t_n,T)} \int_0^\infty \int_0^\infty \frac{2\beta_1^n}{(1+\beta_1)^3} \cdot \frac{2}{(1+\beta_2)^3} \cdot \frac{[\prod_{i=1}^n t_i]^{\beta_1}}{K_1} d\beta_1 d\beta_2 \tau d\tau, \\
B = & \int_{[0,t_1)} \int_0^\infty \int_0^\infty \frac{2}{(1+\beta_1)^3} \cdot \frac{2\beta_2^n}{(1+\beta_2)^3} \cdot \frac{[\prod_{i=1}^n t_i]^{\beta_2}}{K_1} d\beta_1 d\beta_2 d\tau \\
& + \sum_{j=1}^{n-1} \int_{[t_j,t_{j+1})} \int_0^\infty \int_0^\infty \frac{2\beta_1^j}{(1+\beta_1)^3} \cdot \frac{2\beta_2^{n-j}}{(1+\beta_2)^3} \cdot \frac{q_j}{K_1} d\beta_1 d\beta_2 d\tau \\
& + \int_{[t_n,T)} \int_0^\infty \int_0^\infty \frac{2\beta_1^n}{(1+\beta_1)^3} \cdot \frac{2}{(1+\beta_2)^3} \cdot \frac{[\prod_{i=1}^n t_i]^{\beta_1}}{K_1} d\beta_1 d\beta_2 d\tau.
\end{aligned}$$

4 Simulation Results

Each simulation dataset consists of two parts for different shape parameter values of the intensity function of the power law process. We expect to detect a changepoint around the point where the shape parameter value changes. The hyperparameters for the prior of the shape parameter are fixed as $(\xi, \nu) = (1, 1)$.

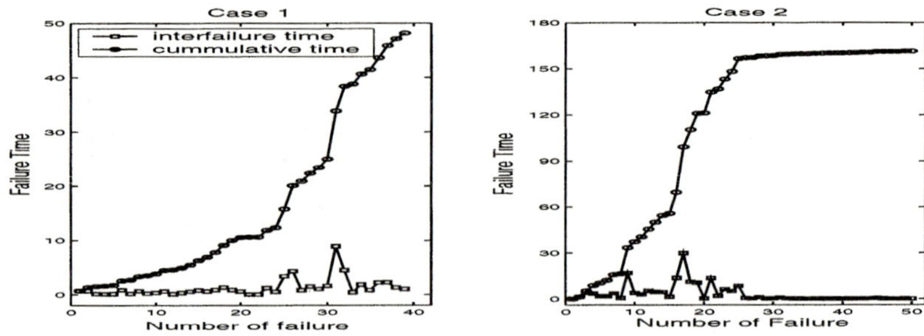

Fig. 1. The plots of simulated datasets with a single changepoint

Case 1. We fix the scale parameter as $\alpha = 0.5$. The datasets are generated with $\beta = 1.0$ and $\beta = 0.6$, respectively. Each dataset size is 20. Since the maximum likelihood estimates of (α, β) are $(0.471, 0.951)$ and $(0.507, 0.624)$ respectively, it seems that the data are fairly well generated. The time interval of the entire data is $(0, 48.185]$. We detect a changepoint using Bayes factor. For the complete data in $(0, 48.185]$, the Bayes factor is 1.902 by (3) and the estimated changepoint is 12.218 by (4), where it is located between the 23th and 24th observation. We want to build a detection system for sequential data. Thus we assume to detect a change at the 25th observation. From the first to the 25th observation, Bayes factor is 4.620, and a changepoint is 11.986 in $(0, 15.734]$. Note that the position of the changepoint estimated with only the 25 incomplete observations is still same as that with the complete data. This demonstrates that the proposed approach can promptly and correctly detect the change.

Case 2. Scale parameter $\alpha = 1.0$, shape parameter $\beta = 0.6$ and 2.0, complete data size: 50, MLEs of (α, β): $(1.160, 0.616)$, $(1.103, 1.887)$, time interval of complete data: $(0, 161.471]$, Bayes factor: 1.591×10^{21}, changepoint: 156.633 (between the 24th and 25th observation), change detection: 30th observation, B_{10} and $\hat{\tau}$ with the first to the 30th observation: 10.683 and 152.863 (between the 24th and 25th observation).

5 Conclusion

In this paper we have proposed a Bayesian approach for detecting changepoint with the power law process of the NHPP. As new data added sequentially, we immediately detect changepoint as soon as Bayes factor becomes greater than 1.

Computer simulation verified the effectiveness and correctness of the proposed approach. This procedure of changepoint detection is applicable when the complete data is unavailable.

References

1. Crow, L. H., "Reliability Analysis of Complex Repairable System," *Reliability and Biometry*, Poroschan,nF. and Serfling, R. J., (Eds.), SIAM, Philadelphi, 379-410 (1974).
2. Hassan Hajji, B. H. Far, and Jingde Cheng, "Detection of network faults and performance problems," *Proceedings of the Internet Conference*, Osaka, Japan., November 2001.
3. Yang, T. Y. and Lynn, K., "Bayesian Binary Segmentation Procedure for a Poisson Process with Multiple Changepoints," *Journal of Computational and Graphical Statistics*, **10**, No. 10, 772-785 (2001).

Author Index

Lecture Notes in Computer Science

For information about Vols. 1–3397

please contact your bookseller or Springer

Vol. 3525: A.E. Abdallah, C.B. Jones, J.W. Sanders (Eds.), Communicating Sequential Processes. XIV, 321 pages. 2005.

Vol. 3517: H.S. Baird, D.P. Lopresti (Eds.), Human Interactive Proofs. IX, 143 pages. 2005.

Vol. 3516: V.S. Sunderam, G.D.v. Albada, P.M.A. Sloot, J.J. Dongarra (Eds.), Computational Science – ICCS 2005, Part III. LXIII, 1143 pages. 2005.

Vol. 3515: V.S. Sunderam, G.D.v. Albada, P.M.A. Sloot, J.J. Dongarra (Eds.), Computational Science – ICCS 2005, Part II. LXIII, 1101 pages. 2005.

Vol. 3514: V.S. Sunderam, G.D.v. Albada, P.M.A. Sloot, J.J. Dongarra (Eds.), Computational Science – ICCS 2005, Part I. LXIII, 1089 pages. 2005.

Vol. 3510: T. Braun, G. Carle, Y. Koucheryavy, V. Tsaoussidis (Eds.), Wired/Wireless Internet Communications. XIV, 366 pages. 2005.

Vol. 3508: P. Bresciani, P. Giorgini, B. Henderson-Sellers, G. Low, M. Winikoff (Eds.), Agent-Oriented Information Systems II. X, 227 pages. 2005. (Subseries LNAI).

Vol. 3503: S.E. Nikoletseas (Ed.), Experimental and Efficient Algorithms. XV, 624 pages. 2005.

Vol. 3501: B. Kégl, G. Lapalme (Eds.), Advances in Artificial Intelligence. XV, 458 pages. 2005. (Subseries LNAI).

Vol. 3500: S. Miyano, J. Mesirov, S. Kasif, S. Istrail, P. Pevzner, M. Waterman (Eds.), Research in Computational Molecular Biology. XVII, 632 pages. 2005. (Subseries LNBI).

Vol. 3498: J. Wang, X. Liao, Z. Yi (Eds.), Advances in Neural Networks – ISNN 2005, Part III. L, 1077 pages. 2005.

Vol. 3497: J. Wang, X. Liao, Z. Yi (Eds.), Advances in Neural Networks – ISNN 2005, Part II. L, 947 pages. 2005.

Vol. 3496: J. Wang, X. Liao, Z. Yi (Eds.), Advances in Neural Networks – ISNN 2005, Part II. L, 1055 pages. 2005.

Vol. 3495: P. Kantor, G. Muresan, F. Roberts, D.D. Zeng, F.-Y. Wang, H. Chen, R.C. Merkle (Eds.), Intelligence and Security Informatics. XVIII, 674 pages. 2005.

Vol. 3494: R. Cramer (Ed.), Advances in Cryptology – EUROCRYPT 2005. XIV, 576 pages. 2005.

Vol. 3492: P. Blache, E. Stabler, J. Busquets, R. Moot (Eds.), Logical Aspects of Computational Linguistics. X, 363 pages. 2005. (Subseries LNAI).

Vol. 3489: G.T. Heineman, J.A. Stafford, H.W. Schmidt, K. Wallnau, C. Szyperski, I. Crnkovic (Eds.), Component-Based Software Engineering. XI, 358 pages. 2005.

Vol. 3488: M.-S. Hacid, N.V. Murray, Z.W. Raś, S. Tsumoto (Eds.), Foundations of Intelligent Systems. XIII, 700 pages. 2005. (Subseries LNAI).

Vol. 3483: O. Gervasi, M.L. Gavrilova, V. Kumar, A. Laganà, H.P. Lee, Y. Mun, D. Taniar, C.J.K. Tan (Eds.), Computational Science and Its Applications – ICCSA 2005, Part IV. XXVII, 1362 pages. 2005.

Vol. 3482: O. Gervasi, M.L. Gavrilova, V. Kumar, A. Laganà, H.P. Lee, Y. Mun, D. Taniar, C.J.K. Tan (Eds.), Computational Science and Its Applications – ICCSA 2005, Part III. LXVI, 1340 pages. 2005.

Vol. 3481: O. Gervasi, M.L. Gavrilova, V. Kumar, A. Laganà, H.P. Lee, Y. Mun, D. Taniar, C.J.K. Tan (Eds.), Computational Science and Its Applications – ICCSA 2005, Part II. LXIV, 1316 pages. 2005.

Vol. 3480: O. Gervasi, M.L. Gavrilova, V. Kumar, A. Laganà, H.P. Lee, Y. Mun, D. Taniar, C.J.K. Tan (Eds.), Computational Science and Its Applications – ICCSA 2005, Part I. LXV, 1234 pages. 2005.

Vol. 3479: T. Strang, C. Linnhoff-Popien (Eds.), Location- and Context-Awareness. XII, 378 pages. 2005.

Vol. 3477: P. Herrmann, V. Issarny (Eds.), Trust Management. XII, 426 pages. 2005.

Vol. 3475: N. Guelfi (Ed.), Rapid Integration of Software Engineering Techniques. X, 145 pages. 2005.

Vol. 3468: H.W. Gellersen, R. Want, A. Schmidt (Eds.), Pervasive Computing. XIII, 347 pages. 2005.

Vol. 3467: J. Giesl (Ed.), Term Rewriting and Applications. XIII, 517 pages. 2005.

Vol. 3465: M. Bernardo, A. Bogliolo (Eds.), Formal Methods for Mobile Computing. VII, 271 pages. 2005.

Vol. 3463: M. Dal Cin, M. Kaâniche, A. Pataricza (Eds.), Dependable Computing - EDCC 2005. XVI, 472 pages. 2005.

Vol. 3462: R. Boutaba, K. Almeroth, R. Puigjaner, S. Shen, J.P. Black (Eds.), NETWORKING 2005. XXX, 1483 pages. 2005.

Vol. 3461: P. Urzyczyn (Ed.), Typed Lambda Calculi and Applications. XI, 433 pages. 2005.

Vol. 3460: Ö. Babaoglu, M. Jelasity, A. Montresor, C. Fetzer, S. Leonardi, A. van Moorsel, M. van Steen (Eds.), Self-star Properties in Complex Information Systems. IX, 447 pages. 2005.

Vol. 3459: R. Kimmel, N.A. Sochen, J. Weickert (Eds.), Scale Space and PDE Methods in Computer Vision. XI, 634 pages. 2005.

Vol. 3456: H. Rust, Operational Semantics for Timed Systems. XII, 223 pages. 2005.

Vol. 3455: H. Treharne, S. King, M. Henson, S. Schneider (Eds.), ZB 2005: Formal Specification and Development in Z and B. XV, 493 pages. 2005.

Vol. 3454: J.-M. Jacquet, G.P. Picco (Eds.), Coordination Models and Languages. X, 299 pages. 2005.

Vol. 3453: L. Zhou, B.C. Ooi, X. Meng (Eds.), Database Systems for Advanced Applications. XXVII, 929 pages. 2005.

Vol. 3452: F. Baader, A. Voronkov (Eds.), Logic for Programming, Artificial Intelligence, and Reasoning. XI, 562 pages. 2005. (Subseries LNAI).

Vol. 3450: D. Hutter, M. Ullmann (Eds.), Security in Pervasive Computing. XI, 239 pages. 2005.

Vol. 3449: F. Rothlauf, J. Branke, S. Cagnoni, D.W. Corne, R. Drechsler, Y. Jin, P. Machado, E. Marchiori, J. Romero, G.D. Smith, G. Squillero (Eds.), Applications of Evolutionary Computing. XX, 631 pages. 2005.

Vol. 3448: G.R. Raidl, J. Gottlieb (Eds.), Evolutionary Computation in Combinatorial Optimization. XI, 271 pages. 2005.

Vol. 3447: M. Keijzer, A. Tettamanzi, P. Collet, J.v. Hemert, M. Tomassini (Eds.), Genetic Programming. XIII, 382 pages. 2005.

Vol. 3444: M. Sagiv (Ed.), Programming Languages and Systems. XIII, 439 pages. 2005.

Vol. 3443: R. Bodik (Ed.), Compiler Construction. XI, 305 pages. 2005.

Vol. 3442: M. Cerioli (Ed.), Fundamental Approaches to Software Engineering. XIII, 373 pages. 2005.

Vol. 3441: V. Sassone (Ed.), Foundations of Software Science and Computational Structures. XVIII, 521 pages. 2005.

Vol. 3440: N. Halbwachs, L.D. Zuck (Eds.), Tools and Algorithms for the Construction and Analysis of Systems. XVII, 588 pages. 2005.

Vol. 3439: R.H. Deng, F. Bao, H. Pang, J. Zhou (Eds.), Information Security Practice and Experience. XII, 424 pages. 2005.

Vol. 3437: T. Gschwind, C. Mascolo (Eds.), Software Engineering and Middleware. X, 245 pages. 2005.

Vol. 3436: B. Bouyssounouse, J. Sifakis (Eds.), Embedded Systems Design. XV, 492 pages. 2005.

Vol. 3434: L. Brun, M. Vento (Eds.), Graph-Based Representations in Pattern Recognition. XII, 384 pages. 2005.

Vol. 3433: S. Bhalla (Ed.), Databases in Networked Information Systems. VII, 319 pages. 2005.

Vol. 3432: M. Beigl, P. Lukowicz (Eds.), Systems Aspects in Organic and Pervasive Computing - ARCS 2005. X, 265 pages. 2005.

Vol. 3431: C. Dovrolis (Ed.), Passive and Active Network Measurement. XII, 374 pages. 2005.

Vol. 3429: E. Andres, G. Damiand, P. Lienhardt (Eds.), Discrete Geometry for Computer Imagery. X, 428 pages. 2005.

Vol. 3427: G. Kotsis, O. Spaniol (Eds.), Wireless Systems and Mobility in Next Generation Internet. VIII, 249 pages. 2005.

Vol. 3423: J.L. Fiadeiro, P.D. Mosses, F. Orejas (Eds.), Recent Trends in Algebraic Development Techniques. VIII, 271 pages. 2005.

Vol. 3422: R.T. Mittermeir (Ed.), From Computer Literacy to Informatics Fundamentals. X, 203 pages. 2005.

Vol. 3421: P. Lorenz, P. Dini (Eds.), Networking - ICN 2005, Part II. XXXV, 1153 pages. 2005.

Vol. 3420: P. Lorenz, P. Dini (Eds.), Networking - ICN 2005, Part I. XXXV, 933 pages. 2005.

Vol. 3419: B. Faltings, A. Petcu, F. Fages, F. Rossi (Eds.), Constraint Satisfaction and Constraint Logic Programming. X, 217 pages. 2005. (Subseries LNAI).

Vol. 3418: U. Brandes, T. Erlebach (Eds.), Network Analysis. XII, 471 pages. 2005.

Vol. 3416: M. Böhlen, J. Gamper, W. Polasek, M.A. Wimmer (Eds.), E-Government: Towards Electronic Democracy. XIII, 311 pages. 2005. (Subseries LNAI).

Vol. 3415: P. Davidsson, B. Logan, K. Takadama (Eds.), Multi-Agent and Multi-Agent-Based Simulation. X, 265 pages. 2005. (Subseries LNAI).

Vol. 3414: M. Morari, L. Thiele (Eds.), Hybrid Systems: Computation and Control. XII, 684 pages. 2005.

Vol. 3412: X. Franch, D. Port (Eds.), COTS-Based Software Systems. XVI, 312 pages. 2005.

Vol. 3411: S.H. Myaeng, M. Zhou, K.-F. Wong, H.-J. Zhang (Eds.), Information Retrieval Technology. XIII, 337 pages. 2005.

Vol. 3410: C.A. Coello Coello, A. Hernández Aguirre, E. Zitzler (Eds.), Evolutionary Multi-Criterion Optimization. XVI, 912 pages. 2005.

Vol. 3409: N. Guelfi, G. Reggio, A. Romanovsky (Eds.), Scientific Engineering of Distributed Java Applications. X, 127 pages. 2005.

Vol. 3408: D.E. Losada, J.M. Fernández-Luna (Eds.), Advances in Information Retrieval. XVII, 572 pages. 2005.

Vol. 3407: Z. Liu, K. Araki (Eds.), Theoretical Aspects of Computing - ICTAC 2004. XIV, 562 pages. 2005.

Vol. 3406: A. Gelbukh (Ed.), Computational Linguistics and Intelligent Text Processing. XVII, 829 pages. 2005.

Vol. 3404: V. Diekert, B. Durand (Eds.), STACS 2005. XVI, 706 pages. 2005.

Vol. 3403: B. Ganter, R. Godin (Eds.), Formal Concept Analysis. XI, 419 pages. 2005. (Subseries LNAI).

Vol. 3402: M. Daydé, J.J. Dongarra, V. Hernández, J.M.L.M. Palma (Eds.), High Performance Computing for Computational Science - VECPAR 2004. XI, 732 pages. 2005.

Vol. 3401: Z. Li, L.G. Vulkov, J. Waśniewski (Eds.), Numerical Analysis and Its Applications. XIII, 630 pages. 2005.

Vol. 3400: J.F. Peters, A. Skowron (Eds.), Transactions on Rough Sets III. IX, 461 pages. 2005.

Vol. 3399: Y. Zhang, K. Tanaka, J.X. Yu, S. Wang, M. Li (Eds.), Web Technologies Research and Development - APWeb 2005. XXII, 1082 pages. 2005.

Vol. 3398: D.-K. Baik (Ed.), Systems Modeling and Simulation: Theory and Applications. XIV, 733 pages. 2005. (Subseries LNAI).